Radiogenic Isotope Geology

Third Edition

Alan P. Dickin
McMaster University

CAMBRIDGE
UNIVERSITY PRESS

CAMBRIDGE
UNIVERSITY PRESS

University Printing House, Cambridge CB2 8BS, United Kingdom

One Liberty Plaza, 20th Floor, New York, NY 10006, USA

477 Williamstown Road, Port Melbourne, VIC 3207, Australia

314-321, 3rd Floor, Plot 3, Splendor Forum, Jasola District Centre, New Delhi - 110025, India

79 Anson Road, #06-04/06, Singapore 079906

Cambridge University Press is part of the University of Cambridge.

It furthers the University's mission by disseminating knowledge in the pursuit of
education, learning and research at the highest international levels of excellence.

www.cambridge.org
Information on this title: www.cambridge.org/9781107099449
DOI: 10.1017/9781316163009

First published 1995
Second edition published 2005
Third edition published 2018

Printed in the United States of America by Sheridan Books, Inc. 2018

A catalogue record for this publication is available from the British Library

Library of Congress Cataloging-in-Publication data
Names: Dickin, Alan P., author.
Title: Radiogenic isotope geology / Alan P. Dickin, McMaster University.
Description: [2018 edition]. | Cambridge : Cambridge University Press, 2018. |
    Includes bibliographical references and index.
Identifiers: LCCN 2017023083 | ISBN 9781107099449 (alk. paper)
Subjects: LCSH: Isotope geology. | Radioactive dating. | Geochemistry. |
    Paleoclimatology. | Environmental archaeology.
Classification: LCC QE501.4.N9 D53 2017 | DDC 551.9–dc23
LC record available at https://lccn.loc.gov/2017023083

ISBN 978-1-107-09944-9 Hardback
ISBN 978-1-107-49212-7 Paperback

Additional resources for this publication at www.cambridge.org/dickin3

Radiogenic Isotope Geology

Third Edition

The third edition of *Radiogenic Isotope Geology* examines revolutionary changes in geochemical thinking that have occurred over the past 15 years. Extinct nuclide studies on meteorites have called into question fundamental geochemical models of the Earth, while new dating methods have challenged conventional views of Earth history. At the same time, the problem of global warming has raised new questions about the causes of past and present climate change. In the new edition, these and other recent issues are evaluated in their scholarly and historical context, so readers can understand the development of current ideas. Controversial theories, new analytical techniques, classic papers, and illustrative case studies all come under scrutiny in this book, providing an accessible introduction for students and critical commentary for researchers.

ALAN P. DICKIN is Professor of Geology at McMaster University.

"The Dickin text provides an excellent introduction to radiogenic isotope geochemistry. I read a previous edition cover-to-cover during preparation for the general knowledge exams in graduate school, and I still suggest that graduate students do the same in preparation for their exams. It continues to be a key reference for teaching and in the classroom and in the laboratory."

– *Matthew Jackson, University of California, Santa Barbara*

"Isotope geochemistry is hugely influential in the development of new approaches and ideas in the Earth sciences. New data challenge models for the formation of the Earth, the evolution of the continental crust, and climate change. An understanding of the basic principles of isotope geology is important in a wide range of the sciences, and this welcome third edition of *Radiogenic Isotope Geology* builds on the success of the previous editions. It is scholarly and accessible, and it combines an all too rare historical context with a comprehensive introduction to a wide range of radiogenic isotope techniques. Written by one of the world's most respected authors in this field, this textbook will be invaluable for undergraduate and graduate courses, and it is an excellent reference text for scientists in other fields."

– *Chris Hawkesworth, University of Bristol*

"For teachers and students in both low- and high-temperature geochemistry who need ready access to geochemical concepts and techniques, Alan Dickin offers an up-to-date, well-written medium-level textbook on isotope geochemistry. A pleasant, handy, and useful book for your shelf."

– *Francis Albarède, Ecole Normale Supérieure de Lyon*

To Margaret
and to the memory of Stephen Moorbath,
isotope pioneer

Contents in Brief

Preface and Acknowledgements   *page* xvii

1 Nucleosynthesis and Nuclear Decay 1

2 Mass Spectrometry 13

3 The Rb–Sr Method 40

4 The Sm–Nd Method 67

5 Lead Isotopes 99

6 Isotope Geochemistry of Oceanic Volcanics 134

7 Isotope Geochemistry of Continental Rocks 167

8 Osmium Isotopes 194

9 Lu–Hf, Ba–La–Ce and K–Ca Systems 218

10 K–Ar, Ar–Ar and U–He Dating 240

11 Noble Gas Geochemistry 274

12 U-Series Dating 306

13 U-Series Geochemistry of Igneous Systems 333

14 Cosmogenic Nuclides 363

15 Extinct Radionuclides 407

16 Fission-Track Dating 444

Appendix 1: Chart of the Nuclides   465
Appendix 2: Meteorite Types   468
Index   471

Contents

Preface and Acknowledgements   *page* xvii

1 Nucleosynthesis and Nuclear Decay 1
  1.1 The Chart of the Nuclides 1
  1.2 Nucleosynthesis 2
     1.2.1 *Stellar Evolution* 3
     1.2.2 *Stages in the Nucleosynthesis of Heavy Elements* 4
  1.3 Radioactive Decay 6
     1.3.1 *Isobaric Decay* 7
     1.3.2 *Alpha and Heavy Particle Decay* 8
     1.3.3 *Nuclear Fission and the Oklo Natural Reactor* 8
  1.4 The Law of Radioactive Decay 10
     1.4.1 *Uniformitarianism* 10

2 Mass Spectrometry 13
  2.1 Chemical Purification 13
     2.1.1 *Ion Exchange Separation* 14
     2.1.2 *Sm–Nd* 15
     2.1.3 *Lu–Hf* 15
     2.1.4 *Lead* 15
     2.1.5 *Analytical Blank* 16
  2.2 Ion Sources 16
     2.2.1 *Thermal Ionization* 16
     2.2.2 *Plasma Source Mass Spectrometry* 17
  2.3 Mass-dependent Fractionation 19
     2.3.1 *Mass Fractionation in TIMS* 19
     2.3.2 *Mass Fractionation in MC–ICP–MS* 21
  2.4 Magnetic Sector Analysis 22
     2.4.1 *Ion Optics* 22
     2.4.2 *Detectors* 24
     2.4.3 *Data Collection* 25
  2.5 Isotope Dilution 26
     2.5.1 *Analysis Technique* 26
     2.5.2 *Double Spiking* 27
     2.5.3 *Pb–Tl Double Spiking* 28
  2.6 MC–ICP–MS Solution-based Applications 28
     2.6.1 *Hf–W* 29
     2.6.2 *Lu–Hf* 29
     2.6.3 *U–Th* 29
     2.6.4 *Pb–Pb* 29
     2.6.5 *Sm–Nd* 29
  2.7 LA–ICP–MS 30
     2.7.1 *U–Pb* 30
     2.7.2 *Lu–Hf* 31
  2.8 Isochron Regression Line Fitting 32
     2.8.1 *Types of Regression Fit* 32
     2.8.2 *Regression Fitting with Correlated Errors* 32
     2.8.3 *Errorchrons* 33
     2.8.4 *Probability of Fit* 34
     2.8.5 *Isoplot* 35

2.9 Probability Density 35
    2.9.1 *Detrital Zircon Distributions* 35
    2.9.2 *Isochron Data Distributions* 36

3 The Rb–Sr Method 40
3.1 The Rb Decay Constant 40
3.2 Dating Igneous Crystallization 41
    3.2.1 *Sr Model Ages* 42
    3.2.2 *The Isochron Diagram* 42
    3.2.3 *Erupted Isochrons* 43
    3.2.4 *Meteorite Chronology* 44
3.3 Dating Metamorphic Systems 45
    3.3.1 *Mineral and Whole-Rock Isochrons* 45
    3.3.2 *Blocking Temperatures* 47
3.4 Dating Ore Deposits 48
3.5 Dating Sedimentary Systems 49
    3.5.1 *Shales* 50
    3.5.2 *Glauconite* 51
3.6 Seawater Evolution 52
    3.6.1 *Measurement of the Curve* 52
    3.6.2 *The Cretaceous–Tertiary Seawater Curve* 54
    3.6.3 *Seawater Sr and Glacial Cycles* 55
    3.6.4 *Modelling the Fluxes* 56
    3.6.5 *Quantifying the Hydrothermal Flux* 58
    3.6.6 *The Effects of Himalayan Erosion* 60
    3.6.7 *Glacial Cycles* 61
    3.6.8 *Stable Sr Isotopes in Seawater* 62

4 The Sm–Nd Method 67
4.1 Sm–Nd Isochrons 67
    4.1.1 *Meteorites* 67
    4.1.2 *Precambrian Mafic Rocks* 68
    4.1.3 *High-Grade Metamorphic Rocks* 70
    4.1.4 *Garnet Geochronology* 70
4.2 Nd Isotope Evolution and Model Ages 72
    4.2.1 *Chondritic Model Ages* 72
    4.2.2 *Depleted Mantle Model Ages* 73
    4.2.3 *Nd Isotope Mapping* 75
4.3 Model Ages and Crustal Processes 78
    4.3.1 *Sedimentary Systems* 79
    4.3.2 *Meta-Sedimentary Systems* 79
    4.3.3 *Meta-Igneous Systems* 80
    4.3.4 *Partially Melted Systems* 81
4.4 The Crustal Growth Problem 82
    4.4.1 *Crustal Accretion Ages* 82
    4.4.2 *Sediment Provenance Ages* 83
    4.4.3 *Archean Depleted Mantle* 84
    4.4.4 *Early Archean Crustal Provinces* 86
4.5 Nd in the Oceans 88
    4.5.1 *Modern Seawater Nd* 88
    4.5.2 *The Oceanic Nd Paradox* 89
    4.5.3 *Ancient Seawater Nd* 90
    4.5.4 *Tertiary Seawater Nd* 91
    4.5.5 *Quaternary Seawater Nd* 92

5 Lead Isotopes 99
  5.1 U–Pb Isochrons 99
    5.1.1 *U–Pb Isochrons and Decay Constants* 100
    5.1.2 *Uranium Isotope Composition* 100
    5.1.3 *U–Pb Isochrons and Timescale Calibration* 101
  5.2 U–Pb Concordia Dating 102
    5.2.1 *Lead Loss Models* 103
    5.2.2 *Air Abrasion and Direct Evaporation* 104
    5.2.3 *Chemical Abrasion and Annealing* 105
    5.2.4 *Concordia Ages and Decay Constants* 107
    5.2.5 *Inherited Zircon* 108
    5.2.6 *In Situ Analysis* 109
    5.2.7 *Alternative U–Pb Dating Materials* 111
  5.3 Pb–Pb Dating 112
    5.3.1 *The Age of the Earth and Pb Paradox* 113
    5.3.2 *Meteorite Dating and the Total Pb Isochron* 115
  5.4 Pb (Galena) Model Ages 118
    5.4.1 *The Holmes–Houtermans Model* 118
    5.4.2 *Conformable Leads* 119
    5.4.3 *Open-System Pb Evolution* 120
    5.4.4 *Plumbotectonics* 121
  5.5 Whole-Rock Pb and Crustal Evolution 122
    5.5.1 *Archean Crustal Evolution* 122
    5.5.2 *Paleo-Isochrons and Metamorphic Disturbance* 124
    5.5.3 *Proterozoic Crustal Evolution* 124
  5.6 Environmental Pb 125
    5.6.1 *Anthropogenic Pb* 126
    5.6.2 *Pb as an Oceanographic Tracer* 127
    5.6.3 *Paleo-Seawater Pb* 129

6 Isotope Geochemistry of Oceanic Volcanics 134
  6.1 Isotopic Tracing of Mantle Structure 134
    6.1.1 *Contamination and Alteration* 134
    6.1.2 *Disequilibrium Melting* 135
    6.1.3 *Mantle Plumes* 136
    6.1.4 *Plum Pudding Mantle* 137
    6.1.5 *Marble Cake Mantle* 138
    6.1.6 *Mantle Convection and Viscosity* 139
  6.2 The Nd–Sr Isotope Diagram 140
    6.2.1 *The Mantle Array and OIB Sources* 140
    6.2.2 *Box Models for the MORB Source* 141
    6.2.3 *Nd–142 and Early Earth Differentiation* 143
  6.3 Pb Isotope Geochemistry 144
    6.3.1 *Pb–Pb Isochrons and the Lead Paradox* 145
    6.3.2 *The Kappa Conundrum* 146
    6.3.3 *The Third Lead Paradox* 149
  6.4 Mantle Reservoirs in Isotopic Multispace 150
    6.4.1 *The Mantle Plane* 151
    6.4.2 *The Mantle Tetrahedron* 151
  6.5 Identification of Mantle Components 154
    6.5.1 *Depleted OIB Sources* 155
    6.5.2 *EMII* 155
    6.5.3 *EMI* 156
    6.5.4 *HIMU* 158
    6.5.5 *The DUPAL Anomaly* 159

6.6 Island Arcs and Mantle Evolution 159
    6.6.1 *Two-Component Mixing Models* 159
    6.6.2 *Three-Component Mixing Models* 161

7 Isotope Geochemistry of Continental Rocks 167
  7.1 Mantle Xenoliths 167
    7.1.1 *Mantle Metasomatism* 168
  7.2 Crustal Contamination 170
    7.2.1 *Two-Component Mixing Models* 171
    7.2.2 *Melting in Natural and Experimental Systems* 172
    7.2.3 *Inversion Modelling of Magma Suites* 174
    7.2.4 *Phenocrysts as Records of Magma Evolution* 178
    7.2.5 *Lithospheric Mantle Contamination* 178
  7.3 Petrogenesis of Continental Magmas 179
    7.3.1 *Kimberlites, Carbonatites and Lamproites* 179
    7.3.2 *Alkali Basalts* 181
    7.3.3 *Flood Basalts* 183
    7.3.4 *Precambrian Granitoids* 185
    7.3.5 *Phanerozoic Batholiths* 187

8 Osmium Isotopes 194
  8.1 Osmium Analysis 194
  8.2 The Re–Os and Pt–Os Decay Schemes 195
    8.2.1 *The Re Decay Constant* 195
    8.2.2 *Meteorite Isochrons* 196
    8.2.3 *Dating Ores and Rocks* 197
    8.2.4 *Os Normalization and the Pt–Os Decay Scheme* 198
  8.3 Mantle Osmium 199
    8.3.1 *The Bulk Silicate Earth* 199
    8.3.2 *Lithospheric Mantle* 200
    8.3.3 *Primitive Upper Mantle* 201
    8.3.4 *Asthenospheric Mantle* 202
    8.3.5 *Enriched Mantle Plumes* 204
    8.3.6 *Subduction Zones* 205
    8.3.7 *The Core Osmium Signature* 206
  8.4 Petrogenesis and Ore Genesis 206
    8.4.1 *The Bushveld Complex* 207
    8.4.2 *The Stillwater Complex* 208
    8.4.3 *The Sudbury Igneous Complex* 209
  8.5 Seawater Osmium 210
    8.5.1 *Osmium Isotope Evolution* 210
    8.5.2 *Os Fluxes and Residence Times* 212
    8.5.3 *Quaternary Seawater Osmium* 213

9 The Lu–Hf, Ba–La–Ce and K–Ca Systems 218
  9.1 Lu–Hf Geochronology 218
    9.1.1 *The Lu Decay Constant and CHUR Composition* 218
    9.1.2 *Dating Metamorphism* 220
  9.2 Modern Mantle Reservoirs 221
    9.2.1 *Depleted Mantle* 221
    9.2.2 *Enriched Mantle* 223
  9.3 Ancient Hf Evolution 226
    9.3.1 *Early Work* 226
    9.3.2 *Detrital Zircon* 227
    9.3.3 *Hf Model Ages* 228
    9.3.4 *Archean Depleted Mantle* 229
  9.4 Seawater Hafnium 230

9.5 The La–Ce and La–Ba Systems 232
    9.5.1 *La–Ba Geochronology* 232
    9.5.2 *La–Ce Geochronology* 232
    9.5.3 *Ce Isotope Geochemistry* 233
    9.5.4 *Seawater Cerium Geochemistry* 234
9.6 The K–Ca System 235

10 K–Ar, Ar–Ar and U–He Dating 240
10.1 The K–Ar Dating Method 240
    10.1.1 *Analytical Techniques* 240
    10.1.2 *Inherited Argon and the K–Ar Isochron Diagram* 242
    10.1.3 *Argon Loss* 244
10.2 The ^{40}Ar–^{39}Ar Dating Method 244
    10.2.1 *^{40}Ar–^{39}Ar Measurement* 244
    10.2.2 *Irradiation Corrections* 245
    10.2.3 *Step Heating* 246
    10.2.4 *Argon Loss Events* 247
    10.2.5 *Excess Argon* 249
    10.2.6 *^{39}Ar Recoil* 250
    10.2.7 *Dating Paleomagnetism* 251
    10.2.8 *Laser Microprobe Dating* 252
10.3 Timescale Calibration 254
    10.3.1 *Magnetic and Astronomical Timescales* 254
    10.3.2 *Intercalibration of Decay Constants* 257
10.4 Thermochronometry 259
    10.4.1 *Arrhenius Modelling* 259
    10.4.2 *Complex Diffusion Models* 261
    10.4.3 *K-Feldspar Thermochronometry* 265
10.5 U–Th–He Dating 268
    10.5.1 *Production and Analysis* 268
    10.5.2 *Annealing Behaviour* 269
    10.5.3 *Cosmogenic Helium Paleothermometry* 269

11 Noble Gas Geochemistry 274
11.1 Helium 274
    11.1.1 *Mass Spectrometry* 274
    11.1.2 *Helium Production in Nature* 275
    11.1.3 *Terrestrial Primordial Helium* 277
    11.1.4 *The 'Two-Reservoir' Model* 279
    11.1.5 *Helium Box Models* 281
    11.1.6 *Crustal and Mantle Helium* 282
    11.1.7 *Oceanic Sediments and Interplanetary Dust* 283
11.2 Neon 285
    11.2.1 *Neon Production* 285
    11.2.2 *Primordial Neon in the Earth* 285
    11.2.3 *Sub-Solar Neon* 287
    11.2.4 *Atmospheric Neon* 288
    11.2.5 *Nucleogenic Neon* 289
11.3 Argon 289
    11.3.1 *Terrestrial Primordial Argon* 290
    11.3.2 *Atmospheric Contamination* 291
    11.3.3 *Argon-38* 293
11.4 Krypton 294
11.5 Xenon 295
    11.5.1 *Iodogenic Xenon* 295
    11.5.2 *Fissiogenic Xenon* 296

11.5.3 *Radiogenic Xenon Reservoirs* 298
11.5.4 *Non-Radiogenic Xenon* 299
11.5.5 *The Barium–Xenon System* 300

12 U-Series Dating 306
12.1 Secular Equilibrium and Disequilibrium 306
12.2 Analytical Methods 307
    12.2.1 *Early Work* 308
    12.2.2 *Mass Spectrometry* 308
    12.2.3 *Half-Lives* 309
12.3 Daughter-Excess Methods 309
    12.3.1 ^{234}U *Dating of Carbonates* 309
    12.3.2 ^{234}U *Dating of Fe–Mn Crusts* 311
    12.3.3 ^{230}Th *Sediment Dating* 313
    12.3.4 ^{230}Th–^{232}Th 314
    12.3.5 ^{230}Th *Sediment Stratigraphy* 315
    12.3.6 ^{231}Pa–^{230}Th 317
    12.3.7 ^{210}Pb 319
12.4 Daughter-Deficiency Methods 321
    12.4.1 ^{230}Th: *Theory* 321
    12.4.2 ^{230}Th: *Applications* 322
    12.4.3 ^{230}Th: *Dirty Calcite* 324
    12.4.4 ^{231}Pa 326
12.5 U-Series Dating of Open Systems 326
    12.5.1 ^{231}Pa–^{230}Th 326
    12.5.2 *ESR–*^{230}Th 328

13 U-Series Geochemistry of Igneous Systems 333
13.1 Geochronology of Volcanic Rocks 334
    13.1.1 *The U–Th Isochron Diagram* 334
    13.1.2 *U–Th (Zircon) Model Ages* 335
    13.1.3 *Ra–Th Isochron Diagrams* 336
    13.1.4 *Ra–Th Model Ages* 337
13.2 Magma Chamber Evolution 337
    13.2.1 *The Th Isotope Evolution Diagram* 338
    13.2.2 *Short-Lived Species in Magma Evolution* 341
13.3 Mantle Melting Models 342
    13.3.1 *Melting Under Ocean Ridges* 343
    13.3.2 *The Effect of Source Convection* 344
    13.3.3 *The Effect of Melting Depth* 347
    13.3.4 *The Effect of Source Composition* 348
    13.3.5 *Crustal Melting and Contamination* 350
13.4 Short-Lived Species and Melting Models 351
    13.4.1 ^{226}Ra *and Melting Models* 351
    13.4.2 ^{231}Pa *and Melting Models* 353
    13.4.3 *Sources of Continental Magmas* 354
13.5 Subduction Zone Processes 355
    13.5.1 *U–Th in Arc Magmas* 355
    13.5.2 *Ra–Th in Arcs* 357
    13.5.3 *U–Pa in Arcs* 358

14 Cosmogenic Nuclides 363
14.1 Carbon-14 363
    14.1.1 *Early Work* 363
    14.1.2 *Closed-System Assumption* 365
    14.1.3 *Initial Ratio Assumption* 366
    14.1.4 *Dendrochronology* 367

14.1.5 *Bayesian Analysis* 369
14.1.6 *Pre-Holocene Calibration* 369
14.2 Radiocarbon and Climate Change 372
14.2.1 *Radiocarbon in the Modern Oceans* 372
14.2.2 *Glacial/Holocene Ventilation Ages* 373
14.2.3 *Causes of Climate Change* 376
14.3 Accelerator Mass Spectrometry 378
14.3.1 *Principles of Accelerator Mass Spectrometry* 378
14.3.2 *Radiocarbon Dating by AMS* 379
14.4 Beryllium-10 380
14.4.1 *^{10}Be in the Atmosphere* 380
14.4.2 *^{10}Be in the Oceans* 381
14.4.3 *^{10}Be in Snow and Ice* 384
14.4.4 *^{10}Be Production and Climate Cycles* 385
14.4.5 *^{10}Be in Soil Profiles* 386
14.4.6 *^{10}Be in Magmatic Systems* 387
14.5 Chlorine-36 390
14.6 Iodine-129 393
14.7 *In Situ* Cosmogenic Nuclides 394
14.7.1 *Meteorite Terrestrial Residence Ages* 394
14.7.2 *Al–Be Terrestrial Exposure Ages* 395
14.7.3 *Al–Be Burial Ages* 396
14.7.4 *Al–Be–Ne Ages* 397
14.7.5 *Chlorine-36 Exposure Ages* 398

15 **Extinct Radionuclides** 407
15.1 Introduction 407
15.1.1 *Nuclide Production and Decay* 407
15.1.2 *Celestial Objects and Ages* 407
15.1.3 *Parent–Daughter Pairs* 409
15.2 Stable Isotopes 409
15.2.1 *Cosmic Building Blocks of the Earth* 410
15.2.2 *Solar System Isotope Heterogeneity* 410
15.3 Extant Actinides 411
15.4 Iodine–Xenon 412
15.4.1 *The Xe–Xe Correlation Diagram* 412
15.4.2 *The Determination of 'Delta'* 413
15.4.3 *Pu–Xe* 414
15.4.4 *I–Xe Chronology* 415
15.5 Al–Mg 415
15.5.1 *^{26}Al in the Allende Meteorite* 416
15.5.2 *Determination of Delta* 416
15.5.3 *Al–Mg Early Nebular Chronometry* 417
15.5.4 *Testing the 'Canonical' Model* 418
15.6 Short-Lived Species in Planetary Differentiation 419
15.6.1 *Pd–Ag* 419
15.6.2 *Mn–Cr* 420
15.6.3 *Fe–Ni* 421
15.6.4 *Hf–W* 423
15.6.5 *Hf–W Solar System Chronometry* 424
15.6.6 *Revisiting the Giant Impact Model* 426
15.7 The Sm–Nd System 427
15.7.1 *Early Work* 427
15.7.2 *Chondrites and the Bulk Earth* 429
15.7.3 *The ^{142}Nd Conundrum* 429
15.7.4 *SCHEM or Chondritic Moon* 430

15.7.5 ^{142}Nd, Core Sulphide and E-Chondrites 432

15.7.6 ^{142}Nd in the Archean Earth 433

15.8 The Curium–Uranium–(Nd) System 433

15.9 Spallogenic Extinct Nuclides 435

15.9.1 Be-10 435

15.9.2 Ca–K 436

15.10 Conclusions 437

16 Fission-Track Dating 444

16.1 Track Formation 444

16.2 Track Etching 446

16.3 Counting Techniques 446

16.3.1 Population Method 447

16.3.2 External Detector Method 447

16.3.3 Re-Etching and Re-Polishing 448

16.3.4 LA–ICP–MS 449

16.3.5 Automated Track Counting 449

16.4 Detrital Populations 449

16.5 Track Annealing 451

16.6 Uplift and Subsidence Rates 452

16.7 Track Length Measurements 454

16.7.1 Projected Tracks (Semi-Tracks) 455

16.7.2 Confined Tracks 456

16.7.3 Track Widths 457

16.7.4 c Axis Projection 458

16.7.5 Forward and Inverse Modelling 459

16.8 Pressure Effects 462

Appendix 1: Chart of the Nuclides   465

Appendix 2: Meteorite Types   468

Index   471

Preface and Acknowledgements

The past fifteen years have seen a quiet revolution in isotope geochemistry, as a once-arcane field involving 'extinct' radionuclides in meteorites has called into question fundamental geochemical models of the Earth itself. At the same time, increasing public awareness of the problem of anthropogenic global warming has focused attention on the role of isotope geochemistry in monitoring past and present influences on climate change.

The third edition of *Radiogenic Isotope Geology* attempts to place these and other recent developments in scientific thinking in their overall scholarly context.

The approach to the subject matter is historical, for three main reasons. Firstly, to give an impression of the development of thought in the field so that the reader can understand the origin of present ideas; secondly, to explain why past theories have had to be modified; and thirdly, to present 'fall back' positions lest current models be refuted at some future date. This approach embodies the scholarly principle that knowledge of the classic work in the field is the starting point for current research.

The text is also particularly focussed on three types of literature. Firstly, it attempts to give accurate attribution of new ideas or methods; secondly, it reviews classic papers which have become standards in their field; and thirdly, it presents case studies that have evoked controversy in the literature, as examples of alternative data interpretations.

The organization of the book allows each chapter to be a relatively free-standing entity covering one segment of the field of radiogenic isotope geology. However, the reader may benefit from an understanding of the thread, which, in the author's mind, links these chapters together.

Chapter 1 introduces radiogenic isotopes by discussing the synthesis and decay of nuclides within the context of nuclear stability. Decay constants and the radioactive decay law are introduced.

Chapter 2 provides an experimental background to many of the chapters that follow by discussing the details of mass spectrometric analysis (TIMS and ICP–MS), along with a discussion of isochron regression fitting.

The next three chapters introduce the three pillars of lithophile isotope geology, comprising the Sr, Nd and Pb isotope methods. Emphasis is placed on their applications to geochronology and their evolution in terrestrial systems. Chapter 3 covers the Rb–Sr system, since this is one of the simplest and most basic dating methods. Chapter 4 covers the Sm–Nd system, including the use of Nd model ages to date crustal formation. Chapter 5 examines U–Pb geochronology and introduces the complexities of terrestrial Pb isotope evolution in a straightforward fashion. Each chapter ends with an examination of these isotopes as environmental tracers, focussing particularly on the oceans.

Chapters 6 and 7 apply Sr, Nd and Pb, as geochemical tracers, to the study of oceanic and continental igneous rocks. This is appropriate, because these isotopes are some of the basic tools of the isotope geochemist, which together may allow understanding of the complexities of mantle processes and magmatic evolution. These methods are supplemented in Chapters 8 and 9 by insights from the Re–Os, Lu–Hf and other lithophile isotope systems, which arise from their distinct chemistry.

Chapter 10 completes the panoply of long-lived isotopic dating systems by introducing the K–Ar, Ar–Ar and U–He methods, including their applications to magnetic and thermal histories. This leads us naturally in Chapter 11 to the consideration of rare gases as isotopic tracers, which give unique insights into the de-gassing history of the Earth.

Chapter 12 introduces the short-lived isotopes of the uranium decay series, covering classical and recent developments in the dating of Quaternary-age sedimentary rocks. This prepares us for the complexities of Chapter 13, which examines U-series isotopes as tracers in igneous systems. Short-lived processes in mantle melting and magma evolution are the focus of attention here.

Chapter 14 examines the most important of the cosmogenic isotopes. These represent a vast and growing field of chronology and isotope chemistry, which is especially pertinent to environmental geoscience. In particular, the radiocarbon method is a vital dating tool in archaeology and a tracer of the ocean–atmosphere system involved in climate change.

Chapter 15 represents a comprehensive review of the 'extinct nuclide' systems in meteorites that have recently raised questions about the cosmic context of terrestrial geochemistry. This overview deals with all of the major extinct nuclide pairs, and discusses their significance for the origins of the solar system and the Earth.

Lastly, Chapter 16 examines the specialized field of (radiogenic) fission track analysis, originally developed as a regular dating method, but increasingly applied to thermal history analysis.

The text is gathered around a large number of diagrams, many of which are classic figures from the literature. I gratefully acknowledge the many authors whose original data and diagrams form the basis for these figures. Author acknowledgement for all figure sources is given within individual figure captions, and corresponding titles, journal names, volumes and pages are contained in the list of cited references at the end of each chapter.

Alan P. Dickin
McMaster University

Chapter 1

Nucleosynthesis and Nuclear Decay

1.1 The Chart of the Nuclides

In the field of isotope geology, neutrons, protons and electrons can be regarded as the fundamental building blocks of the atom. The composition of a given type of atom, called a nuclide, is described by specifying the number of protons (atomic number, Z) and the number of neutrons (N) in the nucleus. The sum of these is the mass number (A). By plotting Z against N for all of the nuclides that have been known to exist (at least momentarily), the chart of the nuclides is obtained (Fig. 1.1). In this chart, horizontal rows of nuclides represent the same element (constant Z) with a variable number of neutrons (N). These are isotopes.

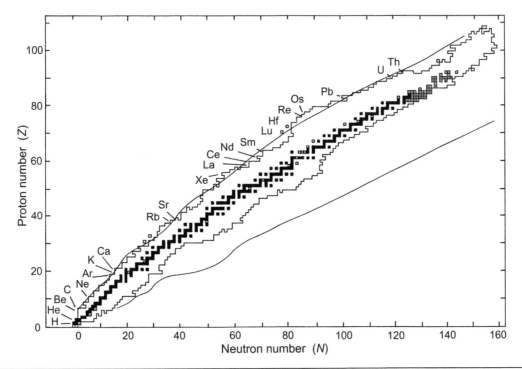

Fig. 1.1 Chart of the nuclides in coordinates of proton number Z, against neutron number N. (■) = stable nuclides; (□) = unstable nuclides; (◉) = naturally occurring long-lived unstable nuclides; (▫) = naturally occurring short-lived unstable nuclides. Some geologically useful radionuclides are marked. Smooth envelope = theoretical nuclide stability limits. For a more detailed nuclide chart, see Appendix 1.

A botal of 264 stable nuclides are known, which have not been observed to decay with available detection equipment. These define a central 'path of stability', coloured black in Fig. 1.1. On either side of this path, the zig-zag outline defines the limits of experimentally known unstable nuclides compiled by Hansen (1987). These species tend to undergo increasingly rapid decay as one moves out on either side of the path of stability. The smooth outer envelopes are the theoretical limits of nuclide stability ('drip lines') beyond which prompt decay occurs. This means that the synthesis and decay of an unstable nuclide occurs in a single particle interaction, giving it a zero effective lifetime.

As work progresses, the domain of experimentally known nuclides should approach the theoretical envelope, as has already occurred for nuclides with $Z < 20$ (Thoennessen, 2013). This has been achieved over the past 60 years using heavy ion accelerators (e.g. Darmstadt) to make exotic species by collision. Because of the curvature of the path of stability (Fig. 1.1), it was relatively easy to populate the proton-rich side of the path of stability, since these species can be made by fusion of lighter elements. Species on the neutron-rich side are made by bombarding target material with ^{238}U, creating unstable heavy atoms which immediately undergo fission to produce very neutron-rich products (e.g. Geissel et al., 2003). Knowledge about these unstable nuclei will improve our understanding of the nucleosynthetic r-process which occurs in supernovae (Thoennessen and Sherrill, 2011).

A small number of unstable nuclides have sufficiently long half-lives that they have not entirely decayed to extinction since the formation of the solar system. A few other short-lived nuclides are either continuously generated in the decay series of uranium and thorium, or produced by cosmic ray bombardment of stable nuclides. These nuclides, and one or two extinct short-lived isotopes, plus their daughter products, are the realm of radiogenic isotope geology. Those with half-lives over 0.5 Ma are marked in Fig. 1.2. Nuclides with half-lives over 1000 Ga decay too slowly to be geologically useful. Observation shows that all of the other long-lived isotopes either have been or are being applied in geology.

1.2 Nucleosynthesis

A realistic model for the nucleosynthesis of the elements must be based on empirical data for their 'cosmic abundance'. True cosmic abundances can be derived from stellar spectroscopy or by chemical analysis of galactic cosmic rays. However, such data are difficult to measure at high precision, so cosmic abundances are normally approximated by solar-system abundances. These can be determined by solar spectroscopy or by direct analysis of the most 'primitive' meteorites, carbonaceous chondrites. A comparison of the latter two sources of data (Ross and Aller, 1976) demonstrates good agreement for most elements (Fig. 1.3). Exceptions are the volatile elements, which have been lost from meteorites, and the Li–Be–B group, which are unstable in stars.

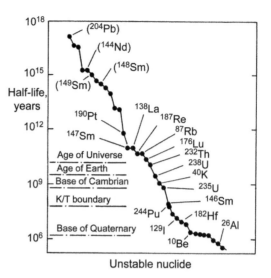

Fig. 1.2 Unstable nuclides with half-lives ($t_{1/2}$) over 0.5 Ma, in order of decreasing stability. Geologically useful parent nuclides are marked. Some very long-lived radionuclides with no geological application are also marked in brackets.

It is widely believed (e.g. Weinberg, 1977) that about 30 minutes after the 'big bang', the matter of the universe (in the form of protons and neutrons) consisted mostly of ^1H and 22–28% by mass of ^4He, along with traces of ^2H (deuterium) and ^3He. Hydrogen is still by far the most abundant element in the universe (88.6% of all nuclei) and with helium, makes up 99% of its mass, but naturally occurring heavy nuclides now exist up to atomic weight 254 or beyond

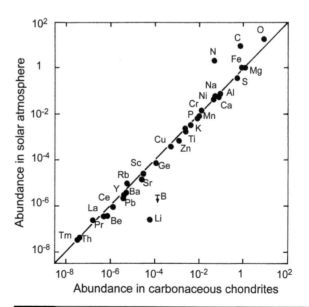

Fig. 1.3 Comparison of solar system abundances (relative to silicon) determined by solar spectroscopy and by analysis of carbonaceous chondrites. After Ringwood (1979).

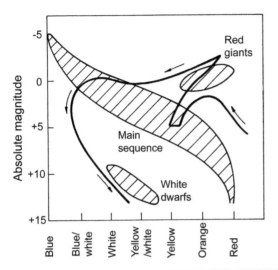

Fig. 1.4 Plot of absolute magnitude against spectral class of stars. Hatched areas show distributions of the three main star groups. The postulated evolutionary path of a star of solar mass is shown.

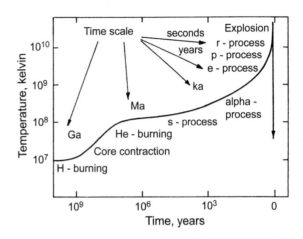

Fig. 1.5 Schematic evolution of a large star showing nucleosynthetic processes along its accelerating life-history in response to increasing temperature. Time is measured backwards from the end of the star's life on the right. After Burbidge *et al.* (1957).

(Fig. 1.1). These heavier nuclei must have been produced by nucleosynthetic processes in stars, and not in the big bang, because stars of different ages have different compositions which can be detected spectroscopically. Furthermore, stars at particular evolutionary stages may have compositional abnormalities, such as the presence of ^{254}Cf in supernovae. If nucleosynthesis of the heavy elements had occurred in the big bang then their distribution would be uniform about the universe.

1.2.1 Stellar Evolution

Present day models of stellar nucleosynthesis are based heavily on a classic review paper by Burbidge *et al.* (1957), in which eight element-building processes were identified (hydrogen burning, helium burning, α, e, x, r, s and p). Different processes were invoked to explain the abundance patterns of different groups of elements. These processes are, in turn, linked to different stages of stellar evolution. It is therefore appropriate at this point to summarize the life-history of some typical stars (e.g. Iben, 1967). The length of this life-history depends directly on the stellar mass, and can be traced on a plot of absolute magnitude (brightness) against spectral class (colour), referred to as the Hertzsprung–Russell or H–R diagram (Fig. 1.4).

Gravitational accretion of a star of solar mass from cold primordial hydrogen and helium would probably take about 1 Ma to raise the core temperature to ca. 10^7 K, when nuclear fusion of hydrogen to helium can begin (Atkinson and Houtermans, 1929). This process is also called 'hydrogen burning'. The star spends most of its life at this stage, as a 'main sequence' star, where an equilibrium is set up between energy supply by fusion and energy loss in the form of radiation. For the Sun, this stage will probably last ca. 10 Ga, but

a very large star with 15 times the Sun's mass may remain in the main sequence for only 10 Ma.

When the bulk of hydrogen in a small star has been converted into ^{4}He, inward density-driven forces exceed outward radiation pressure, causing gravitational contraction. However, the resulting rise in core temperature causes expansion of the outer hydrogen-rich layer of the star. This forms a huge low-density envelope whose surface temperature may fall to ca. 4000 K, observed as a 'red giant'. This stage lasts only one tenth as long as the main sequence stage. When core temperatures reach 1.5×10^7 K, a more efficient hydrogen-burning reaction becomes possible if the star contains traces of carbon, nitrogen and oxygen inherited from older generations of stars. This form of hydrogen burning is called the C–N–O cycle (Bethe, 1939).

At some point during the red giant stage, core temperatures may reach 10^8 K, when helium fusion to carbon is ignited (the 'helium flash'). Further core contraction, yielding a temperature of ca. 10^9 K, follows as helium becomes exhausted. At these temperatures an endothermic process of α-particle emission can occur, allowing the building of heavier nuclides up to mass 40. However, this quickly expends the remaining burnable fuel of the star, which then cools to a white dwarf.

More massive stars (of several solar masses) have a different life-history. In these stars, greater gravitationally induced pressure–temperature conditions allow the fusion of helium to begin early in the red giant stage. This is followed by further contraction and heating, allowing the fusion of carbon and successively heavier elements. However, as lighter elements become exhausted, gravitationally induced contraction and heating occur at an ever increasing pace (Fig. 1.5), until the implosion is stopped by the attainment of neutron-star density. The resulting shock wave causes a

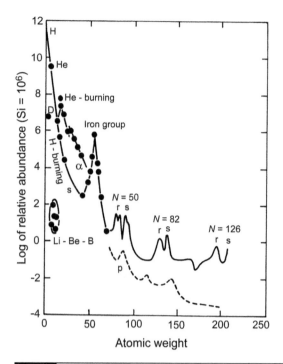

Fig. 1.6 Schematic diagram of the cosmic abundances of the elements, highlighting the nucleosynthetic processes responsible for forming different groups of nuclides. After Burbidge et al. (1957).

supernova explosion which ends the star's life (e.g. Burrows, 2000).

In the minutes before explosion, when temperatures exceed 3×10^9 K, very rapid nuclear interactions occur. Energetic equilibrium is established between nuclei and free protons and neutrons, synthesizing elements like Fe by the so-called e-process. The supernova explosion itself lasts only a few seconds, but is characterized by colossal neutron fluxes. These very rapidly synthesize heavier elements, terminating at ^{254}Cf, which undergoes spontaneous fission. Products of the supernova explosion are distributed through space and later incorporated in a new generation of stars.

1.2.2 Stages in the Nucleosynthesis of Heavy Elements

A schematic diagram of the cosmic abundance chart is given in Fig. 1.6. We will now see how different nucleosynthetic processes are invoked to account for its form.

The element-building process begins with the fusion of four protons to one ^{4}He nucleus, which occurs in three stages:

$$^1\mathrm{H} + {}^1\mathrm{H} \to {}^2\mathrm{D} + e^+ + \nu \qquad (Q = +1.44\,\mathrm{MeV},\ t_{1/2} = 14\,\mathrm{Ga})$$

$$^2\mathrm{D} + {}^1\mathrm{H} \to {}^3\mathrm{He} + \gamma \qquad (Q = +5.49\,\mathrm{MeV},\ t_{1/2} = 0.6\,\mathrm{s})$$

$$^3\mathrm{He} + {}^3\mathrm{He} \to {}^4\mathrm{He} + 2\,{}^1\mathrm{H} + \gamma \quad (Q = +12.86\,\mathrm{MeV},\ t_{1/2} = 1\,\mathrm{Ma}),$$

where Q is the energy output and $t_{1/2}$ is the reaction time of each stage (the time necessary to consume one half of the reactants) for the centre of the Sun. The long reaction time for the first step explains the long duration of the hydrogen-burning (main sequence) stage for small stars like the Sun. The overall reaction converts four protons into one helium nucleus, two positrons and two neutrinos, plus a large output of energy in the form of high-frequency photons. Hence the reaction is very strongly exothermic. Although deuterium and ^{3}He are generated in the first two reactions above, their consumption in the third accounts for their much lower cosmic abundance than ^{4}He.

If heavier elements such as carbon and nitrogen are present in a star, the catalytic C–N–O sequence of reactions can occur, which also combines four protons to make one helium nucleus:

$$^{12}\mathrm{C} + {}^1\mathrm{H} \to {}^{13}\mathrm{N} + \gamma \qquad (Q = +1.95\,\mathrm{MeV},\ t_{1/2} = 13\,\mathrm{Ma})$$

$$^{13}\mathrm{N} \to {}^{13}\mathrm{C} + e^+ + \nu \qquad (Q = +2.22\,\mathrm{MeV},\ t_{1.2} = 7\,\mathrm{min})$$

$$^{13}\mathrm{C} + {}^1\mathrm{H} \to {}^{14}\mathrm{N} + \gamma \qquad (Q = +7.54\,\mathrm{MeV},\ t_{1/2} = 3\,\mathrm{Ma})$$

$$^{14}\mathrm{N} + {}^1\mathrm{H} \to {}^{15}\mathrm{O} + \gamma \qquad (Q = +7.35\,\mathrm{MeV},\ t_{1/2} = 0.3\,\mathrm{Ma})$$

$$^{15}\mathrm{O} \to {}^{15}\mathrm{N} + e^+ + \nu \qquad (Q = +2.70\,\mathrm{MeV},\ t_{1/2} = 82\,\mathrm{s})$$

$$^{15}\mathrm{N} + {}^1\mathrm{H} \to {}^{12}\mathrm{C} + {}^4\mathrm{He} \quad (Q = +4.96\,\mathrm{MeV},\ t_{1/2} = 100\,\mathrm{ka}).$$

The C–N–O elements have greater potential energy barriers to fusion than hydrogen, so these reactions require higher temperatures to operate than the simple proton–proton (p–p) reaction. However, the reaction times are much shorter than for the p–p reaction. Therefore the C–N–O reaction contributes less than 10% of hydrogen-burning reactions in a small star like the Sun, but is overwhelmingly dominant in large stars. This explains their much shorter lifespan in the main sequence.

Helium burning also occurs in stages:

$$^4\mathrm{He} + {}^4\mathrm{He} \leftrightarrow {}^8\mathrm{Be} \qquad (Q = +0.09\,\mathrm{MeV})$$

$$^8\mathrm{Be} + {}^4\mathrm{He} \leftrightarrow {}^{12}\mathrm{C}^* \qquad (Q = -0.37\,\mathrm{MeV})$$

$$^{12}\mathrm{C}^* \to {}^{12}\mathrm{C} + \gamma \qquad (Q = +7.65\,\mathrm{MeV})$$

The ^{8}Be nucleus is very unstable ($t_{1/2} < 10^{-15}$ s) and in the core of a red giant the Be/He equilibrium ratio is estimated at 10^{-9}. However its life is just long enough to allow the possibility of collision with another helium nucleus. (Instantaneous three-particle collisions are very rare.) The energy yield of the first stage is small, and the second is actually endothermic, but the decay of excited ^{12}C* to the ground state is strongly exothermic, driving the equilibrium to the right.

The elements Li, Be and B have low nuclear binding energies, so that they are unstable at the temperatures of 10^7 K and above found at the centre of stars. They are therefore bypassed by stellar nucleosynthetic reactions, leading to low cosmic abundances (Fig. 1.6). The fact that the five stable isotopes ^{6}Li, ^{7}Li, ^{9}Be, ^{10}B and ^{11}B exist at all has been attributed to fragmentation effects (spallation) of heavy cosmic rays (atomic nuclei travelling through the galaxy at relativistic speeds) as they hit interstellar gas atoms (Reeves, 1974). This is termed the x-process.

Problems have been recognized in the x-process model for generating the light elements Li, Be and B, since cosmic ray spallation cannot explain the observed isotope ratios of these elements in solar system materials. However, Casse *et al.* (1995) proposed that carbon and oxygen nuclei ejected from supernovae can generate these nuclides by collision with hydrogen and helium in the surrounding gas cloud. This process is believed to occur in regions such as the Orion nebula. The combination of supernova production with spallation of galactic cosmic rays can explain observed solar system abundances of Li, Be and B.

Following the synthesis of carbon, further helium-burning reactions are possible, to produce heavier nuclei:

$$^{12}C + {}^4He \ \rightarrow \ ^{16}O + \gamma \quad (Q = +7.15\,\mathrm{MeV})$$

$$^{16}O + {}^4He \ \rightarrow \ ^{20}Ne + \gamma \quad (Q = +4.75\,\mathrm{MeV})$$

$$^{20}Ne + {}^4He \rightarrow {}^{24}Mg + \gamma \quad (Q = +9.31\,\mathrm{MeV}).$$

Intervening nuclei such as ^{13}N can be produced by adding protons to these species, but are themselves consumed in the process of catalytic hydrogen burning mentioned above.

In old red giant stars, carbon-burning reactions can occur:

$$^{12}C + {}^{12}C \rightarrow {}^{24}Mg + \gamma \quad (Q = +13.85\,\mathrm{MeV})$$

$$\rightarrow {}^{23}Na + {}^1H \quad (Q = +2.23\,\mathrm{MeV})$$

$$\rightarrow {}^{20}Ne + {}^4He \quad (Q = +4.62\,\mathrm{MeV}).$$

The hydrogen and helium nuclei regenerated in these processes allow further reactions which help to fill in gaps between masses 12 and 24.

When a small star reaches its maximum core temperature of 10^9 K the endothermic α-process can occur:

$$^{20}Ne + \gamma \rightarrow {}^{16}O + {}^4He \, (Q = -4.75\,\mathrm{MeV}).$$

The energy consumption of this process is compensated by strongly exothermic reactions such as:

$$^{20}Ne + {}^4He \rightarrow {}^{24}Mg + \gamma \, (Q = +9.31\,\mathrm{MeV}),$$

so that the overall reaction generates a positive energy budget. The process resembles helium burning, but is distinguished by the different source of ^{4}He. The α-process can build up from ^{24}Mg through the sequence ^{28}Si, ^{32}S, ^{36}Ar and ^{40}Ca, where it terminates, owing to the instability of ^{44}Ti.

The maximum temperatures reached in the core of a small star do not allow substantial heavy element production. However, in the final stages of the evolution of larger

stars, before a supernova explosion, the core temperature exceeds 3×10^9 K. This allows energetic equilibrium to be established by very rapid nuclear reactions between the various nuclei and free protons and neutrons (the e-process). Because ^{56}Fe is at the peak of the nuclear binding energy curve, this element is most favoured by the e-process (Fig. 1.6). However, the other first-series transition elements V, Cr, Mn, Co and Ni in the mass range 50 to 62 are also attributed to this process.

During the last few million years of a red giant's life, a slow process of neutron addition with emission of gamma rays (the s-process) can synthesize many additional nuclides up to mass 209 (see Fig. 1.7). Two possible neutron sources are:

$$^{13}C + {}^4He \ \rightarrow \ ^{16}O + n + \gamma$$

$$^{21}Ne + {}^4He \rightarrow {}^{24}Mg + n + \gamma.$$

The ^{13}C and ^{21}Ne parents can be produced by proton bombardment of the common ^{12}C and ^{20}Ne nuclides.

Because neutron capture in the s-process is relatively slow, unstable neutron-rich nuclides generated in this process have time to decay by β emission before further neutron addition. Hence the nucleosynthetic path of the s-process climbs in many small steps up the path of greatest stability of proton/neutron ratio (Fig. 1.7) and is finally terminated by the α decay of ^{210}Po back to ^{206}Pb and ^{209}Bi back to ^{205}Tl.

The 'neutron capture cross-section' of a nuclide expresses how readily it can absorb incoming thermal neutrons, and therefore determines how likely it is to be converted to a higher atomic mass species by neutron bombardment. Nuclides with certain neutron numbers (e.g. 50, 82 and 126) have unusually small neutron capture cross-sections, making them particularly resistant to further reaction and giving rise to local peaks in abundance at masses 90, 138 and 208. Hence, $N = 50, 82$ and 126 are empirically referred to as neutron 'magic numbers'.

In contrast to the s-process, which may occur over periods of millions of years in red giants, r-process neutrons are added in very rapid succession to a nucleus before β decay is possible. The nuclei are therefore rapidly driven to the neutron-rich side of the stability line, until they reach a new equilibrium between neutron addition and β decay, represented by the hatched zone in Fig. 1.7. Nuclides move along this r-process pathway until they reach a configuration with low neutron capture cross-section (a neutron magic number). At these points a 'cascade' of alternating β decays and single neutron additions occurs, indicated by the notched ladders in Fig. 1.7. Nuclides climb these ladders until they reach the next segment of the r-process pathway. Nuclides with neutron magic numbers build to excess abundances, as with the s-process, but they occur at proton-deficient compositions relative to the s-process stability path. Therefore, when the neutron flux falls off and nuclides on the ladders undergo β decay back to the stability line, the r-process local abundance peaks are displaced about 6–12 mass units below

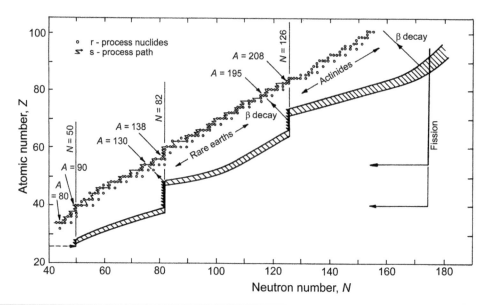

the s-process peaks (Fig. 1.6). The r-process is terminated by neutron-induced fission at mass 254, and nuclear matter is fed back into the element-building processes at masses of ca. 108 and 146. Thus, cycling of nuclear reactions occurs above mass 108.

Because of the extreme neutron flux postulated for the r-process, its occurrence is probably limited to supernovae. However, Blake and Schramm (1976) proposed the existence of a process that occurred at intermediate neutron fluxes between the s- and r-processes, which they called the 'n-process'. This could occur when neutron addition only slightly exceeds rates of β decay. Although neglected for many years, phenomena similar to the n-process have received consideration in some recent modelling of supernova outflows (Meyer, 2005; Wanajo, 2007; Panov and Janka, 2009).

The effects of r- and s-process synthesis of typical heavy elements may be demonstrated by an examination of the chart of the nuclides in the region of the light rare earths (Fig. 1.8). The step-by-step building of the s-process contrasts with the 'rain of nuclides' produced by β decay of r-process products. Some nuclides, such as ^{143}Nd to ^{146}Nd are produced by both r- and s-processes. Some, such as ^{142}Nd are s-only nuclides 'shielded' from the decay products of the r-process by intervening nuclides. Others, such as ^{148}Nd and ^{150}Nd are r-only nuclides which lie off the s-process production pathway.

Several heavy nuclides from ^{74}Se to ^{196}Hg lie isolated on the proton-rich side of the s-process growth path (e.g. ^{144}Sm in Fig. 1.8), and are also shielded from r-process production. In order to explain the existence of these nuclides it is nec-

essary to postulate a p-process by which normal r- and s-process nuclei are bombarded by protons at very high temperature ($>2 \times 10^9$ K), probably in the outer envelope of a supernova.

1.3 Radioactive Decay

Nuclear stability and decay is best understood in the context of the chart of nuclides. It has already been noted that naturally occurring nuclides define a path in the chart of the nuclides, corresponding to the greatest stability of

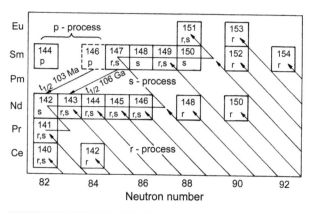

Fig. 1.8 Part of the chart of the nuclides in the area of the light rare earths to show p-, r- and s-process product nuclides. After O'Nions et al. (1979).

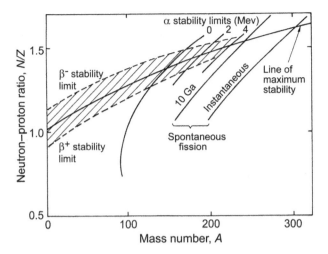

Fig. 1.9 Theoretical stability limits of nuclides illustrated on a plot of N/Z against mass number (A). Lower limits for α emission are shown for α energies of 0, 2 and 4 MeV. Stability limits against spontaneous fission are shown for half-lives of 10 Ga and zero (instantaneous fission). After Hanna (1959).

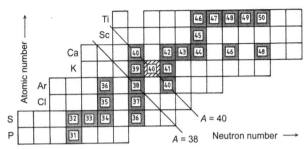

Fig. 1.10 Part of the chart of the nuclides, in coordinates of atomic number (Z) against neutron number (N) in the region of potassium. Stable nuclides are shaded; the long-lived unstable nuclide ^{40}K is hatched. Diagonal lines are isobars (lines of constant mass number, A).

proton/neutron ratio. For nuclides of low atomic mass, the greatest stability is achieved when the numbers of neutrons and protons are approximately equal (N = Z), but as atomic mass increases, the stable neutron/proton ratio increases until N/Z = 1.5. Theoretical stability limits are illustrated on a plot of N/Z against mass number (A) in Fig. 1.9 (Hanna, 1959).

The path of stability is in fact an energy 'valley' into which the surrounding unstable nuclides tend to fall, emitting particles and energy. This constitutes the process of radioactive decay. The nature of particles emitted depends on the location of the unstable nuclide relative to the energy valley. Unstable nuclides on either side of the valley usually decay by 'isobaric' processes. That is, a nuclear proton is converted to a neutron, or vice versa, but the mass of the nuclide does not change significantly (except for the 'mass defect' consumed as nuclear binding energy). In contrast, unstable nuclides at the high end of the energy valley often decay by emission of a heavy particle (e.g. α particle), thus reducing the overall mass of the nuclide.

1.3.1 Isobaric Decay

Different decay processes indicated on Fig. 1.9 can best be understood by looking at example sections of the chart of nuclides. Figure 1.10 shows a part of the chart around the element potassium. The diagonal lines indicate isobars (nuclides of equal mass) which are displayed on energy sections in Fig. 1.11 and Fig. 1.12.

Nuclides deficient in protons decay by transformation of a neutron into a proton and an electron. The latter is then expelled from the nucleus as a negative 'β' particle (β⁻), along with an anti-neutrino (ν̄). The energy released by the

transformation is divided between the β particle and the anti-neutrino as kinetic energy (Fermi, 1934). The observed consequence is that the β particles emitted have a continuous energy distribution from nearly zero to the maximum decay energy. Low-energy β particles are very difficult to separate from background noise in a detector, making the β decay constant of nuclides such as ^{87}Rb very difficult to determine accurately by direct counting (Section 3.1).

In many cases the nuclide produced by β decay is left in an excited state which subsequently decays to the ground state nuclide by a release of energy. This may either be lost as a γ ray of discrete energy, or may be transferred from the nucleus to an orbital electron, which is then expelled

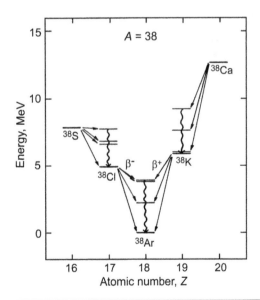

Fig. 1.11 A simple energy section through the chart of nuclides along the isobar A = 38 showing nuclides and isomers. Data from Lederer and Shirley (1978).

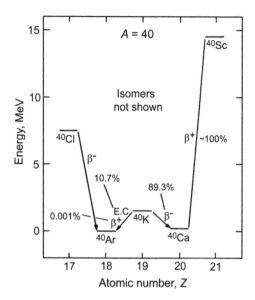

from the atom. In the latter case, nuclear energy emission in excess of the binding energy of the electron is transferred to the electron as kinetic energy, which is superimposed as a line spectrum on the continuous spectrum of the β particles. The meta-stable states, or 'isomers' of the product nuclide are denoted by the superscript 'm', and have half-lives from less than a pico-second up to 241 years (in the case of ^{192m}Ir). Many β emitters have complex energy spectra involving a ground state product and more than one short-lived isomer, as shown in Fig. 1.11. The decay of ^{40}Cl can yield 35 different isomers of ^{40}Ar (Lederer and Shirley, 1978), but these are omitted from Fig. 1.12 for the sake of clarity.

Nuclides deficient in neutrons, e.g. ^{38}K (Fig. 1.11), may decay by two different processes: positron emission and electron capture. Both processes yield a product nuclide which is an isobar of the parent, by transformation of a proton to a neutron. In positron emission a positively charged electron (β^+) is emitted from the nucleus along with a neutrino. As with β^- emission, the decay energy is shared between the kinetic energy of the two particles. After having been slowed down by collision with atoms, the positron interacts with an orbital electron, whereby both are annihilated, yielding two 0.511 MeV γ rays (this forms part of the decay energy of the nuclear transformation).

In electron capture decay (EC) a nuclear proton is transformed into a neutron by capture of an orbital electron, usually from one of the inner shells, but possibly from an outer shell. A neutrino is emitted from the nucleus, and an outer orbital electron falls into the vacancy produced by electron capture, emitting a characteristic X-ray. The product nucleus may be left in an excited state, in which case it decays to the ground state by γ emission.

When the transition energy of a decay route is less than the energy equivalent of the positron mass ($2m_eC^2 = 1.022$ MeV), decay is entirely by electron capture. Thereafter, the ratio β^+/EC increases rapidly with increasing transition energy (Fig. 1.12), but a small amount of electron capture always accompanies positron emission even at high transition energies.

It is empirically observed (Mattauch, 1934) that adjacent isobars cannot be stable. Since ^{40}Ar and ^{40}Ca are both stable species (Fig. 1.10), ^{40}K must be unstable, and exhibits a branched decay to the isobars on either side (Fig. 1.12).

1.3.2 Alpha and Heavy Particle Decay

Heavy atoms above bismuth in the chart of nuclides often decay by emission of an α particle, consisting of two protons and two neutrons (He^{2+}). The daughter product is not an isobar of the parent, and has an atomic mass reduced by four. The product nuclide may be in the ground state, or remain in an excited state and subsequently decay by γ emission. The decay energy is shared between kinetic energy of the α particle and recoil energy of the product nuclide.

The U and Th decay series are shown in Fig. 12.1. Because the energy valley of stable proton/neutron ratios in this part of the chart of the nuclides has a slope of less than unity, α decays tend to drive the products off to the neutron-rich side of the energy valley, where they undergo β decay. In fact β decay may occur before the corresponding α decay.

At intermediate masses in the chart of the nuclides, α decay may occasionally be an alternative to positron or electron capture decay for proton-rich species such as ^{147}Sm. However, α decays do not occur at low atomic numbers because the path of nuclear stability has a Z/N slope close to unity in this region (Fig. 1.1). Any such decays would simply drive unstable species along (parallel to) the energy valley.

An exotic mode of radioactive decay was discovered in the ^{235}U to ^{207}Pb decay series (Rose and Jones, 1984), whereby ^{223}Ra decays by emission of ^{14}C directly to ^{209}Pb with a decay energy of 13.8 MeV. However this mode of decay occurs with a frequency of less than 10^{-9} of the α decay of ^{223}Ra.

1.3.3 Nuclear Fission and the Oklo Natural Reactor

The nuclide ^{238}U (atomic no. 92) undergoes spontaneous fission into two product nuclei of different atomic number, typically ca. 40 and 55 (Zr and Cs), along with various other particles and a large amount of energy. Because the heavy parent nuclide has a high neutron/proton ratio, the daughter products have an excess of neutrons and undergo isobaric decay by β emission. Although the frequency of spontaneous fission of ^{238}U is less than 2×10^{-6} that of α decay, in heavier transuranium elements spontaneous fission is the principal mode of decay. Other nuclides, such as ^{235}U, may undergo fission if they are struck by a neutron. Furthermore, since fission releases neutrons which promote further fission reactions, a chain reaction may be established. If the concentration of fissile nuclides is high enough, this leads to a thermonuclear explosion, as in a supernova or atomic bomb.

In special cases where an intermediate heavy-element concentration is maintained, a self-sustaining but non-explosive chain reaction may be possible. This depends largely on the presence of a 'moderator'. Energetic 'fast' neutrons produced by fission undergo multiple elastic collisions with atoms of the moderator. They are decelerated into 'thermal' neutrons, having velocities characteristic of the thermal vibration of the medium, the optimum velocity for promoting fission reactions in the surrounding heavy atoms. One natural case of such an occurrence is known, termed the Oklo natural reactor (Cowan, 1976; Naudet, 1976).

In May 1972, ^{235}U depletions were found in uranium ore entering a French processing plant and traced to an ore deposit at Oklo in the Gabon Republic of central Africa. In spite of its apparent improbability, there is overwhelming geological evidence that the ^{235}U depletions were caused by the operation of a natural fission reactor at around 1.8 Ga. It appears that in the Early Proterozoic, conditions were such that the series of coincidences needed to create a natural fission reactor were achieved more easily than at the present day.

Uranium dispersed in granitic basement was probably eroded and concentrated in stream-bed placer deposits. It was immobilized in this environment as the insoluble reduced form due to the nature of prevailing atmospheric conditions. With the appearance of blue-green algae, the first organisms capable of photosynthesis, the oxygen content of the atmosphere, and hence river water, probably rose, converting some reduced uranium into more soluble oxidized forms. These were carried down-stream in solution. When the soluble uranium reached a river delta it must have encountered sediments rich in organic ooze, creating an oxygen deficiency which again reduced and immobilized uranium, but now at a much higher concentration (up to 0.5% uranium by weight).

After burial and compaction of the deposit, it was subsequently uplifted, folded and fractured, allowing oxygenated ground waters to re-mobilize and concentrate the ores into veins over 1 m wide of almost pure uranium oxide. Hence the special oxygen fugacity conditions obtaining in the Proterozoic helped to produce a particularly concentrated deposit. However, its operation as a reactor depended on the greater ^{235}U abundance (3%) at that time, compared with the present day level of 0.72%, reduced by α decay in the intervening time (half-life = 700 Ma).

In the case of Oklo, light water (H_2O), must have acted as a moderator, and the nuclear reaction was controlled by a balance between hot water loss by convective heating or boiling, and replacement by cold groundwater influx. In this

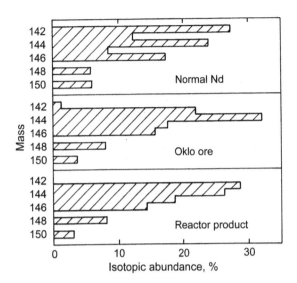

Fig. 1.13 Bar charts of the isotope composition in normal Nd, Oklo ore and reactor fission product waste. Data from Cowan (1976).

way the estimated total energy output (15 000 mega-watt years, representing the consumption of six tons of ^{235}U) was probably maintained at an average of only 20 kW for about 0.8 Ma.

Geochemical evidence for the occurrence of fission is derived firstly from the characteristic elemental abundances of fission products. For example, excess concentrations of rare earths and other immobile elements such as Zr are observed. Alkali metal and alkaline earths were probably also enriched, but have subsequently been removed by leaching. Secondly, the characteristic isotope abundances of some elements can only be explained by fission (Raffenach et al., 1976).

The Nd isotope composition of the Oklo ore is very distinctive (Fig. 1.13). ^{142}Nd is shielded from isobaric decay of the neutron-rich fission products (Fig. 1.8) so that its abundance indicates the level of normal Nd. After correction for an enhanced abundance of ^{144}Nd and ^{146}Nd due to neutron capture by the large-cross-section nuclides ^{143}Nd and ^{145}Nd, Oklo Nd has an isotopic composition closely resembling that of normal reactor fission product waste (Fig. 1.13).

Evidence for a significant neutron flux is also demonstrated by the isotope signatures of actinide elements. For example, the abundant isotope of uranium (^{238}U) readily captures fast neutrons to yield an appreciable amount of ^{239}U, which decays by β emission to ^{239}Np and then ^{239}Pu (Fig. 1.14). The latter decays by α emission with a half-life of 24 ka to yield more ^{235}U, contributing an extra 50% to the 'burnable' fuel, as in a 'fast' breeder reactor ('fast' refers to the speed of the neutrons involved). Because the fission products of ^{239}Pu and ^{235}U have distinct isotopic signatures, it is determined that very little ^{239}Pu underwent neutron-induced

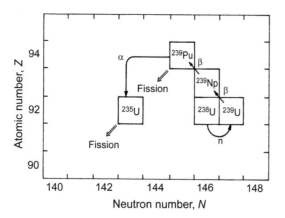

Fig. 1.14 Nuclear reactions leading to 'breeding' of transuranium element fuel in the Oklo natural reactor.

fission before decaying to ^{235}U. Hence, the low flux and prolonged lifetime of the natural reactor are deduced.

1.4 The Law of Radioactive Decay

The rate of decay of a radioactive parent nuclide to a stable daughter product is proportional to the number of atoms, n present at any time t (Rutherford and Soddy, 1902):

$$-\frac{dn}{dt} = \lambda n \qquad [1.1]$$

where λ is the constant of proportionality, which is characteristic of the radionuclide in question and is called the decay constant (expressed in units of reciprocal time). The decay constant states the probability that a given atom of the radionuclide will decay within a stated time. The term dn/dt is the rate of change of the number of parent atoms, and is negative because this rate decreases with time. Rearranging equation [1.1], we obtain:

$$\frac{dn}{dt} = -\lambda n \qquad [1.2]$$

This expression is integrated from $t = 0$ to t, given that the number of atoms present at time $t = 0$ is n_0.

$$\int_{n_0}^{n} \frac{dn}{n} = -\lambda \int_{t=0}^{t} dt \qquad [1.3]$$

Hence:

$$\ln \frac{n}{n_0} = -\lambda t \qquad [1.4]$$

which can also be written as:

$$n = n_0 e^{-\lambda t} \qquad [1.5]$$

A useful way of referring to the rate of decay of a radionuclide is the 'half-life', $t_{1/2}$, which is the time required for half of the parent atoms to decay. Substituting $n = n_0/2$ and $t = t_{1/2}$

into equation [1.5], and taking the natural log of both sides, we obtain:

$$t_{1/2} = \frac{\ln 2}{\lambda} = \frac{0.69315}{\lambda} \qquad [1.6]$$

The number of radiogenic daughter atoms formed, D^*, is equal to the number of parent atoms consumed:

$$D^* = n_0 - n \qquad [1.7]$$

but $n_0 = n\, e^{\lambda t}$ (from equation [1.5]); so substituting for n_0 in equation [1.7] yields:

$$D^* = n\, e^{\lambda t} - n \qquad [1.8]$$

$$D^* = n\left(e^{\lambda t} - 1\right) \qquad [1.9]$$

If the number of daughter atoms at time $t = 0$ is D_0, then the total number of daughter atoms after time t is given as:

$$D = D_0 + n\left(e^{\lambda t} - 1\right) \qquad [1.10]$$

This equation is the fundamental basis of geochronological dating tools.

In the uranium series decay chains, the daughter products of radioactive decay (other than the three Pb isotopes) are themselves radioactive. Hence the rate of decay of such a daughter product is given by the difference between its production rate from the parent and its own decay rate:

$$dn_2/dt = n_1 \lambda_1 - n_2 \lambda_2 \qquad [1.11]$$

where n_1 and λ_1 are the abundance and decay constant of the parent, and n_2 and λ_2 correspond to the daughter.

But equation [1.5] can be substituted for n_1 in equation [1.11] to yield:

$$dn_2/dt = n_{1,\,\text{initial}}\, e^{-\lambda_1 t}\lambda_1 - n_2\lambda_2 \qquad [1.12]$$

This equation is integrated for a chosen set of initial conditions, the simplest of which sets $n_2 = 0$ at $t = 0$. Then:

$$n_2\lambda_2 = \frac{\lambda_1}{\lambda_2 - \lambda_1} n_{1,\,\text{initial}}\left(e^{-\lambda_1 t} - e^{-\lambda_2 t}\right) \qquad [1.13]$$

This type of solution was first demonstrated by Bateman (1910) and is named after him. Recently, Catchen (1984) examined more general initial conditions for these equations, leading to more complex solutions.

1.4.1 Uniformitarianism

When using radioactive decay to measure the age of rocks we must apply the classic principle of uniformitarianism (Hutton, 1788), by assuming that the decay constant of the parent radionuclide has not changed during the history of the Earth. It is therefore important to outline some evidence that this assumption is justified.

The decay constant of a radionuclide depends on nuclear constants, such as a (= elementary charge^2/Plank's constant/velocity of light). Shlyakhter (1976) argued that the neutron capture cross-section of a nuclide is very sensitively dependent on nuclear constants. Because neutron absorbers (such as ^{143}Nd and ^{145}Nd) in the 1.8 Ga Oklo natural reactor

Table 1.1 Summary of decay constants and half-lives of long-lived nuclides.

Nuclide	S & J 1977	'Best value'	Half-life	Reference
^{40}K (^{40}Ar)	5.81 E-11	5.755 E-11	12.04 Ga	Renne et al. (2010)
^{40}K (^{40}Ca)	4.962 E-10	4.974 E-10	1393 Ma	
^{40}K (total)	5.543 E-10	5.549 E-10	1249 Ma	
^{87}Sr	1.42 E-11	1.397 E-11	49.6 Ga	Villa et al. (2015)
^{147}Sm		6.54 E-12	106.0 Ga	Lugmair and Marti (1978)
^{176}Lu		1.86 E-11	37.3 Ga	Nir-El and Lavi (1998)
^{186}Re		1.666 E-11	41.6 Ga	Smoliar et al. (1996)
^{190}Pt		1.477 E-12	469.3 Ga	Brandon et al. (1999)
^{238}U	1.55125 E-10	unchanged	4468.0 Ma	Jaffey et al. (1971)
^{235}U	9.8485 E-10	9.8544 E-10	703.4 Ma	*Mattinson (2010)
^{234}U		2.822 E-6	245.62 ka	Cheng et al. (2013)
^{230}Th		9.1705 E-6	75.58 ka	Cheng et al. (2013)
^{231}Pa		2.116 E-5	32.76 ka	Robert et al. (1969)
^{232}Th		4.9475 E-11	14.01 Ga	Jaffey et al. (1971)

* corrected to ^{238}U/^{235}U = 137.82

give rise to the expected abundance increases in the product isotopes (Fig. 1.13), this constrains nuclear constants to have remained more or less invariant over the last 2 Ga.

The possibility that physical conditions (e.g. pressure and temperature) could affect radionuclide decay constants must also be examined. Because radioactive decay is a property of the nucleus, which is shielded from outside influence by orbital electrons, it is very unlikely that physical conditions influence α or β decay, but electron capture decay could be affected. Hensley et al. (1973) have demonstrated that the electron capture decay of ^{7}Be to ^{7}Li is increased by 0.59% when BeO is subjected to 270 ± 10 kbars pressure in a diamond anvil. This raises the question of whether the electron capture decay of ^{40}K to ^{40}Ar could be pressure dependent, affecting K–Ar ages. In fact this is very unlikely, because at high pressure–temperature conditions at depth in the Earth, K–Ar systems will be chemically open and unable to yield ages at all, while at crustal depths the pressure dependence of λ will be negligible compared with experimental error.

There are a few ways in which the invariance of decay constants has been experimentally verified. Uranium series dates have been calibrated against coral growth bands back to one thousand years ago (Section 12.4.2), the radiocarbon method has been calibrated against tree rings (dendrochronology) back to 12 ka (Section 14.1.4), and K–Ar ages have been calibrated against sea floor spreading rates over periods of several Ma (Heirtzler et al., 1968). In addition, age agreements between systems with very different decay constants also provide supporting evidence.

There have been many attempts to test the consistency of different decay constants by comparing ages for geologically 'well-behaved' systems using different dating methods (e.g. Begemann et al., 2001). As well as verifying the uniformitarian assumption, such geological decay constant comparisons allow the values for poorly known decay schemes to be optimized by comparison with better-known decay schemes. At present, the uranium decay constants are the most well established, and therefore the basis for most comparisons. However, a promising new calibration is the 'astronomical timescale', based on the tuning of glacial cycles to the Earth's orbital motions (Sections 5.1.3, 10.3.1).

The best approach to the improvement of decay constants is new and better measurement. Where this is not possible, a useful alternative is acceptance by the geological community of a 'recommended value'. The most successful application of this procedure was the IUGS Subcommission on Geochronology (Steiger and Jager, 1977). The recommendations served as a useful standard for 20 years but have now been mostly superseded. Therefore, Table 1.1 presents these values where applicable, along with a compilation of the most high-quality decay constant values since that time, by various methods. Full references and further details are given in the appropriate chapters.

References

Atkinson, R. and Houtermans, F. G. (1929). Zur frage der aufbaumoglichkeit der elemente in sternen. Z. Physik **54**, 656–65.

Bateman, H. (1910). Solution of a system of differential equations occurring in the theory of radio-active transformations. Proc. Cambridge Phil. Soc. **15**, 423–7.

Begemann, F., Ludwig, K. R., Lugmair, G. W., et al. (2001). Call for an improved set of decay constants for geochronological use. Geochim. Cosmochim. Acta **65**, 111–21.

Bethe, H. A. (1939). Energy production in stars. Phys. Rev. **55**, 434–56.

Blake, J. B. and Schramm, D. N. (1976). A possible alternative to the r-process. Astrophys. J. **209**, 846–9.

Brandon, A. D., Norman, M. D., Walker, R. J. and Morgan, J. W. (1999). ^{186}Os–^{187}Os systematics of Hawaiian picrites. *Earth Planet. Sci. Lett.* **174**, 25–42.

Burbidge, E. M., Burbidge, G. R., Fowler, W. A. and Hoyle, F. (1957). Synthesis of the elements in stars. *Rev. Mod. Phys.* **29**, 547–647.

Burrows, A. (2000). Supernova explosions in the Universe. *Nature* **403**, 727–33.

Casse, M., Lehoucq, R. and Vangloni-Flam, E. (1995). Production and evolution of light elements in active star-forming regions. *Nature* **373**, 318–19.

Catchen, G. L. (1984). Application of the equations of radioactive growth and decay to geochronological models and explicit solution of the equations by Laplace transformation. *Isot. Geosci.* **2**, 181–95.

Cheng, H., Edwards, R. L., Shen, C. C., *et al.* (2013). Improvements in ^{230}Th dating, ^{230}Th and ^{234}U half-life values, and U-Th isotopic measurements by multi-collector inductively coupled plasma mass spectrometry. *Earth Planet. Sci. Lett.* **371**, 82–91.

Cowan, G. A. (1976). A natural fission reactor. *Sci. Amer.* **235** (1), 36–47.

Fermi, E. (1934). Versuch einer theorie der β-strahlen. *Z. Physik* **88**, 161–77.

Geissel, H., Weick, H., Winkler, M., *et al.* (2003). The Super-FRS project at GSI. *Nucl. Instrum. Meth. in Phys. Res. B.* **204**, 71–85.

Hanna, G. C. (1959). *Alpha-radioactivity*. In: Segre, E. (Ed.) *Experimental Nuclear Physics*, Vol. 3, Wiley, pp. 54–257.

Hansen, P. G. (1987). Beyond the neutron drip line. *Nature* **328**, 476–7.

Heirtzler, J. R., Dickson, G. O., Herron, E. M., Pitman, W. C. and LePichon, X. (1968) Marine magnetic anomalies, geomagnetic field reversals, and motions of the ocean floor and continents. *J. Geophys. Res.* **73**, 2119–36.

Hensley, W. K., Basset, W. A. and Huizenga, J. R. (1973). Pressure dependence of the radioactive decay constant of beryllium – 7. *Science* **181**, 1164–5.

Hutton, J. (1788). Theory of the Earth; or an investigation of the laws observable in the composition, dissolution, and restoration of land upon the globe. *Trans. Roy. Soc. Edin.* **1**, 209–304.

Iben, I. (1967). Stellar evolution within and off the Main Sequence. *Ann. Rev. Astron. Astrophys.* **5**, 571–626.

Jaffey, A. H., Flynn, K. F., Glendenin, L. E., Bentley, W. T. and Essling, A. M. (1971). Precision measurement of half-lives and specific activities of U-235 and U-238. *Phys. Rev. C.* **4**, 1889.

Lederer, C. M. and Shirley, V. S. (1978). *Table of Isotopes* (7th Edn), Wiley.

Lugmair, G. W. and Marti, K. (1978). Lunar initial $^{143}Nd/^{144}Nd$: differential evolution of the lunar crust and mantle. *Earth Planet. Sci. Lett.* **39**, 349–57.

Mattauch, J. (1934). Zur systematiek der isotopen. *Z. Physik* **91**, 361–71.

Mattinson, J. M. (2010). Analysis of the relative decay constants of ^{235}U and ^{238}U by multi-step CA-TIMS measurements of closed-system natural zircon samples. *Chem. Geol.* **275**, 186–98.

Meyer, B. S. (2005). Synthesis of short-lived radioactivities in a massive star. In: Krot, A.N. *et al.* (Eds) *Chondrites and the Protoplanetary Disk*, **341**, 515.

Naudet, R. (1976). The Oklo nuclear reactors: 1800 million years ago. *Interd. Sci.* **1**, 72–84.

Nir-El, Y. and Lavi, N. (1998). Measurement of the half-life of ^{176}Lu. *Applied Rad. Isot.* **49**, 1653–5.

O'Nions, R. K., Carter, S. R., Evensen, N. M. and Hamilton P. J. (1979). Geochemical and cosmochemical applications of Nd isotope analysis. *Ann. Rev. Earth Planet. Sci.* **7**, 11–38.

Panov, I. V. and Janka, H. T. (2009). On the dynamics of proto-neutron star winds and r-process nucleosynthesis. *Astron. Astrophys.* **494**, 829–44.

Raffenach, J. C., Menes, J., Devillers, C., Lucas, M. and Hagemann, R. (1976). Etudes chimiques et isotopiques de l'uranium, du plomb et de plusieurs produits de fission dans un echantillon de mineral du reacteur naturel d'Oklo. *Earth Planet. Sci. Lett.* **30**, 94–108.

Reeves, H. (1974). Origin of the light elements. *Ann. Rev. Astron. Astrophys.* **12**, 437–69.

Renne, P. R., Mundil, R., Balco, G., Min, K. and Ludwig, K. R. (2010). Joint determination of ^{40}K decay constants and $^{40}Ar^*/^{40}K$ for the Fish Canyon sanidine standard, and improved accuracy for $^{40}Ar/^{39}Ar$ geochronology. *Geochim. Cosmochim. Acta*, **74**, 5349–67.

Ringwood, A. E. (1979). Composition and origin of the Earth. In: McElhinny, M. W. (Ed.) *The Earth: its Origin, Structure and Evolution*. Academic Press, pp. 1–58.

Robert, J., Miranda, C. F. and Muxart, R. (1969). Mesure de la periode du protactinium-231 par microcalorimetrie. *Radiochimica Acta* **11**, 104–8.

Rose, H. J. and Jones, G. A. (1984). A new kind of radioactivity. *Nature* **307**, 245–7.

Ross, J. E. and Aller, L. H. (1976). The chemical composition of the Sun. *Science* **191**, 1223–9.

Rutherford, E. and Soddy, F. (1902). The radioactivity of thorium compounds II. The cause and nature of radioactivity. *J. Chem. Soc. Lond.* **81**, 837–60.

Seeger, P. A., Fowler, W. A. and Clayton, D. D. (1965). Nucleosynthesis of heavy elements by neutron capture. *Astrophys. J. Supp.* **11**, 121–66.

Shlyakhter, A. I. (1976). Direct test of the constancy of fundamental nuclear constants. *Nature* **264**, 340.

Smoliar, M. I., Walker, R. J. and Morgan, J. W. (1996). Re-Os ages of group IIA, IIIA, IVA, and IVB iron meteorites. *Science* **271**, 1099.

Steiger, R. H. and Jager, E. (1977). Subcommission on geochronology: convention on the use of decay constants in geo- and cosmo-chronology. *Earth Planet. Sci. Lett.* **36**, 359–62.

Thoennessen, M. (2013). Current status and future potential of nuclide discoveries. *Rep. Prog. Phys.* **76**, 056301.

Thoennessen, M. and Sherrill, B. (2011). From isotopes to the stars. *Nature* **473**, 25–6.

Villa, I. M., De Bièvre, P., Holden, N. E. and Renne, P. R. (2015). IUPAC-IUGS recommendation on the half life of 87 Rb. *Geochim. Cosmochim. Acta* **164**, 382–5.

Wanajo, S. (2007). Cold r-process in neutrino-driven winds. *Astrophys. J. Lett.* **666**, L77.

Weinberg, S. (1977). *The First Three Minutes*. Andre Deutsch, 190 pp.

Chapter 2

Mass Spectrometry

In order to use radiogenic isotopes as dating tools or tracers, they must be separated from non-radiogenic isotopes in a 'mass spectrometer'. The concept of separating positive ions according to their mass was invented by J. J. Thompson, who proposed the use of this technique in chemical analysis (Thompson, 1913). The first type of mass spectrometer to be invented was a 'magnetic sector' instrument, and one of the earliest such instruments was described by Dempster (1918). A cloud of positive ions was accelerated and collimated, then passed through a sector-shaped magnetic field, which separated the ions by mass. Aston (1927) used such an instrument to make the first isotope ratio measurements on lead.

Other methods of mass separation are also possible, such as the 'quadrupole' instrument (Section 2.2.2). A fore-runner of this design was actually the first instrument to be named a mass spectrometer (Smyth and Mattauch, 1932). However, the first high-precision mass spectrometer (Nier, 1940), was based on the magnetic sector approach. Nier's design pioneered so many of the features of modern mass spectrometers (Fig. 2.1) that these have often been called 'Nier-type' instruments.

The normal method of ionizing a sample in a mass spectrometer is simply to heat it under vacuum, either from a gaseous source, as in the case of the rare gases He, Ne, Ar and Xe, or from a purified solid sample loaded on a metal filament. Hence the method is termed thermal ionization mass spectrometry (TIMS). However, thermal ionization is only effective if the sample is first purified by extraction from the matrix that makes up most of the rock sample. Therefore, the normal starting point of precise isotopic measurements by TIMS is chemical separation of the element to be analysed.

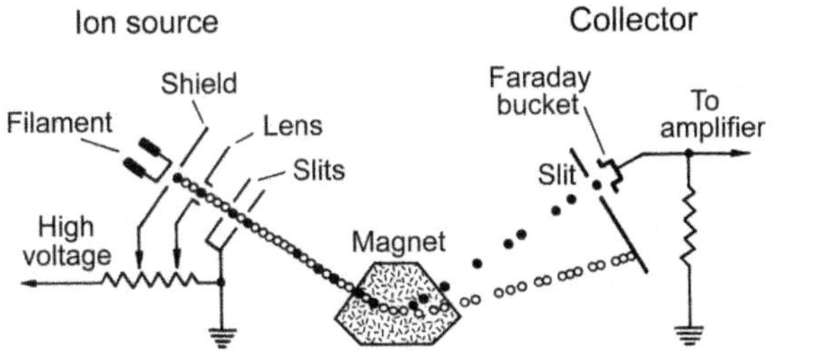

More recently, the use of an inductively coupled plasma (ICP) source has relaxed some of the requirements for chemical purification before mass spectrometry (Section 2.6). In particular, minerals with relatively high concentrations of radiogenic elements may be analysed by *in situ* laser ablation. However, in matrices such as whole rocks, where the concentration of radiogenic isotopes is often low, it is still necessary to pre-concentrate most samples, after which they are introduced into the plasma in aqueous solution. Therefore, methods of dissolution and chemical extraction will now be examined.

2.1 Chemical Purification

Geological samples, which are commonly silicates, are routinely dissolved in concentrated hydrofluoric acid (HF), although some laboratories use perchloric acid as well. Most rock-forming minerals will dissolve in hot concentrated HF at 100 °C (atmospheric pressure). However, certain resistant minerals, such as zircon, must be dissolved in a pressure vessel (hydrothermal bomb), in order to achieve temperatures of up to 200 °C which are necessary for decomposition. The beakers and bomb liners used for dissolution are almost universally made of poly-fluorinated ethylene (PFE). One of the principal manufacturers (Savillex) describes their main product as a variety called PFA (per-fluoro-alkoxy), which has a slightly lower softening point (260 °C) than Teflon ® PTFE (poly-tetra-fluoro-ethylene), but unlike PTFE is translucent.

The conventional bomb dissolution technique for zircons is described by Krogh (1973). A development of this technique for single zircon analysis is to place several 'micro-capsules' carrying different samples into one larger bomb (Parrish, 1987). The micro-capsules are open to vapour transfer of HF,

Fig. 2.1 Schematic illustration of the basic features of a 'Nier type' magnetic sector mass spectrometer. Solid and open circles represent the light and heavy isotopes of an element. After Faul (1966).

but Pb blanks are very low (less than 10 pg), showing that volatile transfer of Pb between samples does not occur.

A major problem which may be encountered after HF dissolution is the formation of fluorides which are insoluble in other 'mineral' acids (e.g. hydrochloric acid, HCl). Refluxing with nitric acid (HNO_3) helps to convert these into soluble forms. Experiments by Croudace (1980) suggested that this process was promoted if additional nitric acid was added before complete evaporation of the HF stage.

If at some stage complete dissolution is not achieved, it may be necessary to decant off the solution and return the undissolved fraction to the previous stage of the process for a second acid attack (Patchett and Tatsumoto, 1980). When complete dissolution has been achieved, the solution may need to be split into weighed aliquots so that one fraction can be 'spiked' with an enriched isotope for isotope dilution analysis (Section 2.5.1) while another is left 'unspiked' for accurate isotope ratio analysis (Section 2.4.3). Following dissolution, the sample is often converted to a chloride for elemental separation, which is normally performed by ion exchange between a dilute acid (eluent) and a resin (stationary phase) contained in quartz or PFE columns.

2.1.1 Ion Exchange Separation

The use of ion exchange as a separation method was developed in support of the Manhattan Project for separating plutonium from fission-product waste (Tompkins *et al.*, 1947). The method uses ion exchange resins, typically in the form of small beads of cross-linked polystyrene with attached sulphonate functional groups that allow the reversible attachment of ions. The partition of cations onto the resin depends on the pH of the solvent and the charge density of the cation. Therefore, when a dissolved rock sample is eluted through an ion exchange 'column' in a solvent such as dilute acid, the matrix elements are eluted in a fixed sequence (Fig. 2.2).

The separation of Rb and Sr is often performed on a cation exchange column eluted with dilute HCl (e.g. Aldrich *et al.*, 1953; Crock *et al.*, 1984; Fig. 2.2a). Columns are normally calibrated in advance of use by means of test solutions, and before each new separation the resin is cleaned by passing sequential volumes of 50% acid and water.

To perform the separation, a small volume of the rock solution is loaded into the column, washed into the resin bed carefully with eluent, and then washed through with more eluent until a fraction is collected when the desired element is released from the resin. For TIMS analysis, the sample is evaporated to dryness, ready to load onto a metal filament. Elements are eluted from the cation column by HCl in roughly the following order: Fe, Na, Mg, K, Rb, Ca, Sr, Ba, rare earth elements (REE) (Crock *et al.*, 1984). This series is defined by increasing partition coefficient onto the solid phase (resin), requiring increasing volumes of eluent to release successive elements.

In the ion exchange separation of Sr, it is critical to remove the 'major elements' of the rock, e.g. Na, K and Ca,

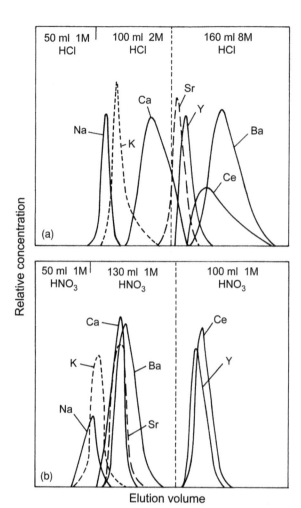

Fig. 2.2 Elution curves for various elements from cation exchange columns: (a) with hydrochloric acid; (b) with nitric acid. Modified after Crock *et al.* (1984).

from the Sr cut. Nitric acid is not effective for this purpose because it does not separate Sr from Ca (Fig. 2.2b). Rb must also be eliminated from Sr because [87]Rb is a direct isobaric interference onto [87]Sr. Small levels of Rb are not a problem in an otherwise clean sample because the Rb burns off before Sr data collection begins. However, the presence of significant Ca in the Sr cut prevents Rb burn-off, causing major interference problems.

The classical ion exchange methods described above are still in widespread use for routine separation of several elements of geochemical interest. However, because they rely on small variations in the partition of each element onto the column, relatively large resin volumes are required. Subsequently, more highly selective resins have been developed that allow the extraction of a particular element such as Sr using a much smaller resin volume (e.g. Horwitz *et al.*, 1992).

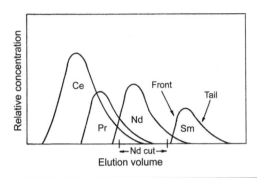

2.1.2 Sm–Nd

Rare earth elements may be separated as a group using cation exchange resin. The preferred medium is dilute nitric acid (Fig. 2.2b), which washes Ba off the column ahead of the rare earth elements (Crock et al., 1984). This is necessary for TIMS analysis because Ba suppresses the ionization of trivalent REE ions. Alternatively, REE-selective resins, such as TRUspec, can be used (Pin et al., 1994).

Because the chemical properties of individual rare earths are so similar, more refined techniques are usually necessary for separations within the REE group. This is particularly important for TIMS analysis because there are several isobaric interferences (e.g. ^{144}Sm interferes onto ^{144}Nd). A few methods are therefore described for separating between the rare earths:

(1) Hexyl di-ethyl hydrogen phosphate (HDEHP)-coated Teflon powder (stationary phase) with dilute mineral acid eluent (Richard et al., 1976). In this technique, which is often called the 'reverse phase' method, light REE are eluted first, whereas in other techniques, heavy REE are eluted first. The reverse phase method yields sharp elution fronts but long tails (Fig. 2.3). It is very effective for removing the Sm interference from ^{144}Nd, and is therefore popular for conventional Sm–Nd analysis. However, substantial Ce is usually present in the Nd cut, so ^{142}Nd cannot be measured accurately. Similarly, the separation between light REE is not good enough for Ce isotope analysis.

(2) Cation exchange resin with hydroxy isobutyric acid (HIBA) eluent (Eugster et al., 1970; Dosso and Murthy, 1980). This method requires more work in preparation of eluent, whose pH must be carefully controlled. It is therefore less popular than (1) for conventional Sm–Nd, but very effective for Ce (Tanaka and Masuda, 1982; Dickin et al., 1987).

(3) For the high-precision analysis of ^{142}Nd, it is necessary to have very low levels of cerium in the purified Nd sample. This can be achieved using the distinct chemistry of the oxidized Ce IV ion, which allows it to be removed from a bulk REE solution by solvent extraction (Rehkamper et al., 1996). The sample is reacted with a strong oxidizing agent such as sodium bromate, after which the cerium can be extracted using HDEHP. Additional ion exchange steps are necessary to remove the added sodium and to separate between Nd and Sm.

(4) An alternative approach, described by Cassidy and Chauvel (1989), is to use high-pressure liquid chromatography (HPLC).

2.1.3 Lu–Hf

A major challenge in the purification of Hf is its separation from Zr, which is much more abundant than Hf and has very similar chemistry. Ti also poses a problem in mafic rocks. Removal of these elements was critical for early Hf analysis by TIMS, because they strongly suppress its thermal ionization. Hence, the method developed by Patchett and Tatsumoto (1980) required three stages of ion exchange column separation. It was also necessary to carry out much of the procedure in an HF medium, due to the risk of deposition of insoluble fluorides.

Although Hf isotope analysis is now performed by MC–ICP–MS, it is still desirable to purify Hf before analysis in order to get good data. A revised method developed by Barovich et al. (1995) also involves three stages of (anion) column chemistry, but only the first is eluted with HF. The second and third columns were eluted principally with H_2SO_4. A modification of this process by David et al. (1999) was able to use only two columns by having a multi-step elution process with several changes in acid type (e.g. H_2SO_4–H_2O–HF–HCl).

2.1.4 Lead

Lead (Pb) and uranium (U) have normally been separated from zircons by elution with HCl on anion exchange columns (Catanzaro and Kulp, 1964). However, this method is not able to separate Pb from the large quantities of Fe in whole-rock samples, which then causes unstable Pb emission during mass spectrometry. A widely used alternative is to elute all elements except Pb from a miniature anion column with dilute hydrobromic acid (Chen and Wasserburg, 1981). The distribution coefficient for Pb onto the resin has a maximum value at just under 1M HBr, and falls off sharply on either side (Fig. 2.4). Elution with HBr effectively strips most elements, including Fe, from the column. However, an alternative is to elute with a mixture of 0.5M HBr and 0.5M HNO_3, which removes transition metals such as Zn more efficiently than HBr alone (e.g. Kuritani and Nakamura, 2002). Finally, Pb is collected with 6M HCl or water. (Pb also has a distribution coefficient maximum onto the solid phase at 2–3M HCl, which falls off in more dilute and more concentrated acid.)

The purity of Pb samples is improved by a second pass through an ion exchange column. This could involve a second pass through the same column (after cleaning). However, a smaller clean-up column gives better results. For example, Kuritani and Nakamura (2002) used a clean-up column

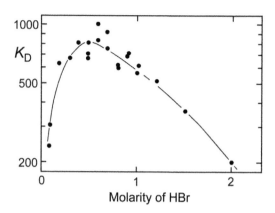

Fig. 2.4 Plot showing the distribution coefficient (K_D) of Pb from dilute HBr onto anion exchange resin as a function of molarity. The curve represents a best fit to the data points. After Manton (1988).

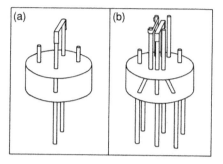

Fig. 2.5 The arrangement of filament ribbons on commonly used single- and triple-filament bead assemblies. Note that only one side filament is shown attached to the 'triple' bead.

with only 10 μl of anion resin. Alternatively, Manton (1988) showed that the distribution coefficient for Pb onto anion resin in dilute HBr is large enough for the absorption of small Pb samples onto a single large resin bead. The Pb is subsequently back-extracted into water. Yields from zircon samples were about 50%, after an equilibration period (with stirring) of about 8 h. Alternatively, the process can be accelerated by agitation in an ultrasonic bath. The somewhat low efficiency may be outweighed by the high purity of the product.

2.1.5 Analytical Blank

Levels of environmental contamination introduced during laboratory procedures are determined by the analysis of 'blanks'. These are measured by taking an imaginary sample through the whole chemical separation procedure, after which the amount of introduced extraneous contamination is measured by isotope dilution (Section 2.5.1). Blank levels must be minimized in all of the chemical procedures described above, but particularly strenuous efforts are necessary to limit Pb contamination in U–Pb analysis, due to the small sample size (e.g. single zircons) and its relatively high concentration in the environment.

The minimum laboratory requirements to maintain low blanks would be an overpressure air system, sub-boiling distillation of all reagents in quartz or PFE stills, and evaporation of all samples under filtered air. For typical terrestrial whole-rock samples, acceptable total chemistry blanks would normally be less than a nanogram (ng = 10^{-9}g) for Pb, Sr and Nd. This is necessary because the samples to be analysed often contain less than 1 microgram (μg = 10^{-6}g) of the element of interest. However, in the analysis of very small samples (e.g. single zircons) blanks of a few picograms (pg = 10^{-12}g) are necessary (e.g. Roddick et al., 1987), since the Pb sample itself may weigh less than 1 ng.

2.2 Ion Sources

As noted above, the traditional start of mass spectrometric analysis is to heat the sample under vacuum, leading to thermal ionization. With the exception of the rare gases, thermal ionization is normally achieved by loading a solid deposit of the sample onto a metal filament, which can then be subjected to resistive heating. However, the manner by which the sample is deposited and then heated has a major effect on the efficiency of the analysis. Therefore these procedures will be discussed in some detail.

2.2.1 Thermal Ionization

For some elements such as Sr, stable emission of metal ions is achieved from a salt deposited directly onto a single metal filament (Fig 2.5a), usually tantalum (Ta). The loading procedure involves evaporating the salt solution onto the filament before insertion into the vacuum system. The sample is often loaded in phosphoric acid, which seems to (a) displace all other anion species to yield a uniform salt composition, (b) destroy organic residues (such as ion exchange resin) mixed with the sample and (c) glue the sample to the filament. During mass spectrometric analysis, the filament current is raised by means of a stabilized power supply to yield a temperature where simultaneous volatilization and ionization of the sample occurs.

However, for many elements, stable volatilization and ionization of metal species does not occur at the same temperature. This problem was noted by Ingram and Chupka (1953), who first proposed the use of multiple source filaments (Fig. 2.5b). In this configuration, one or more filaments bearing the sample load can be heated to the optimum temperature for stable volatilization, while another hotter filament can be used to ionize the atomic cloud by bombarding it with electrons.

This method is particularly effective for REE analysis, where the sample is usually loaded onto one or both of the Ta side filaments of a triple-filament bead. These are held at

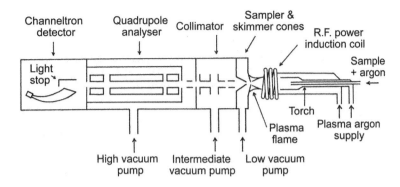

Fig. 2.6 Schematic illustration of an ICP–MS instrument with quadrupole analyser. Modified after Houk (1986).

a moderate temperature (ca. 1400 °C) where REE volatilization is most stable. The centre filament (usually Re) is held at a much higher temperature (ca. 2000 °C), which promotes ionization of the metal vapour. To some extent the ratio of metal to oxide species can be controlled by the centre filament temperature, which may help to suppress REE isobaric interferences. The properties of the REE under such conditions vary from light to heavy rare earths. La and Ce tend to form oxides unless extremely high centre filament temperatures are used, while heavier REE tend to form the metal species (Hooker *et al.*, 1975; Thirlwall, 1982).

Uranium and thorium may also be analysed by the triple-filament technique. Again, the temperature of the centre filament controls the metal/oxide ratio of the emitted ions (Li *et al.*, 1989). The triple-filament method was also used in the first successful analysis of Hf (Patchett and Tatsumoto, 1980). However, Hf analysis is now exclusively performed by ICP–MS, and this method is increasingly used for U and Th (Section 2.6.3).

An alternative to TIMS analysis with multiple filaments is to use special conditions to control the evaporation–ionization behaviour from a single filament. For example, in the case of Pb, the sample is usually loaded on a rhenium filament in a suspension of silica gel (Cameron *et al.*, 1969). This is thought to form a blanket over the sample which effectively retards Pb volatilization so that the filament can be raised to a higher temperature (where Pb fractionation is more reproducible) without burning off the sample uncontrollably.

Early Nd isotope determinations (Lugmair *et al.*, 1975; DePaolo and Wasserburg, 1976) were made using NdO^+ ions, whose emission from a single filament source was promoted by relatively high oxygen pressures in the mass spectrometer source housing. Some workers have continued to use this method, rather than the more popular multiple-filament method, since it can yield higher efficiency. Oxygen may be bled into the source in minute amounts to increase oxide emission. Alternatively, loading with silica gel may achieve the same objective without degrading source vacuum (Thirlwall, 1991a).

2.2.2 Plasma Source Mass Spectrometry

Technically, plasma source mass spectrometry is a kind of thermal ionization mass spectrometry, because it relies on heating in a gas plasma to achieve ionization of the sample. However, for practical purposes, the term 'TIMS' is usually restricted to the case where thermal ionization is achieved by a heated filament under vacuum. Therefore, plasma source mass spectrometry is regarded as a distinct technique.

Plasma source mass spectrometry was invented in the late 1970s when the inductively coupled plasma (ICP) was first attached to a mass spectrometer (MS) to produce the ICP–MS (Houk *et al.*, 1980). The ICP source consists of a plasma torch, made by using a radio frequency (RF) generator to induce intense eddy currents in a stream of ionized argon gas. The RF generator transmits about a kilo-watt of power into the plasma, raising its temperature to about 5000 °C and causing very efficient ionization of most elements (Houk, 1986). Furthermore, the extreme temperature of the plasma ensures that nearly all of these ions are monatomic ions, ideal for mass spectrometric analysis.

When the ICP–MS was first conceived (Dawson, 1976), it was found most convenient to use a quadrupole mass spectrometer as the analyser (Fig. 2.6). This analyser has four rods which are arranged parallel to the path of the ion beam. When alternating voltages with variable frequencies are applied to these rods, the analyser can be tuned to allow a certain mass of ions to pass through to the collector, while all other masses are de-focussed. By rapidly changing the frequencies on the rods, the quadrupole analyser can be made to rapidly 'scan' the mass spectrum from mass zero to 250, allowing the detector (usually an ion multiplier) to measure the relative abundance of every mass in the 'spectrum'. A detailed review is given by Potts (1987).

The main technical breakthrough in the development of ICP–MS was the physical feat of feeding a plasma at 5000 °C and atmospheric pressure into a mass spectrometer whose analyser pressure is ca. 10^8 times lower (10^{-5} mbar). This was achieved by firing the plasma at a two-stage water-cooled orifice ('sampler' and 'skimmer' cones), with continuous pumping of the intermediate space by a mechanical pump

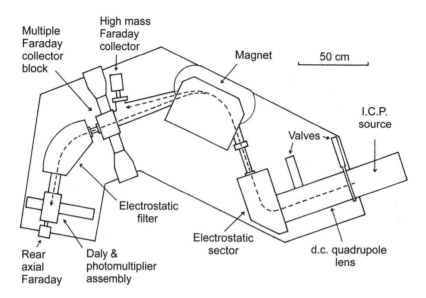

Fig. 2.7 Simplified plan view of the VG Elemental 'Plasma 54' instrument. After Halliday *et al.* (1998).

(Fig. 2.6). Subsequent technical developments in ICP–MS over the next 20 years have mainly involved greatly increased efficiency in the sampling of the plasma by the mass spectrometer. This has allowed ICP–MS to reach remarkable sensitivity, with detection limits as low as parts per trillion (pico-grams per gram).

Plasma sources have an intrinsic instability that causes a slight 'flicker' in the intensity of the plasma. This causes instability in the ion beam, which limits the precision of isotope ratio measurements. Since the quadrupole analyser is restricted to single collector analysis, the best precision that can be achieved is around 1%, which is insufficient for most radiogenic isotope systems. Therefore, to apply ICP–MS to isotope ratio measurement it was necessary to introduce multiple collection techniques to cancel out the instability of the source.

To perform multiple collector mass spectrometry (MC–MS) with a plasma ion source, it was necessary to link the ICP with a magnetic sector analyser of the type normally used in TIMS instruments (Section 2.4.1). This involved several technical challenges (Walder and Freedman, 1992). These technical issues will be described here, whereas the application of MC–ICP–MS to specific isotope systems will be described below (Sections 2.6, 2.7).

The first challenge was an intensification of the pressure differential problem encountered with conventional ICP–MS. Since the magnetic sector analyser has a much longer ion path, it requires a much better vacuum than the quadrupole analyser. This requires additional pumping stages to gradually step the pressure down from the plasma at atmospheric pressure to an analyser approaching 10^{-9} mbar (10^{-12} atmospheres).

A second problem is that magnetic sector mass spectrometry requires a large accelerating voltage to raise ions to the high velocities where magnetic separation is efficient. In order to maintain the normal practice of keeping the analyser assembly at electrical ground, the ion source must be at a few kV positive potential. For MC–ICP–MS, this means holding the sampler and skimmer cones at high voltage. However, the plasma itself need not be at high voltage, because analyte ions are carried from the plasma into the vacuum in a supersonic gas flow that is largely neutral. Only when passing through the skimmer are electrons stripped away from the gas flow, generating a residual beam of positive ions (Niu and Houk, 1996).

A third problem is that sampling of the plasma by the sampler and skimmer cones generates an ion beam with a circular cross-section. This must be converted to a beam with rectangular cross-section to fit the entrance slits of the Faraday collectors. This is typically achieved using a DC quadrupole lens between the plasma source and the beam collimator (e.g. Fig. 2.7).

A final problem is the large range of ion energies generated by the ICP source. Since the magnetic sector is really a momentum analyser rather than purely a mass analyser, this would cause the magnet to de-focus the ion beam for each mass. This is most often overcome using a double-focussing instrument with electrostatic and magnetic sectors (Nier–Johnson geometry, Fig. 2.7). The electrostatic sector is set to cancel out the dispersion of the magnetic sector, so that each ion beam comes back into sharp focus at the collector array (Burgoyne and Hieftje, 1996). To accommodate multiple Faraday collectors, the double-focussing analyser must have 'forward' geometry (electrostatic before magnetic). However, the instrument shown in Fig. 2.7 has a second electrostatic filter to reduce scattering from large ion beams onto small beams measured using the Daly multiplier detector (see below).

MC–ICP–MS instruments were developed by three separate companies, VG Elemental, Micromass and Nu Instruments; the first two of which were later acquired by Thermo-Scientific. The early instruments had various distinctive features which illustrate some of the alternative strategies for dealing with the issues outlined above. For example, the VG Elemental machine (Fig. 2.7) had an intermediate focussing lens between the electrostatic and magnetic sectors. This had the effect of inverting the ion beam, requiring an S-shaped layout of the electrostatic and magnetic sectors (Walder and Freedman, 1992). In contrast, Nu instruments introduced an electrical zoom lens before the collectors, so that their double-focussing geometry had the usual C-shaped layout (Halliday *et al.*, 2000).

An alternative solution to the ion energy problem, offered by Micromass, was to immerse the hexapole lens of the extraction system in a gas reservoir, forming a collision cell that acted to homogenize the kinetic energy of the beam. This avoided the need for an electrostatic analyser, and is useful for destroying molecular ions formed between analyte elements and major plasma ions such as argon, oxygen or nitrogen (Halliday *et al.*, 2000). This is significant for the analysis of isotope systems with masses <80 a.m.u. (atomic mass units).

2.3 Mass-dependent Fractionation

Instrumental mass-dependent fractionation occurs in the ion sources of both TIMS and ICP–MS instruments, leading to similar analytical artefacts. However, it appears that this process has different causes in the two kinds of instrument, leading to different types of behaviour when studied in detail. The following discussion will examine the similarities and differences between these fractionation processes.

2.3.1 Mass Fractionation in TIMS

The process of volatilization and ionization involved in TIMS requires the breaking of chemical bonds, but the strength of these bonds is mass dependent. Therefore, excitation of the sample leads to mass-dependent fractionation, which can be understood by approximating the chemical bond between two atoms as a harmonic oscillator.

The energy of a molecule (or part of an ionic lattice) decreases with decreasing temperature, but at absolute zero it has a certain finite value called the zero point energy, equal to 0.5 $h\nu$ (where h is Plank's constant and ν is the vibrational frequency of the bond). A bond involving the light isotope of an element has a higher vibration frequency and hence a higher zero point energy than one involving a heavier isotope, as illustrated in Fig. 2.8. The difference in bond energies diminishes as temperature rises, but still persists. Because the potential energy well of the bond involving the

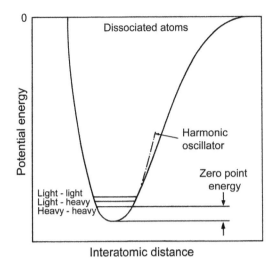

Fig. 2.8 Schematic diagram of potential energy against bond length for a hypothetical molecule made of two isotopes, based on the 'harmonic oscillator' model.

lighter isotope is always shallower than for the heavier, the bond with the lighter isotope is more readily broken. Hence the lighter isotope is preferentially released from the hot filament, causing isotopic fractionation.

As the continual process of fractionation acts on a solid sample loaded on a hot filament, it starts to 'use up' the lighter isotope on the filament. This causes the isotopic composition of the remaining sample to get progressively heavier, termed the 'reservoir effect'. Eberhardt *et al.* (1964) showed that this process follows a Rayleigh fractionation law, and can lead to changes in the measured isotope ratio that are even bigger than the original fractionation effect (Fig. 2.9). This could yield totally unacceptable errors of up to 1% in measured isotope ratio. Therefore corrections for instrumental fractionation are clearly necessary. For elements with two or more non-radiogenic isotopes, an internal normalization for such mass-dependent fractionation can be performed (see below).

This is not the case for Pb isotope analysis by TIMS, or in the isotope dilution analysis of Rb, so an external correction must be used. This depends on achieving uniform fractionation behaviour between standards and unknowns, so that an across-the-board correction can be made to all runs. In the case of Pb, the use of a silica gel blanket on the filament achieves this objective by reducing the magnitude of fractionation processes drastically, from a previous between-lab variation of ca. 3% to a present variation of ca. 2 per mil. This improvement is partly due to the higher filament temperatures possible using silica gel. Because bond energy levels become closer with increasing temperature, the magnitude of isotopic fractionation falls with increasing temperature. However, for this technique to work well, the Pb sample has to be well purified during the chemical separation (Section 2.1.4). In the analysis of uranium, fractionation effects may

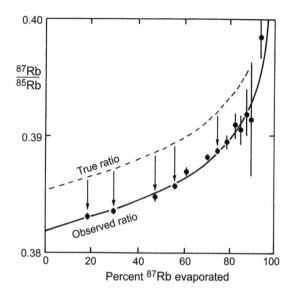

Fig. 2.9 Effect of within-run fractionation, over time, on a sample of natural rubidium undergoing isotopic analysis. Points are observed ratios; dashed line indicates schematically the actual composition of Rb on the filament. Data from Eberhardt *et al.* (1964).

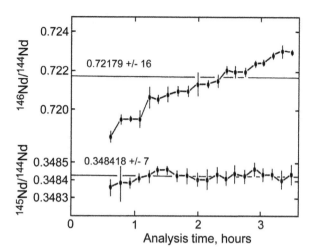

Fig. 2.10 Plot of raw $^{146}Nd/^{144}Nd$ ratios and fractionation-corrected $^{145}Nd/^{144}Nd$ ratios (normalized to $^{146}Nd/^{144}Nd = 0.7219$) for a single mass spectrometer run. Each point is a mean of 10 scans of the mass spectrum, while horizontal lines are grand means. After Noble (pers. comm.).

similarly be reduced by running at high temperature as the oxide. Alternatively, analysis as the metal ion produces larger but relatively consistent degrees of fractionation, which can be corrected by comparison with standard runs.

Elements with two or more non-radiogenic isotopes allow an internal mass fractionation correction. For strontium, the fractionation of $^{87}Sr/^{86}Sr$ is monitored using the $^{86}Sr/^{88}Sr$ ratio, which is known to be constant throughout the Earth, and is taken to be 0.1194 by international convention. This value cannot be measured absolutely, but was originally estimated from the average beam composition half-way through very many TIMS runs. The deviation of observed $^{86}Sr/^{88}Sr$ from 0.1194 at each point through the run is divided by the difference between the two masses ($\Delta_{mass} = 2.003$) in order to calculate a fractionation factor per mass unit:

$$F = \frac{\frac{(^{86}Sr/^{88}Sr)_{obs}}{0.1194} - 1}{\Delta_{mass}} \qquad [2.1]$$

This fractionation factor is then used to correct the observed (raw) $^{87}Sr/^{86}Sr$ ratio, for which $\Delta_{mass} = 1.003$:

$$\left(\frac{^{87}Sr}{^{86}Sr}\right)_{true} = \left(\frac{^{87}Sr}{^{86}Sr}\right)_{obs} \cdot (1 + F\,\Delta_{mass}) \qquad [2.2]$$

This has the effect of improving the within-run precision of the $^{87}Sr/^{86}Sr$ ratio from ca. 1% to better than 0.01%. Neodymium metal analyses are similarly normalized for fractionation (Fig. 2.10) using an internationally agreed value of $^{146}Nd/^{144}Nd = 0.7219$ (O'Nions *et al.*, 1979). However, Nd oxide analyses are normalized to different values (Wasser-

burg *et al.*, 1981) which are incompatible with the Nd metal normalizing value.

The fractionation correction described above is usually called the linear law, but the power law (Wasserburg *et al.*, 1981; Thirlwall, 1991b) is effectively identical. Both of these laws assume that fractionation is proportional to mass difference only, and is independent of the absolute masses of the fractionating species. In other words, fractionation per mass unit is constant. However, this is only an approximation to the real evaporation process, where fractionation per mass unit must vary inversely with the absolute masses of the evaporating species. Russell *et al.* (1978) first observed a break-down of the linear law in isotopic analysis of the 'light' element Ca. To remedy this, they introduced an 'exponential' law, where the fractionation factor depends also on the mass of the evaporating species. This gave a better fit to Ca isotope data than the linear law (Fig. 2.11).

These problems are much less severe for Sr and Nd isotope analysis because of their heavier masses. However, Thirlwall (1991b) found small deviations from linear law behaviour in a large data set of Sr standard analyses. This is revealed by a correlation between normalized $^{87}Sr/^{86}Sr$ and average observed $^{86}Sr/^{88}Sr$ ratios for complete runs (Fig. 2.12). Thirlwall found that he could eliminate the correlation by retrospectively applying an exponential law correction to the data. This is described as follows:

$$\left[\frac{(^{87}Sr/^{86}Sr)_{norm}}{(^{87}Sr/^{86}Sr)_{corr}}\right]^{\ln(86/88)} = \left[\frac{(^{86}Sr/^{88}Sr)_{obs}}{0.1194}\right]^{\ln(87/86)} \qquad [2.3]$$

where 'norm' and 'corr' refer to products of linear normalization and exponential correction, and where 86, 87 and 88 are the actual masses of evaporating ions.

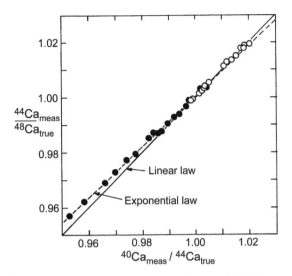

Fig. 2.11 Plot of measured/true $^{44}Ca/^{48}Ca$ versus $^{40}Ca/^{44}Ca$ ratios showing fit of linear and exponential fractionation laws to typical data from two runs. After Russell et al. (1978).

This model assumes that strontium evaporates from the filament as the species 'Sr'. On the other hand, Habfast (1983) suggested that a strontium sample on a rhenium filament might evaporate as a species such as $SrReO_4$ rather than atomic Sr. In that case, the 'apparent mass' of ^{88}Sr which should be used in the exponential correction would be ca. 330 rather than 88. Under such conditions, application of the exponential law with a mass of 88 might produce a worse fit

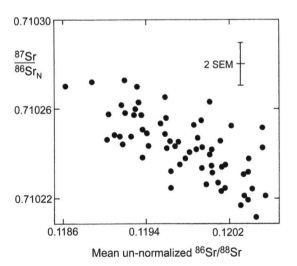

Fig. 2.12 Plot of fractionation-corrected $^{87}Sr/^{86}Sr$ ratios against mean un-corrected $^{86}Sr/^{88}Sr$ ratios for measurements of the SRM987 standard over a period of several months. Error bar indicates average within-run precision. After Thirlwall (1991b).

to the true fractionation behaviour than the linear law. However, Thirlwall's data suggest that evaporation from a tantalum filament does occur as the metal species, so the exponential model is an improvement over the linear model.

2.3.2 Mass Fractionation in MC–ICP–MS

The very high temperature of the ICP source leads to effectively 100% atomization and ionization of most elements (Houk, 1986) so there is no basis for mass-dependent fractionation in the plasma. On the other hand, strong massdependent matrix effects were seen in early quadrupole ICP–MS studies (e.g. Gillson et al., 1988). These effects were first observed between high and low mass species of different elements, but they appeared to be largely physical (mass dependent) rather than chemical in origin. This conclusion is supported by the observation that ICP–OES (optical emission spectrometry) has very low matrix effects.

Various proposals have been made for the site of these effects, but they all concern the *sampling* of the plasma by the mass spectrometer and its shaping into an ion beam. Gillson et al. (1988) suggested that mass fractionation is caused by space-charge effects that begin as the beam passes through the skimmer and loses its electrons, generating a residual beam of positive ions. As a result, the high density of argon ions and analyte ions in this region leads to mutual repulsion between them. This causes the beam to spread out, but the effect is much larger for lighter ions due to their lower momentum. This space-charge effect is translated into a mass-dependent fractionation effect when the outer edges of the diffuse beam are clipped by beam-confining windows in the ion extraction system (Marechal et al., 1999).

In view of the very different origins of mass fractionation in TIMS and ICP–MS analysis, it is surprising that the effects on isotopic analysis can be quite similar. Mass fractionation in ICP–MS analysis may be up to an order of magnitude larger than TIMS (ca. 1% per mass unit rather than 0.1%), but the reproducibility of the effects is comparable in the two techniques. However, because mass fractionation in MC–ICP–MS reflects physical processes during ion extraction rather than chemical processes during volatilization, the fractionation monitor does not necessarily have to be the same element as the analyte. This applies particularly to Pb isotope analysis by MC–ICP–MS, where the two isotopes of thallium can be used to correct Pb isotope data (Section 2.5.3).

Detailed studies of mass fractionation during MC–ICP–MS analysis have shown that the process obeys the same exponential law as in TIMS analysis (e.g. White et al., 2000; Vance and Thirlwall, 2002). In addition, both methods can give rise to variations outside predicted within-run precision for fractionation-corrected standard data. This phenomenon is well illustrated in the analysis of Nd isotopes, for which the two methods yield comparable precision. For example, Thirlwall (1991b) demonstrated a correlation between fractionation-corrected $^{143}Nd/^{144}Nd$ and $^{142}Nd/^{144}Nd$ measured by TIMS over a period of months (Fig. 2.13). Subsequent work on MC–ICP–MS showed similar behaviour (Vance

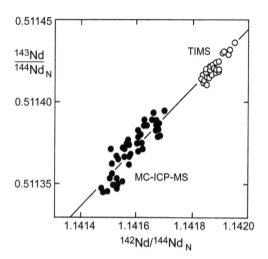

Fig. 2.13 Plot of fractionation-corrected $^{143}Nd/^{144}Nd$ against $^{142}Nd/^{144}Nd$ for analyses of a laboratory standard over a period of several months using normalization to $^{146}Nd/^{144}Nd = 0.7219$. (○) = TIMS data; (●) = MC–ICP–MS data. After Vance and Thirlwall (2002).

and Thirlwall, 2002), but because mass fractionation in ICP–MS is an order of magnitude larger, the effect is somewhat bigger.

Vance and Thirlwall showed that this problem could be solved using two alternative approaches. One is to apply a post analysis correction to $^{143}Nd/^{144}Nd$ ratios using the information from the $^{142}Nd/^{144}Nd$ ratio, as in Fig. 2.13. An alternative approach is to choose masses for the fractionation monitor that bracket the masses to be corrected. For example, the mean mass of ^{143}Nd and ^{144}Nd is 143.5, which is the same as the mean mass of ^{142}Nd and ^{145}Nd. Therefore, $^{142}Nd/^{145}Nd$ represents the ideal monitor ratio for fractionation correction of $^{143}Nd/^{144}Nd$ ratios. This approach is only possible if Ce interferences on ^{142}Nd are low, and is therefore not feasible for TIMS analysis using the popular 'reverse phase' Nd separation (Section 2.1.2). However, for most TIMS analyses the fractionation problem can be avoided by collecting data near a $^{146}Nd/^{144}Nd$ ratio of 0.7219. For Nd analysis by MC–ICP–MS it is necessary to take account of this problem in order to achieve reproducibilities comparable with TIMS. However, Ce isobaric interferences can be accurately corrected in MC–ICP–MS analysis, so ^{142}Nd can be used as one of the fractionation monitors.

2.4 Magnetic Sector Analysis

In a typical 'magnetic sector' instrument (Fig. 2.1) the nuclides to be separated are ionized under vacuum and accelerated through a high potential (V) before passing between the poles of a magnet. A uniform magnetic field (H) acting

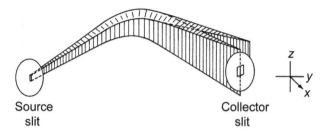

Fig. 2.14 Schematic diagram of the envelope of a double nuclide beam from the source slit to the collector slit in a Nier-type mass spectrometer.

on particles in the ion beam bends them into curves of different radius (r) according to the following equation:

$$r^2 = \frac{m}{e} \cdot \frac{2V}{H^2} \qquad [2.4]$$

where m/e = mass/charge for the ion in question. Since most of the ions produced by TIMS or ICP sources are single charged, the different isotopes in the sample will be separated into a simple spectrum of masses. The relative abundance of each mass is then determined by its corresponding ion current, captured by a Faraday bucket or a multiplier detector.

2.4.1 Ion Optics

The ion optic properties of an instrument determine how the cloud of ions generated in the source is accelerated, focussed into a beam, separated by the magnetic field, and collected for measurement. Correct ion optic alignment is essential to obtain reliable results, because if part of the ion beam hits an obstruction, different masses may be affected to different degrees, leading to a bias in the results.

Most older mass spectrometers followed Nier's (1940) design (Fig. 2.1). From the filament (several kV positive), the beam traversed a series of focussing source plates ('collimator stack'). These plates were at progressively lower potential and brought the beam to a principal focus at the source slit (zero volts). Thereafter, the beam slowly diverged (Fig. 2.14). In the y direction, the magnet brought each nuclide (isotope) beam back into focus at the primary collector slit, which was wide enough to let the whole of one nuclide beam through into the collector. This focussing effect of a sector-shaped magnet was already understood by Aston (1919).

To bring a heavier nuclide into the collector, the magnetic field may be increased, or the momentum of the ions may be reduced by lowering the high voltage (HV) potential across the collimator stack. If the whole of one nuclide beam is focussed into the collector slit in the y direction, an apparently flat-topped peak is produced when magnetic field or HV is varied to sweep the mass spectrum across the collector slit (e.g. Fig. 2.15). In practice, the magnetic field rather than the HV is normally switched to bring different nuclide

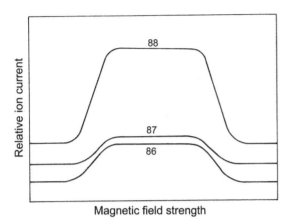

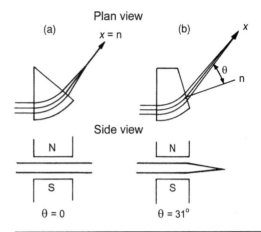

Fig. 2.15 The appearance of flat-topped 'peaks' when strontium 88, 87 and 86 nuclide beams are swept across a triple (Faraday) collector system by varying magnetic field strength.

Fig. 2.16 The effect of different shaped pole faces in generating fringing fields which cause focussing of the ion beam in the y and z planes. (a) Exit pole piece perpendicular to beam yields short focal length in the x direction but no focussing in z. (b) Normal to exit pole face (n) is at an angle θ to the beam direction, yielding longer focal length in x, and also z focussing.

beams into the collector, because the field can be more precisely monitored and controlled, using a 'Hall probe'. This is used to sense the field strength and adjust the magnet power supply in a feed-back loop ('field control'). The magnet (whose pole pieces are perpendicular to the beam) does not focus in the z direction, and in this direction the beam is 'clipped' by baffle plates.

This type of magnet design has a disadvantage, because the process of switching from one mass to another changes 'fringing fields' which are generated by the ends of the magnet poles. The change in these fringing fields may cause slight convergence or divergence of the beam, so that different amounts of different nuclide beams are clipped by the collector slit, and a slight bias is introduced to the beam current reaching the collector according to whether the magnet is switching 'up-mass' or 'down-mass'.

Instruments built since 1980 feature refinements in the design of the magnet pole pieces, based on theoretical work by Cotte (1938). If the pole pieces are set at a slightly oblique angle to the beam, the fringing fields generated by the magnet have the effect of focussing the beam in the z direction (Fig. 2.16). However, the focussing effect in the y direction is weakened, so the distance from the magnet exit pole to the principal focus in the y plane is increased. Therefore this design is referred to as 'extended geometry'. Ion optics in this type of machine are shown in Fig. 2.17.

The extended geometry configuration has three advantages:

1. Because the whole of the z-focussed beam can pass through the collector slit, the transmission of the machine is improved (defined as the number of ions in the source required to yield each ion at the collector).
2. Small variations in fringing field do not cause signal bias, so accuracy is improved.

3. Extended geometry increases the distance between nuclide beams at the collector, so that multiple Faraday buckets can be more easily accommodated. Hence a magnet with 30 cm radius yields a beam separation equivalent to a magnet with 54 cm radius.

If one of the magnet pole faces (e.g. the entry pole) is made slightly convex, this changes the normally oblique focal plane of nuclide beams in the y direction into a flat plane perpendicular to the beams (Fig. 2.18). This facilitates the installation of multiple collectors, whose spacing can then be adjusted to fit ion beams one atomic mass unit apart in any part of the spectrum. However, a more complex adjustable multiple collector configuration can be constructed on the oblique focal plane. At higher mass numbers the spacing of the collectors becomes closer and closer until their outer grounded screens are actually touching during uranium analysis.

A very high vacuum throughout the ion path is essential, otherwise the ion beam becomes scattered, particularly

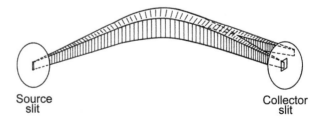

Source slit

Collector slit

Fig. 2.17 Schematic diagram of the ion optics of an extended geometry machine between the source and collector slits (compare with Fig. 2.14).

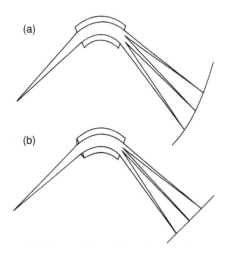

Fig. 2.18 Effect of magnet entry pole face shape. (a) Flat pole face yields oblique curved focal plane at collector; (b) convex pole face yields flat focal plane perpendicular to flight path.

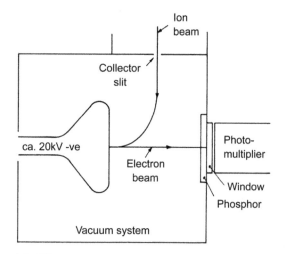

Fig. 2.19 Schematic diagram of a Daly detector showing the means of amplification of an incoming positive ion beam. After Daly (1960).

towards the low mass side, by inelastic collisions with air molecules. Such beam scattering becomes serious at analyser pressures $>10^{-8}$ mbar. This causes the formation of a tail from one peak which may interfere with the adjacent nuclide. The problem is particularly severe in the case of a small peak down-mass from a very large peak. For example, interference by ^{232}Th onto ^{230}Th may be severe in silicate rocks with 232/230 ratios approaching 10^6. The magnitude of interference by a peak on a position one a.m.u. lower is called the 'abundance sensitivity' of the instrument (measured in ppm of the peak size). A typical specification for a single-focussing TIMS machine with analyser vacuum $<5 \times 10^{-9}$ mbar is 2 ppm at 1 a.m.u. from ^{238}U.

If a very high abundance sensitivity is essential, it can be obtained by adding a type of ion energy filter between the magnet and collector, thereby creating a double-focussing machine. The ion energy filter has the effect of removing ions with unusually high or low energy (= velocity). Thus, ions which have suffered a collision, and therefore lost energy, should be weeded out. Three types of filter which have been applied to this task are electrostatic, quadrupole and ion retardation types. They typically result in a 10- to 100-fold improvement in abundance sensitivity.

2.4.2 Detectors

Ion beams in mass spectrometry normally range up to ca. 10^{-10} amps. For beams as small as 10^{-13} amps, the most suitable detector is the Faraday bucket. This is connected to electrical ground (Fig. 2.1) via a large resistance (e.g. 10^{11} ohm). Electrons travel from ground through this output resistor to neutralize the ion beam, and the potential across the resistor is then amplified and converted into a digital signal. A typical ion beam of 10^{-11} amps then generates a potential of 1 volt, converted to, say, 100 000 digital counts. Traditionally,

an indefinite lifetime has been assumed for Faraday buckets. However, the very narrow buckets in modern multi-collector arrays were found to quickly become coated inside with sample debris if large beams were analysed. This degraded ion beam measurements by allowing stray beams to escape out from the Faraday. This was solved by putting absorbant charcoal blocks in the buckets.

For ion beams smaller than ca. 10^{-13} amps, the electrical noise of the Faraday amplifier becomes significant relative to the signal size, so that some form of signal multiplication is necessary. One of the most useful approaches was pioneered by Daly (1960). In the Daly detector, ions passing into the collector are attracted by a large negative potential (e.g. 20 kV). Collision of each ion with the polished electrode surface (Fig. 2.19) yields a secondary electron shower. When this impinges on a phosphor, the resulting light pulse is amplified by a photo-multiplier (situated behind a glass window, outside the vacuum system). In the analogue mode this system can have a gain (i.e. amplification) about 100 times the Faraday cup. Because ions do not strike the multiplier directly, the detector has a long lifetime.

The Daly detector can only be used with positive ion beams. On the other hand, channel electron multipliers (CEM) can be used to amplify either positive or negative ion beams. These devices are therefore used for Re and Os analysis by negative molecular-ion TIMS (Section 8.1). The negative ion enters the orifice of the CEM at a potential near zero, releasing electrons when it strikes the semi-conducting channel wall. These electrons are multiplied during further collisions, as they are attracted to a positive HV collector. Because the collector is at high voltage, a negative ion signal cannot be amplified directly, but a pulsed ion-counting signal can be transmitted through an isolating capacitor to low voltage pulse-counting electronics (e.g. Kurz, 1979). The drawback of CEM detectors is their tendency to suffer damage when struck by heavy ions. Therefore signal sizes should be minimized to prolong their life.

All ion multiplication systems have some degree of non-linearity in their response, typically including a bias effect and a mass discrimination effect. These non-linearities must be carefully corrected if high quality data are to be obtained (e.g. Shen *et al.*, 2012). An alternative approach is to extend the range of sensitivity of Faraday collectors. One approach is to process larger samples, which can be analysed by MC–ICP–MS without the loss of ionization efficiency that would be encountered in TIMS analysis (Cheng *et al.*, 2013). Another approach is the use of a 10^{13} ohm resistor in the amplifier, which can extend Faraday analysis to smaller beam sizes (Trinquier and Komander, 2016). These techniques will undoubtedly become more important in the future for precise analysis of the minor isotopes in U-series dating.

2.4.3 Data Collection

Depending on the size of the ion beam, it is necessary to measure each nuclide signal for up to an hour to achieve high-precision data. To achieve this in a single collector TIMS machine, the magnetic field is 'switched' to cycle round a sequence of peak positions. On switching to a new peak there is a waiting period of 1–2 s to allow the output resistor and amplifier to reach a steady state in response to the new ion current. Then the ion beam is measured for a few seconds. In practice, each peak measurement must be corrected for incomplete decay of the signal from previous peaks (termed 'dynamic zero', 'tau' or 'resistor memory' correction). The computer then cycles round and round a series of peaks, baseline/background(s) and interference monitor position(s), interpolating between successive measurements of the same peak to correct for growth or decay of the beam size. A simple linear time interpolation may be used (Fig. 2.20), but Dodson (1978) developed a more sophisticated 'double interpolation' algorithm which can make better allowance for non-linear beam growth or decay.

Before isotope ratios can be determined from the different signals, background electronic noise must be subtracted in order to determine net peak heights. In TIMS analysis this is done by measuring a baseline position in each collector channel at approximately 0.4 a.m.u. above a whole mass position, sometimes a few a.m.u. away from the masses of interest. In plasma source analysis, backgrounds are often measured under each peak ('on peak zeros') before the analysis starts, in order to remove the effect of previous sample memory from the nebulizer.

From a single cycle of time-interpolated net peak heights, a set of net peak ratios is extracted. These are often collected in blocks of ten scans. The cycle time round a set of peaks may be shortened by the measurement of backgrounds and interferences 'between blocks' rather than 'within-scan'. For a single collector TIMS instrument, collecting 200 scans in about three hours might give a statistical within-run precision of 0.004% ($2\sigma = 2$ standard errors on the mean). That

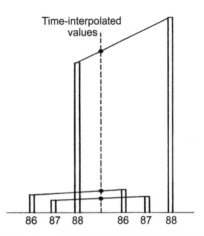

Fig. 2.20 Schematic illustration of the principle of linear time interpolation for a strontium ion beam growing at an (immense) rate of 30% per scan.

is, the scatter of data around the mean suggests that one can be 95% confident that the 'correct' answer lies within 0.004% (40 ppm) on either side of the mean. For $^{87}Sr/^{86}Sr$ this is equivalent to 0.71000 ± 0.00003 (2σ).

Occasional signal 'spikes' and other perturbations outside of normal random error are inevitable in a long mass spectrometer run. It is generally regarded as acceptable to run through the data a few times and test for outliers which are more than a certain number of population standard deviations (SD) from the mean (Pierce and Chauvenet, 1868; in Crumpler and Yoe, 1940). The cut-off level should depend on the size of data set, so that only a minimal number of outliers resulting from normal random variation are rejected. In practice the cut-off would normally be between 2 SD and 3 SD.

Ideally, a multiple collector machine can analyse isotope ratios in a 'static' mode without peak jumping. However, this may be limited, firstly by the extent to which each of the Faraday buckets is identical in terms of beam transmission characteristics; and secondly by the extent to which the gain of each bucket's amplifier system can be calibrated. Until the 1990s, these problems did not allow static analysis to achieve the highest levels of analytical precision, necessary for Sr or Nd isotope analysis. However, the quality of static analysis has steadily improved so that it can now yield data of very high quality. The alternative approach for high-precision analysis is double- or triple-collector peak jumping (multi-dynamic analysis). The simplest (double-collector) method is given below:

High collector:	87	88	91.4	–
Low collector:	86	87	90.4	85
Place in sequence	1	2	3	4

After background subtraction, the ^{85}Rb monitor is used to correct both ^{87}Sr peaks. The following algorithm represents an approximation, assuming unit mass differences between the isotopes:

$$\frac{87}{86}_{\text{true}} = \left[\frac{87}{86_1} \cdot \frac{87}{88_2} \cdot \frac{1}{0.1194} \right]^{0.5} \qquad [2.5]$$

where suffixes denote places in the scan sequence. This equation cancels out beam growth or decay and amplifier bias, as well as performing a power law mass fractionation correction, all in a single calculation. To use the exponential law for Sr evaporation as metal ions, the function above is raised to the power 0.5036 (Thirlwall, 1991b). With both of these methods, within-run precision should reach 0.002% (2σ) after three hours. Triple-collector analysis allows a further improvement in efficiency. In this method, two double-collector determinations on adjacent collectors are averaged to yield a more precise result. Ludwig (1997a) argued that the theoretical limits on precision by multi-dynamic and static analysis should be similar, assuming that the instrument is properly designed and optimized for each type of analysis. For a less skilled operator, multi-dynamic analysis is a more robust method because it makes lesser demands on the hardware performance and calibration.

A more recent development in multi-collector analysis (the Thermo-Scientific 'Triton' instrument) allows the signal from each Faraday to be switched in turn to each amplifier channel. This cancels out the electrical gains of the different amplifiers, allowing reproducibilities of better than 5 ppm to be achieved on isotope ratio measurements (Caro et al., 2006). Provided that the ion optic bias of each collector can be accurately calibrated, this system offers a new level of precision and accuracy in isotope ratio mass spectrometry. It should be particularly useful for applications such as seawater Sr, ^{186}Os and ^{142}Nd, where isotopic variations are at the limits of analytical measurement.

2.5 Isotope Dilution

Isotope dilution is generally agreed to be the supreme analytical method for very accurate concentration determinations. In this technique, a sample containing an element of natural isotopic composition is mixed with a 'spike' solution, which contains a known concentration of the element, artificially enriched in one of its isotopes. When known quantities of the two solutions are mixed, the resulting isotopic composition (measured by mass spectrometry) can be used to calculate the concentration of the element in the sample solution. The element in question must normally have two or more naturally occurring isotopes, one of which can be enriched on a mass separator. However, in some cases a long-lived artificial isotope is used.

2.5.1 Analysis Technique

Before use, the isotopic composition of a 'spike' must be accurately determined by mass spectrometry. This measurement cannot be normalized for fractionation, because there is no

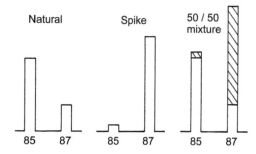

Fig. 2.21 Summation of spike (hatched) and natural ion beams to generate aggregate mixed peaks, as illustrated by rubidium isotope dilution.

'known' ratio to use as a fractionation monitor. Therefore, several long runs are generally made, from which the average midpoint of the run is taken to be the actual spike composition. The concentration of the spike is generally determined by isotope dilution against standard solutions (of natural isotopic composition) whose concentrations are themselves calculated gravimetrically. Metal oxides are generally weighed out, but if these are hygroscopic (e.g. Nd_2O_3) then accurate weighing requires part of a metal ingot.

A simple example of isotope dilution analysis is the determination of Rb concentration for the Rb–Sr dating method. Typical mass spectra of natural, spike and mixed solutions are shown in Fig. 2.21.

In the mixture, each isotope peak is the sum of spike (S) and natural (N) material. Hence,

$$\frac{^{87}\text{Rb}}{^{85}\text{Rb}} = R = \frac{\text{moles } 87_N + \text{moles } 87_S}{\text{moles } 85_N + \text{moles } 85_S} \qquad [2.6]$$

But the number of moles of an isotope is equal to the number of moles of the element as a whole, multiplied by the isotopic abundance. If the total number of moles of natural and spike Rb are represented by M_N and M_S, then

$$R = \frac{M_N \cdot \%87_N + M_S \cdot \%87_S}{M_N \cdot \%85_N + M_S \cdot \%85_S} \qquad [2.7]$$

where percentages indicate the isotopic abundances in the spike and natural solutions. This equation is rearranged in the following steps:

$$R \left(M_N \cdot \%85_N + M_S \cdot \%85_S \right) = M_N \cdot \%87_N + M_S \cdot \%87_S$$

$$R \cdot M_N \cdot \%85_N + R \cdot M_S \cdot \%85_S = M_N \cdot \%87_N + M_S \cdot \%87_S$$

$$R \cdot M_N \cdot \%85_N - M_N \cdot \%87_N = M_S \cdot \%87_S - R \cdot M_S \cdot \%85_S \qquad [2.8]$$

$$M_N \left(R \cdot \%85_N - \%87_N \right) = M_S \left(\%87_S - R \cdot \%85_S \right)$$

$$M_N = M_S \cdot \frac{\left(\%87_S - R \cdot \%85_S \right)}{\left(R \cdot \%85_N - \%87_N \right)}$$

If we insert figures for the isotopic abundance of natural Rb, and the isotopic abundance of a typical spike, such that 27.83/72.17 is the natural ^{87}Rb/^{85}Rb ratio, and 99.4/0.6 is the

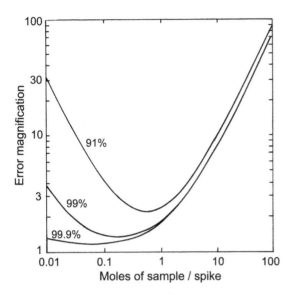

Fig. 2.22 Estimates of error magnification in isotope dilution analysis for different sample/spike mixtures. Cases are shown for 91, 99 and 99.9% spike isotope enrichment, mixed with a natural sample with a 50–50 isotopic abundance ratio. After DeBievre and Debus (1965).

spike $^{87}Rb/^{85}Rb$ ratio, then the number of moles of the natural Rb is given by

$$M_N = \frac{(99.4 - R \cdot 0.6)}{(R \cdot 72.17 - 27.83)} \cdot M_S \qquad [2.9]$$

where R is the measured isotope ratio.

But number of moles = molarity × mass, so the molarity of the natural sample is given by

$$Molarity_N = \frac{(99.4 - R \cdot 0.6)}{(R \cdot 72.17 - 27.83)} \cdot \frac{wt_S}{wt_N} \cdot Molarity_S \qquad [2.10]$$

Molarity is then multiplied by atomic weight to yield concentration:

$$Conc_N = At.\, wt_N. \frac{(99.40 - R.0.6)}{(R.72.17 - 27.83)} \cdot \frac{wt_S}{wt_N} \cdot \frac{Conc._S}{At.wt._S} \qquad [2.11]$$

Because there are only two isotopes of Rb, no internal correction for fractionation is possible in the measurement of $^{87}Rb/^{85}Rb$. However, in the isotope dilution analysis of Sr, fractionation correction based on 88/86 measurement is possible, which allows a much more accurate 84/86 (spike Sr / natural Sr) measurement to be made.

Isotope dilution is potentially a very high-precision method. However, error magnification may occur if the proportions of sample to spike which are mixed are far from unity (Fig. 2.22). Consequently, it is generally believed that the analysed peaks in an isotope dilution mixture should have an abundance ratio close to unity. In actual fact, the ideal composition of the mixture is half-way between that of the natural and spike compositions. However, the precision

normally required in an isotope dilution analysis is ca. 1 per mil (0.1%), which is nearly two orders of magnitude worse than the precision normally achieved in Sr or Nd isotope ratio measurements. Hence, significantly non-ideal spike–natural mixtures can be tolerated in normal circumstances.

The only other sources of error in isotope dilution analysis are incomplete homogenization between sample and spike solution, and weighing errors. The first of these can be overcome by centrifuging the sample solution to check for any undissolved material, and repeating the dissolution steps as necessary until complete solution is achieved. Given sufficient care, including the use of non-hygroscopic standard material and correct balance calibration, spike solutions can be calibrated to 0.1% accuracy (Wasserburg et al., 1981). The use of mixed spikes (e.g. Sm–Nd, Rb–Sr) then eliminates further weighing errors in the analysis of these ratios in sample material. The result is that isotope dilution accuracy can exceed 1% with ease, and 0.1% if necessary. This compares very favourably with all other methods of concentration determination.

2.5.2 Double Spiking

As its name suggests, double spiking uses a spike solution containing two enriched isotopes of the element of interest. Because the ratio of these two isotopes is known, it allows a full correction of mass-dependent fractionation during mass spectrometry of an unknown sample. This does not require any assumptions about the isotopic composition of the unknown, but it does normally require two runs – spiked and un-spiked. The theory of double spiking was first investigated in detail by Dodson (1963). The calculations may be made iteratively (e.g. Compston and Oversby, 1969) or algebraically (e.g. Gale, 1970). However, it is not usually possible to calculate 'absolute' values of isotopic abundance because there is not normally any absolute standard to calibrate the double spike.

There are two main applications of double spiking in geochemistry. One is to systems that have enough non-radiogenic isotopes for an internal fractionation correction, but we want to avoid making any assumptions about these ratios. The second is to systems such as Pb where there are insufficient non-radiogenic isotopes for an internal fractionation normalization.

Selection of an arbitrary non-radiogenic ratio for fractionation normalization (e.g. $^{88}Sr/^{86}Sr = 8.37521$) results in no loss of information for terrestrial samples. However, for meteorites, use of such a procedure means that the true isotopes responsible for certain anomalous isotope ratios cannot be uniquely identified. For example, several alternative nucleosynthetic models could explain the abnormal Sr isotope signatures of the EK 1–4-1 and C1 inclusions in the Allende chondritic meteorite (Clayton, 1978). However, by fully correcting for instrumental mass fractionation, double spiking allows comparison of all the isotope ratios in a suite of samples, including those used for fractionation normalization (Patchett, 1980).

Several workers have investigated the use of double Pb spikes to allow within-run mass fractionation correction of Pb isotope ratios. Most of these studies utilized double stable isotope spikes such as [207]Pb–[204]Pb (Compston and Oversby, 1969; Hamelin *et al.*, 1985), which necessitate two separate mass spectrometer runs. In contrast, the use of a [202]Pb/[205]Pb double spike allows both concentration determination and a correction for analytical mass fractionation to be made on a single Pb mass spectrometer run (Todt *et al.*, 1996). These isotopes are difficult to manufacture, but the procedure for [205]Pb was described by Parrish and Krogh (1987). Because the spike is very expensive (ca. $1 million per milligram), this method is only applicable to very small samples, such as single zircon grains. Hence, the rare spike isotopes must be measured on a multiplier detector.

More recent work on double spike TIMS Pb analysis has been done by Powell *et al.* (1998), Galer (1999) and Thirlwall (2000). However, the advent of plasma source mass spectrometry has offered an alternative approach for fractionation correction in the analysis of common Pb (see below). On the other hand, MC–ICP–MS has opened up the possibility of 'stable isotope' analysis of a wide range of elements, for which double spiking may be useful.

2.5.3 Pb–Tl Double Spiking

Because mass fractionation in ICP-MS instruments occurs largely in the ion extraction process rather than the plasma source, it is almost completely independent of the chemistry of the species involved. Therefore, the double spiking method can be applied using a different 'spike' element from that of the analyte. For example, the two stable isotopes of thallium (202 and 205) can be used to make accurate fractionation corrections for Pb (Walder and Furuta, 1993). Because there is no isobaric interference from this double spike onto the Pb isotopes, this avoids the need for separate spiked and un-spiked runs which are necessary when a double Pb spike such as 207–204 is used. However, because Pb and Tl are different elements, small discrepancies between their fractionation behaviour may be observed.

Experiments by White *et al.* (2000) showed small differences in the mass fractionation behaviour of Pb and Tl over a period of months, forming slightly different Pb-Tl fractionation lines (Fig. 2.23). Thirlwall (2002) suggested that these small variations in fractionation behaviour may be due to subtle solution chemistry effects in the nebulizer. This requires these empirical calibration lines to be established frequently, in order to make accurate corrections for Pb fractionation. However, it is difficult to establish these calibration lines from standard analyses, because the standards do not adequately cover the range of Pb–Tl mass fractionation observed in unknown samples.

To solve this problem, Woodhead (2002) suggested that the standard solution could be deliberately 'contaminated' with a Pb-free matrix solution, which will induce variable fractionation behaviour without affecting the Pb signature of the standard. This approach is demonstrated in Fig. 2.24, where open symbols represent analysis of 'pure' SRM 981 Pb

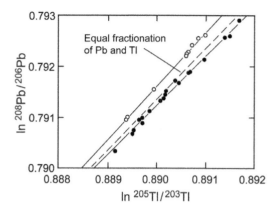

Fig. 2.23 Log–log plot of Pb versus Tl data to show small variations in the location of mass fractionation lines. After White *et al.* (2000).

standard, whereas solid symbols represent standards deliberately contaminated with a rock matrix solution. Woodhead argued that this technique could improve the reproducibility of the thallium double spike method to yield errors as low as 100 ppm for Pb isotope ratios in geological samples.

2.6 MC–ICP–MS Solution-based Applications

MC–ICP–MS is a more expensive technology than conventional TIMS, both to build and to operate. This is because the instruments have several complex engineering systems, involving an RF generator, the torch assembly with its argon supply, and the ion extraction system with its pumps, in addition to the magnetic sector analyser. Hence, MC–ICP–MS is most effectively applied to isotope systems where TIMS has encountered technical difficulties. In this section,

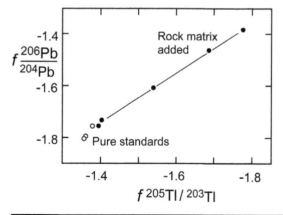

Fig. 2.24 Fractionation factors for Pb and Tl data based on pure (○) and matrix-contaminated standards data (●). Modified after Woodhead (2002).

applications of MC-ICP–MS to isotope analysis by solution nebulization are described. Laser ablation ICP–MS for *in situ* analysis will be described in the following section.

2.6.1 Hf–W

[182]W is the decay product of the short-lived nuclide [182]Hf ($t_{1/2} = 9$ Ma). As an extinct nuclide, [182]Hf can place important constraints on the origin and early history of the solar system, but the very high ionization potential of tungsten (7.98 eV, similar to osmium) precludes its analysis as a positive ion by TIMS. The first high-precision W isotope measurements were made by negative ion TIMS, but MC-ICP–MS proved ideally suited to the analysis of this element. Following establishment of the method (Lee and Halliday, 1995), these authors made numerous W isotope analyses of meteoritic and planetary samples, throwing new light on the timing of terrestrial core formation and the formation of the solar nebula (Section 15.6.4).

2.6.2 Lu–Hf

The Lu-Hf isotope system was first exploited using positive TIMS analysis (Section 9.1), but the high ionization potential of hafnium (6.65 eV) has always rendered TIMS Hf analysis difficult, requiring a large sample size and high purity ion exchange separation to achieve good results. In contrast, the ability of MC-ICP–MS to achieve excellent ionization on impure sample solutions makes the method ideal for Hf isotope analysis. Whole-rock samples containing as little as 100 ng Hf can be analysed after pre-concentration only (Blichert-Toft *et al.*, 1997). Hence, for all of these reasons, MC-ICP–MS has rendered TIMS analysis of hafnium obsolete. The *in situ* Hf isotope analysis of zircons by laser ablation MC–ICP–MS is discussed in Section 2.7.

2.6.3 U–Th

Thorium is another element with a high ionization potential (6.31 eV). It was routinely analysed by TIMS, but the poor ionization efficiency was a significant problem, particularly for samples with a large common thorium ([232]Th) content (Edwards *et al.*, 1987). In such cases, the large total size of the sample markedly reduced the ionization efficiency for TIMS. In contrast, ionization efficiency for ICP–MS is almost perfect, irrespective of the sample size.

Samples with large isotope ratios are well suited to simultaneous analysis on a Faraday and Daly collector (Section 2.4.2). In this approach, the small [230]Th beam is analysed on a Daly or multiplier collector, while the large [232]Th beam is measured simultaneously on a Faraday collector (e.g. Luo *et al.*, 1997). In order to calibrate the relative gain of the Daly and Faraday collectors, Luo *et al.* added uranium of known isotopic composition directly to the Th sample. The [235]U/[238]U ratio measured on two Faraday collectors is used for mass fractionation of both U and Th, while the [234]U/[235]U ratio is used to calibrate the Daly/Faraday collector gain.

More recently, it was shown by Cheng *et al.* (2013) that even the small [234]U and [230]Th signals can be measured using Faraday collectors, by taking advantage of the capability of

ICP-MS to run at high sample concentrations to achieve very large signal size. This allowed precision on these isotopes to reach a few epsilon units for clean carbonate samples.

2.6.4 Pb–Pb

Pb isotope analysis has long been performed by TIMS, but the existence of only one stable non-radiogenic isotope has always made mass fractionation the principal source of analytical error. Because of the different source of mass fractionation in ICP–MS, this allows fractionation correction using a different element (thallium), as described above. This method can achieve accuracies of Pb isotope analysis approaching those produced by the much more experimentally demanding double spiking methods using TIMS (Section 2.5.2). Therefore it can be expected that MC-ICP–MS will take over from TIMS as the preferred method for Pb isotope analysis where large samples suites must be analysed to high precision. This would apply to studies of igneous petrogenesis, and especially to Pb isotope analysis as an environmental tracer, such as in the study of seawater evolution from the analysis of ferromanganese crusts (Christensen *et al.*, 1997).

2.6.5 Sm–Nd

Experiments to test the effectiveness of MC-ICP–MS for Nd isotope analysis were reported by Walder *et al.* (1993b) and Luais *et al.* (1997). Most analyses were made in static multi-collection mode, but a few multi-dynamic analyses gave similar results to static analysis. It was shown that Sm interference corrections could be made very accurately, so that Nd isotope ratio determinations could be made on bulk REE separates from a cation column, without secondary clean-up to remove Sm. In addition, the typical analysis time was only 20 min per sample.

Disadvantages of the ICP method were the longer instrument setup times, the need to analyse more standards and the slightly lower external (between-run) reproducibility. This was found to be ca. 30 ppm over short periods, but 60 ppm over a period of a year. However, a more recent study (Vance and Thirlwall, 2002) suggested that this lower reproducibility was due to the inadequacy of the exponential law to correct for the large fractionation factors encountered with ICP, which are more consistent than TIMS fractionation effects, but usually more than an order of magnitude larger. Vance and Thirlwall showed that these problems could be solved using information from the fractionation of additional non-radiogenic ratios, therefore allowing MC-ICP–MS to generate reproducibilities comparable with TIMS (Section 2.3.2).

This procedure was verified for high-precision [143]Nd analysis by Albarede *et al.* (2004). However, for the very high precision and accuracy approaching 1 ppm, required for [142]Nd analysis, the Thermo-Fisher Triton TIMS instrument with switchable amplifier channels was found to be superior to MC-ICP–MS (for example, compare Boyet *et al.*, 2003; Boyet and Carlson, 2006; Caro *et al.*, 2006).

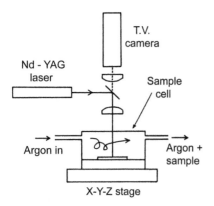

Fig. 2.25 Schematic illustration of the procedure for feeding laser ablated material into the ICP source. After Halliday *et al.* (1998).

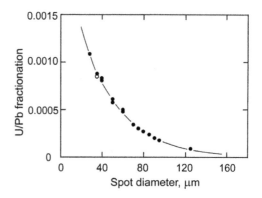

Fig. 2.26 Plot of the slope of the time-dependent U/Pb fractionation process against spot size for excimer laser ablation. Solid symbol = zircon standard; open symbol = solution standard. After Horn *et al.* (2000).

2.7 LA–ICP–MS

There are many applications of *in situ* analysis by laser ablation (LA)–ICP–MS. However, according to a review of publication rates by Schoene (2014), LA–ICP–MS is largely responsible for the 'explosive' growth in U–Pb age dating since the year 2000. Therefore, this review will focus on U–Pb dating by laser ablation, followed by the related subject of zircon Hf isotope analysis by LA–ICP–MS.

2.7.1 U–Pb

The idea of using laser ablation ICP–MS analysis of zircon as a dating tool was demonstrated by Feng *et al.* (1993). They used a pulsed Nd–YAG infra-red laser to ablate cylindrical pits 30–60 μm wide. The ablation was carried out in a quartz cell, and the resulting vapour was conveyed to the ICP source in the argon carrier gas (Fig. 2.25). Pb isotope ratios were measured using a quadrupole instrument adjusted to maximize sensitivity for Pb. Feng *et al.* determined $^{207}Pb/^{206}Pb$ ages on several zircons previously analysed by TIMS, and obtained excellent ages for zircons known to be concordant. Within-run precision as low as 1% was achieved, with a normal range of 2–3% (1σ). U/Pb ratios were also measured, but the difficulty of calibrating the relative ablation efficiencies of U and Pb prevented the acquisition of accurate U/Pb data.

The development of MC–ICP–MS offered the possibility of improving the precision of Pb isotope measurements beyond the 1% capability of quadrupole ICP–MS. Walder *et al.* (1993a) used a glass standard to calculate the efficiency of laser ablation MC–ICP–MS for ion production. They found that the efficiency was comparable with ion microprobe analysis using the SHRIMP (Section 5.2.6). However, because the volume of sample excavated by laser ablation can be larger than with the SHRIMP, laser ablation MC–ICP–MS can generate ion beams large enough for analysis by multiple Faraday collectors (Halliday *et al.*, 1998). Hence the instrument is

potentially capable of generating ages with higher precision than the SHRIMP. The instrument described by Halliday *et al.* (1998) was fitted with a wide flight tube to permit simultaneous analysis of U and Pb.

Unfortunately, the promise of high-precision U–Pb dating by LA-MC-ICP-MS was not quickly realized, due to the problem of accurately calibrating the relative efficiency of U and Pb ablation. In early work on this problem using a frequency-quadrupled Nd-YAG UV laser (λ = 266 nm), Hirata and Nesbitt (1995) found that ablation efficiency was extremely sensitive to the focus depth of the laser on the sample, which changed as the sample was ablated. For example, using a spot size of 15 μm diameter on a glass standard, 60 s of ablation typically caused the Pb signal to decrease by a factor of 25, and U by a factor of 100. However, Hirata (1997) showed that these problems could be somewhat reduced by ramping up the laser power slowly during each spot analysis.

In subsequent work using a 193 nm excimer laser, Horn *et al.* (2000) showed that U/Pb fractionation effects were very sensitive to the spot size. Therefore, increasing the spot size above 35 μm could reduce the problem of variable U/Pb ablation yields to manageable levels (Fig. 2.26). Horn *et al.* (2000) used zircon standards to calibrate the ablation process for a given laser power, spot size and depth of excavation. They then corrected instrumental mass fractionation by introducing an enriched isotope 'spike' to the gas stream from the sample. Using this approach, they claimed to match the dating precision of ion microprobes such as the SHRIMP (Section 5.2.6).

Most other workers have preferred to use alternating analysis of standards and unknowns to correct for both ablation efficiency and instrumental mass fractionation in a single step. Using this procedure, many labs have claimed analytical errors of less than 1%. However, inter-comparison of zircon standards indicated a reproducibility nearer $\pm$3% (Klotzli *et al.*, 2009; Kosler *et al.*, 2013). This led to renewed attempts to find the cause of these errors.

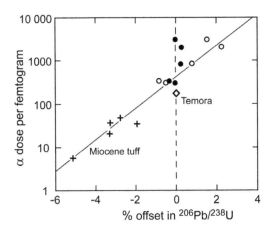

Fig. 2.27 Relationship between calculated alpha dose and offset in measured U/Pb ratio in various zircon standards relative to the Temora standard. (+, ∘) = un-annealed; (•) = annealed. Modified after Allen and Campbell (2012).

By sequentially analysing a set of zircon standards under constant operating conditions, Allen and Campbell (2012) showed that the age variations were due to a breakdown in the correction for U/Pb ablation efficiency. They also showed that the offset in the measured U/Pb ratio (relative to the Temora zircon standard) was strongly correlated with the lifetime alpha dose of each zircon, calculated from its U–Th concentration and age (open circles in Fig. 2.27).

The relationship in Fig. 2.27 was extended to samples with relatively low alpha doses by analysing young (10 Ma) zircons from the Cougar Point tuff (crosses in Fig. 2.27). Allen and Campbell deduced that the ablation process must be affected by the degree of lattice damage. Therefore, they adopted the zircon annealing procedure of Mattinson (2005) in an attempt to remove these effects (Section 5.2.3). Annealing all standards for 48 hours at 850 °C caused the U/Pb ratio of other zircon standards to move into agreement with Temora (solid symbols in Fig. 2.27). This suggests that annealing can increase the reproducibility of the ablation process, and hence improve dating accuracy.

The effectiveness of the annealing procedure in LA–ICP–MS has been tested in several subsequent studies. Marillo-Sialer et al. (2014) initially expressed some reservations about the effectiveness of the technique for increasing U–Pb reproducibility. However, the method was strongly endorsed by Crowley et al. (2014), von Quadt et al. (2014), Solari et al. (2015) and also Marillo-Sialer et al. (2016). These positive findings suggest that in the future, U–Pb dating by LA–ICP–MS may surpass SHRIMP analysis in precision and accuracy, since the latter method also suffers from U/Pb fractionation during sputtering (Section 5.2.6).

2.7.2 Lu–Hf

Hf isotope analysis of zircon follows naturally from U–Pb dating, firstly because Hf is one of the most abundant trace elements in zircon, and secondly because U–Pb ages are a prerequisite for calculating initial Hf isotope ratios for zircons. *In situ* Hf isotope analysis of zircon was first demonstrated by Thirlwall and Walder (1995), and has since grown hugely in importance. Nevertheless, there are some major analytical issues with the method which must be overcome to obtain reliable data. Unfortunately, much of the early *in situ* LA–ICP–MS data were erroneous (Section 9.3.1). However, most of these errors have been attributed to two principal problems (Woodhead et al., 2004; Fisher et al., 2014a, b).

The first of these problems is isobaric interference on the radiogenic isotope ^{176}Hf. Both Lu and Yb form potential interferences, but the latter is much more serious. For the case of Lu the problem is not severe, firstly because Lu levels in zircon are low, and secondly because the monitor isotope, ^{175}Lu, is 37 times more abundant than the interference. This allows very precise correction.

The case of Yb interference is much more severe, since Yb is more abundant in zircon, and the monitor/interference ratio is nearer unity. Hence, the Yb correction can easily be in the hundreds of epsilon units (Fisher et al., 2014b). To achieve an accurate correction on Hf, it is first necessary to correct for mass fractionation of Yb. Early work tended to measure one Yb peak, and then use the Hf fractionation monitor to correct for Yb, in a manner analogous to the thallium–lead double spike method. However, Woodhead et al. (2004) showed that, just as for thallium–lead (Fig. 2.23), the fractionation behaviour of the two elements is slightly different. Therefore, for an accurate Yb correction on Hf, it is necessary to measure two other Yb isotopes in order to correct the interfering isotope, ^{176}Yb, for fractionation.

The second major problem with *in situ* zircon Hf analysis is age correction to initial εHf ratios. The problem is not age correcting the mineral itself (since Lu/Hf in zircon is low), but age correcting the chondritic growth curve to the age of the zircon in order to calculate an accurate εHf value. However, when the data are plotted in epsilon notation, with the CHUR line horizontal, it *appears* that the initial Hf ratio of the zircon is changing with time (Fig. 2.28). This means that if the zircon contains zones with different ages, it is very easy to use an erroneous U–Pb age to determine a highly erroneous εHf value (Section 9.3.1).

To avoid this problem, it is preferable to make simultaneous U–Pb and Hf isotope determinations during ablation, as the laser may drill through zones of different age in a single zircon grain. Two principal strategies have been used to achieve simultaneous or quasi-simultaneous measurement. The first approach is made possible by the zoom lens ion optics of the Nu Instruments MC–ICP–MS (Belshaw et al., 1998). This allows rapid changes in collector configuration so that Hf and Pb isotopes can be measured in rapid succession (Woodhead et al., 2004). An alternative is actually to use two separate ICP–MS instruments for fully simultaneous analysis, termed 'laser ablation split-stream' (LASS) by Fisher et al. (2014a).

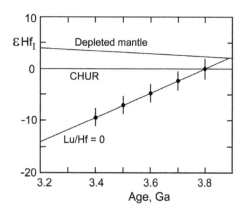

Fig. 2.28 Plot showing the apparent change of initial Hf isotope ratio due to changes in the apparent U–Pb age of a zircon. After Fisher *et al.* (2014b).

2.8 Isochron Regression Line Fitting

The 'isochron' diagram was invented by Nicolaysen (1961) as a way of analysing error systematics in Rb–Sr dating (Section 3.2.2). The Rb–Sr method has now been eclipsed in nearly all dating applications by zircon U–Pb and various other specialist dating methods. However, it remains a useful vehicle to demonstrate many of the principles of isochron fits.

When radiogenic isotope ratios for a cogenetic sample suite (e.g. $^{87}Sr/^{86}Sr$) are plotted against the parent/daughter ratio in those samples (e.g. $^{87}Rb/^{86}Sr$), the points should ideally define a perfect straight line, showing that the suite is 'isochronous'. Typically, the slope of this isochron line can then be used to determine the age of the system. However, since both quantities involved are measured experimentally, experimental errors are inevitable. Hence, these errors must be considered when calculating the slope of a best-fit line though the data.

A straight line fitted to an array of points is termed a 'linear regression'. One of the best approaches involves minimizing the sum of the squares of the distances that data points lie away from a line drawn through the points, and hence is called a least squares fit. This typically involves an iterative solution, which is readily performed by computer.

2.8.1 Types of Regression Fit
In simple linear regression programs, one ordinate is defined as 'free of error' and the regression line is calculated to minimize the misfit of points in the other ordinate (Fig. 2.29a, b). If the data define a very gentle slope and are somewhat scattered, disastrous fits can be produced by regressing onto the wrong ordinate. Where errors are present in both ordinates, as in the case of the isochron fit, a two-error regression treatment must be used (Fig. 2.29c, d). Several methods of this type have been presented in the literature (e.g. McIntyre *et al.*, 1966; York, 1966, 1967; Brooks *et al.*, 1972;

Ludwig, 1997b), sometimes including a ready-to-run computer program.

In cases where the actual deviations of the data points from the regression line are equal to or less than those expected from experimental error, all regression treatments effectively give the same isochron age and initial ratio. In such cases, the only matter of debate is the manner in which experimental errors are assigned.

Ideally, the analysed error in $^{87}Sr/^{86}Sr$ and $^{87}Rb/^{86}Sr$ (for example) would be determined by measuring the reproducibility of an almost infinite number of duplicates (Brooks *et al.*, 1972). Since this is very time consuming, the best empirical estimate is probably the long-term reproducibility of standard analyses. Within-run precision of sample analyses is almost certainly an under-estimate of error, since it is typically about 50% of the reproducibility error. For $^{87}Rb/^{86}Sr$, quoted accuracies must include an estimate for sample weighing errors, spike calibration errors etc., as well as mass spectrometry errors (in the case of isotope dilution).

While some of the regression programs in use provide the facility for weighting each data point according to its precision of measurement, this may sometimes be detrimental, as it tends to 'destabilize' the fit. In practice, $^{87}Rb/^{86}Sr$ and $^{87}Sr/^{86}Sr$ errors are probably best assigned as a blanket percentage (e.g. 0.5% and 0.002% 1σ respectively). If one point has a particularly bad precision, it is better to re-analyse it than give it less weight in the regression.

2.8.2 Regression Fitting with Correlated Errors
In conventional isochron analysis (e.g. Rb–Sr), analytical errors in the two ordinates (isotope ratio and elemental abundance ratio) are effectively uncorrelated. However, in the lead isotope dating methods, this is far from the case. In common Pb–Pb dating, correlated errors are found between $^{207}Pb/^{204}Pb$ and $^{206}Pb/^{204}Pb$, due to greater analytical uncertainties on the small ^{204}Pb peak (which is common to both ratios) and due to the uncertainty of mass fractionation. These two correlation lines have different slopes, and while the former may be important for very small Pb beam sizes, the latter is normally dominant. Data for the NBS 981 standard, shown in Fig. 2.30, yield a correlation coefficient of 0.94 (Ludwig, 1980).

In U–Pb zircon dating, errors may show a much stronger correlation. This is because errors in $^{206}Pb/^{238}U$ and $^{207}Pb/^{235}U$ are mainly attributable to the elemental U/Pb ratio, which may be five or more times less reproducible than the $^{206}Pb/^{207}Pb$ ratio (Davis, 1982). This difference arises from the analytical errors inherent in isotope dilution, and uncertainties in the common Pb correction (Section 5.2.1). Regression treatments for correlated errors using the least squares technique were given by York (1969) and Ludwig (1980). Davis (1982) used the alternative 'maximum likelihood' method, and showed that the two approaches yielded similar estimates of error using test data (see also Titterington and Halliday, 1979).

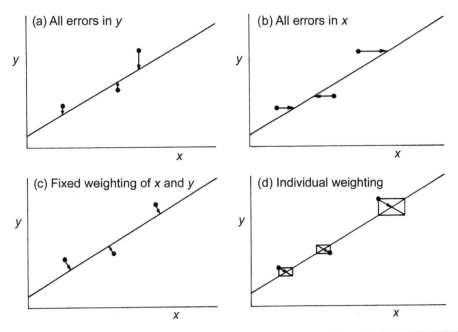

Fig. 2.29 Schematic illustration of least squares regression analysis with different conditions of weighting: (a) infinite weighting of x (all errors in y); (b) infinite weighting of y; (c) fixed weighting of x versus y; (d) individual weighting of each point, inversely proportional to squares of standard deviations. After York (1967).

2.8.3 Errorchrons

Based on statistical principles, Brooks *et al.* (1972) argued that,

A line fitted to a set of data that display a scatter about this line in excess of the experimental error is simply not an isochron.

They proposed that regression fits with excess or 'geological' scatter (McIntyre *et al.*, 1966) should be called 'errorchrons' and treated with a high degree of suspicion. This raises the question of how to detect the presence of geological scatter, bearing in mind the fact that analytical errors are only probabilities.

The sum of the squares of the misfits of each point to the regression line (= squared residuals; York, 1969) or sum of χ^2 (Brooks *et al.*, 1968, 1972), may be divided by the degrees of freedom (number of data points minus two) to yield the mean squared weighted deviates (MSWD), which is the most convenient expression of scatter. If the scatter of data points is, on average, exactly equivalent to that predicted from the analytical errors, the calculation will yield MSWD = 1. Excess scatter of data points yields MSWD > 1, while less scatter than predicted from experimental errors yields MSWD < 1.

Problems may arise with the interpretation of these MSWD values, since the analytical errors input to the program are only *estimates* of error. To address this problem, Brooks *et al.* (1972) constructed an F distribution probability table (Table 2.1) to distinguish between errorchrons and isochrons from their MSWD values. They established a 'rule

of thumb' that on average if MSWD < 2.5 then the data define an isochron, and if >2.5 an errorchron. Unfortunately this rule of thumb was much abused over subsequent years, because the original objectives of Brooks *et al.* (1972) were misunderstood. They set up the MSWD = 2.5 cut-off in order to reject errorchrons with a 95% certainty of excess scatter

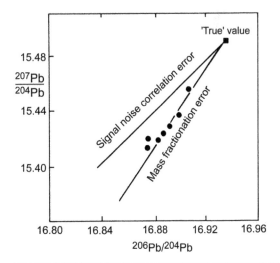

Fig. 2.30 Results of seven analytical runs on the NBS 981 Pb standard performed with large beam sizes at varying filament temperatures. Data cluster near the mass fractionation line. Solid square = 'true value'. After Ludwig (1980).

Table 2.1	MSWD values indicating 95% confidence of an errorchron.								
	Number of samples regressed								
Number of duplicates	3	4	5	6	8	10	12	14	26
10	4.96	4.10	3.71	3.48	3.22	3.07	2.98	2.91	2.74
20	4.35	3.49	3.10	2.87	2.60	2.45	2.35	2.28	2.08
30	4.17	3.32	2.92	2.69	2.42	2.27	2.16	2.09	1.89
40	4.08	3.23	2.84	2.61	2.34	2.18	2.08	2.00	1.79
60	4.00	3.15	2.76	2.53	2.25	2.10	1.99	1.92	1.70
120	3.92	3.07	2.68	2.45	2.18	2.02	1.91	1.83	1.61

Numbers underlined just exceed MSWD = 2.5 cut-off.

over analytical error. This corresponds to only 5% confidence that a fit with MSWD = 2.5 is an isochron (e.g. Wendt and Carl, 1991). However, many workers wrongly assumed that if MSWD was less than 2.5 then there was a high degree of confidence that the data suite formed a true isochron (where analytical errors express most or all of the error on the age). In actual fact, MSWD must be near or below unity in order to give a high degree of confidence that the data represent a true isochron.

Because the number of errorchrons will continually increase as analytical errors decrease, the suggestion that errorchrons be rejected outright is unhelpful. Therefore, other workers have looked for ways of quantifying geological scatter in terms of an error on the age result. Although these methods were developed for the Rb–Sr method, they largely apply to any dating method.

McIntyre et al. (1966) emphasized that statistical error estimation of errorchrons cannot be properly meaningful unless the geological reasons for the misfit are understood. Therefore, they suggested four alternative approaches for error handling. These are as follows:

(1) No excess scatter above predicted analytical errors (= true isochron).
(2) All excess scatter is attributed to Rb/Sr, equivalent to assuming small differences between the ages of the samples.
(3) All excess scatter is attributed to $^{87}Sr/^{86}Sr$, equivalent to assuming variation in the initial isotopic ratio of samples.
(4) Excess scatter is attributed to some combination of models 2 and 3.

The program of York (1966) allowed the analytical errors on x and y ordinates (e.g. $^{87}Rb/^{86}Sr$ and $^{87}Sr/^{86}Sr$) to be multiplied in equal and uniform proportion by an error factor ($\sqrt{MSWD}$) until the expanded errors equal geological scatter (MSWD = 1). The error on the calculated age will be magnified by this process to give a reasonable estimate of uncertainty which includes both geological and analytical scatter.

Some form of error expansion must *always* be performed if a meaningful geological error estimate is to be given for a

data set with MSWD > 1, because this is a *definite* indicator that excess scatter of some form is present. The only uncertainty is whether the excess scatter is geological or analytical. The York (1966) procedure is the most common method of dealing with excess scatter, but it is an arbitrary procedure which takes no account of geological processes and their resulting contribution to errors. Where initial ratio variability is suspected, option 3 of McIntyre (above) is preferable (amplification of isotope ratios only). However, this can lead to misinterpretation of a data set if all points are not identical in age.

2.8.4 Probability of Fit

An alternative way to describe the quality of an isochron is the 'probability of fit' index. This approach was introduced by Krogh (1982) and Davis (1982), and has become the most widely used index for describing the fit of U–Pb discordia lines. Probability of fit is reported alongside MSWD by the Isoplot program (see below), and was defined by Ludwig (2000) as,

The probability that, if the only reason for scatter from a straight line is the analytical errors assigned to the data points, the scatter of the data points will exceed the amount observed for your data.

In other words, it represents the probability that a regression data set is a true isochron. For a high quality data set with well-determined analytical errors and a large number of points, a high probability of fit is achieved fairly easily (equivalent to the bottom right corner of Table 2.1). However, most U–Pb ages are based on quite small data sets, so that a high probability of fit (high confidence of a true isochron) is only achieved if the data are less scattered than predicted by the analytical errors. For example, using four-point test data sets, typical values for MSWD and probability of fit are compared in Table 2.2.

Following a procedure similar to Brooks et al. (1972), and admitting that this was somewhat arbitrary, Ludwig (2000) proposed that 15% probability of fit could be used as a cut-off value for a good quality isochron. These relatively low percentages demonstrate the limitation of probability-based indices in geochronology, since researchers cannot afford

Table 2.2	Typical goodness-of-fit indices for four-point U–Pb data sets.

MSWD	Probability of fit (%)
0.0004	98
0.23	80
0.91	40
1.3	27
2.1	12
2.3	10

to waste resources by producing data sets with significant degrees of redundancy. On the other hand, MSWD remains the simplest description of excess scatter because it states the actual relationship between estimated analytical errors and observed scatter.

2.8.5 Isoplot

By the 1990s, free-standing least-squares programs written in *Fortran* or *Basic* were becoming less attractive for isochron calculation. Therefore, Ludwig (1991) developed a DOS-based program 'Isoplot' incorporating these concepts, later revised by Ludwig (1997b). However, the quality control engineer Dorian Shainin had previously developed a correlation plot with the name Isoplot ®, so the 'Isoplot for geochronology' carries a disclaimer to this effect.

With the adoption of Microsoft Excel as a near-universal spreadsheet package, Ludwig (2000) adapted Isoplot as a Windows Excel macro, and later updated it for the 2003 version of Excel (Isoplot version 3). However, the macro is not compatible with the 'new' and much more restrictive 2007 version of Excel. Although Ludwig has introduced Isoplot version 4 to work under the new Excel, he has recommended that users continue to run the 2003 version, which can be installed alongside the 2007 version (Ludwig, 2012).

Among several types of isochron fit, Isoplot provides York-type fits under three categories similar to those listed above:

(1) Fits based on individually assigned analytical errors. Errors on the age are calculated: (a) based on analytical errors only (applicable if MSWD ≤ 1); and (b) by equal expansion of assigned analytical errors. A warning is given that the latter approach can give rise to serious errors if the assigned analytical errors are significantly variable between individual points.

(2) A fit based on expansion of assigned errors to encompass the scatter, but all points have equal weight. If the assigned errors are uniform, fit 1b is the same as fit 2.

(3) A fit based on model 3 of McIntyre, with excess scatter absorbed by expanding initial ratios only.

With the development of U–Pb dating as the 'gold standard' in geochronology (Section 5.1.3), conventional isochron diagrams of the Rb–Sr type are less used. Therefore, the most commonly used Isoplot output is probably that for U–Pb dating, including Ludwig's 'concordia band'

output that he developed to take account of uncertainties in uranium decay constants (Section 5.2.4). In this connection, it should also be noted that MSWD and probability of fit can also both be used as a measure of the *concordance* of zircon analyses (e.g. Ludwig, 2012; Solari *et al.*, 2015). Concordance is particularly important in detrital zircon analysis, because there is limited contextual information from other zircons in the sample for the construction of discordia lines to determine upper intercept ages. Another feature offered by Isoplot relevant to detrital zircons data is the probability density plot (see below).

2.9 Probability Density

The histogram is a widely used univariate plot that is useful for portraying many different types of frequency distributions. However, the method has two limitations. Firstly, data must be grouped into 'bins' whose size and limits may be arbitrary relative to the raw data. Secondly, the histogram takes no account of the uncertainties of individual values within the data set.

The probability density plot is a development of the histogram that attempts to solve these problems. The uncertainty on each measurement can be assigned individually or as a blanket error to a whole data set. However, the requirement is that the error is symmetrically distributed (Gaussian) for each point. The probability distributions for each data point are then summed to yield an overall curve of probability density defined by the area under the curve.

2.9.1 Detrital Zircon Distributions

The probability density plot is ideally suited to large data sets with complex frequency distributions. Detrital zircon ages are a good example of such data, because they usually comprise a data set of tens or hundreds of ages, often with a complex age distribution.

The method was first applied to detrital fission-track ages (Section 16.4), but has become much more popular for detrital zircon U–Pb ages. Dodson *et al.* (1988) first presented a probability density plot for SHRIMP U–Pb ages (Section 5.2.6), but its usage took off with the development of LA–ICP–MS. It is available in the Isoplot package (Ludwig, 2000), where it is called the cumulative Gaussian (probability) plot.

There has been some discussion concerning the appropriateness of the probability density plot (PDP) versus other methods of graphical representation such as the kernel density estimator (KDE, Vermeesch, 2012). The latter approach creates a distribution based on the frequency distribution alone, without any error weighting. However, the result is highly dependent on the chosen bandwidth (which roughly equates to the bin size of a histogram).

The pros and cons of the two methods can be illustrated using a test data set, presented by Vermeesch (2012), in which the KDE bandwith was chosen automatically based on the data distribution. Using a data set of 145 ages, the two

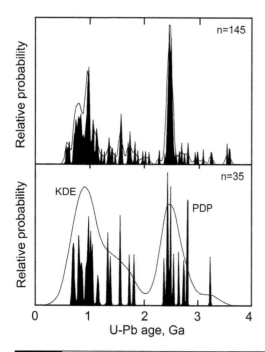

Fig. 2.31 Comparison of detrital zircon data presentation for a suite of 145 points and a random subset of 35. Black envelope = probability density; white envelope = kernel density. Modified after Vermeesch (2012).

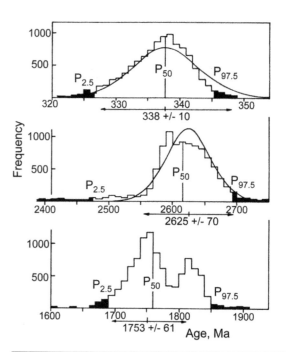

Fig. 2.32 Frequency distribution of 10 000 selection permutations from three sets of errorchron data. 95% (2σ) confidence limits of the 'bootstrap' age determination are indicated ($P_{2.5}$ and $P_{97.5}$). Arrows represent symmetrical 2σ confidence limits resulting from expansion of analytical errors until MSWD = 1. After Kalsbeek and Hansen (1989).

methods yield similar results (Fig. 2.31, upper plot). However, when 35 points were randomly selected from the larger data set, the results were very different (Fig. 2.31, lower plot). Vermeesch argued that the PDP data set is 'under-smoothed'. However, if the PDP data are accurately reflecting provenance ages within the sediment source, a sedimentologist might prefer the un-smoothed data set, while recognizing its incompleteness.

2.9.2 Isochron Data Distributions

Misinterpretation of errorchrons is usually due to a failure to properly visualize the distribution of data and attendant errors. This can be particularly problematical where geological scatter creates a non-normal distribution of data around the regression line. Since the least-squares fit assumes a normal distribution, a non-normal distribution can seriously bias an age calculation.

One way of assessing this problem is the 'bootstrap method' (Kalsbeek and Hansen, 1989). In this method a set of errorchron data is randomly re-sampled statistically to see how stable the regression line is to the application of a greater weighting to different points. What will normally happen is that a few points are selected more than once, while others are omitted. By repeating this process a few thousand times, a probability distribution is set up which portrays the stability of the best-fit line to the influence of certain sub-sets of the data suite (Fig. 2.32).

If geological errors are randomly distributed, the frequency histogram derived from the data set will have a symmetrical (Poisson) distribution. In this case the result is identical to expanding analytical errors by √MSWD. A true isochron should always yield a Poisson distribution, because analytical errors are assumed to be random. However, if geological scatter is uneven, the probability histogram of an errorchron may be skewed or even bimodal (Fig. 2.32), and hence highly suspect in terms of age assignment. This diagram therefore represents an excellent visual test for isochron data quality, and could help to avoid misinterpretation of problematical data sets.

A more recent form of the bootstrap approach was described by Powell et al. (2002). The objective of their method was to downplay the significance of extreme data points that lie outside a Poisson distribution by amplifying the errors on these data points more than the main data set. This approach was intended to generate more 'robust' age and error estimates by avoiding the distortion of an isochron age that could occur by full weighing of 'aberrant' points, while also reducing the 'temptation' to completely exclude such points from the calculation, as many workers do. The ideal approach would be to collect more data in order to resolve the non-Gaussian data more clearly, but in the real world this may not be possible.

References

Albarede, F., Telouk, P., Blichert-Toft, J. *et al.* (2004). Precise and accurate isotopic measurements using multiple-collector ICPMS. *Geochim. Cosmochim. Acta*, **68**, 2725–44.

Aldrich, L. T., Doak, J. B. and Davis, G. L. (1953). The use of ion exchange columns in mineral analysis for age determination. *American J. Sci.* **251**, 377–87.

Allen, C. M. and Campbell, I. H. (2012). Identification and elimination of a matrix-induced systematic error in LA–ICP-MS ^{206}Pb/^{238}U dating of zircon. *Chem. Geol.* **332**, 157–65.

Aston, F. W. (1919). A positive ray spectrograph. *Philos. Mag.* **38**, 707–14.

Aston, F. W. (1927). The constitution of ordinary lead. *Nature* **120**, 224.

Barovich, K. M., Beard, B. L., Cappel, J. B. *et al.* (1995). A chemical method for hafnium separation from high-Ti whole-rock and zircon samples. *Chem. Geol. (Isot. Geosci. Sect.)* **121**, 303–8.

Belshaw, N. S., Freedman, P. A., O'nions, R. K., Frank, M. and Guo, Y. (1998). A new variable dispersion double-focusing plasma mass spectrometer with performance illustrated for Pb isotopes. *Int. J. Mass Spec.* **181**, 51–8.

Blichert-Toft, J., Chauvel, C. and Albarede, F. (1997). Separation of Hf and Lu for high-precision isotope analysis by magnetic sector-multiple collector ICP–MS. *Contrib. Mineral. Petrol.* **127**, 248–60.

Boyet, M., Blichert-Toft, J., Rosing, M. *et al.* (2003). ^{142}Nd evidence for early Earth differentiation. *Earth Planet. Sci. Lett.* **214**, 427–42.

Boyet, M. and Carlson, R. W. (2006). A new geochemical model for the Earth's mantle inferred from ^{146}Sm–^{142}Nd systematics. *Earth Planet. Sci. Lett.* **250**, 254–68.

Brooks, C., Hart, S. R. and Wendt, I. (1972). Realistic use of two-error regression treatments as applied to rubidium–strontium data. *Rev. Geophys. Space Phys.* **10**, 551–77.

Brooks, C., Wendt, I. and Harre, W. (1968). A two-error regression treatment and its application to Rb–Sr and initial Sr^{87}/Sr^{86} ratios of younger Variscan granitic rocks from the Schwarzwald massif, Southwest Germany. *J. Geophys. Res.* **73**, 6071–84.

Burgoyne, T. W. and Hieftje, G. M. (1996). An introduction to ion optics for the mass spectrograph. *Mass Spec. Rev.* **15**, 241–59.

Cameron, A. E., Smith, D. H. and Walker, R. L. (1969). Mass spectrometry of nanogram-size samples of lead. *Anal. Chem.* **41**, 525–6.

Caro, G., Bourdon, B., Birck, J. L. and Moorbath, S. (2006). High-precision ^{142}Nd/^{144}Nd measurements in terrestrial rocks: constraints on the early differentiation of the Earth's mantle. *Geochim. Cosmochim. Acta* **70**, 164–91.

Cassidy, R. M. and Chauvel, C. (1989). Modern liquid chromatographic techniques for the separation of Nd and Sr for isotopic analyses. *Chem. Geol.* **74**, 189–200.

Catanzaro, E. J. and Kulp, J. L. (1964). Discordant zircons from the Little Butte (Montana), Beartooth (Montana) and Santa Catalina (Arizona) Mountains. *Geochim. Cosmochim. Acta* **28**, 87–124.

Chen, J. H. and Wasserburg, G. J. (1981). Isotopic determination of uranium in picomole and sub-picomole quantities. *Anal. Chem.* **53**, 2060–7.

Cheng, H., Edwards, R. L., Shen, C. C. *et al.* (2013). Improvements in ^{230}Th dating, ^{230}Th and ^{234}U half-life values, and U–Th isotopic measurements by multi-collector inductively coupled plasma mass spectrometry. *Earth Planet. Sci. Lett.* **371**, 82–91.

Christensen, J. N., Halliday, A. N., Godfrey, L. V., Hein, J. R. and Rea, D. K. (1997). Climate and ocean dynamics and the lead isotopic records in Pacific ferromanganese crusts. *Science* **277**, 913–18.

Clayton, D. D. (1978). On strontium isotopic anomalies and odd-A p-process abundances. *Astrophys. J.* **224**, L93–5.

Compston, W. and Oversby, V. M. (1969). Lead isotopic analysis using a double spike. *J. Geophys. Res.* **74**, 4338–48.

Cotte, M. (1938). Recherches sur l'optique electronique. *Ann. Physique* **10**, 333–405.

Crock, J. G., Lichte, F. E. and Wildeman, T. R. (1984). The group separation of the rare-earth elements and yttrium from geological materials by cation-exchange chromatography. *Chem. Geol.* **45**, 149–63.

Croudace, I. W. (1980). A possible error source in silicate wet-chemistry caused by insoluble fluorides. *Chem. Geol.* **31**, 153–5.

Crowley, Q. G., Heron, K., Riggs, N. *et al.* (2014). Chemical abrasion applied to LA–ICP-MS U-Pb zircon geochronology. *Minerals* **4**, 503–18.

Crumpler, T. B. and Yoe, J. H. (1940). *Chemical Computations and Errors*, Wiley, pp. 189–90.

Daly, N. R. (1960). Scintillation type mass spectrometer ion detector. *Rev. Sci. Instrum.* **31**, 264–7.

David, K., Birch, J. L., Telouk, P. and Allegre, C. J. (1999). Application of isotope dilution for precise measurement of Zr/Hf and ^{176}Hf/^{177}Hf ratios by mass spectrometry (ID–TIMS/ID–ICP–MS). *Chem. Geol.* **157**, 1–12.

Davis, D. W. (1982). Optimum linear regression and error estimation applied to U–Pb data. *Can. J. Earth Sci.* **19**, 2141–9.

Dawson, P. H. (1976). (Ed.) *Quadrupole Mass Spectrometry and its Applications.* Elsevier, 349 p.

DeBievre, P. J. and Debus, G. H. (1965). Precision mass spectrometric isotope dilution analysis. *Nucl. Instrum. Meth.* **32**, 224–8.

Dempster, A. J. (1918). A new method of positive ray analysis. *Phys. Rev.* **11**, 316–24.

DePaolo, D. J. and Wasserburg, G. J. (1976). Nd isotopic variations and petrogenetic models. *Geophys. Res. Lett.* **3**, 249–52.

Dickin, A. P., Jones, N. W., Thirlwall, M. and Thompson, R. N. (1987). A Ce/Nd isotope study of crustal contamination processes affecting Palaeocene magmas in Skye, NW Scotland. *Contrib. Mineral. Petrol.* **96**, 455–64.

Dodson, M. H. (1978). A linear method for second-degree interpolation in cyclical data collection. *J. Phys. E (Sci. Instrum.)* **11**, 296.

Dodson, M. H. (1963). A theoretical study of the use of internal standards for precise isotopic analysis by the surface ionisation technique: Part I – General first-order algebraic solutions. *J. Sci. Instrum.* **40**, 289–95.

Dodson, M. H., Compston, W., Williams, I. S. and Wilson, J. F. (1988). A search for ancient detrital zircons in Zimbabwean sediments. *J. Geol. Soc. Lond.* **145**, 977–83.

Dosso, L. and Murthy, V. R. (1980). A Nd isotope study of the Kerguelen islands: inferences on enriched oceanic mantle sources. *Earth Planet. Sci. Lett.* **48**, 268–76.

Eberhardt, A., Delwiche, R. and Geiss, Z. (1964). Isotopic effects in single filament thermal ion sources. *Z. Natur.* **19a**, 736–40.

Edwards, R. L., Chen, J. H. and Wasserburg, G. J. (1987). ^{238}U)^{234}U)^{230}Th)^{232}Th systematics and the precise measurement of time over the past 500,000 years. *Earth Planet. Sci. Lett.* **81**, 175–92.

Eugster, O., Tera, F., Burnett, D. S. and Wasserburg, G. J. (1970). The isotopic composition of gadolinium and neutron capture effects in some meteorites. *J. Geophys. Res.* **75**, 2753–68.

Faul, H. (1966). *Ages of Rocks, Planets, and Stars.* McGraw-Hill, 109 pp.

Feng, R., Machado, N. and Ludden, J. (1993). Lead geochronology of zircon by LaserProbe-Inductively coupled plasma mass spectrometry (LP-ICPMS). *Geochim. Cosmochim. Acta* **57**, 3479–86.

Fisher, C. M., Vervoort, J. D. and DuFrane, S. A. (2014a). Accurate Hf isotope determinations of complex zircons using the "laser ablation split stream" method. *Geochem. Geophys. Geosys.* **15**, 121–39.

Fisher, C. M., Vervoort, J. D. and Hanchar, J. M. (2014b). Guidelines for reporting zircon Hf isotopic data by LA-MC–ICPMS and potential pitfalls in the interpretation of these data. *Chem. Geol.* **363**, 125–33.

Gale, N. H. (1970). A solution in closed form for lead isotopic analysis using a double spike. *Chem. Geol.* **6**, 305–10.

Galer, S. J. G. (1999). Optimal double and triple spiking for high precision lead isotopic measurement. *Chem. Geol.* **157**, 255–74.

Gillson, G. R., Douglas, D. J., Fulford, J. E., Halligan, K. W. and Tanner, S. D. (1988). Nonspectroscopic interelement interferences in inductively coupled plasma mass spectrometry. *Anal. Chem.* **60**, 1472–4.

Habfast, K. (1983). Fractionation in the thermal ionization source. *Int. J. Mass Spectrom. Ion Phys.* **51**, 165–89.

Halliday, A. N., Christensen, J. N., Der-Chuen, L. *et al.* (2000). Multiple-collector inductively coupled plasma mass spectrometry. *Practical Spectroscopy Series* **23**, 291–328.

Halliday, A. N., Lee, D.-C., Christensen, J. N. *et al.* (1998). Applications of multiple collector- ICPMS to cosmochemistry, geochemistry, and paleoclimatology. *Geochim. Cosmochim. Acta* **62**, 919–40.

Hamelin, B., Manhes, G., Albarede, F. and Allegre, C. J. (1985). Precise lead isotope measurements by the double spike technique: a reconsideration. *Geochim. Cosmochim. Acta* **49**, 173–82.

Hirata, T. (1997). Ablation technique for laser ablation–inductively coupled plasma mass spectrometry. *J. Anal. Atom. Spec.* **12**, 1337–42.

Hirata, T. and Nesbitt, R. W. (1995). U–Pb isotope geochronology of zircon: Evaluation of the laser probe-inductively coupled plasma mass spectrometry technique. *Geochim. Cosmochim. Acta* **59**, 2491–500.

Hooker, P., O'Nions, R. K. and Pankhurst, R. J. (1975). Determination of rare-earth elements in U.S.G.S. standard rocks by mixed-solvent ion exchange and mass spectrometric isotope dilution. *Chem. Geol.* **16**, 189–96.

Horn, I., Rudnick, R. L. and McDonough, W. F. (2000). Precise elemental and isotope ratio determination by simultaneous solution nebulization and laser ablation-ICP-MS: application to U–Pb geochronology. *Chem. Geol.* **164**, 281–301. (Erratum = vol. **167**, 405–25.)

Horwitz, E. P., Chiarizia, R. and Dietz, M. L. (1992). A novel strontium-selective extraction chromatographic resin. *Solvent Extract. Ion Exchange* **10**, 313–36.

Houk, R. S. (1986). Mass spectrometry of inductively coupled plasmas. *Anal. Chem.* **58**, 97A–105A.

Houk, R. S., Fassel, V. A., Flesch, G. D. *et al.* (1980). Inductively coupled argon plasma for mass spectrometric determination of trace elements. *Anal. Chem.* **52**, 2283–9.

Ingram, M. G. and Chupka, P. (1953). Surface ionisation source using multiple filaments. *Rev. Sci. Instrum.* **24**, 518–20.

Kalsbeek, F. and Hansen, M. (1989). Statistical analysis of Rb)Sr isotope data by the 'bootstrap' method. *Chem. Geol. (Isot. Geosci. Sect.)* **73**, 289–97.

Klotzli, U., Klotzli, E., Gunes, Z. and Kosler, J. (2009). Accuracy of laser ablation U–Pb zircon dating: Results from a test using five different reference zircons. *Geostand. Geoanal. Res.* **33**, 5–15.

Kosler, J., Slama, J., Belousova, E. *et al.* (2013). U–Pb detrital zircon analysis-results of an inter-laboratory comparison. *Geostand. Geoanal. Res.* **37**, 243–59.

Krogh, T. E. (1973). A low contamination method for hydrothermal decomposition of zircon and extraction of U and Pb for isotopic age determination. *Geochim. Cosmochim. Acta* **37**, 485–94.

Krogh, T. E. (1982). Improved accuracy of U–Pb zircon ages by the creation of more concordant systems using the air abrasion technique. *Geochim. Cosmochim. Acta* **46**, 637–49.

Kuritani, T. and Nakamura, E. (2002). Precise isotope analysis of nanogram-level Pb for natural rock samples without use of double spikes. *Chem. Geol.* **186**, 31–43.

Kurz, E. A. (1979). Channel electron multipliers. *Amer. Lab.* **11** (3), 67–74.

Lee D.-C. and Halliday, A. N. (1995). Precise determinations of the isotopic compositions and atomic weights of molybdenum, tellurium, tin and tungsten using ICP source magnetic sector multiple collector mass spectrometry. *Int. J. Mass Spec. Ion Process.* **146/147**, 35–46.

Li, W. X., Lundberg, J., Dickin, A. P. *et al.* (1989). High-precision mass spectrometric uranium-series dating of cave deposits and implications for paleoclimate studies. *Nature* **339**, 534–6.

Luais, B., Telouk, P. and Albarede, F. (1997). Precise and accurate neodymium isotopic measurements by plasma-source mass spectrometry. *Geochim. Cosmochim. Acta* **61**, 4847–54.

Ludwig, K. R. (1980). Calculation of uncertainties of U–Pb isotope data. *Earth Planet. Sci. Lett.* **46**, 212–20.

Ludwig, K. R. (1991). ISOPLOT for MS-DOS, version 2.50. *US Geol. Surv. Open-File Report*, (88–557), 1–64.

Ludwig, K. R. (1997a). Optimization of multicollector isotope-ratio measurement of strontium and neodymium. *Chem. Geol.* **135**, 325–34.

Ludwig, K. R. (1997b). *Isoplot. Program and documentation, version 2.95. Revised edition of U.S. Geol. Surv. Open-File Report*, 91–445.

Ludwig, K. L. (2000). User's manual for Isoplot/Ex version 2.2: A geochronological toolkit for Microsoft Excel. *Berkley Geochronology Center Spec. Pub.* 1a, 1–53.

Ludwig, K. (2012). User's manual for Isoplot version 3.75-4.15: A geochronological toolkit for Microsoft Excel. *Berkley Geochronological Center Spec. Pub.* 5, 1–75.

Lugmair, G. W., Scheinin, N. B. and Marti, K. (1975). Search for extinct 146Sm, 1. The isotopic abundance of ^{142}Nd in the Juvinas meteorite. *Earth Planet. Sci. Lett.* **27**, 79–84.

Luo, X., Rehkamper, M., Lee, D.-C. and Halliday, A. N. (1997). High precision ^{230}Th/^{232}Th and ^{234}U/^{238}U measurements using energy-filtered ICP magnetic sector multiple collector mass spectrometry. *Int. J. Mass Spec. Ion Process.* **171**, 105–17.

Marechal, C. N., Telouk, P. and Albarede, F. (1999). Precise analysis of copper and zinc isotopic compositions by plasma-source mass spectrometry. *Chem. Geol.* **156**, 251–73.

Marillo-Sialer, E., Woodhead, J., Hanchar, J. M. *et al.* (2016). An investigation of the laser-induced zircon 'matrix effect'. *Chem. Geol.* **438**, 11–24.

Marillo-Sialer, E., Woodhead, J., Hergt, J. *et al.* (2014). The zircon 'matrix effect': evidence for an ablation rate control on the accuracy of U–Pb age determinations by LA-ICP-MS. *J. Anal. Atom. Spec.* **29**, 981–9.

Mattinson, J. M. (2005). Zircon U–Pb chemical abrasion ("CA-TIMS") method: combined annealing and multi-step partial dissolution analysis for improved precision and accuracy of zircon ages. *Chem. Geol.* **220**, 47–66.

McIntyre, G. A., Brooks, A. C. Compston, W and Turek, A. (1966). The statistical assessment of Rb)Sr isochrons. *J. Geophys. Res.* **71**, 5459–68.

Manton, W. I. (1988). Separation of Pb from young zircons by single-bead ion exchange. *Chem. Geol. (Isot. Geosci. Sect.)* **73**, 147–52.

Nicolaysen. L. O. (1961). Graphic interpretation of discordant age measurements on metamorphic rocks. *Ann. N. Y. Acad. Sci.* **91**, 198–206.

Nier, A. O. (1940). A mass spectrometer for routine isotope abundance measurements. *Rev. Sci. Instrum.* **11**, 212–16.

Niu, H. and Houk, R. S. (1996). Fundamental aspects of ion extraction in inductively coupled plasma mass spectrometry. *Spectrochim. Acta B.* **51**, 779–815.

O'Nions, R. K., Carter, S. R., Evensen, N. M. and Hamilton P. J. (1979). Geochemical and cosmochemical applications of Nd isotope analysis. *Ann. Rev. Earth Planet. Sci.* **7**, 11–38.

Parrish, R. R. (1987). An improved micro-capsule for zircon dissolution in U–Pb geochronology. *Chem. Geol. (Isot. Geosci. Sect.)* **66**, 99–102.

Parrish, R. R. and Krogh, T. E. (1987). Synthesis and purification of ^{205}Pb for U)Pb geochronology. *Chem. Geol. (Isot. Geosci. Sect.)* **66**, 103–10.

Patchett, P. J. (1980). Sr isotopic fractionation in Ca)Al inclusions from the Allende meteorite. *Nature* **283**, 438–41.

Patchett, P. J. and Tatsumoto, M. (1980). A routine high-precision method for Lu)Hf isotope geochemistry and chronology. *Contrib. Mineral. Petrol.* **75**, 263–7.

Pin, C., Briot, D., Bassin, C. and Poitrasson, F. (1994). Concomitant separation of strontium and samarium–neodymium for isotopic analysis in silicate samples, based on specific extraction chromatography. *Analytica Chimica Acta*, **298**, 209–17.

Potts, P. J. (1987). *Handbook of Silicate Rock Analysis*. Blackie, 622 pp.

Powell, R., Hergt, J. and Woodhead, J. (2002). Improving isochron calculations with robust statistics and the bootstrap. *Chem. Geol.* **185**, 191–204.

Powell, R., Woodhead, J. and Hergt, J. (1998). Uncertainties on lead isotope analyses: deconvolution in the double-spike method. *Chem. Geol.* **148**, 95–104.

Rehkamper, M., Gartner, M., Galer, S. J. G. and Goldstein, S. L. (1996). Separation of Ce from other rare-earth elements with application to Sm–Nd and La–Ce chronometry. *Chem. Geol.* **129**, 201–8.

Richard, P., Shimizu, N. and Allegre, C. J. (1976). ^{143}Nd/^{146}Nd, a natural tracer: an application to oceanic basalts. *Earth Planet. Sci. Lett.* **31**, 269–78.

Roddick, J. C., Loveridge, W. D. and Parrish, R. R. (1987). Precise U/Pb dating of zircon at the sub-nanogram Pb level. *Chem. Geol. (Isot. Geosci. Sect.)* **66**, 111–21.

Russell, W. A., Papanastassiou, D. A. and Tombrello, T. A. (1978). Ca isotope fractionation on the Earth and other solar system materials. *Geochim. Cosmochim. Acta* **42**, 1075–90.

Schoene, B. (2014). 4.10 U–Th–Pb Geochronology. *Treatise on Geochemistry, Second Edn.* Elsevier, 341–78.

Shen, C. C., Wu, C. C., Cheng, H. *et al.* (2012). High-precision and high-resolution carbonate 230Th dating by MC-ICP-MS with SEM protocols. *Geochim. Cosmochim. Acta* **99**, 71–86.

Smyth, W. R. and Mattauch, J. (1932). A new mass spectrometer. *Phys. Rev.* **40**, 429.

Solari, L. A., Ortega-Obregón, C. and Bernal, J. P. (2015). U-Pb zircon geochronology by LAICPMS combined with thermal annealing: achievements in precision and accuracy on dating standard and unknown samples. *Chem. Geol.* **414**, 109–23.

Tanaka, T. and Masuda, A. (1982). The La–Ce geochronometer: a new dating method, *Nature* **300**, 515–18.

Thirlwall, M. F. (1982). A triple-filament method for rapid and precise analysis of rare-earth elements by isotope dilution. *Chem. Geol.* **35**, 155–66.

Thirlwall, M. F. (1991a). High-precision multicollector isotopic analysis of low levels of Nd as oxide. *Chem. Geol. (Isot. Geosci. Sect.)* **94**, 13–22.

Thirlwall, M. F. (1991b). Long-term reproducibility of multicollector Sr and Nd isotope ratio analyses. *Chem. Geol. (Isot. Geosci. Sect.)* **94**, 85–104.

Thirlwall, M. F. (2000). Inter-laboratory and other errors in Pb isotope analyses investigated using a ^{207}Pb–^{204}Pb double spike. *Chem. Geol.* **163**, 299–322.

Thirlwall, M. F. (2002). Multicollector ICP-MS analysis of Pb isotopes using a ^{207}Pb–^{204}Pb double spike demonstrates up to 400 ppm/amu systematic errors in Tl-normalization. *Chem. Geol.* **184**, 255–79.

Thirlwall, M. F. and Walder, A. J. (1995). In situ hafnium isotope ratio analysis of zircon by inductively coupled plasma multiple collector mass spectrometry. *Chem. Geol.* **122**, 241–7.

Thompson J. J. (1913). *Rays of Positive Electricity and their Application to Chemical Analysis*, Longmans, Green and Co. Ltd.

Titterington, D. M. and Halliday, A. N. (1979). On the fitting of parallel isochrons and the method of maximum likelihood. *Chem. Geol.* **26**, 183–95.

Todt, W., Cliff, R. A., Hanser, A. and Hofmann, A. W. (1996). Evaluation of a ^{202}Pb–^{205}Pb double spike for high-precision lead isotopic analysis. In: Basu, A. and Harts, S. R. (Eds) *Earth Processes: Reading the Isotopic Code. Geophys. Monograph* **95**, American Geophysical Union, pp. 429–37.

Tompkins, E. R., Khym, J. X. and Cohn, W. E. (1947). Ion-Exchange as a separation method. I. The separation of fission-produced radioisotopes, including individual rare earths, by complexing elution from Amberlite resin. *J. American Chem. Soc.* **69**, 2769–77.

Trinquier, A. and Komander, P. (2016). Precise and accurate uranium isotope analysis by modified total evaporation using 1013 ohm current amplifiers. *J. Radioanal. Nucl. Chem.* **307**, 1927–32.

Vance, D. and Thirlwall, M. (2002). An assessment of mass discrimination in MC-ICPMS using Nd isotopes. *Chem. Geol.* **185**, 227–40.

Vermeesch, P. (2012). On the visualisation of detrital age distributions. *Chem. Geol.* **312**, 190–4.

Von Quadt, A., Gallhofer, D., Guillong, M. *et al.* (2014). U-Pb dating of CA/non-CA treated zircons obtained by LA-ICP-MS and CA-TIMS techniques: impact for their geological interpretation. *J. Anal. Atom. Spec.* **29**, 1618–29.

Walder, A. J., Abell, I. D., Freedman, P. A. and Platzner, I. (1993a). Lead isotopic ratio measurement of NIST 610 glass by laser ablation-inductively coupled plasma-mass spectrometry. *Spectrochim. Acta* **48B**, 397–402.

Walder, A. J. and Freedman, P. A. (1992). Isotopic ratio measurement using a double focusing magnetic sector mass analyser with an inductively coupled plasma as an ion source. *J. Anal. Atom. Spec.* **7**, 571–5.

Walder, A. J. and Furuta, N. (1993). High precision lead isotope ratio measurement by inductively coupled plasma multiple collector mass spectrometry. *Anal. Sci.* **9**, 675–80.

Walder, A. J., Platzner, I. and Freedman, P. A. (1993b). Isotope ratio measurement of lead, neodymium and neodymium–samarium mixtures, hafnium and hafnium-lutetium mixtures with a double focussing multiple collector inductively coupled plasma mass spectrometer. *J. Anal. Atomic. Spectrom.* **8**, 19–23.

Wasserburg, G. J., Jacobsen, S. B., DePaolo, D. J. McCulloch, M. T. and Wen, T. (1981). Precise determination of Sm/Nd ratios, Sm and Nd isotopic abundances in standard solutions. *Geochim. Cosmochim. Acta* **45**, 2311–23.

Wendt, I. and Carl, C. (1991). The statistical distribution of the mean squared weighted deviation. *Chem. Geol. (Isot. Geosci. Sect.)* **86**, 275–85.

White, W. M., Albarede, F. and Telouk, P. (2000). High-precision analysis of Pb isotope ratios by multi-collector ICP-MS. *Chem. Geol.* **167**, 257–70.

Woodhead, J. (2002). A simple method for obtaining highly accurate Pb isotope data by MC-ICP-MS. *J. Anal. Atom. Spec.* **17**, 1381–5.

Woodhead, J., Hergt, J., Shelley, M., Eggins, S. and Kemp, R. (2004). Zircon Hf-isotope analysis with an excimer laser, depth profiling, ablation of complex geometries, and concomitant age estimation. *Chem. Geol.* **209**, 121–35.

York, D. (1966). Least-squares fitting of a straight line. *Can. J. Phys.* **44**, 1079–86.

York, D. (1967). The best isochron. *Earth Planet. Sci. Lett.* **2**, 479–82.

York, D. (1969). Least-squares fitting of a straight line with correlated errors. *Earth Planet. Sci. Lett.* **5**, 320–4.

Chapter 3

The Rb–Sr Method

The Rb–Sr isotope method was developed in the mid twentieth century as one of the first relatively precise and accurate radiometric dating techniques. It has since been superseded in most areas, but remains significant as a demonstration of scientific principles, and as one of the most important 'conservative' oceanographic tracers (Section 3.6).

Rubidium, a group 1 alkali metal, has two naturally occurring isotopes, ^{85}Rb and ^{87}Rb, whose abundances are 72.17% and 27.83% respectively. These figures yield an atomic abundance ratio of ^{85}Rb/^{87}Rb = 2.593 (Catanzaro et al., 1969), which is a constant throughout the Earth, moon and most meteorites due to isotopic homogenization in the solar nebula. ^{87}Rb is radioactive, and decays to the stable isotope ^{87}Sr by emission of a β particle and antineutrino ($\bar{v}$) . The decay energy (Q) is shared as kinetic energy by these two particles.

$$^{87}_{37}Rb \rightarrow {}^{87}_{38}Sr + \beta^- + \bar{v} + Q$$

3.1 The Rb Decay Constant

The low decay energy for the β decay of ^{87}Rb (275 keV) has always caused problems in the laboratory determination of the decay constant. This makes the problem a particularly good example of the alternative approaches that have been taken to overcome these difficulties.

Because the decay energy is divided between the β particle and anti-neutrino, the β particles have a smooth distribution of kinetic energy from the total energy down to zero. When attempting to accurately determine the decay constant by direct counting, the low-energy β particles cause great problems because they may be absorbed by surrounding Rb atoms before they ever reach the detector. For example, in a thick (>1 μm deep) solid Rb sample, attenuation is so severe that a false frequency maximum is generated at ca. 10 keV (Fig. 3.1).

One way to avoid the attenuation problem is to use a photo-multiplier with a liquid scintillator solution doped with Rb. The β particles will be absorbed by molecules of

the scintillator (emitting light flashes) before they can be absorbed by other Rb atoms. The major problem with this method is that a low-energy cut-off at ca. 10 keV must be applied to avoid the high background noise associated with liquid scintillation. The consequent extrapolation of count-rate curves down to zero energy leads to a large uncertainty, with half-life determinations from 47 to 52 Ga (Flynn and Glendenin, 1959; Brinkman et al., 1965).

Another approach to direct counting is to make measurements with progressively thinner solid Rb sources using a proportional counter. The results are then extrapolated to a theoretical source of zero thickness to remove the effect of self-absorption. The proportional counter has a much lower noise level, so the energy cut-off can be set as low as 0.185 keV. Rb films with thicknesses down to 1 μm were measured by Neumann and Huster (1974), and extrapolated to zero

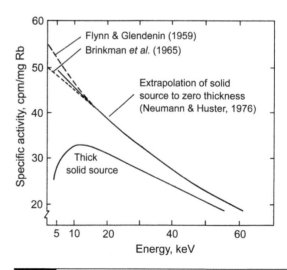

Fig. 3.1 Plot of activity against kinetic energy for β particles generated by ^{87}Rb decay. Solid lines = solid Rb sources; dashed lines = liquid scintillator measurements. After Neumann and Huster (1976).

thickness by Neumann and Huster (1976) to derive an ^{87}Rb half-life of 48.8 ± 0.8 Ga (equivalent to a decay constant of 1.42×10^{-11} a^{-1}).

An alternative approach to determine the decay constant is to measure the amount of ^{87}Sr produced by decay of a known quantity of ^{87}Rb in the laboratory over a known period of time. This method was first attempted by McMullen et al. (1966) on a rubidium sample they had purified in 1956, and was repeated on the same sample batch by Davis et al. (1977). Unfortunately, McMullen et al. omitted to measure the small but significant level of residual ^{87}Sr present in their rubidium before they put it away on the shelf. Hence, the accuracy of their determination was compromised. However, this problem contributes less than 1% uncertainty to the later determination of Davis et al. (1977). Their proposed decay constant ($1.42 \pm 0.01 \times 10^{-11}$ a^{-1}) supported the value of Neumann and Huster (1976) and was adopted by international convention the following year (Steiger and Jager, 1977).

The last approach to determining the decay constant is to date geological samples whose ages have also been determined by other methods with more reliable decay constants. This method has the disadvantage that it involves geological uncertainties, such as whether all isotopic systems closed at the same time and remained closed. However, it provides a useful check on the direct laboratory determinations. The best determination was made by Minster et al. (1982) using very precise U–Pb and Rb–Sr isochrons for chondritic meteorites, and gave a lower decay constant of $1.402 \pm 0.008 \times 10^{-11}$ a^{-1}, equivalent to a half-life of 49.4 ± 0.3 Ga.

The subject was reviewed by Begemann et al. (2001), who suggested that re-analysis of the data sets of Neumann and Huster (1976) and Davis et al. (1977) supported a downward revision of their decay constants towards the value of Minster et al. (1982). However, they also called for new work, which appeared over the next few years (Fig. 3.2), beginning with a new counting determination by Kossert (2003). This gave the lowest decay constant yet determined ($1.396 \pm 0.009 \times 10^{-11}$ a^{-1}), but because the error was larger than that of Minster et al., the two values overlapped.

A new geological determination by Nebel et al. (2011) was based on three intrusions with U–Pb ages of 229, 1166 and 2060 Ma. These gave decay constants of 1.398, 1.391 and 1.390×10^{-11} a^{-1} respectively. However, the two lower values might decrease slightly further if their concordia ages are corrected for a probable error in the ^{235}U decay constant (Section 5.1.1).

Finally, a new measurement of the 'laboratory shelf' sample of Davis was made by Rotenberg et al. (2012). With an additional 30 years of Sr accumulation and more precise measurements, a much more precise decay constant of $1.397 \pm 0.003 \times 10^{-11}$ a^{-11} was determined (equal to a half-life of 49.62 Ga), which lies within the range of geological determinations by Nebel et al. (2011). Hence this value was used as the main basis of a new IUPAC–IUG recommendation by Villa et al. (2015), as quoted in Section 1.4.

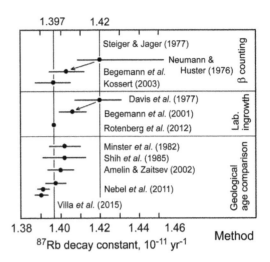

Fig. 3.2 Values of the ^{87}Rb decay constant discussed in the text. Vertical lines indicate values chosen by convention in 1977 and 2015. Modified after Nebel et al. (2011).

3.2 Dating Igneous Crystallization

The Rb–Sr method has largely been superseded as a tool for dating igneous rocks. However, the method provides a good illustration of the principles of isotope dating, and will therefore be reviewed here to demonstrate those principles. This application begins from the general equations for radioactive decay (Section 1.4). Hence, the number of ^{87}Sr daughter atoms produced by decay of ^{87}Rb in a rock or mineral since its formation t years ago is given by substituting into decay equation [1.10]:

$$^{87}\mathrm{Sr} = {}^{87}\mathrm{Sr_I} + {}^{87}\mathrm{Rb} \cdot \left(e^{\lambda t} - 1\right) \qquad [3.1]$$

where ^{87}Sr$_I$ is the number of ^{87}Sr atoms present initially. However, it is difficult to measure precisely the absolute abundance of a given nuclide. Therefore it is more convenient to convert this number to an isotope ratio by dividing through by ^{86}Sr (which is not produced by radioactive decay and therefore remains constant with time). Hence we obtain

$$\left(\frac{^{87}\mathrm{Sr}}{^{86}\mathrm{Sr}}\right)_P = \left(\frac{^{87}\mathrm{Sr}}{^{86}\mathrm{Sr}}\right)_I + \frac{^{87}\mathrm{Rb}}{^{86}\mathrm{Sr}} \cdot (e^{\lambda t} - 1) \qquad [3.2]$$

The present day Sr isotope ratio (P) is measured by mass spectrometry, and the atomic ratio ^{87}Rb/^{86}Sr is calculated from the weight ratio of Rb/Sr. If the initial ratio (^{87}Sr/^{86}Sr)$_I$ is known or can be estimated then t can be determined, subject to the assumption that the system has been closed to Rb and Sr mobility from time t until the present:

$$t = \frac{1}{\lambda} \ln \left\{ 1 + \frac{^{86}\mathrm{Sr}}{^{87}\mathrm{Rb}} \left[\left(\frac{^{87}\mathrm{Sr}}{^{86}\mathrm{Sr}}\right)_P - \left(\frac{^{87}\mathrm{Sr}}{^{86}\mathrm{Sr}}\right)_I \right] \right\} \qquad [3.3]$$

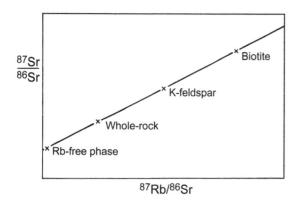

Fig. 3.3 Schematic Rb–Sr isochron diagram for a suite of comagmatic igneous minerals, as conceived by Nicolaysen (1961).

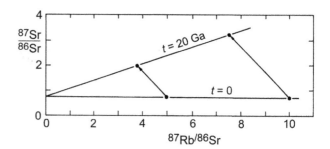

Fig. 3.4 Rb–Sr isochron diagram on axes of equal magnitude showing production of ^{87}Sr as ^{87}Rb is consumed in two hypothetical samples.

3.2.1 Sr Model Ages

When the Rb–Sr method was first used in geochronology, the poor precision attainable with mass spectrometry limited the technique to the dating of Rb-rich minerals such as lepidolite. These minerals develop such high ^{87}Sr/^{86}Sr ratios over geological time that a uniform initial ^{87}Sr/^{86}Sr ratio of 0.712 could be assumed in all dating studies without introducing significant errors. Such determinations are called 'model ages' because the initial ratio is predicted by a model rather than measured directly.

Subsequently, the Rb–Sr method was extended to less exotic rock-forming minerals such as biotite, muscovite and K-feldspar, with lower Rb/Sr ratios. However, discordant dates were often generated, by assuming an initial ratio of 0.712 when the real initial ratio was higher. This problem was first recognized by Compston and Jeffery (1959), and overcome by the invention of the isochron diagram (Nicolaysen, 1961). Model ages subsequently re-appeared in more specialized aspects of Rb–Sr dating such as meteorite chronology, and as an important approach in the Sm–Nd system (Section 4.2).

3.2.2 The Isochron Diagram

An examination of equation [3.2] shows that it is equivalent to the equation for a straight line,

$$y = c + x\,m \qquad [3.4]$$

This led Nicolaysen (1961) to develop a new way of treating Rb–Sr data, by plotting ^{87}Sr/^{86}Sr (y) against ^{87}Rb/^{86}Sr (x). The intercept (c) is then the initial ^{87}Sr/^{86}Sr ratio of the system. On this diagram, a suite of comagmatic minerals having the same age and initial ^{87}Sr/^{86}Sr ratio and which have since remained as closed systems define a line termed an 'isochron'. The slope of this line, m ($= e^{\lambda t} - 1$), yields the age of the minerals. If one of the minerals is very Rb-poor then this may yield the initial ratio directly (Fig. 3.3). Otherwise,

the initial ratio is determined by extrapolating back to the y axis a best-fit line through the available data points. Because $\lambda\,^{87}$Rb is so small, for geologically young rocks the slope may be quite accurately approximated by λt. Such an approximation does not hold for nuclides with shorter half-lives, such as K and U.

The isotopic evolution of a suite of hypothetical minerals in the isochron diagram is illustrated in Fig. 3.4. At the time of crystallization of the rock, both of the minerals have the same ^{87}Sr/^{86}Sr ratio, and plot as points on a horizontal line. After each mineral has become a closed system (effectively at the same instant for the minerals in a high-level, fast-cooled intrusion) isotopic evolution begins. On a diagram where the two axes have the same scale (Fig. 3.4), the points move up straight lines with a slope of -1 as each ^{87}Rb decay increases ^{87}Sr/^{86}Sr and reduces ^{87}Rb/^{86}Sr by the same amount. Each mineral composition remains on the isochron as its slope increases with time. In practice, the y axis is usually very much expanded to display rocks of geological age in a suitable format, and the growth lines are then nearly vertical.

Another development of the Rb–Sr method (Schreiner, 1958) was the analysis of co-genetic whole-rock sample suites, as an alternative to separate minerals. To be effective, a whole-rock suite must display variation in modal mineral content, such that samples display a range of Rb/Sr ratios, without introducing any variation in initial Sr isotope ratio. In actual fact, perfect initial ratio homogeneity may not be achieved, especially in rocks with a mixed magmatic parentage. However, if the spread in Rb/Sr ratios is sufficient, any initial ratio variations are swamped and an accurate age can be determined. Initial ratio heterogeneity is a greater problem in Sm–Nd isochrons, and is therefore discussed under that heading (Section 4.1.2). Schreiner's proposal actually preceded the invention of the Rb–Sr isochron diagram, but some of his data are presented on an isochron diagram in Fig. 3.5 to demonstrate the method.

Graphical calculation of isochron ages was superseded in the 1960s by the application of least squares regression techniques (Section 2.8), but the isochron diagram remains a

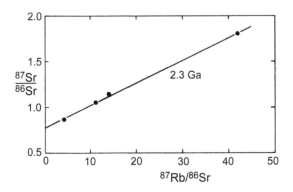

Fig. 3.5 Rb–Sr whole-rock isochron for the 'red granite' of the Bushveld complex, using the data of Schreiner (1958).

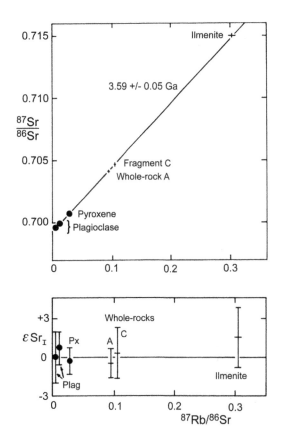

Fig. 3.6 Rb–Sr data for Lunar Mare sample 10017 plotted: (a) on a conventional isochron diagram; and (b) on a diagram of ε against Rb/Sr. After Papanastassiou et al. (1970).

very useful tool for assessing the distribution of data points about an isochron. However, Papanastassiou *et al.* (1970) found that the vertical scale of the isochron diagram was too compressed to allow clear portrayal of the experimental error bars on their data points. To overcome this problem they developed the ε notation, which they defined as the relative deviation of a data point from the best-fit isochron in parts per 10^4. This is given by

$$\varepsilon = \left[\frac{\left(^{87}\mathrm{Sr}/^{86}\mathrm{Sr}\right)_{\mathrm{measured}}}{\left(^{87}\mathrm{Sr}/^{86}\mathrm{Sr}\right)_{\mathrm{best-fit}}} - 1 \right] \times 10^4 \qquad [3.5]$$

Figure 3.6 shows a combined mineral isochron diagram and εSr diagram for an Apollo 11 sample from the Sea of Tranquillity. A limitation of the ε diagram is that the vertical error bars only describe errors in $^{87}\mathrm{Sr}/^{86}\mathrm{Sr}$, whereas errors in Rb/Sr ratio can also cause points to deviate from the line. However, if Rb/Sr ratios are measured by isotope dilution (Section 2.5.1), analytical errors in this variable are normally subordinate to errors in measured isotope ratio.

3.2.3 Erupted Isochrons

A primary basic magma should inherit the isotopic composition of its mantle source, providing that melting occurs in equilibrium conditions. Tatsumoto (1966) first suggested, based on U–Pb data, that primitive basic magmas could also inherit the parent/daughter ratio of their mantle source. If different magma batches were to sample the elemental and isotopic composition of different source domains, this might lead to the eruption of an 'isochron' suite whose slope would yield the time over which these sources were isolated. This concept was examined for the Rb–Sr system in ocean island basalts by Sun and Hansen (1975).

Alkali basalts from 14 different ocean island suites were plotted on an isochron diagram (Fig. 3.7). The data are fairly scattered, but form a positive correlation with a slope age of ca. 2 Ga. Individual ocean islands may also define arrays with positive slope, but usually with more scatter. Sun and

Hansen attributed the positive correlations between Rb/Sr and isotopic composition to mantle heterogeneity, suggesting that the apparent ages represented the time since mantle domains were isolated from the convecting mantle. These were termed 'mantle isochrons' by Brooks *et al.* (1976a).

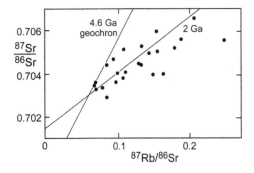

Fig. 3.7 Rb–Sr isochron diagram for young volcanic rocks (mostly alkali basalts) from ocean islands. After Sun and Hansen (1975).

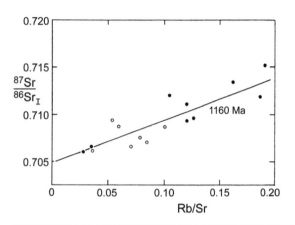

Fig. 3.8 Pseudo-isochron for quartz and olivine norites from Haddo House (●) and Arnage (○) in NE Scotland, yielding an apparent age of 1160 ± 420 Ma (2σ) prior to intrusion. After Brooks *et al.* (1976b).

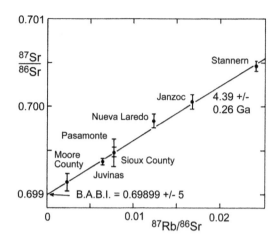

Fig. 3.9 Rb–Sr isochron diagram for whole-rock samples of basaltic achondrites showing the determination of 'BABI'. After Papanastassiou and Wasserburg (1969).

The mantle isochron concept was extended to continental igneous rocks by Brooks *et al.* (1976b). Because these are often ancient (unlike most ocean island basalts), it was necessary to correct measured $^{87}Sr/^{86}Sr$ ratios back to their calculated initial ratios at the time of magmatism, before plotting against Rb/Sr (e.g. Fig. 3.8). Hence Brooks *et al.* termed these plots 'pseudo-isochron' diagrams. They listed 30 examples from both volcanic and plutonic continental igneous rock suites where the data formed a roughly linear array. The controversial aspect of this work was that Brooks *et al.* rejected the possibility that these pseudo-isochrons were mixing lines produced by crustal contamination of mantle-derived basic magmas. Instead they believed them to date mantle differentiation events which established domains of different Rb/Sr ratio in the sub-continental lithosphere.

It is a fundamental assumption of the mantle isochron model that neither isotope nor elemental ratios are perturbed during magma ascent through the crust. However, it is now generally accepted that this assumption is not upheld with sufficient reliability to attribute age significance to erupted isochrons. For example, the Haddo House norites of NE Scotland (Fig. 3.8) are known to contain pelitic xenoliths, so this array must document crustal assimilation. Breakdown of the mantle isochron model can also be caused by low degrees of melting in the mantle source, leading to fractionation between Rb, an ultra-incompatible, and Sr, a moderately incompatible element. Hence, it is concluded that only isotope–isotope erupted isochrons (such as provided by the Pb isotope system) can reliably be interpreted as dating the ages of mantle differentiation events.

3.2.4 Meteorite Chronology
Meteorites have been the subject of numerous Rb–Sr dating studies, but some of the most important Rb–Sr results on meteorites are initial ratio determinations. These have

significance, both as a reference point for terrestrial Sr isotope evolution, and as a model-age dating tool for estimating the relative condensation times of solar system bodies.

The first accurate measurement of meteorite initial ratios was made by Papanastassiou and Wasserburg (1969) on basaltic achondrites. These differ from chondritic meteorites in showing evidence of differentiation after their accretion from the solar nebula. However, they might not have participated in the full planetary differentiation process which generated iron meteorites. Their low Rb/Sr ratios have resulted in only limited radiogenic Sr production since differentiation, so an accurate initial ratio determination is possible.

In order to make this determination, Papanastassiou and Wasserburg analysed whole-rock samples from seven different basaltic achondrites, yielding an isochron (Fig. 3.9) without any excess scatter over analytical error. An age of 4.39 ± 0.26 Ga was calculated using a decay constant of 1.39×10^{-11} a^{-1}. The initial ratio of 0.69899 ± 5 was referred to by Papanastassiou and Wasserburg as the 'basaltic achondrite best initial' or BABI. This value represents a bench-mark to which other meteorite initial ratios may be compared. Birck and Allegre (1978) repeated this study with the addition of separated minerals from Juvinas and Ibitira, yielding an identical initial ratio, but an improved age determination of 4.57 ± 0.13 Ga (with the same decay constant). However, reliable Rb–Sr mineral isochrons are not possible for other achondrites due to later disturbance.

The determination of precise initial ratios for chondritic meteorites is more problematical due to their much higher Rb/Sr ratios than basaltic achondrites. However, by separating out low-Rb/Sr phosphate minerals, Wasserburg *et al.* (1969) and Gray *et al.* (1973) were able to determine good initial ratios for the chondrites Guarena and Peace River.

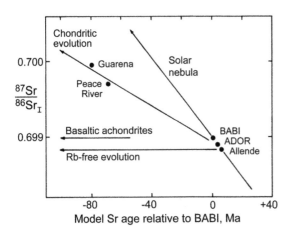

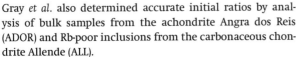

Fig. 3.10 Plot of initial Sr isotope composition for selected meteorites against model ages for condensation or differentiation–metamorphism, based on assumed Rb/Sr ratios in major reservoirs. ADOR = Angra dos Reis. After Gray et al. (1973).

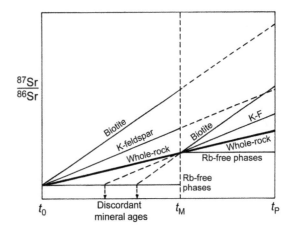

Fig. 3.11 Plot of Sr isotope ratio against time to model the effect of a metamorphic event which opens Rb–Sr mineral systems, but not the whole-rock system. t_0 = age of rock; t_M = age of metamorphism; t_P = present. After Fairbairn et al. (1961).

Gray et al. also determined accurate initial ratios by analysis of bulk samples from the achondrite Angra dos Reis (ADOR) and Rb-poor inclusions from the carbonaceous chondrite Allende (ALL).

These initial ratios were translated into a relative chronology for meteorite condensation (Fig. 3.10) by assuming a homogeneous Rb/Sr ratio in the solar nebula (Papanastassiou and Wasserburg, 1969). The results are only 'model' ages because they depend on an assumed composition for the source reservoir (solar nebula), and they would be rendered invalid if it did not evolve as a homogeneous reservoir. The estimate for Rb/Sr in the solar nebula was based on cited spectroscopic measurements from the Sun, yielding a value of 0.65, which is capable of generating an increase in $^{87}Sr/^{86}Sr$ of ca. 0.0001 in 4 Ma.

If we assume a homogeneous Sr isotope distribution in the solar nebula, the Allende data suggest this to be the oldest known object in the solar system, predating the condensation of basaltic achondrites by ca. 10 Ma (Fig. 3.10). Similarly, Angra dos Reis has a Sr model age ca. 5 Ma older than BABI. Application of the same model to the high initial ratios of Guarena and Peace River would imply unduly late condensation from the solar nebula (25–30 Ma after BABI). Therefore, Gray et al. interpreted these as metamorphic ages produced by re-distribution of Rb and Sr between mineral phases within chondritic bodies (see chondritic evolution line in Fig. 3.10). However, basaltic achondrites and ADOR are themselves products of planetary differentiation. Therefore, a better interpretation (Tilton, 1988) is that the entire model chronology really indicates times of differentiation and metamorphism, rather than condensation. Subsequent work has confirmed this interpretation (e.g. Halliday and Porcelli, 2001).

3.3 Dating Metamorphic Systems

The Rb–Sr method was one of the first to be applied as a general dating tool for metamorphic minerals, but has now been superseded by several different methods with more specialized attributes, some chosen for their greater sensitivity to low-temperature processes (e.g. K–Ar, Ar–Ar, fission-tracks), others for high-temperature applications (e.g. Sm–Nd and Lu–Hf). The Rb–Sr method provides classic case studies that illustrate important principles in the application of isotope systems to metamorphic rocks. However, open whole-rock systems will not be discussed in detail, since Rb–Sr dating of igneous crystallization is no longer considered a useful method in high-grade metamorphic terranes (e.g. Field and Raheim, 1979a, b).

3.3.1 Mineral and Whole-Rock Isochrons

Mineral and whole-rock Rb–Sr systems may respond differently to metamorphic events. ^{87}Sr generated by Rb decay occupies unstable lattice sites in Rb-rich minerals and tends to migrate out of the crystal if subjected to a thermal pulse, even of a magnitude well below the melting temperature. However, if fluids in the rock remain static, Sr released from Rb-rich minerals such as mica and K-feldspar will tend to be taken up by the nearest Sr sink, such as plagioclase or apatite.

The idea of using whole-rock analysis to see back through a metamorphic event which disturbs mineral systems was first conceived by Compston and Jeffery (1959). The model was illustrated graphically by Fairbairn et al. (1961) on a plot of isotope ratio against time (Fig. 3.11). After the formation of a rock at time t_0, different minerals move along

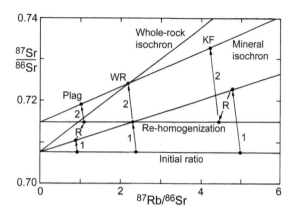

Fig. 3.12 Hypothetical behaviour of a partially disturbed mineral–whole-rock isochron. Evolution lines: 1 = period from igneous crystallization to metamorphism; R = metamorphic re-homogenization; 2 = period from metamorphism to present day.

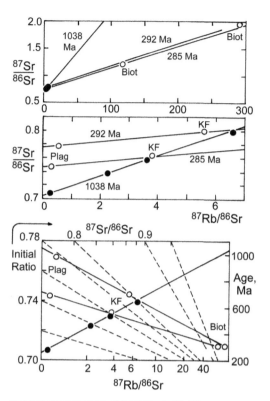

Fig. 3.13 Rb–Sr isochrons for the Baltimore gneiss showing 1038 Ma 'plutonic' and 285–292 Ma metamorphic ages defined by Rb–Sr whole-rock systems (●) and minerals (○). The data are also shown in the isochron diagram of Provost (1990) for comparison. After Provost (1990).

different growth lines, whose steepness corresponds to their Rb/Sr ratio. Isotopic evolution continues until the minerals are homogenized by a thermal event at time t_M. Thereafter, isotopic evolution again continues along different growth lines to the present day (t_P). Individual minerals in this model are open systems during the metamorphism. Therefore, a mineral isochron yields the age of cooling from the thermal event, when each mineral again became a closed system. However, a whole-rock domain of a certain minimum size may remain an effectively closed system during the thermal event, and could therefore be used to date the initial crystallization of the rock.

The effects of metamorphism on mineral and whole-rock systems can also be demonstrated on the isochron diagram, Fig. 3.12 (Lanphere *et al.*, 1964). All systems start on a horizontal cord. Isotopic evolution then occurs along near-vertical parallel paths (due to the extreme amplification of the *y* axis). During the thermal event, isotope ratios are homogenized to the whole-rock value. If this only involved ^{87}Sr, then vertical vectors would be produced. However, a possible complication, illustrated in Fig. 3.12, involves limited Rb re-mobilization. Rb-rich minerals tend to suffer some Rb loss, while Rb-poor phases may be contaminated by growth of Rb-rich alteration products, leading to somewhat unpredictable vectors (R). After the event, whole-rock evolution continues undeflected, while mineral systems define an isochron whose slope yields the age of metamorphism.

A practical example of dating plutonism and metamorphism by whole-rock and mineral analysis of the same body was provided by the work of Wetherill *et al.* (1968) on the Baltimore gneiss (Fig. 3.13). Several mineral isochrons all yield ages of ca. 290 Ma, interpreted as the time of closure of mineral systems after isotopic homogenization associated with

the Appalachian orogeny. The good fit of points to the mineral isochrons is evidence that complete isotopic homogenization on a mineralogical scale was achieved during the metamorphic event. In contrast, whole-rock samples define an isochron whose slope corresponds to an age of 1050 ± 100 Ma. This was interpreted as the time of crystallization of igneous precursors of the gneiss. The age was recalculated to 1038 ± 58 Ma by Provost (1990) using a more recent decay constant, but the original value was actually more accurate.

Many subsequent studies have shown that whole-rock Rb–Sr systems are often opened during high-grade regional metamorphism (e.g. Field and Raheim, 1979a, b). The unit sampled from the Baltimore Gneiss was a coarse-grained biotite–K-feldspar granite that has subsequently been SHRIMP-dated (Section 5.2.6) to 1075 ± 15 Ma (2σ) by Aleinikoff *et al.* (2004). However, the sample also contained older inherited zircons from the surrounding felsic gneisses, which were SHRIMP-dated to 1.25 Ga. Hence, it appears that in this case the whole-rock Rb–Sr system gave the correct igneous crystallization age of the granite.

3.3.2 Blocking Temperatures

After Rb–Sr mineral systems have been opened in the thermal pulse of a regional metamorphic event, there must come a time when they are again closed to element mobility. By dating the closure or 'blocking' of different mineral systems, Rb–Sr ages give information about the cooling history of metamorphic terranes. This was first demonstrated by Jager et al. (1967) and Jager (1973), working on the Central European Alps.

Jager et al. found that in rocks of low metamorphic grade around the exterior of the Central Alps, Hercynian Rb–Sr ages (>200 Ma) were preserved in both biotites and muscovites. However, on moving to a higher metamorphic grade, characterized by the appearance of stilpnomelane (which Jager et al. believed to be equivalent to a temperature of 300 ± 50 °C), Rb–Sr biotite ages of 35–40 Ma were measured. Jager et al. attributed these younger ages to Rb–Sr biotite systems opened at the peak of Lepontine metamorphism. They argued that the 300 °C temperature at which biotites were just opened at the peak of metamorphism would correspond to the temperature at which biotites would re-close up to several Ma after suffering a higher peak temperature (e.g. >500 °C within the central staurolite isograd). In other words, Jager et al. concluded that biotite had a blocking temperature of 300 ± 50 °C for the Rb–Sr system.

The blocking temperature of white mica (muscovite and phengite) was similarly constrained to 500 ± 50 °C by the first resetting of the white mica Rb–Sr ages 'somewhat outside the staurolite–chloritoid boundary' (Purdy and Jager, 1976). However, unlike biotite, white micas can undergo primary crystallization below the Rb–Sr blocking temperature, so that ages as low as 35–40 Ma have been obtained even from the outer zones of low grade Alpine metamorphism. These young ages are argued to date new mica growth at the peak of metamorphism (Hunziker, 1974). This makes the muscovite Rb–Sr system a more problematical tool than biotite for studying post-orogenic cooling processes.

Jager et al. (1967) obtained biotite ages of ca. 12–16 Ma from the Simplon and Gotthard areas of the Central Alps, and ages ca. 8 Ma older in coexisting muscovites. Clark and Jager (1969) used these data to make two different estimates of cooling rate for the Central Alps. Firstly, the age difference between muscovite and biotite closure (200 °C) leads to a cooling rate of ca. 25 °C/Ma between 500 and 300 °C. Secondly, the biotite ages yield cooling rates of ca. 20–25 °C/Ma between 300 and 0 °C (average surface temperature at the present day). Division of these results by an estimated geothermal gradient (25–40 °C/km) allows the calculation of uplift rates between 0.5 and 1.0 km/Ma for the Central Alps, which compare well with modern uplift rates of 0.4–0.8 mm/year from geodetic measurements. More recent calculations of past uplift rate make use of combined Rb–Sr, K–Ar and fission-track cooling ages (Section 16.6).

Purdy and Jager (1976) recognized that the 300 ± 50 °C blocking temperature for biotite might need to be revised if new experimental data for stilpnomelane stability were obtained. Most workers continue to use a value of 300 °C; however, experimental work (e.g. Brown, 1971) points to an upper stilpnomelane stability limit of 440–480 °C at ca. 4 kb, implying a biotite Rb–Sr blocking temperature over 400 °C. This would be consistent with evidence from SW Norway, where biotites subjected to temperatures over 400 °C in the Caledonian orogeny nevertheless preserve Sveco-Norwegian (800 Ma) ages (Verschure et al., 1980).

A more direct method of determining blocking temperatures is to measure mineral ages in deep boreholes. Del Moro et al. (1982) determined biotite–whole-rock Rb–Sr ages at depths of up to 3.8 km in the Sasso 22 well in the Larderello geothermal field, Italy. All of the biotites show almost complete retention of ^{87}Sr at directly measured in-hole temperatures up to nearly 380 °C, supporting a biotite closure temperature of ca. 400 °C. However, Cliff (1985) has argued that in active geothermal systems, convective heat transport could generate localized thermal pulses whose duration is too short to allow significant diffusional Sr loss, thus yielding an anomalously high blocking temperature.

Blocking temperatures can also be determined theoretically, based on calculations of the temperature dependence of volume diffusion processes (Dodson, 1973, 1979). Ideally, closure of the Rb–Sr system represents an instantaneous transition from a time when Rb and Sr were completely mobile to when they were completely immobile. In a fast cooling igneous body the moment of crystallization is a good approximation to this ideal. However, in a slow cooling regional metamorphic terrane there is a continuous transition from a high-temperature regime, when radiogenic ^{87}Sr escapes from crystal lattices by diffusion as fast as it is produced, to low-temperature conditions when there is negligible ^{87}Sr escape (Fig. 3.14).

In such a system, the apparent age of a mineral such as biotite corresponds to a linear extrapolation of the low-temperature ^{87}Sr growth line back into the x axis. The temperature prevailing in the system at the apparent age of the mineral is then defined as the blocking temperature of the mineral in question (Dodson, 1973). This blocking temperature is dependent on cooling rate, since the slower the cooling, the longer will be the time during which partial loss of daughter product may occur, and the lower will be the apparent age (Fig. 3.14).

If a mineral is in contact with a fluid phase that can remove radiogenic Sr from its surface, then the rate of loss of ^{87}Sr depends on the rate of volume diffusion across a certain size of lattice. In the case of biotite, this diffusion will be predominantly parallel to cleavage planes rather than across them. Assuming that the Arrhenius law is obeyed, Dodson (1979) calculated blocking temperatures (at a cooling rate of 30 °C/Ma) of 300 °C for the Rb–Sr system in biotites of 0.7 mm diameter. This was based on experimental work for argon diffusion in biotite (Hofmann and Giletti, 1970), because the two elements are thought to have similar diffusional behaviour in crystal lattices.

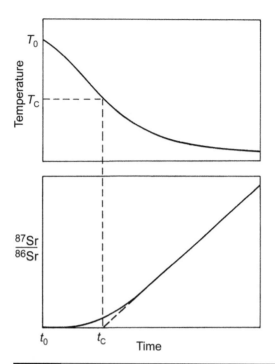

Fig. 3.14 Schematic diagram to show variation of temperature and Sr isotope ratio with time in a mineral cooling from a regional metamorphic event. T_0 = peak metamorphic temperature; T_C = closure or 'blocking' temperature; t_C = apparent closure age. After Dodson (1973).

A problem with the volume diffusional control of blocking temperature is that large (30 cm) fissure-filling biotites in the Central Alps have the same ages, and hence apparent blocking temperatures, as small (<1 mm) groundmass biotites in adjacent gneisses. Dodson (1979) suggested three possible explanations:

(1) Diffusion geometry is independent of grain size. This could be due to the effects of stress on the crystal lattice.
(2) Sr loss is controlled by the rate at which radiogenic atoms leave the site in which they were formed.
(3) Blocking temperature is not kinetically controlled, but depends on a change in the biotite lattice at the blocking temperature.

The susceptibility of Sr to mobilisation by fluids increases complexity in the interpretation of Sr blocking temperatures. Such problems do not arise for argon, because it is an inert gas. Therefore the latter element is a more reliable tool for studies of 'thermochronology'. This subject is discussed in detail in Section 10.4.

3.4 Dating Ore Deposits

Metallic ore deposits have always been notoriously difficult to date reliably. The most common approach to dating such

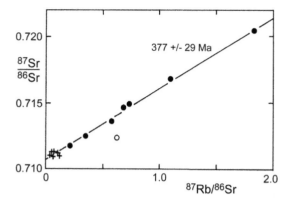

Fig. 3.15 Rb–Sr isochron diagram for sphalerite grains from the Coy mine, Tennessee: (•) = sphalerite host; (+) = fluid inclusions. Open symbol was excluded from age calculation. After Nakai et al. (1990).

deposits is to analyse gangue mineralization and hope that this material was deposited in the same episode as the associated metallic ores. An alternative approach has been to date fluid inclusions believed to form part of a hydrothermal ore-forming system. One or two successful attempts at this technique have been reported (e.g. Sheppard and Darbyshire, 1981), but fluid inclusion populations may represent more than one stage in the evolution of a hydrothermal system, leading to complex mixtures that have no age significance.

Typically, the large ion lithophile (LIL) elements which comprise most of the long-lived decay systems do not partition into metal sulphides, preventing direct dating of such ores. However, LIL elements may partition onto some sulphide ores in just sufficient abundance to allow analysis. One sulphide ore mineral which has been successfully dated by this means is sphalerite.

Nakai et al. (1990) made the first successful Rb–Sr isochron determination on sphalerite samples from a Mississippi Valley type (MVT) lead–zinc deposit from Tennessee. The sphalerite grains were found to have low Sr contents averaging only 1 ppm. As a result, fluid inclusions in the sphalerite grains, estimated to make up only 300 ppm by weight of the host mineral, actually contained more Sr than the host. Therefore, it was necessary to remove these inclusions by crushing the samples and leaching with deionized water before dissolving the sphalerite host for analysis.

This procedure gave a suitable range of Rb/Sr ratios and generated the 'errorchron' shown in Fig. 3.15. One outlier, believed to have been disturbed during a deformation event, was excluded from the data set, after which the remaining seven points gave an age of 377 ± 29 Ma (2σ). This is a 'scatter error', determined by expanding the analytical errors to reduce the original MSWD of 62.6 to unity (Section 2.8.3). The fluid inclusions leached during crushing were also analysed, and were found to lie on the isochron defined by the host phase (Fig. 3.15); however, these analyses were not included

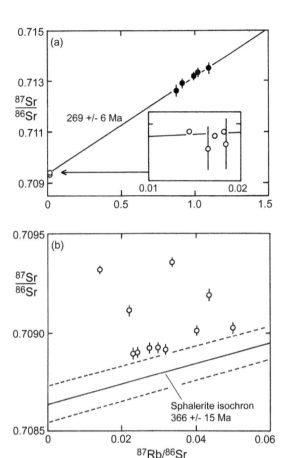

Fig. 3.16 Rb–Sr isochron analysis of sphalerite host (•) and extracted fluid inclusions (○) from MVT lead–zinc deposits: (a) West Hayden, Wisconsin; (b) Polaris, arctic Canada. After Brannon *et al.* (1992) and Christensen *et al.* (1995a).

in the isochron calculation. The age of 377 ± 29 Ma suggested that MVT mineralization occurred during the Acadian orogeny (380–350 Ma ago), which caused the expulsion of basin brines from strata within the deformation zone of the Appalachians. These fluids were then transported to the west, causing ore deposition when they mixed with other fluids during their return to the surface.

Brannon *et al.* (1992) applied this method to other MVT deposits. However, the range of Rb/Sr ratios in the ores themselves was not sufficient for the determination of a precise isochron. Therefore, it was necessary to combine analyses of the host sulphide with fluid inclusions (Fig. 3.16a). This procedure yielded a precise age (269 ± 6 Ma, 2σ), but was effectively a 'two-point' isochron, raising fears that if the host sulphide and the inclusions were not co-genetic, the calculated age might be geologically meaningless.

Further studies by Nakai *et al.* (1993) revealed two more examples (from the Pine Point MVT deposit in Canada, and the Immel mine in east Tennessee) where the inclusions

lay on a well-defined host isochron. However, analysis of the Polaris MVT deposit in arctic Canada provided an example where the inclusions lay off a well-defined host isochron (Christensen *et al.*, 1995a). In this case the host (ore minerals) gave an age of 366 ± 15 Ma, in good agreement with the age of the formation from paleomagnetic evidence, whereas the inclusions defined a cloud of points above the isochron (Fig. 3.16b). Seven of these leachate samples lay just outside the error of the sphalerite isochron, whereas four were more radiogenic, suggesting that the inclusion population included primary inclusions which were co-genetic with the ores, along with more radiogenic secondary inclusions.

To avoid possible complications arising from mixing between host ores and fluid inclusions, Christensen *et al.* (1995b) tested for mixing relationships when they dated sphalerites from the Canning Basin MVT deposit of Western Australia. They found that Sr concentrations in the sphalerite host grains showed no correlation with $^{87}Rb/^{86}Sr$, suggesting that the sphalerite residues after crushing and leaching were not significantly contaminated by unopened inclusions (Fig. 3.17a). On the other hand, Rb contents were found to be strongly correlated with $^{87}Rb/^{86}Sr$. Since the fluid inclusions contain negligible Rb, these Rb contents must have originated from the host sulphide ore. Therefore the isochron must also date the sulphide ore itself.

Pettke and Diamond (1996) used a similar approach to test the possibility of mixing in the sphalerite-inclusion Rb–Sr isochron of Brannon *et al.* (1992). They plotted the data on a graph of Sr isotope ratio against the reciprocal of Sr concentration, on which mixing processes generate straight lines (Fig. 3.17b). On this graph, the fluid inclusions have Sr contents that are essentially infinite (relative to the low abundances in the host), so that they are plotted on the y axis. The results of this analysis showed that one of the isochrons determined by Brannon *et al.* (1992) was probably a mixing line generated by sampling of sphalerite grains with a few un-released inclusions (sample 58-B). Therefore, this age determination is only meaningful if the host and the fluid were cogenetic. However, the other isochron (sample 10-C) does not show the mixing effect, so this age is more reliable. Since both isochrons gave results within error (269 ± 6 and 270 ± 4 Ma), it was concluded that this is a reasonable estimate of the age of ore deposition.

Another sulphide mineral successfully used to date ore deposition is the mercury sulphide, galkhaite. This hydrothermal mineral was found associated with Carlin-type gold mineralization in Nevada, and was used to estimate a date of 39 ± 2 Ma for gold mineralization at the Getchell deposit in northern Nevada (Tretbar *et al.*, 2000).

3.5 Dating Sedimentary Systems

Absolute dating of the time of deposition of sedimentary rocks is an important problem, but one that is very difficult

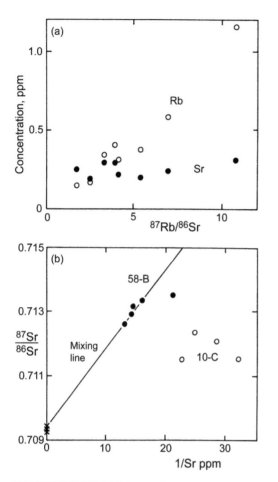

Fig. 3.17 Plots to test for mixing relationships between sphalerite hosts and inclusions: (a) Rb and Sr concentrations versus $^{87}Rb/^{86}Sr$ in the Canning Basin MVT deposit, Australia; (b) Sr isotope ratio versus 1/Sr from the West Hayden MVT deposit, Wisconsin. After Pettke and Diamond (1996).

to solve directly. Accurate dates depend on thorough re-setting of isotopic clocks. In the case of Rb–Sr dating of sediments, this rests on the assumption that Sr isotope systematics in the rock were homogenized during deposition or early diagenesis, and thereafter remained as a closed system until the present day. However, these two requirements may be mutually exclusive.

In principle, sedimentary rocks may be divided into two groups, according to the nature of the Rb-bearing phase present. Allogenic (detrital) minerals are moderately resistant to open-system behaviour during burial metamorphism, but problems arise from inherited isotopic signatures. Authigenic minerals are deposited directly from seawater and hence display good initial Sr isotope homogeneity. However, they are highly susceptible to recrystallization after burial and may not remain closed systems.

In practice, the two distinct dating approaches associated with these sediment types have tended to converge. Analysis of detrital sediments has moved towards the analysis of fine-grained, almost authigenic minerals, such as illite, to escape the effects of the detrital component. In contrast, analysis of authigenic minerals has been focussed on the sub-authigenic mineral glauconite, since the truly authigenic Rb-bearing evaporite minerals are too susceptible to burial metamorphism to be viable geochronometers.

3.5.1 Shales

Detrital Rb-bearing minerals (mica, K-feldspar, clay minerals etc.) can be expected to contain inherited old radiogenic Sr. Therefore, dating of such material should give an average of the provenance ages of the sedimentary constituents. However, if sufficiently fine-grained shales are sampled, it appears that the constituent minerals (mainly illite) often suffer substantial Sr exchange during post-depositional diagenesis. In this case they may develop an almost homogeneous initial Sr isotope composition soon after deposition, thereafter remaining effectively closed systems until the present day.

Compston and Pidgeon (1962) pioneered whole-rock Rb–Sr dating of shales, and found that in some circumstances (e.g. the State Circle shale from SE Australia) the above conditions were closely approached. However, in other cases (e.g. the Cardup shale of W. Australia), gross inherited $^{87}Sr/^{86}Sr$ variations remained, preventing the calculation of a meaningful age. Compston and Pidgeon attributed this to un-decomposed detrital micas, probably sericite. In contrast, the carbonaceous shales of the Cardup unit contained much less detrital mica and, taken alone, gave a tentative depositional age of 660 Ma.

Subsequent work on the dating of shales sought to avoid problems of contamination with detrital micas and feldspars by analysing separated clay-mineral fractions, whose purity is checked by X-ray diffraction (XRD). XRD analysis of illites can also yield information about the nature and origin of clay minerals in a shale which is to be dated.

The 'illite crystallinity index' (Kubler, 1966) is defined as the width of the (001) XRD peak at half its height. A well-crystallized illite, characteristic of a relatively high-temperature history, has sharp peaks, and therefore a low index, while low-temperature illites are more disordered, and have irregular peaks with large indices. In addition to this discriminant, illite has high-temperature (2M) and low-temperature (1M) polymorphs which can also be distinguished by XRD (Dunoyer, 1969). '1M' illites with a large crystallinity index are characteristic of low-temperature growth and recrystallization in the sedimentary–diagenetic regime, while '2M' illites with a small index are indicative of temperatures of zeolite-facies metamorphism or above. The latter reflect a detrital component, or post-diagenetic metamorphism.

A comparison of Rb–Sr whole-rock and clay mineral analysis of a Precambrian shale from Mauritania (W. Africa) is

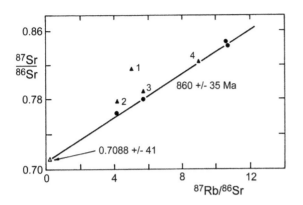

Fig. 3.18 Rb–Sr isochron diagram for whole-rock shales (▲); separated illites (●); and a carbonate sample (△) from Mauritania. Numbered whole-rock samples are discussed in the text. After Clauer (1979).

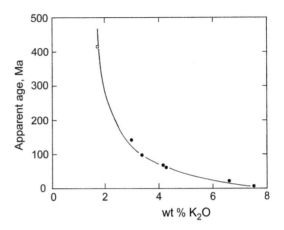

Fig. 3.19 Rb–Sr model ages of Holocene (zero-age) glauconies as a function of potassium content. Open symbol indicates clay fraction. Modified after Clauer et al. (1992).

shown in Fig. 3.18 (Clauer, 1979). Four clay fractions were analysed, containing smectite and the 1M illite polymorph with a crystallinity index over 6 (very low-grade metamorphism is characterized by an index below 5.75). These define a linear array which is colinear with associated dolomites, yielding an age of 860 ± 35 Ma and an initial ratio of 0.7088, typical of late Precambrian seawater. A whole-rock sample (4) shown by XRD to be free of detrital feldspar also lay on the isochron. However, two whole rocks (2 and 3) with traces of microcline lay slightly above it, while one with 15% microcline (1) was displaced well above the isochron.

It appears from the above example that whole-rock Rb–Sr dating of shales is an unreliable geochronometer, but that analysis of separated illite fractions may give meaningful ages of diagenesis or low-grade metamorphism. However, there is always a danger that the detrital component may not be completely eliminated from the illite fraction.

An important example of this problem is provided by the dating of the Sinian–Cambrian boundary. Rb–Sr analysis of shales was attempted for these rocks, but the resulting ages were later shown to be ca. 60 Ma too old (relative to U–Pb ages). Some of these results, summarized by Cowie and Johnson (1985) and Odin et al. (1985) appeared to support an age of ca. 600 Ma for the base of the Cambrian. However, the analysis of fine-grained clay fractions almost invariably gave ages significantly lower than whole-rock or coarse clay fractions. This suggests that a diagenetic event affected the rocks some time after deposition, so that the 600 Ma ages are probably mixed ages between inherited and diagenetic components, rather than depositional ages. Therefore, Rb–Sr dating of shales cannot be considered a reliable technique for dating sedimentary deposition. The more reliable approach is to use U–Pb or Ar–Ar ages on volcanic tuffs intercalated within the sedimentary succession (Sections 5.1.3, 10.3).

3.5.2 Glauconite

The mineral glauconite offers an attractive possibility for direct dating of sedimentary rocks due to its high Rb content, easy identification, and widespread stratigraphic distribution. Glauconite is a micaceous mineral similar to illite which is best developed in macroscopic pellets (called 'glauconies' by Odin and Dodson, 1982). These are probably formed by the alteration of a very fine-grained clay precursor intermixed with organic matter in a faecal pellet. Glauconies form near the sediment–water interface in the marine environment. However, by studying pellets on the present day ocean floor, Odin and Dodson (1982) have shown that 'glauconitization' is a slow process which may take hundreds of thousands of years to reach completion. During this process, the potassium content of the pellet increases, and this can therefore be used to monitor the maturation of the pellet.

Rb–Sr analysis of Holocene glauconies (Clauer et al., 1992) shows that Sr isotope equilibrium with seawater is achieved only slowly as the potassium content increases. The Rb–Sr data can be used to calculate a model Sr age for the pellet by making the initial ratio equal to the isotopic composition of seawater Sr at the estimated time of sedimentation (see below). A zero-age pellet starts with a high apparent model age due to a large content of Sr in detrital mineral phases. However, as it matures, the pellet homogenizes with seawater so that the model age falls to zero in a fully equilibrated pellet (Fig. 3.19). Analysis of the potassium content of glauconies therefore provides an essential screening procedure, in order to select only fully mature material for dating.

Cretaceous and younger glauconies often yield ages concordant with other dating methods (e.g. Harris, 1976), but Paleozoic glauconies commonly give ages that are 10–20% younger than anticipated. Early workers (e.g. Hurley et al., 1960) attributed this to post-depositional uptake of K and Rb during diagenesis. However, Morton and Long (1980) attributed the young ages to ^{87}Sr loss from the expandable layers of the clay lattice, by some form of ion exchange process with circulating brines.

Morton and Long calculated model ages for a series of glauconite separates, based on the assumed initial ^{87}Sr/^{86}Sr ratio of seawater at the time of deposition (see Section 3.6.1). They showed that in some cases erroneous glauconite model ages could be increased to near to the stratigraphic age by leaching with ammonium acetate, which is thought to remove excess loosely bound Rb from the expandable layers of the lattice. In contrast, leaching with acetic acid, HCl etc. had unpredictable effects on the glauconite age, probably due to removal of some tightly bound Sr.

Similar experiments were performed on glauconites from the 525 Ma-old Bonneterre Formation (Missouri) by Grant *et al.* (1984). Eight un-leached glauconite pellets gave model ages in the range 413–440 Ma. However, the most radiogenic sample (model age = 426 Ma) converged only slightly on the true age when subjected to ammonium acetate leaching (437 Ma). Therefore, more rigorous criteria are needed to determine whether old glauconites have suffered open-system behaviour, prior to a dating attempt. Until such criteria are developed, glauconite dating in the Paleozoic must be regarded as a monitor of diagenetic processes, rather than a viable dating tool for stratigraphic correlation.

3.6 Seawater Evolution

Biogenic carbonates fulfil two of the requirements of a sedimentary dating tool: they are fairly resistant to diagenetic alteration, and since they are secreted directly from seawater by the organism, they contain no detrital fraction. The negligible Rb content of carbonates precludes conventional Rb–Sr dating. However, calibration of the seawater Sr isotope evolution path would allow the 'initial' ^{87}Sr/^{86}Sr isotope ratios of carbonates to be used as an indirect dating tool. In the following section we will assess the realization of this concept, as well as the application of Sr isotopes as a conservative oceanographic tracer (a species with a long seawater residence time).

3.6.1 Measurement of the Curve

Interest in the strontium isotope composition of seawater dates back to Wickman (1948). He argued that decay of ^{87}Rb to ^{87}Sr in crustal rocks over geological time, and its subsequent release into the hydrosphere by erosion, should lead to a 25% increase in seawater Sr isotope composition over the last 3 Ga. This model was tested by Gast (1955), who analysed carbonates of different ages as a means of characterizing the evolution of seawater through geological time. However, he found that any natural variations were of the same order as the analytical errors of ^{87}Sr/^{86}Sr analysis pertaining at that time (ca. 0.004), thus refuting Wickman's model. Evidently the average crustal Rb/Sr ratio assumed by Wickman was an over-estimate.

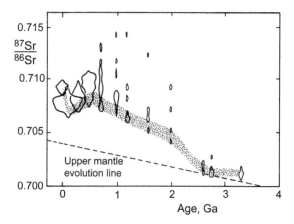

Fig. 3.20 Sr isotope composition of marine carbonates over the last 3.5 Ga, from which the isotopic evolution of seawater is deduced (shaded band). After Veizer and Compston (1976).

Resolution of the actual extent of seawater Sr isotope variation through time had to wait 15 years for the advent of more precise mass spectrometry. Peterman *et al.* (1970) measured the ^{87}Sr/^{86}Sr composition of macro-fossil shell carbonates with an order of magnitude improvement in precision ($\pm$ 0.0005, 2σ). They found a total isotopic range of 0.0022 (4 $\times$ analytical error), which would have been imperceptible using earlier equipment. Peterman *et al.* showed that, contrary to Wickman's prediction, the seawater Sr isotope ratio actually *decreased* during the Paleozoic, reaching a minimum during the Mesozoic before rising quickly to a maximum at the present day.

In order to avoid the effects of post-depositional alteration, Peterman *et al.* rejected any recrystallized shell material, which they claimed to be able to recognize visually. The possibility of Sr exchange between matrix and unrecrystallized shells was rendered unlikely by the good compositional agreement between different shells in a bed. A mixture of mollusc types was used (belemnites, bivalves and brachiopods). Since no variation was seen between such classes in modern samples, they were assumed to behave in the same way as fossils.

Additional data were collected by Dasch and Biscaye (1971) and Veizer and Compston (1974) from different types of sample material. Dasch and Biscaye used Cretaceous-to-Recent pelagic foraminifera, while Veizer and Compston (1974) analysed 'sedimentary carbonate' (in other words not macro-fossil carbonate) to test its reliability for the determination of seawater Sr isotope ratios. Both studies found general agreement with the data of Peterman *et al.* (1970). This implies global homogenization of seawater Sr, which can be attributed to the very long residence time of Sr in seawater (ca. 2.5 Ma; Hodell *et. al.*, 1990) compared to the average mixing time of oceanic water (ca. 1.6 ka; Section 14.2.1). However, Veizer and Compston recognized that 'sedimentary carbonate' is more susceptible to post-depositional

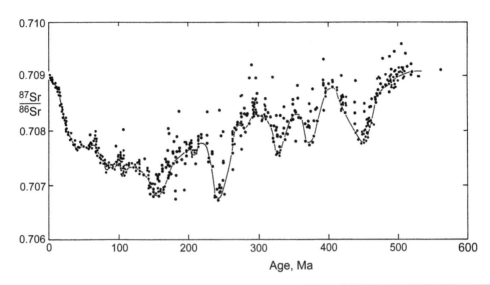

Fig. 3.21 Sr isotope data for Phanerozoic carbonates. Solid line indicates the lower bound of most of the data, which is the most probable seawater Sr composition. After Burke *et al.* (1982).

exchange with pore waters. They argued that since detrital grains would normally have radiogenic Sr isotope signatures, post-depositional exchange would normally be expected to raise $^{87}Sr/^{86}Sr$ ratios. Therefore the minimum Sr isotope ratio found at any given time should be the most reliable guide to contemporaneous seawater composition.

While the analysis of whole-rock carbonate provides fewer constraints on post-depositional processes, it provides more opportunity for sampling, and is essential for Precambrian carbonates. Using the principles outlined above, Veizer and Compston (1976) made a reconnaissance study of the Sr isotope evolution of Precambrian seawater. They found uniformly unradiogenic Sr isotope ratios in Archean carbonates, with values only slightly elevated over contemporaneous upper mantle (Fig. 3.20). However, there was a substantial rise in Sr isotope ratio during the Proterozoic, reaching a maximum in the early Cambrian which was similar to the present day composition.

A major expansion of the seawater Sr data set was achieved by Burke *et al.* (1982), who presented 786 isotopic analyses of marine carbonates, phosphates and evaporites, with good coverage of all of Phanerozoic time except the Lower Cambrian (Fig. 3.21). In addition, work by Derry *et al.* (1989), Asmerom *et al.* (1991) and Kaufman *et al.* (1993) extended the curve back to the Late Proterozoic. In the absence of fossil material, the latter studies were made principally on whole-rock carbonates, which are susceptible to contamination by fluid-borne Sr during post-depositional alteration. Therefore, bulk carbonates were dissolved in dilute acetic acid to reduce the amount of contamination by detrital phases containing radiogenic Sr.

Following the wide-ranging study of Burke *et al.* (1982), subsequent work was generally devoted to improving preci-

sion on small segments of the curve. This requires material to be well-dated stratigraphically, and carefully screened before analysis to exclude post-depositional alteration. In Paleozoic rocks, this screening is best achieved chemically. Brand and Veizer (1980) showed that open-system diagenesis of carbonates is accompanied by a decrease in Sr/Ca ratio and increase in Mn content (Fig. 3.22). Since Mn-enriched calcite can be detected by cathodoluminescence, sections of shell can be screened for alteration before analysis. Popp *et al.* (1986) showed that samples of brachiopod shell prepared in this way gave more reliable results than whole

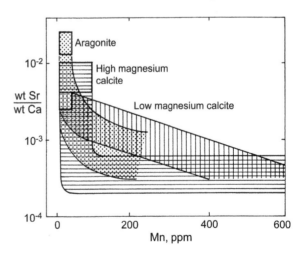

Fig. 3.22 Summary of diagnostic chemical changes which occur during the diagenetic alteration of carbonates. Boxes represent primary fields. After Brand and Veizer (1980).

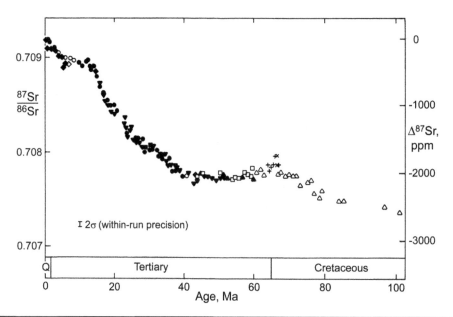

Fig. 3.23 Plot of Sr isotope ratio against age for forams from eight DSDP holes (distinguished by symbol shape). Solid symbols and crosses indicate most reliable data; open symbols may be slightly disturbed. After Hess et al. (1986).

brachiopod shells (which were sometimes contaminated by unradiogenic Sr) and whole-rock carbonates (which were usually contaminated by radiogenic Sr).

The use of high-quality brachiopod shells and belemenites from around the world allowed Veizer et al. (1999) to present a complete Sr evolution curve for the Mesozoic and Paleozoic, based on 1450 new analyses. They utilized the interior shell layers from brachiopods and single laminae of belemnites. In addition, much of their material showed excellent preservation of textures on a sub-micron scale, as demonstrated by examination under the scanning electron microscope (SEM). This study therefore represents a worthy successor to that of Burke et al. (1982) in giving an overview of seawater Sr evolution between 100 and 500 Ma ago.

3.6.2 The Cretaceous–Tertiary Seawater Curve

Construction of a very precise seawater Sr evolution curve for the past 100 Ma was made easier by the availability of numerous Deep Sea Drilling Project (DSDP) cores. These cores provide overlapping continuous sections with well-preserved microfossils such as foraminifera ('forams'). Relatively constant sedimentation rates in these sections are used to interpolate between biostratigraphic and magnetostratigraphic calibration points. This avoids the age uncertainty involved in correlating stratigraphic sections from different localities.

A detailed survey was carried out by Hess et al. (1986), based on over 130 hand-picked whole foram tests. These were screened for secondary alteration by SEM examination and chemical analysis (e.g. Mn and Sr content). Figure 3.23 shows

data from eight partially overlapping DSDP sections. Slight scatter is seen, but much of this can be attributed to analytical error rather than diagenetic effects. In selected samples from two sites, pore-waters had very similar isotope ratios to forams. In one other site, pore-waters were somewhat more radiogenic, but there is no evidence that the foram data have been perturbed. Most subsequent studies have also employed hand-picked forams. Since less than 50 ng of Sr is now needed for a precise analysis, this may be possible on a few or even a single foram. As an additional precaution, Martin and Macdougall (1991) were able to break open large Cretaceous forams to examine them by SEM for internal calcite growth.

The high-precision seawater Sr isotope evolution curve can be used as a stratigraphic dating tool, with a (conservative) precision as good as 0.5 Ma for periods of rapid Sr isotope evolution, but as bad as 2 Ma during periods of slow isotopic evolution. This precision cannot compete with biostratigraphic dating in the Cretaceous and Tertiary periods, but it may be useful for calibration of un-fossiliferous borehole sections (e.g. Rundberg and Smalley, 1989).

An interesting observation by Hess et al. (1986) in their Cretaceous–Tertiary data set was a 'spike' in seawater Sr isotope ratio at the 'K–T' boundary (Fig. 3.23). They speculated as to whether a meteorite impact could release sufficient Sr, either from the bolide or the terrestrial impact ejecta, to explain this peak. If the spike in ^{87}Sr is attributed to a meteorite, it is critical to demonstrate that it occurred at exactly the correct stratigraphic level.

Martin and Macdougall (1991) collected data from four widely spaced localities around the world which appeared to

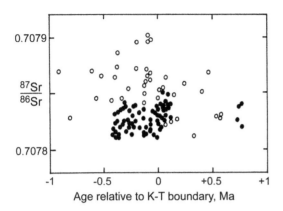

Fig. 3.24 Variation of Sr isotope ratio in the vicinity of the K–T boundary, showing data of McArthur *et al.* (•) in comparison with data from several previous studies (o). After McArthur *et al.* (1998).

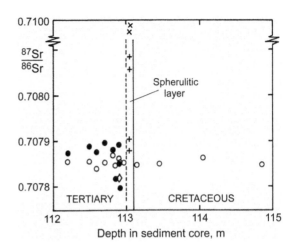

Fig. 3.25 Variation in Sr isotope ratio in different sample types at the K–T boundary, from a DSDP site off Florida: (o) = Cretaceous foram taxon; (•) = Tertiary foram taxon; (◊) = dolomite rhomb overgrowths; (+) = bulk carbonate; (×) = silicate. After MacLeod *et al.* (2001).

support the model. However, more detailed analysis of sample suites close to the K–T boundary in Denmark and Antarctica failed to find such a peak (McArthur *et al.*, 1998). Instead, the data of McArthur *et al.* (1998) fitted the lower bound of the more scattered older data (Fig. 3.24), suggesting that the elevated values in these older studies reflect diagenetic alteration, despite the precautions taken to exclude this effect.

To further examine this problem, MacLeod *et al.* (2001) studied Sr isotope variation in a K–T boundary section from a DSDP hole off Florida. Because this section is close to the impact site, the spherulitic K–T boundary horizon here is 10 cm thick, and can therefore be studied in more detail. Bulk carbonates from the spherulitic layer itself are radiogenic, due to Sr exchange with silicate phases (Fig. 3.25). However, to test for global seawater Sr variations, MacLeod *et al.* analysed the isotopic composition of two different species of foram in the vicinity of the boundary. The results (Fig. 3.25) showed that a foram species found only above the boundary (Tertiary taxon) generally had more radiogenic Sr than a species that straddled above and below the boundary (Cretaceous taxon).

MacLeod *et al.* considered two alternative explanations for these observations. One is that the Tertiary taxon (solid symbols in Fig. 3.25) was more susceptible to contamination by overgrowths because it has a thinner test. However, an analysis of such overgrowths (Fig. 3.25) suggested that this might lower rather than increase $^{87}Sr/^{86}Sr$. The other explanation is that the Tertiary strata were contaminated by Cretaceous forams reworked elsewhere from below the boundary and then carried into the section as clastic sediment. The fact that the drill hole comes from part-way down the continental slope makes this a significant possibility. This would imply that the Tertiary taxon best represents the composition of seawater Sr after the impact, and hence that a very small (0.000 03) increase in seawater $^{87}Sr/^{86}Sr$ occurred across the boundary. This suggests that the impact event at the K–T boundary did have a small effect on seawater Sr. However, this evidence needs to be tested at a site less susceptible to sedimentary reworking.

3.6.3 Seawater Sr and Glacial Cycles

Neogene seawater evolution has provided a challenge to geochemists to find the shortest-period variations in Sr isotope evolution which can be reliably documented. In early work on this problem, Dia *et al.* (1992) and Clemens *et al.* (1993) claimed to observe changes in $^{87}Sr/^{86}Sr$, correlated with $\delta^{18}O$, with a periodicity of about 0.1 Ma (Fig. 3.26 a, b, c). However, subsequent work by the same research groups (Henderson *et al.*, 1994; Clemens *et al.*, 1995) failed to reproduce these cycles in three drill cores (including two used in the original work). Instead, the new data fell on the linear evolution path defined by other studies (e.g. Hodell *et al.*, 1990). Hence, the apparent periodicity in the earlier work is attributed to analytical artefacts and does not reflect seawater Sr isotope evolution.

For the data of Dia *et al.* (1992), the analytical artefact was apparently a breakdown in the accuracy of the fractional correction. Thus, Clemens *et al.* (1995) were able to reproduce the temporal periodicity using a linear-law fractionation correction, but this also generated a positive correlation between $^{88}Sr/^{86}Sr$ and fractionation-corrected $^{87}Sr/^{86}Sr$ ratios, indicative of a fractionation bias (Section 2.3.1). After correction of this bias, the periodicity disappeared (Fig. 3.26d). The data of Clemens *et al.* (1993) were not subject to this bias, since the more accurate exponential law was used. However, Henderson *et al.* (1994) showed that only three out of 75 samples analysed by Clemens *et al.* (1993) lay outside of 2σ (95%) confidence limits from the linear evolution path of Hodell *et al.*

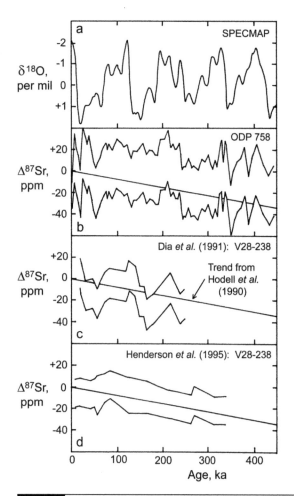

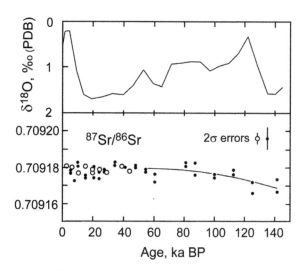

Fig. 3.27 Comparisons of oxygen and strontium isotope data for the last glacial cycle. Modified after Ando *et al.* (2010).

Fig. 3.26 Comparison between seawater Sr and oxygen isotope data for the past 400 ka: (a) oxygen isotope record; (b) data of Clemens *et al.* (1993) expressed by 2σ error limits; (c, d) data of Dia *et al.* (1992) and Henderson *et al.* (1994) on the same drill core. Modified after Henderson *et al.* (1994).

(1990). Since four outliers would be expected at this confidence limit, the apparent periodicity in this data set is probably not statistically significant (Fig. 3.26b).

Farrell *et al.* (1995) carried out a study with similar sampling density and analytical precision to the above work, but using 455 samples extending over the past 6 Ma. These data constrain the seawater evolution curve to an average confidence limit of ± 0.000 02 (2σ). The curve shows undulations with a 1–2 Ma periodicity which are realistic reflections of changing Sr fluxes, given a 2.5 Ma residence time of Sr in the ocean system.

The development of a new generation of ultra-high-precision TIMS instruments (Section 2.4.3) allowed improvements in the precision of Sr isotope analysis to better than 10 ppm (Ando *et al.*, 2010; Mokadem *et al.*, 2015). This allows tighter constraints to be applied to Quaternary seawater Sr evolution, but the subject is still in its infancy. Data of the

highest precision are only presently available for the past 40 ka (Mokadem *et al.*, 2015), but these show no evidence for variations in seawater composition (open symbols in Fig. 3.27). In contrast, slightly lower precision data for the past 200 ka (Ando *et al.*, 2010) show hints of variability which might be related to glacial cycles (solid symbols in Fig. 3.27). Modelling that takes account of periodic glacial erosion predicts variations of approximately this magnitude (Section 3.6.7).

3.6.4 Modelling the Fluxes

The first model for the Sr isotope composition of seawater was constructed by Faure *et al.* (1965) to explain the present day Sr signature of the North Atlantic. They proposed a balance between three components: unradiogenic Sr from erosion of young volcanics; radiogenic Sr from old crustal rocks; and Sr of intermediate composition from the erosion of carbonates.

This model was adopted by Peterman *et al.* (1970), using relative variations of these fluxes to explain the rise and fall of seawater $^{87}Sr/^{86}Sr$ during the Phanerozoic. Armstrong (1971) supplemented this model, suggesting that peaks in seawater Sr isotope ratio during the Carboniferous and Tertiary periods were due to enhanced glacial erosion of old shields with elevated ^{87}Sr contents (Fig. 3.28). However, in other ways the model of Peterman *et al.* remained largely unchallenged.

A major advance in modelling seawater Sr evolution was the proposal of Spooner (1976) that the unradiogenic Sr flux was due to submarine hydrothermal exchange with basaltic crust, rather than sub-aerial erosion of basic rock. Spooner calculated that the hydrothermal flux must be six times the magnitude of the river water Sr flux. However, this was based on high estimates of the isotopic composition of continental run-off (0.716) and hydrothermally buffered water (0.708).

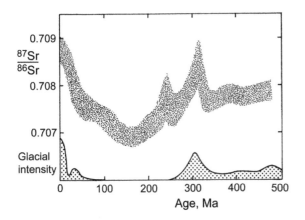

Fig. 3.28 Illustration of a glacial–erosional model to explain the seawater Sr evolution curve of Peterman *et al.* (shaded band). After Armstrong (1971).

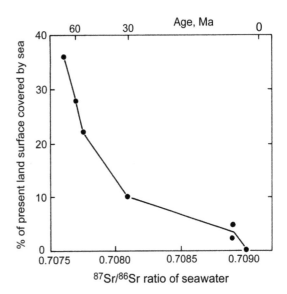

Fig. 3.29 Plot of seawater Sr isotope composition over the past 85 Ma against % continental flooding (relative to the present land area). After Spooner (1976).

Subsequent analysis of hydrothermal vent waters from the East Pacific Rise (Albarede *et al.*, 1981) indicated much less radiogenic compositions. Albarede *et al.* estimated the flux of hydrothermally recycled Sr as less than a quarter of the flux from continental run-off. This model predicted an average Sr isotope composition of between 0.710 and 0.711 for run-off, in good agreement with major rivers such as the Amazon (Brass, 1976).

The recognition of competing riverine and hydrothermal fluxes raises the question of how these fluxes interacted in the past to cause variations in seawater isotope ratio over time. Spooner (1976) assumed that the hydrothermal Sr flux was fairly constant with time. Therefore, he attributed the Tertiary increase in $^{87}Sr/^{86}Sr$ largely to an increase in continent exposure (and hence Sr run-off) over the last 85 Ma (Fig. 3.29). In contrast, Albarede *et al.* (1981) argued that a drop in the ocean ridge Sr exchange flux from a Mesozoic value nearly four times higher was more important than a rise in continental run-off. However, these two effects are difficult to separate, since they are bound together as a system. A drop in spreading rate causes ridge collapse and consequent sealevel fall. Therefore, continental exposure should increase as hydrothermal buffering of seawater decreases.

Based on the premise that the ocean-ridge hydrothermal flux balances the riverine Sr flux, various attempts have been made to reconcile these fluxes. Early studies estimated a range from high to very low ocean-ridge hydrothermal fluxes (Edmond *et al.*, 1979; Morton and Sleep, 1985), which were then used to predict the corresponding composition of riverine Sr (Fig. 3.30).

Other studies (Goldstein and Jacobsen, 1987; Palmer and Edmond, 1989) took the opposite approach, attempting to place tighter constraints on riverine Sr, and hence determining a ridge-crest hydrothermal flux. The first of these studies proposed a relatively low riverine $^{87}Sr/^{86}Sr$ of 0.711, implying a relatively low hydrothermal flux. However, Palmer and Edmond determined a higher riverine $^{87}Sr/^{86}Sr$ value (dis-

cussed in more detail below), which has largely been supported in later work. Balancing this flux implies a much larger hydrothermal flux. However, this has not been supported by detailed studies of sea floor alteration (e.g. Davis *et al.*, 2003), which imply an ocean ridge flux at least three times too small to balance the riverine flux. A later study by Vance *et al.* (2009) implied an even greater mismatch of up to eight times between the hydrothermal flux and present day riverine flux.

Two main alternatives have been proposed to explain this discrepancy. One involves searching for missing fluxes that may have been omitted from the unradiogenic hydrothermal Sr flux. The other attributes the mismatch to rapid changes

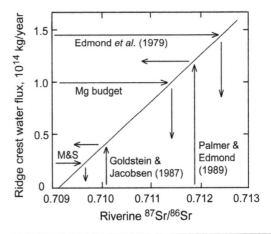

Fig. 3.30 Inter-relationships between ridge-crest and riverine Sr fluxes, assuming a steady state condition. M&S = Morton and Sleep (1985). After Palmer and Edmond (1989).

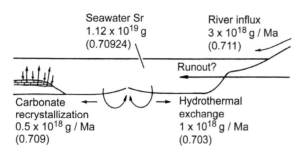

Seawater Sr
1.12×10^{19} g
(0.70924)

River influx
3×10^{18} g / Ma
(0.711)

Runout?

Carbonate
recrystallization
0.5×10^{18} g / Ma
(0.709)

Hydrothermal
exchange
1×10^{18} g / Ma
(0.703)

Fig. 3.31 Simplified circulation model for the present day seawater Sr budget. Modified after DePaolo (1987).

in the riverine flux over time. These alternatives will be briefly examined below.

An additional flux that is widely accepted, although small in size, is the Sr released from ocean floor carbonates by diagenetic recrystallization (Elderfield and Gieskes, 1982). This is estimated at about 10% of the run-off flux, but its isotopic composition is similar to present day seawater Sr. Therefore, its main effect is to dampen isotopic fluctuations, since it recycles old seawater Sr.

Another proposed flux is the sub-surface outflow of continental groundwater, from below the water table, into the sea (Fig. 3.31). This flux was termed 'run-out' by Chaudhuri and Clauer (1986), who proposed that it could explain seawater Sr isotope fluctuations that are not in harmony with variations in sea level. For example, run-out would be affected by the length of the continental perimeter as well as the extent of continental uplift, so plate tectonic configurations which form super-continents would be characterized by low run-out, while fragmented continents (at the present day) should be characterized by high run-out. This model attributes the rising Sr isotope ratio during the early Cretaceous (despite high sea level) to progressive continental break-up at this time.

Chaudhuri and Clauer suggested that the run-out (continental groundwater) Sr flux could be almost as large as the riverine run-off flux. This proposal received little attention for many years, but was re-examined in a study of the Bengal Fan by Basu *et al.* (2001). These authors found groundwater Sr concentrations an order of magnitude higher than the associated Ganges–Brahmaputra river system. Hence they suggested that the groundwater flux could be equal in magnitude to the riverine flux, even at a flow rate ten times less. However, Harvey (2002) questioned whether a flux of this magnitude was possible, given the lack of topography across the delta to provide a hydraulic gradient. Furthermore, the Sr isotope ratio of the groundwater overlaps strongly with river water, and is much more radiogenic than seawater. Hence this flux, if significant, would make the global Sr isotope imbalance even worse.

More recent studies of groundwater Sr fluxes were conducted by Rahaman and Singh (2012) and Beck *et al.* (2013) on various river systems around the world. These studies

revealed Sr concentrations in meteoric groundwater several times greater than river water, with $^{87}Sr/^{86}Sr$ ratios slightly above or below seawater. However, the isotopic differences are not large enough to have any significant effect on seawater Sr isotope composition. Beck *et al.* also found Sr concentrations another order of magnitude higher in brackish groundwater, but recognized that this largely consisted of recycled seawater.

Taking a different approach, Allegre *et al.* (2010) argued that the groundwater flux from ocean islands and island arcs (mainly in the Pacific) had been under-estimated. Based on the high groundwater Sr contents of a few ocean islands, they argued that this flux was nearly three times the magnitude of the global ocean-ridge hydrothermal flux, thus balancing the global Sr isotope budget. The proposed magnitude of this flux has not been supported by subsequent work (e.g. Jones *et al.*, 2014). However, Jones *et al.* suggested that some of the global Sr isotope imbalance could be explained if the suspended sediment flux from ocean islands were also taken into account. These suspended particles were shown to rapidly exchange Sr with seawater in an Icelandic estuary, suggesting that this process could go some way to explaining the global flux imbalance.

In addition to a ridge-crest flux, another possible flux of unradiogenic Sr to the oceans is due to diffuse exchange with sea floor basaltic rocks over large areas on the flanks of ridges. The relatively high heat flux on these flanks implies a substantial hydrothermal water flux there, prompting Hess *et al.* (1991) to test the magnitude of hydrothermal Sr fluxes in these areas. Three off-axis ridge sites were studied in the Pacific Ocean, where warm hydrothermal waters were emitted from the sea floor. However, all measured Sr isotope ratios for these fluids were close to seawater values. In contrast, Elderfield *et al.* (1999) observed relatively unradiogenic fluid compositions ($^{87}Sr/^{86}Sr = 0.707$) on the flanks of the Juan de Fuca ridge. Butterfield *et al.* (2001) attempted to determine the significance of these results for the global seawater Sr balance, but the Juan de Fuca ridge flank has abnormal signatures for other species such as Mg and Ca. It is therefore difficult to extrapolate these Sr data to the whole ocean floor.

3.6.5 Quantifying the Hydrothermal Flux

To quantify the total extent of sea floor Sr exchange (both axial and off-axis), Davis *et al.* (2003) studied hydrothermal alteration effects in analysed sections of ophiolites and oceanic drill cores. They found that hydrothermal Sr exchange was much greater in the ophiolites than *in situ* oceanic crust (Fig. 3.32). They attributed this effect to the abnormal genetic environment of ophiolite crust, which is believed to form in back-arc basins rather than ocean basins. Davis *et al.* argued that while the degree of alteration in ophiolites would be sufficient to balance the riverine Sr flux, the alteration observed in oceanic crustal sections was not.

Coogan (2009) inverted this argument, suggesting that the difference between the 90 Ma Troodos ophiolite and

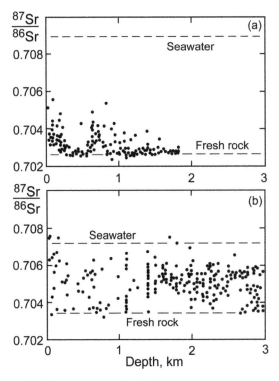

Fig. 3.32 Sr isotope ratios in bulk samples of altered basaltic crust from (a) deep sea drill core (ODP site 504B, off Costa Rica) and (b) the Troodos ophiolite. Modified after Davis *et al.* (2003).

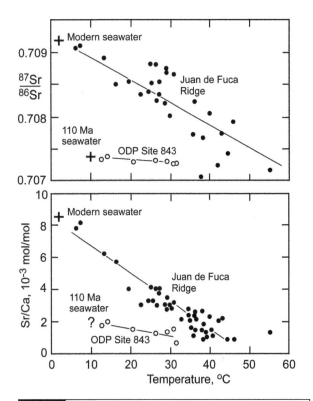

Fig. 3.33 Plots of (a) Sr isotope ratio and (b) Sr/Ca ratios against calculated deposition temperatures of sea floor hydrothermal vein carbonates. After Coggon *et al.* (2010).

modern sea floor reflected a real change in the degree of Sr exchange between seawater and oceanic crust through time. He attributed this effect to enhanced Sr concentrations in the Cretaceous oceans, which would introduce greater amounts of seawater Sr into oceanic crust. However, other groups working on hydrothermal veins in sea floor samples did not find evidence of higher Sr in Cretaceous seawater.

For example, Coggon *et al.* (2010) used the oxygen isotope compositions of sea floor hydrothermal veins to calculate mineral deposition temperatures, which were shown to be strongly correlated with both Sr isotope ratio and Sr/Ca ratio (Fig. 3.33a, b). In both cases, the Juan de Fuca correlation line correctly 'points' to the modern seawater composition, while paleo-seawater $^{87}Sr/^{86}Sr$ is consistent with the correlation line for 110 Ma oceanic crust at ODP site 843. Based on this consistency, the Sr/Ca correlation line for site 843 can be used to estimate the paleo seawater Sr/Ca ratio, suggesting that this ratio dramatically *increased* over the past 40 Ma.

This conclusion is the opposite to that inferred from Sr/Ca measurements on biogenic carbonates, such as foram tests (Steuber and Veizer, 2002; Lear *et al.*, 2003). However, Coggon *et al.* argued that the foram determinations are unreliable, due to the difficulty of correctly determining the Sr/Ca parti-

tion coefficients between seawater and biogenic carbonates. Subsequently, other recent studies of sea floor hydrothermal veins (Rausch *et al.*, 2013) showed corresponding increases in the Sr/Ca and Mg/Ca ratio of seawater, implying that the major change over the past 50 Ma is a marked *decrease* in seawater Ca concentration, whereas Sr remained fairly constant.

Coogan and Dosso (2015) offered an alternative explanation for the reduced degree of alteration in modern oceanic crust compared with 100 Ma ophiolites, arguing that a reduction in ocean bottom temperatures in the Tertiary period led to reduced intensity of sea floor hydrothermal alteration. They also proposed that this change was the driving force behind the global increase in seawater Sr isotope ratios over the past 40 Ma. However, the model assumes a direct relationship between ocean bottom temperature and Sr exchange, whereas data for sea floor hydrothermal veins (Fig. 3.34) show that *average* carbonate deposition temperatures in the cores decreased almost twice as rapidly as ocean bottom temperatures. This suggests that the reduced alteration in modern sea floor veins can be better explained by decreasing ridge heat output over the past 150 Ma. Therefore, enhanced global hydrothermal alteration fluxes 100 Ma ago probably reflected a higher intensity of sea floor spreading at that time, which

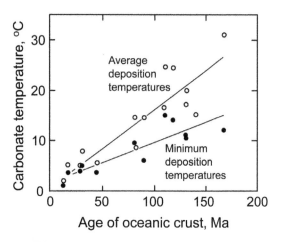

Fig. 3.34 Plot of hydrothermal vein deposition temperatures against crustal age, showing a correlation between minimum (•) and average (○) temperatures of carbonate deposition. Data from Coogan and Dosso (2015).

was also responsible for high sea level stands in the Mesozoic (e.g. Becker *et al.*, 2009).

3.6.6 The Effects of Himalayan Erosion

An alternative or additional cause of the marked increase in seawater Sr isotope ratio over the past 40 Ma is tectonic uplift of the Himalayas, Tibet and the Andes. Raymo *et al.* (1988) proposed that this uplift could have caused a substantial increase in the supply of radiogenic Sr to the oceans, since the rivers which rise in these regions (Ganges–Brahmaputra, Yangtze and Amazon) together supply 20% of the riverine solid load to the oceans.

More detailed constraints on this model came from a comprehensive study of riverine Sr budgets by Palmer and Edmond (1989). This work revealed an important inverse relationship between Sr isotope ratio and concentration (plotted as reciprocal Sr in Fig. 3.35), attributed to mixing between radiogenic Sr from silicate weathering and less radiogenic Sr from carbonate weathering. The Ganges and Brahmaputra, which drain the Himalayan uplift, lie off the general trend. However, within the drainage basin of the Ganges, its tributaries themselves display a similar mixing line, but with a steeper slope than other rivers (Fig. 3.35b).

Further examination of these data (Palmer and Edmond, 1992) showed that the mixing line for the Ganges system also had a more elevated *intercept* (Sr isotope ratio) than other world rivers. Palmer and Edmond attributed this pattern to the presence of carbonate rocks with abnormally radiogenic Sr in the Ganges watershed. They speculated that these carbonates had become enriched in radiogenic Sr by exchange with the surrounding, very radiogenic silicate rocks.

Subsequent to this work, more detailed studies have been made of the rivers draining the High Himalayas, which are tributaries to the Indus and Ganges–Brahmaputra river systems. For example, Blum *et al.* (1998) analysed river water, rock outcrops and river bed sands from the Raikot watershed in northern Pakistan. They showed that stream and river waters define a positive trend of Sr isotope ratio against Ca/Sr ratio (Fig. 3.36). This trend runs from the composition of marbles and marble sands at the unradiogenic end, to a radiogenic end-member with a much higher Ca/Sr ratio than silicate rocks. Therefore, Blum *et al.* speculated that this end-member might be vein calcite, which inherited radiogenic Sr during hydrothermal alteration of the surrounding silicate rocks.

Further study (Jacobson and Blum, 2000) identified radiogenic calcite interstitially within silicates, at grain boundaries and in fracture fillings. This calcite makes up less than 0.5% of the orthogneissic rocks in the Raikot watershed, but appears to dominate the Sr budget of streams draining this

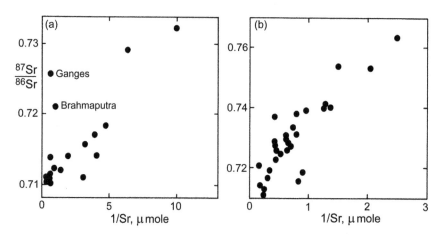

Fig. 3.35 Plot of Sr isotope ratio against reciprocal of Sr concentration: (a) for the world's major rivers; (b) for tributaries of the Ganges (note different axis scales). After Palmer and Edmond (1989).

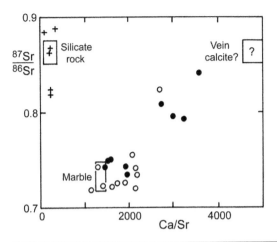

Fig. 3.36 Plot of Sr isotope ratio against Ca/Sr ratio for samples from the Raikot river watershed, northern Pakistan. (•) = waters; (+) = silicate rocks; (○) = carbonate rocks; boxes = proposed end-members. After Blum et al. (1998).

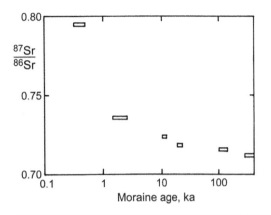

Fig. 3.37 Sr isotope ratio in ammonium acetate soil leachates plotted against the geological age of the moraines on which the soils were formed. After Blum and Erel (1995).

terrain. Similar conclusions about the role of carbonate dissolution were reached by English et al. (2000), based on a study of the Seti watershed in western Nepal. In addition, carbonates in the Seti watershed have also been proposed as a source of radiogenic osmium in Himalayan rivers (Pierson-Wickmann et al., 2002).

3.6.7 Glacial Cycles

Another explanation for recent increases in the Sr isotope ratio of seawater is the onset of Tertiary glaciation, as originally proposed by Armstrong (1971). Hodell et al. (1990) revived this model, attempting to link inflections in the Tertiary Sr isotope evolution path to glacial advances and retreats, and this idea has been developed in several subsequent papers. The basis of the model is that glaciation creates rock flour, which is then more susceptible to chemical weathering than in situ crystalline rocks.

Blum and Erel (1995) attempted to quantify the amount of radiogenic Sr that could be released by glacial erosion. In order to do this, they used ammonium acetate leaching to determine the isotopic composition of exchangeable Sr in glacial moraines. Weathered soils from six moraines in the Wind River Range, Wyoming, displayed a negative correlation between the isotopic composition of leachable Sr and the age of the soil (Fig. 3.37). Most notably, a very radiogenic $^{87}Sr/^{86}Sr$ ratio of 0.795 was obtained by leaching soil from the youngest (400 year old) moraine. Blum and Erel used these data to argue that a spike of radiogenic Sr is released by weathering of moraines immediately after glaciation. Modelling of this spike suggested that it could yield an incremental increase in $^{87}Sr/^{86}Sr$ of 5 ppm (0.000 003) for each 100 ka glacial cycle of the Quaternary period.

This theoretical model was further developed by Vance et al. (2009), leading to the predicted seawater growth curve shown by the heavy dashed line in Fig. 3.38. In contrast, a steady state evolution line with much steeper slope would result from the present day riverine budget, which is believed to be around eight times too great to explain long-term seawater evolution. Based on the age of glacial moraines, Vance et al. argued that peak radiogenic Sr release would have occurred at around 10 ka BP, after which the weathering rate would have decreased to 50% of the maximum value. To make this model fit the overall evolution of seawater Sr over the past 400 ka, Vance et al. invoked enhanced release of less radiogenic Sr (ca. 0.7084) from continental shelves during glacial low sea level stands, as previously suggested by Stoll and Schrag (1998). The composite model was then able to explain most of the data of Henderson et al. (1994) within 2σ confidence limits (Fig. 3.38).

More recent higher precision Sr isotope data were presented by Ando et al. (2010) and Mokadem et al. (2015). One

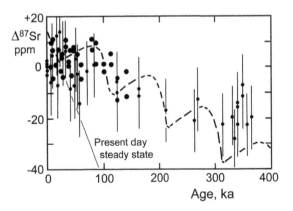

Fig. 3.38 Modelled trend of seawater Sr isotope evolution over the past 400 ka (Vance et al., 2009) compared with foram data. Samples with error bars = Henderson et al. (1994); (•) = Ando et al. (2010) and Mokadem et al. (2015).

problematical issue is that the modern seawater $^{87}Sr/^{86}Sr$ ratio determined in these newer studies (average = 0.709 178) is 0.000 02 (ca. 30 ppm) higher than the value of Henderson *et al.* (1994). This off-set requires paleo-seawater compositions to be presented as Δ values (Fig. 3.38) normalized to the present day seawater Sr ratio ($\Delta^{87}Sr = 0$).

The new high-precision data of Mokadem *et al.* (2015) suggest that seawater Sr has been constant to within ca. 5 ppm over the last 40 ka. This is a problem for the glacial erosion model, because Sr released from glacial rock flour should have increased seawater $^{87}Sr/^{86}Sr$ by more than 10 ppm over the past 20 ka. This seems to cast doubt on the model, although a possible explanation would be an error in modelling the timing of the deglacial Sr spike, so that the present day radiogenic riverine Sr flux has not yet affected seawater Sr. On the other hand, there is some evidence in Fig. 3.38 for a step increase in seawater $^{87}Sr/^{86}Sr$ during the penultimate deglaciation, ca. 120 ka BP. Therefore, more detailed study of that transition is urgently needed to test the glacial model.

3.6.8 Stable Sr Isotopes in Seawater

In the past, all terrestrial Sr isotope ratios have been fractionation-normalized to a constant $^{86}Sr/^{88}Sr$ ratio of 0.1194 ($^{88}Sr/^{86}Sr = 8.375\ 21$). This procedure corrects for the effects of instrumental mass fractionation, but also removes any effects from mass fractionation in nature. It is possible to recover information about natural fractionation effects using the double spiking technique (Section 2.5.2). This has been applied to meteorite studies for many years, to study mass fractionation effects and nucleosynthetic anomalies in the solar nebula. However the method has only recently been applied to study seawater Sr.

To test the usefulness of stable Sr isotopes as a fractionation monitor in seawater, Fietzke and Eisenhauer (2006) studied the temperature dependence of stable Sr isotope variations in inorganic and coral aragonite grown at different temperatures. The inorganic aragonite was grown in lab experiments, whereas the natural coral came from a previous temperature calibration suite collected from the Galapagos Islands. Fietzke and Eisenhauer found that the temperature dependence of isotopic fractionation in natural coral was six times that for inorganic aragonite precipitation (Fig. 3.39).

This behaviour is consistent with that previously observed for stable calcium isotopes, where natural coral shows a temperature-dependent kinetic fractionation effect around 15 times stronger than inorganic aragonite (Gussone *et al.*, 2003). The relatively low fractionation in non-biogenic carbonate is attributed to the formation of complex ions by the addition of a 'hydrate shell' around Ca^{++} and Sr^{++} ions, which increases the apparent mass of these cations to ca. 600 a.m.u. In contrast, most organisms remove the cation from its complex before they secrete aragonite. The effect is approximately twice as large for calcium because its mass is half that of strontium.

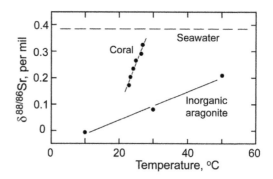

Fig. 3.39 Temperature dependence of stable Sr isotope fractionation for inorganic and coralline aragonite. Ratios are reported relative to the NBS (NIST) 987 Sr standard. After Fietzke and Eisenhauer (2006).

Given this evidence for stable Sr isotope variations in corals, the technique might be useful for understanding the global seawater Sr system. Therefore, Krabbenhoft *et al.* (2010) analysed all of the major terrestrial Sr reservoirs in order to examine their isotopic relationships (Fig. 3.40). They found relatively small variations of $^{88}Sr/^{86}Sr$ in river water and hydrothermal solutions, which therefore form a nearly horizontal mixing line of Sr fluxes to the ocean in Fig. 3.40. However, in the ocean, precipitation of isotopically light Sr in inorganic and biogenic carbonates causes seawater to develop an isotopically heavy $^{88}Sr/^{86}Sr$ signature (vertical arrow in Fig. 3.40). These results were verified by Pearce *et al.* (2015).

Krabbenhoft *et al.* (2010) also speculated that weathering of shelf carbonates exposed by low sea-level stands during glacial periods could release a flux of isotopically *light*

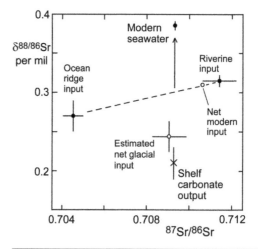

Fig. 3.40 Plot of stable versus radiogenic Sr isotope ratios for major terrestrial reservoirs showing mass-balance relationships. Modified after Krabbenhoft *et al.* (2010).

strontium to the oceans. However, this fractionation process has a relatively weak effect on radiogenic ^{87}Sr/^{86}Sr ratios (Fig. 3.39). Hence, these data do not markedly affect arguments concerning the balance of global Sr fluxes.

References

Albarede, F., Michard, A., Minster, J. F. and Michard, G. (1981). ^{87}Sr/^{86}Sr ratios in hydrothermal waters and deposits from the East Pacific Rise at 21 °N. *Earth Planet. Sci. Lett.* **55**, 229–36.

Aleinikoff, J. N., Horton, J. W., Drake, A. A. et al. (2004). Deciphering multiple Mesoproterozoic and Paleozoic events recorded in zircon and titanite from the Baltimore Gneiss, Maryland: SEM imaging, SHRIMP U-Pb geochronology, and EMP analysis. In: Tollo, R. P., Corriveau, L., McLelland, J. and Bartholomew, M. J. (Eds) *Proterozoic tectonic evolution of the Grenville orogen in North America: An introduction: Geol. Soc. America Mem.* **197**, 411–34.

Allegre, C. J., Louvat, P., Gaillardet, J. et al. (2010). The fundamental role of island arc weathering in the oceanic Sr isotope budget. *Earth Planet. Sci. Lett.* **292**, 51–6.

Amelin, Y. and Zaitsev, A. N. (2002). Precise geochronology of phoscorites and carbonatites: The critical role of U-series disequilibrium in age interpretations. *Geochim. Cosmochim. Acta* **66**, 2399–419.

Ando, A., Nakano, T., Kawahata, H., Yokoyama, Y. and Khim, B. K. (2010). Testing seawater Sr isotopic variability on a glacial–interglacial timescale: An application of latest high-precision thermal ionization mass spectrometry. *Geochem. J.* **44**, 347–57.

Armstrong, R. L. (1971). Glacial erosion and the variable isotopic composition of strontium in sea water. *Nature Phys. Sci.* **230**, 132–3.

Asmerom, Y., Jacobsen, S. B., Knoll, A. H., Butterfield, N. J. and Swett, K. (1991). Strontium isotopic variations of Neoproterozoic seawater: implications for crustal evolution. *Geochim. Cosmochim. Acta* **55**, 2883–94.

Basu, A. R., Jacobsen, S. B., Poreda, R. J., Dowling, C. B. and Aggarwal, P. K. (2001). Large groundwater strontium flux to the oceans from the Bengal Basin and the marine strontium isotope record. *Science* **293**, 1470–3.

Beck, A. J., Charette, M. A., Cochran, J. K., Gonneea, M. E. and Peucker-Ehrenbrink, B. (2013). Dissolved strontium in the subterranean estuary-Implications for the marine strontium isotope budget. *Geochim. Cosmochim. Acta* **117**, 33–52.

Becker, T. W., Conrad, C. P., Buffett, B. and Müller, R. D. (2009). Past and present seafloor age distributions and the temporal evolution of plate tectonic heat transport. *Earth Planet. Sci. Lett.* **278**, 233–42.

Begemann, F., Ludwig, K. R., Lugmair, G. W. et al. (2001). Call for an improved set of decay constants for geochronological use. *Geochim. Cosmochim. Acta* **65**, 111–21.

Birck, J. L. and Allegre, C. J. (1978). Chronology and chemical history of the parent body of basaltic achondrites studied by the ^{87}Rb)^{87}Sr method. *Earth Planet. Sci. Lett.* **39**, 37–51.

Blum, J. D. and Erel, Y. (1995). A silicate weathering mechanism linking increases in marine ^{87}Sr/^{86}Sr with global glaciation. *Nature* **373**, 415–18.

Blum, J. D., Gazis, C. A., Jacobsen, A. D. and Chamberlain, C. P. (1998). Carbonate versus silicate weathering in the Raikot watershed within the High Himalayan Crystalline Series. *Geology* **26**, 411–4.

Brand, U. and Veizer, J. (1980). Chemical diagenesis of a multicomponent carbonate system – 1: Trace elements. *J. Sed. Petrol.* **50**, 1219–36.

Brannon, J. C., Podosek, F. A. and McLimans, R. K. (1992). Alleghenian age of the Upper Mississippi Valley zinc-lead deposit determined by Rb-Sr dating of sphalerite. *Nature* **356**, 509–11.

Brass, G. W. (1976). The variation of the marine ^{87}Sr/^{86}Sr during Phanerozoic time: interpretation using a flux model. *Geochim. Cosmochim. Acta* **40**, 721–30.

Brinkman, G. A., Aten, A. H. W. and Veenboer, J. T. (1965). Natural radioactivity of K-40, Rb-87 and Lu-176. *Physica* **31**, 1305–19.

Brooks, C., Hart, S. R., Hofmann, A. and James, D. E. (1976a). Rb-Sr mantle isochrons from oceanic regions. *Earth Planet. Sci. Lett.* **32**, 51–61.

Brooks, C., James, D. E. and Hart, S. R. (1976b). Ancient lithosphere: its role in young continental volcanism. *Science* **193**, 1086–94.

Brown, E. H. (1971). Phase relations of biotite and stilpnomelane in the green-schist facies. *Contrib. Mineral. Petrol.* **31**, 275–99.

Burke, W. H., Denison, R. E., Hetherington, E. A. et al. (1982). Variations of seawater ^{87}Sr/^{86}Sr throughout Phanerozoic time. *Geology* **10**, 516–19.

Butterfield, D. A., Nelson, B. K., Wheat, C. G., Mottl, M. J. and Roe, K. K. (2001). Evidence for basaltic Sr in midocean ridge-flank hydrothermal systems and implications for the global oceanic Sr isotope balance. *Geochim. Cosmochim. Acta* **65**, 4141–53.

Catanzaro, E. J., Murphy, T. J., Garner, E. L. and Shields, W. R. (1969). Absolute isotopic abundance ratio and atomic weight of terrestrial rubidium. *J. Res. NBS* **73A**, 511–16.

Chaudhuri, S. and Clauer, N. (1986). Fluctuations of isotopic composition of strontium in seawater during the Phanerozoic eon. *Chem. Geol. (Isot. Geosci. Sect.)* **59**, 293–303.

Christensen, J. N., Halliday, A. N., Leigh, K. E., Randell, R. N. and Kesler, S. E. (1995a). Direct dating of sulfides by Rb-Sr: a critical test using the Polaris Mississippi Valley-type Zn-Pb deposit. *Geochim. Cosmochim. Acta* **59**, 5191–7.

Christensen, J. N., Halliday, A. N., Vearncombe, J. R. and Kesler, S. E. (1995b). Testing models of large-scale fluid flow using direct dating of sulfides: Rb-Sr evidence for early dewatering and formation of Mississippi Valley-type deposits, Canning Basin, Australia. *Econ. Geol.* **90**, 877–84.

Clark, S. P. C. and Jager, E. (1969). Denudation rate in the Alps from geochronologic and heat flow data. *Amer. J. Sci.* **267**, 1143–60.

Clauer, N. (1979). A new approach to Rb)Sr dating of sedimentary rocks. In: Jager, E. and Hunziker, J. C. (Eds) *Lectures in Isotope Geology*. Springer, pp. 30–51.

Clauer, N., Keppens, E. and Stille, P. (1992). Sr isotopic constraints on the process of glauconitization. *Geology* **20**, 133–6.

Clemens, S. C., Farrell, J. W. and Gromet, L. P. (1993). Synchronous changes in seawater strontium isotope composition and global climate. *Nature* **363**, 607–10.

Clemens, S. C., Gromet, L. P. and Farrell, J. W. (1995). Artifacts in Sr isotope records. *Nature* **373**, 201.

Cliff, R. A. (1985). Isotope dating in metamorphic belts. *J. Geol. Soc. Lond.* **142**, 97–110.

Coggon, R. M., Teagle, D. A., Smith-Duque, C. E., Alt, J. C. and Cooper, M. J. (2010). Reconstructing past seawater Mg/Ca and Sr/Ca from mid-ocean ridge flank calcium carbonate veins. *Science* **327**, 1114–17.

Compston, W. and Jeffery, P. M. (1959). Anomalous common strontium in granite. *Nature* **184**, 1792–3.

Compston, W. and Pidgeon, R. T. (1962). Rubidium–strontium dating of shales by the total-rock method. *J. Geophys. Res.* **67**, 3493–502.

Coogan, L. A. (2009). Altered oceanic crust as an inorganic record of paleoseawater Sr concentration. *Geochem. Geophys. Geosys.* **10** (4), 1–11.

Coogan, L. A. and Dosso, S. E. (2015). Alteration of ocean crust provides a strong temperature dependent feedback on the geological carbon cycle and is a primary driver of the Sr-isotopic composition of seawater. *Earth Planet. Sci. Lett.* **415**, 38–46.

Cowie, J. W. and Johnson, M. R. W. (1985). Late Precambrian and Cambrian geological time-scale. In: Snelling, N. J. (Ed.) The chronology of the geological record. *Mem. Geol. Soc. Lond.* **10**, 47–64.

Dasch, E. J. and Biscaye, P. E. (1971). Isotopic composition of strontium in Cretaceous-to-Recent, pelagic foraminifera. *Earth Planet. Sci. Lett.* **11**, 201–4.

Davis, A. C., Bickle, M. J. and Teagle, D. A. (2003). Imbalance in the oceanic strontium budget. *Earth Planet. Sci. Lett.* **211**, 173–87.

Davis, D. W., Gray, J. and Cumming, G. L. (1977). Determination of the ^{87}Rb decay constant. *Geochim. Cosmochim. Acta* **41**, 1745–9.

Del Moro, A., Puxeddu, M. and Villa, I. M. (1982). Rb-Sr and K-Ar ages on minerals at temperatures of 300–400 °C from deep wells in the Larderello geothermal field (Italy). *Contrib. Mineral. Petrol.* **81**, 340–9.

DePaolo, D. J. (1987). Correlating rocks with strontium isotopes. *Geotimes* **32**, 16–18.

Derry, L. A., Keto, L. S., Jacobsen, S. B., Knoll, A. H. and Swett, K. (1989). Sr isotopic variations in Upper Proterozoic carbonates from Svalbard and East Greenland. *Geochim. Cosmochim. Acta* **53**, 2331–9.

Dia, A. N., Cohen, A. S., O'Nions, R. K. and Shackleton, N. J. (1992). Seawater Sr isotope variation over the past 300 ka and influence of global climate cycles. *Nature* **356**, 786–8.

Dodson, M. H. (1973). Closure temperature in cooling geochronological and petrological systems. *Contrib. Mineral. Petrol.* **40**, 259–74.

Dodson, M. H. (1979). Theory of cooling ages. In: Jager, E. and Hunziker, J. C. (Eds) *Lectures in Isotope Geology*. Springer, pp. 194–202.

Dunoyer de Segonzac, G. (1969). Les mineraux argileux dans la diagenese. Passage au metamorphisme. *Mem. Serv. Carte Geol. Alsace Lorraine* **29**, 320 pp.

Edmond, J. M., Measures, C., McDuff, R. E. *et al.* (1979). Ridge crest hydrothermal activity and the balances of the major and minor elements in the ocean: the Galapagos data. *Earth Planet. Sci. Lett.* **46**, 1–18.

Elderfield, H. and Gieskes, J. M. (1982). Sr isotopes in interstitial waters of marine sediments from Deep Sea Drilling Project cores. *Nature* **300**, 493–7.

Elderfield, H., Wheat, C. G., Mottl, M. J., Monnin, C. and Spiro, B. (1999). Fluid and geochemical transport through oceanic crust: a transect across the eastern flank of the Juan de Fuca Ridge. *Earth Planet. Sci. Lett.* **172**, 151–65.

English, N. B., Quade, J., DeCelles, P. G. and Garzione, C. N. (2000). Geologic control of Sr and major element chemistry in Himalayan rivers, Nepal. *Geochim. Cosmochim. Acta* **64**, 2549–66.

Fairbairn, H. W., Hurley, P. M. and Pinson, W. H. (1961). The relation of discordant Rb-Sr mineral and rock ages in an igneous rock to its time of subsequent Sr^{87}/Sr^{86} metamorphism. *Geochim. Cosmochim. Acta* **23**, 135–44.

Farrell, J. W., Clemens, S. C. and Gromet, L. P. (1995). Improved chronostratigraphic reference curve of late Neogene seawater $^{87}Sr/^{86}Sr$. *Geology* **23**, 403–6.

Faure, G., Hurley, P. M. and Powell, J. L. (1965). The isotopic composition of strontium in surface water from the North Atlantic Ocean. *Geochim. Cosmochim. Acta* **29**, 209–20.

Field, D. and Raheim, A. (1979a). Rb-Sr total rock isotope studies on Precambrian charnockitic gneisses from South Norway: evidence for isochron resetting during a low-grade metamorphic-deformational event. *Earth Planet. Sci. Lett.* **45**, 32–44.

Field, D. and Raheim, A. (1979b). A geological meaningless Rb-Sr total rock isochron. *Nature* **282**, 497–9.

Fietzke, J. and Eisenhauer, A. (2006). Determination of temperature-dependent stable strontium isotope ($^{88}Sr/^{86}Sr$) fractionation via bracketing standard MC-ICP-MS. *Geochem. Geophys. Geosys.* **7** (8), 1–6.

Flynn, K. F. and Glendenin, L. E. (1959). Half-life and β spectrum of Rb^{87}. *Phys. Rev.* **116**, 744–8.

Gast, P. W. (1955). Abundance of Sr^{87} during geologic time. *Bull. Geol. Soc. Amer.* **66**, 1449–64.

Goldstein, S. J. and Jacobsen, S. B. (1987). The Nd and Sr isotopic systematics of river-water dissolved material: Implications for the sources of Nd and Sr in seawater. *Chem. Geol.: Isot. Geosci. Sect.* **66**, 245–72.

Grant, N. K., Laskowski, T. E. and Foland, K. A. (1984). Rb-Sr and K-Ar ages of Paleozoic glauconites from Ohio–Indiana and Missouri, USA. *Isot. Geosci.* **2**, 217–39.

Gray, C. M., Papanastassiou, D. A. and Wasserburg, G. J. (1973). The identification of early condensates from the solar nebula. *Icarus* **20**, 213–39.

Gussone, N., Eisenhauer, A., Heuser, A. *et al.* (2003). Model for kinetic effects on calcium isotope fractionation ($\delta\,^{44}Ca$) in inorganic aragonite and cultured planktonic foraminifera. *Geochim. Cosmochim. Acta* **67**, 1375–82.

Halliday, A. N. and Porcelli, D. (2001). In search of lost planets – the paleocosmochronology of the inner solar system. *Earth Planet. Sci. Lett.* **192**, 545–59.

Harris, W. B. (1976). Rb-Sr glauconite isochron, Maestrichtian unit of Peedee Formation, North Carolina. *Geology* **4**, 761–2.

Harvey, C. F. (2002). Groundwater flow in the Ganges Delta. *Science* **296**, 1563.

Henderson, G. M., Martel, D. J., O'Nions, R. K. and Shackleton, N. J. (1994). Evolution of seawater $^{87}Sr/^{86}Sr$ over the last 400 ka: the absence of glacial/interglacial cycles. *Earth Planet. Sci. Lett.* **128**, 643–51.

Hess, J., Bender, M. and Schilling, J. G. (1991). Assessing seawater/basalt exchange of strontium isotopes in hydrothermal processes on the flanks of mid-ocean ridges. *Earth Planet. Sci. Lett.* **103**, 133–42.

Hess, J., Bender, M. L. and Schilling, J. G. (1986). Evolution of the ratio of strontium-87 to strontium-86 in seawater from Cretaceous to present. *Science* **231**, 979–84.

Hodell, D. A., Mead, G. A. and Mueller, P. A. (1990). Variation in the strontium isotopic composition of seawater (8 Ma to present): implications for chemical weathering rates and dissolved fluxes to the oceans. *Chem. Geol. (Isot. Geosci. Sect.)* **80**, 291–307.

Hofmann, A. W. and Giletti, B. J. (1970). Diffusion of geochronologically important nuclides under hydrothermal conditions. *Eclogae Geol. Helv.* **63**, 141–50.

Hunziker, J. C. (1974). Rb-Sr and K-Ar age determination and the Alpine tectonic history of the Western Alps. *Mem. Inst. Geol. Min. Univ. Padova* **31**, 1–54.

Hurley, P. M., Cormier, R. F., Hower, J., Fairbairn, H. W. and Pinson, W. H. (1960). Reliability of glauconite for age measurement by K-Ar and Rb-Sr methods. *Amer. Assoc. Pet. Geol. Bull.* **44**, 1793–808.

Jacobson, A. D. and Blum, J. D. (2000). Ca/Sr and $^{87}Sr/^{86}Sr$ geochemistry of disseminated calcite in Himalayan silicate rocks from Nanga Parbat: influence on river-water chemistry. *Geology* **28**, 463–6.

Jager, E. (1973). Die Alpine orogenese im lichte der radiometrischen altersbestimmung. *Eclogae Geol. Helv.* **66**, 11–21.

Jager, E., Niggli, E. and Wenk, E. (1967). Rb)Sr altersbestimmungen an glimmern der Zentralalpen. *Beitr. Geol. Karte Schweiz N. F.* **134**, 1–67.

Jones, M. T., Gislason, S. R., Burton, K. W. *et al.* (2014). Quantifying the impact of riverine particulate dissolution in seawater on ocean chemistry. *Earth Planet. Sci. Lett.* **395**, 91–100.

Kaufman, A. J., Jacobsen, S. B. and Knoll, A. H. (1993). The Vendian record of Sr and C isotopic variations in seawater: implications for tectonics and paleoclimate. *Earth Planet. Sci. Lett.* **120**, 409–30.

Kossert, K. (2003). Half-life measurements of ^{87}Rb by liquid scintillation counting. *Applied Rad. Isot.* **59**, 377–82.

Krabbenhoft, A., Eisenhauer, A., Bohm, F. et al. (2010). Constraining the marine strontium budget with natural strontium isotope fractionations (^{87}Sr/^{86}Sr$*$, δ $^{88/86}$Sr) of carbonates, hydrothermal solutions and river waters. Geochim. Cosmochim. Acta **74**, 4097–109.

Kubler, B. (1966). La cristallinite d'illite et les zones tout a fait superieures du metamorphisme. Colloque. sur les Etages Tectoniques. Univ. Neuchatel, pp. 105–22.

Lanphere, M. A., Wasserburg, G. J., Albee, A. L. and Tilton, G. R. (1964). Redistribution of strontium and rubidium isotopes during metamorphism, World Beater complex, Panamint Range, California. In: Craig, H., Miller, S. L. and Wasserburg, G. J. (Eds) Isotopic and Cosmic Chemistry. North Holland Pub., pp. 269–320.

Lear, C. H., Elderfield, H. and Wilson, P. A. (2003). A Cenozoic seawater Sr/Ca record from benthic foraminiferal calcite and its application in determining global weathering fluxes. Earth Planet. Sci. Lett. **208**, 69–84.

MacLeod, K. G., Huber, B. T. and Fullagar, P. D. (2001). Evidence for a small (~0.000 030) but resolvable increase in seawater ^{87}Sr/^{86}Sr ratios across the Cretaceous–Tertiary boundary. Geology **29**, 303–6.

Martin, E. E. and Macdougall, J. D. (1991). Seawater Sr isotopes at the Cretaceous/Tertiary boundary. Earth Planet. Sci. Lett. **104**, 166–80.

McArthur, J. M., Thirlwall, M. F., Engkilde, M., Zinsmeister, W. J. and Howarth, R. J. (1998). Strontium isotope profiles across K/T boundary sequences in Denmark and Antarctica. Earth Planet. Sci. Lett. **160**, 179–92.

McMullen, C. C., Fritze, K. and Tomlinson, R. H. (1966). The half-life of rubidium-87. Can. J. Phys. **44**, 3033–8.

Minster, J-F., Birck, J-L. and Allegre, C. J. (1982). Absolute age of formation of chondrites studied by the ^{87}Rb–^{87}Sr method. Nature **300**, 414–19.

Mokadem, F., Parkinson, I. J., Hathorne, E. C. et al. (2015). High-precision radiogenic strontium isotope measurements of the modern and glacial ocean: Limits on glacial–interglacial variations in continental weathering. Earth Planet. Sci. Lett. **415**, 111–20.

Morton, J. P. and Long, L. E. (1980). Rb)Sr dating of Palaeozoic glauconite from the Llano region, central Texas. Geochim. Cosmochim. Acta **44**, 663–72.

Morton, J. L. and Sleep, N. H. (1985). A mid-ocean ridge thermal model: Constraints on the volume of axial hydrothermal heat flux. J. Geophys. Res. **90** (B13), 11345–53.

Nakai, S., Halliday, A. N., Kesler, S. E. and Jones, H. D. (1990). Rb–Sr dating of sphalerites from Tennessee and the genesis of Mississippi Valley type ore deposits. Nature **346**, 354–7.

Nakai, S., Halliday, A. N., Kesler, S. E. et al. (1993). Rb–Sr dating of sphalerites from Mississippi Valley-type (MVT) ore deposits. Geochim. Cosmochim. Acta **57**, 417–27.

Nebel, O., Scherer, E. E. and Mezger, K. (2011). Evaluation of the ^{87}Rb decay constant by age comparison against the U–Pb system. Earth Planet. Sci. Lett. **301**, 1–8.

Neumann, W. and Huster, E. (1974). The half-life of ^{87}Rb measured as a difference between the isotopes of ^{87}Rb and ^{85}Rb. Z. Physik **270**, 121–7.

Neumann, W. and Huster, E. (1976). Discussion of the ^{87}Rb half-life determined by absolute counting. Earth Planet. Sci. Lett. **33**, 277–88.

Nicolaysen. L. O. (1961). Graphic interpretation of discordant age measurements on metamorphic rocks. Ann. N. Y. Acad. Sci. **91**, 198–206.

Odin, G. S. and Dodson, M. H. (1982). Zero isotopic age of glauconies. In: Odin, G. S. (Ed.) Numerical Dating in Stratigraphy. Wiley, pp. 277–305.

Odin, G. S., Gale, N. H. and Dore, F. (1985). Radiometric dating of Late Precambrian times. In: Snelling, N. J. (Ed.) The chronology of the geological record. Mem. Geol. Soc. Lond. **10**, 65–72.

Palmer, M. R. and Edmond, J. M. (1989). The strontium isotope budget of the modern ocean. Earth Planet. Sci. Lett. **92**, 11–26.

Palmer, M. R. and Edmond, J. M. (1992). Controls over the strontium isotope composition of river water. Geochim. Cosmochim. Acta **56**, 2099–111.

Papanastassiou, D. A. and Wasserburg, G. J. (1969). Initial strontium isotopic abundances and the resolution of small time differences in the formation of planetary objects. Earth Planet. Sci. Lett. **5**, 361–76.

Papanastassiou, D. A., Wasserburg, G. J. and Burnett, D. S. (1970). Rb–Sr ages of lunar rocks from the Sea of Tranquillity. Earth Planet. Sci. Lett. **8**, 1–19.

Pearce, C. R., Parkinson, I. J., Gaillardet, J. et al. (2015). Reassessing the stable ($\delta^{88/86}$Sr) and radiogenic (^{87}Sr/^{86}Sr) strontium isotopic composition of marine inputs. Geochim. Cosmochim. Acta **157**, 125–46.

Peterman, Z. E., Hedge, C. E. and Tourtelot, H. A. (1970). Isotopic composition of strontium in sea water throughout Phanerozoic time. Geochim. Cosmochim. Acta **34**, 105–20.

Pettke, T. and Diamond, L. W. (1996). Rb–Sr dating of sphalerite based on fluid inclusion–host mineral isochrons: a clarification of why it works. Econ. Geol. **91**, 951–6.

Pierson-Wickmann, A.-C., Reisberg, L. and France-Lanord, C. (2002). Impure marbles of the Lesser Himalaya: another source of continental radiogenic osmium. Earth Planet. Sci. Lett. **204**, 203–14.

Popp, B. N., Podosek, F. A., Brannon, J. C., Anderson, T. F. and Pier, J. (1986). ^{87}Sr/^{86}Sr ratios in Permo-Carboniferous sea water from the analyses of well-preserved brachiopod shells. Geochim. Cosmochim. Acta **50**, 1321–8.

Provost, A. (1990). An improved diagram for isochron data. Chem. Geol. (Isot. Geosci. Sect.)**80**, 85–99.

Purdy, J. W. and Jager, E. (1976). K)Ar ages on rock-forming minerals from the Central Alps. Mem. Inst. Geol. Mineral. Univ. Padova **30**, 3–31.

Rahaman, W. and Singh, S. K. (2012). Sr and ^{87}Sr/^{86}Sr in estuaries of western India: Impact of submarine groundwater discharge. Geochim. Cosmochim. Acta **85**, 275–88.

Rausch, S., Böhm, F., Bach, W., Klügel, A. and Eisenhauer, A. (2013). Calcium carbonate veins in ocean crust record a threefold increase of seawater Mg/Ca in the past 30 million years. Earth Planet. Sci. Lett. **362**, 215–24.

Raymo, M. E., Ruddiman, W. F. and Froelich, P. N. (1988). Influence of late Cenozoic mountain building on ocean geochemical cycles. Geology **16**, 649–53.

Rotenberg, E., Davis, D. W., Amelin, Y., Ghosh, S. and Bergquist, B. A. (2012). Determination of the decay-constant of ^{87}Rb by laboratory accumulation of ^{87}Sr. Geochim. Cosmochim. Acta **85**, 41–57.

Rundberg, Y. and Smalley, P. C. (1989). High-resolution dating of Cenozoic sediments from northern North Sea using ^{87}Sr/^{86}Sr stratigraphy. AAPG Bull. **73**, 298–308.

Schreiner, G. D. L. (1958). Comparison of the Rb-87/Sr-87 age of the Red granite of the Bushveld complex from measurements on the total rock and separated mineral fractions. Proc. Roy. Soc. Lond. A. **245**, 112–17.

Sheppard, T. J. and Darbyshire, D. P. F. (1981). Fluid inclusion Rb–Sr isochrons for dating mineral deposits. Nature **290**, 578–9.

Shih, C. Y., Nyquist, L. E., Bogard, D. D. et al. (1985). Chronology and petrogenesis of a 1.8 g lunar granitic clast: 14321, 1062. Geochim. Cosmochim. Acta **49**, 411–26.

Spooner, E. T. C. (1976). The strontium isotopic composition of seawater, and seawater–oceanic crust interaction. Earth Planet. Sci. Lett. **31**, 167–74.

Steiger, R. H. and Jager, E. (1977). Subcommission on geochronology: convention on the use of decay constants in geo- and cosmo-chronology. Earth Planet. Sci. Lett. **36**, 359–62.

Steuber, T. and Veizer, J. (2002). Phanerozoic record of plate tectonic control of seawater chemistry and carbonate sedimentation. *Geology* **30**, 1123–6.

Stoll, H. M. and Schrag, D. P. (1998). Effects of Quaternary sea level cycles on strontium in seawater. *Geochim. Cosmochim. Acta* **62**, 1107–18.

Sun, S. S. and Hansen, G. N. (1975). Evolution of the mantle: geochemical evidence from alkali basalt. *Geology* **3**, 297–302.

Tatsumoto, M. (1966). Genetic relationships of oceanic basalts as indicated by lead isotopes. *Science* **153**, 1094–101.

Tilton, G. R. (1988). Age of the solar system. In: Kerridge, J. F. and Matthews, M. S. (Eds) *Meteorites and the Early Solar System*, Univ. Arizona Press, pp. 259–75.

Tretbar, D. R., Arehart, G. B. and Christensen, J. N. (2000). Dating gold deposition in a Carlin-type gold deposit using Rb/Sr methods on the mineral galkhaite. *Geology* **28**, 947–50.

Vance, D., Teagle, D. A. and Foster, G. L. (2009). Variable Quaternary chemical weathering fluxes and imbalances in marine geochemical budgets. *Nature* **458**, 493–6.

Veizer, J. and Compston, W. (1974). ^{87}Sr/^{86}Sr composition of seawater during the Phanerozoic. *Geochim. Cosmochim. Acta* **38**, 1461–84.

Veizer, J. and Compston, W. (1976). ^{87}Sr/^{86}Sr in Precambrian carbonates as an index of crustal evolution. *Geochim. Cosmochim. Acta* **40**, 905–14.

Veizer, J. and 14 others. (1999). ^{87}Sr/^{86}Sr, δ^{13}C and δ^{18}O evolution of Phanerozoic seawater. *Chem. Geol.* **161**, 59–88.

Verschure, R. H. Andriessen, P. A. M., Boelrijk, N. A. I. M. *et al.* (1980). On the thermal stability of Rb-Sr and K-Ar biotite systems: evidence from co-existing Sveconorwegian (ca. 870 Ma) and Caledonian (ca. 400 Ma) biotites in S.W. Norway. *Contrib. Mineral. Petrol.* **74**, 245–52.

Villa, I. M., De Bièvre, P., Holden, N. E. and Renne, P. R. (2015). IUPAC–IUGS recommendation on the half life of ^{87}Rb. *Geochim. Cosmochim. Acta* **164**, 382–5.

Wasserburg, G. J., Papanastassiou, D. A. and Sanz, H. G. (1969). Initial strontium for a chondrite and the determination of a metamorphism or formation interval. *Earth Planet. Sci. Lett.* **7**, 33–43.

Wetherill, G. W., Davis, G. L. and Lee-Hu, C. (1968). Rb-Sr measurements on whole rocks and separated minerals from the Baltimore Gneiss, Maryland. *Geol. Soc. Amer. Bull.* **79**, 757–62.

Wickman, F. E. (1948). Isotope ratios: a clue to the age of certain marine sediments. *J. Geol.* **56**, 61–6.

Chapter 4

The Sm–Nd Method

Sm is a rare earth element with seven naturally occurring isotopes, including the long-lived unstable nuclide [147]Sm, with a half-life of 106 Ga. This undergoes α decay to [143]Nd, and the resulting Sm–Nd decay scheme is the main focus of the present chapter. On the other hand, the relatively long-lived *extinct* isotope, [146]Sm, decays to [142]Nd. This yields important information about the early history of the solar system and the Earth, and is discussed in Section 15.7. Two other samarium isotopes ([148]Sm and [149]Sm) are also radioactive, but have such long half-lives (ca. 10^{16} years) that they are not capable of producing measurable variations in the daughter isotopes of [144]Nd and [145]Nd, even over cosmological intervals (10 Ga).

The half-life of [147]Sm is sufficiently short to produce small but measurable differences in [143]Nd abundance over periods of several million years, thus providing the basis for the Sm–Nd dating method. This half-life, equivalent to a decay constant of 6.54×10^{-12} a^{-1}, is the weighted mean of several counting determinations, and yields ages consistent with Pb–Pb dating (Lugmair and Marti, 1977, 1978). More recent radioactive counting determinations, summarized by Boehnke and Harrison (2014), suggested a 0.4% increase in the half-life. However, the best geological determinations, from comparisons with U–Pb and Pb–Pb ages, support the conventional half-life (Section 4.1 below).

The Sm–Nd method is one of the most powerful dating methods in geology, but some errors of interpretation in the early development of the method have discredited it in the eyes of many geologists. However, Nd isotope data give unique insights into crustal evolution that simple chronometers of igneous crystallization, such as U–Pb, cannot provide. These issues will be discussed at some length in this chapter.

4.1 Sm–Nd Isochrons

Considering a given system, such as an igneous rock or mineral, we can write the following equation to describe the build-up of [143]Nd from decay of [147]Sm:

$$^{143}\text{Nd} = {}^{143}\text{Nd}_\text{I} + {}^{147}\text{Sm}\left(e^{\lambda t} - 1\right) \qquad [4.1]$$

where I signifies the initial abundance and t is the age of the system. In view of the possibility of [142]Nd variation (due to [146]Sm decay), it is convenient to divide through by [144]Nd, the second most abundant isotope of Nd. Thus we obtain

$$\frac{^{143}\text{Nd}}{^{144}\text{Nd}} = \left(\frac{^{143}\text{Nd}}{^{144}\text{Nd}}\right)_\text{I} + \frac{^{147}\text{Sm}}{^{144}\text{Nd}} \cdot \left(e^{\lambda t} - 1\right) \qquad [4.2]$$

This equation has the same form as that for Rb–Sr (Section 3.2) and can be plotted as an isochron diagram. However, because Sm and Nd have very similar chemical properties (unlike Rb and Sr), large ranges of Sm/Nd in whole-rock systems are rare, and in particular, low Sm/Nd ratios near the y axis are very rare. Therefore, because of the difficulty of obtaining a wide range of Sm/Nd ratios from a single rock body, and because of the greater technical demands of Nd isotope analysis, the Sm–Nd isochron method was initially applied to problems where Rb–Sr isochrons had proved to be unsatisfactory. Many of these applications were also made before the U–Pb zircon method had reached its present-day level of development (Sections 5.1, 5.2). Therefore, most of these bodies have subsequently been dated to greater accuracy and precision by the U–Pb method. The case studies reviewed here are either significant as developments in understanding, or involve applications (such as dating garnet growth) where the Sm–Nd isochron method continues to provide unique insights.

4.1.1 Meteorites

Chondritic meteorites were readily dated in early Rb–Sr work, but achondrites were more problematical. Bulk samples usually have low Rb/Sr ratios, yielding ages of low precision, while separated minerals give ages below 4.5 Ga due to disturbance. The Sm–Nd system in separated minerals is more resistant to resetting, and the first high-precision Sm–Nd mineral isochron (Lugmair et al., 1975) gave an age of 4560 ± 80 Ma (2σ) for the eucrite achondrite Juvinas. Subsequent work on the angrite achondrite Lewis Cliff 86010 (Lugmair and Galer, 1992) achieved a mineral isochron with greater spread in Sm/Nd ratio (Fig. 4.1). This gave a more precise age

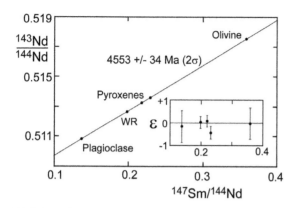

Fig. 4.1 Sm–Nd mineral isochron for the angrite achondrite Lewis Cliff 86010. Nd isotope ratios reflect the normalizing factor used for mass fractionation. After Lugmair and Galer (1992).

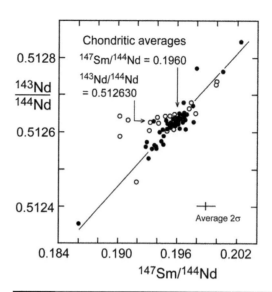

Fig. 4.2 Sm–Nd isochron for unequilibrated (•) and equilibrated (○) chondrite whole-rock samples, showing new average values. After Bouvier et al. (2008).

of 4553 ± 34 Ma (2σ), in excellent agreement with Hf–W and Pb–Pb dating of angrites (Section 15.6.5).

Sm–Nd dating of chondritic meteorites was not a high priority, due to the success of other methods. However, the isotopic *composition* of the chondrites is a critical benchmark for terrestrial evolution, since chondrites are believed to sample the primordial solar nebula. DePaolo and Wasserburg (1976a) coined the acronym CHUR (chondritic uniform reservoir) for this benchmark, but in the absence of isotopic data for chondrites, they used Lugmair et al.'s (1975) initial ratio for Juvinas as representative of the solar system. This value was tested by direct Sm–Nd analysis of chondrites by Jacobsen and Wasserburg (1980) and Wasserburg et al. (1981), with values clustering closely around the original Juvinas measurement.

All of this early work was done using Nd isotope analysis as the oxide species. However, most subsequent workers have analysed Nd as the metal species (Section 2.2.1), thus avoiding the need to made oxide interference corrections. O'Nions et al. (1977) established a different normalizing ratio for mass fractionation correction during Nd metal analysis ($^{146}Nd/^{144}Nd = 0.7219$), leading to CHUR values of $^{143}Nd/^{144}Nd = 0.512\,638$ and $^{147}Sm/^{144}Nd = 0.1966$ (Hamilton et al., 1983).

Bouvier et al. (2008) made minor revisions to the chondritic Nd value in a study mainly aimed at improving the chondritic Hf isotope value. Hf was found to be strongly affected by metamorphic disturbance in what are normally referred to as 'equilibrated' chondrites. These are bodies of the petrographic categories 4–6, signifying metamorphic disturbance. Exclusion of these types dramatically improved the consistency of chondritic Hf isotope signatures, but had a lesser effect on Nd (Fig. 4.2). Note that the four unequilibrated outliers in Fig. 4.2 are all carbonaceous chondrites, while the large apparent scatter is due to the magnified axis scales.

4.1.2 Precambrian Mafic Rocks

The long half-life of ^{147}Sm makes it most useful for dating in the Precambrian. Therefore, most early Sm–Nd studies aimed to determine crystallization ages for Archean igneous rocks. In such suites, the Rb–Sr or K–Ar methods had often shown open-system behaviour, while precise U–Pb dates were not yet available. The Stillwater Complex (DePaolo and Wasserburg, 1979) provides a good example of such an application.

Rb–Sr data for separated minerals from a single adcumulus unit of the Stillwater layered series form a scatter that does not define an isochron (Fig. 4.3a). This was attributed to open-system behaviour of Rb–Sr in minerals. However, Sm–Nd data on the same minerals defined an excellent isochron array (Fig. 4.3b), from which DePaolo and Wasserburg calculated an age of 2701 ± 8 Ma (2σ). The analysis of separated minerals allowed a much greater range of Sm/Nd ratios than whole-rock samples, but raised the possibility that Sm–Nd mineral systems might have been opened by the event that disturbed Rb–Sr. To test for this, DePaolo and Wasserburg also analysed six whole-rock samples with a wide range of plagioclase/pyroxene abundances from different levels in the intrusion. Sm–Nd data for these samples fell within analytical uncertainty of the mineral isochron (Fig. 4.3c), suggesting that the mineral isochron yields a true crystallization age for the intrusion, and that the magma had a homogeneous initial Nd isotope composition.

The Sm–Nd mineral age has subsequently been corroborated by U–Pb dating of the Stillwater Complex. For example, Nunes (1981) obtained an age of 2713 ± 3 Ma (2σ) from the chilled margin, while Premo et al. (1990) obtained an age of 2305 ± 4 Ma (2σ) on the lower banded series. Since these

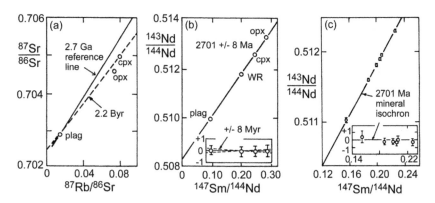

Fig. 4.3 Isochron diagrams for the Stillwater Complex. (a) Rb–Sr diagram showing scatter of mineral data; (b) Sm–Nd mineral isochron; (c) whole-rock data with reference isochron from (b). After DePaolo and Wasserburg (1979).

U–Pb concordia ages would be reduced by 2 to 3 Ma using a revised ^{235}U decay constant (Section 5.1), the agreement with the Sm–Nd mineral age must be considered to be excellent, arguing against any revision of the Sm decay constant.

Unlike the early data, subsequent Sm–Nd data for whole-rock samples did not lie on the mineral isochron. The new samples, from a wider stratigraphic range in the intrusion, indicate larger variations of initial Nd isotope ratio in the complex (Lambert et al., 1989). This is not surprising, since the initial ratio of DePaolo and Wasserburg fell below the CHUR evolution line at 2.7 Ga, with an εNd value of −2.8, best explained by contamination of the magma by old crustal Nd from the Wyoming craton.

The Stillwater data emphasize the importance of combined mineral and whole-rock isochrons to verify the accuracy of Sm/Nd ages. However, this approach is not possible for fine-grained Archean rocks such as basalts and komatiites. In these situations, whole-rock Sm–Nd analysis has often been used alone, but subtle changes to the slopes of whole-rock isochrons can be caused by analysing samples with slight variations in the degree of crustal contamination. This is a particular hazard for mafic–ultramafic rocks emplaced through older crustal basement. A good example is provided by the Kambalda volcanics of Western Australia.

McCulloch and Compston (1981) determined a composite Sm–Nd isochron on a suite of rocks comprising the ore-bearing Kambalda ultramafic unit, the footwall and hanging wall basalts and an 'associated' sodic-granite and felsic porphyry. Although the whole suite yielded a good isochron age of 2790 ± 30 Ma, the basic and ultra-basic samples alone gave an older apparent age of 2910 ± 170 Ma.

The danger of constructing a 'composite' Sm–Nd isochron of acid, basic and ultra-basic rocks which might not be co-magmatic was pointed out by Claoue-Long et al. (1984). These workers attempted to date the Kambalda lavas by the Sm–Nd method without using felsic rocks. However, they were forced to combine analyses from komatiites and basalts in order to achieve a good spread of Sm/Nd ratios (Fig. 4.4). After

the exclusion of an altered komatiite from Kambalda and a suite of basalt lavas from Bluebush (40 km south of Kambalda), ten data points gave an age of 3262 ± 44 Ma (2σ), interpreted as the age of eruption.

Chauvel et al. (1985) challenged this interpretation on the basis that Pb–Pb dating of the Kambalda volcanics and associated igneous sulphide mineralization gave an age of 2726 ± 34 Ma, which they argued to be resistant to resetting by later events. They attributed the 3.2 Ga apparent Sm–Nd age to either variable crustal contamination of the magma suite by older basement, or possibly a heterogeneous mantle source. U–Pb ages of ca. 3.4 Ga in zircon xenocrysts from one of the hanging-wall basalts subsequently confirmed the contamination model (Compston et al., 1985).

There is a danger that such erroneous Sm–Nd ages can discredit Nd isotope analysis as a reliable means of crustal age determination. However, the problem was not with the Sm–Nd method itself, but its application to mafic rocks

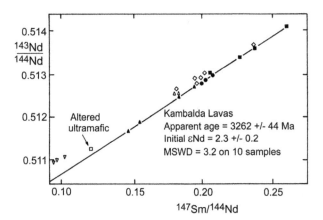

Fig. 4.4 Sm–Nd isochron diagram for whole-rock samples of Kambalda volcanics. (■) = komatiites; (▲) = hanging-wall basalts; (◇) = Bluebush lavas; (△) = 'ocelli' basalts; (▽) = granites. Modified after Claoue-Long et al. (1984).

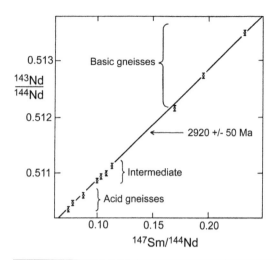

Fig. 4.5 Sm–Nd isochron for a mixed suite of granitic, tonalitic and layered basic gneisses from the Lewisian complex of NW Scotland, yielding an age of 2920 Ma. After Hamilton *et al.* (1979).

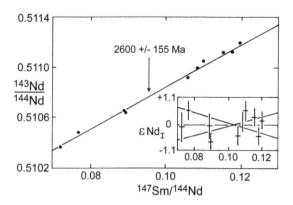

Fig. 4.6 Sm–Nd 'errorchron' for Lewisian tonalitic gneisses, defining an age of 2600 Ma, attributed to granulite-facies metamorphism. After Whitehouse (1988).

emplaced through older crustal basement. These issues will be discussed further below.

4.1.3 High-Grade Metamorphic Rocks

Most dating systems, including U–Pb zircon, can be reset during high-grade metamorphic events. However, the Sm–Nd method provides an opportunity to determine igneous protolith ages in high-grade metamorphic gneiss terranes where other systems are reset. An example is provided by dating work on the Lewisian gneisses of NW Scotland.

Whole-rock Rb–Sr, whole-rock Pb–Pb and U–Pb zircon ages on granulite-facies and amphibolite-facies Lewisian gneisses are concordant at 2630 ± 140, 2680 ± 60 and 2660 ± 20 Ma (2σ), respectively (Pidgeon and Bowes, 1972; Moorbath *et al.*, 1975; Chapman and Moorbath, 1977). However, these gneisses are generally very Rb- and U-depleted, suggesting that even large whole-rock samples were probably open systems for these elements during the depletion event.

A suite of whole-rock samples was dated by the Sm–Nd method (Hamilton *et al.*, 1979) to see whether this system had remained undisturbed during the Badcallian metamorphic event that the other systems are presumed to date. An older age of 2920 ± 50 Ma (2σ) suggested that the gneisses had remained closed systems for Sm–Nd during granulite-facies metamorphism (Fig. 4.5). Hamilton *et al.* therefore interpreted this age as the time of protolith formation, which occurred 200–300 Ma before the closure of U–Pb zircon and whole-rock Rb–Sr and Pb–Pb systems following metamorphism.

In spite of the good quality of the Sm–Nd isochron, there are two problems with the sample selection. Firstly, the sample suite combined amphibolite- and granulite-facies gneisses, and secondly, it contained a bimodal petrological

suite, including tonalitic gneisses and basic rocks from the Drumbeg layered complex. Nevertheless, because the slope ages of the tonalites and mafic gneisses are very similar, the samples as a whole display good linearity, with an MSWD value of only 1.3 (using 1σ errors of 0.1% for Sm/Nd, and the individual within-run isotopic errors).

More detailed investigation by Whitehouse (1988) showed that the Drumbeg layered basic rocks retain a 2.91 Ga isochron age, but Sm–Nd whole-rock systems in intermediate to acid rocks have been reset to the same age as the U–Pb zircon and other whole-rock systems. Ten samples of the latter suite define an errorchron with MSWD = 5.7, yielding an age (with estimate of geological error) of 2600 ± 155 Ma (2σ), shown in Fig. 4.6. Therefore, the isochron of Hamilton *et al.* (1979) apparently does correctly date the time of protolith formation, but only the basic rocks remained closed systems during the Badcallian event. This work shows that even whole-rock Sm–Nd isochrons can be perturbed by granulite-facies metamorphism. However, it will be shown below that Sm–Nd *model ages* can largely preserve the protolith ages of the intermediate gneisses, even though the isochron is disturbed (Section 4.3.3). These model ages agree with the isochron age for the Drumbeg basic pluton.

4.1.4 Garnet Geochronology

One application where the Sm–Nd isochron method is particularly useful is the dating of high-grade metamorphic minerals. This work has focussed particularly on garnets, which typically grow under peak metamorphic conditions, and whose high Sm/Nd ratios allow precise age determinations.

Much of the early work was on dating eclogites, often using garnet–cpx mineral pairs (e.g. Griffin and Brueckner, 1980). However, Mork and Mearns (1986) showed that the development of garnet–omphacite mineralogy may not completely homogenize the Sm–Nd system in the rock. For example, some Caledonian eclogites from western Norway

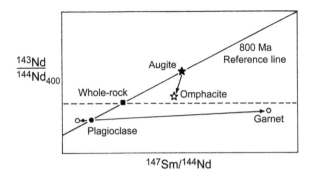

Fig. 4.7 Schematic illustration of the process of Sm–Nd remobilization during the replacement of gabbro by an eclogite mineralogy. Modified after Mork and Mearns (1986).

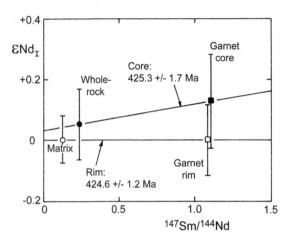

Fig. 4.8 Sm–Nd isochrons (using ε notation) for whole-rock–garnet-core, and matrix–garnet-rim pairs from a meta-pelite. Error bars indicate within-run precision. After Burton and O'Nions (1991).

developed an eclogite mineralogy, but retained a relict igneous texture. Examination of age-corrected Sm-Nd data for these rocks revealed isotopic disequilibrium (Fig. 4.7) and suggested that the main obstacle to isotopic homogenization was the cpx phase. Because the transformation of augite to omphacite requires relatively minor cation exchange, complete resetting of the Sm-Nd system in this mineral rarely occurs. In contrast, major chemical exchange and structural reorganization is required to replace plagioclase with garnet, so complete resetting is more likely. Hence, garnet–whole-rock isochrons are more reliable for dating metamorphic mineral growth than garnet–cpx pairs.

Vance and O'Nions (1990) showed that Sm-Nd dating of garnets is particularly useful for dating prograde metamorphism, in contrast to other methods, such as Rb-Sr and Ar-Ar, which date metamorphic cooling (Sections 3.3, 10.4). Garnets are widely distributed in meta-pelitic rocks and develop in response to the changing *P–T* conditions of prograde metamorphism. Their chemistry (including the Sm-Nd system) is usually preserved during cooling because cation diffusion rates in garnet are very low. The chemical composition of garnets can be used to calculate the *P–T* conditions of their growth, which, combined with age data, provide a method of determining progradational *P–T–time* paths for high-grade metamorphic terranes.

An application of this technique was demonstrated by Burton and O'Nions (1991) in a study of Caledonian regional metamorphism of a Proterozoic supracrustal sequence at Sulitjilma, North Norway. An example is shown in Fig. 4.8 for a case where garnet rims and cores are distinct. The rims yield a slightly younger age, as would be expected. Note that the core is regressed with the whole-rock composition, while the rim is regressed with the matrix of the rock only, since this is the only part of the rock with which the rims were in diffusional contact at the time of their growth.

This approach was further developed by Pollington and Baxter (2010), who dated twelve concentric growth bands in a large (6 cm diameter) garnet from the Tauern window of the Austrian Alps. The garnet was from a highly meta-

somatized shear zone, thought to have been active during exhumation of the Austrain Alps from ca. 30–20 Ma. Successive samples were obtained by milling grooves into a slice sawn through the middle of the garnet, enabling a series of concentric layers to be isolated and crushed (Pollington and Baxter, 2011). Each age determination (Fig. 4.9) represents a two-point isochron between one of the concentric garnet bands and a whole-rock sample of the rock matrix. However, it was necessary to clean the crushed garnet material using a sequential pre-leaching technique to remove the numerous inclusions that are typical of garnet porphyroblasts (Baxter *et al.*, 2002). The results show two periods of

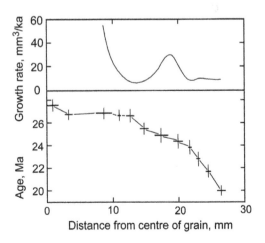

Fig. 4.9 Plot of calculated garnet growth rate against radial distance, based on Sm–Nd dating of concentric sample bands in a large garnet from the Austrian Alps. Modified after Pollington and Baxter (2010).

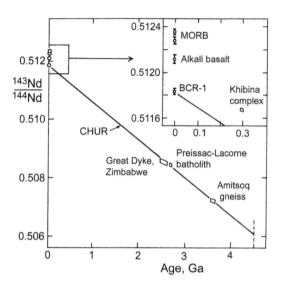

Fig. 4.10 Diagram of ^{143}Nd/^{144}Nd against time showing initial Nd isotope ratios determined in early work on terrestrial igneous rocks, relative to the chondritic growth line. BCR-1 = Columbia River basalt. After DePaolo and Wasserburg (1976a).

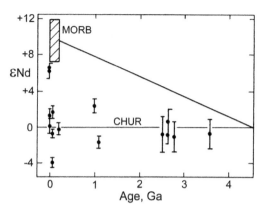

Fig. 4.11 Diagram of Nd isotope evolution against time in the form of deviations from the chondritic evolution line in ε units. After DePaolo and Wasserburg (1976b).

rapid growth, around 26.8 Ma and 24.6 Ma, interpreted as periods of enhanced fluid infiltration that facilitated garnet growth.

4.2 Nd Isotope Evolution and Model Ages

DePaolo and Wasserburg (1976a) made the first whole-rock Nd isotope determinations on terrestrial igneous rocks, plotting their ages and initial ^{143}Nd/^{144}Nd ratios on a diagram of Nd isotope evolution against time (Fig. 4.10). They found that Archean plutons had initial ratios remarkably consistent with the isotopic evolution of the Juvinas achondrite, used as a proxy for the composition of chondritic meteorites (Section 4.1.1). Because chondrites have very consistent Sm/Nd ratios, DePaolo and Wasserburg used them to define a solar system benchmark which they termed the chondritic uniform reservoir (CHUR). Whether or not the Earth has an exactly chondritic signature will be discussed elsewhere (Sections 6.2.3, 15.7). However, even if this were not the case, the CHUR composition would remain a vital benchmark of solar system evolution. The Nd isotope evolution of CHUR is normally drawn as a straight line, but in fact is a very gentle curve, due to the finite half-life of ^{147}Sm (ca. 106 Ga).

Because Sm and Nd are rare earth elements (REE) with atomic numbers only two units apart, their chemical properties are very similar, and they undergo only slight relative fractionation during crystal–liquid processes. This means that in terrestrial rocks, departures of ^{143}Nd/^{144}Nd from the CHUR evolution line are small relative to the steepness of the

line (Fig. 4.12). DePaolo and Wasserburg therefore developed a notation whereby initial ^{143}Nd/^{144}Nd isotope ratios could be represented as parts per 10^4 deviations from the CHUR evolution line, termed epsilon units (εNd). Mathematically, this notation is defined as

$$\varepsilon Nd\,(t) = \left[\frac{\left(^{143}Nd/^{144}Nd\right)_{sample}(t)}{\left(^{143}Nd/^{144}Nd\right)_{CHUR}(t)} - 1\right] \times 10^4 \qquad [4.3]$$

where t indicates the time at which εNd is calculated. The ε notation makes it much easier to compare the initial Nd isotope ratios of bodies of different ages, and also removes the effects of different fractionation corrections applied to Nd analysis as the metal or oxide species. More recently, the chondritic composition of the Bulk Earth has been questioned by some researchers (as noted above). However, the epsilon notation (relative to CHUR) remains the most convenient way of expressing past Nd isotope compositions.

Using this notation, DePaolo and Wasserburg (1976b) presented a larger data set of Nd isotope analyses on a diagram of εNd against time (Fig. 4.11). They noted that continental igneous rocks through time had εNd values very close to zero. Indeed, for Archean rocks, the error bars overlapped with zero, suggesting that continental igneous rocks were 'derived from a reservoir with a chondritic REE pattern, which may represent primary material remaining since the formation of the Earth.' It is now believed that most or all of these samples were derived from a more depleted mantle source, and that their apparently chondritic signatures were either caused by contamination by older crustal material (Section 4.1.2), or by open-system behaviour during high-grade metamorphism (Section 4.1.3).

4.2.1 Chondritic Model Ages
DePaolo and Wasserburg (1976b) argued that if the CHUR evolution line defines the initial ratios of continental igneous rocks through time, measurement of ^{143}Nd/^{144}Nd and ^{147}Sm/^{144}Nd in any crustal rock would yield a model age

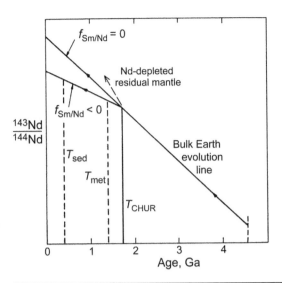

Fig. 4.12 Schematic Nd isotope evolution diagram showing the theory of model ages. T_{met} = age of metamorphic event; T_{sed} = age of erosion–sedimentation event; f = fractionation of sample Sm/Nd relative to Bulk Earth. Dashed vector shows the evolution of the depleted source as a result of crustal extraction. After McCulloch and Wasserburg (1978).

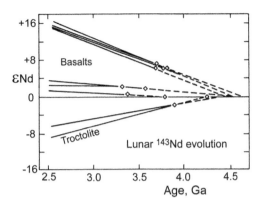

Fig. 4.13 Nd isotope evolution diagram for lunar rocks indicating very early Sm/Nd fractionation between lunar reservoirs. After Lugmair and Marti (1978).

for the formation of that rock (or its precursor) from the chondritic reservoir. For such an age to be meaningful, there must have been sufficient Nd/Sm fractionation during the process of crustal extraction to give a reasonable divergence between the chondritic and crustal evolution lines (Fig. 4.12), and hence a precise intersection. Based on the measured present day ratios, the model age is then given as

$$T_{CHUR} = \frac{1}{\lambda} \times \ln \left[1 + \frac{\left(\frac{143Nd}{144Nd}\right)^0_{sample} - \left(\frac{143Nd}{144Nd}\right)^0_{CHUR}}{\left(\frac{147Sm}{144Nd}\right)^0_{sample} - \left(\frac{147Sm}{144Nd}\right)^0_{CHUR}} \right] \quad [4.4]$$

DePaolo and Wasserburg argued that if the Sm/Nd ratio of a rock sample has not been disturbed since its separation from the chondritic reservoir (taken to be the mantle source), then T_{CHUR} may provide a 'crustal formation' age for a wide variety of rocks. Elemental investigations have pointed to the general immobility of REE on a whole-rock scale during the processes of weathering and low-temperature metamorphism associated with sedimentary rock formation (e.g. Haskin et al., 1966), and even during high-grade metamorphism (Green et al., 1969). This immobility is illustrated schematically in Fig. 4.12 by the lack of deflection in the evolution line of a crustal rock sample during metamorphic and sedimentary events. Hence, Nd model ages may be able to date crustal formation in rocks that have been subjected to high-grade metamorphism and even cycles of erosion–sedimentation.

These premises were applied by McCulloch and Wasserburg (1978) in a model age study aimed at measuring the crustal formation ages of several cratonic rock bodies,

mainly from the Canadian Shield. McCulloch and Wasserburg found Nd model ages within the range 2.5–2.7 Ga for composite samples of the Superior, Slave, and Churchill structural provinces. In the first two areas, previously determined K–Ar and Rb–Sr ages had given the same results, but the 2.7 Ga model age for the Churchill province was 0.8 Ga older than the previously determined K–Ar age, which had presumably been reset by more recent metamorphism. These data supported a model of episodic continental growth, by showing the period from 2.7–2.5 Ga to be a time of remarkably widespread crustal formation. In contrast, a Grenville Province composite yielded a model age of 0.8 Ga which did not reveal any Archean component, suggesting it to be an addition of more recent crust to the pre-existing shield. However, this sample was by no means representative of the Grenville Province as a whole, which actually contains comparatively little juvenile Grenvillian-age crust (Dickin et al., 2010).

4.2.2 Depleted Mantle Model Ages

While observing the good fit of Archean plutons to the CHUR Nd isotope evolution line, DePaolo and Wasserburg (1976b) also noted that young mid-ocean ridge basalts (MORB) lay +7 to +12 epsilon units above the CHUR evolution line (Fig. 4.11). They recognized that Archean continental igneous rocks that fell within error of the CHUR evolution line could conceivably lie on a depleted mantle evolution line characterized by progressively increasing Sm/Nd and $^{143}Nd/^{144}Nd$. Such a source could be formed as a residue from magma extraction, as shown in Fig. 4.12. However, DePaolo and Wasserburg rejected this model in favour of a chondritic source for continental igneous rocks on the basis of a comparison with lunar Nd isotope evolution.

Lunar basalts and troctolites with ages of 3.3–4.3 Ga show a wide range of initial $^{143}Nd/^{144}Nd$ ratios (Fig. 4.13), equivalent to a variation from +7 to −2 ε units relative to CHUR (Lugmair and Marti, 1978). This spread shows that very early

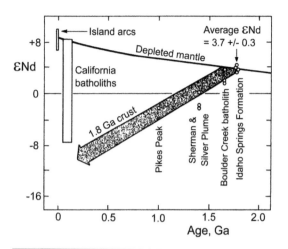

Fig. 4.14 Plot of εNd against time showing Colorado data relative to a model depleted-mantle evolution curve. After DePaolo (1981).

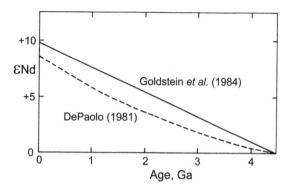

Fig. 4.15 Plot of εNd against time showing two of the most widely used depleted mantle evolution models. Dashed curve: DePaolo (1981); solid line: Goldstein et al. (1984).

Sm/Nd fractionation occurred in the Moon, and that there was no long-lived uniform magma source with a chondritic Sm/Nd ratio. In contrast, none of the Archean terrestrial rocks analysed by 1976 showed any dispersion outside error of CHUR, which led DePaolo and Wasserburg (1976b) to conclude that the Earth did not undergo early differentiation, or if it did, that this was re-mixed by convection.

The paucity of Nd isotope data for the Proterozoic was a serious weakness in this model, since it left a gap between the Archean CHUR data and the recent MORB data, attributed to a 'depleted mantle' source. (A mantle source depleted in large ion lithophile elements will have an elevated Sm/Nd ratio, and hence develop more radiogenic $^{143}Nd/^{144}Nd$ over time.)

An important stage in filling this gap was a study on Proterozoic metamorphic basement from the Colorado Front Range (DePaolo, 1981). Four meta-volcanics and two charnockitic granulites from the Idaho Springs Formation were dated by the Sm–Nd isochron method. In addition, Nd isotope and Sm/Nd determinations were made on three plutons previously dated by the Rb–Sr whole-rock method (the Boulder Creek, Silver Plume and Pikes Peak granitoids). The initial $^{143}Nd/^{144}Nd$ ratios of all these samples are plotted on an εNd versus time diagram in Fig. 4.14.

The Idaho Springs meta-igneous rocks cluster at εNd (t) = +3.7 ± 0.3, showing them to be derived from a depleted mantle reservoir with respect to CHUR at 1.8 Ga. Boulder Creek samples also have positive εNd (+1.7 to +3.5), while the Silver Plume and Pikes Peak granites have progressively lower εNd values which lie on the $^{143}Nd/^{144}Nd$ evolution line of average Idaho Springs crust, suggesting that they contain a large fraction of re-melted 1.8 Ga-old basement.

DePaolo was able to fit a quadratic curve to Idaho Springs and modern island arc data (Fig. 4.14), representing the Nd isotope evolution of a progressively depleted mantle reser-

voir that was the source for calc-alkaline (subduction-related) magmatism. This curve begins on the CHUR evolution line in the Early Archean, but diverges progressively to the present day. The composition of this depleted reservoir at time T, relative to CHUR, is given as

$$\varepsilon Nd\ (T) = 0.25\,T^2 + 3\,T + 8.5 \qquad [4.5]$$

Sm–Nd model ages calculated using this depleted mantle curve are denoted T_{DM}. DePaolo argued that T_{DM} model ages would be a more accurate indication of 'crustal formation ages' than T_{CHUR} ages for studies of continental evolution. For example, an anomalously low T_{CHUR} age of 0.8 Ga in McCulloch and Wasserburg's Grenville composite (Section 4.2.1) is revised to a T_{DM} age of 1.3 Ga, consistent with more recent Nd model ages determined for the nearby Central Metasedimentary Belt of the Grenville Province (Dickin and McNutt, 2007).

It is important to emphasize that the DM model of DePaolo (1981) is empirically based. Therefore, it is not affected by arguments about large-scale mantle structure or the overall composition of the Bulk Silicate Earth, whether these arguments are based on experimental data or theoretical models (Sections 6.2, 15.7). However, subsequent to the discovery of Proterozoic depleted mantle by DePaolo (1981), several alternative proposals have been made for the composition of depleted-mantle reservoirs used to calculate Nd model ages.

One of the most important alternative models, proposed by Goldstein et al. (1984), assumes linear depletion of the mantle from εNd = zero at 4560 Ma to +10 for present day MORB (Fig. 4.15). Unlike the model of DePaolo (1981), this is a theoretical model whose basis for calculating crustal residence ages was not empirically justified. The model provides a good fit to Early Proterozoic greenstones from the SW United States (Nelson and DePaolo, 1984), which probably represent flood basalts erupted in back-arc rifting environments. However, it is not the most appropriate mantle model for calculating crustal extraction ages of tonalitic

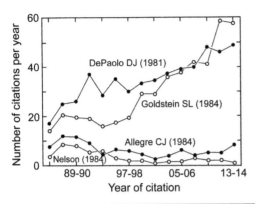

Fig. 4.16 Plot of annual citation rates for four papers that introduced new depleted mantle evolution models for Nd. Data from the Science Citation Index, averaged over two-year intervals.

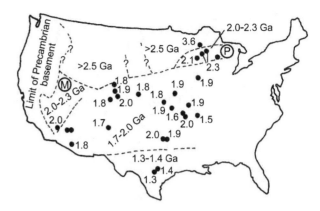

Fig. 4.17 Map of the Unites States showing Nd model age provinces, including the controversial Mojavia (M) and Penokean (P) terranes. After Bennett and DePaolo (1987).

crust-forming rocks generated in arc settings, which at the present day have less depleted Nd isotope signatures than spreading ridges.

There has been a tendency for proliferation of depleted mantle models, as new data for different geographical areas has become available. However, an examination of the literature suggests that the models of DePaolo (1981) and Goldstein et al. (1984) have had the widest application by other workers. This is illustrated in Fig. 4.16 by a comparison of citation rates for these two studies, compared with two control papers: Nelson and DePaolo (1984), discussed above, and Allegre and Rousseau (1984), who proposed a curved mantle evolution line similar to DePaolo (1981). The durability of citations for DePaolo (1981) and Goldstein et al. (1984) indicates the wide usefulness of these mantle models, which therefore provide an important basis for the comparison of different magma suites, even if the absolute values of the model ages are slightly in error. Hence it is desirable that the T_{DM} and T_{CR} notations should be restricted to the models of DePaolo (1981) and Goldstein et al. (1984), while other acronyms can be used to denote different models.

The importance of consistent use of terminology in this field is illustrated by a more recent comment on these issues by Dhuime et al. (2011). These authors suggested that 'the Hf (and Nd) isotope ratios in island arcs are on average lower than the present-day value for the depleted mantle ... primarily because of contributions from subducted sediment'. As a result, Dhuime et al. proposed a new source evolution line termed 'new crust' (NC), which is less depleted than the source of mid ocean ridge basalt (MORB). This was an important point to make, except that Dhuime et al. omitted to mention that their NC model is based on exactly the same premises as the DM model of DePaolo (1981). That model was *explicitly* based on new arc crust, since it was constrained to fit the Nd isotope compositions of Early Proterozoic and present day island arcs.

Many disciplines in science have inherited terminology with historical origins that were largely accidental. A good example is the so-called negative charge of the electron, which is purely a convention. In the case of model ages, DePaolo (1981) clearly defined the concept of depleted mantle model ages (T_{DM}) in comparison with T_{CHUR} ages. Doubtless this was an oversimplification of the complexities of mantle evolution, but it must be seen in its historical context. The onus is on subsequent researchers to state their position in relation to the earlier published record in order to avoid misunderstandings. This is particularly important in the case of Nd model ages, which are probably the most misunderstood major dating technique (Dickin, 2015).

4.2.3 Nd Isotope Mapping

Evidence that intra-crustal melting causes relatively minor perturbations in the Nd isotope evolution of crustal systems encouraged the use of granitic plutons to determine crustal formation ages on associated country-rocks (assuming that the granites are the products of anatexis of those country-rocks). The approach has the advantage of allowing basement mapping of large areas with a minimal number of analyses, since each pluton can be expected to have averaged the composition of a large volume of crust. It was used to great effect by Nelson and DePaolo (1985) to map the crustal extraction ages of huge belts in the central United States (Fig. 4.17). The method is appropriate for this application because Phanerozoic cover obscures most of the central US basement, which can only be dated from drill core or drill chips.

Weaknesses in this approach were revealed, however, when model age results did not correspond to known events represented by igneous crystallization ages. The 2.0–2.3 Ga model ages in the 'Penokean' and 'Mojavia' terranes of Bennet and DePaolo (1987) exemplify this problem. It is likely that they represent Proterozoic mantle-derived magmas which mixed with large quantities of re-melted Archean

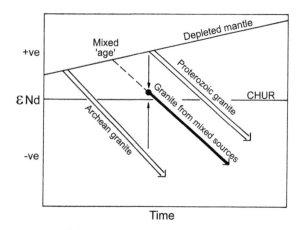

Fig. 4.18 Schematic illustration of magma mixing as a mechanism capable of generating mixed provenance ages which do not date any real geological event. After Arndt and Goldstein (1987).

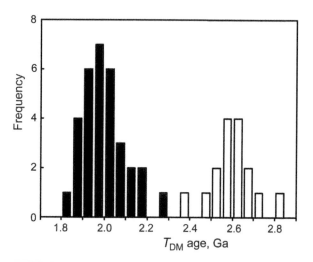

Fig. 4.19 Histogram of T_{DM} ages determined on 1.7–1.9 Ga granitoids intruded on either side of an Early Proterozoic suture in northern Sweden. Data from Wilson et al. (1985) and Skiold et al., 1988).

crust to generate mixed model ages (Fig. 4.18), which have no meaning as crustal formation ages.

In a paper whose title 'Use and abuse of crust-formation ages' has become notorious, Arndt and Goldstein (1987) criticized the interpretation of Nd model ages as crustal formation ages, in cases they considered lacked adequate geological constraint. They suggested that the Nd signatures of continental rocks were often produced by mixing between old crust and younger mantle-derived magmas, giving meaningless mixed ages (Fig. 4.18).

Perhaps there was some over-enthusiastic interpretation of small Nd data sets in the early work, but the warning of Arndt and Goldstein (1987) was taken out of all proportion by large segments of the geological community, who became obsessed with the idea that all Nd model ages represent meaningless mixed ages. However, this is far from the case. Although ensialic arcs are characterized by mixing of crustal components extracted from the mantle at different ages, there are many juvenile terranes that were extracted from the mantle in island arcs and give T_{DM} crustal formation ages consistent with other geological age evidence. This can be demonstrated using a few case studies, including one of the examples cited by Arndt and Goldstein (1987).

One of the studies most strongly criticized by Arndt and Goldstein (1987) was a paper by Reymer and Schubert (1986) which attempted to make the case for very high rates of Early Proterozoic crustal growth using published Nd model age data from Sweden (Wilson et al., 1985), supplemented by an abstract by Patchett et al. (1984). Reymer and Schubert argued that five million square km of Svecokarelian crust had been generated from 2.2 to 1.9 Ga, based on Nd model age evidence. Arndt and Goldstein's claim that this study 'exemplifies dubious use of Nd model ages' is based on taking the arguments of Reymer and Schubert out of context. In fact, it is clear from the Reymer and Schubert paper that these

authors were establishing a *minimum* rate of crustal growth, and therefore that the quoted 300 Ma time span was an approximate overall limit.

In spite of their relatively large analytical errors, these early Nd isotope studies of the Swedish craton display the power of Nd isotope mapping very effectively. Suites of 1.7 to 1.9 Ga plutons from the Proterozoic Skellefte district (to the south) and from the Archean craton (to the north) display distinct distributions of T_{DM} ages (Fig. 4.19), despite being located only 100 km apart on either side of an Early Proterozoic terrane boundary (Wilson et al., 1985; Skiold et al., 1988).

In their critique, Arndt and Goldstein (1987) argued that Nd model ages should only be interpreted as true crustal formation ages 'if they coincide with U–Pb zircon ages or other independent evidence of an orogenic event.' However, this is a misleading claim that has led to much confusion about the correct interpretation of model ages. An orogenic event is most often a collisional event, and may not be a significant crust-forming event at all. In the Swedish example given here, accretion of the Proterozoic arc terrane is thought to have occurred at around 1.92–1.91 Ga, and was followed by intensive granitoid plutonism over the following 200 Ma (e.g. Guitreau et al., 2014). However, the oldest U–Pb crystallization age from the Skellefte district (1959 ± 14 Ma) significantly precedes the estimated accretion age, and is only 40 Ma younger than the average T_{DM} age of 2.0 Ga determined by Wilson et al. (1985).

This example shows the typical history of geological research in orogenic belts. The oldest U–Pb ages determined in early work were more than 100 Ma younger than the average T_{DM} age, whereas further study narrowed the gap to only 40 Ma. However, we should not necessarily *expect* that U–Pb and T_{DM} ages will coincide. The typical granitoid rocks that

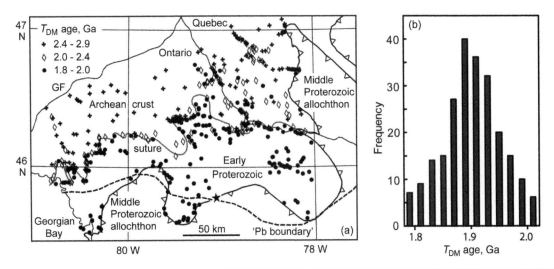

Fig. 4.20 (a) Map of the western Grenville Province adjacent to Georgian Bay (Lake Huron), showing ranges of T_{DM} ages relative to alternative crustal boundaries. (b) Histogram of T_{DM} ages from the Penokean terrane (note that the data set includes some samples outside Fig. 4.20a). Star = locality of quartzite sample used for zircon dating by Culshaw *et al.* (2016).

make up the bulk of the continental crust (tonalites and granodiorites) cannot be direct products of the mantle. When Nd model ages are determined on such material, they are probably dating the mafic crustal precursors, which may be millions of years older than the crystallization age of dated plutons. Since it may take 50 Ma to create a mature island arc, we should expect that the Nd model ages for juvenile terranes will typically be 50 Ma older than the *oldest* U–Pb ages, which therefore provide only a *minimum* age for crustal formation. Therefore, if TDM model ages fall within 100 Ma or so of U–Pb zircon ages, this provides good evidence that a given crustal block is a juvenile terrane.

Having said this, we must remember that Nd model ages can only provide accurate crustal formation ages if the depleted mantle model is an accurate representation of the real source of subduction-related magmatism. But in fact, the depleted mantle model of DePaolo (1981) has been shown in numerous cases to be a surprisingly good representation of such sources (e.g. Dickin *et al.*, 2010, 2016).

In the Swedish case, the early Nd model age data have been spectacularly vindicated by recent Hf isotope analysis of the oldest igneous units from the Skellefte arc terrane. These gave depleted mantle Hf model ages of 2.0–2.1 Ga (Guitreau *et al.*, 2014), leading these authors to suggest that this arc may have grown between ca. 2.2 and 2.0 Ga. If we recognize that additional crustal extraction occurred during the period of ensialic arc magmatism that followed collisional accretion, we can see that Reymer and Schubert's ball-park estimate of 300 Ma to grow Swedish Paleoproterzoic continental crust was quite realistic after all.

This example demonstrates the use of Nd model ages to extend estimates of crustal formation to earlier times than

indicated by U–Pb ages. However, Nd model ages are just as important for placing *upper* limits on the age of crustal formation for an orogenic terrane. Furthermore, placing *geographical* limits on the extent of old crust in complex orogenic belts is one of the unique capabilities of Nd isotope mapping which no other method can provide. This is demonstrated in another Precambrian orogenic terrane with a long history of crustal evolution, located within the Grenville Province of Ontario.

Using Nd model age mapping, Dickin and McNutt (1989) identified the edge of the Archean craton within the younger Grenville orogenic belt of Ontario, Canada. This boundary was interpreted as an Early Proterozoic crustal suture separating gneisses with T_{DM} ages > 2 Ga to the north from those with T_{DM} ages < 2 Ga to the south (Fig. 4.20a). The younger terrane was interpreted as a juvenile Early Proterozoic arc, probably accreted during the 1.85 Ga Penokean orogeny. However, the establishment of a long-lived ensialic arc on the new margin caused severe magmatic reworking that largely erased geological evidence for the Penokean arc terrane.

DeWolf and Mezger (1994) questioned the use of Nd isotope mapping to constrain the extent of Archean crust in the Grenville Province of Ontario, arguing that most of the analysed Early Proterozoic Nd model ages resulted from Mid-Proterozoic reworking of Archean basement (following the mixing model of Arndt and Goldstein, 1987). Based on the Pb analysis of a mixed suite of lithologies (gneiss, granite, pegmatite, amphibolite, pelite and marble), DeWolf and Mezger argued that Pb isotope signatures were a more sensitive detector of the presence of Archean crust than Nd model ages. Hence, they proposed a 'Pb boundary' representing the limit of 'Archean Pb influence' within the Grenville Province of Ontario (dashed line in Fig. 4.20a). This zone extends 50–100 km south of the mapped Archean–Proterozoic suture (Moore and Dickin, 2011), but has no relationship with mapped geological terrane boundaries (solid lines in Fig. 4.20a). Hence, it appears that in this case the claimed greater sensitivity of Pb isotopes for basement mapping is spurious rather than reliable.

The use of Pb isotopes for terrane mapping is discussed in Section 5.5. However, it is important to point out here why Nd isotope analysis is a more reliable method for mapping crustal formation ages than Pb. This goes back to the original scientific basis for Nd model ages (McCulloch and Wasserburg, 1978), which is that Sm/Nd ratios are significantly fractionated by mantle melting, but not significantly affected by intra-crustal differentiation processes. The opposite is true for U/Pb ratios, which are *not* strongly fractionated during crustal formation, but are strongly affected by intra-crustal differentiation processes. This is the reason why Pb model ages are not generally useful for geochronology, even though they are based on the double uranium to lead decay scheme. The exceptions, which will be discussed in Section 5.5.1, generally rely on $^{207}Pb/^{204}Pb$ ratios, which varied very rapidly during early Earth history, and are therefore sensitive to the presence of Early Archean crust.

Contrary to the assertion of Arndt and Goldstein (1987), consistency of Nd model ages within a crustal terrane *does* provide important evidence for the validity of crustal formation ages, provided these are based on adequate Nd data sets. In contrast to the three Nd model ages that were used to suggest a possible 2.3 Ga crustal formation event in Wisconsin (Nelson and DePaolo, 1985), the proposed Penokean crustal formation event in Ontario is based on over 200 published Nd model ages with a strong symmetrical distribution, yielding an average T_{DM} age of 1.9 Ga (Fig. 4.20b). Geological evidence that this dates a real crustal formation event came initially from cited Penokean deformation north of the Grenville Front (Zolnai *et al.*, 1984). More recently, the Penokean crustal formation age has received support from U–Pb dating of detrital zircons in a local quartzite (star in Fig. 4.20). Penokean (1.85 Ga) zircons form the most abundant age peak (Culshaw *et al.*, 2016), as expected if juvenile Penokean crust is a major local sediment source.

In contrast to the homogeneity of model ages within the accreted Penokean arc, the area north of the mapped suture zone (Fig. 4.20) displays marked age heterogeneity, with a mixture of Archean and Early Proterozoic TDM ages. This is clearly an area of Nd isotope mixing, caused by reworking of Archean crust by Proterozoic ensialic arc magmatism. This crustal reworking event is recorded by 1.75 Ga U–Pb ages on both sides of the suture, whereas Archean U–Pb ages are only seen on the north side. However, mapping the suture based on the extent of inherited Archean U–Pb ages may not be reliable, because magmatic zircons may not quantitatively sample source rocks. In contrast, whole-rock Nd isotope analysis of orthogneisses provides a representative sampling of

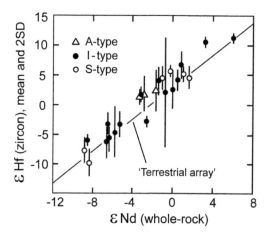

Fig. 4.21 Correlation between average εHf composition of zircons and bulk rock εNd in granitoids from SE Australia with mixed sources. Error bars show two population standard deviations of zircon εHf (*t*) values for each pluton. After Kemp *et al.* (2009).

the bulk crust, and therefore provides a reliable *quantitative* estimate of the extraction age of the crust.

In cases where granitoid rocks are the result of complex mixing of sources of different ages, hafnium isotope analysis of zircons (Section 9.3) offers the potential of a more detailed analysis of magma sources, compared with whole-rock Nd data. The A-type, I-type and S-type granitoid suites of SE Australia represent a good example of plutons with mixed magma sources (Section 7.3.5). Surprisingly, Kemp *et al.* (2009) found that the average εHf signatures of zircons from each pluton were quite well correlated with their whole-rock εNd signatures (Fig. 4.21). Therefore, when deciding what technique to apply, it is important to consider the scientific objectives of the study. If the objective is to understand complex petrogenetic processes, Hf analysis of zircon will usually provide more detail than whole-rock Nd analysis. However, if the objective is detailed *geographical* analysis of age provinces, Nd isotope mapping is the most cost-effective and reliable method.

4.3 Model Ages and Crustal Processes

As outlined above, one of the principal uses of the Sm–Nd model age method is to determine what are usually called 'crustal formation' or 'crustal extraction' ages. However, the model age method has often applied when a long or complicated geological history precludes a more direct method of determining crustal age. One of the strengths of the Sm–Nd model age method, as applied to whole-rock systems, is that it provides the opportunity to see back through erosion, sedimentation, high-grade metamorphism and even crustal melting events which may reset other dating tools. However,

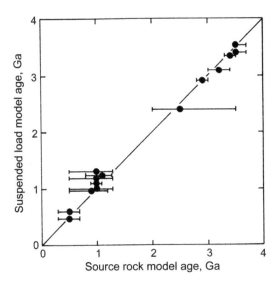

Fig. 4.22 Plot of Nd model ages for river particulates against the area-weighted average crustal residence age of rocks within the watershed. Data are shown for igneous–metamorphic drainage basins only. After Goldstein and Jacobsen (1988).

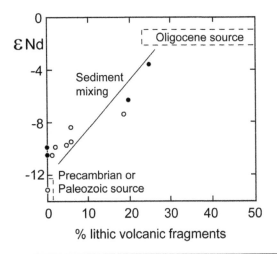

Fig. 4.23 Plot of εNd against modal % lithic volcanic fragments to show petrographic dependence of the Sm–Nd system in sedimentary basins with mixed provenance. (●) = Hagar basin; (○) = Espanola basin. After Nelson and DePaolo (1988).

these processes may cause complications in the interpretation of model ages. Hence, it is important to examine Sm–Nd systematics in well-constrained examples in order to estimate the reliability of model ages in more complex environments.

4.3.1 Sedimentary Systems

The behaviour of the Sm–Nd system during erosion can be examined by comparing the calculated model ages of river-borne particulates with the average geological age of sediment sources in the watershed. Goldstein and Jacobsen (1988) performed such a study on particulates in American rivers. They found that rivers draining primary igneous rocks carried sediment which accurately reflected the crustal residence age of the source (Fig. 4.22). Rivers draining sedimentary watersheds were not properly testable, since the crustal residence age of their sources had not been adequately quantified.

Behaviour of the Sm–Nd system during sedimentation can be further tested by comparing Nd model ages on different size fractions of sediment. An early study on bottom sediment from the Amazon River (Goldstein *et al.*, 1984) found that different size fractions yield only a small range of crustal residence ages (1.54–1.64 Ga), despite a large range in total Nd contents (17–47 ppm). Similar agreements in model age were found by Awwiller and Mack (1991) on mud and sand grade sediments from the Rio Grande and Mississippi rivers. However, the bottom sediments of large rivers may be atypical in displaying good chemical homogeneity.

Nelson and DePaolo (1988) tested the effects of mixed sediment provenance on Sm–Nd systematics in two small basinal systems. In both cases, the different sediment sources were petrographically and geochemically well characterized. In order to quantify the mixing process, Nelson and DePaolo plotted εNd against a petrographic index (percentage of lithic volcanic fragments). The good correlation observed between the end-members and various mixtures (Fig. 4.23) attests to the 'immobile' behaviour of Nd during erosion and sedimentation. This does not *avoid* the problem of mixed provenance, but it shows that coupled isotopic and petrological analysis of a suite of samples can be used to detect and quantify the mixing process.

The work described above was done before the availability of laser ablation U–Pb and Hf isotope data on detrital zircons (Sections 2.7, 9.3). These techniques clearly yield more detailed provenance information than a single whole-rock Nd analysis can provide. However, Nd isotope analysis can still complement these other techniques by providing information about bulk sediment provenance, and specifically concerning the clay fraction in large river systems.

4.3.2 Meta-Sedimentary Systems

Many studies have been undertaken to assess the mobility of REE, and specifically Sm–Nd, under different diagenetic and metamorphic conditions. Paradoxically, the evidence suggests that the REE may be more mobile during diagenesis and low-grade metamorphism than during high-grade metamorphism and partial melting. This may be because of a paucity of mineral phases growing under low-grade metamorphic conditions into which REE are strongly partitioned.

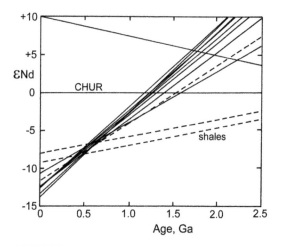

Fig. 4.24 Nd isotope evolution diagram for Middle Ordovician turbidites from eastern New York State showing an average provenance age of 1.8 Ga, and disturbed samples with anomalously old model ages. (——) = sandstone; (– – –) = shale. After Bock et al. (1994).

In contrast, there are several igneous and high-grade metamorphic minerals into which REE are strongly partitioned.

Stille and Clauer (1986) and Bros et al. (1992) demonstrated that in carbonaceous (black) shales, Sm–Nd systematics in the microscopic clay-mineral fraction can be reset by diagenesis. They showed that in some cases, sub-micron sized particles could yield Sm–Nd isochrons, which they interpreted as dating diagenesis. The accuracy of such ages remains to be proven, given the evidence that Rb–Sr dating of clay minerals can be upset by detrital inheritance (Section 3.5.1).

Diagenetic mobilization of REE on a mineralogical scale does not necessarily imply open Sm–Nd systems on a whole-rock scale. A suggestion that such open-system behaviour *could* occur was made by Awwiller and Mack (1991) based on Sm–Nd analysis of sedimentary borehole samples from Texas. Weak positive correlations were observed between depth in the bore hole and depleted mantle model age, which these authors attributed to diagenetic loss of radiogenic Nd, as well as minor increases of Sm/Nd ratio with depth. However, the study was based on very small 'whole-rock' samples (less than 10 g), and variations in sediment provenance could not be ruled out, so the evidence was equivocal.

Additional evidence for diagenetic disturbance of Sm–Nd systems was obtained by Bock et al. (1994), based on sampling turbiditic sandstones and shales from eastern New York State, deposited during the Taconic orogeny (ca. 470 Ma). Nd isotope analysis was performed to determine sediment provenance, but the model age results displayed more scatter than could be explained by variations in provenance alone. This can be seen in Fig. 4.24, where two shales gave impossibly old ages, while two other samples gave ages somewhat older than the remainder of the suite. Furthermore, these four samples were found to be moderately or severely depleted in light REE relative to the other samples. Since Nd isotope evolution lines converged at around 500 Ma, it was suggested that the isotope system in some samples was disturbed, probably during early diagenesis. Unfortunately, the size of samples analysed in this work was not reported.

The lack of sample information in the previous study was rectified in later work by Cullers et al. (1997) on Silurian pelitic schists from western Maine. Whole-rock samples averaging more than 1 kg in weight were collected from pelitic units within lithologically variable formations consisting of finely interbedded quartzite and pelitic schist. Most samples gave very consistent depleted mantle model ages of 1.8 ± 0.1 Ga, but a few samples from the Perry Mountain Formation yielded abnormally old ages, from 2.5 to 5.3 Ga. The samples which gave these old ages were again found to be light-REE-depleted, yielding abnormally large $^{147}Sm/^{144}Nd$ ratios, from 0.15–0.19. These disturbances were attributed to leaching of light REE from shales during diagenesis, and the lack of suitable minerals locally to take up the released REEs. Based on comparison with an earlier study of REE mobility in similar carbonaceous shales from central Wales (Mildowski and Zalasiewicz, 1991), it was suggested that the REEs released from shale layers may have been incorporated into phosphates which grew in more arenaceous layers.

Collectively, these studies show that caution must be exercised when using Nd isotope data to determine sediment provenance ages, especially on carbonaceous shales. One way of dealing with this kind of material is to use a 'two-stage' model age calculation (e.g. Keto and Jacobsen, 1987). In this approach, the measured Sm/Nd ratio of the sample is used to calculate an 'initial' Nd isotope ratio at the estimated time of disturbance, beyond which an average crustal Sm/Nd ratio is used to estimate the provenance age. This approach may have some validity, but there is no substitute for the analysis of a large sample suite containing a variety of rock types. It is then possible to detect and screen out samples that have been subjected to diagenetic disturbance, allowing accurate provenance ages to be determined for the formation as a whole.

4.3.3 Meta-Igneous Systems

Mafic and ultramafic rocks cannot be used to determine accurate crustal formation ages because they have Sm/Nd evolution lines sub-parallel to the chondritic evolution line. However, this same property allows the determination of precise initial Nd isotope ratios, which have been widely used to determine the degree of mantle depletion in early Earth history (Section 4.4.3).

A study by Lahaye et al. (1995) has important implications for this type of work because it implies that the initial Nd isotope signatures of many komatiites may have been disturbed by subsequent alteration. Lahaye et al. compared calculated initial isotope compositions ($\epsilon Nd[t]$) for whole-rock samples and separated pyroxenes in five komatiite flows from the Abitibi and Barberton belts. Many whole-rocks showed small (1–2 ϵ unit) deviations from the pyroxenes, but a few

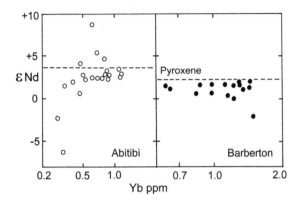

Fig. 4.25 Calculated initial Nd isotope ratios for whole-rock samples of komatiites compared with separated pyroxenes (horizontal dashed lines). Data are plotted against Yb concentration. After Lahaye *et al.* (1995).

had much larger deviations, up to +5 and −10 ε units (Fig. 4.25). In view of this evidence, Nd isotope data on komatiites should be based on a combination of whole-rock and mineral analyses in order to determine reliable initial ratios.

In contrast to the evidence for disturbance of Sm–Nd systems in meta-basic rocks, most granitoid rocks show much greater resistance to resetting. For example, Barovich and Patchett (1992) demonstrated that whole-rock Sm–Nd systems in granitic rocks can remain undisturbed even during severe metamorphic deformation. They studied a 60 m-wide Mesozoic ductile shear zone cutting the Mid Proterozoic Harquahala granite of Arizona. Samples of increasingly deformed granite were found to yield a narrow range of T_{CHUR} ages of around 1.58 Ga in two different traverses to within 1 m of the thrust plane (Fig. 4.26). Closed-system behaviour

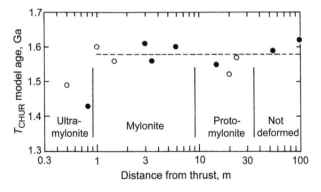

Fig. 4.26 Plot of T_{CHUR} model ages for samples of the Harquahala granite as a function of distance from the Harquahala thrust. Solid and open symbols indicate samples from two different traverses. Approximate boundaries between deformation zones are shown. Data from Barovich and Patchett (1992).

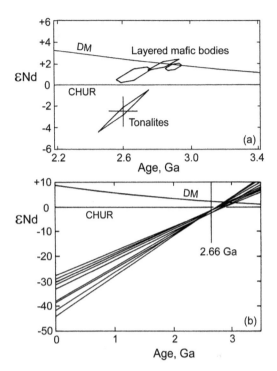

Fig. 4.27 Nd isotope evolution diagrams for the Lewisian Complex of NW Scotland: (a) showing initial ratios for layered mafic bodies and a suite of granulite facies tonalitic gneisses; (b) showing Sm–Nd evolution lines for individual tonalitic gneisses. After Whitehouse (1988).

was preserved even in samples showing widespread sericitization of plagioclase and significant epidote growth. Only in ultra-mylonites less than 1 m from the main thrust was a reduction in model age of up to 150 Ma observed, possibly due to a high fluid flux which caused calcite veining and intense alteration in the immediate vicinity of the thrust.

The resistance of whole-rock Sm–Nd model ages to significant resetting, even during granulite facies metamorphism, is demonstrated by Lewisian granulite facies gneisses from NW Scotland (Whitehouse, 1988). A ten-point Sm–Nd isochron for tonalitic gneisses (Section 4.1.3) yields an age of 2.60 Ga and an initial ratio (ε [*t*]) of −2.4 relative to CHUR (Fig. 4.27a). This isochron is argued to date the metamorphic event. However, T_{DM} model ages for these same gneisses fall in the range 2.84–3.04 Ga, with an average value of 2.93 Ga (Fig. 4.27b). These ages have been slightly scattered by metamorphism, but still yield an average value very close to the undisturbed 2.91 Ga isochron age of the Drumbeg mafic complex.

4.3.4 Partially Melted Systems

Nelson and DePaolo (1985) attempted to place upper limits on the disturbance of model ages under conditions of intracrustal re-working by considering the limiting case of crustal anatexis. From crustal melting models (Hanson, 1978), they

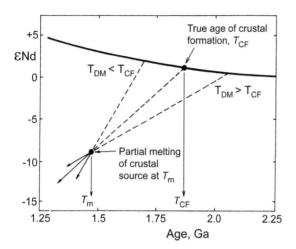

Schematic diagram of Nd isotope systematics to show possible errors in model age arising from Sm/Nd fractionation during intra-crustal melting. After Nelson and DePaolo (1985).

estimated that the maximum amount of Sm/Nd fractionation likely to arise by intra-crustal melting processes (Δ) was 20% of the pre-existing fractionation between sample Sm/Nd and CHUR Sm/Nd. This fractionation factor f was defined by Nelson and DePaolo (1985):

$$f_{Sm/Nd} = \frac{^{147}Sm/^{144}Nd_{sample}}{0.1967} - 1 \qquad [4.6]$$

Using this notation, the error in a T_{DM} age introduced by an intra-crustal fractionation event is given by

$$\mathrm{Err}\,T_{DM} = \Delta\, f_{Sm/Nd.}\,(T_{CF} - T_m) \qquad [4.7]$$

where T_{CF} is the true crustal formation age and T_m is the age of the partial melting event. This error propagation is illustrated schematically in Fig. 4.28. The problem can be minimized by analysing samples with melting ages fairly close (<300 Ma ?) to their formation age.

To avoid possible Sm/Nd fractionation during melting of crustal protoliths, some authors have advocated calculating Nd model ages on granitoids using a two-stage evolution model (e.g. DePaolo et al., 1991). The two-stage model uses the measured Sm/Nd ratio of the rocks to calculate initial Nd isotope ratios at the time of melting (assuming that U–Pb data are available on the same sample), and then uses a 'typical crustal' Sm/Nd ratio to extrapolate the εNd growth line back to the depleted mantle curve. This method is necessary for calculating the Hf isotope evolution of igneous protoliths in order to calculate Hf model ages from zircons (Section 9.3.3). However, Nd model ages are determined on whole-rock samples, for which Sm/Nd ratios are not significantly fractionated by most crustal processes. Therefore, a two-stage model age may actually increase the uncertainty on Nd model ages.

4.4 The Crustal Growth Problem

The question of whether the crust has grown progressively over geological time, or maintained an approximately constant volume, is one of the most fundamental in geology, but has proved hard to answer conclusively. A review of the 'crustal growth' model by its most persistent critic (Armstrong, 1991), shows that Nd isotope data provide critical tests for alternative models. Hence, three of the most important lines of Nd isotope evidence will be examined here.

4.4.1 Crustal Accretion Ages

The ability of the Sm–Nd method to 'see back' through younger thermal events and measure the crustal formation ages of continental rocks makes the method ideally suited to chart the present day age distribution of crustal basement. This yields an apparent profile of continental growth through time which does not take into account crustal recycling into the mantle. Nevertheless, it is an appropriate starting point for this subject.

Attempts to map the age structure of the continents were begun using the Rb–Sr method (Hurley et al., 1962), before the development of Sm–Nd analysis. Hurley et al. compared Rb–Sr isochron ages with Sr model ages calculated assuming a mantle $^{87}Sr/^{86}Sr$ ratio of 0.708. On average the two values were correlated, leading them to suppose that the isochron ages dated the time of crustal extraction from a basic source. This was a good approach, although we now know that the Rb–Sr system is too easily reset to yield reliable crustal extraction ages for old terranes. (Also the mantle growth curve is less radiogenic than 0.708.) Hurley et al. applied the method to the North American continent in order to calculate the approximate area of crustal basement attributed to different age provinces (Fig. 4.29).

Hurley and Rand (1969) extended this approach to include two-thirds of the land area of the world (excluding Russia and China, for which data were not available). K–Ar data were used to geographically extrapolate from the more limited set of Rb–Sr data, bearing in mind the tendency for the former to be reset. Rb–Sr model ages were calculated using an improved mantle $^{87}Sr/^{86}Sr$ growth curve, yielding values now somewhat older than apparent crystallization ages. Hurley and Rand's data are presented on a plot of cumulative crustal age distribution against time (Fig. 4.30). From these data it appeared that crustal growth was accelerating somewhat with time. However, more recent studies have yielded different shaped curves.

A study of comparable sweep to Hurley et al. (1962) was performed by Nelson and DePaolo (1985), who used Nd model ages to map the age structure of the basement of the United States. Nelson and DePaolo found Nd model ages substantially older than igneous crystallization ages, leading to a greatly increased estimate of the rate of Early Proterozoic crustal growth in the mid-continent. These data, along with recently published ages on the Canadian Shield, led to a

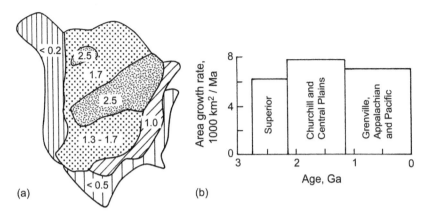

Fig. 4.29 Estimated area of North American crustal basement attributable to different Rb–Sr age provinces. (a) Map showing provinces of different ages in Ga; (b) histogram of growth rate against time. After Hurley *et al.* (1962).

dramatic increase in the estimated average age of the North American craton, compared with that of Hurley *et al.* (1962). This picture was reinforced by Patchett and Arndt (1986), who further amplified the estimated area of newly accreted Early Proterozoic (1.9 Ga) crustal basement in North America. This has generated a 'sigmoidal' curve of crustal formation against time which suggests that the greatest rates of new crustal accretion occurred in the middle of Earth history (Fig. 4.30).

As more detailed model age mapping is performed in areas of old crustal basement it is likely that further increases in average crustal extraction age will be found. The Superior and Slave provinces of Canada provide good examples of this. Hurley *et al.* (1962) mapped the Superior craton as 2.7 Ga in age, a concept still widely held today. However, U–Pb geochronology has shown the presence of large areas of 3 Ga-old crust in the NW Superior Province (reviewed by Thurston *et al.*, 1991). Similarly, the Slave province is largely a 2.7 Ga craton, but a small belt of tonalitic gneisses with

Early Archean zircons yield T_{CHUR} model ages of up to 4.1 Ga (Bowring *et al.*, 1989). Therefore, continued mapping of old cratons will probably fill the dip in the Early to Mid Archean segment of the sigmoid (Fig. 4.30), yielding a linear or concave-downwards apparent crustal growth curve with time.

4.4.2 Sediment Provenance Ages

In response to proponents of the 'continental growth' model (e.g. Moorbath, 1976), Armstrong (1981) argued that the record of continental age provinces documented by various methods (as above) did not prove that the continental area had actually grown over geological time. Armstrong argued that a model in which the continental area was approximately the same 4.5 Ga ago as it is today could also generate *apparent* continental growth with time, provided that the rate of crustal recycling into the mantle (by sediment subduction) was equal to the rate of new crustal formation above subduction zones.

Undoubtedly, crustal recycling back into the mantle does occur on a significant basis. Old crustal terranes may be shortened by orogeny, then flattened again by erosion and sediment subduction. However, some sediment should be expected to escape the recycling process and provide a record of the old, recycled terrane. Therefore, the search for evidence of constancy or growth in the continental mass turned to the sedimentary record. The ability of Nd model ages to 'see back' through erosion and sedimentation to an original crustal extraction event made them ideal for these studies.

To study this problem, it is convenient to portray sediment data on a diagram of Nd model age (crustal residence age) against the stratigraphic depositional age of the sediment in question (Fig. 4.31). Sediments eroded from juvenile mantle-derived sources will have $T_{CR} = T_{STRAT}$ and lie on a 'concordia' line (Allegre and Rousseau, 1984). In contrast, reworking of older sediments without any input of juvenile material will displace compositions to the right along

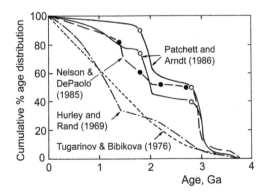

Fig. 4.30 Estimated continental growth rates on a cumulative basis, based on different crustal age data sets. After Jacobsen (1988).

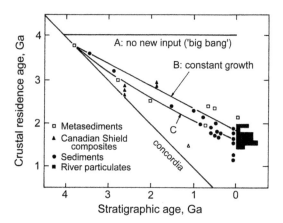

Fig. 4.31 Model age *versus* stratigraphic age diagram showing a compilation of early 1980s data from several clastic sediment studies. Crustal growth models A, B and C are discussed in the text. Data from O'Nions (1984) and Allegre and Rousseau (1984).

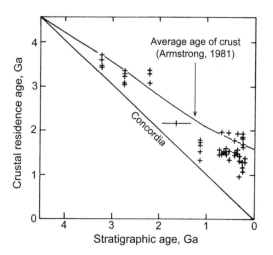

Fig. 4.32 Model age *versus* stratigraphic age diagram showing data of Dia *et al.* (1990) for clastic sedimentary rocks from South Africa. Curve shows provenance ages predicted by the crustal model of Armstrong (1981). After Armstrong (1991).

horizontal vectors. A compilation of data from several sources is shown in Fig. 4.31, including clastic sediments (Hamilton *et al.*, 1983; O'Nions *et al.*, 1983; Taylor *et al.*, 1983; Allegre and Rousseau, 1984) and particulates from major river systems at the present day (Goldstein *et al.*, 1984).

Allegre and Rousseau (1984) compared the sediment data with various theoretical models for continental evolution involving different rates of continental growth through time (Fig. 4.31). A 'big bang' model (A), whereby the whole continental mass was extracted at ca. 4 Ga or before, was ruled out. Allegre and Rousseau argued that a model involving uniform growth of the continents from 3.8 Ga to the present (B) was a better fit to the data, but that the best fit was produced by a curved line (C), representing decreasing growth of the crust through time.

Unfortunately, this diagram is not as conclusive as it may appear, due to the great difficulty of determining a global average sediment provenance age at any given time, from the very variable provenance ages in individual geological provinces. This makes the data very susceptible to sampling bias. One source of such bias is preferential recycling of old sediments relative to erosion of more juvenile crustal basement. This will exaggerate the slowing down of continental growth with time, appearing to favour models of type C over type B. Another source of bias is the neglect of young orogenic belts such as the accreted terranes of the Canadian Cordillera (Samson *et al.*, 1989). The inclusion of such data in Fig. 4.30 would favour linear evolution models (type B), suggesting that crustal growth has *not* slowed significantly in the Phanerozoic.

The interpretation of Fig. 4.31 is also heavily influenced by assumptions about the degree of re-cycling of sediment into the mantle. The so-called 'big bang' model shown in Fig. 4.31 involves no recycling of crustal material into the man-

tle. This does not correspond to Armstrong's model, which involves constant recycling of old crust into the mantle and replacement by an equal volume of juvenile crust. Armstrong (1991) claimed that his model gave rise to a curved evolution line (Fig. 4.32) which looks remarkably like the steady growth model in Fig. 4.31. It is clear then that young sediments provide much too loose a constraint on crustal growth models. Therefore, the argument must focus on the provenance ages of the oldest surviving sediments.

Isua supracrustals from western Greenland, which are the oldest clastic sediments analysed, yield identical stratigraphic and Nd model ages of 3750 Ma, indicating that they did not incorporate a significant amount of older reworked crust. However, the data of Dia *et al.* (1990) from South Africa show surprisingly old provenance ages for Mid to Late Archean sediments. On balance, the sediment data seem to favour a crustal growth model, but ultimately the argument rests on a null hypothesis (no sediments with very old provenance are yet seen, therefore none exist). This is an inherently weak argument upon which to base such an important conclusion. This weakness comes from the need for representative sampling of old crust using a sediment data set that is inherently very noisy.

4.4.3 Archean Depleted Mantle

An alternative route to assessing the volume of crust at a given time in Earth history is to measure the composition of the depleted reservoir which balances the enriched crustal reservoir, namely the Upper Mantle (Section 6.2.2). Because the upper mantle is stirred by convection, we can expect to sample this reservoir (in ancient volcanism) in a much more representative fashion than ancient sediments sample the enriched reservoir. Hence, the problem of crustal growth

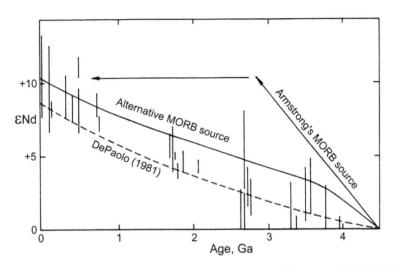

Fig. 4.33 Initial εNd for terrestrial rocks (vertical bars), compiled by Armstrong (1991), compared with his 'big bang' MORB evolution line. The solid curve is an alternative MORB depletion line for a crustal growth model. Note that this is not expected to agree with the dashed arc-source model of DePaolo (1981).

may be soluble by studying the extent of mantle depletion in early Earth history. If there was a large volume of continental crust in the early Earth, there should be evidence for strong mantle depletion.

In the mid 1980s, several studies pointed to initial Nd isotope data for Early and Mid Archean rocks that lay well above the chondritic evolution line, and in some cases above the depleted mantle evolution line of Goldstein *et al.* (1984). Smith and Ludden (1989) argued that some of the strongly positive εNd values calculated for early mafic rocks were in error due to incorrect age assignments. The Kambalda example has already been mentioned (Section 4.1.2), and doubtless there are problems with some of the other data. However, they concluded that there are enough depleted mantle compositions in the Early Archean for the phenomenon to be real.

Such evidence for very early depletion of the upper mantle presented a problem for the model by which continental crust grew gradually at the beginning of Earth history. On the other hand, Armstrong (1991) argued that these data supported his model of no crustal growth. In order to examine this claim, the data compilation of Armstrong (1991) is shown in Fig. 4.33, along with an evolution line for the MORB source which he claimed was a product of his 1981 model. However, most of the available Nd data can be satisfied by a less extreme evolution line in Fig. 4.32 (solid line), which is sub-parallel to DePaolo's curve since 4 Ga ago. The solid evolution line represents the composition of the most depleted mantle sources, whereas DePaolo's line represents the source of arc magmatism, which is generally less depleted.

The gradual depletion of the upper mantle, which is portrayed by the solid line in Fig. 4.33, can only be reconciled with a constant crustal volume model if the average *composition* of the crust changed over geological time. In principle this requirement is met in a model where the Earth begins its evolution with a thick basaltic ('oceanic') crust, which is gradually replaced by continental crust over geological time. This involves a non-plate tectonic model for Archean crustal evolution (e.g. West, 1980). A similar model was also supported by Galer and Goldstein (1991), who proposed that a thick, long-lived alkali basalt crust was built up in the Archean by small degree melting in the deep mantle. However, as evidence mounts for earlier and earlier operation of plate tectonic processes in Earth history (e.g. Turner *et al.*, 2014; Nutman *et al.*, 2015; Hastie *et al.*, 2016), there are limits on the duration of a pre-plate-tectonic era.

Chase and Patchett (1988) proposed that accelerated early mantle depletion is in fact consistent with plate tectonic processes. They postulated that the storage of subducted oceanic crust in the lower mantle, before re-homogenization with the depleted mantle (by convection), would give rise to a hidden enriched reservoir in the deep mantle to balance early depleted mantle. According to this model, the amount of 'stored' subducted oceanic crust has grown over Earth history, although gradual cooling of the earth prevents the system from reaching a steady state, by increasing the lifetime of subducted crust over geological time.

More recently, the creation of an early enriched layer at the base of the mantle has been invoked to explain ^{142}Nd data (Section 15.7.3). Such an enriched layer could have formed by subduction of mafic proto-crust that solidified from an early magma ocean (resulting from the giant Moon-forming impact). Fe-rich meteorites could have accumulated on this proto-crust during the 'late heavy (meteorite) bombardment' of the Earth and Moon (Tolstikhin and Hofmann,

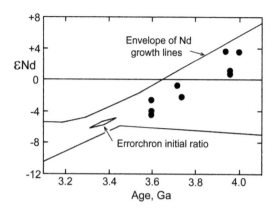

Fig. 4.34 Nd isotope evolution diagram showing initial εNd values calculated at the various U/Pb ages of the Acasta gneisses (•) compared with the initial ratio of a 3.3 Ga old best-fit errorchron. The outer envelope of Nd isotope growth curves is shown for reference. After Moorbath *et al.* (1997).

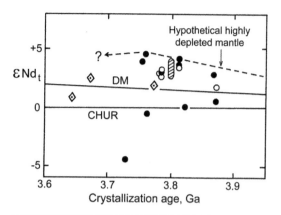

Fig. 4.35 Nd isotope evolution plot showing Nd values of Bennett *et al.* (1993) at the ages determined by U–Pb analysis. (•) = Amitsoq gneiss; (○) = Akilia enclaves in Amitsoq gneisses; shaded zone = Isua supracrustals. Large diamonds indicate ages and initial ratios for three Sm–Nd errorchrons of Moorbath *et al.* (1997).

2005). Such dense material could have enhanced the preservation of a hypothetical enriched reservoir at the base of the mantle.

4.4.4 Early Archean Crustal Provinces

Evidence for open-system behaviour of Sm-Nd in komatiites (Section 4.3.3) cast doubt on some of the evidence for strong mantle depletion in early Earth history. However, new evidence for strong early depletion of the mantle was provided by Bennett *et al.* (1993) and Bowring and Housh (1995), based on the analysis of granitoid orthogneisses from Western Greenland and the Slave Province of northern Canada. Since these rock types are generally very resistant to metamorphic disturbance, the new evidence for highly depleted Early Archean mantle appeared much stronger. However, this new evidence was itself challenged by Moorbath and Whitehouse (1996) and Moorbath *et al.* (1997). Since this discussion has critical implications for crust–mantle evolution, it will be examined here in some detail, beginning with the data from the Acasta gneisses of the Slave Province.

Bowring and Housh (1995) used SHRIMP U–Pb ages (see Section 5.2.6 for method) to calculate Nd initial ratios for a variety of rock types from the Early Archean Acasta gneisses. Based on U–Pb ages of 3.6–4.0 Ga, individual samples gave initial εNd values as high as +4 and as low as −5 (Fig. 4.34). However, Moorbath and Whitehouse (1996) observed that most of the sample suite analysed by Bowring and Housh lay on an Sm-Nd 'errorchron' with an age of ca. 3.3 Ga, which they attributed to an intense metamorphic event that partially homogenized whole-rock Sm-Nd systems at that time. This result was later confirmed (Moorbath *et al.*, 1997) by the analysis of 20 new samples, yielding a combined errorchron age of 3370 ± 60 Ma and initial ratio (εNd) of −5.6 (Fig. 4.34). Hence, they argued that εNd values calculated at the U–Pb

crystallization ages of 3.6 to 4.0 Ga were not accurate measures of the depleted mantle composition at those times.

Bowring and Housh (1996) argued in reply that the 3.3 Ga errorchron age could itself be a mixing line with no age significance. However, Moorbath *et al.* (1997) showed that there was no correlation between Nd isotope ratios and Nd concentrations in the Acasta gneisses, as would be expected from a mixing line. Such a mixing line was seen in Archean lavas from the Abitibi Belt of Ontario which had been contaminated with crustal material (Cattell *et al.*, 1984). Moorbath *et al.* also showed that the relatively low MSWD value for their own Acasta samples (8.8) could not be explained by a fortuitous combination of short segments of 3.8 Ga isochrons, since this would yield a much higher MSWD value of several hundred. From this evidence, it appears that the 3.3 Ga old errorchron may date a real geological event (or series of events) which caused homogenization of Sm-Nd systems in the Acasta gneisses. Since this event postdates the oldest zircon ages by up to 600 Ma, it is concluded that reliable initial εNd values for the mantle source cannot be calculated from these samples.

Early Archean rocks from Western Greenland represent the other principal source of evidence about the composition of the Early Archean mantle. Evidence for strongly depleted mantle sources was first found in the Isua supracrustal sequence (e.g. Hamilton *et al.*, 1983), and was supported by analysis of Amitsoq gneisses and by mafic enclaves (in these gneisses) named the Akilia suite (Bennett *et al.*, 1993). The upper envelope of initial εNd in these three suites defines an evolution line for highly depleted mantle in the Early Archean (Fig. 4.35), which would project to an εNd value of around +25 at the present day. This led Bennett *et al.* to propose a model of two-stage evolution in the early Earth, in which early intense mantle depletion was followed by a period of mixing with deeper less depleted mantle, causing an inflection in the depleted mantle evolution line ('?' in Fig. 4.35).

However, Moorbath *et al.* (1997) showed that the Akilia, Amitsoq and Isua suites all yield Sm–Nd errorchrons with ages significantly younger than SHRIMP U–Pb zircon ages. They also yielded initial ratios near the normal DM line (dimonds in Fig. 4.35). This suggested to Moorbath *et al.* that the Sm–Nd systems in many of these rocks had been reset in a manner similar to the Acasta gneisses. This critique was itself the subject of a scientific discussion (Bennett and Nutman, 1998; Kamber *et al.*, 1998), after which further debate was continued by Kamber and Moorbath (1998), Whitehouse *et al.* (1999) and Nutman *et al.* (2000). Since space is limited, only a brief summary of the debate will be given.

The belt of Early Archean rocks in Western Greenland runs in a northeasterly direction parallel to Godhabsfjord, from Amitsoq on the coast, to Isua at the edge of the inland ice field. Based on detailed SHRIMP U–Pb analysis, it now appears that this belt (termed the Itsaq gneiss complex) was created in two major events. Near Isua in the north, most U–Pb ages cluster around 3.8 Ga, which appears to be the earliest crust-forming event in the area (Nutman *et al.*, 2000). However, near Amitsoq in the south, most U–Pb ages cluster round 3.65 Ga, but zircons sometimes have cores up to 3.8 Ga in age. Furthermore, in the latter area, whole-rock Rb–Sr, Pb–Pb and Sm–Nd errorchrons all give ages around 3.65 Ga (Whitehouse *et al.*, 1999). This suggests that most of the crust in the south is 3.65 Ga-old, but contains inherited fragments of 3.8 Ga material. We can therefore infer that any fragments of 3.8 Ga-old rocks in the south probably had their Nd isotope systems reset at 3.65 Ga, but the extent to which the rocks in the north preserve accurate 3.8 Ga-old initial ratios is unclear.

The Amitsoq gneisses analysed by Bennett *et al.* (1993) came from both ends and the middle of the Itsaq gneiss complex. On the Sm–Nd isochron diagram (Moorbath *et al.*, 1997) these samples were much more scattered than 26 Amitsoq gneisses from the southern end of the belt, which defined an errorchron age of 3640 ± 120 Ma (MSWD = 10) with an initial εNd value of +0.9 ± 1.4 (Fig. 4.36). Unfortunately, there were not enough Nd isotope analyses from the northern area of the gneiss complex, where old U–Pb ages predominate, to see if this part of the complex had consistently different Nd isotope signatures from the southern part.

Moorbath *et al.* suggested that the generally increased scatter in the suite analysed by Bennett *et al.* (1993) was probably due to partial metamorphic disturbance during Late Archean or Mid Proterozoic events. They were not able to prove that resetting had occurred, but Hf isotope analysis (Section 9.3.4) suggested that this might be the case, because the highly positive εNd values determined by Bennett *et al.* were not matched by similarly positive εHf values.

When Moorbath *et al.* (1997) examined Nd data for five samples of the Akilia mafic enclaves analysed by Bennett *et al.* (1993), they discovered a strong isochron array with an

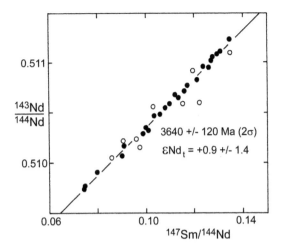

Fig. 4.36 Sm–Nd isochron diagram showing Amitsoq gneisses of the Itsaq gneiss complex. (•) = southern suite of gneisses which define a 3640 Ma errorchron. (○) = more scattered data of Bennett *et al.* (1993). After Moorbath *et al.* (1997).

age of 3675 ± 48 Ma and an initial εNd of +2.6 ± 0.4 (Fig. 4.37). The low MSWD of 2.1 for the regression makes it statistically an isochron, but U–Pb ages for the gneissic host rocks range from 3784 to 3872 Ma. Moorbath *et al.* interpreted the Sm–Nd age as an isotopic homogenization event associated with engulfing of the enclaves by Amitsoq magmas. However, Bennett and Nutman (1998) countered that these samples came from too wide an area to be attributed to metamorphic homogenization. Therefore, Kamber *et al.* (1998) reinterpreted the Sm–Nd age for the enclaves as intrusive, and the

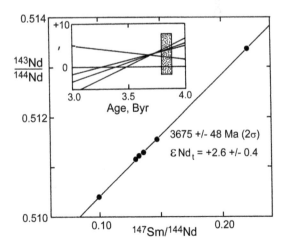

Fig. 4.37 Sm–Nd isochron for Akilia enclaves in the Amitsoq gneisses. Inset shows Nd isotope evolution lines for four samples. Shaded box shows range of U–Pb ages in the enclosing gneisses. After Moorbath *et al.* (1997).

older U–Pb ages in the host gneisses as inherited. This also seems unlikely (Horie *et al.*, 2010). However, Whitehouse *et al.* (2009) argued that the age of the enclosing gneisses cannot be used to date the Akilia enclaves, because field relationships between them are deformational rather than intrusive.

Although these questions remain contentious, it seems clear that arguments for the composition of early depleted mantle must be based on directly dated samples. This is most reliably satisfied by *in situ* Hf isotope analysis of zircon. Therefore, this is the field where questions about the degree of early mantle depletion are now largely centred (Section 9.3.4).

4.5 Nd in the Oceans

The abundance of Nd in seawater is about a million times lower than in rocks, at ca. 3 parts per trillion (Goldberg *et al.*, 1963; Piepgras *et al.*, 1979). This can be attributed to effective scavenging of rare earths from seawater by particulate matter. In contrast, ions such as sodium have similar abundances in rocks and seawater. This led Goldberg *et al.* (1963) to propose that Nd has a very short residence time in seawater, possibly less than 300 years, and less than the turnover time of water in the oceans. As a result, we can expect Nd isotope systematics in seawater to be quite different from Sr (Section 3.6.4), which has an ocean residence time of more than 2 Ma.

4.5.1 Modern Seawater Nd

The very low Nd concentrations in seawater present significant analytical difficulties. In contrast, ferro-manganese nodules, which are believed to precipitate directly from seawater, have Nd contents of up to hundreds of ppm. Consequently the early studies of O'Nions *et al.* (1978) and Piepgras *et al.* (1979) focussed principally on this material. Significant Nd isotopic variations were found between Fe–Mn nodules in different ocean basins (Fig. 4.38b) and attributed by Piepgras *et al.* to real variations in the isotopic composition of seawater.

Piepgras *et al.* justified their interpretation on the grounds that Fe–Mn nodules from a wide geographical area within each ocean mass had distinct but reproducible Nd isotope compositions. This was confirmed by the direct analysis of filtered ocean water samples (Fig. 4.38a), which were shown to be consistent with the isotopic composition of sea floor nodules from the same ocean basin. Direct Nd isotope analyses of four water samples from the Pacific (totalling 10 to 20 litres in size) were presented by Piepgras *et al.* (1979), while analyses of Atlantic Ocean water were presented by Piepgras and Wasserburg (1980).

Comparison of seawater isotope compositions with possible source reservoirs (Fig. 4.38c) suggested that Nd in Atlantic seawater is primarily continental in origin. This is consistent with a large riverine discharge into the Atlantic. In contrast, about 50% of Nd in Pacific seawater appears to be

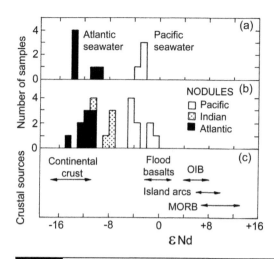

Fig. 4.38 Histograms showing the ranges of εNd displayed by: (a) seawater; and (b) Fe–Mn nodules from different ocean basins; relative to (c) major crustal reservoirs. After Piepgras and Wasserburg (1980).

derived from mafic crustal sources, either from erosion of basaltic arcs or from some form of exchange with ocean floor basalts.

A more detailed study of five vertical sections in the North Atlantic (Piepgras and Wasserburg, 1987) revealed isotopic stratification of water masses and in line with salinity and temperature (Fig. 4.39), consistent with long-established oceanographic observations (Wust, 1924). Surface water at mid latitudes (SW, Fig. 4.39b) has εNd consistent with the dissolved Nd budget of major rivers such as the Amazon and Mississippi (Piepgras and Wasserburg, 1987; Goldstein and Jacobsen, 1987). Outflow of water from the Mediterranean also has a similar composition (Piepgras and Wasserburg, 1983). In contrast, the major water body of the ocean, North Atlantic Deep Water (NADW) has very uniform unradiogenic Nd (ε = −13.5). A major contributor to NADW is Arctic Intermediate Water (AIW), which was shown by Stordal and Wasserburg (1986) to have εNd as low as −25 in Baffin Bay. This mixes with mid latitude surface water, which has become saline due to evaporation. The dense mixture sinks and flows south towards the equator, where it becomes sandwiched between two tongues of water with intermediate εNd, Antarctic Intermediate and Bottom water (AAIW, AABW, Fig. 4.39b).

Although the continental origin of Atlantic seawater is well established, the origin of the radiogenic Nd endmember in Pacific seawater is more problematical. The most obvious source is hydrothermal alteration of ocean floor basalts. However, the low REE contents of hydrothermal vent fluids (Michard *et al.*, 1983) rule out the simple hydrothermal origin proposed for Sr and Pb (Sections 3.6.4 and 5.6.2). A more significant source may be volcanic dust from circum-Pacific volcanoes (Albarede and Goldstein, 1992).

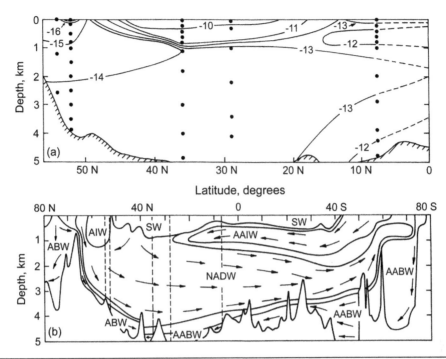

Fig. 4.39 Schematic longitudinal sections through the Atlantic Ocean to show: (a) contoured Nd isotope variations in the North Atlantic; (b) oceanographically established water masses for the whole Atlantic (with vertical profiles in part (a) shown by dashed lines). After Piepgras and Wasserburg (1987).

Wind-blown sediment has also been proposed as a major source for the continental Nd signatures in Atlantic Ocean water. This is supported by the regionality of seawater Nd signatures, which can be tied to prevailing wind directions. For example, unradiogenic Nd in the eastern North Atlantic can be traced to windblown particulates from the Sahara desert (Chester *et al.*, 1979). Similar effects occur in the eastern North Pacific, which receives wind-blown particulates from the deserts of the American southeast (Albarede and Goldstein, 1992).

4.5.2 The Oceanic Nd Paradox

The behaviour of Nd in the modern oceans has turned out to be more complex than anticipated, as indicated by the term 'Nd paradox'. This has been used to express an apparent contradiction between the oceanic Nd residence times implied by isotopic and concentration evidence. However, the term has also been used in other less precise ways.

The Nd isotope variations between the Atlantic and Pacific oceans imply a short Nd residence time, which must be less than the 1500 year circulation time of the oceans. In contrast, Nd concentration data display a nutrient-like behaviour similar to silicon, with apparently low concentrations at the surface but much higher values at depth. If the low surface concentration measurements reflect a low Nd flux to the deep oceans, this would imply a long oceanic Nd residence of up to 10–15 ka (Bertram and Elderfield, 1993).

However, this 'mass balance paradox' may underestimate the effective Nd concentrations in surface waters that are available in an exchangeable form.

To solve this problem, Bertam and Elderfield (1993) proposed a model involving Nd exchange between the suspended and dissolved Nd budgets in seawater. However, they did not take full account of atmospheric Nd inputs to surface water. In contrast, measurement of oceanic Nd depth profiles off western Africa revealed a large flux of Nd to the oceans from the dissolution of clastic particles, largely in the form of Saharan wind-borne particulates (Tachikawa *et al.*, 1999). However, Tachikawa *et al.* also showed that this input flux must be balanced by an equal output flux of Nd *adsorbed* onto sinking biogenic and authigenic particulates produced in the water mass. This creates a process of isotopic exchange that can change the isotopic composition of local ocean water without affecting its Nd concentration. The process has been termed 'reversible scavenging,' and was first invoked to explain the oceanic behaviour of thorium (Bacon and Anderson, 1982). Experimental evidence for this type of behaviour comes from laboratory observations of Nd dissolution (Pearce *et al.*, 2013). Nd released into solution causes seawater to be supersaturated relative to Nd phosphate, so that Nd must precipitate as the mineral rhabdophane (or be adsorbed onto particulates).

Because particulate inputs to the oceans are largely concentrated near continental margins, the exchange process

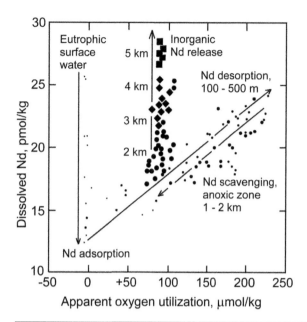

Fig. 4.40 Sequence of Nd scavenging processes on a plot of seawater Nd concentration against apparent oxygen utilization, a measure of remineralization. Symbol size is approximately proportional to water depth. Modified after Stichel *et al.* (2015).

has been called 'boundary exchange' (e.g. Lacan and Jeandel, 2005; Jeandel and Oelkers, 2015). However, because these boundary processes are local, they are notoriously difficult to convert into global flux estimates. For example, Johannesson and Burdige (2007) argued that the flux of riverine and airborne Nd to the oceans was inadequate to support the Nd concentration of deep water, and therefore that an additional subsurface groundwater flux to the oceans was necessary (previously called 'run-out' as a contributor to oceanic Sr budgets). This effect has been justified on the basis of local observations (e.g. Kim and Kim, 2014), but its global significance may be over-estimated.

On the other hand, recent studies have claimed successful modelling of global Nd behaviour based on oceanic box models involving reversible scavenging (e.g. Stichel *et al.*, 2015). This process can be conveniently summarized on a graph of Nd concentration against oxygen utilization (Fig. 4.40). This plot summarizes a sequence of processes that were observed on the continental margin of Western Africa (Stichel *et al.*, 2015). The sequence begins in the shallow water eutrophic zone, where Nd is adsorbed onto organic matter. This is followed by the second step in the sub-photic mixing zone, where desorption of Nd occurs with the breakdown of organic matter, leading to the highest oxygen consumption at around 500 m depth. However, below 1000 m depth, Nd comes back out of solution as organic matter again accumulates. Finally, below 2000 m, Nd concentrations increase without any change in oxygen usage, showing that at these depths, Nd is being released from inorganic material, perhaps Fe–Mn coatings on clastic particles laterally advected into the study area.

The dramatic effect of boundary exchange in decoupling Nd isotope signatures from concentration has also been seen at several other continental margins. For example, Antarctic Intermediate Water, approaching Papua New Guinea, contains 1 part per trillion Nd with an εNd value of −8 (Lacan and Jeandel, 2005). However, as this water passes the island, the εNd value changes dramatically to −2.8, without any significant change in concentration. This type of behaviour suggests that regional Nd isotope signatures in the oceans can be 'quasi-conservative' through time (Siddall *et al.*, 2008). However, this regionality can occur precisely because Nd is *not* a chemically conservative element in seawater. This might seem like a paradox, but it is only a semantic one.

4.5.3 Ancient Seawater Nd

Following the successful characterization of the Nd isotope budget of the modern oceans, Shaw and Wasserburg (1985) evaluated different types of material as indicators of the Nd isotope composition of the paleo-oceans. They found that carbonate and phosphate in living organisms was very low in Nd (part per billion range), but that fossil carbonates and phosphates had concentrations in the tens to hundreds of ppb and ppm respectively. Shaw and Wasserburg attributed the elevated Nd contents of fossil carbonates largely to diagenetic remobilization of detrital Nd, but they attributed the high Nd contents of ancient phosphates (conodonts, fish debris, lingulid brachiopods and inorganic phosphorites) to scavenging directly from seawater (after death). Several Nd isotope studies on this kind of material were made in the late 1980s, allowing a general understanding of the evolution of seawater Nd through time.

Keto and Jacobsen (1988) collated conodont and phosphorite Nd data with analyses of fish teeth (Staudigel *et al.*, 1985), Fe–Mn coatings on forams (Palmer and Elderfield, 1986) and conodonts and lingulids (Keto and Jacobsen, 1987, 1988) to construct a paleo-seawater Nd curve for the Phanerozoic. Because the Pacific Ocean (and its predecessor the Panthalassan Ocean) dominate the world ocean system, this was used to justify the existence of a global seawater Nd evolution curve that was relatively less influenced by boundary effects (Fig. 4.41a). Attempts were then made to extend this curve into the Precambrian by analysis of Archean and Proterozoic banded iron formations (BIF), argued to sample the Nd isotope composition of Precambrian seawater.

Miller and O'Nions (1985) obtained somewhat variable initial Nd ratios in BIFs, consistent with clastic sediments of the same age. They attributed this pattern to the control of seawater Nd by continental run-off, as seen in the modern oceans. In contrast, Jacobsen and Pimentel-Klose (1988) obtained more consistently positive εNd initial ratios (Fig. 4.41b), which they attributed to the control of Archean seawater Nd by mid-ocean ridge hydrothermal circulation, possibly reflecting higher Archean heat flow. Hence,

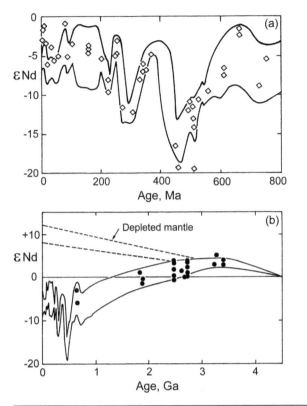

Fig. 4.41 Proposed global seawater evolution curve: (a) for the Phanerozoic, based on phosphate samples from the Pacific–Panthalassa Ocean; (b) for the Precambrian, based on banded iron formations. After Keto and Jacobsen (1988); Jacobsen and Pimentel-Klose (1988).

which presented less of an impediment to circulatory mixing of the oceans. In contrast, other more variable Nd signals have been attributed to continental control of local seawater compositions, as observed at the present day. However, in order to distinguish genuine variations in Precambrian seawater Nd from the effects of metamorphic disturbance, it is necessary to have an *a priori* test for seawater signatures. Shimizu *et al.* (1990) used the consistency of REE profiles to distinguish BIFs with genuine seawater signatures from BIFs with disturbed signatures (and variable REE profiles). This type of screening has been adopted in subsequent studies. For example, consistent REE profiles in the late Archean Mozaan BIF of South Africa led Alexander *et al.* (2009) to attribute unradiogenic Nd signatures to boundary exchange adjacent to a continental margin, rather than metamorphic disturbance.

4.5.4 Tertiary Seawater Nd

Some of the most powerful applications of Nd isotope analysis to oceanography have involved more detailed studies of seawater evolution during the Tertiary epoch, paralleling the detailed study of seawater Sr for this period (Section 3.6.2). These studies have been revolutionized by the ability to measure continuous secular variations of seawater Nd composition from Fe–Mn crusts.

Ferromanganese crusts grow on any exposed surface in the deep ocean, at a rate of about 1–3 mm/Ma. Because of their very slow growth rate, Fe–Mn crusts are easily swamped by sedimentation. However, on elevated areas such as seamounts and volcanic plateaux, crusts can grow unimpeded for more than 20 Ma (Ling *et al.*, 1997). Furthermore, Fe–Mn crusts growing on these features can sample the isotopic composition at different water depths, from as shallow as 850 m to abyssal depths of 5 km (Reynolds *et al.*, 1999). However, in order to use Fe–Mn crusts as an inventory of past seawater Nd signatures, it is necessary to measure their growth rates accurately. Consequently, this has been a major focus of research.

The most precise growth rates for Fe–Mn crusts are obtained from U-series isotopes (Section 12.3.2). Because internal checks can be made using different U-series methods, these are also the most accurate data. However, these methods cannot reach beyond 400 ka, whereas many crusts have grown for more than 20 Ma. An alternative approach attempted in early work was to use Sr isotope stratigraphy (Section 3.6.2). This method was investigated by Ingram *et al.* (1990) and VonderHaar *et al.* (1995), and appeared to give reasonable growth rates on one or two Atlantic crusts (e.g. Burton *et al.*, 1997). However, more detailed studies (e.g. Ling *et al.*, 1997; O'Nions *et al.*, 1998) revealed inconsistencies with other dating techniques, presumably due to open-system behaviour of Sr. Therefore the method has now been abandoned.

An alternative dating technique with a range up to 10 Ma is the cosmogenic isotope [10]Be (Section 14.4.2). This method has proved to be quite reliable, especially when [10]Be

Shimizu *et al.* (1990) and Frei *et al.* (1999) attributed initial εNd outside this range to later metamorphic disturbance, to which Nd-poor BIFs are very susceptible.

Frei and Polat (2007) made a more detailed analysis of iron and silica-rich layers from the Isua iron formation (Western Greenland) and found that initial εNd correlated with the Fe content of the samples. Hence they identified an iron-rich end-member (εNd = +3) which they attributed to local hydrothermal fluids, and a silica-rich end-member (εNd = +1.2) which they identified with Early Archean ocean water. The latter value falls within the expected range for seawater at that time (Fig. 4.41), which was attributed by Frei and Polat to the release of Nd from weathering of mafic arc crust, believed to dominate the limited crustal land masses of the time. A similar model was also proposed by Kamber (2010), based on weathering of emergent mafic volcanic plateaux, argued to have been much more important for the Archean plate tectonic regime than at the present day.

In spite of the lack of reliable paleogeographic information for the Early Archean, it may be justifiable to assume worldwide homogenization of Nd in seawater at that time, due to the smaller continental mass during the Archean, abundances are normalized to ^{9}Be (e.g. Ling *et al.*, 1997). Beyond 10 Ma, the only method that proved to be reliable was cobalt dating. This is based on the assumption that Fe–Mn crusts receive a constant input of cobalt with time, so that lower cobalt concentrations imply a faster growth rate, and vice versa. Hence, Frank *et al.* (1999a) showed that growth rates of three long-lived crusts, based on cobalt abundances, were consistent with growth rates extrapolated from the ^{10}Be/^{9}Be chronometer.

More recently, osmium isotope stratigraphy has been applied as a method for dating crusts beyond the range of ^{10}Be (Section 8.5.1). This is based on the same premise as the failed Sr isotope stratigraphy method, making use of the rapid variations of seawater osmium isotope ratio in the Tertiary period. However, unlike Sr, osmium shows good closed-system behaviour in crusts, and has given results consistent with ^{10}Be and cobalt dating. In addition, the osmium method can detect depositional hiatuses not visible with the cobalt dating method. Hence osmium stratigraphy has the best promise for future studies of older material.

One of the most interesting observations from Fe–Mn crusts, based initially on the analysis of one crust from the North Atlantic and one from the Central Pacific, was an apparent change in Nd isotope composition in both ocean masses at around 4 Ma BP. This period had previously been identified from oxygen isotope evidence as demonstrating increased salinity in the Caribbean, due to closure of the 'Panama Gateway' that once linked the Pacific Ocean to the Caribbean. Hence, Burton *et al.* (1997) attributed inflections in seawater Nd isotope profiles at around 4 Ma to changes in the global ocean circulation pattern prompted by closure of the Gateway (Fig. 4.42).

Unfortunately, the Atlantic data used in these reconstructions were dated by the Sr-stratigraphy method, which was subsequently shown to give ages around double the true value. However, there were actually two inflections in the curve, so that after recalibration of the profile to cosmogenic ^{10}Be ages, the Nd data continue to support a change in Atlantic Ocean Nd signatures at around 4 Ma, probably due to the closure of the Gateway (Fig. 4.42). Later work (Reynolds *et al.*, 1999) suggested that this closure may have been progressive, reflecting a gradual shallowing of the Gateway starting at 8 Ma, but finally completed at around 4 Ma. In addition, some other Atlantic Fe–Mn crusts analysed by Reynolds *et al.* showed inflections at different times. Therefore, closure of the Panama Gateway was probably not the only factor which led to changes in ocean circulation patterns over the past 10 Ma.

4.5.5 Quaternary Seawater Nd

The slow growth rates of ferromanganese crusts preclude their use to study short-term changes in seawater Nd isotope signatures, such as might be found during Quaternary glacial cycles. Therefore, other types of material capable of reliably recording short-term variations were sought. Foram suites from drill-cores are an attractive prospect because they

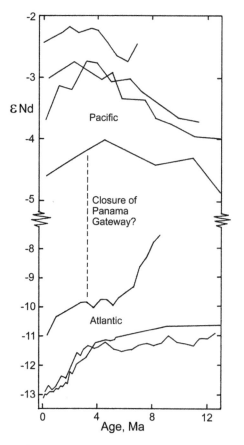

Fig. 4.42 Comparison of several Fe–Mn profiles for the Pacific and Atlantic oceans, relative to the time of closure of the Panama Gateway. After Frank *et al.* (1999b).

are widely distributed, their rapid rates of sedimentation can yield high-resolution profiles, and they are already linked to glacial cycles by stable isotope measurements. A major problem is that forams become coated with Fe–Mn deposits after accumulation on the sea floor, and these coatings have Nd concentrations much higher than the forams themselves. However, Vance and Burton (1999) showed that after removal of these coatings by leaching with a strong reducing agent, the original Nd isotope compositions of forams could be successfully recovered.

Burton and Vance (2000) applied this method to the analysis of forams in a 3 m core from the northern Indian Ocean, covering the past 150 ka. Several tests were done to check that the measured Nd isotope ratios were original rather than secondary. These included analysis of Mn/Ca ratios as a monitor of the effectiveness of the cleaning procedure, comparisons between two different fossil foram species, and comparisons of Nd content and isotope ratio with modern Indian Ocean forams. All these tests gave confidence that the method was recovering original seawater Nd signatures. When the resulting down-hole record of Nd isotope composition was examined (Fig. 4.43), an almost perfect mirror

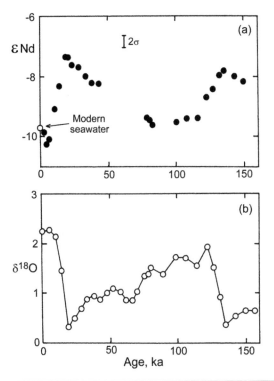

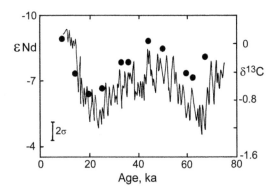

Fig. 4.43 Plot of (a) Nd isotope ratio, and (b) oxygen isotope
ratio (per mil relative to PDB), in planktonic forams from a drill
core in the northern Indian Ocean. After Burton and Vance
(2000).

image of the oxygen isotope record was seen for the last
glacial cycle. This suggests that the Nd isotope record reflects
climatic processes associated with the glacial cycle. Burton
and Vance attributed these isotopic fluctuations to a balance
between the supply of radiogenic Nd from the main body of
the Indian Ocean and the supply of unradiogenic Nd from
Himalayan erosion. Hence, if the monsoon was attenuated
during the last glacial maximum, the result would be more
radiogenic Indian Ocean Nd signatures, as seen in Fig. 4.43.

A somewhat surprising observation made by Burton and
Vance (2000) was that uncleaned forams with ferromanganese coatings had the same isotopic signatures as cleaned
forams, and even bulk sediment samples gave a profile that
was parallel to the cleaned samples, but offset $1-2 \, \varepsilon$ units
above it. This suggests that even dispersed ferromanganese
oxides in sediment cores may be a viable record of past seawater Nd signatures. Based on this assumption, Rutberg et
al. (2000) extracted Nd from bulk sediment cores by leaching
with a strong reducing agent. By this means they examined
an 80 ka record of seawater Nd in a sediment core from the
southeast Atlantic Ocean, in the Cape Basin off South Africa.
This is a critical location for understanding the behaviour
of the ocean circulation system, because many tracer studies
have shown that North Atlantic Deep Water (NADW) is mixed
with Antarctic Bottom Water in the southern ocean to form

Circum Polar Water, which is then exported to the Pacific.
This forms the so-called 'ocean conveyor belt' (Section 14.2).

Based on radiocarbon evidence (Section 14.2.2), it is
expected that the ocean conveyor belt was 'turned off'
or reduced during the last glacial maximum (ca. 20 ka
ago). However, this model was challenged by evidence from
U-series isotopes (Pa/Th activities, Section 12.3.6), which
implied that the conveyor continued unabated during the
glacial maximum. Nevertheless, Pa/Th activity ratios are susceptible to disturbance by changes in biological production,
whereas the Nd isotope system is less susceptible to this
kind of disturbance. Therefore, Nd isotope data may help to
resolve this conflict.

The study of Rutberg et al. (2000) provided preliminary
data to address this problem, assuming that the observed isotopic variations were original and not diagenetic. Evidence
in support of their validity as original seawater compositions came from the preservation of typical seawater Sr isotope signatures in the analysed leachates, despite the presence of radiogenic Sr in coexisting detrital phases. Given this
assumption, variations of Nd isotope ratio can be attributed
to variations in the supply of NADW to the Southern Ocean.
The fact that these variations are in step with climatically
controlled carbon isotope variations (Fig. 4.44) provides evidence to support changes in the strength of the conveyer belt
between glacial and interglacial periods.

More detailed evidence of the relationship between Nd
isotopes and carbon isotopes was obtained by a closer examination of a South Atlantic core over the deglacial period
(Piotrowski et al., 2005). This work showed a large change in
[13]C for only a small change in εNd during early deglaciation
('Oldest Dryas' in Fig. 4.45a). This was attributed to major climatic amelioration accompanied by only minor changes in
ocean circulation. In contrast, the later stages of deglaciation
(Younger Dryas to Holocene) apparently involved only minor
additional climatic amelioration, but involved large changes
in ocean circulation. Piotrowski et al. also showed that

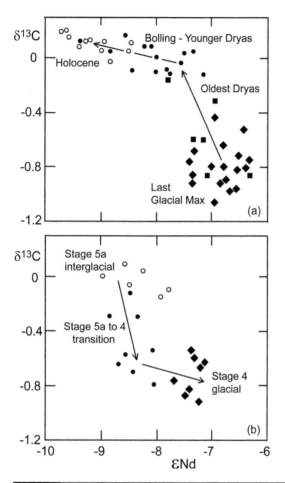

Fig. 4.45 Plots of $\delta^{13}C$ against εNd to show the sequence of isotopic changes: (a) during deglaciation; and (b) during the onset of the last glaciation. (♦) = glacial; (■, •) = transitional; (○) = interglacial. Modified after Piotrowski et al. (2005).

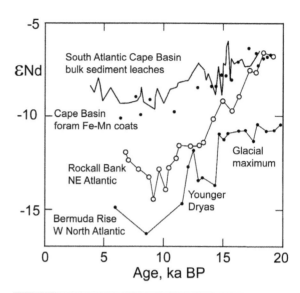

Fig. 4.46 Comparison of εNd profiles for three deep Atlantic Ocean cores during Quaternary deglaciation. Symbols = foram coats; plain line = bulk sediment leaches. Modified after Piotrowski et al. (2012).

converse changes occurred at the *onset* of the last glacial period (stages 5a–4 = ca. 120 ka). Here, a rapid change in $\delta^{13}C$ preceded the change in εNd (Fig. 4.45b), again showing that a climatic change preceded changes in ocean circulation. This suggests that ocean circulation *responds to* rather than precipitates major glacial cycles. However, the evidence for short-term climatic excursions is less clear.

The record of deglaciation is of particular interest for studies of modern climate change, since it records a change from the last glacial maximum (LGM) to modern interglacial conditions. Hence, this period has been the focus of several more recent studies. For example, Piotrowski et al. (2012) made detailed comparisons between cleaned forams, foram Fe–Mn coatings and bulk sediment leachates from Atlantic Ocean bottom cores. They found that in some cases the methods agreed (e.g. for the Cape Basin in Fig. 4.46), while in other cases the cleaned forams and Fe–Mn coatings were consistent but the bulk sediment leachates were unreliable.

Even larger changes in εNd were observed by Roberts *et al.*, 2010 for the NE Atlantic and Burmuda Rise (Fig. 4.46). These profiles suggest that there was a dramatic weakening in the entry of radiogenic Southern Component Water into the deep Atlantic as deglaciation developed, and an intensification in the unradiogenic NADW signal. Unlike the NE Atlantic, the Bermuda Rise core also shows an effect for the Younger Dryas glacial re-advance, attributed to its greater sensitivity to shallow water effects.

References

Albarede, F. and Goldstein, S. L. (1992). World map of Nd isotopes in sea-floor ferromanganese deposits. *Geology* **20**, 761–3.

Alexander, B. W., Bau, M. and Andersson, P. (2009). Neodymium isotopes in Archean seawater and implications for the marine Nd cycle in Earth's early oceans. *Earth Planet. Sci. Lett.* **283**, 144–55.

Allegre, C. J. and Rousseau, D. (1984). The growth of the continents through geological time studied by Nd isotope analysis of shales. *Earth Planet. Sci. Lett.* **67**, 19–34.

Armstrong, R. L. (1981). Radiogenic isotopes: the case for crustal recycling on a near steady-state no-continental-growth Earth. *Phil. Trans. Roy. Soc. Lond.* **A301**, 443–72.

Armstrong, R. L. (1991). The persistent myth of crustal growth. *Aust. J. Earth Sci.* **38**, 613–30.

Arndt, N. T. and Goldstein, S. L. (1987). Use and abuse of crust-formation ages. *Geology* **15**, 893–5.

Awwiller, D. N. and Mack, L. E. (1991). Diagenetic modification of Sm–Nd model ages in Tertiary sandstones and shales, Texas Gulf Coast. *Geology* **19**, 311–14.

Bacon, M. P. and Anderson, R. F. (1982). Distribution of thorium isotopes between dissolved and particulate forms in the deep sea. *J. Geophys. Res.* **87** (C3), 2045–56.

Barovich, K. M. and Patchett, P. J. (1992). Behaviour of isotopic systematics during deformation and metamorphism: a Hf, Nd and Sr isotopic study of mylonitized granite. *Contrib. Mineral. Petrol.* **109**, 386–93.

Baxter, E. F., Ague, J. J. and DePaolo, D. J. (2002). Prograde temperature-time evolution in the Barrovian type-locality constrained by Sm/Nd garnet ages from Glen Clova, Scotland. *J. Geol. Soc.* **159**, 71–82.

Bennett, V. C. and DePaolo, D. J. (1987). Proterozoic crustal history of the western United States as determined by neodymium isotopic mapping. *Geol. Soc. America Bull.* **99**, 674–85.

Bennett, V. C. and Nutman, A. P. (1998). Extreme Nd-isotope heterogeneity in the early Archean – fact or fiction? Case histories from northern Canada and West Greenland – Comment. *Chem. Geol.* **148**, 213–17.

Bennett, V. C., Nutman, A. P. and McCulloch, M. T. (1993). Nd isotopic evidence for transient, highly depleted mantle reservoirs in the early history of the Earth. *Earth Planet. Sci. Lett.* **119**, 299–317.

Bertram, C. J. and Elderfield, H. (1993). The geochemical balance of the rare earth elements and Nd isotopes in the oceans. *Geochim. Cosmochim. Acta* **57**, 1957–86.

Bock, B., McLennan, S. M. and Hanson, G. N. (1994). Rare earth element redistribution and its effects on the neodymium isotope system in the Austin Glen Member of the Normanskill Formation, New York, USA. *Geochim. Cosmochim. Acta* **58**, 5245–53.

Boehnke, P. and Harrison, T. M. (2014). A meta-analysis of geochronologically relevant half-lives: what's the best decay constant?. *Int. Geol. Rev* **56**, 905–14.

Bouvier, A., Vervoort, J. D. and Patchett, P. J. (2008). The Lu-Hf and Sm-Nd isotopic composition of CHUR: constraints from unequilibrated chondrites and implications for the bulk composition of terrestrial planets. *Earth Planet. Sci. Lett.* **273**, 48–57.

Bowring, S. A. and Housh, T. (1995). The Earth's early evolution: *Science* **269**, 1535–40.

Bowring, S. A. and Housh, T. (1996). Sm-Nd isotope data and Earth's evolution: Reply. *Science* **273**, 1878–9.

Bowring, S. A., King, J. E., Housh, T. B., Isachsen, C. E. and Podosek, F. A. (1989). Neodymium and lead isotope evidence for enriched early Archean crust in North America. *Nature* **340**, 222–5.

Bros, R., Stille, P., Gauthier-Lafaye, F., Weber, F. and Clauer, N. (1992). Sm-Nd isotopic dating of Proterozoic clay material: an example from the Francevillian sedimentary series, Gabon. *Earth Planet. Sci. Lett.* **113**, 207–18.

Burton, K. W., Ling, H.-F. and O'Nions, R. K. (1997). Closure of the Central American Isthmus and its effect on deep-water formation in the North Atlantic. *Nature* **386**, 382–5.

Burton, K. W. and O'Nions, R. K. (1991). High-resolution garnet chronometry and the rates of metamorphic processes. *Earth Planet. Sci. Lett.* **107**, 649–71.

Burton, K. W. and Vance, D. (2000). Glacial–interglacial variations in the neodymium isotope composition of seawater in the Bay of Bengal recorded by planktonic foraminifera. *Earth Planet. Sci. Lett.* **176**, 425–41.

Cattell, A., Krogh, T. E. and Arndt, N. T. (1984). Conflicting Sm-Nd whole rock and U-Pb zircon ages for Archean lavas from Newton Township, Abitibi Belt, Ontario. *Earth Planet. Sci. Lett.* **70**, 280–90.

Chapman, H. J. and Moorbath, S. (1977). Lead isotope measurements from the oldest recognised Lewisian gneisses of north-west Scotland. *Nature* **268**, 41–2.

Chase, C. G. and Patchett, P. J. (1988). Stored mafic/ultramafic crust and early Archean mantle depletion. *Earth Planet. Sci. Lett.* **91**, 66–72.

Chauvel, C., Dupre, B. and Jenner, G. A. (1985). The Sm–Nd age of Kambalda volcanics is 500 Ma too old! *Earth Planet. Sci. Lett.* **74**, 315–24.

Chester, R., Griffiths, A. G. and Hirst, J. M. (1979). The influence of soil-sized atmospheric particulates on the elemental chemistry of deep sea sediments of the northeastern Atlantic. *Marine Geol.* **32**, 141–54.

Claoue-Long, J. C., Thirlwall, M. F. and Nesbitt, R. W. (1984). Revised Sm-Nd systematics of Kambalda greenstones, Western Australia. *Nature* **307**, 697–701.

Compston, W., Williams, I. S., Campbell, I. H. and Gresham, J. J. (1985). Zircon xenocrysts from the Kambalda volcanics: age constraints and direct evidence for older continental crust below the Kambalda-Norseman greenstones. *Earth Planet. Sci. Lett.* **76**, 299–311.

Cullers, R. L., Bock, B. and Guidotti, C. (1997). Elemental distributions and neodymium isotopic compositions of Silurian metasediments, western Maine, USA: Redistribution of rare earth elements. *Geochim. Cosmochim. Acta* **61**, 1847–61.

Culshaw, N., Foster, J., Marsh, J., Slagstad, T. and Gerbi, C. (2016). Kiosk domain, Central Gneiss Belt, Grenville Province, Ontario: A Labradorian palimpsest preserved in the ductile deep crust. *Precamb. Res.* **280**, 249–78.

DePaolo, D. J. (1981). Neodymium isotopes in the Colorado Front Range and implications for crust formation and mantle evolution in the Proterozoic. *Nature* **291**, 193–7.

DePaolo, D. J., Linn, A. M. and Schubert, G. (1991). The continental crustal age distribution: Methods of determining mantle separation ages from Sm-Nd isotopic data and application to the southwestern United States. *J. Geophys. Res.* **96** (B2), 2071–88.

DePaolo, D. J. and Wasserburg, G. J. (1976a). Nd isotopic variations and petrogenetic models. *Geophys. Res. Lett.* **3**, 249–52.

DePaolo, D. J. and Wasserburg, G. J. (1976b). Inferences about magma sources and mantle structure from variations of $^{143}Nd/^{144}Nd$. *Geophys. Res. Lett.* **3**, 743–6.

DePaolo, D. J. and Wasserburg, G. J. (1979). Sm-Nd age of the Stillwater complex and the mantle evolution curve for neodymium. *Geochim. Cosmochim. Acta* **43**, 999–1008.

DeWolf, C. P. and Mezger, K. (1994). Lead isotope analyses of leached feldspars: constraints on the early crustal history of the Grenville Orogen. *Geochim. Cosmochim. Acta* **58**, 5537–50.

Dhuime, B., Hawkesworth, C. and Cawood, P. (2011). When continents formed. *Science*, **331**, 154–5.

Dia, A., Allegre, C. J. and Erlank, A. J. (1990). The development of continental crust through geological time: the South African case. *Earth Planet. Sci. Lett.* **98**, 74–89.

Dickin, A. P. (2015). Model Ages (Sm-Nd). In: Rink, W. J. and Thompson, J. W. (Eds) *Encyclopedia of Scientific Dating Methods*, Springer, pp. 573–6.

Dickin, A., Hynes, E., Strong, J. and Wisborg, M. (2016). Testing a back-arc 'aulacogen' model for the Central Metasedimentary Belt of the Grenville Province. *Geol. Mag.* **153**, 681–95.

Dickin, A. P. and McNutt, R. H. (1989). Nd model age mapping of the southeast margin of the Archean foreland in the Grenville province of Ontario. *Geology* **17**, 299–302.

Dickin, A. P. and McNutt, R. H. (2007). The Central Metasedimentary Belt (Grenville Province) as a failed back-arc rift zone: Nd isotope evidence. *Earth Planet. Sci. Lett.* **259**, 97–106.

Dickin, A. P., McNutt, R. H., Martin, C. and Guo, A. (2010). The extent of juvenile crust in the Grenville Province: Nd isotope evidence. *Geol. Soc. America Bull.* **122**, 870–83.

Frank, M., O'Nions, R. K., Hein, J. R. and Banakar, V. K. (1999a). 60 Ma records of major elements and Pb-Nd isotopes from hydrogenous ferromanganese crusts: Reconstruction of seawater paleochemistry. *Geochim. Cosmochim. Acta* **63**, 1689–1708.

Frank, M., Reynolds, B. C. and O'Nions, R. K. (1999b). Nd and Pb isotopes in Atlantic and Pacific water masses before and after closure of the Panama gateway. *Geology* **27**, 1147–50.

Frei, R., Bridgwater, D., Rosing, M. and Stecher, O. (1999). Controversial Pb–Pb and Sm–Nd isotope results in the early Archean Isua (West Greenland) oxide iron formation: Preservation of primary signatures versus secondary disturbances. *Geochim. Cosmochim. Acta* **63**, 473–88.

Frei, R. and Polat, A. (2007). Source heterogeneity for the major components of ~3.7 Ga banded iron formations (Isua Greenstone Belt, Western Greenland): tracing the nature of interacting water masses in BIF formation. *Earth Planet. Sci. Lett.* **253**, 266–81.

Galer, S. J. G. and Goldstein, S. L. (1991). Early mantle differentiation and its thermal consequences. *Geochim. Cosmochim. Acta* **55**, 227–39.

Goldberg, E. D., Koide, M., Schmidt, R. A. and Smith, R. H. (1963). Rare earth distributions in the marine environment. *J. Geophys. Res.* **68**, 4209–17.

Goldstein, S. L. and Jacobsen, S. B. (1987). The Nd and Sr isotopic systematics of river-water dissolved material: implications for the sources of Nd and Sr in seawater. *Chem. Geol. (Isot. Geosci. Sect.)* **66**, 245–72.

Goldstein, S. L. and Jacobsen, S. B. (1988). Nd and Sr isotopic systematics of river water suspended material: implications for crustal evolution. *Earth Planet. Sci. Lett.* **87**, 249–65.

Goldstein, S. L., O'Nions, R. K. and Hamilton, P. J. (1984). A Sm–Nd isotopic study of atmospheric dusts and particulates from major river systems. *Earth Planet. Sci. Lett.* **70**, 221–36.

Green, T. H., Brunfeldt, A. O. and Heier, K. S. (1969). Rare earth element distribution in anorthosites and associated high grade metamorphic rocks, Lofoten-Vesteraalen, Norway. *Earth Planet. Sci. Lett.* **7**, 93–8.

Griffin, W. L. and Brueckner, H. K. (1980). Caledonian Sm–Nd ages and a crustal origin for Norwegian eclogites. *Nature* **285**, 319–20.

Guitreau, M., Blichert-Toft, J. and Billström, K. (2014). Hafnium isotope evidence for early-Proterozoic volcanic arc reworking in the Skellefte district (northern Sweden) and implications for the Svecofennian orogen. *Precam. Res.* **252**, 39–52.

Hamilton, P. J., O'Nions, R. K., Evensen, N. M. and Tarney, J. (1979). Sm–Nd systematics of Lewisian gneisses: Implications for the origin of granulites. *Nature* **277**, 25–8.

Hamilton, P. J., O'Nions, R. K., Bridgwater, D. and Nutman, A. (1983). Sm–Nd studies of Archean metasediments and metavolcanics from West Greenland and their implications for the Earth's early history. *Earth Planet. Sci. Lett.* **62**, 263–72.

Hanson, G. N. (1978). The application of trace elements to the petrogenesis of igneous rocks of granitic composition. *Earth Planet. Sci. Lett.* **38**, 26–43.

Haskin, L. A., Frey, F. A., Schmidt, P. A. and Smith, R. H. (1966). Meteoritic, solar and terrestrial rare-earth distributions. *Phys. Chem. Earth* **7**, 167–321.

Hastie, A. R., Fitton, J. G., Bromiley, G. D., Butler, I. B. and Odling, N. W. (2016). The origin of Earth's first continents and the onset of plate tectonics. *Geology* **44**, 855–8.

Horie, K., Nutman, A. P., Friend, C. R. and Hidaka, H. (2010). The complex age of orthogneiss protoliths exemplified by the Eoarchaean Itsaq Gneiss Complex (Greenland): SHRIMP and old rocks. *Precamb. Res.* **183**, 25–43.

Hurley, P. M., Hughes, H., Faure, G., Fairbairn, H. W. and Pinson, W. H. (1962). Radiogenic strontium-87 model of continent formation. *J. Geophys. Res.* **67**, 5315–34.

Hurley, P. M. and Rand, J. R. (1969). Pre-drift continental nuclei. *Science* **164**, 1229–42.

Ingram, B. L., Hein, J. R. and Farmer, G. L. (1990). Age determinations and growth rates of Pacific ferromanganese deposits using strontium isotopes. *Geochim. Cosmochim. Acta* **54**, 1709–21.

Jacobsen, S. B. (1988). Isotopic constraints on crustal growth and recycling. *Earth Planet. Sci. Lett.* **90**, 315–29.

Jacobsen, S. B. and Pimentel-Klose, M. R. (1988). Nd isotopic variations in Precambrian banded iron formations. *Geophys. Res. Lett.* **15**, 393–6.

Jacobsen, S. B. and Wasserburg, G. J. (1980). Sm–Nd isotopic evolution of chondrites. *Earth Planet. Sci. Lett.* **50**, 139–55.

Jeandel, C. and Oelkers, E. H. (2015). The influence of terrigenous particulate material dissolution on ocean chemistry and global element cycles. *Chem. Geol.* **395**, 50–66.

Johannesson, K. H. and Burdige, D. J. (2007). Balancing the global oceanic neodymium budget: evaluating the role of groundwater. *Earth Planet. Sci. Lett.* **253**, 129–42.

Kamber, B. S. (2010). Archean mafic–ultramafic volcanic landmasses and their effect on ocean–atmosphere chemistry. *Chem. Geol.* **274**, 19–28.

Kamber, B. S. and Moorbath, S. (1998). Initial Pb of the Amîtsoq gneiss revisited: implication for the timing of early Archaean crustal evolution in West Greenland. *Chem. Geol.* **150**, 19–41.

Kamber, B. S., Moorbath, S. and Whitehouse, M. J. (1998). Extreme Nd-isotope heterogeneity in the early Archean - fact or fiction? Case histories from northern Canada and West Greenland - Reply. *Earth Planet. Sci. Lett.* **148**, 219–24.

Kemp, A. I. S., Hawkesworth, C. J., Collins, W. J., Gray, C. M. and Blevin, P. L. (2009). Isotopic evidence for rapid continental growth in an extensional accretionary orogen: The Tasmanides, eastern Australia. *Earth Planet. Sci. Lett.* **284**, 455–66.

Keto, L. S. and Jacobsen, S. B. (1987). Nd and Sr isotopic variations of Early Paleozoic oceans. *Earth Planet. Sci. Lett.* **84**, 27–41.

Keto, L. S. and Jacobsen, S. B. (1988). Nd isotopic variations of Phanerozoic paleo-oceans. *Earth Planet. Sci. Lett.* **90**, 395–410.

Kim, I. and Kim, G. (2014). Submarine groundwater discharge as a main source of rare earth elements in coastal waters. *Marine Chem.* **160**, 11–17.

Lacan, F. and Jeandel, C. (2005). Neodymium isotopes as a new tool for quantifying exchange fluxes at the continent–ocean interface. *Earth Planet. Sci. Lett.* **232**, 245–57.

Lahaye, Y., Arndt, N., Byerly, G. *et al.* (1995). The influence of alteration on the trace-element and nd isotopic composition of komatiites. *Chem. Geol.* **126**, 43–64.

Lambert, D. D., Morgan, J. W., Walker, R. J. *et al.* (1989). Rhenium–osmium and samarium–neodymium isotopic systematics of the Stillwater Complex. *Science* **244**, 1169–74.

Ling, H. F., Burton, K. W., O'Nions, R. K. *et al.* (1997). Evolution of Nd and Pb isotopes in Central Pacific seawater from ferromanganese crusts. *Earth Planet. Sci. Lett.* **146**, 1–12.

Lugmair, G. W. and Marti, K. (1977). Sm-Nd-Pu timepieces in the Angra dos Reis meteorite. *Earth Planet. Sci. Lett.* **35**, 273–84.

Lugmair, G. W. and Galer, S. J. G. (1992). Age and isotopic relationships among the angrites Lewis Cliff 86010 and Angra dos Reis. *Geochim. Cosmochim. Acta* **56**, 1673–94.

Lugmair, G. W. and Marti, K. (1978). Lunar initial ^{143}Nd/^{144}Nd: differential evolution of the lunar crust and mantle. *Earth Planet. Sci. Lett.* **39**, 349–57.

Lugmair, G. W., Scheinin, N. B. and Marti, K. (1975). Search for extinct ^{146}Sm, I. The isotopic abundance of ^{142}Nd in the Juvinas meteorite. *Earth Planet. Sci. Lett.* **27**, 79–84.

McCulloch, M. T. and Compston, W. (1981). Sm–Nd age of Kambalda and Kanowna greenstones and heterogeneity in the Archean mantle. *Nature* **294**, 322–7.

McCulloch, M. T. and Wasserburg, G. J. (1978). Sm–Nd and Rb–Sr chronology of continental crust formation. *Science* **200**, 1003–11.

Michard, A., Albarede, F., Michard, G., Minster, J. F. and Charlou, J. L. (1983). Rare-earth elements and uranium in high-temperature solutions from East Pacific Rise hydrothermal vent field (13 °N). *Nature* **303**, 795–7.

Mildowski, A. E. and Zalasiewicz, J. A. (1991). Redistribution of rare earth elements during diagenesis of turbidite/hemipelagite mudstone sequences of Llandovery age from central Wales. In: Morton, A. C. *et al.* (Eds) Developments in Sedimentary Provenance Studies. *Geol. Soc. Spec. Pap.* **56**, 789–95.

Miller, R. G. and O'Nions, R. K. (1985). Source of Precambrian chemical and clastic sediments. *Nature* **314**, 325–33.

Moorbath, S. (1976). Age and isotope constraints for the evolution of Archaean crust. In: Windley, B. F. (Ed.) *The Early History of the Earth*, Wiley, pp. 351–60.

Moorbath, S., Powell, J. L. and Taylor, P. N. (1975). Isotopic evidence for the age and origin of the grey gneiss complex of the southern Outer Hebrides, Scotland. *J. Geol. Soc. Lond.* **131**, 213–22.

Moorbath, S. and Whitehouse, M. J. (1996). Sm–Nd isotope data and Earth's evolution: Comment. *Science* **273**, 1878.

Moorbath, S., Whitehouse, M. J. and Kamber, B. S. (1997). Extreme Nd-isotope heterogeneity in the early Archean – fact or fiction? Case histories from northern Canada and West Greenland. *Chem. Geol.* **135**, 213–31.

Moore, E. S. and Dickin, A. P. (2011). Evaluation of Nd isotope data for the Grenville Province of the Laurentian shield using a geographic information system. *Geosphere* **7**, 415–28.

Mork, M. B. E. and Mearns, E. W. (1986). Sm–Nd isotopic systematics of a gabbro–eclogite transition. *Lithos* **19**, 255–67.

Nelson, B. K. and DePaolo, D. J. (1984). 1,700-Ma greenstone volcanic successions in southwestern North America and isotopic evolution of Proterozoic mantle. *Nature* **312**, 143–6.

Nelson, B. K. and DePaolo, D. J. (1985). Rapid production of continental crust 1.7 to 1.9 b.y. ago: Nd isotopic evidence from the basement of the North American mid-continent. *Geol. Soc. Amer. Bull.* **96**, 746–54.

Nelson, B. K. and DePaolo, D. J. (1988). Application of Sm–Nd and Rb–Sr isotope systematics to studies of provenance and basin analysis. *J. Sed. Petrol.* **58**, 348–57.

Nunes, P. D. (1981). The age of the Stillwater complex: a comparison of U–Pb zircon and Sm–Nd isochron systematics. *Geochim. Cosmochim. Acta* **45**, 1961–3.

Nutman, A. P., Bennett, V. C. and Friend, C. R. (2015). Proposal for a continent 'Itsaqia' amalgamated at 3.66 Ga and rifted apart from 3.53 Ga: Initiation of a Wilson Cycle near the start of the rock record. *American J. Sci.* **315**, 509–36.

Nutman, A. P., Bennett, V. C., Friend, C. R. L. and McGregor, V. R. (2000). The early Archean Itsaq Gneiss Complex of southern West Greenland: The importance of field observations in interpreting age and isotopic constraints for early terrestrial evolution. *Geochim. Cosmochim. Acta* **64**, 3035–60.

O'Nions, R. K. (1984). Isotopic abundances relevant to the identification of magma sources. *Phil. Trans. Roy. Soc. Lond.* A **310**, 591–603.

O'Nions, R. K., Carter, S. R., Cohen, R. S., Evensen, N. M. and Hamilton, P. J. (1978). Pb, Nd and Sr isotopes in oceanic ferromanganese deposits and ocean floor basalts. *Nature* **273**, 435–8.

O'Nions, R. K., Frank, M., Von Blanckenburg, F. and Ling, H. F. (1998). Secular variation of Nd and Pb isotopes in ferromanganese crusts from the Atlantic, Indian and Pacific Oceans. *Earth Planet. Sci. Lett.* **155**, 15–28.

O'Nions, R. K., Hamilton, P. J. and Evensen, N. M. (1977). Variations in ^{143}Nd/^{144}Nd and ^{87}Sr/^{86}Sr in oceanic basalts. *Earth Planet. Sci. Lett.* **34**, 13–22.

O'Nions, R. K., Hamilton, P. J. and Hooker, P. J. (1983). A Nd isotope investigation of sediments related to crustal development in the British Isles. *Earth Planet. Sci. Lett.* **63**, 229–40.

Palmer, M. R. and Elderfield, H. (1986). Rare earth elements and neodymium isotopes in ferromanganese oxide coatings of Cenozoic foraminifera from the Atlantic Ocean. *Geochim. Cosmochim. Acta* 50, 409–17.

Patchett, P. J. and Arndt, N. T. (1986). Nd isotopes and tectonics of 1.9–1.7 Ga crustal genesis. *Earth Planet. Sci. Lett.* **78**, 329–38.

Patchett, J., Gorbatschev, R., Kuovo, O. and Todt, W. (1984). Origin of continental crust of 1.9–1.7 Ga age: Nd isotopes in the Svecokarelian terrain of Sweden and Finland. *Geol. Soc. America, Abstr. with Prog.* **16**.

Pearce, C. R., Jones, M. T., Oelkers, E. H., Pradoux, C. and Jeandel, C. (2013). The effect of particulate dissolution on the neodymium (Nd) isotope and rare earth element (REE) composition of seawater. *Earth Planet. Sci. Lett.* **369**, 138–47.

Pidgeon, R. T. and Bowes, D. R. (1972). Zircon U/Pb ages of granulites from the central region of the Lewisian, north western Scotland. *Geol. Mag.* **109**, 247–58.

Piepgras, D. J. and Wasserburg, G. J. (1980). Neodymium isotopic variations in seawater. *Earth Planet. Sci. Lett.* **50**, 128–38.

Piepgras, D. J. and Wasserburg, G. J. (1983). Influence of the Mediterranean Outflow on the isotopic composition of neodymium in waters of the North Atlantic. *J. Geophys. Res.* **88**, 5997–6006.

Piepgras, D. J. and Wasserburg, G. J. (1987). Rare earth element transport in the western North Atlantic inferred from Nd isotopic observations. *Geochim. Cosmochim. Acta* **51**, 1257–71.

Piepgras, D. J., Wasserburg, G. J. and Dasch, E. J. (1979). The isotopic composition of Nd in different ocean masses. *Earth Planet. Sci. Lett.* **45**, 223–36.

Piotrowski, A. M., Galy, A., Nicholl, J. A. L. *et al.* (2012). Reconstructing deglacial North and South Atlantic deep water sourcing using foraminiferal Nd isotopes. *Earth Planet. Sci. Lett.* **357**, 289–97.

Piotrowski, A. M., Goldstein, S. L., Hemming, S. R. and Fairbanks, R. G. (2005). Temporal relationships of carbon cycling and ocean circulation at glacial boundaries. *Science* **307**, 1933–8.

Pollington, A. D. and Baxter, E. F. (2010). High resolution Sm–Nd garnet geochronology reveals the uneven pace of tectonometamorphic processes. *Earth Planet. Sci. Lett.* **293**, 63–71.

Pollington, A. D. and Baxter, E. F. (2011). High precision microsampling and preparation of zoned garnet porphyroblasts for Sm–Nd geochronology. *Chem. Geol.* **281**, 270–82.

Premo, W. R., Helz, R. T., Zientek, M. L. and Langston, R. B. (1990). U–Pb and Sm–Nd ages for the Stillwater Complex and its associated sills and dikes, Beartooth Mountains, Montana: Identification of a parent magma? *Geology* **18**, 1065–8.

Reymer, A. and Schubert, G. (1986). Rapid growth of some major segments of continental crust. *Geology* **14**, 299–302.

Reynolds, B. C., Frank, M. and O'Nions, R. K. (1999). Nd- and Pb-isotope time series from Atlantic ferromanganese crusts: implications for changes in provenance and paleocirculation over the last 8 Ma. *Earth Planet. Sci. Lett.* **173**, 381–96.

Roberts, N. L., Piotrowski, A. M., McManus, J. F. and Keigwin, L. D. (2010). Synchronous deglacial overturning and water mass source changes. *Science* **327**, 75–8.

Rutberg, R. L., Hemming, S. R. and Goldstein, S. L. (2000). Reduced North Atlantic Deep Water flux to the glacial Southern Ocean inferred from neodymium isotope ratios. *Nature* **405**, 935–8.

Samson, S. D., McClelland, W. C., Patchett, P. J., Gehrels, G. E. and Anderson, G. (1989). Evidence from neodymium isotopes for mantle contributions to Phanerozoic crustal genesis in the Canadian Cordillera. *Nature* **337**, 705–9.

Shaw, H. F. and Wasserburg, G. J. (1985). Sm-Nd in marine carbonates and phosphates: implications for Nd isotopes in seawater and crustal ages. *Geochim. Cosmochim. Acta* **49**, 503–18.

Shimizu, H., Umemoto, N., Masuda, A. and Appel, P. W. U. (1990). Sources of iron-formations in the Archean Isua and Malene supracrustals, West Greenland: Evidence from La-Ce and Sm-Nd isotopic data and REE abundances. *Geochim. Cosmochim. Acta* **54**, 1147–54.

Siddall, M., Khatiwala, S., van de Flierdt, T. *et al.* (2008). Towards explaining the Nd paradox using reversible scavenging in an ocean general circulation model. *Earth Planet. Sci. Lett.* **274**, 448–61.

Skiold, T., Ohlander, B., Vocke, R. D. and Hamilton, P. J. (1988). Chemistry of Proterozoic orogenic processes at a continental margin in northern Sweden. *Chem. Geol.* **69**, 193–207.

Smith, A. D. and Ludden, J. N. (1989). Nd isotopic evolution of the Precambrian mantle. *Earth Planet. Sci. Lett.* **93**, 14–22.

Staudigel, H., Doyle, P. and Zindler, A. (1985). Sr and Nd isotope systematics in fish teeth. *Earth Planet. Sci. Lett.* **76**, 45–56.

Stichel, T., Hartman, A. E., Duggan, B. *et al.* (2015). Separating biogeochemical cycling of neodymium from water mass mixing in the Eastern North Atlantic. *Earth Planet. Sci. Lett.* **412**, 245–60.

Stille, P. and Clauer, N. (1986). Sm-Nd isochron-age and provenance of the argillites of the Gunflint Iron Formation in Ontario, Canada. *Geochim. Cosmochim. Acta* **50**, 1141–6.

Stordal, M. C. and Wasserburg, G. J. (1986). Neodymium isotopic study of Baffin Bay water: sources of REE from very old terranes. *Earth Planet. Sci. Lett.* **77**, 259–72.

Tachikawa, K., Jeandel, C. and Roy-Barman, M. (1999). A new approach to the Nd residence time in the ocean: the role of atmospheric inputs. *Earth Planet. Sci. Lett.* **170**, 433–46.

Taylor, S. R., McLennan, S. N. and McCulloch, M. T. (1983). Geochemistry of loess, continental crustal composition and crustal model ages. *Geochim. Cosmochim. Acta* **47**, 1897–1905.

Thurston, P. C., Osmani, I. A. and Stone, D. (1991). Northwest Superior province: review and terrane analysis. In: Thurston, P. C., Williams, H. R., Sutcliffe, R. H. and Stott, G. M. (Eds) *Geology of Ontario. Ontario Geol. Surv. Spec. Vol.* **4**, 81–139.

Tolstikhin, I. and Hofmann, A. W. (2005). Early crust on top of the Earth's core. *Phys. Earth Planet. Int.* **148**, 109–30.

Tugarinov, A. I. and Bibikova, Y. V. (1976). Evolution of the chemical composition of the Earth's crust. *Geokhimiya* **1976**, (8) 1151–9.

Turner, S., Rushmer, T., Reagan, M. and Moyen, J. F. (2014). Heading down early on? Start of subduction on Earth. *Geology* **42**, 139–42.

Vance, D. and Burton, K. (1999). Neodymium isotopes in planktonic foraminifera: a record of the response of continental weathering and ocean circulation rates to climate change. *Earth Planet. Sci. Lett.* **173**, 365–79.

Vance, D. and O'Nions, R. K. (1990). Isotopic chronometry of zoned garnets: growth kinetics and metamorphic histories. *Earth Planet. Sci. Lett.* **97**, 227–40.

VonderHaar, D. L., Mahoney, J. J. and McMurtry, G. M. (1995). An evaluation of strontium isotopic dating of ferromanganese oxides in a marine hydrogeneous ferromanganese crust. *Geochim. Cosmochim. Acta* **59**, 4267–77.

Wasserburg, G. J., Jacobsen, S. B., DePaolo, D. J., McCulloch, M. T. and Wen, T. (1981). Precise determination of Sm/Nd ratios, Sm and Nd isotopic abundances in standard solutions. *Geochim. Cosmochim. Acta* **45**, 2311–23.

West, G. F. (1980). Formation of continental crust. In: Strangway, D. W. (Ed.) *The Continental Crust and its Mineral Deposits. Geol. Assoc. Canada Spec. Pap.* **8**, 117–48.

Whitehouse, M. J. (1988). Granulite facies Nd-isotopic homogenisation in the Lewisian complex of northwest Scotland. *Nature* **331**, 705–7.

Whitehouse, M. J., Kamber, B. S. and Moorbath, S. M. (1999). Age significance of U-Th-Pb zircon data from early Archean rocks of west Greenland – a reassessment based on combined ion-microprobe and imaging studies. *Chem. Geol.* **160**, 201–24.

Whitehouse, M. J., Myers, J. S. and Fedo, C. M. (2009). The Akilia controversy: field, structural and geochronological evidence questions interpretations of >3.8 Ga life in SW Greenland. *J. Geol. Soc.* **166**, 335–48.

Wilson, M., Hamilton, P. J., Fallick, A. E., Aftalion, M. and Michard, A. (1985). Granites and early Proterozoic crustal evolution in Sweden: evidence from Sm Nd, U Pb and O isotope systematics. *Earth Planet. Sci. Lett.* **72**, 376–88.

Wust, G. (1924). Florida und Antillenstrom. *Veroffentl. Inst. Meeresh. Univ. Berlin* **12**, 1–48.

Zolnai, A. I., Price, R. A. and Helmstaedt, H. (1984). Regional cross section of the Southern Province adjacent to Lake Huron, Ontario: implications for the tectonic significance of the Murray Fault Zone. *Can. J. Earth Sci.* **21**, 447–56.

Chapter 5

Lead Isotopes

Of the four stable isotopes of lead, only ^{204}Pb is non-radiogenic. The other lead isotopes are the final decay products of three complex decay chains from uranium (U) and thorium (Th). Table 5.1 shows the ultimate parent–daughter decay pairs in this system, of which the highest atomic weight parent (^{238}U) decays to the lowest atomic weight daughter (^{206}Pb) and vice versa. It will be noted that the ^{238}U half-life is comparable with the age of the Earth, whereas the ^{235}U half-life is much shorter, so that almost all primordial ^{235}U in the Earth has now decayed to ^{207}Pb. The ^{232}Th half-life is comparable with the age of the universe. The intermediate members of each series are relatively short-lived, so they have usually been ignored in pre-Tertiary Pb isotope dating studies. With advances in analytical precision, these intermediates have become more significant, but will be discussed later (Section 5.2.7).

If we consider a system of age t (e.g. a granite intrusion which crystallized from a magma), we can write an equation for the nuclides involved in each decay scheme, derived from the general decay equation [1.10]:

$$^{206}\text{Pb}_\text{P} = {}^{206}\text{Pb}_\text{I} + {}^{238}\text{U}\,(e^{\lambda.238\,t} - 1) \qquad [5.1]$$

$$^{207}\text{Pb}_\text{P} = {}^{207}\text{Pb}_\text{I} + {}^{235}\text{U}\,(e^{\lambda.235\,t} - 1) \qquad [5.2]$$

$$^{208}\text{Pb}_\text{P} = {}^{208}\text{Pb}_\text{I} + {}^{232}\text{Th}\,(e^{\lambda.232\,t} - 1) \qquad [5.3]$$

where P indicates the abundance of a given nuclide at the present and I indicates its initial abundance. It is convenient to divide throughout by ^{204}Pb to obtain equations containing isotope ratios rather than absolute nuclide abundances. ^{204}Pb is chosen because it is the only non-radiogenic isotope. Hence for the two uranium decay schemes we obtain

$$\left(\frac{^{206}\text{Pb}}{^{204}\text{Pb}}\right)_\text{P} = \left(\frac{^{206}\text{Pb}}{^{204}\text{Pb}}\right)_\text{I} + \frac{^{238}\text{U}}{^{204}\text{Pb}}\;(e^{\lambda.238\,t} - 1) \qquad [5.4]$$

$$\left(\frac{^{207}\text{Pb}}{^{204}\text{Pb}}\right)_\text{P} = \left(\frac{^{207}\text{Pb}}{^{204}\text{Pb}}\right)_\text{I} + \frac{^{235}\text{U}}{^{204}\text{Pb}}\;(e^{\lambda.235\,t} - 1) \qquad [5.5]$$

Table 5.1	Ultimate parent-daughter pairs of uranium and thorium.		
Decay route		$t_{1/2}$, Ga	Decay const. λ, yr^{-1}
^{238}U $\rightarrow$ ^{206}Pb		4.47	1.55125×10^{-10}
^{235}U $\rightarrow$ ^{207}Pb		0.703	9.8544×10^{-10}
^{232}Th $\rightarrow$ ^{208}Pb		14.01	0.49475×10^{-10}

Data from Section 1.4.1.

5.1 U–Pb Isochrons

In principle, the decay equations [5.4] and [5.5] can be used to construct U–Pb isochron diagrams, and hence to date rocks in a manner analogous to the Rb–Sr system (Section 3.2.2). U–Pb isochrons are subject to similar assumptions as for Rb–Sr, the most critical of which is that the samples remained closed to U and Pb mobility during the lifetime of the system being dated. Unfortunately, the U–Pb system rarely stays closed in silicate rocks, due to the high mobility of U, and to a lesser extent Pb, under conditions of low-grade metamorphism and superficial weathering. For example, in a case study from the Granite Mountains batholith, Wyoming (Fig. 5.1) whole-rock samples suffered disastrous U losses, displacing the data points far from a reference line defined by the intrusive age.

The mobility of uranium greatly limits the application of simple U–Pb isochron dating. However, the unique properties of the U–Pb system, involving two separate decay schemes with common parent and daughter nuclides, mean that age information can be obtained even from disturbed systems. Three dating techniques exploit this situation, the U–Pb 'concordia' method, the common Pb–Pb method and the galena model age method. These methods will be discussed in the subsequent sections of the chapter.

However, recent work has shown that for some Phanerozoic rocks and a very few Precambrian examples, the

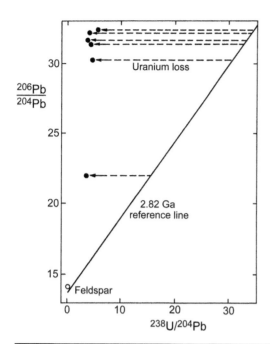

Fig. 5.1 U–Pb isochron diagram for the Granite Mountains batholith showing displacement of whole-rock data points far to the left of the 2.82 Ga reference line, due to disastrous uranium losses. After Rosholt and Bartel (1969).

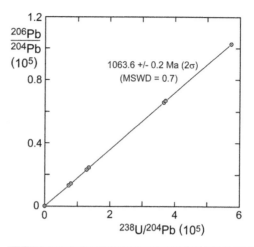

Fig. 5.2 U–Pb isochron diagram for zircons from a late pegmatite in the Grenville Province, Canada, using the decay constant of Jaffey et al. (1971). Errors are comparable with the dots in the centres of the circles. Data from Schoene et al. (2006).

accessory mineral zircon *can* sometimes come close enough to ideal closed-system behaviour that classical U–Pb isochrons can be used as dating tools. A key advance in achieving this degree of ideality is the 'chemical abrasion' method (described in detail in Section 5.2.3). This method can be used to remove domains of a zircon crystal that have suffered open-system behaviour, without disturbing the remainder of the crystal.

5.1.1 U–Pb Isochrons and Decay Constants

As with any other radiometric dating method, U–Pb isochron ages are only as good as the ^{238}U and ^{235}U decay constants. These were measured to high precision by Jaffey et al. (1971) using laboratory counting experiments, and remained unchallenged for many years. However, with improvements in analytical precision, dating of ideal samples began to suggest consistent offsets between U–Pb isochron ages using the two separate decay systems (Ludwig, 1998).

Although ^{235}U has six times the activity of ^{238}U, its 138-fold lower present day abundance is expected to lead to larger uncertainties in the measurement of its decay constant. This was borne out when Mattinson (2010) re-examined the raw data of Jaffey et al. (1971). A new laboratory-based measurement of the ^{238}U decay constant is needed, but in the meantime, the existing value can be used to test the ^{235}U decay constant. After preliminary work by Mattinson (2000), more detailed studies were carried out

by Schoene et al. (2006) and Mattinson (2010) using suites of 'ideal' zircons (e.g. Fig. 5.2). Both of these studies made use of the chemical abrasion method of Mattinson (2005) to remove any traces of zircon with disturbed U–Pb behaviour (Section 5.2.3).

Figure 5.2 shows a ^{238}U–^{206}Pb isochron for zircon from a late Grenville pegmatite with no evidence of U–Pb disturbance. The very high measured ^{238}U–^{204}Pb ratios (ca. 80 000–600 000) can be attributed entirely to U and Pb analytical blanks of ca. 0.1 and 1 pg (10^{-12} g) respectively (Section 2.1.5). In other words, the zircons incorporated essentially zero common Pb during crystallization, and the point near the origin is the composition of the analytical blank. The slope-age of the isochron (1063.6 ± 0.2 Ma, 2σ) can be compared with the analogous result for the ^{235}U–^{207}Pb decay scheme (1064.5 ± 0.2 Ma, 2σ). The difference between the two ages is 0.9 Ma, or 0.08% older for the ^{235}U–^{207}Pb system. Similar offsets were seen for most of the samples analysed by Schoene et al. (2006) and Mattinson (2010), suggesting that an increase in the ^{235}U decay constant is necessary to correct the offset. However, the isotopic composition of uranium is another issue that must be examined.

5.1.2 Uranium Isotope Composition

When calculating a ^{235}U–^{207}Pb age from equation [5.5] it is normal to determine the abundance of ^{235}U from that of ^{238}U using the standard ^{238}U/^{235}U ratio of 137.88 recommended by Steiger and Jager (1977). The recommended value was based on the measured uranium isotope composition of the NBS (NIST) standard 950a (238/235 wt ratio = 139.65), even though this was known at the time to be an outlier from the main

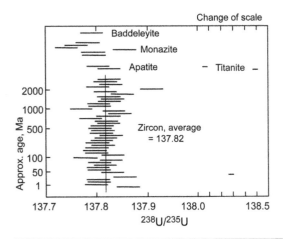

Fig. 5.3 Uranium isotope compositions in zircon and titanite (sphene) samples used for geochronology. After Hiess *et al.* (2012).

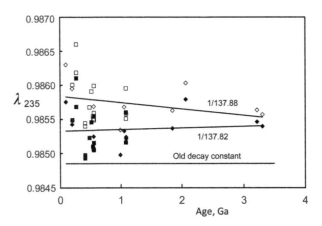

Fig. 5.4 Plot of inferred ^{235}U decay constant against age for 'ideal' zircons, assuming concordance between the two U–Pb decay systems. Diamonds = Schoene *et al.* (2006); squares = Mattinson (2010). Open symbols use the 'old' uranium isotope ratio of 138.88; solid symbols use the average value of 137.818 from Hiess *et al.* (2012).

distribution of ^{238}U/^{235}U ratios in uranium ores (Cowan and Adler, 1976).

Cowan and Adler (1976) had shown that the average composition of 'uranium feed' to the Oak Ridge enrichment plant from several countries was 137.80. This value was supported by gas-source UF$_6$ mass spectrometry by Richter *et al.* (1999) and adopted by the International Union of Pure and Applied Chemistry (IUPAC) four years later (deLaeter *et al.*, 2003). The new value was also confirmed by more precise TIMS analysis using a double ^{233}U–^{236}U spike to correct for instrumental mass fractionation (Richter *et al.*, 2008). However, this knowledge did not reach the isotope geology community until it was 'rediscovered' by Brennecka *et al.* (2010), again based on double-spike TIMS analysis (Sections 2.5.2, 15.8).

It cannot be assumed that zircons will have the same uranium isotope ratio as uraninite. Therefore, this question was tested in an investigation of the uranium isotope composition of accessory minerals used for U–Pb geochronology (Hiess *et al.*, 2012). This study showed that most zircon samples fall close to a mean ratio of 137.818 (Fig. 5.3). However, a few samples, including titanite from the Fish Canyon tuff, had much higher ratios. The fact that Fish Canyon zircon was normal suggests that uranium fractionation can occur in magmatic systems.

It is obviously critical to asses the effects of the uranium isotope composition on ^{235}U–^{207}Pb ages, since this will determine the amount of age discordance (if any) that must be attributed to errors in the decay constant. As pointed out by Mattinson (2010), the two effects are additive, but in a non-linear fashion, because one variable is inside the exponent term and the other is outside. If we rearrange equation [5.5] in terms of t, then we can see this explicitly:

$$t = \frac{1}{\lambda} \cdot \ln\left[\tfrac{^{207}Pb}{^{235}U} + 1 \right] \qquad [5.6]$$

However, if we now rearrange λ and t, we can use this relation to test alternative values of the decay constant by forcing concordance of ^{235}U–^{207}Pb ages with ^{238}U–^{206}Pb ages on 'perfect' zircon samples. This is shown in Fig. 5.4 for the data of Schoene *et al.* (2006) and Mattinson (2010). The data are plotted assuming the 'old' value of the uranium composition (open symbols) and the 'new' value (closed symbols). Using the new uranium composition implies a ^{235}U decay constant of 0.98544×10^{-9} a^{-1}, intermediate between the old value (0.98485×10^{-9}) and the value of 0.98574×10^{-9} a^{-1} proposed by Mattinson (2010). To the extent that a uranium isotope ratio of 137.818 yields a flat (age independent) regression for the decay constant against age, this supports the new ratio. However, the main basis for the new value is its documented widespread occurrence (see above).

5.1.3 U–Pb Isochrons and Timescale Calibration

Until recently, U–Pb systems showing ideal closed-system behaviour were too rare to allow the use of the U–Pb isochron method for timescale calibration. However, the chemical abrasion method of Mattinson (2005), described in detail in Section 5.2.3, now allows nearly ideal zircons to be 'cleaned up' in order to produce ^{238}U–^{206}Pb isochron ages for many Cenozoic and Mesozoic rocks. Because these ages rely only on the ^{238}U decay constant, believed to be the most accurately known of any of the decay constants, they can potentially produce the most accurate ages for timescale calibration. In particular, this method allows the possibility of direct calibration with the 'astrochronology' timescale.

The astrochronology timescale is based on Milankovich theory (Section 12.4.2), which proposes that climatic cycles are based on changes of solar insolation on the Earth due to

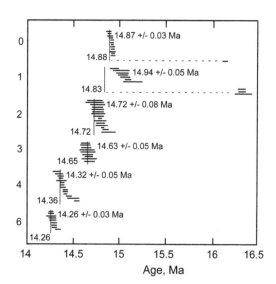

Fig. 5.5 ^{238}U–^{206}Pb zircon isochron ages (horizontal bars) for six tuff horizons in a Miocene sedimentary section, compared with astrochronology ages (vertical lines). U–Pb ages with errors are given for the youngest zircon grain in each horizon. After Wotzlaw et al. (2014).

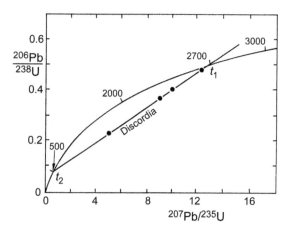

Fig. 5.6 U–Pb concordia diagram showing the concordia line calibrated in Ma, and a discordia line generated by variable Pb loss from U-rich minerals. Ages t_1 and t_2 are upper and lower intersection ages. After Wetherill (1956).

small variations in the Earth's orbit. Since these orbital variations can be back-calculated, this allows the possibility of absolute dating based on proxy records of past climate cycles (such as stable isotope signals). Conventionally, calibration of the astrochronology timescale has been based on Ar–Ar ages (Section 10.3.1). However, these ages must themselves be calibrated against other dated standards, whereas U–Pb isochron ages allow the possibility of direct dating without a calibration step.

The preferred approach (Fig. 5.5) is to date zircons from tuff horizons within an astronomically dated stratigraphy. However, zircons may have grown in a magma chamber for tens of ka before the eruptive age. In an attempt to screen out older zircon xenocrysts, a suite of single zircons is dated from each ash horizon, and the consensus of youngest ages is compared with the astronomical age (vertical lines in Fig. 5.5). In the example shown (Wotzlaw et al., 2014), the two methods gave concordant ages except for one ash layer (#1), for which all zircons ages were too old.

5.2 U–Pb Concordia Dating

Samples showing perfect closed-system behaviour of U and Pb are fairly rare in nature, so the normal challenge is to extract accurate ages from less than perfect systems. In the U–Pb 'concordia' method this involves using the coupled U–Pb decay systems to understand the behaviour of imperfect systems, relative to the ideal of perfectly concordant ages.

In early work on this problem, Ahrens (1955) suggested that although ^{235}U–^{207}Pb and ^{238}U–^{206}Pb ages often appeared to underestimate the true age of a uranium-rich mineral, they might 'converge' on the true age if displayed graphically. In order to do this, he simplified the above equations [5.1] and [5.2] by removing the initial Pb terms. This was reasonable for the very U-rich minerals (such as uraninite) first used for U–Pb dating. Thus, for the ^{238}U–^{206}Pb decay scheme, equation [5.1] is modified to yield the following:

$$^{206}\text{Pb}^* = {}^{238}\text{U} \cdot (e^{\lambda.238\,t} - 1) \qquad [5.7]$$

where Pb* represents radiogenic lead only. Taking ^{238}U to the left side places all of the measured quantities together:

$$\frac{^{206}\text{Pb}^*}{^{238}\text{U}} = (e^{\lambda.238\,t} - 1) \qquad [5.8]$$

This is equivalent to the slope of a two-point U–Pb isochron (Fig. 5.2) between the analysed mineral and the common Pb component. In addition, a similar equation can be derived for the ^{235}U–^{207}Pb system from [5.2] above:

$$\frac{^{207}\text{Pb}^*}{^{235}\text{U}} = (e^{\lambda.235\,t} - 1) \qquad [5.9]$$

Ahrens (1955) plotted ^{238}U/^{206}Pb against ^{235}U/^{207}Pb (the left-hand sides of equations [5.8] and [5.9] above) for uraninite samples from Zimbabwe with discordant U–Pb ages. The result was a good linear array, whose significance was explained by Wetherill (1956). The discordant array is extrapolated until it intersects with the hypothetical compositions of ideal samples with concordant U–Pb ages (Fig. 5.6). These ideal samples define the 'concordia' curve, and the upward extension of the discordant Pb array ('discordia line') intersects with concordia at the true crystallization age of the samples (t_1).

The concordia curve can be drawn by substituting decay constants and successive values of t into the right-hand side of equations [5.8] and [5.9], and plotting the results for each value of t. Hence this plot is termed the concordia diagram.

The slope of discordia lines generally has no age significance. However, assuming a known uranium isotope ratio, a discordia line through the *origin* yields the radiogenic ^{206}Pb/^{207}Pb ratio. This is normally quoted in reciprocal form as the ^{207}Pb/^{206}Pb ratio, equivalent to 1/slope in Fig. 5.6 for a two-point isochron passing through the origin. The resulting '207/206' ages are a useful tool for comparison with upper intersection ages, and will be discussed further below.

Uraninite and monazite were the first minerals used in U–Pb geochronology, in view of their tendency to incorporate large concentrations of uranium but very little initial (non-radiogenic) lead. However, their limited distribution restricts their usefulness. Therefore, the more widely distributed accessory mineral zircon quickly took over as the mineral of choice in U–Pb dating (Silver and Deutsch, 1963). A brief review of alternative dating materials is made in Section 5.2.7.

Because only radiogenic Pb can be plotted on the concordia diagram, a correction must be made for any component of initial ('common') Pb that was incorporated by the zircon when it grew. This is done by measuring the amount of (initial) ^{204}Pb in the zircon and then using the ^{206}Pb/^{204}Pb and ^{207}Pb/^{204}Pb ratios of the whole-rock to estimate the amount of initial ^{206}Pb and ^{207}Pb incorporated in the mineral. This initial Pb is subtracted from the present day ^{206}Pb and ^{207}Pb to yield the radiogenic fraction. For zircons with very low common Pb contents, an adequate correction may be possible by estimating common Pb from a general terrestrial Pb evolution model (e.g. Stacey and Kramers, 1975; Section 5.4.3), rather then by direct analysis of the whole-rock sample.

5.2.1 Lead Loss Models

Early dating work on U-rich minerals soon revealed that most samples yield discordant ^{206}Pb/^{238}U and ^{207}Pb/^{235}U ages. Holmes (1954) attributed this discordance to Pb loss from the crystal lattice. Since that time, much research in U–Pb dating has been devoted to studying the mechanism of lead loss, hopefully to determine accurate ages on samples that have suffered lead loss.

Ahrens (1955) suggested that variable Pb loss from each mineral was due to some kind of continuous diffusional process. This model was elaborated by Russell and Ahrens (1957), who postulated that intermediate members of the uranium decay series were ejected into micro-fissures in the mineral lattice (pitchblende in this case) by the 'recoil energy from α-particle emission.' These nuclides or their decay products must subsequently have been removed by diffusion or by some kind of leaching process.

Wetherill (1956) proposed an alternative explanation, now called the episodic lead loss model. He agreed that the

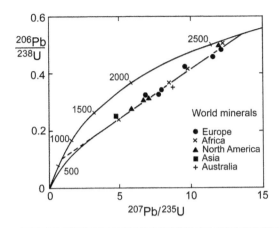

Fig. 5.7 Concordia diagram for Archean U-rich minerals from five continents showing common discordia lower intercept at ca. 600 Ma. Curved line shows expected effect of extreme diffusional lead loss. Dotted line shows extrapolation to an apparent episodic lead loss event. After Tilton (1960).

upper intersection of the discordia line with concordia corresponds to the time of formation of the minerals (t_1). However, he argued that the lower intersection of the discordia and concordia (t_2) also had age significance, representing the time of a thermal event which caused lead loss from the minerals. For the example shown (Fig. 5.6), these episodes are 2700 and 500 Ma respectively. Wetherill supported his model by citing 500 Ma Rb–Sr and K–Ar ages on lepidolite as evidence for a thermal event at that time.

However, Tilton (1960) showed that U-rich minerals with Archean ages from five different continents all lay near a single discordia line with a lower intersection at ca. 600 Ma (Fig. 5.7). Under the episodic lead loss model this would imply a worldwide metamorphic event at 600 Ma, but geological evidence for such an event is lacking. Instead, Tilton proposed that the minerals had undergone continuous diffusional lead loss over geological time, yielding a curve on Fig. 5.7 which for much of its length closely resembles a straight line, only curving downwards to an intersection at the origin for relatively recent lead loss.

Goldrich and Mudrey (1972) developed this diffusional lead loss model by arguing that radiation damage of a U-rich mineral was responsible for the formation of a micro-capillary network in the crystal which would become fluid filled. Pb which diffused into these fluids would be lost from the mineral when uplift of basement rock caused the mineral to dilate and expel the capillary filling fluids. Evidence in support of this 'dilatancy' model was provided by the agreement of various lower intersection ages from North America with times of basement uplift derived from paleogeographical evidence.

Further understanding of lead loss mechanisms came from consideration of the nature of the zircon crystal lattice. For example, Kober (1986) showed that when Pb is evaporated

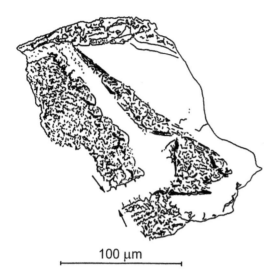

100 μm

Fig. 5.8 Drawing of a metamict zircon showing inward advance of alteration fronts (arrows at left). Unaltered material is white. From a photograph by van Breemen *et al.* (1986).

in situ from zircon grains in the mass spectrometer, discordant Pb can be driven off at low filament temperatures (less than 1350 °C), whereas the concordant Pb fraction is usually emitted between ca. 1400 and 1500 °C (Section 5.2.2).

Based on the high temperatures of concordant Pb emission, Kober (1987) argued that the concordant radiogenic Pb fraction is substituted into the zircon lattice itself, rather than filling defects and voids in the lattice. A stable lattice site would be difficult to envisage for the Pb^{2+} ion, which has an ionic radius (1.18–1.29 Å), much too large to allow substitution for Zr^{4+} (0.72–0.84 Å) or Hf^{4+} (0.71–0.83 Å). However, Pb^{4+} has an ionic radius of only 0.78–0.94 Å, making it a possible candidate for admission into the lattice. Kober (1987) suggested that the emission of β particles during radioactive decay, and the transformation of emitted He^{2+} (α particles) to neutral He, can cause this oxidation.

The microscopic examination of analysed zircon grains now suggests that lead loss from zircons is a fairly 'black and white' process. In other words, unaltered zircon lattices lose very little or no lead, while altered zircon (promoted by metamictization) loses lead very readily. Any given zircon crystal may contain both kinds of material. For example, Fig. 5.8 shows alteration fronts advancing through the metamict U-rich parts of a zircon. In reality, the exact mechanism of lead loss from altered zircon may be different in different circumstances. Hence it is concluded that the lower intersection of a U–Pb discordia should only be attributed age significance if the age is supported by other geological evidence. However, the interpretation of the upper intersection as the age of formation of the zircons is unaffected.

Silver and Deutsch (1963) made a pioneering case study of lead loss from different zircon fractions in a single rock sample. They found that zircons with low uranium contents lost less lead than high-U zircons. This effect was attributed to the greater radiation damage suffered by U-rich grains. In addition to losing lead, metamict zircons tend to incorporate impurities, including iron. Hence magnetic separation of zircons can yield fractions with variable discordance.

In order to obtain the best discordia intersection with the concordia, it is desirable to analyse several zircon fractions with variable discordance, and perform a linear regression on the results. This regression cannot be solved algebraically to yield upper and lower intersection ages; hence these ages are usually calculated iteratively by computer. In order to calculate a regression fit to an array of data displaying some geological scatter, it is common practice to expand analytical error bars to encompass the scatter (Section 2.8.3). However, if lead loss processes have operated at different times in the history of a zircon, the resulting discordia array may fan out somewhat from the upper intercept as the points become more discordant. Therefore, Davis (1982) suggested that, rather than expanding all errors equally to encompass geological scatter, the error bars should instead be magnified in proportion to their discordance.

5.2.2 Air Abrasion and Direct Evaporation

Krogh (1982a,b) argued that instead of refining the mathematical treatment of lead loss models to obtain an accurate upper intersection from discordant zircons, it would be better to remove discordant Pb from the sample before analysis. In early experiments, Krogh and Davis (1975) attempted to remove altered parts of the zircon by hydrofluoric acid leaching prior to analysis. However, they found that Pb was also leached from other parts of the grain. This behaviour has also been seen in more recent leaching studies (e.g. Corfu, 2000; Davis and Krogh, 2000). Therefore, it was concluded that physical rather than chemical methods should be used to remove discordant zones of the zircon crystal.

A technique of physical separation tested by Krogh (1982a) was the use of a very-high-flux magnetic separator, which removes all but the least metamict grains. This was found to be relatively successful (Fig. 5.9), but the most successful approach was to abrade the zircons in a pneumatic mill (Krogh, 1982b). This procedure removes the outer layers of the crystals, which are usually the most U-rich, and hence metamict. Spectacular increases in concordance were obtained in this manner (Fig. 5.9), and the technique has become a standard procedure in zircon geochronology.

The effect of the air abrasion process on prismatic zircon crystals is shown in Fig. 5.10. Cracked and other defective grains (such as that in Fig. 5.8) must be excluded by hand picking under a microscope, since the air abrasion technique cannot remove defective material from the interior of grains.

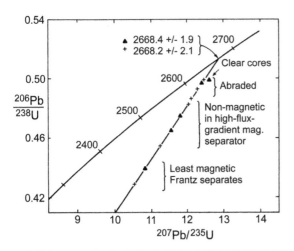

Fig. 5.9 The effect of selecting very non-magnetic zircons, and of abrading off the outer rims, to increase concordance. Symbols (+, ▲) indicate two different rock samples. After Krogh (1982b).

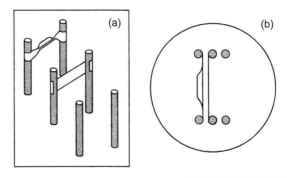

Fig. 5.11 Arrangement of a triple-filament bead for Pb–Pb dating of zircon by the two-stage evaporation method: (a) exploded view; (b) plan view.

An alternative approach to achieving concordancy was pursued by Kober (1986, 1987), using direct evaporation of Pb from a zircon crystal in the mass spectrometer. Kober's method is a two-stage process which represents an improvement on techniques previously tried by other workers (e.g. Gentry et al., 1982). A zircon is wrapped in the side filament of a multiple-filament bead, and the temperature of this filament is raised until Pb evaporates directly from the zircon. Some of this lead is re-deposited on the centre filament of the bead assembly, mounted in front of the evaporation filament (Fig. 5.11). After a deposition period of 5–10 minutes, the side filament is turned off and the centre filament is heated to

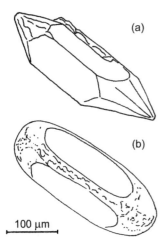

Fig. 5.10 Drawings of zircon grains (a) before and (b) after abrasion. From photographs by Aleinikoff et al. (1990).

re-emit the deposited lead. It is thought that other species evaporated from the zircon (mainly SiO_2) may form a blanket which holds Pb on the centre filament in a manner similar to the silica gel method for direct Pb analysis (Chapman and Roddick, 1994). When the deposited Pb is exhausted, a new deposition step is made (if possible) at a higher side filament temperature.

Kober's Pb evaporation method is based on the premise that discordant lead is contained in less stable lattice sites than concordant lead, and is therefore driven off at comparatively low temperatures. Experiments by Chapman and Roddick (1994) suggested that concordant Pb is released at higher temperatures as a reaction front migrates into the grain, converting zircon into zirconium oxide (baddeleyite). Because no U/Pb concentration measurement is possible in the Kober method, Pb is assumed to be perfectly concordant, yielding only a 207/206 age. If the 207/206 age of each evaporation step is plotted against filament temperature, the ages should increase until a plateau is reached (Fig. 5.12).

The existence of a plateau suggests that Pb emission from the high-temperature steps represents a single phase of lead, rather than mixtures of concordant and discordant lead. The 207/206 ratio should then yield the crystallization age of the zircon. Analysis of very old zircons previously dated by the SHRIMP ion microprobe (Section 5.2.6) gave similar old ages (Kober et al., 1989), but with significant uncertainty. Since only 207/206 ages can be determined, the method is not suitable for metamorphic zircons with a complex Pb-loss history.

5.2.3 Chemical Abrasion and Annealing

Earlier experiments with HF leaching (chemical abrasion) finally led to a reliable dating technique when Mattinson (2005) conceived the idea of 'annealing' zircons in a furnace before leaching. The annealing process heals the radiation damaged lattice in the same way that fission tracks are annealed (Section 16.5).

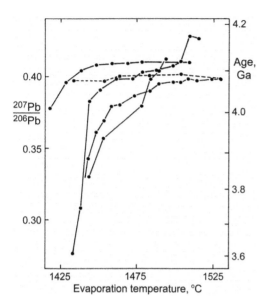

Fig. 5.12 Plot of measured $^{207}Pb/^{206}Pb$ ratios (corresponding to apparent age) against evaporation temperature for five Mount Narryer zircons. Low-temperature domains have suffered Pb loss but high-temperature plateaux are argued to be concordant. After Kober *et al.* (1989).

Mattinson demonstrated the effectiveness of this treatment by performing HF step-leaching experiments on zircons annealed for different periods at different temperatures. Results are shown in Fig. 5.13 in terms of the radiogenic $^{238}U-^{206}Pb$ age (Section 5.1.1) as a function of the

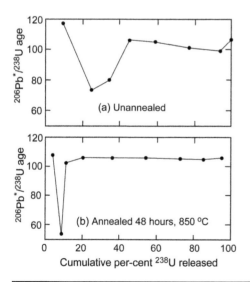

Fig. 5.13 Results of HF leaching experiments on un-annealed and annealed zircons, to show improvements in age reproducibility. After Mattinson (2005).

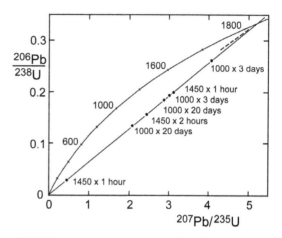

Fig. 5.14 Strongly discordant U–Pb data from the Sudbury felsic norite after CA treatment involving annealing and brief HF leaching before dissolution. After Das and Davis (2010).

percentage of uranium released in each leaching step. The figure shows the results of leaching on two aliquots of Californian Sierra zircons, the first un-annealed, and the second annealed at 850 °C for 48 hours. The effect of annealing is to greatly reduce the amount of uranium released in the early leaching stages, and thereafter generate a 'plateau' of reproducible ages. This shows that the annealing process has stabilized partially damaged parts of the crystal lattice to prevent non-stoichiometric Pb leaching in the earlier stages of the chemical abrasion process. Further experiments showed that annealing for 48 hours at 1000–1100 °C gave the best results.

While the chemical abrasion (CA) method has produced excellent results for relatively concordant Phanerozoic zircons, the treatment of metamorphosed Precambrian rocks is more difficult. Badly damaged zircons from the Sudbury igneous complex represent an extreme example, but may help us to better understand the annealing/leaching process (Das and Davis, 2010). Single zircons were annealed at either 1000 or 1450 °C, and then briefly leached with HF before dissolution. The results displayed very strong Pb loss, but fell consistently close to a single discordia line with a zero age lower intercept (Fig. 5.14). As a result, the analyses yielded relatively consistent 207/206 ages, despite their strong discordance. In contrast, previous conventional TIMS analysis yielded a discordia line with a ca. 400 Ma lower intercept. Das and Davis concluded from this study that there is a threshold of radiation damage above which concordance cannot be restored using the chemical abrasion procedure.

Another example of chemical abrasion applied to ancient rocks (Augland and David, 2015) involved analysis of a felsic (meta-volcanic) schist from the Nuuvuagittuq supracrustal belt of Northern Quebec, known for its ^{142}Nd anomalies (Section 15.7.6). A previous attempt to date the schist using conventional TIMS analysis had yielded somewhat scattered

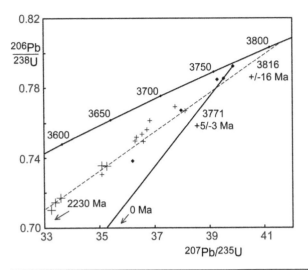

Fig. 5.15 Concordia diagram for conventional and CA–TIMS zircon analyses for a felsic schist from the Nuuvuagittuq supracrustal belt. (+) = conventional data, (♦) = CA data. The dashed line was the original regression age (large crosses). Data from David *et al.* (2009) and Augland and David (2015).

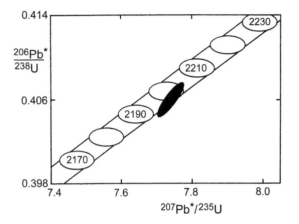

Fig. 5.16 Part of the concordia diagram, showing a nearly concordant data point (solid ellipse) relative to a concordia band that takes decay constant uncertainties into account. After Ludwig (1998).

zircon analyses (crosses in Fig. 5.15) from which the authors (David *et al.*, 2009) had selected a subset claimed to be the 'best quality' zircons (large crosses) with an upper intercept age of 3817 ± 16 Ma (dashed line in Fig. 5.15).

New CA analysis yielded an array with a steeper slope, of which the two most radiogenic zircons lay on a zero-age discordia (solid line), yielding a 207/206 age of 3771 ± 2 Ma (Augland and David, 2015). There is a temptation to adopt this as the age uncertainty of the unit, based on the most concordant point. However, the evidently erroneous 3817 Ma age derived from the earlier TIMS data should warn of the danger of selecting certain points out of a scatter of analyses. Therefore, a conservative treatment recognizes the scatter of the CA data and assigns a wider error limit of 3771 +5/−3 Ma. This is important, because it is difficult to determine what is going on inside a metamorphosed zircon sample during the HF leaching process. To acknowledge the scatter in CA analysis of metamorphosed zircon is to recognize the uncertainties of the process and avoid discrediting the method.

5.2.4 Concordia Ages and Decay Constants

It has often been said over the past 20 years that U–Pb dating represents the 'gold standard' of geochronology, not least because of the high precision of the decay constant determinations of Jaffey *et al.* (1971). However, with the greatly improved quality of upper intercept ages using the physical and chemical abrasion methods, age uncertainties have reached the same range as decay constant uncertainties (0.11% and 0.14%, 2σ for ^{238}U and ^{235}U respectively). Hence, Ludwig (1998, 2000) argued that these uncertainties should

no longer be ignored, and that the quoted error on a typical U/Pb upper intercept age might easily double if they were taken into account.

Many workers have taken the contrary view that provided U–Pb ages are compared only with other U–Pb ages, it is legitimate to ignore decay constant errors. Since U–Pb ages are usually more precise than other absolute dating methods (with a few exceptions, such as Ar–Ar dating in the Tertiary period), this argument is usually reasonable. However, Ludwig (1998, 2000) pointed out that decay constant uncertainties have slightly different effects on age errors using $^{206}Pb/^{238}U$, $^{207}Pb/^{235}U$, $^{207}Pb/^{206}Pb$ and upper intersection ages. He therefore proposed that the concordia line should be represented as a band, whose width takes the decay constant uncertainties into account. The effect of these uncertainties is shown in Fig. 5.16 for a nearly concordant high-precision zircon analysis. Use of the concordia band implies a moderate chance (26%) that the analytical error ellipse is concordant, whereas the probability of concordance falls to only 4% if the concordia is drawn as a single line.

These suggestions were made before development of the chemical abrasion method had improved the quality of zircon samples available for timescale calibration. Since that time, the two high-precision calibration studies summarized in Section 5.1 have been carried out (Schoene *et al.*, 2006; Mattinson, 2010). In addition, evidence has solidified that the widely used uranium isotope ratio (137.88) is wrong. The calibration studies suggest that the concordia curve should be moved significantly to the right, whereas the uranium isotope studies suggest a much smaller movement to the left. Correcting these effects would reduce the calculated $^{207}Pb–^{235}U$ age of a sample by about 0.1%. However, the effect on 207/206 ages is significantly larger. As shown in Fig. 5.17, Late Archean 207/206 ages (ca. 3 Ga) have probably

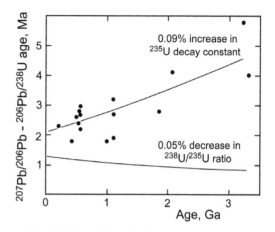

Fig. 5.17 Plot showing the mismatch between 207/206 ages and $^{206}Pb/^{238}U$ ages for ideal samples believed to show no disturbance (data points and best-fit line), along with the offset due to erroneous uranium isotope ratios. (After Mattinson, 2010).

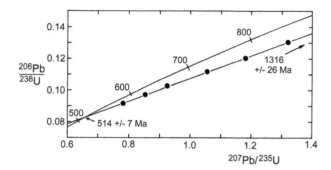

Fig. 5.18 Concordia diagram for Ben Vuirich granite (Scotland) showing mixing between new Caledonian Pb and inherited Mid Proterozoic Pb. After Pankhurst and Pidgeon (1976).

been over-estimated by nearly 5 Ma due to a combination of these effects (Mattinson, 2010). Upper intersection ages may decrease even more, depending on the age of the lower intercept, because they have a more glancing intersection with concordia.

5.2.5 Inherited Zircon

In all of the above studies, it was assumed that the upper intersection with concordia gave the age of igneous crystallization. However, if a magma is derived by partial melting of the crust, or assimilates crustal material, older zircons may be entrained into the magma. These 'inherited' zircons are expected to dissolve in per-alkaline magmas, which have a high Zr saturation level. However, they may survive in per-aluminous melts, especially if cool and dry, due to the low Zr saturation levels of such magmas (Watson and Harrison, 1983). Inherited zircon xenocrysts tend to lose much of their old Pb, and may be overgrown by a new zircon crystal. However, they may still retain a significant proportion of their original Pb. This may be useful as a tracer of source provenance, but it can also complicate interpretation of the crystallization age of the host magma.

A good example is provided by Caledonian-age granites from Scotland, such as the Ben Vuirich granite (Pankhurst and Pidgeon, 1976). In this case, most zircon analyses were concentrated near the lower intercept of a discordia line (Fig. 5.18), whose intersection age (514 ± 7 Ma) was interpreted as the age of intrusion. On the other hand, extrapolation to the upper concordia intersection (1316 ± 26 Ma) gave an approximate age of assimilated old crustal material.

This study was extended by Pidgeon and Aftalion (1978) to include U–Pb analysis of 24 Caledonian granites from Scotland and northern England. Of this suite, 17 plutons with

inherited zircon lay within the Scottish Highlands, while only one, with S-type chemistry, lay to the south (the Eskdale granite). In contrast, all six granites without inherited zircon lay south of the Highland Boundary Fault. In view of the similar chemistry of granites on both sides of the fault, Pidgeon and Aftalion concluded that there must be a fundamental difference in granite source rocks between the Scottish Highlands and crust to the south; a conclusion supported by whole-rock Nd isotope analysis (Halliday, 1984).

The early study of Pankhurst and Pidgeon (1976) made use of bulk zircon separates (total quantity of zircon separated was 8 g!). In an attempt to refine and test the old determination, Rogers et al. (1989) re-dated the pluton using modern techniques of miniature sample analysis and air abrasion. The results (Fig. 5.19) were startlingly different. Abraded and non-abraded needle-shaped grains (# 1–3) defined a Pb-loss

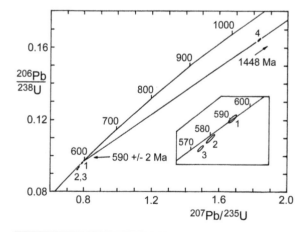

Fig. 5.19 Concordia diagram for Ben Vuirich granite showing a discordia between needle-shaped magmatic zircons (1–3) and stubby inherited zircons (4). Inset shows a Pb-loss line that defines the intrusive age. 1 & 4 = strongly abraded; 2 = slightly abraded; 3 = unabraded, to control Pb-loss line. After Rogers et al. (1989).

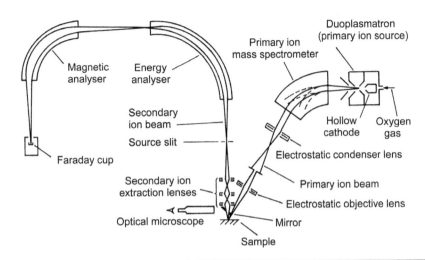

Fig. 5.20 Schematic illustration of a secondary ion mass spectrometer (SIMS) showing the components of the negative ion gun and double focussing analyser. After Potts (1987).

line with an upper intercept of 590 ± 2 Ma interpreted as the age of intrusion (76 Ma older then the previous determination). In contrast, abraded stubby grains defined an inherited Pb discordia, with a more precise upper intercept of 1448 ± 7 Ma (132 Ma older than the previous determination).

The younger ages determined from the earlier study must be attributed to the effects of secondary lead loss after intrusion, from a system that already represented a two-component mixing line. This caused rotation of the apparent discordia, yielding erroneously young ages for both upper and lower intercepts. This study is typical of recent work showing the dangers of bulk zircon analysis in rocks with complex geological histories. Such samples can yield discordia of high statistical quality which nevertheless yield erroneous ages. Therefore, samples showing indications of complex inheritance processes are best analysed by *in situ* methods (see below).

5.2.6 *In Situ* Analysis

In the 1980s, U–Pb dating of zircons with complex histories was revolutionized by *in situ* analysis using an ion microprobe. The general configuration of such an instrument is shown in Fig. 5.20. A beam of light ions (e.g. O^-) is used to bombard and sputter a polished section of a zircon grain to yield a secondary beam of Pb ions (hence the term secondary ion mass spectrometry or 'SIMS'). Pb ions are analysed in a double (electrostatic and magnetic) focussing mass spectrometer (Section 2.2.2). The electrostatic analyser is necessary because emitted secondary ions have a range of energies which would yield bad peak tails in the mass spectrum if not re-focussed in the energy analyser (Fig. 5.20).

A major problem in SIMS analysis is the interference of sputtered molecular ions on the masses of atomic species. In the case of Pb isotope analysis of zircon, this is caused by species such as HfO_2^+, which have almost exactly the same mass as the Pb isotopes, causing isobaric interference (Hinton and Long, 1979). To overcome this problem, the instrument (including magnet) must have a very large physical size, allowing a large spatial separation between the different masses. A resolution of one mass unit in several thousand allows the separation of Pb from molecular ion interferences based on their 'mass defect' (Fig. 5.21). This is the phenomenon by which small variations in atomic mass result from the varying nuclear binding energies of different atoms. The most successful example of a SIMS instrument used for U–Pb dating is the 'sensitive high-resolution ion microprobe' or 'SHRIMP' developed at the Australian National University (Compston *et al.*, 1984).

In addition to molecular ion interferences, another problem in accurate U–Pb dating by SIMS is U/Pb fractionation during the sputtering process. This is caused by the conversion of most of the emitted U^+ ions into the UO^+ species. However, Compston *et al.* (1984) observed a good correlation between the Pb/U and UO^+/U^+ ratios during the analysis of zircon standards of given age (Fig. 5.22). Hence, if we assume that unknown zircons experience the same fractionation effect, the correlation line for the standard can be used with the measured UO^+/U^+ ratio of the unknown to correct the U/Pb ratio of the unknown. This procedure, although empirical, was found to yield reliable results on a variety of standards, thus validating its accuracy (Williams and Claesson, 1987).

An important example of the SHRIMP's capability as a dating tool was provided by the reconnaissance search for vestiges of the early Earth, preserved in ancient sediments (Froude *et al.*, 1983). Zircons were selected from a formation of Archean quartzites surrounded by 3.6 Ga gneisses at Mount Narryer in Western Australia. These metasediments were therefore considered to be possible candidates to contain a very old component derived from a pre-3.6 Ga source.

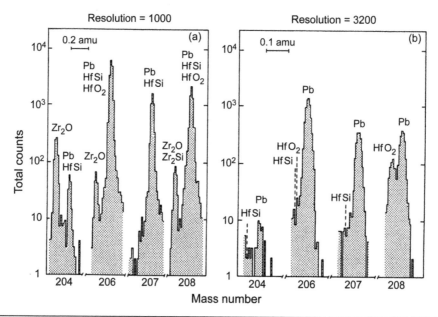

Fig. 5.21 Use of a high-resolution mass spectrometer to separate Pb from interfering molecular ion signals. (a) Low resolution, ca. 1000; (b) high resolution, ca. 3200. After Hinton and Long (1979).

Different areas of single zircon crystals were analysed using the SHRIMP, yielding relatively concordant results. Many ages were in the range 3–3.8 Ga, but a few grains gave ages of 4.1–4.2 Ga (Fig. 5.23).

Some of the Mount Narryer zircon spots fell above the concordia line, displaying what is termed 'reverse discordance'. This phenomenon is common if whole-rock compositions are plotted on a concordia diagram. This is sometimes done for uranium ore deposits, and in that case it is usually caused by uranium loss. Froude *et al.* considered whether a process such as U loss could have caused the data to migrate back up the concordia to yield a spuriously old age. In theory, U loss from 3.7 Ga zircons dur-

ing a Late Archean metamorphic episode could have caused points to move to the right and above concordia. This would have to be followed by recent Pb loss (bringing them back down onto the concordia). However, Froude *et al.* argued that the scatter of data points was too small to be consistent with this model. In addition, Isua zircons were analysed to test the reliability of ion microprobe analyses for complex metamorphic terranes. These gave the expected age of 3.8 Ga.

An ion microprobe study on zircons from Mount Sones, Enderby Land, Antarctica, helps to explain the phenomenon

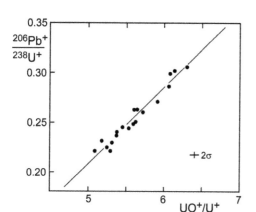

Fig. 5.22 Plot of measured $^{206}Pb/^{238}U$ ratio against UO^+/U^+ for a Sri Lankan zircon standard, to allow calibration of U/Pb ratios in unknown samples. After Compston *et al.* (1984).

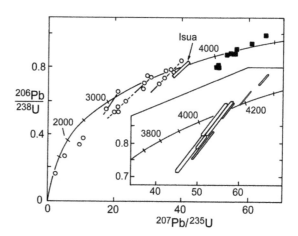

Fig. 5.23 Concordia diagram for ion microprobe analyses of zircon from Mount Narryer quartzite. (■) = very old zircons (inset shows error ellipses). Isua zircons (box) were analysed as a calibration check. After Froude *et al.* (1983).

of reverse discordance in ion microprobe analyses (Williams *et al.*, 1984). Uranium concentrations were found to vary quite smoothly as ion sputtering deepened the analysis spots, whereas lead concentrations varied erratically, giving rise to sudden variations in Pb/U ratio. Hence Williams *et al.* suggested that reverse discordance is due to migration of radiogenic Pb between different regions within a zircon crystal, rather than U loss. Subsequent to these studies, zircons with even older ages (4.40 Ga) have been found in the Jack Hills conglomerate of Western Australia, this time with a range from concordant to normally discordant compositions (Wilde *et al.*, 2001; Valley *et al.*, 2014).

The development of laser ablation as an *in situ* analysis tool for use with inductively coupled plasma–mass spectrometry (ICP–MS) allows this technique to also be used for U–Pb dating (Section 2.7.1). This method has suffered from similar problems of U/Pb fractionation during ablation, limiting its reproducibility to ca. ±3% on dated standards. However, recent work has shown that the zircon annealing process of Mattinson (2005) may greatly reduce these problems. With further improvements in precision expected from multiple collector analysis (Section 2.2.2), U–Pb dating by LA-ICP-MS has now reached the point where it may supersede the SHRIMP in precision and accuracy.

5.2.7 Alternative U–Pb Dating Materials

Zircon has been the mineral of choice for most U–Pb dating work since the earliest studies. However, other minerals may yield valuable U–Pb age data that complement U–Pb zircon ages. The most important of these other minerals are monazite, sphene (titanite) and baddeleyite. In addition, other minerals such as garnet can be used for U–Pb dating in particular circumstances.

Monazite is a light rare earth element (LREE) phosphate that also incorporates significant Th and minor U. It is found in relatively Ca-poor and Al-rich granitoids, and in high-grade metamorphic rocks. It co-exists with zircon, but not with sphene, in many of these rock types. Monazite can show similar Pb inheritance and Pb loss behaviour to zircon (Copeland *et al.*, 1988). However, its distinct chemistry means that these types of behaviour often occur under different conditions to those for co-existing zircon. Thus, monazite has a lower blocking temperature than zircon (Dahl, 1997), which makes any inherited monazites in a crustal melt tend to lose their Pb during the melting event. Because this resets the age clock, monazites can be useful to date aluminous granitoids whose zircon age signatures are complicated by inheritance. On the other hand, despite its lower blocking temperature, monazite seems to be more resistant to Pb loss during low-temperature events. This is probably because monazite, unlike zircon, undergoes annealing at relatively low temperatures, thus healing previous radiation damage to the lattice (Smith and Giletti, 1997).

These properties of monazite were applied by Scharer (1984) to date the Himalayan Makalu granite, whose zircon systematics are complicated by a combination of inherited

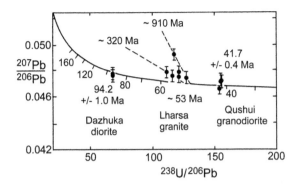

Fig. 5.24 'Tera–Wasserburg' concordia diagram on axes of $^{207}Pb/^{206}Pb$ against $^{238}U/^{206}Pb$ showing data for Himalayan granites. After Scharer *et al.* (1984).

Pb and Pb loss. Monazites in this granite were not affected by these problems. The data are shown in an alternative form of the concordia diagram pioneered by Tera and Wasserburg (1973, 1974), where $^{238}U/^{206}Pb$ is plotted directly against $^{207}Pb/^{206}Pb$ (Fig. 5.24). This concordia has a different curvature to the conventional presentation, and is preferred by workers dating young rocks, because it displays young discordia lines more clearly than the conventional concordia diagram. The Tera–Wasserburg diagram also avoids the issue of error correlation seen in the conventional diagram, caused by the better precision of Pb isotope ratios relative to U/Pb ratios. The correlation of errors is taken into account when fitting discordia regression lines (Section 2.8.2), but is largely avoided in the Tera–Wasserburg diagram.

A complication in the U–Pb dating of monazite arises from its affinity for Th, causing uptake of a significant content of the short-lived U-series isotope ^{230}Th. This decays to ^{206}Pb (Section 12.1), causing an excess abundance of this isotope (Ludwig, 1977). Scharer demonstrated that a correction for this excess production caused apparently discordant analyses to fall properly on the concordia line (Fig. 5.25), yielding a precise age of 24 ± 1 Ma for crystallization of the Makalu pegmatitic granite. Other monazite dating applications were described by Parrish (1990) and Foster *et al.* (2002).

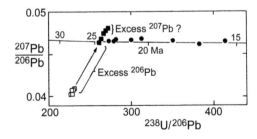

Fig. 5.25 Tera–Wasserburg concordia diagram for the Makalu leucogranite, Himalayas, showing zircon (•) which has lost Pb, and monazite (□, ■) before and after correction for inherited U/Th disequilibrium. After Scharer (1984).

The large Th content of many monazites also allows the possibility of Th–Pb dating. This has been quite widely applied, using the electron microbe to determine total Th–Pb 'chemical age dates' (e.g. Montel *et al.*, 1996). This method is based on the principle that Th abundances in monazite are so high that radiogenic ^{208}Pb totally dominates over uranogenic and non-radiogenic Pb. Chemical dating can only be used as a reconnaissance technique because significant errors arise from uranogenic Pb. However, more accurate Th–Pb dating of monazite can be carried out using the ion microprobe.

Unlike U–Pb dating, the Th–Pb dating method does not allow for an internal correction for Pb loss events. However, *in situ* depth profiling of monazite grains by ion microprobe can allow cooling curves to be determined, based on Pb loss from the grain surface. Grove and Harrison (1999) demonstrated this technique on Tertiary monazites from the hanging wall of the Himalayan Main Central Thrust (MCT). By matching a model for diffusional Pb loss from the grain surface with the variation of Th–Pb ages against depth, Grove and Harrison were able to model the cooling history of the hanging wall since 12 Ma, the average Th–U age derived from the interiors of monazite grains. Additional applications of *in situ* monazite analysis were described by Catlos *et al.* (2002).

Sphene is a titanium silicate (hence often called titanite) with similar properties to zircon and monazite. It has a somewhat lower blocking temperature (ca. 625 °C) than monazite (ca. 715 °C) and zircon (ca. 900 °C; Dahl, 1997). Therefore, sphene may remain open to Pb diffusion during high-temperature cooling of metamorphic terranes. However, it is much less susceptible to *low-temperature* Pb loss than zircon, because it easily recrystallizes, allowing annealing of radiation damage. Sphene was first applied as a dating tool by Tilton and Grunenfelder (1968), and has since been applied widely to date thermal events in poly-metamorphic belts (e.g. Tucker *et al.*, 1987). A good example is the use of combined zircon and sphene ages to date both the formation age and the Caledonian metamorphic age of gneisses from western Norway (Fig. 5.26).

Mafic igneous rocks have very low contents of zircon, monazite and sphene, and therefore have always been difficult to date accurately. However, Krogh *et al.* (1987) showed that the zirconium oxide mineral baddeleyite could be used as a dating tool in these rock types. (In pronouncing 'baddeleyite', it should be remembered that the mineral is named after Baddeley.) This method has since been widely applied to U–Pb dating of mafic rocks (e.g. Heaman and LeCheminant, 1993). In addition, French *et al.* (2002) have shown that total U–Pb analysis of baddeleyite using the electron microprobe can be used as a reconnaissance dating tool for dyke swarms, when supported by conventional U–Pb dating of selected samples.

Garnet is an important mineral in the geothermometry and barometry of metamorphic rocks. Therefore, the direct dating of this material would allow constraints to be

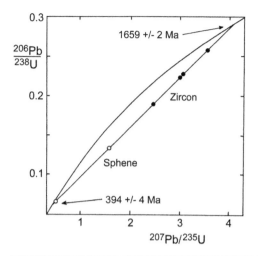

Fig. 5.26 Concordia diagram for migmatite of the Western Gneiss Region (Norway), showing a discordia line defined by zircons (•) which have suffered partial Pb loss, with a lower intercept anchored at the time of Caledonian metamorphism by sphene (○). After Tucker *et al.* (1987).

placed on heating and cooling rates during regional metamorphism. Both the Sm–Nd and U–Pb systems have potential for dating metamorphic garnets, but they have different strengths and weaknesses. Garnets grown under amphibolite facies conditions (ca. 550 °C) can give concordant ages of prograde garnet growth from the two methods (Section 4.1.4). However, Mezger *et al.* (1992) argued that the Sm–Nd system was opened at ca. 600 °C (upper amphibolite facies) in all but very large inclusion-free grains. Therefore, they suggested that garnet Sm–Nd ages usually date cooling rather than prograde mineral growth.

Mezger *et al.* (1991) proposed a higher closure temperature of ca. 800 °C for the U–Pb system in garnet. However, this system suffers from the tendency for uranium to be concentrated in minute inclusions, rather than in the garnet lattice. For example, in the first study of this type, on the Pikwitonei granulite terrane in northern Manitoba, Mezger *et al.* (1989) attempted to use the U–Pb system to date prograde garnet growth during prolonged Late Archean metamorphism. After correction for a small common Pb component, radiogenic ^{206}Pb*/^{238}U and ^{207}Pb*/^{235}U ages were calculated using equations [5.8] and [5.9]. Unfortunately, most samples gave discordant ages, indicative of open-system behaviour. A later study (DeWolf *et al.*, 1996) showed that this was due to Pb loss from micron-sized monazite inclusions in the garnet grains.

5.3 Pb–Pb Dating

The corcordia method is only applicable to systems with negligible common Pb. However, there are many systems containing a common Pb component for which the two U–Pb decay systems can still be usefully combined to derive age information. The first step in this process is to rearrange the U–Pb decay equations [5.5] and [5.4] to bring the initial Pb/Pb terms to the left-hand side:

$$\left(\frac{^{207}Pb}{^{204}Pb}\right)_P - \left(\frac{^{207}Pb}{^{204}Pb}\right)_I = \frac{^{235}U}{^{204}Pb}\ (e^{\lambda_{235}t} - 1) \quad [5.10]$$

$$\left(\frac{^{206}Pb}{^{204}Pb}\right)_P - \left(\frac{^{206}Pb}{^{204}Pb}\right)_I = \frac{^{238}U}{^{204}Pb}\ (e^{\lambda_{238}t} - 1) \quad [5.11]$$

Nier *et al.* (1941) showed that if these two equations refer to the same system, equation [5.10] can be divided by [5.11], and the ^{204}Pb terms in the right-hand side of the equations cancelled, leaving the term $^{235}U/^{238}U$ (now recognized to have a relatively constant value of 1/137.8). This yields the simplified equation

$$\frac{\left(\frac{^{207}Pb}{^{204}Pb}\right)_P - \left(\frac{^{207}Pb}{^{204}Pb}\right)_I}{\left(\frac{^{206}Pb}{^{204}Pb}\right)_P - \left(\frac{^{206}Pb}{^{204}Pb}\right)_I} = \frac{1}{137.8} \cdot \frac{(e^{\lambda_{235}t} - 1)}{(e^{\lambda_{238}t} - 1)} \quad [5.12]$$

This equation is now applicable to dating a variety of systems with common Pb components, without the need for determining the U/Pb ratios of the samples. If we consider a number of samples that have the same age and initial isotopic composition (e.g. whole-rock samples of granite), then it can be seen from equations [5.10] and [5.11] that they will develop different Pb isotope compositions at the present day, according to their individual U/Pb ratios. Therefore, if the present day Pb isotope compositions of this suite are plotted (left-hand side of equation [5.12]), they should form a straight-line array, provided that they have remained as closed systems. The slope of this array, first termed an 'isochrone' by Houtermans (1947), depends only on t, and does not require any knowledge of the U and Pb concentrations in the samples. It should be noted that the isochron equation [5.12] is 'transcendental'. In other words the term on the right-hand side (equal to the slope) cannot be solved algebraically to yield the age, t, but must therefore be solved iteratively.

Since the closed U–Pb system requirement remains, it might be wondered what advantage this method offers over the discredited whole-rock U–Pb isochron method (Section 5.1), in view of the known high mobility of uranium. This question can be answered empirically. Figure 5.27 shows a whole-rock Pb–Pb isochron diagram for the Granite Mountains, Wyoming (Rosholt and Bartel, 1969), which gives a geologically correct age of 2.82 Ga. However, it was shown in Fig. 5.1 that these samples had suffered disastrous uranium losses. This can be explained by the fact that U–Pb whole-rock systems were effectively closed from the time of formation of the intrusion until very near the present day, when uranium was lost in recent weathering processes. This invalidates the U–Pb isochron method, but since the Pb isotope ratios in the rock reflect the pre-weathering U concentrations, they are not upset by the recent alteration event, and therefore yield the true age of the rock.

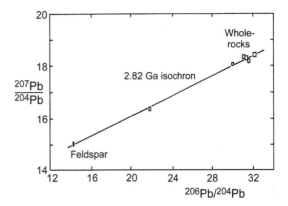

Fig. 5.27 Pb–Pb isochron diagram for whole-rock and mineral samples of the Granite Mountains batholith. After Rosholt and Bartel (1969).

5.3.1 The Age of the Earth and Pb Paradox

The first application of the common Pb–Pb dating technique was actually on meteorites rather than terrestrial rocks, but it also provided the first accurate estimate of the age of the Earth. In this study, Patterson (1956) calculated a Pb–Pb age of 4.55 ± 0.07 Ga on a suite including two chondrites, an achondrite, and two iron meteorites (Fig. 5.28). The least radiogenic of these samples was troilite (FeS) from the Canyon Diablo 1AB iron meteorite, responsible for Meteor Crater, Arizona. The U/Pb ratio measured on this sample (0.025) was so low that Patterson concluded that 'no observable change in the isotopic composition of lead could have resulted from radioactive decay after the meteorite was formed.' Hence, Canyon Diablo troilite is taken as the primordial Pb isotope composition of the solar system, and

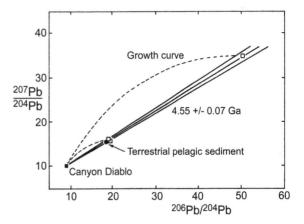

Fig. 5.28 Pb–Pb isochron diagram for iron and stony meteorites (■, □) and a 'Bulk Earth' sample of oceanic sediment (●), showing that the Earth lies on the meteorite isochron, therefore also called the 'geochron'. After Patterson (1956).

provides an important benchmark for terrestrial Pb isotope evolution. This result was recently superseded by Pb analysis of another 1AB iron meteorite, Nantan (Blichert-Toft et al., 2010). However, for most purposes, these results can be considered to be identical.

Patterson's major breakthrough was in using the meteorite isochron to provide an age for the Earth, a problem that had perplexed geochronologists for decades. A sample of recent oceanic sediment, regarded as the best estimate of the Bulk Earth Pb isotope composition, lay on the meteorite isochron, and furthermore had the appropriate U and Pb concentrations to be generated by radiogenic Pb growth from the Canyon Diablo composition in 4.55 Ga. This therefore provided evidence that the Earth has both the same age and ultimate origin as meteorites. The meteorite isochron was therefore termed the geochron.

Subsequent work has shown that the interpretation of pelagic sediment as a Bulk Earth composition is only a rough approximation to the complexities of terrestrial Pb isotope evolution (see below). Similarly, its fortuitous U/Pb ratio, which appeared to support this Pb isotope composition since 4.5 Ga, should be considered a fluke.

In reality, most modern Pb isotope signatures lie to the right of the geochron, just as modern Nd isotope signatures lie off the chondritic evolution line (Section 4.2). The mismatch between modern terrestrial Pb and the geochron (which should represent the Bulk Earth), has been called the 'lead paradox'. It has been studied from several viewpoints, including galena ores (Section 5.4) and modern oceanic volcanics (Section 6.3.1). Several explanations have been proposed, more than one of which could be correct. However, a key to solving the Pb paradox is to examine the *origins* of the Earth, which cannot easily be achieved by studying modern Pb isotope signatures that result from complex mixing processes over Earth history. Therefore, the focus here will be on Earth's oldest rocks.

A key piece of evidence for this problem was a galena Pb analysis from some of the oldest terrestrial rocks in the Isua Supracrustal Belt of Western Greenland (Appel et al., 1978). This analysis can be combined with data for younger galena ores (Section 5.4.2) to reconstruct the terrestrial Pb isotope growth history. However, the curved arrays inherent in the Pb–Pb system make it hard to evaluate the goodness of fit of terrestrial Pb isotope evolution models. Therefore, to simplify the data presentation, Albarede and Juteau (1984) analysed each of the U–Pb systems (and Th–Pb) on a separate diagram of Pb isotope ratio against time, as practised for Nd (Section 4.2). However, for the U–Pb system (unlike Nd), the time dimension must be presented as an exponent (Fig. 5.29) in order to achieve linear evolution lines for the relatively short half-lives of U relative to the age of the Earth.

Fitting a linear growth line to the Pb data in Fig. 5.29 is equivalent to assuming a constant U/Pb ratio (or more precisely a $^{238}U/^{204}Pb$ or 'μ' value of around 9). This causes the terrestrial evolution line to intersect the Canyon Diablo Pb composition at an apparent age of 4.4 Ga (in close agreement

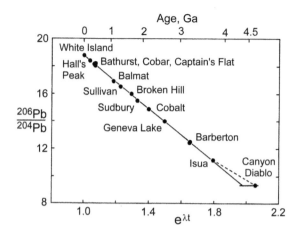

Fig. 5.29 Exponential plot of $^{206}Pb/^{204}Pb$ evolution against time, to test the fit of galena sources to a linear isotopic evolution trend. After Albarede and Juteau (1984).

with earlier calculations by Doe and Stacey, 1974 and Manhes et al., 1979). This is seen most clearly for ^{206}Pb (Fig. 5.24) but also less strongly for ^{207}Pb. However, terrestrial accretion at a date as late as 4.4 Ga is inconsistent with the 4.55 Ga age of differentiated meteorites (Section 3.2.4). Therefore, an Early Archean reservoir with a lower μ value must be postulated. This argument does not rule out other mechanisms to explain more recent Pb isotope redistribution in the Earth, but it means that these other models must be built on top of a viable early-Earth Pb evolution model.

The problem was re-examined by Allegre et al. (1995), who pointed out that bulk chondrites, which are normally regarded as equivalent to the Bulk Earth, have much lower U/Pb ratios than Earth's mantle. When these ratios are expressed as $^{238}U/^{204}Pb$ (μ values), chondrites have a μ value of only 0.15, compared with the Bulk Silicate Earth (BSE) value of around 9 deduced from the Pb compositions of galenas. Allegre et al. suggested that this difference could be due to Pb loss as a volatile during terrestrial accretion, or to Pb partition into the Earth's core.

Neither of these models can be constrained directly, so they are usually approached by comparing Pb with other volatile and siderophile elements. Based on such comparisons, Galer and Goldstein (1996) proposed a two-stage Pb fractionation model. They argued that terrestrial Pb was first depleted by a factor of about 5 (relative to chondrites) by volatile loss from the accreting Earth, yielding a μ value of around 0.7. It was then further depleted by a factor of 13 by Pb partition into the core, yielding a final μ value of 9. Hence they suggested that the ca. 4.45 Ga apparent Pb age of the Earth corresponds to the end of the process of core formation, and that the Pb paradox was a natural outcome of gradual core growth over ca. 80 Ma.

The Hf–W extinct nuclide system was applied to this problem in the late 1990s, but with confusing results

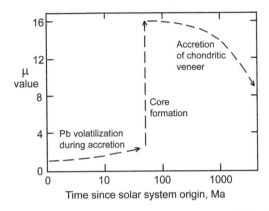

Fig. 5.30 Possible three-stage evolution of μ values in the early Earth: (1) Pb loss by volatilization during accretion; (2) Pb partition into the core; (3) Pb enrichment by late accretion of chondritic material. After Lagos et al. (2008).

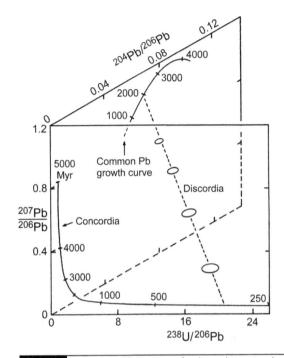

Fig. 5.31 Schematic illustration of a three-dimensional total-Pb/U isochron. The discordia intersects with the concordia curve at its radiogenic end, and with the common Pb growth curve at its non-radiogenic end. After Ludwig (1998).

(Section 15.6.4). Initially, this was due to erroneous W isotope analyses, which implied late core formation after Hf became extinct. After these analyses were corrected, early core formation was implied, around 30 Ma after the origins of the solar system. However, subsequent work has shown that the Hf–W age of core formation is much more model dependent than originally appreciated. The Hf–W data can accommodate a final phase of core formation up to 100 Ma after initial Earth accretion, due to the proposed giant Moon-forming impact (Section 15.6.6). In this scenario, the Pb–Pb age of the Earth can be associated with a final transfer of Pb from the mantle to core during the giant impact (Wood and Halliday, 2005).

Such modelling has been further complicated by proposed late volatile accretions to the Earth by intense meteorite bombardment (Lagos et al., 2008). According to this model, the Moon-forming impact may have led to a mantle μ value as high as 16 (Fig. 5.30), which was subsequently reduced by the later addition of low-μ chondritic material. It would require even greater Pb fractionation into the core to reconcile these two processes. Lagos et al. argued that this was not possible because Pb partition coefficients are too low. However, Wood and Halliday (2010) argued that the experiments of Lagos et al. were biased by carbon-saturated conditions, lowering the apparent Pb partition into the metal phase by nearly two orders of magnitude. Hence they argued that Pb partition into the core is adequate to increase the mantle μ value by the necessary amount to explain the Pb paradox.

Although most workers now agree that a late bombardment added a veneer of volatile-rich material, the magnitude of this process is still in dispute. Albarede (2009) argued that the late bombardment contributed 99% of mantle Pb, and therefore that the Earth's Pb–Pb age dates the late bombardment itself. However, Wood et al. (2010) argued that the bombardment only had a moderate effect on the Pb inventory of the mantle, so that the apparent Pb–Pb age of the

Earth dates the last phase of core formation during the giant impact. This question remains unresolved.

5.3.2 Meteorite Dating and the Total Pb Isochron

Because ^{235}U was relatively abundant in the early solar system, and because of its relatively short half-life of 704 Ma, Pb–Pb dating on meteorites can provide very precise ages for the early evolution of the solar system. However, for a proper appreciation of error sources in this method, it is necessary to understand the relationship between the Pb–Pb and U–Pb dating methods. This was achieved when Wendt (1984) developed the Tera–Wasserburg concordia plot into a three-dimensional U–Pb diagram, by adding an axis of ^{204}Pb/^{206}Pb to represent the common Pb component in a dating suite. Examples of the application of the method to partially open terrestrial systems were provided by Carl and Dill (1985) and Carl et al. (1989). However, its application to meteorite dating is probably more significant.

Ludwig (1998) termed the three-dimensional U–Pb diagram the 'total-Pb/U isochron' diagram, and argued that its main usefulness is to allow a more explicit understanding of error sources. For example, a schematic three-dimensional view of a total-Pb/U isochron in Fig. 5.31 shows that it has two anchor points. The radiogenic end of the isochron

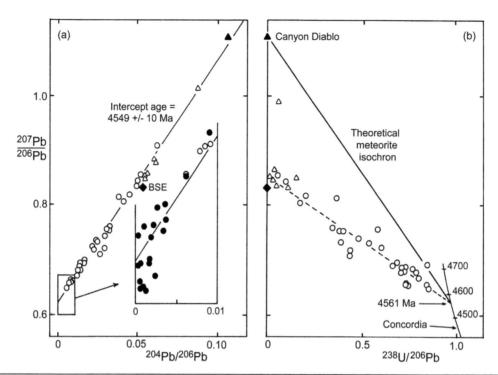

Fig. 5.32 Meteorite mineral data on (a) Pb–Pb isotope and (b) Tera–Wasserburg diagrams, representing the back and front faces of the total-Pb/U isochron diagram. The lower (radiogenic) end of the meteorite discordia shows the effect of Pb loss, while the upper (unradiogenic) end shows the effect of terrestrial contamination. (o) = whole-rock chondrules; (•) = phosphates; triangles = irons. After Tera and Carlson (1999).

intersects the Tera–Wasserburg concordia to define the age of a set of uranium-rich samples, while the other end of the isochron describes the non-radiogenic component of the samples, which should lie close to a reasonable terrestrial (or solar) system common Pb growth curve.

When U–Pb ages are calculated on a concordia diagram with a correction for common Pb, this is equivalent to forcing the three-dimensional isochron through a point on the common Pb growth curve. For samples with low common Pb contents, such a forced fit may actually be more reliable than a 'free fit'. However, when common Pb contents are large, forcing the isochron through an assumed common Pb composition may introduce errors if an incorrect common Pb point is used. A good example comes from the analysis of U–Pb data for 'whole-rock' chondrules from chondritic meteorites (Tera and Carlson, 1999). In this example, raw U–Pb isotope data, without any common Pb correction, are presented on two separate diagrams (Fig. 5.32a, b), based on the x and z axes of the total-Pb/U isochron diagram in Fig. 5.31. This is done to ease the plotting of the data, but the conclusions are the same as for the true 3D diagram.

The Pb–Pb face of the total U–Pb isochron diagram (Fig. 5.32a) has been called the inverse Pb/Pb isochron diagram. Pure radiogenic points fall on the y axis, and their inter-

cept gives the 207/206 age of the analysed meteorite samples. Hence, the most precise ages are given by Pb data points near the y axis. However, the inset shows the effects of Pb loss from meteorite phosphate grains (solid circles) at the radiogenic end of the array. This open-system behaviour limits the precision of the age, and shows that phosphates (which are the result of early secondary alteration) cannot be used to date the oldest objects.

On the Tera–Wasserburg diagram (Fig. 5.32b) whole-rock chondrules define an array with a concordia intersection age of 4561 Ma. This age approximates the correct age of chondrule formation (Section 15.6), but many points in the array project towards a common Pb component similar to present day terrestrial Pb rather than the expected primordial solar system composition. The conclusion from this analysis is that the chondrules must have been partially contaminated by terrestrial Pb. Therefore, to minimize these effects, it was recognized that chondrules should be acid leached before analysis to reduce this contamination effect.

These insights are pertinent to understanding attempts to date the oldest solar system objects, calcium–aluminium inclusions (CAIs) in carbonaceous chondrites. Inclusions and chondrules both have high U/Pb ratios, but they do not contain zircon, and they are often too fine grained for the

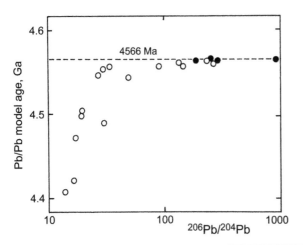

Fig. 5.33 Plot of 207/206 ages for Allende calcium–aluminium inclusions (CAIs), showing an inverse correlation with common Pb content. (○) = bulk samples; (●) = leached. After Allegre *et al.* (1995).

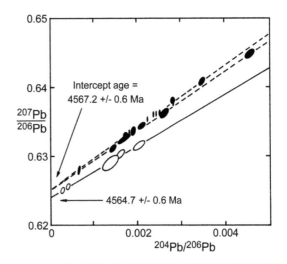

Fig. 5.34 Alternative Pb–Pb isochron diagram showing 207/206 (intercept) ages for acid-washed chondrules (open ellipses) and calcium–aluminium inclusions (= CAIs, solid ellipses). After Amelin *et al.* (2002).

physical separation of other uranium-rich minerals. A significant common Pb component rules out direct application of the U–Pb method. Hence, several Pb–Pb dating studies have been performed, of which a study by Chen and Wasserburg (1981) provided the first precise age from inclusions alone, with a Pb–Pb age of 4568 ± 5 Ma.

Two-point isochrons can be calculated between Canyon Diablo and any individual inclusion. These are termed model 207/206 ages because they rely on the assumption that initial lead in the inclusion was the same as Canyon Diablo. However, Allegre *et al.* (1995) showed that the 207/206 ages of the inclusions were correlated with their $^{206}Pb/^{204}Pb$ ratio, which measures the amount of common Pb in each sample (Fig. 5.33). This suggests that the inclusions were contaminated with extraneous common Pb from outside the chondrules that did not match Canyon Diablo Pb. Therefore, Allegre *et al.* utilized a progressive leaching procedure to remove the common Pb component. The results of this procedure gave 207/206 ages within error of the most radiogenic data of Chen and Wasserburg, with an improved age of 4566 +2/−1 Ma.

In order to use Pb–Pb ages to measure small time differences in the formation of the oldest solar system objects, Amelin *et al.* (2002) argued that it was necessary to make a Pb/Pb isochron determination on each object to be dated, so that accurate common Pb corrections could be made. They used a sequential acid leaching technique to achieve progressively more radiogenic data points on whole-rock chondrules from the Acfer chondrite. The data are shown in Fig. 5.34 on the inverse Pb–Pb isochron, where the intercept indicates the 207/206 age for an infinitely radiogenic sample. The slopes have no age significance, but indicate mixed sources of

common Pb that are not identical to Canyon Diablo. Because the samples analysed by Amelin *et al.* were very radiogenic, they gave an excellent intercept, corresponding to an age of 4564.7 ± 0.6 Ma for chondrule formation (MSWD = 0.5). Acid leached whole-rock fragments of two CAIs from the Efremovka chondrite also gave excellent Pb–Pb isochrons (MSWD = 0.9 and 1.1) with an average intercept age of 4567.2 ± 0.6 Ma, which is 2.5 Ma older than the chondrite isochron, consistent with ages from extinct nuclide systems (Section 15.5.3).

Further refinements in the Pb–Pb ages of the oldest objects followed, but this progress was disrupted when effects from variable uranium isotope compositions were discovered (Brennecka *et al.*, 2010). Using a double $^{233}U-^{236}U$ spike to correct for instrumental mass fractionation, Brennecka *et al.* (2010) made a uranium isotope study of CAIs in search of the signature of extinct curium in the early solar system (Section 15.8). Uranium isotope variations attributed to extinct curium are correlated with the abundance of its analogue species, Nd. However, Brenecka *et al.* also discovered large non-correlated uranium isotope variations in other meteorite types. Although these variations cannot be attributed to extinct curium, they nevertheless have a significant effect on the calculated Pb–Pb ages of these meteorites, as shown in Fig. 5.35.

Angrites show a fairly consistent $^{238}U/^{235}U$ ratio of 137.78, which also represents the bulk solar system value. This requires a subtraction of almost exactly 1 Ma from the previously determined Pb–Pb ages of angrites (Brennecka and Wadhwa, 2012). CAIs have the largest variations, with $^{238}U/^{235}U$ ratios from 137.4 to 137.9 (Fig. 5.35). However, there

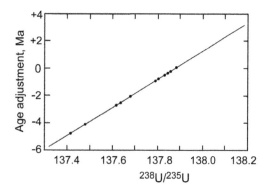

Fig. 5.35 Effect of variable uranium isotope composition (●) on calculated Pb–Pb ages of CAIs. After Brennecka *et al.* (2010).

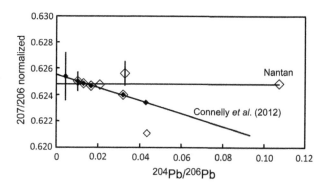

Fig. 5.36 Inverse Pb–Pb isochron plot for acid leach steps from Allende chondrule C30, normalized to a regression through Nantan primordial Pb. The line of Connelly *et al.* (2012) was regressed through the solid points only. Error bars are shown for less precise data.

is a clustering of points near the solar system average. After correcting for these U isotope variations, four CAIs gave very consistent ages of 4567.3 ± 0.15 Ma (Amelin *et al.*, 2010; Connelly *et al.*, 2012).

A more controversial result, reported by Connelly *et al.* (2012), was an identical age of 4567.3 ± 0.4 Ma for an Allende chondrule. This implies that some chondrules are as old as CAIs, in disagreement with the Mg–Al extinct nuclide system (Section 15.5.3). If verified, this result would have profound implications for the origins of the solar system, overturning the 'canonical' model for relatively homogeneous distribution of ^{26}Mg in the solar nebula. However, Kita *et al.* (2015) suggested that the old apparent age could be caused by unresolved common Pb components.

This issue is explored in Fig. 5.36, where Pb signatures of acid leaching steps on whole-rock chondrule C30 are displayed relative to a regression line through the Nantan primordial Pb composition (Blichert-Toft *et al.*, 2010). The best-fit line of Connelly *et al.*, achieved by rejecting three out of nine data points, has a significantly lower slope. Because the high-precision data on this sample are ten times less radiogenic than the best data in Fig. 5.34, the change in slope has a large effect on the age, increasing it by up to 0.5 Ma (remembering that the age is determined from the intercept). Hence this examination suggests that the old chondrule age is not yet proven. It also undermines the claim of Connelly *et al.* (2012) to have detected U/Pb isotope evolution in chondrites that preceded the primordial Pb seen in the 1AB iron meteorites Canyon Diablo and Nantan.

As a final comment on the Pb–Pb dating of meteorites, it should be noted that all of these ages are subject to a possible downward revision of ca. 4 Ma if the ^{235}U half-life is revised according to the evidence presented in Section 5.1. This does not affect comparisons with extinct nuclide systems, because all Pb–Pb ages will essentially change by the same amount. However, it does mean that the true age of the solar system may be only 4564 Ma.

5.4 Pb (Galena) Model Ages

As discussed in Section 5.1, the existence of two parallel U–Pb decay routes offers unique opportunities for obtaining age information from partially disturbed geological systems. The Pb model age method was an ingenious early attempt to utilize these properties to obtain age information from the Pb ore mineral galena (PbS). Since this mineral contains no uranium, there is no problem of U loss after the formation of the mineral. Unfortunately, the method was a failure, due to the great complexities of Pb evolution in the galena source. However, in the process of understanding why the method failed, important insights were obtained on terrestrial Pb isotope evolution.

5.4.1 The Holmes–Houtermans Model

Since there is no U decay in a galena, we are not measuring its age directly back from the present day, but measuring the age of the galena source from the formation of the Earth until the isolation of the galena. This approach was independently conceived by Holmes (1946) and Houtermans (1946). They divided the isotopic evolution of galena Pb into two 'stages'. The first stage was assumed to be a rock system, which needed to be closed to U and Pb movement from the formation of the Earth until galena separation (although we now realize that this is impossible). The second stage was in the galena itself, which must contain no significant amounts of uranium. This model for terrestrial Pb isotope evolution may be summarized as follows:

$$\underset{\text{age of Earth}}{\text{T}} \xrightarrow{\text{U decay in rock}} \underset{\text{age of galena}}{\text{t}} \xrightarrow{\text{no U decay in galena}} \underset{\text{present.}}{\text{P}}$$

Given this model, the basic decay equation for ^{207}Pb is

$$^{207}\text{Pb}_t = {}^{207}\text{Pb}_\text{T} + {}^{235}\text{U}\,(e^{\lambda 235\,T} - e^{\lambda 235\,t}) \qquad [5.13]$$

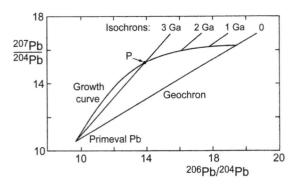

Fig. 5.37 Pb–Pb isochron diagram showing present day composition (P) of galena extracted from a Bulk Earth reservoir at 3 Ga. After Russell and Farquhar (1960).

This decay equation is more complex than [5.1] because 't' is not zero. Each term is next divided through by ^{204}Pb and rearranged. The same procedure is applied to the corresponding equation for ^{206}Pb to yield the following result:

$$\left(\frac{^{207}\text{Pb}}{^{204}\text{Pb}}\right)_t - \left(\frac{^{207}\text{Pb}}{^{204}\text{Pb}}\right)_T = \frac{^{235}\text{U}}{^{204}\text{Pb}}\ (e^{\lambda 235 T} - e^{\lambda 235 t}) \quad [5.14]$$

$$\left(\frac{^{206}\text{Pb}}{^{204}\text{Pb}}\right)_t - \left(\frac{^{206}\text{Pb}}{^{204}\text{Pb}}\right)_T = \frac{^{238}\text{U}}{^{204}\text{Pb}}\ (e^{\lambda 238 T} - e^{\lambda 238 t}) \quad [5.15]$$

equation [5.14] is now divided through by equation [5.15] and the result is simplified as follows:

(1) ^{204}Pb terms are cancelled on the right-hand side of the equation. This leaves a factor for the U isotope ratio at the present day, which is a constant with the value 1/137.8.

(2) $(^{207}\text{Pb}/^{204}\text{Pb})_t$ and $(^{206}\text{Pb}/^{204}\text{Pb})_t$ represent the present day compositions, since galena incorporates no U.

(3) The Pb isotope compositions at time 'T' represent the composition of the solar nebula; which is the primordial composition of the Earth, generally represented by troilite from the Canyon Diablo iron meteorite (C.D.).

The equation can then be written as

$$\frac{\left(\frac{^{207}\text{Pb}}{^{204}\text{Pb}}\right)_P - \text{C.D.}}{\left(\frac{^{206}\text{Pb}}{^{204}\text{Pb}}\right)_P - \text{C.D.}} = \frac{1}{137.8} \cdot \frac{(e^{\lambda 235\,T} - e^{\lambda 235\,t})}{(e^{\lambda 238 T} - e^{\lambda 238\,t})} \quad [5.16]$$

If the isotope ratios on the left-hand side of the equation represent a sample extracted from the mantle at time t, then the term on the right-hand side corresponds to the slope of an 'isochron' line joining it to the solar nebula composition (Fig. 5.37).

To apply the Holmes–Houtermans model, the galena source rock is assumed to be a closed system with a 'single-stage' Pb isotope history. A growth curve is then constructed for the galena source, which runs from the primordial Pb composition to that of the analysed galena, and is calibrated

for different values of t. (Since this is a transcendental curve, t cannot be solved by direct algebra starting with a composition on the left-hand side of equation [5.16].) The shape of the growth curve is determined by the two uranium decay constants, and its trajectory by the ^{238}U/^{204}Pb or 'μ' value of the closed-system galena source. For the single-stage model described, it is called the μ_1 value and would normally be between 7 and 9. According to the Holmes–Houtermans model, not every galena source rock need have the same growth curve, defined by the same μ value. Galena ore bodies were expected to have concentrated the metal from local continental basement in the vicinity. However, this presupposes that the basement in question has been in existence since near the formation of the Earth, which is now regarded as very unlikely (e.g. see Section 4.4).

Major problems with the Holmes–Houtermans model became apparent as more galenas were analysed and found to scatter more and more widely on the Pb–Pb isochron diagram (Fig. 5.38). Some of the ages determined were clearly erroneous, since they were in the future. Others, which were outliers to the main trend, often gave ages which could be shown to be geologically impossible. Since galenas of these two types contradicted the Holmes–Houtermans model, they were called 'anomalous leads'. However, the crucial problem with this situation was the lack of an *a priori* test which could be performed to predict whether a galena would be anomalous, in the absence of other evidence of its age.

5.4.2 Conformable Leads

Given the complexity of Earth evolution, it was realized even in the 1950s that the country-rock source of a given galena ore was unlikely to have been a closed system since the formation of the Earth. Alpher and Herman (1951) attempted to overcome this problem by attributing Pb isotope evolution in the galena source rock to a single worldwide homogeneous reservoir, regarded by Russell (1956) as the Earth's mantle. As an explanation for the observed galena Pb isotope variation, this model is quite obviously inadequate. However, it was the basis of a more geologically realistic model proposed by Stanton and Russell (1959).

A certain class of Pb ores was found by Stanton and Russell which lay on a single closed-system growth curve. These were sulphides associated with sediments and volcanics in greenstone belts and island arcs, which were structurally conformable with the host rocks (in contrast to cross-cutting veins). Stanton and Russell regarded these ores as being formed by syngenetic deposition in sedimentary basins associated with volcanic centres, and therefore as representing galena derived directly from the upper mantle without crustal contamination.

Stanton and Russell selected nine deposits of different ages which met these criteria, and fitted a single-stage (upper mantle) growth curve with a μ_1 value of 9.0 (Fig. 5.39). These ores were termed 'conformable' leads because of their structural occurrence, and all galenas that didn't fit this curve were by inference 'anomalous'. Anomalous leads were

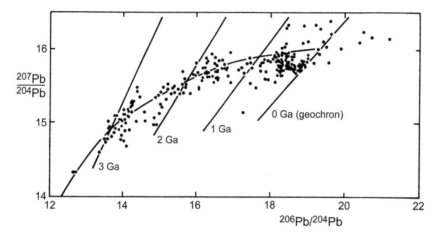

Fig. 5.38 Pb–Pb isochron diagram showing a compilation of many analysed galenas from different environments. After Stanton and Russell (1959).

divided into groups such as 'J-type' leads after a deposit at Joplin, Missouri which gave ages in the future, and some other types such as the 'B' type, which gave ages in the past.

5.4.3 Open-System Pb Evolution

As early as 1956, Russell considered the possibility that the mantle might not have been a closed system to U and Pb, but might have had a variable μ value over time, due to some kind of mantle differentiation process. The relative success of the conformable lead model in explaining uncontaminated galena compositions with a closed-system mantle militated against such complications. However, this situation was not to last.

In the early 1970s, new measurements of the uranium decay constants (Jaffey *et al.*, 1971) and a better estimate of primordial Pb from Canyon Diablo necessitated a re-examination of the conformable Pb model. For example, using the new values, a curve calculated to yield a reason-

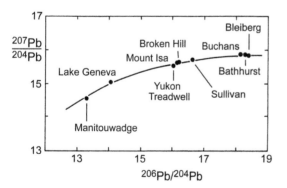

Fig. 5.39 Pb–Pb isochron diagram showing galena ores that form the basis of the 'conformable' Pb model. After Stanton and Russell (1959).

able fit to conformable galenas gave a low apparent age for the Earth of 4.43 Ga (Doe and Stacey, 1974). Alternatively, a terrestrial age of 4.57 Ga based on Pb–Pb dating of meteorites (Tatsumoto *et al.*, 1973), caused the geochron to lie to the left of most Phanerozoic galenas and young oceanic volcanics. This problem became known as the 'Pb paradox' and meant that single-stage Pb models gave 'future ages' up to 1 Ga in error for Phanerozoic rocks.

To rectify these problems, Oversby (1974) proposed a model for an evolving (mantle) source of galena Pb with a progressive increase in μ value with time (approximated by a series of small increments in μ). This model was elaborated upon by Cumming and Richards (1975), who modelled a galena source with a linear increase in μ value. Surprisingly perhaps, Cumming and Richards regarded the galena source as a regional average of the crust. However, this may not be as strange as it sounds, since later work would show that mantle and crustal Pb evolution are in fact coupled together, and upper mantle Pb is largely buffered by the crust (Section 6.3.2). The model of Cumming and Richards yields a good fit to the ages of selected galena data, but still implies a young apparent age for the Earth of 4.50 Ga.

An alternative solution to this problem, proposed by Stacey and Kramers (1975), was to break terrestrial Pb isotope evolution into two stages. Stacey and Kramers used Canyon Diablo Pb and average modern Pb (from a mixture of manganese nodules, ocean sediments and island arc rocks) to anchor the ends of a composite growth curve. This curve was produced by two closed systems (1 and 2) with different μ values (μ_1 and μ_2), separated in time by a worldwide differentiation event. The closed systems consisted of a combination of the upper mantle and upper crust (lower crust, lower mantle and core being isolated). The model gave the best fit to a selection of conformable galenas (dated by the enclosing sediments) when $\mu_1 = 7.2$, $\mu_2 = 9.7$, and the event was at 3.7 Ga (Fig. 5.40). This time was regarded as a peak of crust

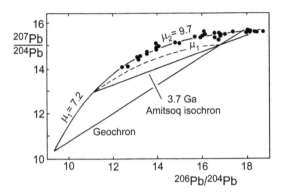

Fig. 5.40 Pb isotope diagram showing a two-stage lead isotope evolution model proposed for the source of galenas (•). After Stacey and Kramers (1975).

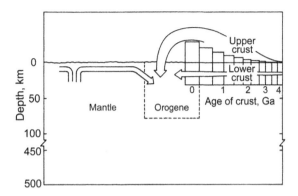

Fig. 5.41 Schematic illustration of the operation of the 'plumbotectonics' model, showing mixing of crustal and mantle reservoirs into the orogene (galena source) reservoir. After Doe and Zartman (1979).

forming events, an interpretation made particularly attractive by the 3.7 Ga age determined for the Amitsoq gneisses of western Greenland (Section 5.5.1). However, Stacey and Kramers noted that their model was only an approximation of Pb isotope evolution in the real Earth. For example, the discrete 3.7 Ga event in the model might actually represent a slow change in the Earth's evolution during the Early Archean.

5.4.4 Plumbotectonics

The above observations suggested that the galena source evolved for the past 3.8 Ga along a higher μ growth curve than the geochron, but the reason for this behaviour was not clear. Armstrong (1968) and Russell (1972) argued that elevated mantle μ values could be explained by recycling (bi-directional transport) of Pb between the crust and mantle. This concept was developed further by Doe and Zartman (1979) and presented as a computer model, 'Plumbotectonics', which modelled the Pb isotope evolution of the Earth.

Doe and Zartman defined three reservoirs: upper crust, lower crust and upper mantle (<500 km depth). Based on evidence that continental accretion began at ca. 4 Ga, and that frequent orogenies mixed mantle and crustal sources to yield differentiated crustal blocks, they modelled orogenies at 400 Ma intervals, with a decreasing mantle contribution through time. Crustal contributions represented erosion and continental foundering. Orogenies instantaneously extracted U, Th and Pb from the three sources, mixed them, and redistributed them back to the sources (Fig. 5.41). U fractionation into the upper crust represented granulite-facies metamorphism.

The orogene composition generated by the Plumbotectonics model was constrained empirically to fit galena ores, and consequent growth curves generated for the other reservoirs are shown in Fig. 5.42. The upper crust develops radiogenic Pb, which is balanced by the development of an unradiogenic lower crustal reservoir, due to preferential retention of Pb relative to U during granulite-facies metamorphism of

the lower crust. The calculated upper mantle μ value is similar to that for the total crust, but recycling of radiogenic upper crustal Pb into the mantle yields an *apparent* increase in mantle μ value with time. In spite of this effect, it is important to note that the Plumbotectonics model was not actually designed to solve the Pb paradox. Thus, in the first and second models (Zartman and Doe, 1981), Pb evolution started with an arbitrary isotope composition at 4 Ga, while in the fourth model (Zartman and Haines, 1988), Pb evolution began at 4.45 Ga from the Canyon Diablo composition (100 Ma after terrestrial accretion).

The Plumbotectonics IV model (Zartman and Haines, 1988) introduced another degree of complexity into the model, based on constraints from the $^{232}Th/^{238}U$ ratio kappa (approximately equal to the Th/U weight ratio). Discrepancies between the estimated present day κ value of the upper mantle and the time-integrated κ value determined from Pb isotope ratios have been called the 'second' Pb paradox, or

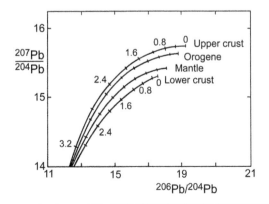

Fig. 5.42 Pb–Pb isochron diagram showing isotopic evolution of the four reservoirs computed by the plumbotectonics model. After Doe and Zartman (1979).

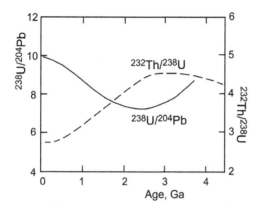

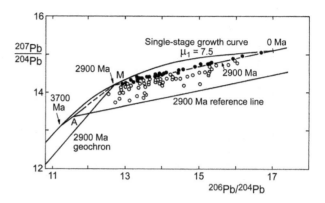

the 'kappa conundrum' (Section 6.3.2). To solve this conundrum, Zartman and Haines proposed an increased crustal U flux into the mantle in the Early Proterozoic, causing a substantial rise in the upper mantle μ value over the past 2 Ga, from a minimum of 7 at the end of the Archean to 10 at the present day (Fig. 5.43). This model was further developed by Kramers and Tolstikhin (1997), who proposed that enhanced uranium recycling was caused by mobilization of uranium in the sedimentary environment as a result of gradual oxygenation of the atmosphere.

Like the earlier plumbotectonics models, the model of Kramers and Tolstikhin (1997) assumed direct recycling of crustal material into the MORB source and no exchange of material with the lower mantle. However, a recent model by Kumari *et al.* (2016) directs a fraction of recycled crust into an 'isolated reservoir' in the lower mantle, where it is aged for 1 Ga before being transferred into the lower mantle source of plumes. Hence this model should begin to approach the true open-system Pb isotope evolution of the mantle.

5.5 Whole-Rock Pb and Crustal Evolution

Although the Pb–Pb whole-rock method has largely been superseded as a dating tool for igneous crystallization, its unique properties continue to make it a useful tool for studies of crustal evolution. However, making the best use of the method involves recognizing its strengths and weaknesses. A major strength is the existence of the two parallel decay schemes, allowing age information to be obtained for crustal reservoirs showing complex mixing relationships. However, the variable chemical behaviour of U and Pb during crustal extraction and reworking can be a major weakness, especially for Proterozoic systems. This is why Pb isotope mapping is generally a much weaker constraint on crustal formation

ages than Sm–Nd (Section 4.2.3). Nevertheless, the Pb–Pb system has one unique advantage, which is the much shorter half-life of ^{235}U (ca. 700 Ma) relative to long-lived species such as ^{87}Rb, ^{176}Lu and ^{147}Sm. This means that Pb isotopes are a particularly powerful tracer for early crustal evolution, as demonstrated by some of the following examples.

5.5.1 Archean Crustal Evolution

Western Greenland contains one of the most important Early Archean crustal provinces, but is also the site of Late Archean magmatism, represented by the Nuk gneisses. For example, Nuk gneisses from Fiskanaesset, Nordland and Sukkertopen in western Greenland have Pb compositions (filled circles in Fig. 5.44) which fall on a reference line with slope age of 2900 Ma. If a single-stage mantle growth curve is calculated to fit this isochron, it yields a μ_1 value of 7.5, which is a typical value for the mantle source of juvenile Archean gneiss terranes (Moorbath and Taylor, 1981). This single-stage model mantle composition is not expected to represent the *real* Earth, since this was shown above to be an oversimplification; however, the value provides a convenient yardstick for comparison between different crust-forming events.

Nuk gneisses from near Godthaab (open circles in Fig. 5.44) fall in a scatter below the 2900 Ma isochron, attributed by Taylor *et al.* (1980) to contamination with the local crust, represented by Early Archean (3700 Ma) Amitsoq gneisses. Because of their low U contents resulting from high-grade metamorphism, the Amitsoq gneisses barely changed in Pb isotope ratio from 3700 to 2900 Ma, resulting in Pb isotope signatures quite distinct from 2900 Ma mantle. Therefore, making use of the dual U–Pb decay schemes, Taylor *et al.* calculated initial ratios for the Nuk gneisses at 2900 Ma by projecting Pb compositions back parallel to the 2900 regression line. The resulting initial ratios lie on a

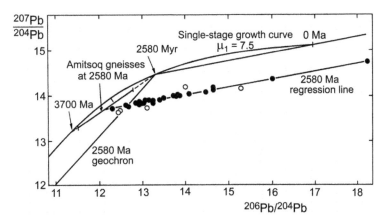

Fig. 5.45 Pb–Pb isochron diagram for Qorqut granite samples showing the coincidence of their initial ratio with the average Amitsoq gneiss composition at 2580 Ma (open symbols omitted from regression). After Moorbath and Taylor (1981).

mixing line between 2900 Ma mantle (M), and the Amitsoq gneisses (A). Therefore, the distance down the mixing line from 'M' to 'A' indicates the fraction of crustal Pb incorporated into the magma. The variable pattern of crustal Pb contamination suffered by Late Archean Nuk magmas is consistent with the known extent of Early Archean crust. Thus, while Godthaab is known to lie on Amitsoq gneiss basement, such rocks are not exposed near Fiskanaesset, Nordland or Sukkertopen.

The Qorqut granite is also exposed within the Amitsoq gneiss terrane near Godthaab (Moorbath *et al.*, 1981). Whole-rock samples of this body define a linear array whose slope corresponds to an age of 2580 Ma (Fig. 5.45). However, if an attempt is made to fit a single-stage mantle growth curve to this data, an impossibly low μ_1 value of 6.23 is obtained, showing that the Qorqut granite cannot be a mantle-derived melt. In fact, the initial Pb isotope ratio of the Qorqut granite coincides closely with the average composition of analysed Amitsoq crust at 2580 Ma, indicating that the Qorqut is probably a partial melt of Amitsoq gneiss. It therefore approximates to three-stage Pb isotope evolution: stage 1 = mantle; stage 2 = Amitsoq crust; stage 3 = Qorqut granite. The initial $^{87}Sr/^{86}Sr$ ratio of the Qorqut granite (0.7083 ± 4) supports this model (Section 7.3.4).

As noted above, the short half-life of ^{235}U makes the Pb isotope systems particularly sensitive for studies of Early Archean crustal evolution. Kamber and Moorbath (1998) used this approach to test the formation age of Early Archean crust in western Greenland. In the coastal Godthabsfjord area south of Nuk, 83 Amitsoq gneiss samples were analysed, yielding a strong Pb–Pb regression age (Fig. 5.46). Because of the large size of the data set, a relatively precise age of 3.65 ± 0.07 Ga (2σ) was obtained. However, the regression gave a large MSWD of 18, suggesting either initial ratio heterogeneity or metamorphic disturbance. Both of these effects could have subtly influenced the errorchron slope to produce a meaningless age. Therefore, to test this possibility, the

Amitsoq Pb–Pb regression line was compared with an open-system mantle growth to determine a Pb model age for the rocks.

The growth curve used was that of Kramers and Tolstikhin (1997), but the curve of Stacey and Kramers (1975) gives almost identical results, due to the convergence of mantle growth curves for the Early Archean. The regression line was shown to intersect the growth curve at a point corresponding to a model Pb age of 3.66 Ga, in excellent agreement with the regression age. The inclusion of leached feldspar analyses in the data set makes the model age particularly robust because these are very close to initial Pb isotope ratios. Hence, these data confirm that the Pb–Pb regression age gives the true age of crustal formation from a typical mantle Pb source. This suggests that the coastal Godthabsfjord area

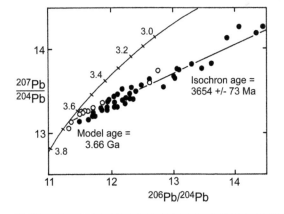

Fig. 5.46 Pb–Pb isochron for Amitsoq gneisses from the Godthabsfjord area of western Greenland, showing intersection with an open-system mantle growth curve at a model Pb age of 3.66 Ga, in good agreement with the regression age. (●) = whole-rocks; (○) = feldspar. After Kamber and Moorbath (1998).

of western Greenland has no significant crustal prehistory before 3.66 Ga. However, this does not rule out such a prehistory for Amitsoq gneisses further inland near Isua (Section 4.4.4).

5.5.2 Paleo-Isochrons and Metamorphic Disturbance

Because of the strong mobility of uranium in crustal systems, it was recognized that Pb–Pb arrays of the type represented by the Amitsoq and Nuk magmas might embody both magmatic differentiation and high-grade metamorphism. Hence these events might be separated by one or more short stages of Pb isotope evolution. For example, Taylor *et al.* (1980) proposed that there might be two short Pb growth stages, representing firstly the time between basalt extraction from the mantle and re-melting to form tonalitic magmas, and secondly the time between tonalite emplacement in the crust and high-grade metamorphism. The whole process was termed a crustal accretion–differentiation super-event or CADS by Moorbath and Taylor (1981).

The relatively short duration of these intermediate stages (<200 Ma) minimizes their effect on long term Pb isotope evolution. However if the period between the two events is substantial, then spurious ages may be obtained. A good example is provided by the Vikan gneiss complex from Lofoten-Vesteralen in NW Norway. If we assume that these rocks behaved as closed systems after their generation from an isotopically homogeneous (mantle?) source, we determine a slope age of 3410 ± 70 Ma (Taylor, 1975). However, Nd model age dating yields ages of ca. 2.4–2.7 Ga (Jacobsen and Wasserburg, 1978).

Subsequent examination of present day U/Pb ratios in the gneisses (Griffin *et al.*, 1978) revealed that they were uniformly far too low to 'support' the observed range of Pb isotope compositions. Therefore, it is now believed that the Pb data reflect a 2680 Ma igneous protolith which suffered high-grade metamorphism at ca. 1760 Ma. To illustrate this interpretation, Pb isotope compositions for the protolith are shown as a paleo-isochron at the time of metamorphism (Fig. 5.47). If the rocks were depleted in U to a nearly uniform level at 1760 Ma, subsequent U decay would yield a 'transposed paleo-isochron' (Griffin *et al.*, 1978; Moorbath and Taylor, 1981) which is almost parallel to the original paleo-isochron.

The slope of the transposed paleo-isochron approximates Pb evolution from time T (protolith age) to t (metamorphic age). This is described by an equation which is analogous to [5.16] for galena evolution:

$$\frac{\left(\frac{^{207}\text{Pb}}{^{204}\text{Pb}}\right)_P - \left(\frac{^{207}\text{Pb}}{^{204}\text{Pb}}\right)_I}{\left(\frac{^{206}\text{Pb}}{^{204}\text{Pb}}\right)_P - \left(\frac{^{206}\text{Pb}}{^{204}\text{Pb}}\right)_I} = \frac{1}{137.8} \frac{(e^{\lambda_{235} T} - e^{\lambda_{235} t})}{(e^{\lambda_{238} T} - e^{\lambda_{238} t})} \quad [5.17]$$

In contrast, the simple Pb/Pb isochron equation [5.12], describing evolution from t to the present, yields too large an age because it is based on the lower $^{235}\text{U}/^{238}\text{U}$ prevailing at the present day, compared to 1760 Ma.

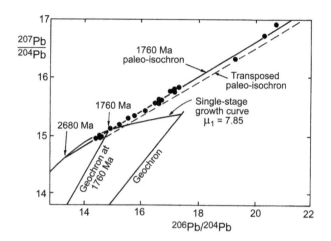

Fig. 5.47 Pb–Pb isochron diagram showing a 'transposed paleo-isochron' in Vikan gneisses of NW Norway. These rocks were formed from 2680 Ma precursors that experienced a granulite-facies uranium depletion event at 1760 Ma. After Moorbath and Taylor (1981).

Transposed paleo-isochrons can be detected by checking concordancy of Pb with Sr or Nd ages and by checking that observed Pb isotope compositions are adequately supported by the U/Pb ratio in the samples. Another example of this phenomenon was found in upper amphibolite-facies gneisses of the Outer Hebrides, NW Scotland, by Whitehouse (1990). By substituting the 2660 Ma (Badcallian) Pb homogenization event as T in equation [5.17], and assuming uniform U/Pb ratios after the second event, he was able to estimate the timing (t) of this second event. The calculated age of 1880 ± 270 Ma was consistent with the timespan of the Laxfordian metamorphic event.

5.5.3 Proterozoic Crustal Evolution

In the light of the above examples, it is clear that the mobility of uranium in crustal processes makes the interpretation of Pb–Pb data generally more complex than Nd data. However, when used correctly, the whole-rock Pb isotope system can also provide unique insights. Demonstrations of successful and unsuccessful applications are found in the Grenville Province of eastern North America.

Like most collisional orogenies, the Grenvillian orogeny was primarily a crustal reworking event. Negligible amounts of new crust were extracted from the mantle during the 1.1 Ga collision (Dickin *et al.*, 2010). However, the high grade of metamorphism resulting from crustal burial obscured many of the geological boundaries between older terranes. Therefore, isotopic tracers have played an important role in understanding pre-Grenvillian crustal evolution. The most important tool for this work is Nd isotope mapping, which yields reliable estimates for crustal formation ages in high-grade terranes (Section 4.2.3).

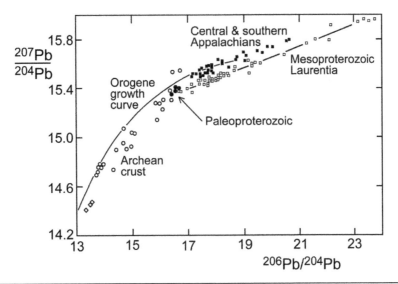

Fig. 5.48 Pb–Pb isotope diagram showing Grenvillian Paleoproterozoic gneisses (●) with signatures overlapping between Archean (○) and Mesoproterozoic Laurentia (□), but distinct from Appalachian terranes (■). (◇) = Superior Province. Growth curve from Doe and Zartman (1979). Data from DeWolf and Mezger (1994) and Fisher *et al.* (2010).

To explore the usefulness of Pb isotopes for mapping Proterozoic terranes, a summary of isotopic data for Eastern North America is shown in Fig. 5.48. Circular symbols are those of DeWolf and Mezger (1994), who attempted to use Pb isotopes to map the edge of the Archean craton within the Grenville Province of Ontario. However, samples with Paleoproterozoic crustal formation ages (solid circles) lie in an area of overlap between Archean and Mesoproterozoic Pb isotope signatures. Pb isotope mapping fails to resolve these suites because U/Pb ratios are not significantly fractionated during crustal extraction, so that crustal rocks often have similar U/Pb ratios to the mantle. In this case, the existence of two coupled U–Pb decay schemes does not help to overcome the problem of variable crustal U/Pb fractionation behaviour.

In contrast, square symbols in Fig. 5.47 are from Fisher *et al.* (2010), who successfully used Pb isotope mapping in the Appalachians to distinguish basement of Amazonian and Laurentian affinity. On this plot, open squares represent Laurentian samples from the Granite–rhyolite Province, the Adirondacks and the Texas Grenville Province, which define a colinear array (solid line). On the other hand, samples from the central and southern Appalachians lie on a parallel array with higher $^{207}Pb/^{204}Pb$ ratios resembling the Pb signature of Amazonia (not shown). This offset in $^{207}Pb/^{204}Pb$ is indicative of a fundamental difference in the Pb evolution of these crustal blocks, because $^{207}Pb/^{204}Pb$ variations were created early in Earth history when the parent isotope was still abundant. Therefore, in this case the coupled U–Pb decay schemes are important in distinguishing the two parallel arrays, allowing their geological histories to be distinguished.

5.6 Environmental Pb

Interest in the isotopic composition of Pb in environmental systems arose from attempts to date the age of the Earth by the Pb/Pb method. In order to determine a Bulk Earth composition for this dating work, Patterson investigated the composition of pelagic sediments, which were thought to provide an average composition of the whole crust. However, the analysis of pelagic sediments led to considerations about the distribution of Pb in the oceanic system.

A primary necessity in attempting to understand the distribution of Pb in the oceans was accurate measurement of the Pb concentration of seawater. However, the very low levels of Pb in seawater presented a considerable analytical challenge. The first problem was to find an analytical method with detection limits as low as one part per billion (ppb). The only method that can routinely achieve these kinds of detection limits is isotope dilution (Section 2.5), which allows the measurement of Pb isotope compositions at the same time. The second problem is anthropogenic contamination of the samples during analysis, referred to as 'blank' (Section 2.1.5). This was to pose a particular problem for Pb, because almost all laboratory materials and equipment had higher Pb levels than seawater.

The first workers to successfully overcome both of these problems and achieve accurate analysis of the Pb content of seawater were Tatsumoto and Patterson (1963). They went to extreme lengths to minimize Pb contamination during analysis, and demonstrated the effectiveness of these measures by using the same analytical procedure to analyse seawater samples of different sizes. Since the amount of

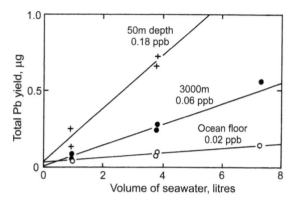

Fig. 5.49 Plot of analytical Pb yield (in micrograms) against volume of seawater (in litres), allowing the Pb content of seawater to be determined. After Tatsumoto and Patterson (1963).

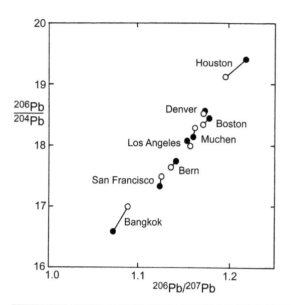

Fig. 5.50 Correspondence between lead ore compositions (•) and petrol (○) from different countries on a Pb/Pb isotope plot. After Chow (1970).

contamination is determined by the procedure, the application of an identical procedure to samples of different sizes should give rise to a constant Pb blank, whereas the total amount of Pb detected is dependent on the sample size. Hence, the two quantities can be separated (Fig. 5.49). This procedure showed that the Pb content of seawater varied from 0.02–0.18 ppb (μg/l), whereas the analytical blank was about 50 ng (0.05 μg). Previous Pb determinations on seawater had been ten to fifty times higher (2–8 ppb), which must be attributed to analytical error.

The accurate measurement of Pb concentrations for different water depths resulted in some surprising observations. These showed that Pb concentrations in surface ocean water (Fig. 5.48) were an order of magnitude higher than in deep ocean water (2000–4000 m). This behaviour was the opposite to that observed for many natural tracers, but resembled the distribution of nuclear fallout in the oceans. Hence, Tatsumoto and Patterson (1963) argued that the principal input of Pb to the oceans at the present day is anthropogenic. Following this discovery, the investigation of Pb in the oceans began to focus on the origins and distribution of anthropogenic Pb in different near-surface environments.

5.6.1 Anthropogenic Pb

The first use of Pb isotopes to trace the sources and distribution of anthropogenic Pb was made by Chow and Johnstone (1965). Based on an observation by Tatsumoto and Patterson (1963) that snow from Lassen Peak National Park had (relatively) very elevated Pb contents (1.6 ppt), Chow and Johnstone made Pb isotope measurements on the snow for comparison with possible sources in Californian leaded petrol (gasoline). They found that the Pb isotope signature of Lassen snow was almost identical to atmospheric particulates recovered from their clean-lab filter in Pasadena (Los Angeles) and also fell within the range of Pb isotope compositions of local petrol. This was due to the practice of adding tetra-ethyl lead to petrol to prevent engine pre-ignition.

Chow (1970) followed up this work with a study of the worldwide compositional variations of leaded petrol. He found large isotopic variations, attributed to the varying geological age of the Pb ores used for making tetra-ethyl lead in different countries. He then compared these Pb signatures with the isotopic composition of locally collected pollutant leads, either from air filters or soil samples. The results were presented on a graph that has often been used since (with minor variations) to compare the signatures of pollutant leads (Fig. 5.50). The data showed a very strong correlation between the Pb isotope composition of petrol Pb and local pollutant Pb, conclusively demonstrating that petrol additives were the principal source of pollutant lead in the environment.

By the late 1960s, steps were under way to convert American cars to lead-free petrol. Ironically, this was not to avoid poisoning the human population but to avoid poisoning catalytic converters that were being fitted to car exhaust systems to control pollution haze (Harrison and Laxen, 1981). As a result of this policy, the use of Pb in American petrol peaked in 1970, and by 1990 had fallen to less than 5% of the peak level (Wu and Boyle, 1997). Therefore, after 1970, Pb isotope tracer studies were devoted to assessing the relative contributions of various Pb pollution sources to the environment.

In an early example of this kind of work, Chow and Earl (1972) showed that atmospheric Pb pollution derived from the combustion of coal could be distinguished from leaded petrol by the more radiogenic Pb signature in coal.

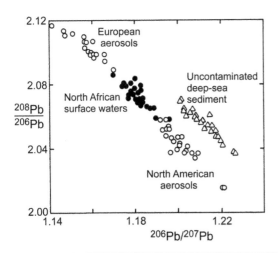

Mixing of anthropogenic Pb components in the North Atlantic, shown on a Pb/Pb isotope diagram. After Hamelin *et al.* (1997).

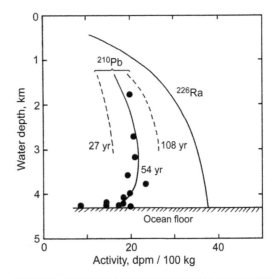

Profiles of ^{210}Pb and ^{226}Ra activity against depth in the central Pacific Ocean, showing good agreement between measured ^{210}Pb activity (●) and a water column model with a Pb residence time of 54 yr. After Craig *et al.* (1973).

This is because Pb in the sedimentary system (where coal is deposited) is more radiogenic than Pb in basement rocks, which are the source of most Pb ore deposits. In a later study, Sturges and Barrie (1987) showed that the isotopic composition of atmospheric Pb pollution from Canadian and American sources could be distinguished, allowing the tracing of cross-border air pollution.

A final example demonstrates the use of Pb isotopes to trace North American and European anthropogenic contributions to the Pb inventory of North Atlantic surface water (Fig. 5.51). In this study, American Pb sources were found to dominate the Pb isotope composition of North Atlantic surface waters off the North African coast between 1990 and 1992, despite the earlier phase-out of American leaded petrol. This Pb was carried across the Atlantic, via the Sargasso Sea, by eastward-moving surface water currents. Hence, this study demonstrates that the signature of leaded petrol will remain in environmental systems for many years to come, as anthropogenic Pb changes from a deadly health hazard to a useful marker of 1970s-age components in hydrological and sedimentary systems.

5.6.2 Pb as an Oceanographic Tracer

The extent of anthropogenic Pb contamination of ocean water is so great that direct Pb isotope measurements of ocean water cannot give information about natural Pb circulation. Therefore, studies of Pb as a natural oceanic tracer must be based on inventories of past oceanic Pb, recorded in ferromanganese nodules and pelagic sediments, as well as on the behaviour of ^{210}Pb, a short-lived isotope in the U-series decay chain (Section 12.1).

The use of Pb isotope analysis in oceanography was pioneered in the studies of Chow and Patterson (1959) on manganese nodules, and Chow and Patterson (1962) on

pelagic sediments. These studies revealed a general distinction between the Pb isotope signatures of Atlantic and Pacific samples, but within each ocean basin manganese nodules and pelagic sediments gave relatively consistent results. Based on these observations, Chow and Patterson concluded that Pb had a relatively short residence time in seawater, and that the distinct Pacific and Atlantic Ocean signatures reflected the average Pb isotope composition of the continents surrounding each ocean basin.

Based on these inter-oceanic variations, Chow and Patterson estimated a dissolved Pb residence time in seawater of ca. 10 ka. However, Craig *et al.* (1973) showed that ^{210}Pb can be used to determine a much more accurate Pb residence time, based on comparison with the relatively long-lived isotope ^{226}Ra, which acts as the parent of ^{210}Pb in seawater. They showed that ^{210}Pb was severely depleted relative to ^{226}Ra in deep ocean water, and hence that Pb must be very rapidly scavenged from seawater by adsorption onto particulate matter.

Based on the degree of ^{210}Pb depletion in a vertical section through the North Pacific off Guadalupe, Craig *et al.* calculated a deep water Pb residence time of only 50 yr (Fig. 5.52). This figure was confirmed as the average oceanic Pb residence time by a recent compilation of oceanic ^{210}Pb data, coupled with a general ocean circulation model (Henderson and Maier-Reimer, 2002). However, residence variations of up to an order of magnitude were observed between areas of high and low biological productivity.

In the 1960s, interest in environmental Pb switched from studies of natural to anthropogenic Pb, and relatively few studies were made of Pb isotopes as oceanic tracers until the 1990s. However, a study by Reynolds and Dasch (1971) led to a better understanding of the sources of dissolved oceanic Pb. Reynolds and Dasch were able to obtain more accurate Pb isotope data than Chow and Patterson due to advances in mass spectrometry, including use of the double spiking technique to correct for instrumental mass fractionation (Section 2.5.2). They showed that Mn nodules from the Atlantic Ocean had Pb signatures consistent with a source from continental erosion, whereas Mn nodules from the Pacific Ocean appeared to contain a mixture of Pb from continental sources and submarine volcanic activity.

Analysis of metal-rich sediments from near the East Pacific Rise revealed very large Pb contents (ca. 200 ppm) and Pb isotope signatures that overlap the composition of Pacific MORB (Dasch *et al.*, 1971). This suggests that submarine hydrothermal activity is a major local source of Pb, but its role as a globally significant flux is less clear. For example, Mn nodules not in the immediate vicinity of the ocean ridge were shown to have more radiogenic Pb signatures, indicative of mixing with continental sources (Reynolds and Dasch, 1971). Heterogeneous Pb isotope signatures in Mn nodules were also found by O'Nions *et al.* (1978). They argued that because the oceanic residence of Pb is so short, the isotopic composition of seawater Pb at any one point is essentially a dynamic equilibrium between continental and hydrothermal Pb fluxes.

The role of submarine hydrothermal activity as a source of seawater Pb was played down by von Blankenburg *et al.* (1996), who argued that the high Pb contents of metal-rich sediments near ridges showed that this Pb was *not* entering the oceans as a whole, precisely because it was being locally precipitated. Therefore, to explain the MORB-like composition of Pacific Pb isotope signatures, von Blankenburg *et al.* invoked volcanic arcs as a significant source of sedimentary Pb that could exchange Pb with seawater.

Other workers (e.g. Duce *et al.*, 1991) emphasized the role of aeolian inputs to explain seawater Pb. This model was supported by Jones *et al.* (2000), who performed sequential leaching experiments on Chinese loess to test the suitability of this component for influencing North Pacific seawater Pb. They showed that the Pb component in loess that was leachable with acetic acid corresponded closely to central Pacific seawater sampled by Fe–Mn crusts. In contrast, marginal Pacific samples were better explained by sedimentary Pb derived from volcanic arcs, but also by local sea floor hydrothermal influences. These relationships are particularly clear using thorogenic Pb (Fig. 5.53), due to the unusually radiogenic $^{208}Pb/^{204}Pb$ signatures of loess.

Much oceanographic evidence shows that Pacific and Atlantic water masses communicate via the circum-polar (Antarctic) Ocean (e.g. Section 14.2.1). Therefore, a good understanding of present day oceanic Pb can be obtained from circum-polar Mn nodules. These were studied by Abouchami and Goldstein (1995), who found evidence for

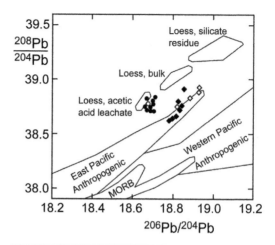

Fig. 5.53 Comparison of Pb isotope signatures for Pacific Fe–Mn crusts with arc sediment sources and leached fractions of Chinese loess. (•) = central Pacific Fe–Mn crusts, (♦) = marginal Pacific. Open symbols = local hydrothermal influence. Modified after Jones *et al.* (2000).

major mixing of Pb between different water masses. This mixing can be seen in a dynamic fashion by plotting the Pb isotope ratio of circum-polar Mn nodules against longitude (Fig. 5.54). This plot reveals two principal trends, involving progressive reduction in the Pb isotope ratio of Antarctic water across the south Pacific, and a progressive increase across the southern Atlantic–Indian oceans. These variations occur in response to mixing of these water masses with the eastward moving circumpolar current, which has a circulation time of ca. 30 yr.

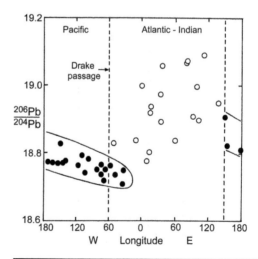

Fig. 5.54 Plot of Pb isotope ratio against longitude, showing the changing composition of circum-polar water due to mixing with water masses of the Pacific Ocean (•) and Atlantic–Indian Ocean (○). Modified after Abouchami and Goldstein (1995).

The Atlantic–Indian Pb trend in Fig. 5.54 is correlated with εNd, which Abouchami and Goldstein attributed to the southerly current that carries North Atlantic Deep Water (NADW) into the Antarctic Ocean. Here, NADW mixes with the Antarctic Bottom Water in which the Mn nodules grow. In contrast, the Pacific Pb trend does not correlate with εNd, which Abouchami and Goldstein attributed to many isolated mixing events between Pacific and Antarctic water as the circumpolar current moves across the southernmost Pacific. However, when this circum-polar water (CPW) reaches the Drake Passage between South America and Antarctica, the circumpolar current carries this water, with its distinct isotopic signature, half way across the south Atlantic, until it suddenly meets the southerly moving NADW.

Within the Atlantic, Indian and Pacific oceans, the Pb isotope compositions of Mn nodules are more homogeneous than the circumpolar ocean (von Blankenburg *et al.*, 1996). However, recent work has revealed isotopic provinciality in the Mn nodules from these oceans, with complex mixing relationships between Pb from different water bodies. For example, the southern Indian Ocean is influenced by NADW and CPW, as discussed above, but the northern Indian Ocean carries a Pb signature from Himalayan erosion (Frank and O'Nions, 1998; Vlastelic *et al.*, 2001).

5.6.3 Paleo-Seawater Pb

Because of their slow growth over millions of years, ferromanganese nodules and crusts preserve a record of past variations of seawater Pb isotope composition, as well as recording geographic variations at the present time. Paleo-seawater reconstructions depend on good dating methods, of which the most precise and accurate, applicable over the last 10 Ma, is cosmogenic ^{10}Be dating.

In view of the very short residence time of Pb in the oceans, and the multitude of Pb sources discussed above, it might be expected that the records of past oceanic Pb isotope composition would show rapid changes. Such changes were indeed observed in the northwest Atlantic over the past 2 Ma (Fig. 5.55). These were originally attributed to the closure of the Panama gateway (Burton *et al.*, 1997). However, the observation of similar changes in the Arctic Ocean suggested that they probably result from the input of very radiogenic Pb from the Canadian and Greenland shields, due to intensified glacial erosion beginning around 3 Ma (Frank *et al.*, 1999). Similarly, the northern Indian Ocean has shown moderately large long-term Pb isotope variations over the past 25 Ma, attributed by Frank and O'Nions (1998) to the effects of Himalayan uplift and erosion. In contrast to these large variations, the central Pacific has maintained a practically constant Pb isotope composition over the past 30 Ma (Ling *et al.*, 1997).

Beyond the range of the ^{10}Be method, other techniques for dating Fe–Mn crusts have been problematical (Section 4.5.4). However, the more recently developed method of

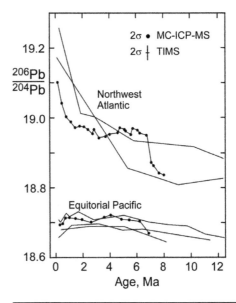

Fig. 5.55 Pb isotope variations in ferromanganese crusts from the Atlantic and Pacific over the past 10 Ma, dated by ^{10}Be. After Frank *et al.* (1999).

osmium isotope stratigraphy has yielded accurate ages for the early Tertiary Period (Section 8.5.1). Using this method, Klemm *et al.* (2007) showed that previously incoherent Pb isotope profiles for Pacific crusts now gave consistent records, with nearly constant signatures for the past 50 Ma (Fig. 5.56). Considering the wide range of Pb compositions for various Pb reservoirs, the very homogeneous Pb signatures of Pacific Fe–Mn crusts are surprising. However, Klemm *et al.* argued that these homogeneous signatures could be explained by a Pb source from aerosols that were well mixed in the atmosphere. Over the past 50 Ma, this appears to be from Chinese loess, whereas earlier Aeolian dust was probably dominated by circum-Pacific volcanism.

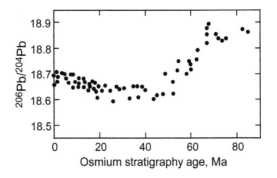

Fig. 5.56 Pb isotope profiles for central Pacific Fe–Mn crust over the past 80 Ma. Modified after Klemm *et al.* (2007).

References

Abouchami, W. and Goldstein, S. L. (1995). A lead isotope study of Circum-Antarctic manganese nodules. *Geochim. Cosmochim. Acta* **59**, 1809–20.

Ahrens, L. H. (1955). Implications of the Rhodesia age pattern. *Geochim. Cosmochim. Acta* **8**, 1–15.

Albarede, F. (2009). Volatile accretion history of the terrestrial planets and dynamic implications. *Nature* **461**, 1227–33.

Albarede, F. and Juteau, M. (1984). Unscrambling the lead model ages. *Geochim. Cosmochim. Acta* **48**, 207–12.

Aleinikoff, J. N., Winegarden, D. L. and Walter, M. (1990). U)Pb ages of zircon rims: a new analytical method using the air-abrasion technique. *Chem. Geol. (Isot. Geosci. Sect.)* **80**, 351–63.

Allegre, C. J., Manhes, G. and Gopel, C. (1995). The age of the Earth. *Geochim. Cosmochim. Acta* **59**, 1445–56.

Alpher, R. A. and Herman, R. C. (1951). The primeval lead isotopic abundances and the age of the Earth's crust. *Phys. Rev.* **84**, 1111–14.

Amelin, Y., Krot, A. N., Hutcheon, I. D. and Ulyanov, A. A. (2002). Lead isotopic ages of chondrules and calcium–aluminum-rich inclusions. *Science* **297**, 1678–83.

Amelin, Y., Kaltenbach, A., Iizuka, T. *et al.* (2010). U–Pb chronology of the Solar System's oldest solids with variable $^{238}U/^{235}U$. *Earth Planet. Sci. Lett.* **300**, 343–50.

Appel, P. W. U., Moorbath, S. and Taylor, P. N. (1978). Least radiogenic terrestrial lead from Isua, west Greenland. *Nature* **272**, 524–6.

Armstrong, R. L. (1968). A model for Sr and Pb isotope evolution in a dynamic Earth. *Rev. Geophys.* **6**, 175–99.

Augland, L. E. and David, J. (2015). Protocrustal evolution of the Nuvvuagittuq Supracrustal Belt as determined by high precision zircon Lu–Hf and U–Pb isotope data. *Earth Planet. Sci. Lett.* **428**, 162–71.

Blichert-Toft, J., Zanda, B., Ebel, D. S. and Albarede, F. (2010). The solar system primordial lead. *Earth Planet. Sci. Lett.* **300**, 152–63.

Brennecka, G. A. and Wadhwa, M. (2012). Uranium isotope compositions of the basaltic angrite meteorites and the chronological implications for the early Solar System. *Proc. Nat. Acad. Sci.* **109**, 9299–303.

Brennecka, G. A., Weyer, S., Wadhwa, M. *et al.* (2010). $^{238}U/^{235}U$ variations in meteorites: Extant ^{247}Cm and implications for Pb-Pb dating. *Science* **327**, 449–51.

Burton, K. W., Ling, H.-F. and O'Nions, R. K. (1997). Closure of the Central American Isthmus and its effect on deep-water formation in the North Atlantic. *Nature* **386**, 382–5.

Carl, C. and Dill, H. (1985). Age of secondary uranium mineralization in the basement rocks of the north eastern Bavaria F. R. G. *Chem. Geol. (Isot. Geosci. Sect.)* **52**, 295–316.

Carl, C., Wendt, I. and Wendt, J. I. (1989). U/Pb whole-rock and mineral dating of the Falkenburg granite in northeast Bavaria. *Earth Planet. Sci. Lett.* **94**, 236–44.

Catlos, E. J., Gilley, L. D. and Harrison, T. M. (2002). Interpretation of monazite ages obtained via *in situ* analysis. *Chem. Geol.* **188**, 193–215.

Chapman, H. J. and Roddick, J. C. (1994). Kinetics of Pb release during the zircon evaporation technique. *Earth Planet. Sci. Lett.* **121**, 601–11.

Chen, J. H. and Wasserburg, G. J. (1981). The isotopic composition of uranium and lead in Allende inclusions and meteoritic phosphates. *Earth Planet. Sci. Lett.* **52**, 1–15.

Chow, T. J. (1970). Isotopic identification of industrial pollutant lead. In: *2nd Int. Clean Air Congress*, pp. 348–52.

Chow, T. J. and Earl, J. L. (1972). Lead isotopes in North American coals. *Science* **176**, 510–11.

Chow, T. J. and Johnstone, M. S. (1965). Lead isotopes in gasoline and aerosols of Los Angeles Basin, California. *Science* **147**, 502–3.

Chow, T. J. and Patterson, C. C. (1959). Lead isotopes in manganese nodules. *Geochim. Cosmochim. Acta* **17**, 21–31.

Chow, T. J. and Patterson, C. C. (1962). The occurrence and significance of lead isotopes in pelagic sediments. *Geochim. Cosmochim. Acta* **26**, 263–308.

Compston, W., Williams, I. S. and Meyer, C. (1984). U)Pb geochronology of zircons from lunar breccia 73217 using a sensitive high mass-resolution ion microprobe. *Proc. 14th Lunar and Planet. Sci. Conf., J. Geophys. Res.* **89** Supp., B525–34.

Connelly, J. N., Bizzarro, M., Krot, A. N. *et al.* (2012). The absolute chronology and thermal processing of solids in the solar protoplanetary disk. *Science* **338**, 651–5.

Copeland, P., Parrish, R. R. and Harrison, T. M. (1988). Identification of inherited radiogenic Pb in monazite and its implications for U–Pb systematics. *Nature* **333**, 760–3.

Corfu, F. (2000). Extraction of Pb with artificially too-old ages during stepwise dissolution experiments on Archean zircon. *Lithos* **53** 279–91.

Cowan, G. A. and Adler, H. H. (1976). The variability of the natural abundance of ^{235}U. *Geochim. Cosmochim. Acta* **40**, 1487–90.

Craig, H., Krishnaswami, S. and Somayajulu, B. L. K. (1973). $^{226}Pb–^{226}Ra$: radioactive disequilibrium in the deep sea. *Earth Planet. Sci. Lett.* **17**, 295–305.

Cumming, G. L. and Richards, J. R. (1975). Ore lead isotope ratios in a continuously changing earth. *Earth Planet. Sci. Lett.* **28**, 155–71.

Dahl, P. S. (1997). A crystal-chemical basis for Pb retention and fission-track annealing systematics in U-bearing minerals, with implications for geochronology. *Earth Planet. Sci. Lett.* **150**, 277–90.

Das, A. and Davis, D. W. (2010). Response of Precambrian zircon to the chemical abrasion (CA-TIMS) method and implications for improvement of age determinations. *Geochim. Cosmochim. Acta* **74**, 5333–48.

Dasch, E. J., Dymond, J. R. and Heath, G. R. (1971). Isotopic analysis of metalliferous sediment from the East Pacific Rise. *Earth Planet. Sci. Lett.* **13**, 175–80.

David, J., Godin, L., Stevenson, R., O'Neil, J. and Francis, D. (2009). U-Pb ages (3.8–2.7 Ga) and Nd isotope data from the newly identified Eoarchean Nuvvuagittuq supracrustal belt, Superior Craton, Canada. *Geol. Soc. Amer. Bull.* **121**, 150–63.

Davis, D. W. (1982). Optimum linear regression and error estimation applied to U–Pb data. *Can. J. Earth Sci.* **19**, 2141–9.

Davis, D. W. and Krogh, T. E. (2000). Preferential dissolution of ^{234}U and radiogenic Pb from alpha-recoil-damaged lattice sites in zircon: implications for thermal histories and Pb isotopic fractionation in the near surface environment. *Chem. Geol.* **172**, 41–58.

deLaeter, J. R., Böhlke, J. K., De Bièvre, P. *et al.* (2003). Atomic weights of the elements. Review 2000 (IUPAC Technical Report). *Pure App. Chem.* **75**, 683–800.

DeWolf, C. P. and Mezger, K. (1994). Lead isotope analysis of leached feldspars: constraints on the early crustal history of the Grenville Orogen. *Geochim. Cosmochim. Acta* **58**, 5537–50.

DeWolf, C. P., Zeissler, C. J., Halliday, A. N., Mezger, K. and Essene, E. J. (1996). The role of inclusions in U-Pb and Sm–Nd garnet geochronology: stepwise dissolution experiments and trace uranium mapping by fission track analysis. *Geochim. Cosmochim. Acta* **60**, 121–34.

Dickin, A. P., McNutt, R. H., Martin, C. and Guo, A. (2010). The extent of juvenile crust in the Grenville Province: Nd isotope evidence. *Geol. Soc. Amer. Bull.* **122**, 870–83.

Doe, B. R. and Stacey, J. S. (1974). The application of lead isotopes to the problems of ore genesis and ore prospect evaluation: a review. *Econ. Geol.* **69**, 757–76.

Doe, B. R. and Zartman, R. E. (1979). *Plumbotectonics*. In: Barnes, H. L. (Ed.) *Geochemistry of Hydrothermal Ore Deposits*. Wiley, pp. 22–70.

Duce, R. A., Liss, P. S., Merrill, J. T. *et al.* (1991). The atmospheric input of trace species to the world ocean. *Global Biogeochem. Cyc.* **5**, 193–259.

Fisher, C. M., Loewy, S. L., Miller, C. F. *et al.* (2010). Whole-rock Pb and Sm-Nd isotopic constraints on the growth of southeastern Laurentia during Grenvillian orogenesis. *Geol. Soc. Amer. Bull.* **122**, 1646–59.

Foster, G., Gibson, H. D., Parrish, R. *et al.* (2002). Textural, chemical and isotopic insights into the nature and behaviour of metamorphic monazite. *Chem. Geol.* **191**, 183–207.

Frank, M. and O'Nions, R. K. (1998). Sources of Pb for Indian Ocean ferromanganese crusts: a record of Himalayan erosion? *Earth Planet. Sci. Lett.* **158**, 121–30.

Frank, M., O'Nions, R. K., Hein, J. R. and Banakar, V. K. (1999). 60 Myr records of major elements and Pb-Nd isotopes from hydrogenous ferromanganese crusts: reconstruction of seawater paleochemistry. *Geochim. Cosmochim. Acta* **63**, 1689–708.

French, J. E., Heaman, L. M. and Chacko, T. (2002). Feasibility of chemical U-Th-total Pb baddeleyite dating by electron microprobe. *Chem. Geol.* **188**, 85–104.

Froude, D. O., Ireland, T. R., Kinny, I. S., Williams, I. S. and Compston, W. (1983). Ion microprobe identification of 4,100)4,200 Myr-old terrestrial zircons. *Nature* **304**, 616–18.

Galer, S. J. G. and Goldstein, S. L. (1996). Influence of accretion on lead in the Earth. In: Basu, A. and Hart, S. R. (Eds) *Earth Processes: Reading the Isotopic Code. Geophys. Monograph* **95**, pp. 75–98. American Geophysical Union.

Gentry, R. V., Sworski, T. J., McKown, H. S. *et al.* (1982). Differential lead retention in zircons: implications for nuclear waste containment. *Science* **216**, 296–7.

Goldrich, S. S. and Mudrey, M. G. (1972). Dilatancy model for discordant U-Pb zircon ages. In: Tugarinov, A. I. (Ed.) *Contributions to Recent Geochemistry and Analytical Chemistry*. Moscow Nauka Publ. Office, 415–18.

Griffin, W. L., Taylor, P. N., Hakkinea, J. W. *et al.* (1978). Archaean and Proterozoic crustal evolution in Lofoten-Vesteraalen, Norway. *J. Geol. Soc. Lond.* **135**, 629–47.

Grove, M. and Harrison, T. M. (1999). Monazite Th-Pb age depth profiling. *Geology* **27**, 487–90.

Halliday, A. N. (1984). Coupled Sm-Nd and U-Pb systematics in Late Caledonian granites and the basement under northern Britain. *Nature* **307**, 229–33.

Hamelin, B., Ferrand, J. L., Alleman, L., Nicolas, E. and Veron, A. (1997). Isotopic evidence of pollutant lead transport from North America to the subtropical North Atlantic gyre. *Geochim. Cosmochim. Acta* **61**, 4423–8.

Harrison, R. M. and Laxen, D. P. H. (1981). *Lead Pollution: Causes and Control*. Chapman and Hall.

Heaman, L. M. and LeCheminant, A. N. (1993). Paragenesis and U-Pb systematics of baddeleyite (ZrO_2). *Chem. Geol.* **110**, 95–126.

Henderson, G. M. and Maier-Reimer, E. (2002). Advection and removal of ^{226}Pb and stable Pb isotopes in the oceans: a general circulation model study. *Geochim. Cosmochim. Acta* **66**, 257–72.

Hiess, J., Condon, D. J., McLean, N. and Noble, S. R. (2012). $^{238}U/^{235}U$ systematics in terrestrial uranium-bearing minerals. *Science* **335**, 1610–14.

Hinton, R. W. and Long, J. V. P. (1979). High-resolution ion-microprobe measurement of lead isotopes: variations within single zircons from Lac Seul, Northwest Ontario. *Earth Planet. Sci. Lett.* **45**, 309–25.

Holmes, A. (1946). An estimate of the age of the Earth. *Nature* **157**, 680–4.

Holmes, A. (1954). The oldest dated minerals of the Rhodesian Shield. *Nature* **173**, 612–17.

Houtermans, F. G. (1946). Die isotopen-haufigkeiten im naturlichen blei und das alter des urans. *Naturwissenschaften* **33**, 185–7.

Houtermans, F. G. (1947). Das alter des urans. *Z. Naturforsch* **29**, 322–8.

Jacobsen, S. B. and Wasserburg, G. J. (1978). Interpretation of Nd, Sr and Pb isotope data from Archaean migmatites in Lofoten Vesteraalen, Norway. *Earth Planet. Sci. Lett.* **41**, 245–53.

Jaffey, A. H., Flynn, K. F., Glendenin, L. E., Bentley, W. C. and Essling, A. M. (1971). Precision measurement of the half-lives and specific activities of U^{235} and U^{238}. *Phys. Rev. C* **4**, 1889–1907.

Jones, C. E., Halliday, A. N., Rea, D. K. and Owen, R. M. (2000). Eolian inputs of lead to the North Pacific. *Geochim. Cosmochim. Acta* **64**, 1405–16.

Kamber, B. S. and Moorbath, S. (1998). Initial Pb of the Amitsoq gneiss revisited: implication for the timing of early Archean crustal evolution in West Greenland. *Chem. Geol.* **150**, 19–41.

Kita, N. T., Tenner, T. J., Ushikubo, T. *et al.* (2015, July). Why do U-Pb ages of chondrules and CAIs have more spread than their ^{26}Al ages? *78th Ann. Meet. Meteoritical Soc.* Abstract #5360.

Klemm, V., Reynolds, B., Frank, M., Pettke, T. and Halliday, A. N. (2007). Cenozoic changes in atmospheric lead recorded in central Pacific ferromanganese crusts. *Earth Planet. Sci. Lett.* **253**, 57–66.

Kober, B. (1986). Whole-grain evaporation for $^{207}Pb/^{206}Pb$-age investigations on single zircons using a double-filament ion source. *Contrib. Mineral. Petrol.* **93**, 482–90.

Kober, B. (1987). Single-zircon evaporation combined with Pb^+ emitter bedding for $^{207}Pb/^{206}Pb$-age investigations using thermal ion mass spectrometry, and implications to zirconology. *Contrib. Mineral. Petrol.* **96**, 63–71.

Kober, B., Pidgeon, R. T. and Lippolt, H. J. (1989). Single-zircon dating by stepwise Pb-evaporation constrains the Archean history of detrital zircons from the Jack Hills, Western Australia. *Earth Planet. Sci. Lett.* **91**, 286–96.

Kramers, J. D. and Tolstikhin, I. N. (1997). Two terrestrial lead isotope paradoxes, forward transport modelling, core formation and the history of the continental crust. *Chem. Geol.* **139**, 75–110.

Krogh, T. E. (1982a). Improved accuracy of U-Pb zircon dating by selection of more concordant fractions using a high gradient magnetic separation technique. *Geochim. Cosmochim. Acta* **46**, 631–5.

Krogh, T. E. (1982b). Improved accuracy of U-Pb zircon ages by the creation of more concordant systems using the air abrasion technique. *Geochim. Cosmochim. Acta* **46**, 637–49.

Krogh, T. E., Corfu, F., Davis, D. W. *et al.* (1987). Precise U-Pb isotopic ages of diabase dykes and mafic to ultramafic rocks using trace amounts of baddeleyite and zircon. In: Halls, H. C. and Fahrig, W. F. (Eds) *Mafic Dyke Swarms. Geol. Assoc. Canada Spec. Pap.* **34**, 147–52.

Krogh, T. E. and Davis, G. L. (1975). Alteration in zircons and differential dissolution of altered and metamict zircon. *Carnegie Inst. Washington Year Book* **74**, 619–23.

Kumari, S., Paul, D. and Stracke, A. (2016). Open system models of isotopic evolution in Earth's silicate reservoirs: Implications for crustal growth and mantle heterogeneity. *Geochim. Cosmochim. Acta* **195**, 142–57.

Lagos, M., Ballhaus, C., Münker, C. *et al.* (2008). The Earth's missing lead may not be in the core. *Nature* **456**, 89–92.

Ling., H. F., Burton, K. W., O'Nions, R. K., *et al.* (1997). Evolution of Nd and Pb isotopes in Central Pacific seawater from ferromanganese crusts. *Earth Planet. Sci. Lett.* **146**, 1–12.

Ludwig, K. R. (1977). Effect of initial radioactive daughter disequilibrium on U-Pb isotope apparent ages of young minerals. *J. Res. U. S. Geol. Surv.* **5**, 663–7.

Ludwig, K. R. (1998). On the treatment of concordant uranium-lead ages. *Geochim. Cosmochim. Acta* **62**, 665–76.

Ludwig, K. R. (2000). Decay constant errors in U-Pb concordia-intercept ages. *Chem. Geol.* **166** 315–18.

Manhes, G., Allegre, C. J., Dupre, B. and Hamelin, B. (1979). Lead–lead systematics, the 'age of the Earth' and the chemical evolution of our planet in a new representation space. *Earth Planet. Sci. Lett.* **44**, 91–104.

Mattinson, J. M. (2000). Revising the "gold standard"– the uranium decay constants of Jaffey *et al.*, 1971. *EOS, Trans. Amer. Geophys. Union* **81**, S444.

Mattinson, J. M. (2005). Zircon U–Pb chemical abrasion ("CA-TIMS") method: combined annealing and multi-step partial dissolution analysis for improved precision and accuracy of zircon ages. *Chem. Geol.* **220**, 47–66.

Mattinson, J. M. (2010). Analysis of the relative decay constants of ^{235}U and ^{238}U by multi-step CA–TIMS measurements of closed-system natural zircon samples. *Chem. Geol.* **275**, 186–98.

Mezger, K., Essene, E. J. and Halliday, A. N. (1992). Closure temperatures of the Sm–Nd system in metamorphic garnets. *Earth Planet. Sci. Lett.* **113**, 397–409.

Mezger, K., Hanson, G. N. and Bohlen, S. R. (1989). U–Pb systematics in garnet: dating the growth of garnet in the Late Archean Pikwitonei granulite domain at Cauchon and Natawahunan Lakes, Manitoba, Canada. *Contrib. Mineral. Petrol.* **101**, 136–48.

Mezger, K., Rawnsley, C. M., Bohlen, S. R. and Hanson, G. N. (1991). U–Pb garnet, sphene, monazite, and rutile ages: implications for the duration of high-grade metamorphism and cooling histories, Adirondack Mts., New York. *J. Geol.* **99**, 415–28.

Montel, J.-M., Foret, S., Veschambre, M., Nicollet, C. and Provost, A. (1996). Electron microprobe dating of monazite. *Chem. Geol.* **131**, 37–53.

Moorbath, S., Taylor, P. N. and Goodwin, R. (1981). Origin of granite magma by crustal remobilisation: Rb–Sr and Pb/Pb geochronology and isotope geochemistry of the late Archaean Qorqut Granite complex of southern West Greenland. *Geochim. Cosmochim. Acta* **45**, 1051–60.

Moorbath, S. and Taylor, P. N. (1981). Isotopic evidence for continental growth in the Precambrian. In: Kroner, A. (Ed.) *Precambrian Plate Tectonics.* Elsevier, pp. 491–525.

Nier, A. O., Thompson, R. W. and Murphy, B. F. (1941). The isotopic constitution of lead and the measurement of geological time III. *Phys. Rev.* **60**, 112–17.

O'Nions, R. K., Carter, S. R., Cohen, R. S., Evensen, N. M. and Hamilton, P. J. (1978). Pb, Nd and Sr isotopes in oceanic ferromanganese deposits and ocean floor basalts. *Nature* **273**, 435–8.

Oversby, V. M. (1974). A new look at the lead isotope growth curve. *Nature* **248**, 132–3.

Parrish, R. R. (1990). U–Pb dating of monazite and its application to geological problems. *Can. J. Earth Sci.* **27**, 1431–50.

Patterson, C. C. (1956). Age of meteorites and the Earth. *Geochim. Cosmochim. Acta* **10**, 230–7.

Pankhurst, R. J. and Pidgeon, R. T. (1976). Inherited isotope systems and the source region pre-history of early Caledonian granites in the Dalradian series of Scotland. *Earth Planet. Sci. Lett.* **31**, 55–68.

Pidgeon, R. T. and Aftalion, M. (1978). Cogenetic and inherited zircon U–Pb systems in granites: Palaeozoic granites of Scotland and England. In: Bowes, D. R. and Leake, B. E. (Eds) *Crustal Evolution in Northwestern Britain and Adjacent Regions. Geol. Soc. Spec. Issue* **10**, 183–220.

Potts, P. J. (1987). *Handbook of Silicate Rock Analysis.* Blackie. 622 pp.

Reynolds, P. H. and Dasch, E. J. (1971). Lead isotopes in marine manganese nodules and the ore-lead growth curve. *J. Geophys. Res.* **76**, 5124–9.

Richter, S., Alonso, A., De Bolle, W., Wellum, R. and Taylor, P. D. P. (1999). Isotopic "fingerprints" for natural uranium ore samples. *Int. J. Mass Spec.* **193**, 9–14.

Richter, S., Alonso-Munoz, A., Eykens, R. *et al.* (2008). The isotopic composition of natural uranium samples – Measurements using the new n(^{233}U)/n(^{236}U) double spike IRMM-3636. *Int. J. Mass Spec.* **269**, 145–8.

Rogers, G., Dempster, T. J., Bluck, B. J. and Tanner, P. W. G. (1989). A high precision U–Pb age for the Ben Vuirich granite: implications for the evolution of the Scottish Dalradian Supergroup. *J. Geol. Soc. Lond.* **146**, 789–98.

Rosholt, J. N. and Bartel, A. J. (1969). Uranium, thorium and lead systematics in Granite Mountains, Wyoming. *Earth Planet. Sci. Lett.* **7**, 141–7.

Russell, R. D. (1956). Lead isotopes as a key to the radioactivity of the Earth's mantle. *Ann. N. Y. Acad. Sci.* **62**, 435–48.

Russell, R. D. (1972). Evolutionary model for lead isotopes in conformable ores and in ocean volcanics. *Rev. Geophys. Space Phys.* **10**, 529–49.

Russell, R. D. and Ahrens, L. H. (1957). Additional regularities among discordant lead–uranium ages. *Geochim. Cosmochim. Acta* **11**, 213–18.

Russell, R. D. and Farquhar, R. M. (1960). *Lead Isotopes in Geology.* Interscience Pub., 243 pp.

Scharer, U. (1984). The effect of initial ^{230}Th disequilibrium on young U–Pb ages: the Makalu case, Himalaya. *Earth Planet. Sci. Lett.* **67**, 191–204.

Scharer, U., Xu, R. H. and Allegre, C. J. (1984). U–Pb geochronology of Gangdese (Transhimalaya) plutonism in the Zhasa-Xigaze region, Tibet. *Earth Planet. Sci. Lett.* **69**, 311–20.

Schoene, B., Crowley, J. L., Condon, D. J., Schmitz, M. D. and Bowring, S. A. (2006). Reassessing the uranium decay constants for geochronology using ID–TIMS U–Pb data. *Geochim. Cosmochim. Acta* **70**, 426–45.

Silver, L. T. and Deutsch, S. (1963). Uranium–lead isotopic variations in zircons: a case study. *J. Geol.* **71**, 721–58.

Smith, H. A. and Giletti, B. J. (1997). Lead diffusion in monazite. *Geochim. Cosmochim. Acta* **61**, 1047–55.

Stacey, J. S. and Kramers, J. D. (1975). Approximation of terrestrial lead isotope evolution by a two-stage model. *Earth Planet. Sci. Lett.* **26**, 207–21.

Stanton, R. L. and Russell, R. D. (1959). Anomalous leads and the emplacement of lead sulphide ores. *Econ. Geol.* **54**, 588–607.

Steiger, R. H. and Jager, E. (1977). Subcommission on Geochronology: Convention on the use of decay constants in geochronology and cosmochronology. *Earth Planet. Sci. Lett.* **36**, 359–62.

Sturges, W. T. and Barrie, L. A. (1987). Lead 206/207 isotope ratios in the atmosphere of North America as tracers of US and Canadian emissions. *Nature* **329**, 144–6.

Tatsumoto, M., Knight, R. J. and Allegre, C. J. (1973). Time differences in the formation of meteorites as determined from the ratio of lead-207 to lead-206. *Science* **180**, 1279–83.

Tatsumoto, M. and Patterson, C. C. (1963). The concentration of common lead in sea water. In: Geiss, J. and Goldberg, E. D. (Eds) *Earth Science and Meteoritics.* North-Holland Pub. Co., pp. 74–89.

Taylor, P. N. (1975). An early Precambrian age for migmatitic gneisses from Vikan i Bo, Vesteraalen, North Norway. *Earth Planet. Sci. Lett.* **27**, 35–42.

Taylor, P. N., Moorbath, S., Goodwin, R. and Petrykowski, A. C. (1980). Crustal contamination as an indicator of the extent of early Archaean continental crust: Pb isotopic evidence from the late Archaean gneisses of West Greenland. *Geochim. Cosmochim. Acta* **44**, 1437–53.

Tera, F. and Carlson, R. W. (1999). Assessment of the Pb-Pb and U-Pb chronometry of the early solar system. *Geochim. Cosmochim. Acta* **63**, 1877–89.

Tera, F. and Wasserburg, G. J. (1973). A response to a comment on U-Pb systematics in lunar basalts. *Earth Planet. Sci. Lett.* **19**, 213–17.

Tera, F. and Wasserburg, G. J. (1974). U-Th-Pb systematics on lunar rocks and inferences about lunar evolution and the age of the Moon. *Proc. 5th Lunar Sci. Conf.* (Supp. **5**), *Geochim. Cosmochim. Acta* (Vol 2), 1571–99.

Tilton, G. R. (1960). Volume diffusion as a mechanism for discordant lead ages. *J. Geophys. Res.* **65**, 2933–45.

Tilton, G. R. and Grunenfelder, M. H. (1968). Sphene: uranium–lead ages. *Science* **159**, 1458–61.

Tucker, R. D., Raheim, A., Krogh, T. E. and Corfu, F. (1987). Uranium–lead zircon and titanite ages from the northern portion of the Western Gneiss Region, south-central Norway. *Earth Planet. Sci. Lett.* **81**, 203–11.

Valley, J. W., Cavosie, A. J., Ushikubo, T. *et al.* (2014). Hadean age for a post-magma-ocean zircon confirmed by atom-probe tomography. *Nature Geosci.* **7**, 219–23.

van Breemen, O., Davidson, A., Loveridge, W. D. and Sullivan, R. W., (1986). U-Pb zircon geochronology of Grenville tectonites, granulites and igneous precursors, Parry Sound, Ontario. In: Moore, J. M., Davidson, A. and Baer, A. J. (Eds) *The Grenville Province. Geol. Assoc. Canada Spec. Pap.* **31**, 191–207.

Vlastelic, I., Abouchami, W., Galer, S. J. G. and Hofmann, A. W. (2001). Geographical control on Pb isotope distribution and sources in Indian Ocean Fe–Mn deposits. *Geochim. Cosmochim. Acta* **65**, 4303–19.

von Blankenburg, F., O'Nions, R. K. and Hein, J. R. (1996). Distribution and sources of pre-anthropogenic lead isotopes in deep ocean water from Fe–Mn crusts. *Geochim. Cosmochim. Acta* **60**, 4957–63.

Watson, E. B. and Harrison, T. M. (1983). Zircon saturation revisited: temperature and composition effects in a variety of crustal magma types. *Earth Planet. Sci. Lett.* **64**, 295–304.

Wendt, I. (1984). A three-dimensional U-Pb discordia plane to evaluate samples with common lead of unknown isotopic composition. *Isot. Geosci.* **2**, 1–12.

Wetherill, G. W. (1956). An interpretation of the Rhodesia and Witwatersrand age patterns. *Geochim. Cosmochim. Acta* **9**, 290–2.

Whitehouse, M. (1990). Isotopic evolution of the southern Outer Hebridean Lewisian gneiss complex: constraints on Late Archean source regions and the generation of transposed Pb–Pb palaeoisochrons. *Chem. Geol.* (*Isot. Geosci. Sect.*) **86**, 1–20.

Wilde, S. A., Valley, J. W., Peck, W. H. and Graham, C. M. (2001). Evidence from detrital zircons for the existence of continental crust and oceans on the Earth 4.4 Gyr ago. *Nature* **409**, 175–8.

Williams, I. S. and Claesson, S. (1987). Isotopic evidence for the Precambrian provenance and Caledonian metamorphism of high grade paragneisses from the Seve Nappes, Scandinavian Caledonides. *Contrib. Mineral. Petrol.* **97**, 205–17.

Williams, I. S., Compston, W., Black, L. P., Ireland, T. R. and Foster, J. J. (1984). Unsupported radiogenic Pb in zircon: a cause of anomalously high Pb-Pb, U-Pb and Th-Pb ages. *Contrib. Mineral. Petrol.* **88**, 322–7.

Wood, B. J. and Halliday, A. N. (2005). Cooling of the Earth and core formation after the giant impact. *Nature* **437**, 1345–8.

Wood, B. J. and Halliday, A. N. (2010). The lead isotopic age of the Earth can be explained by core formation alone. *Nature* **465**, 767.

Wood, B. J., Halliday, A. N. and Rehkamper, M. (2010). Volatile accretion history of the Earth. *Nature* **467**, E6-E7.

Wotzlaw, J. F., Hüsing, S. K., Hilgen, F. J. and Schaltegger, U. (2014). High-precision zircon U-Pb geochronology of astronomically dated volcanic ash beds from the Mediterranean Miocene. *Earth Planet. Sci. Lett.* **407**, 19–34.

Wu, J. and Boyle, E. A. (1997). Lead in the western North Atlantic Ocean: completed response to leaded gasoline phase-out. *Geochim. Cosmochim. Acta* **61**, 3279–83.

Zartman, R. E. and Doe, B. R. (1981). Plumbotectonics – the model. *Tectonophys.* **75**, 135–62.

Zartman, R. E. and Haines, S. M. (1988). The plumbotectonic model for Pb isotopic systematics among major terrestrial reservoirs – a case for bi-directional transport. *Geochim. Cosmochim. Acta* **52**, 1327–39.

Chapter 6

Isotope Geochemistry of Oceanic Volcanics

Some of the most important questions in geology concern the processes which operate in the Earth's mantle. Mantle convection is clearly the driving force behind plate tectonics (e.g. Turcotte and Oxburgh, 1967), but the details of its operation are still unclear. The depth of mantle convection cells, the fate of subducted lithosphere and the source of upwelling mantle plumes are all questions that remain poorly understood. Isotope geochemistry may help to answer these questions by revealing the progress of mantle differentiation into different reservoirs and the extent to which these reservoirs are re-mixed by convective stirring.

The inaccessibility of the mantle presents a severe problem for geochemical sampling. However, mantle-derived basic magmas provide a prime source of evidence about the chemical structure of the mantle. Isotopic tracers represent a particularly powerful tool for such studies because, unlike elemental concentrations, isotope ratios are not affected by crystal fractionation. However, isotope ratios are susceptible to contamination in the continental lithosphere. Therefore the simplest approach to studying mantle chemistry through basic magmas is to analyse oceanic volcanics, which are expected to have suffered minimal contamination in the thin oceanic lithosphere.

Isotopic analysis of oceanic basalts can be used both to probe the structure of the mantle and to model its evolution over time. The approach taken here will be to examine the constraints on mantle structure from single isotopic systems (mainly Sr and Pb), then to examine the constraints on mantle evolution from multiple isotopic systems (Sr–Nd), (U–Th–Pb), (Sr–Nd–Pb). Evidence from other systems will be examined in later chapters.

6.1 Isotopic Tracing of Mantle Structure

Isotopic analysis of ocean island basalts (OIB) was first used to demonstrate the existence of mantle heterogeneity (Faure and Hurley, 1963; Gast et al., 1964). Subsequently, variations were found between the isotope compositions of mid ocean ridge basalts (MORB) and OIB (Tatsumoto, 1966). Some of the key evidence for mantle structure comes from comparing isotope heterogeneity within and between these groups, and especially from case studies of ocean islands located on spreading ridges.

6.1.1 Contamination and Alteration

Before oceanic volcanics can be used to deduce mantle compositions, we must examine and quantify the amounts of alteration and contamination that could occur during magma transport and eruption on ocean islands or the ocean floor.

Sub-solidus alteration of analysed samples could result from hydrothermal interaction with seawater, in the case of submarine basalts, or sub-aerial weathering, in the case of ocean island basalts. For example, Dasch et al. (1973) found a positive correlation between $^{87}Sr/^{86}Sr$ and water content in dredged oceanic basalts of various ages (Fig. 6.1). Samples with over 1% H_2O had almost invariably suffered Sr contamination from seawater, but those with less than 1% alteration appeared to be uncontaminated.

Sub-solidus alteration can be avoided in submarine samples by analysing 100% fresh MORB glasses (Cohen et al., 1980). Where crystalline rock must be analysed (e.g. White et al., 1976), alteration can be avoided by analysing fresh material dredged from the median valley of ocean ridges, where very young, unmetamorphosed basalts outcrop. Alternatively, leaching of crystalline samples before analysis may remove contaminated alteration minerals, also yielding results that are consistent with glasses (Dupre and Allegre, 1980). Unaltered ocean island basalts are easily obtained by sampling only fresh lavas.

Once sub-solidus alteration of samples has been excluded, the next possibility that must be considered is contamination in the oceanic lithosphere. Although this is normally much thinner than the continental lithosphere, some ocean islands could be located on micro-continents or some other kind of abnormal lithosphere.

In their early work on Ascension and Gough islands, Gast et al. (1964) considered the possibility of contamination of the analysed lavas by a crustal micro-plate. They tested this

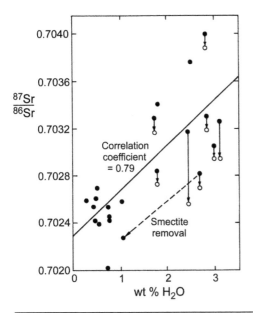

Plot of strontium isotope ratio against water content in ocean floor basalts. Vertical arrows show the effect of leaching before analysis. Dashed arrow shows the effect of smectite removal from an altered sample. After Dasch et al. (1973).

possibility by analysing a range of lavas at variable degrees of magmatic differentiation (Fig. 6.2). The lack of any correlation in all but the most evolved rocks was argued to rule out crustal contamination. High Sr isotope ratios in the highly evolved rocks were attributed to radioactive growth after eruption, since these rocks have very high Rb/Sr ratios. No age corrections could be applied to these lavas since their ages were unknown. Similar problems have been encoun-

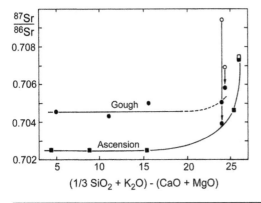

Fig. 6.2 Sr isotope ratios in lavas from Gough and Ascension islands plotted against an index of magmatic differentiation. Radiogenic Sr in highly evolved lavas (open symbols) is attributed to radioactive growth since eruption. Arrows show estimated age corrections. After Gast et al. (1964).

tered in more recent studies of Ascension lavas (Harris et al., 1983). However, most oceanic basalts require no age correction since they have very low Rb/Sr ratios.

Some workers, most notably O'Hara, suggested that isotopic variations in MORB and OIB could be explained by fractionation or contamination processes affecting magmas during their ascent through oceanic crust. In his early papers on the subject, O'Hara (1973, 1975) suggested that $^{87}Sr/^{86}Sr$ variations could be generated by physical fractionation of the isotopes during magmatic differentiation. This was a misconception, since $^{87}Sr/^{86}Sr$ ratios are always fractionation-corrected to the standard $^{88}Sr/^{86}Sr$ ratio of 8.37521 (Section 2.2.1) to eliminate both natural and analytical mass-dependent fractionation. Subsequently, O'Hara and Mathews (1981) argued that large ion lithophile (LIL) elements (including strontium) could be perturbed by contamination with altered oceanic crust in a 'periodically tapped, periodically re-filled, long-lived magma chamber'. However, substantial residence times in ocean ridge magma chambers are ruled out by evidence from U-series isotopes, which allow only a brief period between generation and eruption of ocean floor basalt. This limits the ability of open-system magma chambers to overprint the source isotopic signatures in the erupted products (Section 13.3).

More recently, there has been renewed attention paid to the possibility that some OIB may have been contaminated in the mantle lithosphere. In the case of Hawaii, it was proposed some time ago (Chen and Frey, 1983) that late-stage magmatism on Oahu involved partial melting of the LIL-depleted oceanic lithosphere, and that these melts were variably mixed with melts from a plume source. More recently, it was proposed that several hot-spots surrounding the African continent might have derived their distinctive isotopic signatures from African subcontinental lithosphere that was delaminated during rifting and left as scattered remnants under new oceanic crust. This model was applied to Grand Comore, the Canaries and Sao Miguel in the Azores (e.g. Widom et al., 1997, 1999; Class et al., 1998). However, more recent work has cast doubt on this model (e.g. Elliott et al., 2007; Class et al., 2009). The case of Sao Miguel will be discussed further below.

6.1.2 Disequilibrium Melting

Following the discovery of 'mantle heterogeneity' under the oceans, various workers (e.g. Harris et al., 1972; O'Nions and Pankhurst, 1973; Flower et al., 1975) suggested that mantle temperatures might not be high enough to ensure diffusional homogenization of Sr isotope ratios between different mantle minerals. In that case, grains with higher Rb/Sr ratios (such as the magnesian mica, phlogopite) could develop more radiogenic $^{87}Sr/^{86}Sr$ compositions over geological time. 'Disequilibrium melting' of such phases could then bias the isotopic composition of a melt towards higher $^{87}Sr/^{86}Sr$ compositions. Small degree partial melts would tend to be enriched in Rb/Sr and $^{87}Sr/^{86}Sr$ relative to large degree

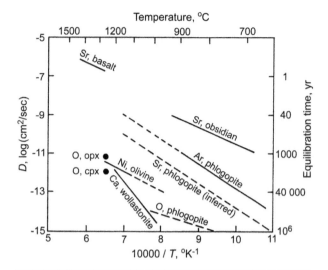

Fig. 6.3 Plot of diffusivity against 1/temperature, showing experimental results for the diffusion of Sr, Ar, Ni, Ca and O in different types of material. Times for effective equilibration are based on 1 cm grain size. Modified after Hofmann and Hart (1978).

partial melts, due to the tendency of high Rb/Sr phases such as phlogopite to enter the melt first.

Harris *et al.* (1972) argued in favour of disequilibrium melting during basalt genesis, based on evidence of isotopic disequilibrium in mantle xenoliths carried to the surface in basic magmas. Isotopic disequilibrium in ultramafic xenoliths is very widespread (Section 7.1), but such cases represent samples of the solid lithosphere. It is questionable whether these observations can be extrapolated to the higher temperature environment of basaltic magma genesis in the convecting asthenosphere.

Hofmann and Hart (1978) examined data for the diffusion of Sr in mantle silicates in order to determine the rates at which isotopic disequilibrium could be eradicated at different temperatures. In Fig. 6.3, values of diffusivity (*D*) are used to calculate times for effective equilibration of a species between a 1 cm diameter sphere and an infinite reservoir such as a slowly moving melt. These times are roughly those taken for diffusion over a 'characteristic transport distance' of 0.25 cm, using the equation $X = (Dt)^{1/2}$.

Using the lower of the measured diffusivities, it would take millions of years to eradicate Sr isotope heterogeneity between large grains of phlogopite and clinopyroxene in solid lithospheric mantle at, say, 600 °C. Even in a solid mantle at 1000 °°C, equilibration could take millions of years if the phlogopite and clinopyroxene grains were separated by intervening olivine or orthopyroxene, which effectively contain no Sr but lengthen the diffusion pathways between phlogopite and cpx. However, as soon as a melt is present, the surface of each crystal is in diffusional contact with nearby (ca. 2 cm distant) grains over a period of a few years.

Therefore isotopic disequilibrium between phlogopite and cpx could be eradicated in a few thousand years at temperatures above the basalt solidus (ca. 1000–1200 °C). Nevertheless, diffusion over long distances, even in a partially molten mantle, is still slow.

Hofmann and Hart (1978) concluded that the evidence favoured 'local equilibrium in a partially molten mantle, local disequilibrium in a completely crystalline mantle, and regional disequilibrium in any mantle that convects only slowly in large convection cells.' This suggests that disequilibrium melting does not preferentially sample mantle isotopic heterogeneity at the mineralogical scale. However, it might well sample heterogeneity between different petrological source types, even if these are streaked out by convection into thin bands (see below).

6.1.3 Mantle Plumes

Following acceptance of the plate tectonic model, it was recognized that the tectonic setting of basic volcanism is a crucial factor in determining the nature of the mantle source being tapped, and hence magma chemistry. Morgan (1971) proposed that the different chemistry of MORB and OIB could be explained if the former were derived directly from the asthenospheric upper mantle, while the latter were generated by upwelling plumes from the lower mantle. Evidence in support of this model was provided by elemental analysis of Icelandic basalts (Schilling, 1973). These data suggested a region of mixing between a mantle plume (OIB source) and the LIL-depleted upper mantle (MORB source) on the Reykjanes Ridge south of Iceland. Sr isotope data for the Reykjanes Ridge (Hart *et al.*, 1973) were slightly more equivocal, since they showed a step-like feature in the data (Fig. 6.4). However, Pb isotope data (Sun *et al.*, 1975) strongly supported the mixing model, thus providing strong evidence for the existence of a mantle plume under Iceland.

White *et al.* (1976, 1979) extended the Reykjanes Ridge Sr data set by analysing dredged samples from the axial valley of the Mid Atlantic Ridge (MAR) between 29 and 63 °N, and by sampling across the Azores platform. Isotopic data are plotted against latitude down the MAR in Fig. 6.4, and longitude across the Azores Plateau in Fig. 6.5. There are large variations in the strontium isotope ratio of MORB samples along the Mid Atlantic Ridge, but where MORB and OIB are erupted alongside each other (the Azores Plateau), they have very similar isotope ratios (with the exception of Sao Miguel). Because tholeiitic (MORB) and alkaline (OIB) magmas are attributed to different degrees of mantle melting, the overlap of their compositions across the Azores Plateau is evidence against sampling of isotopic heterogeneities on a mineralogical scale.

A central question for the plume model was to explain the spatial location of the plume source in the mantle. Anderson (1981) proposed a very shallow source, within the low-velocity zone, but this model has received very little support. Hofmann and White (1982) argued instead that mantle plumes originate from recycled crustal material

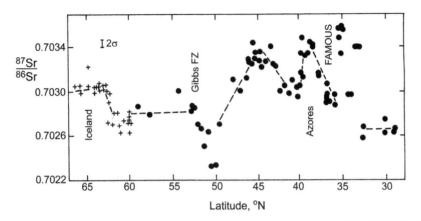

Fig. 6.4 Plot of Sr isotope ratio against latitude for basalts from the Mid Atlantic Ridge. (+) = Iceland – Reykjanes ridge. Age correction of Sr isotope data is unnecessary, due to the low Rb/Sr ratios and young ages of analysed material. After White *et al.* (1976).

stored at the base of the mantle (Section 6.2.1). They argued that subducted oceanic crust would be transformed into dense eclogitic material that would descend to the core–mantle boundary, where it would be stored and reheated for 1–2 Ga before returning to the surface in a plume (Fig. 6.6).

Ringwood (1982) advocated an alternative storage site at the 670 km phase transition. Seismic evidence for depression of the 670 km mantle phase boundary under subduction zones provided some support for this model by suggesting that the descending slab might be deflected horizontally at this level (Shearer and Masters, 1992). However, other evidence suggests that the density contrast is too small

to impede convective transport across this boundary and prevent slab penetration into the lower mantle (Morgan and Shearer, 1993). Furthermore, there is an increasing amount of tomographic evidence showing that some slabs sink to the bottom of the mantle (van der Hilst *et al.*, 1997) and that some plumes rise from the same place (e.g. Bijwaard and Spakman, 1999).

6.1.4 Plum Pudding Mantle

Many workers have questioned whether there might be an intermediate scale of mantle heterogeneity between rare large plumes and mineralogical disequilibrium. Even in their early elemental studies of the Faeroes mantle 'plume', Schilling and Noe-Nygaard (1974) recognized that this structure need not be a continuous column, but could have the

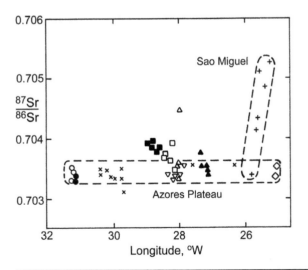

Fig. 6.5 Plot of strontium isotope ratio against longitude for basalt samples from the Azores Plateau. (×) = dredged basalts. Other symbols represent individual islands. After White *et al.* (1979).

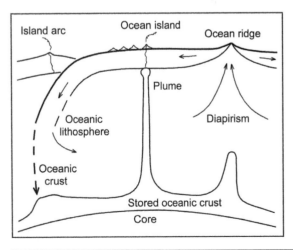

Fig. 6.6 Kinematic model for the origin of mantle plumes at the core–mantle boundary from recycled oceanic crust. After Hofmann and White (1982).

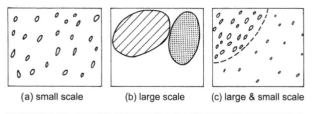

(a) small scale (b) large scale (c) large & small scale

Fig. 6.7 Hypothetical scales of mantle heterogeneity. After Allegre et al. (1980).

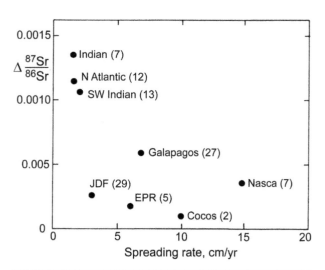

Fig. 6.8 Total ranges of Sr isotope ratio (Δ) for MORB glasses or leached whole-rocks from a given ridge, plotted against spreading rate on that ridge (JDF = Juan de Fuca; EPR = East Pacific Rise). Figures in brackets indicate number of analyses. After Batiza (1984).

form of a 'train of blobs'. Later workers (e.g. Allegre et al., 1980) developed the idea that such trains of blobs need not simply pass in streams from a (hypothetical) lower mantle reservoir through the asthenosphere, but could be part of the convecting asthenosphere itself. Hence Allegre et al. identified three alternative models for 'blob heterogeneity' of the asthenosphere (Fig. 6.7).

In an analysis of basaltic glasses from the major ocean basins, Cohen and O'Nions (1982) showed that the (comparatively) very large range of Pb isotope variation seen in Atlantic MORB was not equalled on the East Pacific Rise. Rather than attributing these differences to a smaller degree of mantle heterogeneity beneath the Pacific, Cohen and O'Nions argued that approximately equal degrees of heterogeneity in the Atlantic and Pacific upper mantle were homogenized in the large magma chambers associated with fast-spreading Pacific ridges. Support for this model came from the observation by Zindler et al. (1984) that seamounts near the East Pacific Rise showed much more variation than the adjacent ridge.

Batiza (1984) confirmed the inverse effect of ridge spreading rate on isotopic heterogeneity by plotting total ranges of $^{87}Sr/^{86}Sr$ (Δ) for different mid-ocean ridges against their spreading rate (Fig. 6.8). He attributed the small range of compositions on the fast-spreading ridges to homogenization, during the melting process, of a mantle 'ubiquitously heterogeneous on a small scale'. Low isotopic variation on some slow-spreading ridges (e.g. Juan de Fuca) was attributed to either their short length or limited sampling. Batiza adopted the more gastronomically elegant term of 'plum pudding mantle' to describe this blob-bearing asthenosphere. Allegre et al. (1984) also found an inverse correlation between ridge spreading rate and isotopic variation, but argued that homogenization must be primarily by (solid state) mantle convection rather than magma mixing.

In order to express the idea that plume and plum pudding models should not be thought of as mutually exclusive, but rather as a continuum of phenomena, Sun (1985) coined the term 'plume pudding mantle' (sic). Plums and plumes might originate from a variety of phenomena. However, this question cannot effectively be answered by the application of single isotope systems, and will be discussed below on the basis of co-variations in multiple isotope systems.

6.1.5 Marble Cake Mantle

Fluid dynamic modelling of the convecting asthenosphere (Richter and Ribe, 1979; McKenzie, 1979) suggested that discrete structures in the mantle (e.g. blobs, plums, etc.) cannot remain undeformed for long periods of time in the convecting asthenosphere. They will tend to be elongated and sheared, until eventually they are physically homogenized with the depleted reservoir. Polve and Allegre (1980) argued that evidence of this process was provided in orogenic lherzolites (Fig. 6.9), which contain alternating bands of (depleted) lherzolite and (enriched) pyroxenite. They suggested that this banding might have been generated by convective 'stirring' and stretching of a two-part sandwich of oceanic crust and underlying residual lherzolite/harzburgite, which is recycled back into the mantle by subduction. Allegre and Turcotte (1986) coined the term 'marble cake mantle' to describe this concept, and argued that it is representative of the structure of much of the upper mantle.

Prinzhofer et al. (1989) argued that random mixing between partial melts of pyroxenite and peridotite in a marble cake mantle could generate the large ranges of incompatible element concentrations and the moderate range of radiogenic isotope ratios seen in lavas from a small (40 × 10 km) area of the East Pacific Rise. However, mixing in the magma chamber is not capable of explaining the length dependence of large-scale isotopic anomalies on ridges (Kenyon, 1990). For example, the isotopic 'texture' of the South Atlantic Ridge requires convective homogenization over distances up to 1000 km (Fig. 6.10). This is too large for a magma chamber, since it is more than the length of ridge segments between transform faults. Hence it follows

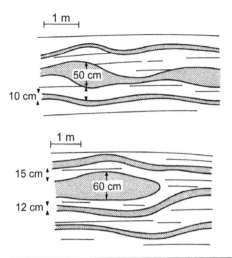

Fig. 6.9 Schematic illustration of 'marble cake' mantle consisting of pyroxenite (shaded) and lherzolite layers in the Beni Bousera peridotite of Morocco. After Allegre and Turcotte (1986).

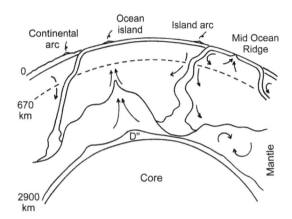

Fig. 6.11 Cartoon to show a hypothetical mantle with 'intrinsically dense' material at its base, upon which recycled slab material would collect. After Kellogg et al. (1999).

that homogenization must be at a deeper level, either by solid state convection of the marble cake mantle, or during magma ascent from the partial melting zone under the ridge (e.g. Section 13.3).

6.1.6 Mantle Convection and Viscosity

The proposed marble cake structure of the mantle, coupled with the existence of mantle plumes, raises questions about the overall convective structure of the mantle. Lithophile isotope and noble gas evidence pointing to a relatively undepleted and undegassed lower mantle led many geochemists in the 1980s to adopt a two-layer model for the convective structure of the mantle (e.g. Allegre, 1982, 1997). In contrast,

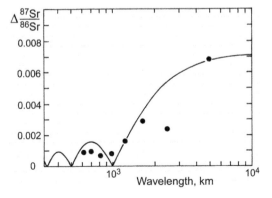

Fig. 6.10 Curve-fit for mixing of isotopic heterogeneity, compared with empirical data for amplitude *versus* wavelength of Sr isotope variation on the South Atlantic Ridge. After Kenyon (1990).

geophysicists have long preferred models involving a single layer of mantle convection (e.g. Davies, 1977).

Over the past couple of decades, geophysical evidence for single-layer convection has become stronger, whereas the geochemical evidence for two-layer convection has weakened (Section 6.2.3). However, evidence that mantle viscosity increases markedly with depth seems to imply a convective regime somewhere between the two extremes of single- and double-layer convection. In principle, the lower mantle is part of the main convective system of the mantle, but in practice its high viscosity may isolate large bodies of it from the more rapidly convecting upper mantle (Bunge *et al.*, 1996).

A model involving a heterogeneous lower mantle was supported by new tomographic evidence for the gross structure of the mantle (van der Hilst and Karason, 1999). Variations in seismic properties suggest that the mid mantle is relatively homogeneous, whereas the upper and lower mantle are more heterogeneous. In the upper mantle, this is attributed to phase changes with depth and to the large temperature variations between plumes, slabs and asthenospheric mantle. In the lower mantle, it is presumed to correspond to chemical heterogeneity between different domains of primordial, recycled and mixed material.

The model of van der Hilst and Karason (1999) was developed by Kellogg *et al.* (1999) to reinstate a sharper division between upper and lower mantle domains. The division between these domains was placed at 1600 km depth, at the top of the heterogeneous lower mantle domain identified by van der Hilst and Karason. The main feature of the new model was a hypothetical reservoir of 'intrinsically dense' material at the bottom of the mantle, perhaps reflecting Fe enrichment. This accumulated dense material was suggested to overlie the D″ (D-double-prime) layer that represents the core–mantle transition zone (Fig. 6.11). However, other workers have proposed a smaller dense layer that corresponds directly with the D″ layer itself (Tolstikhin and Hofmann,

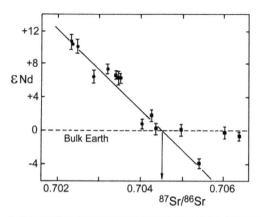

Fig. 6.12 Plot of εNd against Sr isotope ratio for ocean floor, ocean island and continental basalts analysed before 1976. Arrow shows estimated Bulk Earth strontium. After DePaolo and Wasserburg (1976).

2005). Subducted slabs might penetrate the dense material, but would normally collect on its upper surface.

Kellogg et al. suggested that the dense layer originated early in Earth history, perhaps from a magma ocean or by recycling of particularly mafic oceanic crust in the Archean. Because this layer would be enriched in incompatible elements relative to the upper mantle, it would also supply large amounts of heat. However, this presents a problem, because heating of the basal layer would reduce its density, therefore risking entrainment into the convective system of the overlying mantle. However, Tolstikhin and Hofmann (2005) proposed that dense meteoritic material from the 'late heavy bombardment' could accumulate on early mafic crust prior to its subduction, thus stabilizing the early enriched layer against density inversion and preserving it through Earth history. This issue will be discussed further below.

6.2 The Nd–Sr Isotope Diagram

In the mid 1970s, studies of the origins of mantle heterogeneity were revolutionized by the application of Nd isotope analysis to young volcanic rocks (DePaolo and Wasserburg, 1976; Richard et al., 1976). DePaolo and Wasserburg plotted $^{143}Nd/^{144}Nd$ isotope ratios, in the form of εNd (Section 4.2) against $^{87}Sr/^{86}Sr$, and found a negative correlation between them in oceanic and some continental igneous rocks (Fig. 6.12). Based on this evidence, they suggested that the formation of basaltic magma sources in the mantle involved the coupled fractionation of Sm/Nd and Rb/Sr, while some continental samples (which lay to the right of the main correlation line) could have been contaminated by radiogenic Sr in the crust.

On the basis that the 'Bulk Earth' has a chondritic Sm/Nd ratio, DePaolo and Wasserburg (1976) used the intersection of the chondritic (CHUR) composition (εNd = 0) with the mantle Nd–Sr correlation line to calculate an unfractionated mantle (= Bulk Earth) $^{87}Sr/^{86}Sr$ ratio of 0.7045 (Fig. 6.12). On the other hand, more recent evidence from the extinct nuclide system $^{146}Sm–^{142}Nd$ has suggested that the Bulk Earth might not be chondritic after all (Section 15.7.2). This question will be discussed further below.

O'Nions et al. (1977) extended the $^{143}Nd/^{144}Nd$ versus $^{87}Sr/^{86}Sr$ correlation line in oceanic volcanics by analysing a larger suite of ocean island basalts (OIB). This included two samples from Tristan da Cunha with $^{143}Nd/^{144}Nd$ ratios lower than the Bulk Earth, indicative of a mantle source which is slightly enriched in light rare earths relative to Bulk Earth. O'Nions et al. argued that enrichment of some mantle sources in Nd/Sm and Rb/Sr (and depletion of others such as MORB) could be explained by trace element partition between solid and liquid silicate phases. In view of the long half-lives of Rb and Sm, they concluded that these heterogeneities had existed for long periods of time.

6.2.1 The Mantle Array and OIB Sources

The Nd–Sr isotope correlation in oceanic rocks was first referred to as the 'mantle array' by DePaolo and Wasserburg (1979). They attributed the OIB which form most of this array to a chondritic lower mantle source contaminated by mixing with melts from the LIL-depleted MORB source during ascent. Little attention was given to the problem of generating enriched oceanic mantle, since Tristan da Cunha was regarded as more or less representing a primitive mantle composition similar to the Bulk Earth (Allegre et al., 1979; O'Nions et al., 1980).

The first persuasive evidence against simple mixing between Bulk Earth and MORB-source mantle was provided by the extension of the mantle array into the 'enriched' lower right quadrant of the Nd–Sr isotope diagram. This was convincingly demonstrated in a study of the Kerguelen Islands (Dosso and Murthy, 1980), shown in Fig. 6.13. More recent work has shown that part of the Kerguelen Plateau is underlain by a fragment of continental lithosphere, which has imparted enriched isotopic signatures to some of the magmas erupted through it (Weis et al., 2001; Neal et al., 2002). However, the lavas of the Kerguelen Islands themselves are not affected by this phenomenon, and are still attributed by most workers to an enriched source in the deep mantle (e.g. Weis and Frey, 2002).

Another enriched mantle source was sampled by alkali basalts from Sao Miguel in the Azores (Hawkesworth et al., 1979). However, in this case the data trended to enriched $^{87}Sr/^{86}Sr$ compositions off to the right of the mantle array. The Sao Miguel trend was subsequently extended by data from Samoa and the Society Islands (White and Hofmann, 1982), breaking the simple Nd–Sr isotope correlation in OIB into a 'mantle disarray' (White, 1981). Further disarray is caused by the existence of other ocean islands with compositions to the left of the mantle array, such as St Helena.

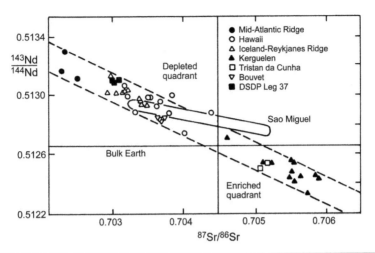

Fig. 6.13 Plot of Nd versus Sr isotope compositions for oceanic volcanics showing extension of the 'mantle array' into the 'enriched' quadrant relative to Bulk Earth, based on Kerguelen data (▲), and the extension of Sao Miguel data into the upper right quadrant. Modified after Dosso and Murthey (1980).

One model proposed to explain the isotopic and trace element characteristics of the Azores data is mantle metasomatism. Hawkesworth *et al.* (1979) suggested that this caused LIL-element enrichment of the source a few tens of Ma before generation of the Azores magmas. This can explain how a mantle source with a long-term depletion in light REE relative to Bulk Earth (as indicated by Nd isotope compositions, Fig. 6.13) can nevertheless be enriched in Sr isotopes and LIL trace elements. More recent work (Elliott *et al.*, 2007) has supported the metasomatism model for the mantle source of Sao Miguel, but also suggested that this was an unusual enrichment process that is not typical of OIB sources in general.

An alternative enrichment model proposed by Hofmann and White (1980, 1982) involved recycling of ancient oceanic crust into the OIB source. This could explain variably enriched trace element and isotopic compositions within the mantle array, while the deviation of the Azores, Samoa and Society Islands to the right of the mantle array could be explained by the addition of subducted sediment to the recycled oceanic crust (Fig. 6.14). This has become the 'standard model' to explain mantle evolution by crustal recycling (e.g. Christensen and Hofmann, 1994; Hofmann, 1997). However, it has more recently been challenged in its ability to generate the isotope signatures of enriched OIB sources, as discussed in Section 6.5.4.

6.2.2 Box Models for the MORB Source

The observed LIL-depleted nature of the MORB source relative to CHUR has very important implications for the evolution of the mantle, and was attributed to the extraction of the continental crust from a chondritic Bulk Earth composition. This was first modelled by Jacobsen and Wasserburg (1979) and O'Nions *et al.* (1979) using calculations commonly termed

'box models'. In such models, the Earth is divided into chemical reservoirs which may exchange matter, grow, shrink, etc., and whose evolution is simulated over the Earth's 4.5 Ga history. Typical reservoirs or 'boxes' are the crust, mantle and core, although these may be subdivided, e.g. into upper and lower mantle. The evolution of the silicate Earth is portrayed in some alternative box models in Fig. 6.15, which will be briefly discussed.

O'Nions *et al.* (1979) examined two alternative models of mantle differentiation and crustal growth (1a and 1b in Fig. 6.15). These models were based on the numerical solution of upward and downward transport coefficients for several elements in 90 steps, each corresponding to 50 Ma of Earth history. The model was constrained by boundary conditions

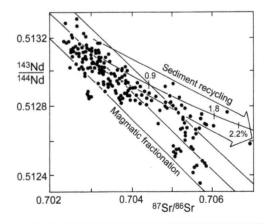

Fig. 6.14 Plot of Nd versus Sr isotope composition for oceanic volcanics, attributed to recycling of oceanic crust and sea floor sediment into the mantle. After Hofmann and White (1982).

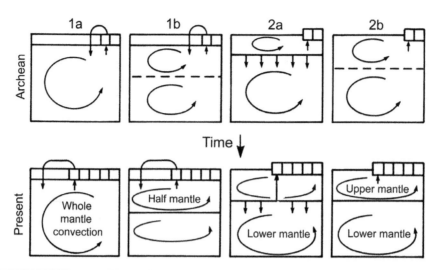

Fig. 6.15 Some box models for mantle evolution. Models 1a and 1b correspond to either whole- or half-mantle depletion due to crustal extraction. Models 2a and 2b show alternative evolution of a depleted mantle reservoir by either increasing its size (2a) or its *degree of depletion* (2b).

in the form of the composition of the primitive chondritic mantle at 4.55 Ga and the estimated composition of the outermost 50 km of the Earth (including the continental and oceanic crust) at the present day. In Fig. 6.16 the results for $^{87}Sr/^{86}Sr$ evolution are shown for cases where (a) the whole mantle is depleted by the extraction of the upper 50 km layer; and (b) only the upper half of the mantle is depleted. (These scenarios correspond to models 1a and 1b in Fig. 6.15.) The conclusion was that model (b) yields a much better fit to the present day $^{87}Sr/^{86}Sr$ ratio of the depleted (MORB) source.

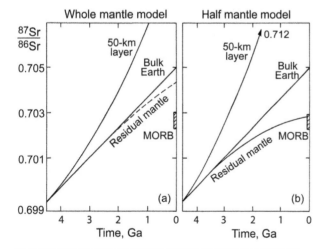

Fig. 6.16 Plots of Sr isotope evolution against time to compare the effects of (a) whole-mantle or (b) half-mantle convection on the degree of depletion predicted for the residual (MORB) reservoir. Hatched area is the present day composition of MORB. After O'Nions *et al.* (1979).

Jacobsen and Wasserburg (1979) used similar box models to examine another aspect of global differentiation (2a and 2b in Fig. 6.15). They simplified their treatment by considering only unidirectional transport of species from the mantle to generate the crust continuously over geological time, and solved the transport equations algebraically. In model 2a (Fig. 6.15) melts are extracted from the primitive mantle and generate the continental crust and a depleted mantle, both of whose *volumes* grow over geological time. However, the elemental *composition* of the depleted mantle remains constant through time. Mass balance calculations based on Sm)Nd data led Jacobsen and Wasserburg to calculate that only 33% of the mantle needs be depleted to generate the continental crust, corresponding to the formation of a depleted MORB reservoir occupying approximately the upper 650 km of the mantle.

In Jacobsen and Wasserburg's second model (2b in Fig. 6.15), the crust is extracted from a fixed *volume* of mantle which therefore becomes more and more depleted through geological time. The mass of this depleted mantle needed to generate the crust was calculated as only 25% of the total mantle. In this model, the isotopic composition of new continental crust will reflect a derivation from depleted mantle, whereas in model 2a new continental crust would have a chondritic (primitive mantle) isotopic signature. On the basis of Nd isotope data available to them at the time (Section 4.2.1), Jacobsen and Wasserburg preferred model 2a. However, more recent Nd isotope evidence (Section 4.2.2) strongly favours model 2b. The different estimates of O'Nions *et al.* (1979) and Jacobsen and Wasserburg (1979) for the volume of the depleted mantle reflect the uncertainties involved in estimating the trace element and isotopic composition of the crust.

DePaolo (1980) studied a model similar to 2b, but with the possibility of crustal recycling into the mantle, and again concluded that only 25–50% of the mantle need be depleted to generate the continental crust. It appears that modest amounts of recycling of continental crust have relatively little effect on mantle Nd–Sr isotope systematics, but a large effect on Pb (Section 5.4.3).

As noted in Section 6.1.6, convective models for the mantle developed over the past 20 years or so have tended to invoke the storage of large volumes of dense subducted oceanic crust at the base of the mantle. The existence of this reservoir will clearly have drastic effects on box model calculations, but its size is difficult to constrain.

By extrapolating the present day production of MORB over an estimated 1 to 1.5 Ga residence time in a basal enriched reservoir, Campbell (2002) estimated that the basal reservoir might be around five times the mass of the continental crust, or 2.5% of the mantle. However, this appears to be a low estimate, since the total production of oceanic crust over Earth history would be around 7% of mantle mass. At the other end of the scale, it seems questionable whether the large basal reservoir (20% of the mantle) proposed by Kellogg et al. (1999) could have sufficient excess density to stabilize all of it against entrainment into the overlying convective mantle. Alternatively, Tolstikhin and Hofmann (2005) proposed a basal layer of intermediate size, comprising ca. 5% of mantle mass. As noted above, they argued that this material had a large density excess over the depleted mantle, due to the weighting of subducted oceanic crust by a dense regolith of meteoritic material, representing the accreted 'veneer' from Earth's intense meteorite bombardment (Section 8.3.1).

The box model of Workman and Hart (2005) was one of the first to explicitly consider the effects of a deep enriched reservoir on mantle depletion models. They calculated that without a basal enriched layer, only 33% of the mantle could be depleted to produce the MORB reservoir, leaving 65% of the mantle undepleted (consistent with the early box models discussed above). At the other extreme, if no undepleted material is allowed to remain, 25% of the mantle must be enriched (including subducted oceanic crust and other OIB sources) in order to balance a MORB source reservoir occupying the remaining 75% of the mantle.

6.2.3 Nd–142 and Early Earth Differentiation

The box models examined above were revolutionized by the discovery that the accessible mantle has a non-chondritic ^{142}Nd signature (Boyet and Carlson, 2005). This implied that either the Earth was built from non-chondritic material, or that an elemental Sm/Nd fractionation event occurred very early in Earth history. However, both alternatives create major difficulties.

Boyet and Carlson (2005, 2006) preferred the early fractionation model. However, the Sm/Nd fractionation required to harmonize ^{142}Nd signatures between Earth and chondrites also affects terrestrial ^{143}Nd evolution. Because of the short half-life of the ^{146}Sm parent of ^{142}Nd, the Nd signatures

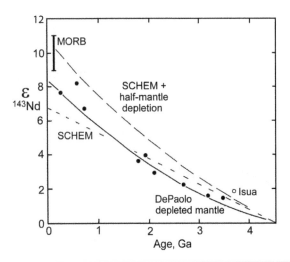

Fig. 6.17   Effects of the SCHEM model, plus extraction of enriched crustal and mantle sources, relative to mafic igneous rocks (●). The depleted mantle of DePaolo (1980) is shown for reference (solid line). Modified after Caro and Bourdon (2010) and Dickin (2016).

could only be satisfied if the fractionation event occurred so early in Earth history that it preceded the giant Moon-forming impact (Section 15.6.6). To many researchers, this seemed an almost impossible requirement.

Other geochemists (e.g. Caro et al., 2008) proposed that the Bulk Earth accreted with a fractionated Sm/Nd ratio relative to chondrites, hence termed the supra-chondritic Earth model (SCHEM). However, this model also creates great difficulties for terrestrial ^{143}Nd evolution. The required Bulk Earth ^{147}Sm/^{144}Nd ratio (ca. 0.206), leads to a present day ε^{143}Nd value of around 6.9 (Fig. 6.17). However, the extraction of the enriched continental crust *and* enriched OIB source reservoirs must cause substantial *further* Sm/Nd fractionation in the depleted mantle, generating a mantle source (SCHEM + half-mantle depletion) that is too LIL-depleted (hence εNd too radiogenic) to generate the signatures of mafic igneous rocks over earth history (solid circles in Fig. 6.17).

Recent ^{142}Nd data for the Moon allow a reconciliation between these models, based on evidence that the Moon represents a sample of the very early Earth's mantle (Dickin, 2016). The Moon has a ^{142}Nd signature within error of enstatite chondrites, showing that it could have accreted without Sm/Nd fractionation relative to the chondritic reservoir (CHUR). On the other hand, the Earth's accessible mantle has a ^{142}Nd signature ca. 6 ppm more radiogenic than the Moon, implying that the Earth *did* undergo early Sm/Nd fractionation (Section 15.7). Nevertheless, the discrepancy in ^{142}Nd signatures is three times less than originally anticipated by Boyet and Carlson (2005), allowing a fractionation event up to 150 Ma after the origin of the solar system. This means that the fractionation event could correspond to the

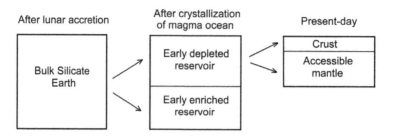

Fig. 6.18 Box model showing mantle evolution in two main stages, comprising the early fractionation event and subsequent separation of crustal and OIB sources. Modified after Korenaga (2009).

formation of the Earth's first mafic crust, generated by cooling of the terrestrial magma ocean that resulted from the giant impact. But although the timing of the event is postponed, the magnitude of the geochemical fractionation, and the size of the resulting early enriched reservoir, remain largely as proposed by Boyet and Carlson (2005, 2006). This leads to box models of the type shown in Fig. 6.18.

In the box model proposed by Boyet and Carlson (2006), the early Sm/Nd fractionation event produces an early enriched reservoir (EER) that must never be significantly sampled by mantle plumes. This is a problem, in view of the high LIL-content of this layer, which should contain 20% of the heat-producing elements of the Bulk Silicate Earth. These heat-producing elements would be expected to reduce its density, causing entrainment into the overlying mantle. However, the model of Tolstikhin and Hofmann (2005) proposes that the basal enriched layer remains at the core–mantle boundary due to its high intrinsic density, attributed partly to its content of meteorite regolith.

If the early enriched layer remains out of circulation, the early depleted reservoir must give rise to the continental crust and the accessible mantle. The latter must by definition include depleted MORB-source mantle (DMM) and enriched OIB source reservoirs. Workman and Hart (2005) proposed a very depleted composition for DMM, based on analyses of abyssal oceanic peridotites. Other workers (e.g. Donnelly *et al.*, 2004; Boyet and Carlson, 2006) have argued that this type of MORB source is too depleted, because it ignores enriched components that make up a substantial fraction of the upper mantle. However, this problem can be solved by attributing much of the accessible mantle to a 'prevalent mantle' reservoir (also called FOZO) that is less depleted than DMM. This issue is discussed in more detail in Section 6.4 and Section 6.5.

Seismic tomography has provided evidence for distinctive bodies of material at the base of the mantle that may contain the proposed early enriched material. This work began in the 1980s, and led to the identification of low-velocity zones below the Western Pacific and Southern Africa (e.g. Dziewonski, 1984; Garnero and Helmberger, 1995; Wen and Helmberger, 1998). These have been termed 'Large Low

Shear-wave Velocity Provinces' (LLSVPs, Fig. 6.19). More recently, it was further suggested that these areas may have significant topographic relief (e.g. Lay and Garnero, 2004), possibly stabilized by their more Fe-rich chemistry and by a phase transition to a denser 'post-perovskite' mineral lattice (Murakami *et al.*, 2004; Lay *et al.*, 2006). Younger cold subducted material may collect at the base of the mantle beside the LLSVPs before being reheated and rising to form plumes (Lay *et al.*, 2006; Burke *et al.*, 2008).

6.3 Pb Isotope Geochemistry

Pb isotopes are a powerful tool in studies of mantle and crustal evolution because the three different radiogenic isotopes are generated from parents with a wide span of half-lives, two of which are isotopes of the same element. By using these isotope tracers in combination, it is not only possible to identify the nature of differentiation events, but also to place constraints on their timing. Nevertheless, mantle Pb isotope systematics are substantially decoupled from

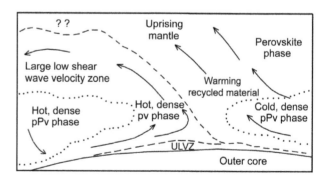

Fig. 6.19 Cross-section of a 'thermochemical pile' comprising dense Fe-rich material (left), beside which more recently subducted oceanic crust (right) may reside at the base of the mantle before rising in a plume. ULVZ = ultra-low-velocity zone; Pv = perovskite structure; pPv = post-perovskite structure. After Lay *et al.* (2006).

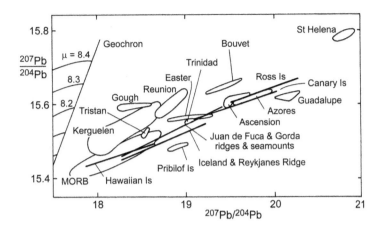

Fig. 6.20 Pb–Pb 'isochron' diagram showing linear arrays of data defined by oceanic volcanics. After Sun (1980).

other lithophile isotope systems, so it is appropriate to consider the evidence provided by this complex system before combining Pb isotope evidence with the other radiogenic tracers.

6.3.1 Pb–Pb Isochrons and the Lead Paradox

Early inferences about the Pb isotope evolution of the mantle were based on the analysis of galenas. These continue to give important information about Archean Pb isotope evolution. However, younger galena ores are plagued by the complex evolutionary history implied by the formation of ore deposits, involving both mantle and crustal residence times. As analytical methods improved, it was possible to analyse more direct mantle-derived samples such as basic magmas. Early work (e.g. Gast et al., 1964; Tatsumoto, 1966) showed substantial Pb isotope heterogeneity in ocean island basalt (OIB), but the data could not be put into a meaningful geochemical framework until the new uranium decay constants were established in the 1970s. Subsequent work (e.g. Sun et al., 1975; Tatsumoto, 1978) showed that oceanic volcanics define a series of arrays to the right of the geochron on the Pb–Pb 'isochron' diagram (Fig. 6.20).

The lack of overlap with the geochron contrasts with the Sr–Nd diagram, where the mantle array overlaps the Bulk Earth point (Section 6.2). This is the opposite of the expected behaviour, since geochemical evidence suggests that U is more incompatible than Pb during mantle melting (e.g. Tatsumoto, 1966, 1978), and should therefore generate low U/Pb ratios in the mantle residue. This problem was termed the 'lead paradox' by Allegre et al. (1980), who argued that the solution was Pb partition into the core, early in Earth history. This is another facet of the problem of the young apparent Pb–Pb age of the Earth. However, that model is best constrained by the study of Archean Pb evolution (Section 5.3.1). Hence, the focus here will be more on the *relative* distribution

of ocean Pb signatures, rather than their position compared to the Bulk Earth (geochron).

The other major feature of the data in Fig. 6.20 is the existence of Pb–Pb arrays with slope ages between 1 and 2.5 Ga. These can be interpreted in three principal ways: as the products of two-component mixing processes; as resulting from discrete mantle differentiation events; or resulting from continuous evolution of reservoirs with changing μ values. Each of these models may be applicable to different magmatic suites.

The mixing model was championed by Sun et al. (1975), who showed that the array of Pb isotope compositions in Reykjanes Ridge basalts was best explained by two-component mixing of 'plume' and 'low-velocity zone' (upper mantle) components under Iceland. They suggested that the linear Pb isotope arrays generated by several other ocean islands might be explained by the same mechanism. However, since these arrays have different slopes, a mixing model can only work if each array is attributed to mixing of the MORB reservoir with a different enriched source (Sun, 1980). Therefore, the problem of explaining the origin of these radiogenic sources still remains.

A model of continuous mantle evolution with a changing μ value was adopted by Dupre and Allegre (1980) to explain the Pb isotope composition of leached basalts dredged from the Mid Atlantic Ridge. The data define a linear array to the right of the geochron, whose slope yields an apparent Pb–Pb isochron of 1.7 Ga age. However, this result was interpreted, not as a worldwide mantle differentiation event, but as an average age for continuous differentiation from ca. 3.8 Ga to the present. This can occur by the mixing of enriched components with the depleted mantle in numerous small events. This model is most applicable to the MORB source, which is now recognized as a mixing zone between many other components. This subject will be dealt with in more detail below.

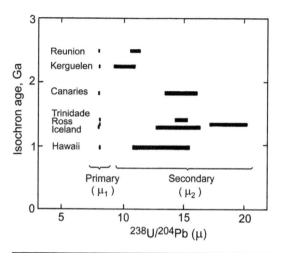

Fig. 6.21 Range of μ values required to explain OIB sources using a two-stage Pb evolution model. Parental mantle (μ_1) undergoes differentiation events at different times to yield discrete OIB source domains (μ_2). Modified after Chase (1981).

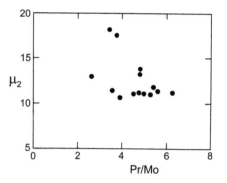

Fig. 6.22 Pb isotope data for OIB, expressed in terms of the μ_2 value of a two-stage evolution model. These show no correlation with a trace element index which measures possible fractionation of siderophile elements into the Earth's core. After Newsome et al. (1986).

The differentiation model was advocated by Chase (1981), who evaluated OIB data in terms of a two-stage Pb isotope evolution model. This allowed $^{238}U/^{204}Pb$ values to be calculated for a 'primary' mantle reservoir (μ_1) and for the secondary sources (μ_2) which yield OIB Pb–Pb arrays. Chase found that values of μ_2 are variable within each island group and between groups, but the calculated μ_1 value was remarkably constant (7.84–7.96) for all of the data (Fig. 6.21). He therefore concluded that ocean islands are derived from separate OIB sources of variable age, but that these in turn were derived from a single long-lived primary reservoir.

Many different explanations have been proposed for the radiogenic Pb signatures in OIB, of which the most extreme compositions are appropriately labelled the 'Hi-Mu' source (Zindler and Hart, 1986). Possible origins of this component by mantle recycling models are discussed in Section 6.5.4. However, an alternative mechanism to create mantle reservoirs with high μ values involves *prolonged* Pb partition into the Earth's core (Vidal and Dosso, 1978; Allegre, 1982). Early Pb fractionation into the core has often been proposed as a solution to the Pb paradox (as mentioned above), but its specific application to OIB sources was based on a model of late core growth, causing continued Pb fractionation through Earth history.

Newsome *et al.* (1986) argued that this model could be tested by examining the distribution of other elements such as Mo and W that have much higher distribution coefficients from a lithophile (mantle) to siderophile (core) phase than Pb. Hence if late Pb partitioning into the core is invoked to explain radiogenic OIB sources (e.g. St Helena) then these sources should be very depleted in Mo and W. However, allowance must be made for the behaviour of Mo during

solid–liquid partitioning in OIB magma genesis. This is done by comparing Mo with a lithophile element with similar bulk partition coefficients, such as the light rare earth element Pr. However, Pr-normalized Mo abundances show no correlation with radiogenic Pb isotope ratios in OIB (Fig. 6.22), suggesting that late fractionation of Pb into the core cannot explain radiogenic OIB arrays.

6.3.2 The Kappa Conundrum

One of the most important questions concerning the origins of Pb signatures in oceanic volcanics is the degree to which their Pb isotope compositions are 'supported' by the actual U/Pb ratios in their sources. Another way of expressing this question is whether the calculated present day or 'instantaneous' μ value of the source is consistent with its 'time-integrated' μ value, meaning the U/Pb ratio over geological time necessary to generate present day Pb isotope ratios. However, because of trace element partitioning during the melting process, it is difficult to recover the instantaneous μ value of the source from erupted products (e.g. Tatsumoto, 1966, 1978). An attempt to answer this question is made in Section 6.3.3, but here we focus on the kappa values ($^{232}Th/^{238}U$ ratios) of magma sources. These are less affected by the melting process, due to the similar partition coefficients of U and Th, and are therefore easier to determine.

In early work, Tatsumoto (1978) estimated an instantaneous κ value of ca. 2 in the MORB source from elemental Th/U ratios in lavas. However, it was not clear that such low values were representative of the mantle source until it was shown, in 1981, that Th/U ratios could be recovered from $^{232}Th/^{230}Th$ activity ratios in oceanic volcanics (Section 13.3).

The alternative time-integrated κ value of a mantle source is determined from the abundances of the U and Th daughter products, ^{206}Pb and ^{208}Pb. The radiogenic $^{208}Pb^*/^{206}Pb^*$ ratio of the source is first determined by approximating a single-stage model between the Earth's formation (T) and its composition at the time of magma extraction (t). Hence (following

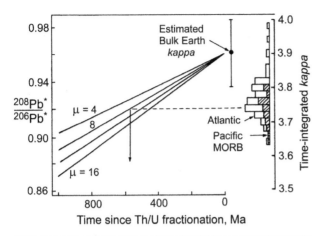

Fig. 6.23 Evolution of time-integrated κ values as a function of residence time in the MORB reservoir, starting at a value of 3.9. Histogram on the right indicates the deduced time-integrated κ values for the sources of Pacific (hatched) and Atlantic MORB. After Galer and O'Nions (1985).

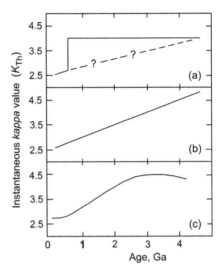

Fig. 6.24 Alternative evolution lines for the instantaneous kappa value of the MORB source. The area under each curve yields time-integrated kappa (κ_t). Modified after Elliott *et al.* (1999).

Allegre *et al.*, 1986) we can define the radiogenic ^{208}Pb/^{206}Pb ratio of a mantle reservoir as

$$\frac{^{208}\text{Pb}^*}{^{206}\text{Pb}^*} = \frac{\left(\frac{^{208}\text{Pb}}{^{204}\text{Pb}}\right)_t - \left(\frac{^{208}\text{Pb}}{^{204}\text{Pb}}\right)_T}{\left(\frac{^{206}\text{Pb}}{^{204}\text{Pb}}\right)_t - \left(\frac{^{206}\text{Pb}}{^{204}\text{Pb}}\right)_T} \quad [6.1]$$

The Th/U ratio (κ value) of a closed-system reservoir can be calculated from ^{208}Pb*/^{206}Pb* by solving U–Th and U–Pb decay equations for values T and t (Chapter 5). However, we can also calculate the average or 'time-integrated' Th/U ratio of an open system from time T to t. It should be noted that in the following discussion, any ancient Th/U values are normalized for subsequent radioactive decay and presented in terms of their 'present day equivalent' κ value.

Galer and O'Nions (1985) made the first explicit comparison between instantaneous and time-integrated κ values for the MORB source. They showed that the time-integrated κ value (3.75) is much higher than the instantaneous value (2.5), and only slightly less than the estimated Bulk Earth value (3.9). This discrepancy has been termed the 'second Pb paradox' (Kramers and Tolstikhin, 1997) or the 'kappa conundrum' (Elliott *et al.*, 1999). To solve this problem, Galer and O'Nions proposed that the MORB source is an open system for Pb, which is buffered over geological time by a less depleted reservoir. Hence they argued that Pb only had a relatively short residence time in the MORB reservoir, and spent most of Earth history in a reservoir with a κ value near Bulk Earth (Fig. 6.23).

Galer and O'Nions calculated that a residence time of 600 Ma in a Th-depleted MORB reservoir with $\kappa = 2.5$ and a 4 Ga residence period in a reservoir with $\kappa = 3.9$ would give the time-integrated κ value of 3.75 needed to explain MORB lead isotope compositions (Fig. 6.23). They examined three pos-

sible locations for their proposed Bulk Earth κ lead source: upper continental crust, sub-continental lithosphere and lower mantle. However, the short upper mantle residence time for Pb calculated using their model was a severe test of the ability of any of these sources to buffer MORB Pb. Upper crust was excluded on the grounds that its high ^{207}Pb/^{204}Pb ratio would excessively perturb this ratio in the MORB source, while lithospheric mantle is too small to buffer the MORB source. Lower mantle buffering was able to explain the data, but required exchange with anywhere between one-quarter and one-half of its mass over geological time. This might be possible if the depleted upper mantle only constitutes the upper 670 km, but more recent evidence for a larger upper mantle MORB source makes it impossible to adequately buffer this reservoir with a Pb residence time of only 600 Ma.

A major omission from the treatment of Galer and O'Nions (1985) was any explanation of how or why the κ value of the MORB source has decreased over geological time. To emphasize the importance of this issue, the model of Galer and O'Nions is presented in Fig. 6.24a in the format used by Elliott *et al.* (1999), where the instantaneous κ value is plotted against time. The time-integrated κ value is then proportional to the area under the curve.

Also shown in Fig. 6.24 are two other models for the evolution of κ values against time. The second model (Fig 6.24b) approximates to that of Allegre *et al.* (1986). Although they did not explicitly consider the variation of instantaneous κ values against time, Allegre *et al.* proposed a gradual decrease of time-integrated κ values over time, which therefore implies model (b). This model requires a starting κ value of around 4.5 in the early mantle. However, Bulk Earth κ

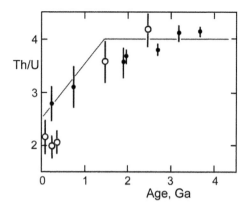

Fig. 6.25 Plot of mantle κ values over geological time derived from Th/U values for mafic rocks (•) and kimberlitic zircon (○), compared to the uranium mobilization model of Elliott *et al.* (1999). Results with large errors are omitted. Data from Collerson and Kamber (1999) and Zartman and Richardson (2005).

values above 4 are rendered very unlikely by new meteorite evidence. Since most meteorites can be considered effectively closed systems over geological time, their instantaneous and time-integrated κ values should be the same, and the latter can be precisely defined. For example, Blichert-Toft *et al.* (2010) obtained a precise time-integrated κ value of 3.876 ± 0.016 from the Pb isotope analysis of a variety of meteorites, supporting the Bulk Earth κ value of 3.9 used by Galer and O'Nions (1985).

A third kappa evolution model (Fig. 6.24c) was presented by Zartman and Haines (1988), in the form of a revised version of the plumbotectonics model (Section 5.4.4). This model was designed to produce the correct time-integrated κ value of the MORB source, but in this case entirely by crustal recycling, with no communication between the MORB reservoir and the lower mantle. Although this model involved some buffering of Pb in the MORB source by crustal recycling, the amount was not enough to perturb the $^{207}Pb/^{204}Pb$ ratio.

In this model (Zartman and Haines, 1988), the main mechanism for controlling the κ value of the MORB source was a variable U flux to the mantle. This was achieved by progressive U/Th differentiation of the crust, in which only U-rich upper crust was recycled into the mantle. However, this strong control of the Th/U ratio of the MORB source by crustal recycling was only possible by restricting the MORB reservoir to the upper 670 km of the mantle, and by assuming a high value of 4.3 for the Bulk Earth κ value.

To solve the kappa conundrum with a lower Bulk Earth κ value of 3.9, Kramers and Tolstikhin (1997) proposed that uranium recycling to the upper mantle was delayed until 1.5 Ga (Fig. 6.25). They argued that any uranium released into the sedimentary system during the Archean was locked up in insoluble reduced forms, due to the low oxygen content

of the atmosphere. This prevented the uranium from entering the upper mantle and lowering its κ value. Oxygenation of the atmosphere released this uranium into the sea in the Mid Proterozoic, from where it entered oceanic crust by sea floor hydrothermal alteration. During subduction, this uranium was released into the upper mantle, causing a rapid lowering of κ in the MORB source (Fig. 6.25). Because the lowering of the upper mantle Th/U ratio is delayed until 1.5 Ga, the model can satisfy the time-integrated κ value (area under the line) despite a lower Bulk Earth κ.

The delayed crustal recycling model was tested by the analysis of mafic igneous rocks through time (Kamber and Collerson, 1999) and by analysing Th/U ratios in zircon from kimberlites (Zartman and Richardson, 2005). The latter data set required a correction for the ten-fold greater partition of U relative to Th into zircon, in order to calculate the κ value of the host magma. However, the results of the two studies were in overall agreement (Fig. 6.25), and support the delayed uranium mobilization model. This model overturns the argument of Galer and O'Nions (1985) that the $^{208}Pb*/^{206}Pb*$ ratio of the MORB source (and hence time-integrated κ) is unsupported by its Th/U ratio (instantaneous κ). This relaxes the requirement for a very short Pb residence time in the MORB source, but still implies relatively short residence (<1.5 Ga). Other lines of evidence for mantle Pb residence times will be discussed below.

Recent support for the delayed uranium recycling model has come from new uranium isotope data for MORB, OIB and arc related samples (Andersen *et al.*, 2015). The data were reported as $\delta^{238}U$ (per mil deviations from the CRM 145 standard), and showed that North Atlantic OIB sources (Azores, Canaries, Cape Verde, Iceland) all had $^{238}U/^{235}U$ ratios within error of Bulk Earth (Juvinas meteorite). In contrast, arc basalts were depleted in ^{238}U, while MORB samples were enriched (Fig. 6.26). The array of arc lavas trends approximately towards the $\delta^{238}U$ signature of the upper part of a Pacific sea floor drill-core, implying that this signal represents U extracted from the upper part of the subducted crust. On the other hand, the residue from this fluid extraction process should be *enriched* in ^{238}U, which would subsequently be subducted into the MORB source.

In contrast to MORB, Andersen *et al.* argued that the lack of a ^{238}U-enriched signature in OIB suggests that OIB source reservoirs were derived from crustal rocks subducted before the *complete* oxygenation of the atmosphere at the beginning of the Phanerozoic. Nevertheless, some analysed OIB did show a variable reduction in Th/U ratio, suggesting that they were derived from crustal units recycled after the *beginning* of atmospheric oxygenation at the end of the Archean. In other words, the lack of correlation between Th/U and ^{238}U in OIB suggests that the OIB source reservoir was derived by crustal recycling before atmospheric oxygenation could cause uranium isotope fractionation. However, the model does not preclude buffering of the MORB source Pb by OIB sources. Such mixing would look the same as mixing between altered oceanic crust and the Bulk Earth.

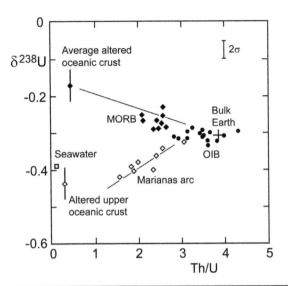

Fig. 6.26 Plot of Th/U ratios against $\delta^{238}U$ relative to CMB 145 standard ($^{238}U/^{235}U = 137.837$, Richter et al., 2010). Bulk Earth = 137.8 (Juvinas). Modified after Andersen et al. (2015).

6.3.3 The Third Lead Paradox

In addition to the puzzling behaviour of Pb isotope systems, leading to the first and second Pb paradoxes, Pb concentrations in MORB also display unexpected behaviour, leading Hart and Gaetani (2006) to label this problem a 'third Pb paradox'. The problem can be investigated on a 'Hofmann plot' (Hart and Gaetani, 2006; Hofmann et al., 1986) where trace element ratios in oceanic volcanics are plotted against elemental abundances (e.g. Fig. 6.27). If such a plot yields a horizontal data array, the trace elements in question are deemed to be equally incompatible, meaning that they have equal bulk distribution coefficients between mantle mineral phases and basaltic magma during partial melting.

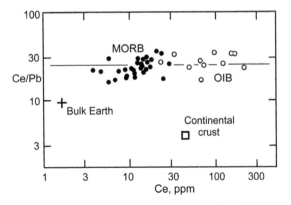

Fig. 6.27 'Hofmann plot' used to estimate the degree of incompatibility of Pb during mantle melting. After Hofmann et al. (1986).

For the case of Pb, Hofmann et al. (1986) demonstrated a horizontal array in MORB and OIB, when ratioed against the light rare earth element Ce (Fig. 6.27). This implies that Ce and Pb are not fractionated by magmatic processes, and hence that the measured Ce/Pb ratios of the volcanics should be representative of their mantle source. Therefore, since the oceanic mantle and continental crust have Ce/Pb ratios on opposite sides of the estimated primitive mantle composition (Bulk Silicate Earth), Hofmann et al. deduced that both the MORB and OIB source reservoirs are residues of continental crust extraction.

It is now clear that MORB and OIB should not be lumped together in Fig. 6.27, and that taken alone, MORB samples show a slight positive slope. More recent data sets have supported this distribution (e.g. Jenner and O'Neill, 2012), suggesting that Pb is actually slightly less incompatible than Ce during MORB melting. However, the revised slope is very gentle, suggesting only a slight reduction in the Ce/Pb ratio of the MORB source, from 25 to 20.

This excess Ce/Pb ratio in MORB relative to primitive mantle is the so-called 'third Pb paradox' of Hart and Gaetani (2006). However, a related puzzle is why Pb has a bulk distribution coefficient as large as the light REE. Experimental partition coefficients (e.g. Hauri et al., 1994) show that Pb is ten times more incompatible than Ce between basaltic melt and cpx, the main light-REE-bearing mantle phase. This implies that another phase must be holding back Pb during melting to increase the bulk distribution coefficient of Pb and explain the sub-horizontal Ce/Pb array in the Hofmann plot (Fig. 6.27). This other phase was argued by Meijer et al. (1990) and Hart and Gaetani (2006) to be mantle sulphide. However, because mantle sulphide must have very low U/Pb ratios, its presence as a long-lived mantle phase would also imply unradiogenic Pb isotope signatures. This is important, because unradiogenic Pb in mantle sulphides could explain the first Pb paradox, as well as the behaviour of Pb during MORB melting.

The first strong evidence for this phenomenon came from Pb analysis of the Horoman peridotite massif in northern Japan (Malaviarachchi et al., 2008). Whole-rock samples of plagioclase lherzolite displayed a very wide range of Pb isotope ratios, with $^{206}Pb/^{204}Pb$ from ca. 16.5 to 18.7 (Fig. 6.28). These were the least radiogenic Pb signatures found in mantle rocks, but could possibly represent samples of subcontinental lithospheric mantle. However, such an explanation was not possible for sulphide phases found by Burton et al. (2012) in an abyssal peridotite from the axial graben of the Mid Atlantic Ridge. This sample contained two different kinds of sulphide phases. Interstitial sulphides showed a trend towards a modern seawater Pb composition. In contrast, sulphides forming inclusions within silicate minerals yielded a Pb–Pb age of 1.83 ± 0.23 Ga (Fig. 6.28), which was supported by a Re–Os isochron age of 2.06 ± 0.26 Ga and Os model ages also around 2 Ga (Harvey et al., 2006).

The Pb inclusion data of Burton et al. (2012) were also colinear with the compositions of nearby MORB samples,

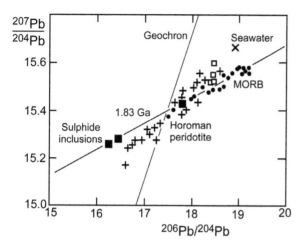

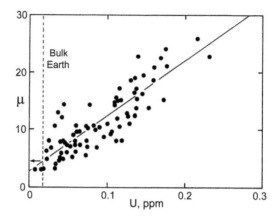

suggesting that the apparent Pb/Pb age of the sulphide inclusions and MORBs corresponds to a real mantle differentiation event at ca. 2 Ga. Burton *et al.* suggested that the analysed sulphide inclusions must have been shielded from the melting process, so that they did not contribute to the MORB liquids. When this evidence is coupled with the data for the Horoman peridotite, it suggests that unradiogenic Pb in refractory or 'shielded' inclusions could represent a substantial mantle Pb reservoir, and therefore can contribute a solution for the first Pb paradox, possibly along with other unradiogenic Pb reservoirs such as the core and lower crust.

Notwithstanding the slightly puzzling behaviour of Pb in comparison with Ce, a much more paradoxical finding concerns the μ value (U/Pb ratio) of the MORB source. It was already noted above that this value is hard to determine, due to the unpredictable behaviour of Pb during melting. However, White (1993) showed that an estimate of the μ value of the MORB source could be determined using the Hofmann plot.

Analysis of U and Pb in 82 glasses from the Atlantic, Pacific and Indian oceans revealed a positive correlation between $^{238}U/^{204}Pb$ and U content (Fig. 6.29). Not surprisingly, this suggests that U is more incompatible than Pb during MORB melting. However, White (1993) argued that even though these elements have different bulk distribution coefficients, the correlation line can still be used to obtain a reliable estimate for the μ value of the upper mantle. Since U is incompatible, the U content of the Bulk Silicate Earth estimated from chondrites (0.018 ppm) should be an upper limit for U in the depleted mantle (MORB source). Applying this value to the μ versus U correlation line leads to a maximum instantaneous μ value of 4.5 in the MORB source (Fig. 6.29). A very similar value is obtained from the larger data set of Jenner and O'Neill (2012).

Following Galer and O'Nions (1985), White (1993) argued that if the instantaneous μ value of the MORB source (from Fig. 6.29) is much lower than the time-integrated μ value (from Pb isotopes), then the radiogenic Pb sampled by MORB magmatism must have a relatively short (ca. 1 Ga) residence time in the upper mantle. He proposed that the longer lived reservoir supplying the MORB source was the lower mantle plume source sampled by OIB. This material generally has radiogenic Pb, and can therefore explain the high time-integrated μ value of MORB. However, the discovery of unradiogenic sulphide inclusions in upper mantle minerals reveals a hidden low-μ component that balances the radiogenic Pb sampled in melting. This hidden component is evidently old and was probably generated by Pb mobilization in ancient subduction zones. How this happened may be revealed by isotopic and geochemical investigation of subduction-related magmatism (Section 6.6). It is concluded that the unradiogenic low-μ sulphide component holds a major fraction of upper mantle Pb, but only makes a minor contribution to MORB magmatism.

6.4 Mantle Reservoirs in Isotopic Multispace

The degree of coherence within the Sr–Nd and Pb–Pb systems described above (Sections 6.2, 6.3) breaks down when Pb isotope ratios are plotted against other isotope tracers on bivariate diagrams (e.g. Fig. 6.30). This indicates that the isotope systematics of the mantle cannot be explained by a two-component mixing model. An exception to this general observation is provided by Pb–Sr isotope systematics on the North Atlantic, which *do* define a coherent positive correlation (Dupre and Allegre, 1980). However, this can be attributed to coincidental contamination of the MORB

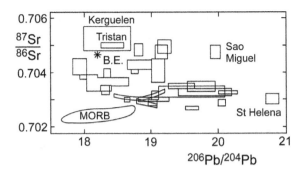

Fig. 6.30 Diagram to show the decoupling of Pb and Sr isotope systems in oceanic volcanics. After Sun (1980). B.E. = Bulk Earth.

reservoir in this area with a single compositional type of enriched plume material.

6.4.1 The Mantle Plane

To explain the scatter of data on Fig. 6.30, Zindler *et al.* (1982) argued that the Pb–Sr–Nd isotope compositions of oceanic volcanics must be caused by (solid state?) mixing of three mantle components. The proposed end-members were a pristine chondritic mantle with a Pb composition on the geochron, a MORB source depleted by continental crustal extraction, and a reservoir containing recycled MORB. This made Kerguelen the best candidate for a primitive mantle source, while St Helena was regarded as having the greatest

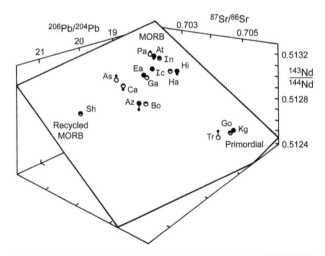

Fig. 6.31 Three-component mantle mixing model for MORB and OIB sources. Solid and open symbols indicate points respectively above and below the mantle plane. Pa, At, In = Pacific, Atlantic and Indian MORB. Hi = Hiva Oa, Ha = Hawaii, Ic = Iceland, Ea = Easter, Ga = Galapagos, As = Ascension, Ca = Canaries, Az = Azores, Bo = Bouvet, Sh = St Helena, Tr = Tristan, Go = Gough, Kg = Kerguelen. After Zindler *et al.* (1982).

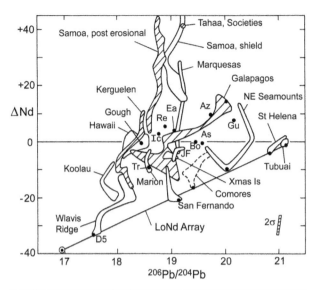

Fig. 6.32 Plot of Δ Nd (part-per-10^5 deviation in $^{143}Nd/^{144}Nd$ ratio from the mantle plane of Zindler *et al.*, 1982) against Pb isotope ratio. OIB compositions are plotted both as fields and discrete points. JF = Juan Fernando, Re = Reunion, Gu = Guadeloupe. Other abbreviations as in Fig. 6.31. After Hart *et al.* (1986).

amount of recycled MORB material in its source. Zindler *et al.* argued that average isotopic compositions of ocean ridges and ocean islands displayed relatively little scatter away from a plane containing the three end-member components (Fig. 6.31).

Zindler *et al.* justified their three-component model by the high correlation coefficient of 0.98 calculated for their data set. However, such a limited scatter was achieved by excluding several ocean islands. For example, Sao Miguel was not included in the Azores average. However, the Sao Miguel signature was argued by White (1985) to be part of a much wider compositional field, including data from the Society Islands, Samoa, and Marquesas, which extend 'above' the mantle plane of Zindler *et al.* (1982) to more radiogenic Sr–Nd signatures. Furthermore, the averaging process also obscured components that lay below the mantle plane, such as Walvis Ridge. Therefore, at least one additional component must be invoked to explain the data.

6.4.2 The Mantle Tetrahedron

Hart *et al.* (1986) suggested that the mantle plane of Zindler *et al.* (1982) could really be a 'co-incidence of similar mixing proportions' of end-members with more extreme compositions, rather than a real entity in its own right. This is illustrated in Fig. 6.32, where samples are plotted in terms of part-per-10^5 deviations in Nd isotope ratio from the mantle plane (Δ Nd), against Pb isotope composition.

Hart *et al.* proposed that the lower bound of $^{143}Nd/^{144}Nd$ compositions on the Nd–Pb isotope diagram (Fig. 6.32) might

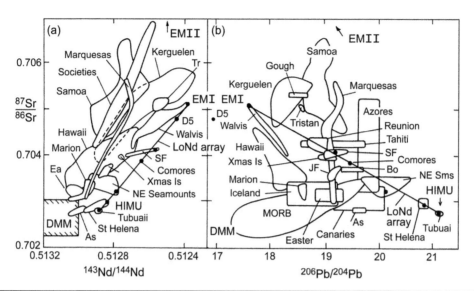

Fig. 6.33 Plots of: (a) Sr versus Nd isotope ratio, and (b) Sr versus Pb isotope ratio, showing the proposed end-members of a four-component mixing system: DMM, HIMU, EMI and EMII (= EM2). Dots are compositions argued to lie on an array between the HIMU and EMI end-members, termed the LoNd array. After Hart *et al.* (1986).

be a more fundamental topological structure, which they termed the 'LoNd' array. The same samples which define the lower bound of the Nd–Pb isotope distribution also define the lower bound of OIB data on the Nd–Sr isotope plot (Fig. 6.33a). In contrast, these samples define a line across the middle of the Sr–Pb isotope plot (Fig. 6.33b), but they remain as a relatively coherent array ($^{208}Pb/^{204}Pb$ ratios in these samples are also coherent with the three other isotope systems).

The LoNd array was itself interpreted as a mixing line between 'HIMU' (high U/Pb) and 'EMI' (enriched mantle one) end-members (Zindler and Hart, 1986). Other important end-members were defined by the most extreme composition of the MORB field (DMM) and the Societies (EMII). In addition, Zindler and Hart (1986) suggested that two other components might be located inside the four-component mixing space in Fig. 6.33. The first of these components is a possible relic of the Bulk Silicate Earth, whose lithophile isotope composition might be represented by Gough–Tristan. The noble gas signature of this component (exemplified by Loihi seamount) was later named 'primitive helium mantle' or 'PHEM' (Farley *et al.*, 1992). The second additional reservoir, termed 'prevalent mantle' or 'PREMA', was justified on the grounds that mixing of discrete components may have reached such a stage of completeness that this mixture itself became a recognizable entity. This component was later referred to as a focus zone of mantle mixing (FOZO) or a 'common mantle component' (C, Hanan and Graham, 1996).

One of the characteristics of the LoNd array (Hart, 1988) is that island groups are not generally elongated *along* the proposed mixing line, but often trend obliquely off the line. This was used as evidence that mixing within the LoNd array occurred a long time ago, before secondary mixing with

other components lying off the array. In addition, Hart *et al.* (1986) argued that the straightness of the proposed LoNd mixing line places tight constraints on the nature of the two mixing end-members, by requiring them to have similar Nd–Sr–Pb ratios and an intimately related environment of formation. Since they believed that such conditions would not be expected between recycled crustal and mantle components, Hart *et al.* argued that the two end-members must have resulted from different metasomatic enrichment processes in the sub-continental lithosphere.

Hart (1988) identified another two-component mixing line within the OIB data set, by using an upper $^{87}Sr/^{86}Sr$ cut-off of 0.703 to exclude all samples with an enriched mantle component. On a diagram of $^{143}Nd/^{144}Nd$ against $^{206}Pb/^{204}Pb$ (Fig. 6.34), these island groups with low $^{87}Sr/^{86}Sr$ define a so-called 'no EM' array between the HIMU and DMM end-members. The straightness of this array again suggests that the end-members had similar Nd/Pb ratios, and hence that DMM, HIMU and EMI all have similar Nd/Pb ratios. However, the geochemical relationship between DMM and HIMU cannot easily be attributed to spatial proximity, as was the EMI–HIMU relationship, because the depleted mantle is a distinct reservoir. This therefore weakens Zindler and Hart's argument for an intimate genetic relationship between the end-members of the LoNd array. Instead, a more general relationship is possible, whereby the three components are generated by similar mantle melting *processes*, but in different locations.

In contrast to the linear mixing lines described above, mixing with the EMII component tends to generate elongate curved arrays within island groups, as shown in Figs. 6.32 and 6.33. This suggests that elemental ratios between EMII

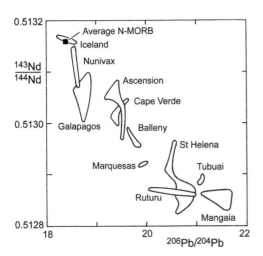

Nd versus Pb isotope diagram showing the linear array of OIB samples with $^{87}Sr/^{86}Sr$ below 0.703, attributed to a 'No EM' mixing line. After Hart (1988).

and the other mantle domains were far from unity, consistent with a model in which DMM, HIMU and EMI are generated by mantle differentiation processes, but EMII represents recycled continental crust with a very different trace element signature. Hart (1988) went further in his distancing of EMII from the other components, suggesting that mixing with this end-member was a late phenomenon which occurred after other mixing processes. However, Staudigel et al. (1991) found strong evidence for mixing between HIMU and EMII in the South Pacific Isotopic and Thermal Anomaly (SOPITA), particularly on the Sr–Pb isotope diagram (not shown here). In view of the intimate geographical association of HIMU and EMII in the SOPITA case, it is likely that

this array was formed prior to mixing with MORB, and it may constitute one of a family of curved 'HiNd' mixing lines analogous to the LoNd array.

There is considerable danger in looking at isotope variations in a number of two-component systems, since arrays are projected onto these surfaces from a multi-dimensional mixing polygon, and in this process the true trends of the arrays may be misunderstood. In order to analyse the data in a more objective fashion, Allegre et al. (1987) ran a principal component analysis on a large set of $^{87}Sr/^{86}Sr$, $^{143}Nd/^{144}Nd$, $^{206}Pb/^{204}Pb$, $^{207}Pb/^{204}Pb$ and $^{208}Pb/^{204}Pb$ data for MORB and OIB samples. This was also performed on an updated sample set by Hart et al. (1992).

Principal component analysis resolves the oceanic data set into five eigenvectors, representing directions in multicomponent space which show the greatest percentage of variance in the data. The magnitudes of these vectors (in the calculation by Hart et al.) are approximately 56%, 37%, 4%, 2% and 1%. The pre-eminence of the first two vectors demonstrates the largely planar form of the data set, as emphasized by Zindler et al. (1982). However, there is enough residual scatter in the data that a third vector is necessary to properly represent the mixing process. The sum of these three vectors is 97.5% in Hart's analysis, and 99.2% in Allegre's analysis. Hence Hart et al. argued that a three-dimensional (four-component) analysis is appropriate to analyse the data with a fairly high degree of reliability. However, the eigenvectors are so divorced from the familiar isotope ratios that it becomes difficult to understand the data. Therefore, Hart et al. presented the data in the form of a three-dimensional isotope plot (of $^{143}Nd/^{144}Nd$, $^{87}Sr/^{86}Sr$ and $^{206}Pb/^{204}Pb$), but projected in such a way as to approximate the eigenvector directions (Fig. 6.35).

This projection led Hart et al. (1992) to argue that the most common feature in the isotope data from plume sources

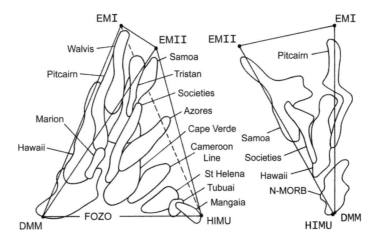

Views of a three-dimensional mantle tetrahedron representing the mixing relationships of four isotopically proposed mantle components seen in oceanic volcanics, along with a proposed 'focus zone', FOZO. Modified after Hart et al. (1992) and Hauri et al. (1994).

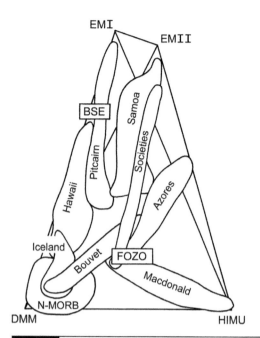

Fig. 6.36 View of the mantle tetrahedron, showing OIB arrays that converge on the revised composition of FOZO from many different directions. After Hauri *et al.* (1994).

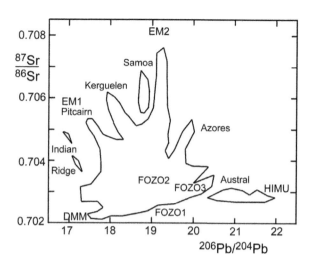

Fig. 6.37 Sr versus Pb isotope signatures of oceanic volcanics to show groups of distinct and overlapping data arrays, and alternative FOZO compositions. Modified after Stracke *et al.* (2005); Stracke (2012).

was a tendency to form linear arrays which appeared to fan out from a 'focus zone' at the base of the mantle tetrahedron towards a variety of enriched mantle end-members. Hence Hart *et al.* named this common component FOZO. The proposed composition of FOZO was on the edge of the tetrahedron between the depleted mantle (DMM) and HIMU (Fig. 6.35). However, it was clearly distinct from DMM. Therefore, Hart *et al.* proposed that it was a lower mantle component that was entrained around enriched mantle plumes rising from the core–mantle boundary.

Subsequent work showed that the location of FOZO as originally proposed was not satisfactory, as several island arrays (e.g. the Macdonald seamounts) trended from HIMU or DMM towards the middle of the mantle tetrahedron. Therefore, Hauri *et al.* (1994) revised the concept of FOZO to a somewhat less depleted signature bearing a strong resemblance to PREMA (Fig. 6.36).

The focussing of data points at the lower corners of the mantle tetrahedron (Fig. 6.36) provides evidence that some of the proposed components are real entities, rather than merely theoretical end-members. This is exemplified by the intersection of the LoNd and No-EM arrays, which provide a relatively strong constraint on the composition of HIMU, suggesting that the 'pure end-member' has a composition very similar to the most radiogenic Pb already analysed, from the island of Mangaia. This conclusion is supported by close agreement between the composition of widely separated HIMU islands from the South Atlantic and South Pacific. In contrast, the density of samples lying at the EMI

and EMII end-member compositions is much lower, and may suggest that these are the most extreme products yet sampled of enrichment *processes*, rather than significant mantle reservoirs in their own right (Barling and Goldstein, 1990).

6.5 Identification of Mantle Components

Since the study of Hart *et al.* (1986), major efforts have been devoted to identifying the proposed mantle components in geological terms, and explaining how they have interacted to generate OIB sources. To a large extent, the debate has been polarized between those who have invoked metasomatic enrichment models (e.g. Hart *et al.*, 1986) and those who invoked crustal recycling models to explain the enriched components (e.g. Weaver, 1991).

Before these arguments are examined for the well-known mantle end-members, it is worth noting that additional analysis of oceanic volcanics has actually broken the trends towards enriched mantle end-members into a number of lobes, so that the Pitcairn/Walvis (EMI) and Society/Samoa (EMII) trends are not unique (Fig. 6.37). Firstly, a few Indian Ocean MORBs define a weak trend towards radiogenic Sr but less radiogenic Pb than Pitcairn, whose meaning is unclear. Secondly, the Samoan samples actually break into two subarrays. Meanwhile Kerguelen continues to form an array between EMI and EMII, while the Azores define a unique trend to radiogenic Sr and Pb. Finally, HIMU actually forms a separate field that is not connected to the densest data distribution at the base of the EM trends. This suggests that

HIMU may not actually have direct genetic links with any of the other components, although classically the Lo-Nd array implied such a link between EMI and HIMU, as discussed above.

It should also be noted here that systems that were once believed to show coherent behaviour, such as radiogenic $^{208}Pb^*/^{206}Pb^*$ versus Sr and Nd isotopes (Allegre *et al.*, 1986), have now been shown to break into a series of trends to different enriched end-members (Stracke, 2012). Hence it is now clear that the mantle is a very complex, partially mixed multi-component system.

6.5.1 Depleted OIB Sources

When Zindler and Hart (1986) integrated enriched mantle signatures into a simple model, they also suggested the existence of a large reservoir of intermediate composition, which they called prevalent mantle (PREMA). Zindler and Hart also noticed that the group of islands exhibiting this signature included Hawaii and Iceland, with enriched helium isotope signatures, but with lithophile isotope signatures depleted relative to Bulk Earth. As a result of these observations, Zindler and Hart suggested two possible alternative origins for PREMA. It could either be a result of mixing of all the other mantle sources, or it could be a kind of depleted mantle formed in early Earth history, before two-layered mantle convection established the existence of the MORB reservoir.

Subsequently, Hart *et al.* (1992) emphasized the tendency for mixing between the prevalent lower mantle composition and individual OIB sources, forming linear arrays in the mantle tetrahedron that fan out from a 'focus zone' at its base. Hence, the term FOZO essentially replaced PREMA. However, as noted in Section 6.4.2, the proposed FOZO composition (FOZO1 in Fig. 6.37) was too depleted to effectively fulfil this role. Therefore, Hauri *et al.* (1994) revised it to the location shown as FOZO2. On the other hand, an alternative much closer to HIMU was proposed by Stracke *et al.* (2005), shown as FOZO3 in Fig. 6.37. However, of these alternatives, FOZO2 is the only one that acts as a credible focus zone to most of the enriched mantle arrays in Fig. 6.37.

Questions about the identity of FOZO were revolutionized by evidence from ^{142}Nd that the 'accessible mantle' has a LIL-depleted composition (hence high Sm/Nd ratio) relative to chondrites. This implied either that the Bulk Earth itself has a suprachondritic Sm/Nd ratio (Caro *et al.*, 2008), or that the mantle underwent a very early differentiation event that created a hidden enriched reservoir and an early depleted mantle (Boyet and Carlson, 2005, 2006). The ^{142}Nd composition of the Moon is critical in choosing between these alternatives, because the Moon is believed to have formed from the early Earth's mantle during a giant impact (Section 15.7). Recent data suggest that the Moon has a distinct ^{142}Nd signature from the accessible Earth, but in common with Mars and enstatite chondrites (Dickin, 2016). Therefore it appears that Earth's mantle underwent early LIL depletion from a common chondritic composition shared by most of the inner solar system (Section 15.7.3).

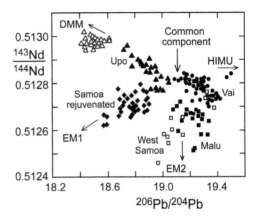

Fig. 6.38 Arrays of OIB data for the SW Pacific that converge on a common composition with a primordial helium signature that closely corresponds to FOZO. After Jackson *et al.* (2014).

This early depleted mantle experienced further LIL depletion over Earth history by extraction of the continental crust and by the formation and storage of large quantities of oceanic crust. Box model calculations show that these events can only be reconciled with the observed Nd isotope composition of MORB if nearly the whole mantle experienced depletion. Again, FOZO2 in Fig. 6.37 is the only composition that fulfils the requirements of a relatively LIL-depleted component that can be representative of most of the mantle, except for the specifically enriched sources that resulted from recycling of continental or oceanic crust.

A good example of a mantle focus zone on a more local scale comes from the Samoan hot-spot, which shows trends from a common component with primordial helium towards all of the major recognized mantle components (Fig. 6.38). This 'local FOZO' may have a slightly more enriched composition than 'global FOZO' due to mixing with other components in the plume. However, it provides an excellent example of a mantle focus zone in action.

6.5.2 EMII

The case for EMII as subducted continental material is almost universally agreed, since this end-member is squarely located on mixing lines between depleted mantle and marine sediments. This model was further strengthened by evidence from peridotite xenoliths in Samoan lavas (Hauri *et al.*, 1993). Trace element data for these xenoliths point to an origin from carbonate-rich melts within the Samoan plume, and the isotopic compositions of the xenoliths are therefore taken as indicative of the EMII mantle component. These xenoliths extend the EMII array directly into the field of marine sediments (Fig. 6.39) and thus provide a compelling case for this material as the source of the EMII component. Similar xenoliths from Tubuai also support the concept of a discrete HIMU component, as previously observed in lavas

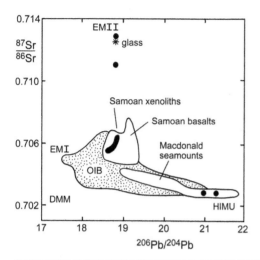

Fig. 6.39 Plot of Sr versus Pb isotope data for cpx grains (•) and glass inclusions (✳) in peridotite xenoliths from Savaii (Samoa) and Tubuai (Austral Islands), indicating affinity with the EMII and HIMU mantle end-members. After Hauri et al. (1993).

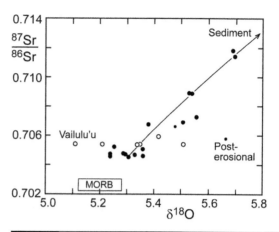

Fig. 6.40 Positive correlation between strontium and oxygen isotope ratios in the main shield-building volcanic phases in Samoa. Curve = mantle–sediment mixing model. After Workman et al. (2008).

from Mangaia, Tubuai and the nearby Macdonald seamount chain.

Further evidence for sediment recycling into the EMII source comes from oxygen isotope measurements, which have consistently shown elevated signatures in these plumes relative to MORB values. Other enriched mantle reservoirs have also appeared in the past to display oxygen isotope variations outside the range of MORB values, based on the analysis of whole-rock basalts or basaltic glasses (see review by Harmon and Hoefs, 1995). However, more recent analysis of olivine phenocrysts from a variety of plume sources showed a much more restricted range (Eiler et al., 1997), suggesting that most of the earlier variations were due to shallow contamination effects, either at the magmatic stage by oceanic crust, or under sub-solidus conditions after eruption. In contrast, phenocryst analyses from EMII plumes continue to show elevated $\delta^{18}O$ ratios correlated with $^{87}Sr/^{86}Sr$ (Workman et al., 2008), considered to be a strong indicator of sediment recycling (Fig. 6.40).

6.5.3 EMI

McKenzie and O'Nions (1983) suggested that sub-continental lithosphere has a greater density than the underlying Fe-depleted asthenosphere, so that over-thickening during continental collision might cause some of the lithosphere to constrict off and fall into the upper mantle convection system. If this material was sampled within a few hundred million years, it might yield OIB magmas before being homogenized into the MORB source by convection.

In their early synthesis on the nature of enriched mantle sources, Hart et al. (1986) argued that both HIMU and EMI were derived from recycled subcontinental lithosphere.

However, Weaver proposed instead that EMI represents subducted pelagic ocean floor sediment (in contrast to the terrigenous source of EMII). With the increasing acceptance of oceanic crust/lithosphere as the origin of HIMU, there has been widespread support for Weaver's model.

Stable isotope support for the sediment recycling model came from elevated $\delta^{18}O$ signatures in submarine glasses from the Pitcairn seamounts (Woodhead et al., 1993). This evidence was significant, because Pitcairn samples display the most extreme EMI signatures for several isotope systems. However, later work by Eiler et al. (1995) failed to find evidence for elevated $\delta^{18}O$ signatures in separated minerals from Pitcairn Island. All olivines had $\delta^{18}O$ values close to +5.2 per mil (relative to the SMOW standard), while plagioclase had $\delta^{18}O$ values near +6.1, consistent with mass fractionation effects at magmatic temperatures. Therefore, it was concluded that the variations seen by Woodhead et al. were probably due to local contamination of the glasses, either before or after solidification. Local contamination effects are also believed to explain oxygen isotope data from Iceland that lie well below the range of MORB values, although some slightly elevated $\delta^{18}O$ signatures and osmium isotope ratios could originate from a plume source (Section 8.3.5).

Much stronger evidence for elevated $\delta^{18}O$ signatures was found in olivine phenocrysts from Hawaiian lavas (Eiler et al., 1996). In this study, oxygen isotope ratios were correlated with lithophile isotope tracers such as Nd (in addition to osmium and helium), consistent with the involvement of three end-members in the Hawaiian plume (Fig. 6.41). Loihi, with a $\delta^{18}O$ signature similar to MORB, has helium signatures indicative of a primordial lower mantle or core component (Section 11.1.3). The Kea component, with a depleted $\delta^{18}O$ signature relative to MORB, is attributed to melting of recycled oceanic lithosphere. Finally, the Koolau component

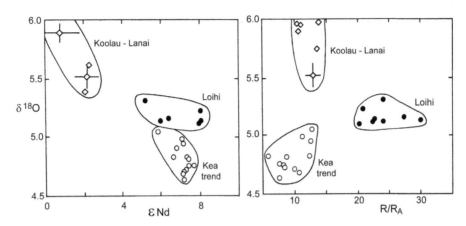

Fig. 6.41 Plots of $\delta^{18}O$ against εNd and He isotope data (R/R$_A$) for olivine phenocrysts in lavas from selected Hawaiian volcanoes. Error bars represent suites where different tracers were determined on different samples, so the symbol shows the mean and standard deviation of the suite. Modified after Eiler *et al.* (1996).

has unradiogenic Pb and Nd signatures characteristic of EMI, along with enriched $\delta^{18}O$ values possibly indicative of a component of recycled sediment.

Although this evidence provides strong support for a sediment signature in Hawaii, it remains possible that this signature was carried to the plume source via the lithospheric mantle wedge. Elevated $\delta^{18}O$ values have been measured in peridotite xenoliths from subduction zones (Liu *et al.*, 2014). This supports the proposal of Eiler *et al.* (1998, 2000, 2007) that elevated $\delta^{18}O$ signatures can be extracted from subducting slabs by metasomatism and carried by fluids into the overlying wedge. This lithospheric mantle would be very susceptible to subsequent subduction erosion, thus facilitating its transport to the lower mantle. By this process, the EMI end-member could have developed elevated $\delta^{18}O$ indicative of an *apparent* sedimentary signature.

Recent support for a more direct sediment recycling model has come from stable sulphur isotope data in different types of OIB samples. For example, Labidi *et al.* (2013) demonstrated a positive correlation between $\delta^{34}S$ and Sr isotope ratio in basaltic glasses with OIB affinities dredged from near the South Atlantic Ridge (Fig. 6.42). Radiogenic isotope signatures of these glasses (Douglass *et al.*, 1999) display affinities with the spectrum from EMI to HIMU, although not approaching the end-member compositions. The sulphur isotope signatures in the glasses were argued to be indicative of the mantle source, because the glasses were in equilibrium with sulphide phases (and hence not fractionated by oxidation during magma evolution). The positive correlation extended from the MORB field (average $\delta^{34}S = -1.5$ per mil) towards post-Archean sediment (deposited in an oxidizing environment). Hence, Labidi *et al.* proposed that the EM plume source was contaminated by post-Archean sediment.

Sulphur isotope analysis on sulphide phases from olivine and plagioclase crystals in Pacific samples with EMI–HIMU affinities gave somewhat different results (Cabral *et al.*, 2013;

Delavault *et al.*, 2016). In this case, all samples had negative $\delta^{34}S$ values below the MORB composition, suggestive of contamination by sediments deposited in reduced conditions. In these studies, $\delta^{34}S$ values were weakly correlated $\Delta^{33}S$, forming a non-mass-dependent fractionation trend that has been attributed to photo-dissociation of SO and SO$_2$ in the early Earth (Farquhar and Wing, 2003). The problematical aspect of these studies is that the Atlantic samples with EMI–HIMU affinities are giving results completely different from the Pacific samples. However, because the sample material was different in the two cases, it cannot be proven that both data sets are recovering true plume-source compositions. Therefore, these data sets must be regarded as somewhat equivocal for the moment. However, the origins of EMI are discussed further below, in relation to the HIMU component. This work points to subducted sedimentary carbonate as a significant mantle component.

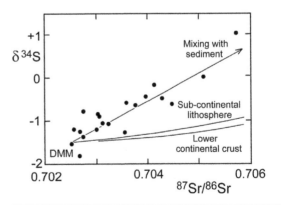

Fig. 6.42 Plot of $\delta^{34}S$ versus $^{87}Sr/^{86}Sr$ for dredged basaltic glasses with EMI–HIMU affinities from the South Atlantic. After Labidi *et al.* (2013).

6.5.4 HIMU

The extreme Pb isotope composition of the HIMU component has always been a challenge to explain geochemically. The 'standard model' of Hofmann and White (1980, 1982) and Chase (1981) is that radiogenic OIB reservoirs can be generated by subduction of U-enriched oceanic crust. The attraction of this model is the certain fact that vast amounts of this material *are* subducted back into the mantle, along with the fact that the oceanic crust is necessarily enriched in many incompatible elements relative to its depleted mantle source. However, the U/Pb ratios of normal MORB are not high enough to explain the composition of the most radiogenic OIB sources. Therefore, several mechanisms have been proposed to elevate the U/Pb ratios of subducted oceanic crust.

One proposed mechanism was the addition of ocean floor sediment to subducted oceanic crust. However, the U/Pb ratios of typical ocean floor sediment (White *et al.*, 1985) are not large enough to explain the radiogenic Pb signatures of most OIB sources. Seawater alteration has also been invoked as a possible mechanism to elevate U/Pb ratios in oceanic crust (Michard and Albarede, 1985), but this might also elevate Rb/Sr ratios, generating more radiogenic strontium than is seen in the HIMU component.

A better model (Weaver, 1991) is to invoke preferential extraction of Pb, relative to U, from the down-going slab in subduction zones. This model is supported by U/Pb ratios nearly an order of magnitude lower in island arc tholeiites than in MORB (Sun, 1980). Mobilization of Pb from the slab in a fluid phase could also explain the surprising degree of Pb isotope homogeneity in arc-related 'conformable' galena deposits (Section 5.4.2). This model was further developed by Chauvel *et al.* (1995), who suggested that Pb is removed from subducting oceanic crust by metasomatism and deposited in the overlying mantle wedge. It would then be incorporated into the continental crust via arc magmatism.

A case against this 'standard model' for the origins of HIMU was made by Niu and O'Hara (2003), who argued that subducted oceanic crust would be too dense to generate rising plumes from the deep mantle, and that the magnesium content of many OIB magmas was too high to be generated by melting of a basaltic source. Hence, this would require plumes to be derived from a mixture of subducted oceanic crust and more buoyant Mg-rich oceanic lithosphere (e.g. Chauvel *et al.*, 1992). On the other hand, Stracke *et al.* (2005) pointed out that there is a gap between the Pb isotope ratios of HIMU and the rest of the OIB compositional field (Fig. 6.37). This implies that relatively 'pure' sampling of the HIMU reservoir is actually quite rare.

One type of 'special' process that has been proposed to account for the HIMU component is enrichment in carbonate. It had already been observed (Bell and Tilton, 2001) that carbonatites of the East African Rift span between the EMI and HIMU end-members (Section 7.3.1). This was taken to show that the isotopic signatures of carbonatites are derived from deep mantle plumes. However, Collerson *et al.* (2010) inverted this argument, proposing that carbonate enrich-

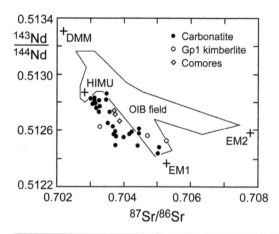

Fig. 6.43 Nd versus Sr isotope plot showing a close relationship between carbonatites and the EMI–HIMU couple. After Weiss *et al.* (2016).

ment was also an essential feature involved in generating the oceanic EMI–HIMU domains. Hence this can explain why the African carbonatites are particularly associated in isotope space with the 'Lo-Nd' array between EMI and HIMU (Fig. 6.43). A carbonate metasomatized peridotite source for the origin of HIMU was also supported by the observation of unusually high Ca/Al ratios in olivine grains in HIMU lavas (Weiss *et al.*, 2016).

This association of HIMU and EMI with carbonate metasomatism seems to be a major step forward in understanding the origins of these end-members, but the site of this process remains unclear. Collerson *et al.* (2010) proposed that the process occurred in the lower mantle, where elevated carbon abundances would act as a flux to lower the melting point of peridotite. Hence they modelled trace element partition during melting of (proposed) lower mantle phases in the presence of carbonate. For example, 5% melting produced a carbonatite melt with trace element ratios that could yield the EMI component, while the residue could generate compositions approaching HIMU after storage in the mantle for a few Ga.

Collerson *et al.* envisaged the carbon in the lower mantle to be of primordial origin. However, other researchers have proposed recycling of sedimentary carbonate into the lower mantle. For example, Castillo (2015) argued that isotopic enrichment in the HIMU source was due to the recycling of marine carbonates to the lower mantle. On the other hand, Weiss *et al.* (2016) argued that the site of carbonate enrichment was in the mantle wedge above subduction zones, and this was also proposed as a site of carbon enrichment by Kelemen and Manning (2015). Metasomatized subcontinental lithosphere could eventually be delaminated from the base of the continent and sink to the lower mantle, entering the plume source.

Additional evidence for a sedimentary contribution to carbonatite genesis is discussed in Section 7.3.1. Based on

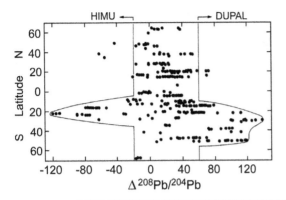

Fig. 6.44 Plot of $\Delta\,^{208}Pb/^{204}Pb$ (deviation from the 'Northern Hemisphere reference line') against latitude to show the geographical distribution of the Dupal and associated HIMU components. After Hart (1988).

the close relationship between carbonatites and EMI–HIMU plumes discussed above, this further strengthens the evidence for the recycling of sedimentary carbonate into these plume sources. However, it remains unclear whether sub-continental lithosphere plays a role as an intermediary in this process.

A final question arising from the above discussion concerns the fate of subducted oceanic crust, if this does not generate the HIMU source. Olson (1984) argued that the process of mantle convection naturally tends to streak out any heterogeneities into narrow schlieren (Olson, 1984), which would then be too small to source large volumes of enriched OIB magmas. However, it has been widely suggested that recycled oceanic crust could form the pyroxenite fraction that appears to be inter-leaved with peridotite in many physical samples of mantle material. This material might in fact represent the 'FOZO3' component in Fig. 6.37.

6.5.5 The DUPAL Anomaly

The subduction of oceanic crust, along with marine sediment or eroded sub-continental lithosphere, may give rise to large-scale isotopic structure in the mantle. Hart (1984) argued that recycling into the mantle was responsible for generating a Pb and Sr isotope anomaly of global scale which he observed to form a small circle of approximately constant latitude encircling the southern hemisphere. He named it the 'Dupal' anomaly because its characteristic signature was first described in Indian Ocean volcanics by Dupre and Allegre (1983). The spatial distribution of the Dupal anomaly is shown in Fig. 6.44, in the form of the variation of Δ 208/204 (per mil $^{208}Pb/^{204}Pb$ deviations from the North Atlantic correlation line) against latitude. Very similar results are obtained for $^{207}Pb/^{204}Pb$, which led Hart to suggest that the recycling process must have occurred fairly early in Earth history, so that the resulting enriched reservoirs, probably located in the lower mantle, could evolve to enhanced Δ 207/204 values.

Around the same time, seismic tomography of the lower mantle was revealing p-wave velocity minima centred under the West Pacific and Africa (Dziewonski, 1984). Castillo (1988) showed that these low-velocity areas in the lower mantle were correlated with the worldwide distribution of mantle plumes; and further, that this distribution was also correlated with the distribution of the Dupal anomaly. Hence, he argued that the Dupal anomaly represents the signature of enriched mantle sources entrained in plumes from the lower mantle. On the other hand, Staudigel et al. (1991) emphasized the particular significance of the isotope anomaly in the southwest Pacific, referring to this as the South Pacific Isotope and Thermal Anomaly (SOPITA).

After the early seismic velocity minima under the West Pacific and Africa evolved into the presently recognized Large Low Shear-wave Velocity Provinces (LLSVPs), it was realized that these represent large structures located on the core–mantle boundary (Section 6.1.6). However, it was argued by Lay et al. (2006) and Burke et al. (2008) that mantle plumes do not sample the material *within* the LLSVPs themselves, but subducted material that has accumulated around their margins. Hence, White (2015) concluded that the Dupal anomaly really consists of two separate bodies originating from the margins of the LLSVPs under the South Pacific and Africa. Of these, the former (corresponding to SOPITA) is dominated by EMII, whereas the latter is dominated by EMI.

6.6 Island Arcs and Mantle Evolution

Island arcs are central to the understanding of mantle evolution because they represent the site where lithospheric material of various types may be returned to the deep mantle. Island-arc magmatism may allow us to sample this material which is in the process of being recycled. Dewey (1980) showed that the volcanic front is always established about 100 km above the descending slab, whatever the angle of subduction. This shows that de-watering of the slab, triggered by pressure, is central to the operation of island-arc magmatism. However, the petrology of island-arc basalts (IAB) precludes their genesis by fusion of subducted oceanic crust (since this would require nearly 100% melting). Therefore, they must be dominantly produced by melting of the 'mantle wedge' overlying the subduction zone (e.g. Wyllie, 1984). Hence, the central problem in interpreting island-arc basalts is to identify which signatures are derived from the slab (and subducted sediment) and which are derived from the overlying wedge. We will therefore examine this problem in terms of two-component mixing between the slab and wedge.

6.6.1 Two-Component Mixing Models

Island-arc basalts have enhanced levels of $^{87}Sr/^{86}Sr$ relative to MORB. However, the origin of these differences is only discernible in the context of other isotope evidence. The first study using combined Sr and Nd isotope data was made by

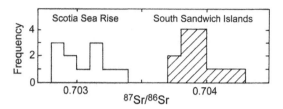

Fig. 6.45 Histograms of Sr isotope ratio for basalts from the South Sandwich arc and the (back-arc) Scotia Sea Rise. After Hawkesworth *et al.* (1977).

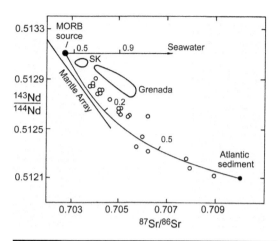

Fig. 6.46 Sr–Nd isotope diagram showing extreme isotopic variation in Martinique lavas (○) compared with Grenada and St Kitts (SK). Mixing lines model the effects of contamination by sediments or seawater. After Davidson (1983).

Hawkesworth *et al.* (1977) on island-arc and back-arc tholeiites from the Scotia Sea (South Sandwich Islands). Analysis of back-arc material provides a control condition because it samples a mantle segment which should be similar to the wedge, but without any slab component.

Hawkesworth *et al.* found that both island-arc and back-arc samples from the Scotia Sea had identical ^{143}Nd/^{144}Nd, overlapping with MORB. However, the island-arc samples had significantly higher ^{87}Sr/^{86}Sr ratios (Fig. 6.45), which could not be explained by subaerial weathering. Therefore, Hawkesworth *et al.* suggested that the enhanced Sr isotope ratios of the island-arc basalts were a product of subducted ^{87}Sr from seawater or (alternatively) oceanic sediments. Possible processes considered were the direct partial melting of altered and subducted oceanic crust, or alternatively, metasomatic contamination of the mantle wedge with elements derived from the oceanic crust. It is now generally accepted that the latter model is correct for the Scotia arc (e.g. Pearce, 1983).

The Scotia arc provides an example of the role of slab-derived fluids in an arc with depleted chemistry. However, in arcs with less depleted chemistry, material contributions from the slab and wedge are more difficult to resolve. The Lesser Antilles (Caribbean) arc provides a test case for the behaviour of arcs with more enriched signatures, since the chemistry of the arc changes along its length. This may help in resolving the origin of enriched components.

A study by Davidson (1983) revealed that St Kitts, situated at the northern end of the arc, has a very small range of isotopic composition close to MORB, while Martinique from the centre of the arc has an extremely large range of isotope composition (Fig. 6.46). If such variations were inherited from the mantle wedge, then 'gross heterogeneity on a scale of kilometres is implied.' Davidson rejected this model, and initially ascribed the isotopic variations at Martinique to contamination of the mantle source with subducted sediment. This model was also more recently proposed by Labanieh *et al.* (2010). However, oxygen isotope analysis of whole-rock samples (Davidson, 1987) and separated minerals (Bezard *et al.*, 2014) revealed positive correlations between Sr isotope ratio, oxygen isotope ratio and silica content in lavas from Martinique and St. Lucia. These correlations are indicative of crustal contamination of ascending magma in the arc

crust, which is thickest in the central region of the arc near Martinique and St Lucia (Bezard *et al.*, 2015). These types of crustal contamination processes will not be detailed here, since they will be discussed in the next chapter.

White and Dupre (1986) presented Pb isotope data for representative samples from the whole length of the Lesser Antilles arc, showing that they were generally intermediate between MORB and sediment compositions. There is no evidence that these signatures are derived from magma contamination in the arc crust. For example, sedimentary xenoliths in Grenada lavas actually have unradiogenic Pb, inherited from an earlier location of the arc to the west, above the subducting Farallon plate. In contrast, Atlantic ocean floor sediments in front of the present day arc have radiogenic Pb signatures.

White and Dupre found a general increase in the Pb isotope ratio of Atlantic floor sediment when going southwards in front of the Lesser Antilles subduction zone, probably reflecting sediment carried onto the sea floor at the south end of the arc by the Orinoco River. This trend was matched by the composition of Lesser Antilles volcanics, suggesting the presence of a subducted sediment component in the arc magmas. This model is supported by the covariation of Pb and Nd isotope data in the volcanics (Fig. 6.47). Two-component mixing between a MORB source and average Atlantic sediment can therefore explain the observed Pb–Nd isotope systematics of Lesser Antilles magmas, avoiding the need to invoke an enriched mantle wedge (Ellam and Hawkesworth, 1988).

Rare earth concentration data may present a problem for this model, since light REE enrichment in some arc volcanics may be too great to be explained by simple mixing between a MORB source and subducted sediment (Hawkesworth *et al.*, 1991). This problem is illustrated in Fig. 6.48 on a plot of

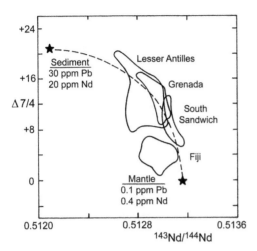

Fig. 6.47 Assessment of a sediment–asthenosphere mixing model for Lesser Antilles volcanics, in terms of Pb and Nd isotope systematics. Δ 7/4 indicates the $^{207}Pb/^{204}Pb$ deviation above the 'Northern Hemisphere reference line' of Hart (1984). After Ellam and Hawkesworth (1988).

Ce/Yb (= REE profile slope) against Sr isotope ratio. LREE enriched basalts and andesites from Grenada, the Sunda arc and the Aeolian arc of southern Italy (triangles) fall above the mixing line between depleted arcs and a typical sediment represented by 'post-Archean average shale' (PAAS). However, White and Dupre (1986) argued that the Pb isotope evidence for sediment involvement in arc magma genesis was so conclusive that it overrides these trace element problems. Given this constraint, the very steep REE profiles must be due to some feature of the melting process. For example, *partial melting of sediment in the presence of residual garnet* could elevate light REE abundances in the melt while depressing the heavy REE.

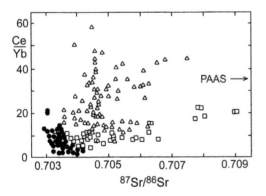

Fig. 6.48 Plot of Ce/Yb ratio against Sr isotope ratio for island-arc basalts and andesites. (●) = normal arc volcanics; (△) = LREE enriched; (□) = Martinique lavas, contaminated during magma ascent. PAAS (post-Archean average shale) is a typical sediment composition. After Hawkesworth *et al.* (1991).

Most workers now accept the supremacy of Pb isotope evidence for sediment involvement in IAB genesis. For example, Ben Othman *et al.* (1989) observed perfect matching of Pb isotope systematics between the West Sunda arc and ocean floor sediment in front of the arc. Since the Pb contents of arc volcanics are nearly an order of magnitude lower than typical sediments, it is unlikely that the sediment signature is itself controlled by erosion of arc volcanics. Therefore, it is most likely that the reverse relationship applies: arc volcanic Pb is controlled by subducted sediment. Further evidence was provided by McDermott *et al.* (1993), who observed Pb isotopic variations along the North Luzon (Philippine) arc that were correlated with the composition of sediment cores from the South China Sea, in front of the trench.

6.6.2 Three-Component Mixing Models

In the above examples, two-component mixing between slab and wedge was examined for cases where the slab-derived component (SDC) was dominantly either a fluid or melted sediment. However, it is clear that in some cases both of these slab-derived components must be present. This requires expansion of the two-component models described above into a three-component mixing model, involving contamination of the depleted-mantle source of IAB with partial melts of subducted sediment *and* large-ion lithophile element enriched slab-derived fluids.

While isotope tracers are very sensitive to the component of melted sediment, the most sensitive tracers for the slab-derived fluid are ratios of low field strength large ion lithophile elements (LILE) such as Ba or Sr, against high field strength elements (HFSE) such as REE. One of the first demonstrations of this approach was by McCulloch and Perfit (1981) on basalts and andesites from the Aleutian arc. When Nd isotope compositions of these lavas were plotted against their Ba/La ratio, most samples lay on a mixing line between depleted mantle and North Pacific sea floor sediment (Fig. 6.49). However, lavas from Bogoslof Island showed strong enrichments in Ba/La that were attributed to metasomatism of the mantle source by slab-derived fluids. The presence of an unusual amount of hydrous fluids in the source of these lavas is supported by the presence of amphibole phenocrysts.

Another element pair expected to show strong fractionation in slab-derived fluids is U/Th, since the former is a very soluble LIL element, while the latter is a high field strength element. This tracer is particularly powerful because these elements are also involved in the short-lived U-series decay chains (Section 13.5). Because the Th–U isotope system is sensitive to the time since U/Th fractionation, it can distinguish contamination by sea floor sediments (generally in secular equilibrium) from slab-derived fluids (out of secular equilibrium). This has provided evidence for recent mobilization of uranium from the subducted slab, which is clearly important for mantle U/Th and U/Pb budgets. Thus, analysis of arc basalts may constrain the effect of subduction zone recycling on mantle Pb isotope budgets.

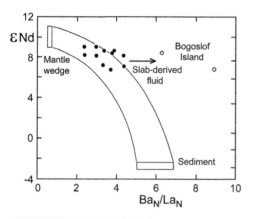

Fig. 6.49 Plot of εNd against chondrite-normalized Ba/La ratio for Aleutians arc volcanics, distinguishing the effects of sediment contamination and slab-derived fluid. Solid and open symbols indicate basalts and andesites respectively. After McCulloch and Perfit (1981).

To test the magnitude of U/Th and U/Pb fractionation in subduction zones, Kelley *et al.* (2005) plotted Pb/U against Th/U for arc volcanics (Fig. 6.50). They found that different arc systems defined separate arrays with negative slopes, but that these arrays converge on the y-axis at a Pb/U ratio of 30, equivalent to a μ value of 2.4. Each array is therefore interpreted as a mixing line between low-Pb melted sediment with variable Th/U ratios and a U- and Pb-rich slab-derived fluid released from sulphide breakdown in altered oceanic crust. However, it is inferred that some of the slab-derived fluid escapes the arc-basalt melting zone and is returned to the upper mantle, where it lowers both the κ and μ values of the MORB source. Kelley *et al.* then calculated the magnitudes of these fluxes to demonstrate that they are capable of controlling the U–Th–Pb systematics of the upper mantle.

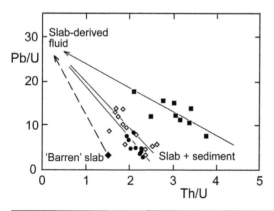

Fig. 6.50 Trace element ratio plot for arc-related volcanics, revealing mixing process in arc magma sources. (◇) = Marianas; (•) = Aleutians; (■) = Honshu. After Kelley et al. (2005).

Measurements of the oxidation state of ferric/total Fe in subduction-related magmas suggest that the oxidation state in the arc magma source is close to the oxygen fugacity of the sulphide–sulphate transition zone (Kelley and Cottrell, 2009; Jego and Dasgupta, 2014). This suggests that oxidation in the subduction zone environment may have been important in releasing Pb from sea floor sulphides, and hence for Pb recycling. Therefore, the rise in atmospheric oxygen during the Proterozoic may have increased Pb as well as U recycling to the upper mantle, causing a coupled reduction of upper mantle μ and κ.

References

Allegre, C. J. (1982). Chemical geodynamics. *Tectonophys.* **81**, 109–32.

Allegre, C. J. (1997). Limitation on the mass exchange between the upper and lower mantle: the evolving convection regime of the Earth. *Earth Planet. Sci. Lett.* **150**, 1–6.

Allegre, C. J., Ben Othman, D., Polve, M. and Richard, P. (1979). The Nd–Sr isotopic correlation in mantle materials and geodynamic consequences. *Phys. Earth Planet. Inter.* **19**, 293–306.

Allegre, C. J., Brevart, O., Dupre, B. and Minster, J. F. (1980). Isotopic and chemical effects produced by a continuously differentiating convecting Earth mantle. *Phil. Trans. Roy. Soc. Lond.* A **297**, 447–77.

Allegre, C. J., Dupre, B. and Lewin, E. (1986). Thorium/uranium ratio of the Earth. *Chem. Geol.* **56**, 219–27.

Allegre, C. J., Hamelin, B. and Dupre, B. (1984). Statistical analysis of isotopic ratios in MORB: the mantle blob cluster model and the convective regime of the mantle. *Earth Planet. Sci. Lett.* **71**, 71–84.

Allegre, C. J., Hamelin, B., Provost, A. and Dupre, B. (1987). Topology in isotopic multispace and origin of mantle chemical heterogeneities. *Earth Planet. Sci. Lett.* **81**, 319–37.

Allegre, C. J. and Turcotte, D. L. (1986). Implications of a two-component marble-cake mantle. *Nature* **323**, 123–7.

Andersen, M. B., Elliott, T., Freymuth, H. *et al.* (2015). The terrestrial uranium isotope cycle. *Nature* **517**, 356–9.

Anderson, D. L. (1981). Hotspots, basalts, and the evolution of the mantle. *Science* **213**, 82–9.

Barling, J. and Goldstein, S. L. (1990). Extreme isotopic variations in Heard Island lavas and the nature of mantle reservoirs. *Nature* **348**, 59–62.

Batiza, R. (1984). Inverse relationship between Sr isotope diversity and rate of oceanic volcanism has implications for mantle heterogeneity. *Nature* **309**, 440–1.

Bell, K. and Tilton, G. R. (2001). Nd, Pb and Sr isotopic compositions of East African carbonatites: evidence for mantle mixing and plume inhomogeneity. *J. Petrol.* **42**, 1927–45.

Ben Othman, D., White, W. M. and Patchett, J. (1989). The geochemistry of marine sediments, island arc magma genesis, and crust–mantle recycling. *Earth Planet. Sci. Lett.* **94**, 1–21.

Bezard, R., Davidson, J. P., Turner, S. *et al.* (2014). Assimilation of sediments embedded in the oceanic arc crust: myth or reality? *Earth Planet. Sci. Lett.* **395**, 51–60.

Bezard, R., Schaefer, B. F., Turner, S., Davidson, J. P. and Selby, D. (2015). Lower crustal assimilation in oceanic arcs: insights from an osmium isotopic study of the Lesser Antilles. *Geochim.t Cosmochim. Acta* **150**, 330–44.

Bijwaard, H. and Spakman, W. (1999). Tomographic evidence for a narrow whole mantle plume below Iceland. *Earth Planet. Sci. Lett.* **166**, 121–6.

Blichert-Toft, J., Zanda, B., Ebel, D. S. and Albarede, F. (2010). The solar system primordial lead. *Earth Planet. Sci. Lett.* **300**, 152–63.

Boyet, M. and Carlson, R. W. (2005). ^{142}Nd evidence for early (>4.53 Ga) global differentiation of the silicate Earth. *Science* **309**, 576–81.

Boyet, M. and Carlson, R. W. (2006). A new geochemical model for the Earth's mantle inferred from ^{146}Sm–^{142}Nd systematics. *Earth Planet. Sci. Lett.* **250**, 254–68.

Bunge, H. P., Richards, M. A. and Baumgardner, J. R. (1996). Effect of depth-dependent viscosity on the planform of mantle convection. *Nature* **379**, 436–8.

Burke, K., Steinberger, B., Torsvik, T. H. and Smethurst, M. A. (2008). Plume generation zones at the margins of large low shear velocity provinces on the core–mantle boundary. *Earth Planet. Sci. Lett.* **265**, 49–60.

Burton, K. W., Cenki-Tok, B., Mokadem, F. *et al.* (2012). Unradiogenic lead in Earth's upper mantle. *Nature Geosci.* **5**, 570–3.

Cabral, R. A., Jackson, M. G., Rose-Koga, E. F. *et al.* (2013). Anomalous sulphur isotopes in plume lavas reveal deep mantle storage of Archaean crust. *Nature* **496**, 490–3.

Campbell, I. H. (2002). Implications of Nb/U, Th/U and Sm/Nd in plume magmas for the relationship between continental and oceanic crust formation and the development of the depleted mantle. *Geochim. Cosmochim. Acta* **66**, 1651–61.

Caro, G. and Bourdon, B. (2010). Non-chondritic Sm/Nd ratio in the terrestrial planets: consequences for the geochemical evolution of the mantle–crust system. *Geochim. Cosmochim. Acta* **74**, 3333–49.

Caro, G., Bourdon, B., Halliday, A. N. and Quitté, G. (2008). Super-chondritic Sm/Nd ratios in Mars, the Earth and the Moon. *Nature* **452**, 336–9.

Castillo, P. (1988). The Dupal anomaly as a trace of the upwelling lower mantle. *Nature* **336**, 667–70.

Castillo, P. R. (2015). The recycling of marine carbonates and sources of HIMU and FOZO ocean island basalts. *Lithos* **216**, 254–63.

Chase, C. G. (1981). Oceanic island Pb: Two-stage histories and mantle evolution. *Earth Planet. Sci. Lett.* **52**, 277–84.

Chauvel, C., Goldstein, S. L. and Hofmann, A. W. (1995). Hydration and dehydration of oceanic crust controls Pb evolution in the mantle. *Chem. Geol.* **126**, 65–75.

Chauvel, C., Hofmann, A. W. and Vidal, P. (1992). HIMU–EM: the French Polynesian connection. *Earth Planet. Sci. Lett.* **110**, 99–119.

Chen, C. Y. and Frey, F. A. (1983). Origin of Hawaiian tholeiite and alkalic basalt. *Nature* **302**, 785–9.

Christensen, U. R. and Hofmann, A. W. (1994). Segregation of subducted oceanic crust in the convecting mantle. *J. Geophys. Res.* **99** (B10), 19 867–84.

Class, C., Goldstein, S. L., Altherr, R. and Bachelery, P. (1998). The process of plume– lithosphere interactions in the ocean basins – the case of Grande Comore. *J. Petrol.* **39**, 937–52.

Class, C., Goldstein, S. L. and Shirey, S. B. (2009). Osmium isotopes in Grande Comore lavas: a new extreme among a spectrum of EM-type mantle end-members. *Earth Planet. Sci. Lett.* **284**, 219–27.

Cohen, R. S., Evensen, N. M., Hamilton, P. J. and O'Nions, R. K. (1980). U–Pb, Sm–Nd and Rb–Sr systematics of ocean ridge basalt glasses. *Nature* **283**, 149–53.

Cohen, R. S. and O'Nions, R. K. (1982). Identification of recycled continental material in the mantle from Sr, Nd and Pb isotope investigations. *Earth Planet. Sci. Lett.* **61**, 73–84.

Collerson, K. D. and Kamber, B. S. (1999). Evolution of the continents and the atmosphere inferred from Th U Nb systematics of the depleted mantle. *Science* **283**, 1519–22.

Collerson, K. D., Williams, Q., Ewart, A. E. and Murphy, D. T. (2010). Origin of HIMU and EM-1 domains sampled by ocean island basalts, kimberlites and carbonatites: The role of CO_2-fluxed lower mantle melting in thermochemical upwellings. *Phys. Earth Planet. Inter.* **181**, 112–31.

Dasch, E. J., Hedge, C. E. and Dymond, J. (1973). Effect of seawater alteration on strontium isotope composition of deep-sea basalts. *Earth Planet. Sci. Lett.* **19**, 177–83.

Davidson, J. P. (1983). Lesser Antilles isotopic evidence of the role of subducted sediment in island arc magma genesis. *Nature* **306**, 253–6.

Davidson, J. P. (1987). Crustal contamination *versus* subduction zone enrichment: examples from the Lesser Antilles and implications for mantle source compositions of island arc volcanic rocks. *Geochim. Cosmochim. Acta* **51**, 2185–98.

Davies, G. F. (1977). Whole-mantle convection and plate tectonics. *Geophys. J. Int.* **49**, 459–86.

Delavault, H., Chauvel, C., Thomassot, E., Devey, C. W. and Dazas, B. (2016). Sulfur and lead isotopic evidence of relic Archean sediments in the Pitcairn mantle plume. *Proc. Nat. Acad. Sci.* **113**, 12 952–6.

DePaolo, D. J. (1980). Crustal growth and mantle evolution: inferences from models of element transport and Nd and Sr isotopes. *Geochim. Cosmochim. Acta* **44**, 1185–96.

DePaolo, D. J. and Wasserburg, G. J. (1976). Inferences about magma sources and mantle structure from variations of ^{143}Nd/^{144}Nd. *Geophys. Res. Lett.* **3**, 743–6.

DePaolo, D. J. and Wasserburg, G. J. (1979). Petrogenetic mixing models and Nd)Sr isotopic patterns. *Geochim. Cosmochim. Acta* **43**, 615–27.

Dewey, J. (1980). Episodicity, sequence and style at convergent plate boundaries. In: Strangway, D. W. (Ed.) *The Continental Crust and its Mineral Deposits.* Geol. Assoc. Canada Spec. Pap. **8**, pp. 553–73.

Dickin, A. P. (2016). The chondritic moon: a solution to the ^{142}Nd conundrum and implications for terrestrial mantle evolution. *Geol. Mag.* **153**, 548–55.

Donnelly, K. E., Goldstein, S. L., Langmuir, C. H. and Spiegelman, M. (2004). Origin of enriched ocean ridge basalts and implications for mantle dynamics. *Earth Planet. Sci. Lett.* **226**, 347–66.

Douglass, J., Schilling, J. G. and Fontignie, D. (1999). Plume–ridge interactions of the Discovery and Shona mantle plumes with the southern Mid-Atlantic Ridge (40°–55° S). *J. Geophys. Res.* **104** (B2), 2941–62.

Dosso, L. and Murthy, V. R. (1980) A Nd isotope study of the Kerguelen islands: inferences on enriched oceanic mantle sources. *Earth Planet. Sci. Lett.* **48**, 268–76.

Dupre, B. and Allegre, C. J. (1980). Pb–Sr–Nd isotopic correlation and the chemistry of the North Atlantic mantle. *Nature* **286**, 17–22.

Dupre, B. and Allegre, C. J. (1983). Pb–Sr isotope variation in Indian Ocean basalts and mixing phenomena. *Nature* **303**, 142–6.

Dziewonski, A. M. (1984). Mapping the lower mantle: determination of lateral heterogeneity in P velocity up to degree and order 6. *J. Geophys. Res.* **89** (B7), 5929–52.

Eiler, J. M., Crawford, A., Elliott, T. I. M. *et al.* (2000). Oxygen isotope geochemistry of oceanic-arc lavas. *J. Petrol.* **41**, 229–56.

Eiler, J. M., Farley, K. A., Valley, J. W. *et al.* (1997). Oxygen isotope variations in ocean island basalt phenocrysts. *Geochim. Cosmochim. Acta* **61**, 2281–93.

Eiler, J. M., Farley, K. A., Valley, J. W., Hofmann, A. W. and Stolper, E. M. (1996). Oxygen isotope constraints on the sources of Hawaiian volcanism. *Earth Planet. Sci. Lett.* **144**, 453–68.

Eiler, J. M., Farley, K. A., Valley, J. W. *et al.* (1995). Oxygen isotope evidence against bulk recycled sediment in the mantle source of Pitcairn Island lavas. *Nature* **377**, 138–41.

Eiler, J. M., McInnes, B., Valley, J. W., Graham, C. M. and Stolper, E. M. (1998). Oxygen isotope evidence for slab-derived fluids in the sub-arc mantle. *Nature* **393**, 777–81.

Eiler, J. M., Schiano, P., Valley, J. W., Kita, N. T. and Stolper, E. M. (2007). Oxygen-isotope and trace element constraints on the origins of silica-rich melts in the subarc mantle. *Geochem. Geophys. Geosys.* **8** (9), 1–21.

Ellam, R. M. and Hawkesworth, C. J. (1988). Elemental and isotopic variations in subduction related basalts: evidence for a three component model. *Contrib. Mineral. Petrol.* **98**, 72–80.

Elliott, T., Blichert-Toft, J., Heumann, A., Koetsier, G. and Forjaz, V. (2007). The origin of enriched mantle beneath Sao Miguel, Azores. *Geochim. Cosmochim. Acta* **71**, 219–40.

Elliott, T., Zindler, A. and Bourdon, B. (1999). Exploring the kappa conundrum: the role of recycling in the lead isotope evolution of the mantle. *Earth Planet. Sci. Lett.* **169**, 129–45.

Farley, K. A., Natland, J. H. and Craig, H. (1992). Binary mixing of enriched and undegassed (primitive?) mantle components (He, Sr, Nd, Pb) in Samoan lavas. *Earth Planet. Sci. Lett.* **111**, 183–99.

Farquhar, J. and Wing, B. A. (2003). Multiple sulfur isotopes and the evolution of the atmosphere. *Earth Planet. Sci. Lett.* **213**, 1–13.

Faure, G. and Hurley, P. M. (1963). The isotopic composition of strontium in oceanic and continental basalt. *J. Petrol.* **4**, 31–50.

Flower, M. F. J., Schmincke, H. U. and Thompson, R. N. (1975). Phlogopite stability and the $^{87}Sr/^{86}Sr$ step in basalts along the Reykjanes Ridge. *Nature* **254**, 404–6.

Galer, S. J. G. and O'Nions, R. K. (1985). Residence time of thorium, uranium and lead in the mantle with implications for mantle convection. *Nature* **316**, 778–82.

Garnero, E. J. and Helmberger, D. V. (1995). A very slow basal layer underlying large-scale low-velocity anomalies in the lower mantle beneath the Pacific: evidence from core phases. *Phys. Earth Planet. Inter.* **91**, 161–76.

Gast, P. W., Tilton, G. R. and Hedge, C. (1964). Isotopic composition of lead and strontium from Ascension and Gough Islands. *Science* **145**, 1181–5.

Hanan, B. B. and Graham, D. W. (1996). Lead and helium isotope evidence from oceanic basalts for a common deep source of mantle plumes. *Science* **272**, 991–5.

Harmon, R. S. and Hoefs, J. (1995). Oxygen isotope heterogeneity of the mantle deduced from global ^{18}O systematics of basalts from different geotectonic settings. *Contrib. Mineral. Petrol.* **120**, 95–114.

Harris, C., Bell, J. D. and Atkins, F. B. (1983). Isotopic composition of lead and strontium in lavas and coarse-grained blocks from Ascension Island, South Atlantic – an addendum. *Earth Planet. Sci. Lett.* **63**, 139–41.

Harris, P. G., Hutchison, R. and Paul, D. K. (1972). Plutonic xenoliths and their relation to the upper mantle. *Phil. Trans. Roy. Soc. Lond.* A **271**, 313–23.

Hart, S. R. (1984). A large-scale isotope anomaly in the Southern Hemisphere mantle. *Nature* **309**, 753–7.

Hart, S. R. (1988). Heterogeneous mantle domains: signatures, genesis and mixing chronologies. *Earth Planet. Sci. Lett.* **90**, 273–96.

Hart, S. R. and Gaetani, G. A. (2006). Mantle Pb paradoxes: the sulfide solution. *Contrib. Mineral. Petrol.* **152**, 295–308.

Hart, S. R., Gerlach, D. C. and White, W. M. (1986). A possible new Sr–Nd–Pb mantle array and consequences for mantle mixing. *Geochim. Cosmochim. Acta* **50**, 1551–7.

Hart, S. R., Hauri, E. H., Oschmann, L. A. and Whitehead, J. A. (1992). Mantle plumes and entrainment: isotopic evidence. *Science* **256**, 517–20.

Hart, S. R., Schilling, J-G. and Powell, J. L. (1973). Basalts from Iceland and along the Reykjanes Ridge: Sr isotope geochemistry. *Nature Phys. Sci.* **246**, 104–7.

Harvey, J., Gannoun, A., Burton, K. W. *et al.* (2006). Ancient melt extraction from the oceanic upper mantle revealed by Re–Os isotopes in abyssal peridotites from the Mid-Atlantic ridge. *Earth Planet. Sci. Lett.* **244**, 606–21.

Hauri, E. H., Shimizu, N., Dieu, J. J. and Hart, S. R. (1993). Evidence for hotspot-related carbonatite metasomatism in the oceanic upper mantle. *Nature* **365**, 221–7.

Hauri, E. H., Whitehead, J. A. and Hart, S. R. (1994). Fluid dynamic and geochemical aspects of entrainment in mantle plumes. *J. Geophys. Res.* **99**, 24275–300.

Hawkesworth, C. J., Hergt, J. M., McDermott, F. and Ellam, R. M. (1991). Destructive margin magmatism and the contributions from the mantle wedge and subducted crust. *Aust. J. Earth Sci.* **38**, 577–94.

Hawkesworth, C. J., Norry, M. J., Roddick, J. C. and Vollmer, R. (1979). $^{143}Nd/^{144}Nd$ and $^{87}Sr/^{86}Sr$ ratios from the Azores and their significance in LIL element enriched mantle. *Nature* **280**, 28–31.

Hawkesworth, C. J., O'Nions, R. K., Pankhurst, R. J., Hamilton, P. J. and Evensen, N. M. (1977). A geochemical study of island-arc and back-arc tholeiites from the Scotia Sea. *Earth Planet. Sci. Lett.* **36**, 253–62.

Hofmann, A. W. (1997). Mantle geochemistry: the message from oceanic volcanism. *Nature* **385**, 219–29.

Hofmann, A. W., Jochum, K. P., Seufert, M. and White, W. M. (1986). Nb and Pb in oceanic basalts: new constraints on mantle evolution. *Earth Planet. Sci. Lett.* **79**, 33–45.

Hofmann, A. W. and Hart, S. R. (1978). An assessment of local and regional isotopic equilibrium in the mantle. *Earth Planet. Sci. Lett.* **38**, 44–62.

Hofmann, A. W. and White, W. M. (1980). The role of subducted oceanic crust in mantle evolution. *Carnegie Inst. Washington Yearbook* **79**, 477–83.

Hofmann, A. W. and White, W. M. (1982). Mantle plumes from ancient oceanic crust. *Earth Planet. Sci. Lett.* **57**, 421–36.

Jackson, M. G., Hart, S. R., Konter, J. G. *et al.* (2014). Helium and lead isotopes reveal the geochemical geometry of the Samoan plume. *Nature* **514**, 355–8.

Jacobsen, S. B. and Wasserburg, G. J. (1979). The mean age of mantle and crustal reservoirs. *J. Geophys. Res.* **84**, 7411–27.

Jego, S. and Dasgupta, R. (2014). The fate of sulfur during fluid-present melting of subducting basaltic crust at variable oxygen fugacity. *J. Petrol.* **55**, 1019–50.

Jenner, F. E. and O'Neill, H. S. C. (2012). Analysis of 60 elements in 616 ocean floor basaltic glasses. *Geochem. Geophys. Geosys.* **13** (2), 1–11.

Kamber, B. S. and Collerson, K. D. (1999). Origin of ocean island basalts: a new model based on lead and helium isotope systematics. *J. Geophys. Res.* **104**, 25, 479–91.

Kelemen, P. B. and Manning, C. E. (2015). Reevaluating carbon fluxes in subduction zones, what goes down, mostly comes up. *Proc. Nat. Acad. Sci.* **112**, E3997–4006.

Kelley, K. A. and Cottrell, E. (2009). Water and the oxidation state of subduction zone magmas. *Science* **325**, 605–7.

Kelley, K. A., Plank, T., Farr, L., Ludden, J. and Staudigel, H. (2005). Subduction cycling of U, Th, and Pb. *Earth Planet. Sci. Lett.* **234**, 369–83.

Kellogg, L. H., Hager, B. H. and van der Hilst, R. D. (1999). Compositional stratification in the deep mantle. *Science* **283**, 1881–4.

Kenyon, P. M. (1990). Trace element and isotopic effects arising from magma migration beneath mid-ocean ridges. *Earth Planet. Sci. Lett.* **101**, 367–78.

Korenaga, J. (2009). A method to estimate the composition of the bulk silicate Earth in the presence of a hidden geochemical reservoir. *Geochim. Cosmochim. Acta* **73**, 6952–64.

Kramers, J. D. and Tolstikhin, I. N. (1997). Two terrestrial lead isotope paradoxes, forward transport modelling, core formation and the history of the continental crust. *Chem. Geol.* **139**, 75–110.

Labanieh, S., Chauvel, C., Germa, A., Quidelleur, X. and Lewin, E. (2010). Isotopic hyperbolas constrain sources and processes under the Lesser Antilles arc. *Earth Planet. Sci. Lett.* **298**, 35–46.

Labidi, J., Cartigny, P. and Moreira, M. (2013). Non-chondritic sulphur isotope composition of the terrestrial mantle. *Nature* **501**, 208–11.

Lay, T. and Garnero, E. J. (2004). Core–mantle boundary structures and processes. In: Sparks, R. S. J. and Hawkesworth, C. J. (Eds) *The State of the Planet: Frontiers and Challenges in Geophysics. Geophys. Monograph Series*, American Geophys. Union, pp. 25–41.

Lay, T., Hernlund, J., Garnero, E. J. and Thorne, M. S. (2006). A post-perovskite lens and D" heat flux beneath the central Pacific. *Science* **314**, 1272–6.

Liu, C. Z., Wu, F. Y., Chung, S. L. *et al.* (2014). A 'hidden' ^{18}O-enriched reservoir in the sub-arc mantle. *Scientific Reports* **4** (4232), 1–6.

Malaviarchchi, S. P., Makishima, A., Tanimoto, M., Kuritani, T. and Nakamura, E. (2008). Highly unradiogenic lead isotope ratios from the Horoman peridotite in Japan. *Nature Geosci.* **1**, 859–63.

McCulloch, M. T. and Perfit, M. R. (1981). ^{143}Nd/^{144}Nd, ^{87}Sr/^{86}Sr and trace element constraints on the petrogenesis of Aleutian island arc magmas. *Earth Planet. Sci. Lett.* **56**, 167–79.

McDermott, F., Defant, M. J., Hawkesworth, C. J., Maury, R. C. and Joron, J. L. (1993). Isotope and trace element evidence for three component mixing in the genesis of the North Luzon arc lavas (Philippines). *Contrib. Mineral. Petrol.* **113**, 9–23.

McKenzie, D. (1979). Finite deformation during fluid flow. *Geophys. J. Roy. Astr. Soc.* **58**, 689–715.

McKenzie, D. P. and O'Nions, R. K. (1983). Mantle reservoirs and ocean island basalts. *Nature* **301**, 229–31.

Meijer, A., Kwon, T. T. and Tilton, G. R. (1990). U-Th-Pb partitioning behavior during partial melting in the upper mantle: Implications for the origin of high Mu Components and the "Pb Paradox". *J. Geophys. Res.* **95** (B1), 433–48.

Michard, A. and Albarede, F. (1985). Hydrothermal uranium uptake at ridge crests. *Nature* **317**, 244–6.

Morgan, J. P. and Shearer, P. M. (1993). Seismic constraints on mantle flow and topography of the 660-km discontinuity: evidence for whole-mantle convection. *Nature* **365**, 506–11.

Morgan, W. J. (1971) Convection plumes in the lower mantle. *Nature* **230**, 42–3.

Murakami, M., Hirose, K., Kawamura, K., Sata, N. and Ohishi, Y. (2004). Post-perovskite phase transition in MgSiO$_3$. *Science* **304**, 855–8.

Neal, C. R., Mahoney, J. J. and Chazey III, W. J. (2002). Mantle sources and the highly variable role of continental lithosphere in basalt petrogenesis of the Kerguelen Plateau and Broken Ridge LIP: results from ODP Leg 183. *J. Petrol.* **43**, 1177–1205.

Newsome, H. E., White, W. M., Jochum, K. P. and Hofmann, A. W. (1986). Siderophile element abundances in oceanic basalts, Pb isotope evolution and growth of the Earth's core. *Earth Planet. Sci. Lett.* **80**, 299–313.

Niu, Y. and O'Hara, M. J. (2003). Origin of ocean island basalts: A new perspective from petrology, geochemistry, and mineral physics considerations. *J. Geophys. Res.* **108** (B4) 2209, ECV5, 1–19.

O'Hara, M. J. (1973). Non-primary magmas and dubious mantle plume beneath Iceland. *Nature* **243**, 507–8.

O'Hara, M. J. (1975). Is there an Icelandic mantle plume? *Nature* **253**, 708–10.

O'Hara, M. J. and Mathews, R. E. (1981). Geochemical evolution in an advancing, periodically replenished, periodically tapped, continuously fractionated magma chamber. *J. Geol. Soc. Lond.* **138**, 237–77.

Olson, P. (1984). Mixing of passive heterogeneities by mantle convection. *J. Geophys. Res.* **89**, B425–36.

O'Nions, R. K., Evensen, N. M. and Hamilton, P. J. (1979). Geochemical modelling of mantle differentiation and crustal growth. *J. Geophys. Res.* **84** 6091–101.

O'Nions, R. K., Hamilton, P. J. and Evensen, N. M. (1977). Variations in ^{143}Nd/^{144}Nd and ^{87}Sr/^{86}Sr ratios in oceanic basalts. *Earth Planet. Sci. Lett.* **34**, 13–22.

O'Nions, R. K., Evensen, N. M. and Hamilton, P. J. (1980). Differentiation and evolution of the mantle. *Phil. Trans. Roy. Soc. Lond.* A **297**, 479–93.

O'Nions, R. K. and Pankhurst, R. J. (1973). Secular variation in the Sr-isotope composition of Icelandic volcanic rocks. *Earth Planet. Sci. Lett.* **21**, 12–21.

Pearce, J. (1983). The role of sub-continental lithosphere in magma genesis at destructive plate margins. In: Hawkesworth, C. J. and Norry, M. J. (Eds) *Continental Basalts and Mantle Xenoliths*. Shiva, pp. 230–49.

Polve, M. and Allegre, C. J. (1980). Orogenic lherzolite complexes studied by ^{87}Rb–^{87}Sr: a clue to understanding the mantle convection process? *Earth Planet. Sci. Lett.* **51**, 71–93.

Prinzhofer, A., Lewin, E. and Allegre, C. J. (1989). Stochastic melting of the marble cake mantle: evidence from local study of the East Pacific Rise at 12° 50' N. *Earth Planet. Sci. Lett.* **92**, 189–206.

Richard, P., Shimizu, N. and Allegre, C. J. (1976). ^{143}Nd/^{144}Nd, a natural tracer: an application to oceanic basalts. *Earth Planet. Sci. Lett.* **31**, 269–78.

Richter, S., Eykens, R., Kühn, H. *et al.* (2010). New average values for the n (238 U)/n (235 U) isotope ratios of natural uranium standards. *Int. J. Mass Spec.* **295**, 94–7.

Richter, F. M. and Ribe, N. M. (1979). On the importance of advection in determining the local isotopic composition of the mantle. *Earth Planet. Sci. Lett.* **43**, 212–22.

Ringwood, A. E. (1982). Phase transformations and differentiation in subducted lithosphere: implications for mantle dynamics, basalt petrogenesis, and crustal evolution. *J. Geol.* **90**, 611–43.

Schilling, J-G. (1973). Iceland mantle plume: geochemical study of Reykjanes Ridge. *Nature* **242**, 565–71.

Schilling, J-G. and Noe Nygaard, A. (1974). Faeroe–Iceland plume; rare-earth evidence. *Earth Planet. Sci. Lett.* **24**, 1–14.

Shearer, P. M. and Masters, T. G. (1992). Global mapping of topography on the 660-km discontinuity. *Nature* **355**, 791–6.

Staudigel, H., Park, K-H., Pringle, M. *et al.* (1991). The longevity of the South Pacific isotopic and thermal anomaly. *Earth Planet. Sci. Lett.* **102**, 24–44.

Stracke, A. (2012). Earth's heterogeneous mantle: A product of convection-driven interaction between crust and mantle. *Chem. Geol.* **330**, 274–99.

Stracke, A., Hofmann, A. W. and Hart, S. R. (2005). FOZO, HIMU, and the rest of the mantle zoo. *Geochem. Geophys. Geosys.* **6** (5), 1–20.

Sun, S. S. (1985). Ocean islands – plums or plumes? *Nature* **316**, 103–4.

Sun, S. S. (1980). Lead isotopic study of young volcanic rocks from mid-ocean ridges, ocean islands and island arcs. *Phil. Trans. Roy. Soc. Lond.* A **297**, 409–45.

Sun, S. S., Tatsumoto, M. and Schilling, J-G. (1975). Mantle plume mixing along the Reykjanes ridge axis: lead isotopic evidence. *Science* **190**, 143–7.

Tatsumoto, M. (1966). Genetic relations of oceanic basalts as indicated by lead isotopes. *Science* **153**, 1094–101.

Tatsumoto, M. (1978). Isotopic composition of lead in oceanic basalt and its implication to mantle evolution. *Earth Planet. Sci. Lett.* **38**, 63–87.

Tolstikhin, I. and Hofmann, A. W. (2005). Early crust on top of the Earth's core. *Phys. Earth Planet. Inter.* **148**, 109–30.

Turcotte, D. L. and Oxburgh, E. R. (1967). Finite amplitude convective cells and continental drift. *J. Fluid. Mech.* **28**, 29–42.

van der Hilst, R. D. and Karason, H. (1999). Compositional heterogeneity in the bottom 1000 kilometers of Earth's mantle: toward a hybrid convection model. *Science* **283**, 1885–8.

van der Hilst, R. D., Widiyantoro, S. and Engdahl, E. R. (1997). Evidence for deep mantle circulation from global tomography. *Nature* **386**, 578–84.

Vidal, P. and Dosso, L. (1978). Core formation: catastrophic or continuous? Sr and Pb isotope geochemistry constraints. *Geophys. Res. Lett.* **5**, 169–72.

Weaver, B. L. (1991). The origin of ocean island basalt end-member compositions: trace element and isotopic constraints. *Earth Planet. Sci. Lett.* **104**, 381–97.

Weis, D. and Frey, F. A. (2002). Submarine basalts of the northern Kerguelen Plateau: Interaction between the Kerguelen Plume and the Southeast Indian Ridge revealed at ODP Site 1140. *J. Petrol.* **43**, 1287–1309.

Weis, D., Ingle, S., Damasceno, D. *et al.* (2001). Origin of continental components in Indian Ocean basalts: Evidence from Elan Bank (Kerguelen Plateau, ODP Leg 183, Site 1137). *Geology* **29**, 147–50.

Weiss, Y., Class, C., Goldstein, S. L. and Hanyu, T. (2016). Key new pieces of the HIMU puzzle from olivines and diamond inclusions. *Nature* **537**, 666–70.

Wen, L. and Helmberger, D. V. (1998). Ultra-low velocity zones near the core–mantle boundary from broadband PKP precursors. *Science* **279**, 1701–3.

White, W. M. (1981). *European Colloquium of Geochronology, Cosmochronology and Isotope Geology* VII, meeting abstract.

White, W. M. (1985). Sources of oceanic basalts: radiogenic isotopic evidence. *Geology* **13**, 115–18.

White, W. M. (1993). $^{238}U/^{204}Pb$ in MORB and open system evolution of the depleted mantle. *Earth Planet. Sci. Lett.* **115**, 211–26.

White, W. M. (2015). Probing the Earth's deep interior through geochemistry. *Geochem. Perspectives* **4** (2), 95–247.

White, W. M. and Dupre, B. (1986). Sediment subduction and magma genesis in the Lesser Antilles: isotopic and trace element constraints. *J. Geophys. Res.* **91**, 5927–41.

White, W. M., Dupre, B. and Vidal, P. (1985). Isotope and trace element geochemistry of sediments from the Barbados Ridge–Demerara Plain region, Atlantic Ocean. *Geochim. Cosmochim. Acta* **49**, 1875–86.

White, W. M. and Hofmann, A. W. (1982). Sr and Nd isotope geochemistry of oceanic basalts and mantle evolution. *Nature* **296**, 821–5.

White, W. M., Tapia, M. D. M. and Schilling, J-G. (1979). The petrology and geochemistry of the Azores islands. *Contrib. Mineral. Petrol.* **69**, 201–13.

White, W. M., Schilling, J-G. and Hart, S. R. (1976). Evidence for the Azores mantle plume from strontium isotope geochemistry of the Central North Atlantic. *Nature* **263**, 659–63.

Widom, E., Carlson, R. W., Gill, J. B. and Schmincke, H.-U. (1997). Th-Sr-Nd-Pb isotope and trace element evidence for the origin of the Sao Miguel, Azores, enriched mantle source. *Chem. Geol.* **140**, 49–68.

Widom, E., Hoernle, K. A., Shirey, S. B. and Schmincke, H. U. (1999). Os isotope systematics in the Canary Islands and Madeira: lithospheric contamination and mantle plume signatures. *J. Petrol.* **40**, 279–96.

Woodhead, J. D., Greenwood, P., Harmon, R. S. and Stoffers, P. (1993). Oxygen isotope evidence for recycled crust in the source of EM-type ocean island basalts. *Nature* **362**, 809–13.

Workman, R. K., Eiler, J. M., Hart, S. R. and Jackson, M. G. (2008). Oxygen isotopes in Samoan lavas: Confirmation of continent recycling. *Geology* **36**, 551–4.

Workman, R. K. and Hart, S. R. (2005). Major and trace element composition of the depleted MORB mantle (DMM). *Earth Planet. Sci. Lett.* **231**, 53–72.

Wyllie, P. J. (1984). Constraints imposed by experimental petrology on possible and impossible magma sources and products. *Phil. Trans. Roy. Soc. Lond.* A **310**, 439–56.

Zartman, R. E. and Haines, S. M. (1988). The plumbotectonic model for Pb isotopic systematics among major terrestrial reservoirs – a case for bi-directional transport. *Geochim. Cosmochim. Acta* **52**, 1327–39.

Zartman, R. E. and Richardson, S. H. (2005). Evidence from kimberlitic zircon for a decreasing mantle Th/U since the Archean. *Chem. Geol.* **220**, 263–83.

Zindler, A. and Hart, S. R. (1986). Chemical geodynamics. *Ann. Rev. Earth Planet. Sci.* **14**, 493–571.

Zindler, A., Jagoutz, E. and Goldstein, S. (1982). Nd, Sr and Pb isotopic systematics in a three-component mantle: a new perspective. *Nature* **298**, 519–23.

Zindler, A., Staudigel, H. and Batiza, R. (1984). Isotope and trace element geochemistry of young Pacific seamounts: Implications for the scale of upper mantle heterogeneity. *Earth Planet. Sci. Lett.* **70**, 175–95.

Chapter 7

Isotope Geochemistry of Continental Rocks

Oceanic volcanics, erupted through thin, young lithosphere, represent a window on the asthenosphere and deep mantle. In contrast, continental basalts and mantle xenoliths, emplaced through thick, old lithosphere, may tell us about the nature of the deep crust and the lithospheric mantle, as well as the evolution of magmas during their ascent to the surface. Isotopic data represent a powerful tool for such studies, firstly because of their ability to date geological events, and secondly because of their usefulness as tracers of complex mixing processes.

Unfortunately, continental igneous rocks are difficult to interpret. This is because they can derive an enriched elemental and isotopic signature from three possible sources: mantle plumes, sub-continental lithosphere and the crust. Resolving these components from one another in continental volcanics and plutons has been a major subject of discussion in geochemistry for several decades. Much progress has been made, but the large number of variables tends to make each case a unique example; or as Read (1948) put it, there are 'granites and granites'. This makes a generalized approach to continental magmas difficult, and forces us to adopt a case study approach as an attempt to illustrate underlying principles.

Mantle xenoliths provide a more direct means of sampling the sub-continental lithosphere. Their texture provides evidence of a solid source, while the peridotite (= lherzolite) petrology of the commonest types is readily distinguished from crustal xenoliths (which will not be dealt with here). Therefore, our approach in this chapter will be firstly to study the lithospheric mantle by means of xenoliths, secondly to examine crustal contamination processes and lastly to look at some classic case studies in the genesis and evolution of continental igneous rocks.

7.1 Mantle Xenoliths

The sub-continental lithosphere is distinguished from the underlying asthenosphere by its non-convecting, rigid state. Hence it was termed the 'tectosphere' by Jordan (1975, 1978).

Jordan argued from seismic and heat-flow evidence that this tectosphere was 200–300 km thick under shield areas. Evidence for 3.5 Ga garnet inclusions in diamonds (Richardson *et al.*, 1984) suggests a similar thickness of continental lithosphere in the Archean.

Alkaline magmas, kimberlites and carbonatites in many continental areas bring up peridotite xenoliths (also called nodules) from great depths. On the basis of their mineral chemistry, these must be samples of the mantle rather than the crust. Maaloe and Aoki (1977) analysed the major element composition of numerous such xenoliths in an attempt to estimate the bulk upper mantle composition. They recognized compositional differences between spinel lherzolite xenoliths, derived from Proterozoic and younger lithosphere, and garnet lherzolites, derived from Archean cratons. Both xenolith types had overlapping ranges of MgO content, but the (Archean) garnet peridotites had distinctly lower FeO contents. In view of their more exotic history, we will direct our main attention to this group.

The world's classic mantle xenolith suites come from the Kaapvaal Craton of South Africa, where they are obtained as by-products of diamond mining. Within this collection, two main textural types are observed; 'granular' and 'sheared'. Harte (1983) proposed that the former were samples of the lithosphere, while the latter, which are more often found around the margins of the Kaapvaal Craton, were derived from the convecting asthenosphere. Of the granular types, Harte further sub-divided samples showing obvious or 'modal' metasomatism (indicated by hydrous or other exotic minerals) from the more normal garnet peridotites. The latter samples come from the centre of the Kaapvaal Craton, at Northern Lesotho and Bultfontein (hence NLB-type), and were regarded as typical samples of the mantle lithosphere.

Various explanations have been proposed to account for the differing FeO contents of spinel and garnet peridotites, but the most satisfactory was developed by Richter (1988). He proposed that garnet peridotites were residues of komatiite extraction in the Archean, and that the large degrees of melting associated with this process caused FeO depletion. This in turn lowered the density of the residuum, relative to

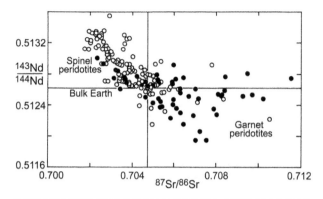

Fig. 7.1 Nd versus Sr isotope diagram showing the largely distinct compositional fields of spinel peridotite (○), and garnet peridotite (●). After Hawkesworth *et al.* (1990).

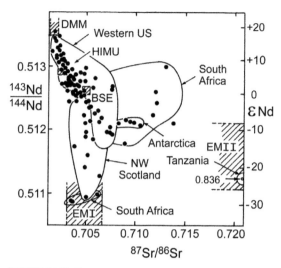

Fig. 7.2 Nd versus Sr isotope diagram showing compositional fields for xenolith suites from different provinces, relative to enriched mantle components identified in OIB sources (hatched fields). After Menzies (1989).

fertile mantle, and allowed its stabilization as sub-continental lithosphere. This material reached sufficient thickness (>150 km) for diamond crystallization to occur at its base. In contrast, Proterozoic lithosphere was stabilized only by conductive cooling of the upper mantle (a mechanism that would not have been possible in the hotter Archean mantle).

Proterozoic mantle lithosphere may be residual from basalt extraction, or may not be depleted by melt extraction at all; hence it has higher levels of FeO and other fertile components. The thickness of lithosphere formed in this way is insufficient for diamond stability, while its high density makes it susceptible to delamination from the base of the crust during orogenic shortening of the lithosphere.

The major element compositional differences between garnet and spinel peridotite xenoliths, described above, are paralleled by isotopic differences. Figure 7.1 shows a compilation of Sr and Nd isotope data for the two groups (Hawkesworth *et al.*, 1990), which define fairly distinct fields. Spinel peridotite data are derived mainly from separated clinopyroxene (cpx), but garnet peridotite data are based on a combination of separated mineral and whole-rock analyses. The latter are less reliable because they are susceptible to contamination by the host magma (usually kimberlite in the case of garnet peridotite xenoliths).

Menzies (1989) presented a summary of Nd–Sr isotope data for xenolith suites analysed to that date (Fig. 7.2), and interpreted them in relation to the DMM, EMI and EMII end-members proposed for OIB sources by Zindler and Hart (Section 6.4.2). There is a strong distribution of data between the HIMU and EMI components in Fig. 7.2, and Zindler and Hart did in fact propose that these end-members (forming the LoNd array) were derived from recycled mantle lithosphere. Additional support for the existence of 'EM1-type' lithosphere with unradiogenic Sr signatures comes from xenolith studies on Eastern Africa (Cohen *et al.*, 1984). This may therefore be a real lithospheric end-member.

The trend towards the EM2 end-member in Fig. 7.2 is somewhat weaker, but it is possible that similar processes

are at work to enrich lithospheric and plume sources. In the latter, the EM2 component is strongly linked to sediment subduction (Section 6.5.2). Subducted sediment may also contribute to enrichment of the mantle lithosphere, by releasing slab-derived fluids that contaminate the overlying mantle wedge (Section 6.5.3). However, such processes have probably given rise to a whole range of variably enriched signatures (e.g. Ackerman *et al.*, 2009), rather than a distinct lithospheric end-member.

Secondary enrichment of the mantle can also be achieved by other processes, some of which will be examined below in examples from the Kaapvaal Craton.

7.1.1 Mantle Metasomatism

Spinel peridotite data in Fig. 7.1 are generally depleted relative to the Bulk Earth composition. Therefore, they may represent 'frozen' samples of asthenospheric upper mantle. However, garnet peridotites generally fall in the enriched quadrant relative to Bulk Earth, despite the fact that they are interpreted as residues of komatiite extraction. This demands a secondary enrichment process, which could be caused by either silicate melts, or by hydrous or carbonate-rich fluids. Only the latter two are examples of metasomatism in the strict sense, but typically, mantle enrichment is regarded as more or less synonymous with mantle metasomatism.

Dawson and Smith (1977) described a suite of mafic xenoliths from kimberlites such as Bultfontein, whose hydrous mineralogy marked them as relics of ancient metasomatising fluids. These nodules, sometimes described as glimmerite, are characterized by the presence of phlogopite mica, along with various other hydrous minerals. These are regarded as 'primary' hydrous minerals, in the sense that

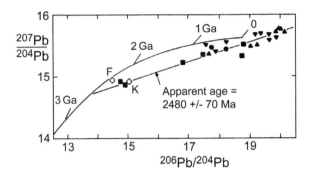

Fig. 7.3 Pb–Pb isochron diagram for nodules from South African kimberlites. (◊) = sulphide inclusions in diamonds (F = Finsch mine, K = Kimberly); filled symbols: cpx from peridotite and cpx megacrysts (different symbols signify different mines). After Kramers (1979).

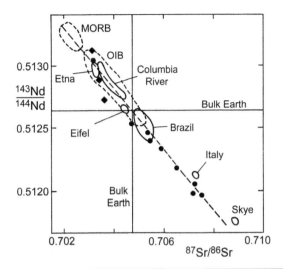

Fig. 7.4 Plot of Nd versus Sr isotope ratios for diopsides from South African kimberlite nodules (•), relative to the mantle array of oceanic basalts. (♦) = whole-rock peridotites. After Menzies and Murthy (1980).

they were generated by high-temperature processes at depth, and are not the result of high-level secondary alteration after kimberlite emplacement.

Dawson and Smith distinguished an important sub-group of these nodules with a characteristic mineral assemblage of mica–amphibole–rutile–ilmenite–diopside, which they dubbed the 'MARID' suite. They suggested that these MARID xenoliths might have crystallized from a pegmatitic magmatic fluid, chemically similar to kimberlite, which would be capable of metasomatising its peridotite wall rocks.

This model was developed by Jones *et al.* (1982), who suggested that peridotite nodules from Bultfontein had been metasomatized by a fluid which, although not exactly like the parent of the MARID suite, was related to it in some way. This metasomatic process is recorded by different peridotite lithologies which form a series. Starting from garnet peridotite, this series progresses through garnet–pargasite peridotite and phlogopite peridotite to phlogopite–K-richterite peridotite in a suite represented as GP–GPP–PP–PKP (Erlank *et al.*, 1987).

Having established the role of mantle metasomatism in generating the incompatible element enrichments of peridotite xenoliths, another important question is the timing of this process. Kramers (1979) analysed the Pb isotope composition of sulphide inclusions in diamonds (and also cpx from eclogite and peridotite xenoliths) in several Cretaceous kimberlite pipes from South Africa. Both inclusion and cpx data lay close to a 2.5 Ga Pb/Pb regression (Fig. 7.3), implying that diamonds and xenoliths are co-genetic, and that mineralogical heterogeneity was preserved in the South African sub-continental lithosphere since the Archean. In particular, the very unradiogenic composition of the diamonds, which yield Pb model ages of over 2 Ga, would be very difficult to explain by any recent metasomatic event. In contrast, Pb isotope compositions of 'fertile peridotites' and cpx megacrysts were interpreted as evidence of fairly recent disturbance.

Menzies and Murthy (1980) analysed the Sr and Nd isotope compositions of diopside in micaceous garnet lherzolite

nodules from South African kimberlite pipes (Bultfontein and Kimberley). The diopsides showed a strong inverse correlation on the Sr–Nd isotope diagram (Fig. 7.4). This was attributed by Menzies and Murthy to gross mantle heterogeneity, randomly sampled by kimberlite magmas. They suggested that these signatures were generated by an ancient metasomatic event, probably related to an upwelling mantle plume, which caused LIL element enrichment of the mantle lithosphere.

Hawkesworth *et al.* (1983) estimated from Nd isotope data that the ancient enrichment event postulated by Menzies and Murthy probably occurred ca. 1–4 Ga ago. However, the Rb/Sr ratios of the analysed diopsides (and indeed any mantle diopsides) are much too low to 'support' their observed $^{87}Sr/^{86}Sr$ compositions (i.e. generate the required extra amount of ^{87}Sr by *in situ* ^{87}Rb decay in the required time). This is demonstrated by the clustering of these points near the *y* axis of the Rb–Sr isochron diagram (Fig. 7.5). Therefore, Hawkesworth *et al.* argued that the diopsides must have crystallized in a recent event, presumably during secondary metasomatism of the enriched mantle generated by the ancient metasomatic event.

In contrast, whole-rock analyses of (garnet-free) phlogopite-bearing and K-richterite-bearing peridotites (PP and PKP) have more radiogenic Sr isotope ratios, and define a linear array with a slope age of 150 Ma. However, this array cannot be generated by contamination with the host kimberlite magma itself, because this has unradiogenic ^{87}Sr.

Kimberlites are actually known to have two distinct isotopic signatures, termed group I and group II. The (group I) kimberlite host of the peridotite nodules was ruled out as the metasomatizing agent of the nodules because of its unradiogenic Sr signature. Therefore, Erlank *et al.* (1987) considered

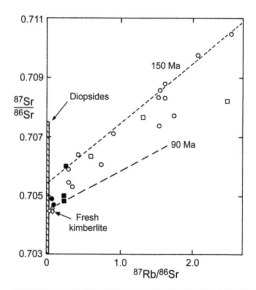

Fig. 7.5 Rb–Sr isochron diagram for South African kimberlite nodules, showing diopside field (hatched) relative to kimberlite host (◊) and nodules of different lithology: (●) = garnet peridotite; (■) = garnet–pargasite peridotite; (□) = phlogopite peridotite; (○) = phlogopite–K-richterite peridotite. After Hawkesworth et al. (1983).

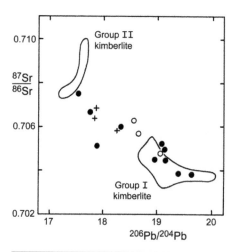

Fig. 7.6 Plot of initial Sr isotope ratio at 90 Ma (kimberlite emplacement age) against Pb isotope ratio for PKP whole-rocks (+), minerals from MARID xenoliths (○) and minerals from other Kimberly peridotites (●), compared to the fields for group I and II kimberlites. After Erlank et al. (1987).

the possibility that the Sr signatures found in the peridotite nodules could be generated by metasomatic fluids related to the more radiogenic group II kimberlites. A compilation of Sr and Pb isotope data for kimberlites, MARID xenoliths and metasomatized peridotite shows that all three suites form a single isotope array with negative slope, which could be a mixing line (Fig. 7.6).

Support for the mixing hypothesis came from the discovery of a new xenolith type (Gregoire et al., 2002) lacking richterite (amphibole) and with low rutile. Gregoire et al. named these 'PIC' xenoliths (phlogopite–ilmenite–cpx), although 'MID' (mica–ilmenite–diopside) would have been more consistent with the established term MARID. However, the important point is that these xenoliths have isotope signatures resembling group I kimberlites.

Gregoire et al. concluded that mantle metasomatism in the Kimberly area, leading to two different xenolith suites, was caused by fluids related to the two recognized kimberlite magma groups. This grouping is supported by Sr–Nd isotope evidence (Fig. 7.7), which also shows that these suites are not related to Karroo basalt magmatism (a model preferred by Erlank et al., 1987).

These metasomatic enrichments caused by kimberlite magmatism are only the final stage in the long geological history of the Kaapvaal Craton. Other stages in the mineralogical and geochemical evolution of the Kaapvaal lithosphere were explored by Simon et al. (2007). In addition, the origins of the kimberlite magma itself will be discussed below (Section 7.3.1).

7.2 Crustal Contamination

Many continental igneous rocks have enriched chemical and isotopic signatures similar to those discussed in the previous section. However, the critical question is whether these signatures were inherited from the mantle or the crust. In principle, isotopic methods represent an ideal tool to solve this problem, since they are not upset by the crystal fractionation processes which affect most magmas during ascent and emplacement. However, the high degrees of enrichment which can occur in plume or lithospheric mantle sources may generate isotopic signatures similar to the crust. Hence it has been argued (e.g. Thirlwall and Jones, 1983; Hawkesworth et al., 1984) that mantle and crustal sources cannot be distinguished simply on the basis of

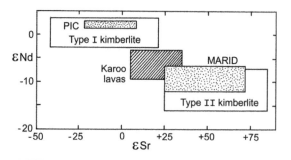

Fig. 7.7. Plot of εNd versus εSr at 90 Ma to show resemblances between the isotope signatures of different types of kimberlite magmas and glimmerite nodules. After Gregoire et al. (2002).

'isotopic discriminant diagrams' in which each component has a unique field. Instead, crustal or mantle contributions to magmatism must be recognized by observing the variable products of *processes* such as magma mixing or crustal assimilation.

Philosophically, one can examine contamination processes in two ways: a predictive model (e.g. DePaolo, 1981a) or an inversion technique (e.g. Mantovani and Hawkesworth, 1990). In the former, we set conditions and then examine consequences. In the latter, we examine products and attempt to reconstruct the original conditions. The predictive model is well suited to two-component mixing processes, such as progressive contamination of a single magma batch by wall-rock assimilation. Some examples of such models will be examined below, followed by an examination of crustal melting processes pertinent to crustal contamination models.

In contrast, volcanic lava piles often involve multi-component mixing. These processes are more difficult to examine using predictive models, because of the plethora of possible mixing scenarios. Therefore, it is more effective to model complex magma suites using the inversion approach, bearing in mind the predictive models already developed for single magma batches. This approach will be illustrated using the classic British Tertiary Igneous Province as a case study.

7.2.1 Two-Component Mixing Models

In its simplest form, contamination of mantle-derived magma by the continental crust can be regarded as a process of two-component mixing. However, magma–crust mixing processes usually have more than one degree of freedom (such as the compositions and proportions of mixed components). Therefore, to adequately evaluate mixing relations it is usually necessary to apply two or more measured variables to the problem. These variables are usually isotope ratio, elemental ratio and elemental abundance. In the context of isotope geology, it is logical to begin by examining the behaviour of isotopic tracers as a function of the elemental concentration of the same element. Therefore, we will begin by studying initial $^{87}Sr/^{86}Sr$ ratios as a function of Sr concentration.

Mixing of components with different isotopic and elemental compositions yields a hyperbolic curve on a diagram of initial $^{87}Sr/^{86}Sr$ against Sr concentration (Fig 7.8a). Ideally, initial Sr isotope ratios should be plotted against ^{86}Sr abundance, since the total concentration of strontium is slightly perturbed by variations in ^{87}Sr. This is particularly true for old rocks, where there is a large age correction to obtain the initial ratio. However, the decay constant of Rb is so low that ^{88}Sr makes up the bulk of strontium in most rocks. Therefore ^{86}Sr abundance can be approximated by total Sr without introducing significant errors. (This is not so for Pb in old rocks, where radiogenic Pb can easily swamp the non-radiogenic component.)

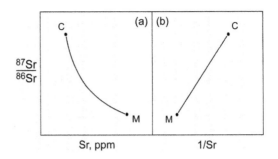

Fig. 7.8 Schematic illustration of two-component mixing on plots of Sr isotope ratio against (a) Sr concentration, and (b) 1/Sr. C = crustal end-member; M = mantle-derived end-member.

A bivariate diagram for two ratios with common denominators must yield linear mixing lines. Therefore the hyperbolic mixing curve of Fig. 7.8a can be transformed into a straight line (Fig. 7.8b) by plotting initial $^{87}Sr/^{86}Sr$ against 1/Sr (approximating $1/^{86}Sr$). Briquet and Lancelot (1979) used this format to examine contamination and fractionation processes in a 'selective contamination' model (Fig. 7.9), which envisaged two-component mixing between a primary basic magma and a hypothetical Sr-rich extract from the crust. Following the contamination process, plagioclase fractionation may cause the Sr content of the magma to fall as it evolves to dacitic and then rhyolitic compositions (Fig. 7.9a). If these contamination and fractionation steps were repeated sequentially, they would create the effect seen in Fig. 7.9b. If the steps become very small, the result is simultaneous fractionation and contamination (Fig. 7.9c). However, Briquet and Lancelot's 'selective' model is probably not the most realistic for magma contamination, since Nd isotope evidence suggests that most contamination is by crustal melts (e.g. Thirlwall and Jones, 1983).

Crustal melting and assimilation is an endothermic process. If the magma is on or below the liquidus, it can only obtain heat to power wall-rock melting by itself undergoing fractional crystallization. Hence, we may expect these two processes to be coupled into a mechanism which DePaolo (1981a) termed 'assimilation fractional crystallisation' (AFC). In this model, the effect of fractionation on the mixing trajectory will depend on the relative importance of assimilation and fractional crystallization, and also on the crystal–liquid distribution coefficient (D) pertaining at the time.

To illustrate these effects, Fig. 7.10 shows calculated mixing lines for different D_{Sr} values at increasing mass fractions of assimilate (M_a) relative to initial magma (M_m), for a fixed proportion of assimilation relative to crystallization (M_a/M_c). (A smaller amount of fractionation relative to assimilation will cause a lesser deviation from the simple mixing line, and a larger relative amount of fractionation will cause greater deviation.) As plagioclase joins the crystallizing assemblage,

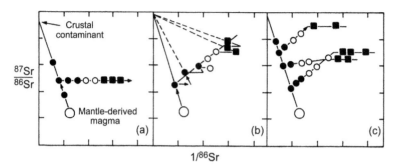

Fig. 7.9 Schematic modelling of selective Sr contamination and fractionation of magmas on plots of Sr isotope ratio against 1/Sr: (a) contamination followed by fractionation; (b) sequential contamination and fractionation events; (c) simultaneous contamination and fractionation, followed by pure fractionation. After Briquet and Lancelot (1979).

this will have a very dramatic effect on D_{Sr} values, changing strontium from an incompatible element ($D_{Sr} < 1$) to a compatible element ($D_{Sr} > 1$) in the crystallizing material. This may cause a magma to follow the inflected curve in Fig. 7.10 during its evolution.

The Sr versus Nd isotope diagram provides a useful tool for assessing crustal contamination models. DePaolo and Wasserburg (1979a) showed that simple two-component mixing on this diagram gives rise to hyperbolae whose trajectories depend on the relative Sr/Nd concentration ratio in the two end-members (Fig. 7.11). For the special case where the Sr/Nd ratio is the same in both end-members, the mixing line is straight. When the mantle-derived component has a higher Sr/Nd ratio, Nd compositions are more readily affected by contamination than Sr, yielding a concave upwards curve (K > 1). This is the normal situation when the mantle-derived component is more basic than the crustal end-member (whose Sr content has been lowered by plagioclase fractionation in its previous history). However, contamination by very plagioclase-rich crust could yield a convex-upward curve (K < 1).

7.2.2 Melting in Natural and Experimental Systems

The isotopic composition of a mantle-derived magma undergoing crustal contamination may be fairly predictable, but the composition of contaminating crustal melts is much more poorly constrained. Therefore a number of studies have been made on crustal melting, both in the laboratory and in 'natural laboratories' in the field. Most of these studies have involved the melting of granitoid rocks, since this is believed to be the most important component available for melting in the continental crust. A few of these studies will be reviewed here.

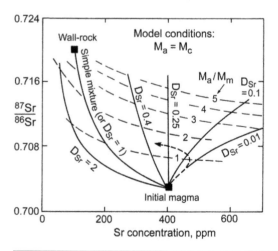

Fig. 7.10 Plot of Sr isotope ratio against concentration to show the effect of different solid/liquid bulk distribution coefficients (D_{Sr}) during the process of assimilation–fractional crystallization (AFC) by a basic magma. For discussion, see text. After DePaolo (1981a).

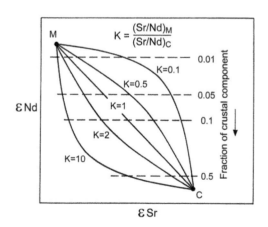

Fig. 7.11 Schematic illustration of two-component mixing on a plot of, Nd versus Sr. M and C are mantle-derived and crustal end-members. K = Sr/Nd ratio in mantle-derived relative to crustal end-member. Normally K is between 2 and 10. After DePaolo and Wasserburg (1979a).

One of the first modern studies of this problem was made by Maury and Bizouard (1974) on partially melted biotite gneiss xenoliths in a basanitic melt from southern France. One of the most important findings was that more than one initial melt composition was present (represented by quenched glasses). The two most important melt compositions were a colourless rhyolitic glass resulting from melting on quartz–feldspar grain boundaries and a brown latite glass resulting from melting on biotite–feldspar grain boundaries. Most subsequent studies have confirmed these findings, but since the colourless and brown glasses tend to mix if the melting interval is prolonged, the 'starting compositions' have often not been found in sufficient quantity for geochemical analysis. For example, two studies on melting of granite from the Sierra Nevada batholith at Rattlesnake Gulch, by a 12 Ma-old trachyandesite plug (Kaczor *et al.*, 1988; Tommasini and Davies, 1997) both identified pale brown and dark brown melt glasses, but were unable to fully separate the *initial* colourless and brown glasses for geochemical analysis.

A laboratory-based melting study was made by Hammouda *et al.* (1996) on a synthetic mixture of plagioclase and phlogopite (= Mg biotite). Preferential melting of the phlogopite was observed at above 1200 °C. Since the phlogopite had been doped with radiogenic Sr to simulate the effects of Rb decay in an old granite, the melt glasses were much more radiogenic than the bulk rock. This experiment simulated the melting behaviour of a tonalitic crustal rock, suggesting that partial melting of such material could cause 'selective contamination' of mafic magmas with radiogenic Sr. However, it was previously suggested by Thompson *et al.* (1982) that the small amounts of fusible *granitic* rock in a crustal section would be more important in promoting crustal contamination of basaltic magmas than the relatively refractory tonalite component.

A combined field-based and laboratory-based melting study of granitic rocks was made by Knesel and Davidson (1999). The field component involved melting of the Sierra Nevada granite in the vicinity of a Pleistocene-age olivine basalt at Tungsten Hills, whereas the laboratory-based component involved melting relatively large 45 g cubes of a 1200 Ma-old Precambrian granite. Sr isotope results from the latter study (Fig. 7.12) showed a progressive evolution in the composition of brown and colourless melt glasses as the melting temperature was increased (for a fixed 24 h duration). The isotope ratios of the two melts were initially very distinct but evolved towards the whole-rock composition of the source. In contrast, the Sr *abundances* of the two melts started relatively close to the whole-rock value, but evolved away from it, reflecting increasing Sr enrichment as melting progressed. This was attributed to the evolution of the restite towards a strontium-free quartz residue.

A somewhat different picture was obtained when the bulk composition of the experimental glasses was calculated, based on the composition and abundance of the two components. When the Sr isotope ratio of the bulk glass was plot-

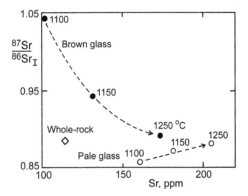

Fig. 7.12 Plot of Sr isotope ratio against concentration, showing the effect of increased melting temperature (1100–1150–1250 °C) on colourless and brown melt glasses, generated in laboratory experiments on a Precambrian granite. After Knesel and Davidson (1999).

ted against melt fraction, a monotonic decrease in (initial) Sr isotope ratio was observed as a function of melt fraction (Fig. 7.13a). Such a pattern was also observed in the field-based data from Tungsten Hills and Rattlesnake Gulch (Tommasini and Davies, 1997; Knesel and Davidson, 1999), shown in Fig. 7.13b, c.

The laboratory experiments described above were all performed at atmospheric pressure (and therefore anhydrous conditions), which do not accurately represent crustal melting in the deep crust. Therefore, Knesel and Davidson (2002) repeated these experiments on the same granite sample at a confining pressure of 600 MPa (6 kb), equivalent to a depth of about 20 km in the Earth's crust. These conditions permitted runs of longer duration (up to 2 months), and also allowed the melt products to be 'extracted' from the source into a vacant pore space created by a bed of industrial diamonds at one end of the sample charge. The sample itself consisted of finely crushed granite (75–100 microns) that was intended to preserve the mineralogical proportions of the original rock.

Results of this experiment are shown in Fig. 7.14 for three different melting temperatures from 850–950 °C. The surprising thing about these results is that, unlike the previously reported field and laboratory experiments, the initial melt was *less* radiogenic than the source, although it eventually reached Sr isotope equilibrium with the whole-rock composition. In contrast, an experiment at 1000 °C (not shown) gave results similar to the previous experiments, with an initial melt more radiogenic than the whole-rock. The unradiogenic Sr composition of the low temperature melts was attributed to the melting of plagioclase, in a reaction involving the dehydration of a small amount of muscovite in the sample. In contrast, biotite breakdown was the most important reaction at above 950 °C. These results are interesting because they suggest that crustal contaminants are not necessarily enriched in radiogenic Sr relative to the source rock. However, the crushing of the original sample may have

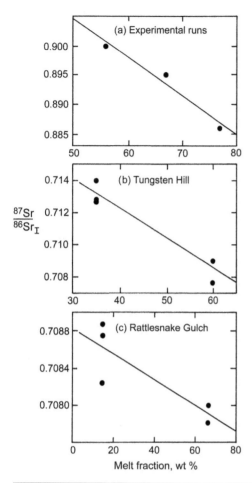

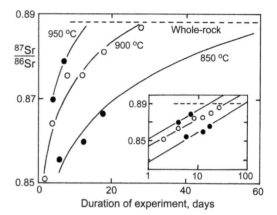

Fig. 7.14 Plot of Sr isotope ratio of granitic melts generated in piston-cylinder experiments over different time intervals and at different temperatures. (Pressure = 6 kbar). Inset shows experimental duration on a log scale. After Knesel and Davidson (2002).

Fig. 7.13 Plot of initial Sr isotope composition of bulk melts as a function of melt fraction in three studies of granitic melting described in the text. After Knesel and Davidson (1999).

created artificial mineral contacts that do not accurately represent the original rock. Therefore, more experiments are needed to test the behaviour of this material and other source compositions under similar melting conditions.

The general conclusion from all of these melting experiments is that the isotopic composition of a crustal melt evolves slowly towards the bulk composition of the rock as melting progresses. However, since the temperature and duration of melting are different for any given crustal contamination event, it is not possible to make general predictions about the extent of disequilibrium melting to be expected in crustal contaminants. Each case must be investigated in its own context.

7.2.3 Inversion Modelling of Magma Suites

The above modelling has considered the evolution of single magma batches during melting, assimilation and/or fractionation processes. However, a suite of analysed lavas may

represent magma batches that reached different stages of differentiation (and hence had different trace element contents) before contamination. Just as different bivariate plots can be used to model progressive contamination of a single magma, the same variety of plots can be used to examine the evolution of magma suites.

The Tertiary volcanic province of NW Scotland represents a good natural laboratory in which to examine some of these processes for two main reasons. Firstly, magma–crust interaction was relatively intense, due to the volatile-poor nature of the magmas. This prevented them from punching through the crust quickly. Secondly, isotopic contrasts between mantle and crustal end-members are well developed, because old lithospheric mantle had been melted away from under the Tertiary volcanic centres by earlier magmatism.

An example of the co-variation of Sr isotope ratio and Sr concentration is provided by Tertiary basic-to-intermediate lavas from the Isle of Skye, NW Scotland (termed the Skye Main Lava Series). Moorbath and Thompson (1980) found a weak negative correlation between Sr isotope ratio and concentration in this suite, forming a hyperbolic trend (Fig. 7.15). However, any individual mixing line between a hypothetical mantle-derived precursor and the estimated crustal component has a slope perpendicular to the observed trend. Such a trajectory is displayed by a small suite of low-potassium (low-K) basalts in Fig. 7.15.

To explain the main data set, Moorbath and Thompson proposed that crystal fractionation had occurred in the upper mantle to yield a series of magmas with variable Sr contents. These were then subjected to similar degrees of contamination with radiogenic crustal Sr, so that those with high Sr contents were less affected than those with low Sr contents; yielding a hyperbolic pattern for the suite as a whole. The scatter in the data probably results from

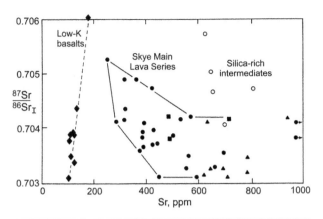

Fig. 7.15 Plot of initial Sr isotope ratio against concentration for Tertiary lavas from Skye, NW Scotland. Skye Main Lava Series: (●) = basalt; (▲) = hawaiite; (■) = mugearite–benmoreite. Other lavas: (○) = silica-oversaturated intermediates; (♦) = Low-K basalts. After Moorbath and Thompson (1980).

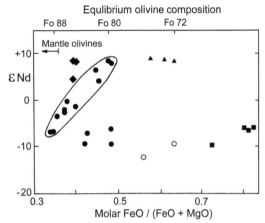

Fig. 7.17 Plot of εNd against 'F/M' ratio for Tertiary lavas from Skye, showing more intense crustal contamination in more magnesian basalts. Symbols as in Fig. 7.15. After Thirlwall and Jones (1983).

somewhat variable degrees of contamination in different magma batches.

Thirlwall and Jones (1983) made Nd isotope determinations on the same suite of Skye lavas. The data are shown (Fig. 7.16) on a plot of Nd isotope ratio against 1/concentration. Most of the basalts define an approximate linear array (equivalent to a hyperbola on a plot of $^{143}Nd/^{144}Nd$ against Nd concentration). However, this linear array does not have the trajectory expected for two-component mixing (steep vector in Fig. 7.16). Instead, it is attributed to contamination of a magma series with variable Nd contents, in which the most 'primitive' magmas, with lowest Nd contents, show the greatest effects of contamination. On the other hand, a few

basalts, along with silica-rich intermediate lavas, show the effects of an AFC process, in which Nd contents rise rapidly as contamination progresses (Fig. 7.16).

Thirlwall and Jones confirmed this interpretation (Fig. 7.17) using a plot of Nd isotope ratio against the major element differentiation index (FeO/FeO + MgO). They showed that the 'F/M' ratios of the Skye lavas must have been generated by fractionation at the base of the crust, since they were too high in most of the rocks to have been in equilibrium with mantle olivines. It follows that the strong correlation of εNd with F/M must be the result of a subsequent process, i.e. contamination in the crust. The most primitive basalts (lowest F/M) were the most contaminated, consistent with their lower Nd contents, which rendered them more sensitive to contamination. Again, the linear array in this diagram does not correspond to a two-component mixing line. Crustal contaminants have low Fe and Mg concentrations, so that they do not affect the F/M ratio of the contaminated magma. Hence, sub-vertical mixing vectors are generated in Fig. 7.17. The formation of the array of lava compositions at an oblique angle to these vectors can be ascribed to a regular and predictable contamination mechanism affecting a suite of related differentiates.

Huppert and Sparks (1985) attributed the type of contamination process seen in the Skye lavas to thermal erosion of wall-rocks by turbulently flowing magma during its ascent through the crust. The more magnesian magmas were hotter and less viscous, therefore enhancing the turbulent flow of these magmas. This prevented the formation of a chilled margin by continually bringing fresh, hot magma into contact with the conduit walls, and thereby allowing more wall-rock erosion. Huppert and Sparks imagined this process occurring in dykes, but a more probable site for such wall-rock assimilation may be sill complexes in the crust,

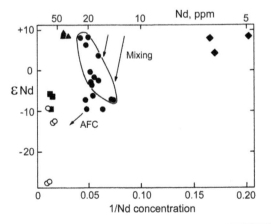

Fig. 7.16 Plot of initial Nd isotope ratio (εNd), against reciprocal Nd content in Skye lavas. Symbols as in Fig. 7.15. Arrows show the effects of contamination by magma mixing and by AFC. After Thirlwall and Jones (1983).

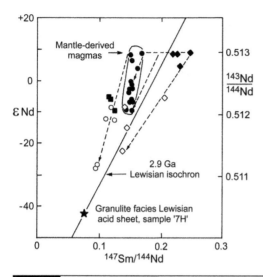

Fig. 7.18 Sm–Nd pseudo-isochron diagram for Tertiary lavas from Skye, showing proposed contamination vectors in comparison with the array of contaminated basalts. Symbols as in Fig. 7.15. After Dickin et al. (1984).

where the longer magma residence time would allow more opportunity for contamination.

An example of the possible effects of turbulent wall-rock assimilation was described by Kille et al. (1986) from the Hebridean island of Mull, where inclined intrusive sheets are intruded into metasedimentary units of the Moine series. Large embayments were seen in the more fusible units of the sedimentary sequence, suggesting 'excavation' by turbulently flowing magma. Subsequently, additional evidence was found from the Mull lava pile for wall-rock assimilation during turbulent magma ascent (Kerr et al., 1995). This study showed that lavas of the Mull Plateau Group (MPG) had similar patterns of εNd against elemental concentration to those previously seen in Skye. Hence, it appears that this process is of quite widespread occurrence, prompting Kerr et al. to coin the expression 'assimilation during turbulent ascent' (ATA).

Isochron diagrams are a particular example of a bivariate plot involving isotope ratios and trace element ratios, and may therefore be useful for studying crustal contamination processes. For old rock suites, initial isotope ratios are plotted on a pseudo-isochron diagram. Because the denominator on both axes is the same, two-component mixing must give rise to products which lie on a straight line between the end-members. However, a magma suite may again generate a data array which does not project to the mixing end-members. The Tertiary lavas from Skye provide a good example of this problem also.

Thirlwall and Jones (1983) found a linear array of εNd versus Sm/Nd ratios in Skye basalts (Fig. 7.18). They interpreted this array as a mixing line between a mantle-derived magma with constant Sm/Nd ratio and a partial melt of intermediate (tonalitic) Lewisian gneiss. The projection of the mixing line

onto the Lewisian isochron then indicates an εNd value (at 60 Ma) of ca. −15. However, Dickin et al. (1984) argued that the basalt array was not a single mixing line, but was generated by a series of obliquely angled mixing lines involving mantle-derived magmas with different Sm/Nd ratios. These trajectories point to a crustal end-member with εNd of ca. −40, corresponding to Lewisian granitic (acid) gneiss. This controversy serves to reiterate the importance of distinguishing between individual mixing lines and magma evolution trends on all plots where contamination models are considered.

Since the Sr and Nd isotope compositions of a contaminated lava suite may be a complex function of Sr and Nd concentrations, depending on the differentiation history of the suite, these factors must be borne in mind when interpreting the Nd versus Sr isotope diagram for a magma suite. For the data from Skye and Mull (Fig. 7.16), most samples define an array with a negative slope, implying coupled behaviour of Sr and Nd in the mixing process. However, it has been shown that contamination effects for each isotope system (^{87}Sr/^{86}Sr and ^{143}Nd/^{144}Nd) are individually controlled by the Sr and Nd concentrations of the differentiating magmas. Furthermore, these trace elements often do not behave coherently during magma differentiation, since Nd is always incompatible, whereas Sr becomes a compatible element once plagioclase crystallization begins. Therefore, we should expect some scatter in the Sr–Nd isotope correlation in order to reflect these complexities.

A further problem with the interpretation of Skye and Mull data on the Sr–Nd isotope diagram is that the correlation line of Tertiary lavas trends half-way between the fields for Rb-depleted granulite-facies and Rb-rich amphibolite-facies Lewisian gneisses, argued to represent the lower and upper parts respectively of the present day crust under Skye. This leads to ambiguity in the interpretation of Sr–Nd isotope data. However, the relative contributions of these crustal components can be resolved using Pb isotope data, as shown below.

Pb isotopes are a powerful tool for studies of mantle and crustal evolution, because the three different radiogenic isotopes are generated from parents with a wide span of half-lives, two of which are a common element. By using ^{206}Pb/^{204}Pb and ^{207}Pb/^{204}Pb ratios in conjunction, it is not only possible to measure the importance of crustal contamination, but also the age of the crustal component. On the other hand, by using ^{206}Pb/^{204}Pb and ^{208}Pb/^{204}Pb ratios in conjunction, it is sometimes possible to locate the depth of the crustal contaminant, since the crust may develop a stratified signature of these isotopes in response to high-grade metamorphism. Both of these possibilities are illustrated by the Tertiary magmatism of Skye and Mull.

Moorbath and Welke (1969) found that both acid and basic Tertiary igneous rocks from Skye lay on a strong linear array on the ^{207}Pb/^{204}Pb versus ^{206}Pb/^{204}Pb diagram, with a slope age of ca. 3 Ga. They interpreted the linear array as a mixing line between radiogenic mantle-derived Pb

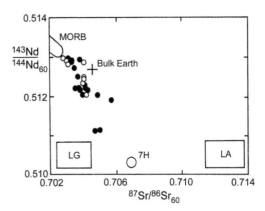

Fig. 7.19 Plot of Nd versus Sr isotope ratio for Skye (•) and Mull (○) lavas, showing a trend towards local crustal units: LG = Lewisian granulite-facies gneiss; LA = Lewisian amphibolite-facies gneiss; 7H = Archean granite pegmatite sheet. After Kerr et al. (1995).

and very unradiogenic Archean (Lewisian) crustal Pb. Dickin (1981) repeated this study with more modern techniques and found a mixing line with a slope age of 2920 ± 70 Ma (Fig. 7.20a), the same as the Sm–Nd age of the Lewisian complex (see Section 4.1.3).

By plotting $^{208}Pb/^{204}Pb$ versus $^{206}Pb/^{204}Pb$ ratios (Fig. 7.20b), it is possible to resolve three components in the Skye Tertiary igneous rocks. The lavas are interpreted as mantle-derived magmas that had suffered strong contamination in the granulite-facies lower crust (solid squares in Fig. 7.20), while gabbros of the Cuillins layered complex are attributed to contamination in amphibolite-facies upper crust (solid diamonds). Finally, the Skye granites (open squares) are attributed to differentiated basic magmas that suffered contamination in the lower crust, followed by further differentiation and contamination in the upper crust.

In this model, the crustal end-members were based on average compositions of gneisses from NW Scotland, supported by evidence from crustal xenoliths carried up in a Tertiary intrusion from Skye. The lower crustal rocks were depleted in both U and Th relative to Pb during the 2.7 Ga-old Scourian granulite-facies metamorphism, while the present day upper crust contains rocks that were depleted in U but not Th (relative to Pb) in the Archean middle crust. The original upper crust, enriched in U and Th relative to Pb, has largely been removed by erosion.

The combination of Pb with Sr isotope evidence allows additional constraints to be applied to the evolution of contaminated magma suites. For example, the shallow slope of the Sr–Pb isotope correlation line in Fig. 7.21 is consistent with contamination of the Skye and Mull lavas by granulite-facies lower crust. However, it must be remembered that (unlike the Pb–Pb isotope diagram) mixing lines on the Sr–Pb isotope diagram can be strongly hyperbolic. Therefore, the lavas were probably contaminated by the most felsic com-

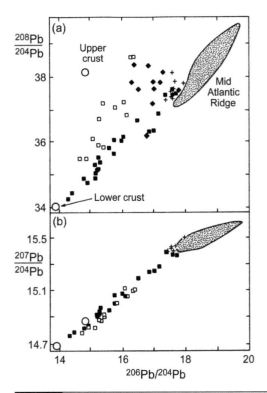

Fig. 7.20 Plot of initial Pb isotope ratios for Tertiary igneous rocks from Skye, showing evidence for three-component mixing. (■) = main lava series; (□) = granites; (+) = low-K basalts; (♦) = layered gabbros. Mid Atlantic Ridge approximates the local mantle composition. Modified after Thompson (1982).

ponents in the granulite-facies basement, which were more Rb-enriched than the bulk lower crust. Felsic minor intrusions were argued by Thompson et al. (1982) to be the most likely fraction in the crust to melt, leading to a kind of 'selective contamination' mechanism due to melting of fusible

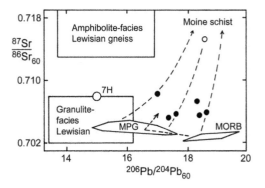

Fig. 7.21 Plot of initial Sr versus Pb isotope ratios in lava suites from Mull, showing the distinct contamination histories of the Mull Plateau Group (MPG) and the early erupting Staffa Magma Type (•). After Morrison et al. (1985).

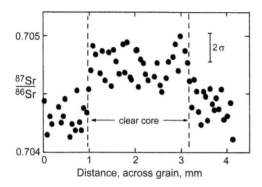

Fig. 7.22 Sr isotope profile along the length of a zoned plagioclase with a clear core and a patchy outer rim zone reflecting two stages of feldspar growth. The width of each data point indicates the approximate size of each ablation pit in the scan. After Davidson et al. (2001).

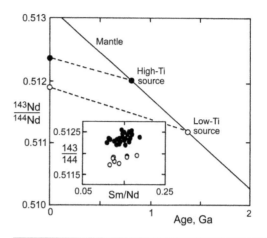

Fig. 7.23 Nd isotope evolution diagram showing predicted evolution lines of the lithospheric sources of high-Ti (●) and low-Ti (○) alkali mafic magmas, with ages of ca. 0.8 and 1.4 Ga respectively. Inset shows the lava data on which the evolution lines are based. After Gibson et al. (1996).

rock types. One such rock (sample '7H') is shown in Fig. 7.19, where it is found to explain the Sr–Nd correlation line in the lavas.

A contrasting type of behaviour was seen in a distinct magma type from SW Mull, which was important in the very early eruptive history of the complex (Morrison et al., 1985). These lavas of the so-called Staffa Magma Type lie far off the Sr–Pb correlation line formed by the later lavas (Fig. 7.21), and were therefore attributed to a two-stage contamination process. This began at the base of the crust, as seen in the other lavas, but was followed by a residence period in the uppermost crust, where the magmas were contaminated by supracrustal metasediments of the Moine series. Thus, as more isotope tracers have become available, the application of multiple tracers to magma suites has allowed more complex magma evolution histories in the crust to be modelled and understood.

7.2.4 Phenocrysts as Records of Magma Evolution

Inversion modelling is unavoidable when attempting to reconstruct the evolution of large magma suites during emplacement through the crust. However, when this is based only on the whole-rock composition of the final products, it may overlook internal mineralogical evidence that could help to constrain contamination models. Therefore, some recent studies have investigated the internal Sr isotope heterogeneity of feldspar phenocrysts in volcanic lavas in order to reconstruct near-surface magma plumbing.

Early work (e.g. Davidson and Tepley, 1997) used a microdrilling technique, followed by conventional ion exchange chemistry, to study Sr zoning in plagioclase phenocrysts from three volcanic systems. However, the advent of MC–ICP–MS (Section 2.2.2) has allowed in situ Sr isotope analysis of feldspar phenocrysts by laser ablation (Figs. 7.22 and 7.23, Davidson et al., 2001).

Figure 7.22 shows a typical isotope profile from the study of Davidson et al. (2001), measured along the length of a plagioclase phenocryst from the El Chichon volcano, Mexico. The data are in good agreement with micro-drill results from the same crystal (Davidson and Tepley, 1997), and clearly show two phases of magma evolution recorded by the crystal. The data are attributed to successive injections of mantle-derived magma with unradiogenic Sr into a magma chamber in the crust. Similar plagioclase zoning has been observed at Stromboli volcano, due to new magma influx into an old magma chamber (Morgan et al., 2007). However, it was also shown by Francalanci et al. (2005) that this volcano has a very complex record of Sr isotope evolution through time. This occurs when new magma entrains plagioclase from more than one generation of older crystal mush magmas with different Sr isotope ratios (Francalanci et al., 2012).

7.2.5 Lithospheric Mantle Contamination

Isotopic tracers have been widely used to monitor crustal contamination of continental magmas during their ascent, and to some extent this process can now be quantified. In contrast, the relative importance of lithospheric and asthenospheric mantle sources continues to be a matter of debate. Thermal constraints (McKenzie and Bickle, 1988) suggest that the high melting rates necessary to erupt flood basalt provinces can only be satisfied by melting in mantle plumes. On the other hand, lithospheric extension will cause small-volume melting of metasomatized lithosphere, generating potassic mafic magmas that may have extreme isotopic compositions. Several workers recognized that these processes may act together, leading to contamination of asthenospheric magmas by the mantle lithosphere, as well as by the overlying crust.

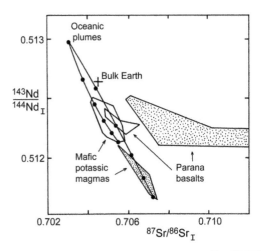

Fig. 7.24 Nd–Sr isotope diagram showing fields for high-Ti and low-Ti Parana basalts. Genesis of each field is explained by two-stage contamination of plume magmas, firstly in the sub-continental lithosphere and secondly in the crust. Mixing lines are marked in 20% increments. After Gibson *et al.* (1996).

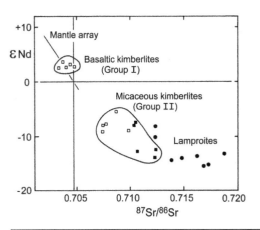

Fig. 7.25 Plot of εNd against Sr isotopic compositions of basaltic and micaceous kimberlites from South Africa (□) and western Australia (■), along with Australian lamproites (•). After DePaolo (1988).

Geochemical provinciality provides one line of evidence for lithospheric control of magma chemistry. An example is the identification of high-Ti and low-Ti flood basalt provinces in the Parana Basin of South America (Hawkesworth *et al.*, 1984). If the same provinciality was found in mafic potassic rocks, this would support the lithospheric contamination model for flood basalts. High-Ti alkali mafic rocks in the centre of the basin were found by Hawkesworth *et al.* (1992), while low-Ti alkali mafic rocks on the flanks were later found by Gibson *et al.* (1996).

Comparison of the Nd isotope systematics of high-Ti and low-Ti alkali mafic rocks from the Parana Basin showed distinct Nd isotope compositions but similar Sm/Nd ratios (Fig. 7.23). This suggests that the lithospheric mantle underlying high-Ti and low-Ti sub-provinces has distinct trace element enrichment ages, possibly reflecting the geographical extent of a Late Proterozoic crustal re-working event. Therefore, this confirms that the high-Ti and low-Ti suites reflect provinciality in the mantle lithosphere.

If the isotopic signatures of the most enriched alkali mafic suites are used as end-members in a mixing model, the isotopic compositions of low-Ti flood basalts can be explained by ca. 20% contamination of a primitive plume end-member in the lithospheric mantle, followed by extensive contamination in the crust (Fig. 7.24). On the other hand, the high-Ti flood basalts require ca. 50% contamination in the mantle lithosphere, but less crustal contamination. Such multi-stage models for the interaction of plumes with the mantle lithosphere are likely to be a continuing focus in geochemical studies of continental basalts.

7.3 Petrogenesis of Continental Magmas

It is impossible here to attempt a comprehensive review of continental magma suites. Instead, a few case studies will be examined for different magma types which illustrate a variety of approaches to problems of petrogenetic interpretation. The general order will be from those showing mantle-derived to crustal features in their genesis.

7.3.1 Kimberlites, Carbonatites and Lamproites

Kimberlites, carbonatites and lamproites are highly incompatible-element-enriched magmas that may be genetically related. Experimental evidence suggests that they are all products of very small-degree partial melting in the deep mantle, and that CO_2 plays an important role in their genesis (e.g. Wendlandt and Mysen, 1980). The volatile-rich nature of these magmas causes rapid ascent through the crust. This, coupled with their high incompatible-element concentrations, renders these magmas very resistant to isotopic modification by crustal contamination.

Early petrological work on South African kimberlites led Nixon *et al.* (1981) to propose a magma mixing model to explain their chemistry, whereby an asthenospheric 'proto-kimberlite' mixed with a melt from the base of the sub-continental lithosphere. However, this model was complicated by the discovery that the two distinct petrological kimberlite types (basaltic and micaceous, Dawson, 1967) also had distinct Sr–Nd isotope signatures (Smith, 1983). Basaltic (group I) kimberlites have isotopic compositions that cluster just within the depleted quadrant relative to Bulk Earth, whereas micaceous (group II) kimberlites fall well inside the enriched quadrant (Fig. 7.25). This trend was further extended by the analysis of micaceous kimberlites

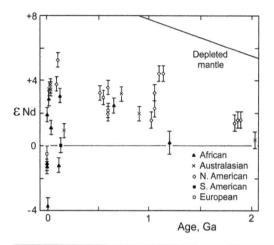

Fig. 7.26 Nd isotope evolution diagram in terms of εNd against time, showing carbonatite derivation from variably depleted (or mixed) mantle sources. After Nelson *et al.* (1988).

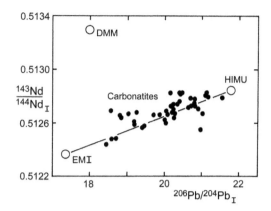

Fig. 7.27 Nd–Pb isotope plot showing carbonatites from the East African Rift (•) relative to the end-members invoked to explain the compositions of OIB. After Bell and Tilton (2001).

and lamproites from Western Australia (McCulloch *et al.*, 1983). Although the implications of the isotopic data were uncertain, McCulloch *et al.* argued that they were consistent with the mixing model of Nixon *et al.* (1981). However, other workers (e.g. Fraser *et al.*, 1985) attributed group II kimberlites and lamproites primarily to a lithospheric source.

Additional light can be thrown on this question by a study of carbonatite genesis. A lithospheric mantle source was invoked to explain the fairly homogeneous isotopic signatures in Canadian carbonatites (Bell and Blenkinsop, 1987). However, Nelson *et al.* (1988) observed a wide variation of εNd values from depleted to enriched signatures in a worldwide data set of carbonatites of various ages (Fig. 7.26). This militated against an origin in the sub-continental lithosphere, and instead favoured a plume origin similar to ocean island basalts. Hence Nelson *et al.* proposed that very small-degree partial melts originate in comparatively 'fertile' asthenospheric sources (similar to the Group 1 kimberlite signature), and are subsequently contaminated to different degrees in the LIL-enriched but refractory sub-continental lithosphere (similar to the Group 2 signature).

This mixing model places kimberlites, carbonatites and lamproites on the same footing as other within-plate magmas. These magmas are now generally thought to originate from plume sources, and subsequently undergo variable degrees of contamination in the lithosphere. Additional support for the importance of plume sources in carbonatite magmatism is provided by noble gas evidence from carbonatites of the Kola Peninsula. For example, neon isotope data provide clear evidence for the involvement of a plume-type lower mantle source in their genesis (Section 11.2.2).

One geological province that has always been important for the study of carbonatites and ultra-alkaline magmas is the East African Rift. Bell and Simonetti (1996) showed that the magmas of this province, including the active carbonatite volcano Oldoinyo Lengai, lie very close to the

HIMU–EMI mixing line (Fig. 7.27) identified in Ocean Island Basalts (OIB). Bell and Simonetti (1996) initially attributed the mixing line to magmas from a HIMU type plume that were subsequently contaminated by subcontinental lithospheric mantle with an isotopic signature resembling EMI (Section 6.5.3). However, subsequent work on a larger number of carbonate and nephelinite volcanic centres from the East African Rift (Bell and Tilton, 2001) showed that this mixing line was a widespread phenomenon.

The discovery that the mixing line originally observed at Oldoinyo Lengai was also seen in carbonatites from elsewhere in the East African rift caused Bell and co-workers to change their ideas about the origin of these components. The widespread occurrence of relatively coherent two-component mixing throughout the rift made a source in the lithosphere less likely, since this is expected to be laterally heterogeneous. Therefore, Bell and Tilton (2001) proposed that the signature of HIMU–EMI mixing in East African carbonatites reflects heterogeneity in the plume source itself, as seen in OIB.

As noted in the introduction to this section, the general relatedness of kimberlites and carbonates has been recognized for a long time (e.g. Dawson, 1971). However, geochemical study of the Udachnaya kimberlite in Siberia provided new evidence supporting a direct link between carbonatites and kimberlites.

The diamondiferous Udachnaya kimberlite is uniquely unaltered, with water contents as low as 0.2%, relative to typical kimberlite water contents of 5 to 10% (e.g. Kamenetsky *et al*,. 2009b). The Udachnaya kimberlite also has much higher sodium contents (up to 6%), reflecting the presence of halite minerals. These were initially attributed to crustal contamination of the kimberlite by sedimentary brines (Pavlov and Ilupin, 1973). However, this interpretation was refuted by isotopic analysis of an aggregate of halite crystals, along with whole-rock samples leached with water or dilute acid (Maas *et al.*, 2005). It was found that these halite mineral signatures were almost identical with the leaching residue, and

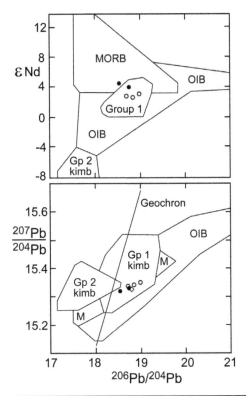

Isotopic analyses of water and acid leaches of bulk kimberlite (○), along with leaching residue (●) and an aggregate of salt crystals (◊), compared with established fields. After Maas *et al.* (2005).

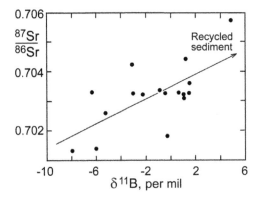

Plot of strontium versus boron isotope ratios for carbonatites of various ages, showing a positive correlation attributed to recycled marine carbonates. After Hulett *et al.* (2016).

lay within the field of typical Group 1 kimberlites (Fig. 7.28). Therefore, Maas *et al.* concluded that the halites were mantle-derived, and characteristic of the original kimberlite magma.

Based on these findings, Kamenetsky *et al.* (2008) proposed that kimberlites are derived from an original sodium-rich carbonatite magma that was subsequently contaminated with silicate phases such as opx and olivine. This model has been supported by numerous subsequent textural studies of kimberlite petrography (e.g. Kamenetsky *et al.*, 2009a, 2014; Brett *et al.*, 2009, 2015), all of which support the model of silicate mineral assimilation by an original carbonatite magma.

The 'unification' of these magma types, and the suggestion of a common origin with the oceanic EMI and HIMU components, means that evidence from carbonatites may also be relevant to the origins of oceanic volcanics, and vice versa. One model that has gained favour recently is an origin involving sedimentary carbonate material carried by subduction or lithospheric delamination into the deep mantle (Section 6.5.4). However, the involvement of surficial crustal/sedimentary material in the formation of diamonds has been advocated for several decades (e.g. Chaussidon *et al.*, 1987; Eldridge *et al.*, 1991).

The most sensitive tracers for the involvement of surficial material in diamond, kimberlite or carbonatite genesis are stable isotope systems, which undergo much more exten-

sive mass fractionation under low-temperature surface conditions than the deep Earth. In addition to early work on sulphur isotopes by the workers cited above, studies involving carbon and oxygen isotopes also indicated the involvement of recycled surficial material in diamond genesis (Kirkley *et al.*, 1991; Ickert *et al.*, 2013; Burnham *et al.*, 2015), although this has sometimes been disputed (Cartigny *et al.*, 1998).

With the recognition that carbonatites form the parental magma of diamond-bearing kimberlites, additional stable isotope studies on carbonatites can throw more light on kimberlite genesis. Boron is a particularly sensitive tracer for detecting recycling of surficial material into plume sources, because marine sediments contain over two orders of magnitude more boron than typical mantle (Hemming and Hanson, 1992; Gurenko and Chaussidon, 1997). There is also a large difference in boron isotope ratio, with a $^{11}B/^{10}B$ ratio ($\delta^{11}B$) of +22 per mil in marine carbonates, compared to a $\delta^{11}B$ value around −10 per mil in typical upper mantle.

In comparison with these end-members, Hulett *et al.* (2016) found $\delta^{11}B$ values from −8 to +5 in carbonatites of various ages, with a positive correlation against Sr isotope ratio (Fig. 7.29). This was attributed to contamination of the lower mantle carbonatite source with the seawater isotope signature carried by marine carbonates. Hence this supports the involvement of sedimentary carbonate in the lower mantle source of these related magma types.

7.3.2 Alkali Basalts

Alkali basalts have been widely studied from oceanic hotspots in order to obtain information about deep mantle plumes. However, the Cameroon Line of West Africa is unique in straddling an ocean–continent boundary, and should therefore allow comparisons between basalt petrogenesis in these two domains.

The Cameroon Line is composed dominantly of alkali basalts, with subordinate tholeiites, and stretches from the Atlantic island of Pagalu (700 km SW of the Niger delta) to the Biu Plateau (800 km inland). In spite of the fact that the

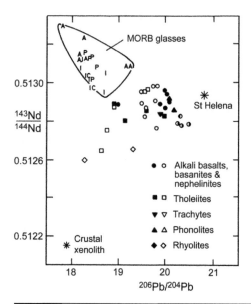

Fig. 7.30 Plot of Nd versus Pb isotope compositions for Cameroon Line volcanics. Solid symbols: continental lavas; open symbols: oceanic; half-filled: continental edge. A, I and P = Atlantic, Indian and Pacific MORB. After Halliday et al. (1988).

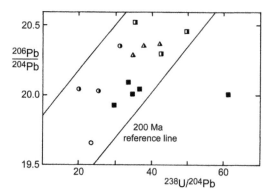

Fig. 7.31 U–Pb isochron diagram for young Cameroon line lavas with over 4% MgO, yielding an apparent age of ca. 200 Ma. Half-open symbols denote individual volcanos from the continental edge. Other symbols indicate continental (o) and oceanic segments (■). After Halliday et al. (1990).

volcanic chain is situated half on young oceanic crust and half on ancient continental crust, trace element contents and Sr isotope ratios are identical in the two sections of the line (Fitton and Dunlop, 1985). Given that oceanic and continental lithosphere would be expected to have different signatures, Fitton and Dunlop argued that the magma source must lie below the lithosphere.

The Cameroon Line shows no evidence of age progression along its length, and must therefore represent a 'hot zone' rather than a hot-spot trail generated by plate motion over a small plume. Fitton and Dunlop argued that because there is no evidence for migration of the area of volcanism over its 65 Ma history (despite movement of the African plate), the mantle source must be coupled to the lithosphere rather than originating from a deep mantle plume that tracked across the overlying plate. This was a surprising result, since the only other explanation (at that time) for the observed range of isotope values was disequilibrium melting (Section 6.1.2).

To further explore this problem, Halliday et al. (1988) performed a more detailed isotopic study on the Cameroon Line, including Pb and Nd isotope measurements (Fig. 7.30). A few samples showed evidence of contamination by continental basement; however, after excluding these samples, the most distinctive feature of the data was the very radiogenic Pb isotope compositions displayed by basic lavas from the continent–ocean boundary, about half-way along the volcanic chain. These compositions approach those of the St Helena hot-spot, but volcanics on either side (within the oceanic and continental segments) are less radiogenic.

Halliday et al. (1988) attributed these features to the 'impregnation' of the upper mantle under the Cameroon Line by material from the St Helena plume. This plume played a major role in promoting the initial opening of the South Atlantic in around 120 Ma. It was probably responsible for the actual location of rifting, which subsequently became the continental edge. With time, the African plate moved away from the St Helena plume, but a 'blob' of hot plume material probably became incorporated into the lithospheric mantle under Cameroon, as the continental margin cooled after the rifting event. As the plume component was gradually dispersed laterally along the volcanic chain, its compositional effect was seen at volcanic centres progressively further from the continental edge.

Halliday et al. (1990) revised this model as a result of additional observations on the Pb isotope data. These revealed that the radiogenic $^{206}Pb/^{204}Pb$ signatures at the continental edge were not accompanied by high enough $^{207}Pb/^{204}Pb$ ratios to represent direct mixing with the St Helena plume. Instead, a positive correlation was observed between $^{206}Pb/^{204}Pb$ and U/Pb ratio (Fig. 7.31) which Halliday et al. interpreted as a 200 Ma erupted isochron. However, it was shown in Fig. 7.30 that $^{206}Pb/^{204}Pb$ correlates with the Nd isotope ratio, which cannot develop large variations from Sm decay over periods of only 200 Ma. Therefore the arrays of Pb isotope ratio are probably mixing lines. The radiogenic end-member must represent young lithosphere whose high $^{206}Pb/^{204}Pb$ signature was generated by magmas with high U/Pb ratios (μ) at the time of continental rifting. Mixing between this component and local asthenospheric upper mantle can explain the isotopic mixing process, which also satisfies more recent U-series evidence from the Cameroon Line (Yokoyama et al., 2007).

These observations have two significant implications. Firstly, continental rifting episodes can replace old subcontinental lithosphere with young lithosphere which may

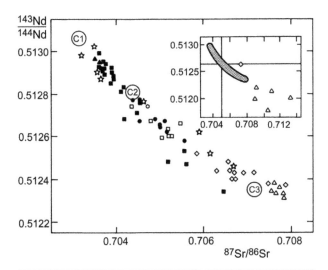

Fig. 7.32 Plot of Nd versus Sr isotope ratio for basalts from the northwestern USA. ($\blacksquare$) = Grand Ronde; ($\bullet$) = Picture Gorge; ($\square$) = Wanapum; ($\blacktriangle$) = Steens Mtn; ($\triangle$) = Saddle Mountains; ($\star$) = HAOT; ($\diamond$) = SROT. C1 to C3 are possible sources discussed in the text. Data from the main diagram define the shaded field on the inset. After Carlson and Hart (1988).

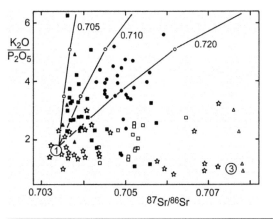

Fig. 7.33 Plot of K_2O/P_2O_5 against Sr isotope ratio for basalts from the northwestern USA, showing mixing models between C-1 magmas and three crustal contaminants with different Sr isotope ratios. Symbols as in Fig. 7.32. After Carlson and Hart (1988).

have an exotic composition. Secondly, U is more incompatible than Pb in magmatic processes, so that preferential extraction of Pb from the mantle cannot be invoked to explain the lead paradox (Section 6.3.1).

7.3.3 Flood Basalts

The northwestern USA displays one of the world's major flood basalt provinces, forming the Columbia River Basalt Group (CRBG). Controversies about the petrogenesis of these lavas serve very well to illustrate the complexities of modelling the genesis of flood basalts.

Early Nd isotope data on the Columbia River basalts clustered near εNd = 0, leading DePaolo and Wasserburg (1976) to propose a chondritic source for these magmas, in contrast to the depleted mantle source of MORB (εNd = +10). At this early stage in the development of the Nd isotope method, other continental igneous rocks had also yielded initial Nd ratios near ε = 0, and it was still believed that continental magmas might sample an undepleted chondritic source that was distinct from the oceanic upper mantle.

More detailed work on several basaltic suites from the northwestern USA presented a rather different picture (Carlson et al., 1981), which was developed by Carlson and Hart (1988). These data display a very strong, almost continuous curved trend, starting from a depleted mantle composition and fanning out somewhat in the enriched quadrant of the Nd–Sr isotope plot (Fig. 7.32). This distribution seemed to imply a relatively simple mixing process, such as crustal contamination, in the genesis of the lavas (Carlson et al., 1981).

However, it has been argued that radiogenic isotopes alone may not be able to distinguish between enriched mantle and crustal sources. Incompatible element ratios and stable isotope data may be needed to assist in this distinction.

Carlson and Hart (1988) suggested that the ratio of a highly incompatible element against a high-field-strength element (e.g. K_2O/P_2O_5) can be used as an index of (specifically) crustal contamination. This index is plotted against Sr isotope ratio in Fig. 7.33. Some Picture Gorge and Grand Ronde basalts of the Columbia River Basalt Group (CRBG), along with Steens Mountain basalts from the Oregon Plateau, have quite elevated K_2O/P_2O_5 ratios, despite low to intermediate Sr isotope ratios. Carlson and Hart attributed this pattern to contamination of magmas from a 'C-1' mantle source by crustal units with a variety of ages. The C-1 source was identified as typical asthenospheric upper mantle, whose melting was probably caused by mantle convection behind the Cascades arc. In contrast to the above lavas, some Saddle Mountains CRBG flows, high-Al olivine tholeiites (HAOT) from the Oregon Plateau and Snake River olivine tholeiites (SROT) have $^{87}Sr/^{86}Sr$ ratios up to 0.708, but low K_2O/P_2O_5. Carlson and Hart attributed these signatures to a lithospheric mantle source ('C-3').

A plot of $\delta^{18}O$ against Sr isotope ratio supports this model (Fig. 7.34). Steep vectors result from contamination of basaltic magmas by typical crustal units. In contrast, subhorizontal vectors could be produced by mixing with old ^{87}Sr-enriched mantle, or possibly by recent contamination of the Sr-poor mantle source by subducted sediment. Such a distinction between source and magma contamination vectors on the oxygen–strontium isotope diagram was explored in detail by Taylor (1980) for granitic rocks (Section 7.3.5).

Pb isotope data revealed another level of complexity in this picture. Basalts with $^{87}Sr/^{86}Sr$ below 0.708 display a

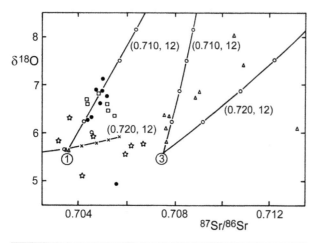

Fig. 7.34 Plot of $\delta^{18}O$ against Sr isotope ratio for basalts from the northwestern USA. Curves show effects of mixing with crust of a given Sr and ^{18}O composition. Steep mixing lines model contamination of magmas ($\circ = 10\%$ increments); shallow mixing line models contamination of MORB-type source with subducted sediment ($\times = 1\%$ increments). Symbols as in Fig. 7.32. After Carlson and Hart (1988).

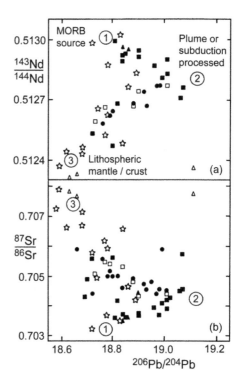

Fig. 7.35 Isotope compositions of basalts from the northwestern USA: (a) Nd versus Pb isotope plot; (b) Sr versus Pb isotope plot. Three distinct mantle sources are resolved (C-1 to C-3). Symbols as in Fig. 7.32. After Carlson and Hart (1988).

triangular distribution on plots of Sr or Nd isotope ratio against $^{206}Pb/^{204}Pb$ (Fig. 7.35). On this diagram, many Grand Ronde lavas trend towards an end-member (C-2) with radiogenic Pb which is distinct from the C-1 and C-3 mantle end-members recognized from other evidence. Carlson and Hart speculated that the C-2 source may have been derived by contamination of C-1 depleted mantle by subducted sediment.

Carlson and Hart did not invoke any lower mantle plume source for the Columbia River basalts, which therefore make this an unusual model for a flood basalt province. Not surprisingly, their model was challenged by DePaolo (1983), who presented a volume-weighted histogram of the major eruptive groups of the Columbia River Basalt Province (Fig. 7.36). This showed that the volume of strongly contaminated Saddle Mountain lavas was insignificant relative to the major basalt groups. Omitting this small group implies that the major eruptive groups could be derived by mixing between depleted upper mantle and chondritic lower mantle melts. Furthermore, the sharp cut-off of the Grande Ronde suite at $\varepsilon Nd = 0$ appears to support this argument, although the abundance peak at $\varepsilon = 0$ was partly a function of double sampling of some flows by combining the data of DePaolo and Wasserburg (1976, 1979b) with that of Carlson $et\ al.$ (1981).

Several subsequent studies of the Columbia River Basalt Province have invoked a plume as one of the end-members. The first of these studies (Brandon and Goles, 1988) involved trace element analysis only, whereas three subsequent studies used radiogenic isotope tracers. However, the analytical results obtained in these isotope studies were largely the same as those of Carlson and Hart ... only the conclusions

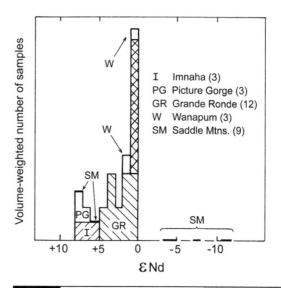

Fig. 7.36 Histogram of εNd compositions for Columbia River basalts, weighted according to eruptive volume. Double-hatched data are from DePaolo and Wasserburg (1976, 1979b). Modified after DePaolo (1983).

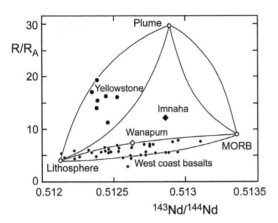

Fig. 7.37 Helium versus Nd isotope data for the Yellowstone/Snake River plume relative to Imnaha and Wanapum basalts of the CRBG. After Graham *et al.* (2009).

differed. Therefore, these models will be discussed using the same isotope data in Fig. 7.35.

Studies by Hooper and Hawkesworth (1993) and Brandon and Goles (1995) used multiple isotope tracers (Pb–Sr–Nd) in their investigation, as well as elemental data. They broadly agreed with Carlson and Hart (1988) that C-1 represents asthenospheric upper mantle and C-3 is lithospheric mantle. However, both groups reinterpreted C-2 as a plume source, identified as part of the track of the present day Yellowstone plume. In contrast, a study by Chamberlain and Lambert (1994) using Pb isotope only, divided both C-1 and C-2 into two sub-component reservoirs (R1–R2 and R4–R5 respectively). However, the main difference was that they identified the C-3 component (their R3) as the plume source and C-2 component (= R4–R5) as crustal.

These competing models were tested by Dodson *et al.* (1997) using helium and neon isotope evidence. Two samples were analysed, from the Imnaha and Wanapum groups. However, these are best understood within the context of helium data for the Yellowstone/Snake River plume (Graham *et al.*, 2009). Contrary to assertions by some geophysicists (Christiansen *et al.*, 2002), the high ^{3}He signatures of the Yellowstone/Snake River samples are clearly indicative of a mantle plume that interacted with the lithosphere (Fig. 7.37). Of the Columbia River samples, only the Imnaha basalt clearly contains a plume component, but this has mixed with both MORB-source and lithospheric components. This work vindicates to some extent the early studies of DePaolo and Wasserburg, by showing that the Columbia River basalts do indeed contain a plume component, but they also demonstrate the need for multiple isotope tracers to distinguish between sources in a complex multi-component system (see also Wolff and Ramos, 2013).

The application of osmium isotope data provides a different perspective on this problem (Chesley and Ruiz, 1998). These data do not distinguish between alternative sources

in the upper and lower mantle (C-1 and C-2), but they throw additional light on the identity of the C-3 source. Contrary to the arguments of Carlson and Hart (1988) based on trace element and oxygen isotope data, Chesley and Ruiz argued that the C-3 end-member was represented by crustal rather than mantle lithosphere. After all of the argumentation above, this seems a surprising conclusion. However, Chesley and Ruiz claimed that initial osmium signatures in lavas from the Grande Ronde, Wanapum and Saddle Mountains groups were so radiogenic that only a crustal source could explain them (see Section 8.3 for background).

Additional evidence for this interpretation comes from dykes of the Wanapum group that are intruded across a suture boundary between Precambrian and Mesozoic terranes. Analyses of lithophile isotope ratios showed no change across this boundary, but initial ^{187}Os/^{188}Os ratios changed from 0.2 on the Mesozoic side to around 3 on the Precambrian side. Hence this provides strong support for a relatively shallow contamination process, attributed by Chesley and Ruiz to mafic lower crust. This material would have low values of δ^{18}O and possibly also low ratios of K_2O/P_2O_5, thus confounding the evidence from these tracers. This should serve as a warning that the composition of the lower continental crust is still not well understood, even for isotope tracer systems which have supposedly reached a mature stage of understanding.

7.3.4 Precambrian Granitoids

One of the most fundamental questions about the continental crust is the extent to which any given block of sialic basement is the product of juvenile separation from the mantle or re-working of older cratonic material. Sr isotope data were originally applied to this problem on the grounds that crustal reservoirs, which have high Rb/Sr ratios, develop higher ^{87}Sr/^{86}Sr ratios over geological time than the low-Rb/Sr mantle. Calculation of the initial Sr isotope composition of a plutonic crustal segment should then indicate whether it has a mantle or crustal source. The evolution line for Sr in the depleted mantle is constructed by drawing a growth curve from the 'basaltic achondrite best initial' (BABI) value of 0.69899 ± 5 (Section 3.2.4) to the ^{87}Sr/^{86}Sr composition of recent ocean ridge basalts in the range 0.702–0.704. Data for specific crustal provinces can then be compared with this evolution line to assess their petrogenesis.

A classic example of the application of the Sr isotope evolution diagram to the provenance of crustal basement is provided by studies of the Archean and Proterozoic gneisses of West Greenland by Moorbath and Pankhurst (1976). Average growth lines are drawn in Fig. 7.38 for 3.7 Ga Amîtsoq gneisses from four localities, 2.8–2.9 Ga Nûk gneisses from five localities, 1.8 Ga-old Ketilidian gneisses from two localities (in South Greenland) and the 2.52 Ga-old Qôrqut granite. The initial ratios of these terranes are compared in Fig. 7.38 with a hypothetical linear upper mantle growth line drawn between BABI and MORB.

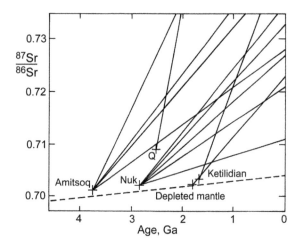

Fig. 7.38 Sr isotope evolution diagram showing the development of four crustal suites relative to the depleted mantle evolution line. Qorqut granite (Q) is attributed to crustal anatexis. After Moorbath and Taylor (1981).

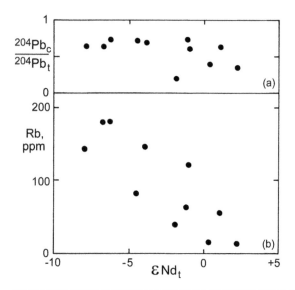

Fig. 7.39 Variation of εNd in Nuk gneisses, compared with (a) the fraction of isotopic contamination by Amitsoq Pb (from Section 5.5); and (b) Rb content. After Taylor et al. (1984).

Moorbath and Pankhurst argued that the Nuk (and Ketilidian) gneisses could not be derived by re-working of older (e.g. Amitsoq) gneiss, since the growth lines of the Amitsoq samples are much too steep to generate products with initial ratios of only 0.702–0.703. Instead they concluded that the igneous precursors of the Nuk gneisses represented a massive addition of juvenile calc-alkaline crust to the Archean basement of West Greenland. The slight elevation of the calculated initial ratios above the upper mantle evolution line was attributed to a period of crustal Sr isotope evolution, lasting perhaps 100–200 Ma, between the separation of the igneous precursors from the mantle and their subjection to granulite-facies metamorphism (see Section 5.5). In contrast, Moorbath and Pankhurst recognized the Qorqut granite as a good candidate for a pluton derived by re-working of older crust, since its initial ratio of 0.709 ± 0.007 is within error of the compositions of Amitsoq gneisses at that time.

Lead isotope analysis of the Nuk gneisses revealed a more complex picture than the Sr isotope data alone, by demonstrating that Nuk magmas emplaced into areas of Amitsoq crust suffered significant contamination with old crustal Pb (Section 5.5). In view of the lack of obvious crustal Sr contamination, selective contamination by Pb was invoked to explain these observations (Taylor et al., 1980). In this situation, the application of Nd isotope analysis provides an ideal tool to test petrogenetic models for the Nuk gneisses.

Taylor et al. (1984) analysed a selection of both Pb-contaminated and Pb-uncontaminated gneisses for Nd isotope composition (Fig. 7.39). The data indicate a good correlation between εNd and Rb, an incompatible trace element expected to be enriched in the Amitsoq gneisses. Taylor et al. attributed these results to contamination of mantle-derived Nuk magmas by partial melts of Amitsoq gneiss in the lower crust. However, εNd does not correlate well with

the degree of Pb isotopic contamination (represented by the index ^{204}Pb contaminant/^{204}Pb total). This suggests the possibility of additional selective Pb contamination, due to Pb-enriched fluids generated by crustal dehydration.

The paradox whereby substantial Pb and Nd contamination of the Nuk magmas was not accompanied by observable Sr isotope disturbance must be attributed to the stratified nature of the Amitsoq crust. Taylor et al. argued that the deep crust responsible for contamination must have had lower Rb/Sr ratios than those analysed from the surface outcrops, presumably due to flushing out of Rb from the lower crust during granulite-facies metamorphism. Hence, this crust did not develop elevated Sr isotope ratios over geological time. It is concluded from this evidence and other studies that Sr isotope data often cannot readily distinguish between mantle and lower crustal source regions. In this situation, Nd isotopes are a more powerful petrogenetic tool because Sm/Nd is fractionated during crustal extraction from the mantle but is not significantly fractionated by intra-crustal processes (Section 4.3).

The Pb isotope system comprises two coupled dating systems, whereas Sm–Nd normally offers only one. Therefore Pb–Pb data should theoretically provide more control than Nd on mixing processes between crustal reservoirs of different ages. However, U–Pb isotope systems are susceptible to open-system behaviour over geological history, whereas whole-rock Sm–Nd systems are very resistant to such effects. Therefore, when Pb–Pb data are used to study ancient mixing events, the measured isotope ratios cannot normally be corrected to unique initial ratios. Instead, model initial ^{207}Pb/^{204}Pb ratios are usually determined by projecting the Pb–Pb data back along an isochron line corresponding to the

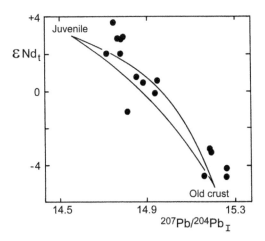

Fig. 7.40 Plot of calculated initial $^{143}Nd/^{144}Nd$ versus $^{207}Pb/^{204}Pb$ compositions of Late Archean granitoid magmas, explained by mixing of juvenile and ancient crustal end-members. After Davis *et al.* (1996).

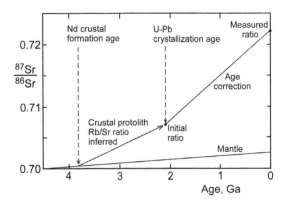

Fig. 7.41 Determination of protolith Rb/Sr ratios from samples with known crystallization and crustal formation ages. After Dhuime *et al.* (2015).

age of the mixing event (e.g. Section 5.5.1). Hence, we are left with a single isotopic tracer, analogous to $^{143}Nd/^{144}Nd$.

Davis *et al.* (1996) compared the use of initial $^{207}Pb/^{204}Pb$ and $^{143}Nd/^{144}Nd$ ratios in late Archean granitoids as tracers of the extent of Early–Mid Archean basement in the Slave Province of NW Canada. They found a relatively good correlation between the two tracers (Fig. 7.40), supporting a model of two-component mixing between juvenile (recently mantle-derived) and old crustal end-members. The mixing line is somewhat curved, indicating a Pb/Nd ratio in the old crustal end-member about three times higher than the juvenile end-member. In principal, this makes Pb a more sensitive tracer of hidden crust than Nd. However, the Greenland evidence discussed above suggests that Pb may behave less reproducibly than Nd. Hence, the scatter in Fig. 7.40 is probably due to non-stoichiometric mixing of Pb.

Dhuime *et al.* (2015) attempted to reinstate the use of Sr isotopes as a useful tracer in crustal evolution studies, by using Sr isotope data to calculate time-integrated Rb/Sr ratios for juvenile crustal suites of different ages. Since the Rb/Sr ratio should correlate with the silica content of granitoid rocks, this offers an opportunity to study variations in crustal composition through time. A calculation of time-integrated Rb/Sr ratio requires the (U–Pb) crystallization age and (Nd) crustal formation age to be known. The initial Sr isotope ratio of the sample is calculated, and the Rb/Sr ratio of the protolith can then be determined from the point of intersection with the Sr mantle growth curve (Fig. 7.41).

Dhuime *et al.* performed this calculation on over 13 000 published analyses of all geological ages, and found a relationship between the Rb/Sr age of the protolith and its crustal formation age (Fig. 7.42). Based on an empirically observed relationship between Rb/Sr and the silica content of igneous rocks, Dhuime *et al.* inferred that the continental

crust became more silica-rich at around 3 Ga. From this, they deduced that plate tectonic processes may not have begun until ca. 3 Ga, prior to which the crust was dominantly mafic.

In contrast, other workers have cited various lines of geological evidence for plate tectonic processes in the Early Archean. For example, Nutman *et al.* (2015) argued that the 'Itsaquia' terrane of western Greenland was assembled by a plate collision, before breaking apart due to the effects of a deep mantle plume. Using a different approach, Smart *et al.* (2016), argued that the stable isotope compositions of South African diamonds are indicative of active subduction in the Early Archean. Therefore, a possible alternative explanation of the low Rb/Sr ratios deduced for Early Archean rocks is that these rocks were more affected by high-grade metamorphism, thereby lowering their Rb contents (see above).

7.3.5 Phanerozoic Batholiths

The Sierra Nevada batholith of Southern California is a classic example of a batholith system, and therefore represents

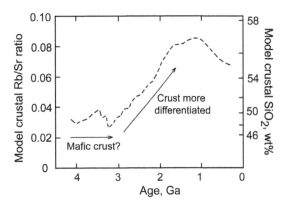

Fig. 7.42 Plot of calculated Rb/Sr ratio for crustal protoliths against their crustal formation age, showing variation through time. After Dhuime *et al.* (2015).

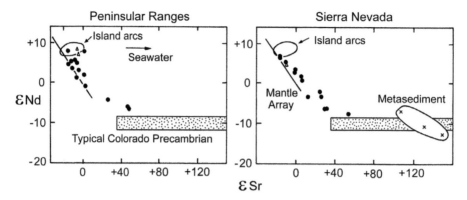

Fig. 7.43 Plots of εNd against εSr for granitoids (•) from the Peninsular Ranges and the Sierra Nevada. Compositions of crustal reservoirs and the effect of seawater alteration are also shown. After DePaolo (1981b).

an ideal case study for granitoid petrogenesis in Phanerozoic arc systems. Hurley *et al.* (1965) made an early Sr isotope study of the Sierra Nevada batholith and concluded that most of the intrusive bodies making up the batholith had initial $^{87}Sr/^{86}Sr$ ratios of ca. 0.7073, intermediate between expected upper mantle and Precambrian crustal signatures. However they were unable to determine from Sr isotope evidence alone whether the Sierra Nevada batholith represented mantle-derived magmas subsequently contaminated by the crust, or simply partial melting of geosynclinal sediments and volcanics.

DePaolo (1981b) made a combined Sr and Nd isotope study of both the Sierra Nevada and Peninsular Ranges batholiths in a further attempt to resolve the genesis of these bodies. The data define hyperbolic arrays on εSr versus εNd diagrams (Fig. 7.43), running from the island-arc basalt field towards the composition of nearby Precambrian schists. The latter were regarded as representative of the source area which yielded the Paleozoic–Mesozoic geosynclinal sediments into which the batholiths are intruded. DePaolo recognized that some of the western Peninsular Ranges samples closely conformed with the Sr–Nd mantle array and that they could therefore be products of a heterogeneous mantle without crustal contamination. However, in the context of the Sierra Nevada data, crustal contamination of magmas originating within the island-arc field seems much more likely.

This interpretation is supported by a comparison of strontium and oxygen isotope data (Taylor and Silver, 1978; DePaolo, 1981b), which together form another powerful tool for studies of granite petrogenesis. (For background to stable isotope geology, see Hoefs, 2008.) Sierra Nevada and Peninsular Range granitoids form a hyperbolic array on the εSr versus $\delta^{18}O$ diagram (Fig. 7.44), between mantle-derived and Paleozoic sediment end-members. The shape of the hyperbola is determined by the relative strontium/oxygen concentrations in the two end-members, and is consistent with a simple mixture of high $\delta^{18}O$ sedimentary crustal melts with basic magmas.

Three alternative models can all be ruled out because they would cause vertical vectors in Fig. 7.44, in which Sr isotope increases would not be accompanied by appreciable change in $\delta^{18}O$. These models are:

1. Sr (and Nd) isotopic enrichment of a mantle source along the mantle array in Fig. 7.43.
2. Contamination of the mantle source by sediment subduction. The much lower strontium content of the mantle, relative to basic magmas, would make it much more susceptible to contamination by subducted sedimentary or seawater Sr, whereas the oxygen content of the mantle and of basic magmas is the same. In other words, the mantle has a lower Sr/O ratio than basic magma,

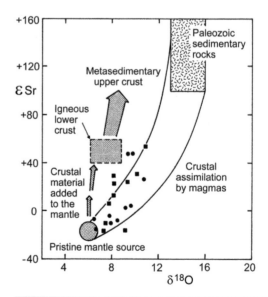

Fig. 7.44 Plot of εSr against $\delta^{18}O$ showing data for the Peninsular Ranges (•) and Sierra Nevada (■) batholiths, relative to various possible magma sources and models. After DePaolo (1981b).

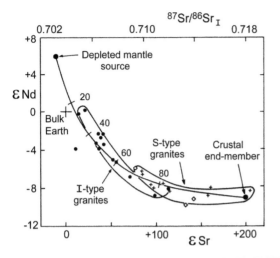

Fig. 7.45 Diagram of εNd against εSr for 'I type' ($\bullet$) and 'S type' ($+$) granites and crustal xenoliths ($\Diamond$) from SE Australia. A best-fit mixing line is shown between hypothetical crustal and mantle-derived end-members. After McCulloch and Chappell (1982).

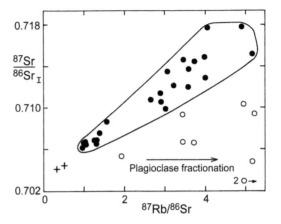

Fig. 7.46 Rb–Sr pseudo-isochron diagram for granites of SE Australia ($\bullet$) showing possible 'mixing fan'. ($+$) = gabbros; ($\circ$) = granites argued to have fractionated plagioclase after contamination. After Gray (1984).

which would yield a mixing hyperbola of steep slope in Fig. 7.44.

3. Contamination with a hypothetical lower crust of Precambrian basement which would have low $\delta^{18}O$.

Neither of the above discriminant diagrams (Figs. 7.43 and 7.44) can distinguish between genesis of the Peninsular Ranges batholith as a direct mantle-derived differentiate or a re-melt of young basic igneous rock at the base of the crustal geosyncline. However, this question has been investigated in some detail for the Paleozoic granitoids of SE Australia.

As important products of Phanerozoic crustal evolution, the California batholiths are paralleled by the Berridale and Kosciusko batholiths of the Lachlan fold belt in SE Australia. However, the genesis of these granitoid suites has proved particularly controversial. This debate began when Chappell and White (1974) distinguished two major granite types there, on the basis of chemical and mineralogical criteria. 'S-type' granites with low Ca contents and a tendency to per-aluminous character ($Al_2O_3/[Na_2O+K_2O+CaO]$ > 1.05) were regarded as partial melts of sedimentary rocks, whereas 'I-type' granites with high Ca contents and $Al_2O_3/[Na_2O+K_2O+CaO] < 1.05$ were interpreted as partial melts of young igneous crustal rocks.

McCulloch and Chappell (1982) tested this model by performing Sr and Nd isotope analysis on a suite of samples from the Berridale and Kosciusko batholiths. The data formed two overlapping fields, which together define a hyperbolic array in the lower right quadrant of the εNd–εSr diagram (Fig. 7.45). McCulloch and Chappell interpreted these data as supporting the crustal melting model of Chappell and White (1974).

In contrast, Gray (1984, 1990) attributed these data to contamination of mantle-derived basic magmas, by mixing with a sedimentary crustal component. Possible end-members are represented by young basic rocks with a mantle-like signature, and Ordovician flysch with a model Nd age of ca. 1400 Ma. The left end of the array projects back to a depleted mantle-like end-member with εNd of +6. The existence of 'rare gabbros' found in the vicinity of the batholiths demonstrates that such magmas were available in the crust, while their rarity at the surface can be attributed to the 'density problem' of raising basic magma through felsic continental crust. On the other hand, the crustal end-member is well represented by the Cooma granodiorite, which displays strong structural evidence of being an *in situ* melt of Ordovician flysch.

Gray supported his model using major element variation diagrams, and by examining Sr isotope compositions on a Rb–Sr isochron diagram (Fig. 7.46). Average initial $^{87}Sr/^{86}Sr$ and Rb/Sr ratios were plotted for two gabbros and for several plutons from the major 'S-type' and I-type batholiths. Most of the data form a cone-shaped array which Gray argued to represent mixing between a low Rb/Sr basaltic or andesitic end-member ($^{87}Sr/^{86}Sr$ = ca. 0.703–0.704), and a somewhat heterogeneous crustal end-member, typified by the crustally derived Cooma granodiorite. Compositions to the right of this array were attributed to plagioclase fractionation subsequent to mixing, which would yield horizontal displacements.

McCulloch and Chappell (1982) and Chappell and White (1992) acknowledged that the isotopic data for the Lachlan fold belt could be explained by two-component mixing between mafic and sedimentary end-members. However, they rejected this model on the grounds that mixing of basic igneous and greywacke components could not explain the

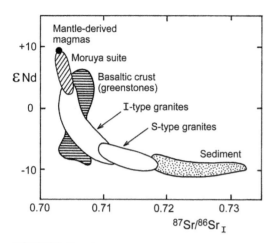

Fig. 7.47 Plot of εNd versus initial Sr isotope ratio for granitoids of the Lachlan Fold Belt, along with mantle-derived and crustal end-members (shaded) involved in a three-component mixing model. After Keay *et al.* (1997).

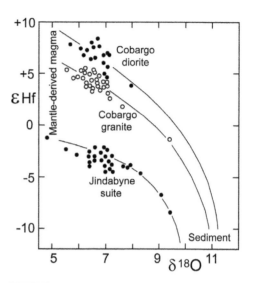

Fig. 7.48 Plot of $\delta^{18}O$ against εHf for Australian I-type plutonic rocks, showing evidence for mixing between mantle-derived magmas and sedimentary crustal units.

major element signatures of the rocks (White and Chappell, 1988).

More light was thrown on this problem by a detailed study of the most mantle-like granitoids of the Lachlan Fold Belt, comprising the Moruya granitoid suite (Keay *et al.*, 1997). A variety of rock types from this suite define an elongated distribution with εNd values from +8 to +4 (Fig. 7.47), lying between the fields for Upper-Paleozoic mantle and Cambrian greenstones. The latter rock type is believed to be an important deep crustal component of the Lachlan fold belt.

Keay *et al.* attributed the 'mafic' Moruya Suite granitoids (diagonal shading in Fig. 7.47) to mixing between mantle-derived basic magmas and tonalitic melts of the Cambrian greenstones. These mixed magmas then provide one end-member for an additional mixing process involving melts of Ordovician turbidites, leading overall to a three-component mixing model. This model also explains the relatively greater abundance of the S-type granites (containing a large sedimentary component) in the western part of the belt, since the turbidite sequence is thickest in the west and dies out eastwards (Collins, 1998).

The mixing model was tested by an oxygen versus Hf isotope study of zircons from three I-type granitoid plutons and associated mafic rocks (Kemp *et al.*, 2007). This study revealed correlated variations between $\delta^{18}O$ and εHf at the time of intrusion. However, distinct hyperbolic mixing lines pointed to different mantle-derived end-members in three distinct bodies (Fig. 7.48). For example, the Jindabyne and Cobargo zircons formed convex-upwards mixing lines, attributed by Kemp *et al.* to contamination of mantle-derived magma by an Hf-poor crustal melt (leaving behind a zircon-bearing restite).

It is somewhat ironic that the conclusion that arises from this work is that the classic I-type granitoids of the Lach-

lan Fold Belt are actually mixed S-type magmas. However, it does not follow that the true I-type granitoids that form most Precambrian continental crust have such an origin. It was shown in Section 4.2.3 that juvenile Precambrian crustal suites typically have homogeneous Nd isotope signatures that are supported by U–Pb crystallization ages. In contrast, the huge range of εNd in the granitoids of the Lachlan Fold Belt immediately points to crust–mantle magma mixing processes.

References

Ackerman, L., Jelínek, E., Medaris, G. *et al.* (2009). Geochemistry of Fe-rich peridotites and associated pyroxenites from Horní Bory, Bohemian Massif: insights into subduction-related melt–rock reactions. *Chem. Geol.* **259**, 152–67.

Bell, K. and Blenkinsop, J. (1987). Archean depleted mantle: evidence from Nd and Sr initial isotopic ratios of carbonatites. *Geochim. Cosmochim. Acta* **51**, 291–8.

Bell, K. and Simonetti, A. (1996). Carbonatite magmatism and plume activity: implications from the Nd, Pb and Sr isotope systematics of Oldoinyo Lengai. *J. Petrol.* **37**, 1321–39.

Bell, K. and Tilton, G. R. (2001). Nd, Pn and Sr isotopic compositions of East African carbonatites: evidence for mantle mixing and plume heterogeneity. *J. Petrol.* **42**, 1927–45.

Brandon, A. D. and Goles, G. G. (1988). A Miocene subcontinental plume in the Pacific Northwest: geochemical evidence. *Earth Planet. Sci. Lett.* **88**, 273–83.

Brandon, A. D. and Goles, G. G. (1995). Assessing subcontinental lithospheric mantle sources for basalts: Neogene volcanism in the Pacific Northwest, USA as a test case. *Contrib. Mineral. Petrol.* **121**, 364–79.

Brett, R. C., Russell, J. K. and Moss, S. (2009). Origin of olivine in kimberlite: Phenocryst or impostor? *Lithos* **112**, 201–12.

Brett, R. C., Russell, J. K., Andrews, G. D. M. and Jones, T. J. (2015). The ascent of kimberlite: Insights from olivine. *Earth Planet. Sci. Lett.* **424**, 119–31.

Briquet, L. and Lancelot, J. R. (1979). Rb–Sr systematics and crustal contamination models for calc-alkaline igneous rocks. *Earth Planet. Sci. Lett.* **43**, 385–96.

Burnham, A. D., Thomson, A. R., Bulanova, G. P. *et al.* (2015). Stable isotope evidence for crustal recycling as recorded by superdeep diamonds. *Earth Planet. Sci. Lett.* **432**, 374–80.

Carlson, R. W. and Hart, W. K. (1988). Flood basalt volcanism in the northwestern United States. In: MacDougall, J. D. (Ed.) *Continental Flood Basalts.* Kluwer, pp. 35–62.

Carlson, R. W., Lugmair, G. W. and MacDougall, J. D. (1981). Columbia River volcanism: the question of mantle heterogeneity or crustal contamination. *Geochim. Cosmochim. Acta* **45**, 2483–99.

Cartigny, P., Harris, J. W. and Javoy, M. (1998). Eclogitic diamond formation at Jwaneng: no room for a recycled component. *Science* **280**, 1421–4.

Chamberlain, V. E. and Lambert, R. St. J. (1994). Lead isotopes and the sources of the Columbia River Basalt Group. *J. Geophys. Res.* **99**, 11 805–17.

Chappell, B. W. and White, A. J. R. (1974). Two contrasting granite types. *Pacific Geol.* **8**, 173–4.

Chappell, B. W. and White, A. J. R. (1992). I- and S-type granites in the Lachlan Fold Belt. *Trans. Roy. Soc. Edin.: Earth Sci.* **83**, 1–26.

Chaussidon, M., Albarede, F. and Sheppard, M. F. (1987). Sulphur isotope heterogeneity in the mantle from ion microprobe measurements of sulphide inclusions in diamonds. *Nature* **330**, 242–4.

Chesley, J. T. and Ruiz, J. (1998). Crust–mantle interaction in large igneous provinces: implications from Re–Os isotope systematics of the Columbia River flood basalts. *Earth Planet. Sci. Lett.* **154**, 1–11.

Christiansen, R. L., Foulger, G. R. and Evans, J. R. (2002). Upper-mantle origin of the Yellowstone hotspot. *Geol. Soc. Amer. Bull.* **114**, 1245–56.

Cohen, R. S., O'Nions, R. K. and Dawson, J. B. (1984). Isotope geochemistry of xenoliths from East Africa: implications for development of mantle reservoirs and their interaction. *Earth Planet. Sci. Lett.* **68**, 209–20.

Collins, W. J. (1998). Evaluation of petrogenetic models for Lachlan Fold belt granitoids: implications for crustal architecture and tectonic models. *Australian J. Earth Sci.* **45**, 483–500.

Davidson, J. P. and Tepley, F. J. (1997). Recharge in volcanic systems; evidence from isotopic profiles of phenocrysts. *Science* **275**, 826–9.

Davidson, J., Tepley, F., Palacz, Z. and Meffan-Main, S. (2001). Magma recharge, contamination and residence times revealed by *in situ* laser ablation isotopic analysis of feldspar in volcanic rocks. *Earth Planet. Sci. Lett.* **184**, 427–42.

Davis, W. J., Gariepy, C. and van Breemen, O. (1996). Pb isotopic composition of late Archean granites and the extent of recycling early Archean crust in the Slave Province, northwest Canada. *Chem. Geol.* **130**, 255–69.

Dawson, J. B. (1971). Advances in kimberlite geology. *Earth Sci. Rev.* **7**, 187–214.

Dawson, J. B. and Smith, J. V. (1977). The MARID (mica-amphibole-rutile-ilmenite-diopside) suite of xenoliths in kimberlite. *Geochim. Cosmochim. Acta* **41**, 309–23.

Dawson, J. B. (1967). A review of the geology of kimberlite. In: Wyllie, P. J. (Ed.) *Ultramafic and Related Rocks.* Wiley, pp. 241–51.

DePaolo, D. J. (1981a). Trace elements and isotopic effects of combined wallrock assimilation and fractional crystallisation. *Earth Planet. Sci. Lett.* **53**, 189–202.

DePaolo, D. J. (1981b). A neodymium and strontium isotopic study of the Mesozoic calc-alkaline granitic batholiths of the Sierra Nevada and Peninsular Ranges, California. *J. Geophys. Res.* **86**, 10470–88.

DePaolo, D. J. (1983). Comment on 'Columbia River volcanism: the question of mantle heterogeneity or crustal contamination' by R. W. Carlson, G. W. Lugmair and J. D. Macdougall. *Geochim. Cosmochim. Acta* **47**, 841–4.

DePaolo, D. J. (1988). *Neodymium Isotopes in Geology.* Springer-Verlag, 187 pp.

DePaolo, D. J. and Wasserburg, G. J. (1976). Nd isotopic variations and petrogenetic models. *Geophys. Res. Lett.* **3**, 249–52.

DePaolo, D. J. and Wasserburg, G. J. (1979a). Petrogenetic mixing models and Nd–Sr isotopic patterns. *Geochim. Cosmochim. Acta* **43**, 615–27.

DePaolo, D. J. and Wasserburg, G. J. (1979b). Neodymium isotopes in flood basalts from the Siberian Platform and inferences about their mantle sources. *Proc. Nat. Acad. Sci. USA* **76**, 3056–60.

Dhuime, B., Wuestefeld, A. and Hawkesworth, C. J. (2015). Emergence of modern continental crust about 3 billion years ago. *Nature Geosci.* **8**, 552–5.

Dickin, A. P. (1981). Isotope geochemistry of Tertiary igneous rocks from the Isle of Skye, N. W. Scotland. *J. Petrol.* **22**, 155–89.

Dickin, A. P., Brown, J. L., Thompson, R. N., Halliday, A. N. and Morrison, M. A. (1984). Crustal contamination and the granite problem in the British Tertiary Volcanic Province. *Phil. Trans. Roy. Soc. Lond. A* **310**, 755–80.

Dodson, A., Kennedy, B. M. and DePaolo, D. J. (1997). Helium and neon isotopes in the Imnaha Basalt, Columbia River Basalt Group: evidence for a Yellowstone plume source. *Earth Planet. Sci. Lett.* **150**, 443–51.

Eldridge, C. S., Compston, W., Williams, I. S., Harris, J. W. and Bristow, J. W. (1991). Isotope evidence for the involvement of recycled sediments in diamond formation. *Nature* **353**, 649–53.

Erlank, A. J., Waters, F. G., Hawkesworth, C. J. *et al.* (1987). Evidence for mantle metasomatism in peridotite nodules from the Kimberly pipes, South Africa. In: Menzies, M. A. and Hawkesworth, C. J. (Eds) *Mantle Metasomatism.* Academic Press, pp. 221–311.

Fitton, J. G. and Dunlop, H. M. (1985). The Cameroon line, West Africa, and its bearing on the origin of oceanic and continental alkali basalt. *Earth Planet. Sci. Lett.* **72**, 23–38.

Francalanci, L., Avanzinelli, R., Nardini, I. *et al.* (2012). Crystal recycling in the steady-state system of the active Stromboli volcano: a 2.5-ka story inferred from in situ Sr-isotope and trace element data. *Contrib. Mineral. Petrol.* **163**, 109–31.

Francalanci, L., Davies, G. R., Lustenhouwer, W. *et al.* (2005). Intra-grain Sr isotope evidence for crystal recycling and multiple magma reservoirs in the recent activity of Stromboli volcano, southern Italy. *J. Petrol.* **46**, 1997–2021.

Fraser, K. J., Hawkesworth, C. J., Erlank, A. J., Mitchell, R. H. and Scott-Smith, B. H. (1985). Sr, Nd and Pb isotope and minor element geochemistry of lamproites and kimberlites. *Earth Planet. Sci. Lett.* **76**, 57–70.

Gibson, S. A., Thompson, R. N., Dickin, A. P. and Leonardos, O. H. (1996). High-Ti and low-Ti mafic potassic magmas: key to plume–lithosphere interactions and continental flood-basalt genesis. *Earth Planet. Sci. Lett.* **141**, 325–41.

Graham, D. W., Reid, M. R., Jordan, B. T. *et al.* (2009). Mantle source provinces beneath the northwestern USA delimited by helium isotopes in young basalts. *J. Volcanol. Geotherm. Res.* **188**, 128–40.

Gray, C. M. (1984). An isotopic mixing model for the origin of granitic rocks in southeastern Australia. *Earth Planet. Sci. Lett.* **70**, 47–60.

Gray, C. M. (1990). A strontium isotopic traverse across the granitic rocks of southeastern Australia: petrogenetic and tectonic implications. *Aust. J. Earth Sci.* **37**, 331–49.

Gregoire, M., Bell, D. R. and Le Roex, A. P. (2002). Trace element geochemistry of phlogopite-rich mafic mantle xenoliths: their classification and their relationship to phlogopite-bearing peridotites and kimberlites revisited. *Contrib. Mineral. Petrol.* **142**, 603–25.

Gurenko, A. A. and Chaussidon, M. (1997). Boron concentrations and isotopic composition of the Icelandic mantle: evidence from glass inclusions in olivine. *Chem. Geol.* **135**, 21–34.

Halliday, A. N., Dickin, A. P., Fallick, A. E. and Fitton, J. G. (1988). Mantle dynamics: a Nd, Sr, Pb and O isotopic study of the Cameroon line volcanic chain. *J. Petrol.* **29**, 181–211.

Halliday, A. N., Davidson, J. P., Holden, P. *et al.* (1990). Trace-element fractionation in plumes and the origin of HIMU mantle beneath the Cameroon line. *Nature* **347**, 523–8.

Hammouda, T., Pichavant, M. and Chaussidon, M. (1996). Isotopic equilibration during partial melting: an experimental test of the behaviour of Sr. *Earth Planet. Sci. Lett.* **144**, 109–21.

Harte, B. (1983). Mantle peridotites and processes – the kimberlite sample. In: Hawkesworth, C. J. and Norry, M. J. (Eds) *Continental Basalts and Mantle Xenoliths*. Shiva, pp. 46–91.

Hawkesworth, C. J., Erlank, A. J., Marsh, J. S., Menzies, M. A. and van Calsteren, P. W. C. (1983). Evolution of the continental lithosphere: evidence from volcanics and xenoliths in Southern Africa. In: Hawkesworth, C. J. and Norry, M. J. (Eds) *Continental Basalts and Mantle Xenoliths*. Shiva, pp. 111–38.

Hawkesworth, C. J., Gallagher, K., Kelley, S. *et al.* (1992). Parana magmatism and the opening of the South Atlantic. *Geol. Soc. Lond. Spec. Pub.* **68**, 221–40.

Hawkesworth, C. J., Kempton, P. D., Rogers, N. W., Ellam, R. M. and van Calsteren, P. W. C. (1990). Continental mantle lithosphere, and shallow level enrichment processes in the Earth's mantle. *Earth Planet. Sci. Lett.* **96**, 256–68.

Hawkesworth, C. J., Rogers, N. W., van Calsteren, P. W. C. and Menzies, M. A. (1984). Mantle enrichment processes. *Nature* **311**, 331–3.

Hemming, N. G. and Hanson, G. N. (1992). Boron isotopic composition and concentration in modern marine carbonates. *Geochim. Cosmochim. Acta* **56**, 537–43.

Hoefs, J. (2008). *Stable Isotope Geology*. 6th Edn, Springer-Verlag. 286 pp.

Hooper, P. R. and Hawkesworth, C. J. (1993). Isotopic and geochemical constraints on the origin and evolution of the Columbia River basalt. *J. Petrol.* **34**, 1203–46.

Hulett, S. R., Simonetti, A., Rasbury, E. T. and Hemming, N. G. (2016). Recycling of subducted crustal components into carbonatite melts revealed by boron isotopes. *Nature Geosci.* **9**, 904–8

Huppert, H. E. and Sparks, R. S. J. (1985). Cooling and contamination of mafic and ultramafic magmas during ascent through continental crust. *Earth Planet. Sci. Lett.* **74**, 371–86.

Hurley, P. M., Bateman, P. C., Fairbairn, H. W. and Pinson, W. H. (1965). Investigation of initial Sr^{87}/Sr^{86} ratios in the Sierra Nevada plutonic province. *Bull. Geol. Soc. Amer.* **76**, 165–74.

Ickert, R. B., Stachel, T., Stern, R. A. and Harris, J. W. (2013). Diamond from recycled crustal carbon documented by coupled $\delta^{18}O–\delta^{13}C$ measurements of diamonds and their inclusions. *Earth Planet. Sci. Lett.* **364**, 85–97.

Jones, A. P., Smith, J. V. and Dawson, J. B. (1982). Mantle metasomatism in 14 veined peridotites from Bultfontein Mine, *South Africa. J. Geol.* **90**, 435–53.

Jordan, T. H. (1975). The continental tectosphere. *Rev. Geophys. Space Phys.* **13** (3), 1–12.

Jordan, T. H. (1978). Composition and development of the continental tectosphere. *Nature* **274**, 544–8.

Kaczor, S. M., Hanson, G. N. and Peterman, Z. E. (1988). Disequilibrium melting of granite at the contact with a basic plug: a geochemical and petrographic study. *J. Geol.* **96**, 61–78.

Kamenetsky, V. S., Golovin, A. V., Maas, R. *et al.* (2014). Towards a new model for kimberlite petrogenesis: Evidence from unaltered kimberlites and mantle minerals. *Earth Sci. Rev.* **139**, 145–67.

Kamenetsky, V. S., Kamenetsky, M. B., Sobolev, A. V. *et al.* (2008). Olivine in the Udachnaya-East kimberlite (Yakutia, Russia): types, compositions and origins. *J. Petrol.* **49**, 823–39.

Kamenetsky, V. S., Kamenetsky, M. B., Sobolev, A. V. *et al.* (2009a). Can pyroxenes be liquidus minerals in the kimberlite magma? *Lithos* **112**, 213–22.

Kamenetsky, V. S., Maas, R., Kamenetsky, M. B. *et al.* (2009b). Chlorine from the mantle: magmatic halides in the Udachnaya-East kimberlite, Siberia. *Earth Planet. Sci. Lett.* **285**, 96–104.

Keay, S., Collins, W. J. and McCulloch, M. T. (1997). A three-component Sr-Nd isotopic mixing model for granitoid genesis, Lachlan fold belt, eastern Australia. *Geology* **25**, 307–10.

Kemp, A. I. S., Hawkesworth, C. J., Foster, G. L. *et al.* (2007). Magmatic and crustal differentiation history of granitic rocks from Hf–O isotopes in zircon. *Science* **315**, 980–3.

Kerr, A. C., Kempton, P. D. and Thompson, R. N. (1995). Crustal assimilation during turbulent magma ascent (ATA); new isotopic evidence from the Mull Tertiary lava succession, N. W. Scotland. *Contrib. Mineral. Petrol.* **119**, 142–54.

Kille, I. C., Thompson, R. N., Morrison, M. A. and Thompson, R. F. (1986). Field evidence for turbulence during flow of a basalt magma through conduits from southwest Mull. *Geol. Mag.* **123**, 693–7.

Kirkley, M. B., Gurney, J. J., Otter, M. L., Hill, S. J. and Daniels, L. R. (1991). The application of C isotope measurements to the identification of the sources of C in diamonds: a review. *Applied Geochem.* **6**, 477–94.

Knesel, K. M. and Davidson, J. P. (1999). Sr isotope systematics during melt generation by intrusion of basalt into continental crust. *Contrib. Mineral. Petrol.* **136**, 285–95.

Knesel, K. M. and Davidson, J. P. (2002). Insights into collisional magmatism from isotopic fingerprints of melting reactions. *Science* **296**, 2206–8.

Kramers, J. D. (1979). Lead, uranium, strontium, potassium and rubidium in inclusion-bearing diamonds and mantle-derived xenoliths from southern Africa. *Earth Planet. Sci. Lett.* **42**, 58–70.

Maaloe, S. and Aoki, K. (1977). The major element composition of the upper mantle estimated from the composition of lherzolites. *Contrib. Mineral. Petrol.* **63**, 161–73.

Maas, R., Kamenetsky, M. B., Sobolev, A. V., Kamenetsky, V. S. and Sobolev, N. V. (2005). Sr, Nd, and Pb isotope evidence for a mantle origin of alkali chlorides and carbonates in the Udachnaya kimberlite, Siberia. *Geology* **33**, 549–52.

Mantovani, M. S. M. and Hawkesworth, C. J. (1990). An inversion approach to assimilation and fractional crystallisation processes. *Contrib. Mineral. Petrol.* **105**, 289–302.

Maury, R. C. and Bizouard, H. (1974). Melting of acid xenoliths into a basanite: an approach to the possible mechanisms of crustal contamination. *Contrib. Mineral. Petrol.* **48**, 275–86.

McCulloch, M. T. and Chappell, B. W. (1982). Nd isotopic characteristics of S- and I-type granites. *Earth Planet. Sci. Lett.* **58**, 51–64.

McCulloch, M. T., Jaques, A. L., Nelson, D. R. and Lewis, J. D. (1983). Nd and Sr isotopes in kimberlites and lamproites from Western Australia: an enriched mantle origin. *Nature* **302**, 400–3.

McKenzie, D. and Bickle, M. J. (1988). The volume and composition of melt generated by extension of the lithosphere. *J. Petrol.* **29**, 625–79.

Menzies, M. A. (1989). Cratonic, circumcratonic and oceanic mantle domains beneath the Western United States. *J. Geophys. Res.* **94**, 7899–915.

Menzies, M. A. and Murthy, V. R. (1980). Enriched mantle: Nd and Sr isotopes in diopsides from kimberlite nodules. *Nature* **283**, 634–6.

Moorbath, S. and Pankhurst, R. J. (1976). Further rubidium–strontium age and isotope evidence for the nature of the late Archean plutonic event in West Greenland. *Nature* **262**, 124–6.

Moorbath, S. and Taylor, P. N. (1981). Isotopic evidence for continental growth in the Precambrian. In: Kroner, A. (Ed.) *Precambrian Plate Tectonics*. Elsevier, pp. 491–525.

Moorbath, S. and Thompson, R. N. (1980). Strontium isotope geochemistry and petrogenesis of the early Tertiary lava pile of the Isle of Skye, Scotland and other basic rocks of the British Tertiary Province: an example of magma crust interaction. *J. Petrol.* **21**, 217–31.

Moorbath, S. and Welke, H. (1969). Lead isotope studies on igneous rocks from the Isle of Skye, Northwest Scotland. *Earth Planet. Sci. Lett.* **5**, 217–30.

Morgan, D. J., Jerram, D. A., Chertkoff, D. G. *et al.* (2007). Combining CSD and isotopic microanalysis: magma supply and mixing processes at Stromboli Volcano, Aeolian Islands, Italy. *Earth Planet. Sci. Lett.* **260**, 419–31.

Morrison, M. A., Thompson, R. N. and Dickin, A. P. (1985). Geochemical evidence for complex magmatic plumbing during development of a continental volcanic center. *Geology* **13**, 581–4.

Nelson, D. R., Chivas, A. R., Chappell, B. W. and McCulloch, M. T. (1988). Geochemical and isotopic systematics in carbonatites and implications for the evolution of ocean-island sources. *Geochim. Cosmochim. Acta* **52**, 1–17.

Nixon, P. H., Rogers, N. W., Gibson, I. L. and Grey, A. (1981). Depleted and fertile mantle xenoliths from southern African kimberlites. *Ann. Rev. Earth Planet. Sci.* **9**, 285–309.

Nutman, A. P., Bennett, V. C. and Friend, C. R. (2015). Proposal for a continent 'Itsaqia' amalgamated at 3.66 Ga and rifted apart from 3.53 Ga: Initiation of a Wilson Cycle near the start of the rock record. *Amer. J. Sci.* **315**, 509–36.

Pavlov, D. I. and Ilupin, I. P. (1973). Halite in Yakutian kimberlite, its relations to serpentine and the source of its parent solutions. *Transactions (Doklady) Russian Acad. Sci.* **213**, 178–80.

Read, H. H. (1948). Granites and granites. In: Gilluly, J. (Ed.) *Origin of Granite. Geol. Soc. Amer. Mem.* **28**, 1–19.

Richardson, S. H., Gurney, J. J., Erlank, A. J. and Harris, J. (1984). Origin of diamonds in old enriched mantle. *Nature* **310**, 198–202.

Richter, F. M. (1988). A major change in the thermal state of the Earth at the Archean)Proterozoic boundary: consequences for the nature and preservation of continental lithosphere. *J. Petrol. Spec. Vol.*, 39–52.

Simon, N. S., Carlson, R. W., Pearson, D. G. and Davies, G. R. (2007). The origin and evolution of the Kaapvaal cratonic lithospheric mantle. *J. Petrol.* **48**, 589–625.

Smart, K. A., Tappe, S., Stern, R. A., Webb, S. J. and Ashwal, L. D. (2016). A review of the isotopic and trace element evidence for mantle and crustal processes in the Hadean and Archean: implications for the onset of plate tectonic subduction. *Nature Geosci.* **9**, 255–9.

Smith, C. B. (1983). Pb, Sr and Nd isotopic evidence for sources of southern African kimberlites. *Nature* **304**, 51–4.

Taylor, H. P. (1980). The effects of assimilation of country rocks by magmas on $^{18}O/^{16}O$ and $^{87}Sr/^{86}Sr$ systematics in igneous rocks. *Earth Planet. Sci. Lett.* **47**, 243–54.

Taylor, H. P. and Silver, L. T. (1978). Oxygen isotope relationships in plutonic igneous rocks of the Peninsular Ranges Batholith, southern and Baja California. *US Geol. Surv. Open File Rep.* **79–701**, 423–6.

Taylor, P. N., Jones, N. W. and Moorbath, S. (1984). Isotopic assessment of relative contributions from crust and mantle sources to the magma genesis of Precambrian granitoid rocks. *Phil. Trans. Roy. Soc. Lond.* A **310**, 605–25.

Taylor, P. N., Moorbath, S., Goodwin, R. and Petrykowski, A. C. (1980). Crustal contamination as an indicator of the extent of Early Archean continental crust: Pb isotopic evidence from the Late Archean gneisses of West Greenland. *Geochim. Cosmochim. Acta* **44**, 1437–53.

Thirlwall, M. F. and Jones, N. W. (1983). Isotope geochemistry and contamination mechanisms of Tertiary lavas from Skye, northwest Scotland. In: Hawkesworth, C. J. and Norry, M. J. (Eds) *Continental Basalts and Mantle Xenoliths*. Shiva, pp. 186–208.

Thompson, R. N. (1982). Magmatism of the British Tertiary Volcanic Province. *Scott. J. Geol.* **18**, 49–107.

Thompson, R. N., Dickin, A. P., Gibson, I. L. and Morrison, M. A. (1982). Elemental fingerprints of isotopic contamination of Hebridean Palaeocene mantle derived magmas by Archean sial. *Contrib. Mineral. Petrol.* **79**, 159–68.

Tommasini, S. and Davies, G. R. (1997). Isotope disequilibrium during anatexis: a case study of contact melting, Sierra Nevada, California. *Earth Planet. Sci. Lett.* **148**, 273–85.

Wendlandt, R. F. and Mysen, B. O. (1980). Melting phase relations of natural peridotite + CO_2 as a function of degree of partial melting at 15 and 30 kbar. *Amer. Mineral.* **65**, 37–44.

White, A. J. R. and Chappell, B. W. (1988). Some supracrustal (S-type) granites of the Lachlan Fold Belt. *Trans. Roy. Soc. Edin.: Earth Sci.* **79**, 169–81.

Wolff, J. A. and Ramos, F. C. (2013). Source materials for the main phase of the Columbia River Basalt Group: Geochemical evidence and implications for magma storage and transport. In: Reidel, S. P. *et al.* (Eds) *The Columbia River Flood Basalt Province, Geol. Soc. Amer. Spec. Pap.* **497**, 273–91.

Yokoyama, T., Aka, F. T., Kusakabe, M. and Nakamura, E. (2007). Plume-lithosphere interaction beneath Mt. Cameroon volcano, West Africa: Constraints from $^{238}U–^{230}Th–^{226}Ra$ and Sr–Nd–Pb isotope systematics. *Geochim. Cosmochim. Acta* **71**, 1835–54.

Chapter 8

Osmium Isotopes

Osmium is the least abundant member of the group of six elements called the PGE (platinum group elements). Like lead, osmium is an element with siderophile–chalcophile affinities, but unlike lead, osmium appears to be a strongly 'compatible' element during melting in silicate systems (meaning that it is strongly retained in the mantle source mineralogy). These geochemical properties mean that osmium can be used as a dating tool and a tracer in different ways from 'lithophile' isotope systems such as Sr, Pb and Nd, providing unique insights that complement these other systems.

Osmium has seven naturally occurring isotopes, two of which (^{187}Os and ^{186}Os) are the decay products of long-lived radioactive isotopes, ^{187}Re and ^{190}Pt. Of these two decay schemes, the Re–Os method has been widely used as a dating tool and geochemical tracer, since its parent, ^{187}Re, has a half-life of ca. 42 Ga and makes up 62% of natural rhenium. Rhenium is a typical chalcophile element which behaves like molybdenum in magmatic and ore-forming systems.

The Pt–Os method was developed more recently, since the radioactive parent (^{190}Pt) makes up only 0.013% of natural platinum, and has an extremely long half-life of ca. 470 Ga. This means that natural variations in ^{186}Os abundance are extremely small and hard to measure. However, in combination with the Re–Os couple the Pt–Os system provides unique information that justifies the effort of its analysis. Technically, ^{186}Os is itself radioactive, but the half-life is so long that it can be considered to be stable for geological purposes.

8.1 Osmium Analysis

In spite of its great potential as a geochemical tool, analytical difficulties have hindered the development of the osmium isotope method. The chief of these difficulties is the high ionization potential of Os (ca. 9 eV) which prevents the formation of positive osmium ions at temperatures attainable in conventional thermal ionization mass spectrometry (TIMS). Alternative methods of excitation therefore had to be sought.

Hirt et al. (1963) analysed osmium isotopes as the gaseous species OsO_4, but precision was low ($\pm10\%$ on a 200 ng sam-

ple of pure radiogenic osmium). This was probably due to dissociation of OsO_4 during thermal ionization of the molecule. Consequently, this method was not pursued for over 25 years. Instead, subsequent work focussed on the enhanced production of atomic osmium ions using more energetic ion sources. One approach was secondary ion mass spectrometry (SIMS), in which a beam of light negative ions (e.g. O^-) is used to sputter and ionize a solid osmium metal sample (e.g. Allegre and Luck, 1980). An alternative method was ICP–MS (Russ et al., 1987), which uses the high temperature of the argon plasma to ionize osmium in solution. However, these excitation methods for atomic osmium ions were rendered largely obsolete by the discovery that a solid osmium sample could yield negative Os molecular ions by conventional thermal ionization (Volkening et al., 1991). This N–TIMS method allows levels of precision over an order of magnitude better than the positive ion techniques described above.

In the N–TIMS method, Os is measured as the species OsO_3^- using platinum filaments. These are coated with a barium salt to lower the work function of the filament, which enhances the emission of negative ions relative to electrons. Formation of the oxide species may also be enhanced by bleeding oxygen into the source (Walczyk et al., 1991). The same N–TIMS method may be used to perform isotope dilution analysis of other PGE, as well as rhenium, which forms the ReO_4^- species (Fig. 8.1). This method can generate beams large enough for analysis by Faraday detector from a few ng of osmium, while multiplier analysis allows picogram size samples to be analysed (Creaser et al., 1991). This approach has now brought the osmium isotope system to the same wide range of applications as the Sr, Nd and Pb isotope methods.

In addition to the difficulties of osmium ionization, another major problem with Re–Os analysis has been the chemical behaviour of osmium in solution, due to the existence of multiple oxidation states, including the volatile tetroxide species. The volatility of osmium tetroxide allowed Luck et al. (1980) to establish a chemical extraction method in which samples were oxidized after dissolution, allowing separation by distillation. However, the variable oxidation states of osmium have continually plagued its isotope

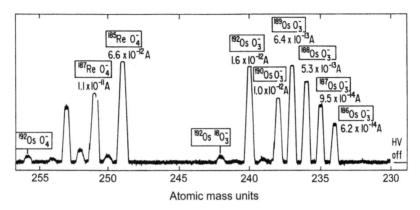

Fig. 8.1 Mass spectrum of Re and Os molecular ions produced from a Ba-doped Pt filament at 770 °C (ca. 2 A), loaded with 5 ng Os and 3 ng Re. After Creaser *et al.* (1991).

dilution analysis, by preventing complete homogenization between sample and spike osmium.

This problem was finally resolved by introduction of the somewhat hazardous 'Carius tube' digestion method (Shirey and Walker, 1995). In this technique, samples are dissolved in sealed glass ampoules under high temperature and pressure. Reagents are sulphuric acid and chromium oxide, and the samples are heated to 180 °C. Outer metal safety jackets are used, but the pressure is retained entirely by the sealed glass tube, which may quite often break! After a successful reaction, the products are frozen before the vial is broken to release the sample.

The development of multiple collector ICP–MS reopened the possibility of making high-precision osmium isotope analyses with the plasma source. However, the principal advantage of MC–ICP–MS is the ability to perform *in situ* analysis by laser ablation. Hirata *et al.* (1998) first demonstrated this method by performing *in situ* analysis of the platinum group metal (PGM) phase osmiridium. The rarity of this material limits the usefulness of the technique. However, Pearson *et al.* (2002) showed that laser ablation MC–ICP–MS could be used to make *in situ* measurements of osmium isotope ratio and Re/Os ratios in sulphide inclusions within mantle olivines. Since most of the osmium inventory from mantle rocks is probably in sulphide inclusions, the ability to perform *in situ* analyses on this material offers a powerful technique for understanding the behaviour of the Re–Os system in the mantle.

8.2 The Re–Os and Pt–Os Decay Schemes

8.2.1 The Re Decay Constant

^{187}Re decays to ^{187}Os by β emission, but the decay energy of 2.65 keV is extremely low, even compared with ^{87}Rb (275 keV). This makes measurement of the decay constant by direct

counting very difficult. Conventional counting of solid samples is impossible, due to absorption of β particles by surrounding Re atoms. Alternative techniques have used either a gaseous Re compound to replace the gas filling of a proportional counter, or a liquid Re compound in a scintillation detector. However, it was difficult to find Re compounds with suitable properties, and both methods had large errors.

Ironically, the very low decay energy of ^{187}Re makes this transition useful for measuring the mass of a neutrino that is emitted with the β decay. This method uses microcalorimetry near absolute zero temperature to measure the energy of each β particle absorbed in the detector (Gatti *et al.*, 2006). The neutrino mass (in eV) can be determined by subtraction from the total decay energy, but the Re half-life is determined as a by-product of the experiment. Even this method cannot detect β particles with energies below 60 eV, requiring assumptions about the numbers of very low energy emissions. This introduces systematic errors, so that the best current half-life determination (42.1 ± 1.1 Ga) has moderate uncertainty. However, future results should be more precise.

The difficulty with counting determinations encouraged alternative measurements of the ^{187}Re half-life based on growth of the ^{187}Os daughter product, either in the laboratory or in geological samples. One of the most successful was by Lindner *et al.* (1986), who used the 'laboratory shelf' technique to make an independent half-life determination. A 1 kg sample of purified perrhenic acid (HReO$_4$) was spiked with two different non-radiogenic Os isotopes (190 and 192), set aside for two years to allow radiogenic Os growth and then sampled for Os isotope composition over a further two-year interval. Os isotope measurements by LAMMA and ICP–MS gave results in good agreement, although the two spikes gave results that differed by 2%. Unfortunately, the starting material used by Lindner *et al.* (1986) had a non-zero level of initial radiogenic Os, as indicated by the positive intercept in Fig. 8.2. Hence the first two years of storage were effectively wasted. However, a later result by Lindner *et al.* (1989) gave

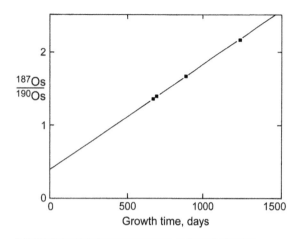

Fig. 8.2 Least squares growth line of $^{187}Os/^{190}Os$ as a function of time 'on the shelf' for a Re stock solution. Note non-zero initial ratio. After Lindner et al. (1986).

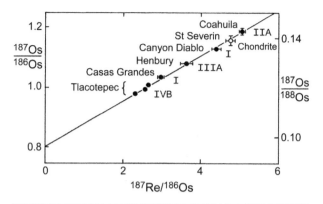

Fig. 8.3 Re–Os isochron diagram showing best-fit regression for five iron meteorites (notation indicates sub-group) and one chondrite (○). For $t = 4.55$ Ga, $\lambda = 1.62 \times 10^{-11}$ yr^{-1} ($t_{1/2} = 42.8$ Ga). After Luck et al. (1980).

a good half-life value of 42.3 ± 1.3 Ga, equivalent to a decay constant of 1.64×10^{-11} yr^{-1}.

Geological measurements of the half-life are an attractive alternative to laboratory experiments, but these have also encountered technical difficulties. In view of the low concentrations of Re in normal rocks, early determinations (Hirt et al., 1963) used molybdenite (MoS_2), which strongly concentrates Re at ca. 10–50 ppm. Unfortunately, the data were scattered, yielding an imprecise half-life of 43 ± 5 Ga. Further determinations on molybdenite were made by Luck and Allegre (1982), who selected a variety of samples with known ages and analysed Re and Os concentrations by isotope dilution. Since insignificant amounts of common osmium were found, no isotope ratio determination was necessary. However, the resulting Re–Os ages were often too old, implying Re loss during alteration. This discouraged further Re–Os dating of molybdenite, since it appeared to be an unreliable geochronometer. However, subsequent work (discussed below) has yielded more reliable data.

8.2.2 Meteorite Isochrons

In view of these problems, other attempts at geological half-life determination were focussed on iron meteorites. These have moderately large Re and Os contents, commonly in the high ppb (parts per billion) to low ppm range, and also have Re/Os ratios large enough to yield a precise age. However, the age must be calculated on an isochron diagram, since these samples contain significant initial osmium. Luck et al. (1980) ratioed radiogenic ^{187}Os against ^{186}Os, following Hirt et al. (1963). However, most other workers have normalized osmium isotope ratios to ^{188}Os because ^{186}Os can itself show small variations in nature (see below). This leads to the following Re–Os isochron equation:

$$\left(\frac{^{187}Os}{^{188}Os}\right)_P = \left(\frac{^{187}Os}{^{188}Os}\right)_I + \frac{^{187}Re}{^{188}Os}\left(e^{\lambda t} - 1\right) \quad [8.1]$$

Luck et al. (1980) determined a good isochron on 'whole-rock' (bulk) samples of five iron meteorites from different petrographic groups (Fig. 8.3). The good fit suggested that all iron meteorites, except group IVA, were formed during a narrow time interval. Based on extinct nuclide chronometers, their age is believed to be about the same as angrites (Section 15.6). The Re half-life is then calculated by substituting this age t into equation [8.1].

Unfortunately, meteorite half-life determinations have also been dogged by analytical problems, principally involving spike calibration. For example, Luck et al. (1980) determined a half-life of 42.8 ± 2.4 Ga from the isochron in Fig. 8.3, but Luck and Allegre (1983) retracted this value on the grounds that isotope dilution analysis of Os in their 1980 data set was upset by a change of osmium species in the spike solution subsequent to its calibration. However, subsequent work showed the original determination to be more nearly correct.

Problems with spike calibration were also encountered by Walker and Morgan (1989), but this time involving an over-estimate of the Re content of the spike, as later demonstrated by Morgan et al. (1992). However, Walker and Morgan also encountered problems with inadequate homogenization between sample and spike. For example, analysis of seven chondrites gave results that did not lie on the isochron through iron meteorites (using the same spike calibration for both types of meteorite). This discrepancy was probably caused by the very different chemistry of the two types of meteorite during dissolution, since more recent work places the chondrites on the iron meteorite isochron, although with a large degree of scatter attributed to later disturbance of Re–Os systems in chondrites (Walker et al., 2002).

To avoid problems of inadequate spike homogenization, the Carius tube technique was used in subsequent studies, aimed at more accurate dating of meteorites. In the most detailed of these studies, Smoliar et al. (1996) determined

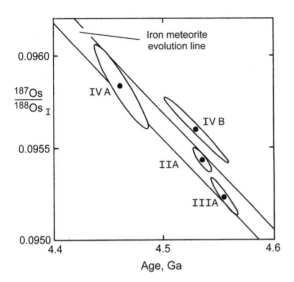

Fig. 8.4 Osmium isotope evolution diagram for ages and initial ratios of iron meteorite isochrons, used to reconstruct the evolution of the solar nebula. After Smoliar et al. (1996).

Re–Os isochrons for four different classes of iron meteorites, of which the IIIA group is the oldest. Assuming an age of 4558 Ma (from argrites) the Re–Os isochron slope was used to determine a half-life of 41.6 ± 0.15 Ga, equivalent to a decay constant of 1.666 ± 0.005 × 10^{-11} yr^{-1}.

This result has been supported by additional, more consistent work on molybdenite (Section 8.2.3), yielding a decay constant of 1.667 ± 0.003 × 10^{-11} yr^{-1}. However, this result and the iron meteorite determination are both anchored by the U–Pb method. If the ^{235}U half-life is revised in line with recent chemical abrasion dating work (Section 5.1), this will lower the accepted age of the solar system by nearly 5 Ma (0.1%) which in turn affects all other geological half-life determinations.

Using the 41.6 Ga half-life, other iron meteorite groups (except IVB) yield ages and initial ratios on a common osmium evolution curve (Fig. 8.4). The slope of this evolution curve represents the Re/Os ratio for the iron meteorite source, and is within error of the H chondrite Re/Os ratio. Hence, Smoliar et al. suggested that this source is probably the evolving solar nebula. Shen et al. (1996) also made a precise isochron determination on group IIAB irons, yielding a slope which was identical to the IIAB isochron of Smoliar et al. (within analytical error). The spread of Re–Os ratios in the other groups did not allow precise ages to be determined.

8.2.3 Dating Ores and Rocks

Very few radiometric methods have shown much success in the dating of ore deposits, despite the importance of such studies to economic geology. The difficulties arise because most dating schemes involve lithophile elements that are not stable in ore minerals. In contrast, Re, and to some extent

Os, display chalcophile chemistry, so both parent and daughter should occupy relatively stable lattice sites in sulphide minerals. Hence, many attempts have been made at Re–Os dating of ore deposits. However, following the problems encountered by Luck and Allegre (1982) in dating molybdenite, there has been continuing disagreement over the susceptibility of sulphide minerals to open-system behaviour.

Suzuki et al. (1993) claimed that recent Re–Os dates on molybdenite were usually concordant with other methods, and suggested that errors in the earlier work might be due to poor sample–spike homogenization rather than geological disturbance. However, other workers such as McCandless et al. (1993) maintained that open-system behaviour is a major problem, at least in old molybdenites. They suggested that a combination of microprobe analysis, electron back-scatter imaging and X-ray diffraction should be used to screen samples for alteration prior to analysis.

More success has been achieved in dating young ore deposits with molybdenite. For example, Selby et al. (2002) achieved perfect Re/Os isochrons (MSWD < 1) for molybdenites associated with Mesozoic gold deposits from Alaska. The ages were close to the time of igneous intrusion (from U–Pb dating) and older than Ar–Ar dates on hydrothermal muscovite and sericite. More recently, Selby et al. (2007) demonstrated concordant Re–Os and U–Pb ages on molybdenite and zircon from 11 ore deposits with a much wider range of ages from 90 to 2700 Ma.

Other studies have also generated good isochrons from sulphide-poor material. One example is a dating study on the Deccan basalts (Allegre et al., 1999). A suite of ten whole-rock basalt samples formed an excellent Re–Os isochron with a good spread of data points. Using a decay constant of 1.663 × 10^{-11} yr^{-1}, the isochron gave a precise age of 65.6 ± 0.3 Ma, in excellent agreement with previous K–Ar and Ar–Ar ages, averaging 64.5 ± 1.5 Ma. This study shows that young whole-rock suites are capable of generating Re–Os isochrons, although older material may be more problematical.

Another recent dating study (Kirk et al., 2002) successfully determined a Re–Os isochron age for gold samples from the Witwatersrand Supergroup of South Africa (Fig. 8.5). Gold from the Vaal Reef had moderate Re/Os ratios, and despite the authigenic appearance of some of the grains, gave an age of 3033 ± 21 Ma (MSWD = 1.06) which was older than the maximum age of deposition. This age provides powerful evidence that the Witwatersrand gold is of detrital origin, and was not introduced by later hydrothermal fluid circulation.

In contrast, Re–Os dating on other types of sulphide has continued to show some evidence for disturbance. Most Fe–Ni–Cu sulphide ores analysed by Dickin et al. (2000) from the Sudbury nickel deposit in Ontario appeared to be relatively undisturbed, but some pyrrhotite ores gave impossible (negative) initial ratios, indicative of major disturbance. This behaviour was confirmed by (Morgan et al., 2002). Suites of sulphide ores from two different mines gave errorchrons with large MSWD values, although the resulting ages were

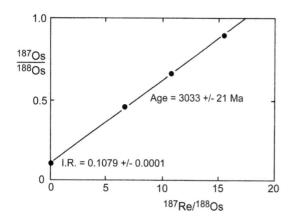

Fig. 8.5 Re–Os isochron defined by gold samples from the Vaal Reef in the Witwatersrand Supergroup of South Africa. After Kirk et al. (2002).

within error of the known age of the complex from U–Pb dating.

Laboratory experiments (Brenan et al., 2000) also confirmed the susceptibility of pyrrhotite to resetting by diffusional gain or loss of osmium. An Arrhenius relationship was observed in high-temperature diffusion experiments. This allowed a pyrrhotite 'blocking' or closure temperature to be calculated (Sections 3.3.2, 10.4). The resulting estimates of blocking temperature for a variety of grain size were between 300 and 400 °C, similar to the blocking temperature for Rb–Sr and K–Ar in biotite. This means that pyrrhotite as a dating tool is easily reset by metamorphic events. In contrast, Brenan et al. found diffusion rates over an order of magnitude lower in pyrite, implying a blocking temperature over 500 °C.

A final example of the potential of the Re–Os dating method is its application to dating the depositional age of organic-rich sediments. Sedimentary deposition is one of the most difficult events to date, and attempts to apply the Rb–Sr method have met with mixed success (Section 3.5). However, it was shown by Ravizza and Turekian (1992) that carbonaceous shales could be used to recover seawater osmium signatures (Section 8.5), and more recent work (e.g. Cohen et al., 1999) has shown that this material can also be used for dating sedimentary deposition.

Evidence for open Re–Os systems in sulphides suggested that open-system behaviour might also be a problem in dating black shales. However, a study of Paleozoic black shales from the Western Canada sedimentary basin (Creaser et al., 2002) suggested that Re–Os systems remain closed over a variety of degrees of hydrocarbon maturation. On the other hand, contamination of some samples with detrital osmium led to a large scatter on the isochron diagram (MSWD = 103). Therefore, Creaser et al. selected samples with total organic carbon over 5% in order to sample only seawater-derived (hydrogenous) osmium. These samples formed a

much tighter array (MSWD = 1.8) with an age of 358 ± 9 Ma. This suggests that the hydrogenous osmium component in the rock can be successfully isolated to determine depositional ages.

8.2.4 Os Normalization and the Pt–Os Decay Scheme

Hirt et al. (1963) established the convention of ratioing ^{187}Os data against ^{186}Os, and was followed in this practice by Luck and Allegre (1983), who normalized Os data for within-run fractionation to a ^{192}Os/^{188}Os value of 3.0827. However, ^{186}Os is itself the α decay product of the rare long-lived unstable isotope ^{190}Pt. This is not a significant problem in most geological applications, since ^{190}Pt makes up only 0.013% of total platinum. However, in view of the growing importance of the Pt–Os decay scheme for understanding mantle evolution (Section 8.3.7) it now seems best to use an alternative normalizing isotope for both decay schemes, and most workers are now using ^{188}Os for this purpose. Hence, in this book, ^{188}Os will be used in all cases as the normalizing isotope, and isotope ratios previously quoted as ^{187}Os/^{186}Os ratios are now quoted in terms of ^{187}Os/^{188}Os. However, ^{187}Os/^{186}Os ratios are shown on some figures where ^{186}Os was the original normalizing isotope. ^{187}Os/^{186}Os ratios can be converted to ^{187}Os/^{188}Os by multiplying by 0.12034.

The Pt–Os decay scheme was first applied by Walker et al. (1991), who analysed a suite of Pt-rich Fe–Cu–Ni sulphide ores from the Strathcona mine of the Sudbury nickel deposit. They constructed a Pt–Os isochron by substituting into the general decay equation as follows:

$$\left(\frac{^{186}\text{Os}}{^{188}\text{Os}}\right)_P = \left(\frac{^{186}\text{Os}}{^{188}\text{Os}}\right)_I + \frac{^{190}\text{Pt}}{^{188}\text{Os}}\left(e^{\lambda t} - 1\right) \qquad [8.2]$$

By using the known age of the complex, Walker et al. were able to use the isochron for a rough determination of the ^{190}Pt decay constant, obtaining a value similar to a counting determination of 1×10^{-12} yr^{-1} (Macfarlane and Kohman, 1961).

A much more precise determination of the ^{190}Pt decay constant was made by Walker et al. (1997) by analysis of Fe–Ni sulphide ores with very high Pt/Os ratios from the Noril'sk Complex in Siberia. Using the isochron shown in Fig. 8.6 and a published U–Pb age of 251.2 ± 0.3 Ma, Walker et al. determined a precise value for the decay constant. After modification of the ^{190}Pt abundance to 0.01296% (Brandon et al., 1999), this gave a decay constant of 1.477 ± 0.002 × 10^{-12} yr^{-1}, equivalent to a half-life of 469 Ga.

An alternative osmium isotope notation proposed by Walker et al. (1989a) is in the form of percentage deviations (γ) from a chondritic reference point. However, the Bulk Silicate Earth does not have an exactly chondritic Os signature (see below), so the CHUR reference point is less powerful for Os than in the Sm–Nd system (Section 4.2). Nevertheless, if we wish to compare Os isotope data of different ages, we may need to use this notation, so it will be used here to a limited extent. The present day average chondrite reference

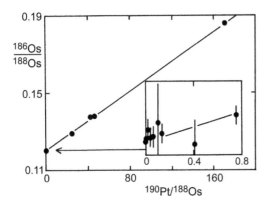

Fig. 8.6 Pt–Os isochron diagram for Fe–Ni sulphide ores from the Noril'sk Complex, Siberia. Inset shows data for samples with low Pt/Os ratios. After Walker et al. (1997).

values chosen by Walker et al. were $^{187}Os/^{188}Os = 0.127$ and $^{187}Re/^{188}Os = 0.402$ (after conversion to the new normalization). However, more recent values are slightly higher (see below).

8.3 Mantle Osmium

Based on the siderophile chemistry of Re and Os, most of the inventory of these elements originally accreted to the Earth must have been partitioned into the core. Since it is very unlikely that Re and Os would have identical partition coefficients between the mineralogy of the mantle and core, we should expect to see Re/Os fractionation during core formation. Therefore, if Re and Os in the mantle represent the residue from core formation, we would not expect them to display a Bulk Earth (= chondritic) ratio. Thus, many recent Os isotope studies of mafic igneous rocks have been devoted to establishing a mantle osmium evolution line.

8.3.1 The Bulk Silicate Earth

The first determination of the terrestrial osmium evolution line was made by Allegre and Luck (1980) on placer samples of the platinoid alloy osmiridium, whose crystallization age could be estimated. Not only can osmiridium be analysed directly on the ion probe, but since its Re content is zero, no age correction was necessary to obtain initial ratios. These analyses were combined with the initial Os isotope ratio of the iron meteorite isochron to estimate an evolution line for the Bulk Silicate Earth. The data appeared to define a straight evolution line for all samples except the Urals (Fig. 8.7), implying that the mantle was not progressively depleted in Re/Os due to crustal extraction of the crust, as seen for Nd. Allegre and Luck inferred that the Os budget of the crust was too low to affect the Re/Os budget of the mantle. However, the linearity of the evolution line depended largely on the Witwatersrand point, which was later discredited (Hart

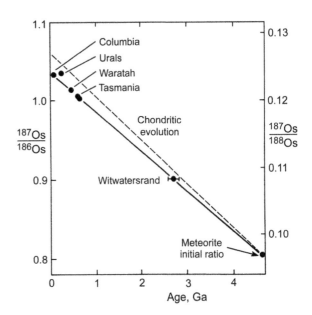

Fig. 8.7 Os isotope evolution diagram showing the first determination of the mantle growth curve, compared with the evolution line for average chondrites. The Witwatersrand point has subsequently been discredited. After Allegre and Luck (1980).

and Kinloch, 1989). Therefore, such a linear evolution model should no longer be assumed.

The osmium data of Allegre and Luck (1980) can be used to calculate a Re/Os ratio for the mantle. This ratio was compared with data for different meteorite types by Morgan (1985). The data showed a very close match to the Re/Os ratios of most chondrite groups, which was surprising, in view of the expectation of mantle Re/Os fractionation by partition into the core. To explain this co-incidence, Morgan (1985) suggested that the Re–Os budget of the mantle was generated by late accretion of chondritic material to the Earth, after core formation. This model was supported by more recent osmium isotope analysis of chondrites (Walker and Morgan, 1989). These yield an average $^{187}Os/^{188}Os$ ratio of 0.1296 for ordinary chondrites, and a slightly lower ratio of 0.128 for enstatite chondrites (Brandon et al., 2006).

Relative to the chondritic reference line (Fig. 8.7), the placer ores analysed by Allegre and Luck (1980), and similar samples from Phanerozoic ophiolites (Hattori and Hart, 1991) imply Re depletion of the modern mantle. However, it was important to analyse dated ancient samples of mantle-derived materials to chart the osmium evolution of the mantle over time. Several studies were made to examine this question (Puchtel et al., 1999, 2001; Bennett et al., 2002), as summarized in Fig. 8.8. The data show a certain amount of scatter, which reflects metamorphic disturbance of some samples. However, there is a general clustering of Os

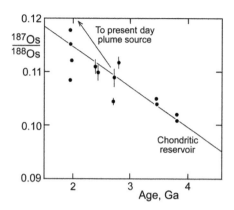

Fig. 8.8 Ranges of initial Os isotope ratio in Archean and Early Proterozoic mantle-derived samples relative to the chondritic evolution line. After Bennett *et al.* (2002).

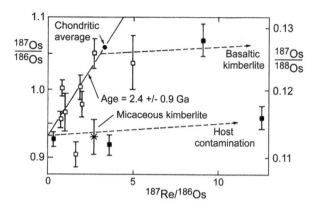

Fig. 8.9 Re–Os isochron diagram for South African peridotite xenoliths. ($\square$) = high-temperature peridotite; ($\blacksquare$) = low-temperature peridotite; ($*$) = Group II (micaceous) kimberlite. After Walker *et al.* (1989a).

initial ratios around or below the chondritic evolution line throughout the Archean.

These observations lead to two important conclusions. The first of these is that the addition of the accreted chondritic veneer was completed by the Early Archean, and largely homogenized into the mantle. (However, elemental PGE data have been used to argue for later homogenization by Maier *et al.*, 2009.) The second conclusion is that the mantle has undergone Re depletion over Earth history from an early chondritic composition. However, this Re fractionation may result from the extraction of mafic crust that has been subducted and stored in the lower mantle, rather than extraction of the continental crust. These issues will be discussed further below.

8.3.2 Lithospheric Mantle

Because of the possibility that mantle-derived magmas sample enriched plums or plumes in the asthenospheric mantle, the most detailed understanding of the Re–Os evolution of the upper mantle may be obtained from detailed studies of lithospheric mantle peridotite suites. Because these mantle samples have been preserved in the solid state for long periods, they offer the opportunity of studying ancient upper mantle differentiation events, as well as the effects of more recent mantle enrichment.

The first osmium isotope determinations on ancient subcontinental lithosphere were made by Walker *et al.* (1989a), who analysed peridotite xenoliths from South African kimberlites. The Re–Os isochron diagram (Fig. 8.9) allowed the effects of contamination by the host (Group I type) basaltic kimberlite to be assessed. Walker *et al.* argued that the process of contamination caused samples to move along a subhorizontal vector towards high Re/Os ratios characteristic of the host magma. Samples argued not to have suffered this effect define a steep array in Fig. 8.9, corresponding to an age of 2.4 ± 0.9 Ga.

Walker *et al.* attributed the relatively coherent Re–Os isotope systematics of the peridotite xenoliths to separation of South African lithosphere from the convecting asthenosphere in the Archean. Their unradiogenic compositions, relative to Allegre and Luck's upper mantle $^{187}Os/^{188}Os$ ratio of 0.127 ($^{187}Os/^{186}Os = 1.06$), suggested that these peridotites represent residues from partial melting, consistent with gross incompatible-element depletion of the sub-continental lithosphere (Jordan, 1978). Hence, Walker *et al.* suggested that the Re–Os system could provide insights into mantle depletion events in a distinct manner from LIL tracers such as Sr and Nd, which commonly chart secondary *enrichment* of the sub-continental lithosphere by metasomatism.

Because the Re–Os system can effectively date mantle depletion events, as well as the mantle enrichment events normally dated by lithophile isotope systems, osmium model ages may be particularly useful for dating sub-continental lithospheric mantle. Samples of depleted mantle can be used to calculate conventional osmium model ages, based on the intersection of sample and Bulk Earth evolution lines (Fig. 8.10). However, the Re/Os ratio of mantle samples may be disturbed by metasomatic rhenium addition after the initial crust-forming Re depletion event. This may cause conventional Re–Os 'mantle extraction' ages to yield too old an age.

An alternative approach proposed by Walker *et al.* is to assume that the depletion event generated Re/Os ratios of zero in the samples. So-called 'rhenium-depletion' model ages (T_{RD}) can then be calculated by projecting a horizontal line back to the Os mantle growth curve (Fig. 8.10). However, if the mantle differentiation event did not reduce Re contents to zero, the Re depletion age will under-estimate the true age. These alternative strategies will be discussed further below.

The first attempt to determine a Re–Os isochron on a section of exhumed lithospheric mantle was made by

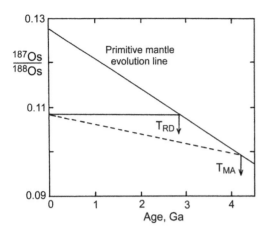

Fig. 8.10 Comparison of a rhenium-depletion model age (T_{RD}) with a conventional 'mantle extraction' age (T_{MA}). Modified after Walker et al. (1989b).

Reisberg et al. (1991) on the Ronda 'orogenic lherzolite' massif in southern Spain. The massif contains both mafic and ultramafic layers, which Reisberg et al. interpreted as relics of old recycled oceanic mantle ('marble cake mantle', Section 6.1.5), later incorporated into the continental lithosphere. Whole-rock samples of mafic and ultramafic layers defined Re–Os errorchrons with similar apparent ages (Fig. 8.11a, b), interpreted as the time when this piece of mantle was accreted to the continental lithosphere. However, the ranges of isotopic composition were very different: ultramafic units displayed both isotopic enrichment and depletion relative to the chondritic composition, whereas the mafic units displayed extreme isotopic enrichment. The significance of the latter component will be discussed further below.

Reisberg and Lorand (1995) attempted to apply the Re–Os isochron technique to date a similar mantle differentiation event in the East Pyrenees lherzolite, but the data were badly scattered on the Re–Os isochron diagram. The scattered data were attributed to open-system behaviour of rhe-

nium during mantle metasomatism. However, Reisberg and Lorand observed good correlations between osmium isotope ratios and aluminium contents of whole-rock peridotite samples. A reduction in scatter on the isochron array was also observed for ultramafic units from Ronda (Fig. 8.11c). Hence, Reisberg and Lorand proposed that these plots are isochron analogues, in which aluminium, an immobile and incompatible element, is a proxy for rhenium.

Because the Re/Al ratio of the original rock before metasomatism is unknown, the slope of an 'aluminochron' cannot be solved directly to yield an age. However, if Re and Al behaved coherently in the original rock, the intercept on the y axis will yield the initial osmium isotope ratio in the usual way. This quantity can be used to date mantle differentiation events using Os model ages. This age has been termed an 'aluminochron extrapolation age' (Lassiter et al., 2014), and should be more accurate than either T_{RD} or T_{MA} model ages because it is not making any assumptions about Re mobility in the sample.

8.3.3 Primitive Upper Mantle

Since mantle depletion events are quantitatively the most important processes affecting the osmium evolution of lithospheric mantle, Meisel et al. (1996) argued that the most fertile (Al-rich) lherzolites from the subcontinental lithosphere could be used to constrain the osmium isotope evolution of the 'primitive upper mantle' (PUM). Mantle xenoliths from Kilburn Hole (New Mexico), West Eifel (Germany) and the Baikal Rift (Mongolia) formed a linear array on a plot of Os isotope ratio against Al_2O_3 (Fig. 8.12). The most Al-rich samples were consistent with an estimated Al_2O_3 content of 4.2% in the PUM, and gave an average $^{187}Os/^{188}Os$ ratio of 0.129 ($^{187}Os/^{186}Os = 1.08$), in good agreement with the most Al-rich of the massive peridotites analysed by Reisberg and Lorand (1995). The PUM composition is in agreement with the average for ordinary chondrites, but is significantly more radiogenic than the mantle value of Allegre and Luck (1980).

The composition of PUM determined by Meisel et al. (1996) was further refined by Meisel et al. (2001) using a much larger data set of spinel- and garnet-lherzolites from around the

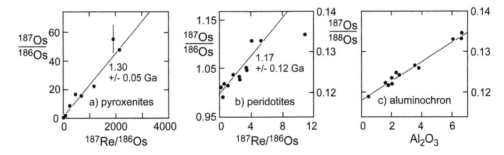

Fig. 8.11 Re–Os isochron diagrams for the Ronda ultramafic complex: (a) mafic units; (b) ultramafic units; (c) 'aluminochron' for ultramafics. After Reisberg et al. (1991).

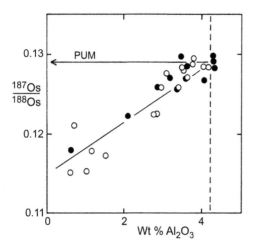

Fig. 8.12 Al–Os aluminochron diagram for mantle xenoliths used to determine the composition of PUM: (○) = older data; (●) = newer data. After Meisel *et al.* (1996).

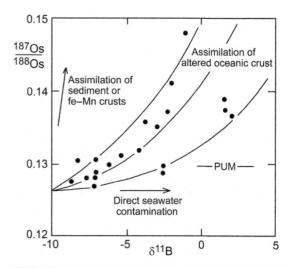

Fig. 8.13 Plot of osmium versus boron isotope ratio, showing the effects of different seawater contamination scenarios for MORB magmas. Modified after Gannoun *et al.* (2007).

world. Again using aluminochrons, they obtained a PUM $^{187}Os/^{188}Os$ signature of 0.1296 ± 0.0008. As noted above, analysis of Early Archean peridotites from western Greenland and western Australia has also given initial osmium ratios close to a PUM evolution line (Bennett *et al.*, 2002), suggesting that the upper mantle has had this composition since very early in Earth history.

An important question about these mantle samples is the mineralogical location of osmium. To investigate this question, Burton *et al.* (1999) made a detailed petrographic and geochemical examination of a spinel lherzolite xenolith from Kilburn Hole, New Mexico. This revealed that most osmium was concentrated in Fe–Ni sulphide minerals, which contain about 4 ppm osmium, two thousand times as much as the bulk xenolith. These findings have been confirmed by subsequent studies (e.g. Burton *et al.*, 2002), showing that osmium is not really a compatible element in mantle minerals like olivine, but is contained in sulphide phases that tend to crystallize at the same time as olivine.

Of the sulphide material in the Kilburn Hole xenoliths, Burton *et al.* found that 80–90% was interstitial between the silicate minerals, while the remainder was mostly in small inclusions between 5 and 10 μm in diameter, enclosed in pyroxenes. Isotopic analysis showed that the interstitial sulphide was in equilibrium with the silicate minerals of the xenolith, and with the host basalt. However, the sulphide inclusions within the pyroxenes were out of isotopic equilibrium with the rest of the system. It was concluded that these inclusions were protected from diffusional communication with the rest of the rock by the high sulphide/silicate partition coefficient of osmium, which tends to keep osmium 'locked up' in the inclusions. Subsequent work showed similar features in abyssal peridotites, interpreted as samples of the asthenospheric mantle (see below).

8.3.4 Asthenospheric Mantle

Roy-Barman and Allegre (1994) made a detailed comparison between ocean ridge basalts and abyssal peridotites from different ridges. These two sample types are expected to have the same osmium isotope ratios, since they are thought to represent (respectively) the eruptive product and the residue from melting. Samples were leached with oxalic acid (and then HBr) to remove ferro-manganese coatings, which have radiogenic seawater osmium signatures. Leached peridotites from the flanks of the Mid Atlantic Ridge (MAR) had a very narrow range of $^{187}Os/^{188}Os$ ratios, in good agreement with previous work. However, leached MORB samples had elevated ratios of 0.128 to 0.133, within the range of North Atlantic OIB. (Roy-Barman and Allegre interpreted the MORB values as magmatic because they found good agreement between leached glass and crystalline basalt compositions.) They proposed two alternative models to explain the differences between MORBs and peridotites: contamination of MAR basaltic magmas by assimilation of hydrothermally altered oceanic crust; or contamination of the MAR magma source with isotopically enriched osmium, either from the Azores plume or from enriched streaks in a marble cake mantle (Roy-Barman *et al.*, 1996).

Schiano *et al.* (1997) supported the latter model, based on even more radiogenic $^{187}Os/^{188}Os$ ratios of up to 0.16 in pristine MORB glasses from various ridges. However, these very radiogenic signatures were not reproduced when replicate analyses were performed by Gannoun *et al.* (2007). Furthermore, all of the samples with radiogenic osmium analysed by Gannoun *et al.* were also found to have elevated boron isotope signatures, indicative of exchange with a seawater component (Fig. 8.13). However, the curvature of the mixing hyperbolae is consistent with assimilation of altered sea floor

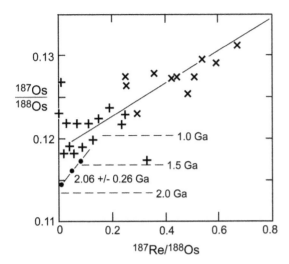

Fig. 8.14 Re–Os isochron diagram for whole-rock abyssal peridotites (+, ×) and rounded sulphide inclusions (•) from the Mid Atlantic Ridge. Horizontal dashed lines are Re depletion model ages. Modified after Harvey et al. (2006).

Additional studies on abyssal peridotites have also shed new light on the depletion history of the MORB source. For example, whole-rock samples of peridotite from the Kane fracture zone and the axial valley of the Mid Atlantic Ridge were shown to form colinear positive arrays on the Re–Os isochron diagram (Alard et al., 2005; Harvey et al., 2006). The scatter of the data is evidence of Re mobility, but the least radiogenic samples yield a rhenium depletion age of ca. 1.3 Ga (Fig. 8.14). Sulphide inclusions from within silicate phases were also analysed. The majority of these inclusions had skeletal textures, low osmium contents and defined a horizontal isotope array attributed to seawater contamination. However, six inclusions with rounded habit and low Re/Os ratio defined a linear array with a slope age of 2.06 ± 0.06 Ga (Fig. 8.14) which was also in good agreement with the Re-depletion age of the isochron intercept (ca. 1.9 Ga). These data suggest that these peridotites suffered an ancient melt-depletion event very similar to sub-continental lithosphere, despite being currently located on the Mid Atlantic Ridge. This points to ancient melt depletion of the MORB source on a mineralogical scale that has evidently not been erased by mantle convection.

Some workers have suggested that these melt depletion signatures in the MORB source are shallow level features not representative of the asthenosphere as a whole. However, Lassiter et al. (2014) showed that the Re–Os signatures of abyssal peridotites are very similar to peridotite xenoliths in ocean island basalts (but not their host magmas). This similarity is demonstrated by the 'aluminochron' diagrams in Fig. 8.15, where the two suites show almost exactly the same initial ratios, yielding 'aluminochron extrapolation' ages of ca. 1.4 Ga for both suites. In contrast, the Al_2O_3 contents of abyssal peridotites are slightly lower, reflecting greater melt extraction from asthenospheric mantle under ridges. Hence it is concluded that both of these suites are sampling the asthenospheric upper mantle, showing this

material by MORB magmas, rather than direct seawater contamination of the samples themselves. After this effect was removed, MORB samples all had $^{187}Os/^{188}Os$ ratios of 0.13 or less, a conclusion that was tested by osmium analysis of sulphide phases within the same MORB glasses. These phases are believed to have separated at an early stage of MORB magma evolution, and therefore to have escaped the assimilation effects seen in the glasses. Consistent with this theory, most MORB sulphides were found to have $^{187}Os/^{188}Os$ ratios below PUM, and those with osmium content over 100 ppb gave a mean $^{187}Os/^{188}Os$ ratio of 0.1263 ± 0.0012 (2σ), within error of abyssal peridotites.

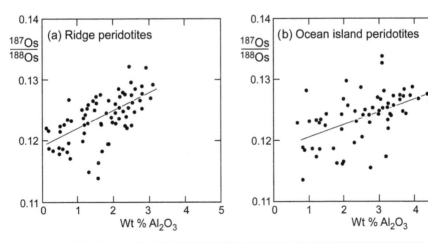

Fig. 8.15 Aluminochrons for (a) abyssal peridotites and (b) OIB xenoliths, showing similar Re–Os signatures. After Lassiter et al. (2014).

to be moderately Re depleted overall relative to primitive mantle.

8.3.5 Enriched Mantle Plumes

The first osmium isotope analyses of ocean island basalts (OIB) were made on samples from Iceland and Hawaii (Martin, 1991). These showed modest enrichment in ^{187}Os relative to the values seen in ophiolites and peridotite xenoliths (see above). However, as seen for lithophile isotope systems (Section 6.5), the best understanding of mantle enrichment processes in OIB was obtained by the analysis of extreme examples. For example, Hauri and Hart (1993) determined initial $^{187}Os/^{186}Os$ ratios as high as 1.25 ($^{187}Os/^{188}Os$ approaching 0.15) in lavas from the HIMU island of Mangaia. These radiogenic compositions are best explained by recycling of enriched crustal material back into the upper mantle. Thus, Hauri and Hart calculated that 16% of re-cycled 2.1 Ga oceanic crust can explain the osmium signature of the HIMU component. On the other hand, 1–2% of recycled continental crust in the EMII source does not markedly affect its osmium isotope signature.

Reisberg *et al*. (1993) found even more radiogenic osmium isotope ratios in the HIMU islands of the Comores and St Helena, with initial $^{187}Os/^{186}Os$ ratios as high as 1.7. However, they cautioned that, because these signatures were carried by osmium-poor lavas (10–30 parts per trillion = 0.01–0.03 ppb), their elevated ^{187}Os signatures could be introduced by processes in the oceanic lithosphere. This could involve contamination of OIB magmas with ocean floor sediments included in the volcanic edifice, or accumulated lithospheric olivine xenocrysts. However, work on the Canaries (Marcantonio *et al*., 1995) established that samples with more than about 30 to 50 parts per trillion (0.03–0.05 ppb) osmium are generally immune to sea floor contamination processes. Such effects were also ruled out in Mauna Kea lavas by strong correlations between Os and Sr, Nd and Pb isotope ratios, showing that the osmium signatures must be derived from primary mantle sources (Hauri *et al*., 1996).

Although sea floor contamination was ruled out for the Canaries samples, they nevertheless gave evidence of complex mixing processes in the plume source, since they had radiogenic osmium signatures, coupled with Pb isotope ratios intermediate between HIMU and EMI (Fig. 8.16). This cannot be explained by simple mixing between DMM and HIMU (Marcantonio *et al*., 1995). Therefore, the Canaries data were tentatively attributed to mixing between DMM and a HIMU–EMI mix. This model is also consistent with lithophile isotope evidence, which has suggested an intimate relationship between HIMU and EMI in plume sources (Section 6.5).

New data for the Koolau volcano, Hawaii (Bennett *et al*., 1996) extended the Hawaiian Os–Sr–Pb correlation line to more enriched compositions, approaching the composition of a Pitcairn sample (Fig. 8.16), thought to approximate the isotopic composition of the EMI mantle component. However, the trajectory of the Hawaiian array in Fig. 8.16 cannot be explained by simple two-component mixing between

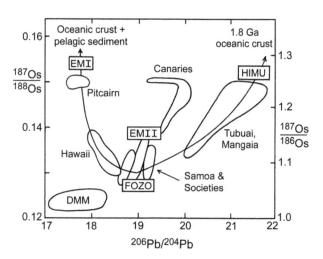

Fig. 8.16 Plot of Os against Pb isotope ratios for various plume sources to show possible mixing lines between a depleted lower mantle end-member (**FOZO**) and distinct enriched end-members. Data were from OIB samples with more than 45 ppt osmium. Hauri *et al*. (1996).

EMI and DMM mantle. Therefore, Hauri *et al*. (1996) identified the other mixing component as entrained lower mantle (FOZO, Section 6.5.1) with unradiogenic osmium. In contrast, post-erosional alkaline mafic rocks of the Honolulu volcanic series define a mixing with a quite different trajectory that does trend towards DMM. This suite is best explained by interaction of the plume source with partial melts of the old oceanic lithosphere under Hawaii (Class and Goldstein, 1997).

An important discovery by Lassiter and Hauri (1998) is that radiogenic osmium in Koolau basalts is associated with elevated $\delta^{18}O$ signatures (Fig. 8.17). Since elevated oxygen isotope compositions are normally only produced in the sedimentary system, these workers attributed the enriched (EMI) signature to a mixture of recycled oceanic crust and subducted pelagic sediment. An osmium–oxygen isotope correlation similar to the Kea-Loa trend was also observed in Iceland basalts (Skovgaard *et al*., 2001). However, these authors attributed the elevated oxygen and osmium signatures to altered oceanic crust, without a sediment component.

It has also been shown more recently that peridotite xenoliths from subduction zones can have $\delta^{18}O$ signatures as high as +8 (Liu *et al*., 2014). Therefore, high $\delta^{18}O$ signatures in plumes could also originate from the metasomatized mantle wedge above subduction zones. The osmium signatures of this material will be examined below (Section 8.3.6).

Comparison between osmium and strontium isotope data for various ocean islands revealed a series of separate subvertical mixing lines that seem to originate from a curved 'baseline' with a hyperbolic shape (Fig. 8.18). This pattern led Class *et al*. (2009) to argue that there is no unique enriched mantle source, but a series of slightly different enrichment

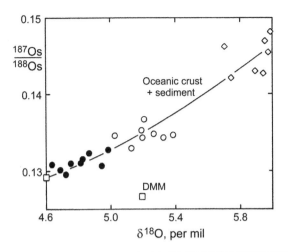

Fig. 8.17 Plot of osmium against oxygen isotope composition for Hawaiian lavas, showing a positive correlation: ($\Diamond$) = Koolau; ($\bullet$) = Mauna Kea; ($\circ$) = Mauna Loa. After Lassiter and Hauri (1998).

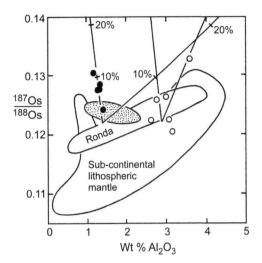

Fig. 8.19 Plot of osmium isotope data for xenoliths from subduction-related magmas against their Al_2O_3 contents: ($\bullet$) = Simcoe; ($\circ$) = Ichinomegata. Mixing lines show the effect of contamination with subduction-related fluids (% increments marked). Shaded field = abyssal peridotites. After Brandon et al. (1996).

processes. However, new analyses from the Samoan plume (Jackson and Shirey, 2011) show a horizontal trend in Fig. 8.18 characterized by increasing Sr isotope ratio without any correlated increase in osmium. This suggests that sediment contamination may not explain the positive correlations between osmium and oxygen seen in Hawaiian lavas (Fig. 8.17), or the steep Os–Sr isotope arrays in Fig. 8.18. Therefore, to explore the possible origin of these correlations, it is necessary to examine subduction-related materials, where these components originate.

8.3.6 Subduction Zones

Most subduction-related magmas have very low osmium contents that are very susceptible to high-level contamination in

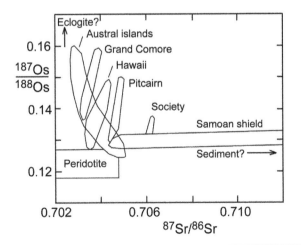

Fig. 8.18 Plot of Os versus Sr isotope data for OIB, showing a series of sub-vertical arrays in different island groups. After Jackson and Shirey (2011).

the arc crust. Therefore, more reliable sampling of the products of subduction-related processes are provided by mantle xenoliths from subduction zones. Brandon et al. (1996) analysed lherzolite and harzburgite xenoliths from the Cascades and Japanese arcs that were believed to sample the metasomatized mantle wedge above these subduction zones. These xenoliths have osmium concentrations of 0.1–2.1 ppb and osmium isotope ratios ranging from a minimum near the field of abyssal peridotites to a maximum $^{187}Os/^{188}Os$ ratio of 0.134 (Fig. 8.19). Evidence for a metasomatic origin for the osmium enrichments is provided by the relatively low Al_2O_3 ratios of the xenoliths (especially those from Simcoe, Washington). These define a vertical trend which is distinct from the diagonal enrichment–depletion trend seen in orogenic lherzolites. Brandon et al. suggested that their results reflected metasomatic osmium transport in a chloride-rich slab-derived fluid or melt.

Suzuki et al. (2011) proposed a method of sampling the osmium isotope composition of primitive subduction-related magmas, by the analysis of chrome spinels from Japanese beach sands. These mineral phases are believed to have crystallized at an early stage of magmatic evolution, and their osmium contents up to 10 ppb make them very resistant to the crustal contamination processes that affect most arc volcanics. Suzuki et al. showed that spinels derived from boninites had unradiogenic $^{187}Os/^{188}Os$ ratios averaging 0.124, in good agreement with whole-rock abyssal peridotites. Therefore, the somewhat elevated osmium signatures in the host magmas must be attributed to high-level crustal contamination, even for lavas with relatively high osmium contents of 100 ppt (Fig. 8.20). In comparison,

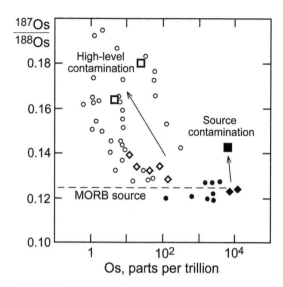

Fig. 8.20 Plot of Os isotope ratio against Os concentration to show compositions of Japanese boninites (◇) and tholeiites (□) and associated spinel sands (◆, ■), in comparison with other arc volcanics (○) and peridotite xenoliths (●). After Suzuki *et al.* (2011).

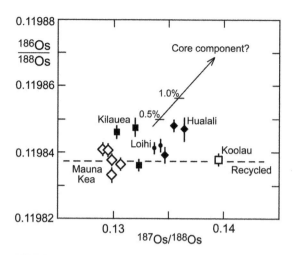

Fig. 8.21 Enrichments of ^{186}Os in Hawaiian picrites relative to normal mantle values (dashed horizontal line), possibly attributable to contamination by a core component. After Ireland *et al.* (2011).

tholeiitic host magmas had much more radiogenic osmium, also attributed to high-level contamination, but in this case the associated chrome spinels also showed elevated ^{187}Os/^{188}Os ratios of 0.143. Hence this provides reliable evidence for some metasomatic contamination of the arc magma source by slab-derived fluids.

8.3.7 The Core Osmium Signature

Walker *et al.* (1995) and Widom and Shirey (1996) proposed an alternative model for isotopic enrichment of some OIB sources that derived their radiogenic osmium signatures directly from the core. They suggested that a metallic core component could be incorporated into plumes at the core–mantle boundary, such that its high osmium content would overprint the isotopic signature of the deep mantle with radiogenic osmium from the outer core. However, the existing isotope tracer evidence was inadequate to properly test this model.

The development of the Pt–Os decay scheme provided an additional isotopic tracer to test the possibility of contamination of plumes by material from the outer core. To provide a basis for interpretation, Walker *et al.* (1997) predicted the behaviour of Re–Os and Pt–Os systems in the core and mantle from known silicate–metal partition coefficients and by comparison with the isotope and PGE systematics of iron meteorites. It is well established that all PGE were strongly partitioned from the mantle into the core during the early history of the Earth. However, the meteorite evidence also suggests that the PGE were partitioned between the *liquid* and *solid* core once the latter had begun to crystallize. It was argued that stronger partitioning of Os into the solid metal

phase would have caused the liquid outer core to be enriched in Pt and Re relative to Os. Over time, this would lead to correlated enrichments of ^{186}Os and ^{187}Os in the outer core.

The first evidence in support of this predictive model came from analysis of the Noril'sk Complex, which showed a significant enrichment of ^{186}Os relative to other terrestrial and meteorite samples (Walker *et al.*, 1997). This is consistent with the derivation of the Noril'sk Complex from a plume source. These findings were supported by the observation of small ^{186}Os enrichments in picrites from several different Hawaiian volcanoes (Brandon *et al.*, 1998). Initially, the most radiogenic ^{186}Os signatures were observed in the Koolau volcano, but these disappeared when subjected to more detailed analysis (Brandon *et al.*, 1999). It was suggested that the original Koolau data were probably perturbed by molecular ion interferences on the ^{186}Os peak.

Brandon *et al.* (1999) did continue to find elevated ^{186}Os signatures in Loihi seamount, which also carries high ^{3}He values, possibly from the outer core (Section 11.1). However, the predicted core contribution should also affect the isotopic composition of ^{182}W, since tungsten is also a highly siderophile element. Scherstén *et al.* (2004) found tungsten anomalies smaller than predicted by the core-contamination model, casting some doubt on the model. In addition, more recent work by Ireland *et al.* (2011) failed to replicate the osmium data for Loihi. On the other hand, ^{186}Os excesses of similar magnitude were observed in picrites from Kilauea and Hualalai (Fig. 8.21). However, it remains to be seen whether these excesses will be replicated in future work.

8.4 Petrogenesis and Ore Genesis

As one of the platinum group elements (PGE), osmium is uniquely suited as a tracer of petrogenetic and ore-forming processes of noble metal deposits. PGE deposits are generally associated with major mafic complexes, and their development has been attributed to mixing between mantle-derived and crustal components, causing the precipitation of PGE-bearing chromite or sulphide phases (e.g. Naldrett, 1989). The strong fractionation between Re and Os in crust-forming processes, which generates very radiogenic osmium in the crust relative to the mantle, makes osmium a powerful tracer for such mixing processes. Several of the world's largest basic–ultrabasic intrusions have been subjected to Re–Os analysis; however, the complexities of their chemistry has typically required several studies to get a reasonable understanding of their genesis.

8.4.1 The Bushveld Complex

The Bushveld Complex is the world's largest layered mafic intrusion and principal PGE producer. Most of these PGEs come from the famous Merensky Reef, but the UG1 and UG2 chromite layers (chromitites) are also major sources. Hart and Kinloch (1989) made an ion probe study of PGE sulphides from the Merensky Reef on the western lobe of the intrusion. They found radiogenic initial $^{187}Os/^{188}Os$ ratios of around 0.175 ($^{187}Os/^{186}Os = 1.45$) for grains of laurite (RuS_2) from the Rustenburg, Union and Amandelbult mining areas. These values are far above the mantle growth line for osmium at the time of intrusion of the Bushveld (at 2.05 Ga), indicating a large crustal component in the ore. However, two grains of erlichmanite (OsS_2) from Rustenburg and Union gave low $^{187}Os/^{186}Os$ ratios, lying on the chondritic evolution line ($^{187}Os/^{188}Os = 0.112$).

The Bushveld laurites can be interpreted as the products of crustally contaminated magmas, as has been proposed to explain Sr isotope data for the Bushveld Complex (Sharpe, 1985). However, the erlichmanite results pose a major problem for this interpretation. They cannot be attributed to open-system perturbation of Re–Os age corrections, since these minerals contain no rhenium. If the osmium isotope variations in the laurites are attributed to magmatic processes, then the erlichmanites seem to represent mantle-derived PGE phases which were somehow carried into the intrusion without contamination (which seems unlikely). Alternatively, osmium isotope variations in the laurites must be hydrothermal in origin, and some component of the Merensky Reef mineralization would therefore be attributed to hydrothermal PGE introduction.

In an attempt to solve this conundrum, Schoenberg *et al.* (1999) analysed separated minerals from the 'critical zone' underlying the Merensky Reef and the unmineralized 'bastard unit' of the main zone, immediately above it (Fig. 8.22). Powerful evidence against the 'metasomatic contamination model' came from four analyses of a poikilitic pyroxenite from the bastard unit. These samples, from 4–20 m above the reef, defined a perfect Re–Os isochron with an age of 2043 ± 11 Ma (MSWD = 0.7). At the time, the best available crystallization age for the complex was an Rb–Sr age of 2061 ± 27 Ma.

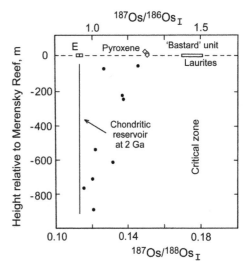

Fig. 8.22 Plot of initial osmium isotope ratios in the Bushveld Complex against stratigraphic height relative to the Merensky Reef: (●) = chromite; E = erlichmanite (sulphide). Modified after Schoenberg *et al.* (1999).

A much more precise crystallization age has since been determined on zircon from the Merensky Reef using the chemical abrasion method (Scoates and Friedman, 2008). The most precise and concordant single grains give a $^{238}U/^{206}Pb$ age of 2054.0 ±1 Ma (2σ), only 0.5 Ma less than the 207/206 age. Hence these ages are unaffected by possible uncertainties in the ^{235}U half-life (Section 5.1). In comparison, Ar–Ar dating on biotite from the UG chromitite gave a slightly younger age of 2042 ± 3 Ma (Nomade *et al.*, 2004), which can be explained by a slight under-estimate of the potassium half-lives (Section 10.3.2). Such good agreement between all of these ages would be very unlikely if pervasive osmium introduction by metasomatism had occurred in the vicinity of the Merensky Reef.

Support for the alternative 'magma mixing model' came from analysis of chromitites from the underlying critical zone (McCandless *et al.*, 1999; Schoenberg *et al.*, 1999). These chromitites were found to have variable initial ratios (Fig. 8.22), including some unradiogenic values near the chondritic evolution line that supported the erlichmanite data of Hart and Kinloch (1989).

Chromite crystallization is attributed to magma mixing that generates a more silicic melt that can no longer carry high dissolved Cr abundances, thus pushing the melt into the field of chromite precipitation. The variable initial Os ratios of the critical zone chromitites suggest that two magmas mixed several times in the history of the magma chamber, in different proportions. The magma mixing process probably also pushed the more silicic product melt into the sulphide immiscibility field, causing the separation of sulphide droplets. The chromite grains and sulphide droplets would then have settled at the same time, forming cumulate

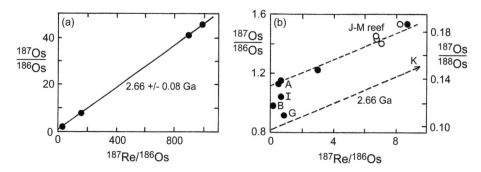

Fig. 8.23 Re–Os isochron diagrams for whole-rock samples from the Stillwater Complex. (a) Rhenium-rich samples used to construct an isochron. (b) Re-poor samples, including the J–M reef (○), chromitite bands A to K, and un-named chromitites (●). After Lambert *et al.* (1989).

chromites with interstitial sulphide. The isotopic composition of the sulphides thus formed might have been quite variable, as they scavenged PGE from different pockets of an isotopically heterogeneous magma. Hence the data of Hart and Kinloch (1989) are reasonably explained within the overall framework of the magma mixing model.

In spite of the success of the magmatic mixing model, Coggon *et al.* (2012) revived the hydrothermal remobilization model on the basis of a Pt–Os isochron age of 1995 ± 50 Ma, determined on PGM from the Merensky reef. They argued that since this age is just outside error of the U–Pb intrusive age, this implies that some PGM were remobilized around 50 Ma after magmatic crystallization. Since Pt/Os ratios were determined by laser ablation microprobe, Coggon *et al.* included 5% uncertainty in Pt/Os ratios (when calculating isochron slopes) to reflect possible elemental fractionation during analysis. However, if Pt/Os ratios have 5% uncertainties due to a possibly systematic fractionation effect, the true uncertainty on the isochron age also cannot be less than 5% (100 Ma). Therefore, it follows that the new Pt–Os age is within error of the intrusive age, and does not provide evidence of hydrothermal PGE remobilization.

8.4.2 The Stillwater Complex

The Stillwater Complex in Montana has similarities to the Bushveld, and also produced Re–Os results which were at first puzzling. As in the Bushveld case, two distinct magmas have been proposed to explain the petrology and chemistry of the pluton. An ultramafic liquid apparently gave rise to the lower Ultramafic Series (UMS), while a magma similar to high-Al basalt formed most of the overlying Banded Series. The PGE-bearing J–M reef is located near the stratigraphic boundary between cumulates from these two different liquids. Hence it is likely that mixing between these magmas caused the segregation of PGE-bearing sulphide droplets to form the reef. However, chromite layers scattered through the ultramafic series may also reflect small influxes of a high-Al basaltic liquid into a magma chamber crystallizing an

ultramafic liquid. These chromitites are identified by letters (A, B, G etc).

In spite of the initial ratio heterogeneity of the complex, Lambert *et al.* (1989) obtained a Re–Os isochron from four Re-rich whole-rock samples, comprising two sulphide-rich cumulates, a bronzite pegmatite and the K-seam chromitite band. The isochron had a low MSWD of 0.03, although this was achieved at the expense of throwing out a fifth data point from a hydrothermally altered harzburgite (Fig. 8.23a). Using a decay constant of 1.64×10^{-11} yr^{-1}, the isochron gave an age of 2.66 ± 0.08 Ga, but recalibration of the spike (Lambert *et al.*, 1994) increased this to 2.83. Nevertheless, the age is revised back down to 2.78 ± 0.08 Ga using the new decay constant of 1.666×10^{-11} yr^{-1}. This result is within error of 2.7 Ga U–Pb and Sm–Nd ages for the intrusion (Section 4.1.2), and suggests that fresh whole-rock samples from the Stillwater Complex generally remained as closed systems for Re and Os during a thermal event which reset Rb–Sr *mineral* systems in the complex.

Analysis of whole-rock samples with low Re/Os ratios by Lambert *et al.* revealed a degree of initial ratio heterogeneity which was similar to the Bushveld (Fig. 8.23b). Chromitite bands from the ultramafic series had a range of initial ^{187}Os/^{188}Os compositions from a late Archean chondritic value of 0.109 to a maximum initial ratio of 0.145 in the J–M reef. Similar results were also obtained on a smaller suite of samples by Martin (1989). Lambert *et al.* attributed the variable initial Os ratios in the chromitites to crustal contamination of mantle-derived magmas by an enriched crustal component, while Martin (1989) attributed isotopic variation in the reefs to variable mixing between chromite cumulates and contaminated intercumulus liquid.

Further study of fresh Stillwater chromites by Marcantonio *et al.* (1993) gave puzzling results. Chromite separates and chromitite whole-rocks from four horizons had initial ^{187}Os/^{188}Os ratios within error of the 'chondritic' mantle ratio of 0.109 at the time of intrusion. On the other hand, samples from the fifth horizon (G chromitite) gave very variable initial ratios, ranging from below the chondritic

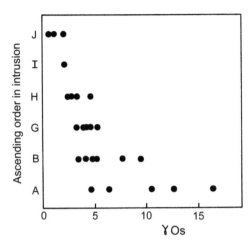

Fig. 8.24 Plot of initial Os isotope ratios (γOs) for massive chromite layers of the Stillwater Complex against stratigraphic height in the ultramafic series (UMS). After Horan et al. (2001).

evolution line to well above it (initial $^{187}Os/^{188}Os = 0.79$) in a molybdenite-bearing sample. The simplest explanation for the latter sample would be later disturbance of the Re–Os system, but the molybdenite from this sample gave a Re–Os age of 2.74 ± 0.08 Ga, suggesting that osmium redistribution occurred soon after emplacement of the complex. Therefore, Marcantonio et al. attributed the elevated initial ratios in this and other chromitite samples to hydrothermal introduction of radiogenic Os immediately after crystallization. They also suggested that the Sm–Nd system might have been similarly upset by hydrothermal remobilization, but this seems extremely unlikely.

Subsequent work by Lambert et al. (1994) and Horan et al. (2001) helped to resolve some of these problems. For example, analysis of additional chromitite samples gave a more coherent picture in which initial osmium isotope ratios of massive chromitites gradually decreased upwards through the ultramafic series (Fig. 8.24). Horan et al. attributed this effect to the mixing of two magmas in changing proportions; an osmium-rich melt with a chondritic isotope signature that was probably derived from a mantle plume, and a second magma probably contaminated by upper crustal rocks. According to this model, the fraction of the second (contaminated) magma must have decreased with time. However, the lack of a complete Nd data set on all samples analysed for Os makes the model speculative.

The J–M reef itself represents a reversal of this trend which is difficult to explain (Lambert et al., 1994). However, as in the Bushveld, it seems most likely that a much larger influx of crustally contaminated melt, mixing with the plume-derived magma, caused major precipitation of immiscible sulphide, which then took on the relatively radiogenic osmium signature of the mixed liquid. Additional Os and Nd analyses on the same samples are needed to test and clarify these models.

8.4.3 The Sudbury Igneous Complex

The Sudbury Igneous Complex (SIC) in Ontario, Canada, is a large mafic body which also hosts the world's largest nickel reserves. However, isotopic evidence indicates a unique origin for the magmas which gave rise to this mineralization. This can be attributed to the genesis of the Sudbury structure in a meteorite impact, a model first proposed by Dietz (1964) and now confirmed by numerous lines of evidence.

Nd isotope data for the silicate rocks of the SIC were presented by Faggart et al. (1985) and Naldrett et al. (1986). Both groups showed that the silicate rocks had a remarkably strong crustal signature, with εNd at 1.85 Ga averaging about −7.5, although Naldrett et al. found a range of ε values from −5 to −9. While Faggart et al. argued that their data could be explained by an exclusively crustal origin for the SIC, Naldrett et al. preferred a model involving gross crustal contamination of a mantle-derived magma. However, because Nd is a lithophile element, this evidence cannot reliably be extrapolated to deduce such an origin for the nickel-bearing sulphide ores of the complex. Furthermore, the enrichment of Nd in crustal relative to mantle-derived melts makes it an insensitive tracer for a small mantle-derived source component, which would tend to be swamped by crustal Nd. In this situation, Os data may be more diagnostic.

Walker et al. (1991) demonstrated approximate agreement between Re–Os isochron ages for sulphide ores and the 1.85 Ga U–Pb age of the silicate rocks (Krogh et al., 1984). This substantiated previous geochemical evidence that the sulphide and silicate melts were co-genetic. However, age correction of measured isotope ratios in ores from the Levack West, Falconbridge and Strathcona mines gave rise to variable initial Os isotope ratios at 1.85 Ga. This was attributed to a heterogeneous magma body, formed by variable mixing between mantle-derived osmium and radiogenic crustal osmium.

In contrast to these results, Dickin et al. (1992) observed relatively good homogeneity of initial osmium isotope ratios for sulphide ores from the Creighton, Falconbridge and Levack West mines. Tails to lower initial ratios in two of these mines, and the large scatter of initial ratios from the Strathcona mine, were attributed to post-intrusive open-system behaviour of the Re–Os system, possibly in response to the Grenville orogeny. The consensus of initial ratios for Sudbury mines falls within the range of estimated crustal compositions at 1.85 Ga, and was attributed by Dickin et al. to an entirely crustal source for osmium in the Sudbury ores. This is consistent with an origin of the SIC as an impact melt sheet (Dietz, 1964). However, no evidence of material contribution from the meteorite itself is seen.

Further work by Cohen et al. (2000), Dickin et al. (2000) and Morgan et al. (2002) has resulted in some convergence between the previous positions. It is now recognized that the complex probably had a 100% crustal origin, but it is also recognized that there is considerable Os isotope heterogeneity in the complex due to incomplete mixing between the melted target rocks. The degree of osmium heterogeneity

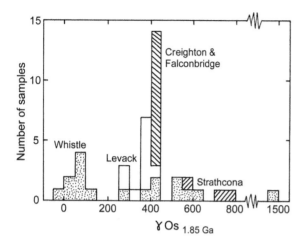

Fig. 8.25 Histogram of initial Os isotopic ratios (γOs) for Sudbury ores at 1.85 Ga. Some mines display homogeneous osmium, while others (e.g. Whistle) are very heterogeneous. After Cohen et al. (2000).

in the complex is demonstrated by the compilation of data in Fig. 8.25. This histogram shows good agreement between the initial ratios of Creighton and Falconbridge ores, but much larger variations in mineralized 'inclusions' from the Whistle mine. These 'inclusions' are mafic–ultramafic xenoliths whose elemental chemistry is indicative of an origin as cumulates from the SIC magma. However, their isotopic signatures are indicative of an origin from locally melted crustal rocks, at least some being of basaltic composition. Hence it appears that the SIC magma was very poorly homogenized at an early stage in its evolution when these cumulates were formed, but became much better homogenized as it cooled and differentiated (Dickin et al., 2000).

The evidence from the Whistle mine for impact melting of mafic as well as felsic crustal rocks is supported by ^{190}Pt–^{186}Os isotope evidence (Morgan et al., 2002). Comparison of initial ^{186}Os and ^{187}Os abundances in three mines revealed a rough inverse correlation, consistent with the mixing of distinct lithologies. Thus, Strathcona samples had radiogenic ^{187}Os but unradiogenic ^{186}Os, indicative of a large component of melted felsic rocks with high Re/Pt ratios; whereas samples from Falconbridge and McCreedy West had less radiogenic ^{187}Os, but more radiogenic ^{186}Os, indicative of a component of melted basic rocks with lower Re/Pt ratios. This mixing model is supported by Sr isotope evidence (e.g. Dickin et al., 2000). However, Pb isotope evidence provides a different slant, by revealing distinct contributions to the impact melt sheet from different crustal depths (Dickin et al., 1996, 2000). Thus, most Pb in North Range ores came from Archean crust, whereas most Pb in South Range ores came from Huronian supracrustals.

Hence it is concluded that the melt sheet formed by the Sudbury meteorite impact was a complex mixture of shock-melted crustal rocks. Mixing of mafic and felsic lithologies probably caused the melt to enter the field of immiscibility between silicate and sulphide melts (Naldrett et al., 1986). This sulphide melt was of crustal origin and probably did not originally contain high levels of PGE. However, the melt must have been in intimate contact with the pool of fused crustal material for a considerable time. During this time, PGE were partitioned from the bulk crustal melt into the sulphide phase in a process analogous to the nickel sulphide fire assay method (e.g. Hofmann et al., 1978). Hence, Sudbury represents 'Nature's largest fire assay'.

8.5 Seawater Osmium

A comparison with Sr isotope systematics suggests that the large differences in osmium isotope composition between crustal and mantle reservoirs should generate large seawater Os isotope variations through time. Seawater itself contains very little osmium, but chemical sediments act to pre-concentrate osmium by scavenging it from seawater, thus reaching quite high abundances. Ravizza and Turekian (1992) showed that this 'hydrogenous' osmium component can be extracted from the substrate by leaching. Hence, they demonstrated that modern seawater has radiogenic $^{187}Os/^{186}Os$ ratios around 8.5 ($^{187}Os/^{188}Os = 1.0$). In contrast, residues from leaching have significantly lower Os isotope ratios due to the presence of a micro-meteorite (cosmic dust) component which is constantly raining down upon the Earth.

It is most convenient to review this subject under the same categories that were used for seawater Sr (Section 3.6). Therefore, we first examine the evidence used to reconstruct a seawater osmium curve, before considering the competing fluxes which cause changes in seawater osmium through time.

8.5.1 Osmium Isotope Evolution

Pegram et al. (1992) analysed leached carbonaceous sediments of various ages in the first study of seawater osmium isotope evolution through the Cenozoic. Osmium was extracted by acid hydrogen peroxide leaching of pelagic black shales from a large piston core recovered from the North Pacific. Pegram et al. interpreted the measured osmium isotope ratios as primary signatures of the sediments, reflecting seawater osmium, and not the product of secondary mixing between re-mobilized terrestrial and meteoritic osmium. Given this assumption, the data implied a sharp increase in seawater osmium isotope composition through the Tertiary period, mimicking the seawater Sr profile for this period (Section 3.6.2).

These preliminary findings were confirmed by Ravizza (1993), who used metalliferous sediments deposited near mid-ocean ridges as recorders of seawater osmium. Because these sediments have greater rates of deposition than the pelagic clays used previously, the fraction of hydrogenous

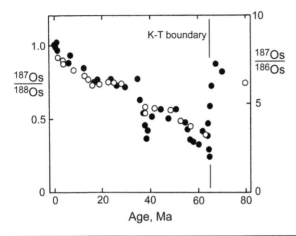

Fig. 8.26 Plot of Os isotope ratio of sediment leachates against age, attributed to increasing seawater Os isotope ratio during the Tertiary. Data sources: (○) = Ravizza (1993) and Peucker-Ehrenbrink *et al.* (1995); (●) = Pegram and Turekian (1999).

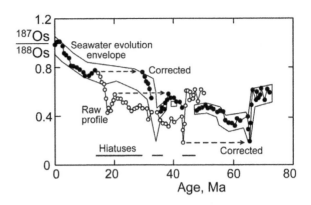

Fig. 8.27 Use of osmium isotope stratigraphy as a dating tool for Fe–Mn crusts, demonstrated for the Tertiary period: (○) = raw data; (●) = corrected for hiatus. After Klemm *et al.* (2005).

(seawater-derived) osmium dominates over the cosmic dust fraction. The results were in good agreement with the leached pelagic clay data, and showed a rapid increase in seawater $^{187}Os/^{188}Os$ over the last 15 Ma, but relatively constant ratios between 18 and 28 Ma (Fig. 8.26). Sr isotope ratios from the same samples were in good agreement with published data, but when compared with osmium, they showed significant decoupling between the two systems. This is not surprising, in view of the very different chemistry of the two elements.

Peucker-Ehrenbrink *et al.* (1995) extended the detailed seawater osmium record to 80 Ma, using a combination of leached and bulk sediment analyses. However, they showed that for slowly deposited pelagic clays, a more gentle leaching procedure was necessary to remove hydrogenous osmium without releasing the cosmic dust fraction. Data from sediments with a relatively high deposition rate were in good agreement between different studies, but data from slowly deposited sediments were less reliable. Leached 80 Ma sediment showed for the first time a radiogenic osmium isotope signature below the K–T boundary (Fig. 8.26). In addition, gently leached samples on either side of the boundary constrained a sharp drop in osmium isotope ratio to the immediate vicinity of the boundary. To explain this dip at the K–T boundary, Peucker-Ehrenbrink *et al.* calculated that a meteorite impact could have released a pulse of dissolved osmium into the oceans equivalent in size to a 5 Ma duration of the global run-off flux (see below).

Data from the above studies were augmented by Pegram and Turekian (1999) in a larger study of the same North Pacific sediment core used by Pegram *et al.* (1992). The K–T boundary horizon was confirmed by an iridium anomaly, and by the presence of impact-derived spherules. Other

points were then interpolated using cobalt accumulation rates (Section 4.5.4), and the accuracy of Os measurements on bulk sediments was improved by correcting for the cosmogenic Os component in the sediment on the basis of ^{3}He measurements. The result was a more detailed profile for the early Tertiary which was largely consistent with earlier work, but also reached to even less radiogenic compositions at the K–T boundary.

The completeness of this core led to its use by Klemm *et al.* (2005) as a dating standard for other Tertiary core sections. They made osmium isotope determinations on a Fe–Mn crust from the central Pacific, which had been dated using the Co accumulation method. They showed that this matched the sedimentary osmium record in Fig. 8.26, except that the Fe–Mn crust had a long hiatus of deposition from ca. 30–15 Ma (Fig. 8.27). Hence they proposed that osmium isotope stratigraphy represents a superior calibration method for dating Fe–Mn crusts beyond the range of the ^{10}Be method. This method promises to improve the calibration of the seawater evolution curve for other isotopic methods such as Nd (Section 4.5.4).

The much shorter seawater residence of osmium relative to strontium (see below) means that seawater osmium is capable of much more rapid variations in response to a variety of environmental impacts. It is not yet possible to produce a detailed osmium isotope stratigraphy for the whole Mesozoic–Tertiary geological record (Fig. 8.28), but detailed studies of osmium isotope stratigraphy have been made in the vicinity of major boundary events or environmental perturbations, a few of which will be mentioned. One of these notable events is the Triassic–Jurassic boundary extinction. This is not known to be associated with a meteorite impact, but to test for possible environmental causes, Cohen and Coe (2002) made a detailed osmium isotope study of a sedimentary succession from SW England that crosses the T–J boundary. They found that a dip in seawater osmium isotope ratio occurred from about 5 m below to 10 m above

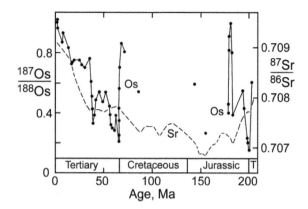

Fig. 8.28 Plot of inferred Os isotope ratios of seawater through the Jurassic to Tertiary periods. T = Triassic. After Cohen (2004).

the period boundary, representing an interval of several Ma. Hence they suggested that this prolonged isotopic minimum was more likely to be due to magmatism marking the onset of Atlantic Ocean rifting, rather than a meteorite impact.

A few Ma later, in the early Jurassic period, an osmium isotope maximum of shorter duration (possibly about 100 ka) is correlated with the Toarcian oceanic anoxic event (Cohen *et al.*, 2004). This maximum (seen in Fig. 8.29) is attributed to a spike in continental weathering rates due to a brief 'greenhouse' event attributed to a tripling of atmospheric CO_2 concentration. An associated very large $\delta^{13}C$ excursion has been attributed to a massive methyl hydrate emission,

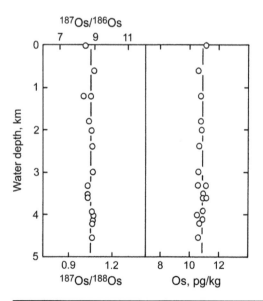

Fig. 8.29 Profiles of (a) osmium isotope ratio, and (b) concentration, in sections through the SW Indian Ocean. After Levasseur *et al.* (1998).

which was rapidly oxidized to CO_2 (Beerling *et al.*, 2002). Enhanced continental weathering apparently acted as a negative feed-back, acting to reduce atmospheric CO_2 and bringing the greenhouse event to an end. Hence, this event represents an important analogue for present day climate perturbations and their possible solution.

Detailed study of the osmium isotope minimum near the Eocene–Oligocene boundary (Fig. 8.29) suggests that this feature actually consists of two different events (Paquay *et al.*, 2014). A very sharp dip at 35.8 Ma (probably lasting less than 100 ka) is attributed to the Popagai impact event, whereas the cause of a more prolonged dip between 34 and 35 Ma remains unknown.

8.5.2 Os Fluxes and Residence Times

In order to understand seawater osmium evolution it is important to quantify the fluxes which control its composition. A comparison between these fluxes and the oceanic osmium budget will then allow calculation of the residence time of osmium in seawater. The principal source of radiogenic osmium is river-borne run-off from old continental crust. On the other hand, possible sources of unradiogenic osmium are low-temperature hydrothermal alteration of ultramafic rocks (Sharma *et al.*, 2000) and the dissolution of cosmic dust. The magnitude of the unradiogenic component is unknown, but its composition is well constrained. Therefore, a major objective has been to constrain the size and composition of the river water budget.

In early work on this subject, Pegram *et al.* (1994) argued that the oxidizing conditions of river water would cause osmium to be adsorbed by ferro-manganese coatings on particulate sediment, rather than remaining in solution. They speculated that when this sediment reached the sea, reducing conditions would break down these Fe–Mn oxides, releasing osmium to seawater. By analysing leaches of river sediment, Pegram *et al.* found very variable $^{187}Os/^{188}Os$ ratios, ranging from 0.17–0.85 in rivers draining ultramafic rocks to values of 1.4–2.8 in more typical drainage basins. However, large rivers such as the Mississippi and Ganges had ratios close to 2.2, a best estimate for the global average. This was substantially more radiogenic than an estimate of 1.26 in average upper crust (Esser and Turekian, 1993).

The first direct measurements of dissolved riverine osmium were made by Sharma and Wasserburg (1997). Because these concentrations are so low, it is convenient to quote them in picogram/kg (10^{-15} g/g). Analysis of four major rivers suggested a concentration range from 2.8 to 8.5 pg/kg and a range of $^{187}Os/^{188}Os$ from 1.2 to 2.0. Hence Sharma and Wasserburg estimated the total riverine supply of dissolved osmium at 320 kg/yr. These results were refined by Levasseur *et al.* (1999), based on 17 of the world's largest rivers. The average dissolved osmium concentration was estimated as 7.9 pg/kg, with a $^{187}Os/^{188}Os$ ratio of 1.5, leading to an estimated global riverine flux of 295 kg/yr, in good agreement with Sharma and Wasserburg (1997). In addition, the flux of non-dissolved Os carried on particulate matter was estimated at less than 25% of the dissolved flux.

The first direct measurements of seawater osmium concentration were made by Sharma et al. (1997), yielding a best estimate of 3.6 pg/kg, with a $^{187}Os/^{188}Os$ ratio of 1.04 ($^{187}Os/^{186}Os = 8.7$). Subsequent work by Levasseur et al. (1998) and Woodhouse et al. (1999) confirmed the mean $^{187}Os/^{188}Os$ ratio of modern seawater as 1.06, but suggested that the concentration determined by Sharma et al. was an underestimate, due to a failure to achieve complete isotopic homogenization between sample and spike osmium. Thus, Levasseur et al. determined a constant concentration of 10.9 pg/kg in a 5 km-deep section from the southwest Indian ridge (Fig. 8.29). A similar experiment by Woodhouse et al. (1999), on a 3 km-deep section from the eastern Pacific, gave identical isotope ratios but suggested some variations in osmium concentration. Below 2 km depth, these ranged from 8.5 to 9.5 pg/kg, but at 500 m depth the concentration dropped as low as 6.5 pg/kg.

A comparison between the isotopic composition of average river water and seawater allows the relative fluxes of riverine osmium and unradiogenic osmium (meteoritic and mantle derived) to be estimated. Estimates of the size of the riverine flux vary from a low of 50% (using the riverine $^{187}Os/^{188}Os$ ratio of 2.2 from Pegram et al., 1994) to a high of 81% (using the value of 1.5 from Sharma and Wasserburg, 1997). The intermediate riverine value of 1.8 from Levasseur et al. (1999) corresponds to a 70% contribution. Ignoring the redissolution of particulate osmium from rivers, this implies a global osmium flux to the oceans (riverine and unradiogenic) of 420 kg/yr.

Assuming that the system is in a steady state condition, comparison of the seawater osmium inventory with the total input flux allows the oceanic residence time of osmium to be calculated. An average seawater concentration of 10 pg/kg, divided by the mass of the oceans (1.4×10^{21} kg) leads to an oceanic osmium inventory of 1.4×10^7 kg. Given a total input flux of 420 kg/yr, this leads to a residence time (τ) of 33 ka. This residence time is near the middle of many estimates made over the past ten years. It is much shorter than that of strontium (Section 3.6.4), but also substantially longer than non-conservative elements such as Nd, Th and Be. In fact it is similar to the time taken for river water to fill the oceans (37 ka).

In spite of the efforts described above to place constraints on global osmium fluxes, there are still major uncertainties. One of these concerns the fate of dissolved osmium in the estuarine and coastal zone. As noted above, Pegram et al. (1994) initially proposed that particulate riverine osmium would be released into solution in estuaries. On the other hand, a study of the Lena River in Siberia (Levasseur et al., 2000) showed that nearly 30% of dissolved riverine osmium was lost from solution in the estuary by adsorption onto suspended particles. If this process occurred on a worldwide scale, the seawater residence time of osmium would have to be increased by 30%. However, a study of the Fly and Sepik

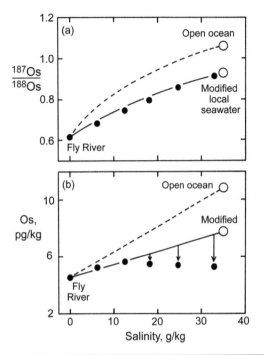

Fig. 8.30 Salinity profiles from the estuary of the Fly River, Papua New Guinea: (a) against isotope ratio; (b) against Os concentration. After Martin et al. (2001).

rivers in Papua New Guinea (Martin et al., 2001) suggested that *both* riverine and seawater were removed by adsorption in the estuary and the near shore zone. This was demonstrated in a 200 km-long 'salinity transect' (Fig. 8.30).

When osmium isotope ratio is plotted against salinity (Fig. 8.30a) it provides evidence of fairly simple mixing between seawater and the unradiogenic osmium carried by the river (reflecting its source in young ultramafic rocks). However, the plot of osmium *concentration* against salinity (Fig. 8.30b) shows that removal of dissolved osmium has occurred, particularly at the distal end of the transect. This implies that seawater osmium may be lost at a similar or greater rate than riverine osmium. The relative magnitude remains unknown, certainly at the global scale, but it suggests that 33 ka may represent a maximum rather than a minimum value for seawater osmium residence.

8.5.3 Quaternary Seawater Osmium

The tentative consensus of seawater osmium residence times at around 30–40 ka was shattered by evidence for very rapid changes in seawater osmium isotope composition during the Quaternary period (Oxburgh, 1998). Two 200 ka sediment cores from the East Pacific Rise showed consistent osmium isotope variations that were also in step with glacial cycles, as represented by oxygen isotope analysis (Fig. 8.31). However, the sharpness of the changes during the last two deglaciations (at ca. 10 and 140 ka BP) implied an extremely short

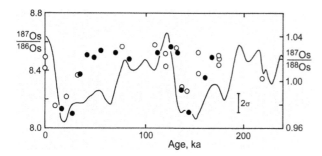

Fig. 8.31 Plot of osmium isotope ratio against age for two sediment cores (●, ○) from the flanks of the East Pacific Rise. These are compared with a record of oxygen isotope variations, representative of glacial cycles. After Oxburgh (1998).

seawater osmium residence time, possibly as low as 3000 yr. Given such a short residence time, the apparent sharp drops in the isotope signature during the last two glacial periods were attributed to a reduction in chemical weathering at these times. However, in order to reconcile these data with the box model estimate of seawater residence, it is necessary to postulate a riverine flux at least three times greater than that estimated by Sharma and Wasserburg (1997) and Levasseur et al. (1999).

More recent data for the Indian Ocean (Burton et al., 2010) gave results largely in agreement with the Pacific data. In addition, most of a record from the Cariaco Basin in the Caribbean was also consistent (Oxburgh et al., 2007). However, the Cariaco Basin gave less radiogenic results from ca. 50 to 70 ka (Fig. 8.32). Based on additional unradiogenic osmium from 10–15 ka, Paquay and Ravizza (2012) attributed unradiogenic osmium data for the Cariaco Basin to a local bias arising from a mafic sedimentary component. They showed that another restricted basin, the Gulf of California,

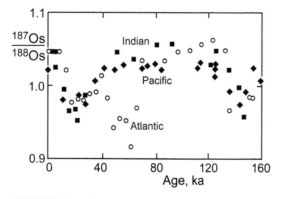

Fig. 8.32 Plot of osmium isotope ratio for three different ocean basins, compared with a record of oxygen isotope variations, representative of glacial cycles. (○) = Atlantic; (♦) = Pacific; (■) = Indian Ocean. After Burton et al. (2010).

also had abnormal osmium isotope signatures, but in this case biased to radiogenic values. However, these perturbations in restricted basins are not surprising if osmium has a short oceanic residence time of the order of 5 ka.

More problematical were new measurements from the East Pacific (Paquay and Ravizza, 2012) that implied a lack of correlation with glacial cycles. These observations led Paquay and Ravizza to suggest that unradiogenic osmium at glacial maxima in the East Pacific data of Oxburgh (1998) might reflect a failure to properly correct for contamination by non-hydrogenous osmium. Because Oxburgh applied these corrections equally through the whole core, she might not have taken account of variations in sediment flux or composition at glacial maxima. Hence, additional data are clearly required from open ocean sites to verify the linkage between osmium and glacial cycles.

References

Alard, O., Luguet, A., Pearson, N. J. et al. (2005). In situ Os isotopes in abyssal peridotites bridge the isotopic gap between MORBs and their source mantle. Nature 436, 1005–8.

Allegre, C. J., Birck, J.-L., Capmas, F. and Courtillot, V. (1999). Age of the Deccan traps using 187Re 187Os systematics. Earth Planet. Sci. Lett. 170, 197–204.

Allegre, C. J. and Luck, J. M. (1980). Osmium isotopes as petrogenetic and geological tracers. Earth Planet. Sci. Lett. 48, 148–54.

Beerling, D. J., Lomas, M. R. and Grocke, D. R. (2002). On the nature of methane gas-hydrate dissociation during the Toarcian and Aptian oceanic anoxic events. Amer. J. Sci. 302, 28–49.

Bennett, V. C., Esat, T. M. and Norman, M. D. (1996). Two mantle-plume components in Hawaiian picrites inferred from correlated Os–Pb isotopes. Nature 381, 221–3.

Bennett, V. C., Nutman, A. P. and Esat, T. M. (2002). Constraints on mantle evolution from 187Os/188Os isotopic compositions of Archean ultramafic rocks from southern West Greenland (3.8 Ga) and Western Australia (3.46 Ga). Geochim. Cosmochim. Acta 66, 2615–30.

Brandon, A. D., Creaser, R. A., Shirey, S. B. and Carlson, R. W. (1996). Osmium recycling in subduction zones. Science 272, 861–4.

Brandon, A. D., Norman, M. D., Walker, R. J. and Morgan, J. W. (1999). 186Os–187Os systematics of Hawaiian picrites. Earth Planet. Sci. Lett. 174, 25–42.

Brandon, A. D., Walker, R. J., Morgan, J. W., Norman, M. D. and Prichard, H. M. (1998). Coupled 186Os and 187Os evidence for core–mantle interaction. Science 280, 1570–3.

Brandon, A. D., Walker, R. J. and Puchtel, I. S. (2006). Platinum–osmium isotope evolution of the Earth's mantle: constraints from chondrites and Os-rich alloys. Geochim. Cosmochim. Acta 70, 2093–103.

Brenan, J. M., Cherniak, D. J. and Rose, L. A. (2000). Diffusion of osmium in pyrrhotite and pyrite: implications for closure of the Re–Os isotopic system. Earth Planet. Sci. Lett. 180, 399–413.

Burton, K. W., Gannoun, A., Birk, J.-L. et al. (2002). The compatability of rhenium and osmium in natural olivine and their behaviour during mantle melting and basalt genesis. Earth Planet. Sci. Lett. 198, 63–76.

Burton, K. W., Gannoun, A. and Parkinson, I. J. (2010). Climate driven glacial–interglacial variations in the osmium isotope composition of seawater recorded by planktic foraminifera. Earth Planet. Sci. Lett. 295, 58–68.

Burton, K. W., Schiano, P., Birck, J.-L. and Allegre, C. J. (1999). Osmium isotope disequilibrium between mantle minerals in a spinel-lherzolite. *Earth Planet. Sci. Lett.* **172**, 311–22.

Class, C. and Goldstein, S. L. (1997). Plume–lithosphere interactions in the ocean basins: constraints from the source mineralogy. *Earth Planet. Sci. Lett.* **150**, 245–60.

Class, C., Goldstein, S. L. and Shirey, S. B. (2009). Osmium isotopes in Grande Comore lavas: a new extreme among a spectrum of EM-type mantle endmembers. *Earth Planet. Sci. Lett.* **284**, 219–27.

Coggon, J. A., Nowell, G. M., Pearson, D. G. *et al.* (2012). The ^{190}Pt–^{186}Os decay system applied to dating platinum-group element mineralization of the Bushveld Complex, South Africa. *Chem. Geol.* **302**, 48–60.

Cohen, A. S. (2004). The rhenium–osmium isotope system: applications to geochronological and palaeoenvironmental problems. *J. Geol. Soc.* **161**, 729–34.

Cohen, A. S., Burnham, O. M., Hawkesworth, C. J. and Lightfoot, P. C. (2000). Pre- emplacement Re–Os ages from ultramafic inclusions in the sublayer of the Sudbury Igneous Complex, Ontario. *Chem. Geol.* **165**, 37–46.

Cohen, A. S. and Coe, A. L. (2002). New geochemical evidence for the onset of volcanism in the Central Atlantic magmatic province and environmental change at the Triassic–Jurassic boundary. *Geology* **30**, 267–70.

Cohen, A. S., Coe, A. L., Bartlett, J. M. and Hawkesworth, C. J. (1999). Precise Re–Os ages of organic-rich mudrocks and the Os isotope composition of Jurassic seawater. *Earth Planet. Sci. Lett.* **167**, 159–73.

Cohen, A. S., Coe, A. L., Harding, S. M. and Schwark, L. (2004). Osmium isotope evidence for the regulation of atmospheric CO_2 by continental weathering. *Geology* **32**, 157–60.

Creaser, R. A., Papanastassiou, D. A. and Wasserburg, G. J. (1991). Negative thermal ion mass spectrometry of osmium, rhenium, and iridium. *Geochim. Cosmochim. Acta* **55**, 397–401.

Creaser, R. A., Sannigrahi, P., Chacko, T. and Selby, D. (2002). Further evaluation of the Re–Os geochronometer in organic-rich sedimentary rocks: a test of hydrocarbon maturations effects in the Exshaw Formation, Western Canada Sedimentary Basin. *Geochim. Cosmochim. Acta* **66**, 3441–52.

Dickin, A. P., Artan, M. A. and Crocket, J. H. (1996). Isotopic evidence for distinct crustal sources of North and South Range ores, Sudbury Igneous Complex. *Geochim. Cosmochim. Acta* 60, 31605–13.

Dickin, A. P., Nguyen, T. and Crocket, J. H. (2000). Isotopic evidence for a single impact melting origin of the Sudbury Igneous Complex. *Geol. Soc. America Spec. Paper* **339**, pp. 361–71.

Dickin, A. P., Richardson, J. M., Crocket, J. H., McNutt, R. H. and Peredery, W. V. (1992). Osmium isotope evidence for a crustal origin of platinum group elements in the Sudbury nickel ore. *Geochim. Cosmochim. Acta* **56**, 3531–7.

Dietz, R. S. (1964). Sudbury structure as an astrobleme. *J. Geol.* **72**, 412–34.

Esser, B. K. and Turekian, K. K. (1993). The osmium isotopic composition of the continental crust. *Geochim. Cosmochim. Acta* **57**, 3093–104.

Faggart, B. E., Basu, A. R. and Tatsumoto, M. (1985). Origin of the Sudbury Complex by meteoritic impact: neodymium isotope evidence. *Science* **230**, 436–9.

Gannoun, A., Burton, K. W., Parkinson, I. J. *et al.* (2007). The scale and origin of the osmium isotope variations in mid-ocean ridge basalts. *Earth Planet. Sci. Lett.* **259**, 541–56.

Gatti, F. *et al.* (2006). *MARE: Microcalorimeter Arrays for a Rhenium Experiment.* 149 pp.

Hart, S. R. and Kinloch, E. D. (1989). Osmium isotope systematics in Witwatersrand and Bushveld ore deposits. *Econ. Geol.* **84**, 1651–5.

Harvey, J., Gannoun, A., Burton, K. W. *et al.* (2006). Ancient melt extraction from the oceanic upper mantle revealed by Re–Os isotopes in abyssal peridotites from the Mid-Atlantic ridge. *Earth Planet. Sci. Lett.* **244**, 606–21.

Hattori, K. and Hart, S. R. (1991). Osmium-isotope ratios of platinum-group minerals associated with ultramafic intrusions: Os-isotopic evolution of the oceanic mantle. *Earth Planet. Sci. Lett.* **107**, 499–514.

Hauri, E. H. and Hart, S. R. (1993). Re–Os isotope systematics of HIMU and EMII oceanic island basalts from the south Pacific Ocean. *Earth Planet. Sci. Lett.* **114**, 353–71.

Hauri, E. H., Lassiter, J. C. and DePaolo, D. J. (1996). Osmium isotope systematics of drilled lavas from Mauna Loa, Hawaii. *J. Geophys. Res.* **101**, 11793–806.

Hirata, T., Hattori, M. and Tanaka, T. (1998). In-situ osmium isotope ratio analyses of iridosmines by laser ablation–multiple collector-inductively coupled plasma mass spectrometry. *Chem. Geol.* **144**, 269–80.

Hirt, B., Tilton, G. R., Herr, W. and Hoffmeister, W. (1963). The half life of 187Re. In: Geiss, J. and Goldberg, E. (Eds) *Earth Science Meteoritics.* North Holland Pub., pp. 273–80.

Hofmann, E. L., Naldrett, A. J., van Loon, J. C., Hancock, R. G. V. and Manson, A. (1978). The determination of all the platinum group elements and gold in rocks and ore by neutron activation analysis after preconcentration by a nickel sulfide fire-assay technique on large samples. *Anal. Chim. Acta* **102**, 157–66.

Horan, M. F., Morgan, J. W., Walker, R. J. and Cooper, R. W. (2001). Re–Os isotopic constraints on magma mixing in the Peridotite Zone of the Stillwater Complex, Montana, USA. *Contrib. Mineral. Petrol.* **141**, 446–57.

Ireland, T. J., Walker, R. J. and Brandon, A. D. (2011). ^{186}Os–^{187}Os systematics of Hawaiian picrites revisited: New insights into Os isotopic variations in ocean island basalts. *Geochim. Cosmochim. Acta* **75**, 4456–75.

Jackson, M. G. and Shirey, S. B. (2011). Re–Os isotope systematics in Samoan shield lavas and the use of Os-isotopes in olivine phenocrysts to determine primary magmatic compositions. *Earth Planet. Sci. Lett.* **312**, 91–101.

Jordan, T. H. (1978). Composition and development of the continental tectosphere. *Nature* **274**, 544–8.

Kirk, J., Ruiz, J., Chesley, J., Walshe, J. and England, G. (2002). A major Archean, gold- and crust-forming event in the Kaapvaal Craton, South Africa. *Science* **297**, 1856–8.

Klemm, V., Levasseur, S., Frank, M., Hein, J. R. and Halliday, A. N. (2005). Osmium isotope stratigraphy of a marine ferromanganese crust. *Earth Planet. Sci. Lett.* **238**, 42–8.

Krogh, T. E., Davis, D. W. and Corfu, F. (1984). Precise U–Pb zircon and baddeleyite ages from the Sudbury area. In: Pye, E. G., Naldrett, A. J. and Giblin, P. E. (Eds) *The Geology and Ore Deposits of the Sudbury Structure. Ont. Geol. Surv. Spec. Pub* Vol. 1, pp. 431–47.

Lambert, D. D., Morgan, J. W., Walker, R. J. *et al.* (1989). Rhenium–osmium and samarium–neodymium isotopic systematics of the Stillwater Complex. *Science* **244**, 1169–74.

Lambert, D. D., Walker, R. J., Morgan, J. W. *et al.* (1994). Re–Os and Sm–Nd isotope geochemistry of the Stillwater Complex, Montana: implications for the petrogenesis of the J–M reef. *J. Petrol.* **35**, 1717–53.

Lassiter, J. C., Byerly, B. L., Snow, J. E. and Hellebrand, E. (2014). Constraints from Os-isotope variations on the origin of Lena Trough abyssal peridotites and implications for the composition and evolution of the depleted upper mantle. *Earth Planet. Sci. Lett.* **403**, 178–87.

Lassiter, J. C. and Hauri, E. H. (1998). Osmium-isotope variations in Hawaiian lavas: evidence for recycled oceanic lithosphere in the Hawaiian plume. *Earth Planet. Sci. Lett.* **164**, 483–94.

Levasseur, S., Birk, J.-L. and Allegre, C. J. (1998). Direct measurement of femtomoles of osmium and the ^{187}Os/^{186}Os ratio in seawater. *Science* **282**, 272-4.

Levasseur, S., Birk, J.-L. and Allegre, C. J. (1999). The osmium riverine flux and the oceanic mass balance of osmium. *Earth Planet. Sci. Lett.* **174**, 7-23.

Levasseur, S., Rachold, V., Birk, J.-L. and Allegre, C. J. (2000). Osmium behaviour in estuaries: the Lena River example. *Earth Planet. Sci. Lett.* **177**, 227-35.

Lindner, M., Leich, D. A., Borg, R. J. et al. (1986). Direct laboratory determination of the ^{187}Re half-life. *Nature* **320**, 246-8.

Lindner, M., Leich, D. A., Russ, G. P., Bazan, J. M. and Borg, R. J. (1989). Direct determination of the half-life of ^{187}Re. *Geochim. Cosmochim. Acta* **53**, 1597-606.

Liu, C. Z., Wu, F. Y., Chung, S. L. et al. (2014). A 'hidden' ^{18}O-enriched reservoir in the sub-arc mantle. *Sci. reports* **4** (4232), 1-6.

Luck, J. M. and Allegre, C. J. (1982). The study of molybdenites through the ^{187}Re-^{187}Os chronometer. *Earth Planet. Sci. Lett.* **61**, 291-6.

Luck, J. M. and Allegre, C. J. (1983). ^{187}Re-^{187}Os systematics in meteorites and cosmochemical consequences. *Nature* **302**, 130-2.

Luck, J. M., Birck, J. L. and Allegre, C. J. (1980). ^{187}Re -^{187}Os systematics in meteorites: early chronology of the solar system and the age of the galaxy. *Nature* **283**, 256-9.

Macfarlane, R. D. and Kohman, T. P. (1961). Natural α radioactivity in medium-heavy elements. *Phys. Rev.* **121**, 1758-69.

Maier, W. D., Barnes, S. J., Campbell, I. H. et al. (2009). Progressive mixing of meteoritic veneer into the early Earth's deep mantle. *Nature* **460**, 620-3.

Marcantonio, F., Zindler, A., Elliot, T. and Staudigel, H. (1995). Os isotope systematics of la Palma, Canary Islands: evidence for recycled crust in the mantle source of HIMU ocean islands. *Earth Planet. Sci. Lett.* **133**, 397-410.

Marcantonio, F., Zindler, A., Reisberg, L. and Mathez, E. A. (1993). Re-Os isotopic systematics in chromitites from the Stillwater Complex, Montana, USA. *Geochim. Cosmochim. Acta* **57**, 4029-37.

Martin, C. E. (1989). Re-Os isotopic investigation of the Stillwater Complex, Montana. *Earth Planet. Sci. Lett.* **93**, 336-44.

Martin, C. E. (1991). Osmium isotopic characteristics of mantle-derived rocks. *Geochim. Cosmochim. Acta* **55**, 1421-34.

Martin, C. E., Peucker-Ehrenbrink, B., Brunskill, G. and Szymczak, R. (2001). Osmium isotope geochemistry of a tropical estuary. *Geochim. Cosmochim. Acta* **65**, 3193-200.

McCandless, T. E., Ruiz, J. R., Adair, B. I. and Freydier, C. (1999). Re-Os isotope and Pd/Ru variations in chromitites from the critical zone, Bushveld Complex, South Africa. *Geochim. Cosmochim. Acta* **63**, 911-23.

McCandless, T. E., Ruiz, J. and Campbell, A. R. (1993). Rhenium behaviour in molybdenite in hypogene and near-surface environments: implications for Re-Os geochronometry. *Geochim. Cosmochim. Acta* **57**, 889-905.

Meisel, T., Walker, R. J. and Morgan, J. W. (1996). The osmium isotopic composition of the Earth's primitive upper mantle. *Nature* **383**, 517-20.

Meisel, T., Walker, R. J., Irving, A. J. and Lorand, J.-P. (2001). Osmium isotopic compositions of mantle xenoliths: a global perspective. *Geochim. Cosmochim. Acta* **65**, 1311-23.

Morgan, J. W. (1985). Osmium isotope constraints on Earth's accretionary history. *Nature* **317**, 703-5.

Morgan, J. W., Walker, R. J. and Grossman, J. N. (1992). Rhenium-osmium isotope systematics in meteorites I: magmatic iron meteorite groups IIAB and IIIAB. *Earth Planet. Sci. Lett.* **108**, 191-202.

Morgan, J. W., Walker, R. J., Horan, M. F., Beary, E. S. and Naldrett, A. J. (2002). ^{190}Pt-^{186}Os and ^{187}Re-^{187}Os systematics of the Sudbury Igneous Complex, Ontario. *Geochim. Cosmochim. Acta* **66**, 273-90.

Naldrett, A. J. (1989). *Magmatic Sulphide Deposits*. Oxford University Press, 186 pp.

Naldrett, A. J., Rao, B. V. and Evensen, N. M. (1986). Contamination at Sudbury and its role in ore formation, In: Gallagher, M. J., Ixer, R. A., Neary, C. R. and Pritchard, H. M. (Eds) *Metallogeny of Basic and Ultrabasic Rocks. Spec. Pub. Inst. Mining & Metall.*, pp. 75-92.

Nomade, S., Renne, P. R. and Merkle, R. K. (2004). ^{40}Ar/^{39}Ar age constraints on ore deposition and cooling of the Bushveld Complex, South Africa. *J. Geol. Soc.* **161**, 411-20.

Oxburgh, R. (1998). Variations in the osmium isotope composition of sea water over the past 200,000 years. *Earth Planet. Sci. Lett.* **159**, 183-91.

Oxburgh, R., Pierson-Wickmann, A. C., Reisberg, L. and Hemming, S. (2007). Climate-correlated variations in seawater ^{187}Os/^{188}Os over the past 200,000 yr: Evidence from the Cariaco Basin, Venezuela. *Earth Planet. Sci. Lett.* **263**, 246-58.

Paquay, F. S. and Ravizza, G. (2012). Heterogeneous seawater ^{187}Os/^{188}Os during the Late Pleistocene glaciations. *Earth Planet. Sci. Lett.* **349**, 126-38.

Paquay, F. S., Ravizza, G. amd Coccioni, R. (2014). The influence of extraterrestrial material on the late Eocene marine Os isotope record. *Geochim. Cosmochim. Acta* **144**, 238-57.

Pearson, N. J., Alard, O., Griffin, W. L., Jackson, S. E. and O'Reilly, S. Y. (2002). In situ measurement of Re-Os isotopes in mantle sulfides by laser ablation multicollector- inductively coupled plasma mass spectrometry: analytical methods and preliminary results. *Geochim. Cosmochim. Acta* **66**, 1037-50.

Pegram, W. J., Esser, B. K., Krishnaswami, S. and Turekian, K. K. (1994). The isotopic composition of leachable osmium from river sediments. *Earth Planet. Sci. Lett.* **128**, 591-9.

Pegram, W. J., Krishnaswami, S., Ravizza, G. E. and Turekian, K. K. (1992). The record of seawater ^{187}Os/^{186}Os variation through the Cenozoic. *Earth Planet. Sci. Lett.* **113**, 569-76.

Pegram, W. J. and Turekian, K. K. (1999). The osmium isotopic composition change of Cenozoic sea water as inferred from deep-sea core corrected for meteoritic contributions. *Geochim. Cosmochim. Acta* **63**, 4053-8.

Peucker-Ehrenbrink, B., Ravizza, G. and Hofmann, A. W. (1995). The marine ^{187}Os/^{186}Os record of the past 80 million years. *Earth Planet. Sci. Lett.* **130**, 155-67.

Puchtel, I. S., Brügmann, G. E. and Hofmann, A. W. (1999). Precise Re-Os mineral isochron and Pb-Nd-Os isotope systematics of a mafic-ultramafic sill in the 2.0 Ga Onega plateau (Baltic Shield). *Earth Planet. Sci. Lett.* **170**, 447-61.

Puchtel, I. S., Brugmann, G. E. and Hofmann, A. W. (2001). ^{187}Os-enriched domain in an Archean mantle plume: evidence from 2.8 Ga komatiites of the Kostomuksha greenstone belt, NW Baltic Shield. *Earth Planet. Sci. Lett.* **186**, 513-26.

Ravizza, G. (1993). Variations of the ^{187}Os/^{186}Os ratio of seawater over the past 28 million years as inferred from metalliferous carbonates. *Earth Planet. Sci. Lett.* **118**, 335-48.

Ravizza, G. E. and Turekian, K. K. (1992). The osmium isotopic composition of organic-rich marine sediments. *Earth Planet. Sci. Lett.* **110**, 1-6.

Reisberg, L. C., Allegre, C. J. and Luck, J. M. (1991). The Re-Os systematics of the Ronda Ultramafic Complex of southern Spain. *Earth Planet. Sci. Lett.* **105**, 196-213.

Reisberg, L. and Lorand, J-P. (1995). Longevity of sub-continental mantle lithosphere from osmium isotope systematics in orogenic peridotite massifs. *Nature* **376**, 159-62.

Reisberg, L. C., Zindler, A., Marcantonio, F. *et al.* (1993). Os isotope systematics in ocean island basalts. *Earth Planet. Sci. Lett.* **120**, 149–67.

Roy-Barman, M. and Allegre, C. J. (1994). ^{187}Os/^{186}Os ratios of mid-ocean ridge basalts and abyssal peridotites. *Geochim. Cosmochim. Acta* **58**, 5043–54.

Roy-Barman, M., Luck, J.-M. and Allegre, C. J. (1996). Os isotopes in orogenic lherzolite massifs and mantle heterogeneities. *Chem. Geol.* **130**, 55–64.

Russ, G. P., Bazan, J. M. and Date, A. R. (1987). Osmium isotopic ratio measurements by inductively coupled plasma source mass spectrometry. *Anal. Chem.* **59**, 984–9.

Schersten, A., Elliott, T., Hawkesworth, C. and Norman, M. (2004). Tungsten isotope evidence that mantle plumes contain no contribution from the Earth's core. *Nature* **427**, 234–7.

Schiano, P., Brick, J.-L. and Allegre, C. J. (1997). Osmium–strontium–neodymium–lead isotopic covariations in mid-ocean ridge basalt glasses and the heterogeneity of the upper mantle. *Earth Planet. Sci. Lett.* **150**, 363–79.

Schoenberg, R., Kruger, F. J., Nagler, T. F., Meisel, T. and Kramers, J. D. (1999). PGE enrichment in chromite layers in the Merensky Reef of the western Bushveld Complex; a Re Os and Rb Sr isotope study. *Earth Planet. Sci. Lett.* **172**, 49–64.

Scoates, J. S. and Friedman, R. M. (2008). Precise age of the platiniferous Merensky Reef, Bushveld Complex, South Africa, by the U–Pb zircon chemical abrasion ID–TIMS technique. *Econ. Geol.* **103**, 465–71.

Selby, D., Creaser, R. A., Hart, C. J. R. *et al.* (2002). Absolute timing of sulfide and gold mineralization: a comparison of Re–Os molybdenite and Ar–Ar mica methods from the Tintina Gold Belt, Alaska. *Geology* **30**, 791–4.

Selby, D., Creaser, R. A., Stein, H. J., Markey, R. J. and Hannah, J. L. (2007). Assessment of the ^{187}Re decay constant by cross calibration of Re–Os molybdenite and U–Pb zircon chronometers in magmatic ore systems. *Geochim. Cosmochim. Acta* **71**, 1999–2013.

Sharma, M., Papanastassiou, D. A. and Wasserburg, G. J. (1997). The concentration and isotopic composition of osmium in the oceans. *Geochim. Cosmochim. Acta* **61**, 3287–99.

Sharma, M. and Wasserburg, G. J. (1997). Osmium in the rivers. *Geochim. Cosmochim. Acta* **61**, 5411–16.

Sharma, M., Wasserburg, G. J., Hofmann, A. W. and Butterfield, D. A. (2000). Osmium isotopes in hydrothermal fluids from the Juan de Fuca Ridge. *Earth Planet. Sci. Lett.* **179**, 139–52.

Sharpe, M. R. (1985). Strontium isotope evidence for preserved density stratification in the main zone of the Bushveld Complex, South Africa. *Nature* **316**, 119–26.

Shen, J. J., Papanastassiou, D. A. and Wasserburg, G. J. (1996). Precise Re–Os determinations and systematics of iron meteorites. *Geochim. Cosmochim. Acta* **60**, 2887–900.

Shirey, S. B. and Walker, R. J. (1995). Carius tube digestion for low-blank Re–Os analyses. *Anal. Chem.* **67**, 2136–41.

Skovgaard, A. C., Storey, M., Baker, J., Blusztajn, J. and Hart, S. (2001). Osmium–oxygen isotopic evidence for a recycled and strongly depleted component in the Iceland mantle plume. *Earth Planet Sci. Lett.* **194**, 259–75.

Smoliar, M. I., Walker, R. J. and Morgan, J. W. (1996). Re–Os ages of group IIA, IIIA, IVA, and IVB iron meteorites. *Science* **271**, 1099–102.

Suzuki, K., Senda, R. and Shimizu, K. (2011). Osmium behavior in a subduction system elucidated from chromian spinel in Bonin Island beach sands. *Geology* **39**, 999–1002.

Suzuki, K., Lu, Q., Shimizu, H. and Masuda, A. (1993). Reliable Re–Os age for molybdenite. *Geochim. Cosmochim. Acta* **57**, 1625–8.

Volkening, J., Walczyk, T. and Heumann, K. G. (1991). Osmium isotope ratio determinations by negative thermal ionization mass spectrometry. *Int. J. Mass Spectrom. Ion Proc.* **105**, 147–59.

Walczyk, T., Hebeda, E. H. and Heumann, K. G. (1991). Osmium isotope ratio measurements by negative thermal ionization mass spectrometry (N–TIMS). *Fres. J. Anal. Chem.* **341**, 537–41.

Walker, R. J., Carlson, R. W., Shirey, S. B. and Boyd, F. R. (1989a). Os, Sr, Nd, and Pb isotope systematics of southern African peridotite xenoliths: implications for the chemical evolution of subcontinental mantle. *Geochim. Cosmochim. Acta* **53**, 1583–95.

Walker, R. J., Horan, M. F., Morgan, J. W. *et al.* (2002). Comparative ^{187}Re–^{187}Os systematics of chondrites: implications regarding early solar system processes. *Geochim. Cosmochim. Acta* **66**, 4187–201.

Walker, R. J. and Morgan, J. W. (1989). Rhenium–osmium isotope systematics of carbonaceous chondrites. *Science* **243**, 519–22.

Walker, R. J., Morgan, J. W. and Horan, M. F. (1995). Osmium-187 enrichment in some plumes: evidence for core–mantle interaction? *Science* **269**, 819–22.

Walker, R. J., Morgan, J. W., Beary, E. S. *et al.* (1997). Applications of the ^{190}Pt–^{186}Os isotope system to geochemistry and cosmochemistry. *Geochim. Cosmochim. Acta* **61**, 4799–807.

Walker, R. J., Morgan, J. W., Naldrett, A. J. and Li, C. (1991). Re–Os isotopic systematics of Ni–Cu sulfide ores, Sudbury Igneous Complex, Ontario: evidence for a major crustal component. *Earth Planet. Sci. Lett.* **105**, 416–29.

Walker, R. J., Shirey, S. B., Hanson, G. N., Rajamani, V. and Horan, M. F. (1989b). Re–Os, Rb–Sr, and O isotopic systematics of the Archean Kolar schist belt, Karnataka, India. *Geochim. Cosmochim. Acta* **53**, 3005–13.

Widom, E. and Shirey, S. B. (1996). Os isotope systematics in the Azores: implications for mantle plume sources. *Earth Planet. Sci. Lett.* **142**, 451–65.

Woodhouse, O. B., Ravizza, G., Falkner, K. K., Statham, P. J. and Peucker-Ehrenbrink, B. (1999). Osmium in seawater: vertical profiles of concentration and isotopic composition in the eastern Pacific Ocean. *Earth Planet. Sci. Lett.* **173**, 223–33.

Chapter 9

The Lu–Hf, Ba–La–Ce and K–Ca Systems

Most of this chapter will be concerned with the Lu–Hf isotope system, which has risen hugely in importance over the past 20 years due to the analytical improvements made possible by MC–ICP–MS (Sections 2.6, 2.7). Brief reviews of the Ba–La–Ce and K–Ca systems complete the toolbox of long-lived lithophile isotope systems used in the earth sciences.

9.1 Lu–Hf Geochronology

Lutetium lies at the end of the lanthanide series as the 'heaviest' of the rare earth elements (REE). It has two isotopes, ^{175}Lu and ^{176}Lu, whose respective abundances are 97.4 and 2.6%. ^{176}Lu displays a branched isobaric decay, by β^- to ^{176}Hf and electron capture to ^{176}Yb. However the latter makes up only a few per cent at most of the total activity and can be more or less ignored (Dixon et al., 1954). ^{176}Hf is left in an excited state after β emission, and decays to the ground state by γ emission. It is one of six isotopes and makes up 5.2% of total hafnium, an element which is not a rare earth but resembles Zr very closely in its crystal chemical behaviour.

The decay scheme

$$^{176}_{71}\text{Lu} \rightarrow {}^{176}_{72}\text{Hf} + \beta^- + v + Q$$

yields a decay equation

$$^{176}\text{Hf} = {}^{176}\text{Hf}_I + {}^{176}\text{Lu}\,(e^{\lambda t} - 1) \qquad [9.1]$$

This is conveniently divided through by ^{177}Hf:

$$\frac{^{176}\text{Hf}}{^{177}\text{Hf}} = \left(\frac{^{176}\text{Hf}}{^{177}\text{Hf}}\right)_I + \frac{^{176}\text{Lu}}{^{177}\text{Hf}}\,(e^{\lambda t} - 1) \qquad [9.2]$$

The first Lu–Hf geochronological measurement was made by Herr et al. (1958), who attempted to determine the half-life of ^{176}Lu by analysing the isotopic composition of Hf in the heavy-REE-rich mineral gadolinite (containing several thousand ppm Lu). However, routine Hf isotope analysis was prevented until 1980 by the difficulties of low blank chemical separation and by the poor ionization efficiency of Hf during thermal ionization mass spectrometry (TIMS). These problems were finally overcome by Patchett and Tatsumoto

(1980a), using a modified form of triple filament analysis with a very hot centre filament (Section 2.2.1). However, TIMS analysis of Hf continued to be limited by poor ionization efficiency, and it has now been superseded by MC–ICP–MS (Section 2.6.2).

9.1.1 The Lu Decay Constant and CHUR Composition

Patchett and Tatsumoto (1980b) presented the first Lu–Hf isochron, based on a suite of eucrite meteorites (achondrites). These have an estimated age of ca. 4.55 Ga, from which Patchett and Tatsumoto were able to make a geological determination of the ^{176}Lu decay constant. Their original ten-point whole-rock isochron was improved by the addition of three extra points (Tatsumoto et al., 1981) to yield a half-life of 35.7 ± 1.2 Ga (equivalent to a decay constant of 1.94×10^{-11} yr^{-1}) and an initial ^{176}Hf/^{177}Hf ratio of 0.27978 ± 9 (2σ), Fig. 9.1. However, it should be noted that the good isochron fit obtained from achondrites was obtained by rejecting one sample, the Antarctic meteorite ALHA (Allan Hills).

The low Hf abundances in chondrites, coupled with the poor efficiency of Hf analysis by TIMS, prevented a direct determination of their Hf isotope composition. Therefore, Patchett and Tatsumoto (1981) determined the composition of the chondritic uniform reservoir (CHUR) from the intersection of the eucrite meteorite isochron and the ^{176}Lu/^{177}Hf ratio of 0.0334 derived from the carbonaceous chondrites Murchison and Allende. This gave a present day chondritic ^{176}Hf/^{177}Hf ratio of 0.28286.

Direct analysis of chondritic Hf was finally made possible by MC–ICP–MS (Blichert-Toft and Albarede, 1997). This yielded a cluster of data points close to the eucrite isochron of Patchett and Tatsumoto, with a somewhat lower CHUR value of 0.28277 ± 3. However, individual chondrite Lu/Hf analyses scattered over a wide range (28%) that was much more than Sm/Nd (4%), leading Bouvier et al. (2008) to suggest that the Lu/Hf system in chondrites was disturbed. This seems likely, since the samples used by Blichert-Toft and Albarede (1997) included 'equilibrated' (metamorphosed) chondrites of petrological types 4 to 6. Therefore, Bouvier

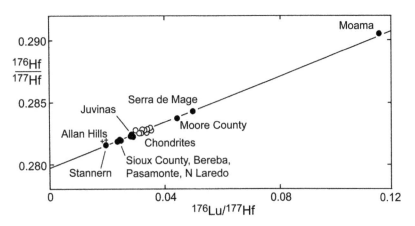

Fig. 9.1 Lu–Hf isochron for eucrite meteorites. Allan Hills samples (+) were omitted from the regression; (○) = chondrites analysed in later work. Modified after Blichert-Toft and Albarede (1997).

et al. made a new study using only un-equilibrated chondrites from petrological types 1 to 3. This gave a narrower range of Lu/Hf ratios, and a more precise CHUR ^{176}Hf/^{177}Hf ratio of 0.282785 ± 11 (2σ).

More recent physical determinations yield significantly higher values of the half-life, averaging 37.3 ± 0.1 Ga (Nir-El and Lavi, 1998), equivalent to a decay constant of 1.86 ± 0.005 × 10^{-11} yr^{-1}. This value was supported by geological half-life determinations based on mineral isochrons on four Proterozoic rock samples (Scherer *et al.*, 2001). Three of these samples were pegmatites, while the fourth was a monazite–xenotime gneiss from Grenville-age rocks of the Hudson Highlands, New York State. All four samples were dated by U–Pb, but the gneissic sample displayed a large amount of geological scatter, so it should probably be excluded from the half-life determination. On the other hand, the three pegmatites gave very consistent results, clustering closely round the decay constant of 1.86 × 10^{-11} yr^{-1} determined by counting experiments. This result was supported by two additional mineral isochrons on mafic minor intrusions (Soderlund *et al.*, 2004), yielding an overall 'terrestrial' decay constant of 1.867 ± 0.013 × 10^{-11} yr^{-1} (2σ).

To test these results, additional work was done on the Lu–Hf systematics of meteorites, but with perplexing results. The first of these studies, by Blichert-Toft *et al.* (2002), re-examined the suite of eucrites analysed by Patchett and Tatsumoto (Fig. 9.1). However, this suite actually comprises two different petrological types, the cumulate eucrites (with Lu/Hf ratios greater than chondrites) and the basaltic eucrites, also called basaltic achondrites (with lower Lu/Hf ratios similar to chondrites). Basaltic achondrites have been successfully dated by Sm–Nd mineral isochrons (Section 4.1.1), but Sm–Nd analysis of the cumulate eucrites shows evidence of major disturbance, especially for Moama, the high Lu/Hf sample that controls the eucrite Lu–Hf isochron in Fig. 9.1. On the other hand, if the Lu–Hf data set is restricted to basaltic eucrites (achondrites), the spread in

Lu/Hf ratios is not sufficient to obtain a precise regression. Therefore, although these data appeared to favour the old half-life, they were not conclusive.

A second study focussed on ordinary chondrites and carbonaceous chondrites, with a few eucrite analyses for comparison (Bizzarro *et al.*, 2003). The ordinary chondrites and basaltic eucrites formed colinear isochrons with a good spread of Lu/Hf ratios, yielding an overall MSWD of 1.04. Using the 'terrestrial' decay constant, this gave an age of 4.84 Ga, 280 Ma older than the accepted age of the solar system (equivalent to a ^{176}Lu decay constant of 1.98 × 10^{-11} yr^{-1}). An even more extreme result was obtained by Thrane *et al.* (2010) on a mineral isochron for the SAH 99555 angrite. This precise isochron (MSWD = 1.3) gave an age of 4.88 Ga using the terrestrial decay constant.

Various alternative explanations have been examined for these puzzling discrepancies. For example, Amelin and Davis (2005) tested whether branched decay of ^{176}Lu to ^{176}Yb (in addition to ^{176}Hf) could explain the old apparent age in meteorites. However they found no evidence for excess ^{176}Yb in old terrestrial minerals.

Alternatively, Albarede *et al.* (2006) suggested that meteorites might have been bombarded with high-energy γ rays in the early solar nebula, causing nuclear transformations of the ^{176}Lu nuclide from the ground state to an excited isomer with an unusual spin state. This spin state prevents the isomer from decaying back to the ground state, and instead it decays by β emission to ^{176}Hf, but with a very short half-life of 3.7 hours, 10^{11} times shorter than the normal half-life of ^{176}Lu. Although this process sounds highly improbable, it has been demonstrated in laboratory experiments, where the half-life of ^{176}Lu was massively shortened by irradiation with 1332 keV γ rays, produced from a ^{60}Co source (Norman *et al.*, 1985). However, Thrane *et al.* (2010) argued that γ rays were not sufficiently penetrative to cause this effect in meteorites, and instead proposed cosmic rays (primarily high-energy protons).

Wimpenny et al. (2015) argued that accelerated decay of [176]Lu should be detectable by creating deficits in the abundance of this isotope relative to stable [175]Lu. Using a new method for Lu isotope analysis by ICP-MS, they presented high-precision Lu data for a variety of terrestrial samples and achondrites, but found no isotope anomalies. Therefore, it was concluded that the accelerated [176]Lu decay model must be rejected.

In fact, Amelin (2005) had already obtained 'normal' Lu–Hf isochrons for phosphate phases (apatite and merrillite) from the Richardton ordinary chondrite and the Acapulco primitive achondrite. Both meteorites formed excellent Lu–Hf mineral isochrons, giving precise Lu decay constants within error of the terrestrial value, when coupled with precise U–Pb ages of 4551 and 4556 Ma on these meteorites. Some workers (e.g. Bizzarro et al., 2012) suggested that the hypothetical [176]Lu irradiation event might have occurred before apatite crystallization, which necessarily represents a late event relative to primary meteorite differentiation. However, Bouvier et al. (2015) and Sanborn et al. (2015) determined relatively normal Lu–Hf isochron ages on several angrites, with some young ages indicative of relatively late metamorphic disturbance. When coupled with the Lu isotope evidence discussed above, this suggests that anomalous Lu–Hf ages on meteorites are due to open-system behaviour rather than enhanced Lu decay.

In addition to their repercussions for the Lu decay constant, the disturbed Lu–Hf systematics in meteorites have had profound effects on the understanding of terrestrial Hf isotope evolution, because they create uncertainty in the initial Hf isotope ratio of the solar system. This problem is in addition to the issues discussed above in determining the present day chondritic composition. These effects are at the level of several epsilon units (Fig. 9.2). Interestingly, the consensus of recent determinations (shown as $\varepsilon = 0$) is very close to the initial ratio originally proposed by Tatsumoto et al. (1981). However, it should be remembered that this was with a different decay constant. Therefore, although the mantle growth curve proposed in the early studies was evidently close to the true value, erroneous age corrections for individual rock samples may have caused them to deviate from chondritic values. These issues will be discussed further below.

9.1.2 Dating Metamorphism

The high Lu/Hf ratios found in garnets make these minerals useful for Lu–Hf dating of metamorphic events in a manner analogous to Sm–Nd. In particular, the spread of Lu/Hf ratios measured in metamorphic garnets is typically greater than for Sm/Nd, and when coupled with the lower half-life of Lu, allows more precise dating of young (Cenozoic) metamorphic events using the Lu–Hf method (Scherer et al., 1997). Applications have included the dating of garnet granulite lower crustal xenoliths (Scherer et al., 1997) and the dating of high-pressure metamorphism in the Alps (Duchene et al., 1997).

Scherer et al. (2000) made a detailed study of the application of Lu–Hf garnet geochronology, including the effects

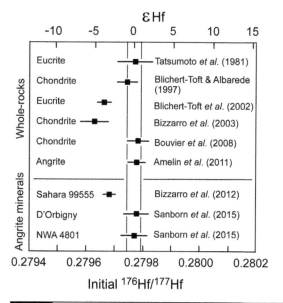

Fig. 9.2 Comparison of solar system initial Hf isotope ratios determined in various studies. Modified after Sanborn et al. (2015).

of Hf-rich accessory minerals and an estimate of the blocking temperature of the Lu–Hf system in garnet. Of the major accessory minerals, apatite, monazite and zircon, only the latter contains high levels of Hf, whereas monazite contains high levels of Nd. This means that comparison of Lu–Hf and Sm–Nd ages can be used to test for the perturbation of garnet–whole-rock isochrons by these minerals. This is important, because zircon may contain as much as 95% of the total Hf inventory of a rock, so that any discordance between zircon and garnet will have a large effect on the Lu–Hf age. Such discordance is quite likely when we are trying to date the growth of metamorphic garnet, because the rock will probably contain pre-metamorphic zircon grains which do not equilibrate with garnet under peak metamorphic conditions.

Two alternative scenarios involving discordant zircon are shown in Fig. 9.3. The first rock (Fig. 9.3a) has zircon grains in the groundmass, but not in the garnet crystals. Comparison between zircon and whole-rock compositions shows that the zircon grains bias the whole-rock point to give an apparent isochron age which is slightly too old. In the other case (Fig. 9.3b), there are zircon inclusions in the garnet as well as zircon in the groundmass. In this case, comparison of leached and unleaded garnets shows that the apparent age will be too young. To avoid these problems, Scherer et al. recommended that inclusion-free garnets should be selected by hand picking. In addition, analysis of other zircon-free minerals such as hornblende or clinopyroxene can be used (instead of the whole-rock point) to avoid the effects of groundmass zircon.

Having applied such corrections, Scherer compared the resulting Lu–Hf ages with Sm–Nd ages on the same samples.

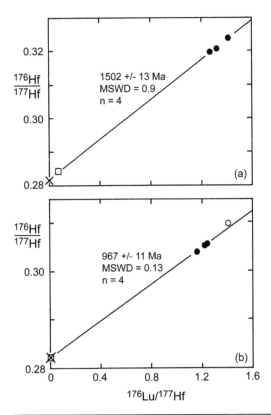

Fig. 9.3 The effects of zircon on Lu–Hf ages of metamorphic garnet growth: (a) involving groundmass zircons only; (b) involving zircon inclusions in garnet grains. (•) = garnet; (○) = leached garnet; (□) = whole-rock; (×) = zircon. After Scherer et al. (2000).

Their published ages should also be corrected for the new decay constant. The result is that the Lu-Hf ages are either within error of or older than the Sm-Nd ages, so the blocking temperature of the Lu-Hf system in garnet appears to be greater or equal to Sm-Nd. However, because metamorphic garnets have large variations in texture and chemistry, both isotope systems will have a fairly wide range of blocking temperatures.

9.2 Modern Mantle Reservoirs

Patchett and Tatsumoto (1980c) made the first Hf isotope measurements on selected MORB and OIB samples previously analysed for $^{143}Nd/^{144}Nd$ and $^{87}Sr/^{86}Sr$. These data (augmented by Patchett, 1983) show that $^{176}Hf/^{177}Hf$ very closely parallels $^{143}Nd/^{144}Nd$ in ocean island basalts (Fig. 9.4). However, MORB samples display a proportionally greater degree of spread in $^{176}Hf/^{177}Hf$ (60% of the total range for oceanic basalts) than $^{143}Nd/^{144}Nd$ (only 30% of the total range). Hence the MORB arrays in Fig. 9.4 are nearly three times steeper than the OIB arrays. Patchett and Tatsumoto attributed these differences to stronger fractionation of Lu/Hf than Sm/Nd and Sr/Rb in very trace-element depleted source regions such as MORB, due to the greater incompatibility displayed by Hf relative to Lu than Nd/Sm or Rb/Sr.

9.2.1 Depleted Mantle

Because Hf is not a true rare earth element its chemistry is not coherent with the heavy rare earths. This is illustrated by the mineral–magma partition coefficients shown in Fig. 9.5. Hence there is an opportunity for more extreme fractionation in Lu/Hf than Sm/Nd when low-degree melting occurs at depths greater than 80 km, within the garnet stability field. Salters and Hart (1991) argued that variable garnet contents in ancient melting events could explain the partial decoupling of Hf-Nd isotope systematics in MORB. When garnet is present in the residue from melting, this residue develops high Lu/Hf ratios, and hence, over time, a radiogenic Hf isotope signature. On the other hand, the melting residues of (garnet-free) spinel peridotites will have lower Lu/Hf ratios, resulting in less radiogenic Hf isotope signatures over geological time.

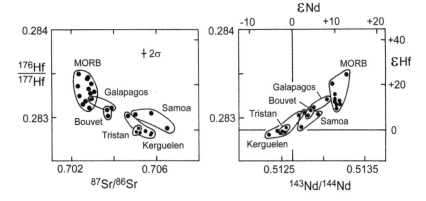

Fig. 9.4 Hf versus Sr and Nd isotope diagrams showing early data for oceanic volcanics. OIB samples define a mantle array, but MORB samples show some decoupling of Hf systematics from Sr and Nd. After Patchett (1983).

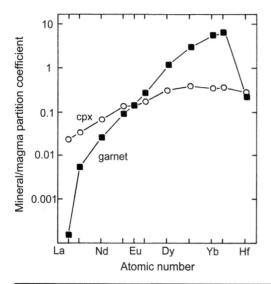

Fig. 9.5 Partition coefficients for REE and Hf between minerals (cpx, garnet) and kimberlite magma. After Fujimaki et al. (1984).

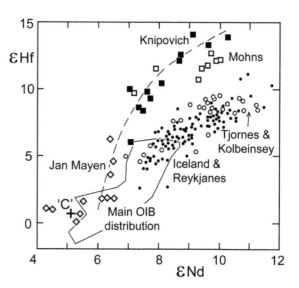

Fig. 9.6 Hf versus Nd isotope plot for MORB suites from North Atlantic ridge segments relative to the common component 'C' of Hanan and Graham (1996). Dashed line = possible mixing line between depleted mantle and plume sources. Modified after Blichert-Toft et al. (2005).

The advent of MC–ICP–MS offered the opportunity to test the earlier work on MORB samples by the analysis of larger sample suites at levels of precision nearly an order of magnitude better. In particular, the analysis of suites of samples from single ridge segments revealed a series of sub-parallel arrays on the Hf–Nd isotope diagram (Blichert-Toft et al., 2005). Some of the clearest evidence comes from the North Atlantic (Fig. 9.6). Several ridge segments near Iceland display 'normally' correlated MORB data within the mantle array, but the Knipovich and Mohns ridges in the Arctic sample much more depleted sources (with radiogenic Hf).

Blichert-Toft et al. suggested that this behaviour might be caused by disequilibrium melting, whereby selective melting of ancient garnet grains could release radiogenic Hf from streaks of old lithosphere, possibly of continental affinity. The association of radiogenic Hf signatures with the slow-spreading 'immature' Knipovich and Mohns ridges makes pollution of the mantle source by lithospheric fragments a credible model. This type of pollution of the mantle source may also explain the very radiogenic Hf signatures observed on another part of the Mid Atlantic Ridge (Hamelin et al., 2013), the so-called 'Lucky Strike' ridge segment west of the Canary Islands. This is reasonable, because some aspects of the chemistry of the Canary Islands have previously been attributed to a fragment of sub-continental lithospheric mantle.

Other workers have not accepted the disequilibrium melting model. For example, Salters et al. (2011) argued that mantle temperatures are too high to preserve such isotopic disequilibrium during ridge melting, and also pointed out that garnet and cpx typically melt together, in approximately modal proportions. Instead they attributed the radiogenic

Hf signatures on certain segments to real changes in the regional isotopic signature of the mantle from one segment to another.

Evidence for systematic variation of mantle Hf signatures on a scale of tens to hundreds of km was demonstrated by Graham et al. (2006). They showed that basalts from the southeast Indian Ridge have bimodal Hf isotope signatures, which they attributed to melting of large streaks of mantle with different time-integrated Lu/Hf ratios which have been attenuated by mantle convection over hundreds of Ma (Fig. 9.7). Because Nd isotope compositions do not show the bimodal distribution of the Hf data, it seems likely that the different histories represented by the bimodal source compositions are related to ancient fractionation events controlled by garnet, as suggested above.

Clues about the origins of Hf isotope variations in MORB can be derived from the study of mantle peridotites. The first evidence for large variations in the Hf isotope composition of mantle peridotites was found by Salters and Zindler (1995) in diopside separates from lzerzolite xenoliths of Salt Lake Crater, from the Hawaiian island of Oahu. These minerals displayed very large variations of Hf isotope ratio (ε up to +80), unaccompanied by significant variations in Nd. More recent work by Bizimis et al. (2007) extended this range of εHf values to +115, and showed a reasonable correlation with Lu/Hf ratio. When plotted on an isochron diagram (Fig. 9.8), most of the data cluster round a ca. 1 Ga reference line, showing that the garnet fractionation event that formed the xenoliths was ancient, and cannot be related to the formation of Hawaiian oceanic lithosphere. Therefore, the radiogenic

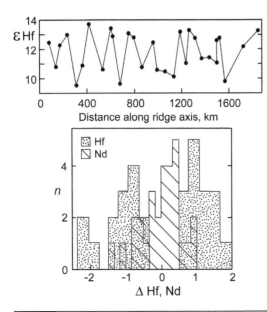

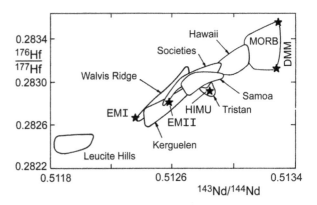

Fig. 9.9 Hf–Nd isotope diagram, showing fields for geochemically important ocean islands, along with the estimated compositions of end-members. The Leucite Hills represent subcontinental lithosphere. Modified after Salters and Hart (1991).

Fig. 9.7 Bimodality of Hf isotope compositions in MORBs from the southeast Indian ridge, seen on a plot of along-ridge distance and in histogram form. After Graham *et al.* (2006).

Hf component is presumed to be derived from the Hawaiian plume.

Whole-rock osmium isotope data on the same samples define unradiogenic signatures in the same range as abyssal peridotites (Section 8.3.4) with rhenium depletion ages up to 2 Ga (average = ca. 1 Ga). Hence these results are consistent with the Hf data in pointing to an ancient fractionation event, in which the xenoliths sampled ancient mantle litho-

sphere, formed as a residue from melting. Bizimis *et al.* suggested that mantle components with such extreme Hf isotope signatures do not give rise to MORB magmas, because they are too depleted to yield magma, except when subjected to very large degree melting. This may occur when large % melting occurs at very shallow depths under ridges, perhaps as seen below the Arctic ridges described above.

The first high-precision Hf isotope determinations on abyssal peridotites were made by Stracke *et al.* (2011). Cpx was separated from abyssal peridotites of the Gakkel Ridge in the Arctic. These gave εHf values of up to +104, again somewhat correlated with Lu/Hf ratio, yielding a mantle differentiation age of ca. 1.2 Ga. These results show that ancient melt-depleted components are present within the upper mantle as well as plumes, although rarely sampled by MORB magmatism.

9.2.2 Enriched Mantle

Hf analyses of OIB samples were presented by Salters and Hart (1991) and Salters and White (1998) in order to establish the locations of the end-member components proposed by Zindler and Hart (Section 6.4.2). The results (Fig. 9.9) showed that EMI and EMII were strongly colinear. On the other hand, the isolation of HIMU below the main trend suggests a possible connection with unradiogenic MORB signatures, attributed by Johnson and Beard (1993) to ancient depleted spinel peridotite. In other words, if we take into account the location of the HIMU field below the main OIB array, the degree of decoupling between Hf and Nd isotope signatures in OIB mirrors that seen in MORB. Hence, the HIMU Hf signature may represent recycled (shallow) oceanic lithosphere.

Another geological environment where Lu/Hf can undergo strong fractionation relative to Sm/Nd is the sedimentary system. Patchett *et al.* (1984) plotted Lu/Hf ratios against Sm/Nd for different marine sediment types

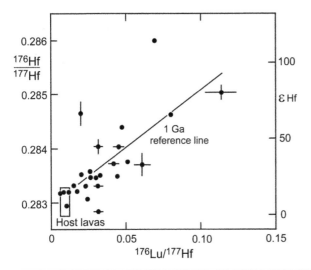

Fig. 9.8 Lu–H isochron diagram for cpx grains separated from mantle xenoliths in Salt Lake Crater, Hawaii. After Bizimis *et al.* (2007).

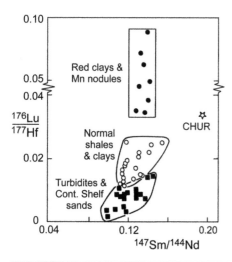

Fig. 9.10 Plot of Lu/Hf versus Sm/Nd ratio in different sediment types, showing that large fractionations in Lu/Hf are not accompanied by significant changes in Sm/Nd. After Patchett *et al.* (1984).

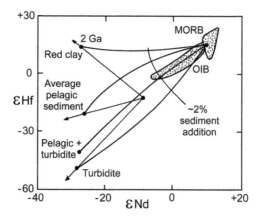

Fig. 9.11 Trends in OIB data predicted to result from the subduction of different types of sediment into the deep mantle, followed by storage for 1–2 Ga. After Patchett *et al.* (1984).

(Fig. 9.10). While ^{147}Sm/^{144}Nd ratios are more or less constant at ca. 0.12–0.14 in most analysed samples, ^{176}Lu/^{177}Hf is strongly fractionated between sandstones and clays. Patchett *et al.* attributed this fractionation to the very strong affinity of Hf for zircon, which, because of its resistance to mechanical and chemical attack, becomes enriched in sand-grade sediments. Hf is correspondingly depleted in the fine-grained clay fraction.

The sorting of marine sediments according to grain size is expected to yield low Lu/Hf sands and turbidites on the continental shelf and continental slope, medium Lu/Hf in shales and clays, and very high Lu/Hf in red clays and Mn nodules deposited in the deep ocean where terrigenous sediment is lacking. These variations in sediment Lu/Hf ratios were in turn predicted to generate large variations in Hf isotope ratio after recycling and storage in deep mantle reservoirs. Hence, Patchett *et al.* (1984) calculated that any individual sediment type subducted (e.g. red clay, average pelagic sediment, or turbidite) would yield distinctive isotopic compositions after 2 Ga of residence in the mantle (Fig. 9.11). Since such dramatic divergences are not seen in OIB samples, Patchett *et al.* concluded that the magnitude of sediment recycling into the mantle must be strictly limited.

Island-arc basalts (IAB) offer a means of monitoring the composition of material actually being recycled into the mantle, in order to test theoretical models such as Patchett *et al.* (1984). Hf–Nd isotope data were presented by White and Patchett (1984) for arc basalts sampling depleted and enriched sources (solid symbols in Fig. 9.12). These data fall within the field of OIB samples, showing that old sedimentary material presently being subducted into the mantle has appropriate Hf–Nd systematics to explain the composition of

OIB magmas. Therefore, contrary to the prediction of Patchett *et al.* (1984), this implies that non-extreme mixtures of the different sediment types *can* explain the composition of OIB sources with moderate ease.

This conclusion was supported by MC–ICP–MS analysis of Hf in a much larger suite of sediment samples by Vervoort *et al.* (1999). This work showed that Hf–Nd isotope systematics in the global sedimentary system were more coherent than previously expected. Apart from a very few extreme samples, the vast majority of sediments (of a wide variety of ages) lie along the same trend as the OIB mantle array (Fig. 9.13). This is particularly true for sediments from active margins, which are much more likely to be recycled into the mantle than passive margin sediments. Therefore, it was concluded that Hf isotope data provide a weaker constraint on sediment recycling into the mantle than originally expected, so that models of terrigenous or pelagic sediment recycling into different OIB reservoirs are no longer ruled out.

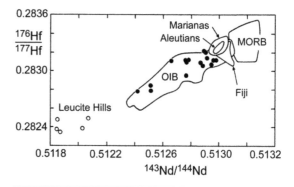

Fig. 9.12 Hf–Nd isotope plot showing the composition of island-arc basalts (solid symbols and marked fields) relative to MORB, OIB and subcontinental lithosphere (represented by the Leucite Hills). After Salters and Hart (1991).

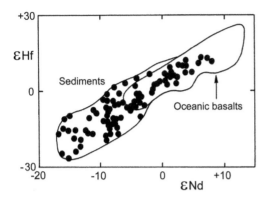

Fig. 9.13 Hf–Nd isotope data for a large suite of sediments of different ages (•) to show variation relative to the field of oceanic volcanics. After Vervoort et al. (1999).

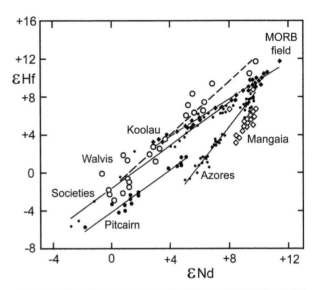

Fig. 9.15 Hf versus Nd isotope plot for several OIB suites showing a series of strong arrays with variable slopes. After Salters et al. (2011).

More recently, a few cases have been found where arc magmas do show Hf isotope variations that are attributed to subduction of variable sediment compositions. One of the strongest examples is the Sunda arc of Indonesia, which is known to span from continental to oceanic basement, moving along the arc from Java to Flores. Handley *et al.* (2011) analysed lavas from the continental segment of the arc in Java, and found variable Hf isotope compositions along this part of the arc. They attributed these variations to changing sediment compositions, with a higher pelagic fraction in the central part of Java (Fig. 9.14). Handley *et al.* also examined the significance of Hf concentration anomalies (Hf/Hf*), which are calculated in a similar way to europium anoma-

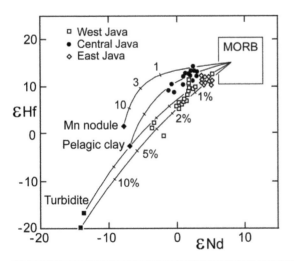

Fig. 9.14 Hf versus Nd isotope plot showing different sedimentary contaminants in arc volcanics in central Java, compared with east and west Java. Mixing lines show % contamination of each sediment type. After Handley *et al.* (2011).

lies. These anomalies have previously been proposed as a possible indicator of subducted sediment type. However, Hf/Hf* was not found to correlate with Hf isotope composition in the Javanese lavas, and is more likely to be controlled by magmatic differentiation processes.

With the advent of more high-precision Hf data sets from MC–ICP–MS, more subtle trends have also been observed within the OIB data set, which can give a better understanding of the origins of OIB sources from different types of recycled material. For example, a detailed Hf isotope study of Hawaiian lavas (Blichert-Toft *et al.*, 1999) revealed a much shallower slope on the Hf–Nd isotope plot than the main mantle array. This shallow array was mainly formed by samples from the Koolau volcano, Oahu, and was attributed to a significant fraction of recycled pelagic sediment in parts of the Hawaiian plume, mixed in different proportions with a component from recycled oceanic lithosphere. Subsequent high-precision Hf work on samples from the Koolau Scientific Drilling Project by Salters *et al.* (2006) confirmed the shallower Hf–Nd slope, but with a less extreme deviation from other OIB suites than originally claimed (Fig. 9.15). Hence Salters *et al.* suggested that this deviation did not require a pelagic sediment component in the Hawaiian plume, but could be explained by recycling of other types of enriched material.

Similar arguments have been made concerning the origins of the EMI mantle component seen at Pitcairn Island. Eisele *et al.* (2002) found a slightly shallower Hf–Nd slope in Pitcairn samples than other OIB (Fig. 9.15), attributing this signature to a subducted pelagic sediment component. This evidence is important, since Pitcairn is the most extreme example of the EMI mantle reservoir (Section 6.5.3). However,

Salters *et al.* (2011) pointed out that Pitcairn actually has the same Hf-Nd isotope slope at the Societies (Fig. 9.15), which are normally attributed to recycled continental sediment. This therefore suggests that shallow Hf-Nd slopes are not a unique signature of a pelagic sediment component. This argument is supported by a comparison with samples from Walvis Ridge, also regarded as an important sampler of the EMI mantle component. The Walvis and Pitcairn arrays converge at an εNd value of around −9 (εHf = −14), similar to the Leucite Hills (Fig. 9.12), suggesting that sub-continental lithosphere is also a viable candidate to give rise to the EMI Hf composition.

9.3 Ancient Hf Evolution

As noted above, the Lu–Hf system displays many similarities to Sm–Nd, and both isotope methods can be used to study ancient crust-mantle processes. Nd analysis is a less demanding method for whole-rock studies, but Hf isotope analysis offers a unique advantage over Sm–Nd for the mineral zircon, which is ideally suited to Hf isotope analysis for several reasons (Patchett *et al.*, 1981):

1. Hf forms an integral part of the zircon lattice, which is therefore very resistant to Hf mobility and contamination.
2. The very high Hf concentrations in zircon (ca. 10 000 ppm) yield very low Lu/Hf ratios and consequently minute age corrections.
3. There are large quantities of zircon separates previously prepared for U–Pb analysis, which yield accurate dates for the same material.
4. Metamorphic overprinting or zircon inheritance from a previous crustal history are clearly revealed by the U–Pb data.

9.3.1 Early Work

Patchett *et al.* (1981) made the first major study of mantle Hf isotope evolution over geological time, using zircon separates to calculate accurate initial ^{176}Hf/^{177}Hf ratios for many rock bodies without making numerous isochron determinations. When these data are plotted on a hafnium isotope evolution diagram, all igneous rocks with a mantle-derived signature lay within error of, or above, the chondrite evolution line (Fig. 9.16). However, only one sample, a metatholeiite dyke cutting the Suomussalmi–Kuhmo greenstone belt in eastern Finland, had an initial ^{176}Hf/^{177}Hf above a linear depleted mantle evolution line drawn from the primordial solar system value to the most radiogenic MORB analysis (Fig. 9.16).

One explanation for the fan of Hf data in Fig. 9.16 is their derivation from a heterogeneous mantle showing variable trace element depletion of Hf relative to Lu through space and time. An alternative would be to derive magmas from a more depleted homogeneous source (such as defined by the

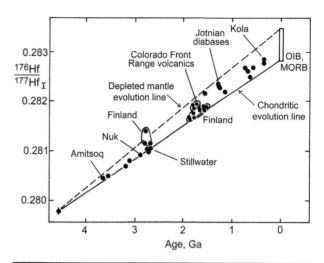

Fig. 9.16 Diagram of Hf isotope evolution over geological time. Initial ratios of uncontaminated mantle-derived magmas show them to be derived from a slightly depleted source relative to chondrites. Data from Patchett *et al.* (1981).

dashed evolution line in Fig. 9.16) and subject them to contamination by older crustal basement. Patchett *et al.* (1981) preferred this model for the 1.4 Ga Silver Plume and 1.0 Ga Pikes Peak batholiths of Colorado, which have Nd *and* Hf initial ratios near the chondritic evolution line, and were argued to contain large fractions of 1.7 Ga crust by DePaolo (Section 4.2.2). Two 1.8 Ga post-tectonic granites intruded into the Archean craton of North Finland had spectacularly low initial ratios, corresponding to εHf values of −10 and −12. These two samples are clearly of crustal derivation on the basis of Pb and Sr isotope data, and were selected to demonstrate the effects of crustal re-working on Hf isotope systematics.

It is important to remember that when this paper was published (1981) the only conclusive Nd isotope evidence for depleted mantle in the Proterozoic was provided by the work of DePaolo on the Front Ranges of Colorado (see Section 4.2.2). With the subsequent determination of strongly depleted εNd signatures for early Archean rocks (Section 4.4.3), another important application of Hf isotope data is to test this Nd evidence.

Nd isotope analysis of Amitsoq gneisses by Bennett *et al.* (Section 4.4.4) had previously implied a wide range of εNd values, based on initial ratios calculated at the U–Pb ages of the rocks. This included some strongly positive epsilon values, suggesting strong mantle depletion in the Early Archean. To test these results, Vervoort *et al.* (1996) analysed zircon separates from a selection of the samples analysed by Bennett *et al.* However, the εHf values fell in a narrower range than εNd, suggesting to Vervoort *et al.* that whole-rock Sm–Nd systems in some of these rocks had probably been disturbed.

In a continuation of this study, Vervoort and Blichert-Toft (1999) analysed Hf in both whole-rock samples and

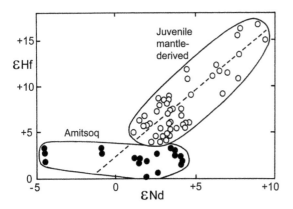

Fig. 9.17 Hf–Nd isotope plot showing initial ratios for Amitsoq gneisses (solid symbols), compared with a variety of younger juvenile samples. After Vervoort and Blichert-Toft (1999).

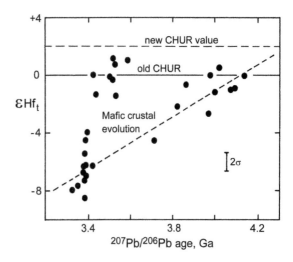

Fig. 9.18 Plot of εHf against 207/206 lead ages for 37 detrital zircon grains from the Jack Hills meta-conglomerate in the Narryer gneiss complex of western Australia. Mafic crustal evolution line has $^{176}Lu/^{177}Hf$ of 0.022. Modified after Amelin et al. (1999).

additional zircon separates. The results bore out the earlier work, showing that whole-rock Hf–Nd isotope systematics in the West Greenland samples depart from the mantle array defined by juvenile samples younger than 3.5 Ga (Fig. 9.17). Hence, the whole-rock Hf isotope data support the earlier data on zircon separates in implying that the Nd isotope system was somewhat disturbed. It should also be noted at this point that these conclusions apply to samples from the heavily reworked southern area of the Itsaq gneiss terrane (as it is now called), as well as the less reworked northern area (see Section 4.4.4).

9.3.2 Detrital Zircon

As noted above, the resistance of zircon to weathering and to metamorphic resetting makes Hf isotope analysis of this mineral a good test for Nd isotope data in studies of early Earth evolution. However, these advantages were particularly enhanced by the development of MC–ICP–MS (Section 2.7.2) which allowed Hf analysis of single zircon grains, which can be individually dated by the U–Pb method. Application of the method will be illustrated on detrital zircons from the Jack Hills meta-conglomerate in western Australia, which have yielded U–Pb ages as old as 4.38 Ga (Holden et al., 2009). These represent the oldest dated terrestrial samples, and offer a window into the earliest evolution of the crust–mantle system.

In the first Hf study of this material, Amelin et al. (1999) analysed 37 whole grains that mostly had less than 10% discordant Pb. The data fell into three groups on a plot of εHf against lead 207/206 ages (Fig. 9.18). However, two of the groups lay on an isotopic growth line (dashed) with a $^{176}Lu/^{177}Hf$ ratio (0.022) typical of mafic rocks. This implies that the zircon grains in these two groups (whose elemental chemistry is indicative of felsic magmas) might be derived from granitoid rocks of various Early Archean ages derived by melting of evolving mafic crust.

Uncertainties in the meteorite Lu–Hf isochrons (Section 9.1.1) had major repercussions for the interpretation of Jack Hills Hf isotope data. These problems do not affect the initial ratios of the analysed zircons, since minimal age corrections are required. However, any errors in the meteorite initial ratio affect the CHUR evolution line for old terrestrial samples. (This does not apply to modern rocks, since the present day chondritic value is well determined.) It now appears that the CHUR initial ratio proposed by Patchett and Tatsumoto (1981) was approximately correct, despite their use of an erroneous decay constant. For comparison, the more recent value of Bouvier et al. (2008) is about two epsilon units higher than the value used by Amelin et al. (Fig. 9.18), while the now discredited CHUR values of Bizzarro et al. (2003, 2012) are about five ε units lower (not shown).

Following the early whole-grain ICP–MS work, another major development in detrital zircon Hf analysis was the introduction of in situ analysis by laser ablation (Harrison et al., 2005). Unfortunately, the first in situ analyses produced many erroneous results, which appeared to provide evidence for very early terrestrial mantle depletion, with radiogenic Hf values far above the expected depleted mantle evolution line (small dots in Fig. 9.19).

Subsequent work (Holden et al., 2009) showed that all of the oldest Jack Hills zircons contained components with multiple ages, such as younger overgrowths on older cores, leading Holden et al. to suggest that these grains represent inherited zircons in Hadean plutonic rocks that were later eroded, sorted and deposited. This makes it difficult to guarantee that the dated age domains are exactly the same age as those analysed for Hf isotope measurement. The Hf data of Harrison et al. (2005) were based on large laser spot sizes,

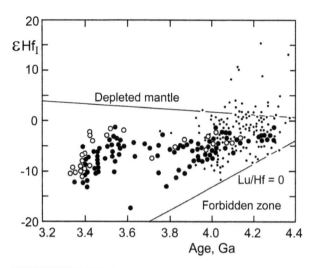

Fig. 9.19 Hf isotope evolution diagram showing Hf analysis for detrital zircons from the Jack Hills, relative to the CHUR value of Bouvier et al. (2008). (○) = Amelin et al. (1999); (•) = 2005–2008 data; (•) = newer in situ data. Modified after Bell et al. (2011).

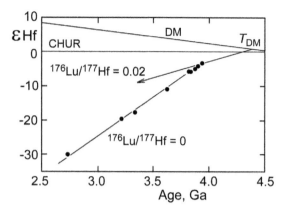

Fig. 9.20 Hf isotope evolution diagram showing calculation of a zircon Hf model age based on a mafic precursor. After Zeh et al. (2014).

which may therefore have sampled Hf from more than one age domain within the analysed grains.

Dissolution analysis of some whole grains appeared to provide support for the large positive Hf anomalies seen in the laser ablation data (Harrison et al., 2005; Blichert-Toft and Albarede, 2008). However, the sequential analysis of Hf and Pb isotope ratios from the same ablation pit (Woodhead et al., 2004; Harrison et al., 2008) yielded only data points less radiogenic than the CHUR evolution line of Bouvier et al. (2008). On the other hand, some of these analyses are also problematical, because they plot on the edge of the forbidden zone (Fig. 9.19), defined by Lu/Hf ratios of zero starting at ca. 4.5 Ga. In contrast, more recent analyses of the oldest zircons fall more centrally between the depleted mantle growth line and the forbidden zone (e.g. Kemp et al., 2010).

9.3.3 Hf Model Ages

Zircon Hf data can be used to calculate crustal formation ages for crustal precursors, but there are several complexities, which can be demonstrated for the Jack Hills zircons. The first issue is the CHUR evolution line. Erroneous values (e.g. Bizzarro et al., 2012) will clearly affect Hf model ages, but the CHUR value of Bouvier et al. (2008) is now generally accepted.

Given this baseline, the next consideration is the depleted mantle evolution line. Many workers (e.g. Bell et al., 2011) have plotted a linear mantle evolution line anchored by the modern DMM composition (depleted MORB mantle). However, it is questionable whether such a depleted mantle model is appropriate for the source of continental crust. Dhuime et al. (2011) argued that the DM line should be anchored by the composition of modern arc basalts, with an

average εHf value of +13.2. They termed this the 'new crust' evolution line, on the basis that this is the mantle composition that generates new arc crust, which is then amalgamated to produce the continents. This argument is very similar to that used by DePaolo (1981) as the basis for T_{DM} Nd model ages, except that DePaolo's DM source is even less depleted over most of Earth history, since it has a concave-upwards form (Section 4.2.2).

Given these parameters, Hf model ages remain more complex than Nd model ages because zircon has a Lu/Hf ratio much lower than the host rock. As a result, simply projecting the Hf growth line of a zircon back to the DM line will give a minimum crustal formation age that normally under-estimates the true age. This is demonstrated by a complexly zoned detrital zircon from a Limpopo Belt quartzite (Zeh et al., 2014). Zones with discordant Pb yielded young 207/206 ages, but similar Hf isotope signatures to other parts of the grain. The resulting Hf growth curve would project to a model age of only 4.1 Ga if extended back to the DM line (Fig. 9.20). In contrast, the most concordant Pb data points, coupled with a $^{176}Lu/^{177}Hf$ of 0.02 characteristic of mafic crust, yield an Hf model age for the crustal precursor of 4.42 Ga, indicative of very early Hadean crust. However, assuming a lower $^{176}Lu/^{177}Hf$ ratio of 0.01 typical of felsic crust would yield a younger Hf model age of around 4.3 Ga.

Ideally, the Lu/Hf ratio of the crust can be constrained by Hf analysis of zircon grains of different ages, forming a crustal evolution line that is then projected back to CHUR or DM. For example, Amelin et al. (1999) showed that most of their concordant whole-grain data were consistent with a high $^{176}Lu/^{177}Hf$ value of 0.022 (Fig. 9.18). When projected back to the new CHUR line, this gives an approximate crustal formation age of 4.4 Ga for the Jack Hills crustal precursor. However, the large number of erroneous Hf initial ratios determined in the early laser ablation work shows the difficulty of achieving accurate U–Pb ages on the same domains analysed for Hf. For example, too young a U–Pb age can lead

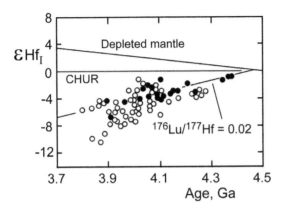

Fig. 9.21 Hf evolution diagram comparing a filtered data set of the least disturbed Jack Hills grains (solid symbols) with more disturbed grains (open). Points with large errors omitted. Modified after Kemp *et al.* (2010).

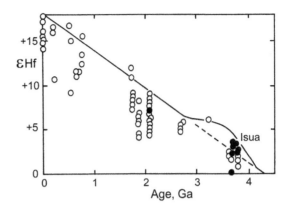

Fig. 9.22 Hf isotope evolution diagram showing inconsistent εHf values in Amitsoq zircons (solid symbols) relative to whole-rock samples. After Vervoort and Blichert-Toft (1999).

to more negative apparent εHf values, implying evolution lines with lower Lu/Hf.

The data set of Kemp *et al.* (2010) allowed a more detailed analysis of this problem. Their entire Jack Hills data of 96 points defined a fairly scattered array on the Hf evolution diagram (Fig. 9.21) with several quite negative εHf values in zircons with younger U–Pb ages. This might imply a low Lu/Hf ratio characteristic of felsic crust. However, zircons with occillatory zoning, believed to indicate the least disturbed igneous grains, gave more consistent results, with a best fit ^{176}Lu/^{177}Hf value of 0.020 (Fig. 9.20). This evolution line gave a T_{CHUR} age for the Jack Hills source of 4.4 Ga. These points fell within error of Hf data for lunar crustal zircons analysed by Taylor *et al.* (2009). Together they represent the oldest crustal nuclei established from cooling of global magma oceans after the giant Moon-forming impact.

9.3.4 Archean Depleted Mantle

With the increasing amount of Hf isotope data for early Archean and Hadean zircons, it is important to revisit the Hf

isotope evidence for early mantle evolution. As noted above (Section 9.3.1), Hf isotope analysis of Amitsoq gneisses from western Greenland helped to refute Nd isotope evidence that had suggested extreme early mantle depletion. On the other hand, early Hf analysis of zircon (Vervoort *et al.*, 1996) also appeared to indicate more depleted mantle signatures than whole-rock Hf analysis (Fig. 9.22). However, much of this discrepancy can be explained by problems with the Lu decay constant. Zircons have Lu/Hf ratios much more divergent from the chondritic evolution line than whole-rock samples, so that zircon εHf values were more affected by errors in the decay constant.

With the adoption of the new decay constant and the CHUR values of Bouvier *et al.* (2008), it is necessary to reexamine this question. New analyses of less altered Amitsoq zircons were also provided by Hiess *et al.* (2009). These new data point to a less depleted Early Archean mantle than the previous work, with a maximum εHf value of around +2, based on laser ablation analysis of the most concordant zones (Fig. 9.23). However, five other samples yielded initial εHf values between +0.5 and −0.1 relative to CHUR, leading Hiess *et al.* to argue that the source of Amitsoq magmas was

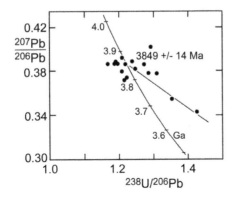

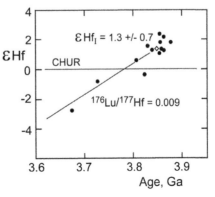

Fig. 9.23 U–Pb and Hf data for an Amitsoq zircon grain. After Hiess *et al.* (2009).

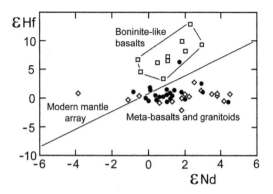

Fig. 9.24 Hf–Nd isotope diagram for metabasic rocks from Isua, western Greenland. After Hoffmann *et al.* (2011).

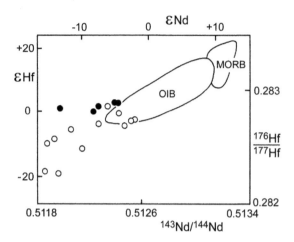

Fig. 9.25 Plot of hafnium isotope compositions against Nd and Sr to show variation on Fe–Mn nodules (•) and marine sediments (○) relative to oceanic volcanics. After White *et al.* (1986).

essentially chondritic. Several other authors (e.g. Kemp *et al.*, 2010; Bell *et al.*, 2011; Vervoort and Kemp, 2016) have likewise down-played the existence of Early Archean depleted mantle Hf signatures.

In contrast to this recent zircon work, Hf analysis of boninite-like metabasalts from the Isua Supracrustal Belt of Western Greenland has yielded very radiogenic Hf compositions far above the DM line (Hoffmann *et al.*, 2011). These authors argued from geochemical indices that these samples were not significantly altered, so that their Hf signatures should be indicative of the composition of real mantle sources. However, the Nd signatures of these samples are relatively normal, reaching εNd values of only +3 (Fig. 9.24). In addition, all tholeiitic metabasalts except one displayed normal εHf signatures clustering on either side of the CHUR line. The fact that one tholeiitic sample displayed elevated Hf is problematical, and suggests that, despite the lack of geochemical indicators, the extreme Hf signatures may be alteration effects. This has been established as the cause of extreme Nd isotope signatures in ancient metabasic samples (Section 4.3.3). In spite of these problems, the tholeiitic metabasalts provide evidence for moderately depleted Hf in the early Earth, consistent with Nd evidence for early mantle depletion.

9.4 Seawater Hafnium

The first investigations of seawater hafnium were made by White *et al.* (1986), based on Hf analysis of four Fe–Mn nodules from the Pacific and one each from the Atlantic and Indian oceans. These were argued to be indicative of the composition of Hf in modern seawater, but in contrast to their wide range of Nd signatures, the nodules were found to have relatively homogeneous εHf, from +0.4 to +3.5 (Fig. 9.25). Because these values were barely outside the large analytical errors of TIMS, White *et al.* attributed them to relatively constant mixing between 'crustal' and 'mantle' sources of dissolved hafnium. Assuming a crustal (riverine) source with

εHf of around −9 and a mantle (island arc or MORB-derived) source with εHf of around +16 implied mixing in approximately equal proportions.

The narrow range of seawater Hf determined by White *et al.* between widely separated end-members implied a long Hf residence time in seawater. This was surprising, since the hafnium concentration in seawater is low, implying a short residence. This enigma could not be solved until the advent of MC–ICP–MS, allowing a 2–4 times improvement in precision on Hf isotope analysis.

Albarede *et al.* (1998) analysed a larger suite of ferromanganese nodules from the Atlantic Ocean, and found that samples with over 7.5 ppm Hf displayed a strong co-variation with Nd isotope composition, strengthening the tentative Hf–Nd correlation previously observed into a clearly defined 'seawater array' (Fig. 9.26). On the other hand, some aberrant signatures in samples with lower Hf contents were attributed to the incorporation of less radiogenic Hf of a detrital or diagenetic origin. The trajectory of the right-hand end of the array is not precisely defined, but it converges on the general field of ocean floor basalts. On the other hand, the left-hand end of the mixing line in Fig. 9.22 points to an end-member distinctly more radiogenic than most sediments.

Based on this work, there was a growing conviction that the left-hand end of the seawater array is controlled by 'incongruent weathering' of crustal material, by which erosion-resistant zircon locks up unradiogenic Hf (reflecting its low Lu/Hf ratios) and hence causes the crustal end-member to have a much more radiogenic Hf signature than the main terrestrial array.

More detailed studies of Fe–Mn crusts forming 'time series' over the last 50 Ma showed a somewhat more complex picture, with less coherent variation of Hf and Nd signatures (Lee *et al.*, 1999; Piotrowski *et al.*, 2000; van de Flierdt

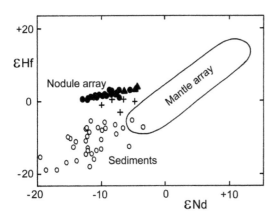

Fig. 9.26 Plot of Hf versus Nd isotope composition for ferromanganese nodules from the Atlantic Ocean (•) and Pacific Ocean (triangles), compared with sedimentary rocks (○) and the oceanic mantle array. (+) = Mn nodules with low Hf contents. After Albarede *et al.* (1998).

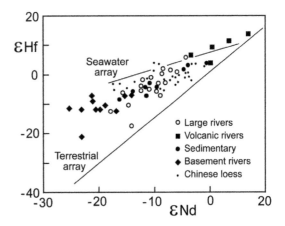

Fig. 9.27 Hf–Nd plot of clay-sized sediment from world rivers relative to the terrestrial and seawater arrays. After Bayon *et al.* (2016).

et al., 2002). However, van de Flierdt *et al.* (2004) emphasized that these effects were subordinate to the overall Hf–Nd correlation in the seawater array, and can be explained by more local effects. For example, glacial erosion causes mechanical breakdown of zircons, releasing a pulse of less radiogenic Hf and temporarily deflecting the riverine Hf component below the main seaweater array (van be Flierdt *et al.*, 2002).

van de Flierdt *et al.* (2004, 2007) argued that the 'mantle' end of the seawater array is also largely 'continental', being derived from the erosion of circum-pacific arc terranes. Because these rocks have much lower zircon contents, weathering of such material would cause much less incongruent Hf erosion, explaining the convergence between the seawater and terrestrial arrays at the right-hand end. This model minimizes the significance of hydrothermal sources of Hf and Nd in ocean water, and also plays down the significance of wind-blown dust, except in the last 3.5 Ma when Chinese loess became a more significant component.

Bau and Koschinsky (2006) attempted to reinstate the hydrothermal source model, arguing that very little of the nominally dissolved Hf budget in seawater is 'truly dissolved'. They accepted that the Nd composition of Fe–Mn crusts is a true reflection of the seawater Nd isotope composition, due to exchange and isotopic homogenization between truly dissolved and colloidal material. However, they argued that Hf is not exchanged in this way, so that Fe–Mn crusts and nodules do not accurately represent the Hf isotope composition of seawater. However, clear evidence that Fe–Mn crusts *are* indicative of seawater Hf was provided by the first Hf isotope analyses of filtered ocean water from the Atlantic and Pacific (Rickli *et al.*, 2009; Zimmermann *et al.*, 2009). These results showed Nd–Hf compositions colinear with the Fe–Mn crust and nodule array, and also extended the crustal end of the array to even less radiogenic compositions.

Based on very low Hf concentrations in Pacific Deep Water (<1 pico-mole/kg), Rickli *et al.* (2009) argued that (unlike Nd) Hf concentrations are not enriched from Atlantic to Pacific along the 'ocean conveyer belt' (Section 14.2). Hence Rickli *et al.* argued that the seawater residence time of Hf is shorter than for Nd, in the range of a few hundred years. However, such a short residence time prevents significant Hf isotope homogenization in the oceans, implying that the narrow range of seawater εHf values reflects isotopically homogenous inputs. This provides strong support for the incongruent weathering model, including the derivation of most Pacific Ocean Hf from arc-like rather than sea floor hydrothermal sources.

In order to understand how incongruent weathering can supply Hf to the oceans, it was necessary to investigate the sediment types that could mediate this process. For example, Rickli *et al.* (2010) analysed the Hf isotope composition of windblown Saharan dust in ocean water of the eastern North Atlantic. They found that bulk sediment Hf was too unradiogenic to give rise to the seawater composition, but argued that partial dissolution could preferentially release more radiogenic Hf. On the other hand, Zhao *et al.* (2015) showed that the finest clay-sized sediment fraction in dust from the Northern Mongolian Plateau forms an array close to the seawater Hf–Nd array, meaning that this could theoretically be a source of radiogenic Hf to the oceans.

To assess the potential of incongruent weathering to supply riverine Hf, Bayon *et al.* (2016) analysed the clay fraction of several major world rivers. They showed that these riverine clays formed a Hf–Nd array close to the seawater array and the Mongolian dust array of Zhao *et al.* (2015), thus demonstrating that fine particulate matter in major rivers can generate the seawater array with only minor additional incongruent Hf dissolution (Fig. 9.27). Notably, the clay-sized array for rivers draining volcanic terranes is colinear with the seawater array, showing that such riverine Hf is fully capable of generating the 'mantle' end of the seawater array.

9.5 The La–Ce and La–Ba Systems

[138]La exhibits branched decay: by β emission to [138]Ce and by electron capture to [138]Ba. The La–Ba–Ce decay scheme is a potentially useful isotopic tracer, and the La–Ba scheme may form a useful geochronometer, but their application has been greatly hindered by the very low abundance of the parent isotope (0.089% of natural lanthanum) and its very long half-life (over 100 Ga, totalled between both routes). Nevertheless, both methods have recently been applied to geological problems.

In addition to the general problems mentioned above, counting determinations of the La decay constants are hampered by the low energy of emitted particles. Hence, early measurements, particularly of the β decay branch, were scattered. To overcome this problem, the counting experiments actually measure the β decay of isomers (excited states) of the product nuclide, rather than the isobaric decay process itself.

A further complication for counting experiments is the hygroscopic nature of La_2O_3, the material usually used in these studies. Transformations to the hydroxide or carbonate result in weight gains of 17% and 2.5% respectively, but in ten out of twelve recent counting determinations on La, no volatile data were reported (Tanaka and Masuda, 1982). However, in spite of a large variation in absolute values, all counting determinations since 1970 yield a ratio for β/electron capture decay constants of near 0.51. In addition, two recent counting experiments on anhydrous La oxide yielded average values for the β decay constant of 2.29 and 2.22 × 10^{-12} yr^{-1} respectively (Sato and Hirose, 1981; Norman and Nelson, 1983). The La–Ce system was the first of the two methods to be applied geologically, but since the La–Ba case is simpler it will be discussed first.

9.5.1 La–Ba Geochronology

[138]Ba, the daughter product of the electron capture decay of [138]La, is also the most abundant isotope of barium, making up 88% of the natural element. In view, therefore, of the very low abundance of the parent nuclide, significant variations in the abundance of [138]Ba are only found in REE-rich and Ba-poor minerals. The first geological measurements were made by Nakai *et al.* (1986) on epidote, allanite and sphene from Precambrian rocks (Fig. 9.28). Nakai *et al.* ratioed [138]Ba against [137]Ba to yield the following decay equation:

$$\frac{^{138}Ba}{^{137}Ba} = \left(\frac{^{138}Ba}{^{137}Ba}\right)_I + \frac{^{138}La}{^{137}Ba} \cdot \frac{\lambda_{E.C.}}{\lambda_{total}} (e^{\lambda\, total\, t} - 1) \quad [9.3]$$

Nakai *et al.* found that all analysed whole-rock samples had [138]Ba/[137]Ba initial ratios within error of 6.3897, attributed to the very low [138]La/[137]Ba ratios of all such materials. Since the Ba isotope ratio of whole-rock systems is effec-

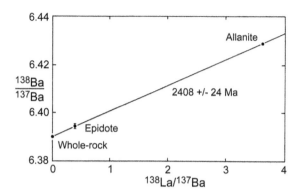

Fig. 9.28 La–Ba isochron diagram for a sample of Amitsoq gneiss, western Greenland. The isochron yields an age of 2408 ± 24 Ma, which dates a metamorphic event rather than the age of the rock. The initial ratio of 6.3897 is invariant in whole-rock systems. After Nakai *et al.* (1986).

tively invariant over time, a La–Ba mineral age can be based simply on the analysis of one or more La-enriched minerals. Using the electron capture decay constant of 4.44 × 10^{-12} yr^{-1} determined by Sato and Hirose (1981), Nakai *et al.* obtained relatively good agreement between the La–Ba and Sm–Nd ages of a pegmatite from Mustikkamaki (Finland) and Amitsoq gneiss from Greenland.

9.5.2 La–Ce Geochronology

The relative harmony between counting and geological determinations of the La electron capture decay branch was not matched by the La β decay route to cerium. This branch is beset by much larger analytical problems, but has more geochemical applications. Tanaka and Masuda (1982) determined the first La–Ce isochron, ratioing [138]Ce against [142]Ce and normalizing to a value of 0.0172 for the [136]Ce/[142]Ce ratio. The decay equation is

$$\frac{^{138}Ce}{^{142}Ce} = \left(\frac{^{138}Ce}{^{142}Ce}\right)_I + \frac{^{138}La}{^{142}Ce} \cdot \frac{\lambda_\beta}{\lambda_{total}} (e^{\lambda\, total\, t} - 1) \quad [9.4]$$

Because the half-lives of both decay branches are so long, they have very little effect on each other. For example, simplifying the equation to

$$\frac{^{138}Ce}{^{142}Ce} = \left(\frac{^{138}Ce}{^{142}Ce}\right)_I + \frac{^{138}La}{^{142}Ce} (e^{\lambda_\beta\, t} - 1) \quad [9.5]$$

only causes a 0.5% over-estimate in age.

Two further technical problems are encountered in Ce isotope analysis. One is the extreme size of the [140]Ce peak relative to the small [136]Ce and [138]Ce peaks (e.g. [140]Ce/[136]Ce = 464.65). Collision of the [140]Ce ion beam with gas molecules in the vacuum system causes down-mass peak tailing whose effect on the small peaks must be carefully corrected, preferably using a polynomial curve to model the shape of the peak tail.

A second major problem is the isobaric interference of ^{138}Ba onto ^{138}Ce. ^{138}Ba is six times more abundant than any other natural Ba isotope that could be used to monitor Ba interference. Therefore any interference correction for ^{138}Ba (even if near zero) will amplify detector noise six-fold. The solution to this problem is to analyse Ce as the oxide species CeO^+. Because barium is divalent, the BaO^+ species is very unfavourable, so, provided that overall Ba levels are kept low by good ion exchange chemistry, the Ba interference can be taken to be zero without correction. Analysing Ce as the oxide introduces other isobaric interference problems, but these are easily overcome by good chemistry (Section 2.1.2). More recent work by Willbold (2007) has also validated TIMS oxide analysis as the best method for Ce isotope analysis.

Tanaka and Masuda (1982) attempted to date separated minerals from the Bushveld pluton, but because of the geological similarity between La and Ce, a limited range of La/Ce ratios was available and the isochron had a large analytical error. Subsequently, Dickin (1987a) was able to determine a La–Ce isochron on a suite of Lewisian whole-rock gneisses from north-west Scotland. Using a decay constant of 2.29×10^{-12} yr^{-1} from a laboratory counting determination (Sato and Hirose, 1981), the La–Ce isochron age of 2.99 Ga was in good agreement with the Lewisian Sm–Nd age of 2.91 Ga (Section 4.1.3). A further geological determination of the La β decay constant (Makishima et al., 1993) has also supported the value of Sato and Hirose. Two La–Ce mineral isochrons were determined on Archean granites from Western Australia, which gave more or less concordant U–Pb zircon and Rb–Sr mineral isochron ages. The granites gave La–Ce mineral ages of 2.76 ± 0.41 and 2.69 ± 0.38 Ga, in good agreement with U–Pb ages of 2.665 and 2.692 Ga, respectively.

Because of the gradual variation of chemical properties along the lanthanide series, chondrite-normalized REE patterns for large rock reservoirs tend to define approximately linear profiles. Tanaka et al. (1987) considered the behaviour of an idealized group of rocks which underwent Ce and Nd isotope evolution starting at the Bulk Earth isotope composition. If these rocks all had linear rare earth profiles, then their Ce–Nd isotope compositions must lie on a linear array whose slope is solely a function of the relative decay constants of the parent nuclides, irrespective of age. Ideally, if such a rock suite were analysed for Ce and Nd isotope composition, the relative decay constants could be calculated without the need to know any concentration or age information.

Tanaka et al. attempted to apply this model in practice, and determine the La β decay constant by the analysis of four unrelated continental rocks from around the world with approximately linear REE profiles. An almost perfect linear array was found, but unfortunately this must be attributed to coincidence, since the calculated initial ratios of these samples are actually *more* dispersed than their present day compositions. Therefore the linearity of this particular data set is coincidental, and its slope cannot be used to determine the La β decay constant. This concept is probably destined to remain a theoretical construct, since the principal difficulty

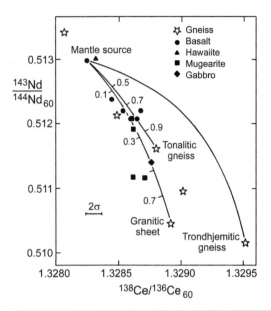

Fig. 9.29   Plot of initial Nd versus Ce isotope composition of Skye lavas (solid symbols), relative to Archean basement gneisses at 60 Ma (stars). The array of lava compositions points to mixing of mantle-derived and granitic crustal melts. After Dickin et al. (1987).

in determining the decay constant is not determination of the La/Ce ratio, but the Ce isotope ratio itself.

9.5.3 Ce Isotope Geochemistry

Since La and Ce are light rare earth elements (LREE), Ce isotope data form a tracer for time-integrated LREE enrichment or depletion of geological reservoirs. Similarly, Nd isotope data are a tracer for time-integrated fractionation between the middle REE. Therefore, a combination of Ce and Nd isotope data should provide a unique control of the time-integrated light-to-middle REE evolution of complex geological reservoirs in the mantle or crust. Unfortunately this has not always been borne out in practice.

A combination of Ce and Nd isotope data was used by Dickin et al. (1987) as a tool to study mixing relations during crustal contamination of continental magmas. Twelve Tertiary igneous rocks from Skye in north-west Scotland were analysed for Ce isotope composition. These were compared with alternative mixing models involving analysed crustal end-members (Fig. 9.29). The mixing line with trondhjemitic gneiss is clearly ruled out by the data, but contamination by tonalitic or granitic gneiss are harder to resolve because the mixing lines are sub-parallel. However, the tonalitic gneiss requires excessive fractions of crustal contamination (see mixing lines), and can therefore be ruled out on these grounds.

A similar approach was used more recently by Bellot et al. (2015) to study crustal contamination processes in Martinique. In this Caribbean island, arc magmas have been

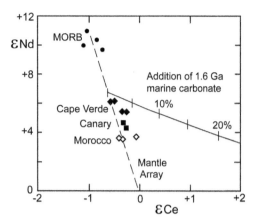

Fig. 9.30 Plot of εNd against εCe (parts per 10⁴ deviation from Bulk Earth) for young oceanic volcanics showing a carbonate contamination model relative to the mantle array. Modified after Doucelance *et al.* (2014).

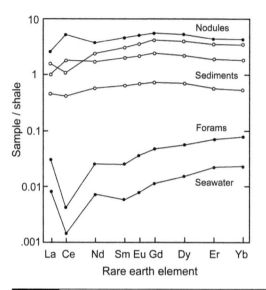

Fig. 9.31 REE profiles for a variety of marine reservoirs. Modified after Elderfield *et al.* (1981).

very strongly contaminated during their ascent through thickened arc crust, leading to well-developed isotopic mixing trends (Section 6.6). These also yield hyperbolic mixing lines for Ce–Nd isotope data.

Attempts were made by Dickin (1987b) and Tanaka *et al.* (1987) to determine the Ce isotope compositions of MORB and OIB samples. However, due to analytical errors, Dickin obtained too shallow a mantle array, while Tanaka *et al.* obtained scattered results. More accurate data for MORB were obtained by Makishima and Masuda (1994), but the Finigan Triton instrument (Section 2.4.3) has since led to ten-times improvements in analytical precision. This has allowed precise analysis of MORB and OIB samples (Doucelance *et al.*, 2014), forming a good quality mantle array (Fig. 9.30). The analysed suite included carbonatites from the Cape Verde Islands, Canaries and Morocco, as well as nearby MORB and Cape Verde basalts. The data were used to test a model of ancient carbonate subduction into the carbonatite magma source (see below).

9.5.4 Seawater Cerium Geochemistry

Cerium has chemical behaviour distinct from the other REE in seawater, which may lead to applications as a tracer of marine sediments. Cerium exists in the 4⁺ state in oxidized terrestrial environments (e.g. Goldberg *et al.*, 1963; Elderfield *et al.*, 1981). The oxidized state is less soluble, causing cerium to be enriched in ferro-manganese crusts and nodules, and correspondingly depleted in seawater and most marine sediments. The resulting 'cerium anomalies' can be portrayed by plotting REE profiles for marine reservoirs (Fig. 9.31). Carbonate sediments show cerium depletion most strongly, because they are directly precipitated from seawater with a minimal clastic component.

Doucelance *et al.* (2014) modelled the effect recycling marine carbonate into the deep mantle, allowing it to age for

1.6 Ga before mixing with recycled oceanic crust. The high La/Ce ratio of the carbonate component leads to radiogenic Ce compositions after storage for long periods of time, producing a Ce–Nd mixing line too shallow to yield the Cape Verde samples. This could imply that recycling of ancient marine carbonates cannot generate the carbonatite source. However, REE analysis of ancient carbonates (Kamber *et al.*, 2014) has shown that these do not display cerium anomalies, due to the coherent behaviour of cerium in seawater prior to oxygenation of the atmosphere. This therefore suggests that Ce–Nd isotope systematics cannot be used to test for ancient carbonate recycling into mantle sources.

The existence of a seawater cerium concentration anomaly does not affect the Ce isotopic composition of modern seawater or marine sediments, because the Ce in these systems has spent most of its history in igneous rocks or minerals, which display smooth REE profiles. Therefore, cerium in the oceans behaves coherently with Nd, and can be used to study sedimentary mixing processes.

Ce–Nd isotope data for Atlantic and Pacific ferromanganese nodules were presented by Tanaka *et al.* (1986) and Amakawa *et al.* (1991). These isotope ratios are argued to be indicative of the composition of the ocean water from which the nodules grew. In the light of Nd and Sr isotope data (Section 4.5), the Ce–Nd data are expected to reflect mixing between continental and MORB-type REE fluxes into the ocean system. However, the data were widely scattered, and did not lie on a single mixing line between reasonable MORB and continental end-members.

New Ce isotope analyses on ocean floor manganese nodules from the Atlantic Ocean (Amakawa *et al.*, 1996) gave values which were much less scattered than previous data. Hence, the scatter of the old data may have been due to

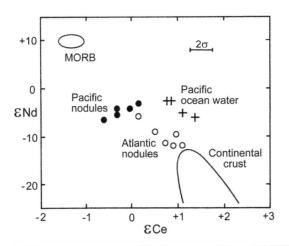

Fig. 9.32 Ce–Nd isotope plot showing data for Pacific (•) and Atlantic (o) manganese nodules attributed to mixing of REE from MORB and continental end-members. + = Pacific seawater. Data from Amakawa *et al.* (1996) and Tazoe *et al.* (2007).

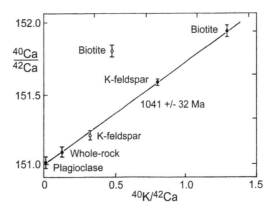

Fig. 9.33 K–Ca isochron plot for separated minerals from the Pikes Peak batholith. Note erroneous results of spiking aliquots (o) rather than the whole dissolution (•). After Marshall and DePaolo (1982).

analytical error. On a plot of εNd versus Ce (Fig. 9.32), the new data fall close to a mixing line between MORB and continental crust, consistent with simple mixing of REE from these sources.

Subsequent analytical advances allowed the first direct Ce isotope analysis of seawater samples by filtering up to 3000 litres of seawater (Tazoe *et al.*, 2007). Excluding one sample from Tokyo Bay, which may show local contamination effects, four samples of Pacific Ocean water showed positive εCe values (+0.9 to +1.4) to the right of the MORB–ontinental mixing trend implied by Fe–Mn nodules (Fig. 9.31). Tazoe *et al.* suggested that these signatures may reflect light REE inputs to seawater from wind-blown dust, whose higher Ce/Nd ratios define a more hyperbolic mixing line.

9.6 The K–Ca System

The K–Ca couple was actually the first isotopic system to be suggested as a geochemical tracer for granite petrogenesis (Holmes, 1932). However, this was on the assumption that the major isotope of potassium, ^{41}K, was the radioactive nuclide. Fortunately this is not really the case or the Earth would have melted from the heat. When it was realized that ^{40}K was actually the radioactive nuclide, the idea of pursuing the K–Ca system was abandoned, since it was anticipated that radiogenic ^{40}Ca would be swamped by the dominant non-radiogenic ^{40}Ca component. The method finally became viable with the development of modern high-precision mass spectrometers, but has not been widely applied.

Russell *et al.* (1978) used Ca isotope analysis to investigate mass-dependent fractionation processes, but the first geochronological application of the method was made by

Marshall and DePaolo (1982). Because of the large relative differences between Ca nuclide masses, isotope ratios must be corrected for natural and instrumental mass fractionation using a more complex procedure than the simple linear law (Section 2.3.1). In practice an exponential mass fractionation correction was used in the two studies mentioned above. Marshall and DePaolo quoted their Ca isotope data as ^{40}Ca/^{42}Ca ratios, corrected to a value of 0.31221 for the non-radiogenic ^{42}Ca/^{44}Ca ratio.

A variety of meteorites, lunar samples and mantle-derived materials was analysed by Russell *et al.* (1978) and Marshall and DePaolo (1982). When age corrected to yield initial Ca isotope ratios at various times between 1.3 and 4.6 Ga, all of the measurements fell within analytical uncertainty of a ^{40}Ca/^{42}Ca ratio of 151.016. This tells us that, because of its very low K/Ca ratio, the Earth's mantle demonstrates negligible growth of radiogenic Ca with time. Rather than quoting raw isotope ratios, Ca isotope compositions can be reported in terms of epsilon units (part per 10^4 deviation from the mantle composition). However this is more a matter of convenience than necessity, in view of the zero Ca isotope evolution of the mantle over time.

Bearing in mind the branched decay of ^{40}K, we can substitute into the general decay equation [1.10] to derive the following isochron equation for the K–Ca system:

$$\frac{^{40}\text{Ca}}{^{42}\text{Ca}} = \left(\frac{^{40}\text{Ca}}{^{42}\text{Ca}}\right)_I + \frac{^{40}\text{K}}{^{42}\text{Ca}} \cdot \frac{\lambda_\beta}{\lambda_{\text{total}}} (e^{\lambda_{\text{total}} t} - 1) \quad [9.6]$$

The branching ratio of β to total decays is 0.8952, and the total decay constant is 5.543×10^{-10} yr^{-1} (Chapter 10).

Marshall and DePaolo tested the K–Ca system as a dating tool by analysing a small suite of separated minerals from the Pikes Peak batholith of Colorado. Plagioclase, whole-rock, K-feldspar and biotite define an isochron array (Fig. 9.33), whose slope yields an age of 1041 ± 32 Ma (2σ). This is within error of other age determinations on this largely

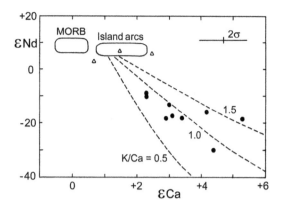

Fig. 9.34 Plot of εNd against εCa showing compositions of Cenozoic granitoids emplaced into young basement (triangles) and old basement (•), relative to island-arc volcanics and MORB. Curves show model K/Ca ratios for basement with $^{147}Sm/^{144}Nd = 0.1$. After Marshall and DePaolo (1989).

un-metamorphosed pluton. The initial ratio of the Pikes Peak batholith (151.024) is within error of the mantle value.

One severe analytical problem that was encountered during this work (other than the mass fractionation behaviour mentioned above) was that samples divided into aliquots before mixing with spike gave erroneous K/Ca ratios. Marshall and DePaolo speculated that this might have been due to partial potassium precipitation from the rock solutions. It is avoided by spiking the whole sample before dissolution.

Marshall and DePaolo (1989) went on to apply the K–Ca method as a petrogenetic tracer in a study of Cenozoic plutons from the western USA. Granites emplaced into Paleozoic crust on the continental margin had a similar range of Ca isotope ratios to island-arc volcanics. The slightly elevated Ca isotope signatures in arc volcanics could reflect Ca-bearing fluids derived from subducted sediment. However, these enrichments are barely outside analytical error (see also Nelson and McCulloch, 1989). In contrast, granites emplaced into Lower Proterozoic basement showed larger Ca isotope enrichments which were correlated with εNd (Fig. 9.34). Hence, the Ca isotope ratios of the plutons must be inherited from the crustal source at depth.

Marshall and DePaolo compared the εCa and εNd compositions of the plutons with crustal evolution models shown by curves in Fig. 9.34. Given a crustal $^{147}Sm/^{144}Nd$ ratio of 0.1 (determined using the known Nd isotope signature of Colorado basement), Ca–Nd isotope evolution curves were drawn for different crustal K/Ca ratios. If we assume the granites to be total crustal melts, then the data in Fig. 9.34 imply a K/Ca ratio of about 1 in the source. However, this is higher than most estimates for bulk crust, so a more likely explanation is that εCa ratios were fractionated during the melting processes by preferential extraction of the most fusible components of the crust.

More recent work on the K–Ca system has shown it to provide a possible complement to Rb–Sr dating of glauconies

(Section 3.5.2) for dating diagenesis (Gopalan, 2008) and as a complement to the Sr isotope analysis of river water (Caro et al., 2010).

References

Albarede, F., Scherer, E. E., Blichert-Toft, J. et al. (2006). γ-ray irradiation in the early Solar System and the conundrum of the ^{176}Lu decay constant. Geochim. Cosmochim. Acta **70**, 1261–70.

Albarede, F., Simonetti, A., Vervoort, J. D., Blichert-Toft, J. and Wafa, A. (1998). A Hf-Nd isotopic correlation in ferromanganese nodules. Geophys. Res. Lett. **25**, 3895–8.

Amakawa, H., Ingri, J., Masuda, A. and Shimizu, H. (1991). Isotopic compositions of Ce, Nd and Sr in ferromanganese nodules from the Pacific and Atlantic Oceans, the Baltic and Barents Seas and the Gulf of Bothnia. Earth Planet. Sci. Lett. **105**, 554–65.

Amakawa, H., Nozaki, Y. and Masuda, A. (1996). Precise determination of variations in the $^{138}Ce/^{142}Ce$ ratios of marine ferromanganese nodules. Chem. Geol. **131**, 183–95.

Amelin, Y. (2005). Meteorite phosphates show constant ^{176}Lu decay rate since 4557 million years ago. Science **310**, 839–41.

Amelin, Y. and Davis, W. J. (2005). Geochemical test for branching decay of ^{176}Lu. Geochim. Cosmochim. Acta **69**, 465–73.

Amelin, Y., Lee, D.-C., Halliday, A. N. and Pidgeon, R. T. (1999). Nature of the Earth's earliest crust from hafnium isotopes in single detrital zircons. Nature **399**, 252–5.

Bau, M. and Koschinsky, A. (2006). Hafnium and neodymium isotopes in seawater and in ferromanganese crusts: the "element perspective". Earth Planet. Sci. Lett. **241**, 952–61.

Bayon, G., Skonieczny, C., Delvigne, C. et al. (2016). Environmental Hf-Nd isotopic decoupling in World river clays. Earth Planet. Sci. Lett. **438**, 25–36.

Bell, E. A., Harrison, T. M., McCulloch, M. T. and Young, E. D. (2011). Early Archean crustal evolution of the Jack Hills Zircon source terrane inferred from Lu-Hf, $^{207}Pb/^{206}Pb$, and $δ^{18}O$ systematics of Jack Hills zircons. Geochim. Cosmochim. Acta **75**, 4816–29.

Bellot, N., Boyet, M., Doucelance, R. et al. (2015). Ce isotope systematics of island arc lavas from the Lesser Antilles. Geochim. Cosmochim. Acta **168**, 261–79.

Bizimis, M., Griselin, M., Lassiter, J. C., Salters, V. J. and Sen, G. (2007). Ancient recycled mantle lithosphere in the Hawaiian plume: osmium–hafnium isotopic evidence from peridotite mantle xenoliths. Earth Planet. Sci. Lett. **257**, 259–73.

Bizzarro, M., Baker, J. A., Haack, H., Ulfbeck, D. and Rosing, M. (2003). Early history of Earth's crust-mantle system inferred from hafnium isotopes in chondrites. Nature **421**, 931–3.

Bizzarro, M., Connelly, J. N., Thrane, K. and Borg, L. E. (2012). Excess hafnium-176 in meteorites and the early Earth zircon record. Geochem. Geophys. Geosys. **13** (3), 1–10.

Blichert-Toft, J., Agranier, A., Andres, M. et al. (2005). Geochemical segmentation of the Mid-Atlantic Ridge north of Iceland and ridge-hot spot interaction in the North Atlantic. Geochem. Geophys. Geosys. **6** (1), 1–27.

Blichert-Toft, J. and Albarede, F. (1997). The Lu-Hf isotope geochemistry of chondrites and the evolution of the mantle–crust system. Earth Planet. Sci. Lett. **148**, 243–58.

Blichert-Toft, J. and Albarede, F. (2008). Hafnium isotopes in Jack Hills zircons and the formation of the Hadean crust. Earth Planet. Sci. Lett. **265**, 686–702.

Blichert-Toft, J., Albarede, F. Rosing, M., Frei, R. and Bridgwater, D. (1999). The Nd and Hf isotopic evolution of the mantle through the Archean. Results from the Isua supracrustals, West Greenland, and from the Birimian terranes of West Africa. *Geochim. Cosmochim. Acta* **63**, 3901–14.

Blichert-Toft, J., Boyet, M., Telouk, P. and Albarede, F. (2002). ^{147}Sm–^{143}Nd and ^{176}Lu–^{176}Hf in eucrites and the differentiation of the HED parent body. *Earth Planet. Sci. Lett.* **204**, 167–81.

Blichert-Toft, J., Frey, F. A. and Albarede, F. (1999). Hf isotope evidence for pelagic sediments in the source of Hawaiian basalts. *Science* **285**, 879–82.

Bouvier, A., Blichert-Toft, J., Boyet, M. and Albarède, F. (2015). ^{147}Sm–^{143}Nd and ^{176}Lu –^{176}Hf systematics of eucrite and angrite meteorites. *Meteoritics Planet. Sci.* **50**, 1896–1911.

Bouvier, A., Vervoort, J. D. and Patchett, P. J. (2008). The Lu-Hf and Sm-Nd isotopic composition of CHUR: constraints from unequilibrated chondrites and implications for the bulk composition of terrestrial planets. *Earth Planet. Sci. Lett.* **273**, 48–57.

Caro, G., Papanastassiou, D. A. and Wasserburg, G. J. (2010). ^{40}K–^{40}Ca isotopic constraints on the oceanic calcium cycle. *Earth Planet. Sci. Lett.* **296**, 124–32.

DePaolo, D. J. (1981). Neodymium isotopes in the Colorado Front Range and crust–mantle evolution in the Proterozoic. *Nature* **291**, 193–6.

Dhuime, B., Hawkesworth, C. and Cawood, P. (2011). When continents formed. *Science* **331**, 154–5.

Dickin, A. P. (1987a). La–Ce dating of Lewisian granulites to constrain the ^{138}La β-decay half-life. *Nature* **325**, 337–8.

Dickin, A. P. (1987b). Cerium isotope geochemistry of ocean island basalts. *Nature* **326**, 283–4.

Dickin, A. P., Jones, N. W., Thirlwall, M. F. and Thompson, R. N. (1987). A Ce/Nd isotope study of crustal contamination processes affecting Palaeocene magmas in Skye, northwest Scotland. *Contrib. Mineral. Petrol.* **96**, 455–64.

Dixon, D., McNair, A. and Curran, S. C. (1954). The natural radioactivity of lutetium. *Phil. Mag.* **45**, 683–4.

Doucelance, R., Bellot, N., Boyet, M., Hammouda, T. and Bosq, C. (2014). What coupled cerium and neodymium isotopes tell us about the deep source of oceanic carbonatites. *Earth and Planetary Science Letters*, **407**, 175–86.

Duchene, S., Blichert-Toft, J., Luais, B., Telouk, P. and Albarede, F. (1997). The Lu-Hf dating of garnets and the ages of the Alpine high-pressure metamorphism. *Nature* **387**, 586–9.

Eisele, J., Sharma, M., Galer, S. J. G. *et al.* (2002). The role of sediment recycling in EM-1 inferred from Os, Pb, Hf, Nd, Sr isotope and trace element systematics of the Pitcairn hotspot. *Earth Planet. Sci. Lett.* **196**, 197–212.

Elderfield, H., Hawkesworth, C. J., Greaves, M. J. and Calvert, S. E. (1981). Rare earth element geochemistry of oceanic ferromanganese nodules and associated sediments. *Geochim. Cosmochim. Acta* **45**, 513–28.

Fujimaki, H., Tatsumoto, M. and Aoki, K. (1984). Partition coefficients of Hf, Zr and REE between phenocryst phases and groundmass. *Proc. 14th Lunar Planet. Sci. Conf., J. Geophys. Res.* **89** (supp.), B662–72.

Goldberg, E. D., Koide, M., Schmitt, R. A. and Smith, R. H. (1963). Rare-earth distributions in the marine environment. *J. Geophys. Res.* **68** (14), 4209–17.

Gopalan, K. (2008). Conjunctive K–Ca and Rb–Sr dating of glauconies. *Chem. Geol.* **247**, 119–23.

Graham, D. W., Blichert-Toft, J., Russo, C. J., Rubin, K. H. and Albarede, F. (2006). Cryptic striations in the upper mantle revealed by hafnium isotopes in southeast Indian ridge basalts. *Nature* **440**, 199–202.

Hamelin, C., Bezos, A., Dosso, L. *et al.* (2013). Atypically depleted upper mantle component revealed by Hf isotopes at Lucky Strike segment. *Chem. Geol.* **341**, 128–39.

Hanan, B. B. and Graham, D. W. (1996). Lead and helium isotope evidence from oceanic basalts for a common deep source of mantle plumes. *Science* **272**, 991–5.

Handley, H. K., Turner, S., Macpherson, C. G., Gertisser, R. and Davidson, J. P. (2011). Hf-Nd isotope and trace element constraints on subduction inputs at island arcs: limitations of Hf anomalies as sediment input indicators. *Earth Planet. Sci. Lett.* **304**, 212–23.

Harrison, T. M., Blichert-Toft, J., Muller, W. *et al.* (2005). Heterogeneous Hadean hafnium: evidence of continental crust at 4.4 to 4.5 Ga. *Science* **310**, 1947–50.

Harrison, T. M., Schmitt, A. K., McCulloch, M. T. and Lovera, O. M. (2008). Early (≥ 4.5 Ga) formation of terrestrial crust: Lu-Hf, δ^{18}O, and Ti thermometry results for Hadean zircons. *Earth Planet. Sci. Lett.* **268**, 476–86.

Herr, W., Merz, E., Eberhardt, P. and Signer, P. (1958). Zur bestimmung der β halbwertszeit des ^{176}Lu durch den nachweis von radiogenem ^{176}Hf. *Z. Natur.* **13a**, 268–73.

Hiess, J., Bennett, V. C., Nutman, A. P. and Williams, I. S. (2009). In situ U–Pb, O and Hf isotopic compositions of zircon and olivine from Eoarchaean rocks, West Greenland: New insights to making old crust. *Geochim. Cosmochim. Acta* **73**, 4489–516.

Hoffmann, J. E., Munker, C., Polat, A., Rosing, M. T. and Schulz, T. (2011). The origin of decoupled Hf–Nd isotope compositions in Eoarchean rocks from southern West Greenland. *Geochim. Cosmochim. Acta* **75**, 6610–28.

Holden, P., Lanc, P., Ireland, T. R. *et al.* (2009). Mass-spectrometric mining of Hadean zircons by automated SHRIMP multi-collector and single-collector U/Pb zircon age dating: the first 100,000 grains. *Int. J. Mass Spec.* **286**, 53–63.

Holmes, A. (1932). The origin of igneous rocks. *Geol. Mag.* **69**, 543–58.

Johnson, C. J. and Beard, B. L. (1993). Evidence from hafnium isotopes for ancient sub-oceanic mantle beneath the Rio Grande rift. *Nature* **362**, 441–4.

Kamber, B. S., Webb, G. E. and Gallagher, M. (2014). The rare earth element signal in Archaean microbial carbonate: information on ocean redox and biogenicity. *J. Geol. Soc.* **171**, 745–63.

Kemp, A. I. S., Wilde, S. A., Hawkesworth, C. J. *et al.* (2010). Hadean crustal evolution revisited: new constraints from Pb-Hf isotope systematics of the Jack Hills zircons. *Earth Planet. Sci. Lett.* **296**, 45–56.

Lee, D.-C., Halliday, A. N., Hein, J. R. *et al.* (1999). Hafnium isotope stratigraphy of ferromanganese crusts. *Science* **285**, 1052–4.

Makishima, A. and Masuda, A. (1994). Ce isotope ratios of N-type MORB. *Chem. Geol.* **118**, 1–8.

Makishima, A., Nakamura, E., Akimoto, S., Campbell, I. H. and Hill, R. I. (1993). New constraints on the ^{138}La β-decay constant based on a geochronological study of granites from the Yilgarn Block, Western Australia. *Chem. Geol. (Isot. Geosci. Sect.)* **104**, 293–300.

Marshall, B. D. and DePaolo, D. J. (1982). Precise age determination and petrogenetic studies using the K–Ca method. *Geochim. Cosmochim. Acta* **46**, 2537–45.

Marshall, B. D. and DePaolo, D. J. (1989). Calcium isotopes in igneous rocks and the origin of granite. *Geochim. Cosmochim. Acta* **53**, 917–22.

Nakai, S., Shimizu, H. and Masuda, A. (1986). A new geochronometer using lanthanum-138. *Nature* **320**, 433–5.

Nelson, D. R. and McCulloch, M. T. (1989). Petrogenetic applications of the ^{40}K–^{40}Ca radiogenic decay scheme – a reconnaissance study. *Chem. Geol. (Isot. Geosci. Sect.)* **79**, 275–93.

Nir-El, Y. and Lavi, N. (1998). Measurement of the half-life of ^{176}Lu. *Appl. Radiat. Isot.* **49**, 1653–5.

Norman, E. B., Bertram, T., Kellogg, S. E., Wong, P. and Gil, S. (1985). Equilibration of ^{176}Lu^{g,m} during the s-process. *Astrophys. J.* **291**, 834–7.

Norman E. B. and Nelson M. A. (1983). Half-life and decay scheme of ^{138}La. *Phys. Rev. C* **27**, 1321–4.

Patchett, P. J. (1983). Hafnium isotope results from Mid-ocean ridges and Kerguelen. *Lithos* **16**, 47–51.

Patchett P. J., Kouvo O., Hedge C. E. and Tatsumoto M. (1981). Evolution of continental crust and mantle heterogeneity: Evidence from Hf isotopes. *Contrib. Mineral. Petrol.* **78**, 279–97.

Patchett P. J. and Tatsumoto M. (1980a). A routine high-precision method for Lu–Hf isotope geochemistry and chronology. *Contrib. Mineral. Petrol.* **75**, 263–7.

Patchett, P. J. and Tatsumoto, M. (1980b). Lu–Hf total-rock isochron for the eucrite meteorites. *Nature* **288**, 571–4.

Patchett P. J. and Tatsumoto M. (1980c). Hafnium isotope variations in oceanic basalts. *Geophys. Res. Lett.* **7**, 1077–80.

Patchett P. J. and Tatsumoto M. (1981). Lu/Hf in chondrites and definition of a chondritic hafnium growth curve. *Lunar Planet. Sci.* **XII**, 822–4.

Patchett P. J., White W. M., Feldmann H., Kielinczuk S. and Hofmann A. W. (1984). Hafnium/rare earth element fractionation in the sedimentary system and crustal recycling into the Earth's mantle. *Earth Planet. Sci. Lett.* **69**, 365–78.

Piotrowski, A. M., Lee, D.-C., Christensen, J. N. *et al.* (2000). Changes in erosion and ocean-circulation recorded in the Hf isotopic compositions of North Atlantic and Indian Ocean ferromanganese crusts. *Earth Planet. Sci. Lett.* **181**, 315–25.

Rickli, J., Frank, M., Baker, A. R. *et al.* (2010). Hafnium and neodymium isotopes in surface waters of the eastern Atlantic Ocean: Implications for sources and inputs of trace metals to the ocean. *Geochim. Cosmochim. Acta* **74**, 540–57.

Rickli, J., Frank, M. and Halliday, A. N. (2009). The hafnium–neodymium isotopic composition of Atlantic seawater. *Earth Planet. Sci. Lett.* **280**, 118–27.

Russell W. A., Papanastassiou D. A. and Tombrello T. A. (1978). Ca isotope fractionation on the Earth and other solar system materials. *Geochim. Cosmochim. Acta* **42**, 1075–90.

Salters, V. J., Blichert-Toft, J., Fekiacova, Z., Sachi-Kocher, A. and Bizimis, M. (2006). Isotope and trace element evidence for depleted lithosphere in the source of enriched Ko'olau basalts. *Contrib. Mineral. Petrol.* **151**, 297–312.

Salters, V. J. and Hart, S. R. (1991). The mantle sources of ocean ridges, islands and arcs: the Hf-isotope connection. *Earth Planet. Sci. Lett.* **104**, 364–80.

Salters, V. J., Mallick, S., Hart, S. R., Langmuir, C. E. and Stracke, A. (2011). Domains of depleted mantle: New evidence from hafnium and neodymium isotopes. *Geochem. Geophys. Geosys.* **12** (8), 1–18.

Salters, V. J. M. and White, W. M. (1998). Hf isotope constraints on mantle evolution. *Chem. Geol.* **145**, 447–60.

Salters, V. J. M. and Zindler, A. (1995). Extreme ^{176}Hf/^{177}Hf in the suboceanic mantle. *Earth Planet. Sci. Lett.* **129**, 13–30.

Sanborn, M. E., Carlson, R. W. and Wadhwa, M. (2015). ^{147,146}Sm–^{143,142}Nd, ^{176}Lu–^{176}Hf, and ^{87}Rb–^{87}Sr systematics in the angrites: Implications for chronology and processes on the angrite parent body. *Geochim. Cosmochim. Acta* **171**, 80–99.

Sato J. and Hirose T. (1981). Half-life of ^{138}La. *Radiochem. Radioanal. Lett.* **46**, 145–52.

Scherer, E. E., Cameron, K. L. and Blichert-Toft, J. (2000). Lu–Hf garnet geochronology: closure temperature relative to the Sm–Nd system and the effects of trace mineral inclusions. *Geochim. Cosmochim. Acta* **64**, 3413–32.

Scherer, E. E., Cameron, K. L., Johnson, C. M. *et al.* (1997). Lu–Hf geochronology applied to dating Cenozoic events affecting lower crustal xenoliths from Kilburn Hole, New Mexico. *Chem. Geol.* **142**, 63–78.

Scherer, E., Munker, C. and Mezger, K. (2001). Calibration of the lutetium-hafnium clock. *Science* **293**, 683–8.

Söderlund, U., Patchett, P. J., Vervoort, J. D. and Isachsen, C. E. (2004). The ^{176}Lu decay constant determined by Lu–Hf and U–Pb isotope systematics of Precambrian mafic intrusions. *Earth Planet. Sci. Lett.* **219**, 311–24.

Stracke, A., Snow, J. E., Hellebrand, E. *et al.* (2011). Abyssal peridotite Hf isotopes identify extreme mantle depletion. *Earth Planet. Sci. Lett.* **308**, 359–68.

Tanaka, T. and Masuda, A. (1982). The La–Ce geochronometer: a new dating method. *Nature* **300**, 515–18.

Tanaka, T., Shimizu, H., Kawata, Y. and Masuda, A. (1987). Combined La–Ce and Sm–Nd isotope systematics in petrogenetic studies. *Nature* **327**, 113–17.

Tanaka, T., Usui, A. and Masuda, A. (1986). Oceanic Ce and continental Nd: multiple sources of REE in oceanic ferromanganese nodules. *Terra Cognita* **6**, 114 (abstract).

Tatsumoto, M., Unruh, D. M. and Patchett, P. J. (1981). U–Pb and Lu–Hf systematics of Antarctic meteorites. *Nat. Inst. Polar Res. Tokyo.* <Au: full reference?>

Taylor, D. J., McKeegan, K. D. and Harrison, T. M. (2009). Lu–Hf zircon evidence for rapid lunar differentiation. *Earth Planet. Sci. Lett.* **279**, 157–64.

Tazoe, H., Obata, H., Amakawa, H., Nozaki, Y. and Gamo, T. (2007). Precise determination of the cerium isotopic compositions of surface seawater in the Northwest Pacific Ocean and Tokyo Bay. *Marine Chem.* **103**, 1–14.

Thrane, K., Connelly, J. N., Bizzarro, M. and Meyer, B. S. (2010). Origin of excess ^{176}Hf in meteorites. *Astrophys. J.* **717**, 861–7.

van de Flierdt, T., Frank, M., Lee, D. C. and Halliday, A. N. (2002). Glacial weathering and the hafnium isotope composition of seawater. *Earth Planet. Sci. Lett.* **201**, 639–47.

van de Flierdt, T., Frank, M., Lee, D. C. *et al.* (2004). New constraints on the sources and behavior of neodymium and hafnium in seawater from Pacific Ocean ferromanganese crusts. *Geochim. Cosmochim. Acta* **68**, 3827–43.

van de Flierdt, T., Goldstein, S. L., Hemming, S. R. *et al.* (2007). Global neodymium–hafnium isotope systematics – revisited. *Earth Planet. Sci. Lett.* **259**, 432–41.

Vervoort, J. D. and Blichert-Toft, J. (1999). Evolution of the depleted mantle: Hf isotope evidence from juvenile rocks through time. *Geochim. Cosmochim. Acta* **63**, 533–56.

Vervoort, J. D. and Kemp, A. I. (2016). Clarifying the zircon Hf isotope record of crust–mantle evolution. *Chem. Geol.* **425**, 65–75.

Vervoort, J. D., Patchett, P. J., Blichert-Toft, J. and Albarede, F. (1999). Relationships between Lu Hf and Sm Nd isotopic systems in the global sedimentary system. *Earth Planet. Sci. Lett.* **168**, 79–99.

Vervoort, J. D., Patchett, P. J., Gehrels, G. E. and Nutman, A. P. (1996). Constraints on early Earth differentiation from hafnium and neodymium isotopes. *Nature* **379**, 624–7.

White, W. M. and Patchett, J. (1984). Hf–Nd–Sr isotopes and incompatible element abundances in island arcs: implications for magma origins and crust–mantle evolution. *Earth Planet. Sci. Lett.* **67**, 167–85.

White, W. M., Patchett, J. and Ben Othman, D. (1986). Hf isotope ratios of marine sediments and Mn nodules: evidence for a mantle source of Hf in seawater. *Earth Planet. Sci. Lett.* **79**, 46–54.

Willbold, M. (2007). Determination of Ce isotopes by TIMS and MC-ICPMS and initiation of a new, homogeneous Ce isotopic reference material. *J. Anal. Atomic Spectrom.* **22**, 1364–72.

Wimpenny, J., Amelin, Y. and Yin, Q. Z. (2015). The Lu isotopic composition of achondrites: closing the case for accelerated decay of ^{176}Lu. *Astrophys. J. Lett.* **812** (L3), 1–5.

Woodhead, J., Hergt, J., Shelley, M., Eggins, S. and Kemp, R. (2004). Zircon Hf-isotope analysis with an excimer laser, depth profiling, ablation of complex geometries, and concomitant age estimation. *Chem. Geol.* **209**, 121–35.

Zeh, A., Stern, R. A. and Gerdes, A. (2014). The oldest zircons of Africa – their U–Pb–Hf–O isotope and trace element systematics, and implications for Hadean to Archean crust–mantle evolution. *Precamb. Res.* **241**, 203–30.

Zhao, W., Sun, Y., Balsam, W. *et al.* (2015). Clay-sized Hf–Nd–Sr isotopic composition of Mongolian dust as a fingerprint for regional to hemispherical transport. *Geophys. Res. Lett.* **42**, 5661–9.

Zimmermann, B., Porcelli, D., Frank, M. *et al.* (2009). The hafnium isotope composition of Pacific Ocean water. *Geochim. Cosmochim. Acta* **73**, 91–101.

Chapter 10

K–Ar, Ar–Ar and U–He Dating

Potassium is one of the eight most abundant chemical elements in the Earth's crust and a major constituent of many rock-forming minerals. However, the radioactive isotope, ^{40}K, makes up only 0.012% of total potassium, so it effectively falls in the low ppm concentration range. ^{40}K exhibits a branched decay scheme to ^{40}Ca and ^{40}Ar. The major branch leads to ^{40}Ca, but in most rocks the daughter product is swamped by common (non-radiogenic) ^{40}Ca, which makes up 97% of total calcium. Because variations in radiogenic ^{40}Ca abundance are very limited in most rock systems, this method has a restricted application as a dating tool (Section 9.6). Only 11% of ^{40}K decays lead to ^{40}Ar, but since this is a rare gas, the radiogenic component is dominant. It makes up 99.6% of atmospheric argon, equal to 0.93% of dry air by volume.

Decay to ^{40}Ar is by three different routes but the only one with a measurable decay constant is by electron capture to an excited state that emits a 1.5 MeV gamma ray. The other two routes are by electron capture or positron emission to the ground-state. However, these make up less than 0.1% of decays, and can be ignored. Therefore, the electron capture (E.C.) decay constant can be taken to represent all of the routes from ^{40}K to ^{40}Ar. The 'conventional' decay constant (Steiger and Jager, 1977) has a value of 0.581×10^{-10} yr^{-1}, equivalent to a half-life of 11.93 Ga. It was based on a weighted mean of the best gamma counting determinations evaluated by Beckinsale and Gale (1969).

Decay to ^{40}Ca is by emission of a β particle, and the β decay constant has a conventional value of 4.962×10^{-10} yr^{-1}, equivalent to a half-life of 1.397 Ga. The sum of the decay constants for the two branches yields the total ^{40}K decay constant of 5.543×10^{-10} yr^{-1}, equivalent to a half-life of 1.25 Ga. The possible need for revision to this value is discussed in Section 10.3, because it pertains to both the K–Ar and the Ar–Ar methods. The U–He dating method is also introduced in this chapter because of its related applications in thermochronometry.

10.1 The K–Ar Dating Method

The fraction of ^{40}K atoms that decay into ^{40}Ar is equal to $\lambda_{EC}/(\lambda_{EC} + \lambda_{\alpha})$. Hence, substituting into the general decay equation [1.10], the growth of ^{40}Ar in a K-bearing rock or mineral can be written as

$$^{40}\text{Ar}_{\text{total}} = {}^{40}\text{Ar}_\text{I} + \frac{\lambda_{EC}}{\lambda_{\text{total}}} \cdot {}^{40}\text{K}(e^{\lambda\,\text{total}\,t} - 1) \qquad [10.1]$$

However, if the system was completely outgassed of Ar at the time of formation, the initial Ar term disappears, and the equation is simplified to

$$^{40}\text{Ar}^* = (\lambda_{EC}/\lambda_{\text{total}}) \cdot {}^{40}\text{K}(e^{\lambda\,\text{total}\,t} - 1), \qquad [10.2]$$

where ^{40}Ar* signifies radiogenic argon only. As will be seen below, this is the usual assumption in K–Ar dating.

10.1.1 Analytical Techniques

The isotopic composition of naturally occurring potassium has been found to be effectively constant in all types of rock throughout the Earth, with a few minor exceptions (e.g. Garner et al., 1976). Therefore the ^{40}K content of a mineral or rock is usually found by straightforward chemical analysis for total potassium, followed by multiplication by 1.2×10^{-4} to derive the concentration of the radioactive isotope. Various methods can be used to determine potassium, including ICP–OES (inductively coupled plasma – optical emission spectrometry), X-ray fluorescence and isotope dilution. In the past, a commonly used method was flame photometry (Vincent, 1960), a form of optical emission spectrometry especially suitable for the alkali metals. This technique was less accurate than isotope dilution but could achieve a precision of ca. 1% and was quick and inexpensive.

Argon trapped in a geological sample is released and purified in an argon extraction line, 'spiked' with an enriched isotope and then fed into a mass spectrometer for isotopic analysis (Fig. 10.1). Samples must have the minimum possible

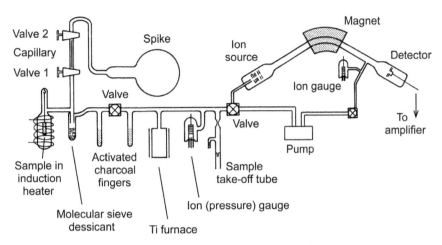

Fig. 10.1 Schematic diagram of an argon extraction line coupled to a static gas mass spectrometer. After Dalrymple and Lanphere (1969).

surface area for absorption of atmospheric argon; therefore mineral separates or whole-rock chips are not powdered. After loading the sample(s) in the extraction line, the whole line, and especially the sample itself, must be baked under vacuum to extract all possible atmospheric argon from the system. Next, after isolating the pump, the sample is manoeuvred into a disposable molybdenum crucible, which is positioned in a radio frequency induction furnace. The crucible is heated to ca. 1400 °C, whereupon the sample melts and releases all of the trapped gases. These consist mostly of H_2O and CO_2, with a very small amount of argon and other rare gases. All gases except the rare gases can be removed by reaction with titanium vapour in a Ti sublimation pump or by using a zeolite 'getter'. Activated charcoal fingers may be used for temporary absorption of gases during their manipulation.

Highly enriched ^{38}Ar spike is usually stored in a large glass reservoir bulb. This is connected to a length of capillary tube of fixed volume between two valves with low deadspace (Fig. 10.1). The capillary is opened to the reservoir while valve 1 is closed. Valve 2 is then closed, and the known volume of spike between the two valves is added to the sample by opening valve 1. Because the reservoir pressure falls with each gas withdrawal, successive spike aliquots contain smaller and smaller fractions of ^{38}Ar. However, aliquots are periodically calibrated by mixing with a known volume of atmospheric argon and performing an isotope dilution analysis. The amount of ^{38}Ar spike added to each sample is determined by noting its order in the sequence and interpolating between the calibration runs (Lanphere and Dalrymple, 1966). The amount of argon released from a typical sample of a few hundred milligrams is very small, generally less than 10^{-6} cc (cm^3) at STP (standard temperature and pressure = 25 °C at 1 atm). For this reason the isotopic analysis is performed statically; in other words, the entire sample is

fed into the mass spectrometer at once, after isolation from the pumps. The ratio ^{40}Ar/^{36}Ar in the air may be measured between unknown samples as a check on the calibration of the machine, and normally has a value of 295.5 ± 0.5.

Two different types of mass spectrometer are in common use. Modern rare gas machines tend to be very similar to solid source TIMS machines, with a high accelerating potential of several kV, and peak switching by changing the magnetic field. The problem with this type of machine is that the high velocities of the ions make them implant into metal components in the vacuum system whenever these are struck by the ion beam. Such ions diffuse back out of the metal surfaces during the next sample analysis, and this memory effect must be carefully corrected. The effect may be reduced by polishing metal components which the beam is likely to strike. Many older instruments used a low accelerating potential of a few hundred volts and a small permanent-field magnet. The accelerating potential was then switched to focus different nuclides into the collector. This type of machine suffered from very little memory effect but was capable of much poorer precision in the measurement of isotope ratios. Since the source is gaseous, there is no problem of mass-dependent fractionation in either type of machine.

A typical argon isotope mass spectrum is shown in Fig. 10.2. The presence of any ^{36}Ar signal shows that common or non-radiogenic argon is present. This is almost inevitable, because of the great difficulty of removing all atmospheric argon from the system. However, if the sample was completely outgassed at the time of its formation, it will not contain any inherited non-radiogenic Ar. In that case, the measured ^{40}Ar peak can be corrected for atmospheric contamination by subtracting 295.5 times the ^{36}Ar peak:

$$^{40}Ar^* = {}^{40}Ar_{total} - 295.5\,^{36}Ar \qquad [10.3]$$

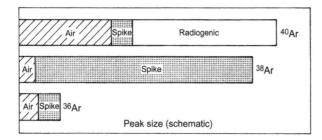

Fig. 10.2 Schematic argon isotope mass spectrum showing fractions of each peak due to radiogenic Ar (white), spike (stipple) and atmospheric contamination (hatched). Size fractions are not shown to scale. After Dalrymple and Lanphere (1969).

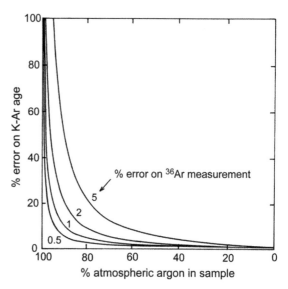

Fig. 10.3 Error magnification in K–Ar dating (y axis) resulting from atmospheric argon contamination. Curves are calculated for 0.5, 1, 2 and 5% errors in the measurement of ^{36}Ar. After Cox and Dalrymple (1967).

A value of 0.063% of atmospheric argon is similarly subtracted from the ^{38}Ar peak. The ^{40}Ar and ^{36}Ar peaks must also be corrected for small fractions of these isotopes in the spike. The amount of radiogenic ^{40}Ar* in the sample is then found by comparison with the size of the net ^{38}Ar peak, formed by a known quantity of spike. (In other words, this is an isotope dilution determination.) Given the abundances of ^{40}Ar and ^{40}K in the sample, the age is calculated by re-arranging equation 10.2:

$$t = \frac{1}{\lambda_{\text{total}}} \ln \left[\frac{^{40}\text{Ar}^*}{^{40}\text{K}} \cdot \frac{\lambda_{\text{total}}}{\lambda_{\text{EC}}} + 1 \right] \qquad [10.4]$$

K–Ar ages depend on closed-system behaviour of the sample for K and Ar throughout its history. In addition, it is necessary to assume that the sample contains no initial argon (usually called 'excess' argon), because this might have a ^{40}Ar/^{36}Ar ratio different from atmospheric argon, leading to a mixture with indeterminate ^{40}Ar/^{36}Ar ratio which could not be corrected for atmospheric contamination.

The ^{36}Ar/^{40}Ar ratio must be analysed to very high precision because the atmospheric Ar correction magnifies any errors in this measurement by nearly 300. The importance of this effect is shown in Fig. 10.3, where the effect of errors in the measurement of ^{36}Ar are shown in terms of the resulting error on the calculated age (Cox and Dalrymple, 1967). Once atmospheric contamination exceeds 70% of total argon, errors in ^{36}Ar have serious consequences on the age measurement. This correction is not a problem for old and/or K-rich samples, but is the principal limitation to dating young material.

Since 1967, great improvements have been made in measurement precision, allowing the dating of very young rocks. However, Mussett and Dalrymple (1968) showed that volcanic rocks contain 'locked-in' atmospheric (non-radiogenic) argon, some of which cannot be removed even by baking in a vacuum. Hence, even with a low-blank analytical system, a small residual atmospheric fraction is almost unavoidable in terrestrial lavas.

10.1.2 Inherited Argon and the K–Ar Isochron Diagram

Since ^{36}Ar is used as a monitor of atmospheric contamination, there is no facility in K–Ar dating to correct for initial argon incorporated into minerals or rocks at the time of crystallization. Hence, it must be assumed to be absent. However, early work by Damon and Kulp (1958) showed the presence of initial or 'excess' argon in beryl, cordierite and tourmaline. Since these minerals all have a ring structure, it was initially assumed that the stacking of rings created channels in which excess argon inherited from fluids could reside. Hence, Damon and Kulp suggested that this problem might also occur in hornblende, where partial vacancy of the alkali-cation site might provide a location for excess argon.

However, excess argon was subsequently also found in pyroxenes by Hart and Dodd (1962). Since the pyroxene structure does not have any suitable voids for argon accommodation, Hart and Dodd argued that it must be located in crystal dislocations and defect structures. This implies that excess argon is a product of the environment of crystallization rather than the host mineral. Hart and Dodd noted that their analysed pyroxenes were from originally deep-seated rocks, unlike the volcanic or shallow intrusive rocks normally used in K–Ar dating. Hence they warned that excess argon might be a common feature in samples from deep-seated (plutonic) environments.

The occurrence of excess argon was extended to submarine lavas by Dalrymple and Moore (1968). They dated glassy pillow rims and whole-rock pillow cores from flows at 500–5000 m depth on the north-east ridge of Kilauea volcano,

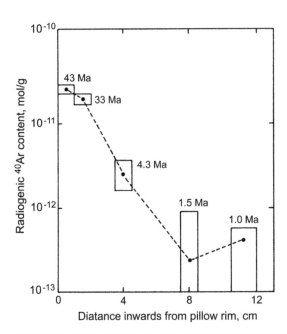

Fig. 10.4 Contents of (excess) radiogenic ^{40}Ar in submarine pillows from Hawaii, plotted against inward distance from the pillow rim. Apparent K–Ar ages for each sample are noted in Ma. After Dalrymple and Moore (1968).

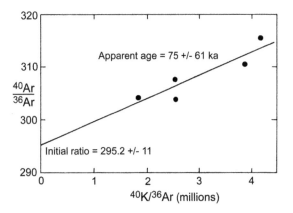

Fig. 10.5 K–Ar isochron plot for a lava of historical age from Mount Wellington, New Zealand, showing a best-fit slope age of 75 ka. After McDougall et al. (1969).

Hawaii. Various geological lines of evidence suggested a historical age for the samples, but K–Ar ages of up to 43 Ma were found. Furthermore, a series of samples from rim to interior of one pillow (from 2590 m depth) showed an inverse correlation of apparent K–Ar age with distance from the rim (Fig. 10.4). The results were attributed to entrapment of initial or excess argon which was inherited from the mantle source by the magmas. Dalrymple and Moore concluded that because these magmas were quenched under substantial hydrostatic pressure, inherited argon was not completely outgassed at the time of eruption, as usually occurs in terrestrial lavas. It was eventually able to escape from the slowly crystallizing core of the pillow, but was retained in the glassy rim.

Even some subaerially erupted lavas were subsequently found to contain inherited argon. For example, McDougall et al. (1969) encountered measurable radiogenic ^{40}Ar contents in historical-age subaerial basalts from New Zealand. Lavas shown by ^{14}C dating of wood inclusions to be less than 1 ka in age nevertheless gave K–Ar ages up to 465 ka. This led McDougall et al. to consider whether such cases of inherited argon could be detected and/or corrected.

They proposed that the raw ^{40}Ar signal (uncorrected for atmospheric contamination) be divided by ^{36}Ar and plotted against the K/Ar ratio to form an isochron diagram analogous to that for Rb–Sr (Fig. 10.5). This is achieved by expanding the initial Ar term in equation [10.1] to include both atmospheric and excess argon components, and by dividing throughout by ^{36}Ar:

$$\left(\frac{^{40}\text{Ar}}{^{36}\text{Ar}}\right)_{\text{total}} = \left(\frac{^{40}\text{Ar}}{^{36}\text{Ar}}\right)_{\text{atm+excess}} + \frac{^{40}\text{K}}{^{36}\text{Ar}} \cdot \frac{\lambda_{\text{EC}}}{\lambda_{\text{EC}} + \lambda_{\beta}} \left(e^{\lambda\,\text{total}\,t} - 1\right)$$

[10.5]

This equation has the form

$$y = c + x \cdot m$$

[10.6]

When a suite of samples is analysed from a single completely outgassed system such as a lava flow, the c term is entirely atmospheric. Therefore, the analysed points, when plotted on an isochron diagram, should define a straight line with an intercept of 295.5, whose slope yields the age of eruption. In fact, this array is merely a mixing line between the samples and atmospheric argon. When the atmospheric correction is performed on a single analysis, we effectively make 295.5 the origin and determine the slope.

In the case studied by McDougall et al. (1969), the lavas are of approximately zero age. Hence, the analyses which make up their 'isochrons' (e.g. Fig. 10.5) represent trapped argon in the magma variably mixed with atmospheric argon. McDougall et al. speculated that the trapped argon might originate from partially digested crustal xenocrysts.

Roddick and Farrar (1971) considered the case of a geologically old sample suite displaying both inherited argon and atmospheric contamination (Fig. 10.6). With inherited and radiogenic argon only, the array ABC is defined, but if variable atmospheric contamination occurs, a scatter (DEF) may result. In principle, a good linear array on the K–Ar isochron diagram should indicate that both the age and initial Ar isotope ratio are meaningful. However, it may be possible for the slope of the line to swing round in a systematic way due to complex mixing processes, so that it yields a good array of meaningless slope. Nevertheless, the isochron diagram is a useful test of K–Ar data where inherited Ar is suspected.

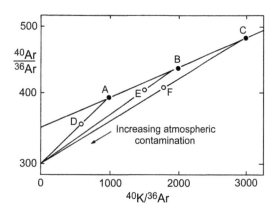

Fig. 10.6 Schematic K–Ar isochron diagram to show the effect of mixing inherited and radiogenic argon (A, B, C), coupled with variable atmospheric contamination (D, E, F). After Roddick and Farrar (1971).

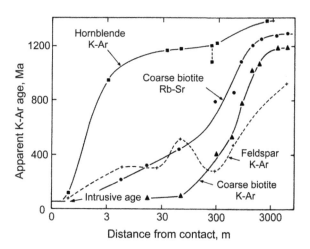

Fig. 10.7 Plot of apparent K–Ar mineral ages against outward distance from the contact of the 60 Ma Eldora stock, Colorado. After Hart (1964).

Lanphere and Dalrymple (1976) drew a distinction between inherited argon and excess argon. They defined the former as argon which 'originates within mineral grains by decay of ^{40}K prior to the rock-forming event'. Hence, this definition includes the examples given above. However, excess argon has a wider definition that also includes sources of extraneous argon that diffuse into a system from outside. Lanphere and Dalrymple (1971, 1976) pioneered the use of the ^{40}Ar/^{39}Ar method to identify excess argon in rocks (Section 10.2.5).

10.1.3 Argon Loss

The K–Ar method is unique amongst the major radiometric dating methods in having a gaseous daughter product. This means that the K–Ar system reacts differently from lithophile isotope systems such as Rb–Sr in response to thermal and hydrothermal events. Because argon is a non-reactive gas, its partition into the fluid phase is limited. Therefore, the K–Ar system may be more resistant than Rb–Sr to hydrothermal metamorphism. On the other hand, no mineral phase preferentially takes up argon when it is lost from the mineral where it was originally produced. This means that in K–Ar dating, whole-rock analysis confers no additional resistance to metamorphic resetting (as it does for the Rb–Sr method). On the contrary, in K–Ar analysis a whole-rock sample is only as resistant to resetting as its *least* retentive phase. Consequently, whole-rock K–Ar analysis is a last resort, when all mineral phases in the rock are too fine-grained for mineral separation.

A good comparison of argon loss from different minerals during a thermal event is provided by contact metamorphism associated with the Eldora stock in the Colorado front ranges (Hart, 1964). The 54 Ma quartz monzonite stock is intruded into ca. 1350 Ma amphibolites and schists. Hart analysed biotite, hornblende and K-feldspar at increasing distances from the intrusive contact (Fig. 10.7), and found that

despite the limited extent of petrographic alteration, K–Ar mineral ages were reset at large distances from the stock. Hornblende displayed good Ar retention properties, with loss of argon confined primarily to within ten feet (ca. 3 m) of the contact. However, coarse biotites were largely reset at distances up to 1000 ft (300 m) from the contact, while K-feldspars had lost a substantial fraction of argon even 20 000 feet (6 km) from the contact. The latter can hardly be said to be within the thermal aureole of the stock, and reflects the now widely accepted view that K-feldspars may lose argon by diffusion even at ambient temperatures. It is now recognized that systems such as these, that have suffered open-system behaviour of argon, must be studied by the ^{40}Ar–^{39}Ar technique (see below).

10.2 The ^{40}Ar–^{39}Ar Dating Method

The very different chemical affinity of potassium and argon causes limitations in the K–Ar dating method. However, these limitations can be overcome by converting ^{39}K to ^{39}Ar in a nuclear reactor, by irradiation with fast neutrons. This causes an n, p (neutron capture, proton emission) reaction:

$$^{39}_{19}K + n \rightarrow\ ^{39}_{18}Ar + p$$

This reaction permits the potassium determination for a K–Ar age to be made as part of the argon isotope analysis, thus opening up many new opportunities for the K–Ar dating technique.

10.2.1 ^{40}Ar–^{39}Ar Measurement

The comparatively long half-life of ^{39}Ar ($t_{1/2} = 269$ yr) means that it can be regarded as a stable isotope for mass spectrometric analysis, which was first applied to ^{40}Ar–^{39}Ar dating by Merrihue and Turner (1966). It is interesting to note, however, that the concept of combined irradiation and mass spectrometric analysis was applied to the I–Xe system in meteorite studies five years earlier (Section 15.4).

The production of ^{39}Ar from ^{39}K during the irradiation is expressed as

$$^{39}\text{Ar} = {}^{39}\text{K} \, \Delta t \int_{\min e}^{\max e} \phi_e \, \sigma_e \, de, \qquad [10.7]$$

where Δt is the irradiation time, ϕ_e is the flux density of neutrons with energy e and σ_e is the capture cross-section of ^{39}K for neutrons of energy e. The production must be integrated over the total range of neutron energies, which is a very difficult calculation in practice. Therefore, the normal procedure is to use a sample of known age as a flux monitor.

Taking the K–Ar decay equation [10.2], which is reproduced here:

$$^{40}\text{Ar}^* = (\lambda_{\text{EC}}/\lambda_{\text{total}}) \cdot {}^{40}\text{K} \, (e^{\lambda \, \text{total} \, t} - 1)$$

and dividing through on both sides by equation [10.7] yields

$$\frac{^{40}\text{Ar}^*}{^{39}\text{Ar}} = \left[\frac{\lambda_{\text{EC}}}{\lambda_{\text{total}}} \cdot \frac{^{40}\text{K}}{^{39}\text{K} \, \Delta t \int \phi_e \, \sigma_e \, de} \right] (e^{\lambda \, \text{total} \, t} - 1) \qquad [10.8]$$

However, the boxed term is the same for sample and standard. Therefore, it is customary to refer to it as a single quantity, whose reciprocal J can be evaluated as a constant (Mitchell, 1968). Hence, for the standard:

$$J = \frac{e^{\lambda t} - 1}{^{40}\text{Ar}^* / ^{39}\text{Ar}} \qquad [10.9]$$

where t is known. Rearranging equation [10.8] for samples of unknown age yields:

$$t = \frac{1}{\lambda} \cdot \ln \left[J \cdot \left(\frac{^{40}\text{Ar}^*}{^{39}\text{Ar}} \right) + 1 \right] \qquad [10.10]$$

In order to obtain an accurate value of J for each unknown sample, several standards need to be run, representing known spatial positions relative to the unknown samples within the reactor core (Mitchell, 1968). Hence, J values for each of the samples can be interpolated.

10.2.2 Irradiation Corrections

During the irradiation of ^{39}K, interfering Ar isotopes are generated by neutron reactions from calcium and other potassium isotopes (Fig. 10.8). Brereton (1970) and Dalrymple and Lanphere (1971) made detailed studies of the magnitude of these effects and their correction. However, it appears in practice that many workers have simply ignored the interferences.

Mitchell (1968) suggested that acceptable results could be obtained without interference correction on minerals over 1 Ma in age, provided that K/Ca was greater than 1. In such circumstances, a simple atmospheric correction may be considered adequate:

$$\frac{^{40}\text{Ar}^*}{^{39}\text{Ar}} = \left(\frac{^{40}\text{Ar}}{^{39}\text{Ar}} \right)_{\text{meas}} - 295.5 \left(\frac{^{36}\text{Ar}}{^{39}\text{Ar}} \right)_{\text{meas}} \qquad [10.11]$$

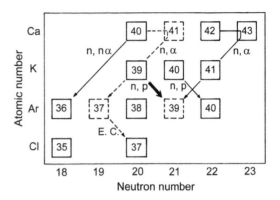

Fig. 10.8 Part of the chart of the nuclides in the region of potassium showing the production reaction (heavy arrow) and major interfering reactions (solid) during neutron activation. Dashed reaction to ^{37}Ar is the interference monitor. Data from Mitchell (1968).

Turner (1971a) showed that Ar interferences could be kept to a minimum by variation of certain irradiation parameters. The principal interferences which must be considered (Fig. 10.9) are:

$$^{40}\text{K n, p} \rightarrow {}^{40}\text{Ar}$$

$$^{40}\text{Ca n, n}\alpha \rightarrow {}^{36}\text{Ar}$$

$$^{42}\text{Ca n, }\alpha \rightarrow {}^{39}\text{Ar}$$

Other interferences occur but may be omitted as insignificant.

The approaches suggested by Turner were:

(1) Optimization of neutron dose according to age (Fig. 10.9a) to maximize ^{39}Ar production, without generating significant artificial ^{40}Ar from ^{40}K. (K content is not considered as a factor because the intended and interfering targets are both K isotopes.)

(2) Optimization of sample size according to age and K content, in order to obtain the total ^{40}Ar and ^{39}Ar yields necessary to achieve the desired counting statistics during mass spectrometric analysis.

(3) The K/Ca ratio in the sample also dictates an optimum neutron dose to generate enough ^{39}Ar without significant interfering ^{36}Ar (Fig. 10.9b). However, the optimum values largely overlap with those prescribed by criterion (1).

Theoretically, very young rocks can be activated with less than 1% interference by following these rules. However, this may require an immense sample size. In practice, a better alternative may be to use more irradiation but apply corrections. The complete correction formula (in terms of

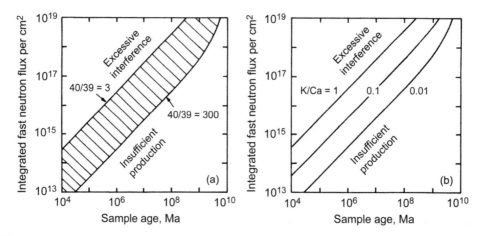

Fig. 10.9 Optimization of neutron dose for (a) K content, and (b) K/Ca ratio. Hatched area in (a) and bold lines in (b) indicate regions of acceptable compromise between sufficient ^{39}Ar production and minimal ^{40}Ar or ^{36}Ar interference in typical rocks. After Turner (1971a).

Ar isotope ratios) is

$$\frac{^{40}\text{Ar}^*}{^{39}\text{Ar}} = \frac{\frac{^{40}\text{Ar}}{^{39}\text{Ar}} - 295.5 \cdot \frac{^{36}\text{Ar}}{^{39}\text{Ar}} + 295.5 \cdot \frac{^{37}\text{Ar}}{^{39}\text{Ar}} \cdot \left(\frac{^{36}\text{Ar}}{^{37}\text{Ar}}\right)_{\text{Ca}} - \left(\frac{^{40}\text{Ar}}{^{39}\text{Ar}}\right)_{\text{K}}}{1 - \frac{^{37}\text{Ar}}{^{39}\text{Ar}} \cdot \left(\frac{^{39}\text{Ar}}{^{37}\text{Ar}}\right)_{\text{Ca}}}$$

[10.12]

where $^{37}\text{Ar}/^{39}\text{Ar}$ is the interference monitor ratio measured for the unknown, which must be corrected for ^{37}Ar decay from the time of irradiation until analysis ($t_{1/2} = 35$ days); and where $(^{36}\text{Ar}/^{37}\text{Ar})_{\text{Ca}}$, $(^{39}\text{Ar}/^{37}\text{Ar})_{\text{Ca}}$ and $(^{40}\text{Ar}/^{39}\text{Ar})_{\text{K}}$ are production ratios of Ar isotopes from the subscripted elements. These production ratios are determined by irradiating pure salts of Ca and K respectively in the reactor of interest, and are characteristic of the neutron flux of that reactor. Values for these production ratios measured by different authors for different reactors have typical ranges of 2.1–2.7, 6.3–30 and 0.006–0.031 respectively (Dalrymple and Lanphere, 1971).

10.2.3 Step Heating

Because the potassium signature of a sample is converted *in situ* to an argon signature by the Ar–Ar technique, it is possible to liberate argon in stages from different domains of the sample and still recover full age information from each step. Merrihue and Turner (1966) demonstrated the effectiveness of this 'step heating' technique in their original Ar–Ar dating study of meteorites, adapting the method from its previous application to I–Xe analysis of meteorites (Section 15.4.1).

The great advantage of the step heating technique over the conventional 'total fusion' technique is that progressive outgassing allows the possibility that anomalous subsystems within a sample may be identified, and, ideally, excluded from an analysis of the 'properly behaved' parts of the sample. This can apply to both separated minerals and whole-rock samples. Most commonly the technique is used

to understand samples which have suffered argon loss, but it may also be of help in interpreting samples with inherited argon.

In the case of partially disturbed systems, the domains of a sample which are most susceptible to diffusional argon loss (such as the rim of a crystal) should be outgassed at relatively low temperatures, whereas domains with tightly bound argon (which are more resistant to disturbance) should release argon at higher temperature. In order to understand the history of disturbed samples, results of the step heating analysis are normally presented in one of two ways: as a K–Ar isochron diagram, analogous to a suite of samples analysed by conventional K–Ar; or as an age spectrum plot.

Step heating results from the meteorite Bjurbole (Merrihue and Turner, 1966) are plotted on an isochron diagram in Fig. 10.10. The straight line array indicates a simple one-stage closed-system history for the meteorite. However, the initial $^{40}\text{Ar}/^{36}\text{Ar}$ ratio may be only partially meaningful, since it is a mixture of initial Ar and atmospheric contamination. The isochron plot can be useful to see the relative amounts of radiogenic argon and atmospheric/inherited argon in the sample. However, it conceals one of the most useful pieces of information about a step heating analysis, which is the position of each argon release step in the overall heating experiment. This information is displayed in the argon spectrum plot.

To construct a spectrum plot, the size of each gas release at successively higher temperature is measured by the magnitude of the ^{39}Ar ion beam produced. Each gas release can then be plotted as a bar, whose length represents its volume as a fraction of the total ^{39}Ar released from the sample, and whose value on the y axis is the corrected $^{40}\text{Ar}/^{39}\text{Ar}$ ratio from equation [10.12]. The latter is proportional to age, which is sometimes plotted on a log scale, and sometimes

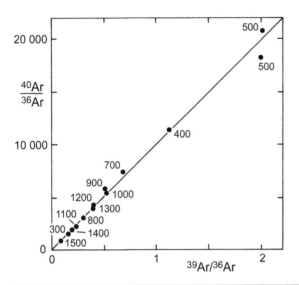

Fig. 10.10 Step heating data for the Bjurbole meteorite presented on the Ar–Ar isochron diagram. Numbers by data points signify temperatures of each release step in °C. After Merrihue and Turner (1966).

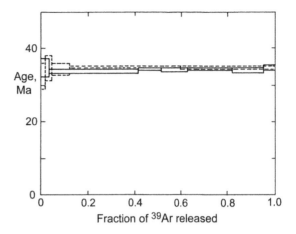

Fig. 10.11 Ideal ^{40}Ar/^{39}Ar age spectra for two Texas tektites, distinguished by solid and dashed boxes. After York (1984).

linear. Determination of a reliable crystallization age from the spectrum plot depends on the identification of an age 'plateau'. A rigorous criterion for a plateau age is the identification of a series of adjacent steps which together comprise more than 50% of the total argon release, each of which yields an age within two standard deviations of the mean (Dalrymple and Lanphere, 1974; Lee *et al.*, 1991). However, plateaus have been 'identified' in many instances on weaker evidence.

The age spectrum plot displays the ideal behaviour of the K–Ar system in tektite glasses. These are objects which were completely melted during flight through the atmosphere and then rapidly quenched on landing. Thus, young tektites which have not been affected by weathering yield perfect plateaus (Fig. 10.11). However, the most useful application of the 40–39 method is on samples with a complex geological history that *does* involve secondary argon loss.

10.2.4 Argon Loss Events

In order to assess the usefulness of Ar–Ar dating on disturbed systems, Turner *et al.* (1966) applied the method to chondritic meteorites. Many of these objects yield conventional K–Ar ages below 4.5 Ga, with U–He ages clustering around 500 Ma (Anders, 1964). For example, step heating results from a whole-rock sample of the Colby meteorite generate a complex age spectrum (Fig. 10.12), which was attributed to argon loss at ca. 500 Ma (Turner *et al.*, 1966; Turner, 1968).

Turner *et al.* suggested that when these meteorites were subjected to an ancient heating event (possibly in a collision between planetessimals), Ar loss occurred from the surface of mineral grains, and its transport within grains was by vol-

ume diffusion. This argon loss model is illustrated schematically in Fig. 10.13. Turner *et al.* argued that step heating analysis of the sample in the vacuum system would mimic the natural thermal event, so that domains near the surface of minerals, which had suffered geological disturbance, would outgas first in the experiment. In contrast, domains near the cores of grains would be resistant to geological disturbance, and would also outgas at the highest temperatures in the laboratory. (Although these meteorites are chondrites, it is the minerals in the chondrules that retain argon, rather than the chondrules themselves.)

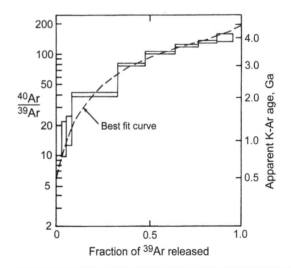

Fig. 10.12 Ar–Ar argon release pattern for the Colby meteorite, showing evidence for disturbance after formation. The best-fit curve is consistent with a model in which 40% of argon was lost during a thermal event (see below). After Turner (1968).

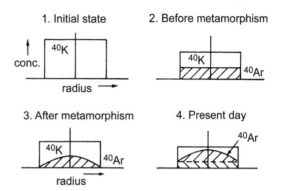

1. Initial state

2. Before metamorphism

3. After metamorphism

4. Present day

Fig. 10.13 Schematic illustration of the geological history of a mineral grain in a partially disturbed meteorite: (1) at 4500 Ma; (2) 500 Ma ago, before thermal event; (3) immediately after the event; (4) present day. After Turner (1968).

In order to test their diffusion model, Turner *et al.* calculated theoretical age spectrum plots, assuming that the meteorites were formed at 4500 Ma and metamorphosed in a single event at 500 Ma. Argon loss was modelled by assuming volume diffusion from spherical K-bearing mineral grains (probably mainly feldspar). The results of two different models are compared with the measured argon release profile of the Bruderheim meteorite in Fig. 10.14. The first model assumed argon loss from spherical grains of uniform size, but this gave a bad fit to the measured profile (curve a). However, thin section analysis of the meteorite revealed the existence of variably sized feldspar grains. Therefore, Turner *et al.* calculated the result of argon diffusion from grains with a log-normal size distribution. This model generates a family of solutions according to the amount of grain size variation allowed. Curve b in Fig. 10.14 shows the best result, assuming a standard deviation (σ[log radius]) of 0.20, equivalent to 4/5 of grains falling within a factor of two from the mean radius.

The shape of the best-fitting curve in Fig. 10.14 is also dependent on the fraction of total Ar that was lost in the heating event. Thus, the concave upwards shape of the curve indicates that Bruderheim suffered more than 80% argon loss. In this case the low-temperature part of the profile yields the age of metamorphism (0.5 Ga) but the formation age is lost, because even the highest temperature release steps give ages well below 4.5 Ga. On the other hand, the shape of the Colby release profile (Fig. 10.12) approximates to ca. 40% argon loss, so that the high-temperature argon release steps still preserve an indication of the age of formation. For samples which lose less than 20% (not shown), the high-temperature plateau still records a good crystallization age but the metamorphic age is badly constrained.

The relatively good agreement between the analysed and the log-normal model pattern in Fig. 10.14 suggests that the diffusional loss mechanism is a good description of the thermal disturbance of meteorites. This might be expected, since both the geological event and laboratory measurements were based on heating of anhydrous phases in a vacuum. Turner (1972) demonstrated a similar good fit to the diffusional model for experimental data from a lunar anorthosite. In contrast, terrestrial Ar–Ar dating generally involves hydrated minerals such as biotite and hornblende. In this case, the diffusional argon loss mechanism may not provide such an accurate model. An assessment can be made by examining Ar–Ar data for the Eldora stock (Berger, 1975), for which conventional K–Ar data have been described above. Figure 10.15 shows age spectrum plots for hornblendes, biotites and feldspars in the vicinity of the stock.

Of the three minerals studied, hornblende (Fig. 10.15a, b) displays the type of pattern most similar to Turner's thermal diffusion degassing model, although this resemblance may be misleading. The most distant sample (not shown) yields an excellent plateau age of ca. 1400 Ma. Samples at 1130, 950, 248 and 34 ft (ca. 350, 290, 75 and 10 m) display serious Ar loss from the outside of grains, but approach the 'true' age in the highest temperature fractions. However, Berger recognized that the pattern of Ar loss might reflect alteration to biotite, rather than diffusional Ar loss from hornblende. This interpretation is supported by dating experiments on synthetic hornblende–biotite mixtures (Rex *et al.*, 1993). Another problematical observation is that the sample 11 ft (3.5 m) from the contact displays an intermediate 'false' plateau of high quality. Finally, the sample 2 ft (0.6 m) from the contact displays a saddle-shaped pattern, in which the lowest-age fraction approaches the age of metamorphism.

Coarse biotite (Fig. 10.15c) behaves somewhat differently. Its maximum age at infinite distance from the stock

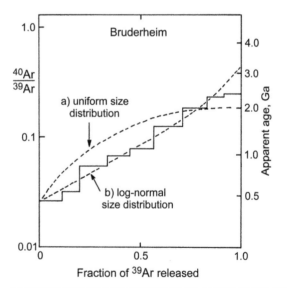

Fig. 10.14 Argon release profile of the Bruderheim meteorite, compared with calculated argon loss profiles from spherical mineral grains formed at 4.5 Ga and disturbed at 0.5 Ga. Curve a: assuming uniform size distribution; curve b: log-normal size distribution with $\sigma = 0.20$. After Turner (1968).

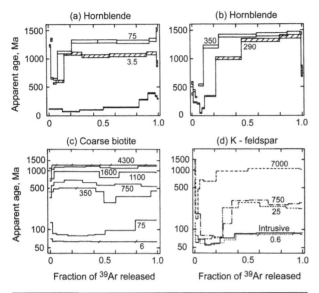

Fig. 10.15 Ar–Ar age spectrum plots for mineral phases at different distances from the Eldora stock. Figures beside age spectra indicate distances in metres. (a) and (b) hornblende, (c) biotite, (d) K-feldspar. Release steps with identical ages are separated by slashes. After Berger (1975).

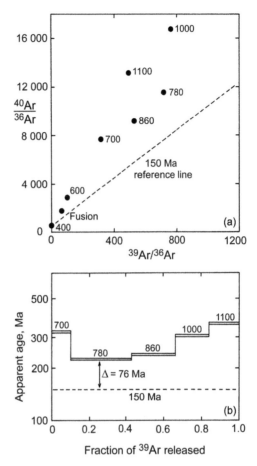

Fig. 10.16 Comparison between the K/Ar isochron plot and age spectrum plot for a 150 Ma biotite grain with excess argon. Note the characteristic 'saddle-shaped' profile. Numbers indicate the temperature of each heating step in °C. F = fusion step. After Lanphere and Dalrymple (1976).

(1250 Ma) is lower than the hornblende age. At intermediate distances the spectra are irregular, but show a general decrease in 'plateau' age as the stock is approached. Hence, it appears that biotites can be partially but uniformly outgassed, possibly because of enhanced diffusion parallel to the cleavage. Finally, K-feldspar suffers irregular and disastrous Ar losses, as is known from conventional K–Ar analysis (Fig. 10.15d).

Berger concluded that hornblendes were able to generate plateaus of high quality which were nevertheless meaningless. This may make hornblende a dangerous material on which to base geological interpretations of age, in the absence of independent confirmatory evidence. On the other hand, partially reset biotites were always identifiable by their irregular patterns, making biotite ages a more reliable tool for age interpretation. The exact meaning of the plateaus in the biotite and hornblende samples distant from the stock is equivocal, since the country-rocks are paragneisses with a long history of thermal events. Subsequent studies have indeed generated many examples of meaningless plateaus in hornblende, and more rarely, in biotite.

10.2.5 Excess Argon

As well as detecting argon loss, step heating analysis can also be used to evaluate cases where excess argon is present in a ^{40}Ar–^{39}Ar analysis. Following early work by Lanphere and Dalrymple (1971), a more detailed examination of this problem was made by Lanphere and Dalrymple (1976). In this study, step heating analyses were made on separated min-

eral phases from several rocks that were known from earlier work to contain excess (inherited) argon. The data were presented on K/Ar isochron diagrams and age spectrum plots for comparison. The example shown in Fig. 10.16(a, b) comes from a sample of Mg-rich biotite separated from a kimberlite dyke that intrudes Devonian sediments in New York State. Previous studies had suggested that these minerals might be xenocrysts, since K–Ar ages were variable.

On the K/Ar isochron diagram, it can be seen that the data scatter badly above a 150 Ma reference line. This line was based on the estimated age of the kimberlite, with an intercept equal to the atmosphere point. The scatter of the data provides some evidence of excess argon but is not strongly diagnostic. However, when the data are plotted on the spectrum plot they form a 'saddle-shaped' pattern that was found to be characteristic of all of the samples with excess argon analysed by Lanphere and Dalrymple (1976). Unfortunately, the minimum age from the saddle does not give the age of intrusion, since it is still above the estimated age. Therefore, such minima must only be regarded as maximum ages for the rock, as argued by Kaneoka (1974).

In seeking an explanation for the saddle-shaped age spectrum associated with excess (inherited) argon, Kelley (2002) suggested that this feature was caused by inclusions of various types. For example, fluid inclusions in mineral grains are expected to release argon at low temperature, whereas mineral inclusions may release argon at high temperature. A special case of the former type is exhibited by anorthoclase grains in a lava from Mt Erebus, Antarctica (Esser et al., 1997). The lava is of zero age, but the anorthosite phenocrysts contain excess argon. ^{40}Ar–^{39}Ar analysis showed that this argon was inherited by melt inclusions in the phenocrysts that were not completely outgassed during eruption. A different type of excess argon observed in some other cases is caused by back-diffusion into a mineral from argon in the surrounding rock. This will be discussed below (Section 10.4.2).

10.2.6 ^{39}Ar Recoil

The Ar–Ar dating technique was found to be particularly useful for dating small whole-rock samples of lunar material, especially fine-grained mare basalts. However, some samples showed either a sharp decrease in apparent age in the high-temperature fractions, or, particularly in fine-grained rocks, a progressive decrease in apparent age over most of the gas release. The latter examples led workers to suspect Ar redistribution within the sample, possibly during the irradiation process.

Turner and Cadogan (1974) calculated that this effect could deplete argon from the surface of a K-bearing mineral to a mean depth of 0.08 μm. In order to test the practical effects of this process, they powdered a sample of medium-grained ferrobasalt to a grain size of 1–10 μm before irradiation. This was thought to bring ca. 10% of K-bearing lattice sites to within 0.1 μm of a grain boundary, whereupon ^{39}Ar could recoil out of the lattice. It was anticipated that the ^{39}Ar released would enter low-K minerals such as plagioclase, pyroxene and ilmenite, leading to an old apparent age during low temperature release (K-bearing minerals) and a young apparent age during high-temperature release.

However, while abnormally old ages were produced at low temperature, the data approached the 'true' plateau age at intermediate temperatures (Fig. 10.17). Therefore, Turner and Cadogan argued that ^{39}Ar released by recoil must have been lost from the sample altogether, rather than absorbed by low-K phases. This is probably due to the fact that adjacent grains are in less intimate contact in a powdered sample than a fine-grained rock sample. The unusually high ages in the highest temperature fraction (Fig. 10.17) were tentatively attributed to an incorrect Ca correction, due to recoil of the monitor isotope ^{37}Ar during the n, α reaction from ^{40}Ca. This transformation should result in four times more recoil than proton emission from ^{39}Ar.

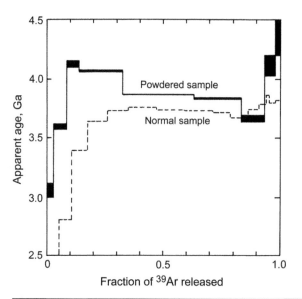

Fig. 10.17 The effect of fine crushing on a 40–39 age spectrum, due to ^{39}Ar recoil. Dashed profile = analysed rock chip of a lunar mare basalt. Solid profile = similar sample activated after fine powdering. After Turner and Cadogan (1974).

Argon recoil has important implications for minerals whose diffusional history is explained in terms of micro domains (Section 10.4.3). The most important examples are feldspars, which have exsolution lamellae about 0.01–0.3 μm thick, but show little evidence for recoil effects in their plateau ages or Arrhenius plots. The lack of any such evidence led McDougall and Harrison (1988) to speculate that a large fraction of ^{39}Ar recoils might occur at low energy, with reduced displacements.

Onstott et al. (1995) re-examined this question using theoretical calculations and ion implantation experiments, but these continued to support a mean ^{39}Ar recoil distance of 0.082 μm. The implications were examined for three minerals showing exsolution of K-rich and K-poor lamellae (amphibole, plagioclase and K-feldspar). In all three cases, calculations indicated that ^{39}Ar concentrations would be significantly homogenized and ^{37}Ar almost totally homogenized between adjacent lamellae (Fig. 10.18). Therefore, Onstott et al. concluded that the lamellae were too small to be the domains controlling volume diffusion of argon in the samples of amphibole and plagioclase studied. The situation for K-feldspar was less clear, but they suggested that some reinterpretation of results might be necessary for the smallest domain sizes of K-feldspar used in thermal history analysis.

The calculations of Onstott et al. were tested by a direct determination of the ^{39}Ar recoil distance by Villa (1997). A thin slab of KCl was sandwiched between two sheets of silica, but on one side the silica layer was shielded by a silicon

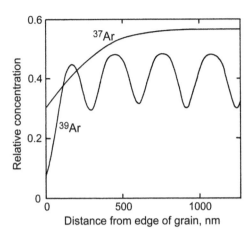

Fig. 10.18 Predicted argon isotope distribution after neutron irradiation of a plagioclase grain. The grain has alternating lamellae of calcic plagiocase (60 nm wide) and 50% plagioclase and K-feldspar (320 nm wide). After Onstott *et al.* (1995).

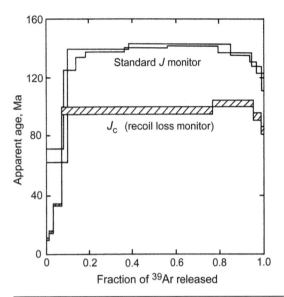

Fig. 10.19 Ar spectrum plot on a 95 Ma glauconite standard showing the results of conventional step heating (top two profiles) as well as a step-heating analysis corrected using the J_C recoil loss monitor (hatched). Modified after Kapusta *et al.* (1997).

coating 95 nm thick. The whole assembly was then irradiated to simulate a 40–39 argon analysis. After irradiation, the ^{39}Ar concentration on the inner surface of each silica sheet was analysed, and from the difference between the shielded and unshielded surface, a mean ^{39}Ar recoil distance of 80 ± 20 nm in silicon was calculated. Based on the relative densities of silicon and K-feldspar, the recoil distance in this mineral was estimated as about 70 nm. This measurement supported the theoretical calculations of Onstott *et al.* (1995), and therefore led Villa (1997, 1998) to question the meaning of K-feldspar thermochronometry using the Micro Diffusion Domain (MDD) model (Section 10.4.3). This question remains controversial.

The problem of ^{39}Ar recoil was found to be particularly severe in attempts to apply Ar–Ar dating to the authigenic sedimentary mineral glauconite (e.g. Foland *et al.*, 1984). This is probably due to the very small grain size of the glauconite crystallites which make up the grains of a pellet. Smith *et al.* (1993) showed that this problem might be overcome by encapsulating glauconite grains in small glass ampoules prior to irradiation. The recoil products can then be collected for analysis, in order to correct the Ar release from the rest of the grain. However, this method is only applicable to a whole-sample degassing analysis (analogous to a conventional K–Ar age) and cannot be used with the step heating method.

Kapusta *et al.* (1997) proposed a new method whereby step-heating experiments could be performed on fine-grained material such as glauconite or clay. This approach involves irradiating two aliquots of the sample to be dated. The first is used for the step heating analysis, while the second is encapsulated and used to determine a total release age (relative to a standard of known age). The standard of known age is used to determine the J value of the irradiation in the usual

way. The total release age of the encapsulated aliquot is then used in turn to calculate a 'J_C' correction for the step-heated aliquot, modified from equation [10.9]:

$$J_C = \frac{e^{\lambda t} - 1}{(^{40}\text{Ar}^* / ^{39}\text{Ar})_{\text{total release}}} \qquad [10.13]$$

The J_C value allows normalization of both the neutron flux and recoil loss, provided it has the same grain size distribution and crystal make-up as the step-heated sample. Kapusta *et al.* (1997) demonstrated the method on a glauconite standard (Fig. 10.19). However, it should be applicable to clay minerals that have suffered a combination of radiogenic ^{40}Ar loss over geological time as well as ^{39}Ar loss during irradiation.

10.2.7 Dating Paleomagnetism

Paleomagnetic measurements are a vital tool in the reconstruction of ancient plate tectonic motions, by comparison of 'apparent polar wander paths' (APWPs) for different continental fragments. One essential step in the construction of an APWP 'track' for a given terrane is to date the time when magnetic remanence was inherited by the rock. However, the magnetic remanence is relatively easily overprinted because it has a comparatively low blocking temperature.

The dating of magnetic remanence took a major step forward when York (1978) showed from theoretical principles that the processes of thermal de-magnetization and argon loss from a mineral grain were related. This is because they are both almost exclusively the products of thermal kinetics, in contrast to ^{87}Sr loss (for example) which may be

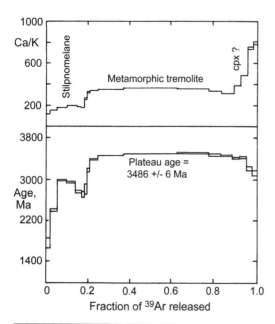

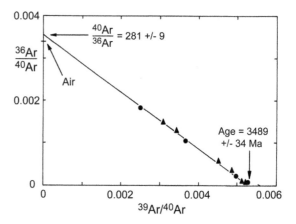

Fig. 10.21 Inverse Ar–Ar isochron plot for two Ar–Ar runs (•, ▲) on the Barberton komatiite B40A. The age is determined from the intersection on the x axis. Modified after Lopez Martinez et al. (1984).

dependent on the presence or absence of aqueous fluids. Hence, the Ar–Ar method is an ideal tool to date paleomagnetic remanence.

A good example of such work is provided by the oldest reliable Ar–Ar age for terrestrial rocks (Lopez Martinez et al., 1984), on the Barberton komatiites. Although the time of eruption was constrained by Sm–Nd dating, knowledge of their subsequent thermal history was required to interpret paleomagnetic data. Analyses were performed on whole-rock powders, which were irradiated alongside the '3 GR' hornblende standard to determine neutron fluxes. Figure 10.20 shows an age spectrum from the best sample analysed, with three separate sections. At low temperatures (600–800 °C) and high temperatures (>1100 °C) argon loss was observed, resulting in low ages. However, at intermediate temperatures (925–1035 °C) a very stable plateau was observed, from which a best age of 3486 ± 6 Ma (2σ) was obtained. The integrated (total fusion) age of 3336 Ma was significantly younger, due to the effects of the low and high-temperature steps.

The top half of Fig. 10.20 reports Ca/K ratios, calculated from the measured $^{37}Ar/^{39}Ar$ ratio (Section 10.2.2), which help to characterize the mineral phases in the sample which gave rise to different parts of the age spectrum. By microprobe analysis, the authors were able to deduce that the mineral giving rise to the age plateau was metamorphic tremolite, while the low temperature, low Ca/K phase was stilpnomelane. The high Ca/K phase may represent pyroxene relics of the original igneous mineralogy. Since the plateau age is identified with a metamorphic mineral, it must be dat-

ing a thermal event which occurred less than 100 Ma after eruption. Hale (1987) tentatively identified this event as the intrusion of the nearby Threespruit granitoid pluton.

A K-Ar isochron diagram was plotted (Fig. 10.21) in order to examine the composition of the non-radiogenic end-member, and test for inherited argon. In this case the isochron diagram was plotted in the alternative form $^{36}Ar/^{40}Ar$ versus $^{39}Ar/^{40}Ar$ (Turner, 1971b). This presentation helps to curtail the strong correlation between the two ordinates which occurs with the conventional K-Ar isochron diagram, making error estimates easier. An initial $^{40}Ar/^{36}Ar$ ratio of 281 ± 18 (2σ) was calculated from the inverse of the normal y axis intercept, after expansion of analytical errors to absorb a small amount of geological scatter. This is within error of the atmospheric value of 295.5, so insignificant initial argon was probably present. These data are from a sample which was stored in vacuum between irradiation and analysis. This was found to be necessary to prevent a strong absorption of atmospheric argon by the sample. The x axis intercept corresponds to 1/slope for a conventional Ar-Ar isochron (Fig. 10.10), yielding an age of 3489 ± 68 Ma (2σ). This is very similar to the plateau age, but the uncertainty is probably a better representation of the true error on the age than the unrealistically low apparent error on the plateau age.

10.2.8 Laser Microprobe Dating

The ability of the Ar–Ar method to obtain age information from argon analysis alone makes it ideally suited to microanalysis by laser ablation. Megrue (1967) pioneered the use of laser ablation for rare gas analysis, but did not apply the method to geochronology until six years later (Megrue, 1973). This study made use of the laser probe in order to date small clasts in a polymict lunar breccia. After activation, spots 100 μm in diameter were irradiated with single pulses

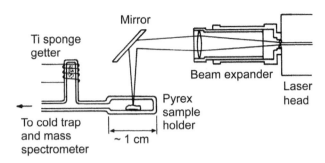

Fig. 10.22 Schematic illustration of laser ablation Ar–Ar dating equipment. After York *et al.* (1981).

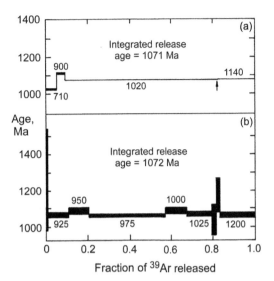

Fig. 10.23 Step heating results for the Hb3GR standard. (a) Conventional; (b) laser single grain. Quoted ages are average (integrated release) ages. Numbers on each release segment are temperatures in °C. After Layer *et al.* (1987).

from a ruby laser. Each pulse ablated a pit about 30 μm deep, equivalent to about 1 μg of rock, representing a miniature total fusion analysis of the exposed surface. The aggregate gas fraction from several nearby spots was gettered and cryogenically trapped, before admission to the mass spectrometer for analysis. Typical equipment is shown in Fig. 10.22. Analysis of ten different clasts revealed two arrays of data on a K–Ar isochron diagram with ages of approximately 3.7 and 2.9 Ga.

York *et al.* (1981) developed the laser microprobe technique by showing that a defocussed continuous wave laser could be used to perform step heating analysis in a manner analogous to conventional Ar–Ar dating. The technique was demonstrated on a whole-rock sample of slate from the Kidd Creek mine, near Timmins, Ontario. The laser beam was focussed to generate a spot 0.6 mm in diameter, which caused progressive argon release from the surface after a few minutes, using a 1 watt power setting. The laser step heating analysis produced results consistent with conventional step heating of the same sample, representing the timing of a thermal event which opened the K–Ar system in the slate.

The low sensitivity of the MS–10 mass spectrometer used by York *et al.* (1981) limited application of the method, but in subsequent development a purpose-built continuous laser system was coupled to a high-sensitivity mass spectrometer. Layer *et al.* (1987) tested this system by analysing the hornblende standard Hb3GR. This is known from previous step heating analysis (Turner, 1971a) to yield a perfect plateau age (Fig. 10.23a). After activation, single grains up to 0.5 mm across were heated within the laser beam for 30 seconds at increasing power levels. After each heating episode, argon was gettered and then analysed. Excellent plateaus were generated (e.g. Fig. 10.23b), and the integrated release ages fell within error of the conventional step heating result.

The laser step heating technique has been shown to be a very effective method for rapid sample analysis, but is subject to the same limitations as conventional step heating analysis. This was demonstrated by a case study on biotite and hornblende grains which had suffered a thermal disturbance long after initial cooling (Lee *et al.*, 1990). The sample consisted of baked Archean gneiss adjacent to an Early

Proterozoic dyke, and both minerals were analysed by three methods: conventional step heating, single grain laser step heating and laser spot dating. Biotite ages for the three methods clustered closely around 2050 Ma, interpreted as the time of dyke intrusion. On the other hand, hornblende produced very different results from the three techniques. Conventional step heating of a multi-grain population and laser spot dating generated very variable ages (Fig. 10.24), while laser step heating generated a good plateau, with an apparent age of 2430 Ma. However, this does not correspond to a known geological event.

Lee *et al.* (1991) speculated that the plateau could result from mixing of argon from different domains in the mineral before release. Heating experiments on the hornblende standard Mmhb–1 showed that argon was released in three principal pulses (Fig. 10.25). The first of these, at 930 °C, was correlated with the onset of structural breakdown at the margins of grains. However, the main phase of breakdown occurred at 1050 °C, forming a strong fabric parallel to cleavage and accompanied by the breakdown of titanite lamellae in the crystal. Finally, at 1130 °C the grains melted. The laser step heating plateau in Fig. 10.24 was formed by argon release between 960 and 1250 °C, suggesting that it may result from argon homogenization in the grain during structural breakdown. Therefore, although laser step heating is a powerful technique, it is necessary to check data from disturbed systems by a second technique such as laser spot dating or laser depth profiling (e.g. Roberts *et al.*, 2001).

Laser probe dating was developed using continuous wave infra-red lasers, either defocussed for step heating, or focussed for spot analysis. These lasers are effective for

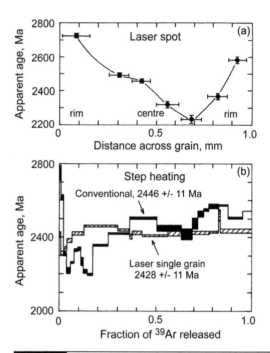

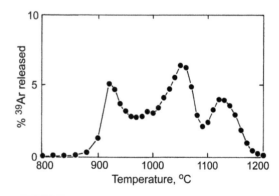

Fig. 10.24 Comparison of spot and step heating ages for a disturbed hornblende sample: (a) profile of laser spot ages across a single grain; (b) laser and conventional step heating profiles. After Lee *et al.* (1990, 1991).

heating most samples, but cannot be focussed below a spot size of 50 μm, and are not effective for analysing pale coloured minerals such as feldspar. To overcome these problems, ultra-violet lasers have been introduced (e.g. Kelley *et al.*, 1994).

Ultra-violet laser light is obtained by frequency doubling, which is only possible using pulsed lasers. The power available with such a system is much lower, but it is effective for spot ablation because the energy is more efficiently con-

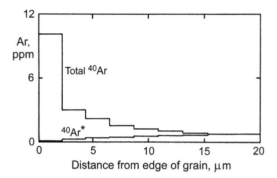

Fig. 10.26 UVLAMP depth profiling measurements of total ^{40}Ar concentration and radiogenic ^{40}Ar in a K-feldspar grain. The data fit a model of radiogenic Ar diffusion out of the grain and atmospheric Ar diffusion into the grain, except for a high value at the surface, probably due to adsorption. After Kelley *et al.* (1994).

centrated in a 10 μm diameter spot. The UV laser ablation microprobe (UVLAMP) offers the opportunity for *in situ* analysis of thin sections with high spatial precision, both laterally and with depth in the sample. Kelley *et al.* demonstrated the depth resolution by using a rastered beam to ablate successive 2 μm-deep steps into the surface of a K-feldspar grain. The resulting isotopic depth profile shows the diffusion of atmospheric argon into the surface of the grain (Fig. 10.26). The method shows great potential for detailed studies of argon diffusion in minerals (Section 10.4.2).

10.3 Timescale Calibration

The high precision that is obtainable in K-Ar dating, particularly using the ^{40}Ar–^{39}Ar step heating method or the laser probe method, makes this dating method very useful for timescale calibration. This applies particularly to the Tertiary period, where Ar-Ar dating can exceed the precision of U-Pb zircon geochronology. The following sections will review some of the most important applications.

10.3.1 Magnetic and Astronomical Timescales
One of the most important applications of the K-Ar method has been to calibrate the magnetic reversal timescale defined by sea floor magnetic anomaly 'stripes'. The amount of ocean floor material recovered which is fresh (unaltered) enough for dating is limited, so most attention has been focussed on dating terrestrial sections (such as basic lavas) which yield a good magnetostratigraphy. Until recently, the K-Ar method was really the only geochronometer capable of dating young volcanic rocks. Since its establishment, the reversal timescale has been subject to almost continuous revision, but a few landmarks are reviewed here.

Pioneering work was performed by *Cox et al.* (1963) on 0–3 Ma lavas from California, and by McDougall and Tarling

Fig. 10.25 Argon release pattern observed in response to heating of the hornblende standard Mmhb–1. After Lee *et al.* (1991).

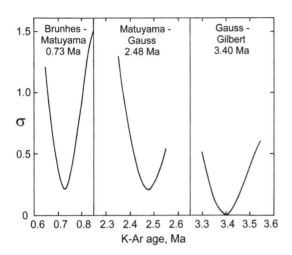

Fig. 10.27 Standard deviation of apparent dating inconsistencies as a function of 'trial' values for polarity boundaries. The best estimate of each boundary age is where error is a minimum. After Mankinen and Dalrymple (1979).

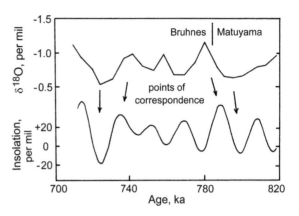

Fig. 10.28 Section of a DSDP core displaying the Brunhes–Matuyama reversal boundary. This is dated by tuning oxygen isotope variations (caused by glacial cycles) to the history of northern hemisphere insolation, calculated from orbital mechanics. Arrows show suggested correlation points. After Johnson (1982).

(1964) on similar-aged lavas from the Hawaiian islands. Cox *et al.* used K–Ar dates on sanidine, obsidian, biotite and whole-rocks, while McDougall and Tarling worked on basalt whole-rocks. Good agreement between the two data sets confirmed that the reversal timescale is due to worldwide changes in the polarity of the Earth's magnetic field, and not due to post-crystallization alteration phenomena, as had been suggested by some workers.

A comprehensive compilation of data for 354 terrestrial lavas (mostly from ocean islands) was used by Mankinen and Dalrymple (1979) to constrain the polarity timescale for the last 5 Ma more precisely, using the new K–Ar decay constants of Steiger and Jager (1977). Not all of the available data were in perfect agreement; therefore Mankinen and Dalrymple used a statistical technique to calculate the most probable ages of the three most recent polarity epoch boundaries, such that the standard deviation of apparent dating inconsistencies was minimized (Fig. 10.27).

An alternative approach to calibrating the magnetic reversal timescale was pioneered by Johnson (1982), based on planetary mechanics (Milankovich forcing, Section 12.4.2). According to this theory, small variations of the Earth's orbit have led to variations in the intensity of solar radiation reaching the Earth, which were responsible for the glacial–interglacial cycles of the Quaternary period. These glacial cycles caused variations in the oxygen isotope composition of seawater which were recorded in fossil forams.

Because the Earth's orbital variations can be projected back into the past very accurately, it is possible to 'tune' the oxygen isotope record in deep sea cores with a precision of better than 1%. Based on coherent patterns from two high-quality cores from the western Pacific, Johnson was able to date the Brunhes–Matuyama magnetic reversal boundary,

which was preserved in sediment 7.25 m below the surface of core V28–239 (Fig. 10.28). He estimated a date of 790 ka for the boundary, with a probable uncertainty of less than 5 ka. However, this work did not receive much attention because of the large discrepancy between this date and the younger value of 730 ka (Fig. 10.28) determined by Mankinen and Dalrymple (1979).

The publication of a new high-resolution oxygen isotope record for the Pleistocene (Shackleton *et al.*, 1990) focussed attention back onto the discrepancy between the astronomical and K–Ar calibrations. A slight revision of the orbital tuning calculation placed the Brunhes–Matuyama boundary at 780 ka and suggested that the date of Mankinen and Dalrymple (1979) might indeed be in error. In fact, a more detailed examination of the data analysis of Mankinen and Dalrymple (1979) by Tauxe *et al.* (1992) showed significant internal disagreement in the data used to constrain the Brunhes–Matuyama (Fig. 10.29). The failure of the error function to reach a zero value in Fig. 10.27 should a have been a warning that the raw data were unreliable, but the scientific community (including the present author) was 'bewitched' by the statistical treatment of the data. In hindsight we can see that the low ages for the reversed (Matuyama) chron were due to some early work by Doell and Dalrymple (1966).

Subsequent work to inter-calibrate the astronomical and geomagnetic timescales has been based almost exclusively on Ar–Ar dating, either of whole-rock basalts, or more commonly on sanidine phenocrysts from volcanic ash beds. The latter have been shown to give homogeneous Ar–Ar ages with little sign of excess argon or argon loss, so that individual grains can be dated by laser total fusion (Fig. 10.30). The individual ages for each grain are stacked, and if the errors bars are variable, the data can be summed using an age

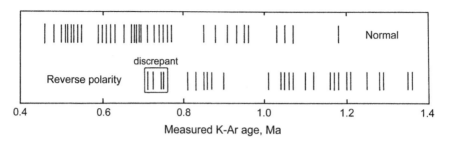

Fig. 10.29 Chart showing raw data used by Mankinen and Dalrymple to date the Brunhes–Matuyama reversal boundary. The best-fit age for the boundary was biased downwards by some early analyses (boxed) that were probably affected by argon loss. Modified after Tauxe *et al.* (1992).

probability curve to calculate the average (Deino and Potts, 1990). Using this technique, Spell and McDougall (1992) dated sanidine crystals from the Valles Caldera, New Mexico, where the anomalously young ages had come from. They obtained ages over 790 ka on reversely magnetized lavas, confirming that the Brunhes–Matuyama boundary should be located at 780 ± 10 ka.

A similar age for the Brunhes–Matuyama reversal boundary was obtained by Baksi *et al.* (1992) using ^{40}Ar–^{39}Ar ages of whole-rock samples of basalt lava, exposed in the caldera wall of Haleakala volcano on the island of Maui. The magnetic signature of the section had previously been studied, and gives such detailed coverage of the reversal boundary that several flows actually have magnetic signatures which are transitional between normal and reversed polarities. Samples were analysed in two different labs, yielding consistent results with an average age for the reversal boundary of 783 ± 11 ka, in excellent agreement with the astronomical calibration.

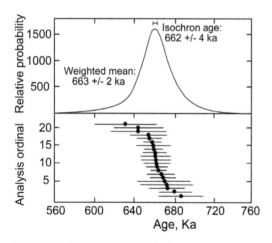

Fig. 10.30 Stacked Ar–Ar ages for individual K feldspar phenocrysts used to determine an age probability curve for a volcanic ash bed. After Deino and Potts (1990).

Use of the astronomical timescale reached a mature stage in 1994 when it was used to back-calibrate the Fish Canyon sanidine standard, one of the primary ^{40}Ar/^{39}Ar dating standards (Renne *et al.*, 1994). This was possible because sanidine crystals from Fish Canyon tuff had been used in several earlier studies to calibrate Ar–Ar ages for six important reversal boundaries. With the advent of a precise astronomical calibration of these boundaries (including a Brunhes–Matuyama boundary age of 0.78 Ma), the Ar–Ar calibration calculation was reversed to determine an 'astronomical' age of 28.03 ± 0.18 Ma for the Fish Canyon tuff.

The Brunhes–Matuyama reversal boundary is the most intensively studied calibration point, but more recent studies have led to somewhat younger astronomical ages for this boundary. For example, an astronomical age of 770 ± 6 ka was estimated for the mid-point to the polarity reversal (Dreyfus *et al.*, 2008), based on a peak of ^{10}Be abundance at the corresponding depth in the Dome C ice core, Antarctica. This age was supported by astronomical dating of North Atlantic sediment cores, yielding a precise boundary age of 773 ± 1 ka (Channell *et al.*, 2010). A lower precision SHRIMP ^{238}U/^{206}Pb age of 770 ± 7 Ma (2σ) also supports a young age (Suganuma *et al.*, 2015). In comparison, a somewhat old Ar–Ar age of 776 ± 2 ka was determined by Coe *et al.* (2004) for transitionally magnetized lavas from Maui, using the conventional half-life. These ages were determined using Taylor Creek and Alder Creek Rhyolite standards, inter-calibrated against a Fish Canyon age of 28.03 Ma (Renne *et al.*, 1998b). However, evidence to be discussed below seems to require an increase in the age of Fish Canyon sanidine, which would further increase the Ar–Ar age for the Maui reversal section to a value well above the new astronomical ages. This problem has not yet been resolved.

The first strong evidence for an older Fish Canyon calibration age came from a Miocene astronomically dated section from Morocco (Kuiper *et al.*, 2008). Ar–Ar ages were determined on tephra horizons in a strongly cyclic 6–7 Ma sedimentary section, and the calibration was then inverted to determine an astronomical age of 28.20 ± 0.05 Ma for Fish Canyon sanidine (using the decay constants of Steiger and Jager, 1977). This represents an increase of nearly 1% in the calibration age. However, a weakness of the study of Kuiper *et al.* (2008) is that the astronomical calibration of the Moroccan section was indirect, using biostratigraphy to calibrate the section from a composite astronomically dated sequence in Spain. Therefore, a more recent study by Rivera *et al.* (2011) used a 6.8 Ma section from Crete with a direct astronomical calibration. This gave a similar age of 28.17 ± 0.03 Ma for Fish Canyon sanidine.

In comparison with these ages, air abrasion U–Pb dating of Fish Canyon zircons by Schmitz and Bowring (2001) gave an average age of 28.48 ± 0.02 Ma. Further analysis by Bachmann *et al.* (2007) gave a slightly older chemical abrasion age of 28.61 ± 0.08 Ma but a younger air abrasion age of 28.37 ± 0.05 Ma. This range of ages implies zircon residence times in the Fish Canyon magma chamber of ca. 200–400 ka, which is slightly higher than the 30–250 ka range of residence ages determined for other high-level magma chambers (Simon *et al.*, 2008), but not unreasonable.

10.3.2 Intercalibration of Decay Constants

In addition to these difficulties with the astronomical timescale, it is observed in older rocks that Ar–Ar ages are consistently slightly younger than U–Pb ages for the same rock unit. This discordance problem is particularly acute for some critical calibration points in the pre-Tertiary Phanerozoic timescale, such as the end of Permian extinction (ca. 250 Ma), and has been called a 'crisis' by some authors (Schmitz and Kuiper, 2013). U–Pb dating is regarded as the 'gold standard' for geochronology (Ludwig, Section 5.2.4), because the uranium decay constants are based on the best laboratory determinations and because the two U–Pb decay schemes provide a built in check for disturbance. Hence this raises the question of whether the decay constant determinations of Beckinsale and Gale (1969), adopted by Steiger and Jager (1977) as recommended values, are significantly in error.

These recommendations also included an improvement in the potassium isotope composition over that used by Beckinsale and Gale (1969), and further small improvements were made by Min *et al.* (2000). A more recent determination by Naumenko *et al.* (2013) improved the precision but did not significantly change the ratio.

Several recent studies have used high-precision measurements on 'geologically ideal' samples to test for concordance between U–Pb and Ar–Ar ages. However, these studies have encountered two difficulties that are specific to the K–Ar system. One of these arises because the Ar–Ar method normally yields higher precision ages than conventional K–Ar dating, but because the efficiency of the activation process is difficult to calculate, Ar–Ar ages are usually determined relative to another age standard. Most of the early age standards used for K–Ar dating were biotite or hornblende, and these became the primary standards for subsequent Ar–Ar dating (e.g. Baksi *et al.*, 1996). However, with the advent of laser step heating, sanidine phenocrysts from Tertiary volcanic ash beds became more popular as dating standards due to their better homogeneity.

A disadvantage of K-feldspar is the difficulty of achieving quantitatively complete extraction of argon for analysis (McDowell, 1983). This is a problem for conventional K–Ar dating, but does not matter for Ar–Ar dating. Therefore, although sanidine makes excellent standards, it cannot be directly calibrated by conventional K–Ar dating. To solve this problem, Renne *et al.* (1998b) calibrated several sanidine standards from volcanic tuffs against the Australian K–Ar biotite standard GA-1550, using potassium concentrations determined by isotope dilution instead of the old flame photometry (atomic absorption) method. The result was an age of 28.02 ± 0.28 Ma for Fish Canyon sanidine, which was refined by Jourdan and Renne (2007), who calibrated Fish Canyon sanidine against four primary standards, yielding a more precise value of 28.03 ± 0.08 Ma (using the conventional decay constants of Steiger and Jager, 1977).

The second major problem specific to K–Ar and Ar–Ar dating is the branched nature of ^{40}K decay, which means that two different decay constants should be considered in any calibration. This means that if lab-based measurements are used, errors in either branch will cause errors in the overall value. Furthermore, the branch leading to argon is theoretically a composite of electron capture and positron decay, although the latter has never definitely been observed experimentally (Renne *et al.*, 2010).

Typically, the two decay constants are combined into a single overall value (Section 10.1), which is then used in comparative dating experiments. However, because the β decay route to calcium affects the amount of potassium believed to be available for decay to argon, the β decay constant has much more effect on old age determinations than young ages. This means that accurate calibration of both branches requires calibration over a range of geological ages, specifically including both old material to constrain the β decay constant and young material to constrain the electron capture branch (Renne *et al.*, 2010). Since geological calibration attempts have often glossed over this problem, this has caused further confusion in the calibration effort.

One of the first of these inter-calibration studies was made on the 1.1 Ga Palisades Rhyolite by Min *et al.* (2000). They obtained an Ar–Ar age of 1088.4 ± 4 Ma, in comparison with a U–Pb concordia age of 1097.6 ± 2 Ma (including analytical errors only). The age difference was used to justify a proposed 1.6% decrease in the β decay constant of ^{40}K to 4.884×10^{-10} yr^{-1}, similar to that proposed in an early counting determination by Endt and Van der Leun (1973). Their recommended values are shown in Table 10.1, and the effect on calculated K–Ar or Ar–Ar ages is shown in Fig. 10.31. However, using a new ^{235}U decay constant (Section 5.1.1), the concordia intercept age of the Palisades Rhyolite would decrease by ca. 5 Ma, bringing the two dating methods within analytical error. This largely undermines this justification for revising the decay constant.

To avoid such problems, Renne *et al.* (2010) made a much more detailed comparison of Ar–Ar and U–Pb ages based on the ^{238}U–^{206}Pb system alone. This avoids uncertainties in the

Author	λ EC	Change	λ β	Change	λ total
Steiger & Jager (1977)	0.581*		4.962*		5.543*
Min *et al.* (2000)	0.580	(−0.17%)	4.884	(−1.6%)	5.370
Renne *et al.* (2010)	0.5755	(−0.95%)	4.974	(+0.2%)	5.549

Table 10.1 Alternative decay constants relative to Steiger and Jager (1977).

* All values are $\times 10^{-10}$ yr^{-1}

location of the concordia line (Section 5.2.4), and was made possible by the development of the chemical abrasion technique, which can remove discordant Pb in relatively pristine zircon (Section 5.2.3). However, there is one age bias that cannot be easily screened for, which is the tendency, noted for the Fish Canyon tuff, for zircons to give U–Pb ages older than the estimated time of eruption, due to a finite magma chamber residence time. Based on previous work, Renne *et al.*, made an across-the-board correction to U–Pb ages of ca. 100 ka (Fig. 10.32). This correction was not large enough to explain the abnormally old U–Pb age for the youngest (17 Ma) sample. Conversely, all other samples in the study are over 100 Ma, and for them the residence effect is negligible.

Based on the age dependence of the bias between U–Pb and Ar–Ar ages (Fig. 10.32), Renne *et al.* (2010) attempted to predict the optimal changes in both ^{40}K decay constants that would fit these data. Ignoring the young sample with excess magma chamber residence, the Phanerozoic samples all fall within error of a 1% correction to the electron capture decay constant. However, one older sample around 1 Ga in age fell below this curve. This led Renne *et al.* to propose that the β decay constant should actually be *increased* (the opposite of the change proposed by Min *et al.*, 2000).

One other data point that is particularly pertinent for modelling the branching ratio of the decay constants is derived from the Ste Marguerite and Forest Vale H4 chondrites (open circle in Fig. 10.32). These were not included in the original data set of Renne *et al.* (2010) because they

can only be dated by the 207/206 method, not the ^{238}U–^{206}Pb method. However, very accurate Pb–Pb ages for these meteorites can be estimated by adjusting the ^{235}U decay constant (Section 5.1.1). This gives an average Pb–Pb age of 4557 Ma, in comparison with an average Ar–Ar age of 4527 (Trieloff *et al.*, 2001), corresponding to a 0.66% bias (Fig. 10.32). Other evidence cited by Trieloff *et al.* (2001) suggests that these particular H chondrites underwent rapid cooling and were never reheated, so their ages should be considered very reliable.

Overall, the data in Fig. 10.32 represent a fairly close fit to a model involving a 1% increase in the electron capture decay branch (compared with Steiger and Jager, 1977). The data also clearly refute the model of Min *et al.* (2000), which should be abandoned. However, a problem with the values of Renne *et al.* (2010) is that all Ar–Ar data in their study were tied to a Fish Canyon age of 28.03 Ma. When the age calculations are run backwards using measured U–Pb ages and the new decay constants to date Fish Canyon sanidine as an 'unknown', the result is an age of 28.35 Ma. The 1% discrepancy between these two calibrations is almost exactly equivalent to the proposed 1% decrease in electron capture decay constant.

An alternative approach would be to continue to use the old decay constants, with a new Fish Canyon calibration age

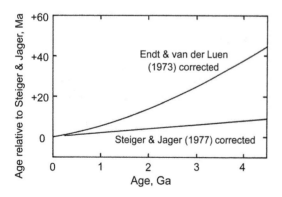

Fig. 10.31 Effect on calculated geological ages of using alternative ^{40}K β decay constants. After Min *et al.* (2000).

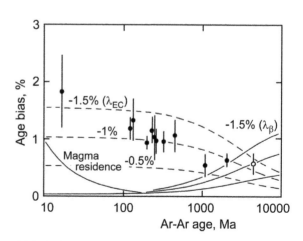

Fig. 10.32 Age biases in U–Pb relative to Ar–Ar ages, compared with alternative corrections to the β and electron capture potassium decay constants. After Renne *et al.* (2010).

of 28.35. This might also require a revision of the β decay constant to prevent H chondrites from giving too old an age. It is also notable that the astronomically tuned Fish Canyon age of 28.17 (Rivera *et al.*, 2011) is intermediate between the two calibration ages mentioned above. However, this age (and that of Kuiper *et al.*, 2008) was determined using the erroneous decay constant of Min *et al.* (2000).

Ultimately, the decay constant is always used together with the age of the standard in an Ar–Ar age determination, so these two quantities essentially form a package for all except the oldest rocks. The calibration of Renne *et al.* (2010) may be provisional, but it gives correct Ar–Ar ages relative to U–Pb for most calibration sections other than the Brunhes–Matuyama boundary (discussed above). For example, their calibration increases the Ar–Ar age of the K–T boundary to 66.0 Ma, consistent with U–Pb dating (Renne *et al.*, 2013).

10.4 Thermochronometry

The thermal history of meteorites was interpreted in Section 10.2.4 in terms of short-lived thermal events, including the initial cooling and subsequent collision of their parent bodies, separated by long periods under cold conditions. However, Turner (1969) recognized that in different circumstances, slow cooling from a single event could yield an age spectrum rather similar to that produced by episodic thermal events.

10.4.1 Arrhenius Modelling
A special feature of the Ar–Ar step heating method is that argon release in the laboratory should be controlled by the same diffusional properties of minerals that cause argon loss in nature. The long timescales of geological events obviously preclude laboratory analysis under natural conditions. However, if diffusion obeys the Arrhenius law, a rapid argon release experiment at high temperature in the laboratory can mimic much slower argon release at lower temperatures in the crust. Hence, step heating data may be used to reconstruct thermal histories such as post-orogenic cooling.

The result of slow cooling in a young intrusive body is illustrated by the age spectrum of a biotite from the La Encrucijada pluton in Venezuela (Fig. 10.33). Unfortunately, this approach can rarely be used on terrestrial rocks because the low-temperature part of the profile, which is critical in the determination of a precise cooling rate, becomes 'corrupted' by minor diffusional loss of argon at ambient temperatures over geological time.

A more useful approach to quantify the cooling history of crustal rocks is the blocking temperature concept (Section 3.3.2), whose theoretical basis was examined by Dodson (1973). Argon loss from a mineral can be described by the thermal diffusion coefficient:

$$D = D_0\, e^{-E/RT} \qquad [10.14]$$

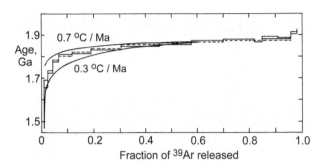

Fig. 10.33 ^{40}Ar/^{39}Ar age spectrum of La Encrucijada biotite, Venezuela, compared to predicted cooling curves based on modelling of Ar diffusion in biotite. After York (1984).

where D_0 is the thermal diffusivity of the mineral, E is the activation energy of argon diffusion, R is the gas constant and T is absolute temperature. The exponent causes D to be a very strong function of temperature. Therefore a small temperature drop can cause a transition from a state where Ar loss by diffusion is rapid to a state where Ar loss by diffusion is very slow. This relatively sharp transition constitutes the process of blocking. The blocking temperature T_B is defined by Dodson (1973) as follows:

$$T_B = \frac{E}{R\,\ln\left(A\,\tau\,D_0/a^2\right)} \qquad [10.15]$$

where A is a geometrical parameter which takes account of the crystal form of the argon-bearing mineral (55, 27 or 9 for a sphere, cylinder or sheet respectively), a is the length of the average diffusion pathway from the interior to the surface of the grain and τ is the cooling time constant. The latter is in turn defined as follows:

$$\tau = R\,T_B^2/E(-C)_B \qquad [10.16]$$

where $(-C)_B$ is the cooling rate at the blocking temperature T_B. Hence, substituting equation [10.15] into [10.14] yields

$$T_B\,R = E/\ln\left[A \cdot \frac{R\,T_B^2}{E(-C)_B} \cdot \frac{D_0}{a^2}\right] \qquad [10.17]$$

A method to calculate blocking temperatures from Ar–Ar spectrum plots was proposed by Buchan *et al.* (1977) and developed by Berger and York (1981a). A plateau age must be available on a mineral from a slowly cooled terrane. For each heating step in the plateau, the volume of radiogenic ^{40}Ar released in a given time is used to calculate D/a^2. For planar minerals such as biotite the diffusion equation has the following general form (e.g. Harrison and McDougall, 1981):

$$\frac{D}{a^2} = \frac{(q\,f)^2}{t} \qquad [10.18]$$

where f is the fractional loss of argon, t is the heating time and q is a geometric factor.

The results are plotted on a log scale against the reciprocal temperature of each step, forming an Arrhenius plot

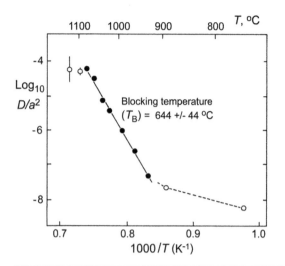

Fig. 10.34 Arrhenius plot for hornblende from a Grenville diorite, Haliburton Highlands, Ontario. The blocking temperature was determined from the array of seven solid data points. Error bars are 1σ. After Berger and York (1981a).

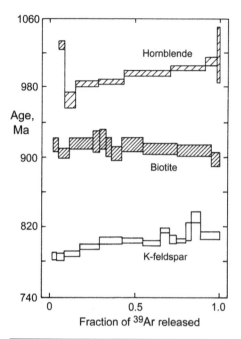

Fig. 10.35 40–39 age spectra for the hornblende, biotite and K-feldspar analyses used to determine blocking temperatures in Figs. 10.34 and 10.36. Modified after Berger and York (1981a).

(e.g. Fig. 10.34). If diffusional Ar loss obeys the Arrhenius law as expected, then the steps in the plateau should define a straight line whose slope is the activation energy E and whose y intercept is the frequency factor D_0/a^2. These values enable equation [10.17] to be solved, provided that the cooling rate $(-C)_B$ at the blocking temperature (T_B) can be estimated. Fortunately, the temperature solution has a weak dependency on cooling rate, such that an order of magnitude change in this value only causes a 10% change in the calculated blocking temperature. Because T_B appears on both sides of equation [10.17], it must be solved iteratively, but it converges quickly. The power of this technique is that the mineral blocking temperature is calculated directly on the dated material, rather than having to depend on generalized blocking temperatures for different mineral types from the literature, which may not be applicable to the specific cooling conditions under study.

Berger and York (1981a) applied this 'thermochronometry' method to a study of post-orogenic cooling in the Grenville Province of southern Ontario, Canada. Plutonic ages in the Grenville belt vary from 1.0 to 2.7 Ga, but most K–Ar dates fall below 1 Ga, and are attributed to uplift and cooling after collisional orogeny (Harper, 1967). Berger and York studied dioritic and gabbroic plutons from the Haliburton Highlands, both to determine a detailed cooling curve for the area and to interpret paleomagnetic data on these rocks.

Typical 40–39 profiles from Haliburton diorites which gave reasonable plateaus are shown in Fig. 10.35. These samples are plotted on Arrhenius plots in Figs. 10.34 and 10.36. The hornblende displays a relatively simple array in Fig. 10.34, although low-temperature and high-temperature

points must be excluded from the regression. K-feldspar displays coherent low-temperature behaviour, but high-temperature data are irregular, possibly due to disruption of the lattice at above 900 °C. The most unusual behaviour was demonstrated by biotite, which in many cases gave rise to two heating pulses which defined sub-parallel arrays (such as in Fig. 10.36). Nevertheless, the blocking temperatures calculated from the two segments were usually within 2 sigma error. Berger and York speculated that the break in regular behaviour was due to structural breakdown.

Results from all of the analysed minerals are shown on a diagram of blocking temperature against age (Fig. 10.37). Points without error bars failed a reliability criterion which required both the age plateau and Arrhenius correlation line to have four or five statistically well-fitting data points. The data show a clear picture of fairly rapid cooling (ca. 5 °C/Ma) from the hornblende blocking temperature of ca. 700 °C at 980 Ma to the biotite blocking temperature of ca. 380 °C at 900 Ma. Thereafter, the data are mainly from plagioclase, which displays considerable scatter. Berger and York's original interpretation (solid line in Fig. 10.39) called for very slow cooling (under 1 °C/Ma) for a further 300 Ma. However, in a study of gabbro from the Hastings Basin of the Grenville (ca. 80 km east of the Haliburton Highlands), Berger and York (1981b) recognized that the apparent slow cooling curve after 900 Ma might really represent a more recent thermal event.

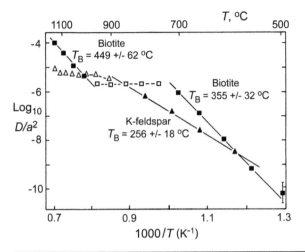

Fig. 10.36 Arrhenius plot for Grenville K-feldspar (triangles) and biotite (squares) dated in Fig. 10.37. Blocking temperature calculations were based on solid data points only. Errors are 1σ. After Berger and York (1981a).

The latter interpretation of the plagioclase data was supported by Hanes *et al.* (1988), based on Ar–Ar dating of the Elzevir and Skootamata plutons of the Hastings basin. Hanes *et al.* analysed three plagioclases which displayed a range of rather mediocre plateau ages between 400 and 600 Ma. Variations in $^{37}Ar/^{39}Ar$ with the fraction of ^{39}Ar released were used as an index of the Ca/K ratio of different domains within the minerals. A pronounced hump in the middle of these profiles indicated that the analysed plagioclases were multi-phase systems. They display two different types of alteration which may be of different age. Hanes *et al.* suggested that scattered coarse epidote and muscovite alteration might have

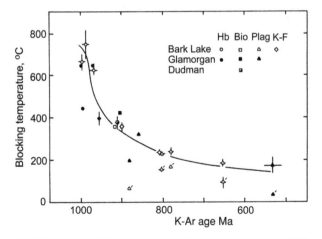

Fig. 10.37 Plot of calculated mineral blocking temperatures against plateau ages to show a model for crustal cooling after the Grenville orogeny. Solid and open symbols indicate minerals from different plutons. After Berger and York (1981a).

formed at high temperatures soon after plutonism, while fine-grained sericitic alteration probably represents an event younger than 400 Ma.

The evidence for structural breakdown in biotite points to a weakness in the thermochronometry method of Berger and York. Because biotite is a hydrous mineral, diffusional loss of Ar during vacuum heating may not accurately mimic Ar loss in nature under (probable) hydrothermal conditions, as suggested by Giletti (1974). This problem was confirmed by Gaber *et al.* (1988), who showed large divergences in argon diffusivity between hydrothermal and vacuum heating experiments on biotite and hornblende.

The susceptibility of these hydrous minerals to structural breakdown during vacuum heating has largely discredited the application of step heating analysis to determine their blocking temperatures. Instead, attention has switched to K-feldspar, the most common anhydrous K-bearing mineral. As an anhydrous mineral, the behaviour of this mineral during vacuum degassing can mimic argon loss in nature with some accuracy (Section 10.4.3). However, studies of diffusional argon loss from hydrous minerals have been continued using laser spot analysis. For example, this method has been applied to the study of complex argon diffusion models in micas and amphiboles.

10.4.2 Complex Diffusion Models

The traditional approach to mineral blocking temperatures is based on 'Fickian' behaviour, which means that argon diffusion in minerals is assumed to occur by volume diffusion, obeying Fick's first law (e.g. McDougall and Harrison, 1999). However, detailed analysis of Ar–Ar ages in minerals has revealed breakdowns in this simple model, requiring more complex models of argon diffusion to be proposed. The advent of the laser probe has allowed the detailed spatial analysis of minerals in order to test these more complex models.

In order to test the laser step heating method on a slowly cooled geological system, Layer *et al.* (1987) analysed biotites from the Trout Lake batholith, NW Ontario. Laser step heating of a small (0.25 mm) biotite grain yielded an age of 2600 Ma, which was identical to a conventional step heating analysis on 13 mg of biotite. However, laser step heating of a large (1 mm) biotite from the same hand specimen yielded a significantly older age. The total release age for this grain (2654 ± 5 Ma) was closer to the U–Pb intrusive age for the batholith of 2699 ± 2 Ma. This shows that the larger biotites probably closed earlier during metamorphic cooling.

Wright *et al.* (1991) developed this study on the Trout Lake batholith by using the laser step heating method to date a range of single biotite grains of different sizes. Only grains having a regular shape similar to a thin cylinder were analysed. After measurement of the grain radii and activation in the reactor, each specimen was subjected to laser step heating analysis, during which the whole grain was bathed in the laser beam at increasing intensities. For samples displaying normal plateaus, their integrated ages were plotted

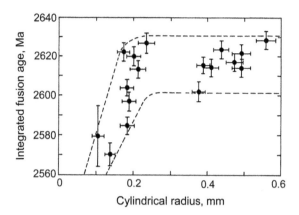

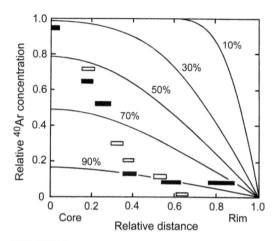

against grain size (Fig. 10.38). For small grains, the results display a positive correlation between grain size (cylindrical radius) and integrated age. This was attributed to diffusional argon loss during the original cooling history of the batholith. Wright et al. speculated that the scatter of data points to the right of the steep array might represent large grains which were either damaged during sample crushing or consisted of natural aggregates of smaller sub-domains.

The positive correlation on the left side of Fig. 10.38 is explained by the larger surface area / volume ratio of smaller grains, resulting in a lower effective blocking temperature than larger grains. Geological determinations of Ar diffusion in biotite (e.g. Onstott et al., 1989) can be used to calculate the size dependence of blocking temperature (0.1 mm = 275 °C; 0.23 mm = 295 °C). Hence, the sloping array in Fig. 10.38 translates into a cooling curve for the batholith of temperature against time. This yields a calculated cooling rate between 295 and 275 °C of about 0.33 °C/Ma. The high-temperature cooling curve of the pluton can be calculated between the older biotite ages and the U–Pb zircon age of 2700 Ma (with a blocking temperature estimated at around 750 °C). This segment of the cooling curve is much steeper, at around 5 °C/Ma.

The data points on the right-hand side of Fig. 10.38 illustrate a well-known phenomenon, that large biotite grains can lose radiogenic strontium or argon more rapidly than predicted by volume diffusion (e.g. Section 3.3.2). This behaviour was also seen in argon diffusion experiments using hydrothermal bombs, and can be explained if biotite has an 'effective diffusion radius' of about 150 μm (e.g. Harrison et al., 1985). This implies that biotite grains consist of domains of ca. 150 μm radius, within which argon moves by volume diffusion, but between which there is a mechanism of enhanced transport along crystallographic defects.

The increased amount of spatial argon data available from the laser probe has allowed complex diffusion mod-

els to be tested using natural mineral systems. Phillips and Onstott (1988) made such a study, based on mantle-derived phlogopites from the Premier kimberlite, South Africa. 'Chrontour' mapping of 75 laser spots revealed 1.2–1.4 Ga apparent ages in the rim of a large grain, rising to a maximum of 2.4 Ga in the core. This pattern was attributed to the loss of inherited mantle argon from the rim of the grain during 1.2 Ga kimberlite emplacement. However, the shape of the argon loss profile (boxes in Fig. 10.39) did not fit a volume diffusion model (curves in Fig. 10.39). Phillips and Onstott attributed the observed patterns to enhanced diffusive loss of argon from a wide band (round the rim of the grain) due to structural defects. However, the nature of these defects was unclear.

Hodges et al. (1994) compared argon diffusion patterns in muscovite and biotite using the laser probe. Single mica grains were analysed from the 1700 Ma Crazy Basin monzogranite of central Arizona, which has been argued to display either very slow cooling or multiple metamorphic events. Laser step heating of both muscovite and biotite gave very similar plateau ages of 1412 and 1410 Ma. However, for muscovite, the gas release steps which form the plateau occurred after melting had begun. Therefore, this plateau undoubtedly results from argon homogenization during melting, and the similarity to the biotite plateau age is probably a coincidence. These findings are consistent with the work described above, where argon homogenization was shown to be a greater problem in laser step heating than conventional step heating (Section 10.2.8).

Laser spot dating revealed very different age patterns in other muscovite and biotite grains from the same sample. Muscovite spot ages ranged from 1270 Ma at the rim to 1650 Ma (the 'true' age) in a small core area, with a roughly

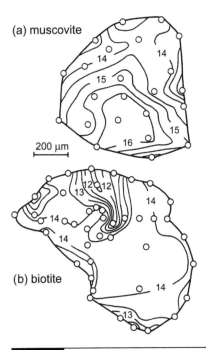

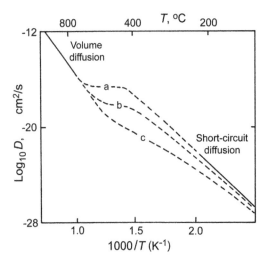

Fig. 10.41 Predicted effect of an 'exchange coefficient' (K2) in changing the dominant diffusive mechanism from volume to short-circuit diffusion. Curves a, b, and c represent exchange coefficients of $-10^{-13}, -10^{-14}$ and -10^{-15} s^{-1} for the transfer of species from the crystal lattice to high-diffusivity paths. After Lee (1995b).

Fig. 10.40 Comparison between 'chrontour' age maps for (a) muscovite and (b) biotite, determined from laser spot dating. The biotite ages cannot be explained by volume diffusion. After Hodges *et al.* (1994).

concentric pattern (Fig. 10.40a). This age distribution fits a simple volume diffusion model for cylindrical geometry with an effective diffusion radius similar to the grain size. On the other hand, laser spot ages in the biotite grain had a quite different distribution, with ages ranging from 1150 Ma at the rim to 1420 Ma in a large core area (Fig. 10.40b). The large area with (reset) 1400 Ma ages, along with an asymmetric zone of young ages approaching the core, was evidence of the operation of fast diffusion pathways. Hodges *et al.* interpreted the data in terms of small domains with an effective diffusion radius of 150 μm, as postulated in hydrothermal experiments.

Lee and Aldama (1992) attempted to explain complex diffusion behaviour by combining mechanisms for diffusion within the crystal lattice with enhanced ('short-circuit') diffusion between lattice domains; thus making a model termed 'multi-path' diffusion. In this model, diffusion in both of the pathways (lattice and short-circuit) is Fickian, but is of very different magnitude in each pathway. In addition, the transfer of species from the crystal lattice to the high diffusivity paths, and vice versa, is attributed to two exchange coefficients, termed K2 and K1 respectively. The magnitude of these coefficients is critical in determining whether multi-path diffusion will occur in a given situation.

The operation of the multi-path diffusion model can be examined on an Arrhenius plot (Fig. 10.41). At high tempera-

tures (>800 °C), volume diffusion in the crystal lattice is the dominant mechanism for argon loss from a mineral. However, if the exchange coefficient (K2) is relatively large, this causes a sudden change from lattice dominated to short-circuit argon transport as the temperature falls, forming a sigmoidally shaped diffusion curve. On the other hand, smaller values of K2 inhibit the role of short-circuit diffusion, maintaining a lattice dominated diffusion mechanism to lower temperatures (<500 °C).

Although the multi-path model is an attempt to create a realistic representation of the physical movement of argon in minerals, McDougall and Harrison (1999) argued that it is of limited use in analysing thermal histories from geological samples, because the postulated exchange coefficients cannot be determined by laboratory measurement. In addition, the model cannot explain the results of 'cycling step heating', a method developed by Lovera *et al.* (1991) for K-feldspar thermochronometry (see below). In this method, the heating schedule in an Ar–Ar degassing experiment is not increased monotonically as in a conventional analysis, but is increased in short 'bursts' with a period in between, when the temperature is either reduced or maintained at a constant level for a relatively long time. An example of such a heating schedule is shown in Fig. 10.42.

According to the multi-path diffusion model, the short-circuit diffusion path should be 'replenished' by volume diffusion out of the lattice during the backward cycling stage of the step heating procedure, so that the subsequent low-temperature forward cycles reproduce the low-temperature behaviour from the beginning of the experiment. However,

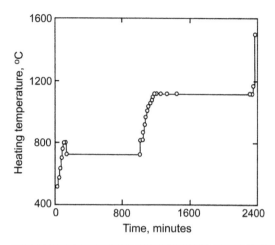

Fig. 10.42 Demonstration of a cyclic heating schedule for Ar–Ar analysis. After Lovera *et al.* (1991); Richter *et al.* (1991).

such behaviour was not seen (McDougall and Harrison, 1999). Instead, they pointed out that if the exchange coefficients are ignored, the multi-path diffusion model is mathematically equivalent to argon diffusion from two different domain sizes in the multi-domain model (Section 10.4.2). In other words, argon extraction over relatively long distances along high-diffusivity pathways is equivalent to volume diffusion from very small lattice domains.

Complex diffusion models can be tested against the results of hydrothermal diffusion experiments on hornblende and biotite at different grain sizes (Fig. 10.43). Lee (1995b) argued that the enhanced diffusivities for large grain sizes favour the multi-path model over simple volume diffusion. However, the data can also be explained by a domain model, where the size of the lattice domains controlling volume diffusion is smaller than the complete crystal.

In the hydrothermal experiments, the behaviour of hornblende and biotite may result from similar styles of alteration, both in the laboratory and natural systems. For example, Onstott *et al.* (1991) suggested that fast diffusion pathways in hydrothermal biotite may be created during the experiment by the breakdown of chlorite layers, whereas Kelley and Turner (1991) attributed fast diffusion pathways in natural hornblende to biotite alteration.

Some recent experiments using the UV laser ablation microprobe (UVLAMP) suggest that in pristine, unaltered grains of biotite, most argon diffusion out of (or into) the grains occurs by volume diffusion. For example, Pickles *et al.* (1997) demonstrated such behaviour in biotite grains of various sizes (50 to 4500 μm diameter) in a pegmatite affected by Alpine metamorphism. During the metamorphic event, excess ^{40}Ar diffused into the grains from the grain boundary fluid and created diffusion profiles. In several grains, diffusion profiles up to 100 μm long were successfully modelled by volume diffusion. In the largest grain, the diffusion profile spanned a distance of nearly 300 μm, although the out-

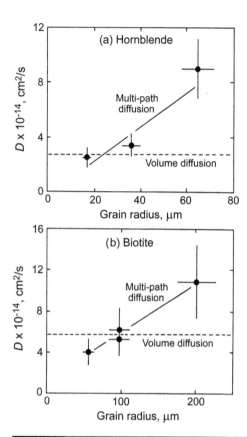

Fig. 10.43 Plot of bulk argon diffusion coefficients against grain radius. Measurements from hydrothermal experiments on hornblende and biotite are compared with the predicted results of volume and short-circuit diffusion models. After Lee (1995b).

ermost 90 μm of the profile was lost due to argon loss during a later alteration event.

Other experiments suggest that short-circuit diffusion is principally a feature of low-temperature argon release. For example, Lo *et al.* (2000) found low-temperature argon diffusivities two to four orders of magnitude faster than the values that would result from extrapolation of hydrothermal experiments to low-temperature conditions. However, they suggested that this low-temperature release might represent argon that had accumulated in radiation induced lattice defects. Evidence for this interpretation comes from the fact that the outgassing of ^{39}Ar usually exceeds ^{40}Ar during very low-temperature argon release (ca. 400 °C), which would be consistent with the preferential release of recoiled ^{39}Ar atoms residing in defect sites. In this case the occurrence of short-circuit diffusion might be to some extent an artefact of the irradiation procedure involved in the Ar–Ar method. Similar doubts have been expressed about the meaning of the lowest temperature emission step from K-feldspars (Section 10.2.6 and below).

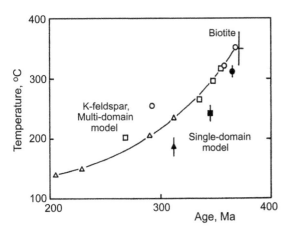

10.4.3 K-Feldspar Thermochronometry

K-feldspars have long been known to show complex diffusion behaviour. Because of the unpredictable effects of perthitic exsolution, K-feldspars have very variable blocking temperatures, which must be determined for individual dated samples by the step heating method. Heizler *et al.* (1988) demonstrated this technique in determining the cooling curve of the Chain of Ponds pluton, NW Maine. They obtained quite distinct blocking temperatures and ages on three feldspar separates, establishing a post-Appalachian cooling curve from 330 °C to 180 °C (solid symbols in Fig. 10.44).

Other workers (e.g. Foland, 1974) suggested that diffusion in K-feldspars is controlled by microstructural domains of variable size, rather than on a whole-grain scale. The variable domain sizes invoked by this model can explain the traditional reputation of K-feldspars for having such poor argon retentivity as to be useless as a dating tool. The smallest of the domains do indeed suffer argon loss at near ambient temperatures. However, the larger domains, sampled in a step heating analysis, can have blocking temperatures as high as biotite. Therefore, if this variation in blocking temperatures can be exploited using the step heating method, K-feldspar can be a powerful tool in thermochronometry. Because this method is based on the assumption of domains of various sizes, it is termed the multiple diffusion domain (MDD) model.

Adopting a MDD model, Lovera *et al.* (1989) reinterpreted the data of Heizler *et al.* (1988) by breaking each Ar–Ar analysis into a series of sub-plateaus with distinct ages and blocking temperatures. They attributed these sub-plateaus to diffusional domains within each feldspar grain, varying in size by two orders of magnitude. This model was tested by com-

paring measured step heating data with model spectra based on different domain size distributions, as applied by Turner *et al.* (1966) to meteorite studies (Section 10.2.4). The variable domain size model was shown to fit the experimental data much better than the uniform model, thus confirming its usefulness. The result of this approach was that each analysis yielded a separate but overlapping cooling curve *segment* (open symbols in Fig. 10.44), rather than a single point on the cooling curve. Further experiments on single feldspar crystals by Lovera *et al.* (1991) showed that domains of varying size are an intrinsic property of alkali feldspars, which therefore cannot be separated by hand-picking of material for analysis.

Thermochronometry is based on linear Arrhenius relationships observed when ^{39}Ar is released from diffusion domains of uniform size and K content, and which obey a simple diffusion law. In their development of the K-feldspar method, Lovera *et al.* (1989) showed that non-linear Arrhenius trends produced by conventional step heating could be resolved into separate linear segments by cycling the heating schedule up and down.

Subsequent work on a variety of samples showed that these line segments were effectively parallel, indicating relative constancy of diffusional activation energies. The vertical separation between different linear segments on the Arrhenius diagram can then be interpreted in terms of relative domain size. If the effective diffusion radius of the domain which generates the first, low-temperature array is set arbitrarily to be r_0 ('*r*' is the equivalent of '*a*' in Section 10.4.1), then the relative diffusion dimension of each subsequent gas release point can be calculated using the following relation:

$$\log\left(\frac{r}{r_0}\right) = \frac{\log\left(\frac{DT}{r^2}\right) - \log\left(\frac{DT}{r_0^2}\right)}{2} \qquad [10.19]$$

Because we are only interested in the relative diffusion dimension, the exact geometry assumed (slab or sphere etc.) has little effect on the calculations. Richter *et al.* (1991) proposed that each value of $\log(r/r_0)$ should be plotted against cumulative ^{39}Ar release in a manner analogous to an age spectrum plot (Fig. 10.45b). This plot therefore bridges the gap between the age spectrum diagram and the Arrhenius diagram.

The working of the 'Richter plot' was demonstrated on Ar–Ar data from the Quxu pluton of the Himalayas. The diagram gave evidence of four plateaus, the three highest temperature of which correspond to plateaus on the age spectrum plot (Fig. 10.45a). The lowest temperature plateau had no corresponding age because the first five steps were perturbed by excess ^{40}Ar, which probably diffused into the grains after cooling of the pluton. However, this did not affect the ^{39}Ar released during this part of the analysis, which can still be used to obtain diffusional data. On the other hand, the highest temperature step gave an age, but no diffusional information, because it occurred during sample melting. After the gas release is broken into plateaus, the

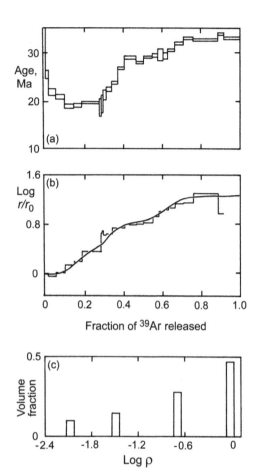

Fig. 10.45 Comparison between different data presentations for K-feldspar thermochronometry: (a) traditional age spectrum plot; (b) r/r_0 plot; (c) resulting solution of volume fraction against relative domain size. After Richter *et al.* (1991).

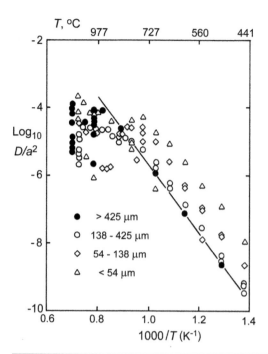

Fig. 10.46 Arrhenius plot for K-feldspar from the Chain of Ponds pluton, Maine. Parallel linear arrays were obtained from the low-temperature gas releases, after crushing to four different grain sizes. After Lovera *et al.* (1993).

volume fraction of each is calculated, and an iterative program is used (Lovera, 1992) to determine the actual radius ratios of the domains which will model the observed profile of relative diffusion dimensions. The solution is shown in Fig. 10.45c (Richter *et al.*, 1991). Finally, a cooling curve is determined (iteratively) which will yield the observed age spectrum.

In several experiments on K-feldspars from different orogens, the radius ratio of largest to smallest domains was determined to be between 100 and 500. Lovera *et al.* (1993) performed step heating experiments on the Chain of Ponds pluton, Maine (sample MH-10) in an attempt to find the absolute sizes of these domains. After crushing, K-feldspar grains were separated into four size fractions, averaging 425, 138, 54 and 42 μm diameter. Following irradiation, a full step heating analysis was performed on each size fraction, and the resulting Arrhenius plots compared (Fig. 10.46). The results showed a dramatic decrease in argon retentivity between the

138 μm and 54 μm size fractions, attributed to the 'breaking open' of the largest diffusion domain in crushing to 54 μm. Hence, the diameter of this domain was inferred as 50–100 μm. Based on relative domain size ratios, the smallest domain size was then estimated at ca. 0.1 μm.

In an accompanying optical and TEM study, FitzGerald and Harrison (1993) attempted to determine the crystallographic identity of the different domain sizes. They tentatively correlated the largest domains with blocks of K-feldspar surrounded by fractured and turbid zones, and the smallest with the 0.1 μm (100 nm) distance between albite exsolution lamellae. However, they were unable to identify any intermediate sized domains.

More recent studies have attempted to achieve a better understanding of argon diffusion in K-feldspar by studying gem-quality crystals, which are expected to display simpler diffusional behaviour. Two of these studies were made on the same gem-quality orthoclase from Madagascar. The first study (Arnaud and Kelley, 1997) primarily used cyclic step heating experiments to investigate the argon release behaviour of the orthoclase. This experiment revealed that Ar release occurred in two stages with different activation energies. The main argon release was consistent with simple volume diffusion from a single domain size. However, the low-temperature release could be explained by either rapid diffusion pathways (e.g. short-circuit diffusion) or by multi-diffusion domains. However, this release stage represented only 0.5% of the total ^{39}Ar inventory of the sample, so Arnaud and Kelley speculated that it actually represented argon release from the surface of the sample, possibly damaged during sample preparation.

The second study of this material used the UV laser ablation microprobe to perform depth profiles of the sample surface (Wartho *et al.*, 1999). This study revealed that argon release from the surfaces of the orthoclase crystal was controlled by volume diffusion, just like the crystal interior. In fact, calculations of the effective diffusion radius for the low-temperature release and the main argon release gave values of 1.8 mm and 1.3 mm respectively, both of which were consistent with the 2–3 mm dimensions of the analysed fragments. Therefore, despite the different activation energies of the different release stages, they both reflected volume diffusion from a single domain, consistent with the unflawed nature of the analysed fragments. Possibly, the low-temperature release represented argon from vacancy sites in the mineral. It was concluded from these experiments that cyclic step heating faithfully measured the diffusion parameters of the sample. However, they could not provide further help in understanding complex metamorphic K-feldspars.

Because the estimated size of the smallest diffusion domains in the MDD model is similar to the ^{39}Ar recoil distance, doubts have been expressed (Section 10.2.6) about the meaning of the lowest temperature argon emission data. In addition, Parsons *et al.* (1999) argued that many K-feldspars display micro-structural complexity that is metastable during step heating experiments. Therefore, they suggested that alkali feldspar thermochronology using the MDD model is a 'mathematical mirage' rather than a method that can be used to recover real thermal histories. However, several empirical demonstrations have been made where cooling curves based on K-feldspar thermochronology are in good agreement with other geochronological evidence. For example, Lee (1995a) demonstrated good agreement of K-feldspar cooling curves with 40–39 muscovite ages and fission track apatite ages in a study of tectonic uplift in the Snake Range, Nevada (Fig. 10.47).

To provide a more objective test of the reliability of K-feldspar thermochronology as a technique to recover real thermal histories, Lovera *et al.* (1997, 2002) analysed the argon release patterns of a suite of nearly 200 basement samples. They found that the most common obstacle to the derivation of thermal histories from step heating results was inherited argon. This has long been recognized as a problem in K–Ar and Ar–Ar dating, and in the study of Lovera *et al.* it became manifest through saddle-shaped profiles (intermediate age minimum) on the age spectrum diagram. On the other hand, a second type of anomalous behaviour became manifest as an intermediate age *maximum*, attributed to low-temperature alteration. Together, these problems affected about half of the total sample suite of 194 K-felspars studied. The remaining half of the sample suite was considered to be 'well behaved', and therefore suitable for analy-

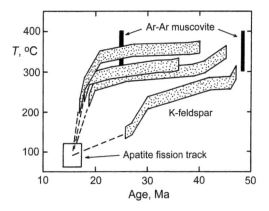

Fig. 10.47 Comparison of cooling data for the Snake Range, Nevada. Shaded fields derived from K-feldspar thermochronometry are consistent with control points from apatite fission track analysis and Ar–Ar muscovite analysis (boxes). After Lee (1995a).

sis by the MDD model. In addition, the activation energies in these samples had a normal distribution (average = 46 ± 6 kcal/mol) so that a standard protocol could be used to analyse all samples, thereby minimizing subjectivity in interpretation of the data.

Lovera *et al.* (2002) argued that the effectiveness of the MDD model in explaining the real thermal history of a sample could be assessed by the goodness of fit between the step heating age spectrum and the pattern of changing effective diffusion radius ratio (r/r_0). If these two profiles matched in a K-feldspar analysis, this was held to be evidence that the laboratory argon release experiment successfully mimicked the diffusional loss of argon over the cooling history of the sample. However, in order to demonstrate the quality of fit between the two profiles, it was necessary to develop quantitative indices for comparing their shape. Two indices were calculated for the middle part of the argon release curve, after low-temperature inherited argon had essentially disappeared, and before the sample started to melt. These indices are compared in Fig. 10.48, which shows the analysis spectrum from two K-feldspars, one showing an ideal argon release pattern and one showing poor quality.

The first index is a correlation coefficient, derived by fitting a polynomial function to each of the two patterns, extracting a series of discrete points at evenly spaced intervals then calculating the correlation coefficient between the two series. The second index is obtained simply by determining the average value of each series, and then determining where this point is achieved in the cumulative argon release pattern (mean point). For the ideal experiment (Fig. 10.48a) the correlation coefficient was 0.98 and the mean points coincided. For the poor experiment the correlation coefficient was very poor (0.5) and the mean points did not coincide. On several other occasions, one or other of the criteria was met, but not both.

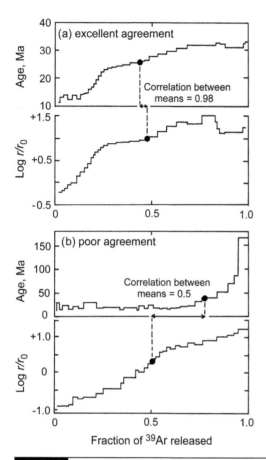

Age spectra and profiles of changing effective diffusion radius ratio (r/r_0) in two step heating experiments on basement samples of K-feldspar: (a) excellent agreement between the two profile shapes; (b) poor agreement between profile shapes. After Lovera et al. (2002).

The result of applying this analysis to the complete sample suite was that 40% of the samples had correlation coefficients above 0.9, leading Lovera et al. (2002) to suggest that most of the alteration-free samples without large inventories of inherited argon were suitable for thermal history analysis by the MDD model. In response to the criticism that the postulated multi diffusion domains cannot be identified structurally, Lovera et al. argued that the empirical results are adequate in themselves to demonstrate the effectiveness of the method for recovering thermal histories, and that understanding the structural basis of the model was of secondary importance. Doubtless, this debate will continue.

10.5 U–Th–He Dating

The uranium–helium dating method was one of the very first to be applied after the discovery of natural radioactivity (e.g.

Strutt, 1905). It is based on the accumulation of ^{4}He produced by the alpha decay of uranium and thorium (and also samarium). In view of the relatively large production rate of helium from these parent nuclides, it should be readily usable to date the time of mineral formation.

Attempts to apply the method were made periodically through the twentieth century, but the extreme diffusivity of helium always presented great problems, causing most uranium–helium ages to be too young (e.g. Hurley, 1954). A renewal of interest in the U–He method was provoked by Zeitler et al. (1987), who pointed out that the low retentivity of helium in minerals could be used as a dating tool for systems displaying low-temperature cooling histories.

For this purpose, the very low closure temperature of the U–He system now becomes an advantage. For example, the U–He system in apatite has the lowest closure temperature of any widely used system (ca. 75 °C), making it one of the most useful for studying recent tectonic uplift, especially in combination with other systems having higher blocking temperatures.

10.5.1 Production and Analysis

The production of helium is mainly from uranium and thorium, and is described by the equation below. The half-life of ^{147}Sm is much longer than the others, but on the other hand Sm can be strongly enriched in some minerals, and may therefore be significant:

$$^4\text{He} = 8 \cdot {}^{238}\text{U}\,(e^{\lambda 238\,t} - 1) + 7 \cdot {}^{235}\text{U}\,(e^{\lambda 235\,t} - 1)$$
$$+ 6 \cdot {}^{232}\text{Th}\,(e^{\lambda 232\,t} - 1) + {}^{147}\text{Sm}\,(e^{\lambda 147\,t} - 1)$$

Zeitler et al. (1987) made exploratory analyses of helium, uranium and thorium in gem-quality apatite from the Durango mine, Mexico. This material is well recognized as a K–Ar and fission-track standard, showing no evidence of disturbance of these systems. The concentrations of U, Th and He measured by Zeitler et al. were approximately consistent with the 31.4 Ma age of Durango apatite, but with relatively large errors (ca. 20%), due to the measurement of helium and U–Th abundances on different aliquots. These measurements were repeated by McDowell et al. (2005), who determined an average (U–Th–Sm)/He age of 31.0 ± 0.4 Ma (2σ), within error of the accepted age. These authors found that helium contributions were 18.5% from U decay, 81.1% from Th decay and 0.4% from Sm decay (the latter ignored by Zeitler et al., 1987).

Typical analytical methods for U–He analysis are similar to conventional K–Ar analysis (e.g. Farley, 2002). Samples are heated in a furnace to release helium, followed by spiking with ^{3}He and collection on a charcoal finger. After removal of other gases, helium is admitted to the mass spectrometer for static gas isotope analysis. The sample residue from heating is recovered, dissolved in acid and spiked with suitable U and Th isotopes to determine concentrations by isotope dilution analysis.

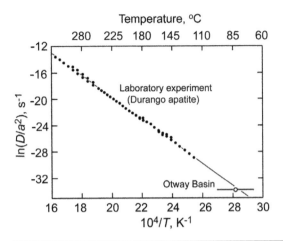

Fig. 10.49 Plot of the diffusion dimension against reciprocal of temperature (Arrhenius plot) for the U–He system in apatite, showing a linear relationship indicative of simple radius-dependent diffusion behaviour. After Farley (2002).

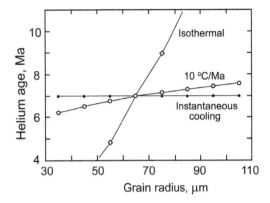

Fig. 10.50 Plot of simulated U–He ages against grain radius, showing the effect of cooling rate on apparent age. After Farley (2002).

10.5.2 Annealing Behaviour

The relative homogeneity of apatite makes it an ideal material for diffusion studies, and particular attention has been focussed on gem-quality apatite from Durango, Mexico, also used to study fission track annealing. The early work of Zeitler *et al.* (1987) revealed non-linear relationships of the diffusion parameter against $1/T$ on the Arrhenius plot, under relatively high-temperature conditions. However, experiments under lower temperature conditions by Farley (2000) revealed very linear relationships (Fig. 10.49), which were also consistent with measurements from Otway Basin drill cores in Australia (House *et al.*, 1999).

The linearity of these Arrhenius relationships suggests that apatite grains were behaving as single diffusion domains during the experiments. This was verified by cyclic heating experiments of the type used by Lovera *et al.* (1989) to reveal variably sized diffusion domains in K-feldspar, but no such effect was seen for U–He in apatite.

The dependence of diffusion on grain radius makes closure temperatures dependent on cooling rate, as long ago demonstrated by Dodson (1973). This relationship is illustrated in Fig. 10.50. The simplest case is for instantaneous cooling, where all sizes of grain yield the same age. The opposite end-member is for isothermal conditions (in this case, samples held at a constant 65 °C), whereby each size of grain reaches its own equilibrium between radiogenic helium production and diffusional loss. The third case is a typical geological cooling rate of 10 °C/Ma, which yields a modest dependence of age on grain radius. In principle, these alternative cooling histories can be resolved by dating grains of different sizes from the same rock sample. However, the grain size effect is only clearly seen in very slowly cooled conditions (Reiners and Farley, 2001).

Wolf *et al.* (1996) determined a U–He blocking temperature for apatite of 75 °C at a cooling rate of 10 °C per Ma, dropping to 60 °C for a cooling rate of 1 °C/Ma. This dependence of blocking temperature on cooling rate gives rise to the so-called 'partial annealing zone' (PAZ) for U–He ages, as previously demonstrated for fission track ages by Naeser (Section 16.6).

10.5.3 Cosmogenic Helium Paleothermometry

In comparison with the production of radiogenic helium from U and Th decay, cosmic ray bombardment produces ^{3}He in exposed rock surfaces by spallation reactions from various light elements (Section 11.1.2). Similar processes generate cosmogenic neon (^{21}Ne), which can be used with other cosmogenic isotopes in cosmogenic exposure/burial dating. The high diffusivity of helium prevents its use in this way, but offers the alternative of cosmogenic noble gas paleothermometry (Tremblay *et al.*, 2014).

In this method, cosmogenic neon is used to monitor the production of cosmogenic ^{3}He in the rock surface. By comparing the measured ^{3}He abundance with this production value, the losses due to thermal diffusion can be determined. Because these losses occur even at ambient temperatures, the method can be used to measure paleotemperatures in cold regions. Tremblay *et al.* demonstrated the method on rocks from the Trans Antarctic Mountains, with an annual mean temperature of −23 °C. ^{3}He exposure ages were measured on a suite of samples from 500–1000 m elevation with a range of ^{10}Be exposure ages from 4 to 10 ka (Fig. 10.51). These ages are thought to date unroofing of Antarctic nunataks due to melting of ice at the end of the last glaciation (Stone *et al.*, 2003). ^{3}He exposure ages were generally less than 50% of the ^{10}Be exposure ages, attributed to diffusional loss of helium from the rocks at sub-zero temperatures.

The analysed data points fell close to a diffusional helium loss curve with an average effective diffusion temperature

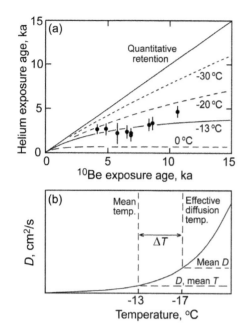

Fig. 10.51 Plot of measured ^{3}He versus ^{10}Be exposure ages for Antarctic nunataks (a) and correction of modelled effective diffusion temperatures for temperature cycles (b). Modified after Tremblay et al. (2014).

of $-13\ ^\circ$C (Fig. 10.51a). However, because diffusional helium loss is a non-linear function of temperature, the effective diffusion temperature must be converted to a mean annual temperature by taking account of the temperature range experienced by the rocks (Fig. 10.51b). Based on temperature fluctuations recorded at a nearby weather station, the correction (ΔT) for this area is ca. 7 °C, leading to a calculated average annual temperature of $-20\ ^\circ$C, which compares well with the nearby station average of $-23\ ^\circ$C. The slight over-estimate can be attributed to *in situ* warming of the rock surface by solar radiation, causing additional diffusional losses on sunny days. Nevertheless, the relatively good agreement suggests that this is a viable method for paleotemperature estimates in polar regions.

References

Anders, E. (1964). Meteorite ages. *Rev. Mod. Phys.* **34**, 287–325.

Arnaud, N. O. and Kelley, S. P. (1997). Argon behaviour in gem-quality orthoclase from Madagascar: experiments and some consequences for ^{40}Ar/^{39}Ar geochronology. *Geochim. Cosmochim. Acta* **61**, 3227–55.

Bachmann, O., Oberli, F., Dungan, M. A. et al. (2007). ^{40}Ar/^{39}Ar and U-Pb dating of the Fish Canyon magmatic system, San Juan Volcanic field, Colorado: Evidence for an extended crystallization history. *Chem. Geol.* **236**, 134–66.

Baksi, A. K., Archibald, D. A. and Farrar, E. (1996). Intercalibration of ^{40}Ar/^{39}Ar dating standards. *Chem. Geol.* **129**, 307–24.

Baksi, A. K., Hsu, V., McWilliams, M. O. and Farrar, E. (1992). ^{40}Ar/^{39}Ar dating of the Brunhes–Matuyama geomagnetic field reversal. *Science* **256**, 356–7.

Beckinsale, R. D. and Gale, N. H. (1969). A reappraisal of the decay constants and branching ratio of ^{40}K. *Earth Planet. Sci. Lett.* **6**, 289–94.

Berger, G. W. (1975). ^{40}Ar/^{39}Ar step heating of thermally overprinted biotite, hornblende and potassium feldspar from Eldora, Colorado. *Earth Planet. Sci. Lett.* **26**, 387–408.

Berger, G. W. and York. D. (1981a). Geothermometry from ^{40}Ar/^{39}Ar dating experiments. *Geochim. Cosmochim. Acta* **45**, 795–811.

Berger, G. W. and York. D. (1981b). ^{40}Ar/^{39}Ar dating of the Thanet gabbro, Ontario: looking through the metamorphic veil and implications for paleomagnetism. *Can. J. Earth Sci.* **18**, 266–73.

Brereton, N. R. (1970). Corrections for interfering isotopes in the ^{40}Ar/^{39}Ar dating method. *Earth Planet. Sci. Lett.* **8**, 427–33.

Buchan, K. L., Berger, G. W., McWilliams, M. O., York, D. and Dunlop, D. J. (1977). Thermal overprinting of natural remanent magnetization and K/Ar ages in metamorphic rocks. *J. Geomag. Geoelectr.* **29**, 401–10.

Channell, J. E. T., Hodell, D. A., Singer, B. S. and Xuan, C. (2010). Reconciling astrochronological and ^{40}Ar/^{39}Ar ages for the Matuyama-Brunhes boundary and late Matuyama Chron. *Geochem. Geophys. Geosys.* **11** (12), 1–21.

Coe, R. S., Singer, B. S., Pringle, M. S. and Zhao, X. (2004). Matuyama-Brunhes reversal and Kamikatsura event on Maui: paleomagnetic directions, ^{40}Ar/^{39}Ar ages and implications. *Earth Planet. Sci. Lett.* **222**, 667–84.

Cox, A. and Dalrymple, G. B. (1967). Statistical analysis of geomagnetic reversal data and the precision of potassium-argon dating. *J. Geophys. Res.* **72**, 2603–14.

Cox, A., Doell, R. R. and Dalrymple, G. B. (1963). Geomagnetic polarity epochs and Pleistocene geochronology. *Nature* **198**, 1049–51.

Dalrymple, G. B. and Lanphere, M. A. (1969). *Potassium-Argon Dating*. Freeman, 258 pp.

Dalrymple, G. B. and Lanphere, M. A. (1971). ^{40}Ar/^{39}Ar technique of K-Ar dating: a comparison with the conventional technique. *Earth Planet. Sci. Lett.* **12**, 300–8.

Dalrymple, G. B. and Lanphere, M. A. (1974). ^{40}Ar/^{39}Ar age spectra of some undisturbed terrestrial samples. *Geochim. Cosmochim. Acta* **38**, 715–38.

Dalrymple, G. B. and Moore, J. G. (1968). Argon 40: excess in submarine pillow basalts from Kilauea Volcano, Hawaii. *Science* **161**, 1132–5.

Damon, P. E. and Kulp, L. (1958). Excess helium and argon in beryl and other minerals. *Amer. Miner.* **43**, 433–59.

Deino, A. and Potts, R. (1990). Single-crystal ^{40}Ar/^{39}Ar dating of the Olorgesailie Formation, Southern Kenya Rift. *J. Geophys. Res.B* **95**, 8453–70.

Dodson, M. H. (1973). Closure temperature in cooling geochronological and petrological systems. *Contrib. Mineral. Petrol.* **40**, 259–74.

Doell, R. R. and Dalrymple, G. B. (1966). Geomagnetic polarity epochs: A new polarity event and the age of the Brunhes–Matuyama boundary. *Science* **152**, 1060–1.

Dreyfus, G. B., Raisbeck, G. M., Parrenin, F. et al. (2008). An ice core perspective on the age of the Matuyama-Brunhes boundary. *Earth Planet. Sci. Lett.* **274**, 151–6.

Endt, P. M. and Van der Leun, C. (1973). Energy levels of A = 21—44 nuclei (V). *Nucl. Phys. A*, **214**, 1–625.

Esser, R. P., McIntosh, W. C., Heizler, M. T. and Kyle, P. R. (1997). Excess argon in melt inclusions in zero-age anorthoclase feldspar from Mt Erebus, Antarctica, as revealed by the ^{40}Ar/^{39}Ar method. *Geochim. Cosmochim. Acta* **61**, 3789–801.

Farley, K. A. (2000). Helium diffusion from apatite: General behavior as illustrated by Durango fluorapatite. *J. Geophys. Res. B* **105**, 2903–14.

Farley, K. A. (2002). (U–Th)/He dating: Techniques, calibrations, and applications. *Reviews in Mineralogy and Geochemistry* **47**, 819–44.

FitzGerald, J. D. and Harrison, T. M. (1993). Argon diffusion domains in K-feldspar I: microstructures in MH–10. *Contrib. Mineral. Petrol.* **113**, 367–80.

Foland, K. A. (1974). Ar-40 diffusion in homogeneous orthoclase and an interpretation of Ar diffusion in K-feldspar. *Geochim. Cosmochim. Acta* **38**, 151–66.

Foland, K. A., Linder, J. S., Laskowski, T. E. and Grant, K. (1984). ^{40}Ar–^{39}Ar dating of glauconies: measured ^{39}Ar recoil loss from well-crystallized specimens. *Chem. Geol. (Isot. Geosci. Sect.)* **46**, 241–64.

Gaber, L. J., Foland, K. A. and Corbato, C. E. (1988). On the significance of argon release from biotite and amphibole during ^{40}Ar/^{39}Ar vacuum heating. *Geochim. Cosmochim. Acta* **52**, 2457–65.

Garner, E. L., Machlan, L. A. and Barnes, I. L. (1976). The isotopic composition of lithium, potassium, and rubidium in some Apollo 11, 12, 14, 15, and 16 samples. *Proc. 6th Lunar Sci. Conf.* Pergamon, pp. 1845–55.

Giletti, B. J. (1974). Diffusion related to geochronology. In: Hofmann, A. W., Giletti, B. J., Yoder, H. S. and Yund, R. A. (Eds) *Geochemical Transport and Kinetics.* Carnegie Inst. Wash., pp. 61–76.

Hale, C. J. (1987). The intensity of the geomagnetic field at 3. 5 Ga: paleointensity results from the Komati Formation, Barberton Mountain Land, South Africa. *Earth Planet. Sci. Lett.* **86**, 354–64.

Hanes, J. A., Clark, S. J. and Archibald, D. A. (1988). An ^{40}Ar/^{39}Ar geochronological study of the Elzevir batholith and its bearing on the tectonothermal history of the southwestern Grenville Province, Canada. *Can. J. Earth Sci.* **25**, 1834–45.

Harper, C. T. (1967). On the interpretation of potassium–argon ages from Precambrian shields and Phanerozoic orogens. *Earth Planet. Sci. Lett.* **3**, 128–32.

Harrison, T. M. (1990). Some observations on the interpretation of feldspar ^{40}Ar/^{39}Ar results. *Chem. Geol. (Isot. Geosci. Sect.)* **80**, 219–29.

Harrison, T. M., Duncan, I. and McDougall, I. (1985). Diffusion of ^{40}Ar in biotite: Temperature, pressure and compositional effects. *Geochim. Cosmochim. Acta* **49**, 2461–8.

Harrison, T. M. and McDougall, I. (1981). Excess ^{40}Ar in metamorphic rocks from Broken Hill, New South Wales: implications for ^{40}Ar/^{39}Ar age spectra and the thermal history of the region. *Earth Planet. Sci. Lett.* **55**, 123–49.

Hart, S. R. (1964). The petrology and isotopic–mineral age relations of a contact zone in the Front Range, Colorado. *J. Geol.* **72**, 493–525.

Hart, S. R. and Dodd, R. T. (1962). Excess radiogenic argon in pyroxenes. *J. Geophys. Res.* **67**, 2998–9.

Heizler, M. T., Lux, D. R. and Decker, E. R. (1988). The age and cooling history of the Chain of Ponds and Big Island Pond plutons and the Spider Lake granite, west-central Maine and Quebec. *Amer. J. Sci.* **288**, 925–52.

Hodges, K. V., Hames, W. E. and Bowring, S. A. (1994). ^{40}Ar/^{39}Ar age gradients in micas from a high-temperature – low-pressure metamorphic terrane: evidence for very slow cooling and implications for the interpretation of age spectra. *Geology* **22**, 55–8.

House, M. A., Farley, K. A. and Kohn, B. P. (1999). An empirical test of helium diffusion in apatite: borehole data from the Otway basin, Australia. *Earth Planet. Sci. Lett.* **170**, 463–74.

Hurley, P. M. (1954). The helium age method and the distribution and migration of helium in rocks. In: Faul, H. (Ed.) *Nuclear Geology.* Wiley, pp. 301–29.

Johnson, R. G. (1982). Brunhes–Matuyama magnetic reversal dated at 790,000 yr B.P. by marine–astronomical correlations. *Quaternary Res.* **17**, 135–47.

Jourdan, F. and Renne, P. R. (2007). Age calibration of the Fish Canyon sanidine ^{40}Ar/^{39}Ar dating standard using primary K-Ar standards. *Geochim. Cosmochim. Acta* **71**, 387–402.

Kaneoka, I. (1974). Investigation of excess argon in ultramafic rocks from the Kola peninsula by the ^{40}Ar/^{39}Ar method. *Earth Planet. Sci. Lett.* **22**, 145–56.

Kapusta, Y., Steinitz, G., Akkerman, A. *et al.* (1997). Monitoring the deficit of ^{39}Ar in irradiated clay fractions and glauconites: modelling and analytical procedure. *Geochim. Cosmochim. Acta* **61**, 4671–8.

Kelley, S. P. (2002). Excess argon in K-Ar and Ar-Ar geochronology. *Chem. Geol.* **188**, 1–22.

Kelley, S. P. and Turner, G. (1991). Laser probe ^{40}Ar–^{39}Ar measurements of loss profiles within individual hornblende grains from the Giants Range granite, northern Minnesota, USA. *Earth Planet. Sci. Lett.* **107**, 634–48.

Kelley, S. P., Arnaud, N. O. and Turner, S. P. (1994). High spatial resolution ^{40}Ar/^{39}Ar investigations using an ultra-violet laser probe extraction technique. *Geochim. Cosmochim. Acta* **58**, 3519–25.

Kuiper, K. F., Deino, A., Hilgen, F. J. *et al.* (2008). Synchronizing rock clocks of Earth history. *Science* **320**, 500–4.

Lanphere, M. A. and Dalrymple, G. B. (1966). Simplified bulb tracer system for argon analysis. *Nature* **209**, 902–3.

Lanphere, M. A. and Dalrymple, G. B. (1971). A test of the ^{40}Ar/^{39}Ar age spectrum technique on some terrestrial materials. *Earth Planet. Sci. Lett.* **12**, 359–72.

Lanphere, M. A. and Dalrymple, G. B. (1976). Identification of excess ^{40}Ar by the ^{40}Ar/^{39}Ar age spectrum technique. *Earth Planet. Sci. Lett.* **32**, 141–8.

Layer, P. W., Hall, C. M. and York, D. (1987). The derivation of ^{40}Ar/^{39}Ar age spectra of single grains of hornblende and biotite by laser step-heating. *Geophys. Res. Lett.* **14**, 757–60.

Lee, J. (1995a). Rapid uplift and rotation of mylonitic rocks from beneath a detachment fault: insights from potassium feldspar ^{40}Ar/^{39}Ar thermochronology, northern Snake range, Nevada. *Tectonics* **14**, 54–77.

Lee, J. K. W. (1995b). Multipath diffusion in geochronology. *Contrib. Mineral. Petrol.* **120**, 60–82.

Lee, J. K. W. and Aldama, A. A. (1992). Multipath diffusion: a general numerical model. *Comput. Geosci.* **18**, 531–55.

Lee, J. K. W., Onstott, T. C., Cashman, K. V., Cumbest, R. J. and Johnson, D. (1991). Incremental heating of hornblende in vacuo: implications for ^{40}Ar/^{39}Ar geochronology and the interpretation of thermal histories. *Geology* **19**, 872–6.

Lee, J. K. W., Onstott, T. C. and Hanes, J. A. (1990). An ^{40}Ar/^{39}Ar investigation of the contact effects of a dyke intrusion, Kapuskasing Structural Zone, Ontario. *Contrib. Mineral. Petrol.* **105**, 87–105.

Lo, C.-H., Lee, J. K. W. and Onstott,. C. (2000). Argon release mechanisms of biotite in vacuo and the role of short-circuit diffusion and recoil. *Chem. Geol.* **165**, 135–66.

Lopez Martinez, M., York, D., Hall, C. M. and Hanes, J. A. (1984). Oldest reliable ^{40}Ar/^{39}Ar ages for terrestrial rocks: Barberton Mountain komatiites. *Nature* **307**, 352–4.

Lovera, O. M. (1992). Computer programs to model ^{40}Ar/^{39}Ar diffusion data from multidomain samples. *Comput. Geosci.* **18**, 789–813.

Lovera, O. M., Grove, M. and Harrison, T. M. (2002). Systematic analysis of K-feldspar ^{40}Ar/^{39}Ar step heating results II: relevance of laboratory argon diffusion properties to nature. *Geochim. Cosmochim. Acta* **66**, 1237–55.

Lovera, O. M., Grove, M., Harrison, T. M. and Mahon, K. I. (1997). Systematic analysis of K-feldspar ^{40}Ar/^{39}Ar step heating results: I. Significance of activation energy determinations. *Geochim. Cosmochim. Acta* **61**, 3171–92.

Lovera, O. M., Heizler, M. T. and Harrison, T. M. (1993). Argon diffusion domains in K-feldspar II: kinetic properties of MH–10. *Contrib. Mineral. Petrol.* **113**, 381–93.

Lovera, O. M., Richter, F. M. and Harrison, T. M. (1989). The ^{40}Ar/^{39}Ar thermochronometry for slowly-cooled samples having a distribution of domain sizes. *J. Geophys. Res.* **94**, 17 917–35.

Lovera, O. M., Richter, F. M. and Harrison, T. M. (1991). Diffusion domains determined by ^{39}Ar released during step heating. *J. Geophys. Res.* **96**, 2057–69.

McDowell, F. W. (1983). K–Ar dating: Incomplete extraction of radiogenic argon from alkali feldspar. *Chem. Geol.* **41**, 119–26.

McDowell, F. W., McIntosh, W. C. and Farley, K. A. (2005). A precise ^{40}Ar–^{39}Ar reference age for the Durango apatite (U–Th)/He and fission-track dating standard. *Chem. Geol.* **214**, 249–63.

McDougall, I. and Harrison, T. M. (1988). *Geochronology and Thermochronology by the ^{40}Ar/^{39}Ar Method.* Oxford University Press, 212 pp.

McDougall, I. and Harrison, T. M. (1999). *Geochronology and Thermochronology by the ^{40}Ar/^{39}Ar Method.* 2nd Edn. Oxford University Press, 269 pp.

McDougall, I., Polach, H. A. and Stipp, J. J. (1969). Excess radiogenic argon in young subaerial basalts from the Auckland volcanic field, New Zealand. *Geochim. Cosmochim. Acta* **33**, 1485–520.

McDougall, I. and Tarling, D. H. (1964). Dating geomagnetic polarity zones. *Nature* **202**, 171–2.

Mankinen, E. A. and Dalrymple, G. B. (1979). Revised geomagnetic polarity time scale for the interval 0 to 5 m.y. B.P. *J. Geophys. Res.* **84**, 615–26.

Megrue, G. H. (1967). Isotopic analysis of rare gases with a laser microprobe. *Science* **157**, 1555–6.

Megrue, G. H. (1973). Spatial distribution of ^{40}Ar/^{39}Ar ages in lunar breccia 14301. *J. Geophys. Res.* **78**, 3216–21.

Merrihue, C. and Turner, G. (1966). Potassium–argon dating by activation with fast neutrons. *J. Geophys. Res.* **71**, 2852–7.

Min, K., Mundil, R., Renne, P. R. and Ludwig, K. R. (2000). A test for systematic errors in ^{40}Ar/^{39}Ar geochronology through comparison with U/Pb analysis of a 1.1-Ga rhyolite. *Geochim. Cosmochim. Acta* **64**, 73–98.

Mitchell, J. G. (1968). The argon-40/argon-39 method for potassium–argon age determination. *Geochim. Cosmochim. Acta* **32**, 781–90.

Mussett, A. E. and Dalrymple, G. B. (1968). An investigation of the source of air Ar contamination in K–Ar dating. *Earth Planet. Sci. Lett.* **4**, 422–6.

Naumenko, M. O., Mezger, K., Nägler, T. F. and Villa, I. M. (2013). High precision determination of the terrestrial ^{40}K abundance. *Geochim. Cosmochim. Acta* **122**, 353–62.

Onstott, T. C., Hall, C. M. and York, D. (1989). ^{40}Ar/^{39}Ar thermochronometry of the Imataca complex, Venezuela. *Precamb. Res.* **42**, 255–91.

Onstott, T. C., Miller, M. L., Ewing, R. C., Arnold, G. W. and Walsh, D. S. (1995). Recoil refinements: implications for the ^{40}Ar/^{39}Ar dating technique. *Geochim. Cosmochim. Acta* **59**, 1821–34.

Onstott, T. C., Phillips, D. and Pringle-Goodell, L. (1991). Laser microprobe measurement of chlorine and argon zonation in biotite. *Chem. Geol.* **90**, 145–68.

Parsons, I., Brown, W. L. and Smith, J. V. (1999). ^{40}Ar/^{39}Ar thermochronology using alkali feldspars: real thermal history or mathematical mirage of microtexture? *Contrib. Mineral. Petrol.* **136**, 92–110.

Phillips, D. and Onstott, T. C. (1988). Argon isotopic zoning in mantle phlogopite. *Geology* **16**, 542–6.

Pickles, C. S., Kelley, S. P., Reddy, S. M. and Wheeler, J. (1997). Determination of high spatial resolution argon isotope variations in metamorphic biotites. *Geochim. Cosmochim. Acta* **61**, 3809–33.

Reiners, P. W. and Farley, K. A. (2001). Influence of crystal size on apatite (U–Th)/He thermochronology: an example from the Bighorn Mountains, Wyoming. *Earth Planet. Sci. Lett.* **188**, 413–20.

Renne, P. R., Deino, A. L., Hilgen, F. J. *et al.* (2013). Time scales of critical events around the Cretaceous–Paleogene boundary. *Science* **339**, 684–7.

Renne, P. R., Deino, A. L., Walter, R. C. *et al.* (1994). Intercalibration of astronomical and radioisotopic time. *Geology* **22**, 783–6.

Renne, P. R., Mundil, R., Balco, G., Min, K. and Ludwig, K. R. (2010). Joint determination of ^{40}K decay constants and ^{40}Ar*/^{40}K for the Fish Canyon sanidine standard, and improved accuracy for ^{40}Ar/^{39}Ar geochronology. *Geochim. Cosmochim. Acta* **74**, 5349–67.

Renne, P. R., Swisher, C. C., Deino, A. L. *et al.* (1998). Intercalibration of standards, absolute ages and uncertainties in ^{40}Ar/^{39}Ar dating. *Chem. Geol.* **145**, 117–52.

Rex, D. C., Guise, P. G. and Wartho, J.-A. (1993). Disturbed ^{40}Ar/^{39}Ar spectra from hornblendes: thermal loss or contamination? *Chem. Geol. (Isot. Geosci. Sect.)* **103**, 271–81.

Richter, F. M., Lovera, O. M., Harrison, T. M. and Copeland, P. (1991). Tibetan tectonics from ^{40}Ar/^{39}Ar analysis of a single K-feldspar sample. *Earth Planet. Sci. Lett.* **105**, 266–78.

Rivera, T. A., Storey, M., Zeeden, C., Hilgen, F. J. and Kuiper, K. (2011). A refined astronomically calibrated ^{40}Ar/^{39}Ar age for Fish Canyon sanidine. *Earth Planet. Sci. Lett.* **311**, 420–6.

Roberts, H. J., Kelley, S. P. and Dahl, P. S. (2001). Obtaining geologically meaningful ^{40}Ar–^{39}Ar ages from altered biotite. *Chem. Geol.* **172**, 277–90.

Roddick, J. C. and Farrar, E. (1971). High initial argon ratios in hornblendes. *Earth Planet. Sci. Lett.* **12**, 208–14.

Schmitz, M. D. and Bowring, S. A. (2001). U–Pb zircon and titanite systematics of the Fish Canyon Tuff: an assessment of high-precision U–Pb geochronology and its application to young volcanic rocks. *Geochim. Cosmochim. Acta* **65**, 2571–87.

Schmitz, M. D. and Kuiper, K. F. (2013). High-precision geochronology. *Elements* **9** (1), 25–30.

Shackleton, N. J., Berger, A. and Peltier, W. R. (1990). An alternative astronomical calibration of the lower Pleistocene timescale based on ODP Site 677. *Trans. Roy. Soc. Edinburgh: Earth Sci.* **81**, 251–61.

Simon, J. I., Renne, P. R. and Mundil, R. (2008). Implications of pre-eruptive magmatic histories of zircons for U–Pb geochronology of silicic extrusions. *Earth Planet. Sci. Lett.* **266**, 182–94.

Smith, P. E., Evensen, N. M. and York, D. (1993). First successful ^{40}Ar/^{39}Ar dating of glauconies: argon recoil in single grains of cryptocrystalline material. *Geology* **21**, 41–4.

Spell, T. L. and McDougall, I. (1992). Revisions to the age of the Brunhes-Matuyama boundary and the Pleistocene geomagnetic polarity timescale. *Geophys. Res. Lett.* **19**, 1181–4.

Steiger, R. H. and Jager, E. (1977). IUGS Subcommission on Geochronology: convention on the use of decay constants in geo- and cosmochronology. *Earth Planet. Sci. Lett.* **36**, 359–62.

Stone, J. O., Balco, G. A., Sugden, D. E. *et al.* (2003). Holocene deglaciation of Marie Byrd land, west Antarctica. *Science* **299**, 99–102.

Strutt, R. J. (1905). On the radio-active minerals. *Proc. Roy. Soc. Lond. A* **76**, 88–101.

Suganuma, Y., Okada, M., Horie, K. *et al.* (2015). Age of Matuyama-Brunhes boundary constrained by U-Pb zircon dating of a widespread tephra. *Geology* **43**, 491–4.

Tauxe, L., Deino, A. D., Behrensmeyer, A. K. and Potts, R. (1992). Pinning down the Brunhes/Matuyama and upper Jaramillo boundaries: a reconciliation of orbital and isotopic time scales. *Earth Planet. Sci. Lett.* **109**, 561–72.

Tremblay, M. M., Shuster, D. L. and Balco, G. (2014). Cosmogenic noble gas paleothermometry. *Earth Planet. Sci. Lett.* **400**, 195–205.

Trieloff, M., Jessberger, E. K. and Fieni, C. (2001). Comment on "^{40}Ar/^{39}Ar age of plagioclase from Acapulco meteorite and the problem of systematic errors in cosmochronology" by Paul. R. Renne. *Earth Planet. Sci. Lett.* **190**, 267–9.

Turner, G. (1968). The distribution of potassium and argon in chondrites. In: Ahrens, L. H. (Ed.) *Origin and Distribution of the Elements.* Pergamon, pp. 387–97.

Turner, G. (1969). Thermal histories of meteorites by the ^{39}Ar–^{40}Ar method. In: Millman, P. M. (Ed.) *Meteorite Research.* Reidel, pp. 407–17.

Turner, G. (1971a). Argon 40–argon 39 dating: the optimisation of irradiation parameters. *Earth Planet. Sci. Lett.* **10**, 227–34.

Turner, G. (1971b). ^{40}Ar/^{39}Ar ages from the lunar maria. *Earth Planet. Sci. Lett.* **11**, 169–91.

Turner, G. (1972). ^{40}Ar–^{39}Ar age and cosmic ray irradiation history of the Apollo 15 anorthosite, 15415. *Earth Planet. Sci. Lett.* **14**, 169–75.

Turner, G. and Cadogan, P. H. (1974). Possible effects of ^{39}Ar recoil in ^{40}Ar/^{39}Ar dating. *Proc. 5th Lunar Sci. Conf.*, pp. 1601–15.

Turner, G., Miller, J. A. and Grasty, R. L. (1966). Thermal history of the Bruderheim meteorite. *Earth Planet. Sci. Lett.* **1**, 155–7.

Villa, I. M. (1997). Direct determination of ^{39}Ar recoil distance. *Geochim. Cosmochim. Acta* **61**, 689–91.

Villa, I. M. (1998). Reply to the comment by T. M. Harrison, M. Grove, and O. M. Lovera on "Direct determination of ^{39}Ar recoil distance". *Geochim. Cosmochim. Acta* **62**, 349.

Vincent, E. A. (1960). Analysis by gravimetric and volumetric methods, flame photometry, colorimetry and related techniques. In: Smales, A. A. and Wager, L. R. (Eds) *Methods in Geochemistry.* Interscience. pp. 33–80.

Wartho, J.-A., Kelley, S. P., Brooker, R. A. *et al.* (1999). Direct measurement of Ar diffusion profiles in a gem-quality Madagascar K-feldspar using the ultra-violet laser ablation microprobe (UVLAMP). *Earth Planet. Sci. Lett.* **170**, 141–53.

Wolf, R. A., Farley, K. A. and Silver, L. T. (1996). Helium diffusion and low-temperature thermochronometry of apatite. *Geochim. Cosmochim. Acta* **60**, 4231–40.

Wright, N., Layer, P. W. and York, D. (1991). New insights into thermal history from single grain ^{40}Ar/^{39}Ar analysis of biotite. *Earth Planet. Sci. Lett.* **104**, 70–9.

York, D. (1978). A formula describing both magnetic and isotopic blocking temperatures. *Earth Planet. Sci. Lett.* **39**, 89–93.

York, D. (1984). Cooling histories from ^{40}Ar/^{39}Ar age spectra: implications for Precambrian plate tectonics. *Ann. Rev. Earth Planet. Sci.* **12**, 383–409.

York, D., Hall, C. M., Yanase, Y., Hanes, J. A. and Kenyon, W. J. (1981). ^{40}Ar/^{39}Ar dating of terrestrial minerals with a continuous laser. *Geophys. Res. Lett.* **8**, 1136–8.

Zeitler, P. K., Herczeg, A. L., McDougall, I. and Honda, M. (1987). U–Th–He dating of apatite: A potential thermochronometer. *Geochim. Cosmochim. Acta* **51**, 2865–8.

Chapter 11

Noble Gas Geochemistry

The elements known as the rare, inert or noble gases possess unique properties which make them important in isotope geology. The low abundance of these rare gases makes them sensitive recorders of several types of nuclear process, even including rare nuclear fission reactions. In contrast, the relatively larger abundance of other fission product nuclides such as the 'rare' earths swamps fissiogenic production. Another property of these gases is their inertness, which allows unique insights into the Earth's interior, because of their lack of interaction with other materials. Finally, as isotopic tracers, they can give information about the degassing history of the mantle, the formation of the atmosphere and about mixing relationships between different mantle reservoirs.

Over the past 20 years, the mass spectrometry of noble gases has made huge advances, which have greatly expanded the number of isotope tracers that can be deployed. This has reached the point where it seems desirable to summarize the isotope couples available (Table 11.1). These are generally expressed as radiogenic/non-radiogenic ratios, or in other forms most commonly used. They are listed in order of atomic weight, which in general leads to increasing complexity, and this is the order in which they are discussed in this chapter.

11.1 Helium

Helium has two isotopes, ^4He and ^3He. The former was recognized by Rutherford (1906) to be the α decay product of actinide elements and hence comprised the first radiometric dating method. However, the great diffusivity of helium made the method very susceptible to thermal disturbance, and it was therefore largely abandoned as a dating tool. Its reinstatement by Zeitler et al. (1987) was as a tool to date low-temperature cooling phenomena (Section 10.5).

Non-radiogenic ^3He was first discovered in nature by Alvarez and Cornog (1939). Being the more 'exotic' isotope, ^3He was usually normalized against ^4He. Because atmospheric helium is universally used as a mass spectrometric

Table 11.1	Noble gas isotope systems.	
Tracer	Cause of variation	Half-life of parent
$^4He/^3He$	alpha decay	Various
$^{20}Ne/^{22}Ne$	cosmic?	
$^{21}Ne/^{22}Ne$	spallation	
$^{38}Ar/^{36}Ar$	primordial?	
$^{40}Ar/^{36}Ar$	^{40}K decay	12 Ga
$^{84}Kr/^{82}Kr$	primordial?	
$^{86}Kr/^{82}Kr$	primordial?	
$^{124}Xe/^{130}Xe$	mass fractionation?	
$^{126}Xe/^{130}Xe$	mass fractionation?	
$^{128}Xe/^{130}Xe$	mass fractionation?	
$^{129}Xe/^{130}Xe$	^{129}I decay	16 Ma
$^{130}Xe/^{131}Xe$	^{129}Ba decay	1000 Ga
$^{131}Xe/^{130}Xe$	^{228}U and ^{244}Pu fission	
$^{133}Xe/^{130}Xe$	^{228}U and ^{244}Pu fission	
$^{134}Xe/^{130}Xe$	^{228}U and ^{244}Pu fission	
$^{136}Xe/^{130}Xe$	^{228}U and ^{244}Pu fission	.

standard, it was convenient to express $^3He/^4He$ ratios in unknown samples relative to the atmospheric ratio in the form $R_{unknown}/R_{air}$ (R/R_A). However, several authors have argued that it is better to express helium isotope ratios in the form $^4He/^3He$ (radiogenic / non-radiogenic) to bring this tracer into line with most other isotope systems. Therefore, this is the format that will be adopted here, with frequent reference to R/R_A values for comparison. This index is normally based on an atmospheric $^4He/^3He$ ratio of 720 000, averaged from the determinations of Mamyrin et al. (1970) and Clarke et al. (1976).

11.1.1 Mass Spectrometry

Mass spectrometric analysis of helium is broadly similar to argon isotope analysis for K–Ar dating (Section 10.1.1). However, there are no 'extra' isotopes available during helium isotope analysis to allow precise corrections for atmospheric contamination. Therefore it is critical to minimize the *extent* of atmospheric contamination during helium extraction and analysis. Uncertainties in the atmospheric 'blank' may contribute the principal error in helium isotope analysis, especially for rock samples. Well-gas samples, being larger, are less susceptible to atmospheric contamination during analysis, but may have come from an open system in the natural environment. In the case of rock analysis, absorbed atmospheric helium is usually driven off by overnight heating at 200–300 °C. The sample gas may then be extracted by melting the rock or by crushing under vacuum. A combination of both techniques (e.g. Kurz and Jenkins, 1981) provides an extra check against the possibility of atmospheric contamination, both in the laboratory and the environment.

Two steps are necessary to reduce blank levels in the mass spectrometer for all noble gas analyses. One is to polish all internal surfaces of a metal instrument to minimize gas absorption onto the vacuum system walls. Another is to reduce the internal surface area of the instrument as much as possible, for example by boring the flight tube out of a solid piece of steel, rather then using welded pipe. A low internal volume also yields better sensitivity for very small samples.

All noble gas analyses are performed in the static gas mode (i.e. with vacuum pumps isolated). As a result, hydrogen tends to build up in the instrument so that its molecular ions HD^+ and H_3^+ cause isobaric interferences onto $^3He^+$. Therefore, the vacuum system in some older machines contains a small titanium 'getter', designed to absorb H_2 released inside the instrument (Clarke *et al.*, 1969). Nevertheless, the peak composed of HD and H_3 may still be much larger than ^3He, and it is essential to separate them by mass. This can be done by making use of the 0.006 atomic mass unit (a.m.u.) difference between ^3He and the other two species (Fig. 11.1), which results from their different nuclear binding energies. In order to achieve this separation at mass 3, a resolution of one mass unit in 600 is necessary, which can be achieved with an instrument of ca. 25 cm radius (Clarke *et al.*, 1969; Kurz and Jenkins, 1981).

In order to measure the very large intensity difference between ^3He and ^4He signals, it is most convenient to measure the former on a multiplier detector and the latter by Faraday detector. These can only be used in the static collection mode if a branched flight tube is available, because of the extreme divergence of the mass 3 and 4 ion beams (Lupton and Craig, 1975). Alternatively, peak switching is performed by changing the accelerating potential or magnetic field (e.g. Clarke *et al.*, 1969; Poreda and Farley, 1992).

11.1.2 Helium Production in Nature

Early work by Alvarez and Cornog (1939) revealed distinct helium signatures in well-gases and the atmosphere. This led Aldrich and Nier (1948) to suggest that there must be independent sources of the two isotopes, one of which could be primordial. However, in order to determine whether primordial helium is an important constituent in the Earth, it was necessary to establish benchmarks, such as the $^4He/^3He$ ratio

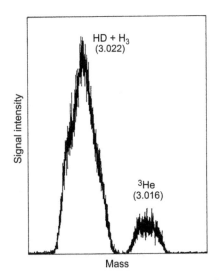

Fig. 11.1 Scan of peaks in the region of mass 3 during helium isotope analysis, showing the separation of molecular interference using high spectral resolution. Masses are quoted relative to $^{12}C = 12.000$. After Lupton and Craig (1975).

of primordial solar system helium and the production ratios in nuclear and cosmogenic processes (Fig. 11.2).

An early indication of the composition of primordial helium was provided by the $^4He/^3He$ ratio of ca. 2500–5000 measured in gas-rich carbonaceous chondrites (Pepin and Signer, 1965). More recently, the helium signature of the solar wind was determined by exposing aluminium foil on the lunar surface during the Apollo missions, yielding an average $^4He/^3He$ ratio of 2350 (Geiss *et al.*, 2004). This value was confirmed by a two-year exposure of foils in space on the Genesis Mission (Grimberg *et al.*, 2008). In contrast, helium analysis of the atmosphere of Jupiter by the Galileo space probe (Mahaffy *et al.*, 1998) gave a much higher $^4He/^3He$ ratio

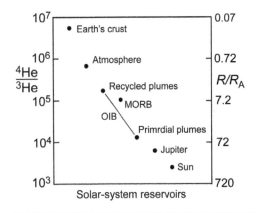

Fig. 11.2 Range of helium isotope ratios in some important reservoirs. Modified after Moreira (2013).

of 6000 ($R/R_A = 120$). This value is interpreted as the primordial composition of the solar system, whereas the solar wind has evolved in composition due to nuclear reactions in the sun (Moreira, 2013). Meanwhile the compositions of gas-rich meteorites can be attributed to modification of a primordial helium signature by impregnation from the solar wind.

Early calculations of the helium production ratio in igneous rocks were made by Morrison and Pine (1955). Radiogenic production of ^{4}He is obvious, since the α particle is synonymous with a ^{4}He nucleus. However, 'nucleogenic' ^{3}He can also be generated by neutron bombardment of light atoms. Radioactive decay of uranium generates a neutron flux in rocks by two mechanisms. Spontaneous fission is a minor source, but by far the dominant source of neutrons is the collision of α particles with the nuclei of light elements. Some of these neutrons reach epithermal energies, where they can induce the (n, α) reaction on lithium. The tritium thus produced decays to ^{3}He:

$$^6\text{Li} + \text{n} \rightarrow {}^3\text{H} + \alpha$$

$$^3\text{H} \rightarrow {}^3\text{He} + \beta (t_{1/2} = 12\,\text{yr}).$$

Kunz and Schintlmeister (1965) calculated that ^{3}He generation by this reaction is at least three orders of magnitude more efficient than all other neutron induced reactions, and this was confirmed by Mamyrin and Tolstikhin (1984). Given the uranium (+ thorium) and lithium content of a rock, the ^{4}He/^{3}He yield can be calculated (Gerling *et al.*, 1971). The results are consistent with values of around 10^7 measured empirically in old granites. It was concluded from these observations and calculations that no nuclear process has been discovered which is capable of generating ^{4}He/^{3}He ratios significantly below 10 million in normal rocks. However, uranium ores generate higher ratios, while Li-rich *minerals* generate abnormally low ratios.

Another mineral in which low ^{4}He/^{3}He ratios have been observed is diamond. Values as low as 3000 were originally interpreted as primordial mantle signatures (Ozima and Zashu, 1983), but have been attributed by later workers to either nucleogenic or cosmogenic ^{3}He production. For example, Lal *et al.* (1987) attributed high ^{3}He/^{4}He ratios in alluvial diamonds from Zaire to cosmogenic production while exposed at the surface. On the other hand, Kurz *et al.* (1987) and Zadnik *et al.* (1987) measured ^{4}He/^{3}He as low as 700 in diamonds mined directly from kimberlite pipes at depths from 26 m to 200 m. Since cosmic rays cannot penetrate to such depths, these helium signatures were attributed to nucleogenic production. This was based on observations of isotopic variability within individual diamonds, and ^{4}He/^{3}He ratios lower than solar in the latter study. In both cases, ^{3}He production was attributed to the (n, α) reaction on lithium. For this process to occur, the diamond and its inclusions must be irradiated by neutrons from outside the crystal, so that radiogenic ^{4}He production in the diamond itself is suppressed.

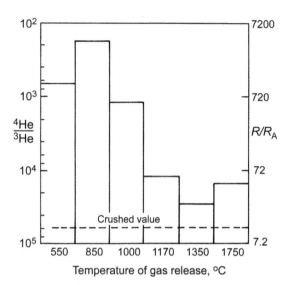

Fig. 11.3 Step heating helium isotope analysis of a surface sample of Haleakala lava, showing a large cosmogenic component (high R/R_A), especially in the low-temperature release steps. However, crushed vesicles yield a typical mantle value. After Kurz (1986a).

In situ cosmogenic helium production in terrestrial rocks was proposed by Jeffrey and Hagan (1969), but was not identified unambiguously until work by Kurz (1986a) and Craig and Poreda (1986). In a detailed helium isotope study of subaerial lavas from Haleakala volcano, Kurz discovered very high ^{3}He signals in gases released by step heating of exposed 0.5–0.8 Ma alkali basalts. Low-temperature gas releases from samples within 0.5 m of the weathered surface gave R/R_A values of up to 1000 (Fig. 11.3). This corresponds to ^{4}He/^{3}He values below 1000, even lower than primordial meteoritic or solar-wind helium.

In contrast, step heating of samples from a similar stratigraphic horizon that were buried under ca. 160 m of younger flows yielded MORB-like helium (^{4}He/^{3}He = 80 000). Helium released by crushing of phenocrysts also gave a MORB signature for both the buried and surface samples. Therefore, Kurz argued that crushing released magmatic helium from vesicles, but step heating of old surface samples released dispersed cosmogenic helium from the rock matrix. Young surface samples such as the 1790 flow on Haleakala do not show these effects, ruling out anthropogenic bomb tritium as the source of the ^{3}He.

Kurz (1986b) examined cosmogenic ^{3}He production as a function of depth below the surface of a lava flow. Spallation reactions caused by cosmogenic neutrons are the dominant source of ^{3}He at the surface, but neutrons are attenuated exponentially downwards. Nevertheless, ^{3}He abundances showed less attenuation with depth than expected. This was attributed to production by cosmic-ray muons, which have a greater penetration depth than neutrons. Muon capture

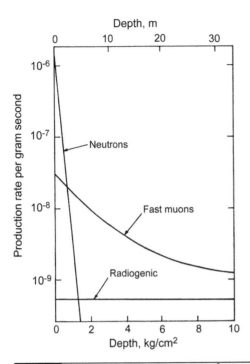

Fig. 11.4 Calculated production rates for ^{3}He by different processes as a function of depth in a rock surface. Depths are expressed as kg/cm^2, which is approximately equal to 0.3 × depth in m. After Lal (1987).

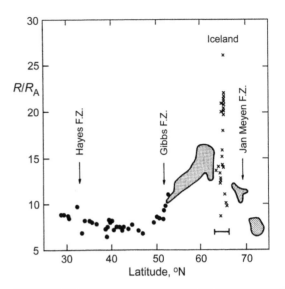

Fig. 11.5 Plot of helium isotope ratios along the Mid Atlantic Ridge, expressed as deviations from the atmospheric value (R/R_A). The primordial ^{3}He signature of the Iceland plume (×) is elevated relative to MORB (●). Shaded fields display mixing of sources. After Kurz et al. (1985).

by nuclei causes neutron emission, which in turn produces ^{3}He via the (n, α) reaction on lithium. The depth dependence of different production routes for ^{3}He is summarized in Fig. 11.4 (Lal, 1987).

Cosmogenic isotopes represent a useful tool for determining exposure ages of rock surfaces (Section 14.7). However, the great diffusivity of helium means that it is not quantitatively retained in quartz, the most widely used material in surface exposure dating (Cerling, 1989). In contrast, ^{21}Ne displays cosmogenic production with an attenuation depth similar to ^{3}He (Sarda et al., 1993), but is not subject to significant diffusive losses. This means that ^{21}Ne is more suitable for direct dating of exposure ages (Section 14.7.4). However, it has recently been proposed that ^{3}He can be used together with ^{21}Ne to study the thermochronometry of exposed rock surfaces (Section 10.5).

11.1.3 Terrestrial Primordial Helium

Alvarez and Cornog (1939) showed that the atmosphere has a ^{3}He/^{4}He ratio about ten times higher than most well-gases, a result supported by Aldrich and Nier (1948). This suggested that the excess ^{3}He abundance in the atmosphere resulted from cosmogenic production. In contrast, Clarke et al. (1969) discovered that deep water from the Pacific Ocean was *enriched* in ^{3}He by up to 20% relative to atmosphere. This implied a possible source of primordial ^{3}He in the Earth's

interior. However, Sheldon and Kern (1972) and Lupton and Craig (1975) hypothesized that this could conceivably be due to a past temporary weakening of the Earth's magnetic field, during which atmospheric ^{3}He was elevated by greater cosmic ray penetration. This might then have modified the composition of deep ocean water.

More convincing evidence of primordial helium in the Earth was provided by Mamyrin et al. (1969), who found ^{3}He/^{4}He signatures ten times the atmospheric ratio in thermal fluids from the Kuril Islands. Subsequently, ^{3}He/^{4}He ratios as high as 20 times atmospheric were found in hot springs from Iceland (Mamyrin et al., 1972), up to 32 times atmospheric in basaltic glass from Loihi Seamount, Hawaii (Kurz et al., 1982) and up to 38 times atmospheric (^{4}He/^{3}He around 20 000) in olivine basalt from the neovolcanic zone in NW Iceland (Hilton et al., 1999). The helium signature of the Iceland plume can also be seen spreading out over a large area of the North Atlantic as the plume head contaminates the asthenospheric upper mantle, leading to mixed isotopic signatures on the Reykjanes Ridge (Fig. 11.5).

The existence of a primordial helium reservoir in the Earth was questioned by Anderson (1993), who attributed the high ^{3}He signatures in oceanic volcanic rocks to the subduction of cosmic (interplanetary) dust particles. These particles were found to accumulate in ocean floor sediments by Merrihue (1964). Cosmic dust has ^{4}He/^{3}He ratios similar to gas-rich meteorites (ca. 3000), but unlike meteorites, these particles can fall to Earth without burning up in the atmosphere (Nier and Schlutter, 1990). Hence, ocean floor sediments develop a 'primordial' helium isotope signature (Fig. 11.6a).

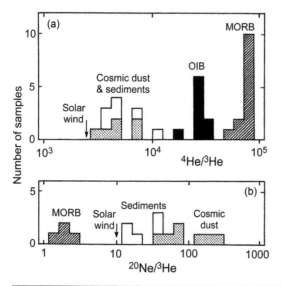

Fig. 11.6 Histograms of (a) $^4He/^3He$ and (b) $^{20}Ne/^3He$ in cosmic dust particles (stipple) and ocean floor sediments (white) compared with the noble gas composition of MORB (hatched) and OIB (black). The Solar wind composition is shown for reference. Data from Allegre et al. (1993).

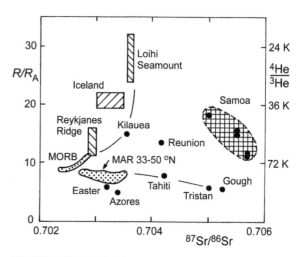

Fig. 11.7 Plot of helium against Sr isotope ratios to show possible mixing between the MORB reservoir and primordial and recycled plume sources. After Lupton (1983).

The noble gases in cosmic dust particles are encapsulated in magnetite grains, which are relatively resistant to thermal degassing (Matsuda et al., 1990). Therefore, the cosmic helium in ocean floor sediments might survive the subduction process and be transported into the deep mantle. In contrast, *atmospheric* noble gases trapped in ocean floor sediments are very susceptible to thermal degassing. Staudacher and Allegre (1988) argued that subduction-related volcanism is at least 98% efficient in scavenging these atmospheric gases from subducted sediments before they can reach the deep mantle.

Because cosmic dust might survive the 'subduction barrier' against atmospheric noble gases (Staudacher and Allegre, 1988), it has the potential to deliver helium with a primordial signature into the deep mantle. However, Allegre et al. (1993) used neon/helium ratios to place upper limits on the amount of cosmic ^3He which can enter plume sources. The $^{20}Ne/^3He$ ratio of cosmic dust is one to two orders of magnitude higher than the upper mantle (Fig. 11.6b). Furthermore, helium has a much greater diffusivity than neon, which would promote its preferential degassing from cosmic dust grains during subduction (Hiyagon, 1994). Therefore it appears that subduction of cosmic dust cannot contribute more than a small fraction of the mantle ^3He budget without causing excessive enrichment of ^{20}Ne in submarine glasses.

Because the helium isotope signature in plumes provides the best evidence for a primordial gas reservoir in the deep Earth, it provides a possible constraint on the location of other mantle components identified in plumes (Section 6.5).

Hence, various attempts have been made to compare helium isotope signatures with other isotope ratios in oceanic volcanics, in order to provide extra constraints on mantle processes.

One such approach is the comparison of helium and strontium isotope data (Kurz et al., 1982; Lupton, 1983). MORBs define a restricted range of compositions on a plot of helium isotope ratio against $^{87}Sr/^{86}Sr$, but ocean islands are widely scattered (Fig. 11.7). While Loihi has one of the most extreme primordial helium signatures, other ocean islands such as Tristan, Gough and the Azores have $^3He/^4He$ ratios *lower* (more radiogenic) than MORB. Similar low ratios have subsequently been found in the HIMU islands of the SW Pacific (Hanyu and Kaneoka, 1997). These low $^3He/^4He$ ratios require a component of radiogenic helium from a long-lived U- or Th-rich source, which can most easily be satisfied by the recycling of oceanic crust and sediments into the mantle, as inferred from lithophile isotope data (Section 6.5).

A third possible type of plume source was exemplified by the data from Samoa (Fig. 11.7), which showed a negative correlation between R/R_A value and Sr isotope ratio. This suggested that primordial helium signatures could be associated with moderately depleted mantle sources identified with lithophile isotope tracers (Section 6.5.1). These sources were tentatively attributed to mixed lower mantle reservoirs such as a 'focus zone' of mantle mixing (FOZO) or a 'common' component (C) in the lower mantle (Hart et al., 1992; Hanan and Graham, 1996; van Keken et al., 2002).

More extreme R/R_A values (around 50) were subsequently determined on Tertiary picrite basalts from Baffin Island (Stuart et al., 2003). Since these samples came from eroding sea cliffs, they were shielded from cosmogenic production, and provide samples of the early Iceland plume at the beginning of opening of the North Atlantic. A puzzling aspect of

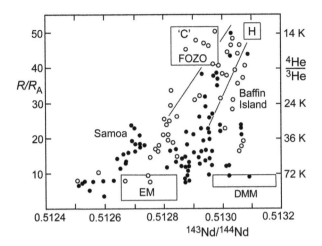

Fig. 11.8 Helium versus Nd isotope compositions in early Baffin Island and Samoan data (•), implying mixing between enriched mantle (EM) and LIL-depleted primordial helium mantle (H), compared with later scattered data (○). Modified from Ellam and Stuart (2004); Starkey et al. (2009).

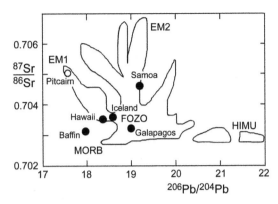

Fig. 11.9 Plot of Sr against Nd isotope ratios, showing the compositions of plumes with R/R_A values > 25 (solid symbols), compared with the total range of Sr–Pb isotopes in oceanic volcanics. Approximate compositions of major mantle components are shown. Modified after Garapic et al. (2015).

the new Baffin data was that the high ^{3}He signal appeared to be correlated with radiogenic Nd (H, Fig. 11.8), implying an association of primordial helium with lithophile isotope signatures *more* depleted than the Iceland plume itself (Ellam and Stuart, 2004). This observation strongly influenced the controversial helium box model of Class and Goldstein (Section 11.1.5), by suggesting that the high ^{3}He source was identified with depleted rather than primordial mantle. However, later work (Starkey et al., 2009) broke this correlation by showing that samples with high R/R_A values could also have lower ^{143}Nd/^{144}Nd isotope ratios near FOZO. Hence this implied that helium might be decoupled from lithophile isotope systems.

Further evidence for such decoupling came from the discovery of much higher R/R_A values in the Samoan plume than previously known (Fig. 11.7). The new data had R/R_A values as high as 34 (Jackson et al., 2007), suggesting that the Samoan, Hawaiian and Iceland data were not pointing towards a single lower mantle source with primordial helium, but separate components with primordial helium and distinct lithophile isotope signals. Jackson et al. labelled these components as FOZO 'A' and FOZO 'B'.

New helium data for the Galapagos Islands further complicated this picture by revealing R/R_A values as high as 29 in this hot-spot (Kurz et al., 2009). Hence, this adds yet another plume with primordial helium but a distinct Pb/Sr/Nd isotope composition. This proliferation of high-^{3}He plumes (R/R_A > 25) with distinct lithophile signatures can be seen on a Sr–Pb isotope plot (solid circles in Fig. 11.9). The spread in lithophile isotope ratios in plumes with primordial noble gases is made even wider if Pitcairn is included (open circle in Fig. 11.9), based on its primordial neon signature

(Section 11.2.2). The conclusion that flows from this evidence is that noble gases are significantly decoupled from lithophile isotope systems. This implies the existence of a deep, helium-rich primordial source capable of releasing noble gases into several different types of mantle reservoirs.

11.1.4 The 'Two-Reservoir' Model

The intermediate helium isotope composition of MORB, between atmospheric and plume sources, can be explained by partial outgassing of primordial helium from the upper mantle, followed by radiogenic helium production. This would have caused the upper mantle to develop a higher ^{4}He/^{3}He composition than a less degassed lower mantle source, where radiogenic production would be swamped by primordial helium.

This partial degassing or 'two-reservoir' model for the mantle was originally proposed to explain argon isotope systematics (Hart et al., 1979), and was applied to helium by Kaneoka and Takaoka (1980). Their ^{3}He-enriched samples from Hawaii were later shown to be contaminated with cosmogenic helium (Kurz, 1986a). However, the widespread evidence for a primordial ^{3}He component in the Earth has led to general acceptance of the two-reservoir model for mantle helium. Nevertheless, the model still faces several challenges.

One problem for the two-reservoir model is to explain the respective *concentrations* of helium and other noble gases in the two-reservoirs. Thus, if OIB come from an undegassed source, we would expect them to contain more helium than MORB glasses from the degassed upper mantle. However, OIB glasses actually have ten times *less* ^{3}He than MORB (Fisher, 1985). This observation has sometimes been called the helium paradox (e.g. Hilton et al., 2000). However, although this evidence is problematical, it is not definitive, due to the poorly constrained behaviour of noble gases

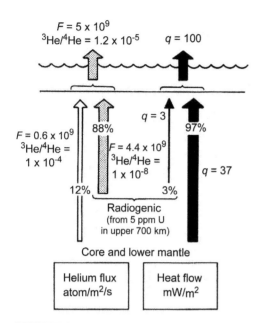

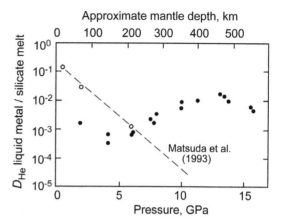

Fig. 11.11 Evidence from high pressure experiments for the $D_{metal/silicate}$ partition coefficient of helium at high pressures. (○) = old data. After Bouhifd et al. (2013).

Fig. 11.10 A comparison of calculated helium (F) and heat fluxes (q) in the oceans. After O'Nions and Oxburgh (1983).

during the melting process. For example, the dynamics of mantle convection and melt segregation under ridges must be different from plumes: ridge magmas probably collect helium from a greater volume of mantle during the melting process (Section 13.3). Hence, most workers have taken the isotopic evidence in favour of the two-reservoir model for helium as definitive, and over-riding any problems involving noble gas abundances. The case for the heavy noble gases will be discussed later.

Another test for the two-reservoir model comes from a comparison of helium and heat fluxes in the earth (O'Nions and Oxburgh, 1983). These fluxes should be related, because the decay of uranium and thorium produces both radiogenic helium (α particles) and also radioactive heating. Taking account of the small amount of heat also derived from ^{40}K decay, O'Nions and Oxburgh (1983) calculated that 10^{12} atoms of ^4He would be generated in the mantle per joule of heat production. They then calculated the concentration of U necessary to generate the observed helium and heat fluxes. The results were somewhat surprising, because the amount of uranium required to generate 88% of the Earth's oceanic helium flux can only produce 3% of the observed oceanic heat flow (Fig. 11.10).

The logical source for some of the remaining heat flux is crystallization of the inner core, which releases heat through the outer core and mantle by convection. However, this convection must operate in such a way that the reservoir of primordial ^3He in the Earth's interior is not completely exhausted. Therefore, O'Nions and Oxburgh (1983) proposed that a boundary layer inhibits upward transport of helium from the primordial reservoir much more effectively than

the transport of heat. They envisaged this boundary layer at 700 km depth, separating the upper and lower mantle, and implying that the whole lower mantle is a primordial helium reservoir. However, this simple model has been undermined by increasing geophysical evidence for whole mantle convection (Section 6.1.6), which militates against simple box models with separate upper and lower mantle reservoirs.

One way of preserving the two-reservoir model in a mantle with single-layer convection is to invoke increasing mantle viscosity with depth. This might cause large lumps of the lower mantle to be preserved intact, without being streaked out and homogenized by convection. This model was tested by 2D numerical modelling of one-layer convection in such a mantle (van Keken and Ballentine, 1998, 1999). However, these workers argued that models which were realistically close to the real Earth in terms of viscosity and phase transformations could not preserve lower mantle domains large enough to act as a primordial helium reservoir.

A second possible location for the primordial helium reservoir is the core (Porcelli and Halliday, 2001). The core–mantle boundary provides an obvious boundary layer for the retention of helium relative to heat. However, early high-pressure experiments (Matsuda et al., 1993) suggested that helium partition into the metal phase decreased significantly as pressure was increased, implying that the core would only have a limited helium budget (Fig. 11.11). Nevertheless, more recent high pressure experiments by Bouhifd et al. (2013) showed no correlation of helium metal/silicate partition coefficients with pressure (Fig. 11.11), yielding a modest $D_{metal/silicate}$ value of around 10^{-2} at 15 GPa (ca. 500 km depth). They also cited theoretical calculations (Zhang and Yin, 2012) implying that a helium $D_{metal/silicate}$ value around 10^{-2} extends to 40 GPa (ca. 1400 km depth), suggesting that similar values might pertain at the core/mantle boundary (ca. 2900 km).

Helium partition into the core could be further enhanced if the core contains a significant fraction of sulphides. In support of this model, a sulphide phase in the core is necessary to explain the observed mass fractionation of stable copper isotopes in the Earth (Savage *et al.*, 2015). The case for xenon partition into the core will be discussed in Section 11.5.

A third possible location for a primordial helium reservoir is the basal layer of the mantle immediately overlying the core, usually termed the D double-prime layer (D″). There is much uncertainty about this layer, whose thickness is poorly defined (tens to hundreds of km) and whose melt content is also poorly defined. For example, some models suggest that a basal magma ocean existed at the bottom of the mantle for much of Earth history (Sections 6.1.6, 15.7).

11.1.5 Helium Box Models

Numerous attempts have been made to model the helium isotope evolution of the Earth using box models, but it has been surprisingly difficult to reach firm conclusions. A recurring issue of disagreement is whether variable ^{4}He/^{3}He ratios in different reservoirs (e.g. OIB and MORB) are primarily due to differential degassing, or could be explained by more constant degrees of degassing, coupled with variable uranium concentrations.

An early extreme form of the two-reservoir model was presented by Allegre *et al.* (1986), in which a strongly degassed upper mantle was separated from a relatively un-degassed lower mantle by a boundary layer at the 670 km seismic discontinuity. Only helium was allowed to diffuse from the lower to the upper mantle (based on its high diffusivity), whereas plumes conveyed noble gases directly from the lower mantle to the atmosphere.

A more balanced box model was presented by Kellogg and Wasserburg (1990), recognizing the substantial flux of helium that was carried from the lower mantle to the upper mantle in plumes. This model still involved a relatively un-degassed lower mantle plume source and a degassed upper mantle MORB source, but a helium flux from the lower mantle caused the upper mantle to be in steady state, with a helium residence time of 1.4 Ga.

The opposite extreme position (to that of Allegre *et al.*) was taken by Anderson (1998, 2001), who argued that the whole mantle was equally degassed, but that so-called 'primitive' helium signatures (low ^{4}He/^{3}He ratios) in plumes such as Hawaii are due to shallow uranium-depleted sources. This theory that 'primordial' ^{3}He signatures arise from the upper mantle has been rejected by most researchers, but early helium isotope data from Baffin Island (Fig. 11.8) caused Class and Goldstein (2005) to revive a version of Armstrong's model.

Class and Goldstein (2005) proposed that the degassed MORB source was the principal helium reservoir in the Earth, and that the OIB source only deviated from it over the last 1.5 Ga of Earth history. Their model involved the ^{4}He/^{3}He ratio of the MORB source evolving through contin-

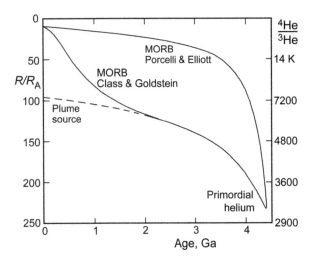

Fig. 11.12 Plot of helium isotope ratios against time to show the consequences of alternative degassing models for the MORB source (solid line) and plume source (dashed) according to the models of Class and Goldstein (2005) and Porcelli and Elliott (2008).

ual degassing, whereas the OIB source was isolated for the last 1.5 Ga at the base of the mantle, so that its ^{4}He/^{3}He evolution was retarded relative to MORB (Fig. 11.12).

However, Porcelli and Elliott (2008) pointed out that the inflection in the MORB helium evolution curve of Class and Goldstein (2005) had not been geologically justified. It was caused by assuming that the rate of ^{4}He degassing from the upper mantle decreased linearly over Earth history, despite the fact that the production of ^{4}He by radioactive decay decreases exponentially. Geological justification of the model would have required extreme uranium enrichment of the upper mantle at around 1.5 Ga, similar to an earlier proposal of Seta *et al.* (2001). However, the evidence from actual uranium measurements of MORB samples (Section 6.3.3) is that the U/Pb ratio (μ value) of the upper mantle has *decreased* with time. Therefore it appears that the helium degassing model of Class and Goldstein (2005) is invalid. On the other hand, Porcelli and Elliott (2008) showed that if the helium degassing rate over Earth history is proportional to concentration, a degassed source such as MORB evolves to radiogenic ^{4}He/^{3}He ratios early in Earth history (Fig. 11.12). Therefore, they concluded that the primordial unradiogenic ^{4}He/^{3}He ratio of the plume source can only be produced from a mantle source that was relatively less degassed early in Earth history.

There may now be general agreement on these conclusions, but there is still disagreement about the origins of individual helium components, just as there have been for lithophile isotope mantle components. A prime example is the Baffin Island helium signature, which was shown by Jackson *et al.* (2010) to be associated with unradiogenic Pb compositions close to the geochron. Hence Jackson *et al.* argued

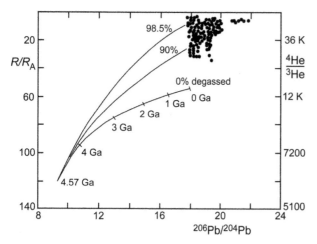

Fig. 11.13   Plot of helium versus lead isotope ratios in OIB (•), compared with mantle growth models with different degrees of helium degassing. Modified after Huang *et al.* (2014).

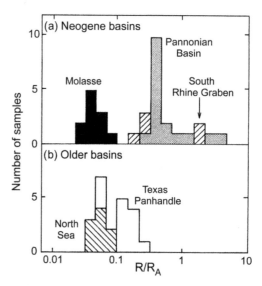

Fig. 11.14   Histograms showing variation in R/R_A values (on a log scale) in different types of sedimentary basin, showing more crustal or mantle type signatures. After Oxburgh *et al.* (1986).

that this source could be an ancient relic of primitive undifferentiated and undegassed mantle.

On the other hand, Huang *et al.* (2014) argued that this signature could be derived from ancient sulphide grains, which inherited unradiogenic He and Pb from ancient subduction zones. They argued that these signatures were preserved in mafic under-plates of arc crust that foundered and settled to the bottom of the mantle, to be sampled by later plumes. Huang *et al.* claimed to observe an overall correlation between helium and lead signatures in OIB (Fig. 11.13), due to storage of such signatures in sulphide-bearing cumulates of different ages. However, it seems clear from the data in Fig. 11.13 that there is no simple relationship between Pb and He systematics in OIB.

The evidence from Fig. 11.9 is that primordial noble gas signatures are associated with a wide variety of lithophile mantle components, pointing to a large degree of decoupling of helium from lithophile isotope tracers. However, if we look past the scatter of OIB data in Fig. 11.13, we recall that Pb partition into the core has been argued as a fundamental process controlling Pb evolution in the mantle as a whole (Section 5.3.1). If sulphide in the core acted as a primordial helium reservoir, this could likewise control the overall evolution of mantle helium.

11.1.6 Crustal and Mantle Helium

99% of the continental helium flux is radiogenic, and can be sustained by a U equivalent concentration of 6 ppm in the upper 8 km of the crust. This can also explain 50% of the continental heat flux. Hence, the other 50% of continental heat flow must be sub-continental, whereas less than 1% (primordial + radiogenic) of the continental helium flux comes from the mantle. Hence it is clear that the continental crust is a boundary layer. Mantle-derived heat can be carried across it

conductively, but mantle-derived helium only leaks through the crust in certain discrete areas. These are normally areas of active magmatism, and are of particular interest in studying crust–mantle helium mixing processes (Fig. 11.14).

Well-gas studies demonstrate the local nature of mantle helium transport through the crust. Oxburgh *et al.* (1986) showed that sedimentary basins which result from crustal loading, such as the Alpine Molasse basin, yield helium with very low R/R_A values around 0.05, whereas sedimentary basins formed by extensional tectonics, such as the Rhine Graben and the Pannonian basin of Hungary, may yield helium with much higher R/R_A values of around unity (Fig. 11.14a). The huge 'Panhandle' gas field in the southern USA is particularly interesting. It is one of the world's largest gas fields, and has helium contents of up to 2%. In the south, the reservoir is draped over uplifted Proterozoic–Paleozoic basement, and in this region R/R_A values as low as 0.06 have been measured (Fig. 11.14b). In contrast, the northern part of the reservoir is in an area of recent igneous activity. Here, R/R_A values of up to 0.2 have been measured, corresponding to 2% MORB type helium (Oxburgh *et al.*, 1986).

It has traditionally been assumed that crust–mantle mixing processes of this type are not important in oceanic volcanics. However, recent work shows that helium isotope signatures in OIB may also be susceptible to high-level contamination processes. For example, Hilton *et al.* (1993) found a strong correlation between R/R_A value and petrology in submarine volcanic glasses from the Lau back-arc basin, situated behind the Tongan arc. Basaltic samples from the centre of the basin had relatively high helium contents (up to 10 µcc/g), and normal MORB-like R/R_A values of 8 (Fig. 11.15). However, more differentiated glasses from just behind the

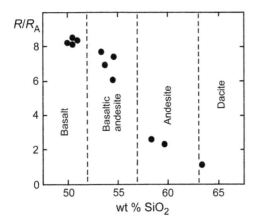

Fig. 11.15 Plot of helium isotope composition (in vesicles) against the silica content of Lau Basin submarine volcanics, showing an inverse correlation. After Hilton et al. (1993).

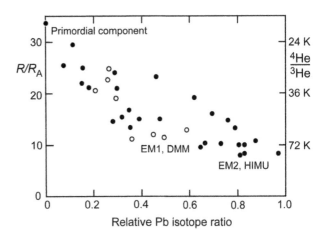

Fig. 11.16 Correlation between helium and lead isotope ratios in three hot-spot trails of the Samoan plume. (•) = EM1, DMM; (○) = EM2, HIMU. After Jackson et al. (2014).

magmatic arc had much lower helium contents (<0.2 μcc/g), and R/R_A values as low as unity. Based on the correlation between $^3He/^4He$, helium content and petrology, Hilton et al. attributed the lower R/R_A values in differentiated glasses to shallow level contamination, probably due to crustal assimilation by magmas which had been largely degassed of mantle helium.

Questions about the relative importance of shallow contamination effects versus deep mantle source signatures are critically important for the Samoan plume, which has a unique combination of elevated R/R_A and radiogenic Sr and Pb (Figs. 11.7, 11.9). Peridotite xenoliths in young Samoan lavas have very radiogenic Sr isotope signatures, attributed to a recycled sediment component. However, helium isotope analysis of fluid inclusions from the xenoliths revealed high R/R_A values of around 12 (Farley, 1995a). This was unexpected, since a recycled sediment component should have radiogenic helium with low R/R_A values.

Because subducted sediment accumulates radiogenic helium from uranium decay, the high R/R_A value of the xenoliths should place limits on the mantle residence time of the sediment since subduction. Based on binary Sr–He mixing calculations, Farley (1995a) estimated a residence time of only 10 Ma, suggesting that the sediments were incorporated into the plume from the nearby Tongan trench (120 km west of Samoa). However, this calculation changed when 5 Ma submarine lavas off Samoa were shown to have similar Sr–He isotope signatures (Jackson et al., 2007). Since the Samoan plume was over 1000 km east of the Tongan trench at the time of eruption, it was no longer possible to invoke recent sediment subduction to explain the enriched isotope signatures of the Samoan plume.

Further work by Jackson et al. (2014) showed that the Samoan plume actually consists of four closely spaced hot-spot trails, each of which displays a mixing line between a common primordial component and distinct depleted/recycled components. Strong mixing lines are displayed between helium and lead isotopes (Fig. 11.16), despite the fact that two suites trend towards distinct *unradiogenic* Pb components (EM1, DMM), while two others trend towards distinct *radiogenic* Pb components (EM2, HIMU). This suggests that a single primordial helium component mixes with different evolved components within the up-welling plume. Since the Pb isotope signature of the primordial Samoan component (Fig. 6.38) is completely different from that in Baffin Island, this demonstrates the decoupling of helium from lithophile mantle components.

11.1.7 Oceanic Sediments and Interplanetary Dust

It is well established that high $^3He/^4He$ ratios in ocean floor sediments reflect the accumulation of inter-planetary dust particles (IDPs). However, the question of temporal variability in the IDP flux has more recently been examined (Takayanagi and Ozima, 1987). These authors studied ^3He variability in a 10 m pelagic clay core from the Central Pacific and a 150 m nanno-fossil ooze core from the South Atlantic. The former spanned 0–3 Ma, while the latter, with generally higher sedimentation rates, spanned 0–40 Ma. Sedimentation rates were determined in both cases by paleomagnetism, supplemented in the 3 Ma core with ^{10}Be data (Section 14.4.2). The observed range of $^3He/^4He$ ratios was attributed to mixing of 0.1–1 ppm of IDPs with terrestrial sediment (Fig. 11.17). However, the ^3He content of IDPs is ten orders of magnitude higher than terrestrial sediment, so the IDPs totally dominate the ^3He budget of the samples.

In both cores studied by Takayanagi and Ozima, ^3He contents were inversely correlated with sedimentation rate. The ^3He deposition flux was therefore determined by multiplying the ^3He content by the sediment mass accumulation rate (mass is used because ocean floor sediments undergo

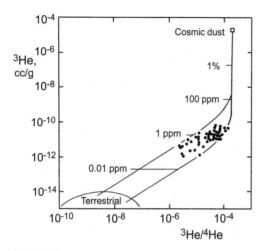

Fig. 11.17 Plot of ^{3}He abundance against isotope ratio in pelagic clays from the Central Pacific, compared with mixing lines between cosmic dust and terrestrial sediments. After Takayanagi and Ozima (1987).

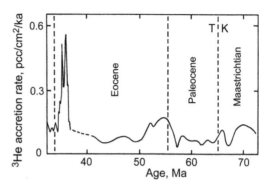

Fig. 11.18 Record of ^{3}He abundances in Cretaceous to Tertiary age sediments, showing a ^{3}He peak in the late Eocene, but no peak at the K–T boundary. After Mukhopadhyay et al. (2001).

compaction after deposition). The results suggested flux variations over time, but did not display any overall trend. The average ^{3}He flux over the past 40 Ma was estimated as 1.5 ($\pm$1) $\times$ 10^{-15} cc/cm^2/yr (at STP).

Generally similar results were obtained by Farley (1995b) on a 22 m core of pelagic clay from the central North Pacific, spanning the past 72 Ma. During the Quaternry, the sedimentation rate was high, yielding a ^{3}He flux of about 1.1 $\times$ 10^{-15} cc/cm^2/yr, in good agreement with Takayanagi and Ozima (1987). However, in the deeper part of the core, the calculated ^{3}He flux was lower, as shown by Mukhopadhyay et al. (2001). It is not clear whether this represents a real variation in the interplanetary dust flux over time, or a reduction in the retentivity of ^{3}He with depth.

Farley (1995b) also observed no ^{3}He peak at the K–T boundary, indicating that the extra-terrestrial signals from iridium and helium are decoupled (Fig. 11.18). This was attributed to impact-induced vaporization and outgassing of the K–T bolide. In contrast, Farley et al. (1998) discovered a spike of ^{3}He in the late Eocene (35–36 Ma ago) which did correlate with the iridium signal (Fig. 11.18). This event was attributed to a comet shower, but there is no ^{3}He support for a comet shower at the Paleocene–Eocene boundary (Schmitz et al., 2004).

A more controversial question concerns the evidence for variation of the ^{3}He flux during the glacial cycles of the Quaternary period. A detailed study in this time range was made by Marcantonio et al. (1995), based on a 4 m core of carbonate-rich sediment from the Central Pacific, spanning the last 200 ka. After correcting for dilution by biogenic carbonate, their ^{3}He/^{4}He data lay on the same mixing line observed by Takayanagi and Ozima between terrigenous and IDP components. However, Marcantonio also determined initial excess ^{230}Th activities on the same samples. Normalization of ^{3}He

to ^{230}Th can remove the effects of variable sediment dilution, because ^{230}Th is constantly produced in seawater from ^{234}U and is rapidly transported to the ocean floor by adsorption onto sinking particulate matter (Section 12.3.3). When plotted against ages from oxygen isotope stratigraphy, ^{3}He and ^{230}Th showed strong covariation, with peak signals during interglacial periods (Fig. 11.19). These peaks were attributed, not to variations of the IDP flux, but to intensified carbonate dissolution during interglacial periods. Hence, based on the ratio of ^{230}Th activity to ^{3}He content, an average ^{3}He deposition flux of 0.96 $\times$ 10^{-15} cc/cm^2/yr was determined for the past 200 ka.

Farley and Patterson (1995) made a similar study of Quaternary ^{3}He variation based on a 9 m core of foram nanofossil ooze from the flank of the Mid Atlantic Ridge, spanning the period 250–450 ka. ^{3}He contents were inversely correlated with δ^{18}O variations, which were interpreted as monitors of

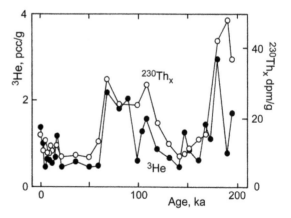

Fig. 11.19 Isotope stratigraphy of a carbonate-rich sediment from the Central Pacific, showing strong co-variation between ^{3}He abundance ($\bullet$) and excess initial ^{230}Th activity ($\circ$). After Marcantonio et al. (1995).

glacial–interglacial cycles. Similar results were obtained by Patterson and Farley (1998), leading them to speculate that the helium isotope data were recording a causal relationship between IDP accumulation and climate. For example, Muller and MacDonald (1995) proposed that glacial cycles reflect variation in Earth's orbital inclination, leading to periodic encounters with a cloud of IDPs which partially block out solar radiation. However, according to this model, interglacial periods should be characterized by the lowest ^{3}He flux, whereas Farley and Patterson found the opposite relationship.

A more likely interpretation (Marcantonio et al., 1995, 1996) is that climatically induced variations in sedimentation rate caused apparent variations in the ^{3}He flux which could not be adequately corrected with the available age data for the core. This explanation was confirmed by ^{3}He data from an early Quaternary (1.35–1.6 Ma) core from the Equatorial Pacific (Winckler et al. (2004). In this core, ^{3}He abundances were strongly correlated with the non-carbonate fraction in the sediment, with a 41 ka periodicity that is related to climate cycles, but unrelated to orbital inclination. Therefore the ^{3}He abundance variation must be attributed to variations in sediment focussing during different climatic periods (Section 12.3.5). Thus it is concluded that the IDP flux has been essentially uniform and constant over the past 2 Ma, and is not responsible for glacial cycles.

Reversing the sense of the above arguments, Farley and Eltgroth (2003) proposed that if the ^{3}He flux is assumed to be constant, the method can be used to calibrate short-term variations in sedimentation rate. They demonstrated the usefulness of the method in studying rapid climatic changes at the Paleocene–Eocene boundary (see also Murphy et al., 2010).

11.2 Neon

Unlike helium, the heavier noble gases accumulate in the atmosphere over geological time, so their atmospheric abundances are much larger relative to rocks. For this reason, atmospheric contamination has always been a huge problem in the isotopic analysis of heavy noble gases (including neon). Improvements in analytical procedures and instrumentation were critical to achieving accurate data, and these new data have largely superseded older low-precision results. For neon, this technical revolution occurred in the late 1980s to early 1990s, when high-precision data distinct from the atmosphere were first obtained for mantle-derived rocks. In this period, neon isotope systematics went from a state of complete confusion to a degree of clarity that has since enabled them to be a kind of 'Rosetta Stone' for understanding the behaviour of other noble gases in the Earth.

11.2.1 Neon Production

Of the three stable neon isotopes, ^{20}Ne is non-radiogenic, whereas ^{21}Ne and ^{22}Ne are generated by nucleogenic

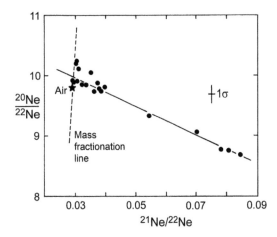

Fig. 11.20 Neon three-isotope correlation diagram showing well-gases from the Alberta basin on a mixing line between atmospheric and nucleogenic neon. After Kennedy et al. (1990).

interactions with α particles and neutrons. The α particles are mostly derived from the U-series decay chains, while the neutrons are mostly produced by secondary reactions from α particles. The principal reactions are n, α reactions on ^{24}Mg and ^{25}Mg, producing ^{21}Ne and ^{22}Ne respectively. Subsidiary pathways are α, n reactions on ^{18}O and ^{19}F which produce ^{21}Ne and ^{22}Na, the latter undergoing β decay to ^{22}Ne.

These reactions were first studied by Wetherill (1954) and have been refined in subsequent work (e.g. see Kennedy et al., 1990). The net result of these reactions is to yield a trend towards lower ^{20}Ne/^{22}Ne and higher ^{21}Ne/^{22}Ne ratios, which is most clearly seen in uranium-rich rocks such as granites. Figure 11.20 shows isotopic data for gas wells from Alberta, Canada, plotted on the commonly used neon three-isotope diagram. The data form a linear array which was attributed to mixing between atmospheric and nucleogenic neon. This is consistent with helium isotope data for these gases, which show a strong radiogenic signature with no mantle-derived component.

Isotopic analysis of exposed terrestrial rocks has also demonstrated the cosmogenic production of ^{21}Ne (Marty and Craig, 1987). This isotope is produced by spallation reactions on Mg, Na, Si and Al, generating a sub-horizontal array on the three-isotope plot. By analysing all three isotopes, the cosmogenic component can be resolved from trapped (magmatic) neon and nucleogenic neon. Coherent behaviour of ^{21}Ne has been demonstrated with ^{10}Be and ^{26}Al (Section 14.7.4), suggesting that neon will be a useful tool in determining cosmic exposure ages of surficial rocks.

11.2.2 Primordial Neon in the Earth

The first evidence for non-atmospheric neon in the mantle was presented by Craig and Lupton (1976) on samples of MORB and volcanic gases. The sample most enriched in ^{20}Ne was a fumarole gas from Kilauea volcano, with a ^{20}Ne/^{22}Ne

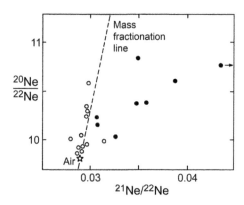

Fig. 11.21 Comparison of mantle samples (•), and mass-fractionated geothermal gases (○), on a neon three-isotope diagram. After Kyser and Rison (1982).

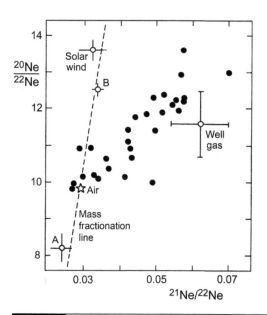

Fig. 11.22 Compilation of MORB neon data (•) on a three-isotope plot. A and B represent the 'planetary' neon compositions seen in some meteorites. After Hiyagon et al. (1992).

ratio of 10.3 ± 0.04. Subsequently, Harding County well-gas was also found to have a composition well removed from atmosphere (Phinney et al., 1978). These authors attributed the ^{20}Ne-enriched components to exotic primordial noble gas components in the Earth, possibly representing solar neon. However, similar neon compositions in geothermal gases from Japan (Nagao et al., 1979) were attributed to mass fractionation due to preferential diffusion of light neon to the sampling sites, through porous rock or soil (Fig. 11.21). Therefore, when Kyser and Rison (1982) found ^{20}Ne excesses in a variety of mantle xenoliths and megacrysts (Fig. 11.21), they attributed these signatures to mass fractionation of neon from an original mantle composition similar to the atmosphere (along with nucleogenic production).

Elevated ^{20}Ne abundances above atmospheric were subsequently found in diamonds by Honda et al. (1987) and Ozima and Zashu (1988, 1991). However, it is not logical to attribute elevated ^{20}Ne in solid mantle samples to a gaseous fractionation model. Therefore Ozima and Zashu reversed the mass fractionation argument of Kyser and Rison, suggesting that diamonds sample a solar neon reservoir in the Earth, whereas the present day atmosphere has been *depleted* in ^{20}Ne by mass fractionation. They argued that bombardment of the early Earth by radiation caused the massive blow-off of a primitive solar type atmosphere, leaving a residue enriched in heavy neon.

Fractionation of the proposed magnitude between mantle and atmosphere should be accompanied by fractionation of the non-radiogenic isotopes of argon, krypton and xenon. Therefore, it should be possible to test the atmospheric fractionation model by isotopic analysis of the heavy noble gases. However, two factors make such a test difficult to perform. Firstly, the heavier noble gases are expected to undergo lesser amounts of mass fractionation, making them less sensitive tracers of this process. Secondly, the heavy noble gases are more likely to be recycled back into the mantle, masking the effect of any mantle–atmosphere fractionation of such gases. This subject will be re-examined below.

Additional insights into terrestrial neon systematics were obtained by neon analysis of larger suites of submarine basaltic glasses. Sarda et al. (1988) demonstrated the existence of a MORB correlation line passing through the atmosphere point, which has been confirmed in several subsequent studies (e.g. Marty, 1989; Hiyagon et al., 1992; Moreira et al., 1998). This array (Fig. 11.22) can be explained by three-component mixing of solar type, atmospheric and nucleogenic neon, as discussed below.

Because of their primordial helium signatures, Loihi glasses were the preferred target for neon analysis of OIB samples, but the early data fell within error of atmosphere (Sarda et al., 1988). Nevertheless, subsequent analyses of submarine basalt glasses from Loihi and Kilauea revealed a wider range of neon isotope ratios, stretching from the atmospheric composition towards ^{20}Ne-enriched compositions (Honda et al., 1991; Hiyagon et al., 1992). The enriched end of this array approaches the solar wind composition, but the array has a slope intermediate between the pure mass fractionation line and the MORB correlation line (Fig. 11.23). Therefore, Honda et al. and Hiyagon et al. attributed all neon in the Earth's interior to mixing between solar and nucleogenic isotopes. The sloping arrays of MORB and OIB were then explained by variable atmospheric contamination of this solar + radiogenic mantle neon, and the Loihi neon samples analysed by Sarda et al. (1988) were attributed to severe atmospheric contamination.

Allegre et al. (1993) proposed alternatively that high solar type ^{20}Ne/^{22}Ne ratios in the mantle could be explained by

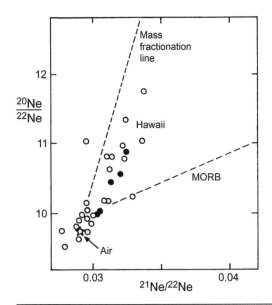

Fig. 11.23 Compilation of Hawaiian neon data on a three-isotope plot. ($\circ$) = Loihi; ($\bullet$) = Kilauea. After Hiyagon *et al.* (1992).

subduction of cosmic dust particles in deep sea sediments. This proposal was somewhat ironic, because it contradicted the 'noble gas subduction barrier' invoked by Staudacher and Allegre (1988). In addition, experimental studies by Hiyagon (1994) suggested that neon would be completely extracted from cosmic dust within three years at 500 °C, which is insufficient to sustain neon subduction. Hence it appears that the noble gas subduction barrier *is* effective for neon. This implies that mantle-derived magmas, once corrected for atmospheric contamination during eruption, sample an essentially primordial neon reservoir. However, the origin of this reservoir has remained in some doubt.

11.2.3 Sub-Solar Neon

In addition to solar neon, with a $^{20}Ne/^{22}Ne$ ratio of ca. 13.8, early work on meteorites (Black, 1972) identified several other distinct and reproducible isotope signatures. Because the identity of these components was unclear, Black designated them by the letters A and B (Fig. 11.22). These components were termed 'planetary' because they were considered to be possible sources of terrestrial neon. Of these components, neon-B, with a 'sub-solar' $^{20}Ne/^{22}Ne$ ratio of 12.7, represents an alternative to solar neon as a source for noble gases in the Earth's interior. However, Raquin and Moreira (2009) argued that neon-B is not so much a discrete solar system component as the result of a process of ion implantation and sputtering of meteoritic or planetary substrates by the solar wind. This is demonstrated by the neon composition of lunar soil, which lies very close to the composition of neon-B because it has been exposed to irradiation by the solar wind for millions of years (Moreira, 2013).

Moreira and Charnoz (2016) modelled this type of signature by simulating solar irradiation of cosmic dust for 0.1 Ma prior to accretion. Hence they argued that the neon composition of the Earth's interior is consistent with its derivation from inner solar system material with neon isotope systematics similar to enstatite chondrites. This material would have been subjected to intense irradiation by the solar wind before its accretion into planetary bodies, thus incorporating the signature of neon-B into bodies such as the Earth.

This model seems much more attractive as an explanation for the Earth's neon signature than the solar neon hypothesis, which requires that the early liquid Earth dissolved solar neon from a dense proto-nebular 'atmosphere' with solar composition (e.g. Harper and Jacobsen, 1996). This neon dissolution model is contradicted by the 'onion skin' model of planet formation (Section 15.1.2), which suggests that small planetary bodies underwent cold accretion, and then started to melt from the interior outwards due to radioactive heating. Hence the exterior of even quite large bodies would have remained solid during the early condensation of the solar system, and would subsequently have given rise to un-processed chondritic meteorites.

Evidence in support of a sub-solar (neon-B) component in the mantle is derived from neon–argon isotope mixing lines. These mixing processes were originally investigated by Farley and Poreda (1993) as a means of correcting for atmospheric contamination of heavier noble gases such as argon (Section 11.3.2). However, because mixing lines between different noble gas tracers typically define hyperbolic shapes, they can also be used to define the composition of end-members. In the case of argon–neon mixing, the atmosphere has much larger argon/neon ratios than the mantle, causing convex-upwards mixing lines that can be used to define the mantle neon composition of the Earth (Fig. 11.24).

The first detailed study of this phenomenon, by Trieloff *et al.* (2000), used argon–neon data for sub-glacially erupted basalt glasses from Iceland and volatile-rich dunite xenoliths from Loihi to constrain the mantle neon signature. These sample suites defined arrays slightly steeper than previous Loihi data on the neon three-isotope plot. In addition, the upper ends of these data arrays ($^{20}Ne/^{22}Ne = 12.5$) fell within error of neon-B on a neon–argon plot, leading Trieloff *et al.* to suggest that this was the composition of 'solar neon' in the Earth.

On the other hand, evidence for $^{20}Ne/^{22}Ne$ ratios *above* neon-B in the Earth was obtained from neon analysis of dunite from an ultramafic–carbonatite complex in the Kola Peninsula of northern Russia (Yokochi and Marty, 2004). However, the relatively large errors on these data mean that even the highest $^{20}Ne/^{22}Ne$ ratio (13.04 ± 0.4 2σ) is within error of the neon-B composition. In addition, further evidence for a sub-solar composition in the OIB neon reservoir was obtained by laser analysis of individual gas vesicles in sub-glacial volcanic glasses from Iceland (Mukhopadhyay, 2012; Colin *et al.*, 2015). Because of the large noble gas contents of these vesicles, they are resistant to laboratory

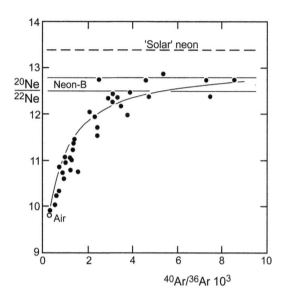

Fig. 11.24 Neon–argon isotope plot for Icelandic subglacial basalts (●) showing hyperbolic mixing line between atmospheric and sub-solar noble gases. Modified after Colin *et al.* (2015).

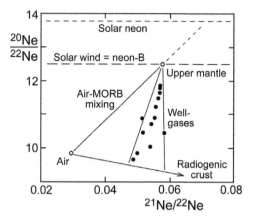

Fig. 11.25 Neon three-isotope plot for well-gases attributed to mixing between nucleogenic crustal neon and a MORB source with sub-solar neon. After Ballentine and Holland (2008).

contamination, and define an excellent argon–neon hyperbola that confirms the sub-solar composition of the source (Fig. 11.24). This conclusion was supported by similar results from Fernandina island in the Galapagos (Peron *et al.*, 2016). In addition, further support of a sub-solar neon composition for the upper mantle is obtained from the comparison of neon with argon-38 data (Section 11.3.3).

A different line of evidence for the composition of upper mantle neon is based on Harding County well-gases from New Mexico (Ballentine *et al.*, 2005; Ballentine and Holland, 2008). These well-gases contain a neon component from a nucleogenic crustal source, which modifies the composition of the atmospheric neon end-member that contaminates the well-gases via circulating ground water. Because the lower end of the mixing line is displaced, it intersects the MORB neon array (also caused by atmospheric contamination) at an oblique angle (Fig. 11.25). This allows the ^{20}Ne/^{22}Ne ratio of the local MORB source to be pinpointed to a value of 12.5, which can be explained by nucleogenic neon growth in a source of original sub-solar (neon-B) composition.

11.2.4 Atmospheric Neon

The origin of atmospheric neon has itself been a subject of extensive debate. Early models (outlined above) attributed the neon signature of Earth's atmosphere to mass fractionation from a solar composition in the interior. This could have been caused by intense bombardment of Earth's early atmosphere by solar ultra-violet radiation, or by atmospheric blow-off in the giant Moon-forming impact (Pepin, 1991, 1997, 2006).

An alternative model proposed by Marty (1989) is that Earth's atmosphere was formed by late accretion of gas-rich meteorites with planetary neon, of which the lower ^{20}Ne/^{22}Ne end-member is neon-A (Fig. 11.26). Dissolution experiments on carbonaceous chondrites have shown that the signature of neon-A originates from micro-diamonds (Tang and Anders, 1988; Huss and Lewis, 1994). Based on the isotopic signatures of several elements, at least some of these diamonds are believed to be relics from before the formation of the solar system (Ott, 2014). These diamonds seem to be responsible for the relatively lower ^{20}Ne/^{22}Ne ratios in bulk

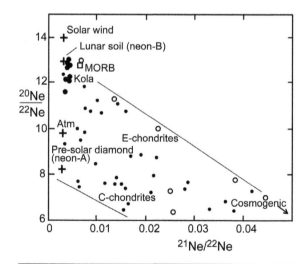

Fig. 11.26 Neon three-isotope plot for carbonaceous and enstatite chondrites (●, ○), showing postulated approach to the compositions of neon-A and neon-B for gas-rich meteorites with a low cosmogenic neon component. Modified after Moreira (2013).

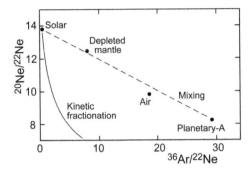

Fig. 11.27 Comparison between alternative mixing and fractionation models to explain neon versus argon/neon data in major reservoirs. After Marty (2012).

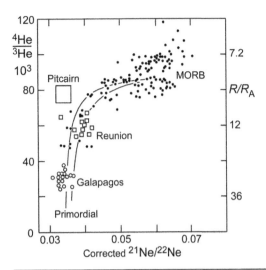

Fig. 11.28 Correlation of helium versus neon isotope signatures in mantle sources, showing a hyperbolic mixing line. Modified after Moreira (2013).

carbonaceous chondrites relative to enstatite chondrites (Fig. 11.26).

The 'late accretion' model of Marty (1989) is supported by osmium isotope data (Section 8.3.1). Because osmium is a siderophile element, early partition into the core should have led to a non-chondritic osmium signature in the overlying mantle. However, since the Earth does in fact have chondritic osmium, this points to late addition of this component after core formation. Since this work, several other lines of isotopic evidence have pointed to the accretion of a 'late veneer' of carbonaceous chondrite material to the Earth, after the giant Moon-forming impact (Section 15.6.6).

Marty (2012) argued that this model is also supported by the neon–argon compositions of solar system reservoirs. These form a relatively linear array (Fig. 11.27), which is better explained by mixing of these reservoirs rather than mass fractionation. For example, kinetic fractionation should yield a hyperbolic curve (Fig. 11.27) which is a bad fit to observed terrestrial compositions relative to solar and planetary neon.

11.2.5 Nucleogenic Neon

With the steepest OIB arrays falling within error of primordial solar-system components on the neon three-isotope plot, nucleogenic/radiogenic neon and helium can now be placed in a unified model. This is possible because helium does not suffer significant atmospheric contamination, while atmospheric contamination of neon can be corrected using the three-isotope plot. Relationships between neon and heavy noble gases will be discussed in the next section.

The fraction of nucleogenic neon in the mantle can be represented in different ways. One approach (Poreda and Farley, 1992) is to force a regression line through the air composition on a three-isotope plot. The gradient of this regression line can then be expressed as the ratio $\delta^{20}Ne/\delta^{21}Ne$. An alternative approach which seems more popular is to project neon data away from the atmosphere point onto the solar or sub-solar $^{20}Ne/^{22}Ne$ ratio, whereupon the neon data are

presented as normalized $^{21}Ne/^{22}Ne$ ratios, which can then be plotted against helium data (Fig. 11.28).

Early He–Ne isotope data from Loihi appeared to lie on a linear mixing line between MORB and a primordial neon source (Porcelli and Wasserburg, 1995b). This would be consistent with the common progenitors U and Th, which produce radiogenic helium and nucleogenic neon. However, Niedermann et al. (1997) found hyperbolic He–Ne mixing lines for MORB samples from the East Pacific Rise, believed to be contaminated by a variable plume-derived component (solid symbols in Fig. 11.28). This was supported by additional work on plume sources such as Reunion and the Galapagos (Hanyu et al., 2001; Hopp and Trieloff, 2005; Kurz et al., 2009). These samples lie on a similar hyperbolic trend (Fig. 11.28), suggesting different He/Ne ratios in upper and lower mantle reservoirs.

A particularly interesting case is represented by Pitcairn, where the EMI lithophile end-member has relatively radiogenic (non-primitive) helium, but non-radiogenic (primordial) neon (Fig. 11.28). Since the evidence suggests that ocean floor neon cannot be subducted, this makes the popular sediment recycling model for this mantle component less attractive. The alternative sub-continental lithospheric source for EMI is a better fit to the neon data, since such a source is not expected to be outgassed of its primordial neon, but could easily have acquired radiogenic helium.

11.3 Argon

Initial, 'excess' or 'inherited' argon is normally regarded as a problem to be avoided in K–Ar and Ar–Ar dating (Section 10.1.2). However, the isotopic composition of initial argon

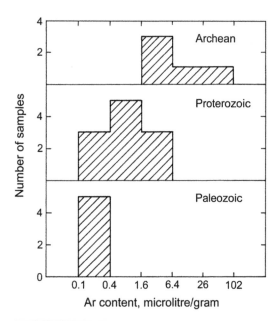

can be used as a powerful geochemical tracer, especially when used alongside other noble gas data.

Atmospheric contamination is a more serious problem in the isotopic analysis of argon than it is for helium and neon. Atmospheric helium has a very low abundance due to its complete escape from the atmosphere, and even neon is believed to have been significantly depleted in the atmosphere by intense solar irradiation. Furthermore, the neon three-isotope plot allows atmospheric contamination to be monitored. On the other hand, the heavy noble gases (argon, krypton and xenon) have accumulated in the atmosphere over Earth history, and corrections for atmospheric contamination are much more difficult. Because of this problem, the first clear evidence for inherited argon was provided by the analysis of beryl, whose ring-type structure accommodates unusually large quantities of initial argon, swamping the effects of atmospheric contamination. 'Excess' argon was first found in beryl by Aldrich and Nier (1948) and studied in more detail by Damon and Kulp (1958). The latter workers discovered Archean beryl containing more than 99% of excess argon and with ^{40}Ar/^{36}Ar ratios as high as 10^5.

11.3.1 Terrestrial Primordial Argon

The ^{40}Ar contents of beryl were observed to decrease over geological time (Fig. 11.11), leading Damon and Kulp to propose extensive early degassing of the Earth in the Archean, decreasing exponentially towards the present. The beryl data shown in Fig. 11.29 were also used by Fanale (1971) to support a more extreme model of catastrophic early degassing of the

Earth. He argued that they were not consistent with models of constant degassing intensity through Earth history, such as proposed by Turekian (1964).

Schwartzman (1973) supported the early degassing model using Ar isotope data from the 2.7 Ga Stillwater Complex. Because this is a mafic–ultramafic intrusion, it can yield more direct information about the Archean mantle than beryl-bearing pegmatites with a potentially large crustal input. A Stillwater pyroxene had an excess ^{40}Ar/^{36}Ar ratio of at least 17 900, corresponding to a calculated maximum ^{36}Ar/silicon ratio of 3×10^{-11} for the 2.7 Ga mantle source. The estimated ratio of (outgassed) atmospheric ^{36}Ar to mantle silicon at the present day is 1×10^{-10}, so the Earth was apparently outgassed to at least 70% of its present extent by the end of the Archean.

Ocean floor basalt glasses are an important source of information about the noble gas budget of the present day mantle because the high water pressure at the site of eruption retains initial magmatic argon in the sample (Section 10.1.2). Furthermore, rapid quenching reduces contamination by atmospheric argon dissolved in seawater. In contrast, the crystalline cores of basalt pillows are largely outgassed of magmatic noble gases and contaminated with atmospheric gases during crystallization (Fisher, 1971).

Hart et al. (1979) used the maximum ^{40}Ar content of 3×10^{-6} ml/g in ocean floor basalt glasses as an estimate of the present day ^{40}Ar concentration in the upper mantle source (UM). By subtracting this value from the average ^{40}Ar concentration in the Bulk Silicate Earth (estimated from its K abundance), and comparing this to the total ^{40}Ar budget of the atmosphere and crust, Hart et al. were able to calculate the mass of mantle which must be outgassed. This calculation is shown in equation [11.1], where square brackets denote concentrations:

$$\text{mass of mantle outgassed} = \frac{^{40}\text{Ar}_{atm} + {}^{40}\text{Ar}_{crust}}{[{}^{40}\text{Ar}_{BSE}] - [{}^{40}\text{Ar}_{UM}]} \quad [11.1]$$

Using this equation, a potassium abundance of 660 ppm in the Bulk Silicate Earth implied that only 25% of the total mass of the mantle need be outgassed.

Hart et al. proposed that the upper mantle was also thoroughly degassed of ^{36}Ar, so that subsequent radiogenic ^{40}Ar production generated high ^{40}Ar/^{36}Ar ratios of up to 16 000. On the other hand, they argued that the lower mantle was not significantly degassed of argon, such that radiogenic ^{40}Ar production was swamped by the primordial component, allowing only a modest rise in ^{40}Ar/^{36}Ar ratio above the atmospheric value. Hart et al. noted the similarity of these model predictions to the ^{40}Ar/^{36}Ar ratios observed in ocean floor glasses from ridges and mantle plumes respectively (Fig. 11.30). This suggested to them that these data were not seriously perturbed by atmospheric contamination.

However, the concept of a relatively less degassed or 'undegassed' mantle reservoir with respect to heavy noble gases was strongly contested by Fisher (1983, 1985). He argued that the low ^{40}Ar/^{36}Ar ratios measured in plume

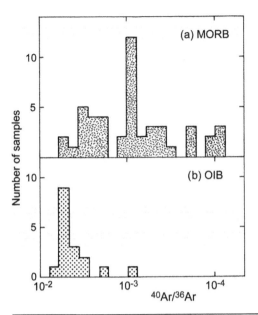

Fig. 11.30 Histograms of early $^{40}Ar/^{36}Ar$ data for MORB and OIB, showing contrasting ranges of isotope composition. After Hart et al. (1979).

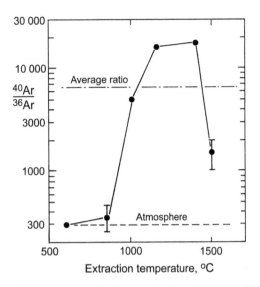

Fig. 11.31 Plot of measured argon isotope ratio against extraction temperature for MORB glass showing probable atmospheric contamination of the low temperature (<1000 °C) fractions. After Staudacher and Allegre (1982).

environments were a result of atmospheric contamination. This argument was initially aimed at data on xenolithic inclusions in lavas (e.g. Kaneoka and Takaoka, 1980). These appeared to contain primordial argon and helium, but were subsequently shown to be contaminated by atmospheric argon and cosmogenic helium (Section 11.1.2). Submarine glasses may be more resistant to such effects, but the interpretation of these data has nevertheless provoked intense controversy.

11.3.2 Atmospheric Contamination

Experience has shown that all mantle-derived noble gas samples are contaminated to some extent by atmospheric gases. Therefore, it is necessary to take stringent experimental precautions to minimize as far as possible the extent of this contamination. Samples and equipment are thoroughly baked before analysis, and frequent blanks are determined to verify the effectiveness of these procedures. It may be possible to extract a mantle signature from a contaminated sample by extracting argon from the sample in separate aliquots, which could sample isotopic heterogeneities within the sample.

One approach is to perform the noble gas analysis by step heating (e.g. Staudacher and Allegre, 1982). This procedure is demonstrated for a MORB sample in Fig. 11.31. Absorbed atmospheric noble gases are released in the low-temperature heating steps, allowing an estimate to be made of the severity of contamination effects. Another approach (Hart et al., 1983) is to do separate experiments by thermal degassing and by crushing. The latter method releases gases from vesicles, which may have undergone less (but sometimes more)

contamination than the rock matrix. Staudacher et al. (1986) used both step heating and crushing to make the most rigorous search for sample contamination.

In other early work on this subject, attempts were made to use elemental noble gas ratios (e.g. $^4He/^{40}Ar$) to distinguish between atmospheric and primordial signatures in MORB and OIB samples (e.g. Hart et al., 1983, 1985). However, it has become apparent that elemental fractionation of noble gases can occur during several phases of the evolution of the samples, including partial melting, magmatic differentiation and solidification. Hence, it appears that isotope signatures are the only reliable discriminant between atmospheric and primordial signatures, and even these can be misleading.

In an attempt to resolve these components in the argon isotope signatures of MORB and OIB samples, Allegre et al. (1983) and Staudacher et al. (1986) compared argon and helium isotope ratios in glasses from ocean ridges and from the Hawaiian plume. Since Loihi Seamount gave the most primordial (lowest $^4He/^3He$) ratios for uncontaminated mantle-derived materials, Allegre et al. interpreted the low argon isotope ratio in these samples as likewise indicative of an undegassed mantle source.

In Fig. 11.32, the MORB field has a forked shape, attributed by Staudacher et al. (1986) to mixing with primordial and atmospheric helium reservoirs. Both mixing branches show very strong curvature, attributed to the very low argon content of the degassed MORB reservoir, relative to both atmosphere and plume reservoirs. However, a dunite xenolith from Loihi lay far off the proposed MORB–plume mixing line. The argon signature of this sample is consistent

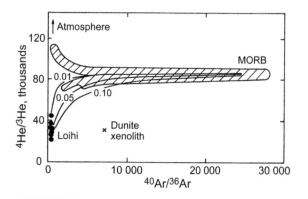

Fig. 11.32 Plot of helium versus argon isotope ratio for submarine glasses from MORB and plume environments. Mixing lines are shown for different $^3He/^{36}Ar$ ratios in Loihi relative to MORB sources. After Staudacher *et al.* (1986).

with a source in oceanic lithosphere. The high ^3He content of this sample cannot be cosmogenic (as for some other xenoliths), since it is submarine; therefore, it was attributed to diffusion of helium from the host magma into the xenolith before eruption. However, this sample could also imply that the plume source has much higher $^{40}Ar/^{36}Ar$ ratios than indicated by the Loihi data.

Patterson *et al.* (1990) pointed out that seawater has between two and four orders of magnitude more ^{36}Ar than Loihi glasses, but two orders of magnitude less ^3He. Hence, argon isotope ratios in Loihi magmas might have been contaminated by seawater without affecting their helium signature. In MORB glasses, variable contamination of this kind generates correlations between $^{40}Ar/^{36}Ar$ ratio and $1/^{36}Ar$ abundances, indicative of simple mixing between atmospheric and mantle argon (Fisher, 1986). However, Loihi data do not display such a correlation. Hence, the atmospheric

contamination model is hard to evaluate critically in plume environments.

Farley and Craig (1994) made a new examination of this problem, based on helium and argon measurements on olivine phenocrysts in a tholeiitic basalt from the Juan Fernandez hot-spot. They demonstrated a positive correlation between ^4He and ^{40}Ar abundances released from fluid inclusions by crushing (Fig. 11.33a). However, this correlation cannot be attributed to atmospheric contamination, because the atmosphere has negligible ^4He. Therefore, the correlation must reflect variable gas inventories sampled from a mantle source with constant $^4He/^{40}Ar$ ratio.

In contrast, there was no correlation between ^4He and ^{36}Ar abundances in these samples (Fig. 11.33b), implying that ^{36}Ar abundances were perturbed by atmospheric contamination. Even gas-rich samples showed this behaviour, suggesting that analytical blank was not the cause. Therefore, the magma itself was probably contaminated by seawater argon in the oceanic crust before eruption. Hence, Farley and Craig argued that the mantle source sampled by the Juan Fernandez hot-spot has a minimum $^{40}Ar/^{36}Ar$ ratio equal to the maximum observed $^{40}Ar/^{36}Ar$ ratio of 7700. However, this does not place strong constraints on the argon isotope ratio of the lower mantle (undegassed reservoir), because the Juan Fernandez plume may have been contaminated with radiogenic argon during its ascent through the MORB source.

If this seawater contamination model is correct, it casts doubt on the reliability of basaltic glasses *in general* as samples of primordial argon from the mantle. However, it does not disprove the 'two-reservoir' model for mantle noble gases; it simply implies that the $^{40}Ar/^{36}Ar$ of the deep mantle cannot be determined directly.

Farley and Poreda (1993) suggested that $^{20}Ne/^{22}Ne$ ratios could be used to monitor and correct atmospheric contamination in heavier noble gases such as argon. However, because the end-members will have different noble

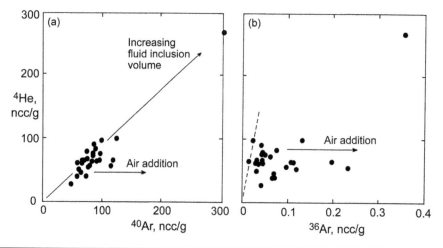

Fig. 11.33 Plots of gas release (10^{-9} cc/gram) during crushing of olivine phenocrysts from a Juan Fernandez basalt: (a) ^4He against ^{40}Ar; (b) ^4He against ^{36}Ar; showing effects of atmospheric contamination. Modified after Farley and Craig (1994).

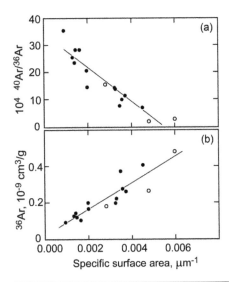

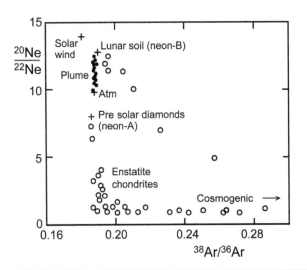

Fig. 11.34 Plot of $^{40}Ar/^{36}Ar$ and ^{36}Ar against the relative surface area of analysed glass sherds, showing strong correlations. Open symbols showed evidence of sea floor contamination. After Stroncik and Niedermann (2016).

Fig. 11.35 Plot of non-radiogenic Ne versus Ar isotope data for the Galapagos plume (●) and E-chondrites (○), compared with other solar system reservoirs (+). Data from Raquin and Moreira (2009) and Ott (2014).

gas abundance ratios (e.g. Ne/Ar), mixing will generate hyperbolic rather than linear arrays, leading to somewhat greater uncertainty in the calculation of uncontaminated end-members. Mixing lines were calculated between the atmosphere point and the best estimate for primordial neon and argon, based on the composition of planetary neon ($^{20}Ne/^{22}Ne = 12.5$) and the highest $^{40}Ar/^{36}Ar$ ratio (near 30 000) measured in a volatile-rich 'popping rock' from the Mid Atlantic Ridge (Staudacher *et al.*, 1989).

More recently, Stroncik and Niedermann (2016) argued that argon contamination of ocean floor samples occurred in two stages: the first involving contamination of the basaltic magma by assimilation of seawater-altered older crust; and the second due to atmospheric contamination of samples after recovery. They suggested that incorporation of altered sea floor can be monitored by measuring Cl/K ratios, which show a weak inverse correlation with Ne/Ar ratio in the more altered samples. However, a much stronger inverse correlation between $^{40}Ar/^{36}Ar$ ratio and the surface area of the analysed mineral separates suggests that air adsorption onto grain surfaces is a more important source of atmospheric contamination (Fig. 11.34a).

The low intercept of the correlation line between surface area and ^{36}Ar (Fig. 11.34b) suggests that even the most radiogenic samples contain significant atmospheric ^{36}Ar contamination. Therefore, it is inferred that the probable $^{40}Ar/^{36}Ar$ ratio of the MORB source is at least 40 000 (Fig. 11.34a). This value is also supported by the hyperbolic mixing line on a plot of Ar versus Ne isotope ratio (not shown). On the other hand, Ne–Ar data for Iceland (Fig. 11.24) point to a $^{40}Ar/^{36}Ar$ ratio of around 10 000 in this plume source, which may

be typical of the lower mantle. These values are somewhat higher than the values first estimated by Farley and Poreda (1993) based on a Ne–Ar mixing hyperbola, but are essentially in the same 'ball-park'. They support the two-reservoir model, but suggest that the lower mantle is significantly degassed.

11.3.3 Argon-38

In principle, ^{38}Ar can be combined with ^{36}Ar and ^{40}Ar data to make a three-isotope system analogous to neon. Unfortunately, ^{38}Ar is about five times less abundant than ^{36}Ar, which is itself a rare isotope. Furthermore, the total variation between atmospheric and solar compositions is only 5%. Therefore, until recently, errors on $^{38}Ar/^{36}Ar$ ratios were too large to clearly distinguish between these sources. For example, Valbracht *et al.* (1997) and Pepin (1998) claimed to find trends towards a solar argon composition in terrestrial rocks, whereas Kunz (1999) and Trieloff *et al.* (2000) did not find any such deviations.

New high-precision data were measured on samples from the Galapagos island of Fernandina, which shows the least nucleogenic neon of any OIB suite (Raquin and Moreira, 2009). These data confirm the 'negative' result of Kunz (1999), showing that OIB samples do *not* trend towards the solar composition. In contrast, the composition of sub-solar argon, as sampled in lunar soil, is a good fit to the Galapagos data. These data are also consistent with analyses of gas-rich enstatite chondrites showing the least amounts of cosmogenic contamination (Fig. 11.35). Hence these data provide support for the enstatite chondrite Earth model (Javoy, 1995). The vertical array of Galapagos data in Fig. 11.35 is also consistent with the late-veneer model for derivation of Earth's

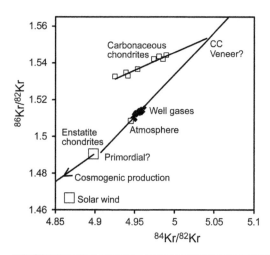

Fig. 11.36 Krypton three-isotope plot showing Harding County well-gases and possible source reservoirs. Data from Eugster *et al.* (1967); Crabb and Anders (1981); Holland *et al.* (2009).

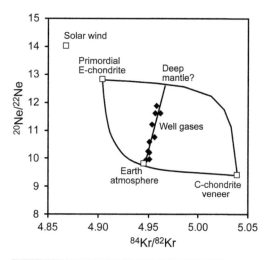

Fig. 11.37 Neon–krypton isotope diagram showing the well-gas array in relation to possible carbonaceous/E-chondrite mixing lines for the atmosphere and mantle. Data sources as in Fig. 11.36.

atmosphere by noble gas addition from carbonaceous chondrites. However the data do not provide any further constraints on this model.

11.4 Krypton

Krypton has six stable isotopes, but unlike the other noble gases, none of them has a significant radiogenic or nucleogenic component. Nevertheless, different solar system objects have significant variations in krypton isotope composition, which are generally attributed to nucleosynthetic variations inherited from the solar nebula.

Carbonaceous chondrites contain relatively large amounts of trapped krypton (orders of magnitude higher than terrestrial rocks). Therefore, they were some of the first samples to be analysed for krypton (Eugster *et al.*, 1967). In contrast, it was not possible to analyse krypton isotopes in any terrestrial materials except the atmosphere. However, the increased precision of multi-collector noble gas mass spectrometry allowed Holland *et al.* (2009) to detect a non-atmospheric krypton component in Harding County well-gases. On a krypton three-isotope plot (Fig. 11.36), these gases define a linear array whose slope is much lower than any reasonable mass fractionation line. After correction for a component from uranium fission in the crust, Holland *et al.* attributed the well-gas array to mixing between atmospheric krypton and an upper mantle component. Extensions of this mixing line are therefore compared in Fig. 11.36 with possible solar system sources.

Since neon and argon isotope data favour a sub-solar source for the Earth's interior, an estimate for the composition of enstatite chondrites is obtained from bulk

E-chondrite analyses by Crabb and Anders (1981). Most E-chondrites define a broadly scattered trend towards light krypton in the three-isotope plot, attributed to a large cosmogenic component. However, an average of four bulk E-chondrites with the heaviest isotope composition (lowest cosmogenic component) lies between the atmosphere and solar wind points in Fig. 11.36, making E-chondrites a plausible source of terrestrial krypton. On the other hand, carbonaceous chondrites (CC) may have contributed a later veneer of material to the Earth after the Moon-forming impact. The linear array of bulk carbonaceous chondrite analyses in Fig. 11.36 (Eugster *et al.*, 1967) probably reflects mixing between primordial and cosmogenic components, so the most likely composition of the accreted CC veneer would lie at the heavy end of the cosmogenic mixing line.

Although mixing between carbonaceous and enstatite chondrites could explain the slope of the well-gas mixing line, there is a major problem with the order of the reservoirs, as originally noted by Holland *et al.* (2009) and emphasized by Halliday (2013). Specifically, the order of the terrestrial krypton components: solar – *atmosphere* – *mantle* – CC is the opposite of the order for neon: solar – *mantle* – *atmosphere* – CC. To explain this reversal, Holland *et al.* suggested that the mantle contains mass-fractionated solar and/or CC krypton, whereas the atmospheric composition is controlled by a very late addition of commentary ices with solar type krypton signatures (Owen *et al.*, 1992).

In fact, the data distribution in Fig. 11.36 suggests that the krypton signatures of both atmosphere and mantle probably produced by mixing of widely separated end-members. This can be seen by plotting a neon–krypton isotope diagram (Fig. 11.37). On this plot, well-gases span almost the whole range of neon compositions between carbonaceous and E-chondrite end-members. This is because the noble gas subduction barrier for neon prevents atmospheric and mantle neon reservoirs from mixing. In contrast, subduction of atmospheric krypton may lead to efficient mixing between these reservoirs (as proposed for xenon, Section 11.5). In such a scenario, small late additions of light cometary krypton to the atmosphere could indeed bias its $^{84}Kr/^{82}Kr$ ratio downwards relative to the mantle. In contrast, these ices would not be expected to contain a significant component of neon, due to its greater volatility (Owen *et al.*, 1992). To test these ideas, it is clearly a priority to compare the Harding County data with other well-gas reservoirs worldwide.

11.5 Xenon

Xenon is a heavy noble gas with nine stable isotopes. Reynolds (1960a) first demonstrated variations in the abundance of ^{129}Xe in meteorites, produced from the extinct nuclide ^{129}I (Section 15.4.1). In addition, the four heaviest xenon isotopes are fission products of both ^{238}U (Wetherill, 1953), and the extinct nuclide ^{244}Pu (Kuroda, 1960). Xenon isotope abundances are normally ratioed against ^{130}Xe, which is non-radiogenic and is also shielded from spallation production. However, exceptions apply when particular processes are under study (see below).

11.5.1 Iodogenic Xenon

Because Xe is a tracer for two extinct nuclides, meteorite 'xenology' is a powerful tool for studying the condensation of the solar system (Section 15.4). However, 'terrestrial xenology' is also a powerful tool for understanding terrestrial differentiation (Staudacher and Allegre, 1982). The first evidence for excess ^{129}Xe in the Earth (relative to atmospheric xenon) was found in CO_2 well-gases from Harding County, New Mexico (Butler *et al.*, 1963). This evidence has such far-reaching implications that numerous subsequent studies have been devoted to Harding County well-gases (e.g. Hennecke and Manuel, 1975; Phinney *et al.*, 1978; Staudacher, 1987).

Studies of xenon isotope data from granitic rocks (Butler *et al.*, 1963) showed that unlike the heavy xenon isotopes, ^{129}Xe is not generated in significant amounts (relative to non-radiogenic xenon) by fission or neutron activation reactions (Fig. 11.38). Therefore Butler *et al.* concluded that the ^{129}Xe excess in mantle-derived gases must have been due to decay, soon after the formation of the Earth, of extinct ^{129}I. The presence of this extinct nuclide ($t_{1/2} = 16$ Ma) in the Earth demonstrates that accretion must have occurred within a few Ma of meteorites, which also commonly display ^{129}Xe anomalies (Section 15.4.1).

Further advances in terrestrial xenology required the well-gas data to be put into the wider perspective of major terrestrial reservoirs. Technical developments, allowing the xenon analysis of submarine glasses, made this possible.

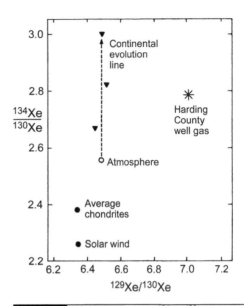

Fig. 11.38 Plot of fissiogenic $^{134}Xe/^{130}Xe$ against iodogenic $^{129}Xe/^{130}Xe$ for granites and well-gases analysed prior to 1978. Modified after Staudacher and Allegre (1982).

In early work, Staudacher and Allegre (1982) found correlated enrichments of ^{129}Xe and other heavy xenon isotopes in MORB glasses. This was confirmed by Staudacher (1987), who showed that the most ^{129}Xe-enriched well-gas analyses lay on the same correlation line as MORB (Fig. 11.39). This provides strong evidence that the Harding County well-gases sample an upper mantle reservoir similar to the depleted MORB source.

Additional evidence for the xenon isotope evolution of the upper mantle has been obtained from the analysis of 'coated' diamonds (Ozima and Zashu, 1991). The coats of these diamonds contain relatively large noble gas contents,

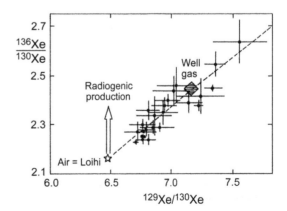

Fig. 11.39 Fissiogenic versus iodogenic xenon for MORB glasses, compared to other terrestrial components. After Staudacher (1987).

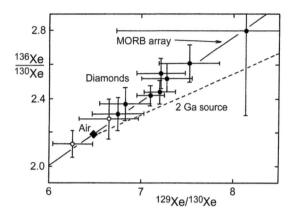

Fig. 11.40 Plot of $^{136}Xe/^{130}Xe$ against $^{129}Xe/^{130}Xe$ for coated diamonds, compared with the MORB correlation line of Staudacher (1987). (●) = coats; (○) = cores. Dashed line shows result of uranogenic production from a 2 Ga source. After Ozima and Zashu (1991).

and are thus suited to isotopic analysis. Ozima and Zashu found xenon isotope ratios identical to the MORB correlation line (Fig. 11.40), suggesting that the same evolution processes gave rise to the mantle sources of MORB and diamonds. In the light of neon and argon isotope evidence (Sections 11.2, 11.3), the MORB array is best attributed to atmospheric contamination of magmas with radiogenic xenon. Similarly, Ozima and Zashu attributed the diamond array to mixing between a radiogenic xenon source and atmospheric xenon contamination (represented by the cores).

Analysis of xenon in glasses from Loihi Seamount and the Reykjanes Ridge gave results within error of atmospheric noble gases, but the origin of this signature was widely disputed. Allegre et al. (1983) and Hart et al. (1983) attributed it to a less degassed mantle source, in which radiogenic ^{129}Xe and ^{40}Ar are swamped by large primordial noble gas contents. Alternatively, Ozima et al. (1985) proposed that atmospheric xenon is recycled into the deep mantle, while Patterson et al. (1990) argued that the xenon isotope ratios of Loihi glasses (along with argon) are due to atmospheric contamination, introduced directly into plume magmas from seawater.

Poreda and Farley (1992) suggested that the relatively shallow ocean depths at which Loihi basalts were erupted allowed much of the magmatic gas content to be lost, making them very susceptible to contamination by noble gases in seawater. To avoid these problems, Poreda and Farley analysed two suites of very volatile-rich harzburgite xenoliths from the Samoan hot-spot, which are more resistant to atmospheric contamination. The helium and neon isotope signatures of these rocks were somewhat more radiogenic than Loihi (Section 11.1.3), but still distinct from MORB. Therefore, these samples clearly contain a large plume-derived component. Xenon isotope ratios were found to be elevated by up to 6% relative to the atmospheric point, providing the first evidence for a non-atmospheric xenon signature from the lower mantle.

Based on this evidence for non-atmospheric xenon in plumes, Porcelli and Wasserburg (1995a, b) extended the steady state helium model of Kellogg and Wasserburg (1990) to xenon. They argued that the radiogenic signature of upper mantle xenon (relative to the atmosphere) is not a residue of early Earth degassing, as previously proposed, but the result of constant input from the lower mantle, along with in situ production from uranium fission. The supply of noble gases from lower to upper mantle was attributed to mass transfer in plumes. A fraction of the plume is degassed at the hot-spot, but the bulk is mixed into the upper mantle. In this mass transfer model, noble gases are not fractionated from one another, so they are all thought to have the same upper mantle residence time of around 1.4 Ga.

If the upper mantle is assumed to be in steady state, it can carry no memory of its early history. Hence, according to the steady state model, $^{129}Xe/^{130}Xe$ variations in the upper mantle (attributed to extinct iodine) must be explained solely by mixing of lower mantle and atmospheric xenon. The atmosphere is at the unradiogenic end of the MORB $^{129}Xe/^{130}Xe$ range, and must therefore have been outgassed from the Earth before a substantial amount of the iodine budget had decayed. However, Porcelli and Wasserburg suggested that some of the atmosphere was probably contributed by late accretion of volatile-rich material after degassing of the deep earth.

This model predicts that plumes from the lower mantle should have a more radiogenic xenon signature than MORB, once corrected for atmospheric contamination. New analyses from Loihi and Iceland were indeed found to yield xenon compositions distinct from atmospheric (Trieloff et al., 2000), but they extended only half-way up the MORB array (Fig. 11.41), apparently ruling out the model of Porcelli and Wasserburg (1995a, b).

It is not clear from the xenon three-isotope plot whether the upper limit of the MORB and OIB arrays is a real mantle component, or whether its composition has been biased by atmospheric contamination. However, by plotting xenon data against neon (Fig. 11.42), the probable compositions of both mantle end-members can be estimated, as previously demonstrated for MORB by Moreira et al. (1998). The MORB and OIB data in Fig. 11.42 lie on different mixing hyperbolae, and the early Iceland and Loihi data have been supported by additional data from Iceland and the Lau basin (Mukhopadhyay, 2012; Peto et al., 2013). Thus, it is confirmed that the lower mantle has a radiogenic xenon signature much closer to atmosphere than the upper mantle.

11.5.2 Fissiogenic Xenon

Variations in fissiogenic xenon are more complex than iodogenic xenon because fission produces a spread of product masses rather than a single radiogenic isotope. The four heavy xenon isotopes are produced from both extinct ^{244}Pu and extant ^{238}U fission, but plutogenic xenon is slightly

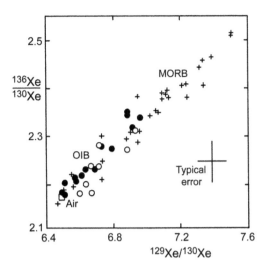

Fig. 11.41 Xenon three-isotope plot showing data for Icelandic glasses (●) and Loihi dunites (○) compared with the MORB array (+). Modified after Trieloff et al. (2000).

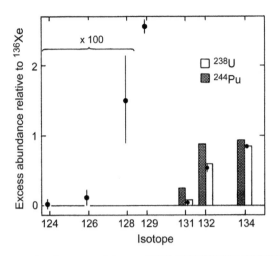

Fig. 11.43 Plot of excess abundances of xenon isotopes in well-gas relative to the atmosphere, ratioed against excess ^{136}Xe. The data are compared with modelled production of fissiogenic xenon from ^{238}U and ^{244}Pu. After Phinney et al. (1978).

enriched in the lighter isotopes relative to uranogenic xenon, allowing these sources to be distinguished (Fig. 11.43). Thus Phinney et al. (1978) were able to show that abundances of ^{131}Xe, ^{132}Xe and ^{134}Xe relative to ^{136}Xe in Harding County well-gases are a better fit to spontaneous fission of uranium, rather than plutonium.

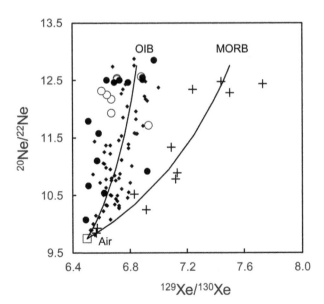

Fig. 11.42 Neon–xenon isotope plot showing early data from Iceland (●) and Loihi (○) and compared with MORB (+) and more recent OIB data (•). Data from Moreira et al. (1998), Trieloff et al. (2000), Mukhopadhyay (2012) and Peto et al. (2013).

Ozima et al. (1985) argued that fissiogenic xenon data can be evaluated in a more quantitative manner on a plot in which different fissiogenic xenon isotopes are ratioed against ^{130}Xe. Any of the heavy isotopes ^{131}Xe, ^{132}Xe or ^{134}Xe can be compared with ^{136}Xe. The latter two isotopes are usually compared because they have the best analytical precision, but for these isotopes, the alternative sources yield only very subtle differences in slope. These isotopes are ratioed in Fig. 11.44, from a re-evaluation of published data by Tolstikhin and O'Nions (1996), but the errors on the MORB data were too large to resolve the alternative models. The well-gas data were clearly shown to be uranogenic, but this could reflect crustal sources of xenon.

More recently, xenon isotope analysis of carbonatites from Brazil and Canada gave a signature that was distinctly uranogenic (Sasada et al., 1997). On the other hand, Kunz et al. (1998) claimed a mixture of uranogenic and plutogenic xenon in new MORB analyses. However, because the slopes of the alternative production vectors are very similar, their analysis was very dependent on the choice of non-radiogenic end-member taken as the starting point for fissiogenic production.

Ratioing other fissiogenic isotopes against ^{132}Xe rather than ^{130}Xe allows plutogenic and uranogenic end-members to be plotted directly, making the analysis less dependent on the choice of non-radiogenic starting point. The greatest separation between the two production vectors is achieved by plotting ^{131}Xe/^{132}Xe against ^{136}Xe/^{132}Xe (Fig. 11.45). In spite of the large scatter, the MORB data (Tucker et al., 2012) largely follow the uranogenic trend seen for the well-gases (Caffee et al., 1999). However, contrary to some optimistic suggestions, even the most recent high-precision data for

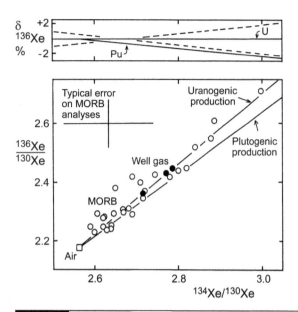

Fig. 11.44 Plot of two fissiogenic xenon isotopes, ratioed against ^{130}Xe, to compare data for well-gas (●) and MORB (○) with alternative uranogenic and plutogenic production routes. The δ ^{136}Xe diagram above shows the 95% confidence envelope for the best-fit regression of MORB data (dashed lines), relative to the alternative production routes. After Tolstikhin and O'Nions (1996).

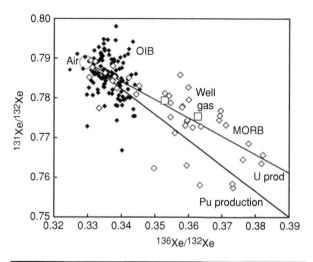

Fig. 11.45 Plot of heavy xenon isotope ratios to show the failure to resolve uranogenic and plutogenic xenon production in OIB (solid diamonds) and MORB (open diamonds). Data from Caffee et al. (1999); Tucker et al. (2012); Mukhopadhyay (2012) and Peto et al. (2013).

OIB samples (Mukhopadhyay, 2012; Peto et al., 2013) cannot clearly resolve the fissiogenic components, because errors in ^{131}Xe/^{132}Xe are too large (this is demonstrated by the large number of 'impossible' ratios above atmospheric). On the other hand, ^{134}Xe/^{132}Xe ratios have smaller errors, but the production vectors have very similar slopes. Hence it is concluded that there is no unequivocal evidence for a plutogenic xenon signature in the mantle. In contrast, the calculated initial plutonium/uranium ratio of 0.008 in the early Earth (Turner et al., 2007), would lead to plutogenic/uranogenic xenon production at around 30 for a closed-system Earth (Caffee et al., 1999). Therefore, it is concluded that the mantle must have lost up to 99.9% of its original xenon inventory (e.g. Tolstikhin et al., 2014).

11.5.3 Radiogenic Xenon Reservoirs

More success has been achieved in resolving iodogenic and fissiogenic xenon in mantle sources, which can be seen by plotting ^{136}Xe/^{132}Xe against ^{129}Xe/^{132}Xe (Fig. 11.46). On this plot, the OIB trend comprises samples from Iceland and the Lau basin, which are significantly different from the MORB trend, as observed by Peto et al. (2013). However, since there is no evidence for plutogenic xenon in MORB and OIB, the different slopes in Fig. 11.46 are best attributed to variations in iodogenic relative to uranogenic contributions. Since ^{129}I is an extinct nuclide, iodogenic xenon sig-

natures can be regarded as primordial relative to later mantle differentiation. It can then be seen that the behaviour of xenon is somewhat analogous to neon (Fig. 11.23), where OIB fall close to a primordial ^{20}Ne/^{22}Ne mixing line, whereas MORB shows the effect of nucleogenic production over Earth history. In a similar way, the larger fissiogenic xenon component in MORB can be attributed to uranium decay in the degassed MORB reservoir over Earth history.

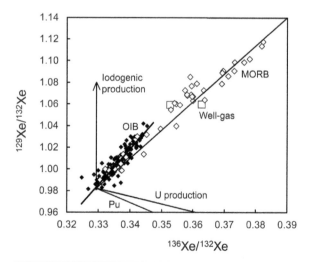

Fig. 11.46 Plot of iodogenic against fissiogenic xenon isotopes to resolve the contributions of these production routes for MORB and OIB source reservoirs. Data sources as in Fig. 11.45.

Evidence for the location of iodogenic xenon in the Earth was recently provided by Armytage *et al.* (2013), who found evidence that iodine is siderophile, implying that the majority of Earth's iodine budget is probably contained in the core. In view of the relatively short half-life of ^{129}I, significant decay must have occurred before core formation, which is dated to around 40 Ma after the origin of the solar system by the extinct Hf–W system (Section 15.6.6). Such an interval represents between two and three half lives of ^{129}I, leading Armytage *et al.* to suggest that 80% of iodine could have been lost from the Earth before core formation was complete. However, almost all the xenon inventory of the mantle may have been outgassed during the giant Moon-forming impact, so that the major reservoir of terrestrial ^{129}Xe at the present day may be located in the core.

Locating iodogenic xenon in the core can help to explain how the Earth appears to contain a much larger budget of the daughter of short-lived ^{129}I than longer-lived ^{244}Pu. If plutogenic xenon was largely degassed from the mantle, this implies that the abundance of iodogenic xenon in the mantle is even lower. However, these species could have been lost from the mantle during the giant impact and subsequent magma ocean without necessarily degassing iodogenic xenon from the core. Since plutonium is lithophile, its abundance in the core must be negligible, so the core cannot contain plutogenic xenon.

11.5.4 Non-Radiogenic Xenon

The non-radiogenic (light) isotopes of xenon are potentially very useful for unravelling the mysteries of terrestrial xenology. These isotopes should be usable in the same way as the non-radiogenic isotopes of other noble gases, to assess the relationship between atmospheric xenon and primordial xenon in the Earth. However, it became apparent with the first xenon isotope analysis of chondrites (Reynolds, 1960a, b) that the isotopic composition of light xenon in Earth's atmosphere is fractionated relative to meteorites in a manner that is clearly mass dependent.

Examination of the relative atmospheric abundances of the lighter noble gases (Ne, Ar and Kr) shows that they also fall on a mass fractionation line (Fig. 11.47). However, unlike ratios involving xenon, the fractionation line for the other noble gas ratios passes through the origin. The coherent behaviour of these gases is explained by very early isotopic fractionation in the nebula, probably during condensation of solid bodies from the gaseous phase. This interpretation is supported by similar abundance ratios in the Earth and meteorites. However, xenon is unique in showing isotopic fractionation in the atmosphere that is distinct from meteorites (Reynolds, 1960b). As a result, xenon abundances are offset on a different fractionation line in Fig. 11.47.

This behaviour is usually attributed to xenon loss from the early Earth, hence termed the 'missing xenon' problem, but its causes are still unclear. Pepin (1991) proposed a 'hydrodynamic escape' model, by which all of the noble gases were largely lost from the Earth (along with an early

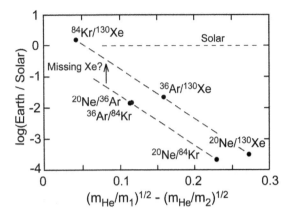

Fig. 11.47 Plot of the relative abundance of all noble gases except helium, showing two stages of mass fractionation behaviour. After Ozima and Podosek (1999).

hydrogen-rich atmosphere), possibly due to intense radiation from the early Sun. However, Tolstikhin and O'Nions (1994) pointed out that this model should cause less mass fractionation for the heaviest noble gas (xenon) than the lighter gases.

A significant constraint on the origin of Earth's fractionated atmospheric xenon signature comes from comparison with Mars (Owen *et al.*, 1992). On an argon/krypton plot normalized against xenon, terrestrial basalts and Shergottite meteorites both lie on the same mixing line between Chassigny (representing the Martian interior) and the terrestrial and Martian atmospheres (Fig. 11.48). This correlation was further strengthened by new high-precision noble gas data recently determined on the Martian atmosphere (Ott, 2014).

The fact that the terrestrial and Martian atmospheres are so similar points to a common source or process, and Owen *et al.* (1992) suggested that this involved late accretion of

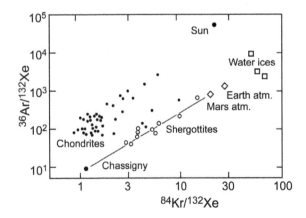

Fig. 11.48 Comparison of solar–chondritic noble gas signatures with terrestrial and Martian signatures, showing common behaviour of the two planets. Modified after Ott and Begemann (1985) and Ott (2014).

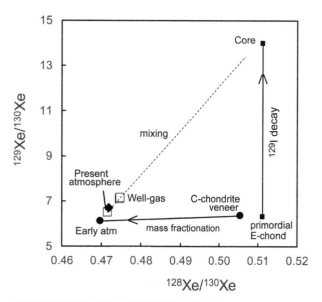

Fig. 11.49 Plot of iodogenic $^{129}Xe/^{130}Xe$ against non-radiogenic $^{128}Xe/^{130}Xe$, showing the possible evolution of mantle and atmospheric xenon and their mixing in well-gases. ($\blacklozenge$) = OIB. Data from Caffee et al. (1999), Mukhopadhyay (2012) and Ott (2014).

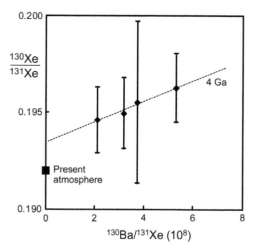

Fig. 11.50 Ba–Xe isochron plot for Archean fluid inclusions from western Australia, shown with a 4 Ga reference line. Error bars are 1σ analytical errors. Data from Pujol et al. (2011).

cometary material on both planets. Comets are believed to carry unique noble gas signatures in their water ices, derived from solar noble gas abundances that were mass fractionated during deposition at above 35 K (Fig. 11.48). This cometary component was previously invoked in Section 11.4, because it may explain the light krypton signature of Earth's atmosphere relative to well-gases.

The mass fractionation processes invoked to explain elemental noble gas ratios should also lead to observable variations in the abundances of non-radiogenic light xenon isotopes. The first evidence for such phenomena in mantle samples was provided by well-gas ^{128}Xe data (Caffee et al., 1988, 1999) which showed a correlation between iodogenic $^{129}Xe/^{130}Xe$ and non-radiogenic $^{128}Xe/^{130}Xe$ in the mantle. This correlation implies that the (upper) mantle reservoir sampled by well-gases was formed by mixing between primordial xenon from the deep Earth, and a component of subducted atmospheric xenon with a mass fractionated composition similar to the present atmosphere (Fig. 11.49). The mixing line projects to a $^{129}Xe/^{130}Xe$ ratio of around 13 in this deep Earth reservoir (argued above to be the core).

As observed by Caffee et al. (1999), well-gases from two different continents (Caroline, Australia; Harding County, USA) lie on the same mixing line between atmosphere and the iodogenic end-member (Fig. 11.49). Although it has a relatively larger error, the average composition of a large number of Icelandic basalt glasses (Mukhopadhyay, 2012) also falls within error of this mixing line (diamond in Fig. 11.49). This linearity implies mixing of only two xenon com-

ponents, whereas other noble gas evidence implies four distinct xenon sources in the Earth: iodogenic xenon (probably from the core), outgassed primordial xenon, xenon from a late carbonaceous chondrite veneer, and very late additions of cometary noble gases. However, the atmosphere provides the most probable site for mixing between the latter three sources, which could then have been subducted into the mantle. If atmospheric xenon was subducted into the plume source at the base of the mantle, this could explain why OIB have relatively less iodogenic xenon signatures than MORB and well-gases.

11.5.5 The Barium–Xenon System

The extreme sensitivity of xenon as a radiogenic tracer is perfectly illustrated by the technetium–xenon and barium–xenon decay schemes. Te–Xe isochron analysis of dated Precambrian technetium ores from several mines yielded a half-life of $7.7 \pm 0.4 \times 10^{24}$ years for the double β decay of ^{128}Te to ^{128}Xe (Bernatowicz et al., 1992). This is the longest half-life ever determined, and was measured in order to place limits on the neutrino mass. However, the initial ratios of such systems can also provide tests for past xenon isotope compositions in the Earth.

Another long-lived route for the production of radiogenic xenon is the double electron capture route from ^{130}Ba to ^{130}Xe, with a half-life estimated at 6.0×10^{20} yr (Pujol et al., 2009). Using this decay scheme, Pujol et al. (2011) attempted to place limits on the isotopic composition of the Archean atmosphere. Fluid inclusions from quartz-filled vesicles in a 3.5 Ga komatiite from 'North Pole' in western Australia defined a Ba–Xe isochron with a ca. 4 Ga slope age (Fig. 11.50). The fluid inclusions are believed to sample Ba and Xe derived from an Archean barite deposit. In this context,

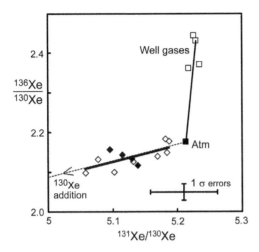

Fig. 11.51 Xenon–xenon isotope plot showing alternative production routes due to ^{130}Ba decay and ^{238}U fission, compared with the distribution of 'North Pole' isochron samples (diamonds). Heavy sub-horizontal line is a best-fit regression to all fluid inclusion analyses. Error bars show typical analytical uncertainty. Data from Pujol et al. (2011).

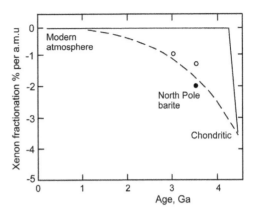

Fig. 11.52 Alternative models for the xenon isotope evolution of the atmosphere due to mass fractionation from an original chondritic/solar composition. Modified after Pujol et al. (2011).

^{131}Xe is regarded as a non-radiogenic isotope, in view of its limited fissiogenic production (to be demonstrated below). The isochron yields an initial ^{130}Xe/^{131}Xe ratio well above atmosphere, which, if verified, would have significant implications for the fractionation history of atmospheric xenon. However, the quoted 1σ analytical errors for each data point also fall within error limits of a regression line through the present day atmospheric composition.

Another way of assessing these data is on a xenon–xenon isotope plot (Fig. 11.51), where alternative production routes due to ^{130}Ba decay and ^{238}U fission are shown, starting from a modern atmosphere point. The complete 'North Pole' array (heavy regression line) is consistent with ^{130}Xe addition to a modern atmospheric composition from ^{130}Ba decay. These data show no tendency towards fissiogenic production, which is illustrated by the trend of well-gas compositions in Fig. 11.50. In contrast, the four North Pole samples selected for the Ba–Xe isochron (solid diamonds in Fig. 11.51) have a distribution almost orthogonal to the larger sample array, showing that these are not representative samples for the construction of a Ba–Xe isochron. Therefore, the evidence for a normal atmospheric xenon composition based on the larger Xe–Xe data set (Fig. 11.51) overrides the evidence from the isochron array (Fig. 11.50).

As noted above, this case is important because it was claimed to provide the strongest evidence for gradual evolution of the xenon isotope composition of the atmosphere over geological time (Fig. 11.52). In contrast, two earlier studies of xenon production in Archean barites (open symbols in Fig. 11.52; Srinivasan, 1976; Pujol et al., 2009) were not aiming to constrain the initial isotope ratio of radiogenic xenon.

Consequently initial ratios calculated from these studies have large errors that do not provide precise constraints on past atmospheric xenon composition. Therefore, it is concluded that the sum of the isotopic evidence favours very early atmospheric xenon fractionation, rather than gradual evolution over geological time.

Zahnle et al. (2007) suggested that the missing xenon problem and xenon isotope fractionation could be explained by xenon's ionization potential, which is unique in the noble gases in being lower than that of hydrogen. Since hydrogen was probably a dominant species in the early atmosphere, it would have soaked up ultra-violet radiation, preventing any species with higher ionization potential from being ionized. Xenon ions could then be preferentially removed by acceleration in the Earth's magnetic field. This model can also explain the unexpected depletion of nitrogen and carbon in Earth's atmosphere, since these elements can be ionized in the form of ammonia and methane (Halliday, 2013).

References

Aldrich, L. T. and Nier, A. O. (1948). The occurrence of He^3 in natural sources of helium. *Phys. Rev.* **74**, 1590–4.

Allegre, C. J., Sarda, P. and Staudacher, T. (1993). Speculations about the cosmic origin of He and Ne in the interior of the Earth. *Earth Planet. Sci. Lett.* **117**, 229–33.

Allegre, C. J., Staudacher, T. and Sarda, P. (1986). Rare gas systematics: formation of the atmosphere, evolution and structure of the Earth's mantle. *Earth Planet. Sci. Lett.* **81**, 127–50.

Allegre, C. J., Staudacher, T., Sarda, P. and Kurz, M. (1983). Constraints on evolution of Earth's mantle from rare gas systematics. *Nature* **303**, 762–6.

Alvarez, L. W. and Cornog, R. (1939). Helium and hydrogen of mass 3. *Phys. Rev.* **56**, 613.

Anderson, D. L. (1993). Helium-3 from the mantle: primordial signal or cosmic dust? *Science* **261**, 170–6.

Anderson, D. L. (1998). The helium paradoxes. *Proc. Nat. Acad. Sci.* **95**, 4822–7.

Anderson, D. L. (2001). A statistical test of the two reservoir model for helium isotopes. *Earth Planet. Sci. Lett.* **193**, 77–82.

Armytage, R. M., Jephcoat, A. P., Bouhifd, M. A. and Porcelli, D. (2013). Metal–silicate partitioning of iodine at high pressures and temperatures: Implications for the Earth's core and ^{129}Xe budgets. *Earth Planet. Sci. Lett.* **373**, 140–9.

Ballentine, C. J. and Holland, G. (2008). What CO_2 well gases tell us about the origin of noble gases in the mantle and their relationship to the atmosphere. *Phil. Trans. Roy. Soc. Lond. A* **366**, 4183–203.

Ballentine, C. J., Marty, B., Sherwood Lollar, B. and Cassidy M. (2005). Neon isotopes constrain convection and volatile origin in the Earth's mantle. *Nature* **433**, 33–8.

Bernatowicz, T., Brannon, J., Brazzle, R. *et al.* (1992). Neutrino mass limits from a precise determination of $\beta\beta$-decay rates of ^{128}Te and ^{130}Te. *Phys. Rev. Lett.* **69**, 2341–4.

Black, D. C. (1972). On the origins of trapped helium, neon and argon isotopic variations in meteorites – I. Gas-rich meteorites, lunar soil and breccia. *Geochim. Cosmochim. Acta* **36**, 347–75.

Bouhifd, M. A., Jephcoat, A. P., Heber, V. S. and Kelley, S. P. (2013). Helium in Earth's early core. *Nature Geosci.* **6**, 982–6.

Butler, W. A., Jeffery, P. M., Reynolds, J. H. and Wasserburg, G. J. (1963). Isotopic variations in terrestrial xenon. *J. Geophys. Res.* **68**, 3283–91.

Caffee, M. W., Hudson, G. B., Velsko, C. *et al.* (1988). Non-atmospheric noble gases from CO_2 well gases. *Lunar Planet. Sci.* **XIX**, 154–5 (abs).

Caffee, M. W., Hudson, G. B., Velsko, C. *et al.* (1999). Primordial noble gases from Earth's mantle: identification of a primitive volatile component. *Science* **285**, 2115–18.

Cerling, T. E. (1989). Dating geomorphologic surfaces using cosmogenic ^{3}He. *Quaternary Res.* **33**, 148–56.

Clarke, W. B., Beg, M. A. and Craig, H. (1969). Excess ^{3}He in the sea: evidence for terrestrial primordial helium. *Earth Planet. Sci. Lett.* **6**, 213–20.

Clarke, W. B., Jenkins, W. J. and Top, Z. (1976). Determination of tritium by mass-spectrometric measurement of ^{3}He. *Int. J. Appl. Rad. Isot.* **27**, 515–22.

Class, C. and Goldstein, S. L. (2005). Evolution of helium isotopes in the Earth's mantle. *Nature* **436**, 1107–12.

Colin, A., Moreira, M., Gautheron, C. and Burnard, P. (2015). Constraints on the noble gas composition of the deep mantle by bubble-by-bubble analysis of a volcanic glass sample from Iceland. *Chem. Geol.* **417**, 173–83.

Crabb, J. and Anders, E. (1981). Noble gases in E-chondrites. *Geochim. Cosmochim. Acta* **45**, 2443–64.

Craig, H. and Lupton, J. E. (1976). Primordial neon, helium, and hydrogen in oceanic basalts. *Earth Planet. Sci. Lett.* **31**, 369–85.

Craig, H. and Poreda, R. J. (1986). Cosmogenic ^{3}He in terrestrial rocks: the summit lavas of Maui. *Proc. Natl. Acad. Sci. USA* **83**, 1970–4.

Damon, P. E. and Kulp, L. (1958). Excess helium and argon in beryl and other minerals. *Amer. Miner.* **43**, 433–59.

Ellam, R. M. and Stuart, F. M. (2004). Coherent He–Nd–Sr isotope trends in high ^{3}He/^{4}He basalts: implications for a common reservoir, mantle heterogeneity and convection. *Earth Planet. Sci. Lett.* **228**, 511–23.

Eugster, O., Eberhardt, P. and Geiss, J. (1967). ^{81}Kr in meteorites and ^{81}Kr radiation ages. *Earth Planet. Sci. Lett.* **2**, 77–82.

Fanale, F. P. (1971). A case for catastrophic early degassing of the Earth. *Chem. Geol.* **8**, 79–105.

Farley, K. A. (1995a). Rapid cycling of subducted sediments into the Samoan mantle plume. *Geology* **23**, 531–4.

Farley, K. A. (1995b). Cenozoic variations in the flux of interplanetary dust recorded by ^{3}He in a deep-sea sediment. *Nature* **376**, 153–6.

Farley, K. A. and Craig, H. (1994). Atmospheric argon contamination of ocean island basalt olivine phenocrysts. *Geochim. Cosmochim. Acta* **58**, 2509–17.

Farley, K. A. and Eltgroth, S. F. (2003). An alternative age model for the Paleocene–Eocene thermal maximum using extraterrestrial 3 He. *Earth Planet. Sci. Lett.* **208**, 135–48.

Farley, K. A., Montanari, A., Shoemaker, E. M. and Shoemaker, C. S. (1998). Geochemical evidence for a comet shower in the late Eocene. *Science* **280**, 1250–3.

Farley, K. A. and Patterson, D. B. (1995). A 100-ka periodicity in the flux of extraterrestrial ^{3}He to the sea floor. *Nature* **378**, 600–3.

Farley, K. A. and Poreda, R. J. (1993). Mantle neon and atmospheric contamination. *Earth Planet. Sci. Lett.* **114**, 325–39.

Fisher, D. E. (1971). Incorporation of Ar in East Pacific basalts. *Earth Planet. Sci. Lett.* **12**, 321–4.

Fisher, D. E. (1983). Rare gases from the undepleted mantle? *Nature* **305**, 298–300.

Fisher, D. E. (1985). Noble gases from oceanic island basalts do not require an undepleted mantle source. *Nature* **316**, 716–18.

Fisher, D. E. (1986). Rare gas abundances in MORB. *Geochim. Cosmochim. Acta* **50**, 2531–41.

Garapic, G., Jackson, M. G., Hauri, E. H. *et al.* (2015). A radiogenic isotopic (He–Sr–Nd–Pb–Os) study of lavas from the Pitcairn hotspot: Implications for the origin of EM-1 (enriched mantle 1). *Lithos* **228**, 1–11.

Geiss, J., Bühler, F., Cerutti, H. *et al.* (2004). The Apollo SWC experiment: results, conclusions, consequences. *Space Sci. Rev.* **110**, 307–35.

Gerling, E. K., Mamyrin, B. A., Tolstikhin, I. N. and Yakovleva, S. S. (1971). Isotope composition of helium in some rocks. *Geokhimiya* **10**, 1209–17.

Grimberg, A., Baur, H., Buhler, F., Bochsler, P. and Wieler, R. (2008). Solar wind helium, neon, and argon isotopic and elemental composition: data from the metallic glass flown on NASA's Genesis mission. *Geochim. Cosmochim. Acta* **72**, 626–45.

Halliday, A. N. (2013). The origins of volatiles in the terrestrial planets. *Geochim. Cosmochim. Acta* **105**, 146–71.

Hanan, B. B. and Graham, D. W. (1996). Lead and helium isotope evidence from oceanic basalts for a common deep source of mantle plumes. *Science* **272**, 991–5.

Hanyu, T., Dunai, T. J., Davies, G. R. *et al.* (2001). Noble gas study of the Reunion hotspot: evidence for distinct less-degassed mantle sources. *Earth Planet. Sci. Lett.* **193**, 83–98.

Hanyu, T. and Kaneoka, I. (1997). The uniform and low ^{3}He/^{4}He ratios of HIMU basalts as evidence for their origin as recycled materials. *Nature* **390**, 273–6.

Harper, C. L. and Jacobsen, S. B. (1996). Noble gases and Earth's accretion. *Science* **273**, 1814–18.

Hart, R, Dymond, J. and Hogan, L. (1979). Preferential formation of the atmosphere–sialic crust system from the upper mantle. *Nature* **278**, 156–9.

Hart, R, Dymond, J., Hogan, L. and Schilling, J. G. (1983). Mantle plume noble gas component in glassy basalts from Reykjanes Ridge. *Nature* **305**, 403–7.

Hart, R., Hogan, L. and Dymond, J. (1985). The closed-system approximation for evolution of argon and helium in the mantle, crust and atmosphere. *Chem. Geol. (Isot. Geosci. Sect.)* **52**, 45–73.

Hart, S. R., Hauri, E. H., Oschmann, L. A. and Whitehead, J. A. (1992). Mantle plumes and entrainment: isotopic evidence. *Science* **256**, 517–20.

Hennecke, E. W. and Manuel, O. K. (1975). Noble gases in CO_2 well gas, Harding County, New Mexico. *Earth Planet. Sci. Lett.* **27**, 346–55.

Hilton, D. R., Gronvold, K., Macpherson, C. G. and Castillo, P. R. (1999). Extreme ^{3}He/^{4}He ratios in northwest Iceland: constraining the common component in mantle plumes. *Earth Planet. Sci. Lett.* **173**, 53–60.

Hilton, D. R., Hammerschmidt, K., Loock, G. and Friedrichsen, H. (1993). Helium and argon isotope systematics of the central Lau Basin and Valu Fa Ridge: evidence of crust/mantle interactions in a back-arc basin. *Geochim. Cosmochim. Acta* **57**, 2819–41.

Hilton, D. R., Thirlwall, M. F., Taylor, R. N., Murton, B. J. and Nichols, A. (2000). Controls on magmatic degassing along the Reykjanes Ridge with implications for the helium paradox. *Earth Planet. Sci. Lett.* **183**, 43–50.

Hiyagon, H. (1994). Retention of solar helium and neon in IDPs in deep sea sediment. *Science* **263**, 1257–9.

Hiyagon, H., Ozima, M., Marty, B., Zashu, S. and Sakai, H. (1992). Noble gases in submarine glasses from mid-ocean ridges and Loihi seamount: constraints on the early history of the Earth. *Geochim. Cosmochim. Acta* **56**, 1301–16.

Holland, G., Cassidy, M. and Ballentine, C. J. (2009). Meteorite Kr in Earth's mantle suggests a late accretionary source for the atmosphere. *Science* **326**, 1522–5.

Honda, M., McDougall, I., Patterson, D. B., Doulgeris, A. and Clague, D. A. (1991). Possible solar noble-gas component in Hawaiian basalts. *Nature* **349**, 149–51.

Honda, M., Reynolds, J. H., Roedder, E. and Epstein, S. (1987). Noble gases in diamonds: occurrences of solar-like helium and neon. *J. Geophys. Res.* **92**, 12507–21.

Hopp, J. and Trieloff, M. (2005). Refining the noble gas record of the Reunion mantle plume source: Implications on mantle geochemistry. *Earth Planet. Sci. Lett.* **240**, 573–88.

Huang, S., Lee, C. T. A. and Yin, Q. Z. (2014). Missing lead and high ^{3}He/^{4}He in ancient sulfides associated with continental crust formation. *Scientific rep.* **4** (5314), 1–6.

Huss, G. R. and Lewis, R. S. (1994). Noble gases in presolar diamonds I: Three distinct components and their implications for diamond origins. *Meteoritics* **29**, 791–810.

Jackson, M. G., Carlson, R. W., Kurz, M. D. *et al.* (2010). Evidence for the survival of the oldest terrestrial mantle reservoir. *Nature* **466**, 853–6.

Jackson, M. G., Hart, S. R., Konter, J. G. *et al.* (2014). Helium and lead isotopes reveal the geochemical geometry of the Samoan plume. *Nature*, **514**, 355–8.

Jackson, M. G., Kurz, M. D., Hart, S. R. and Workman, R. K. (2007). New Samoan lavas from Ofu Island reveal a hemispherically heterogeneous high ^{3}He/^{4}He mantle. *Earth Planet. Sci. Lett.* **264**, 360–74.

Javoy, M. (1995). The integral enstatite chondrite model of the Earth. *Geophys. Res. Lett.* **22**, 2219–22.

Jeffrey, P. M. and Hagan, P. J. (1969). Negative muons and the isotopic composition of the rare gases in the Earth's atmosphere. *Nature* **223**, 1253.

Kaneoka, I. and Takaoka, N. (1980). Rare gas isotopes in Hawaiian ultramafic nodules and volcanic rocks: constraints on genetic relationships. *Science* **208**, 1366–8.

Kellogg, L. H. and Wasserburg, G. J. (1990). The role of plumes in mantle helium fluxes. *Earth Planet. Sci. Lett.* **99**, 276–89.

Kennedy, B. M., Hiyagon, H. and Reynolds, J. H. (1990). Crustal neon: a striking uniformity. *Earth Planet. Sci. Lett.* **98**, 277–86.

Kunz, J. (1999). Is there solar argon in the Earth's mantle? *Nature* **399**, 649–50.

Kunz, J., Staudacher, T. and Allegre, C. J. (1998). Plutonium-fission xenon found in the Earth's mantle. *Science* **280**, 877–80.

Kunz, W. and Schintlmeister, I. (1965). *Tabellen der Atomkerne, teil II, Kernreaktionen*. Akademie–Verlag, 1022 pp.

Kuroda, P. K. (1960). Nuclear fission in the early history of the Earth. *Nature* **187**, 36–8.

Kurz, M. D. (1986a). Cosmogenic helium in a terrestrial rock. *Nature* **320**, 435–9.

Kurz, M. D. (1986b). *In-situ* production of terrestrial cosmogenic helium and some applications to geochronology. *Geochim. Cosmochim. Acta* **50**, 2855–62.

Kurz, M. D., Curtice, J., Fornari, D., Geist, D. and Moreira, M. (2009). Primitive neon from the center of the Galapagos hotspot. *Earth Planet. Sci. Lett.* **286**, 23–34.

Kurz, M. D., Gurney, J. J., Jenkins, W. J. and Lott, D. E. (1987). Helium isotopic variability within single diamonds from Orapa kimberlite pipe. *Earth Planet. Sci. Lett.* **86**, 57–68.

Kurz, M. D. and Jenkins, W. J. (1981). The distribution of helium in oceanic basalt glasses. *Earth Planet. Sci. Lett.* **53**, 41–54.

Kurz, M. D., Jenkins, W. J. and Hart, S. R. (1982). Helium isotopic systematics of oceanic islands and mantle heterogeneity. *Nature* **297**, 43–6.

Kurz, M. D., Meyer, P. S. and Sigurdsson, H. (1985). Helium isotopic systematics within the neovolcanic zones of Iceland. *Earth Planet. Sci. Lett.* **74**, 291–305.

Kyser, T. K. and Rison, W. (1982). Systematics of rare gas isotopes in basaltic lavas and ultramafic xenoliths. *J. Geophys. Res.* **87**, 5611–30.

Lal, D. (1987). Production of ^{3}He in terrestrial rocks. *Chem. Geol. (Isot. Geosci. Sect.)* **66**, 89–98.

Lal, D., Nishiizumi, K., Klein, J., Middleton, R. and Craig, H. (1987). Cosmogenic ^{10}Be in Zaire alluvial diamonds: implications to ^{3}He excess in diamonds. *Nature* **328**, 139–41.

Lupton, J. E., (1983). Terrestrial inert gases: isotope tracer studies and clues to primordial components in the mantle. *Ann. Rev. Earth Planet. Sci.* **11**, 371–414.

Lupton, J. E. and Craig. H. (1975). Excess ^{3}He in oceanic basalts: evidence for terrestrial primordial helium. *Earth Planet. Sci. Lett.* **26**, 133–9.

Mahaffy, P. R., Donahue, T. M., Owen T. C., Niemann, H. B. and Atreya, S. K. (1998) Galileo probe measurements of D/H and ^{3}He/^{4}He in Jupiter's atmosphere. *Space Sci.Rev.* **84**, 251–63.

Mamyrin, B. A., Anufriyev, G. S., Kamenskiy, I. L. and Tolstikhin, I. N. (1970). Determination of the composition of atmospheric helium. *Geochem. Int.* **7**, 498–505.

Mamyrin, B. A. and Tolstikhin, I. N. (1984). *Helium Isotopes in Nature*. Elsevier, 273 pp.

Mamyrin, B. A., Tolstikhin, I. N., Anufriev, G. S. and Kamenskiy, I. L. (1969). Anomalous isotopic composition of helium in volcanic gases. *Dokl. Akad. Nauka SSSR* **184**, 1197–9.

Mamyrin, B. A., Tolstikhin, I. N., Anufriyev, G. S. and Kamenskiy, I. L. (1972). Isotopic composition of helium in Icelandic hot springs. *Geokhimiya* **11**, 1396.

Marcantonio, F., Anderson, R. F., Stute, M. *et al.* (1996). Extraterrestrial ^{3}He as a tracer of marine sediment transport and accumulation. *Nature* **383**, 705–7.

Marcantonio, F., Kumar, N., Stute, M. *et al.* (1995). A comparative study of accumulation rates derived by He and Th isotope analysis of marine sediments. *Earth Planet. Sci. Lett.* **133**, 549–55.

Marty, B. (1989). Neon and xenon isotopes in MORB: implications for the Earth–atmosphere evolution. *Earth Planet. Sci. Lett.* **94**, 45–56.

Marty, B. (2012). The origins and concentrations of water, carbon, nitrogen and noble gases on Earth. *Earth Planet. Sci. Lett.* **313**, 56–66.

Marty, B. and Craig, H. (1987). Cosmic-ray-produced neon and helium in the summit lavas of Maui. *Nature* **325**, 335–7.

Matsuda, J., Murota, M. and Nagao, K. (1990). He and Ne isotopic studies on the extraterrestrial material in deep-sea sediments. *J. Geophys. Res.* **95**, 7111–17.

Matsuda, J., Sudo, M., Ozima, M. *et al.* (1993). Noble gas partitioning between metal and silicate under high pressures. *Science* **259**, 788–90.

Merrihue, C. (1964). Rare gas evidence for cosmogenic dust in modern Pacific red clay. *Ann. N. Y. Acad. Sci.* **119**, 351–67.

Moreira, M. and Charnoz, S. (2016). The origin of the neon isotopes in chondrites and on Earth. *Earth Planet. Sci. Lett.* **433**, 249–56.

Moreira, M., Kunz, J. and Allegre, C. J. (1998). Rare gas systematics in popping rock: isotopic and elemental compositions in the upper mantle. *Science* **279**, 1178–81.

Moreira, M. (2013). Noble gas constraints on the origin and evolution of Earth's volatiles. *Geochem. Perspectives* **2**, 229–403.

Morrison, P. and Pine, J. (1955). Radiogenic origin of the helium isotopes in rocks. *Ann. N. Y. Acad. Sci.* **62**, 69–92.

Mukhopadhyay, S. (2012). Deep mantle neon and xenon preserve a record of early planetary differentiation and heterogeneous volatile accretion. *Nature* **486**, 101–4.

Mukhopadhyay, S., Farley, K. A. and Montanari, A. (2001). A 35 Ma record of helium in pelagic limestones from Italy: implications for interplanetary dust accretion from the early Maastrichtian to the middle Eocene. *Geochim. Cosmochim. Acta* **65**, 653–69.

Muller, R. A. and Macdonald, G. J. (1995). Glacial cycles and orbital inclination. *Nature* **377**, 107–8.

Murphy, B. H., Farley, K. A. and Zachos, J. C. (2010). An extraterrestrial [3]He-based timescale for the Paleocene–Eocene thermal maximum (PETM) from Walvis Ridge, IODP Site 1266. *Geochim. Cosmochim. Acta* **74**, 5098–108.

Nagao, K., Takaoka, N. and Matsubayashi, O. (1979). Isotopic anomalies of rare gases in the Nigorikawa geothermal area, Hokkaido, Japan. *Earth Planet. Sci. Lett.* **44**, 82–90.

Niedermann, S., Bach, W. and Erzinger, J. (1997). Noble gas evidence for a lower mantle component in MORBs from the southern East Pacific Rise: decoupling of helium and neon isotope systematics. *Geochim. Cosmochim. Acta* **61**, 2697–715.

Nier, A. O. and Schlutter, D. J. (1990). Helium and neon in stratospheric particles. *Meteoritics* **25**, 263–7.

O'Nions, R. K. and Oxburgh, E. R. (1983). Heat and helium in the Earth. *Nature* **306**, 429–36.

Ott, U. (2014). Planetary and pre-solar noble gases in meteorites. *Chemie der Erde* **74**, 519–44.

Ott, U. and Begemann, F. (1985). Are all the 'Martian' meteorites from Mars? *Nature* **317**, 509–12.

Owen, T., Bar-Nun, A. and Kleinfeld, I. (1992). Possible cometary origin of heavy noble gases in the atmospheres of Venus, Earth and Mars. *Nature* **358**, 43–6.

Oxburgh, E. R., O'Nions, R. K. and Hill, R. I. (1986). Helium isotopes in sedimentary basins. *Nature* **324**, 632–5.

Ozima, M. and Podosek, F. A. (1999). Formation age of Earth from [129]I/[127]I and [244]Pu/[238]U systematics and the missing Xe. *J. Geophys. Res. B* **104**, 25 493–9.

Ozima, M., Podosek, F. A. and Igarashi, G. (1985). Terrestrial xenon isotope constraints on the early history of the Earth. *Nature* **315**, 471–4.

Ozima, M. and Zashu, S. (1983). Primitive helium in diamonds. *Science* **219**, 1067–8.

Ozima, M. and Zashu, S. (1988). Solar-type Ne in Zaire cubic diamonds. *Geochim. Cosmochim. Acta* **52**, 19–25.

Ozima, M. and Zashu, S. (1991). Noble gas state of the ancient mantle as deduced from noble gases in coated diamonds. *Earth Planet. Sci. Lett.* **105**, 13–27.

Patterson, D. B. and Farley, K. A. (1998). Extraterrestrial [3]He in seafloor sediments: evidence for correlated 100 ka periodicity in the accretion rate of interplanetary dust, orbital parameters, and Quaternary climate. *Geochim. Cosmochim. Acta* **62**, 3669–82.

Patterson, D. B., Honda, M. and McDougall, I. (1990). Atmospheric contamination: a possible source for heavy noble gases in basalts from Loihi Seamount, Hawaii. *Geophys. Res. Lett.* **17**, 705–8.

Pepin, R. O. (1991). On the origin and early evolution of terrestrial planet atmospheres and meteoritic volatiles. *Icarus* **92**, 2–79.

Pepin, R. O. (1997). Evolution of Earth's noble gases: consequences of assuming hydrodynamic loss driven by giant impact. *Icarus* **126**, 148–56.

Pepin, R. O. (1998). Isotopic evidence for a solar argon component in the Earth's mantle. *Nature* **394**, 664–7.

Pepin, R. O. (2006). Atmospheres on the terrestrial planets: Clues to origin and evolution. *Earth Planet. Sci. Lett.* **252**, 1–14.

Pepin, R. O. and Signer, P. (1965). Primordial rare gases in meteorites. *Science* **149**, 253–65.

Peron, S., Moreira, M., Colin, A. *et al.* (2016). Neon isotopic composition of the mantle constrained by single vesicle analyses. *Earth Planet. Sci. Lett.* **449**, 145–54.

Peto, M. K., Mukhopadhyay, S. and Kelley, K. A. (2013). Heterogeneities from the first 100 million years recorded in deep mantle noble gases from the Northern Lau Back-arc Basin. *Earth Planet. Sci. Lett.* **369**, 13–23.

Phinney, D., Tennyson, J. and Frick, U. (1978). Xenon in CO_2 well gas revisited. *J. Geophys. Res.* **83**, 2313–19.

Porcelli, D. and Elliott, T. (2008). The evolution of He isotopes in the convecting mantle and the preservation of high [3]He/[4]He ratios. *Earth Planet. Sci. Lett.* **269**, 175–85.

Porcelli, D. and Halliday, A. N. (2001). The core as a possible source of mantle helium. *Earth Planet. Sci. Lett.* **192**, 45–56.

Porcelli, D. and Wasserburg, G. J. (1995a). Mass transfer of xenon through a steady-state upper mantle. *Geochim. Cosmochim. Acta* **59**, 1991–2007.

Porcelli, D. and Wasserburg, G. J. (1995b). Mass transfer of helium, neon, argon, and xenon through a steady-state upper mantle. *Geochim. Cosmochim. Acta* **59**, 4921–37.

Poreda, R. J. and Farley, K. A. (1992). Rare gases in Samoan xenoliths. *Earth Planet. Sci. Lett.* **113**, 129–44.

Pujol, M., Marty, B. and Burgess, R. (2011). Chondritic-like xenon trapped in Archean rocks: a possible signature of the ancient atmosphere. *Earth Planet. Sci. Lett.* **308**, 298–306.

Pujol, M., Marty, B., Burnard, P. and Philippot, P. (2009). Xenon in Archean barite: weak decay of [130]Ba, mass-dependent isotopic fractionation and implication for barite formation. *Geochim. Cosmochim. Acta* **73**, 6834–46.

Raquin, A. and Moreira, M. (2009). Atmospheric [38]Ar/[36]Ar in the mantle: implications for the nature of the terrestrial parent bodies. *Earth Planet. Sci. Lett.* **287**, 551–8.

Reynolds, J. H. (1960a). Determination of the age of the elements. *Phys. Rev. Lett.* **4**, 8–10.

Reynolds, J. H. (1960b). Isotopic composition of primordial xenon. *Phys. Rev. Lett* **4**, 351–4.

Reynolds, J. H. (1963). Xenology. *J. Geophys. Res.* **68**, 2939–56.<Au: text citation?>

Rutherford, E. (1906). The production of helium from radium and the transformation of matter. In: Rutherford, E. *Radioactive Transformations*. Yale University Press, pp. 187–93.

Sarda, P., Staudacher, T. and Allegre, C. J. (1988). Neon isotopes in submarine basalts. *Earth Planet. Sci. Lett.* **91**, 73–88.

Sarda, P., Staudacher, T., Allegre, C. J. and Lecomte, A. (1993). Cosmogenic neon and helium at Reunion: measurement of erosion rate. *Earth Planet. Sci. Lett.* **119**, 405–17.

Sasada, T., Hiyagon, H., Bell, K. and Ebihara, M. (1997). Mantle-derived noble gases in carbonatites. *Geochim. Cosmochim. Acta* **61**, 4219–28.

Savage, P. S., Moynier, F., Chen, H. *et al.* (2015). Copper isotope evidence for large-scale sulphide fractionation during Earth's differentiation. *Geochem. Perspect. Lett.* **1**, 53–64.

Schmitz, B., Peucker-Ehrenbrink, B., Heilmann-Clausen, C. *et al.* (2004). Basaltic explosive volcanism, but no comet impact, at the Paleocene–Eocene boundary: high-resolution chemical and isotopic records from Egypt, Spain and Denmark. *Earth Planet. Sci. Lett.* **225**, 1–17.

Schwartzman, D. W. (1973). Argon degassing models of the Earth. *Nature Phys. Sci.* **245**, 20–1.

Seta, A., Matsumoto, T. and Matsuda, J.-I. (2001). Concurrent evolution of ^{3}He/^{4}He ratio in the Earth's mantle reservoirs for the first 2 Ga. *Earth Planet. Sci. Lett.* **188**, 211–19.

Sheldon, W. R. and Kern, J. W. (1972). Atmospheric helium and geomagnetic field reversals. *J. Geophys. Res.* **77**, 6194–201.

Srinivasan, B. (1976). Barites: anomalous xenon from spallation and neutron-induced reactions. *Earth Planet. Sci. Lett.* **31**, 129–41.

Starkey, N. A., Stuart, F. M., Ellam, R. M. *et al.* (2009). Helium isotopes in early Iceland plume picrites: Constraints on the composition of high ^{3}He/^{4}He mantle. *Earth Planet. Sci. Lett.* **277**, 91–100.

Staudacher, T. (1987). Upper mantle origin for Harding County well gases. *Nature* **325**, 605–7.

Staudacher, T. and Allegre, C. J. (1982). Terrestrial xenology. *Earth Planet. Sci. Lett.* **60**, 389–406.

Staudacher, T. and Allegre, C. J. (1988). Recycling of oceanic crust and sediments: the noble gas subduction barrier. *Earth Planet. Sci. Lett.* **89**, 173–83.

Staudacher, T., Kurz, M. D. and Allegre, C. J. (1986). New noble-gas data on glass samples from Loihi Seamount and Hualalai and on dunite samples from Loihi and Reunion Island. *Chem. Geol.* **56**, 193–205.

Staudacher, T., Sarda, P., Richardson, S. H. *et al.* (1989). Noble gases in basalt glasses from a Mid-Atlantic Ridge topographic high at 14 °N: geodynamic consequences. *Earth Planet. Sci. Lett.* **96**, 119–33.

Stroncik, N. A. and Niedermann, S. (2016). Atmospheric contamination of the primary Ne and Ar signal in mid-ocean ridge basalts and its implications for ocean crust formation. *Geochim. Cosmochim. Acta* **172**, 306–21.

Stuart, F. M., Lass-Evans, S., Fitton, J. G. and Ellam, R. M. (2003). High ^{3}He/^{4}He ratios in picritic basalts from Baffin Island and the role of a mixed reservoir in mantle plumes. *Nature* **424**, 57.

Takayanagi, M. and Ozima, M. (1987). Temporal variation of ^{3}He/^{4}He ratio recorded in deep-sea sediment cores. *J. Geophys. Res.* **92**, 12 531–8.

Tang, M. and Anders, E. (1988) Isotopic anomalies of Ne, Xe, and C in meteorites. III. Local and exotic noble gas components and their interrelations. *Geochim. Cosmochim. acta* **52**, 1245–54.

Tolstikhin, I., Marty, B., Porcelli, D. and Hofmann, A. (2014). Evolution of volatile species in the earth's mantle: A view from xenology. *Geochim. Cosmochim. Acta* **136**, 229–46.

Tolstikhin, I. N. and O'Nions, R. K. (1994). The Earth's missing xenon: a combination of early degassing and of rare gas loss from the atmosphere. *Chemical geology*, **115**(1–2), 1–6.

Tolstikhin, I. N. and O'Nions, R. K. (1996). Some comments on isotopic structure of terrestrial xenon. *Chem. Geol.* **129**, 185–99.

Trieloff, M., Kunz, J., Clague, D. A., Harrison, D. and Allegre, C. J. (2000). The nature of pristine noble gases in mantle plumes. *Science* **288**, 1036–8.

Tucker, J. M., Mukhopadhyay, S. and Schilling, J. G. (2012). The heavy noble gas composition of the depleted MORB mantle (DMM) and its implications for the preservation of heterogeneities in the mantle. *Earth Planet. Sci. Lett.* **355**, 244–54.

Turekian, K. K. (1964). Outgassing of argon and helium from the Earth. In: Brancazio, P. and Cameron, A. G. W. (Eds) *The Origin and Evolution of Atmospheres and Oceans.* Wiley, pp. 74–83.

Turner, G., Busfield, A., Crowther, S. A. *et al.* (2007). Pu-Xe, U-Xe, U-Pb chronology and isotope systematics of ancient zircons from Western Australia. *Earth Planet. Sci. Lett.* **261**, 491–9.

Valbracht, P. J., Staudacher, T., Malahoff, A. and Allegre, C. J. (1997). Noble gas systematics of deep rift zone glasses from Loihi Seamount, Hawaii. *Earth Planet. Sci. Lett.* **150**, 399–411.

van Keken, P. E. and Ballentine, C. J. (1998). Whole-mantle versus layered mantle convection and the role of a high-viscosity lower mantle in terrestrial volatile evolution. *Earth Planet. Sci. Lett.* **156**, 19–32.

van Keken, P. E. and Ballentine, C. J. (1999). Dynamical models of mantle volatile evolution and the role of phase transitions and temperature-dependent rheology. *J. Geophys. Res.* **104**, 7137–51.

van Keken, P. E., Hauri, E. H. and Ballentine, C. J. (2002). Mantle mixing: the generation, preservation, and destruction of chemical heterogeneity. *Ann. Rev. Earth Planet. Sci.* **30**, 493–525.

Wetherill, G. W. (1953). Spontaneous fission yields from uranium and thorium. *Phys. Rev.* **82**, 907–12.

Wetherill, G. W. (1954). Variations in the isotopic abundances of neon and argon extracted from radioactive materials. *Phys. Rev.* **96**, 679–83.

Winckler, G., Anderson, R. F., Stute, M. and Schlosser, P. (2004). Does interplanetary dust control 100 kyr glacial cycles? *Quaternary Sci. Rev.* **23**, 1873–8.

Yokochi, R. and Marty, B. (2004). A determination of the neon isotopic composition of the deep mantle. *Earth Planet. Sci. Lett.* **225**, 77–88.

Zadnik, M. G., Smith, C. B., Ott, U. and Begemann, F. (1987). Crushing of a terrestrial diamond: ^{3}He/^{4}He higher than solar meteorites. *Meteoritics* **22**, 541–2.

Zahnle, K., Arndt, N., Cockell, C. *et al.* (2007). Emergence of a habitable planet. *Space Sci. Rev.* **127**, 35–78.

Zeitler, P. K., Herczeg, A. L., McDougall, I. and Honda, M. (1987). U-Th-He dating of apatite: A potential thermochronometer. *Geochim. Cosmochim. Acta* **51**, 2865–8.

Zhang, Y. and Yin, Q. Z. (2012). Carbon and other light element contents in the Earth's core based on first-principles molecular dynamics. *Proc. Nat. Acad. Sci.* **109**, 19 579–83.

Chapter 12

U-Series Dating

The intermediate nuclides in the U–Pb and Th–Pb decay series have very short half-lives in comparison to their parents, and are usually ignored in the Pb isotope dating methods. However, their short half-lives make these nuclides useful for dating Pleistocene geological events which are too old to be well resolved by the radiocarbon method and too young to be well resolved by decay schemes with long half-lives. The manner in which U-series nuclides can fill this 'dating gap' is shown in Fig. 12.1. Generally, they are most useful for dating events of similar age to their half-life.

12.1 Secular Equilibrium and Disequilibrium

A distinctive property of the U-series nuclides which sets them apart from other isotope dating schemes is that the radiogenic daughters are themselves radioactive. Hence, in a uranium-bearing system which has been undisturbed for a few million years, a state of 'secular equilibrium' becomes established between the abundances of successive parent and daughter nuclides in the U and Th decay chains, such that the decay rate (or 'activity') of each daughter nuclide in the chain is equal to that of the parent:

$$\text{Activity} = \lambda_0 n_0 = \lambda_1 n_1 = \lambda_2 n_2 = \lambda_N n_N \quad [12.1]$$

where λ_0 is the decay constant and n_0 is the number of atoms of the original parent, λ_1 and n_1 are the decay constant and abundance of the first daughter, and so on. It follows that the abundance of each nuclide will be directly proportional to its half-life (i.e. inversely proportional to its decay constant). The relevant parts of the decay chains are shown in Fig. 12.2.

During geological processes such as erosion, sedimentation, melting or crystallization, different nuclides in the decay series can become fractionated relative to one another, due to variations in their chemistry or the structural site they occupy. This results in a state of secular disequilibrium. Such a situation can be utilized in two different ways as

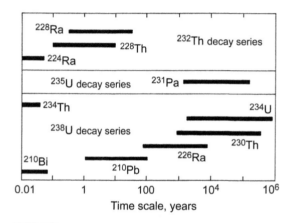

Fig. 12.1 Diagram showing the dating ranges of different nuclides within the three U-series decay chains, to show their utility. After Potts (1987).

a dating tool, called respectively the 'daughter-excess' and 'daughter-deficiency' dating methods.

In the daughter-excess method, a deposit is formed with an excess of the daughter beyond the level which can be sustained by the abundance of its parent nuclide. Over time, the excess or 'unsupported' daughter decays back until secular equilibrium with its parent is restored. If the original fractionation can be estimated, the age of the deposit can be calculated by the progress of decay of the excess.

In the daughter-deficiency method, chemical fractionation during the formation of a deposit causes it to take up a radioactive parent but effectively none of its daughter. The age of the deposit can then be determined by measuring the growth of the daughter, up to the point when its abundance is within error of secular equilibrium of the parent. Using high-precision mass spectrometric data (Section 12.2.2) the useful dating range of U-series nuclides may be up to seven half-lives, but other factors may impose lower limits. Table 12.1 summarizes some of the more important U-series dating methods.

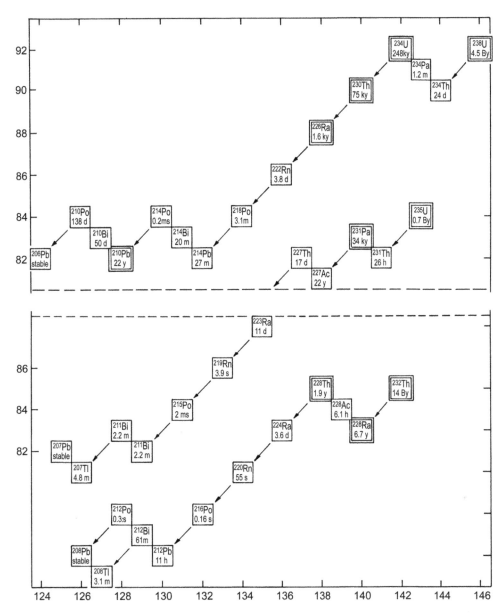

Fig. 12.2 Part of the chart of the nuclides, in terms of Z against N, to show species in the Th- and U-series decay chains and their half-lives. Useful species are indicated by double boxes. In early research on decay series nuclides, some species were given special names (e.g. ^{230}Th = ionium). However, these are now obsolete.

Note that in this chapter, all nuclide abundances are given as activities unless otherwise stated.

12.2 Analytical Methods

As noted above, the atomic abundance of a U-series nuclide in secular equilibrium is proportional to its half-life. Therefore, the very variable half-lives of the U-series radionuclides cause them to have extreme abundance ratios. For many years this discouraged mass spectrometric determination of U-series nuclides for dating purposes. In contrast, species in secular equilibrium have equal activities (by definition), so radioactive counting was an obvious method for their determination. Counting techniques utilizing β and γ particles are not favoured because of the low energies of β transitions and the complexity of γ ray spectra (Yokoyama and Nguyen, 1980). Therefore, the traditional technique for measurement of U-series nuclides was α spectrometry.

Method	Measurement	$t_{1/2}$, ka	Range, ka	Application
Daughter excess				
^{234}U–^{238}U	^{234}U decay	245.6	< 1500	Coral, closed-system test for ^{230}Th
^{230}Th	^{230}Th "	75.6	< 500	Deep sea sedimentation rates
^{231}Pa	^{231}Pa "	32.8	< 200	Deep sea sedimentation rates (advection test for ^{230}Th)
^{210}Pb	^{210}Pb "	0.022	< 0.1	Recent sedimentation, snow accumulation
Daughter deficiency				
^{230}Th–^{234}U	^{230}Th accum.	75.6	< 500	Marine and fresh-water carbonate, volcanics
^{231}Pa–^{235}U	^{231}Pa "	32.8	< 200	Closed-system test for ^{230}Th
^{226}Ra–^{230}Th	^{226}Ra "	1.6	< 10	Historical age volcanic systems

Table 12.1 | U-series dating methods.

12.2.1 Early Work

Because of the very short range of α particles in matter, samples for counting must be made into a thin film and placed under vacuum in a gridded ion chamber (a type of gas ionization chamber with a short dead-time). If the applied potential between cathode and anode is within a certain range, the electrical pulses generated by α particle emission are proportional in size to the kinetic energy of those particles. The output is fed to a multi-channel analyser, which registers count rates as a function of energy level. This allows the α particles from different decay transitions to be distinguished as separate 'peaks' in the energy spectrum. To obtain 1σ counting errors of 1%, total counts of 10^4 are required on each peak ($\sigma = \sqrt{n}$). To achieve this, counting times of at least a week were required for most natural samples.

Because α counting only measures the abundance of nuclides that actually decay during the measurement period, it is a very inefficient measurement technique, especially for long-lived nuclides. In contrast, mass spectrometry offers the opportunity of measuring the abundance of every atom in the sample, and is therefore much more sensitive. In the case of U-series nuclides, mass spectrometry offers at least an order of magnitude improvement in both sensitivity and precision, and has therefore replaced α spectrometry as a measurement technique. However, these advances also throw the emphasis of U-series dating work back onto sample collection and preparation, since open-system behaviour of samples becomes more obvious with improvements in analytical precision. These problems will be discussed below.

Given the low abundances of the U-series nuclides to be measured in natural materials (part per trillion to part per million range), chemical purification is essential for both alpha counting and mass spectrometry. This normally involves dissolution of the sample in HNO_3 (carbonates) or HF (silicates) followed by anion exchange separation (Section 2.1.4). Anion exchange is also used to separate between U and Th. Since chemical extractions are not expected to give a 100% yield, the sample is 'spiked' before chemistry with a known quantity of artificially enriched isotopes, allowing an isotope dilution determination of isotope abundances in the sample (Section 2.5). In alpha counting analysis, short-lived radioactive species were usually chosen as spikes. For example, a widely used U–Th spike was ^{232}U ($t_{1/2} = 70$ yr), which had been allowed to generate its daughter ^{228}Th ($t_{1/2} = 1.9$ yr) naturally. The short half-life of the latter nuclide meant that it reached secular equilibrium with its parent in ca. 20 years (Ivanovich, 1982a). For mass spectrometry, the longer-lived isotopes ^{229}Th and ^{236}U are preferred, with half-lives of ca. 6 ka and 70 ka respectively.

12.2.2 Mass Spectrometry

Uranium-series dating by mass spectrometry was one of the missed opportunities of 1970s isotope geology, since the analytical equipment available at that time was equal to this task, but was not applied until the late 1980s. This omission can be explained by a communication gap between workers in the two fields, and by exaggerated estimates of the problems which might be posed by large nuclide abundance ratios. The gap was closed in two stages, by Chen *et al.* (1986) who made the first precise mass spectrometric analysis on ^{234}U, and by Edwards *et al.* (1987) who made the first ^{230}Th measurements. Using multiplier detectors to measure the small ^{234}U and ^{230}Th beams, these workers showed that mass spectrometric U-series dating offered great improvements in precision over the best α counting determinations.

Edwards *et al.* avoided the difficulty of measuring large ^{238}U/^{234}U ratios by measuring ^{235}U/^{234}U instead. Since ^{238}U/^{235}U has a constant ratio of 137.8 in normal rocks, the conversion is simple. However, Cheng *et al.* (2013) argued that it is preferable to collect the large ^{238}U beam in a Faraday bucket to prevent the beam from reflecting off the inside of the instrument and potentially biasing baseline values. It is also necessary to make an accurate correction for the peak tails of the larger ion beams, which can easily reach 2 mass units in the down-mass direction. These problems can be avoided in the determination of ^{230}Th by analysing pure samples with a low detrital ^{232}Th content (see below). Hence, for corals the ^{232}Th/^{230}Th atomic abundance ratio can be as low as unity, compared to ratios over 250 000 in silicate rocks.

Thorium has a relatively high ionization potential. Therefore thermal ionization mass spectrometry (TIMS) analysis of this element is relatively inefficient. Li *et al.* (1989) used the conventional double-filament technique employed for Nd isotope analysis (Section 2.2.1), with a very hot centre filament to promote the formation of Th metal ions. This method is not very demanding of chemical purity but is relatively inefficient. Edwards *et al.* (1987) loaded both U and Th (separately) on graphite-coated single rhenium filaments, and analysed them as the metal species. This method is more efficient for very small samples, but the ionization efficiency drops rapidly as the size of loaded sample increases, from 0.1% in very small samples to 0.001% in large samples. This is due to a failure to make proper contact with the heated metal filament as the amount of loaded sample increases.

Ionization problems are avoided using an ICP source, which achieves nearly complete ionization of all elements (Section 2.2.2). As a result, MC–ICP–MS has largely superseded TIMS analysis for Th analysis (Section 2.6.3). It also has the capability of performing *in situ* U-series analysis of uranium-rich samples using the laser microprobe (Stirling *et al.*, 2000). Sampling of the plasma by the mass spectrometer is about 1% efficient, so MC–ICP–MS can offer approximately an order of magnitude increase in signal size relative to TIMS. This does not necessarily translate into a corresponding improvement in dating precision, because the accuracy of multiplier detectors is limited to ca. 0.1%. However, transmission efficiency for MC–ICP–MS does not significantly fall with increasing sample size, so this method offers the opportunity of analysing large samples with Faraday detectors (e.g. Cheng *et al.*, 2013). This can then yield an order of magnitude improvement in precision and/or reproducibility to the epsilon level (part per 10 000).

12.2.3 Half-Lives
A pre-requisite to precise and accurate dating with U-series nuclides is the availability of good half-life determinations. However, the attainment of secular equilibrium allows these half-lives to be determined relative to the very well-constrained ^{238}U half-life. For example, the half-life of ^{234}U can be determined very accurately relative to ^{238}U by measurement of the ^{234}U/^{238}U ratio on a sample in secular equilibrium, such as uraninite ore. Using this technique, de Bievre *et al.* (1971) determined a value of 244.6 ± 0.7 ka by α spectrometry, which was revised to 245.3 ± 0.14 ka by TIMS using a multiplier detector (Ludwig *et al.*, 1992). More recently, MC–ICP–MS with a Faraday collector has given a slightly longer half-life of 245.62 ± 0.26 (Cheng *et al.*, 2013).

The ^{230}Th half-life can also be determined by analysis of uraninite in secular equilibrium, using a mixed ^{229}Th–^{236}U spike. Meadows *et al.* (1980) determined a half-life of 75.4 ± 0.6 ka from α counting, which was revised to 75.69 ± 0.23 ka by TIMS (Cheng *et al.*, 2000), and 75.58 ± 0.11 by ICP–MS (Cheng *et al.*, 2013). Fortunately, gravimetric weighting errors cancel out for the ^{230}Th daughter-deficiency method (Section 12.4), which uses both half-lives together. Also, since the two half-life determinations increased by a similar proportion moving from alpha counting to mass spectrometry, the overall effect on ^{230}Th ages has been small.

12.3 Daughter-Excess Methods

As shown in Table 12.1, the most important species for U-series dating are isotopes of uranium and thorium. These elements have very different chemistries in aqueous systems, which leads to elemental fractionation and hence isotopic disequilibrium. In this section we examine systems where elemental fractionation has led to excess abundances of the daughter nuclide of an isotope pair.

12.3.1 ^{234}U Dating of Carbonates
^{238}U decays via two very short-lived intermediates to ^{234}U (Fig. 12.2). Since ^{234}U and ^{238}U have the same chemical properties, it might be expected that they would not be fractionated by geological processes. However, Cherdyntsev and co-workers (1965, 1969) showed that such fractionation does occur. In fact, natural waters show a considerable range in ^{234}U/^{238}U activities from unity (secular equilibrium) to values of ten or more (e.g. Osmond and Cowart, 1982).

Cherdyntsev *et al.* (1961) attributed these fractionations to radiation damage of crystal lattices, caused both by α emission and by recoil of parent nuclides. In addition, radioactive decay may leave ^{234}U in a more soluble +6 charge state than its parent (Rosholt *et al.*, 1963). These processes (termed the 'hot atom' effect) enable preferential leaching of the two very short-lived intermediates and the longer-lived ^{234}U nuclide into groundwater. The short-lived nuclides have a high probability of decaying into ^{234}U before they can be adsorbed onto a substrate, and ^{234}U is itself stabilized in surface waters as the soluble UO_2^{++} ion, due to the generally oxidizing conditions prevalent in the hydrosphere.

The variety of weathering conditions prevailing in the terrestrial environment leads to very variable ^{234}U/^{238}U activity ratios in fresh water systems. However, the long residence time of uranium in seawater (>300 ka, Ku *et al.*, 1977) maintains seawater ^{234}U/^{238}U within narrow limits, corresponding to an activity ratio of ca. 1.14 (Goldberg and Bruland, 1974). This has since been refined to an activity ratio of 1.147 in the open oceans (Stirling and Andersen, 2009). However, in restricted basins such as the Arctic, seawater can show small variations in uranium activity ratio due to incomplete homogenization of freshwater inputs (Andersen *et al.*, 2007).

A major uranium sink in the oceans is calcium carbonate, with which uranium is co-precipitated. This is deposited in shallow water by marine organisms and in deep water as an authigenic mineral (i.e. by direct chemical precipitation). At the time of deposition, this material takes on the 'daughter-excess' ^{234}U/^{238}U activity ratio of seawater, but once isolated, the excess decays away until secular equilibrium with the parent is regained (Fig. 12.3). Given an estimate of the original ^{234}U/^{238}U fractionation, and given subsequent closed-system behaviour, the system can be used as a dating tool until it returns to within analytical error of secular equilibrium.

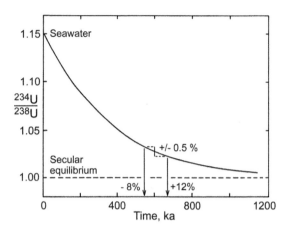

Fig. 12.3 Plot of $^{234}U/^{238}U$ activity against time showing the return to secular equilibrium after isolation from seawater. Arrows show the amplification of analytical errors as the system approaches equilibrium.

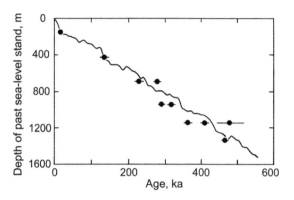

Fig. 12.4 Plot of terrace depth against mass spectrometric ^{234}U age for corals off NW Hawaii, showing the good fit to a cooling subsidence curve (modulated by eustatic variations). After Ludwig et al. (1991).

Unfortunately, many problems are encountered in the practical application of this method. As noted above, the variable uranium isotope fractionations observed in fresh-water systems preclude its application there. In addition, pelagic sediments are ruled out by open-system behaviour of uranium after deposition (Ku, 1965), while mollusc shells also tend to take up uranium after deposition (Kaufman et al., 1971). However, the method has been applied with reasonable success to the dating of corals (e.g. Thurber et al., 1965).

The decay of excess ^{234}U can be expressed by the fundamental decay equation [1.5]. Although this equation was derived in Section 1.4 for atomic abundances, it is also true for activities (by dividing both sides by the decay constant, e.g. $\lambda^{234}U$):

$$\frac{n}{\lambda} = \frac{n_0}{\lambda}e^{-\lambda t} \quad [12.2]$$

$$A = A_0\,e^{-\lambda t} \quad [12.3]$$

In order to date a carbonate sample by the decay of excess ^{234}U (Fig. 12.5), we can substitute into equation [12.3] to yield

$$^{234}U^X_{present} = {}^{234}U^X_{initial}\,e^{-\lambda.234 t} \quad [12.4]$$

where 'x' signifies excess activities above secular equilibrium, and 'initial' signifies the activity at the time of precipitation.

As noted above, *all nuclide quantities in this chapter will be presented in terms of activities* (unless otherwise stated). However, absolute activities are not as readily measurable as activity ratios, so it is convenient to divide through by ^{238}U activities. But because of the very long half-life of ^{238}U, the activity of $^{238}U_{present}$ is the same as $^{238}U_{initial}$. So:

$$\left(^{234}U^X/^{238}U\right)_{present} = \left(^{234}U^X/^{238}U\right)_{initial}e^{-\lambda.234 t} \quad [12.5]$$

Since these quantities are in the form of activities, the excess $^{234}U/^{238}U$ activity is equal to the total activity ratio minus one (that part corresponding to secular equilibrium). So:

$$\left(\frac{^{234}U}{^{238}U}\right)^{total}_{present} - 1 = \left[\left(\frac{^{234}U}{^{238}U}\right)^{total}_{initial} - 1\right]e^{-\lambda.234 t} \quad [12.6]$$

Hence, if we assume that the initial activity ratio of the sample is given by present day seawater, we can calculate the age of a coral simply by measuring the present day activity ratio. Chen et al. (1986) showed that modern seawater in the Pacific and Atlantic oceans has a homogeneous $^{234}U/^{238}U$ activity ratio, with values of 1.143 and 1.144 respectively. Given the >300 ka residence time of uranium in seawater (Ku et al., 1977), this gives us a strong expectation that the activity ratio should have been close to this value within the 1.2 Ma theoretical dating range of the method. This was indeed demonstrated by more recent work (Henderson, 2002), which indicated a $^{234}U/^{238}U$ activity ratio of 1.145 for the last 350 ka. This value is often expressed as δ (excess ^{234}U), with a value of 145 per mil (Fig. 12.3).

Because the seawater $^{234}U/^{238}U$ value is relatively close to secular equilibrium, a small error in the ^{234}U measurement leads to a large error in the calculated age (Fig. 12.3). Hence, the ^{230}Th deficiency method (Section 12.4) yields more precise ages below 500 ka. However, using mass spectrometric analysis, the ^{234}U method allows the possibility of dating back to 1 Ma with tolerable precision. This was demonstrated by Ludwig et al. (1991), who used ^{234}U to date submerged coral terraces off NW Hawaii. Comparison of ^{234}U ages with terrace depth led to a subsidence curve which is approximately linear for the last 500 ka, at a rate of 2.6 mm/yr (Fig. 12.4). Small undulations on the subsidence curve represent the calculated effect of eustatic sea level fluctuations. These cause development of coral terraces by periodically neutralizing subsidence (to create a sea level 'stand') and then exacerbating subsidence, to drown the reef.

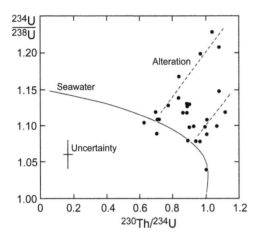

Fig. 12.5 Plot of $^{234}U/^{238}U$ against $^{230}Th/^{234}U$ activities for Barbados coral terraces, showing trend lines above the seawater value resulting from alteration. After Bender et al. (1979).

The good fit of data points to a linear subsidence model in Fig. 12.4 provides evidence of the overall reliability of the $^{234}U/^{238}U$ dating method, including the approximate constancy of the $^{234}U/^{238}U$ ratio of seawater over time. However, Bard et al. (1991) pointed out that multiple samples from the same depth sometimes gave ages well outside of error, such as the samples from 1150 m depth (Fig. 12.4). This could indicate open-system behaviour of the samples, despite their submarine location.

The open-system problem is expected to be much worse for sub-aerially exposed coral terraces, which are subject to groundwater percolation. This has been most extensively documented for the raised terraces on the island of Barbados, which is undergoing net tectonic uplift with time. An early alpha-counting study by Bender et al. (1979) found arrays of points with $^{234}U/^{238}U$ and $^{230}Th/^{234}U$ activity ratios above the seawater activity decay line (Fig. 12.5). This was attributed to open-system behaviour of the corals, which apparently occurred despite the exclusion of any samples showing visible evidence of recrystallization. These observations were confirmed by Gallup et al. (1994) using mass spectrometric analysis. These authors found similar positive trends of activity ratios above the seawater line, which they likewise attributed to open-system behaviour.

These results essentially invalidate use of the $^{234}U/^{238}U$ method to date corals, but the fact that $^{234}U/^{238}U$ and $^{230}Th/^{234}U$ activity ratios are correlated suggests that the $^{230}Th/^{234}U$ daughter-deficiency method (Section 12.4) is also subject to disturbance. However, the apparent coherent behaviour of these systems allows the possibility of using $^{234}U/^{238}U$ activities to monitor the disturbance of $^{230}Th/^{234}U$ ages.

Gallup et al. (1994) felt that the substantial scatter of data points around the $^{234}U-^{230}Th$ enrichment lines precluded the calculation of accurate ^{230}Th ages on samples with elevated $^{234}U/^{238}U$ activity ratios. However, this can mean excluding as much as 90% of analysed samples (e.g. Medina-Elizalde, 2013), since no a priori test has been discovered to allow screening out of disturbed samples before isotopic analysis. Therefore, to avoid this waste, other authors have attempted to develop open-system models to allow ^{230}Th ages on disturbed samples to be corrected.

Similar models were proposed by Thompson et al. (2003) and Villemant and Feuillet (2003), to account for the leaching of ^{234}Th and ^{230}Th from lattice sites damaged by α recoil (the 'hot atom' effect). These species will then be transported in fluids and concentrated elsewhere in the reef structure. Although these models allow for improvement in old ages biased by open-system behaviour, it is questionable whether the corrected ages are accurate enough to usefully enhance the smaller number of ages from pristine samples (with seawater $^{234}U/^{238}U$ activities) that do not require corrections (Scholz and Mangini, 2007).

A very different theory was proposed by Esat and Yokoyama (2006, 2010), who argued that the elevated $^{234}U/^{238}U$ activities in ancient corals reflect true variation in the initial $^{234}U/^{238}U$ ratio of seawater. Hence they attempted to make the case that the correlated enrichments in $^{234}U/^{238}U$ and $^{230}Th/^{234}U$ activity ratios reflected short time-period changes in ocean chemistry. For example, it was argued that sea level rise after a glacial period could release a pulse of ^{234}U to the sea that had been locked up as reduced uranium in organic-rich tidal mudflats during the previous interglacial. This pulse would then gradually diminish over time, yielding the observed $^{234}U/^{238}U - ^{230}Th/^{234}U$ correlation lines. While theoretically possible, this model is not well supported empirically. For example, the activity correlation lines for Barbados coral terraces yield overlapping apparent U–Th ages that do not make sense as real ages, and are better explained by diagenetic processes (e.g. Fig. 12.5 above and Potter et al., 2004).

To avoid these problems, the emphasis in more recent work has been on selecting parts of the coral skeleton that are more resistant to alteration. For example, Obert et al. (2016) showed that denser thecal wall material gave more reliable results for the Atlantic brain coral *Diploria strigosa* than bulk samples. Significantly, not all samples showed a positive trend of $^{234}U/^{238}U$ activity ratio against apparent U–Th age (Fig. 12.6), showing that improvements in sample quality are more effective than attempts at open-system modelling. However, it is also notable that the thecal wall samples themselves define an array with positive slope which intersects with the modern seawater value at its lower end. This suggests that even superior samples may have some alteration problems.

12.3.2 ^{234}U Dating of Fe–Mn Crusts

Another application of ^{234}U is to the dating of ocean floor ferro-manganese crusts. These crusts grow over long periods of time on the ocean floor, and represent very useful

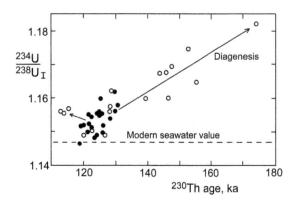

Fig. 12.6 Plot of uranium isotope activity against apparent age to show the effect of diagenesis on bulk coral samples (○) compared with thecal wall material (●). After Obert *et al.* (2016).

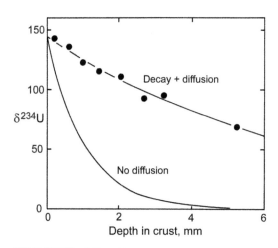

Fig. 12.8 Plot of ^{234}U activity (parts per mil above secular equilibrium) against depth in a Fe–Mn crust from the North Atlantic. The measured ^{234}U 'decay curve' is compared with the predicted decay curve for zero diffusion. After Henderson and Burton (1999).

archives of past seawater chemistry if they can be dated accurately. In the first U-series dating study on this material, Chabaux *et al.* (1995) analysed two crusts dredged from 1900 m depth on a West Pacific seamount. For both crusts, the ^{234}U and ^{230}Th daughter excess methods gave consistent growth rates of ca. 7.8 mm/Ma and 6.6 mm/Ma respectively (Fig. 12.7). This suggested that closed-system conditions were preserved, and that initial uranium and thorium isotope ratios remained constant (within error) during the 150 ka period of deposition.

A problem with the sampling of Fe–Mn crusts is that the outer surface can be abraded during dredging operations,

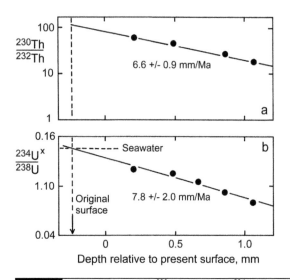

Fig. 12.7 The use of excess ^{230}Th and excess ^{234}U in a Fe–Mn crust to determine the growth rate and the zero-age surface of the crust before abrasion. Modified after Chabaux *et al.* (1995).

preventing determination of the absolute age of the crust from its growth rate. This is specifically a problem with the ^{230}Th method, because the initial thorium isotope ratio of seawater is variable. Therefore, in the absence of an 'initial ratio' determination from the surface of the crust, an absolute date is not possible. On the other hand, the initial $^{234}U/^{238}U$ ratio of seawater is constant in space and time. Therefore, it should be possible to use ^{234}U to determine the original growth surface of the crust by projecting the excess ^{234}U activity back to the known seawater composition (Fig. 12.7b).

Fe–Mn crusts are quite porous and have very slow growth rates. Therefore, there was concern that diffusion could cause open-system behaviour of U-series nuclides in the crust after deposition. The work of Chabaux *et al.* (1995) appeared to allay these fears, but subsequent work by Chabaux *et al.* (1997), Neff *et al.* (1999) and Henderson and Burton (1999) has confirmed that U diffusion is a problem in Fe–Mn crusts. These later studies gave Fe–Mn growth rates from 7 to 19 mm/Ma, based on excess ^{234}U measurements, but only 3 to 4 mm/Ma based on excess ^{230}Th. The discrepancy between these growth rates, especially for the very high value of 19 mm/Ma, is best explained by diffusional redistribution of uranium. By comparing the observed decay profile of ^{234}U with a theoretical decay curve based on excess ^{230}Th dates, Henderson and Burton were able to calculate effective diffusion coefficients for uranium in the different Fe–Mn crusts (Fig. 12.8). These values were in the range 10^{-6} to 5×10^{-8} cm^2/yr.

Henderson and Burton argued that the magnitude of U diffusion in the crust must be controlled by the ability of uranium to be exchanged from the solid crust into pore water

| Table 12.2 | Partition and diffusion coefficients in Fe–Mn crusts. | |

Element	Partition coefficient crust/seawater	Diffusion coefficient in crust, cm^2/yr
Particle reactive species		
Th	2.0×10^9	2×10^{-12}
Nd	2.6×10^8	2×10^{-11}
Pb	1.6×10^8	3×10^{-11}
Be	4.0×10^7	1×10^{-10}
Hf	5.2×10^6	9×10^{-10}
Os	1.7×10^5	3×10^{-8}
Seawater conservative species		
U	4.0×10^3	1×10^{-6}
Sr	2.1×10^2	2×10^{-5}

within the crust. Hence, the diffusion coefficient must be dependent on the partition coefficient of uranium between seawater and the Fe–Mn crust. The solubility of U in surface water has already been discussed. In contrast, thorium is said to be 'particle reactive', meaning that it is readily adsorbed onto the surface of detrital grains and has a very short residence time in natural waters. Thus, uranium has a seawater residence time exceeding 300 ka, whereas thorium has a seawater residence more than 1000 times shorter than this.

Based on these principles, Henderson and Burton estimated the diffusion coefficients of other elements from their relative concentrations in the crust and in seawater. This calculation places these species in two groups (Table 12.2). Uranium, along with strontium has a relatively large diffusion coefficient in crusts, consistent with these elements being non-particle-reactive, and hence conservative in seawater (osmium also falls in this group although its behaviour is less conservative). At the other extreme, thorium has a partition coefficient about six orders of magnitude lower than uranium, consistent with it being the most particle-reactive of its group. In addition, other members of the particle-reactive group also have sufficiently low diffusion coefficients in Fe–Mn crusts that they can be considered immobile. This is encouraging for the use of Fe–Mn crusts as an inventory of the past seawater signatures of these tracers (Sections 4.5, 5.6, 9.4, 14.4). However, it is concluded that the ^{234}U method is not reliable for dating this material.

Another area where excess ^{234}U activity data would be very useful is in the dating of planktonic foraminifera, since these are the basis of the seawater oxygen isotope record (Section 12.4.2). However, forams have low U contents (typically 20 ppb), which tend to be swamped by the U contents of ferro-manganese diagenetic overgrowths. Henderson and O'Nions (1995) showed that dithionite solution (a reducing agent) could be used to clean recent forams in order to recover normal seawater uranium isotope ratios. However, a test on 2 Ma forams showed excess ^{234}U activities above the seawater value, which must have been introduced from pore

waters after sedimentation. This suggests that forams do not remain a closed system for uranium, and therefore cannot be used for dating or to constrain the uranium isotope evolution of seawater.

12.3.3 ^{230}Th Sediment Dating

The differing behaviour of uranium and thorium in seawater causes U/Th fractionation during the formation of different sediment types, leading to systems out of secular equilibrium. As noted above, ^{238}U decays via two very short-lived intermediates to ^{234}U in seawater. This in turn decays to ^{230}Th, but the latter is almost immediately adsorbed onto sinking particulate matter. Because it is preferentially enriched on the sediment surface, relative to its (^{234}U) parent, ^{230}Th is 'unsupported' and out of secular equilibrium. However, after isolation from the sediment–water interface, this unsupported ^{230}Th begins to decay back to secular equilibrium with its parent. Hence, this method should allow the dating of sedimentary deposition.

Thorium adsorption onto detrital grains is so much more effective than uranium adsorption that for young sediments the uranium-supported component (i.e. the component in secular equilibrium) can be effectively ignored. In other words:

$$^{230}Th_{excess} \approx ^{230}Th_{total} \qquad [12.7]$$

Therefore we can use the method as a dating tool by means of the simple decay equation

$$^{230}Th_{present} = ^{230}Th_{initial}e^{-\lambda_{230}\,t} \qquad [12.8]$$

Since the ^{230}Th excess method is used to study sedimentation, it is convenient to formulate t in terms of sediment depth, D (in a core) and sedimentation rate, R:

$$t = D/R \qquad [12.9]$$

If we substitute this into equation [12.8] and take the natural log of both sides, we obtain

$$\ln \left(^{230}Th_P \right) = \ln \left(^{230}Th_I \right) - D \left(\lambda_{230}/R \right) \qquad [12.10]$$

This corresponds to the equation for a straight line:

$$y = c - x\,m \qquad [12.11]$$

Hence, if the natural log of the present day ^{230}Th activity is plotted against depth in the core, the sedimentation rate can be obtained from the reciprocal of the slope (solid line in Fig. 12.9):

$$R = -\lambda_{230} \cdot \frac{D}{\ln\,^{230}Th_P - \ln\,^{230}Th_I} \qquad [12.12]$$

Although the effects of U-supported ^{230}Th may be negligible near the sediment surface, this component becomes increasingly important as the system approaches secular equilibrium with increasing burial depth (dashed line in Fig. 12.9). Two possible sources of U-supported ^{230}Th may be present. Authigenic minerals such as calcite contain no ^{230}Th

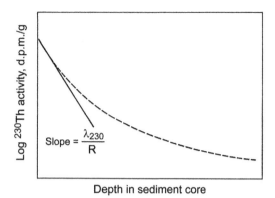

Fig. 12.9 Schematic plot of log ^{230}Th activity (decays per minute/gram) against depth, to show behaviour expected in a core formed by a constant sedimentation rate. Solid line = young sediments; dashed line = older sediments with U-supported ^{230}Th.

at the sediment surface, but their uranium budget generates ^{230}Th until this reaches secular equilibrium with the parent. This fraction is best removed physically by mineral separation. On the other hand, detrital grains contain ^{230}Th which is in secular equilibrium with ^{234}U and ^{238}U, even at the sediment surface. This fraction is removed by subtracting ^{234}U activity (in secular equilibrium with the ^{230}Th daughter) from total ^{230}Th activity (e.g. Ku, 1976). This leaves the 'excess' ^{230}Th activity of the clay fraction:

$$^{230}\text{Th}_{\text{excess}} = {}^{230}\text{Th}_{\text{total}} - {}^{234}\text{U} \qquad [12.13]$$

The corrected (excess) activities determined in this way are substituted into equations [12.10] and [12.12] to determine sedimentation rates. Since the concentration of ^{230}Th in the oceans is expected to be constant through time and the adsorption process is expected to be of constant efficiency, the initial concentration of ^{230}Th in the detrital sediment fraction should be constant. Then, if the bulk sedimentation rate (R) remains constant with time, excess ^{230}Th activity will decrease as a log function with depth. Figure 12.10a shows data from a Caribbean core which fit this model (Ku, 1976), yielding a linear fit (of log activity against depth). The regression slope yields a sedimentation rate R of 25 ± 1 mm/ka for the last 300 ka.

Within the decay chain of ^{235}U, the species ^{231}Pa (protactinium) is another particle-reactive species that behaves very similarly to thorium in seawater. Therefore, it also develops excess activities at the sediment surface relative to its parent isotope of uranium. ^{231}Pa has a half-life of 32.76 ka (Roberts et al., 1969), and is therefore used in an analogous way to ^{230}Th. However, because the parent (^{235}U) has a much lower abundance than ^{238}U, analytical errors are larger. Therefore, ^{231}Pa is usually used only as a concordancy test for ^{230}Th dates, to check that the dating assumptions have been upheld. This application is shown in Fig. 12.10b.

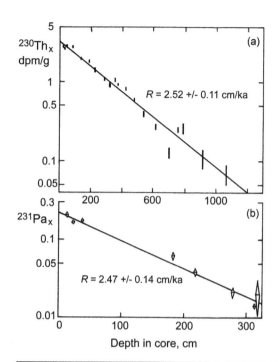

Fig. 12.10 Plots of (a) excess ^{230}Th activity, and (b) excess ^{231}Pa activity, against depth in a sediment core, yielding two independent estimates of average sedimentation rate. After Ku (1976).

12.3.4 ^{230}Th–^{232}Th

Unfortunately, not all cores yield such good results as that in Fig. 12.10, because ^{230}Th and ^{231}Pa are sometimes variably diluted in sediments. Picciotto and Wilgain (1954) suggested that this problem could be avoided by using ^{232}Th as a reference isotope to normalize for variable absolute levels of adsorbed Th. They justified this approach on the basis that ^{230}Th and ^{232}Th ($t_{1/2}$ = 14 Ga) are chemically identical, so they should be removed from seawater at the same rate. Because ^{232}Th has such a long half-life, it suffers no significant decay within the dating range of ^{230}Th. Therefore, if we assume that initial ^{230}Th/^{232}Th activities at the sediment surface remain constant at any given locality through time, we can divide both sides of equation [12.8] by ^{232}Th (where × signifies excess activities):

$$\left(\frac{^{230}\text{Th}^X}{^{232}\text{Th}}\right)_{\text{present}} = \left(\frac{^{230}\text{Th}^X}{^{232}\text{Th}}\right)_{\text{initial}} \cdot e^{-\lambda_{230}t} \qquad [12.14]$$

Applying this to the activity versus depth plot we obtain

$$\ln\left(\frac{^{230}\text{Th}^X}{^{232}\text{Th}}\right)_{\text{P}} = \ln\left(\frac{^{230}\text{Th}^X}{^{232}\text{Th}}\right)_{\text{I}} - D\frac{\lambda_{230}}{R} \qquad [12.15]$$

Picciotto and Wilgain pointed out that, for this method to work effectively, all of the Th in the sediment must have been chemically adsorbed onto the detrital phases, and not been within them. However, 30% or more of the total ^{232}Th

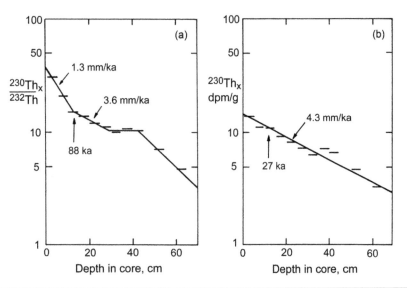

Fig. 12.11 Thorium isotope results from the ZEP 15 core (Mid Atlantic Ridge) showing interpretations of sedimentation history using (a) the ^{230}Th/^{232}Th method and (b) the simple ^{230}Th method. After Ku (1976).

budget in pelagic sediment is normally within the detrital phases (Goldberg and Koide, 1962). Consequently, Ku *et al.* (1972) argued that the effect of dividing by ^{232}Th is similar to the effect of dividing by the detrital (non-carbonate) fraction in the analysed sample. If the detrital fraction in the sediment is constant then this does not cause a problem, but if it varies with depth, this will perturb the initial ^{230}Th/^{232}Th ratios and hence lead to erroneous ages and sedimentation rates. This problem is illustrated in Fig. 12.11 using data for a core from the Mid Atlantic Ridge. The ^{232}Th/^{230}Th plot (Fig. 12.11a) yields an age for the 12 cm deep horizon (arrowed) which is more discordant from the ^{14}C age of 17 ka BP (before present) than the simple ^{230}Th plot (Fig. 12.11b).

In order to reduce the perturbing effect of the detrital component on ^{230}Th/^{232}Th ages, Goldberg and Koide (1962) used a technique by which authigenic minerals and adsorbed Th were leached from the detrital component with hot hydrochloric acid. This led them also to adopt a different correction for U-supported ^{230}Th. On the assumption that no detrital ^{230}Th component was leached, they excluded the component in secular equilibrium. Instead, they corrected for U-supported ^{230}Th in the authigenic (carbonate) component, which is expected to grow with time. This is equivalent to the ^{230}Th daughter-deficiency method, and will be dealt with in detail below (Section 12.4.1). If the immediate parent (^{234}U) is assumed to be in equilibrium with ^{238}U (an approximation), then the growth of U-supported Th is given by equation [12.24]. This is subtracted from total ^{230}Th activity to determine excess ^{230}Th:

$$^{230}\text{Th}_{\text{excess}} = {}^{230}\text{Th}_{\text{total}} - {}^{238}\text{U}\left(1 - e^{-\lambda_{230} t}\right) \quad [12.16]$$

Ku (1976) argued that this method also had drawbacks, since thorium leaks from detrital phases during the acid leach pro-

cess. Hence, it is concluded that normalizing to ^{232}Th can sometimes improve ^{230}Th data, but sometimes has a degrading effect. Therefore, it tends to be used on an *ad hoc* empirical basis.

12.3.5 ^{230}Th Sediment Stratigraphy

In view of the difficulties described above, the ^{230}Th dating method should probably be regarded as semi-quantitative in most circumstances. However, ^{230}Th data may be a powerful tool for stratigraphic correlation of Quaternary sediments. An example of this application is provided by the study of Scholten *et al.* (1990) on a 5 m core from the Norwegian Sea near Jan Mayen (Fig. 12.12). In general, excess ^{230}Th activity data from this study fitted an average decay curve equivalent to a sedimentation rate of 1.9 cm/ka. This is in reasonable agreement with the rate of 1.6 cm/ka calculated from oxygen

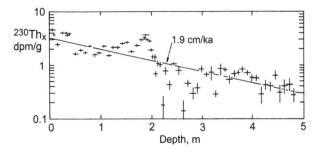

Fig. 12.12 Plot of excess ^{230}Th activity (on a log scale) against depth in core 23059 from the Norwegian Sea. Regression line indicates average sedimentation rate. After Scholten *et al.* (1990).

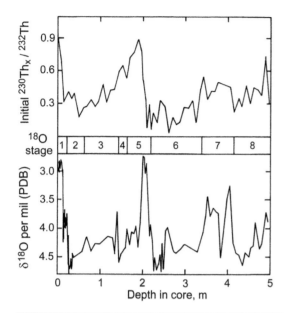

Fig. 12.13 Comparison of the depth dependence of excess initial ^{230}Th/^{232}Th and δ^{18}O in core 23059. Numbered intervals are stages based on ^{18}O stratigraphy. Stages 1 and 5 represent the Holocene and the 120–130 ka interglacials respectively. After Scholten *et al.* (1990).

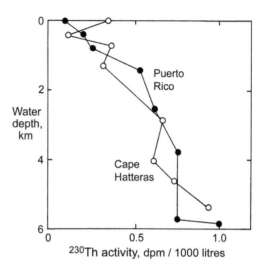

Fig. 12.14 Plot of total ^{230}Th activity as a function of depth in waters off Cape Hatteras (o) and north of Puerto Rico (•). After Cochran *et al.* (1987).

isotope stratigraphy. However, the data display large short-term variations superimposed on the mean decay curve.

Traditionally, variations of this type have been attributed to changes in sedimentation rate. However, this is clearly impossible for some segments of core 23059, which define a positive slope of excess activity against depth (opposite to the effect of radioactive decay). In order to examine these short-term activity variations, Scholten *et al.* corrected the data for radioactive decay since burial (using the mean decay curve), and then ratioed these *initial* (excess) ^{230}Th activities against ^{232}Th to correct for variable carbonate contents. The resulting values display variations with depth which are correlated with δ^{18}O (Fig. 12.13). Scholten *et al.* attributed these variations to the influence of climatic factors on ^{230}Th deposition rate. Climatic changes affect the productivity of plankton, and hence the amount of sinking organic matter relative to total particulate sediment.

Biogenic particle fluxes were argued by Mangini and Diester-Haass (1983) to control the downward flux of radionuclides off NW Africa, and hence ^{230}Th activity variations in sediment cores. Therefore, Scholten *et al.* argued that the low initial excess ^{230}Th/^{232}Th activity ratios in isotope stages 2 and 6 (Fig. 12.13) were due to a widespread reduction of biogenic paleo-productivity during these cold periods. This regional climatic control of radionuclide deposition allows the opportunity of correlating ^{230}Th variations between different sites in an ocean system. This approach can be further developed by comparing ^{230}Th activities with other uranium

series isotopes such as ^{231}Pa (Section 12.3.6), and with the cosmogenic isotope ^{10}Be (Section 14.4.2).

The rapid adsorption of ^{230}Th onto particulate matter makes it a very useful oceanographic tracer. Hence, several studies have been directed at understanding its behaviour in seawater, including its ocean residence time. The activity of ^{230}Th in North Atlantic seawater was determined by Cochran *et al.* (1987) by pumping large volumes of seawater, at different depths, through a filter system designed to scavenge ^{230}Th. Two profiles showed increasing activity with depth, both on particulates and in solution (Fig. 12.14). High levels of dissolved ^{230}Th at depth were attributed to attainment of sorption equilibrium between particulates and seawater. In addition, riverine supply of ^{230}Th causes slight enrichment in shallow seawater off Cape Hatteras, but makes a negligible contribution to the total ^{230}Th inventory.

Yu *et al.* (1996) used these results to make a new estimate of the ^{230}Th residence time (τ) in North Atlantic seawater. This value can be determined from the ^{230}Th inventory per unit volume of water (n = activity / λ_{230}), divided by the supply flux per unit volume. Since riverine supply of ^{230}Th is considered insignificant, the supply flux is equal to oceanic ^{234}U decay. Hence, in terms of activities:

$$\tau_{\text{Th}-230} = \frac{1}{\lambda_{230}} \cdot \frac{^{230}\text{Th}}{^{234}\text{U}} \qquad [12.17]$$

Based on profiles of activity against depth, Yu *et al.* estimated an average ^{230}Th activity of 0.65 d.p.m./m^3 in the North Atlantic at 25 °N. This compares with a ^{234}U activity of 2700 d.p.m./m^3 which is constant throughout the oceans due to the long residence time of uranium. Plugging these values into the above equation gave a τ value of only 26 yr, much shorter than previously estimated.

12.3.6 ^{231}Pa–^{230}Th

Similarities in the chemistry of Pa and Th prompted Sackett (1960) and Rosholt *et al.* (1961) to suggest their use in conjunction as a dating tool. Three factors suggested that the adsorbed initial ^{230}Th/^{231}Pa activity ratio should be a constant (~11) defined by the radiogenic production ratio of the two species. Firstly, the isotope ratio of their parents is relatively constant in seawater (as demonstrated by the concordance of ^{231}Pa and ^{230}Th dates); secondly, they are both adsorbed rapidly compared to their half-lives; and thirdly, the direct river-borne contribution of ^{231}Pa and ^{230}Th to the oceans is negligible (Scott, 1968). In this case, equation [12.8] can be divided by the corresponding equation for protactinium, yielding

$$\left(\frac{^{230}\text{Th}}{^{231}\text{Pa}}\right)^{\text{excess}}_P = \left(\frac{^{230}\text{Th}}{^{231}\text{Pa}}\right)^{\text{excess}}_I e^{-(\lambda 230 - \lambda 231)t} \qquad [12.18]$$

Equation [12.17] can then be solved for t by assuming the initial (production) ratio to be 11. The early work of Sackett (1960) and Rosholt *et al.* (1961) appeared to bear out the assumption. However, subsequent work has yielded variable excess ^{230}Th/^{231}Pa activities at the sediment surface. Sediments often have surface ratios much higher than 11 (e.g. Sackett, 1964), while manganese nodules may have ratios much lower than 11 (e.g. Sackett, 1966). Hence, it is concluded that variable fractionation between ^{231}Pa and ^{230}Th occurs during sedimentation, rendering the method useless as a dating tool.

The variable fractionation between ^{231}Pa and ^{230}Th can now be explained by the different seawater residence times of these species. Because of its extremely particle-reactive behaviour, very little ^{230}Th can be transported laterally (advected) before it is scavenged and sedimented. In contrast, ^{231}Pa can be advected by ocean currents before it is scavenged in locations with a high sinking particle flux. As a result, ^{230}Th/^{231}Pa ratios vary across ocean basins, normally with high ratios in the centre of the basin, where sedimentation rates are low, and low ratios near the margins, where sedimentation rates are high (Yang *et al.*, 1986).

Yu *et al.* (1996) proposed that the different seawater residence times of Pa and Th allow their activity ratios to be used as monitors of ocean circulation. The present-day Atlantic Ocean is dominated by a 'conveyer belt' which transports North Atlantic Deep Water (NADW) southwards to the Antarctic. This phemonenon is also called the Atlantic meridional overturning circulation (AMOC), because northward-moving shallow water becomes more saline and sinks in the North Atlantic before flowing southward again as deep water. Radiocarbon evidence (Section 14.2.1) suggests that NADW has a residence time of 200–300 yr in the Atlantic. Comparison of this value with the ocean residence times of ^{231}Pa and ^{230}Th indicates that about 50% of ^{231}Pa, but only 10% of ^{230}Th produced in the Atlantic will be exported to the Southern Ocean.

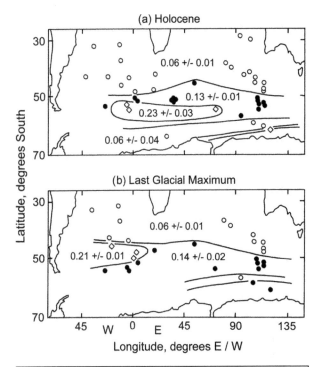

Fig. 12.15 ^{231}Pa/^{230}Th activity ratios for the Southern Ocean at the present day (a); and the last glacial maximum (b). (○) <0.1; (●) 0.1–0.2; (◇) >0.2. After Yu *et al.* (1996).

These predictions were supported by activity measurements on (recent) core tops from ocean floor sediment (Fig. 12.15). These data are presented in the reciprocal form (^{231}Pa/^{230}Th), and show an average activity ratio of only 0.06 in the Atlantic, but 0.17 in the Southern Ocean (relative to a production ratio of 0.09). Yu *et al.* made the critical observation that sediments deposited at the time of the last glacial maximum had exactly the same distribution pattern of ^{231}Pa/^{230}Th activity ratios (Fig. 12.15b) as present day sediments. From this observation they concluded that the ocean conveyor belt operated at a very similar rate during the glacial maximum. This result cast doubt on the widely favoured model in which the conveyer belt was thought to have partially or completely ceased during the last glacial maximum (Section 14.2.2).

Yu *et al.* attributed the high ^{231}Pa/^{230}Th activities in the southern ocean to increased sedimentation in this area, including biogenic opal precipitation, which is particularly effective at scavenging protactinium. These observations have focussed much attention on the influence of boundary scavenging on Pa/Th ratios, especially involving opal precipitation (e.g. Walter *et al.*, 1997; Luo and Ku, 1999; Chase *et al.*, 2002). However, other workers have emphasized that in order to assess changes in global ocean circulation, it is more profitable to examine time series at open-ocean localities far from shore-lines, such as mid-ocean ridges or islands.

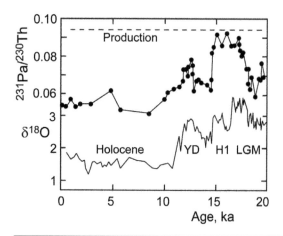

Fig. 12.16 Comparison of $\delta^{18}O$ and $^{231}Pa/^{230}Th$ activity ratios over the past 20 ka from a sediment core at 4.5 km depth on the Bermuda Rise. After McManus *et al.* (2004).

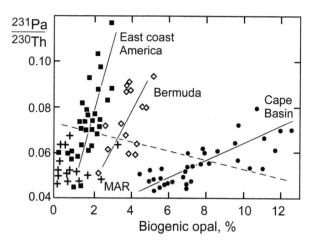

Fig. 12.17 Plot of $^{231}Pa/^{230}Th$ activity ratio against sedimentary opal content for cores from different areas of the Atlantic Ocean, to explore the possible effects of biogenic productivity on Pa/Th signals. Modified after Lippold *et al.* (2016).

Pa/Th deposition fluxes in the North Atlantic region are likely to be the most sensitive to variation in the AMOC. Therefore, to test for short-term variation in the Pa/Th deposition flux, McManus *et al.* (2004) analysed a sediment core covering the past 20 ka from 4.5 km depth on the Bermuda Rise. They found that $^{231}Pa/^{230}Th$ activity ratios during the last glacial maximum (LGM) were moderately reduced relative to the present day (Fig. 12.16), indicative of a ca. 30% reduction in intensity of the AMOC. However, short-period cooling events during deglaciation (most notably the Heinrich event from 17 to 15 ka BP) exhibited sharp increases in $^{231}Pa/^{230}Th$ activity to values near the production ratio of 0.093. This suggested an almost complete shut-down of meridional overturning during the Heinrich event, in response to a catastrophic release of melt-water into the North Atlantic.

This interpretation was challenged by Keigwin and Boyle (2008), based on ^{14}C ventilation ages for the Bermuda Rise (Section 14.2.2), which showed similar water ages for the Heinrich event and the last glacial maximum. Hence, they argued that the Atlantic meridional overturn could not have diminished to the degree proposed by McManus *et al.* (2004) during the Heinrich event, and that the unusually high $^{231}Pa/^{230}Th$ activity ratios for this period must reflect enhanced biogenic opal precipitation. This argument was supported by Lippold *et al.* (2009), who found strong historical correlations between $^{231}Pa/^{230}Th$ activity ratios and diatom abundances in another core from the Bermuda Rise. This suggests that even at an open-ocean site, preferential Pa sedimentation was induced by biogenic opal precipitation, possibly due to an influx of silica-rich water from elsewhere in the Atlantic basin. Therefore, the Pa/Th activity time-series may not be due only to ocean circulation, but may reflect a combination of circulation and biogenic opal precipitation.

Various strategies have been developed to deal with this problem, of which the most basic is to sample from sedimentation sites with low biological productivity, where opal precipitation is not expected to be as significant. The opal flux

can then be monitored, to see if there is any correlation with $^{231}Pa/^{230}Th$ activity ratios. However, this does not necessarily yield straightforward results, as demonstrated by results presented in Fig. 12.17 by Lippold *et al.* (2016). If all of these data are combined, the best-fit regression (dashed line) has a negative slope with very low correlation coefficient. However, several individual cores yield positive regressions with moderately high correlation coefficients, suggesting that there is a relationship between $^{231}Pa/^{230}Th$ activity ratio and sediment opal content. However, it has been suggested that the opal content of many cores may have been advected from elsewhere in the oceans by the very ocean currents under investigation (Deng *et al.*, 2014). Therefore, these positive correlations may not invalidate $^{231}Pa/^{230}Th$ activity ratios as monitors of ocean circulation.

Evidence that the $^{231}Pa/^{230}Th$ data for the Bermuda Rise core may genuinely reflect water body activity ratios (and hence ocean circulation) was provided by a very similar signal from another North Atlantic core (Gherardi *et al.*, 2009). These authors used several North Atlantic sample localities (with different water depths) in an attempt to explore $^{231}Pa/^{230}Th$ time sections as a function of ocean-wide water depth. The results for three time slices, the Holocene, LGM and Heinrich event (H1) are compared in Fig. 12.18. Relative to the production ratio of 0.093, a lower Pa/Th ratio indicates a water body from which Pa is being more rapidly exported by currents. Hence, the Holocene Pa/Th activity profile is consistent with rapid export of NADW from the North Atlantic from 3.5–4.5 km depth. The glacial pattern is almost the opposite, showing that water was being strongly exported at shallower depths (2–3 km, termed Glacial North Atlantic Intermediate Water, GNAIW), but is largely stagnant below 3.5 km depth. Finally, the H1 pattern shows more modest Pa/Th values at intermediate depths, and high Pa/Th ratios

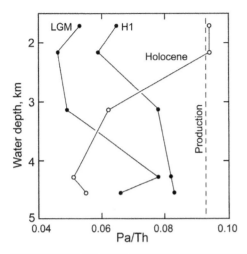

Fig. 12.18 Comparison of ^{231}Pa/^{230}Th activity ratios as a function of water depth in the North Atlantic during the Holocene and LGM. Modified after Gherardi *et al.* (2009).

at all depths below 3 km, consistent with a significant shutdown of the AMOC at this time, as proposed by McManus *et al.* (2004).

To study the operation of AMOC in the South Atlantic, Negre *et al.* (2010) compared Pa/Th activity in a South Atlantic core in the middle of present day southward flow of NADW, with the average of two North Atlantic cores with a similar history of ^{231}Pa/^{230}Th activity (SW Ireland and Bermuda Rise). Negre *et al.* observed almost the opposite temporal trend in ^{231}Pa/^{230}Th activity ratios in the South Atlantic core compared with the two North Atlantic cores (Fig. 12.19). Since ^{231}Pa/^{230}Th activity ratio is believed to increase along the direction of current flow due to the greater oceanic residence

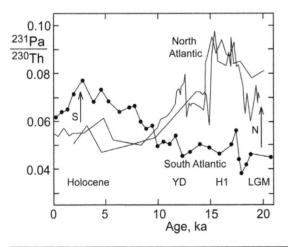

Fig. 12.19 Comparison of South Atlantic (•) and North Atlantic (—) ^{231}Pa/^{230}Th activity ratios against time for the past 20 ka. S and N arrows indicate direction of current flow. Modified after Negre *et al.* (2010).

time of protactinium, Negre *et al.* argued that relative differences in Pa/Th activity ratio across ocean basins could be used to infer the direction of current flow (from low to high Pa/Th ratio). Hence they argued that deep water flow in the Atlantic was largely reversed from the LGM to the Holocene.

This is something of an over-simplification, because the two North Atlantic cores were from much greater depths (4.28 and 4.55 km) than the South Atlantic core (2.44 km). Therefore, these activity ratios cannot be compared directly to infer the direction of north–south water transport without considering the depth ranges of the different water currents. However, it does appear that Southern Component Water reached much further north during the LGM than at the present day, so that the deep North Atlantic may have been fed by intrusion of Antarctic Bottom Water (AABW). At the same time, shallower North Atlantic cores suggest very strong export of GNAIW (Gherardi *et al.*, 2009), and yet this Northern Component Water appears to lose its influence in the Southern Ocean (based on Nd isotope data, Section 4.5.5). Therefore, the water bathing the African margin core may indeed have been changed from north-flowing Southern Component water during the LGM to south-flowing NADW during the Holocene.

Evidence in support of this model was provided by another core from 3 km depth near the centre of the South Atlantic (Jonkers *et al.*, 2015). Because the site analysed in this study is in the centre of a subtropical gyre with low biogenic productivity, it is believed to be unaffected by opal precipitation. Nd isotope signatures in this core record a change in the water body bathing the core site from Southern Component Water during the LGM to NADW in the Holocene. This change is accompanied by a change in ^{231}Pa/^{230}Th activity ratio very similar to that on the African margin. Hence the change from north-flowing Circumpolar Water to south-flowing NADW appears to have been widely experienced in the South Atlantic during deglaciation.

The present state of understanding was summarized by Bohm *et al.* (2015), as shown in Fig. 12.20. This presents a N–S longitudinal section of the Atlantic Ocean in a more stylized form than was shown in Fig. 4.39. At the present day, NADW flow reaches as deep as 4.5 km in the North Atlantic, before rising above intruding AABW (Fig. 12.20a). During the LGM, AABW intruded all of the way north along the bottom of the Atlantic, pushing the southward flow of NADW to shallower depths (Fig. 12.20b). During the Heinrich event (H1), NADW flow may have been significantly attenuated, but probably not completely turned off, as implied in Fig. 12.20c.

12.3.7 ^{210}Pb

Within the ^{238}U decay chain, the daughter product of ^{226}Ra is the rare gas ^{222}Rn. This escapes into the atmosphere from the whole land surface. However, ^{222}Rn has a half-life of only three days, and is followed by four intermediates with half-lives of minutes to seconds, ultimately yielding longer-lived ^{210}Pb. This is estimated to remain in the upper atmosphere for a few days, before the majority returns to earth in precipitation. Thereafter, unsupported ^{210}Pb decays away

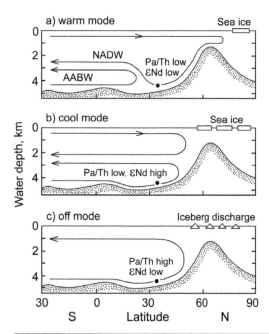

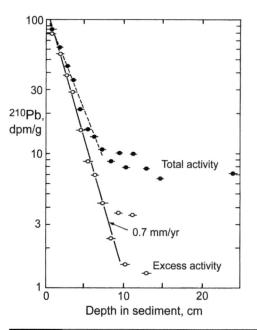

Fig. 12.20 Alternative modes of operation of the AMOC at (a) the present day; (b) during the LGM; (c) an extreme interpretation of the Heirich event. After Bohm *et al.* (2015).

Fig. 12.21 Plot of ^{210}Pb activity against depth in recent sediments from the Santa Monica marine basin. (●) = total ^{210}Pb activity, including ^{226}Ra-supported fraction. (○) = excess ^{210}Pb only. After Bruland *et al.* (1974).

with a half-life of 22.3 yr. The use of ^{210}Pb was first suggested as a tool to date snow accumulation by Goldberg (1963). However, it can also be used to date very recent fresh-water and marine sedimentation (e.g. Krishnaswamy *et al.*, 1971; Koide *et al.*, 1972) because ^{210}Pb has an aqueous residence time of only a year or two before adsorption onto sediment.

If the ^{210}Pb concentration in newly precipitated snow or sediment remains more or less constant with time at a given locality (as expected), then the system will behave exactly the same as the ^{230}Th excess method. We can then use ^{210}Pb activity at the present day surface to determine initial ^{210}Pb, and solve for the age of a buried ice or sediment sample:

$$^{210}\text{Pb} = {}^{210}\text{Pb}_{\text{initial}}\, e^{-\lambda\,210\,t} \qquad [12.19]$$

Then (as with ^{230}Th) if we plot the log of ^{210}Pb activity against depth, the slope yields the sedimentation rate. The first application of the method was to snow chronology (Crozaz *et al.*, 1964). The calculated sedimentation rate of snow at the South Pole in water equivalents (6 ± 1 cm/yr) compared well with a rate determined from yearly 'ice varves'.

The short half-life of ^{210}Pb also makes it ideally suited to the dating of historical-age sediments. For example, the method has become an important tool in studying the history of heavy metal pollution of coastal waters and lakes. Bruland *et al.* (1974) used the method in a study of metal pollution of the Santa Monica basin off Los Angeles. A log plot of total ^{210}Pb activity against depth (Fig. 12.21) yields a linear fit at shallow depths, but the profiles flatten out at ca. 8 cm depth due to the effect of ^{210}Pb supported by ^{226}Ra. However,

this can be corrected by subtracting ^{226}Ra activity, yielding excess ^{210}Pb activities:

$$^{210}\text{Pb}_{\text{excess}} = {}^{210}\text{Pb}_{\text{total}} - {}^{226}\text{Ra} \qquad [12.20]$$

When the data are plotted in this form, the usable range of the method is extended to ca. 150 yr. For the Santa Monica basin the corrected (excess) ^{210}Pb data yield a sedimentation rate of 0.7 mm/yr (Fig. 12.21).

A particularly appropriate application of the ^{210}Pb method is to studies of anthropogenic Pb contamination of sediments. Shirahata *et al.* (1980) applied the method to a remote sub-alpine pond in Yosemite National Park, in order to assess the regional atmospheric fallout of Pb from car exhausts. A sedimentation rate of 0.6 mm/yr was calculated from ^{210}Pb data. Bioturbation of the sediment was ruled out because all bomb-produced radionuclides remained within ca. 2 cm of the sediment surface. Total Pb concentrations in the sediment were found to increase four-fold over the past 100 years, and this change was accompanied by a change in ^{206}Pb/^{207}Pb ratio from a natural local value of 1.15 to an exotic value of 1.2. The latter was typical of the sources of Pb ore used in the United States for the manufacture of leaded petrol (Section 5.6.1).

In spite of these achievements with the ^{210}Pb method, caution must be exercised in the interpretation of data, since some studies (e.g. Santschi *et al.*, 1983; Benoit and Hemond, 1991) have shown that ^{210}Pb can be re-mobilized from the surfaces of sediment grains into sediment pore-waters, and thence into the overlying water column. A recent review of ^{210}Pb accumulation and dating models is given by Sanchez-Cabeza and Ruiz-Fernandez (2012).

12.4 Daughter-Deficiency Methods

By far the dominant daughter-deficiency dating method is ^{230}Th. The protactinium method has similar behaviour, but due to its low abundance, and hence low precision, it is usually used as a closed-system test for ^{230}Th. The radium method will not be discussed here.

12.4.1 ^{230}Th: Theory

The tendency described above for thorium adsorption onto clay minerals leads to low Th levels in groundwaters, in contrast to their moderate U levels. Thus, when biogenic or authigenic calcite is formed, it tends to contain appreciable U concentrations (a few ppm) but negligible Th. This leads to a situation where ^{230}Th is strongly deficient relative to its parent, ^{234}U. The subsequent regeneration of ^{230}Th can then be used as a dating tool.

The first application of this '^{230}Th deficiency' technique was made as early as 1926 by Khlapin, who used short-lived ^{226}Ra as a measure of ^{230}Th activity. Khlapin assumed that the ^{234}U parent taken up by calcite was itself in secular equilibrium with ^{238}U, and that Th uptake was negligible. Under these conditions, we can treat ^{230}Th production from ^{234}U as if it were derived directly from ^{238}U. To calculate net ^{230}Th accumulation, we must then subtract the fraction which has decayed to ^{226}Ra. Substituting into the relevant Bateman equation [1.13], the abundance (not activity) of ^{230}Th after time t is given as follows:

$$n^{230}\mathrm{Th} = \frac{\lambda_{238}}{\lambda_{230} - \lambda_{238}} \cdot n^{238}\mathrm{U_I} \left(e^{-\lambda_{238} t} - e^{-\lambda_{230} t} \right) \quad [12.21]$$

where I signifies the initial ratio. But these abundances may be easily converted into activities by dividing by the relevant decay constants:

$$\frac{^{230}\mathrm{Th}}{\lambda_{230}} = \frac{\lambda_{238}}{\lambda_{230} - \lambda_{238}} \cdot \frac{^{238}\mathrm{U_I}}{\lambda_{238}} \left(e^{-\lambda_{238} t} - e^{-\lambda_{230} t} \right) \quad [12.22]$$

Now, cancelling λ_{238} and multiplying both sides by λ_{230}:

$$^{230}\mathrm{Th} = \frac{\lambda_{230}}{\lambda_{230} - \lambda_{238}} \cdot {^{238}\mathrm{U_I}} \left(e^{-\lambda_{230} t} - e^{-\lambda_{230} t} \right) \quad [12.23]$$

However, because of the very long half-life of ^{238}U relative to the other species, its activity is effectively constant over time. Therefore, ^{238}U initial activity can be approximated by ^{238}U, $e^{-\lambda_{238} t}$ is approximately 1, and $\lambda_{230} - \lambda_{238}$ is approximately λ_{230}, which then cancels to yield

$$^{230}\mathrm{Th} = {^{238}\mathrm{U}} \left(1 - e^{-\lambda_{230} t} \right) \quad [12.24]$$

Finally, dividing through by ^{238}U activity yields the decay equation which can be used for dating:

$$\frac{^{230}\mathrm{Th}}{^{238}\mathrm{U}} = 1 - e^{-\lambda_{230} t} \quad [12.25]$$

However, it was noted above that ^{234}U and ^{238}U activities in natural waters are very rarely in secular equilibrium. This introduces a complication into the decay equation, since there is an extra contribution to ^{230}Th activity by excess ^{234}U, until the latter has decayed away. ^{230}Th production by excess ^{234}U (X) is given by an equation analogous to [12.23]:

$$^{230}\mathrm{Th_X} = \frac{\lambda_{230}}{\lambda_{230} - \lambda_{234}} \cdot {^{234}\mathrm{U_I^X}} \left(e^{-\lambda_{234} t} - e^{-\lambda_{230} t} \right) \quad [12.26]$$

But excess ^{234}U activities can only conveniently be measured as a ratio against ^{238}U. Therefore, we divide both sides of equation [12.26] by ^{238}U activity. This is effectively constant over time due to its long half-life, so that present and initial ^{238}U activities are interchangeable:

$$\left(\frac{^{230}\mathrm{Th}}{^{238}\mathrm{U}} \right)^X = \frac{\lambda_{230}}{\lambda_{230} - \lambda_{234}} \cdot \left(\frac{^{234}\mathrm{U}}{^{238}\mathrm{U}} \right)_I^X \left(e^{-\lambda_{234} t} - e^{-\lambda_{230} t} \right) \quad [12.27]$$

But the excess activity ratio is equal to the total activity ratio minus one (corresponding to secular equilibrium). So:

$$\left(\frac{^{230}\mathrm{Th}}{^{238}\mathrm{U}} \right)^X = \frac{\lambda_{230}}{\lambda_{230} - \lambda_{234}} \cdot \left[\left(\frac{^{234}\mathrm{U}}{^{238}\mathrm{U}} \right)_I - 1 \right] \cdot \left(e^{-\lambda_{234} t} - e^{-\lambda_{230} t} \right) \quad [12.28]$$

We can substitute equation [12.6] into this equation in order to convert initial 234/238 activities to the present day measured activities (P):

$$\left(\frac{^{230}\mathrm{Th}}{^{238}\mathrm{U}} \right)^X = \frac{\lambda_{230}}{\lambda_{230} - \lambda_{234}} \cdot \left[\left(\frac{^{234}\mathrm{U}}{^{238}\mathrm{U}} \right)_P - 1 \right] \cdot \frac{\left(e^{-\lambda_{234} t} - e^{-\lambda_{230} t} \right)}{e^{-\lambda_{234} t}} \quad [12.29]$$

But the final term simplifies to yield

$$\left(\frac{^{230}\mathrm{Th}}{^{238}\mathrm{U}} \right)^X = \frac{\lambda_{230}}{\lambda_{230} - \lambda_{234}} \cdot \left[\left(\frac{^{234}\mathrm{U}}{^{238}\mathrm{U}} \right)_P - 1 \right] \cdot \left(1 - e^{-(\lambda_{230} - \lambda_{234}) t} \right) \quad [12.30]$$

Finally, adding the ^{230}Th production from equilibrium and excess ^{234}U (equations [12.25] and [12.30]), we obtain

$$\frac{^{230}\mathrm{Th}}{^{238}\mathrm{U}} = 1 - e^{-\lambda_{230} t} + \frac{\lambda_{230}}{\lambda_{230} - \lambda_{234}}$$
$$\cdot \left[\frac{^{234}\mathrm{U}}{^{238}\mathrm{U}} - 1 \right] \cdot \left(1 - e^{-(\lambda_{230} - \lambda_{234}) t} \right) \quad [12.31]$$

This equation could be used directly to solve ages, but it has become normal procedure to rearrange it by dividing through by ^{234}U/^{238}U. This yields

$$\frac{^{230}\mathrm{Th}}{^{234}\mathrm{U}} = \frac{1 - e^{-\lambda_{230} t}}{^{234}\mathrm{U}/^{238}\mathrm{U}} + \frac{\lambda_{230}}{\lambda_{230} - \lambda_{234}}$$
$$\cdot \left[1 - \frac{1}{^{234}\mathrm{U}/^{238}\mathrm{U}} \right] \cdot \left(1 - e^{-(\lambda_{230} - \lambda_{234}) t} \right) \quad [12.32]$$

This equation was plotted as an 'isochron' diagram (Fig. 12.22), by Kaufman and Broecker (1965). Technically, it is a true isochron diagram, since the near-vertical curves are lines of equal age. However, this is a misleading term for

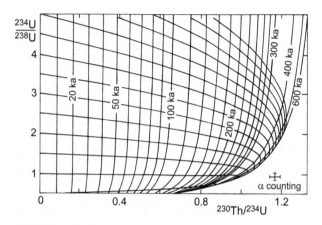

Fig. 12.22 Th–U diagram for systems containing no ^{232}Th. Labelled, steeply dipping lines are isochrons; lateral lines are growth lines. Error bar shows typical uncertainty for α spectrometry. After Kaufman and Broecker (1965).

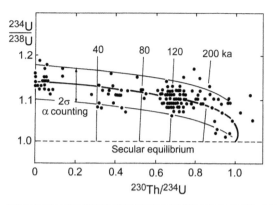

Fig. 12.23 Plot of ^{234}U/^{238}U versus ^{230}Th/^{234}U activity ratio for un-recrystallized coral data with 2σ errors better than 5%. Heavy near-horizontal curve shows the decay path starting from present day seawater. Vertical lines are isochrons. After Veeh and Burnett (1982).

most geologists, who are more familiar with the classical 'Rb–Sr type' of isochron, which will be applied to U–Th dating below.

Effectively, the calibration line for ^{234}U/^{238}U activity = 1 (secular equilibrium) yields the age in terms of ^{230}Th build-up, while the isochron lines apply the correction for non-equilibrium U isotope compositions. As can be seen on the diagram, this correction is unnecessary for samples less than ca. 30 ka in age. The maximum dating range of the ^{230}Th method is ca. 300 ka by α counting, but this may be extended to over 400 ka by mass spectrometry.

12.4.2 ^{230}Th: Applications

The ^{230}Th–^{234}U method is applicable to the dating of any closed-system carbonate which is free from contamination by initial detrital thorium. It can provide far better precision for coral dating than the ^{234}U–^{238}U method alone. This is illustrated in Fig. 12.23 by a compilation of high-precision α spectrometry data for un-recrystallized corals (Veeh and Burnett, 1982). It can be seen that typical measurement errors in ^{234}U/^{238}U ratio lead to age uncertainties of over 100%, whereas errors in ^{230}Th/^{234}U lead to age errors of only 10%.

The precision of ^{230}Th coral dating has been further enhanced by the mass spectrometric method, as demonstrated by the analysis of live reef-forming corals from the Vanuatu arc, east of Australia (Edwards et al., 1988). In these specimens, ^{230}Th ages were compared with historical ages based on yearly growth bands. The latter are about 1 cm wide, and can be accurately counted in specimens of at least 200 years old. ^{230}Th ages were determined with errors as low as ±3 yr (2σ), and were in excellent agreement with the historical age of the corals. ^{230}Th dating of corals between 9000 and 40 000 yr old has been used very effectively to calibrate the radiocarbon timescale (Section 14.1.6).

One of the most important applications of ^{230}Th dating is in the study of Pleistocene (Quaternary age) climatic varia-

tions associated with glacial cycles. These cycles caused periodic variations in the global ice mass at the expense of seawater, and therefore left two types of record. The first is the direct variation in sea level through time, while the second is an indirect record of sea level variations due to the fractionation of oxygen isotopes as different amounts of seawater were converted into ice. These oxygen isotope variations can be seen in marine and terrestrial carbonates, as well as in ice records. The ^{230}Th (daughter-deficiency) method can be used to date two of these climatic records: direct sea level variations, and oxygen isotope variations recorded in terrestrial carbonates. These dates can then be used to test the third main record of Quaternary climatic variation, the so-called SPECMAP record of marine oxygen isotope variations.

The SPECMAP model (Imbrie et al., 1984) attributes Pleistocene glacial cycles, recorded as the oxygen isotope variations in marine forams, to variations in the intensity of solar radiation. This is based on the theory of Milankovitch (1941), who suggested that changes in Pleistocene climate were largely due to changes of 'insolation' in the Northern Hemisphere, caused by variations in the Earth's orbit. Because the Earth's orbital variations can be precisely calculated and projected back in time, it is possible to model insolation variations back through several million years. Then, if the Milankovitch theory is correct, these orbital variations can be used to 'tune' the climatic record of oxygen isotope variations in order to date glacial cycles precisely (termed 'Milankovitch forcing').

Reef-building corals represent a useful record of past sea level variations, because sea level highs during interglacial periods become marked by coral terraces which are stranded when the sea level falls again. In general there is a good correlation between sea level highs and Northern Hemisphere insolation, providing support for the Milankovitch theory. However, dating the onset of the last interglacial, marked by a high sea level stand (stage 5e in the stable isotope record)

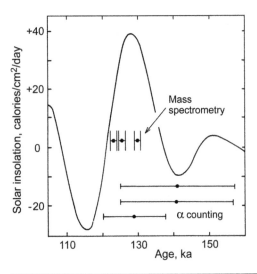

Plot of solar insolation against time, to compare analytical uncertainties for mass spectrometric and α counting U-series ages for coral terraces from the last interglacial (stage 5e). After Edwards *et al.* (1987).

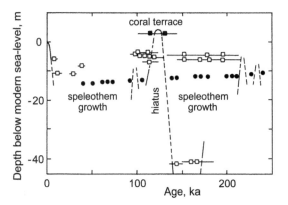

Pleistocene sea level curve for the Bahamas (dashed lines), based on U-series ages on drowned speleothems. (•) = mass spectrometric data. Squares = older speleothem (□) and coral terrace ages (■) by α spectrometry. After Li *et al.* (1989).

has always been problematical. Early α counting ages for stage 5e were spread over a very large range, from 120 to 140 ka BP, due to the large α-counting error bars for events of this age (Fig. 12.24). Mass spectrometric dates on the same samples were in all three cases within 2σ errors of the α-counting determinations, but were centred unequivocally on the insolation high at 127 ka (Edwards *et al.*, 1987).

Another approach for constraining Pleistocene sea level variations is ^{230}Th dating of speleothem (stalactites, stalagmites etc.) from submarine caves. These formations grow during periods of low sea level stand, when they are exposed subaerially to percolating calcareous solutions. When the sea level rises and they become drowned, growth stops, and an erosional hiatus is formed. The densely crystalline form of speleothem deposits is conducive to good closed-system behaviour, so this material is ideal for U-series dating. Therefore, drowned speleothem and coral terraces form a complementary couple for Pleistocene sea level studies. The first mass spectrometric dating study on such material was made by Li *et al.* (1989) on a sample from 12 m depth in a Bahamas 'Blue Hole'. A detailed sequence of U-series age determinations on the 12 cm-thick flowstone showed carbonate deposition over 280 ka (Fig. 12.25). Within this period, there were four internal hiatuses corresponding to sea level stands above −12 m (relative to present day sea level). These data supported the orbitally tuned SPECMAP record in suggesting that the last interglacial began *after* 140 ka BP.

Speleothems can also be used as inventories of past atmospheric oxygen isotope variations, which are also linked to glacial cycles. These records are well suited to dating by the ^{230}Th (daughter-deficiency) method. Hence, climatic variations can be dated by measuring δ^{18}O signatures and U-

series ages on the same cave deposits. For example, Winograd *et al.* (1992) made a combined U-series and stable isotope study on the calcite lining of a water-filled cavern in Nevada (USA) called Devils Hole. The results of this study suggested similar glacial cycles to the SPECMAP oceanic record, but the warming trend associated with the last interglacial (at around 140 ka) appeared to precede the increase in solar insolation which should have driven this warming. This led Winograd *et al.* to question the SPECMAP model.

Similar arguments were made by Henderson and Slowey (2000), who determined U–Th ages on fine-grained carbonate sediments from a Bahamas core, whose stratigraphy was determined by oxygen isotope analysis. Many corrections were required for these samples, including firstly, a washing technique to remove detrital grains containing common thorium; secondly, use of the U–Th isochron technique (Section 12.4.3) to correct for common thorium adsorbed from seawater; and thirdly a recoil correction to compensate for radiogenic thorium loss from the small sediment grains due to the hot atom effect (Section 12.3.3). The calculated ages must also be corrected for bioturbation, which can contaminate the dated horizons with older sediment particles from lower in the section. After these corrections, Henderson and Slowey placed the mid-point of the penultimate deglaciation at 135 ± 2.5 ka BP, 8 ka older than the SPECMAP value. Hence, they also challenged the Milokovitch model for the penultimate deglaciation.

These arguments show that dating the penultimate deglaciation continues to be an important test case for the integrity of the Milankovitch model. Hence, this event has been studied in several new age determinations, based on both coral terraces and core samples. For example, Rohling *et al.* (2008) made a new determination of the oxygen isotope record for the last interglacial using a new core sample from the Red Sea, which is particularly sensitive to climate cycles. They concluded that the penultimate deglaciation

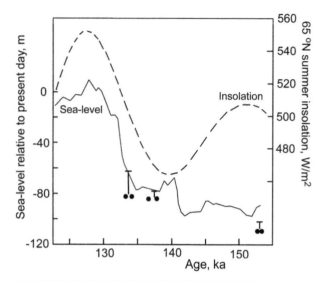

Fig. 12.26 Evidence for the onset of the last interglacial, based on a $\delta^{18}O$ record from the Red Sea and the best Tahiti coral terrace ages (•). Vertical brackets indicate possible coral growth depths. Data from Thomas et al. (2009) and Grant et al. (2012).

the depth uncertainty slightly, but the most robust dates from the penultimate glaciation agree very well with the revised sea level curve based on the Red Sea core (Fig. 12.26). Significantly, the Tahiti data suggest that sea level remained more than 70 m below its present level until 133 ka, well after the rise in northern hemisphere insolation. However, evidence from an older 153 ka terrace, coupled with the Red Sea record, suggests that a much smaller (20 m) rise in sea level occurred at around 10 ka before the main sea level rise. This early event cannot be attributed to variations in northern hemisphere insolation, and implies more than one orbital forcing mechanism. The Devils Hole signal may record this earlier event, which was possibly followed by a glacial readvance similar to the Heinrich (H1) event of the last deglaciation (17 ka BP). Subsequently, deglaciation resumed, with the beginning of the last interglacial (*sensu stricto*) at around 130 ka.

12.4.3 ^{230}Th: Dirty Calcite

Because fossil bones may be encased by subsequent tufa deposits, U-series analysis of such material has been very useful for dating Pleistocene human and animal remains (e.g. Schwarcz and Blackwell, 1991). However, the most interesting tufas are often impure, for the very reason that if they contain bones they will probably contain other detrital material. This introduces initial ^{230}Th, which, if not corrected for, may cause serious errors in calculated ages.

In cases where detrital contamination is minor, the same laboratory technique may be used as for clean material: the sample is leached with dilute nitric acid in an attempt to dissolve the carbonate fraction without disturbing the detrital component. This may diminish the contamination to a level where it is swamped by other errors. However, the detrital component is not usually inert in nitric acid, but often contains a certain fraction of loosely bound uranium and thorium, which is removed by the leaching process. The extent of this leakage may be monitored by measuring the activity of ^{232}Th. If this reaches a level of more than a few % of ^{230}Th activity then it may be necessary to correct the carbonate data for leaching of radionuclides from the contaminating detrital phase (Ku and Liang, 1984).

U-series data for dirty calcites are best visualized on an isochron diagram. The most common form involves ratioing both ^{230}Th and ^{234}U against ^{232}Th (Fig. 12.27). If all U and Th isotopes are leached from the residue with equal efficiency then a cord joining the leachate and residue points can be interpreted as an isochron line. The slope will then yield the ^{230}Th/^{234}U ratio of the carbonate component, which can be used to calculate the sample age in the same way as for clean material (Fig. 12.22).

Przybylowicz et al. (1991) performed leaching experiments on artificial mixtures of pure calcite speleothem and mud in order to test the reliability of the leaching method in dating dirty calcites (e.g. Fig. 12.27). The results show that residues are displaced slightly (occasionally substantially) above the array of leachate compositions. This is probably

was caused by orbital forcing, and this conclusion has been strengthened by improvements in dating the Red Sea core by comparison with a very well-dated Mediterranean core (Grant et al., 2012). The latter was itself dated by correlation of $\delta^{18}O$ records with a U-Th dated speleothem from Soreq cave, Israel (Bar-Matthews et al., 2003). The newly calibrated Red Sea record is in almost perfect step with northern hemisphere insolation (Fig. 12.26).

Direct dating support for this new sea level curve is provided by new U-Th ages on coral terraces from Tahiti (Thomas et al., 2009). These coral terraces were formed under glacial conditions with low sea level stands, and were accessed from up to 150 m below present day sea-level by an Ocean Drilling Program expedition. Because Tahiti is subject to slow subsidence and is far from any isostatic rebound effects, this means that dated coral terraces from Tahiti provide precise determinations of actual past sea level stands. Furthermore, the submarine location of the samples renders them less susceptible to open-system behaviour due to percolating groundwater. As a result, half of the analysed samples had calculated initial ^{234}U/^{238}U ratios between +137 and +151 per mil relative to secular equilibrium, regarded as indicative of a pristine seawater signature.

To test the method, samples from the last glacial period were dated, giving ages of around 30 ka BP (just before the LGM). However, after correction for subsidence, these samples were assigned depths ca. 20 m below other well-defined sea level records of this age. Hence, Thomas et al. determined that these corals probably grew in deeper water than the main reef, as part of the fore-reef. This increases

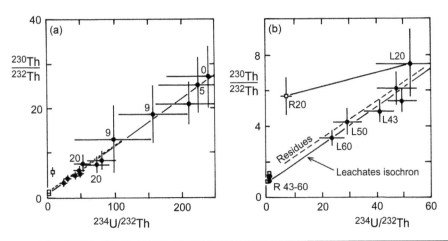

Fig. 12.27 U–Th isochron diagrams showing results from leaching of artificial mixtures of calcite and mud with 5–7 M nitric acid. (•, L) = leachates; (□, R) = residues from leaching; numbers indicate % of mud in the sample. Diagram (b) is a blow-up of the lower left corner of (a). After Przybylowicz *et al.* (1991).

due to slight preferential leaching of uranium relative to thorium from the detrital phase during the leaching process, and may yield apparent ages somewhat below the true value. Schwarcz and Latham (1989) argued that this problem could be diminished by regressing leachate analyses alone. In this case it is no longer necessary to assume a lack of differential isotopic fractionation during the leaching process. Isotopic fractionation is permitted, provided that the amount of such fractionation is the same in all samples. This has the effect of shifting the isochron line sideways in Fig. 12.27b, but not changing its slope. A similar type of correlation diagram can also be used to correct the ^{234}U/^{238}U ratio for detrital contamination. However, this is less important, since the ^{230}Th age is only weakly dependent on the ^{234}U/^{238}U ratio in samples less than 300 ka in age.

One problem with the data presentation in Fig. 12.27 is that the two variables become very highly correlated as the ^{232}Th fraction diminishes in size. Therefore, a regression program utilizing correlated errors should be used to calculate isochron slopes. Similarly, the large error bars should be represented by elongated error ellipses rather then rectangular error boxes. The correct data presentation is shown in Fig. 12.28, involving usage of the 'leach–leach' technique to date contaminated travertines (Schwarcz and Skoflek, 1982). These deposits are an important dating target because they enclose the 'Mousterian cultural layer' at Tata, Hungary, an important deposit of Neanderthal flint tools. Regression of four leachates leads to an age of 101 ± 4 ka for carbonate enclosing the cultural layer, which is bracketed between the ages of 78 ± 5 and 118 ± 37 ka in overlying and underlying clean travertine layers.

In order to achieve a high-precision result from the leach–leach technique it is desirable to leach three or more samples with variable detrital contents from the horizon to be dated. However, under these circumstances it is possible that differ-

ent samples might show variable degrees of isotopic fractionation during leaching. This problem can be avoided by total digestion of a suite of variably contaminated samples from the same deposit. If they all contain the same detrital component (i.e. have the same initial ^{230}Th/^{232}Th ratio), and have remained as closed systems, then they will define a perfect isochron line.

Bischoff and Fitzpatrick (1991) tested the relative performance of the total dissolution, leach–leach and leachate-residue methods on a series of artificial mixtures of natural detritus and carbonate. (They also tested the effect of

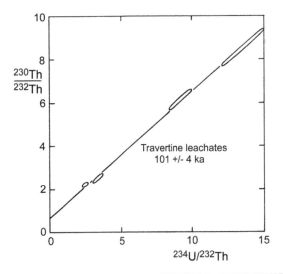

Fig. 12.28 ^{230}Th/^{232}Th versus ^{234}U/^{232}Th isochron diagram for leachates of contaminated travertine from Tata, Hungary. Ellipses portray correlated error limits. Modified after Schwarcz and Latham (1989).

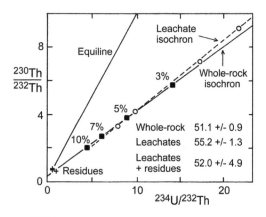

Fig. 12.29 U–Th isochron diagram showing tests of different dating approaches on artificial mixtures of carbonate and detritus. (■) = whole-rock (total dissolution); (○) = leachate; (+) = residue. After Bischoff and Fitzpatrick (1991).

leaching with different acid strengths.) Typical results showed that the total dissolution method was superior to the other two techniques for artificial mixtures (Fig. 12.29). The total dissolution method is also more versatile, in that it can be applied to the dating of any type of Pleistocene material with homogeneous initial ratio and closed-system behaviour (Luo and Ku, 1991). However, it may be difficult to obtain a large enough range of detrital–carbonate variations to define a good regression line. Furthermore, in the pursuit of such a range of mixtures, samples with variable initial ratio may be analysed. Therefore, the total dissolution, leach–leach and leach–residue methods may all be viable alternatives for dating dirty calcite in different circumstances.

An alternative isochron presentation, utilized by Kaufman (1971) but not shown here, involves plotting ^{230}Th/^{234}U against ^{232}Th/^{234}U. On this 'Th–U' isochron diagram, the age of the sample is represented by the intercept on the y axis, and increasing detrital contamination is indicated by displacement away from the y axis. Similar plots have been used more widely in U-series studies of silicate systems (Section 13.1.3). Th/U isochrons offer some advantages over the more popular U/Th isochron presentation because Th/U isochrons minimize the problem of error correlation between the variables.

Another important constraint in dating contaminated travertines is the principle that samples should yield ages consistent with their stratigraphic location in the deposit. Hellstrom (2006) emphasized the use of this method to constrain U-Th isochron ages in cases where the detrital contaminant has a variable isotopic composition, which tends to add statistical noise to the age sequence. This method was recently applied by Roy-Barman and Pons-Branchu (2016) to U-Th dating series in contaminated speleothems.

12.4.4 ^{231}Pa

The build-up of ^{231}Pa in carbonates can be used as a dating tool in a way analogous to ^{230}Th. The immediate parent of ^{231}Pa (^{231}Th) is assumed to be always in equilibrium with its parent (^{235}U) due to its short half-life of 26 hours. Hence, the age relationship is analogous to the simple form of equation [12.26] for ^{230}Th build-up:

$$\frac{^{231}\text{Pa}}{^{235}\text{U}} = 1 - e^{-\lambda_{231} t} \qquad [12.33]$$

Because ^{235}U is so much less abundant than ^{238}U, the abundance of ^{231}Pa or its daughters is twenty times lower than for ^{230}Th. Since counting statistics on minor isotopes are the major source of uncertainty in U-series dating, the ^{230}Th method has been much preferred to ^{231}Pa as a practical dating tool. However, the ^{231}Pa technique is potentially valuable as a concordancy test for ^{230}Th dates, and the two systems can be used together to date partially open systems (Section 12.5.1).

Mass spectrometry potentially offers the same advantages for ^{231}Pa analysis as for ^{230}Th, involving an order of magnitude improvement in precision over α counting (Edwards *et al.*, 1997). However, a problem encountered in this work is the lack of any long-lived nuclides for use as spike isotopes. Apart from ^{231}Pa, with a half-life of 32.76 ka, the second most long-lived isotope of protactinium is ^{233}Pa, with a half-life of 27.4 days. This must be prepared anew every few months and must be repeatedly calibrated so that its concentration is known on the exact day (!) of its analysis.

Two methods of preparing ^{233}Pa are in use. The first, described by Pickett *et al.* (1994), involves periodically 'milking' ^{233}Pa from a solution of ^{237}Np. This is done by ion exchange chemical separation, but it must be done under a stringent radiological protection regime, because the parent isotope is very highly active. An alternative method, described by Bourdon *et al.* (1999), is much less hazardous. This involves periodic neutron activation of ^{232}Th in a reactor, to produce ^{233}Th. This short-lived species (half-life = 22 minutes) then decays into ^{233}Pa. Because the parent (^{232}Th) has low activity, the ion exchange purification can be done under normal lab conditions, after allowing short-lived by-products of the activation process to die away.

12.5 U-Series Dating of Open Systems

There has already been some discussion of open systems in ^{230}Th dating, mainly concerning the dating of corals and other material that are directly deposited from seawater. In this section we are concerned with the coupled use of ^{230}Th and other dating methods to form a 'concordia' method analogous to U–Pb dating.

12.5.1 ^{231}Pa–^{230}Th

Because the ^{231}Pa and ^{230}Th (daughter-deficiency) dating methods share a common parent element (uranium), they

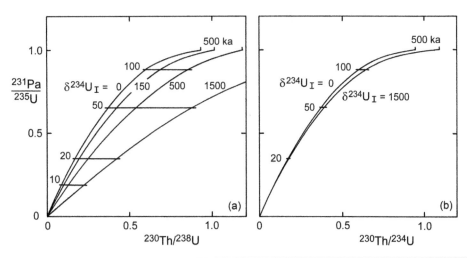

Fig. 12.30 Concordia diagrams for the U–Th–Pa system: (a) against ^{238}U; and (b) against ^{234}U. Concordia curves are shown for a variety of δ^{234}U values. After Cheng et al. (1998).

form a single U–Th–Pa system which is analogous to the U–Pb dating system. This similarity was first discussed by Allegre (1964), who showed that the U–Th–Pa system can be studied using the same 'concordia' diagram used for U–Pb zircon dating (Section 5.2). A combination of U–Th and U–Pa analyses can be used as a concordancy test to check for closed-system behaviour. This approach was adopted by Chiu et al. (2006) in the dating of fossil corals, as an alternative to the use of ^{234}U as a closed-system monitor (Section 12.3.1). However, the U–Th–Pa method can in principle be used to date systems that have been partially opened during their history, as for the U–Pb method.

Because the U–Th–Pa system involves three different elements, rather than two, it is necessarily more complex than the U–Pb system. However, this does not have a large effect on the method in practice, because the daughter nuclides Th and Pa are essentially immobile, thus behaving as if they were the same element. On the other hand, the principal cause of open-system behaviour is the mobility of uranium, which is the common element between the two systems.

In spite of the fact that the U–Th and U–Pa sub-systems have the same parent element, the U–Th–Pa system is also more complex than U–Pb because the parent uranium isotopes (^{234}U and ^{235}U) can undergo fractionation in the environment (Section 12.3.1). As a result, the U–Th–Pa system actually gives rise to a family of curves in a concordia diagram. This complication was ignored in the analysis of Allegre (1964) and in later studies by Kaufman and Ku (1989) and Kaufman et al. (1995), but was treated in the more detailed discussion of Cheng et al. (1998). As a result, two different concordia diagrams can be plotted, one involving the ultimate parent (^{238}U, Fig. 12.30a) and one involving the immediate parent (^{234}U, Fig. 12.30b). On each of these diagrams, concordia curves can then be shown for different 234/238 activities, presented in the form of δ^{234}U. For reference, it

should be remembered that δ^{234}U = 0 signifies secular equilibrium, while δ^{234}U = 150 is the uranium isotope composition of seawater, corresponding to a 234/238 activity ratio of 1.15.

The final complicating factor in the U–Th–Pa system, relative to U–Pb, is the fact that the daughter products are themselves radioactive. Indeed, this is actually the feature which makes the combined system useful for dating partially open systems in the age range 0–200 ka. It is the return of ^{230}Th and ^{231}Pa activities into secular equilibrium with their parents, at different rates, which allows age information to be recovered from partially open systems. As a result, when the concordia diagram is plotted (Fig. 12.30), the curvature is in the opposite sense to that of the U–Pb concordia, with the longer-lived nuclide (^{230}Th) along the x axis and the shorter lived (^{231}Pa) on the y axis.

Uranium gain is probably the most important type of open-system behaviour in U–Th–Pa dating, and causes the data points to move towards the origin in a similar way to Pb loss in the U–Pb system. The simplest scenario is probably a single episode of U gain. This forms a linear array through the origin at the time of open-system behaviour, and the array then rotates as the system ages under closed-system conditions. The result at the present time is a discordia array similar to a U–Pb discordia (Fig. 12.31). The upper intercept with the concordia line then gives the true age of the system. However, the lower intercept only gives the age of U gain for the special case where δ^{234}U = 0 (secular equilibrium of parent uranium isotopes).

The case of continuous open-system uranium gain is more complex, and gives rise to a curved 'discordia' array. However, as in the case of U–Pb dating, the upper end of this continuous U-gain line is relatively straight, and can give a fairly precise upper intercept age. A special case of this scenario is 'linear uptake' of uranium through the life of the system,

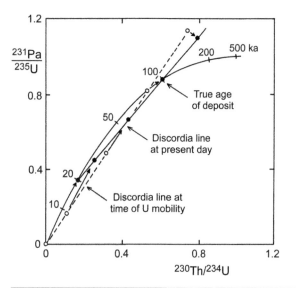

Fig. 12.31 ^{234}U–Pa–Th plot showing the evolution of a discordia line generated by a single episode of uranium mobility. After Cheng et al. (1998).

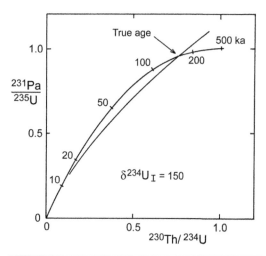

Fig. 12.32 U–Pa–Th plot showing a discordia line generated by continuous uranium gain. After Cheng et al. (1998).

which is shown on the ^{234}U–Pa–Th diagram in Fig. 12.32. As for U–Pb dating, the best estimate for the upper intercept age is obtained by having a suite of samples with variable degrees of U gain, but with some samples near the concordia (minor late U gain). However, a minimum estimate of the upper intercept age can be calculated by dividing the U–Pa equation [12.34] by the U–Th equation [12.33], as suggested by Ivanovich (1982b). The resulting Pa–Th age is analogous to a 207/206 lead age.

12.5.2 ESR–^{230}Th

Some of the most interesting U-series dating applications involve human bones and cultural deposits, but these materials are notorious for their open-system behaviour of U-series isotopes, as well as their small sample size. In the past, the ages of this type of material were based on speleothem deposits which pre-dated and post-dated a 'cultural' layer (e.g. Schwarcz, 1989). However, with the advent of mass spectrometric U-series analysis, ^{230}Th ages can be used in combination with electron spin resonance (ESR) to obtain reliable dates for teeth (Grun et al., 1988; Grun and McDermott, 1994). Human teeth are not dated directly, but human bones are often found in deposits with large numbers of bovoid teeth (from the cow family). The relatively large size of these teeth provides sufficient material for ESR and U-series dating.

Both ^{230}Th (daughter-deficiency) and ESR dating are based on U-series isotopes, but in different ways. ESR measures the accumulation of trapped electrons at defect sites, caused by the time-integrated radiation dose to the sample, mainly derived from ^{238}U (for reviews of ESR dating, see Jonas (1997) and Rink (1997)). On the other hand, ^{230}Th dating measures the return of this nuclide to secular equilibrium with ^{238}U,

via its daughter product ^{234}U. In both cases, a dating signal is derived largely from a common parent (^{238}U), and in both cases the dating signal itself is largely immobile (trapped electrons versus ^{230}Th nuclides). On the other hand, the dating signal has a different half-life in the two methods: infinite for trapped electrons but 72.5 ka for ^{230}Th. Therefore, the combined ESR–^{230}Th method can be used to date systems partially open to uranium mobility in a similar way to the U–Th–Pa dating system.

Actually, ESR–^{230}Th dates are based on a system that is almost completely open to uranium, because the entire uranium inventory of a tooth is acquired after deposition (live teeth have essentially no uranium). However, experience has shown that the U-uptake history of a tooth can be approximated by three types of model, termed early uptake (EU), linear uptake (LU) and recent uptake (RU), as shown in Fig. 12.33.

Of these three models, early uptake is the best scenario for dating because it approximates a case of closed-system evolution, whereby the age of the tooth is essentially the same as the uptake event that occurred soon after burial. In this case, the ESR and U-series (^{230}Th) age will be concordant. On the other hand, a recent uptake scenario is the worst for dating because the U-series method cannot see back any further than the recent uranium enrichment event. ESR dating can only give a useful date in these circumstances if the majority of the radiation dose experienced by the tooth comes from the sedimentary environment *outside* the tooth. Linear uptake is the approximation which represents all scenarios between these two extremes. It generates relatively large uncertainties, but may be susceptible to analysis if a large enough data set is available. In detail, the uptake history of three uranium reservoirs must be considered: tooth enamel, tooth dentine and surrounding sediment. The ESR measurement is made on the enamel and dentine, and its

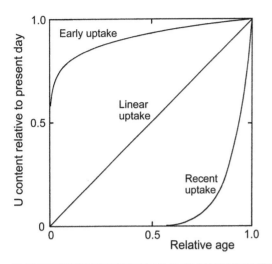

Fig. 12.33 Schematic illustration of different uptake models experienced by buried teeth.

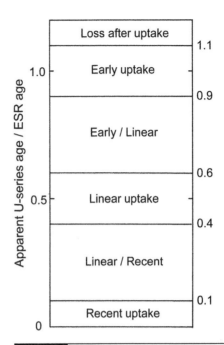

Fig. 12.35 Categories of uranium uptake model based on the ratio between apparent U-series and ESR ages (provisionally assuming early uptake for both systems). After Rink *et al.* (2001).

signal is a time-integrated function of the radiation dose from the three reservoirs.

An example of age concordance between U-series analyses and ESR ages is provided by a study of bovoid dental fragments associated with the skeletons of early modern humans (Fig. 12.34). In this case, agreement between most of the U-series and ESR dates indicated that uranium uptake occurred soon after deposition. Hence the U-series ages confirmed ESR dates for the appearance of early modern humans at least 100 ka BP in Israel (McDermott *et al.*, 1993). Two points in Fig. 12.34 (samples a and b) display markedly lower U-series ages

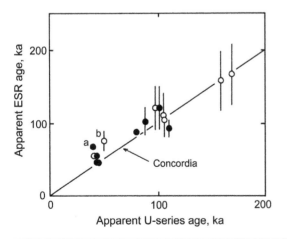

Fig. 12.34 Plot of apparent ESR and U-series ages for enamel (•) and dentine (○) samples of bovoid teeth from Israel, assuming an early uptake model. Error bars are 1σ. After Grun and McDermott (1994).

relative to their ESR ages, outside the limits of error. These are indicative of models approximated by recent and linear uptake respectively (Grun and McDermott, 1994).

Grun *et al.* (1988) proposed a complex parametric calculation to model the uptake history of a tooth based on the relative ESR and U-series ages of open-system samples. They demonstrated the effectiveness of this technique on samples from Hoxne, England, which have very young U-series ages relative to their ESR ages, and are therefore indicative of relatively recent U uptake. However, Grun and McDermott (1994, p. 123) admitted that, 'The mathematical formulation of these steps is very complex and we have not yet been able to establish a rigorous error-calculation procedure'. Therefore, for practical purposes, a simpler method used by Rink *et al.* (2001) may be adequate. These authors categorized the uptake model (Fig. 12.35) based on the ratio between the U-series age and the ESR age calculated from an early uptake model. They then interpolated between the different uptake models to get a best estimate of the age of the tooth.

An alternative approach might be to plot a concordia diagram of the ^{230}Th/^{234}U ratio against the ESR age. The upper intercept of this concordia diagram should give the approximate age of the tooth, in a similar way to the U–Th–Pa and U–Pb concordia diagrams. Unfortunately, the quality of experimental data has not yet been adequate to demonstrate this approach.

References

Allegre, C. (1964). De l'extension de la methode de calcul graphique concordia aux mesures d'ages absolus effectues a l'aide du desequilibre radioactif. *C. R. Acad. Sci.* **259**, 4086–9.

Andersen, M. B., Stirling, C. H., Porcelli, D. *et al.* (2007). The tracing of riverine U in Arctic seawater with very precise $^{234}U/^{238}U$ measurements. *Earth Planet. Sci. Lett.* **259**, 171–85.

Bard, E., Fairbanks, R. G., Hamelin, B., Zindler, A. and Hoang, C. T. (1991). Uranium–234 anomalies in corals older than 150,000 years. *Geochim. Cosmochim. Acta* **55**, 2385–90.

Bar-Matthews, M., Ayalon, A., Gilmour, M., Matthews, A. and Hawkesworth, C. J. (2003). Sea–land oxygen isotopic relationships from planktonic foraminifera and speleothems in the Eastern Mediterranean region and their implication for paleorainfall during interglacial intervals. *Geochim. Cosmochim. Acta* **67**, 3181–99.

Bender, M. L., Fairbanks, R. G., Taylor, F. W. *et al.* (1979). Uranium-series dating of the Pleistocene reef tracts of Barbados, West Indies. *Geol. Soc. Amer. Bull.* **90**, 577–94.

Benoit, G. and Hemond, H. F. (1991). Evidence for diffusive redistribution of ^{210}Pb in lake sediments. *Geochim. Cosmochim. Acta* **55**, 1963–75.

Bischoff, J. L. and Fitzpatrick, J. A. (1991). U-series dating of impure carbonates: an isochron technique using total-sample dissolution. *Geochim. Cosmochim. Acta* **55**, 543–54.

Bohm, E., Lippold, J., Gutjahr, M. *et al.* (2015). Strong and deep Atlantic meridional overturning circulation during the last glacial cycle. *Nature* **517**, 73–6.

Bourdon, B., Joron, J.-L. and Allegre, C. J. (1999). A method for ^{231}Pa analysis by thermal ionization mass spectrometry in silicate rocks. *Chem. Geol.* **157**, 147–51.

Bruland, K. W., Bertine, K., Koide, M. and Goldberg, E. D. (1974). History of metal pollution in Southern California coastal zone. *Envir. Sci. Tech.* **8**, 425–32.

Chabaux, F., Cohen, A. S., O'Nions, R. K. and Hein, J. R. (1995). $^{238}U-^{234}U-^{230}Th$ chronometry of Fe–Mn crusts: growth processes and recovery of thorium isotopic ratios of seawater. *Geochim. Cosmochim. Acta* **59**, 633–8.

Chabaux, F., O'Nions, R. K., Cohen, A. S. and Hein, J. R. (1997). $^{238}U-^{234}U-^{230}Th$ disequilibrium in hydrogenous oceanic Fe–Mn crusts: palaeoceanographic record or diagenetic alteration? *Geochim. Cosmochim. Acta* **61**, 3619–32.

Chase, Z., Anderson, R. F., Fleisher, M. Q. and Kubik, P. W. (2002). The influence of particle composition and particle flux on scavenging of Th, Pa and Be in the ocean. *Earth Planet. Sci. Lett.* **204**, 215–29.

Chen, J. H., Edwards, R. L. and Wasserburg, G. J. (1986). ^{238}U, ^{234}U and ^{232}Th in seawater. *Earth Planet. Sci. Lett.* **80**, 241–51.

Cheng, H., Edwards, R. L., Hoff, J. *et al.* (2000). The half-lives of uranium-234 and thorium-230. *Chem. Geol.* **169**, 17–33.

Cheng, H., Edwards, R. L., Murrell, M. T. and Benjamin, T. M. (1998). Uranium thorium protoactinium dating systematics. *Geochim. Cosmochim. Acta* **62**, 3437–52.

Cheng, H., Edwards, R. L., Shen, C. C. *et al.* (2013). Improvements in ^{230}Th dating, ^{230}Th and ^{234}U half-life values, and U-Th isotopic measurements by multi-collector inductively coupled plasma mass spectrometry. *Earth Planet. Sci. Lett.* **371**, 82–91.

Cherdyntsev, V. V. (1969). Uranium–234. Atomizdat, Moskva. Translated by Schmorak, J. Israel Prog. Sci. Trans. 1971, 234 pp.

Cherdyntsev, V. V., Kazachevskii, I. V. and Kuz'mina, E. A. (1965). Dating of Pleistocene carbonate formations by the thorium and uranium isotopes. *Geochem. Int.* **2**, 794–801.

Cherdyntsev, V. V., Orlov, D. P., Isabaev, E. A. and Ivanov, V. I. (1961). Isotopic composition of uranium in minerals. *Geochemistry* **10**, 927–36.

Chiu, T. C., Fairbanks, R. G., Mortlock, R. A. *et al.* (2006). Redundant $^{230}Th/^{234}U/^{238}U$, $^{231}Pa/^{235}U$ and ^{14}C dating of fossil corals for accurate radiocarbon age calibration. *Quaternary Sci. Rev.* **25**, 2431–40.

Cochran, J. K., Livingston, H. D., Hirschberg, D. J. and Surprenant, L. D. (1987). Natural and anthropogenic radionuclide distributions in the northwest Atlantic Ocean. *Earth Planet. Sci. Lett.* **84**, 135–52.

Crozaz, G., Picciotto, E. and DeBreuck, W. (1964). Antarctic snow chronology with Pb-210. *J. Geophys. Res.* **69**, 2597–604.

de Bievre, P., Lauer, K. F., Le Duigou, Y. *et al.* (1971). In: Hurrell, M. L. (Ed.) *Proc. Int. Conf. Chem. Nucl. Data, Inst. Civil Eng. Lond.*, pp. 221–5.

Deng, F., Thomas, A. L., Rijkenberg, M. J. and Henderson, G. M. (2014). Controls on seawater ^{231}Pa, ^{230}Th and ^{232}Th concentrations along the flow paths of deep waters in the Southwest Atlantic. *Earth Planet. Sci. Lett.* **390**, 93–102.

Edwards, R. L., Chen, J. H. and Wasserburg, G. J. (1987). $^{238}U-^{234}U-^{230}Th-^{232}Th$ systematics and the precise measurement of time over the past 500,000 years. *Earth Planet. Sci. Lett.* **81**, 175–92.

Edwards, R. L., Cheng, H., Murrell, M. T. and Goldstein, S. J. (1997). Protactinium-231 dating of carbonates by thermal ionization mass spectrometry: implications for Quaternary climate change. *Science* **276**, 782–6.

Edwards, R. L., Taylor, F. W. and Wasserburg, G. J. (1988). Dating earthquakes with high-precision thorium-230 ages of very young corals. *Earth Planet. Sci. Lett.* **90**, 371–81.

Esat, T. M. and Yokoyama, Y. (2006). Variability in the uranium isotopic composition of the oceans over glacial–interglacial timescales. *Geochim. Cosmochim. Acta* **70**, 4140–50.

Esat, T. M. and Yokoyama, Y. (2010). Coupled uranium isotope and sea-level variations in the oceans. *Geochim. Cosmochim. Acta* **74**, 7008–20.

Gallup, C. D., Edwards, R. L. and Johnson, R. G. (1994). The timing of high sea levels over the past 200,000 years. *Science* **263**, 796–9.

Gherardi, J. M., Labeyrie, L., Nave, S. *et al.* (2009). Glacial–interglacial circulation changes inferred from $^{231}Pa/^{230}Th$ sedimentary record in the North Atlantic region. *Paleoceanog.* **24** (PA2204), 1–14.

Goldberg, E. D. (1963). Geochronology with Pb-210. In: *Radioactive Dating and Methods of Low-Level Counting, Proc. Symp. I.A.E.A.*, Vienna, pp. 121–31.

Goldberg, E. D. and Bruland, K. (1974) Radioactive geochronologies. In: Goldberg, E. D. (Ed.) *The Sea.* vol. **5**, Wiley Interscience, pp. 451–89.

Goldberg, E. D. and Koide, M. (1962). Geochronological studies of deep sea sediments by the ionium/thorium method. *Geochim. Cosmochim. Acta* **26**, 417–50.

Grant, K. M., Rohling, E. J., Bar-Matthews, M. *et al.* (2012). Rapid coupling between ice volume and polar temperature over the past 150,000 years. *Nature* **491**, 744–7.

Grun, R. and McDermott, F. (1994). Open system modelling for U-series and ESR dating of teeth. *Quaternary Geochron. (Quaternary Sci. Rev.)* **13**, 121–1.

Grun, R., Schwarcz, H. P. and Chadham, J. (1988). ESR dating of tooth enamel: coupled correction for U-uptake and U-series disequilibrium. *Nucl. Tracks Radiat. Meas.* **14**, 237–41.

Hellstrom, J. (2006). U-Th dating of speleothems with high initial ^{230}Th using stratigraphical constraint. *Quaternary Geochron.* **1**, 289–95.

Henderson, G. M. (2002). Seawater $(^{234}U/^{238}U)$ during the last 800 thousand years. *Earth Planet Sci. Lett.* **199**, 97–110.

Henderson, G. M. and Burton, K. W. (1999). Using $(^{234}U/^{238}U)$ to assess diffusion rates of isotopic tracers in ferromanganese crusts. *Earth Planet. Sci. Lett.* **170**, 169–79.

Henderson, G. M. and O'Nions, R. K. (1995). $^{234}U/^{238}U$ ratios in Quaternary planktonic foraminifera. *Geochim. Cosmochim. Acta* **59**, 4685–94.

Henderson, G. M. and Slowey, N. C. (2000). Evidence from U–Th dating against Northern Hemisphere forcing of the penultimate deglaciation. *Nature* **404**, 61–6.

Imbrie, J., Hays, J. D., Martinson, D. G. *et al.* (1984). The orbital theory of Pleistocene climate: support from a revised chronology of the marine $\delta^{18}O$ record. In: Berger, A. L. *et al.* (Eds) *Milankovitch and Climate*, Part 1. Reidel, pp. 269–305.

Ivanovich, M. (1982a). Spectroscopic methods. In: Ivanovich, M. and Harmon, R. S. (Eds) *Uranium Series Disequilibrium Applications to Environmental Problems*. Oxford University Press, pp. 107–44.

Ivanovich, M. (1982b). Uranium series disequilibria applications in geochronology. In: Ivanovich, M. and Harmon, R. S. (Eds) *Uranium Series Disequilibrium Applications to Environmental Problems*. Oxford University Press, pp. 56–78.

Jonas, M. (1997). Concepts and methods of ESR dating. *Radiation Meas.* **27**, 943–73.

Jonkers, L., Zahn, R., Thomas, A. *et al.* (2015). Deep circulation changes in the central South Atlantic during the past 145 kyrs reflected in a combined $^{231}Pa/^{230}Th$, neodymium isotope and benthic record. *Earth Planet. Sci. Lett.* **419**, 14–21.

Kaufman, A. (1971). U-series dating of Dead Sea carbonates. *Geochim. Cosmochim. Acta* **35**, 1269–81.

Kaufman, A. and Broecker, W. S. (1965). Comparison of Th-230 and C-14 ages for carbonate materials from lakes Lahontan and Bonneville. *J. Geophys. Res.* **70**, 4039–54.

Kaufman, A., Broecker, W. S., Ku, T. L. and Thurber, D. L. (1971). The status of U-series methods of mollusc dating. *Geochim. Cosmochim. Acta* **35**, 1155–83.

Kaufman, A. and Ku, T.-L. (1989). The U-series ages of carnotites and implications regarding their formation. *Geochim. Cosmochim. Acta* **53**, 2675–81.

Kaufman, A., Ku, T.-L. and Luo, S. (1995). Uranium-series dating of carnotites: concordance between $^{230}Th–^{231}Pa$ ages. *Chem. Geol. (Isot. Geosci. Sect.)* **120**, 175–81.

Keigwin, L. D. and Boyle, E. A. (2008). Did North Atlantic overturning halt 17,000 years ago? *Paleoceanog.* **23** (PA1101), 1–5.

Khlapin, V. G. (1926). *Dokl. Akad. Nauka SSSR* 178.

Koide, M., Soutar, A. and Goldberg, E. D. (1972). Marine geochronology with Pb-210. *Earth Planet. Sci. Lett.* **14**, 442–6.

Krishnaswamy, S., Lal, D., Martin, J. M. and Meybek, M. (1971). Geochronology of lake sediments. *Earth Planet. Sci. Lett.* **11**, 407–14.

Ku, T. L. (1965). An evaluation of the U^{234}/U^{238} method as a tool for dating pelagic sediments. *J. Geophys. Res.* **70**, 3457–74.

Ku, T. L. (1976). The uranium series methods of age determination. *Ann. Rev. Earth Planet. Sci.* **4**, 347–79.

Ku, T. L., Bischoff, J. L. and Boersma, A. (1972). Age studies of Mid-Atlantic Ridge sediments near 42 °N and 20 °N. *Deep-Sea Res.* **19**, 233–47.

Ku, T. L., Knauss, K. G. and Mathieu, G. G. (1977). Uranium in open ocean: concentration and isotopic composition. *Deep-Sea Res.* **24**, 1005–17.

Ku, T. L. and Liang, Z. C. (1984). The dating of impure carbonates with decay-series isotopes. *Nucl. Instr. Meth. in Phys. Res. A* **223**, 563–71.

Li, W. X., Lundberg, J., Dickin, A. P. *et al.* (1989). High-precision mass-spectrometric uranium-series dating of cave deposits and implications for palaeoclimate studies. *Nature* **339**, 534–6.

Lippold, J., Grutzner, J., Winter, D *et al.* (2009). Does sedimentary $^{231}Pa/^{230}Th$ from the Bermuda Rise monitor past Atlantic meridional overturning circulation? *Geophys. Res. Lett.* **36** (L12601), 1–6.

Lippold, J., Gutjahr, M., Blaser, P. *et al.* (2016). Deep water provenance and dynamics of the (de) glacial Atlantic meridional overturning circulation. *Earth Planet. Sci. Lett.* **445**, 68–78.

Ludwig, K. R., Simmons, K. R., Szabo, B. J. *et al.* (1992). Mass-spectrometric $^{230}Th–^{234}U–^{238}U$ dating of the Devils Hole calcite vein. *Science* **258**, 284–7.

Ludwig, K. R., Szabo, B. J., Moore, J. G. and Simmons, K. R. (1991). Crustal subsidence rate off Hawaii determined from $^{234}U/^{238}U$ ages of drowned coral reefs. *Geology* **19**, 171–4.

Luo, S. and Ku, T. L. (1991). U-series isochron dating: a generalised method employing total-sample dissolution. *Geochim. Cosmochim. Acta* **55**, 555–64.

Luo, S. and Ku, T.-L. (1999). Oceanic $^{231}Pa/^{230}Th$ ratio influenced by particle composition and remineralization. *Chem. Geol.* **167**, 183–95.

Mangini, A. and Diester-Haass, L. (1983). Excess Th-230 in sediments off NW Africa traces upwelling in the past. In: Suess, A. E. and Thiede, J. (Eds) *Coastal Upwelling*. Plenum. Part A, pp. 455–70.

McDermott, F., Grun, R., Stringer, C. B. and Hawkesworth, C. J. (1993). Mass-spectrometric U-series dates for Israeli Neanderthal/early modern hominid sites. *Nature* **363**, 252–5.

McManus, J. F., Francois, R., Gherardi, J. M., Keigwin, L. D. and Brown-Leger, S. (2004). Collapse and rapid resumption of Atlantic meridional circulation linked to deglacial climate changes. *Nature* **428**, 834–7.

Meadows, J. W., Armani, R. J., Callis, E. L. and Essling, A. M. (1980). Half-life of ^{230}Th. *Phys. Rev. C* **22**, 750–4.

Medina-Elizalde, M. (2013). A global compilation of coral sea-level benchmarks: Implications and new challenges. *Earth Planet. Sci. Lett.* **362**, 310–18.

Milankovitch, M. M. (1941). Canon of insolation and the ice age problem. *Koniglich Serbische Akademie, Beograd.* Translation, Israel Prog. Sci. Trans., Washington D. C.

Neff, U., Bollhofer, A., Frank, N. and Mangini, A. (1999). Explaining discrepant depth profiles of $^{234}U/^{238}U$ and $^{230}Th_{exc}$ in Mn-crusts. *Geochim. Cosmochim. Acta* **63**, 2211–18.

Negre, C., Zahn, R., Thomas, A. L. *et al.* (2010). Reversed flow of Atlantic deep water during the Last Glacial Maximum. *Nature* **468**, 84–8.

Obert, J. C., Scholz, D., Felis, T. *et al.* (2016). $^{230}Th/U$ dating of Last Interglacial brain corals from Bonaire (southern Caribbean) using bulk and theca wall material. *Geochim. Cosmochim. Acta* **178**, 20–40.

Osmond, J. K. and Cowart, J. B. (1982). Ground water. In: Ivanovich, M. and Harmon, R. S. (Eds) *Uranium Series Disequilibrium Applications to Environmental Problems*, Oxford University Press, pp. 202–45.

Picciotto, E. G. and Wilgain, S. (1954). Thorium determination in deep-sea sediments. *Nature* **173**, 632–3.

Pickett, D. A., Murrell, M. T. and Williams, R. W. (1994). Determination of femtogram quantities of protoactinium in geological samples by thermal ionization mass spectrometry. *Anal. Chem.* **66**, 1044–9.

Potter, E. K., Esat, T. M., Schellmann, G. *et al.* (2004). Suborbital-period sea-level oscillations during marine isotope substages 5a and 5c. *Earth Planet. Sci. Lett.* **225**, 191–204.

Potts, P. J. (1987). *Handbook of Silicate Rock Analysis*. Blackie, 602 pp.

Przybylowicz, W., Schwarcz, H. P. and Latham, A. G. (1991). Dirty calcites. 2. Uranium-series dating of artificial calcite–detritus mixtures. *Chem. Geol. (Isot. Geosci. Sect.)* **86**, 161–78.

Rink, W. J. (1997). Electron spin resonance (ESR) dating and ESR applications in Quaternary science and archaeometry. *Radiation Meas.* **27**, 975–1025.

Rink, W. J., Schwarcz, H. P., Lee, H. K. *et al.* (2001). Electron spin resonance (ESR) and thermal ionization mass spectrometric (TIMS) $^{230}Th/^{234}U$ dating of teeth in Middle Paleolithic layers at Amud Cave, Israel. *Geoarchaeology* **16**, 701–17.

Roberts, J., Miranda, C. F. and Muxart, R. (1969). Mesure de la periode du protoactinium-231 par microcalorimetrie. *Radiochim. Acta* **11**, 104–8.

Rohling, E. J., Grant, K., Hemleben, C. H. *et al.* (2008). High rates of sea-level rise during the last interglacial period. *Nature Geosci.* **1**, 38–42.

Rosholt, J. N., Emiliani, C., Geiss, J., Koczy, F. F. and Wangersky, P. J. (1961). Absolute dating of deep-sea cores by the Pa-231/Th-230 method. *J. Geol.* **69**, 162–85.

Rosholt, J. N., Shields, W. R. and Garner, E. L. (1963). Isotopic fractionation of uranium in sandstone. *Science* **139**, 224–6.

Roy-Barman, M. and Pons-Branchu, E. (2016). Improved U-Th dating of carbonates with high initial ^{230}Th using stratigraphical and coevality constraints. *Quaternary Geochron.* **32**, 29–39.

Sackett, W. M. (1960). The protoactinium-231 content of ocean water and sediments. *Science* **132**, 1761–2.

Sackett, W. M. (1964). Measured deposition rates of marine sediments and implications for accumulation rates of extraterrestrial dust. *Ann. N. Y. Acad. Sci.* **119**, 339–46.

Sackett, W. M. (1966). Manganese nodules: thorium-230: protoactinium-231 ratios. *Science* **154**, 646–7.

Sanchez-Cabeza, J. A. and Ruiz-Fernandez, A. C. (2012). ^{210}Pb sediment radiochronology: an integrated formulation and classification of dating models. *Geochim. Cosmochim. Acta* **82**, 183–200.

Santschi, P. H., Li, Y. H., Adler, D. M *et al.* (1983). The relative mobility of natural (Th, Pb and Po) and fallout (Pu, Am, Cs) radionuclides in the coastal marine environment: results from model ecosystems (MERL) and Narragansett Bay. *Geochim. Cosmochim. Acta* **47**, 201–10.

Scholten, J. C., Botz, R., Mangini, A. *et al.* (1990). High resolution ^{230}Th$_{ex}$ stratigraphy of sediments from high-latitude areas (Norwegian Sea, Fram Strait). *Earth Planet. Sci. Lett.* **101**, 54–62.

Scholz, D. and Mangini, A. (2007). How precise are U-series coral ages? *Geochim. Cosmochim. Acta* **71**, 1935–48.

Schwarcz, H. P. (1989). Uranium series dating of Quaternary deposits. *Quaternary Int.* **1**, 7–17.

Schwarcz, H. P. and Blackwell, B. (1991). Archaeological applications. In: Ivanovich, M. and Harmon, R. S. (Eds) *Uranium Series Disequilibrium Applications to Environmental Problems.* 2nd Edn, Oxford University Press, pp. 513–52.

Schwarcz, H. P. and Latham, A. G. (1989). Dirty calcites. 1. Uranium-series dating of contaminated calcite using leachates alone. *Chem. Geol. (Isot. Geosci. Sect.)* **80**, 35–43.

Schwarcz, H. P. and Skoflek, I. (1982). New dates for the Tata, Hungary archaeological site. *Nature* **295**, 590–1.

Scott, M. R. (1968). Thorium and uranium concentrations and isotope ratios in river sediments. *Earth Planet. Sci. Lett.* **4**, 245–52.

Shirahata, H., Elias, R. W., Patterson, C. C. and Koide, M. (1980). Chronological variations in concentrations and isotopic compositions of anthropogenic atmospheric lead in sediments of a remote subalpine pond. *Geochim. Cosmochim. Acta* **44**, 149–62.

Stirling, C. H. and Andersen, M. B. (2009). Uranium-series dating of fossil coral reefs: extending the sea-level record beyond the last glacial cycle. *Earth Planet. Sci. Lett.* **284**, 269–83.

Stirling, C. H., Lee, D.-C., Christensen, J. N. and Halliday, A. N. (2000). High-precision *in situ* ^{238}U–^{234}U–^{230}Th isotopic analysis using laser ablation multiple-collector ICP–MS. *Geochim. Cosmochim. Acta* **64**, 3737–50.

Thompson, W. G., Spiegelman, M. W., Goldstein, S. L. and Speed, R. C. (2003). An open-system model for U-series age determinations of fossil corals. *Earth Planet. Sci. Lett.* **210**, 365–81.

Thomas, A. L., Henderson, G. M., Deschamps, P. *et al.* (2009). Penultimate deglacial sea-level timing from uranium/thorium dating of Tahitian corals. *Science* **324**, 1186–9.

Thurber, D. L., Broecker, W. S., Blanchard, R. L. and Potratz, H. A. (1965). Uranium-series ages of Pacific atoll coral. *Science* **149**, 55–8.

Veeh, H. H. and Burnett, W. C. (1982). Carbonate and phosphate sediments. In: Ivanovich, M. and Harmon, R. S. (Eds) *Uranium Series Disequilibrium Applications to Environmental Problems.* Oxford University Press, 459–80.

Villemant, B. and Feuillet, N. (2003). Dating open systems by the ^{238}U–^{234}U–^{230}Th method: application to Quaternary reef terraces. *Earth Planet. Sci. Lett.* **210**, 105–18.

Walter, H. J., Rutgers van der Loeff, M. M. and Hoeltzen, H. (1997). Enhanced scavenging of ^{231}Pa relative to ^{230}Th in the South Atlantic south of the Polar Front: implications for the use of the ^{231}Pa/^{230}Th ratio as a paleoproductivity proxy. *Earth Planet. Sci. Lett.* **149**, 85–100.

Winograd, I. J. (1990). Dating sea level in caves: comment. *Nature* **343**, 217–8.

Winograd, I. J., Coplen, T. B., Landwehr, J. M. *et al.* (1992). Continuous 500,000-year climate record from vein calcite in Devils Hole, Nevada. *Science* **258**, 284–7.

Yang, H.-S., Nozaki, Y., Sakai, H. and Masuda, A. (1986). The distribution of ^{230}Th and ^{231}Pa in the deep-sea surface sediments of the Pacific Ocean. *Geochim. Cosmochim. Acta* **50**, 81–9.

Yokoyama, Y. and Nguyen H. V. (1980). Direct and non-destructive dating of marine sediments, manganese nodules, and corals by high resolution (γ-ray spectrometry. In: Goldberg, E. D., Horibe, Y. and Saruhashi, K. (Eds) *Isotope Marine Chemistry.* Uchida Rokkaku, Ch. 14.

Yu, E.-F. Francois, R. and Bacon, M. P. (1996). Similar rates of modern and last-glacial ocean thermohaline circulation inferred from radiochemical data. *Nature* **379**, 689–94.

Chapter 13

U-Series Geochemistry of Igneous Systems

U-series dating of sedimentary rocks was discussed in the previous chapter. These isotopes can also be used as dating tools for igneous rocks; however, their application as isotopic tracers is probably more important. The short half-lives of these decay series nuclides makes them ideally suited to studies of magma segregation from the mantle and magma evolution in the crust, since these processes operate over similar time periods. With a half-life of 75.6 ka, ^{230}Th is by far the most important of these geological tracers, and will be the main focus of this chapter. However, attention will also be given to other shorter-lived isotopes used in conjunction with thorium.

Note that all isotopic abundances of U-series nuclides referred to in this chapter are expressed as activities, unless specifically stated to be atomic.

Until recently, all U-series measurements on igneous rocks were made by α spectrometry, as for sedimentary rocks (Section 12.2.1). Following the application of mass spectrometry to U-series dating of carbonates, it was quickly applied to igneous systems (Goldstein *et al.*, 1989). However, analysis of ^{230}Th in silicate rocks is made more difficult by the very large atomic ^{232}Th/^{230}Th ratios. For example, a basalt with a typical Th/U concentration ratio of 4, and in secular equilibrium, will have a ^{232}Th/^{230}Th atomic ratio of 240 000. For a single-sector mass spectrometer with an abundance sensitivity of 1 ppm at 2 a.m.u. (Section 2.4.1), the ^{232}Th peak will then generate a peak tail at mass 230 which is one-quarter of the size of the ^{230}Th peak. Normally, the abundance sensitivity is improved using an energy filter (e.g. Cohen *et al.*, 1992; Section 2.4.1). However, McDermott *et al.* (1993) showed that relatively accurate ^{230}Th data can be obtained without a filter if the exponentially curved baseline shape is carefully interpolated under the ^{230}Th peak.

In addition to ^{230}Th, there are five shorter-lived nuclides in the U and Th decay schemes which may be useful in the study of igneous systems. These were shown in Fig. 12.2, but for convenience they are summarized below. The equilibration times shown (t_{Eq}) represent the maximum useful range of each species, based on the assumption that its activity will be within error of secular equilibrium after five half-lives.

Each arrow represents a decay transition, but only nuclides relevant to volcanic systems are shown.

^{238}U $> > > >$ ^{230}Th $>$ ^{226}Ra $> > > > > >$ ^{210}Pb $> > >$ ^{206}Pb

$t_{1/2}$, yr		75 580	1600		22
t_{Eq}, yr		400 000	8000		100

^{235}U $> >$ ^{231}Pa $> > > > > > > >$ ^{207}Pb

$t_{1/2}$, yr	34 300
t_{Eq}, yr	170 000

^{232}Th $>$ ^{228}Ra $> >$ ^{228}Th $> > > > > >$ ^{208}Pb

$t_{1/2}$, yr	5.77	1.91
t_{Eq}, yr	30	10

Disequilibrium between short-lived nuclei in volcanic rocks was discovered in very early work (Joly, 1909), but has only recently been the subject of detailed study. ^{210}Pb was observed to be out of secular equilibrium with ^{230}Th in ocean island lavas by Oversby and Gast (1968). This demonstrated the occurrence of Th/Pb fractionation within 100 years of eruption. However, the chemistry of Pb is so different from the other nuclides that it is difficult to use in petrogenetic interpretations. It will not be discussed further here. ^{228}Ra and ^{228}Th can be used to measure even shorter-period changes in magma chemistry, but have only rarely been found out of isotopic equilibrium.

Of the other short-lived nuclides, ^{226}Ra has been the most widely used. It has traditionally been measured by radioactive counting (sometimes via its shorter-lived decay products), but despite its short half-life, technical advances have allowed its measurement by mass spectrometry (e.g. Cohen and O'Nions, 1991). This permits ^{226}Ra abundances in the femtogram range (10^{-9} ppm) to be determined to better than 1% precision. ^{231}Pa abundances in igneous rocks are too low to measure by α spectrometry because of the low abundance of the parent isotope, ^{235}U. However, with the application of mass spectrometry, ^{231}Pa measurements are also possible in the femtogram range. Hence this nuclide also shows promise as a useful geochemical and chronological tool (Goldstein *et al.*, 1993).

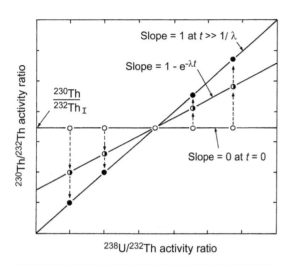

Fig. 13.1 Isotopic evolution of igneous rocks on the ^{230}Th/^{232}Th versus ^{238}U/^{232}Th isochron diagram. Symbols: (○): time $t = 0$; (half-filled): samples after elapsed time t; (●): samples after effectively infinite time ($t \gg 1/\lambda$).

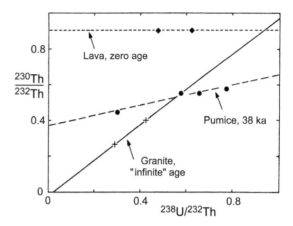

Fig. 13.2 ^{230}Th/^{232}Th versus ^{238}U/^{232}Th isochron diagram for three suites of leachates from whole-rock samples. After Kigoshi (1967).

13.1 Geochronology of Volcanic Rocks

In some ways, U-series systems in igneous rocks are simpler than carbonates, because ^{234}U and ^{238}U are always effectively in secular equilibrium. On the other hand, igneous systems are more complex than pure carbonates in that they invariably contain initial Th at the time of cooling. Hence, a U–Th isochron diagram must normally be used to date igneous rocks by the ^{230}Th method.

13.1.1 The U–Th Isochron Diagram

After time t, the net ^{230}Th activity in a silicate sample is the sum of ^{230}Th growth from U decay and the residue of partially decayed initial ^{230}Th. Thus (reversing the order) we sum equations [12.8] and [12.24] used in sedimentary systems:

$$^{230}\text{Th}_P = {}^{230}\text{Th}_I\, e^{-\lambda 230\, t} + {}^{238}\text{U}\left(1 - e^{-\lambda 230\, t}\right) \quad [13.1]$$

It is convenient to divide through by ^{232}Th, whose activity is effectively constant between t initial and the present:

$$\left(\frac{^{230}\text{Th}}{^{232}\text{Th}}\right)_P = \left(\frac{^{230}\text{Th}}{^{232}\text{Th}}\right)_I e^{-\lambda 230 t} + \frac{^{238}\text{U}}{^{232}\text{Th}}\left(1 - e^{-\lambda 230 t}\right) \quad [13.2]$$

This is the equation for a straight line, and is plotted on a diagram of ^{230}Th/^{232}Th against ^{238}U/^{232}Th (Fig. 13.1) which is analogous to the Rb–Sr isochron diagram. As with the Rb–Sr system, a suite of cogenetic samples of the same age define a linear array whose slope yields the age. However, because the ^{230}Th daughter product is itself subject to decay, this leads to more complex isotope systematics.

The evolution of igneous systems on the U–Th isochron diagram depends on their composition with respect to a state of secular equilibrium. Samples in secular equilibrium must, by definition, have equal ^{230}Th and ^{238}U activities. Hence, they must have equal ^{230}Th/^{232}Th and ^{238}U/^{232}Th activity ratios in Fig. 13.1. Such samples lie on a slope of unity in this diagram, called the 'equiline' by Allegre and Condomines (1976).

Now, considering a suite of rock or mineral samples; at the time of their crystallization they have variable U/Th ratios but a constant (initial) ^{230}Th/^{232}Th activity ratio, forming a horizontal line in Fig. 13.1. The point of intersection of this array with the equiline must by definition remain invariant, since it starts its closed-system evolution already in secular equilibrium. However, all other samples in the suite evolve with time. Those to the right of the invariant point are daughter (^{230}Th) deficient relative to ^{238}U, and ^{230}Th therefore builds up with time. They move vertically upwards until they also reach the equiline (secular equilibrium). Those to the left of the equiline have daughter (^{230}Th) excess relative to the parent. They move vertically downwards with time until they reach the equiline. The more initial thorium that is present, the higher the intersection between the initial composition and the equiline. Conversely, when no initial Th is present (as in pure carbonates), evolution begins along the x axis. Hence, in a cogenetic suite which has not yet reached secular equilibrium, the initial Th isotope ratio is given by the intersection of the isochron array with the equiline. It is not the intercept on the y axis.

The U–Th isochron method was first applied to the dating of igneous minerals by Cerrai *et al.* (1965), and subsequently tested by Kigoshi (1967). Kigoshi used the method to date three igneous rocks of different ages; a Cretaceous granite (effectively of infinite age), a 35.7 ka pumice (dated by ^{14}C on a wood inclusion) and a lava of historical (effectively zero) age. His results (Fig. 13.2) demonstrated the method to be effective. The old granite samples lie on the equiline, the

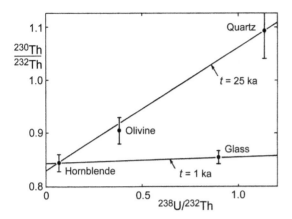

Fig. 13.3 $^{230}Th/^{232}Th$ versus $^{238}U/^{232}Th$ isochron diagram for hornblende–olivine–quartz phenocryst assemblages and glass matrix from a Mono Craters rhyolite (California), showing isotopic discordancy between phenocrysts and glass. After Allegre (1968).

pumice samples yield a U–Th isochron age of 38 ka and the historical lava yields a zero slope.

To achieve a high-precision age from the U–Th method, a reasonable spread of $^{238}U/^{232}Th$ ratios is needed within each sample suite. Kigoshi carried over the leaching techniques of carbonate dating in order to maximize this spread of U/Th ratios. However, this is a potentially dangerous technique, since in the leaching process disequilibrium may be introduced between different parents and daughters in the decay series. For example, ^{230}Th may be preferentially leached relative to ^{238}U from radiation-damaged lattice sites, yielding spuriously old ages. This is called the 'hot atom effect', and is the same process that gives rise to variable $^{234}U/^{238}U$ ratios in natural waters (Section 12.3.1).

Taddeucci et al. (1967) avoided the complexities of the 'hot atom' effect by using conventional physical separation and total dissolution of minerals to date five rhyolitic tuffs from the Mono Craters of California. However, U–Th dates on phenocryst–glass pairs did not display good agreement with other methods. For example, the hornblende–glass pair gave an apparent U–Th age of only 1 ka, whereas K–Ar dating yielded an age of ca. 7 ka, and ^{14}C dating of undisturbed lake sediments near the volcano gave a minimum eruptive age of 2.2 ka.

Allegre (1968) subsequently showed that separated phenocryst phases from one of Taddeucci et al.'s rocks defined an isochron age of 25 ka (Fig. 13.3). The implication of these results is that the different analytical methods are dating different events. U–Th ages on phenocryst minerals probably date their crystallization in a magma chamber, whereas the K–Ar method dates the time of eruption (if outgassing of volatiles during eruption was effective). The hornblende–glass age is meaningless, since these two systems did not close at the same time. The discordance between dat-

ing methods is therefore caused by the relatively long residence period of magma in the chamber, after phenocryst formation.

In contrast to the Mono Craters case, Allegre and Condomines (1976) and Condomines and Allegre (1980) were able to achieve good linearity of phenocryst and whole-rock points in dating studies of the Irazu volcano (Costa Rica) and Stromboli volcano (Italy). This implies that in these systems phenocryst growth only briefly preceded eruption. On the other hand, Capaldi and Pece (1981) claimed to find gross Th isotope disequilibrium between different mineral phases in modern lavas from Etna, Vesuvius and Stromboli. This led Capaldi et al. (1982) to completely write off the U–Th method as a dating tool. However, in repeat analyses of samples from the same Etna and Vesuvius lavas, Hemond and Condomines (1985) were unable to find mineralogical disequilibrium of Th isotope ratios. This suggests that Capaldi et al. (1982) overreacted when they dismissed the method. It is true that there are quite a number of instances where Th isotope disequilibrium has been found on a mineralogical scale (Capaldi et al., 1985). However, if phenocryst phases are screened by petrographic examination to exclude entrained xenocrysts, then many young lavas can yield accurate U–Th crystallization ages (e.g. Condomines et al., 1982).

13.1.2 U–Th (Zircon) Model Ages

The analysis of zircon as a mineral phase for U–Th dating was proposed in very early work (e.g. Cerrai et al., 1965), but not exploited much due to the low abundances of zircons in most rocks. However, the high uranium content of zircons often generates high U/Th ratios, which can yield good isochron fits. Hence, there has been renewed interest in U–Th dating of zircons with the development of mass spectrometric methods for the analysis of small samples.

Condomines (1997) used conventional TIMS analysis to demonstrate the usefulness of U–Th zircon analysis on a young trachytic rock in the Puy de Dome area of France. In this sample, the zircon analysis lay on the same isochron as several other mineral phases, and the isochron slope was consistent with the known age of eruption from other methods. However, in view of the laborious process of mineral separation and analysis, it has become normal just to analyse the zircon and whole-rock points, forming a two-point isochron that is also called a U–Th model age (Reid et al., 1997).

Zircon U–Th analyses can be performed by complete dissolution (TIMS analysis) or using in situ analysis by ion microprobe (SIMS). The former yields more precise ages, but the latter allows a larger suite of gains to be analysed, giving a better estimate of the statistical variability of the model ages, as well as testing for possible internal zoning of ages from core to rim. Good examples are provided by eruptions from the Taupo Volcanic Zone of New Zealand (Charlier et al., 2003). For example, TIMS analysis of zircons from the 62 ka Rotoiti Pumice (solid circles in Fig. 13.4) gave an isochron age of 77 ± 4 ka (2σ), 15 ka older than the age of eruption. Other

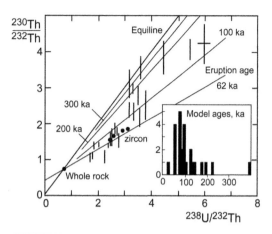

Fig. 13.4 Th–U isochron diagram for zircons from the Rotoiti Pumice, New Zealand, analysed by TIMS (•) and SIMS (error bars). Model ages for SIMS data are also plotted in histogram form. After Charlier et al. (2003).

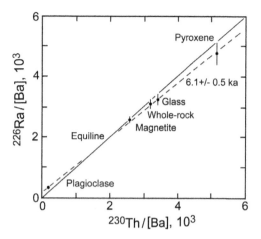

Fig. 13.5 Th–Ra/[Ba] isochron diagram for the 1985 pumice eruption of Nevado del Ruiz volcano, Columbia. Square brackets indicate weights, not activities. After Schaefer et al. (1993).

mineral phases from the pumice also lay on the same isochron, but too close to the whole-rock point to further constrain the age. SIMS analyses of 26 zircon grains are plotted on the same isochron diagram (vertical error bars). When these are plotted in the form of model ages they yield a large scatter, but with a peak centred on the TIMS age.

Guillong *et al.* (2016) showed that data of comparable quality can be obtained by laser ablation ICP-MS, provided that suitable corrections are applied for molecular-ion interferences. Further examples of U–Th model age dating will be given in Section 13.2 dealing with magma chamber evolution.

13.1.3 Ra–Th Isochron Diagrams

In view of its short equilibration time of 8000 yr, ^{226}Ra is useful in studies of geologically rapid magmatic processes. However, a disadvantage is the lack of a longer-lived radium isotope to normalize against, in order to exclude chemical fractionation. Williams *et al.* (1986) proposed that this problem might be overcome by using barium as a proxy for a stable radium isotope. For this to be useful, the two elements must have similar distribution coefficients, so that they behave in the same way during partial melting and crystal fractionation.

If barium is an accurate analogue for stable radium, the Th–Ra/[Ba] method can be used in conventional isochron dating of magma fractionation events. Reagan *et al.* (1992) applied this method to the dating of anorthosite phenocrysts in phonolitic magmas of Mount Erebus volcano, Antarctica. This was the first work to demonstrate the use of the Th–Ra/[Ba] isochron diagram for radium (analogous to the U–Th isochron diagram).

Since the half-life of ^{226}Ra is much less than the parent (^{230}Th), the activity of the latter is considered effectively con-

stant over the time periods under consideration (<10 ka). Therefore, after time t, the net ^{226}Ra activity is the sum of ^{226}Ra growth from ^{230}Th decay and the residue of partially decayed initial ^{226}Ra:

$$\frac{\left(^{226}\mathrm{Ra}\right)_P}{[\mathrm{Ba}]} = \frac{\left(^{226}\mathrm{Ra}\right)_I}{[\mathrm{Ba}]}\, e^{-\lambda\,226\,t} + \frac{^{230}\mathrm{Th}}{[\mathrm{Ba}]}\left(1 - e^{-\lambda\,226\,t}\right) \qquad [13.3]$$

where square brackets denote concentrations. This is analogous to equation [13.2] for the U–Th system.

Because anorthoclase readily takes up divalent but not trivalent ions, it has Th/Ba ratios of effectively zero. The isochron age is then determined by the glass points, yielding crystallization ages of ca. 2.5 ka for samples from two recent eruptions. However, it may be dangerous to rely on two-point phenocryst–glass ages without other supporting evidence (Section 13.1.1). A more complete example of a Th–Ra/[Ba] isochron was provided by Schaefer *et al.* (1993) on the 1985 pumice eruption of Nevado del Ruiz volcano, Columbia (Fig. 13.5). In this case the glass point was colinear with three different mineral phases (and the whole-rock), yielding a best-fit age of 6.1 ± 0.5 ka. This result was interpreted as the average age of an extended period of crystal fractionation, rather than a discrete magmatic differentiation event.

Ra–Th isotope data can also be presented on an 'alternative' isochron diagram that was first used by Kaufman (Section 12.4.3) for dating dirty calcite. We will take the opportunity here to examine this format, and compare it with the conventional isochron plot, using data for MORB glasses from the East Pacific Rise (Rubin and Macdougall, 1990).

The premise of the alternative isochron diagram is to reverse the ratio on the x axis, so that radiogenic samples with zero common radium (in this case represented by zero barium) plot on the y axis. The age equation can then be

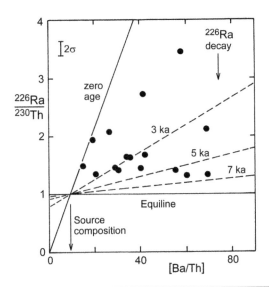

Fig. 13.6 Ra/Th–[Ba/Th] 'isochron' diagram for MORB glasses from the East Pacific Rise. After Rubin and Macdougall (1990).

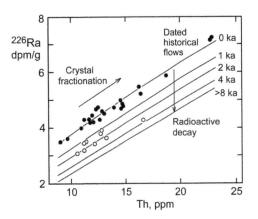

Fig. 13.7 Plot of calculated initial ^{226}Ra activities for previously dated samples from Etna (•), to show a good relationship with common Th concentration, which is then used to calculate Ra model ages for unknown samples (○). After Branca et al. (2015).

simplified by considering only pure radiogenic radium formed during the lifetime of the system. After rearranging, the ^{226}Ra/^{230}Th activity ratio, plotted on the y axis, yields the age:

$$\frac{^{226}\text{Ra}}{^{230}\text{Th}} = 1 - e^{-\lambda 226\, t} \qquad [13.4]$$

If a magma suite is extracted from a homogeneous source that is in secular equilibrium, zero-age lavas should have a constant Ra/Ba ratio, forming a straight line through the origin (Fig. 13.6). After separation from the source, magmas evolve vertically towards the equiline. A theoretical radiogenic sample would then evolve from 0 to 1 (secular equilibrium) on the y axis. However, since real MORB magmas have Ba/Th ratios greater than the source, isotopic evolution is by decay of excess ^{226}Ra. Therefore, it is convenient to show the ages since melt extraction as isochron lines for a given source composition (Fig. 13.6). However, it is clear that the samples are not actually grouped along isochron arrays, but are really a collection of unrelated samples that form two-point isochrons with an assumed source composition. Hence the dashed lines in Fig. 13.6 are actually model ages.

Later work by Volpe and Goldstein (1993) confirmed from other ocean ridges that within a given ridge segment, Ba/Th ratios are constant in lavas of different ages (below 10 ka). Hence, it is valid to determine Ra–Th–Ba model ages for individual MORB samples. However, other geochemical pairs can also be used to model the source composition of a sample suite in order to determine Ra–Th model ages, as shown below.

13.1.4 Ra–Th Model Ages

Since barium is an imperfect analogue for the behaviour of common radium in dating experiments, other geochemical indices can be used to estimate initial ^{226}Ra activities, and hence calculate Ra–Th model ages. This approach was demonstrated by Condomines et al. (1995, 2005) on lavas up to 5 ka in age from the Etna volcano. Using a suite of previously dated lavas, Condomines et al. calculated initial ^{226}Ra activities, and showed that these were well correlated with common thorium contents. Hence, Th concentrations can be used to predict the initial ratio of unknown samples. However, post-1970s lavas did not fit this relation, thus demonstrating the limited empirical basis of the approach. Nevertheless, Condomines et al. demonstrated that Ra–Th model ages for unknown pre-modern samples were in agreement with ages determined from secular variations in paleomagnetic orientation.

This method was used to date many additional lavas from the summit eruptions of Etna (Tanguy et al., 2007), as well as smaller flank eruptions (Branca et al., 2015). These authors applied the same Ra–Th relation to determine initial ^{226}Ra activities for their new samples, whose ages were then determined by the extent of ^{226}Ra decay from the Ra–Th correlation line (Fig. 13.7).

13.2 Magma Chamber Evolution

Just as initial Sr isotope compositions are useful as a geochemical tracer when using the Rb–Sr method, initial Th isotope compositions are also a useful product of the U–Th dating method. One area where they have proved particularly valuable is in studying magma chamber evolution.

In the same way that Th isotope evolution in a volcanic rock can be used to date crystallization, Th isotope evolution between successive eruptions can be used to date the residence of magma in a chamber. We can use the same equation as for the rock system, except that what we input as the 'final composition' on the left-hand side is actually the initial Th activity ratio of a magma at the time of eruption (E), while the Th activity ratio on the right-hand side of the equation is the composition of the magma in the chamber at the time of influx (I), simplistically, from the mantle. The quantity 't' then represents the residence time of the magma batch in the chamber (all nuclide ratios in activities):

$$\left(\frac{^{230}Th}{^{232}Th}\right)_E = \left(\frac{^{230}Th}{^{232}Th}\right)_I e^{-\lambda_{230}t} + \frac{^{238}U}{^{232}Th}\left(1 - e^{-\lambda_{230}t}\right) \quad [13.5]$$

Allegre and Condomines (1976) preferred to refer all times to the present, introducing T as the age of influx into the chamber and t as the time of eruption. It is best to first rearrange equation [13.5] to gather the exponent terms in one place:

$$\left(\frac{^{230}Th}{^{232}Th}\right)_E = \left[\left(\frac{^{230}Th}{^{232}Th}\right)_I - \frac{^{238}U}{^{232}Th}\right] e^{-\lambda_{230}t} + \frac{^{238}U}{^{232}Th} \quad [13.6]$$

Then:

$$\left(\frac{^{230}Th}{^{232}Th}\right)_t = \left[\left(\frac{^{230}Th}{^{232}Th}\right)_T - \frac{^{238}U}{^{232}Th}\right] e^{-\lambda_{230}(t-T)} + \frac{^{238}U}{^{232}Th} \quad [13.7]$$

Just as we make a closed-system assumption in the case of rock dating, so we must assume that the magma in the chamber remains a closed system to U and Th during its evolution and eruption. This assumption should not be upset by Rayleigh crystal fractionation, since the distribution coefficients of both U and Th are so low that both elements normally remain entirely in the liquid.

13.2.1 The Th Isotope Evolution Diagram

The ^{230}Th evolution of magmas can be shown on U–Th isochron diagrams, but it is also convenient to display isotopic evolution on a plot of Th activity ratio against time (Fig. 13.8). This plot is analogous to the evolution diagrams of Sr or Nd isotope composition against time (e.g. Section 4.2.1), but because the ^{230}Th half-life is short relative to the time periods under study, the x axis must be calibrated in log time.

A magma body in secular equilibrium must maintain its ^{230}Th/^{232}Th activity ratio. Hence, such evolution is described by a horizontal line in Fig. 13.8. If the system undergoes U/Th fractionation (horizontal displacement from M_0 to M or M′ on the inset diagrams) then the ^{230}Th/^{232}Th activity ratio of the magma will evolve over time to regain secular equilibrium. A magma enriched in ^{238}U/^{230}Th (to the right of the equiline on the inset, with activity M) defines a line of negative slope on the main diagram. Similarly, a magma depleted in ^{238}U/^{230}Th (activity ratio M′ on the inset) defines an evolution line of positive slope on the main diagram. If we extrapolate the growth lines to $e^{\lambda t} = 0$, then we can see from equation [13.6] that the y ordinate in the main diagram describes the ^{238}U/^{232}Th activity ratio of the evolving magma.

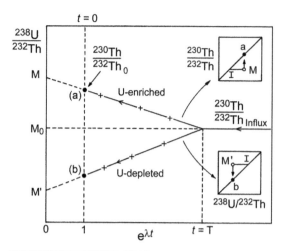

Fig. 13.8 Schematic diagram of Th isotope evolution against time for two closed-system magma chambers: (a) displaying daughter deficiency; and (b) daughter excess. Plotted data are initial Th activities at the time of eruption. Insets show U/Th fractionation and subsequent evolution on U–Th isochron diagrams. After Condomines et al. (1982).

The U–Th isotope system is a very powerful tool for studying magma chamber evolution because the 75.6 ka half-life of ^{230}Th is the same order of magnitude as the time interval between magma chamber events. However, a significant database is needed to unravel the history of most volcanoes, which usually involve repeated magma injection and eruption events. A simple scenario of this type is illustrated in Fig. 13.9. In this case, a primary mantle source in secular equilibrium supplies a series of magma batches over a period of time which have a constant disequilibrium ^{238}U/^{232}Th

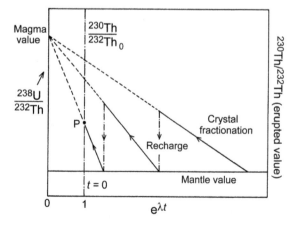

Fig. 13.9 Schematic illustration of the thorium isotope evolution of a periodically tapped and re-filled magma chamber with constant U/Th ratio. After Condomines et al. (1982).

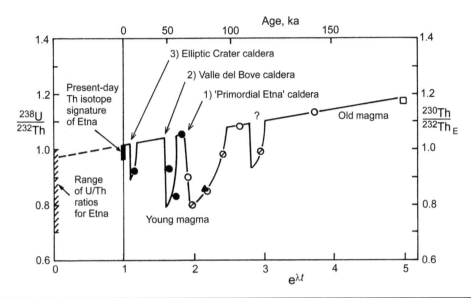

Fig. 13.10 History of the Etna volcano on a thorium isotope evolution diagram to show four magma influx–mixing–eruption events. Symbols indicate erupted products at different stages of volcano evolution. After Condomines et al. (1982).

activity ratio generated by the partial melting process (see below). After a period of magma evolution in a high-level chamber (sloping line), the chamber is emptied by eruption and re-filled, causing a kick back to the starting composition. However, in the real world, mixing of magmas of different ages is likely to occur, and this will generate a more complex pattern of magma evolution.

Early studies (Allegre and Condomines, 1976; Condomines and Allegre, 1980) lacked sufficient data to resolve the magmatic history of long-lived volcanoes; hence, their results were ambiguous. However, a later study by Condomines et al. (1982) provided enough data to interpret the history of the Etna volcano in Sicily. Thirteen mineral–whole-rock U–Th isochrons were determined, along with analyses of recent lavas. The results were plotted on a Th isotope evolution diagram (Fig. 13.10), and an isochron diagram (Fig. 13.11).

On the Th isotope evolution diagram (Fig. 13.10) the Etna data provide evidence for four episodes of eruption and magma replenishment, the last three of which (numbered) tie in with known dates of major caldera collapse events. Between these events, small magma tappings monitor Th isotope evolution in the high-level chamber. However, ^{230}Th/^{232}Th ratios fall too rapidly during these periods to be explained by closed-system evolution, given the observed range of U/Th ratios (hatched band on the left-hand axis). Therefore, Condomines et al. invoked a magma mixing model to explain these steep trends, suggesting that the sub-horizontal evolution line represented a deep, long-lived alkali basalt reservoir which continually supplied magma to higher levels, where a low-^{230}Th/^{232}Th tholeiitic component was added intermittently.

The sub-horizontal evolution line in Fig. 13.10 corresponds to periods where Ra–Th model age dating can be used to date historical age (<5 ka) eruption events, since initial magma ^{226}Ra activities can be accurately predicted during these periods (Section 13.1.4). On the other hand, the steep or vertical evolution lines represent periods of rapid change in magma chemistry where the Ra–Th model age method cannot be used for dating. The eruptive activity since 1970 represents one such period.

Initial ratios of lavas at the time of eruption are plotted on a U–Th isochron diagram in Fig. 13.11. The data suggest

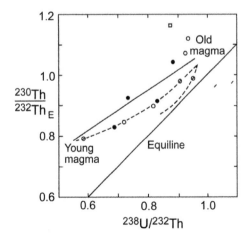

Fig. 13.11 U–Th activity data for Etna lavas at the time of eruption, showing possible mixing processes (symbols as in Fig. 13.10). After Condomines et al. (1982).

mixing between an old magma nearly in secular equilibrium and a young one substantially out of equilibrium. Condomines *et al.* argued that the low ^{230}Th/^{232}Th (young) component could not be a crustal contaminant, since this should be close to secular equilibrium, whereas Fig. 13.10 indicates it to be far from equilibrium. However, this does not exclude the possibility of sediment contamination of the mantle source of these magmas. The straight line (1) represents an instantaneous mixing model whereas the curved lines (2, 3) model progressive mixing over a time interval. Present day ratios (of old lavas) are not plotted on Fig. 13.11 because they do not yield any useful information about magma evolution.

The determination of a mineral isochron for several points on a magma evolution diagram is extremely labour intensive, and in practice many mineral suites do not yield good isochron fits (e.g. Condomines *et al.*, 2003). Assuming that analytical errors or sample alteration are not responsible, the most likely cause is that different minerals in the rock crystallized at different times prior to eruption. Therefore, several recent studies of magma chamber evolution have adopted an alternative approach using U–Th zircon model ages to determine the range of magma chamber residence times in crystal populations as an indicator of magma chamber evolution.

This approach was demonstrated by Reid *et al.* (1997), using ion microprobe (SIMS) analysis of zircons from the Long Valley Caldera, California. Zircons were analysed from two domes, one aged 115 ka and one less than 1 ka, located a few km apart. Each zircon analysis was combined with the whole-rock composition to determine a two-point isochron that Reid *et al.* termed a 'zircon model age'. These model ages were somewhat scattered, due to a combination of relatively large analytical errors and some geological scatter. However, it was observed that the average zircon model age for the two domes was the same, despite their different eruption ages. This suggests that the two domes might be sampling the same relatively long-lived magma chamber, which cooled through the zircon saturation temperature at about 230 ka, causing the crystallization of a crop of zircon grains that subsequently remained entrained in the viscous magma.

This approach has been adopted in several recent studies, but with a more sophisticated analysis of the range of model ages in the data set. For example, Charlier *et al.* (2005) made a combined TIMS and SIMS investigation of several eruptive products from Taupo caldera in the northern island of New Zealand. Data are shown in Fig. 13.12 for the large 26.5 ka Oruanui eruption, with a volume 530 km³. SIMS data produce a large scatter on the U–Th isochron diagram (Fig. 13.12a), but this can be analysed more quantitatively from the slopes of two-point isochrons between each zircon spot and the whole-rock composition (Fig. 13.12b). Axis labels along the top of Fig. 13.12b show that U–Th model ages are a non-linear function of isochron slopes due to the short half-life of ^{230}Th. Hence, as argued by Wilson and Charlier (2009),

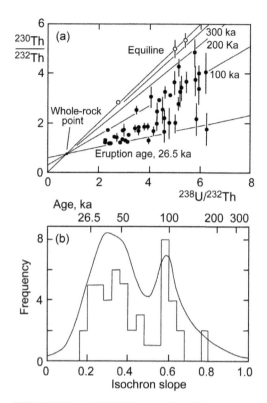

Fig. 13.12 SIMS zircon data for the 26 ka eruption of Oruanui volcano, Taupo Volcanic Zone. (a) U–Th isochron diagram with 1σ errors; (b) histogram and cumulative frequency plot of two-point isochron slopes. Open symbols in (a) are omitted from (b). After Charlier *et al.* (2005).

the isochron slopes are more suitable for statistical analysis, since the errors are symmetrically distributed.

The data are shown in Fig. 13.12b in the form of a histogram and cumulative frequency plot, both methods indicating a bimodal distribution of zircon crystallization ages, one around 100 ka and a broader one beginning about 25 ka before the time of eruption. The second peak corresponds fairly well with the range of TIMS analyses of different-sized zircon fractions, ranging from 32 ± 0.5 ka in the smallest fraction to 40 ± 1 ka in the largest. This suggests zircon growth over a period of 20 ka in a temporary high-level magma storage chamber. On the other hand, the older 100 ka age peak is also seen in previous smaller eruptions of Taupo volcano, suggesting some entrainment of zircons from a crystal mush forming the floor of the high-level chamber. There was little evidence for entrainment of major mineral phases in the Oruanui eruption that were analysed for U–Th (opx, plagioclase, hornblende, magnetite). However, these phases have a small range of ^{238}U/^{232}Th ratios (0.72–0.80) that are very similar to the whole-rock (0.745). This limits their sensitivity to crystal mixing processes.

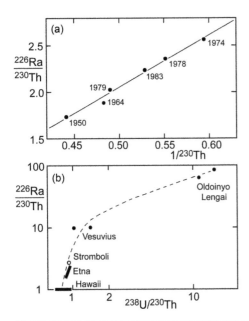

For magma systems exceeding the effective dating range of the U–Th system, U–Pb dating of zircons has been used in a similar way to constrain the magma chamber residence time of zircon populations. Studies of zircon magma chamber residence times in young well-dated eruptions are important to constrain zircon behaviour in older eruptions used for stratigraphic correlation (Section 5.1.3). A good example is the 760 ka Bishop Tuff of the Long Valley Caldera (e.g. Simon and Reid, 2005; Ickert et al., 2015).

13.2.2 Short-Lived Species in Magma Evolution

Shorter-lived species offer the opportunity of more detailed study of magmatic processes such as the magma influx–mixing–eruption events already studied using the U–Th system. Magma replenishment processes may precede eruption events, and a better understanding of their timescale can therefore allow better prediction of impending eruption, and hence enhanced hazard mitigation.

Etna provides one of the best examples for testing this approach, in view of the large number of well-dated lavas from the historical period. Early work was carried out by Capaldi et al. (1976), and was followed by more detailed studies by Condomines et al. (1987, 1995). This work showed that pre-1970 lavas had relatively constant ^{226}Ra/^{230}Th activity ratios, whereas post-1970 lavas showed a rapid rise in ^{226}Ra/^{230}Th activity that was inversely correlated with Th activity (Fig. 13.13a). These results were interpreted by Condomines et al. (1995) as indicative of magma mixing processes between a long-lived hawaiite (differentiated basaltic)

magma and a new magma pulse with a radium-enriched signature.

The radium enrichments seen in the 1974 magma were attributed to selective crustal contamination by Condomines et al. (1995). However, it is now known that such signatures are typical of subduction-related magmas, where radium and uranium enrichments are carried from the subducting crust by slab-derived fluids (Section 13.5). This model is supported by similar radium-enriched signatures in other volcanoes of the Aeolian Arc: Stromboli and Vesuvius.

Much more extreme radium enrichments are seen at Oldoinyo Lengai volcano, in the East African Rift of Tanzania (Fig. 13.13b). In this case, disequilibrium was also seen in the very short-lived species ^{228}Ra ($t_{1/2} = 5.77$ yr) and even ^{228}Th ($t_{1/2} = 1.91$ yr). Hence these systems can provide additional information about the radium enrichment process at Oldoinyo Lengai. In contrast, Condomines et al. (1987, 1995) did not find any samples outside of secular equilibrium for these species at Etna, thus contradicting earlier claims by Capaldi et al. (1976).

The 1960 and 1988 carbonatite eruptions at Oldoinyo Lengai were studied by Williams et al. (1986) and Pyle et al. (1991). Both of these eruptions showed strong disequilibrium between ^{228}Ra and its ultimate parent, ^{232}Th. The 1988 eruption also demonstrated ^{228}Th disequilibrium, but it was not possible to test for this phenomenon in the 1960 eruption, since the samples reached secular equilibrium in the 20 years between sampling and analysis.

It is now generally agreed that carbonatites are formed by the evolution of peralkaline magmas in conditions of strong CO_2 enrichment, probably involving the segregation of immiscible droplets of carbonate magma from a silicate magma host (e.g. Pyle et al., 1991). The discovery of ^{228}Ra disequilibrium in the Oldoinyo carbonatites suggests that the segregation process probably occurred shortly before eruption. Over this timescale, ^{226}Ra ($t_{1/2} = 1620$ yr) can be treated effectively as a stable isotope. Therefore the ^{228}Ra decay equation can be divided by ^{226}Ra to yield an isochron relation analogous to equation [13.2] (Capaldi et al., 1976):

$$\left(\frac{^{228}\text{Ra}}{^{226}\text{Ra}}\right)_{\text{P}} = \left(\frac{^{228}\text{Ra}}{^{226}\text{Ra}}\right)_{\text{I}} e^{-\lambda_{228}t} + \frac{^{232}\text{Th}}{^{226}\text{Ra}}\left(1 - e^{-\lambda_{228}t}\right) \quad [13.8]$$

However, Williams et al. (1986) preferred to use the alternative isochron diagram, where ages are defined by the intercept on the left-hand axis, using the following equation, analogous to [13.4]:

$$\left(\frac{^{228}\text{Ra}}{^{232}\text{Th}}\right)_{\text{P}} = 1 - e^{-\lambda_{228}t} \quad [13.9]$$

This activity ratio is plotted in Fig. 13.14a against ^{226}Ra/^{232}Th activity. The zero-age line passes through the origin and represents a Ra/Th fractionation line. The slope of this line is the ^{228}Ra/^{226}Ra activity ratio, equal in turn to the ^{232}Th/^{238}U activity ratio, since ^{226}Ra and ^{228}Ra are in separate decay chains. For Oldoinyo Lengai, this ratio was found to be unity. Radioactive decay of a system located on the fractionation

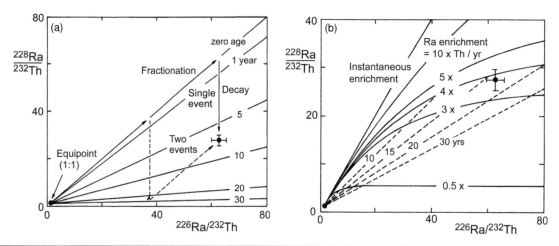

Fig. 13.14 Ra–Th isochron diagrams showing (a) instantaneous and (b) continuous radium enrichment models for the 1960 carbonatite of Oldoinyo Lengai. After Williams *et al.* (1986).

line causes it to move vertically downwards towards the equiline and permits an age to be assigned.

The initial ratio of the 1960 carbonatite is plotted on Fig. 13.14a. If this magma is attributed to a single instantaneous event which caused Ra/Th fractionation, it follows that this event occurred seven years before eruption. However, more complex models are also possible. For example, if segregation occurred in two events, ^{228}Ra formed in the first event might decay before the second event (Fig. 13.14a). In this case, the time from the second event to eruption is less than seven years. Alternatively, carbonatite segregation might have occurred over a period of time. In Fig. 13.14b, this process is modelled by drawing Ra/Th growth curves for different rates of Ra enrichment in the carbonatite magma (relative to the conjugate silicate liquid). These growth curves are then calibrated by determining the time necessary for enrichment of the effectively stable ^{226}Ra isotope. A simple model of constant enrichment rate yields a calculated duration of this process for the 1960 magma of ca. 18 years, if this was occurring immediately prior to eruption (Fig. 13.14b).

The addition of ^{228}Th data can potentially allow selection between short-term differentiation models such as those outlined above, because of its dependence on very recent events. These data were collected by Pyle *et al.* (1991) for the 1988 carbonatite of Oldoinyo Lengai, but due to analytical difficulties, Ra/Th activity ratios could not be measured. Therefore, the advantage of the combined systems was lost.

13.3 Mantle Melting Models

Volcanoes located on continental crust appear to show complex ^{230}Th evolution patterns, probably because mantle-derived magmas become trapped during their rise through low-density sialic basement. In contrast, oceanic volcanoes might be expected to show simpler behaviour, since oceanic crust is easily punctured by rising magma. Therefore, oceanic volcanics provide a window to study Th isotope fractionation processes in the mantle.

The first detailed U-series measurements on oceanic volcanics were made by Oversby and Gast (1968) by α counting on recent ocean island lavas. This study revealed disequilibrium between isotopes of the ultra-incompatible elements, radium and thorium. Oversby and Gast suggested that these fractionations were probably inherited from a melting event in the mantle source of OIB magmas. ^{230}Th activities were observed to be higher than ^{238}U, suggesting that thorium was a more incompatible element than uranium. This conclusion was supported more than ten years later (Condomines *et al.*, 1981) in studies of MORB magma genesis (see below).

In view of the short half-life of ^{226}Ra, Oversby and Gast attributed disequilibrium of this species to rapid ascent of ocean island magmas from the source area (<10 000 yr). This conclusion has also been supported by more recent radium analysis of MORB (Section 13.4.1). In addition, rapid ascent of ocean island magmas has been supported by studies of their Th isotope chemistry over time. For example the volcanoes of Mauna Kea (Hawaii), Marion Island (SW Indian Ocean) and Piton de la Fournaise (Fig. 13.15) all show constant initial ^{230}Th/^{232}Th ratios (within error) over the last 250 ka (Newman *et al.*, 1984; Condomines *et al.*, 1988). This suggests that magma transport from the melting zone to the surface probably occurred within a few ka, without storage in a deep crustal reservoir that would have perturbed the observed ratios. Therefore, the calculated initial Th isotope ratio for each eruption is probably very close to the source value.

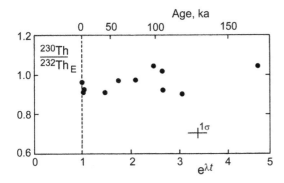

Fig. 13.15 Plot of erupted (initial) Th activity ratios against time for Piton de la Fournaise volcano (Reunion Island) showing constant magma composition over time, within error. After Condomines *et al.* (1988).

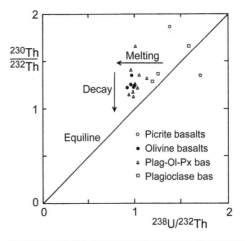

Fig. 13.16 U–Th isochron diagram showing analyses of young lavas from the FAMOUS area of the Mid Atlantic Ridge, to the left of the equiline. Arrows show the effects of partial melting and radioactive decay. After Condomines *et al.* (1981).

13.3.1 Melting Under Ocean Ridges

Mid ocean ridges present the minimum crustal thickness which must be traversed by ascending mantle-derived magmas. Therefore, in this environment we should have the best opportunity to see back through the processes of magma evolution during ascent to study source processes and chemistry. Some workers (e.g. O'Hara and Mathews, 1981) suggested that MORB magmas spend many eruptive cycles in periodically re-filled, periodically tapped magma chambers under the ridge, which would then grossly perturb the incompatible element and isotopic signatures of the product magmas. In this case, they argued, it would be almost impossible to 'invert' the data (see Section 7.2.3) to reconstruct source chemistry.

The short half-life of ^{230}Th provides a powerful tool to test these models of MORB magma evolution. In the first detailed study of MORB samples, Condomines *et al.* (1981) found that fresh, young crystalline basalts and glasses from the 'FAMOUS' area on the Mid Atlantic Ridge (37 °N) had a narrow range of Th isotope and U/Th ratios. Because these samples are all less than 5 ka in age, their present Th isotope compositions can be taken as initial ratios at the time of eruption. When plotted on a U–Th isochron diagram (Fig. 13.16) the FAMOUS data fall well to the left of the equiline, showing them to be far from in isotopic equilibrium.

The U–Th disequilibrium shown in Fig. 13.16 must have been inherited during the melting process, because U and Th are both ultra-incompatible elements and cannot be fractionated from each other during Rayleigh crystallization in a magma chamber. If the residence time of magma in such chambers was more than a few tens of ka, U–Th activities would again reach equilibrium (Fig. 13.16). Because this has not happened, we can deduce that the transport of magma from the melting zone to the surface was relatively rapid, which is not consistent with prolonged evolution in an open-system magma chamber. This conclusion has subsequently

been supported by evidence of isotopic disequilibrium of even shorter-lived U-series nuclides in MORB glasses (Section 13.4).

In addition to the FAMOUS area, Condomines *et al.* (1981) showed that other young ridge basalts and OIB also fell on the left side of the equiline (Fig. 13.17). Hence, in all of these cases, melts were enriched in Th/U relative to the source. This

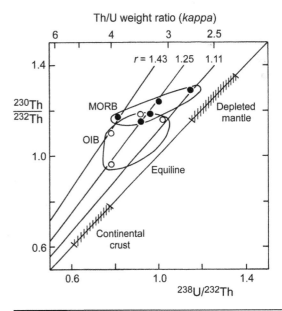

Fig. 13.17 U–Th isochron diagram showing fields for MORB and OIB relative to the Th/U fractionation factor during melting (*r*). Modified after Allegre and Condomines (1982).

implies greater incompatibility of Th over U during melting. Allegre and Condomines (1982) expressed this Th/U fractionation during melting by the factor 'k':

$$k = \frac{\left(^{238}U/^{232}Th\right)_{magma}}{\left(^{238}U/^{232}Th\right)_{source}} \quad [13.10]$$

However, it is more useful to express this ratio by its reciprocal, termed 'r', as used by McKenzie (1985a). In addition to facilitating the algebra, this formulation avoids confusion of 'k' with kappa (the atomic $^{232}Th/^{238}U$ ratio, which is approximately equal to the Th/U weight ratio):

$$r = \frac{\left(^{238}U/^{232}Th\right)_{source}}{\left(^{238}U/^{232}Th\right)_{magma}} \quad [13.11]$$

Since the source is assumed to be on the equiline, its $^{238}U/^{232}Th$ activity is equal to its $^{230}Th/^{232}Th$ activity. Furthermore, if the analysed sample was extracted from the source in less than a few thousand years, the source Th activity ratio is equal to that measured in the magma:

$$\left(\frac{^{238}U}{^{232}Th}\right)_{source} = \left(\frac{^{230}Th}{^{232}Th}\right)_{source} = \left(\frac{^{230}Th}{^{232}Th}\right)_{magma} \quad [13.12]$$

Hence, substituting into equation [13.11], we obtain

$$r = \frac{\left(^{230}Th/^{232}Th\right)_{magma}}{\left(^{238}U/^{232}Th\right)_{magma}} = \left(\frac{^{230}Th}{^{238}U}\right)_{magma} \quad [13.13]$$

This is represented in Fig. 13.17 by the gradient of lines which project from the origin. Hence we can determine U/Th fractionation during melting from a U-series analysis of the magmatic product.

Corroboration of the Mid Atlantic thorium data came from a similar study on the East Pacific Rise by Newman *et al.* (1983). These data also lie to the left of the equiline, and fall within error of unaltered basalts from the FAMOUS area, but with more scatter, particularly in $^{238}U/^{232}Th$ activity. Since U and Th are both ultra-incompatible elements, Newman *et al.* recognized that it was very difficult to generate the required U/Th fractionations at the degrees of melting normally expected for MORB (ca. 10%). Hence, they suggested that under these conditions a U–Th-rich accessory phase might be required to explain the data.

Thompson *et al.* (1984) reversed this problem, arguing that the incompatible element signatures of MORB rocks, including U–Th disequilibrium data, could only be generated by very low degrees of mantle melting. They noted that such an explanation was consistent with observations for other isotope systems on MORB glasses (e.g. Cohen *et al.*, 1980), which showed that Rb/Sr, Sm/Nd and U/Pb ratios measured in MORB samples were fractionated relative to the ratios in the source required to generate the observed isotope signatures.

13.3.2 The Effect of Source Convection

This line of argument was developed by McKenzie (1985a), who performed calculations to determine the maximum percentage of partial melting that was consistent with the

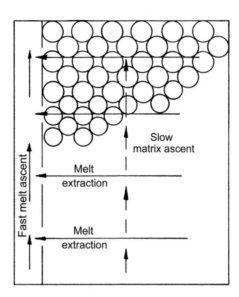

Fig. 13.18 Schematic view of a dynamic melting model for the generation of MORB. See text for discussion. After McKenzie (1985a).

observed U/Th fractionations. For this purpose he assumed that Th was perfectly incompatible (i.e. its bulk distribution coefficient between solid and liquid, D, is zero), and that there was no magma residence time in a sub-ridge chamber before eruption. In order to generate a $^{230}Th/^{238}U$ enrichment (r) of 1.25 using a batch melting model and a D value of 0.005 for uranium, the maximum degree of melting permitted was 2%. However, this result is not consistent with major element considerations, which require MORB to be a large degree (ca. 15%) melt of the mantle. Hence, McKenzie ruled out a simple batch melting model for MORB genesis, and adopted instead the dynamic melting model of Langmuir *et al.* (1977), shown in Fig. 13.18.

In the dynamic melting model, melts are extracted simultaneously from a vertical thickness of perhaps 60 km of mantle (horizontal lines in Fig. 13.18). The melts ascend quickly to the surface in a conduit, mixing as they go (long vertical arrows on the left of Fig. 13.18). Meanwhile the source itself moves slowly upwards through the melting zone (short vertical arrows in the middle). At any given point the source contains less than 2% melt (termed the 'porosity'), but as it moves upwards and melts are tapped off, the source becomes more and more depleted in incompatible elements. If melts mix equally from the whole melting zone, the effect of dynamic partial melting on incompatible *element* abundances is similar to batch melting. This is because (in the extreme case) the source is completely exhausted of these elements by the time it reaches the top of the melting zone. In other words, incompatible element extraction is 100% efficient. However, for short-lived *nuclides*, the two melting models can yield quite different results.

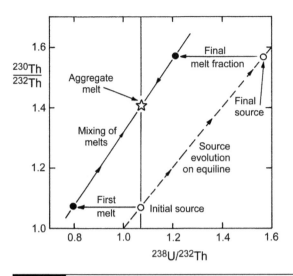

Fig. 13.19 Consequences of slow mantle upwelling under the ridge for the Th isotope systematics of MORB magmas. Note that the rate of *magma* upwelling is very rapid relative to the ^{230}Th half-life.

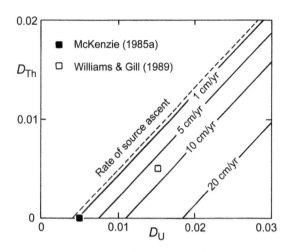

Fig. 13.20 Diagram showing the relationship between calculated mantle upwelling rate (cm/yr) and bulk distribution coefficients (D) for U and Th, assuming a dynamic melting model with 2% porosity and yielding a ^{230}Th/^{238}U activity (r) of 1.2 in MORB. After Williams and Gill (1989).

If the rate of mantle upwelling is rapid relative to the half-life of the nuclide in question (e.g. ^{230}Th), then this nuclide behaves like a stable element. In that case, dynamic melting will yield an aggregate melt similar to batch melting, and the 15% melt fraction necessary to explain major element data cannot satisfy the Th isotope data. However, if the rate of mantle upwelling is very slow relative to the ^{230}Th half-life, then ^{230}Th which is removed from the source at the base of the melting column is replenished in the source as it ascends, by decay from residual uranium (which is less incompatible than thorium). Consequently, as upwelling progresses, the ^{238}U/^{232}Th activity of the source increases, but it remains on the equiline (Fig. 13.19). After extraction of all U and Th from the source (15% total melt) the ^{238}U/^{232}Th activity of the aggregate melt will be the same as the initial source, but the ^{230}Th/^{238}U activity ratio (r) is still the same as in the first increment of melting at the base of the melting zone (Fig. 13.19). Minor variations to this model have been proposed, but they were all classified as 'ingrowth' models by Elliott (1997).

In between these two extremes (simple batch and dynamic melting), it is possible to determine the rate of mantle upwelling which will yield the observed r value in MORB, given the bulk distribution coefficients (D) for U and Th. Williams and Gill (1989) presented these relationships in diagrammatic form, based on the equations developed by McKenzie (1985a). This information is shown in Fig. 13.20. The values of D chosen by McKenzie (1985a) lead to a calculated mantle upwelling rate of only 1 cm/yr. In contrast, Williams and Gill (1989) argued for a much higher D value for uranium (and a lesser increase for thorium, Fig. 13.20). This allows a more rapid rate of upwelling (ca. 7 cm/yr).

These different upwelling rates make very different predictions about mantle processes under the ridge. A value of 1 cm/yr is less than the rate of plate motion, which led McKenzie (1985b) to argue that the melting zone under the ridge is funnel shaped (as sketched in Fig. 13.21a, after Galer and O'Nions, 1986). Trace elements are then extracted from a wide swath of mantle near the base of the funnel, while major elements are dominated by the melt extracted from the apex of the funnel (Fig. 13.21a). Because the trace element extraction zone is larger, this lower domain dominates the U–Th systematics of the melt, which acts like a small-degree batch melt of the original mantle source.

In contrast, the higher upwelling rate of Williams and Gill (1989) results in melt extraction from a vertical slice under the ridge, yielding a result closer to the dynamic melting model. However, the consequence is that the ^{230}Th/^{232}Th activity of the erupted products is substantially higher than the source (Fig. 13.19), while the ^{238}U/^{232}Th activity is similar to the source. O'Nions and McKenzie (1993) pointed out that in this case, the Th/U ratio (kappa value) of the source (Sections 6.3.2, 13.3.4) should be determined, *not* from the Th isotope ratio of MORB (equation [13.12] above), but directly from the elemental U/Pb ratio of MORB. Hence, this model predicts that short-lived isotopes *are* fractionated by the melting process under ridges, but stable incompatible elements are *not* fractionated under these conditions.

The real model may be somewhere between the two extremes described above; however, several lines of evidence tend to support the slow upwelling model. Perhaps the most important piece of evidence comes from the so-called mantle electromagnetic and tomography (MELT) experiment, which used geophysical methods to image the distribution of

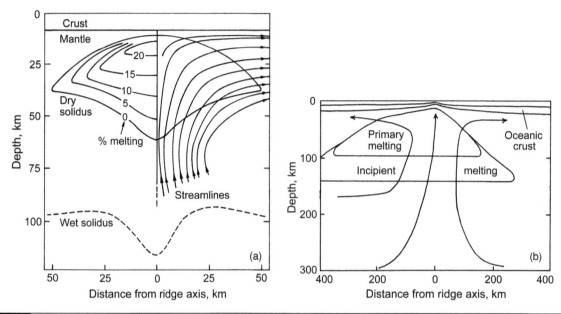

Fig. 13.21 Predicted and observed distribution of melting zones under a mid ocean ridge: (a) model predicted by Galer and O'Nions (1986); (b) cross section of the East Pacific Rise based on results of the MELT seismic experiment. After Forsyth et al. (1998).

partial melt zones under the East Pacific Rise (Forsyth et al., 1998). Seismic evidence collected in this experiment showed that the zone of incipient melting under the ridge extended to depths of over 150 km, and to a width of nearly 800 km, matching the shape of the melting zone predicted in Fig. 13.21a but having an extent three to four times greater in each direction (Fig. 13.21b). Other evidence in support of slow upwelling with an extended zone of incipient melting comes from the continuing need to explain the fractionation of incompatible trace element ratios such as Rb/Sr and Sm/Nd in erupted products, relative to the mantle source composition indicated by Sr and Nd isotope ratios (as discussed above).

The opposite end of this spectrum of upwelling rates is demonstrated by OIB sources, for which U–Th data indicate much larger upwelling rates. One of the best examples of this phenomenon is provided by the Iceland plume, whose products can be sampled over a large geographical area due to its numerous volcanic centres distributed over a wide subaerial extent. A large sample suite analysed by Kokfelt et al. (2003) revealed a strong inverse relationship between ^{230}Th/^{238}U activity ratio and radial distance from the centre of the plume (Fig. 13.22). Hence, Kokfelt et al. argued that mantle upwelling rates in the centre of the plume (ca. 5–20 cm/yr) were about five times greater than at the periphery (ca. 1–4 cm/yr). Similar results were obtained from the Hawaiian Islands (Sims et al., 1999), but the concentration of activity into a few large centres makes the correlation between activity ratio and distance less well defined.

According to the dynamic melting model (mantle) upwelling velocity is inversely correlated with Th disequilibrium, because more Th ingrowth occurs at low upwelling rates. However, Sims et al. (1995) showed that Th disequilibrium in Hawaii is also well correlated with major element compositions (such as silica saturation) that are a function of melt fraction. This correlation is also seen to a lesser extent in Iceland (Kokfelt et al., 2003). Since the rapidly upwelling

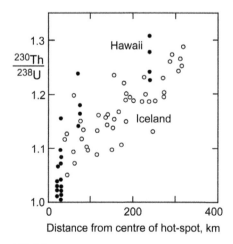

Fig. 13.22 Plot of ^{230}Th/^{238}U activity ratio against radial distances from the centre of two different plumes. (●) = Hawaii; (○) = Iceland. Modified after Bourdon et al. (2005).

centre of a plume has much larger melt fractions than the periphery, the correlation of Th disequilibrium with distance from the plume centre can be explained by variable dilution of small degree melts from the stem of the plume by large degree melts in the plume head as it spreads out under the oceanic lithosphere (Fig. 13.21a). According to this model, ^{230}Th ingrowth might not be required to explain Th disequilibrium in plumes. However, Sims *et al.* (1999) argued that this model cannot explain the degree of radium disequilibrium in Hawaii. This will be discussed below (Section 13.4).

13.3.3 The Effect of Melting Depth

As noted above, ^{230}Th/^{238}U enrichment (increased *r*) is only possible if thorium is more incompatible than uranium. Therefore, to model Th/U fractionation more accurately, it was necessary to refine the determinations of crystal/melt distribution coefficients for U and Th. Since the only mineral phases that host significant inventories of U and Th are garnet and clino-pyroxene (cpx), these minerals have been the focus of attention.

In the first detailed work on this subject, Beattie (1993a) measured solid/liquid partition coefficients (*D*) for cpx that were *greater* for Th than U, implying that melting in the spinel peridotite stability field (above 70 km depth) could not generate the observed Th/U fractionations. On the other hand, Beattie (1993b) showed that garnet has solid/liquid partition coefficients which *are* suitable to generate the observed ^{230}Th/^{238}U enrichments. This was confirmed by LaTourrette *et al.* (1993), who determined D_{Th}/D_U values of 0.1 for garnet. Hence, it was concluded that the Th/U fractionations observed in MORB must originate from melting at greater than 70 km depth, and the resulting liquids must be transported to the surface quickly, before substantial ^{230}Th decay.

These conclusions were questioned by Wood *et al.* (1999) based on modelling of the pressure dependence of cpx/liquid partition coefficients for U and Th. Wood *et al.* predicted that at pressures slightly above 1 GPa (= 10 kbar, equivalent to 35–50 km depth) the D_U/D_{Th} ratio should rise above unity, allowing Th/U excesses to be generated by MORB melting in the spinel peridotite field. Experimental work by Landwehr *et al.* (2001) showed that this claim was true in principle (Fig. 13.23). However, significant ^{230}Th excesses could only be generated by cpx melting at *extremely* low porosity. For example, the excess ^{230}Th values shown in Fig. 13.23 were obtained by assuming a porosity of 10^{-5}, equivalent to a melt fraction of only 0.001%! A somewhat more realistic porosity of 0.1% yields D_{Th}/D_U values for cpx that are 30% lower, and less than half the value obtained in the garnet stability field. Hence, although these results permit limited Th excesses to be obtained in the spinel field, melting in the garnet zone is still likely to be the major source of Th/U disequilibrium.

Further light is thrown on the question of melting depths under ridges by the observation of a relationship between U-series activity ratios and ridge depth. In a study of MORB

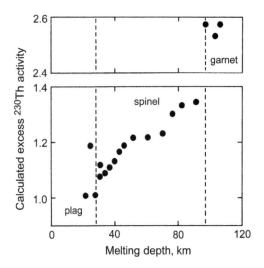

Fig. 13.23 Predicted excess ^{230}Th activity in basaltic magmas as a function of melting depth, based on measured cpx/liquid and garnet/liquid partition coefficients for U and Th, and assuming a porosity of 0.001%. Note the jump in scale on the y axis. Modified after Landwehr *et al.* (2001).

glasses from the Azores plateau, Bourdon *et al.* (1996a) demonstrated an inverse correlation between ^{230}Th/^{238}U activity ratio and water depth. This correlation was subsequently extended by Bourdon *et al.* (1996b) to other ridge segments (Fig. 13.24). This correlation is surprising, in view of the great sensitivity of Th/U disequilibrium to melting porosity. As a result, one would *not* expect to see the greatest disequilibrium in samples from the shallowest part of the ridge, which has the highest upwelling and melting rates.

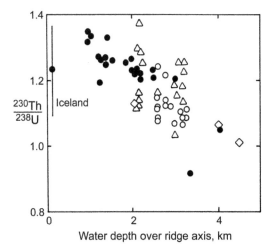

Fig. 13.24 Plot of ^{230}Th/^{238}U activity ('r') for MORB against depth to the ridge axis. (•) = Mid Atlantic; (△) = Gorda, JDF and EPR; (◊) = Tamayo. After Bourdon *et al.* (1996b).

The observation of an inverse correlation between isotope disequilibrium and depth led Bourdon *et al.* (1996a, b) to suggest that the main control on Th/U disequilibrium across the Azores plateau must be the depth at which melting is *initiated*. If melting is initiated at greater depths, the longer melting column leads to greater overall melt fractions, but the longer melting interval within the garnet zone increases Th/U disequilibrium. Bourdon *et al.* attributed the greater melting depth on the Azores plateau to increased heating associated with a mantle plume. An alternative explanation is enhanced contamination of the mantle by eclogite or garnet pyroxenite from the plume (see below). However, Bourdon *et al.* argued that this could not account for the Azores observations, since Th/U excesses do not correlate with isotopic or trace element evidence for source enrichment.

Data from Iceland represent an exception to the generally good correlation between ridge depth and thorium disequilibrium, since they lie well below the main trend. However, Iceland obviously represents an extreme example of plume contamination of a ridge, since the ridge actually emerges above sea level. Therefore, Burdon *et al.* attributed this anomalous behaviour to the much larger upwelling velocity under Iceland (Section 13.3.2). Therefore, in order to further test the depth dependence of Th disequilibrium, it was necessary to examine ridges with large variations in depth without variations in upwelling rate.

The Arctic Ridge segments north of Iceland represent one of the best examples worldwide of such variability (Elkins *et al.*, 2014). Here, water depth varies from 5 to 1 km at the ridge crest, and Elkins *et al.* found that Th disequilibrium is indeed well correlated with ridge depth on these segments. However, other examples of variable ridge depth in the South Indian Ocean and South Atlantic do not show this correlation (Russo *et al.*, 2009; Turner *et al.*, 2015). Instead, they show a larger effect from source variability. Hence it is necessary to examine this phenomenon as a third explanation for Th disequilibrium (in addition to upwelling rate and melting depth).

13.3.4 The Effect of Source Composition

Analogous to the Nd versus Sr mantle of long-lived isotope systems, Condomines *et al.* (1981) demonstrated that MORB and OIB samples form a mantle array on a Th activity ratio versus Sr isotope ratio plot. This suggests that the bulk composition of the mantle source can also lead to variations in activity ratio. The small data set presented by Condomines *et al.* (1981) implied a linear mantle array. However, as further analyses were done, an increasing degree of spread was observed (Fig. 13.25). Some of these displacements off the array can be explained by sea floor alteration due to contamination with ^{230}Th-enriched seawater (in the case of Pacific MORB). In other cases, prolonged magma evolution in the crust allowed ^{230}Th disequilibrium to decay away, as seen in the Canaries and Iceland (Section 13.3.2). However, with time, the number of pristine samples lying outside the linear mantle array also increased. This focussed attention on

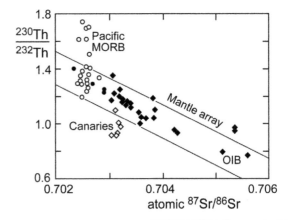

Fig. 13.25 Comparison between Th activity ratios and Sr isotope ratios, showing a negative correlation in most oceanic volcanics. Deviations from the array are attributed to alteration or contamination effects. After Gill and Condomines 1992.

upwelling rate and melting depth as the *principal* factors controlling Th disequilibrium in MORB.

Condomines and Sigmarsson (2000) summarized Th disequilibrium data on MORB and OIB available at the time, with additional data from less depleted sources and more stringent exclusion of altered samples. They showed that ^{230}Th/^{232}Th activity ratio formed a slightly curved correlation against Sr isotope ratios (Fig. 13.26). However, they treated the array as a linear relation for the purpose of calculating correlation coefficients. Significantly, Condomines and Sigmarsson also showed that the correlation of Sr isotope ratio with ^{230}Th/^{232}Th activity (Fig. 13.26a) is stronger than with ^{238}U/^{232}Th activity (Fig. 13.26b). The latter quantity is related to the ^{238}U/^{232}Th atomic ratio (kappa) by multiplying by the ratio of the half-lives (1.34).

Condomines and Sigmarsson suggested that the relative strength of these two correlations (R values in Fig. 13.26) could be used to test the alternative batch melting or Th-ingrowth models. If U/Th *fractionation* during partial melting (in the garnet zone) is the principal origin of excess Th activities, the Th/U (κ) value of the source should best be determined from the ^{230}Th/^{232}Th activity ratio of the products (Fig. 13.26a). On the other hand, if Th isotope ingrowth during source upwelling is the principal origin of excess Th activities, the κ value of the source should best be determined from the ^{238}U/^{232}Th activity ratio of the products (Fig. 13.26b). Since the former defines the best correlation, this supports the batch melting (garnet source) model rather than the Th ingrowth model. However, most of the correlation line is defined by OIB rather than MORB, so the test is only relevant to OIB.

This work was updated by Sims and Hart (2006) with the addition of samples from the EM2 mantle component from Samoa and HIMU from Mt Erebus, Antarctica (the HIMU islands Mangaia and Tubuai do not have young

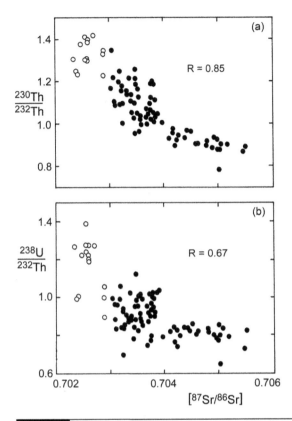

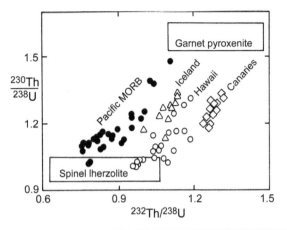

Fig. 13.27 Plot of ^{230}Th/^{238}U activity ('r') against ^{232}Th/^{238}U activity (proportional to the atomic Th/U ratio, κ) to show sub-linear arrays for different ocean ridge and ocean island suites, lying between the fields for garnet–pyroxenite and garnet–peridotite melts. After Sigmarsson *et al.* (1998).

Fig. 13.26 Comparison between the strength of Th–Sr isotope mantle arrays involving (a) ^{238}U/^{232}Th activities, and (b) ^{230}Th/^{232}Th activities, plotted against [atomic] Sr isotope ratio for MORB (o) and OIB (•). Canary Island data are excluded, due to their long pre-eruptive history. After Condomines and Sigmarsson (2000).

lavas suitable for U-series analysis). The Samoan samples significantly extend the distributions in Fig. 13.26 to ^{87}Sr/^{86}Sr values around 0.709, showing that the correlation has a hyperbolic form. Sims and Hart attributed the hyperbola to two-component mixing, but continued to show that Sr isotope ratio is somewhat better correlated with ^{230}Th/^{232}Th activity than ^{238}U/^{232}Th activity. However, departures from the mixing line are larger for MORB samples, suggesting that these are most affected by variations in the melting regime, as described above.

The effect of source composition is expected to be weaker for MORB than OIB. However, Sigmarsson *et al.* (1998) demonstrated strong arrays for both OIB and MORB on a plot of ^{230}Th/^{238}U activity ratio ('r' value) against ^{232}Th/^{238}U activity (related to kappa). This plot (Fig. 13.27) is actually equivalent to the Th/U 'alternative isochron' plot (Section 12.4.3), which was also used for presenting Ra/Th isotope data (Section 13.1.3). On this diagram, MORB and OIB suites define arrays with a slope of unity (the equiline), implying infinite age relative to the ^{230}Th half-life. However, these are believed to

be essentially zero-age volcanic rocks. Therefore, Sigmarsson *et al.* argued that the correlation between r values and κ values is indicative of the source composition, hence implying mixing between different mantle components.

Lundstrom *et al.* (1999) tested this model for lavas of the East Pacific Rise (EPR) in the vicinity of the Siqueiros Fracture Zone, NW of the Galapagos. Based on trace element data and long-lived isotope signatures (Sr–Nd) they identified magmas from an enriched source (E-MORB), a depleted source (D-MORB) and mixtures between these components (N-MORB). These three types of sources also gave rise to distinct differences in Th isotope disequilibrium, with E-MORB showing excess ^{230}Th activities, whereas D-MORBs were essentially in secular equilibrium. Hence, Lundstrom *et al.* concluded that E-MORBs were sampling garnet-bearing veins that were distributed in a source with 'marble cake' structure (Section 6.1.5), whereas D-MORBs sampled the garnet-free peridotite 'host' (Fig. 13.28a). N-MORBs were then attributed to mixing between these two end-members.

This interpretation was challenged by Sims *et al.* (2002), who analysed a suite of samples from the axial graben of the EPR, just north of the Siqueiros Fracture Zone. On the plot of Th isotope disequilibrium 'r' (^{230}Th/^{238}U activity) against Th/U ratio, they found that their axial graben samples lay on the mixing line between the E-MORB and D-MORB of Lundstrom *et al.* (Fig. 13.28a). However, on a plot of Th isotope disequilibrium ('r') against long-lived isotope tracers, they found that the axial graben samples did not lie on mixing lines between the E-MORB and D-MORB end-members defined by the Siqueiros Fracture Zone (Fig. 13.28b). Sims *et al.* therefore concluded that variations of U-series isotope activities were due to polybaric melting of a homogeneous source, and not due to mixing between melts from compositionally distinct sources.

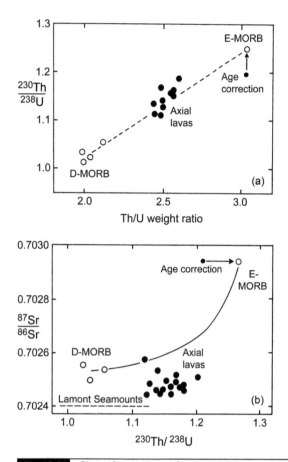

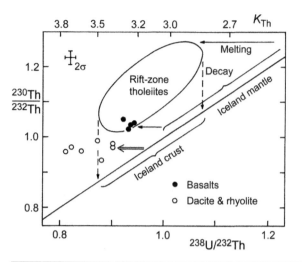

Fig. 13.29 U–Th isochron diagram showing the development of distinct Th activity ratios for Icelandic mantle and crust, relative to felsic magmas from Hekla volcano. After Sigmarsson *et al.* (1991).

Fig. 13.28 Plots of isotope data for the axial graben of the East Pacific Rise (•) compared with D-MORB and E-MORB samples from the Siqueiros Fracture Zone (○). (a) Th activity ratios; (b) Th activity as a function of Sr isotope ratio. Modified after Sims *et al.* (2002).

However, studies of long-lived isotope systems have suggested that homogeneous isotope signatures are due to magma mixing under the ridge rather than sampling of a homogeneous mantle source (Section 6.1.4). In fact, a re-examination of the Sims *et al.* paper suggests that basalts from the Lamont Seamounts (Fornari *et al.*, 1988) are a better indicator of the nature of D-MORB than the basalts of the Siqueiros Fracture Zone. Firstly, the Lamont Seamounts are much closer to the study area than the Siqueiros Fracture Zone, and secondly, Pb isotope data from the axial graben samples actually form an array that falls within the larger envelope of the Lamont Seamount array (not shown here). The Lamont Seamounts have Sr isotope signatures indicative of a more depleted source (Fig. 13.26b) which would better explain the compositions of the axial graben samples by mixing with E-MORB. Hence it appears that the axial graben samples of the East Pacific Rise can be explained by mixing between magmas from a 'marble cake' mantle. However, polybaric melting may also play a role.

13.3.5 Crustal Melting and Contamination

Work on Icelandic lavas by Sigmarsson *et al.* (1991) has shown that Th isotope data can also be used to investigate the genesis of felsic rocks in oceanic environments. Two alternative models for these rocks involve either direct fractionation of mafic magmas or partial melting of pre-existing mafic crust. However, because of the young age of this crust, conventional radiogenic tracers such as Sr and Nd cannot resolve these models. In contrast, the short half-life of ^{230}Th may allow a solution to this problem.

Rift-zone tholeiites from Iceland are displaced to the left of the equiline by partial melting, but after 0.5 Ma (when their excess ^{230}Th activity has decayed back to secular equilibrium) they are displaced downwards to a different range of Th activity ratios on the equiline (Fig. 13.29). Consequently, historical-age felsic melts produced by anatexis of Icelandic crust will have a lower ^{230}Th/^{232}Th activity than direct magmatic differentiates of juvenile basic magma. Sigmarsson *et al.* found that dacitic volcanics from the Hekla volcano in southern Iceland had ^{230}Th/^{232}Th activities too low to be derived from contemporaneous mantle-derived magmas (horizontal single arrows), but consistent with the melting of older crust (double arrow). Hekla rhyolites were then modelled by magmatic fractionation of the dacite melts (Fig. 13.29).

Th isotope studies of volcanic systems from Iceland have also shown the effect of contamination of juvenile basic magmas by crust a few Ma in age (Hemond *et al.*, 1988; Sigmarsson *et al.*, 1992). In major volcanic centres, where the oxygen isotope composition of the crust has been overprinted by meteoric hydrothermal convection, Th activity ratios also correlate with δ^{18}O (Hemond *et al.*, 1988).

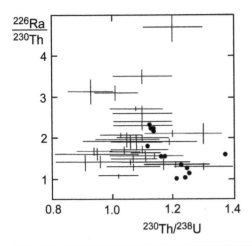

Fig. 13.30 Plot of early radium–thorium disequilibrium data for young MORB glasses believed to represent initial ratios. (+) = α counting; (•) = mass spectrometric data. Modified after Spiegelman and Elliott (1993).

13.4 Short-Lived Species and Melting Models

Observations of disequilibrium in shorter-lived species such as ^{226}Ra and ^{231}Pa place even tighter time constraints on melting models than Th/U data. In view of their short half-lives, it is critical to analyse recent or young dated samples. For OIB, this usually means collection from historical eruptions of known ages. For MORB, this means unaltered glasses collected by submersible (rather than dredge) from the ridge axis. Alternatively, Po–Pb dating can be used to select samples with ages less than 1000 yr (e.g. Rubin *et al.*, 1994). In the following examples, it will be assumed that the activities reported are all initial ratios indicative of magmatic processes.

13.4.1 ^{226}Ra and Melting Models

Early α counting measurements on MORB glasses by Rubin and Macdougall (1988) revealed ^{226}Ra/^{230}Th ratios as high as 2.7 (even higher in some later data), despite the short 8000 yr equilibration time of ^{226}Ra (Fig. 13.30). Because radium is even more incompatible than Th in deep mantle melting, essentially instantaneous extraction would be required to preserve Ra anomalies generated at the base of the melting column (ca. 80 km depth). However, it seems questionable whether these Ra excesses can be preserved during magma ascent to the surface.

In view of these problems, and the very strong ^{226}Ra enrichments seen in the MORB α-counting data, Rubin and Macdougall (1988) speculated that conventional batch or dynamic melting models might not be adequate to explain the Ra data, and that some kind of kinetic process (disequi-

librium melting) might be responsible for the radium enrichments. Such a model would differ from the melting models discussed above, in which the inventory of trace elements within each mineral grain undergoing melting is believed to be homogenized by diffusion before any melt is removed from the surface of the grain (equilibrium melting). In contrast, disequilibrium melting could strip cations out of mineral grains in layers, like the shells of an onion.

Qin (1992) argued that disequilibrium melting is almost unavoidable in the generation of ^{226}Ra excesses, since the secular equilibration time of this nuclide (8000 yr) is comparable with the 10^4 yr diffusional equilibration time of cations in a mineral grain at the temperature of basaltic melting (Section 6.1.2). Qin then took this argument further, suggesting that differing rates of volume diffusion for different cations could cause *diffusional fractionation* of U-series nuclides and other incompatible element couples. However, this overlooks the fact that the source spends a minimum of 100 ka in the melting zone, for a 50 km-deep melting zone upwelling at 5 cm/yr. These figures suggest that disequilibrium melting cannot be maintained over the whole depth of the melting column under ridges, but might occur at the base of the column in such as way as to hold back complete release of ultra-incompatible elements into the melt.

Spiegelman and Elliott (1993) also believed that rates of MORB magma ascent were probably too slow to preserve radium excesses generated by conventional dynamic melting (whose U-series signatures are dominated by the deepest melt fraction). Instead, they proposed that the small melt fractions generated at the base of the melting column do not easily escape from the source, but ascend relatively slowly to the surface by a mechanism they termed equilibrium porous flow (EPF). They proposed that this process was also accompanied by an increase in the porosity of the melting system from a value of zero at the base of the melting column, to a maximum of ca. 0.5% at the top of the melting column. This model has the effect of holding back the early release of Th, which can then decay to Ra as the melt and source ascend together. The ^{226}Ra extracted from shallower levels in the melting column could then reach the surface without undergoing significant decay. This model can therefore be classified as a particular kind of ingrowth model.

One of the best tests of the viability of different melting models is their ability to explain the observed correlations between ^{226}Ra/^{230}Th and ^{230}Th/^{238}U activities. In the case of MORB, the early α counting data for MORB showed no correlation, while mass spectrometric data (Volpe and Goldstein, 1993) suggested a negative correlation (Fig. 13.30). This was confirmed by Lundstrom *et al.* (1999) and Sims *et al.* (2002) for very young samples from the axial graben of the East Pacific Rise (EPR). These samples were collected by submersible, including an eruption that was actually in progress on the sea floor in 1992. The strong inverse correlation (Fig. 13.31) suggests that the EPR magmas were produced by mixing between E-MORB and D-MORB end-members with distinct

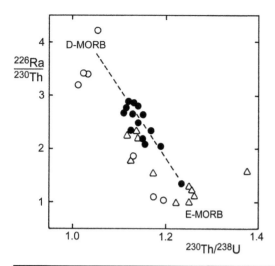

Fig. 13.31 Plot of $^{226}Ra/^{230}Th$ versus $^{230}Th/^{238}U$ activity ratios to show an inverse correlation in MORB, attributed to mixing between D-MORB and E-MORB melting processes. (•) = EPR; (△) = JDF–Gorda; (○) = Siqueiros Fracture Zone. After Sims et al. (2002).

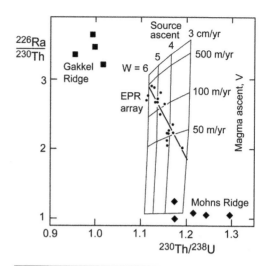

Fig. 13.32 Radium–thorium disequilibrium plot showing an attempt to model EPR magmas by variable mantle upwelling rate (W) and variable magma ascent rate (v) at constant porosity (0.075%). (•) = EPR; (■, ◆) = Arctic Ridge segments. Modified after Stracke et al. (2006).

origins. The former have large ^{230}Th excesses but little excess ^{226}Ra, pointing to an origin from deep melting in the garnet zone, whereas the latter have large ^{226}Ra excesses but little excess ^{230}Th, implying a shallow origin after radium ingrowth.

Lundstrom et al. (1999) suggested that E-MORBs were derived principally by a channel-flow melting process, whereas D-MORBs were generated by a porous-flow melting process. These two process must have been going on at the same time under the ridge, and could simply reflect heterogeneity in the melting process. For example, garnet peridotite veins melt preferentially to form ^{230}Th-enriched E-MORB melts. These may rise to the surface along channels that begin as zones of porous flow, mixing with a ^{226}Ra-enriched D-MORB fraction from their wall-rocks. As melt focussing increases, a few of the zones of porous flow probably evolved into open magma conduits (Kelemen et al., 1997).

The difficulties of bringing radium anomalies from the melting zone to the surface were somewhat alleviated by evidence from subduction-related magmas for very rapid magma ascent (Section 13.5.2). This therefore encouraged a re-examination of more conventional dynamic melting models as an explanation for large excess radium activities. For example, Stracke et al. (2006) made a more generalized examination of these models by specifying the magma ascent rate as an independent variable, in addition to the source ascent rate and the melting porosity.

The inverse correlation between ^{226}Ra and ^{230}Th activities in EPR MORB might be explained by anti-correlated variations of mantle upwelling rate and porosity, but this is antithetical, because these variables should be positively correlated (faster upwelling should yield more melting).

A more realistic explanation was achieved by fixing the porosity and exploring the consequences of a positive correlation between mantle upwelling rate (W) and magma ascent rate (v), shown by the grid of model lines in Fig. 13.32. This modelling showed that ^{230}Th-depleted ridge crest EPR magmas could be explained by rapid magma ascent (ca. 500 m/yr) from a source undergoing rapid upwelling (6 cm/yr). In contrast, ^{230}Th-enriched flank magmas could be explained by slow magma ascent (ca. 30 m/yr) from a source undergoing slow upwelling (3 cm/yr). Thus, the negative correlation of EPR basalts in Fig. 13.32 is effectively achieved by generating high initial ^{226}Ra excess in all melts at very low porosity (0.075%) and allowing ^{226}Ra to decay away during the slow ascent of ridge flank magmas. This turns an initial positive correlation between radium and thorium into the observed negative trend.

As Stracke et al. admitted, this model has problems explaining the more extreme radium disequilibrium seen in D-MORB magmas from the EPR and more recently on the Artic Gakkel Ridge (solid squares in Fig. 13.32). The deep Gakkel Ridge (Elkins et al., 2014) and EPR D-MORB both have large radium disequilibria accompanied by very small Th disequilibria, which cannot be produced by the model parameters in Fig. 13.32. However, the Gakkel Ridge data can readily be explained by melting in the spinel stability field, where Th has partition coefficients similar to U. In contrast, the shallower Mohns and Knipovich ridge segments have low ^{226}Ra but large ^{230}Th activities similar to Pacific E-MORB, and indicative of deep melting in the garnet field. Hence we can see that for MORB, the melting depth is the most important consideration, overriding most others. This favours the equilibrium porous flow model for these types of magmas, since

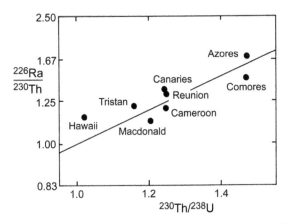

Fig. 13.33 Plot of $^{226}Ra/^{230}Th$ versus $^{230}Th/^{238}U$ activity ratios for OIB, showing a correlation consistent with a simple melting model. Flipped from Chabaux and Allegre (1994).

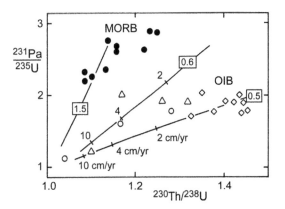

Fig. 13.34 Plot of excess ^{231}Pa versus ^{230}Th activities to compare MORB (•) with different ocean islands: (○) = Hawaii; (△) = Iceland; (◇) = Grande Comore. Lines represent constant $^{231}Pa/^{230}Th$ activity ratios and are marked with upwelling rates in cm/yr, modelled assuming a melting porosity of 0.1% and an integrated melt fraction of 4%. After Bourdon et al. (1998).

the EPF model is sensitive to the pressure at which melting *ends*, when melt is finally extracted from the source (Lundstrom et al., 1995). This model is therefore a better fit than dynamic melting to explain ridges with very slow upwelling rates.

In contrast to MORB, Chabaux and Allegre (1994) showed that radium and thorium disequilibria are positively correlated in OIB, whose small ^{226}Ra excesses may be preserved in OIB due to their volatile-charged alkaline chemistry, causing rapid magma ascent. In the original paper, Chabaux and Allegre presented their data on an 'inverse' plot of $^{230}Th/^{226}Ra$ versus $^{230}Th/^{238}U$ activity ratio. However, this is confusing relative to the plots used by all other workers, and the diagram is therefore flipped to the normal form in Fig. 13.33. The larger degree of disequilibrium in plumes such as the Azores and Comores, relative to Hawaii, is inversely correlated with the diapiric mantle plume upwelling rate, which is proportional to the melting porosity.

According to the dynamic melting model, the greater Ra excesses in slowly upwelling plumes are due to greater ingrowth of radium after the first melts have been tapped off. However, Sims et al. (2008) calculated high maximum porosities of 2–6% to generate the Ra disequilibrium in Samoan basalts, similar to the porosity needed to fractionate Sm/Nd. Hence they deduced that for enriched OIB sources, Ra ingrowth in the melting column was not required. Based on the modelling by Stracke et al. (2006) discussed above, the most important factor for the preservation of large radium disequilibria is rapid magma ascent. Provided that this is achieved, the large radium and thorium disequilibria in the Azores and Camores can be achieved by extraction of melts at low porosity in the garnet stability field.

13.4.2 ^{231}Pa and Melting Models
The first comprehensive Pa isotope study of oceanic volcanics (Pickett and Murrell, 1997) revealed much larger variations

in $^{231}Pa/^{235}U$ activity ratios than for $^{230}Th/^{238}U$, but comparable with the results of radium/thorium analysis discussed above. However, the half-life of ^{231}Pa is more comparable with ^{230}Th than ^{226}Ra, so the large variations in $^{231}Pa/^{235}U$ activity ratio must be attributed primarily to differences in partition coefficient. In other words, protactinium is even more incompatible than thorium during mantle melting processes. Pickett and Murrell also observed that MORB showed much greater $^{231}Pa/^{235}U$ activity ratios than OIB for a given $^{230}Th/^{238}U$ ratio.

These findings were confirmed by Bourdon et al. (1998), who analysed additional OIB samples with large $^{230}Th/^{238}U$ activity ratios from Grand Comore (Fig. 13.34). These $^{231}Pa/^{235}U$ activity ratios cannot have been affected by radioactive decay, since excess activities are also seen in shorter-lived ^{226}Ra. Instead, the lower $^{231}Pa/^{230}Th$ activities in OIB relative to MORB were attributed to a greater fraction of garnet in plume sources. This is shown in Fig. 13.34 by lines of constant $^{231}Pa/^{230}Th$ activity, modelled according to different cpx/garnet abundances in the source. This reflects the fact that significant excess $^{230}Th/^{238}U$ activity ratios can only be generated in the garnet stability field (Section 13.3.3), whereas excess $^{231}Pa/^{235}U$ activity ratios can be generated in the spinel field, because protactinium is twice as incompatible as uranium during cpx melting.

Plume-derived lavas have a larger range of $^{230}Th/^{238}U$ activities in Fig. 13.34, reflecting more variable upwelling rates, even within a single hot-spot. For example, Hawaii and Iceland show very variable upwelling rates, from ca. 2 cm/yr to 30 cm/yr. On the other hand Grande Comore plume shows a more consistent but lower upwelling rate coupled with an enriched source. In a more detailed study of the Hawaiian plume, using combined Pa–Th data, Sims et al. (1999)

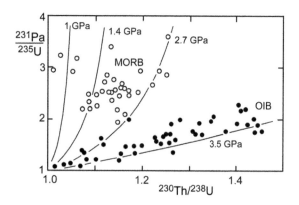

Fig. 13.35 Plot of excess ^{231}Pa versus ^{230}Th activities to compare MORB and OIB signatures with modelled melting depths. Modified after Bourdon and Sims (2003); Bourdon *et al.* (2005).

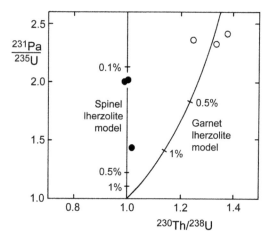

Fig. 13.36 Plot of excess ^{231}Pa activity against excess ^{230}Th activity showing age-corrected data from the south-west USA. Solid symbols = lithospheric melts, open symbols = asthenospheric melts. Curve shows the result of simple batch melting of spinel or garnet peridotite at different melt fractions, using reasonable partition coefficients. After Asmerom *et al.* (2000).

calculated source upwelling rates as high as 20–80 cm/yr for Kilauea and Mauna Loa, but only 2–6 cm/yr in Hualalai and Mauna Kea. They attributed these variations in upwelling velocity to the radial distances of these different volcanoes from the centre of the plume, believed to lie under the SE coast of the main island.

Subsequent work strengthened the ^{231}Pa/^{230}Th correlation line in OIB, but weakened the correlation in MORB (e.g. Bourdon *et al.*, 2005). This is because mantle upwelling rates are the dominant factor controlling thorium disequilibrium in OIB, whereas it was shown above that melting depth and source composition are important factors affecting ^{230}Th/^{238}U activities in MORB. For example, modelling the effect of melting depth (pressure) on ^{231}Pa/^{230}Th activities (Fig. 13.35) shows that OIB can be modelled by a constant melting pressure of 3.5 GPa (ca. 120 km depth), whereas the melting pressure for MORB varies from 1.4 to 2.7 GPa (ca. 50–95 km depth).

13.4.3 Sources of Continental Magmas

Evidence that excess ^{230}Th activities require the presence of garnet except at very low melting porosities (Section 13.3.3) has provided a basis for understanding U-series systematics in young continental lavas. Asmerom and Edwards (1995) demonstrated this approach in comparing young (<10 ka) basalts from the Pincate volcanic field of the Basin and Range province in Mexico, and the 'San Francisco' volcanic field of the Colorado Plateau.

Nd isotope and geochemical data are consistent with an asthenospheric origin for basalts of the Basin and Range, but a lithospheric origin for those of the Colorado Plateau. U-series analysis was performed to see whether basalts from these different sources would possess different ^{230}Th signatures, reflecting melting at different depths. The results (Fig. 13.36) showed that magmas regarded as asthenospheric had excess ^{230}Th activities typical of MORB magmas

(r = 1.2–1.35), whereas those attributed to lithospheric melting had no excess ^{230}Th activity (r = 0.99–1.02). Some possible explanations for equilibrium ^{230}Th/^{238}U activity ratios could be large degrees of mantle melting or slow ascent to the surface. However, both of these possibilities were ruled out by the observation of large excess ^{231}Pa activities in these rocks (^{231}Pa/^{235}U activity = 2.0). Melting in the spinel peridotite field can generate excess ^{231}Pa, but cannot generate significant excess ^{230}Th activity (Section 13.3.3). Therefore Asmerom and Edwards suggested that alkali basalts of the Colorado Plateau were probably generated by shallow melting of subcontinental lithosphere within the spinel peridotite field, whereas Basin and Range magmas were produced by deeper asthenospheric melting in the garnet field.

Subsequent work (Asmerom *et al.*, 2000) widened this study to include two samples from the Rio Grande Rift (RGR). The Zumi Band tholeiite is on the margin of the RGR and is attributed to a lithospheric source under the Colorado Plateau, whereas the Potrillo basalts are located in the southern central part of the RGR and attributed to an asthenospheric source. U-series data (Fig. 13.36) are consistent with these predictions. All of the alkali basalts have large excess ^{231}Pa activities, consistent with small degree melting and rapid magma ascent. In contrast, the tholeiite has a smaller ^{231}Pa excess, either due to a larger degree of melting or slower ascent. On the other hand, excess ^{230}Th activities are again grouped according to source type, with both asthenospheric suites yielding large excess activities, whereas both lithospheric suites yield data within error of the equiline.

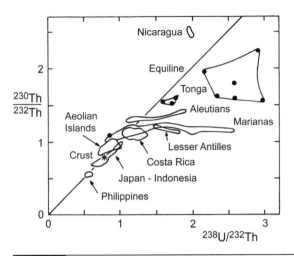

Fig. 13.37 U–Th isochron diagram showing representative data for subduction-related magmas. Star = average crust. After Hawkesworth et al. (1991).

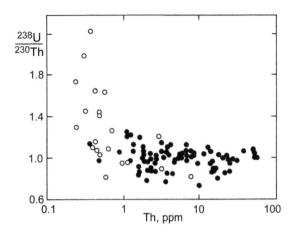

Fig. 13.38 Plot of $^{238}U/^{230}Th$ activity (= U/Th fractionation from the equiline) against total Th content of arc volcanics. Open symbols are from the Tonga and Mariana arcs. After McDermott and Hawkesworth (1991).

13.5 Subduction Zone Processes

Subduction-related magmatism represents one of the most complex environments of magma genesis, since there are numerous potential sources of magmas and fluids in the subducting slab, the mantle wedge and the overlying arc crust (Section 6.6). In this situation, evidence from U-series nuclides can provide additional constraints, especially on the timing of fluid metasomatism and magma genesis.

13.5.1 U–Th in Arc Magmas

The first comprehensive study of Th isotope systematics in subduction-related magmas was undertaken by Hemond (1986). This work showed a distribution clustering near the equiline on the U–Th isochron diagram, but with strong departures from the Th–Sr isotope mantle array seem in MORB and OIB.

The incoherent behaviour of Th–Sr isotope systematics suggested more complex processes at subduction zones, compared to ridges and hot-spots. With additional work (Gill and Williams, 1990; McDermott and Hawkesworth, 1991), systematic variations began to emerge on the U–Th isochron diagram (Fig. 13.37), where continental arcs tend to lie on the U-depleted side of the equiline, while oceanic arcs lie to the U-enriched side. High elemental U/Th ratios (low Th/U) have long been known in the island arc tholeiite series (Jakes and Gill, 1970), so the distribution of these data to the right of the equiline is not surprising. This effect is seen clearly in oceanic arcs such as Tonga and the Marianas, but is small or absent in associated back-arc volcanics. For these reasons, the effect is best explained by U metasomatism from the subducted slab into the melting zone in the overlying mantle wedge.

The slab-derived uranium flux can be seen most clearly when the overlying wedge is LIL-depleted (e.g. Tonga, Marianas), but in arcs with a more enriched wedge, the uranium flux may be swamped by U and Th derived by normal partial melting processes. Hence, continental arcs, underlain by enriched mantle lithosphere, tend to show U/Th behaviour similar to within-plate basalts. This effect can be seen by plotting the $^{238}U/^{230}Th$ activity ratio ($1/r$ in Section 13.3.1) against the total Th content of the rock (Fig. 13.38), thus showing that arcs with strong U/Th enrichment are characterized by low total Th contents (McDermott and Hawkesworth, 1991).

Magmas to the right of the equiline in Fig. 13.37 must reflect U/Th fractionation shortly before eruption, in order to preserve isotopic disequilibrium. Independent evidence for the role of slab-derived fluids in this process comes from correlated $^{10}Be/^{9}Be$ and $^{238}U/^{230}Th$ ratios in the Southern Volcanic Zone of the Andes (Sigmarsson et al., 1990). Because ^{10}Be is a cosmogenic isotope, it can only be introduced into the melting zone of arc magmas by subduction of ocean floor crust and sediment (Section 14.4.6). Therefore, a positive correlation between $^{10}Be/^{9}Be$ and $^{238}U/^{230}Th$ ratios suggests a similar location for uranium enrichment of arc magmas.

Several of the Southern Volcanic Zone samples have $^{238}U/^{230}Th$ ratios close to the equiline (in common with other continental arcs), but these all have elevated cosmogenic beryllium (Fig. 13.39). Therefore, if we project the array in Fig. 13.39 back to zero cosmogenic beryllium, we can estimate the $^{238}U/^{230}Th$ ratio of the wedge-derived component in subduction-related magmas. This has a value of ca. 0.82 ($r = 1.2$) which is typical of asthenosphere-derived magmas from ocean ridges. This implies that the type of decompression melting seen at ridges (due to mantle diapirism) also occurs to some extent under arcs. This model is supported by ^{231}Pa data (Section 13.5.3).

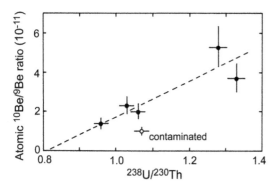

Fig. 13.39 Plot of [10]Be/[9]Be ratio against [238]U/[230]Th activity ratio in the Southern Volcanic Zone of the Andes, showing correlated enrichments. The open circle indicates a sample that has been perturbed by contamination in the continental crust. After Sigmarsson et al. (1990).

A second distinct feature of subduction-related magmas on the U–Th isochron diagram is the extension of some arc suites to [232]Th-enriched compositions towards the origin (e.g. Philippines, Indonesia, Fig. 13.37). Because these suites lie close to the equiline, it appears that U/Th fractionation was relatively ancient. This is confirmed on a plot of κ_{Pb} against κ_{Th} in Fig. 13.40. Correlated increases in these values must reflect ancient U–Th fractionation events, since κ_{Pb} is controlled by long-lived U and Th isotopes. The best explanation of these data is contamination of arc magmas by partial melts of subducted sediments, which have appropriate compositions on both the U–Th isochron diagram and κ–κ dia-

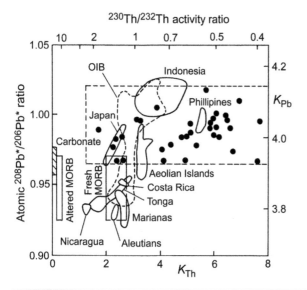

Fig. 13.40 Plot of κ_{Th} against κ_{Pb} to show correlated high values in some arcs and in ocean floor sediments (•). Fields for OIB, MORB, altered MORB and marine sediments are shown for reference. After McDermott and Hawkesworth (1991).

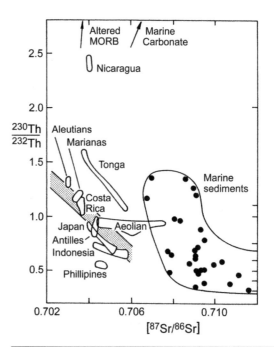

Fig. 13.41 Th–Sr isotope diagram showing possible mixing models to explain the departure of arc magmas from the mantle array of MORB and OIB compositions (shaded). After McDermott and Hawkesworth (1991).

grams (McDermott and Hawkesworth, 1991). Some sediment inevitably escapes the melting process, and is then recycled into the deep mantle.

Processes of fluid metasomatism and sediment contamination in arcs can be summarized on a Th activity versus Sr isotope plot (Fig. 13.41). On this diagram, the Aeolian arc defines an array between normal depleted mantle and subducted sediment. In contrast, Tongan data trend upwards towards altered MORB and marine carbonate, from whence the U-enriched fluid flux is probably derived. A final note must be made regarding the Nicaraguan data. These have very high [230]Th contents, reminiscent of a metasomatized source, but actually fall on the U-depleted side of the equiline (Fig. 13.36). This unusual signature is probably best attributed to a source which suffered U metasomatism some time prior to magma generation.

More recent studies on arc lavas have attempted to put tighter limits on the timing of fluid influx and magma genesis in the mantle wedge. Evidence relevant to the timing of fluid influx came from new studies by Turner et al. (1997) and Elliott et al. (1997) on the Tonga–Kermadec and Marianas arcs. These data sets revealed low positive slopes on the U–Th isochron diagram, especially if the lower bound to each data set was considered (Fig. 13.42). This lower bound is believed to be most indicative of the typical time from fluid metasomatism of the mantle source to the time of eruption. In contrast, individual data points lying above the line are attributed to magma batches that had longer residence times in U-rich high-level magma chambers. The lower

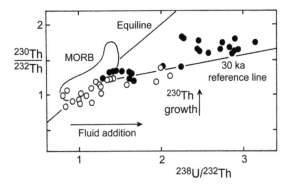

Fig. 13.42 Data arrays from the Tongan arc (•) and Kermadec arc (○) on a U–Th isochron diagram, showing arrays with possible age significance. After Turner et al. (1997).

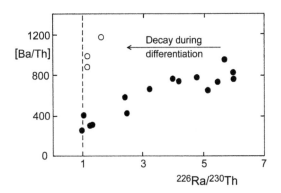

Fig. 13.43 Plot of Ba/Th weight ratio against $^{226}Ra/^{230}Th$ activity, showing a positive correlation between ^{226}Ra enrichment and Ba/Th ratio, except in evolved lavas. (•) = basalts; (○) = evolved lavas. After Turner et al. (2000).

bounds to each data set gave slope ages of 30 ka, 50 ka and 30 ka respectively for the Tonga, Kermadec and Marianas arc, suggesting quite rapid processes of magma genesis and ascent in oceanic arcs. However, because of its half-life of only 1600 years, ^{226}Ra analysis offers the opportunity of seeing even more short-term processes in arc magmas.

13.5.2 Ra–Th in Arcs

Early work on the Ra–Th systematics of arc lavas (Gill and Williams, 1990) revealed large ^{226}Ra excesses in many lavas, with a weak positive correlation between ^{226}Ra and ^{238}U enrichment. This led Gill and Williams to speculate that ^{226}Ra enrichment might be linked to subduction-related fluids. This was confirmed by a study on several volcanic centres of the Lesser Antilles arc (Chabaux et al., 1999), which showed a good correlation between $^{226}Ra/^{230}Th$ and $^{238}U/^{230}Th$ activity ratios, suggesting that both ^{238}U and ^{226}Ra enrichment were caused by subduction-related fluids. This model was confirmed in a study of the Tonga–Kermadec arc (Turner et al., 2000), which found the largest ^{226}Ra excesses in the most depleted basalts with the highest Ba/Th ratios (Fig. 13.43). Turner et al. also found that as the silica contents of the magmas increased, ^{226}Ra excesses declined, reflecting increasing residence times in differentiating magma chambers.

Evidence that ^{226}Ra enrichment is caused by fluid metasomatism is problematical, because the earlier U–Th evidence from the Tongan arc suggested that fluid enrichment occurred at ca. 30 ka. To explain this problem, Turner et al. suggested that multiple fluid injections could have invaded the mantle wedge (Fig. 13.44). According to this model, an earlier event would have introduced both ^{238}U and ^{226}Ra into the wedge, but the latter nuclide soon decayed away. Meanwhile, as ^{226}Ra was decaying in the wedge, it was being replenished in the Th-rich subducting slab. Therefore, when a second episode of fluid metasomatism occurred, the replenished inventory of ^{226}Ra was again released into the mantle wedge. On this second occasion, the metasomatic event was quickly followed by the extraction of a basaltic

melt, which therefore picked up ^{238}U from the first event and ^{226}Ra from the second event.

Further confirmation of the role of fluid metasomatism in causing ^{226}Ra enrichments was provided by the observation of a positive correlation with beryllium isotope ratios in Andean lavas (Fig. 13.45). Since cosmogenic ^{10}Be can only originate from subducted pelagic sediment (Section 14.4.6), the correlation with ^{226}Ra provides strong evidence for the origin of these enrichments in the subduction-related fluids (Sigmarsson et al., 2002). However, Sigmarsson suggested that the data arrays with shallow slopes on the U–Th isochron diagram might be mixing lines produced by a multi-stage extraction of fluids from the slab, rather than isochrons dating the first of a two-phase process of fluid metasomatism.

As well as throwing more light on the timing of fluid metasomatism, ^{226}Ra evidence from arc basalts also provides new constraints on the rates of ascent of basic magmas. This is because the process of fluid metasomatism must occur

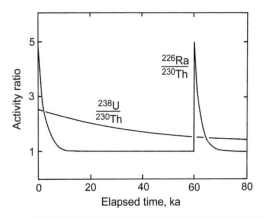

Fig. 13.44 Plot of excess activities of ^{226}Ra and ^{238}U against time to show two-stage metasomatism proposed to explain the enrichment of Tongan lavas in U-series nuclides from subduction-related fluids. After Turner et al. (2000).

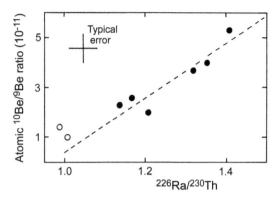

Fig. 13.45 Plot of atomic beryllium isotope ratio against ^{226}Ra/^{230}Th activity ratio for lavas from the Southern Volcanic Zone of the Andes. (○) = lavas which probably lost their excess ^{10}Be and ^{226}Ra during a longer magmatic history in the crust. After Sigmarsson et al. (2002).

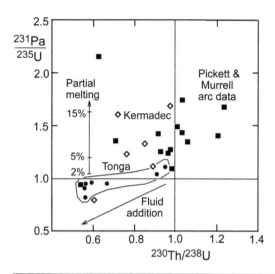

Fig. 13.46 Plot of Pa versus Th disequilibrium for arc volcanics, showing the trend resulting from U addition in slab-derived fluid. Data: (■) = Pickett and Murrell (1997); (●, ◇) = Bourdon et al. (1999).

at nearly 100 km depth, since the fluids are released by pressure-induced breakdown of amphibole in the subducting slab (Section 6.6). In contrast, it has been argued that ^{226}Ra enrichments under ocean ridges occur at quite shallow depths, due to ingrowth in the upwelling mantle source (Section 13.4.1).

To explore these constraints in more detail, Turner et al. (2001) measured Ra–Th activities in seven additional arc systems, for comparison with the Tongan data. It was found that these other arc systems display behaviour similar to the Tongan arc, but less extreme in magnitude. Thus the other arcs also displayed positive correlations of initial ^{226}Ra/^{230}Th activity ratios with ^{238}U/^{230}Th activity and with Ba/Th weight ratio, and inverse hyperbolic relationships with silica content. These findings confirm that ^{226}Ra enrichments in arcs are a general product of metasomatism by slab-derived fluids, but they also place tight constraints on rates of magma ascent, suggesting that primary subduction-related magmas probably rise at rates approaching 1 km/yr from their mantle source at ca. 100 km depth. This implies that these magmas rise through open channel-ways rather than by melt percolation. This is likely to be a common feature in melting zones with large upwelling and magma production rates, but less likely in situations with low upwelling rates such as slow-spreading ridges.

13.5.3 U–Pa in Arcs

Protactinium has chemical behaviour similar to thorium, remaining essentially immobile in aqueous fluids. This makes the U–Pa system a useful complement to U–Th and Th–Ra in subduction-related magmatism, and may further help to disentangle the effects of the slab-derived fluids from melting behaviour in the overlying mantle wedge.

The first ^{231}Pa data for subduction-related volcanic rocks were obtained by Pickett and Murrell (1997), and showed Pa

excesses in all samples (Fig. 13.46) except a Tongan andesite. This analysis was confirmed by a larger suite of Tongan samples presented by Bourdon et al. (1999). Overall, the data define a positive trend of Pa versus Th disequilibrium which can be explained by addition of U to the magma source in a slab-derived fluid. The greater addition of U to the Tongan magma source is consistent with previous evidence (Fig. 13.37). The positive intercept of ^{231}Pa/^{235}U activity for samples with ^{230}Th/^{238}U activity = 1 shows that signatures in apparent Th/U secular equilibrium result from opposing enrichments of U from the slab-derived fluid and Th from melting of the mantle wedge, and are not indicative of an ancient source. On the other hand, the anomalously high ^{231}Pa activity in a Lesser Antilles andesite was attributed to sediment contamination.

Bourdon et al. (1999) used the Tongan data to construct a U–Pa–Nb pseudochron diagram (Fig. 13.47), which provides an estimate of the timing of U enrichment of the arc magma source. In this diagram, niobium is used as a geochemical proxy for protactinium, in a manner similar to the Th–Ra–Ba pseudochron diagram (Fig. 13.5). However, in the present case this is a mantle pseudochron, since the event being dated is uranium metasomatism of the magma source. The age of 60 ka should only be regarded as approximate, since mantle isochrons are based on the principle that parent and daughter (in this case U and Pa) were not fractionated during the melting process, which is known to be a false assumption. Since Fig. 13.46 shows that melting causes Pa-enrichment of the magma, the pseudochron age should yield a maximum value.

Turner et al. (2006) analysed a larger suite of arc samples from seven different subduction zones, and observed

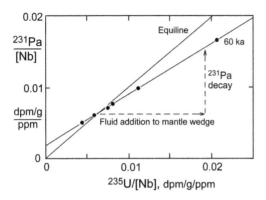

Fig. 13.47 U–Pa–Nb pseudochron diagram for Tongan volcanics, giving an approximate age for U metasomatism of the arc magma source. After Bourdon *et al.* (1999).

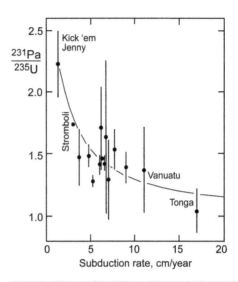

Fig. 13.48 Plot of Pa disequilibrium against subduction rate, showing averages for a variety of island and continental arc magmatic suites. After Huang and Lundstrom (2007).

an inverse correlation between $^{231}Pa/^{235}U$ activity and subduction rate. Huang and Lundstrom (2007) strengthened this correlation by analysing samples from the submarine volcano Kick'em Jenny in the Lesser Antilles arc. Averages for 19 subduction zone segments are compared in Fig. 13.48 with a dynamic melting model in which the subduction rate determines the movement of the mantle wedge into the melting zone in a manner analogous to upwelling under ridges. The difference here is that the mantle wedge is moving *downwards* into the melting zone, during or shortly after the addition of slab-derived fluids, which provide a flux to lower the peridotite solidus (Thomas *et al.*, 2002; Turner *et al.*, 2006).

It should be noted that individual subduction zone segments show large degrees of scatter about the activity-ratio means in Fig. 13.48, showing that rate of subduction is not the only factor affecting $^{231}Pa/^{235}U$ activities in subduction-related magmas. However, it appears to be the most important factor. Unlike ridges and hot-spots, pressure has only a minor effect on U-series distribution coefficients because melting is constrained to be largely in the spinel stability

field by the release of slab-derived fluids at 100 km depth. However, debate continues about the importance of the slab-derived fluid in controlling U-series signatures in arc volcanics.

Huang *et al.* (2011) argued that the majority of all U-series signatures in Kick'em Jenny lavas were derived by ingrowth during the melting process, rather than slab-derived fluids. This interpretation may appear to be in conflict with the evidence cited above that radium enrichments in arc magmas are largely inherited from the slab-derived fluid. To examine this question, Ra/Th activities are compared with U/Th and Pa/U activity ratios for Kick'em Jenny samples in Fig. 13.49. In these plots, solid circles represent samples with excess radium activities, whereas open circles are samples where

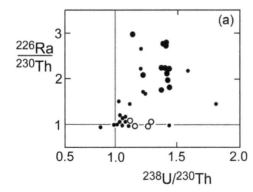

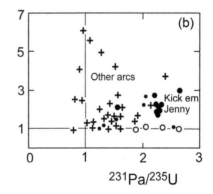

Fig. 13.49 Correlated enrichments of radium in Kick'em Jenny (●, ○); other Lesser Antilles islands (•) and other oceanic arcs (+), with uranium and protactinium enrichments due to fluid addition and dynamic melting. Note different *y* axis scales. Modified after Huang *et al.* (2011).

the radium has decayed away. The radium data are equivocal, because in Fig. 13.49a, $^{226}Ra/^{230}Th$ correlates with $^{238}U/^{230}Th$, which is a signature of fluid addition, whereas in Fig. 13.49b, $^{226}Ra/^{230}Th$ correlates with $^{231}Pa/^{235}U$, which is a signature of dynamic melting. The most logical conclusion is that radium enrichment reflects both processes, but the relative proportions are variable. Since Kick'em Jenny is the extreme example of melting-induced Pa enrichment, it would not be surprising if most Ra in these samples was also a melting feature. On the other hand, Tongan lavas define the opposite extreme, where Pa enrichment is modest, and most radium enrichment is likely to be slab-derived.

References

Allegre, C. J. (1968). ^{230}Th dating of volcanic rocks. *Earth Planet. Sci. Lett.* **5**, 209–10.

Allegre, C. J. and Condomines, M. (1976). Fine chronology of volcanic processes using ^{238}U–^{230}Th systematics. *Earth Planet. Sci. Lett.* **28**, 395–406.

Allegre, C. J. and Condomines, M. (1982). Basalt genesis and mantle structure studied through Th-isotopic geochemistry. *Nature* **299**, 21–4.

Asmerom, Y., Cheng, H., Thomas, R., Hirschmann, M. and Edwards, R. L. (2000). Melting of the Earth's lithospheric mantle inferred from protactinium thorium uranium isotopic data. *Nature* **406**, 293–6.

Asmerom, Y. and Edwards, R. L. (1995). U-series isotope evidence for the origin of continental basalts. *Earth Planet. Sci. Lett.* **134**, 1–7.

Beattie, P. (1993a). The generation of uranium series disequilibria by partial melting of spinel peridotite: constraints from partitioning studies. *Earth Planet. Sci. Lett.* **117**, 379–91.

Beattie, P. (1993b). Uranium–thorium disequilibria and partitioning on melting of garnet peridotite. *Nature* **363**, 63–5.

Bourdon, B., Joron, J.-L., Claude-Ivanaj, C. and Allegre, C. J. (1998). U-Th-Pa-Ra systematics for the Grande Comore volcanics: melting processes in an upwelling plume. *Earth Planet. Sci. Lett.* **164**, 119–33.

Bourdon, B., Langmuir, C. H. and Zindler, A. (1996a). Ridge-hotspot interaction along the Mid-Atlantic Ridge between 37° 30' and 40° 30' N: the U-Th disequilibrium evidence. *Earth Planet. Sci. Lett.* **142**, 175–89.

Bourdon, B. and Sims, K. W. (2003). U-series constraints on intraplate basaltic magmatism. *Reviews in Mineralogy and Geochemistry* **52**, 215–54.

Bourdon, B., Turner, S. and Allègre, C. (1999). Melting dynamics beneath the Tonga-Kermadec island arc inferred from ^{231}Pa-^{235}U systematics. *Science* **286**, 2491–3.

Bourdon, B., Turner, S. P. and Ribe, N. M. (2005). Partial melting and upwelling rates beneath the Azores from a U-series isotope perspective. *Earth Planet. Sci. Lett.* **239**, 42–56.

Bourdon, B., Zindler, A., Elliot, T. and Langmuir, C. H. (1996b). Constraints on mantle melting at mid-ocean ridges from global ^{238}U-^{230}Th disequilibrium data. *Nature* **384**, 231–5.

Branca, S., Condomines, M. and Tanguy, J. C. (2015). Flank eruptions of Mt Etna during the Greek-Roman and Early Medieval periods: New data from ^{226}Ra-^{230}Th dating and archaeomagnetism. *J. Volcanol. Geotherm. Res.* **304**, 265–71.

Capaldi, G., Cortini, M., Gasparini, P. and Pece, R. (1976). Short-lived radioactive disequilibria in freshly erupted volcanic rocks and their implications for the pre-eruption history of a magma. *J. Geophys. Res.* **81**, 350–8.

Capaldi, G., Cortini, M. and Pece, R. (1982). Th isotopes at Vesuvius: evidence for open system behaviour of magma-forming processes. *J. Volc. Geotherm. Res.* **14**, 247–60.

Capaldi, G., Cortini, M. and Pece, R. (1985). On the reliability of the ^{230}Th-^{238}U dating method applied to young volcanic rocks – (reply). *J. Volc. Geotherm. Res.* **26**, 369–76.

Capaldi, G. and Pece, R. (1981). On the reliability of the ^{230}Th-^{238}U dating method applied to young volcanic rocks. *J. Volc. Geotherm. Res.* **11**, 367–72.

Cerrai, E., Dugnani Lonati, R., Gazzarini, F. and Tongiorgi, E. (1965). Il methodo iono-uranio per la determinazione dell'eta dei minerali vulcanici recenti. *Rend. Soc. Mineral. Ital.* **21**, 47–62

Chabaux, F. and Allegre, C. J. (1994). ^{238}U–^{230}Th–^{226}Ra disequilibria in volcanics: a new insight into melting conditions. *Earth Planet. Sci. Lett.* **126**, 61–74.

Chabaux, F., Hemond, C. and Allegre, C. J. (1999). ^{238}U ^{230}Th ^{226}Ra disequilibria in the Lesser Antilles arc: implications for mantle metasomatism. *Chem. Geol.* **153**, 171–85.

Charlier, B. L., Peate, D. W., Wilson, C. J. et al. (2003). Crystallisation ages in coeval silicic magma bodies: ^{238}U–^{230}Th disequilibrium evidence from the Rotoiti and Earthquake Flat eruption deposits, Taupo Volcanic Zone, New Zealand. *Earth Planet. Sci. Lett.* **206**, 441–57.

Charlier, B. L. A., Wilson, C. J. N., Lowenstern, J. B. et al. (2005). Magma generation at a large, hyperactive silicic volcano (Taupo, New Zealand) revealed by U-Th and U-Pb systematics in zircons. *J. Petrol.* **46**, 3–32.

Cohen, A. S., Belshaw, N. S. and O'Nions, R. K. (1992). High precision uranium, thorium and radium isotope ratio measurements by high dynamic range thermal ionisation mass spectrometry. *Int. J. Mass Spec. Ion Proc.* **116**, 71–81.

Cohen, A. S. and O'Nions, R. K. (1991). Precise determination of femtogram quantities of radium by thermal ionization mass spectrometry. *Anal. Chem.* **63**, 2705–8.

Cohen, R. S., Evensen, N. M., Hamilton, P. J. and O'Nions, R. K. (1980). U-Pb, Sm-Nd and Rb-Sr systematics of mid-ocean ridge basalt glasses. *Nature* **283**, 149–53.

Condomines, M. (1997). Dating recent volcanic rocks through ^{230}Th-^{238}U disequilibrium in accessory minerals: example of the Puy de Dome (French Massif Central). *Geology* **25**, 375–8.

Condomines, M. and Allegre, C. J. (1980). Age and magmatic evolution of Stromboli volcano from ^{230}Th-^{238}U disequilibrium data. *Nature* **288**, 354–7.

Condomines, M., Bouchez, R., Ma, J. L. et al. (1987). Short-lived radioactive disequilibria and magma dynamics in Etna volcano. *Nature* **325**, 607–9.

Condomines, M., Gauthier, P. J. and Sigmarsson, O. (2003). Timescales of magma chamber processes and dating of young volcanic rocks. *Reviews in Mineralogy and Geochemistry*, **52**, 125–74.

Condomines, M., Gauthier, P. J., Tanguy, J. C. et al. (2005). ^{226}Ra or $^{226}Ra/Ba$ dating of Holocene volcanic rocks: application to Mt. Etna and Merapi volcanoes. *Earth Planet. Sci. Lett.* **230**, 289–300.

Condomines, M., Hemond, Ch. and Allegre, C. J. (1988). U-Th-Ra radioactive disequilibria and magmatic processes. *Earth Planet. Sci. Lett.* **90**, 243–62.

Condomines, M., Morand, P. and Allegre, C. J. (1981). ^{230}Th-^{238}U radioactive disequilibria in tholeiites from the FAMOUS zones (Mid-Atlantic Ridge, 36° 50' N): Th and Sr isotopic geochemistry. *Earth Planet. Sci. Lett.* **55**, 247–56.

Condomines, M. and Sigmarsson, O. (2000). ^{238}U-^{230}Th disequilibria and mantle melting processes: a discussion. *Chem. Geol.* **162**, 95–104.

Condomines, M., Tanguy, J. C. and Michaud, V. (1995). Magma dynamics at Mt Etna: constraints from U-Th-Ra-Pb radioactive disequilibria and Sr isotopes in historical lavas. *Earth Planet. Sci. Lett.* **132**, 25–41.

Condomines, M., Tanguy, J. C., Kieffer, G. and Allegre, C. J. (1982). Magmatic evolution of a volcano studied by ^{230}Th–^{238}U disequilibrium and trace elements systematics: the Etna case. *Geochim. Cosmochim. Acta* **46**, 1397–416.

Elkins, L. J., Sims, K. W. W., Prytulak, J. *et al.* (2014). Melt generation beneath Arctic Ridges: Implications from U decay series disequilibria in the Mohns, Knipovich, and Gakkel Ridges. *Geochim. Cosmochim. Acta* **127**, 140–70.

Elliott, T. (1997). Fractionation of U and Th during mantle melting: a reprise. *Chem. Geol.* **139**, 165–83.

Elliott, T., Plank, T., Zindler, A., White, W. and Bourdon, B. (1997). Element transport from slab to volcanic front at the Mariana arc. *J. Geophys. Res.* **102**, 14 991–5 019.

Fornari, D. J., Perfit, M. R., Allan, J. F. and Batiza, R. (1988). Small-scale heterogeneities in depleted mantle sources: near-ridge seamount lava geochemistry and implications for mid-ocean-ridge magmatic processes. *Nature* **331**, 511–13.

Forsyth, D. W., Scheirer, D. S., Webb, S. C. *et al.* (1998). Imaging the deep seismic structure beneath a Mid-Ocean Ridge: the MELT experiment. *Science* **280**, 1215–18.

Galer, S. J. G. and O'Nions, R. K. (1986). Magma genesis and the mapping of chemical and isotopic variations in the mantle. *Chem. Geol.* **56**, 45–61.

Gill, J. and Condomines, M. (1992). Short-lived radioactivity and magma genesis. *Science* **257**, 1368–76.

Gill, J. B. and Williams, R. W. (1990). Th isotope and U-series studies of subduction-related volcanic rocks. *Geochim. Cosmochim. Acta* **54**, 1427–42.

Goldstein, S. J., Murrell, M. T. and Janecky, D. R. (1989). Th and U isotopic systematics of basalts from the Juan de Fuca and Gorda Ridges by mass spectrometry. *Earth Planet. Sci. Lett.* **96**, 134–47.

Goldstein, S. J., Murrell, M. T. and Williams, R. W. (1993). ^{231}Pa and ^{230}Th chronology of mid-ocean ridge basalts. *Earth Planet. Sci. Lett.* **115**, 151–9.

Guillong, M., Sliwinski, J. T., Schmitt, A., Forni, F. and Bachmann, O. (2016). U–Th zircon dating by laser ablation single collector inductively coupled plasma–mass spectrometry (LA-ICP-MS). *Geostand. Geoanal. Res.* **40**, 377–87.

Hawkesworth, C. J., Hergt, J. M., McDermott, F. and Ellam, R. M. (1991). Destructive margin magmatism and the contributions from the mantle wedge and subducted crust. *Australian J. Earth Sci.* **38**, 577–94.

Hemond, Ch. (1986). *Geochimie Isotopique du Thorium et du Strontium dans la Serie Tholeiitique d'Islande et dans des Series Calco-alcalines Diverses.* These 3eme Cycle, Universite Paris VII, 151 pp.

Hemond, Ch. and Condomines, M. (1985). On the reliability of the ^{230}Th–^{238}U dating method applied to young volcanic rocks – discussion. *J. Volc. Geotherm. Res.* **26**, 365–9.

Hemond, Ch., Condomines, M., Fourcade, S. *et al.* (1988). Thorium, strontium and oxygen isotopic geochemistry in recent tholeiites from Iceland: crustal influence on mantle-derived magmas. *Earth Planet. Sci. Lett.* **87**, 273–85.

Huang, F. and Lundstrom, C. C. (2007). ^{231}Pa excesses in arc volcanic rocks: Constraint on melting rates at convergent margins. *Geology* **35**, 1007–10.

Huang, F., Lundstrom, C. C., Sigurdsson, H. and Zhang, Z. (2011). U-series disequilibria in Kick'em Jenny submarine volcano lavas: A new view of time-scales of magmatism in convergent margins. *Geochim. Cosmochim. Acta* **75**, 195–212.

Ickert, R. B., Mundil, R., Magee, C. W. and Mulcahy, S. R. (2015). The U–Th–Pb systematics of zircon from the Bishop Tuff: A case study in challenges to high-precision Pb/U geochronology at the millennial scale. *Geochim. Cosmochim Acta* **168**, 88–110.

Jakes, P. and Gill, J. B. (1970). Rare earth elements and the island arc tholeiitic series. *Earth Planet. Sci. Lett.* **9**, 17–28.

Joly., J. (1909). On the radioactivity of certain lavas. *Phil. Mag.* **18**, 577.

Kelemen, P. B., Hirth, G., Shimizu, N., Spiegelman, M. and Dick, H. J. B. (1997). A review of melt migration processes in the adiabatically upwelling mantle beneath spreading ridges. *Phil. Trans. Roy. Soc. Lond.* A **355**, 283–318.

Kigoshi, K. (1967). Ionium dating of igneous rocks. *Science* **156**, 932–4.

Kokfelt, T. F., Hoernle, K. and Hauff, F. (2003). Upwelling and melting of the Iceland plume from radial variation of ^{238}U–^{230}Th disequilibria in postglacial volcanic rocks. *Earth Planet. Sci. Lett.* **214**, 167–86.

Landwehr, D., Blundy, J., Chamorro-Perez, E. M., Hill, E. and Wood, B. (2001). U-series disequilibria generated by partial melting of spinel lherzolite. *Earth Planet. Sci. Lett.* **188**, 329–48.

Langmuir, C. H., Bender, J. F., Bence, A. E. and Hanson, G. N. (1977). Petrogenesis of basalts from the FAMOUS area: Mid-Atlantic Ridge. *Earth Planet. Sci. Lett.* **36**, 133–56.

LaTourrette, T. Z., Kennedy, A. K. and Wasserburg, G. J. (1993). Thorium-uranium fractionation by garnet: evidence for a deep source and rapid rise of oceanic basalts. *Science* **261**, 739–42.

Lundstrom, C. C., Gill, J., Williams, Q. and Perfit, M. R. (1995). Mantle melting and basalt extraction by equilibrium porous flow. *Science* **270**, 1958–61.

Lundstrom, C. C., Sampson, D. E., Perfit, M. R., Gill, J. and Williams, Q. (1999). Insights into mid-ocean ridge basalt petrogenesis: U-series disequilibria from Siqueiros Transform, Lamont Seamounts, and East Pacific Rise. *J. Geophys. Res.* **104**, 13 035–48.

McDermott, F., Elliott, T. R., van Calsteren, P. and Hawkesworth, C. J. (1993). Measurement of ^{230}Th/^{232}Th ratios in young volcanic rocks by single-sector thermal ionisation mass spectrometry. *Chem. Geol. (Isot. Geosci. Sect.)* **103**, 283–92.

McDermott, F. and Hawkesworth, C. (1991). Th, Pb, and Sr isotope variations in young island arc volcanics and oceanic sediments. *Earth Planet. Sci. Lett.* **104**, 1–15.

McKenzie, D. (1985a). ^{230}Th–^{238}U disequilibrium and the melting processes beneath ridge axes. *Earth Planet. Sci. Lett.* **72**, 149–57.

McKenzie, D. (1985b). The extraction of magma from the crust and mantle. *Earth Planet. Sci. Lett.* **74**, 81–91.

Newman, S., Finkel, R. C. and Macdougall, J. D. (1983). ^{230}Th–^{238}U disequilibrium systematics in oceanic tholeiites from 21 °N on the East Pacific Rise. *Earth Planet. Sci. Lett.* **65**, 17–33.

Newman, S., Finkel, R. C. and Macdougall, J. D. (1984). Comparison of ^{230}Th–^{238}U disequilibrium systematics in lavas from three hot spot regions: Hawaii, Prince Edward and Samoa. *Geochim. Cosmochim. Acta* **48**, 315–24.

O'Hara, M. J. and Mathews, R. E. (1981). Geochemical evolution in an advancing, periodically replenished, periodically tapped, continuously fractionated magma chamber. *J. Geol. Soc. Lond.* **138**, 237–77.

O'Nions, R. K. and McKenzie, D. (1993). Estimates of mantle thorium/uranium ratios from Th, U and Pb isotope abundances in basaltic melts. *Phil. Trans. Roy. Soc. Lond.* A **342**, 65–77.

Oversby, V. M. and Gast, P. W. (1968). Lead isotope compositions and uranium decay series disequilibrium in recent volcanic rocks. *Earth Planet. Sci. Lett.* **5**, 199–206.

Pickett, D. A. and Murrell, M. T. (1997). Observations of ^{231}Pa/^{235}U disequilibrium in volcanic rocks. *Earth Planet. Sci. Lett.* **148**, 259–71.

Pyle, D. M., Dawson, J. B. and Ivanovich, M. (1991). Short-lived decay series disequilibria in the natrocarbonatite lavas of Oldoinyo Lengai, Tanzania: constraints on the timing of magma genesis. *Earth Planet. Sci. Lett.* **105**, 378–96.

Qin, Z. (1992). Disequilibrium partial melting model and its implications for trace element fractionations during mantle melting. *Earth Planet. Sci. Lett.* **112**, 75–90.

Reagan, M. K., Volpe, A. M. and Cashman, K. V. (1992). ^{238}U- and ^{232}Th-series chronology of phonolite fractionation at Mount Erebus, Antarctica. *Geochim. Cosmochim. Acta* **56**, 1401–7.

Reid, M. R., Coath, C. D., Harrison, T. M. and McKeegan, K. D. (1997). Prolonged residence times for the youngest rhyolites associated with Long Valley Caldera: ^{230}Th–^{238}U ion microprobe dating of young zircons. *Earth Planet. Sci. Lett.* **150**, 27–39.

Rubin, K. H. and Macdougall, J. D. (1988). ^{226}Ra excesses in mid-ocean-ridge basalts and mantle melting. *Nature* **335**, 158–61.

Rubin, K. H. and Macdougall, J. D. (1990). Dating of neovolcanic MORB using (^{226}Ra/^{230}Th) disequilibrium. *Earth Planet. Sci. Lett.* **101**, 313–22.

Rubin, K. H., Macdougall, J. D. and Perfit, M. R. (1994). ^{210}Po–^{210}Pb dating of recent volcanic eruptions on the sea floor. *Nature* **368**, 841–4.

Russo, C. J., Rubin, K. H. and Graham, D. W. (2009). Mantle melting and magma supply to the Southeast Indian Ridge: The roles of lithology and melting conditions from U-series disequilibria. *Earth Planet. Sci. Lett.* **278**, 55–66.

Schaefer, S. J., Sturchio, N. C., Murrell, M. T. and Williams, S. N. (1993). Internal ^{238}U-series systematics of pumice from the November 13, 1985, eruption of Nevado del Ruiz, Colombia. *Geochim. Cosmochim. Acta* **57**, 1215–19.

Sigmarsson, O., Carn, S. and Carracedo, J. C. (1998). Systematics of U-series nuclides in primitive lavas from the 1730–36 eruption of Lanzarote, Canary island, and implications for the role of garnet pyroxenites during oceanic basalt formations. *Earth Planet. Sci. Lett.* **162**, 137–51.

Sigmarsson, O., Chmeleff, J., Morris, J. and Lopez-Escobar, L. (2002). Origin of ^{226}Ra–^{230}Th disequilibria in arc lavas from southern Chile and implications for magma transfer time. *Earth Planet. Sci. Lett.* **196**, 189–96.

Sigmarsson, O., Condomines, M. and Fourcade, S. (1992). Mantle and crustal contribution in the genesis of Recent basalts from off-rift zones in Iceland: constraints from Th, Sr and O isotopes. *Earth Planet. Sci. Lett.* **110**, 149–62.

Sigmarsson, O., Condomines, M., Morris, J. D. and Harmon, R. S. (1990). Uranium and ^{10}Be enrichments by fluids in Andean arc magmas. *Nature* **346**, 163–5.

Sigmarsson, O., Hemond, Ch., Condomines, M., Fourcade, S. and Oskarsson, N. (1991). Origin of silicic magma in Iceland revealed by Th isotopes. *Geology* **19**, 621–4.

Simon, J. I. and Reid, M. R. (2005). The pace of rhyolite differentiation and storage in an 'archetypical' silicic magma system, Long Valley, California. *Earth Planet. Sci. Lett.* **235**, 123–40.

Sims, K. W., DePaolo, D. J., Murrell, M. T. and Baldridge, W. S. (1995). Mechanisms of magma generation beneath Hawaii and mid-ocean ridges: uranium/thorium and samarium/neodymium isotopic evidence. *Science* **267**, 508–12.

Sims, K. W. W., DePaolo, D. J., Murrell, M. T. *et al.* (1999). Porosity of the melting zone and variations in the solid mantle upwelling rate beneath Hawaii: inferences from ^{238}U–^{230}Th–^{226}Ra and ^{235}U–^{231}Pa disequilibria. *Geochim. Cosmochim. Acta* **64**, 4119–38.

Sims, K. W. W., Goldstein, S. J., Blichert-Toft, J. *et al.* (2002). Chemical and isotopic constraints on the generation and transport of magma beneath the East Pacific Rise. *Geochim. Cosmochim. Acta* **66**, 3481–504.

Sims, K. W. W. and Hart, S. R. (2006). Comparison of Th, Sr, Nd and Pb isotopes in oceanic basalts: implications for mantle heterogeneity and magma genesis. *Earth Planet. Sci. Lett.* **245**, 743–61.

Sims, K. W., Hart, S. R., Reagan, M. K. *et al.* (2008). ^{238}U–^{230}Th–^{226}Ra–^{210}Pb–^{210}Po, ^{232}Th–^{228}Ra, and ^{235}U–^{231}Pa constraints on the ages and petrogenesis of Vailulu'u and Malumalu Lavas, Samoa. *Geochem. Geophys. Geosys.* **9** (4), 1–30.

Spiegelman, M. and Elliott, T. (1993). Consequences of melt transport for uranium series disequilibrium in young lavas. *Earth Planet. Sci. Lett.* **118**, 1–20.

Stracke, A., Bourdon, B. and McKenzie, D. (2006). Melt extraction in the Earth's mantle: constraints from U–Th–Pa–Ra studies in oceanic basalts. *Earth Planet. Sci. Lett.* **244**, 97–112.

Taddeucci, A., Broecker, W. S. and Thurber, D. L. (1967). ^{230}Th dating of volcanic rocks. *Earth Planet. Sci. Lett.* **3**, 338–42.

Tanguy, J. C., Condomines, M., Le Goff, M. *et al.* (2007). Mount Etna eruptions of the last 2,750 years: revised chronology and location through archeomagnetic and ^{226}Ra–^{230}Th dating. *Bull. Volcanol.* **70**, 55–83.

Thomas, R. B., Hirschmann, M. M., Cheng, H., Reagan, M. K. and Edwards, R. L. (2002). (^{231}Pa/^{235}U)–(^{230}Th/^{238}U) of young mafic volcanic rocks from Nicaragua and Costa Rica and the influence of flux melting on U-series systematics of arc lavas. *Geochim. Cosmochim. Acta* **66**, 4287–309.

Thompson, R. N., Morrison, M. A., Hendry, G. L. and Parry, S. J. (1984). An assessment of the relative roles of crust and mantle in magma genesis: an elemental approach. *Phil. Trans. Roy. Soc. Lond.* A **310**, 549–99.

Turner, S., Bourdon, B., Hawkesworth, C. J. and Evans, P. (2000). ^{226}Ra–^{230}Th evidence for multiple dehydration events, rapid melt ascent and the time scales of differentiation beneath the Tonga Kermadec island arc. *Earth Planet. Sci. Lett.* **179**, 581–93.

Turner, S., Evans, P. and Hawkesworth, C. (2001). Ultra-fast source-to-surface movement of melt at island arcs from ^{226}Ra–^{230}Th systematics. *Nature* **292**, 1363–6.

Turner, S., Hawkesworth, C., Rogers, N. *et al.* (1997). ^{238}U–^{230}Th disequilibria, magma petrogenesis, and flux rates beneath the depleted Tonga-Kermadec island arc. *Geochim. Cosmochim. Acta* **61**, 4855–84.

Turner, S., Kokfelt, T., Hauff, F. *et al.* (2015). Mid-ocean ridge basalt generation along the slow-spreading, South Mid-Atlantic Ridge (5–11 S): Inferences from ^{238}U–^{230}Th–^{226}Ra disequilibria. *Geochim. Cosmochim. Acta* **169**, 152–66.

Turner, S., Regelous, M., Hawkesworth, C. and Rostami, K. (2006). Partial melting processes above subducting plates: constraints from ^{231}Pa–^{235}U disequilibria. *Geochim. Cosmochim. Acta* **70**, 480–503.

Volpe, A. M. and Goldstein, S. J. (1993). ^{226}Ra–^{230}Th disequilibrium in axial and off-axis mid-ocean ridge basalts. *Geochim. Cosmochim. Acta* **57**, 1233–41.

Williams, R. W. and Gill, J. B. (1989). Effects of partial melting on the uranium decay series. *Geochim. Cosmochim. Acta* **53**, 1607–19.

Williams, R. W., Gill, J. B. and Bruland, K. W. (1986). Ra–Th disequilibria systematics: timescale of carbonatite magma formation at Oldoinyo Lengai volcano, Tanzania. *Geochim. Cosmochim. Acta* **50**, 1249–59.

Wilson, C. J. N. amd Charlier, B. L. A. (2009). Rapid rates of magma generation at contemporaneous magma systems, Taupo Volcano, New Zealand: insights from U–Th model-age spectra in zircons. *J. Petrol.* **50**, 875–907.

Wood, B. J., Blundy, J. D. and Robinson, J. A. C. (1999). The role of clinopyroxene in generating U-series disequilibrium during mantle melting. *Geochim. Cosmochim. Acta* **63**, 1613–20.

Chapter 14

Cosmogenic Nuclides

The Earth undergoes continuous bombardment by cosmic rays from the galaxy. These are atomic nuclei (mainly protons) travelling through interstellar space at relativistic speeds. The net flux of cosmic ray energy intercepted by the Earth is low, and roughly equivalent in intensity to visible starlight. However, the energy of each particle is very high, averaging several billion electron volts (the kinetic energy of a gas molecule at 10 000 K is about one electron volt). Cosmic rays can therefore interact strongly with matter.

Cosmic rays generate unstable nuclides in two principal ways: by direct bombardment of target atoms (causing atomic fragmentation or 'spallation'), and by the agency of cosmic-ray-generated fast neutrons. The latter are produced by the collision of cosmic rays with target molecules and slowed by further collisions to thermal kinetic energies. These 'thermal' neutrons are able to interact with the nuclei of stable atoms, causing transformations to radioactive nuclei. The 'cosmogenic' nuclides thus produced can be used as dating tools and as radioactive tracers.

Terrestrial cosmogenic nuclides (TCN) are produced in two principal sites. The first is the atmosphere, where cosmic rays interact with nitrogen, oxygen and rare gases. The resulting 'atmospheric cosmogenic nuclides' include radiocarbon, and also other cosmogenic isotopes such as ^{10}Be, ^{36}Cl and ^{129}I, which are useful as environmental tracers. The second site of production occurs in the surface of terrestrial rocks, termed *in situ* production. These nuclides, including ^{26}Al, ^{10}Be and ^{36}Cl, are useful for dating the surface exposure of rocks.

The measurement of cosmogenic nuclides falls into two developmental stages. Early work, almost entirely on ^{14}C, was by radioactive counting. More recently, accelerator mass spectrometry (AMS) has revolutionized the field of cosmogenic nuclides, allowing ^{14}C measurement on very small samples and allowing the utilization of several other cosmogenic nuclides for the first time.

14.1 Carbon-14

The collision of cosmic-ray-produced thermal neutrons with nitrogen nuclei has a reasonable probability of generating radiocarbon by the n, p reaction:

$$^{14}_{7}N + n \rightarrow {}^{14}_{6}C + p$$

Oxidation to carbon dioxide follows rapidly, and this radioactive CO_2 joins the carbon cycle. It may be absorbed photosynthetically by plants, or may exchange with CO_2 in water and ultimately be deposited as carbonate.

^{14}C decays by β emission back to ^{14}N with a half-life of ca. 5700 yr. Hence, atmospheric ^{14}C activity is the result of equilibrium between cosmogenic production, radioactive decay and exchange with other reservoirs. During their lifetime, living tissues will exchange CO_2 with the atmosphere, and hence remain in radioactive equilibrium with it. However, on death this exchange is expected to stop, whereupon ^{14}C in the tissue decays with time. If the initial level of ^{14}C activity in a carbon sample at death (A_0) can be predicted, and if it has subsequently remained a closed system, then by measuring its present level of activity (A), its age (t) can be determined. This can be expressed as the radioactive decay law (from equation [1.5]):

$$A = A_0 e^{-\lambda t} \qquad [14.1]$$

Thus, the radiocarbon method is one of a small minority of dating methods that determines the age of a system from the reduction in abundance of the radioactive parent rather than accumulation of the radiogenic daughter. The latter would be impossible, because radiocarbon decays back into nitrogen gas.

14.1.1 Early Work

The idea of using radiocarbon as a dating tool was conceived by Willard F. Libby, for which he received the Nobel Prize for Chemistry in 1960. The early history of the field is described by Kamen (1963) and a 25-year review was given by Ralph and Michael (1974).

The development of the radiocarbon method went hand in hand with the development of low-level counting techniques. The specific activity of ^{14}C is small, yielding a maximum count rate of 13.6 decays per minute per gram (dpm/g) for modern wood, but only 0.03 dpm/g for a sample of 50 ka age. Furthermore, the maximum β energy is low (156 keV), so

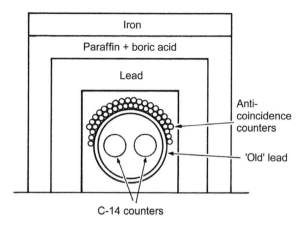

Fig. 14.1 Components in passive and active screening of a CO_2 gas counter. After Mook and Streurman (1983).

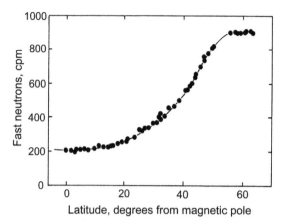

Fig. 14.2 Plot of cosmogenic neutron flux as a function of latitude to show the geographical variation in cosmic ray intensity. After Simpson (1951).

that in a solid source of non-zero thickness a significant fraction of particles would be absorbed by other carbon atoms in the sample. Libby's early determinations of ^{14}C activity were on samples of solid carbon, but this method was soon replaced by the analysis of CO_2 in a gas counter (de Vries and Barendsen, 1953). CO_2 is readily prepared, and in the gas counter there is no risk of losing counts by absorption before the β particles reach the detector.

Unfortunately, the natural background level of activity (cosmic rays and gamma emission from natural materials) is far larger than the level of activity from the sample itself. Hence, two screening techniques have been used (Fig. 14.1). The first is a thick wall of material which itself has a low level of activity (e.g. 'old' lead). The second component is an array of geiger tubes arranged immediately round the gas proportional counter. The geiger tubes are electronically connected in anti-coincidence to the proportional counter. If a high-energy particle such as a cosmic ray enters the shielding, it will trigger the geiger tubes at almost the same time as the proportional counter, and the two signals will cancel out. The dramatic effects of these shielding techniques on the counting background were demonstrated by Ralph (1971) using a counter filled with 'dead' CO_2 made from anthracite coal. Count rates (dpm) were:

No shielding:	1500
Shielded by iron and lead:	400
Shielded and with anti-coincidence counters turned on:	8

The Earth's magnetic field deflects incoming charged particles so that the equatorial cosmic ray flux is four times less than the polar flux (Fig. 14.2). Therefore, one of the first questions which Libby and his co-workers investigated was whether the present day activity of ^{14}C was uniform over the Earth's surface. No latitude dependence was found in modern wood (Anderson and Libby, 1951), and the average specific activity found was 15.3 disintegrations per minute per gram of carbon (dpm/g). Hence, geographical homogenisation of ^{14}C in the atmosphere (before its uptake by plants) appears to be a justifiable assumption.

More recent evidence for the rate of atmospheric ^{14}C homogenization came from atmospheric nuclear tests. Figure 14.3 shows the level of ^{14}C at different locations around the world after the addition of excess ^{14}C from atmospheric explosions (Libby, 1970). Worldwide atmospheric homogenization occurs after only two or three years. The recovery rate of the Mojave Desert samples after 1965 suggests that the timescale for buffering of the atmosphere by surface ocean water is somewhat longer (17 years), but this is still very short relative to the ^{14}C half-life.

Libby (1952) also assumed that the atmosphere had a constant ^{14}C activity through time, as a result of equilibrium between a constant rate of production and decay. Hence, the ^{14}C activity of recent organic tissue was taken to be equal to the 'initial' activity of carbon samples formed in the past. A closed-system assumption was also argued, on the basis that complex organic molecules cannot exchange carbon with the environment after death. (However, such exchange can occur in many carbonates, making them less reliable as dating material.) The above-mentioned assumptions were supported (Arnold and Libby, 1949) by a good concordance between ^{14}C dates and historical ages for a suite of test samples (Fig. 14.4). These ages were based on a ^{14}C half-life of 5568 ± 30 yr obtained from a weighted mean of the four most precise laboratory counting determinations, all of which clustered closely around the mean.

Subsequent to Libby's work, his dating assumptions and half-life value were re-examined. However, it was decided to continue to publish radiocarbon ages using Libby's

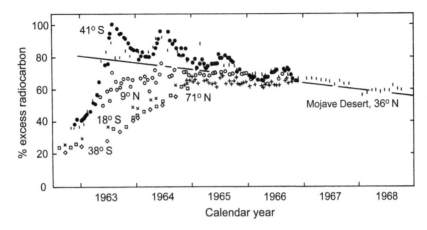

Fig. 14.3 Excess (bomb-produced) atmospheric ^{14}C measured at different localities on the globe, during and after the peak of atmospheric nuclear testing. Localities: (□) 71 °N; (|) Mojave Desert, 36 °N; (○) 9 °N; (×) 18 °S; (+) 21 °S; (◇) 38 °S; (●) 41 °S. After Libby (1970).

atmospheric composition and half-life (Godwin, 1962). These are called 'conventional ages'. Correction factors are subsequently applied to determine a true 'historical' age. These will be discussed below.

14.1.2 Closed-System Assumption

Loss of carbon from a system during its geological lifetime is not usually a problem in radiocarbon dating. However, contamination with extraneous environmental carbon may be

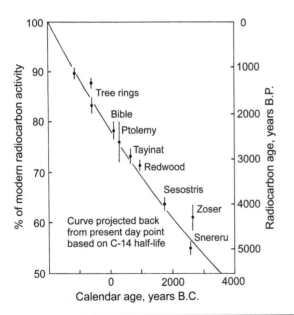

Fig. 14.4 Plot of measured ^{14}C activity (disintegrations per minute per gram of carbon) in archaeological samples of known age against predicted activity based on modern wood and a 5568 yr half-life. After Libby (1952).

a major problem. To exclude such contamination, rigorous sample preparation procedures have been developed.

When dating wood or charcoal for archaeological purposes, the objective is to determine the time when the tree was cut down. Hence, it is only necessary to exclude postmortem exchange with the environment. For this, an acid–alkali–acid leaching treatment referred to as the AAA treatment was found to be effective (Olsson, 1980). The three steps are:

(1) Leach with 4% HCl at 80 °C for 24 hours to remove sugars, resins, soil carbonate and infiltrated humic acids.
(2) Leach with up to 4% NaOH at up to 80 °C for at least 24 hours to remove infiltrated tannic acids (this step also removes part of the lignin).
(3) Repeat step 1 to remove any atmospheric CO_2 absorbed during the alkali step.

The overall process also removes about 50% of the original carbon in the sample.

When dating tree rings for calibration studies (see below), the objective is quite different. In this case it is essential to sample only material laid down in the year of growth corresponding to the annual ring. This requires that all material deposited during the subsequent life of the tree (e.g. lignin) must be leached away. This is accomplished by inserting a step 1a into the above procedure, in which the wood chips are bleached by progressive addition of an almost equal weight of sodium perchlorate powder in dilute acetic acid at 70 °C. The procedure removes up to 75% of the carbon, leaving a residue of pure cellulose for analysis (Mook and Streurman, 1983).

When dating bones, all of the inorganic carbonate fraction must be removed by leaching with very dilute HCl, because this fraction invariably exchanges carbon with groundwater. The organic carbon fraction in the bone is in the form of collagen, which is resistant to post-mortem exchange. Different methods for the treatment of bones are described by Olsson *et al.* (1974). Leaching with acid has also been shown to improve the accuracy of radiocarbon ages on corals (see below).

14.1.3 Initial Ratio Assumption

As radiocarbon measurements became more precise, systematic age discrepancies between historical material and radiocarbon dates began to suggest that the level of ^{14}C activity in the atmosphere had varied with time. The first evidence for such temporal variations in ^{14}C activity was provided by Suess (1955), who found that twentieth-century wood showed a 2% depletion in activity relative to nineteenth-century wood. This was attributed to dilution of radioactive carbon by 'dead' carbon introduced into the atmosphere by burning fossil fuel (nuclear tests later drove the equilibrium in the other direction by adding ^{14}C to the atmosphere). Subsequently, de Vries (1958) found that late-seventeenth-century wood had ca. 2% *higher* activity than nineteenth-century wood. These two 'anomalies' are sometimes called the 'Suess' and 'de Vries' effects.

The discovery of secular variations in ^{14}C activity has provoked various models which attempt to explain these variations. Forbush (1954) observed that the 11-year cycle of sunspot activity was inversely correlated with cosmic ray intensity. This is because high levels of solar activity (marked by increased sunspot activity) cause an increase in the solar wind of ionized particles, which extends the Sun's magnetic field and deflects galactic cosmic rays away from the Earth. Thus, calculations by Oeschger *et al.* (1970) suggest that the stratospheric cosmic ray flux may be nearly doubled at solar minima, relative to maxima.

Because historical records are available for sunspot frequency, this provides a means of calculating past cosmic ray intensity, and hence ^{14}C production, over the past few hundred years. Using this approach, Stuiver (1961) suggested that a sunspot minimum in the late seventeenth century could explain the 'de Vries effect' ^{14}C activity maximum at that time. This was confirmed by Stuiver (1965) using more detailed ^{14}C data, reported as Δ values (per mil) relative to nineteenth-century wood (Fig. 14.5).

Extension of the ^{14}C activity curve to well before the time of Christ revealed large long-term variations, in addition to the short-term effects attributed to changes in solar cosmic ray modulation (Suess, 1965). Elsasser *et al.* (1956) had predicted that if the strength of the Earth's magnetic field displayed secular variations, as suggested by Thellier (1941 and following), then this would have affected the paleo cosmic-ray flux incident on the atmosphere, and hence ^{14}C production. However, strong evidence of a causal relationship with the Earth's field strength was not established until Bucha and Neustupny (1967) provided more extensive paleomagnetic intensity measurements. These data revealed sinusoidal variations in the Earth's magnetic field strength

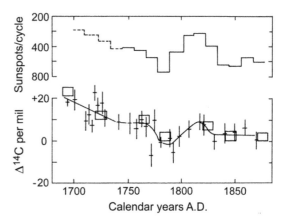

Fig. 14.5 Plots of sunspot activity (on an inverted scale) and relative ^{14}C activity to show anti-correlated variations in the eighteeenth and nineteenth centuries. A best-fit curve is drawn using two ^{14}C data sets (error boxes and error crosses). After Stuiver (1965).

which matched the sinusoidal deviations between radiocarbon and absolute ages.

By modelling the effect of paleomagnetic intensity variations on ^{14}C activity, Bucha and Neustupny were able to match the deviations between tree-ring and radiocarbon timescales almost exactly (Fig. 14.6). A comparison with historically dated wood showed a very similar result, except that this curve was translated upwards by ca. 100 yr. This can be attributed to the average time delay between wood growth and utilization. Because the model of Bucha and Neustupny linked the long time-period deviations between radiocarbon and absolute ages to variations in the global magnetic field, it also implied that the deviations should be of a systematic worldwide nature. Hence it gave grounds for the establishment of very precise calibration sequences, which could then be used for worldwide correction of 'conventional' radiocarbon ages to calendar ages.

In the natural reduction of CO_2 to carbon by photosynthesis, and during laboratory preparation for analysis (e.g. combustion of carbon to CO_2), isotopic fractionation between carbon isotopes can occur. This is due to the weaker bonding, and hence greater reactivity, of the lighter isotope (Section 2.3.1). In order to assess the fractionation between ^{14}C and ^{12}C in natural and laboratory processes, Craig (1954) proposed that the ^{13}C/^{12}C ratio of samples be measured by mass spectrometry. Because fractionation is mass dependent, ^{14}C/^{12}C fractionation will be twice as great as ^{13}C/^{12}C fractionation. The latter is normally expressed relative to the PeeDee belemnite (PDB) standard (Craig, 1957):

$$\delta^{13}C = \left[\frac{(^{13}C/^{12}C)_{sample}}{(^{13}C/^{12}C)_{PDB}} - 1 \right] \cdot 10^3 \qquad [14.2]$$

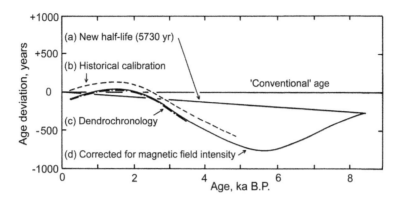

Fig. 14.6 Plot of age deviation between 'conventional' radiocarbon ages (half-life = 5568 yr) and other age determinations: (a) radiocarbon method using 5730 yr half-life; (b) historical timescale; (c) dendrochronology timescale; (d) radiocarbon method using 5730 yr half-life and correction for variations in Earth's magnetic field intensity. After Bucha and Neustupny (1967).

This fractionation factor can be directly converted into a correction to the ^{14}C age using Fig. 14.7 (Mook and Streurman, 1983). In this diagram, normal δ^{13}C compositions for various types of sample are shown. Because 'modern wood' is established as the reference point for calibrating the efficiency of ^{14}C counting equipment, age corrections must be applied relative to this type of material (Fig. 14.7), which has a normal or 'calibration' value of δ^{13}C = −25 per mil (relative to PDB). In marine carbonates, this effect is offset by the 400 yr ^{14}C age of ocean surface water, which must be subtracted from measured ages (the 'reservoir' age, Section 14.2.2).

14.1.4 Dendrochronology

It was quickly realized that the most accurate way to calibrate the 'conventional' ^{14}C timescale for initial ^{14}C variations was to integrate radiocarbon dates with tree-ring chronologies. Great efforts have been expended in this task over the last 50 years.

The longest dendrochronology calibration range has been achieved using the stunted bristlecone pine. When this work began, the species was known as *Pinus aristata*. However, the great longevity of some populations of the bristlecone pine was subsequently recognized by placing these populations in a new species named *Pinus longaeva*. The semi-desert habitat of this tree gives rise to its great longevity and also permits good preservation of the dry wood after death. Thus, Ferguson (1970) erected a continuous master chronology reaching back over 7000 years, based on several living trees and 17 specimens of dead wood from the White Mountains of eastern-central California (Fig. 14.8). This suite now extends nearly 8700 years (to 6700 BC), and includes the oldest living tree at more than 4600 years old! (Ferguson and Graybill, 1983).

Suess (1970) presented a data set of 315 radiocarbon measurements for bristlecone pine from Ferguson's collection, and used this data set to construct a continuous calibration curve from 5200 years BC to the present. One of the

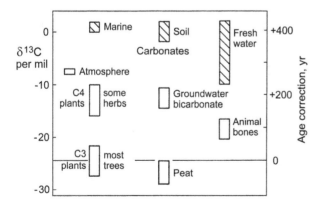

Fig. 14.7 Carbon isotope fractionation effects in different materials, and necessary corrections to calibrated ^{14}C ages for C3 plants (wood). Carbonates are hatched. After Mook and Streurman (1983).

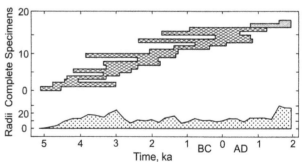

Fig. 14.8 A 'master' tree-ring chronology based on living and dead specimens of Bristlecone pine with overlapping age ranges. Upper chart shows range of each specimen. Lower chart shows total number of radii from which raw data were derived. After Ferguson (1970).

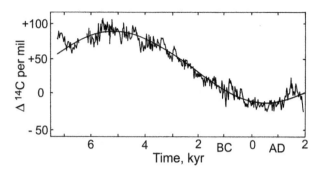

Fig. 14.9 Changes in atmospheric ^{14}C activity in the last 9000 years, presented in the form of isotopic fractionation per mil, based on 'continuous' Bristlecone pine and 'floating' European oak chronologies. After Bruns et al. (1983).

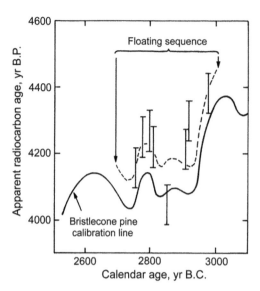

Fig. 14.10 Comparison of ^{14}C data for a wood sample and the calibration curve to show the application of 'wiggle matching'. The dashed line is the proposed fit to the measured data, shown with error bars. After Suess and Strahm (1970).

prominent features of this curve was the presence of numerous 'wiggles' with wavelengths of 100–300 years, superimposed on the longer-term variations discussed above. Suess attracted much criticism because his calibration curve was drawn by eye through the measured points (with 'cosmic schwung'), rather than using a statistical curve fit. Many other workers (as late as Pearson et al., 1977) maintained that the second-order 'wiggles' identified by Suess were an artefact of statistical uncertainties in the data, and had no real meaning. However, this was illogical, since the known 'de Vries effect' wiggles of the seventeenth century AD were of similar magnitude. The reality of the 'Suess' wiggles in the ancient radiocarbon record (3500 years BC) was finally established by De Jong et al. (1979). These wiggles are seen superimposed on long-term ^{14}C variations in the 9000 yr calibration curve shown in Fig. 14.9.

Comparatively large (20 per mil) ^{14}C variations in wood from single sunspot cycles were claimed by some workers (e.g. Baxter and Farmer, 1973; Fan et al., 1986). However, atmospheric ^{14}C variations on this timescale are not consistent with the experimental data of Stuiver and Quay (1981), who modelled small (4 per mil) ^{14}C variations over sunspot cycles which are at the limits of measurement precision. Solar modulation of radiocarbon production is discussed in more detail in Section 14.2.3.

The convoluted shape of the calibration curve introduces ambiguities to ^{14}C dating within many periods, since a single radiocarbon age can correspond to more than one historical age. These ambiguities may sometimes be resolved by applying historical constraints (e.g. Section 14.1.5). Alternatively, they may be avoided in the dating of wood samples if a piece spanning more than about 50 growth rings can be dated. This ring sequence then forms a small 'floating' calibration curve which can be 'wiggle matched' with the known calibration curve to yield a much more accurate timespan for the growth of the sample wood. Suess and Strahm (1970) demonstrated this technique when they dated a floating tree ring sequence from Auvernier (Switzerland) against the

bristlecone pine calibration curve (Fig. 14.10). This procedure allowed the age uncertainties on the Auvernier material to be reduced from hundreds of years to decades.

In order to obtain the highest quality calibration curve, it is desirable to analyse samples representing single annual rings. However, the small size of the bristlecone pine limits the precision which can be obtained, because of the limited amount of sample for analysis. Therefore, other work has been devoted to obtaining a more detailed calibration curve from larger trees (e.g. De Jong et al., 1979).

In Europe, the most important tree for detailed calibration purposes is the oak (Quercus petraea). This is partly because the oak very rarely has missing annual growth rings. In contrast, the widespread alder may lack up to 45% of its annual growth rings (Huber, 1970). The oak is also ideal because it is a long-lived, large tree which displays good resistance to decay after death. In North America, the last 1500 yr of the ^{14}C timescale has been calibrated in great detail (Stuiver and Pearson, 1986) using large trees such as the Douglas fir (Pseudotsuga menziesii) and giant redwood (Sequoia gigantea).

Even more exact dates are possible if the floating chronology comes from an area geographically near to the calibration chronology. Having wiggle-matched the radiocarbon data to obtain historical ages with uncertainties of a few decades, the widths of the tree rings themselves are then matched between the floating and calibration material, to obtain an exact date. However, this procedure is only possible if the two chronologies come from areas with the same weather pattern, thus giving rise to similar growth variations.

Hillam *et al.* (1990) used this procedure to date a Neolithic wooden walkway from Somerset (England) to the probable year of its construction (3806 BC). This age is based on the fact that ten timbers had bark surfaces with ages of 3807/3806 BC, and must therefore have been cut down in that calendar year. A single sample with a bark age of 3800 BC probably represents a later repair to the walkway. This work suggests that as dendrochronologies are completed for more areas of the world, it should increasingly be possible to date large wood samples to the exact age of their felling.

14.1.5 Bayesian Analysis

Wiggle matching annual rings in a sample of wood to a dendro-calibration curve is a special case of a more general problem where a series of radiocarbon ages is available, but the intervals between them are less tightly defined. For example, it should also be possible to create a floating radiocarbon sequence by analysis of several plant macro-fossils (such as seeds) from a soil section. Such a sequence could then be wiggle matched to the dendro-calibration to determine a much more reliable calendar age than is possible from the calibration of a single radiocarbon age. Problems arise however, if such a section has been disturbed by animal bioturbation or human disturbance. In that case some of the data points would be in conflict with their supposed depth in the sequence, and a statistical method would be needed to resolve these conflicts. However, this is difficult for calibrated radiocarbon data, because the symmetrical bell-shaped error function of a conventional radiocarbon age (which basically equals λt) becomes non-symmetrical once it is mapped onto the non-linear calibration curve. Bayesian statistics attempt to deal with this problem.

Bayes' theorem, as published posthumously (Bayes and Price, 1763), attempts to establish the probability of an outcome, given a previously obtained data set concerning the same scenario. It is not just concerned with the likelihood of the outcome, but with putting an error estimate on that likelihood. For example, if a lottery is observed to have a prior success rate of 1 in 11, how confident can we be that the future success rate will be within certain limits (such as between 1 in 9 and 1 in 11) based on a given number of prior observations? This is what standard deviation does for a normally distributed data set, but Bayesian analysis is a more generalized approach that can deal with non-normal distributions. Although credited to Bayes, in its modern formulation it probably goes far beyond his original conception.

In radiocarbon dating, Bayesian analysis is typically used to place error limits on a series of radiocarbon dates that are grouped into sub-sets. These sub-sets could refer to a series of specific events, or to successive time periods. The method was demonstrated by Naylor and Smith (1988) using data from the Danebury Iron Age hill fort in England, and refinements were demonstrated on the same data set by Buck *et al.* (1992). Sixty radiocarbon determinations were available (after rejecting anomalous data) for small samples of charcoal, cereal grain or bone, each of which was closely associ-

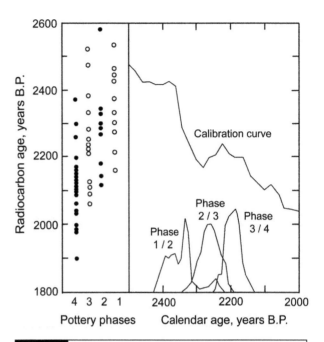

Fig. 14.11 Plot of conventional radiocarbon ages against calibrated ages for pottery phases of the Danebury Iron Age hill fort. Modified after Buck *et al.* (1992).

ated with a pottery fragment. These fragments were categorized into four phases, which are used as prior information in the analysis.

The raw Danebury radiocarbon data are shown on the left-hand side of Fig. 14.11, grouped according to the four pottery phases. Two different models were investigated, in which the pottery phases were either defined as non-overlapping in age, or were allowed to overlap temporally. Based on these models, the data were used to create probability distributions for the ages of the beginning and end of each phase. The assumption of overlapping pottery phases means that the radiocarbon data are essentially unconstrained by the pottery evidence, leading to very large uncertainties on the limits of each phase (ca. 300 years, at a 95% confidence limit, not shown in Fig. 14.11). However, an assumption of non-overlapping phases allows the pottery evidence to tightly constrain the radiocarbon ages, leading to greatly reduced uncertainty on the phase boundaries (right-hand side of Fig. 14.11). Note that the trimodal distribution for the Phase 1/2 boundary is caused by the inflections in the calibration curve, which increases uncertainty on this boundary. This type of Bayesian analysis tool was made widely available to archaeologists in the Windows-based OxCal program (Ramsey, 1995).

14.1.6 Pre-Holocene Calibration

Many attempts have been made to extend the calibrated radiocarbon timescale beyond the limit of dendrochronology. Early work was mainly based on varved lake sediments

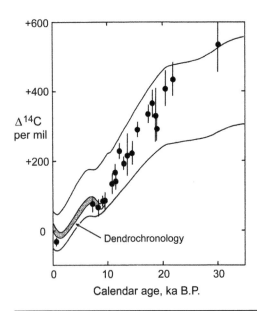

Fig. 14.12 Plot of $\Delta^{14}C$ activity in corals (relative to modern wood) against U–Th ages. Shaded curve = dendrochronology calibration. Solid curves show the envelope of ^{14}C activity predicted from a theoretical cosmogenic model. After Bard et al. (1993).

(e.g. Tauber, 1970) or ice cores (e.g. Hammer et al., 1986). Sediment varves are usually caused by a change in the type of minerals being deposited at different times of the year, but are not as reproducible as tree rings. For example, if sediment from the bottom is stirred up by strong winds and then redeposited, it may be possible for more than one varve layer to be deposited in a year. As a result, many different and conflicting calibration lines were proposed, which largely discredited this approach.

Bard et al. (1990a) took a major step forward in extending the radiocarbon calibration using mass spectrometric U-series analysis (Section 12.2.2). This method was used to assign absolute ages to Barbados corals previously analysed for ^{14}C. In view of uncertainties about closed-system behaviour in carbonates, the method was tested by analysis of samples less than 10 ka in age. These gave ages in good agreement with the dendrochronology timescale, after applying a 400 yr correction for ^{14}C equilibration between atmospheric and surface seawater (the 'reservoir age'). Results for older samples were presented on a plot of $\Delta^{14}C$ activity (relative to modern wood) against U-Th age (Fig. 14.12). Samples in the 10–15 ka range gave $\Delta^{14}C$ activities well within error of those predicted from geomagnetic field strength data. Samples older than 15 ka initially gave more scattered data. However, repeat analysis of the ^{14}C measurements after strong acid leaching gave more consistent results (Bard et al., 1993, 1998).

Because the atmosphere contains only 5% of the carbon budget of the ocean–atmosphere system, climatic changes

might have had a major influence on atmospheric ^{14}C abundances, modifying the effects of cosmogenic radiocarbon production. Large effects are not expected during the Holocene period covered by the dendrochronology timescale, due to its relatively consistent climate. However, much larger climatic effects are expected during the last glacial cycle. Therefore, the use of U-series dating to extend the calibrated timescale further back in time allows a test to be made of these effects on atmospheric ^{14}C abundances.

Mazaud et al. (1991) compared the coral data of Bard et al. (1990a, b) with a ^{14}C production model based on an improved geomagnetic intensity record. The good agreement between the coral data and the predicted ^{14}C activity curve means that long-term activity variations in the atmosphere and hydrosphere can be largely explained by variable cosmogenic production (in response to secular variations in the magnetic field). Hence, climatic effects, which can affect the ^{14}C equilibrium between atmospheric and marine carbonate reservoirs, must play a subordinate role. However, Stuiver et al. (1991) argued that climate could have a second-order effect on atmospheric $^{14}C/^{12}C$ activity ratios by releasing ^{12}C from oceanic carbonate sinks through changes in ocean circulation, possibly during the Younger Dryas event.

The Younger Dryas event was a brief glacial re-advance that occurred between 13 and 11.5 ka BP, and has been a testing ground of high-precision ^{14}C measurements designed to compare atmospheric ^{14}C abundances with production rates. The first of these detailed studies was made by Edwards et al. (1993a) using an 8–14 ka coral record from Papua New Guinea. They found a markedly rapid decrease in ^{14}C activity near the end of the Younger Dryas (12–11 ka PB), which they attributed to dilution of atmospheric ^{14}C by 'dead' carbon as a result of the Younger Dryas event. On the other hand, Goslar et al. (1995) observed a peak of $\Delta^{14}C$ in slightly older (>12 ka) varved sediments from Lake Gosciaz, Poland (followed by a rapid decline). This was attributed to a temporary 'hold-back' in the ventilation of old carbon from the oceans. Hence, it was argued that climatic changes *could* perturb the overall control of the geomagnetic field on atmospheric ^{14}C activity for short periods of time.

Further investigation of this problem was made using two new varved sediment records. The first example was from Lake Suigetsu, near the coast of Central Japan (Kitagawa and van der Plicht, 1998). This small freshwater lake has a water depth of 34 m, but contains a 75 m-deep thickness of sediment at its bottom. The sediment shows annual varves about 1 mm thick, formed by variation from a winter season of dark clay deposition to a spring season of white (siliceous) diatom deposition. The sediments were cored, and the most clearly banded section, from 10 m to 30 m depth, showed 29 000 varves. From within this section, 250 samples of windblown plant debris and insect wings were dated by radiocarbon, allowing this section to be anchored to the tree ring calibration curve between 9 and 11 ka BP.

The second new calibrated varve section was formed by sediments in the marine Cariaco Basin in the Caribbean

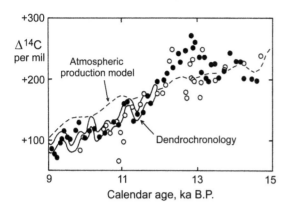

Fig. 14.13 Radiocarbon measurements for the Cariaco basin varve section (•), showing excellent agreement with the dendrochronology calibration line (wide band) and U–Th dates on corals (○). Dashed line = production model. After Hughen et al. (1998).

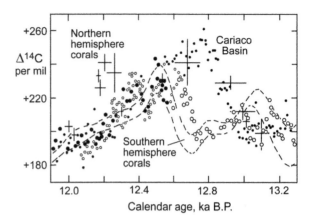

Fig. 14.14 Radiocarbon signatures of analysed reservoirs during the Younger Dryas event. (•) = Cariaco Basin; (●) = dendrochronology calibration; (○, O) = floating dendrochronology; dashed envelope = Pacific corals; + = Atlantic corals. Modified after Hua et al. (2009).

(Hughen et al., 1998). The dated points in this sediment core showed exceptionally good agreement with the tree ring calibration curve and dated coral samples in the range from 9 to 12 ka BP, but had an unexpectedly large peak of $\Delta^{14}C$ activity around 13 ka BP, considerably above the atmospheric production model for this time (Fig. 14.13). This was problematical, because this peak is just beyond the range of the dendrochronology calibration, while U–Th dated coral determinations give more scattered results (open circles in Fig. 14.12). Lake Suigetsu data also show a peak in $\Delta^{14}C$ at this time (not shown in Fig. 14.13), but this is considerably smaller (ca. 220 per mil) than the Cariaco Basin peak (>250 per mil).

A resolution of this problem was achieved by extending the dendrochronology calibration using a floating sequence from Tasmania (Hua et al., 2009). Because the southern hemisphere was less glaciated, Tasmanian and New Zealand data can be used to tie northern hemisphere sequences together where these become fragmented due to ice cover in Europe. A floating Huon Pine sequence from Tasmania was used to tie the Late Glacial Pine (LGP) sequence from central Europe (Kromer et al., 2004) to the main northern hemisphere dendrochronology calibration by wiggle matching (Fig. 14.14). It should be noted that there are small (ca. 40 year) offsets in tree-ring radiocarbon signatures between northern and southern hemispheres, attributed to local buffering with radiocarbon from distinct ocean water masses. These offsets are not constant with time, but they are about an order of magnitude smaller than the offsets in ocean water composition, termed the reservoir effect (see below).

Significantly, the extended dendrochronology calibration is in good agreement with a calibration based on Pacific Ocean corals, whereas Atlantic Ocean corals (mainly from the Caribbean region) show resemblances to the Cariaco Basin (Fig. 14.14). It is also notable that the Cariaco Basin comes back into agreement with the Southern Hemisphere dendrochronology and Pacific coral data at ages over 13 ka. Hence, Hua et al. suggested that the Cariaco Basin varve chronology was maintained correctly through the Younger Dryas event, but its $\Delta^{14}C$ signatures were perturbed by local ocean circulation effects. The fact that the Cariaco Basin records high $\Delta^{14}C$ values ca. 250 years before the atmosphere suggests that the atmosphere was responding to a reorganization of marine radiocarbon caused by changes in ocean circulation. This is consistent with models that attribute the Younger Dryas event to a reduction in the thermohaline circulation due to a sudden release of glacial melt-water into the North Atlantic (Section 12.3.6 and Section 14.2.2 below). This evidence is important as a possible analogue for sudden melting of Greenland glaciers due to anthropogenic global warming.

A new determination of geomagnetic field intensity over the past 200 ka by Guyodo and Valet (1996) suggested for the first time that cosmogenic C-14 production might not be closely in step with the radiocarbon calibration curve beyond 30 ka. However, this question could not be addressed until additional radiocarbon data were available beyond 25 ka. The Lake Suigetsu and Cariaco Basin records provided evidence pertinent to this question, and these records were augmented by new data for a U–Th-dated speleothem from the Bahamas (Beck et al., 2001). The latter data were used to calculate $\Delta^{14}C$ values back to 45 ka (Fig. 14.15).

Because speleothems incorporate significant amounts of dead inorganic carbon from groundwater, radiocarbon ages of this material require comparatively large corrections to bring them into line with the atmospheric calibration derived from other sources (e.g. Suigetsu and Cariaco). In fact, subsequent work (Hoffmann et al., 2010) showed that the contamination problem was worse than originally

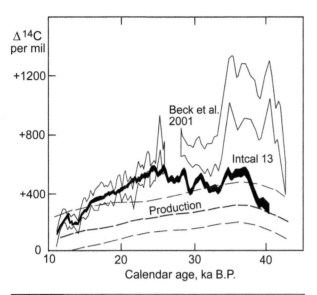

Fig. 14.15 Plot of $\Delta^{14}C$ records (represented by 2σ error envelopes) from a Bahamas speleothem (Beck *et al.*, 2001) and the Intcal 2013 model (Reimer *et al.*, 2013) compared with a ^{14}C production model and its uncertainty limits (dashed lines). Modified after Beck *et al.* (2001).

thought, requiring age corrections of 1000–2500 years. As a result, $\Delta^{14}C$ values above 800 (Fig. 14.15) have not been supported by more recent data. For example, the Intcal 2013 calibration (Reimer *et al.*, 2013) is much lower than Beck *et al.* (2001) beyond 25 ka. However, it should be remembered that according to the six-half-lives rule of thumb, radiocarbon is essentially dead after 35 ka, so that measurements in this range are very demanding.

The $\Delta^{14}C$ records of Beck *et al.* (2001) and Reimer *et al.* (2013) are compared in Fig. 14.15 with a production model in which $\Delta^{14}C$ variations with time are attributed entirely to changes in geomagnetic field strength. This production model is used to calculate $\Delta^{14}C$ values using a box model based on the present day operation of the carbon cycle. The model of Laj *et al.* (2002) is not shown in Fig. 14.15, but is generally elevated above the model of Guyodo and Valet (1996) used by Beck *et al.* (2001). However, even the model of Laj *et al.* is still not able to explain the elevated $\Delta^{14}C$ values of the Intcal 2013 calibration between 20 and 40 ka. This period corresponds approximately to the last glacial maximum, supporting the theory that ventilation of CO_2 with low $\Delta^{14}C$ from the deep oceans was temporarily suppressed during the glacial maximum (Section 14.2.2).

14.2 Radiocarbon and Climate Change

The effect of atmospheric CO_2 concentrations on Earth's climate is one of the most important issues in the earth sci-

ences. Radiocarbon is obviously an important tool to study the effects of CO_2 on climate change because it is directly involved in the carbon cycle. Therefore, in addition to its role as a dating tool, radiocarbon can also be used as a tracer of carbon exchange between reservoirs. The oceans must figure prominently in all such investigations because deep ocean water holds approximately 30 times as much dissolved CO_2 as the pre-industrial atmosphere (Siegenthaler and Sarmiento, 1993).

14.2.1 Radiocarbon in the Modern Oceans

Radiocarbon is a very useful tracer in oceanography because it allows quantitative estimates to be made of the residence times of water at different depths, the mixing between different water bodies, and the magnitude of ocean currents. Radiocarbon evolution in the oceans begins with 'ventilation' of surface water to the atmosphere, which allows this water to reach equilibrium with atmospheric radiocarbon. After a water body moves away from the surface, it can be dated by the decay of this radiocarbon, and hence its flow path and mixing history can be traced.

Studies of the radiocarbon budget of the oceans began in the 1950s at the same time as the first atmospheric nuclear tests, which produced large quantities of C-14 by neutron activation of nitrogen. This 'bomb' radiocarbon complicates the interpretation of natural radiocarbon variations in the oceans, but the entry of this 'spike' of anthropogenic radiocarbon into the oceans also provides a useful tracer of the very recent movement of water bodies. However, because the radiocarbon method was in its infancy when atmospheric nuclear testing began, there was an inadequate data set of pre-bomb measurements on seawater to provide a proper baseline to evaluate the magnitude of the bomb signature. Hence, a full understanding of the early data was not achieved until later studies revealed the composition of pre-bomb radiocarbon inventories. For example, the analysis of corals provides a means of sampling pre-bomb radiocarbon signatures of the surface oceans (Druffel, 1996).

The first major programme for the radiocarbon analysis of ocean water, called GEOSECS (Geochemical Ocean Sections Study) was undertaken in the mid 1970s, near the peak of bomb radiocarbon in the atmosphere. Hence, surface water analyses from this programme provide a dramatic picture of the effects of ocean currents on the bomb radiocarbon signature (Fig. 14.16). The highest values of bomb radiocarbon were found in the sub-tropics, where water has the longest surface residence time. In contrast, Antarctic water was found to have essentially no bomb radiocarbon signature, which was attributed to strong mixing between surface water and deep water (e.g. Nydal, 2000). Comparison of radiocarbon data from the GEOSECS programme with more recent sampling programmes such as WOCE (in the early 1990s) showed a diminution with time of the bomb signature in the surface ocean (Fig. 14.14), but also revealed a concomitant increase of the bomb signature in the deep ocean (e.g. Ostlund and Rooth, 1990).

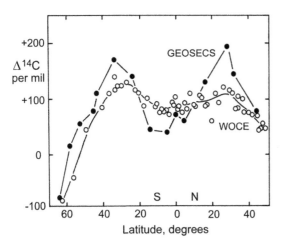

Radiocarbon variations in surface ocean waters of the Pacific as a function of latitude, attributed to atmospheric nuclear tests. (•) = GEOSECS; (○) = WOCE. After Key *et al.* (1996).

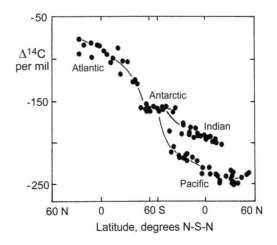

Plot of $\Delta^{14}C$ against latitude, showing the progressive drop in radiocarbon abundance from the North Atlantic, through the Antarctic mixing zone, to the Pacific. After Stuiver *et al.* (1983).

Natural radiocarbon variations in the oceans provide evidence about ocean circulation on a longer timescale. The first detailed radiocarbon study of the Atlantic Ocean was made by Broecker *et al.* (1960). After correction for anthropogenic contamination, tropical waters were found to have a $\Delta^{14}C$ composition of −50 per mil (relative to nineteenth-century wood), equivalent to an apparent ^{14}C age of ca. 400 yr. As they flow north, these waters impart the same apparent ^{14}C age to North Atlantic surface water. After mixing with cold Arctic water, they feed North Atlantic Deep Water with $\Delta^{14}C$ values beginning around −70 parts per mil. However, as this deep water moves southward, ^{14}C activities fall progressively as the water body ages, reaching a Δ value of −160 per mil in Antarctic Bottom Water.

These results are consistent with the established oceanographic model for the Atlantic (Fig. 4.39). In this model, tropical waters are transported (advected) to the North Atlantic, where they cool, and because of their high salinity, sink to form North Atlantic Deep Water (NADW). This body of deep salty water flows southward to the Antarctic, and after mixing with Antarctic Bottom Water, ultimately reaches the Pacific. The role of the Antarctic Ocean as a mixing zone between young Atlantic water (NADW) and older deep water from the Pacific and Indian oceans was confirmed by a large radiocarbon data set from the GEOSECS program (Stuiver *et al.*, 1983), as shown in Fig. 14.17.

If the change in $\Delta^{14}C$ from North Atlantic surface water to Pacific Deep Water was due entirely to radiocarbon decay (at ca. 11 per mil/century), this would imply a total age of ca. 1700 yr since the time when this water was last at the sea surface. However, the 'true' age of a water body, defined as the average time since the water in that package was at the sea surface, is considerably less. This was demonstrated

by the use of a general circulation model (Campin *et al.*, 1999). This modelling successfully recreated the old radiocarbon age of Antarctic Bottom Water (AABW), but the true age of AABW is much lower. The difference is due to the presence of sea ice, which prevents the ventilation of radiocarbon from Antarctic surface water to the atmosphere. Furthermore, the average mixing time in each individual ocean basin is much shorter than the radiocarbon age: ca. 200–300 yr in the Atlantic, 500 yr in the Pacific and only 80 yr in the Antarctic.

14.2.2 Glacial/Holocene Ventilation Ages

The global system of deep water transport from the North Atlantic, via the Antarctic, to the Pacific has been termed the oceanic thermo-haline circulation (THC) or the 'ocean conveyer belt' (Broecker and Denton, 1989). These authors suggested that this thermohaline circulation (Fig. 14.18) played a critical role in controlling climate switches between glacial and interglacial periods. For example, evidence from elemental tracers suggested that the conveyer belt might have been 'turned off' during glacial periods, which would have tended to amplify the cooling effect in the North Atlantic by preventing the export of cold NADW and the import of warm tropical water. Broecker and Denton also speculated that a similar effect occurred during the Younger Dryas event, the temporary glacial re-advance which interrupted the last deglaciation. The significance of this model is that a destabilization of the THC by excessive melting of the Greenland ice sheet could have a dramatic effect on future climate.

Radiocarbon analysis of biogenic carbonates of different ages can be used to study changes in the operation of the ocean conveyer belt by comparing the apparent ages of given water masses at different times. This apparent age is usually

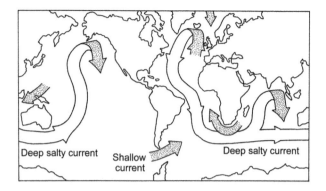

Fig. 14.18 World map showing the Ocean Conveyer Belt connecting major world oceans. After Broecker and Denton (1989).

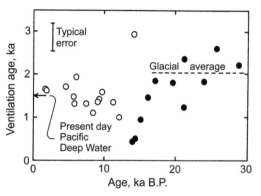

Fig. 14.19 Plot of apparent 'ventilation' age for Pacific Deep Water based on differences in radiocarbon ages between planktonic and benthic forams from a Central Pacific core. (●) = Shackleton et al.; (○) = Andree et al. After Shackleton et al. (1988).

termed a 'reservoir' or 'ventilation' age. These terms mean essentially the same thing, but the former tends to be used when correcting radiocarbon ages for shallow marine samples to bring them into harmony with the dendrochronology calibration, whereas the latter term is more often used when describing deep ocean circulation.

One of the most effective ways of measuring the relative ventilation ages of shallow and deep water bodies is to compare the radiocarbon ages of benthic and planktonic forams in a given sediment deposit. This work was not possible until the advent of accelerator mass spectrometry (see below), because of the small amount of sample material available. The first study was attempted by Andree et al. (1986) but was complicated by the effects of bioturbation because the core under study had a sedimentation rate of only 1.5 cm/ka.

Shackleton et al. (1988) avoided these problems by working on a core from the Central Pacific with a sedimentation rate of 10 cm/ka. For each sampled increment of the core (typically 2–3 cm) the difference between the radiocarbon ages of benthic and planktonic forams was termed the BF–PF age. This difference can be converted into a total ventilation age for Pacific Deep Water by adding the Pacific surface water reservoir age of 650 yr. These data are compared in Fig. 14.19 with the 1500 yr ventilation age of present day Pacific Deep Water, measured in the GEOSECS program. The results of Andree et al. (1986) are also shown for comparison. The ventilation ages measured by Shackleton et al. were quite variable, especially during the period of deglaciation, 18–2 ka BP. However, Shackleton et al. argued that it was possible to identify a 'glacial mean value' that was about 500 years older than the ventilation age of Pacific Deep Water at the present day.

While B–P (benthic–pelagic) ages allow a good approximation of the relative ventilation ages of deep and shallow water masses, Adkins and Boyle (1997) pointed to inaccuracies when atmospheric radiocarbon was changing rapidly in response to changing cosmogenic production. For example, during a period of decreasing atmospheric ^{14}C abundances

(as seen for much of the past 20 ka), the initial radiocarbon activity of an old deep water sample would have been higher (when that water body was at the sea surface) than the activity level in surface water by the end of the period of evolution of the deep water body. This may cause B–P ages to underestimate the true ventilation ages of water bodies.

To avoid this problem, Adkins and Boyle developed the 'projection' age method, which involves projecting the evolution line of the deep water sample back in time until it reaches the atmospheric evolution curve. The ventilation age of deep water is then calculated by subtracting the 'reservoir age' of surface ocean water. However, other workers (e.g. DeVries and Primeau, 2010) have suggested that water body mixing processes could introduce errors in projected ages that are as large as the corrections themselves. They argued that more accurate corrections could be achieved using an ocean 'general circulation model'. However, this can only be done for ancient samples once the circulation model itself is established for the period under investigation.

An alternative approach adopted by other workers is to compare benthic foram radiocarbon ages directly with atmospheric radiocarbon, using an independent age calibration of the sediment core. For example, Bard et al. (1994) used terrestrial plant fossils in a north Atlantic core, while Sikes et al. (2000) and Skinner et al. (2015) used dated volcanic tuff horizons in SW Pacific cores. These approaches have their own pros and cons relative to using a planktonic foram reference. Without planktonic forams the benthic data may be more susceptible to bioturbation errors (Bard et al., 1994). On the other hand, the direct calibration approach avoids the variable reservoir ages of shallow ocean water.

Marchitto et al. (2007) used oxygen isotope stratigraphy to date benthic forams in a core from 700 m depth off Baja California. The benthic foram ages generally paralleled the atmospheric radiocarbon curve, but two very large negative excursions were observed during the Heinrich event

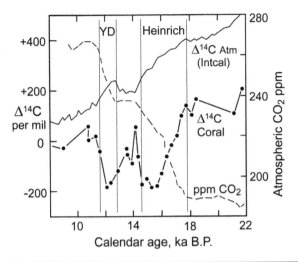

Fig. 14.20 Plot of $\Delta^{14}C$ against age for benthic forams off Baja California compared with the atmospheric $\Delta^{14}C$ calibration line. Dashed line shows rising atmospheric CO_2. Modified after Marchitto et al. (2007).

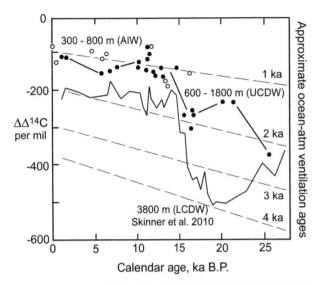

Fig. 14.21 Plot of $\Delta\Delta^{14}C$ and ocean–atmosphere ventilation ages for corals sampling Antarctic Intermediate Water (o = AIW) and Upper Circumpolar Deep Water (• = UCDW) relative to ventilation ages from a deep South Atlantic core. Modified after Burke and Robinson (2012).

(ca. 17–14 ka BP)and the Younger Dryas (Fig. 14.20). Marchitto et al. argued that these excursions reflect large volumes of ancient radiocarbon released into the Pacific Ocean by sudden ventilation of CO_2 from Antarctic Deep Water during deglaciation. This sudden release can explain the rapid rise of atmospheric CO_2 concentrations during deglaciation recorded in Antarctic ice cores (Monnin et al., 2001).

Deep sea corals represent an alternative source of information about the radiocarbon signatures of deep ocean water masses. These corals are slow growing solitary corals that do not rely on symbiotic algae for an energy source, and can therefore grow outside the range of reef-building corals, in the deep seas and in polar regions. These corals can offer high-resolution climatic records because, unlike deep sea sediments, they are not affected by bioturbation. To test the method, Goldstein et al. (2001) compared ventilation ages of very young deep sea corals from the South Pacific with GEOSECS radiocarbon ages. The agreement was very good for a sample from 950 m depth, but less good for samples from 100 m and 350 m depth. However, Goldstein et al. suggested that part of the misfit for the shallowest point might be due to contamination of the coral sample by bomb radiocarbon in the water mass in which it grew.

Burke and Robinson (2012) used a 25 ka record from U–Th dated deep sea corals to reconstruct the deglacial radiocarbon evolution of the southern ocean. Corals were dredged from up to 2 km depth in the Drake Passage between South America and Antarctica, where they are bathed at the present day by Antarctic Intermediate Water (AIW) and Upper Circumpolar Deep Water (UCDW). The results are expressed in Fig. 14.21 in the form of '$\Delta\Delta^{14}C$' meaning the difference between the coral and atmospheric $\Delta^{14}C$ signatures at the measured U–Th age.

$\Delta^{14}C$ values can be used to determine approximate deep water–atmosphere ventilation ages using the projection method (see above), as indicated by the sloping dashed lines in Fig. 14.21. Corals sampling UCDW (solid symbols) show a rapid decrease in ventilation ages during deglaciation, with an evolution path that approximately parallels that derived from BF–PF ventilation ages on a South Atlantic core from 3.8 km depth (Skinner et al., 2010). This water depth (solid line in Fig. 14.21) is believed to sample Lower Circumpolar Deep Water (LCDW). Taken together, the evidence favours ventilation of old radiocarbon from Circumpolar Deep Water during deglaciation.

Temporal variations in the ventilation age of Pacific Deep Water are particularly important for climate, because the Pacific represents 50% by volume of the world's oceans. Since deglaciation was accompanied by a fall in the $\Delta^{14}C$ signature of the atmosphere, it was anticipated that Pacific Deep Water might be a reservoir where old radiocarbon was stored during the last glacial maximum, to be released during deglaciation (Section 14.1.6). However, studies by Broecker et al. (2004, 2008) showing similar BF–PF ages in the glacial and modern Pacific (up to 3 km water depth) created a problem for this model. Nevertheless, the hint of a solution came from the study of a core from 3.6 km depth in the North Pacific (Galbraith et al., 2007). This showed a decrease in BF–PF ages during deglaciation from slightly above present day to slightly below present day Pacific Deep Water values (Galbraith et al., 2007).

To study the depth dependence of Pacific ventilation ages in more detail, Keigwin and Lehman (2015) combined data

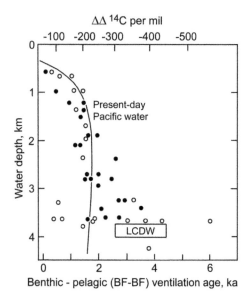

Fig. 14.22 Plot of ¹⁴C data against depth in the Pacific Ocean for two temporal snap-shots, representing the last glacial maximum (● >19.6 ka) and late glacial period (○ <19.6 ka), compared with modern seawater. Glacial LCDW from Fig. 14.21. Modified after Keigwin and Lehman (2015).

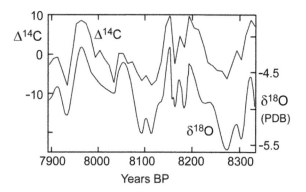

Fig. 14.23 Comparison of $\Delta^{14}C$ and $\delta^{18}O$ records over the period 7.9–8.3 ka BP, showing strong synchroneity. After Neff et al. (2001).

from broader geographical and age suites to compare glacial, late-glacial and present day ventilation ages. In this study, the effect of bioturbation was minimized by sampling from foram abundance peaks, since bioturbation is likely only to smear out such peaks. In addition, effects of variable foram dissolution over time were tested by comparing data from dissolution-resistant species with dissolution-sensitive species. The results (Fig. 14.22) showed a change in the ventilation behaviour of Pacific deep water at around 3 km depth, with a reservoir of deep glacial water with high ventilation ages. This suggests that glacial Pacific Deep Water was the source of large quantities of CO_2 with ancient radiocarbon that was released to the atmosphere during deglaciation, via the Antarctic overturning circulation.

14.2.3 Causes of Climate Change

One of the major questions about modern climate change is the degree to which global warming results from anthropogenic CO_2 release relative to natural climate cycles. This question became particularly important in the early twentyfirst century, with the appearance that late twentieth-century global temperature increases might have temporarily slowed in the first decade or so of the new millennium (e.g. Stocker et al., 2014). Therefore, in order to predict the effect of anthropogenic CO_2 release, it is vital to be able to model natural climate variations into the future (e.g. Mann and Park, 1994).

Based on sunspot observations during the Maunder Minimum (Section 14.1.3) it was established that short-term vari-

ations in atmospheric radiocarbon are controlled by solar cycles, which modulate the intensity of galactic cosmic rays (GCRs). However, Eddy (1977) argued that the radiocarbon record could be used to identify other periods with notably lower solar activity before the record of sunspot observations, by subtracting the geomagnetic modulation of radiocarbon from the dendrochronology timescale (Fig. 14.9). He summarized these anomalies through time and attempted to compare them with climatic cycles. This semi-quantitative analysis suggested that such correlations might exist, but was not conclusive.

Much stronger evidence for a link between GCRs and climate came from the study of speleothem records from caves in Oman (Neff et al., 2001; Fleitmann et al., 2003). Neff et al. made radiocarbon and oxygen isotope measurements on a U–Th dated speleothem from northern Oman, covering an age span from 6.1–9.6 ka BP. A shorter time-section showing the largest variations is shown in Fig. 14.23. Because Oman was on the edge of the Indian Ocean monsoonal region during the early Holocene, this area was particularly sensitive to short-term climatic variations, which are seen to covary with past GCR intensities recorded in the radiocarbon record. These observations were supported by analysis of a longer speleothem record from southern Oman (Fleitmann et al., 2003), showing very similar results for the period 10.3 to 2.7 ka BP.

Given a link between proxy records of GCR intensity and climate, the critical question is whether cosmic rays affect climate directly, or whether the effect is simply an indirect consequence of solar activity variations. Svensmark and Christensen (1997) observed a six-year correlation between GCR intensity and satellite measures of cloud cover between 1984 and 1990. Hence they suggested that GCRs can cool the climate by enhancing cloud cover, and thereby increasing the reflection of solar radiation. However, subsequent cloud cover data degraded the correlation (Sun and Bradley, 2002; Laken et al., 2012) and the model has been rejected by most climate scientists (e.g. Gray et al., 2010). Therefore, a more likely explanation is that GCR intensity and climate are both consequences of solar activity variations (e.g. Bard and Frank, 2006).

Recent satellite measurements have shown that total solar irradiance varies by ca. 0.1% with sunspot cycles (Kelly and Wigley, 1992; Schlesinger and Ramankutty, 1992). Since the Maunder Minimum was a cold period in Europe, this suggests that total solar irradiance at that time may have been significantly lower than the present day. Solar models can be used to extrapolate recent variations of irradiance back to the Maunder minimum (e.g. Lean *et al.*, 1992), but these need to be constrained by proxy records. Sunspot records are an obvious source of evidence, but because these are only seen at peaks of solar activity (Maunder, 1922), they are a non-linear function of total irradiance. On the other hand, cosmogenic isotope data represent another proxy record that can constrain models of solar activity in order to extrapolate recent variations of solar irradiance back to the Maunder minimum.

A summary of modelled total solar irradiance (TSI) for the past 1000 years is shown in Fig. 14.24 (Vieira *et al.*, 2011), derived from three different data sources. The heavy solid line is a solar model extrapolated from satellite observations based on solar property measurements (Krivova *et al.*, 2010). The heavy dashed line was deduced from solar modulation of GCRs, based on subtraction of geomagnetic and climate reservoir exchange effects from the Intcal radiocarbon record (for the historical period, the Intcal trend has not changed significantly in recent iterations).

The light dashed line in Fig. 14.24 was derived in a similar way from ^{10}Be data (Steinhilber *et al.*, 2009), but the latter model was substantially modified in subsequent work (Steinhilber *et al.*, 2012) using a composite of seven ice core records from Greenland and Antarctica (see Section 14.4.3). The new model then gave rise to a ^{10}Be-derived GCR record almost identical to the radiocarbon based record, with a TSI value

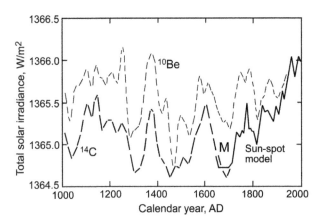

Fig. 14.24 Models of total solar irradiance based on astronomical observations, radiocarbon and ^{10}Be records. M = Maunder Minimum. Modified after Vieira *et al.* (2011).

of 1364.8 W/m^2 during the Maunder Minimum (Steinhilber *et al.*, 2012, supporting information).

The variations of irradiance shown in Fig. 14.24 clearly have much longer periodicities than the sunspot cycle. These long-term variations in solar activity have been known for decades, and have periodicities which can be identified by (Fourier transform) frequency analysis (e.g. Cohen and Lintz, 1974). For example, Abreu *et al.* (2012) determined the Fourier spectrum in Fig. 14.25a using the composite (^{14}C and ^{10}Be) cosmic ray intensity record of Steinhilber *et al.* (2012). Some of these frequency peaks are very well established, and have even been named after pioneers in this field (Fig. 14.25a).

Surprisingly, Abreu *et al.* showed that a very similar frequency spectrum is generated by the torque modulus of the giant planes of the solar system (Fig. 14.25b). This expresses

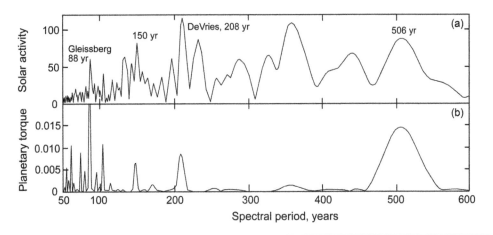

Fig. 14.25 Fourier spectrum showing links between (a) solar activity based on cosmogenic isotope production; (b) calculated torque modulus of the giant planets. After Abreu *et al.* (2012).

the frequency spectrum of tidal forces exerted in the sun by the four largest planets. Although these tidal forces are relatively very small, Abreu *et al.* argued that their effect can be amplified when they act on a non-spherical boundary layer (the tachocline) that separates the sun's non-convecting radiative interior from its convecting outer envelope. This explanation of short-term climate cycles based on planetary dynamics is therefore somewhat analogous to Milankovich forcing of long-term climate (Abreu *et al.*, 2012).

The quality of the frequency matches in Fig. 14.25 have been questioned by some authors, but also supported by others (e.g. McCracken *et al.*, 2014; Scafetta, 2014, 2016). However, it is clear that many of the frequency maxima in Fig. 14.25a remain unexplained by this model. In addition, it was proposed by Scafetta and Willson (2013) that planetary alignment may directly modulate the strength of the solar wind itself. Therefore, it is clear that the subject of planetary influence on solar irradiance is still in its infancy.

Since this model is still largely untested, predicting future solar irradiance still depends on extrapolating proxy records from the past. The Maunder Minimum is usually taken as the type-example of a 'grand minimum' of solar activity (e.g. Tobias, 1996), in comparison with which the high solar activity over the twentieth century has been termed a 'grand maximum' (Abreu *et al.*, 2008). Usoskin *et al.* (2007) argued that these periodicities were chaotic, and therefore could not be predicted, whereas Abreu *et al.* (2008) concluded from studies on Greenland [10]Be records that grand maxima have a statistically predictable duration. They argued that the twentieth-century maximum is likely to end in the near future, but could not predict the 'depth' of any ensuing minimum.

Feulner and Rahmstorf (2010) attempted to model the effects of a future grand minimum on climate change, and concluded that such a scenario would not significantly diminish anthropogenic global warming. However, other workers (e.g. Lockwood, 2010; Maycock *et al.*, 2015) have argued that a grand minimum could have significant effects on *regional* climate, with the possibility of major cooling in northern Europe. Nevertheless, these cooling effects can only be temporary, and the need remains for humanity to adopt a carbon-free economy before catastrophic consequences ensue.

14.3 Accelerator Mass Spectrometry

Mass spectrometry is potentially a powerful alternative to radioactive counting in the determination of cosmogenic nuclides, because it utilizes every atom of the nuclide available in the sample. In contrast, counting determinations utilize only the small number of atoms which decay during the measurement experiment. If decay rates are very high (corresponding to half-lives of less than a thousand years) then counting may be most efficient. However, for longer-lived nuclides, mass spectrometry has the ability to out-perform counting.

14.3.1 Principles of Accelerator Mass Spectrometry

Cosmogenic nuclides are characterized by very low abundances, both relative to other isotopes of the same element and to isobaric interferences from other elements. The first problem is exemplified by the fact that even modern carbon, with the highest $^{14}C/^{12}C$ ratio (1.2×10^{-12}), would yield a ^{14}C peak too small to see above the tail of the very large ^{12}C peak in the mass spectrum of a 'conventional' magnetic sector mass spectrometer used by geologists. Such machines typically have 'abundance sensitivities' (peak tail at one mass unit distance) of 10^{-6} at the uranium mass, which may decrease to ca. 10^{-9} at the mass of carbon.

Abundance sensitivity might be improved sufficiently to measure ^{14}C in a 'conventional' mass spectrometer by increasing the magnet radius, introducing electrostatic filters, and by increasing the accelerating potential and magnet current. These approaches respectively filter and overwhelm the energy spread of ions emitted by the source. Accelerator mass spectrometers usually have all of these features (Fig. 14.26), but they are not central to accelerator mass spectrometry (AMS). In contrast, the three principal attributes of the tandem accelerators used in AMS are the caesium ion sputter source, the charge-exchange process, and the very high ion energies achieved, which allows the use of special detectors.

The essence of the tandem accelerator is the initial acceleration of negative ions by a positive potential in the megavolt range, followed by charge exchange of the ion beam, after which positive ions are accelerated back to zero potential. During the charge-stripping process, isobars of different elements often adopt different charge states, allowing their subsequent separation, while molecular ion isobaric interferences are destroyed.

Charge stripping may be performed by passing the ion beam through an electron-stripping gas (e.g. argon), through a thin graphite film, or (in very high-energy accelerators) a thin metal foil. Experience with carbon has shown that charge stripping to a 3+ state is often most effective, since CH_2 (the principal molecular ion interference) breaks apart rather than forming triple-charged ions (Litherland, 1987). This avoids the need for a high-resolution magnetic analyser to resolve molecular ions by their mass defect. By using only a low-resolution magnetic analyser, the transmission of the instrument for rare isotopes (e.g. ^{14}C) is maximized.

It was suggested by Lal (1988) that the principal impetus for the development of AMS was the fact that accelerators became available for cosmogenic isotope analysis as their applications in nuclear physics diminished. This certainly helped the development of AMS, and the similar ion beam requirements of the two communities facilitated this transition (Litherland *et al.*, 2011). However, the case for continuing

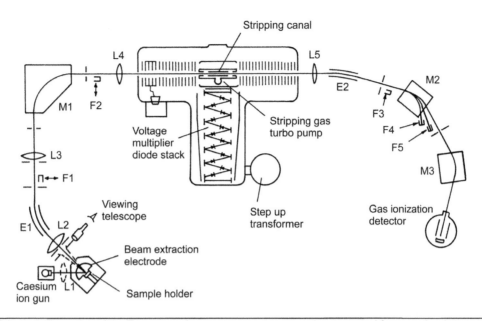

Fig. 14.26 Schematic illustration of the Toronto accelerator mass spectrometer, showing typical features of such instruments. M1–M3: magnetic analysers; E1–E2: electrostatic analysers; L1–L5: electrostatic lenses; F1–F5: Faraday cups; After Kieser *et al.* (1986).

to run these expensive instruments is based on the unique capabilities of AMS, which cannot be achieved by any other means. An alternative method of excluding isobaric interferences is laser induced resonance ionization (e.g. Labrie and Reid, 1981). However, this method has not lived up to its early promise in mass spectrometry.

14.3.2 Radiocarbon Dating by AMS

Most ^{14}C analyses by AMS are performed on solid graphite samples. A typical preparation method is the catalytic reduction of CO_2. The efficiency of AMS radiocarbon measurement is illustrated by the fact that it yields the same count rate from <1 mg of carbon as the β count rate from a whole gram of carbon. Nevertheless, an early instrument transmitting a 2 μA ^{12}C beam produced a ^{14}C count rate of only one ion per minute on a 55 ka carbon sample, corresponding to a $^{14}C/^{12}C$ ratio of 1.2×10^{-15}. In contrast, modern instruments can transmit ^{12}C ion beams above 100 μA, yielding precisions on routine samples of ca. 0.3% (Roberts *et al.*, 2010).

These intense ^{12}C ion beams are achieved using a caesium sputter source, which ejects negative carbon ions by bombarding the sample with Cs^+ from an ion gun. This is particularly suitable for ^{14}C (and ^{26}Al and ^{129}I) analysis because the direct atomic isobars of these species (^{14}N, ^{26}Mg and ^{129}Xe) do not form stable negative ions. Therefore, complete separation from these species occurs in the source (e.g. Purser *et al.*, 1977). However, the sputter source generates an ion beam with variable ion energies. Therefore, this beam must be 'cleaned up' using an electrostatic filter (at a few tens of kV acceleration) before the beam is ready for the mega-volt tandem accelerator. It is also necessary to split the major and

minor ion beams with a magnetic analyser before the accelerator, in order to minimize scattering of the ^{14}C beam by collision of the ^{12}C beam with gas molecules.

Separation of the atomic $^{14}C^-$ ion from the molecular ion $^{12}CH_2{}^-$ depends on the charge stripping stage of the tandem accelerator. The most effective charge stripping medium is provided by a relatively higher gas pressure in the central ultra high voltage 'stripping canal' of the tandem accelerator (Fig. 14.26). Differential pumping of the acceleration tubes at either end of the tandem generator can maintain a pressure 5000 times lower here than in the stripping canal (Litherland, 1987).

The charge stripping process generates a range of charge states in the positive ion beam, such that only ca. 50% of ions have the selected charge. Therefore, the accelerator system must be calibrated against standards of known $^{14}C/^{12}C$ ratio before unknown samples are run. Production of $^{14}C^{3+}$ using a 3 MV accelerator is ideal for radiocarbon measurement, but $^{14}C^{2+}$ ions from a 1.4 MV accelerator can also be used (Lee *et al.*, 1984). Lower accelerating energies have also been used with a +1 charge state (e.g. Jull and Burr, 2006). However, these instruments require higher gas pressures in the stripping channel, which reduces transmission (Suter *et al.*, 2000).

The very high energy of the positive ion beam at the collector end of the instrument (normally >1 MeV) allows the use of ionization counters which can *identify* collected ions as well as measuring their abundance. This is done by measuring the energy loss of the ions in a 'collision cell', in which the ion beam is gradually decelerated by collision with gas molecules. Different kinds of ions lose energy at different

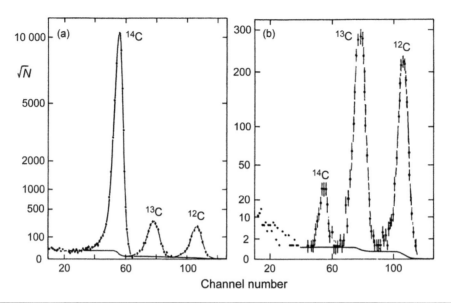

Fig. 14.27 Multi-channel pulse-height (energy-loss) analysis of radiocarbon dating samples from an ionization detector. (a) Modern carbon; (b) 47.4 ka carbon. Typical error bars are shown. After Litherland (1987).

rates in the collision cell. Hence, this provides a final means of resolving any residual ^{12}C molecular ions in the ^{14}C beam (possibly generated by recombination after the accelerator). Figure 14.27 shows that the molecular ion beams of ^{13}C and ^{12}C are barely significant in modern carbon, but dominant in 47 ka carbon. Additional discrimination tools for other elements of interest will be discussed below.

14.4 Beryllium-10

Cosmic rays interact directly with nitrogen and oxygen atoms in the atmosphere, causing spallation (fragmentation) into the light atoms Li, Be and B. Amongst these, one of the nuclides produced is the unstable isotope ^{10}Be, with a half-life of 1.39 Ma (Korschinek *et al.*, 2010; Chmeleff *et al.*, 2010). This half-life was recently revised from an earlier value of 1.51 Ma (Hofmann *et al.*, 1987), and all of the dating studies described here use the old half-life unless specifically mentioned. Cosmogenic ^{10}Be can also be generated in the surface layer of exposed rocks by *in situ* production. However, this subject will be dealt with under Section 14.7.

^{10}Be decays by pure β emission to ^{10}B. It was first observed in naturally occurring material by radioactive counting (Arnold, 1956). However, even at that time Arnold recognized that mass spectrometry might supplant radioactive counting as the best method for measuring ^{10}Be. This is because the relatively long half-life of 1.39 Ma makes counting a very inefficient process for ^{10}Be analysis of natural samples. For example, McCorkell *et al.* (1967) used 1200 tonnes of ice-water to make ^{10}Be (and ^{26}Al) measurements by β counting on Greenland ice. In contrast, Raisbeck *et al.* (1978) made the

first AMS measurement on similar material using only 10 kg of ice-water.

^{10}Be determination by AMS is similar to ^{14}C, but involves some additional complications. Because beryllium does not form stable negative ions, the BeO^- species must be used, upon which the isobaric interference of ^{10}BO is a serious problem. This is overcome by passing the ion beam through an absorber gas (in front of the detector) whose pressure is adjusted to completely stop ^{10}B transmission. The high ion velocity of the ^{10}Be beam generated by AMS allows this species to pass through to the detector, which consists of a gas ionization counter in front of a surface barrier detector which finally absorbs the ion beam. The first detector measures the energy loss (ΔE) of ions as they collide with gas molecules in the chamber. This property of the ions is inversely proportional to their atomic number. The second detector then measures residual ion energy. Using this bivariate discriminant, ^{10}Be ions can be resolved from all other signals (Fig. 14.28) to yield a very low background. ^{10}Be contents of samples are measured relative to added ^9Be spike, and normalized for machine mass discrimination by frequent standard analysis.

14.4.1 ^{10}Be in the Atmosphere

^{10}Be enters the hydrological cycle by attachment to aerosols, from which it is scavenged by precipitation. Consequently, it has a very short (ca. 1 week to 2 yr) residence time in the atmosphere and, unlike ^{14}C, is not homogenized within the atmosphere prior to its fallout.

It was originally assumed (e.g. Raisbeck *et al.*, 1979) that the ^{10}Be analysis of rainwater would allow accurate constraints to be placed on the global production rate.

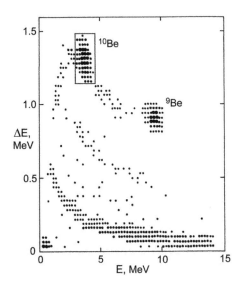

Fig. 14.28 Plot of ΔE against E for a typical geological sample, to show resolution of ^{10}Be from other species. Dot size indicates the number of counts in each bin (smallest = 1 count). After Brown *et al.* (1982).

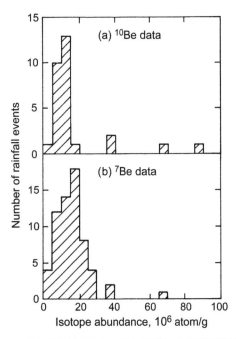

Fig. 14.29 Histograms of ^{10}Be and ^{7}Be concentration respectively in rainfall from Mauna Loa, Hawaii and Bondville, Illinois. The former are weekly rainfall aggregates, the latter represent individual showers. After Brown *et al.* (1989).

However, two factors complicate a determination of the ^{10}Be flux in rainwater. Firstly, there is a tendency for comparatively Be-rich soil particulates to be caught up in the atmosphere and cause secondary contamination of rain (Stensland *et al.*, 1983). Secondly, individual depositional events have very variable ^{10}Be contents, even when the soil contamination effect is removed (Brown *et al.*, 1989).

One way of gauging the effect of soil re-suspension on atmospheric ^{10}Be abundances is to compare them with ^{7}Be data. The latter species has relatively similar atmospheric production rates to ^{10}Be, but much lower levels in soils due to its very short (53 day) half-life. It is measured by γ counting. An alternative way of assessing the effect of soil re-suspension is to compare continental and oceanic ^{10}Be deposition rates. Brown *et al.* (1989) used both of these approaches in an analysis of ^{10}Be precipitation in Hawaii and the continental USA. Average ^{10}Be contents of Hawaiian rain, with negligible soil re-suspension, were very similar to ^{7}Be in rain from Illinois (Fig. 14.29). However, in both cases a few events yield very large relative contents.

The variability of ^{10}Be contents in individual rain showers makes it difficult to determine accurate annual fluxes for mid latitudes. A summary of these data as a function of latitude (Fig. 14.30) shows the variability of these estimates (Brown *et al.*, 1992). However, at tropical latitudes the estimates of annual ^{10}Be flux are more reproducible. The latter are in good agreement with a global ^{10}Be flux estimate of 10^{6} atom/cm^2/yr, based on cosmic ray intensity as a function of latitude (Lal and Peters, 1967). Therefore, the curve in Fig. 14.30 represents a reasonable estimate of the atmospheric ^{10}Be flux.

14.4.2 ^{10}Be in the Oceans

Marine sediments were some of the first materials to be successfully analysed for ^{10}Be, since they have concentrations measurable by β counting. The objective was to use ^{10}Be as a dating tool for oceanic sediments. However, early studies, which simply compared ^{10}Be abundances at various depths

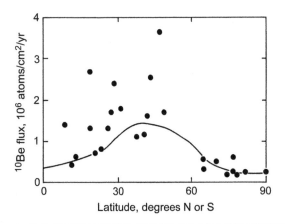

Fig. 14.30 Summary of empirical estimates of annual ^{10}Be flux in rainfall, as a function of latitude. These are compared with a theoretical model based on cosmic ray intensity (shown by the curve). After Brown *et al.* (1992).

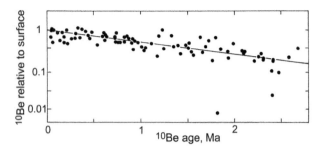

Fig. 14.31 Compilation plot of ^{10}Be activity (normalized relative to the sediment surface) against burial age (depth) for five cores from the North Pacific. After Tanaka and Inoue (1979).

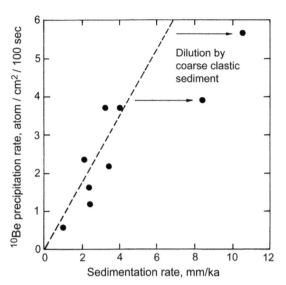

Fig. 14.32 Plot of ^{10}Be precipitation rate against particulate sedimentation rate for the different Pacific Ocean sites. Dashed line represents a constant ^{10}Be concentration in different cores. Points to the right of this correlation line are attributed to dilution by excess (coarse) clastic material. After Tanaka and Inoue (1979).

with theoretical cosmogenic production rates, were unreliable. A more rigorous study was made by Tanaka and Inoue (1979) on paleomagnetically dated sediment cores from the Pacific Ocean. These workers showed that absolute ^{10}Be concentrations were variable from site to site, but values at a given depth relative to the sediment surface were consistent with a theoretical decay path (Fig. 14.31). The good agreement between ^{10}Be activities and the reference decay trajectory suggests that cosmogenic ^{10}Be production has been constant to within about 30% over the last 2.5 Ma.

An important application of ^{10}Be as a dating tool has been to ferromanganese crusts. These deposits represent an ideal archive for particle-reactive species in the ocean system, such as Nd and Pb (Section 4.5). Because the crusts grow into free space, they are resistant to contamination by clastic sediment particles. Hence they can accurately record long-term seawater isotope variations, provided their growth rates can be accurately measured. For the past 500 ka this can be performed using U-series isotopes (Section 12.3.2), but between 0.5 Ma and 10 Ma, ^{10}Be is the best technique. This was first applied by Turekian *et al.* (1979) and is now used as a standard technique in oceanographic paleotracer analysis (Section 4.5.4).

Early studies of the behaviour of stable ^{9}Be in the oceans suggested that it was one of the class of particle-reactive elements that very quickly precipitates from seawater (Merrill *et al.*, 1960). This implies that fine particulates are the principal carrier of ^{10}Be. This model was tested by Tanaka and Inoue (1979) by plotting the ^{10}Be precipitation rate against sedimentation rate at different sites in the Pacific (Fig. 14.32). The good positive correlation displayed by most of the data suggests that the particulate model is valid. Therefore, the net ^{10}Be deposition flux at any given locality is dependent on the sedimentation rate rather than the cosmogenic flux. ^{10}Be deposition rates are actually seen to vary by a factor of three above and below the theoretically predicted flux from the atmosphere. These variations in the ^{10}Be depositional flux at different localities were attributed by Tanaka and Inoue to the lateral transport (advection) of fine particulates by ocean currents.

A detailed understanding of the behaviour of ^{10}Be in the aqueous system requires a consideration of the oceanic residence time. Merrill *et al.* (1960) estimated the residence time of beryllium using equation [14.3] (Goldberg and Arrhenius, 1958):

$$\text{residence time} = \frac{\text{total oceanic inventory}}{\text{total rate of introduction}} \quad [14.3]$$

This equation applies to a steady state (equilibrium) system, which is approximated if the flux is constant for three residence times. For ^{10}Be the equation can conveniently be calculated per unit area:

$$\text{residence time} = \frac{\text{total water column budget/unit area}}{\text{supply flux/unit area}}$$

$$[14.4]$$

Merrill *et al.* determined a residence time for ^{9}Be attached to particulate matter of only 150 yr, but they estimated a longer residence time of 570 yr for the dissolved beryllium budget.

The first estimate of the soluble ^{10}Be budget of the oceans was made by Yokoyama *et al.* (1978) based on ^{10}Be/^{9}Be ratios in manganese nodules. By assuming that these incorporated dissolved beryllium directly from seawater, and using published ^{9}Be abundances in the oceans, they calculated the dissolved ^{10}Be budget as 2×10^{9} atoms/g. Almost identical concentrations were determined by Raisbeck *et al.* (1980) in the first direct ^{10}Be determinations on deep ocean waters,

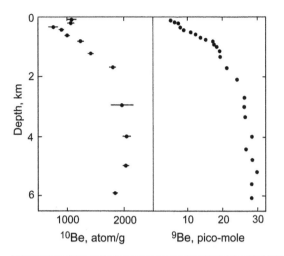

Fig. 14.33 Concentrations of dissolved ^{10}Be (atom/g) and ^{9}Be (picomole) plotted as a function of water depth in the open ocean of the east Pacific. After Kusakabe *et al.* (1987).

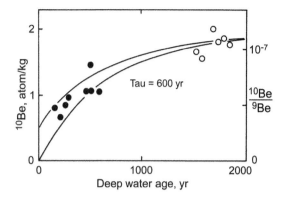

Fig. 14.34 Plot of ^{10}Be concentration against radiocarbon age in oceanic deep water to show build-up of ^{10}Be along the Ocean Conveyer Belt. Solid symbols = NADW; open symbols = Pacific deep water. After von Blankenburg *et al.* (1996).

but their estimated residence time (630 yr) differed markedly from that of Yokoyama *et al.* (1600 yr) due to the use of different cosmogenic flux estimates. Raisbeck *et al.* used their own estimate of the cosmogenic ^{10}Be flux, based on one year's rain from a single locality in France, uncorrected for re-suspension of soil. This can now be seen to be an overestimate. Using the theoretical production rate of Reyss *et al.* (1981), both studies lead to a soluble ^{10}Be residence time in the oceans of ca. 1200 yr.

Arnold (1958) divided the behaviour of elements such as beryllium in the oceanic system into three categories: soluble/sorbed ions; pelagic particulate-controlled ions; and inshore particulate-controlled ions. All three types of behaviour can be seen to control ^{10}Be. In spite of its tendency to be adsorbed onto particulates, dissolved ^{10}Be has a longer residence time in the oceans than similar adsorbable species such as ^{230}Th (Section 12.3.3). This difference can be explained by a nutrient-like behaviour in beryllium (Measures and Edmond, 1982). Profiles of dissolved beryllium concentration in Pacific Ocean water (Fig. 14.33) show strong depletion near the surface where beryllium is adsorbed onto organic matter, but relative enrichment at depth due to the breakdown of dead organic matter (releasing Be) as it falls through the water column (Kusakabe *et al.*, 1987). These Pacific Ocean data yield a deep water ^{10}Be concentration of 2×10^9 atom/g which is in agreement with the earlier results quoted above. However, Atlantic Ocean water displays a different signature which may reflect the large riverine input into this ocean basin.

The behaviour of ^{10}Be in the near-shore environment is very different from the open ocean, as proposed in Arnold's model. Sediment cores from the continental rises off Western Africa and Western North America show ^{10}Be accumulation rates at least an order of magnitude larger than the

theoretical cosmogenic flux (Mangini *et al.*, 1984; Brown *et al.*, 1985). This is attributed to the advection of large quantities of ^{10}Be, by ocean currents, into areas of high deposition. The high deposition rates may be caused by continental erosion or excess biological production, and in these localities the transported ^{10}Be is effectively scavenged and carried to the bottom.

Comparisons between the Be isotope ratio of modern ocean masses were first made by Ku *et al.* (1990). These revealed a consistent global pattern of increasing ^{10}Be/^{9}Be ratio along the ocean conveyer belt, with values ranging from 0.6×10^{-7} in NADW, through 1×10^{-7} in the Antarctic, to values as high as 1.6×10^{-7} in Pacific deep water (Ku *et al.*, 1990). This trend is shown in terms of increasing ^{10}Be content with water age in Fig. 14.34. The positive trend is due to the low particulate sedimentation rate from NADW, which allows ^{10}Be to be advected to the Pacific, and implies a beryllium residence time similar to the circulation age of the ocean water masses. The precise residence time is difficult to constrain because of the different sources of the two isotopes: ^{10}Be from global rain and ^{9}Be from river water. However, the curve shown in Fig. 14.33 was modelled assuming a residence time of 600 yr (von Blankenburg *et al.*, 1996).

Further understanding of the behaviour of beryllium in the oceans can be achieved by comparing ^{10}Be with other isotope tracers. Ku *et al.* (1995) made the first comparison of ^{26}Al and ^{10}Be abundances in ocean water. They found ^{26}Al/^{27}Al ratios in surface water which were consistent with the expected ^{26}Al flux from atmospheric production (by spallation of argon). Therefore, contributions of *in situ* cosmogenic ^{26}Al from cosmic dust or wind-blown continental dust do not appear to be significant in seawater. However, the ^{26}Al/^{10}Be ratio measured in surface water was an order of magnitude lower than the atmospheric production ratio. This was therefore attributed to the much shorter ocean residence time of ^{26}Al.

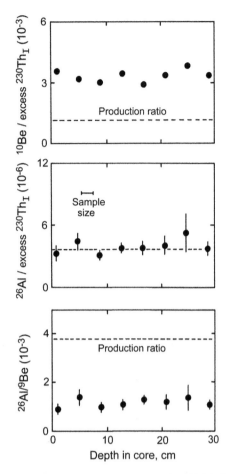

Fig. 14.35 Relative abundances of radionuclides in a sediment core from the North Pacific, showing ratios close to the production ratio for ^{26}Al–^{230}Th, but out of equilibrium with ^{10}Be. After Wang et al. (1996).

Wang et al. (1996) compared ^{26}Al, ^{10}Be and ^{230}Th records in ocean floor sediments. The authigenic (seawater-derived) fractions of these nuclides were leached from core samples using NaOH. ^{10}Be/^{9}Be ratios in the leachates were in good agreement with the composition of overlying (North Pacific) deep water, whereas ^{10}Be/^{9}Be ratios in the bulk sediment were 50% lower, due to a component of detrital continental beryllium. The ^{26}Al budget in the sediment column was also in good agreement with the atmospheric production flux and with ^{26}Al in ocean water (Ku et al., 1995).

Comparison of authigenic ^{26}Al, ^{10}Be and excess initial ^{230}Th revealed ^{26}Al/^{230}Th ratios consistent with the estimated production ratio, but excess abundances of ^{10}Be relative to both of the other nuclides (Fig. 14.35). Hence, it was concluded that ^{26}Al and ^{230}Th have similar (very short) residence times in ocean water, whereas the longer residence time of ^{10}Be allows advection of ^{10}Be into the North Pacific, where it is scavenged by high biogenic production. Similar effects

are seen in the ^{231}Pa/^{230}Th system (Section 12.3.6), but the effect on ^{10}Be/^{26}Al is larger, due to the longer ocean residence of ^{10}Be. On the other hand, ratios of ^{231}Pa/^{230}Th are easier to measure, so both methods are likely to be very useful in paleo-oceanography.

14.4.3 ^{10}Be in Snow and Ice

Atmospheric ^{10}Be accumulates in snow and ice, but its half-life is too long to date such deposits. However, it can be used as a tracer of climatic change and to understand the processes modulating cosmogenic ^{14}C production in the atmosphere. The first detailed study of this type was made by Raisbeck et al. (1981a) on a 906 m-long ice core from the Dome C station, eastern Antarctica. This core had been dated using oxygen isotope variations correlated with ^{14}C-dated marine sediments. A detailed analysis of the top 40 m of core revealed a ^{10}Be maximum at around 1700 AD which correlated with the ^{14}C de Vries effect maximum (Section 14.1.3) and the 'Maunder' sunspot minimum at this time (Eddy, 1976). Consequently these data supported the model of solar modulation of cosmic ray intensity.

Subsequent studies of Greenland ice cores from the Camp Century and Milcent stations (Beer et al., 1984) confirmed the ^{10}Be maximum associated with the Maunder sunspot minimum. In addition, Beer et al. (1985) performed a Fourier transform analysis of very detailed isotopic data from the Milcent core. This revealed second-order ^{10}Be variations with a 9–11 year cycle time equal to the 'sunspot cycle', both before and during the Maunder minimum. Hence, Beer et al. concluded that solar activity continues to vary, even when sunspots are not actually visible.

Studies of the solar modulation of ^{10}Be production over the past 1000 years were re-examined with greater age precision in a more recent study from the South Pole (Bard et al., 1997). The section was precisely dated using 20 impurity layers which were correlated with known volcanic eruptions. The production peak at the Maunder Minimum (1700 AD) is seen very sharply (M, Fig. 14.36), but similar peaks are also visible at other periods, particularly at around 1050 and 1460 AD.

Measurement of ^{10}Be in polar ice is particularly effective for studying solar modulation of cosmogenic production, because the shielding effect of the Earth's magnetic field is reduced at the poles, enhancing the sensitivity of cosmogenic production to the effects of the solar wind. Furthermore, these local production effects are fully transmitted into the ice core record, because ^{10}Be is not well homogenized in the atmospheric system. In contrast, the radiocarbon signal from tree rings shows a strongly damped signal due to buffering of atmospheric radiocarbon abundances by exchange with the oceans. However, Bard et al. used a box model to unravel these effects in the carbon system and calculate a synthetic record of 'un-damped' radiocarbon production. As expected, this was well correlated with the ^{10}Be record (Fig. 14.36).

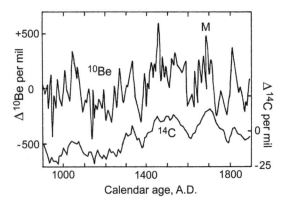

Fig. 14.36 Plot of ^{10}Be variation in the South Pole ice core from ca. 900 to 1900 AD, compared with ^{14}C levels in tree rings, corrected for exchange with other reservoirs. M = Maunder Minimum. After Bard et al. (1997).

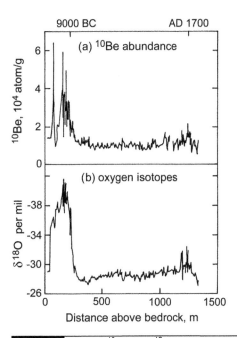

Fig. 14.37 Plot of ^{10}Be and δ^{18}O variations in the Camp Century ice core (Greenland) over a depth of 1400 m, corresponding to ca. 10 ka. After Beer et al. (1988).

In contrast to the successful use of ice core material to study *solar* modulation of cosmogenic nuclide production, attempts to use ice core records of ^{10}Be to chart *geomagnetic* modulation of cosmogenic production were much less successful. For example, a reconnaissance study of the long Dome C core (Raisbeck *et al.*, 1981a) showed no significant correlation with geomagnetic field strength, which reached a minimum value 6000 years ago and is argued to control long-term ^{14}C variations. Instead, a strong correlation was observed with δ^{18}O, which was attributed to the climatic effect of the last ice age. Yiou *et al.* (1985) suggested a partial solution to this problem by attributing the ^{10}Be maximum during the last glaciation to lower precipitation at that time. This would sweep the same amount of ^{10}Be out of the atmosphere, but concentrate it in a lower volume of ice, causing the ^{10}Be record to be compressed.

The long-term Antarctic Dome C data were again matched by results from Camp Century in Greenland, showing a strong correlation of ^{10}Be with δ^{18}O (Fig. 14.37), but a weak correlation with geomagnetic intensity data. Beer *et al.* (1988) acknowledged that at high latitudes, the field strength has a weak influence on atmospheric cosmic ray intensity. However, they concluded that long-term ^{10}Be and ^{14}C activity variations are *not* caused by geomagnetic field changes, but by a complex interplay of climatic effects and solar cosmic ray modulations. This was a questionable proposal even in 1988, and has since been clearly refuted. U–Th calibration of the radiocarbon timescale (Section 14.1.6) shows that ^{14}C variations up to 30 ka in age can *largely* be explained by variations in the Earth's magnetic field. Hence, the 'climate' modulation theory is now largely discredited, except for second-order effects that are discussed in Section 14.1.6.

It has since been shown that dramatic ^{10}Be abundance variations in ice cores over the last glacial cycle are almost entirely a result of variable ice accumulation rates. Indeed, the method has more recently been used to chart paleo-

accumulation rates in central Greenland, with a correction for variable atmospheric production based on geomagnetic field intensity data (Wagner *et al.*, 2001). Such a correction would not be necessary if the ^{10}Be signal was entirely local, since geomagnetic field intensity has little effect on cosmogenic production at the poles. Again, there are second-order climatic effects, such as the transport of ^{10}Be by wind-blown dust (Lal, 2007). As expected, this is much more significant for Greenland than Antarctica, in view of the proximity of Greenland to the North American continent.

14.4.4 ^{10}Be Production and Climate Cycles

In order to study geomagnetic modulation of ^{10}Be production, it is necessary to study equatorial production records from deep sea sediments. These records are subject to perturbation by oceanic currents, but if these effects are quantified, suitable cores can yield reliable records of paleo ^{10}Be production. To overcome the problem of advection, Lao *et al.* (1992) compared ^{10}Be abundances with the U-series nuclides ^{230}Th and ^{231}Pa. These nuclides have similar ocean chemistry to beryllium, but are produced at a constant rate from uranium in solution. Lao *et al.* compared ^{10}Be production at the present day and 20 ka ago (corresponding to the last glacial maximum when the ^{14}C flux was 140% of its present value). They normalized both ^{10}Be and ^{231}Pa fluxes against ^{230}Th. However, based on a comparison of ocean residence times (Section 12.3.6), we can best normalize climatic effects on ^{10}Be deposition by comparing the ^{10}Be/^{231}Pa ratio at the present day and 20 ka ago. After excluding one site with abnormal chemistry, 17 sites from the Pacific gave an average ^{10}Be flux enhancement of 144% during the last glacial maximum, in excellent agreement with ^{14}C.

Comparisons between the geomagnetic field strength and ^{10}Be deposition have also been made by observing their variation over time at a single site. However, the existence of a direct (inverse) relationship between these variables depends on the neutralization of climatic effects. In the absence of U-series data for normalization, this may only happen by chance. Thus, Robinson *et al.* (1995) observed a good inverse relationship between ^{10}Be and magnetic intensity in an 80 ka sediment core from the central North Atlantic. However, Raisbeck *et al.* (1994) observed no such relationship in a 600–800 ka section from the equatorial Pacific Ocean (beyond the range of the ^{230}Th dating method).

To avoid this type of problem, Frank *et al.* (1997) used only ^{230}Th-dated cores to compile a global average of ^{10}Be inventories for the past 200 ka. This was based on 19 long cores which covered most of this age range, supplemented by 18 shorter cores covering the last 25 ka. The global stack of ^{10}Be abundance data was then inverted to determine relative geomagnetic field intensity, which was found to correlate very well with a globally stacked paleointensity record (solid lines in Fig. 14.38a,b). However, Kok (1999) showed that the modelled paleointensity record shows a significant correspondence with the SPECMAP record of oxygen isotope variations (dashed lines in Fig. 14.38). This suggests that the ^{10}Be paleointensity record might still not be completely free of climatic effects, despite use of the ^{230}Th method to calibrate the sections.

Two alternative explanations have since been offered for the ^{10}Be – δ^{18}O correspondence. The first suggestion (Yamazaki and Oda, 2002) is that the geomagnetic field intensity is *itself* modulated by the Earth's orbital eccentricity (which is thought to modulate climatic cycles). In that case, we would expect to see a correlation between variations in climate and the geomagnetic field. However, close examination of Fig. 14.38 shows that the ^{10}Be-derived magnetic paleointensity stack (Fig. 14.38a) shows a stronger correspondence with the SPECMAP δ^{18}O record than the directly measured magnetic record. This suggests that it is the ^{10}Be record, rather than the magnetic record, that contains a component with a climatic linkage.

The second explanation (Sharma, 2002) is the existence of a residual ^{10}Be modulation effect, after the subtraction of geomagnetic modulation. Sharma normalized this to its present day value, and it is plotted in Fig. 14.38c along with the SPECMAP record. He attributed the residual modulation effect to solar magnetic activity, which he suggested was also responsible for the 100 ka glacial climate cycles during this period. This would then provide a 'rival' explanation for glacial cycles, relative to the widely accepted Milankovich model of 'orbital tuning' (Section 12.4.2).

This proposal might be partially true, but it suffers from similar problems to the suggestion of Beer *et al.* (1988) discussed above. Just as polar ice records are not a good basis

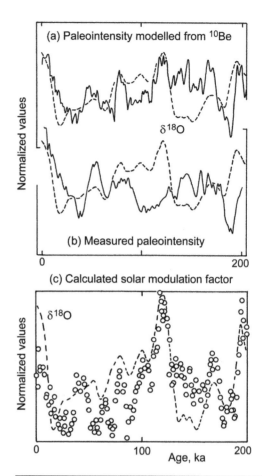

Fig. 14.38 Comparison of paleointensity record with the SPECMAP δ^{18}O record (dashed line): (a) magnetic paleointensity derived from the inversion of ^{10}Be data; (b) directly measured magnetic paleointensity; (c) normalized residual ^{10}Be modulation, attributed to solar field effects. After Kok (1999) and Sharma (2002).

for studying geomagnetic modulation of cosmogenic production, so deep sea sediments are not a good basis for studying solar modulation. A more reliable link between climate and solar intensity is provided by radiocarbon data (Section 14.2.3). These issues are well summarized on a plot comparing solar and geomagnetic modulation with latitude (Fig. 14.39), showing that both modulation effects are highly latitude dependent.

14.4.5 ^{10}Be in Soil Profiles

Beryllium is partitioned very strongly from rain water onto the surface of soil particles such as clay minerals. If we assume that ^{10}Be adsorption is perfect, and that a given soil section is developed by weathering of rock or rock debris without the addition or removal of sediment during the weathering process, then the soil section should contain a

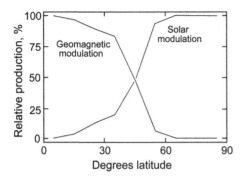

Fig. 14.39 Plot showing the calculated latitude dependence of geomagnetic and solar modulation of cosmogenic isotope production. After Beer *et al.* (2013).

complete inventory of all deposited ^{10}Be that has not yet decayed. This process offers the opportunity of dating a soil profile by measuring the total accumulation of ^{10}Be in the section, but it is apparent that there are a large number of assumptions.

The ^{10}Be inventory of a soil profile in Virginia represents a case where beryllium uptake appears to be nearly 100% efficient (Pavich *et al.*, 1985). ^{10}Be activities display a smooth decay curve against depth (Fig. 14.40a), with a total inventory of 9×10^{11} atom/cm^2. We can compare this value with a theoretical inventory, N, assuming 100% uptake over a given period of time. This is given by the equation

$$N = \frac{q}{\lambda} \left(1 - e^{-\lambda\,t} \right) \qquad [14.5]$$

where q is the input flux from rainfall and t is the accumulation time. If the profile is infinitely old, relative to the 1.4 Ma half-life of ^{10}Be, it will reach saturation, after which the input flux from rainfall is balanced by the rate of decay (λ). Equation [14.3] then simplifies to $N = q/\lambda$. Given an annual deposition flux of 1.3×10^6 atom/cm^2 at this latitude, the

saturation inventory will be 3×10^{12} atom/cm^2, which is three times the observed inventory. The discrepancy can be explained by loss of ^{10}Be-enriched soil from the top of the profile by erosion, and its replacement at the bottom of the profile by weathering of bedrock with no ^{10}Be. Solving equation [14.3] for t (using the observed inventory) yields the residence time of ^{10}Be in the profile, equal to 800 ka.

Very different ^{10}Be behaviour is demonstrated by a soil profile from the Orinoco Basin (Fig. 14.40b), which has a total inventory of only 5×10^9 atom/cm^2. Assuming an annual ^{10}Be flux for this latitude of 0.4×10^6 atom/cm^2, we obtain a ^{10}Be residence time in the soil profile of only 12 ka (because t is short relative to the ^{10}Be half-life, it can be approximated by N/q). This low value is best explained by 'breakthrough' of ^{10}Be from the base of the profile by leaching (Brown *et al.*, 1992).

Differences in ^{10}Be retention between the two cases described above can be understood in the light of laboratory experiments on beryllium partition between soil and water (You *et al.*, 1989). These studies showed that beryllium retention on soil particles is strongly pH dependent, with distribution coefficients of ca. 10^5 in neutral conditions (pH 7), but less than 100 at pH 2. Hence the more acidic conditions in tropical soils are less favourable for beryllium retention.

In alkaline soils, beryllium mobility within the soil profile may be very limited, and in these conditions ^{10}Be may be used as a stratigraphic tool. An example is provided by a ^{10}Be study of Chinese loess, in which carbonate-rich conditions yield a pH value of 8 (Chengde *et al.*, 1992). The profile was dated magnetically back to 800 ka, and represents the products of wind-borne deposition through varying climatic conditions. Chengde *et al.* tuned their profile to the Quaternary climatic record provided by sea floor ^{18}O variations. They concluded that during arid periods, rapid loess deposition was accompanied by high fluxes of ^{10}Be, adsorbed onto wind-blown particles. These sections were interspersed with wetter periods with lower ^{10}Be depositional fluxes.

14.4.6 ^{10}Be in Magmatic Systems

The most important application of ^{10}Be as a petrogenetic tracer is in studies of the relationship between sediment subduction and island arc volcanism. In a reconnaissance study, Brown *et al.* (1982) demonstrated ^{10}Be concentrations of $3–7 \times 10^6$ atom/g in island arc volcanics. These ratios were generally much higher than those in a control group of continental and oceanic flood basalts ($<1 \times 10^6$ atom/g). Brown *et al.* argued against high-level contamination of the analysed volcanics on the grounds that the short half-life of ^{10}Be renders it extinct in all but surficial deposits, while ^{10}Be levels in rainwater are too low to cause the observed enrichments. In contrast, it has long been known (e.g. Arnold, 1956) that pelagic sediments have very large ^{10}Be contents in excess of 10^9 atoms/g. Brown *et al.* therefore attributed their data to the involvement of subducted ocean floor sediment in the genesis of island arc magmatism.

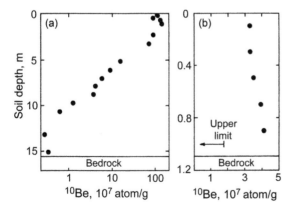

Fig. 14.40 Plots of ^{10}Be activity as a function of depth in two soil profiles: (a) Virginia Piedmont, after Pavich *et al.* (1985); (b) Orinoco Basin, after Brown *et al.* (1992).

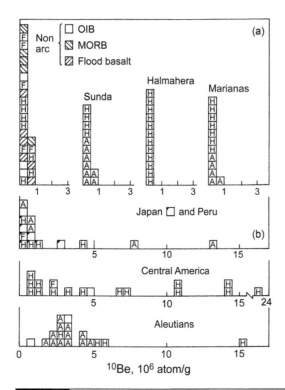

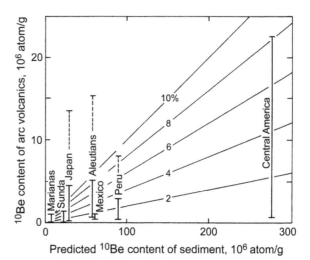

Fig. 14.42 Plot of ^{10}Be data for seven arcs against a model parameter for the efficiency of ^{10}Be supply to arc magma sources. ^{10}Be signals are modelled for different bulk percentage sediment contributions to magmas. After Tera *et al.* (1986).

Fig. 14.41 Histograms of ^{10}Be abundance in volcanic rocks. (a) Non-arc control group and low-^{10}Be arcs; (b) high-^{10}Be arcs. Symbols: A = active volcano; H = historic flow; F = fresh sample collected during or immediately after eruption. After Tera *et al.* (1986).

Subsequent studies (e.g. Tera *et al.*, 1986) confirmed the general observation of high ^{10}Be in island-arc volcanics and low ^{10}Be in the non-arc control group (Fig. 14.41). Detailed studies were also undertaken to assess the effects of weathering on the ^{10}Be contents of lavas. Analyses of material collected during or immediately after eruption were shown to contain the same range of ^{10}Be contents as historical lavas. Similarly, contamination of non-arc samples by rainwater-derived cosmogenic ^{10}Be was only observed in the top 1 cm of severely altered samples of 16 Ma Columbia River basalt. Furthermore, *in situ* cosmogenic production of ^{10}Be in rocks is at a much lower level than atmospheric production.

A surprising result of the detailed study of Tera *et al.* (1986) was the observation that several arcs have ^{10}Be levels as low as the maximum of 1×10^6 atoms/g in the control group. These included all samples from the Mariana, Halmahera (Moluccan) and Sunda arcs. To explain these observations, Tera *et al.* suggested four requirements for a positive ^{10}Be signal in arc volcanics:

(1) adequate ^{10}Be inventory in trench sediments;
(2) subduction rather than accretion of uppermost Be-enriched sediments;

(3) incorporation of sediment in the magma source area;
(4) transport time of <10 Ma from sedimentation to the magma source area.

Failure of any one of these criteria could preclude a positive ^{10}Be signal. However, Tera *et al.* did not observe simple correlations between ^{10}Be and geophysical parameters such as age of the subducting plate.

In an attempt to harmonize their data from different arcs, Tera *et al.* plotted ^{10}Be against a model parameter involving sedimentation rate, sediment thickness, plate velocity, and distance from trench to magma source. This quantity is specified as:

$$\text{Predicted }^{10}\text{Be abundance} = \frac{\eta_0 s}{\lambda h} \exp - \left(\frac{\lambda l}{v} \right) \quad [14.6]$$

where $\eta_0 = {}^{10}$Be abundance in the sediment, s = Plio-Pleistocene sedimentation rate, h = sediment thickness, l = the arc–trench gap and v = plate velocity. Since the volcanic front is always located about 100 km above the seismic plane, the arc–trench gap is inversely proportional to the dip of the Benioff zone. Using this model, the contrast between the ^{10}Be-rich Central American data and other arcs is explained by the high sedimentation rate and steep subduction angle of the former. However, both the Japanese and Aleutian arcs have a single ^{10}Be-rich data point (shown by the dashed ranges in Fig. 14.42) which does not fit the model.

A further step in rationalizing the ^{10}Be systematics of arc volcanics was achieved by considering the data relative to non-cosmogenic (^{9}Be) abundances (Monaghan *et al.*, 1988; Morris and Tera, 1989). Within the different minerals of a single rock sample, ^{10}Be is normally strongly correlated

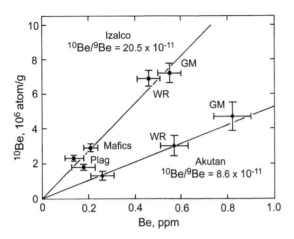

Fig. 14.43 Plot of ^{10}Be against total Be content for separated mineral phases, plus whole-rock (WR) and groundmass (GM) from two young lavas. Samples are from the Izalco volcano (Central America) and Akutan (Aleutians). After Morris and Tera (1989).

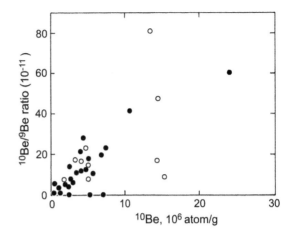

Fig. 14.44 Plot of ^{10}Be/^{9}Be against ^{10}Be abundance for basalts (•) and evolved rocks (o) from different arc and non-arc environments. After Morris and Tera (1989).

with total Be content (Fig. 14.43), implying that radiogenic and non-radiogenic Be were mixed before magmatic differentiation occurred. This further strengthens the arguments against surficial contamination of the lavas by ^{10}Be, and also argues against crustal ^{10}Be assimilation by magmas. Finally, the enriched Be contents of the groundmass, relative to phenocrysts, show that Be behaves as an incompatible element during magmatic differentiation.

In contrast to Be mineral systematics, most whole-rock samples analysed by Morris and Tera did not show a strong correlation between ^{10}Be and total Be. However, they did show a good correlation between ^{10}Be/^{9}Be ratios and absolute ^{10}Be abundances (Fig. 14.44). These findings suggest that most of the rocks analysed, which were basalts, did not have their ^{10}Be contents perturbed by magmatic differentiation. However, some andesites lie significantly to the right of the main trend, including the Japanese and Aleutian samples with abnormally high ^{10}Be contents in Fig. 14.42 (Tera *et al.*, 1986).

Further constraints on the timing of subduction-related processes were obtained by combining ^{10}Be/^{9}Be and ^{238}U/^{230}Th data. Sigmarsson *et al.* (1990) observed that these ratios were correlated in the Southern Volcanic Zone of the Andes (Section 13.5). Based on this correlation, and the much shorter half-life of ^{230}Th than ^{10}Be, Sigmarsson *et al.* suggested that the timescale for dehydration, melting and eruption of these arc magmas was probably less than 20 ka.

A further step in understanding subduction zone processes was achieved by comparing ^{10}Be/^{9}Be and boron/beryllium ratios in arc lavas (Morris *et al.*, 1990). Several arcs display a strong positive correlation between these two variables (Fig. 14.45), despite the fact that the two beryllium isotopes and boron have different distributions in

the subducted slab. ^{10}Be is concentrated in the uppermost sediment layers and diminishes rapidly downwards; ^{9}Be is distributed throughout the sediment column, whereas B is principally concentrated in the hydrothermally altered basaltic crust.

The fact that these species behave coherently in widely separated volcanoes along the length of an arc suggests a very thorough homogenization mechanism for Be and B during the process of subduction-related magma genesis. While such a process might occur in the solid state, it is easiest to conceive of the mixing of fluids that escape from different parts of the subducted slab. The convergence of the ^{10}Be/B correlation lines at the origin points to complete stripping out of all boron from the subduction zone, with no long-term residence of this element in the mantle.

The observed correlation between ^{10}Be/^{9}Be and elemental B/Be ratio suggests that the latter may represent a useful proxy for the former. This is important because it widens the applicability of beryllium data. Firstly, the elemental B/Be ratio can be used as a tracer of the slab component in arcs with low subduction rates where ^{10}Be is extinct by the time of eruption. Secondly, elemental ratios can be measured with less sophisticated analytical equipment such as ICP–MS. These advantages were demonstrated by Edwards *et al.* (1993b) in a study of basaltic lavas from the Indonesian arc. Edwards *et al.* were able to combine B/Be ratios with other radiogenic isotope systems to uniquely specify the Pb, Sr and Nd isotope signatures of the slab-derived component, which was modelled by a 80%–20% mixture of basaltic crust and Indian Ocean sediment. This signature was also distinguishable from enriched and depleted reservoirs in the mantle wedge. The use of elemental B/Be data made these deductions possible despite the fact that ^{10}Be abundances were at baseline, showing this nuclide to be extinct in the analysed lavas.

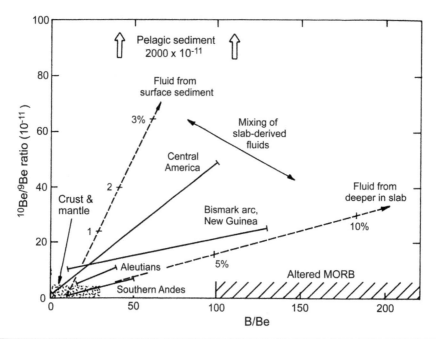

14.5 Chlorine-36

^{36}Cl is analogous to ^{10}Be in its atmospheric production, in this case by spallation of ^{40}Ar rather than ^{14}N, and like ^{10}Be it is quickly swept from the atmosphere by precipitation. However, unlike ^{10}Be, ^{36}Cl is not removed from groundwater by adsorption onto particulates, but remains in the aqueous medium as it travels through geological strata. This fact, coupled with its relatively short half-life of 0.301 Ma, makes ^{36}Cl potentially very useful in the dating or tracing of Quaternary groundwater systems. Cosmogenic ^{36}Cl can also be generated in the surfaces of exposed rocks by *in situ* production. However, this subject will be discussed in Section 14.7.

The principal obstacle in AMS analysis of ^{36}Cl is isobaric interference by ^{36}S. This forms abundant negative ions and is not removed by the charge stripping process. It can be resolved by its lower energy loss in the gas counter, but this is most effective at energy levels above 48 MeV, requiring an accelerator of at least 6 MV potential. This rules out ^{36}Cl analysis with lower-energy (2 MV) tandetrons (Wolfli, 1987). A 'time-of-flight' analyser may also be used before the gas counter (Fig. 14.46) in order to eliminate peak tailing from the relatively very large ^{35}Cl and ^{37}Cl ion beams, which are not adequately resolved by the preceding magnetic and electrostatic analysers in the system. Time-of-flight analysis can only be performed on pulsed ion beams, which are controlled by pulsing the sputter source. This analysis relies on

the fact that lighter masses are accelerated to slightly higher velocities than heavier ones, so that after traversing a distance of a metre or so, they arrive at the detector a few

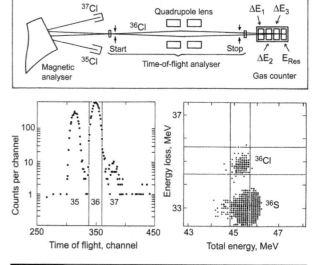

Fig. 14.46 Analyser segment and output data of an AMS instrument designed for ^{36}Cl determination, showing the use of time-of-flight analysis to resolve ^{36}Cl from ^{35}Cl and ^{37}Cl and energy loss detection to resolve from ^{36}S. After Wolfli (1987).

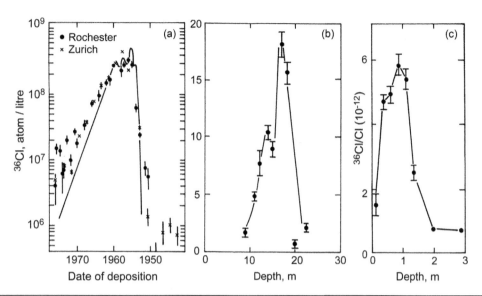

Fig. 14.47 Profiles of anthropogenic ^{36}Cl as a function of depth in different environments. (a) Ice (Dye 3 station, central south Greenland); (b) groundwater (Borden landfill, Ontario); (c) desert soil (New Mexico). After Gove (1987) and Fabryka-Martin *et al.* (1987).

nanoseconds earlier. Hence ^{36}Cl is resolved from both ^{36}S, ^{35}Cl and ^{37}Cl (Fig. 14.46).

The first use of ^{36}Cl as a hydrological tracer was not based on the cosmogenic isotope at all, but on anthropogenic bomb-produced ^{36}Cl. This resulted from seven large nuclear tests conducted on the sea surface from 1952 to 1958, which caused neutron activation of marine chlorine. Profiles of anthropogenic ^{36}Cl against time were determined in a Greenland ice core (Elmore *et al.*, 1982), in Canadian groundwater (Bentley *et al.*, 1982) and in a soil profile from New Mexico (Phillips *et al.*, 1988). All of these measurements showed a very sharp spike in ^{36}Cl, with a duration of 15 to 20 years (Fig. 14.47). It is anticipated that in the near future anthropogenic ^{36}Cl will be a useful hydrological tracer, replacing bomb-produced tritium as the latter becomes extinct.

As seen for other cosmogenic isotopes, the production of natural ^{36}Cl is expected to have varied in the past due to modulation of the cosmic ray flux by the solar wind and Earth's magnetic field. The most easily measured inventories of ^{36}Cl are the ice cores from Greenland and Antarctica, which have also been studied for several other environmental tracers. However, it was shown that ^{36}Cl and ^{10}Be abundances in these cores are better correlated with climatic variations than with past variations in geomagnetic field intensity (e.g. Beer *et al.*, 1988). This result caused considerable puzzlement at the time, but is not really surprising, since the cosmogenic isotope flux in polar snow is largely of local (polar) origin, where cosmic ray intensity is not significantly shielded by the Earth's magnetic field. Therefore, to obtain more representative records of past changes in global ^{36}Cl production it

was necessary to find a suitable inventory from a non-polar source. Such an inventory was discovered by Plummer *et al.* (1997) in the form of fossil packrat urine from Nevada.

Packrats obtain all of their water from the desert plants that they eat, and these plants in turn derive their water from surface-infiltrated rainfall. Therefore the abundant chlorine in packrat urine accurately reflects the ^{36}Cl/Cl ratio of recent rainfall. Furthermore, this urine may be preserved for thousands of years in underground middens, and can be dated by the radiocarbon method. Hence, this material represents an ideal inventory of past cosmogenic ^{36}Cl production. ^{36}Cl/Cl ratios for packrat urine up to 40 ka in age are presented in Fig. 14.48, along with a record of past ^{14}C production compiled from several sources. The two data sets are relatively well correlated, especially at the present day and at the peak of cosmogenic isotope production around 30 ka. Since geomagnetic modulation is the principal cause of past ^{14}C variations, it follows that ^{36}Cl production is subject to the same controls.

Following this demonstration of the geomagnetic modulation of global ^{36}Cl production, it was found that appropriate corrections for the variable accumulation rates of Greenland snow *did* yield a record of past ^{36}Cl variations that was well correlated with geomagnetic field strength (Baumgartner *et al.*, 1998). This suggests that a significant fraction of the precipitation in Greenland is actually derived from more temperature latitudes. However, the large corrections that must be applied for variable snow accumulation rates mean that ice cores are not reliable as prime records of past cosmogenic isotope production. Instead, the known variations in past cosmogenic isotope production may be more useful

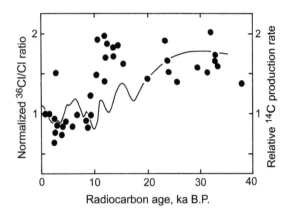

Fig. 14.48 Plot of chlorine isotope variation against age in samples of packrat urine from Nevada, USA. The data are compared with a curve of past radiocarbon abundances compiled from several sources. After Plummer *et al.* (1997).

for calibrating the variable ice accumulation rates in these cores (Wagner *et al.*, 2001).

The most important application of cosmogenic ^{36}Cl (as opposed to anthropogenic ^{36}Cl) is to the dating of ancient groundwater, hundreds of ka in age. For simple sedimentary aquifers this has been quite successful, as demonstrated by studies on the Great Artesian Basin of eastern Australia (Bentley *et al.*, 1986; Torgersen *et al.*, 1991). These studies presented a total of five ^{36}Cl transects across the basin, reaching as far as 800 km from the recharge area. Comparisons were made with average groundwater ages from hydrological modelling, and comparisons were also made between three different ways of calculating the ^{36}Cl age. However, the simplest method is probably the most reliable in most circumstances. This method calculates the groundwater age from the total abundance of ^{36}Cl (above secular equilibrium) in an unknown sample, relative to the total ^{36}Cl abundance above secular equilibrium at the recharge site (where water inters the aquifer):

$$t = -\frac{1}{\lambda_{36}} \cdot \ln \frac{^{36}Cl_{sample} - {}^{36}Cl_{equilib}}{^{36}Cl_{recharge} - {}^{36}Cl_{equilib}} \qquad [14.7]$$

The ^{36}Cl groundwater ages calculated from this equation are shown in Fig. 14.49, grouped into N–S and E–W transects and compared with the average age profile from hydraulic measurements. These results show that the two N–S transects, which are in the westerly part of the basin (open symbols), generally have younger ^{36}Cl ages than predicted from hydraulic measurements. This can be explained by additional water input into the system along the length of the basin, which dilutes old basin water with young recharge water. On the other hand, the E–W transects (solid symbols), which span the easterly half of the basin, generally have older ^{36}Cl ages than predicted. This implies that basin water tends to accumulate in this area and develop older ages.

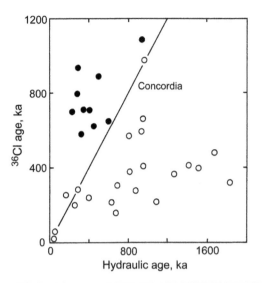

Fig. 14.49 Plot of ^{36}Cl groundwater ages for four transects across the Great Artesian Basin against average age since recharge from hydraulic measurements (solid line). ^{36}Cl ages: (○) = N–S transects; (●) = E–W transects. Data from Torgersen *et al.* (1991).

The ages from the transects were used to calculate a contour diagram of ^{36}Cl groundwater age, shown in Fig. 14.50, and compared with streamlines based on hydraulic measurements. The latter data imply that water flows mainly in a NE to SW direction across the basin from Queensland to South Australia. However, the contour plot shows the presence of old ages in the middle of the basin, implying that water tends to pool here in what is also the deepest part of the basin. When coupled with the evidence for younger groundwaters in the western part of the basin, this implies that a somewhat radial water flow from basin margins to centre is imposed on the general NE–SW flow direction deduced from hydrological modelling. These results have been confirmed by more recent work (e.g. Mahara *et al.*, 2009), showing that the ^{36}Cl method is a reliable method for dating water up to 1 Ma in age in sedimentary aquifers such as the Great Artesian Basin.

The ^{36}Cl method has been more problematical when applied to groundwater ages in igneous rocks, due to interference from local radiogenic ^{36}Cl production. These problems have been evaluated in a case study of the Stripa granite, Sweden, which has unusually high uranium contents of ca. 40 ppm. This generates a substantial neutron flux, which produces ^{36}Cl by the n, γ reaction on ^{35}Cl. The U content of metasedimentary country-rocks (5 ppm) also generates significant, if much lower, levels of radiogenic ^{36}Cl. Analysis of Stripa groundwater yields ^{36}Cl/Cl ratios between the *in situ* radiogenic production of the two dominant rock types (Andrews *et al.*, 1989). These values are so high that they exceed and swamp normal cosmogenic ^{36}Cl/Cl ratios. ^{36}Cl is

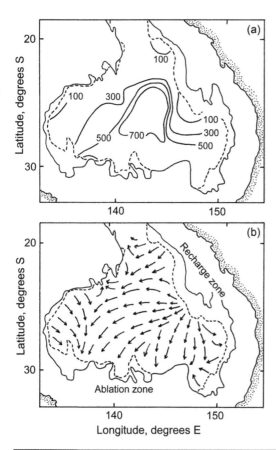

Fig. 14.50 Map of the Great Artesian Basin showing, (a) groundwater age contours from ^{36}Cl dating and, (b) groundwater flow directions based on hydraulic measurements. After Torgersen *et al.* (1991).

therefore only a viable dating method for waters in uranium-poor rocks such as sedimentary aquifers.

14.6 Iodine-129

There are over 100 cosmogenic isotopes with masses over 40 and half-lives over one year, which are therefore potentially useful geochemical tracers or dating tools (Henning, 1987). However, most of these elements are metals, and they are not suited to AMS analysis due to the difficulty of forming negative ions. One of the few heavy isotopes to have found significant application is ^{129}I, which is formed in modest abundance in the atmosphere by spallation of Xe, and as a non-metal, forms good negative ion beams.

^{129}I analysis by AMS is relatively straightforward, since the only isobaric interference (^{129}Xe) does not form stable negative ions (Elmore *et al.*, 1980). The principal interference is ^{127}I, which at isotope ratios above 10^{12} forms a peak tail that must be removed by time-of-flight analysis in addition

to magnetic and electrostatic analysers. The ^{129}I/^{127}I detection limit under these conditions is about 10^{-14}.

Anthropogenic ^{129}I was generated as a fission product by atmospheric nuclear testing, but this source is dwarfed by the amount released from nuclear reprocessing plants in Britain and France (Snyder *et al.*, 2010). In marine sediments from the continental slope off Cape Hatteras (North Carolina), Fehn *et al.* (1986) found ^{129}I/^{127}I levels at the sediment surface that were two orders of magnitude higher than the relatively constant abundances at depth. This has been confirmed by more recent studies. For example, studies of Mississippi delta sediments (Oktay *et al.*, 2000) gave ^{129}I depth profiles very similar to those for ^{36}Cl.

Natural ^{129}I produced in the atmosphere enters the hydrosphere, where it mixes with an approximately equal amount of radiogenic ^{129}I released from rocks (Fabryka-Martin *et al.*, 1985). Iodine is readily soluble in water, but the affinity of iodine for organic matter complicates the residence of iodine in the hydrosphere. For example, marine organic material is enriched by a factor of one hundred relative to seawater, but it releases this iodine back to pore waters within the upper part of the sediment column, from where it may return to seawater (Ullman and Aller, 1980). Hence iodine must be considered a semi-conservative element, and estimates of the marine residence of iodine have varied from above 300 ka (Broecker and Peng, 1982) to 40 ka for the whole ocean (Fabryka-Martin *et al.*, 1985) and as little as 1 ka in the deep ocean (Fabryka-Martin *et al.*, 1985).

Following the early work of Fehn *et al.* (1986), additional studies of pre-anthropogenic iodine in shallow marine sediments gave an average ^{129}I/^{127}I ratio of ca. 1.5×10^{-12} which is regarded as the starting point for attempts to date sedimentary systems (Moran *et al.*, 1998; Fehn *et al.*, 2007a). For a detection limit of ca. 2×10^{-14}, this gives the method a dating range of ca. 100 Ma (Fehn, 2012).

An interesting application of this technique is to subduction-related geothermal fluids, which show excess ^{129}I abundances attributed to subducted sediment. These signatures are analogous to ^{10}Be in subduction-related volcanics (Section 14.4.6), but the longer half-life of ^{129}I allows the original age of deposition of this sediment to be determined. The iodine signatures of arc-related magmas (solid symbols in Fig. 14.51) were obtained from geothermal fumaroles, hot springs and wells above subduction zones (Snyder and Fehn, 2002; Snyder *et al.*, 2002). Based on rates of plate movement, the total age of each slab was calculated at the point where hydrous mineral breakdown releases incompatible element-enriched fluids from subducted sediments and oceanic crust. The inverse correlation between this slab age and the ^{129}I/^{127}I ratio of the volcanic fluids is consistent with iodine decay in each sediment package as it was transported across the ocean floor and down a subduction zone. The two dashed curves in Fig. 14.51 represent modelled ^{129}I decay curves for the oldest sediments at the bottom of the section (minimum ages) and an average for the whole sedimentary section (Fehn, 2012).

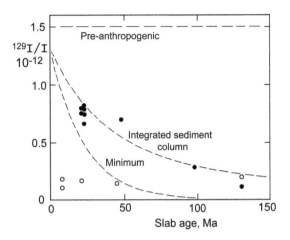

Fig. 14.51 Plot of $^{129}I/^{127}I$ in arc-related rocks compared with the average age of the subducted slab. (•) = arc geothermal fluids; (○) = gas hydrates. Modified after Fehn (2012).

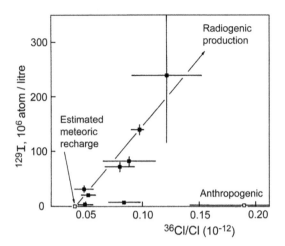

Fig. 14.52 Plot of absolute ^{129}I abundance against the ^{36}Cl/total Cl ratio for groundwaters from depths up to 1200 m in the Stripa mine, Sweden. Cosmogenic, radiogenic and anthropogenic signatures are shown. After Fabryka-Martin et al. (1989).

Open symbols in Fig. 14.51 represent iodine analyses of pore waters associated with gas hydrates on passive margins and in fore-arc regions (Fehn et al., 2007b). The $^{129}I/^{127}I$ ratios in these samples yield ages between 50 and 100 Ma, much older than the age of the enclosing sediments (<5 Ma). Based on the carbonaceous affinity of iodine, Fehn et al. argued that the $^{129}I/^{127}I$ ratios date the organic sources of methane in the gas hydrates, which moved upwards in the sediment column (along with iodine) from Early Tertiary sediments deeper in each sedimentary basin.

The long 15.7 Ma half-life of ^{129}I limits its use in the dating of groundwater, but as a fission product of uranium, ^{129}I can be used to monitor the radiogenic production of ^{36}Cl from wall-rocks. This is illustrated by a study of the Stripa granite of Sweden (Fabryka-Martin et al., 1989), where radiogenic ^{129}I is present at levels two orders of magnitude higher than cosmogenic iodine. When ^{129}I abundances are plotted against $^{36}Cl/Cl$, all samples except one with anthropogenic ^{36}Cl form an array trending from estimated meteoric recharge towards a pure radiogenic component (Fig. 14.52). This array could result from mixing between two end-members, but it could also result from variable, but correlated, production of radiogenic ^{129}I and ^{36}Cl in the granite, since both are controlled by the uranium content of the rock.

14.7 *In Situ* Cosmogenic Nuclides

The term *in situ* cosmogenic nuclides is taken here to refer to short-lived radioactive species generated in rock surfaces. Their most common application is to date the age of un-roofing (or burial) of rock surfaces exposed to cosmic ray bombardment. The species involved are generally similar to atmospheric cosmogenic nuclides, and must therefore be analysed by AMS in most cases. However, analysis of *in situ* cosmogenic nuclides may be more demanding, since these nuclides must first be separated from the rock matrix.

14.7.1 Meteorite Terrestrial Residence Ages

After the break-up of their parent bodies, meteorite fragments are exposed to intense cosmic ray bombardment during their travel through space, causing spallation production of species such as ^{26}Al. Since this nuclide has a half-life of only 0.7 Ma, meteorite fragments with cosmic exposure ages of a few Ma will reach saturation in ^{26}Al abundance. After falling to Earth, atmospheric shielding protects these fragments from significant further ^{26}Al production, and decay of this inventory can then be used to determine the terrestrial residence age of the meteorites.

Because the abundances of ^{26}Al in meteorites are comparatively high, they do not demand measurement by accelerator mass spectrometry. Consequently, these were some of the first cosmogenic exposure determinations, made by putting whole meteorite fragments in a large shielded γ counter. Attenuation of γ particles by the sample itself is corrected by empirical modelling (e.g. Evans et al., 1979).

In the first large-scale survey of Antarctic meteorites, Evans et al. (1979) compared the ^{26}Al activities in these samples with those in 'falls' (observed at impact), which have a zero terrestrial residence. The falls had a moderately well-defined range of ^{26}Al activities for a given compositional class of meteorites (Fig. 14.53). The outlier with low activity is attributed to a failure to reach ^{26}Al saturation, due to the short cosmic ray exposure history of the fragment. In contrast, Antarctic meteorites range to substantially lower activities, indicating significant terrestrial residence ages

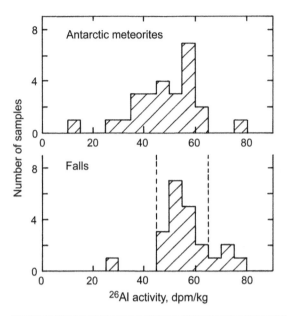

Fig. 14.53 Histogram of ^{26}Al activities in Antarctic meteorites, compared with American 'falls'. After Evans *et al.* (1979).

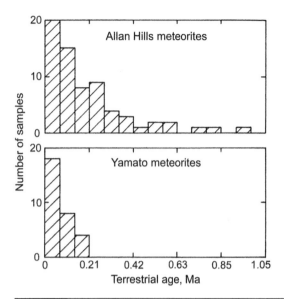

Fig. 14.54 Histograms of calculated terrestrial residence age for Antarctic meteorites from the Allan Hills and Yamato Mountains areas, based on the decay of cosmogenic ^{36}Cl. After Nishiizumi *et al.* (1989a).

in several cases. Unfortunately, these values are only semi-quantitative, due to the relatively long half-life of ^{26}Al, and due to uncertainties in production. The latter problem arises because of the poor penetrative capacity of the low-energy cosmic rays which generate ^{26}Al, making the cosmic production rate susceptible to depth within the fragment.

Although ^{26}Al represents a good reconnaissance tool for meteorite terrestrial residence determinations (e.g. Evans and Reeves, 1987), ^{36}Cl provides a more precise method (Nishiizumi *et al.*, 1979). This arises from its shorter (0.3 Ma) half-life, and from more accurately known saturation values, due to its generation by penetrative high-energy cosmic rays. However, the analysis is more technically demanding, and must be performed by accelerator mass spectrometry. Results of a large ^{36}Cl study of Antarctic meteorite ages are shown in Fig. 14.54 (Nishiizumi *et al.*, 1989a).

The high quality of terrestrial ^{36}Cl ages for Antarctic meteorites has led to their use to determine the half-life of another cosmogenic nuclide, ^{41}Ca. With a half-life of only 0.1 Ma, this shows promise as a precise dating tool, but the formation of negative calcium ions for AMS analysis is very inefficient. The most favourable molecular ion for AMS analysis is calcium hydride, and the triple hydride ion avoids potassium interference, since this does not form triple hydride ions (Fink *et al.*, 1990). Another problem limiting the application of ^{41}Ca has been uncertainty in the half-life. To solve this problem, Klein *et al.* (1991) performed ^{41}Ca analyses on aliquots of Antarctic iron meteorites which had already been dated by ^{36}Cl. The results revealed a strong linear correlation between the abundance of the two species,

whose slope corresponds to the ratio of the half-lives. Taking a ^{36}Cl half-life of 301 ± 4 ka yields a precise ^{41}Ca half-life of 103 ± 7 ka.

14.7.2 Al–Be Terrestrial Exposure Ages

Because of atmospheric attenuation of cosmic rays, most terrestrial materials have ^{26}Al/^{27}Al ratios of less than 10^{-14}. However, in some aluminium-poor minerals such as quartz, the content of (non-cosmogenic) ^{27}Al may be as low as few ppm, so that after a few thousand years of exposure to cosmic rays, ^{26}Al/^{27}Al ratios of 10^{-11} to 10^{-13} may be generated, within the measurement range of AMS. These data can then be used to calculate exposure ages of terrestrial rock surfaces.

The principal obstacle to AMS analysis of ^{26}Al is the formation of sufficient negative Al ions during sputtering, which is only about 25% efficient (Middleton and Klein, 1987). The metal species must be used rather than the oxide because the latter suffers a severe interference from MgO. However, Mg does not form negative metal ions at all, so there are no isobaric interferences on the Al metal ion signal.

It would be possible to use ^{26}Al alone for the determination of rock exposure ages, but in view of the many possible permutations of exposure and erosion history, the use of two nuclides with different half-lives provides a more powerful constraint on these models. The normal choice is to combine ^{26}Al measurements ($t_{1/2} = 0.705$ Ma) with ^{10}Be ($t_{1/2} = 1.39$ Ma). The latter value is the revised half-life of Korschinek *et al.* (2010) and Chmeleff *et al.* (2010). However, most of the examples described below use the earlier value of 1.51 Ma

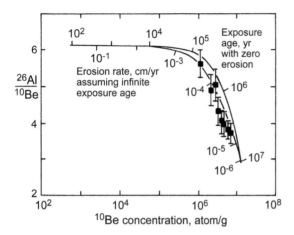

Fig. 14.55 Plot of analysed $^{26}Al/^{10}Be$ against ^{10}Be abundance (corrected to production at sea level) for Allan Hills quartz samples, used as a measure of minimum exposure age and/or maximum erosion rate. After Nishiizumi et al. (1991a).

(Hofmann et al., 1987). The new value will have significant effects on Al–Be exposure ages, including a change in the expected saturation $^{26}Al/^{10}Be$ ratio that is achieved after long exposure.

The atmospheric $^{26}Al/^{10}Be$ production ratio has been determined as about 4×10^{-3} by sampling from high-flying aircraft. In contrast, Nishiizumi et al. (1989b) determined an *in situ* production ratio of 6 in quartz, more recently revised to 6.75 based on new analytical standards (Balco and Rovey, 2008). Because the atmospheric ^{10}Be production rate is comparatively high, great care must be taken to ensure that rock samples to be used for exposure dating are not contaminated by the atmospheric or so-called 'garden variety' of ^{10}Be (Nishiizumi et al., 1986). Because of its resistance to chemical weathering, quartz is comparatively resistant to contamination by 'garden variety' ^{10}Be. This, along with its low ^{27}Al content, makes it an excellent material for exposure dating. In quartz, ^{10}Be is derived from spallation of ^{16}O, while ^{26}Al is produced by spallation of ^{28}Si and mu-meson capture by the same species. Most of this production occurs in the top half-metre of the rock surface, but limited ^{26}Al production can occur at depths of up to 10 m (Middleton and Klein, 1987).

There are two limiting models for the interpretation of surface exposure data (e.g. Nishiizumi et al., 1991a). These are illustrated on a plot of $^{26}Al/^{10}Be$ ratio against absolute ^{10}Be abundance (Fig. 14.55). The upper curve shows the effect of increasing exposure age, for the case where the erosion rate is zero. The lower curve shows the effect of different steady state erosion rates, for the case where exposure age is infinite (relative to the half-lives of ^{26}Al and ^{10}Be). The space between the two curves was called the 'erosion island' by Lal (1991) because continuously eroding objects plot between the two curves. Finally the curves meet (after a few half-lives) at the point of saturation, with a $^{26}Al/^{10}Be$ ratio of 2.88.

Unfortunately, the limited (factor of two) difference between the half-lives of ^{26}Al and ^{10}Be causes only a relatively small separation between the curves for the two end-member models. This places limits on the resolving power of the Al–Be method, given the relatively large analytical errors on AMS measurements.

Application of the Al–Be exposure method is demonstrated in Fig. 14.55 using data from nunataks of the Allan Hills area of Antarctica (Nishiizumi et al., 1991a). The results display a range of Al/Be ratios close to the steady state erosion curve. However, the zero erosion rate (variable exposure age) model cannot be ruled out. The lowest $^{26}Al/^{10}Be$ ratios yield the strongest constraint, representing a minimum exposure age of 1.4 Ma or a maximum erosion rate of 0.24 mm/ka. Samples to the left of the erosion line in Fig. 14.55 may be explained by burial under ice for some period in the past. During times of burial, points move downwards to the left, due to the greater rate of decay of ^{26}Al relative to ^{10}Be.

14.7.3 Al–Be Burial Ages

The combination of these two cosmogenic isotopes can also be used to date sediment burial, as demonstrated by Granger et al. (2001) in a study on Mammoth Cave, Kentucky. Clastic sediments are thought to have entered this cave system a few Ma ago from the Green River, a tributary of the Ohio River, which cut into a karst landscape. There is a very strong likelihood that the sediments were exposed on the land surface prior to their deposition in the cave system. Therefore, quartz grains in the sediment are expected to have reached steady state abundances of cosmogenic Al and Be, with a $^{26}Al/^{10}Be$ ratio of around 6 (Fig. 14.56).

After burial in the cave system, radioactive decay begins, and the samples follow a trend to the lower left in Fig. 14.56. This allows the dating of sediments at different levels in the cave system, so that the down-cutting of the river can be traced against time. Projection of the decay trend back to the steady state line also allows the erosion rate of the surficial sediment source to be estimated at around 3 mm/ka (m/Ma). The combination of ^{26}Al and ^{10}Be is particularly suitable for burial dating because the pre-burial cosmogenic production of the two nuclides occurs under similar conditions. Accurate modelling of isotope production during the surface exposure history of the sediment is a prerequisite to obtaining accurate burial ages (Granger and Muzikar, 2001).

The data in Fig. 14.56 suggest that Mammoth Cave is a relatively ideal case for burial dating. The sediment supply appears to have had a relatively simple cosmogenic exposure history before deposition in the cave, and the case system is so deep that no production occurred after deposition. However, many dating situations are less ideal, prompting the search for a more versatile method. For example, Balco and Rovey (2008) sought to date paleosols developed on top of Quaternary glacial till deposits, in order to date the overlying glacial formation and hence the date of the next glaciation. However, it was recognized that the tills might have complex exposure histories, with $^{26}Al/^{10}Be$ ratios below the

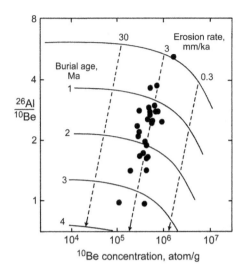

Fig. 14.56 Plot of analysed $^{26}Al/^{10}Be$ against ^{10}Be abundance in sediments from different levels in Mammoth Cave, Kentucky. Vertical displacement allows calculation of the age of different deposits in the cave, while the intersection with the steady state erosion curve indicates the erosion rate of the sediment source at the surface. After Granger et al. (2001).

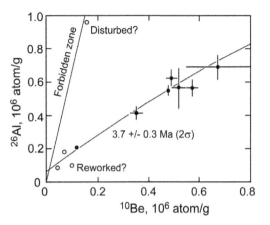

Fig. 14.57 Al–Be isochron plot for samples from Sterkfontein Cave, South Africa, associated with an Australopithecus skull. Solid ellipses = sand; open = chert. Modified after Granger et al. (2015).

production value of 6. To take account of more complex burial models, Balco and Rovey analysed a suite of samples at various depths in the paleosols, presenting the data on Al–Be isochron diagrams. This strategy was quite successful, and Balco and Rovey (2010) were able to date a series of glacial advances by the Laurentide ice sheet from 1.3 to 0.2 Ma.

Since glacial deposits are not very deeply buried, Balco and Rovey (2008, 2010) made a correction to each of their analysed data points for cosmogenic production after burial from cosmic-ray-generated muons, which can still cause significant production at depths up to 30 m (Heisinger et al., 2002). However, other workers suggested that the Al–Be isochron diagram was ideally suited to remove the effect of post-burial production, avoiding the uncertainty of attempting to model the muon induced production.

This use of the Al–Be isochron method was demonstrated by Granger et al. (2015) in dating an Australopithecus skeleton buried in a shallow cave at Sterkfontein, South Africa (Fig. 14.57). Granger et al. analysed five samples of fine sand, two of coarse sand/pebbles and four cherts. Two of the chert analyses were rejected, but the other two were included in the isochron calculation, yielding an overall slope age of 3.7 ± 0.3 Ma (2σ). The offset of the isochron from the zero point is attributed to post-burial production from muons, which are assumed to have affected all samples equally. The new Al–Be age is significantly older than a U–Th age of 2.2 Ma, which is argued to date a later cavity-filling episode rather than the age of deposition. On the other hand, the new age is younger than a previous 4.2 Ma Al–Be age deter-

mined on a single sample, which probably suffered from too large a correction for post-burial muon production.

14.7.4 Al–Be–Ne Ages

Dating surface exposure or burial using the relative decay rates of two short-lived nuclides (^{26}Al and ^{10}Be) is not necessarily the most effective way of using cosmogenic isotopes. It would be more effective to measure the accumulation or decay of ^{26}Al or ^{10}Be in comparison with a stable cosmogenic isotope. Graf et al. (1991) proposed that cosmogenic ^{21}Ne could be used in such a way with ^{26}Al or ^{10}Be, but the method was not widely applied, perhaps because it requires the combination of two completely different analytical methods (AMS and noble gas mass spectrometry). However, several more recent studies have begun to use this approach.

Kober et al. (2007) used the Al–Be–Ne exposure method to study rates of erosion in the Peruvian Andes, comparing the large rates of erosion in the high (wet) Cordillera with the much smaller rates in the lower (dry) Western Escarpment. The Ne–Be data from this work are shown in Fig. 14.58, but the Ne–Al plot is basically similar. It is clear that the Ne–Be plot better resolves alternative erosion models compared with the traditional Al–Be plot (Fig. 14.55). The 'erosion island' is greatly enlarged on the new plot, allowing variable erosion rates to be more clearly resolved. Analytical errors are also smaller (barely larger than the symbol size) since ^{21}Ne can be measured more precisely than ^{26}Al.

Balco and Shuster (2009) showed that the Al–Be–Ne method can also be used effectively to measure burial ages (Fig. 14.59). They tested the method on the Whippoorwill Formation, a bedrock weathering horizon that underlies a till deposited by the Laurentide ice sheet. The data were presented in a reversed form relative to Fig. 14.58, using ^{21}Ne as the denominator of the cosmogenic isotope ratio. The results showed good agreement of Al–Ne and Be–Ne ages with a

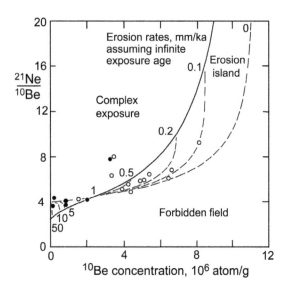

Fig. 14.58 Ne–Be cosmogenic exposure plot showing samples from the wet Cordillera (•) and the dry Western Escarpment of the Andes (○) relative to the 'erosion island'. Numbers indicate erosion rates in mm/ka (m/Ma). Modified after Kober et al. (2007).

previously determined Al–Be age, dating the advance of the Laurentide ice sheet to ca. 2.4 Ma.

14.7.5 Chlorine-36 Exposure Ages

The concept of exposure dating using *in situ* produced cosmogenic ^{36}Cl was suggested by Davis and Schaeffer (1955), but could not be effectively applied until the advent of AMS analysis (Phillips *et al.*, 1986). Although developed after the ^{26}Al and ^{10}Be and methods, ^{36}Cl offers several advantages. Whereas cosmogenic Al–Be dating is largely restricted to

quartz, ^{36}Cl is applicable to a variety of rock types because it is generated from three parents (K, Ca and Cl) with different chemistry. In addition, most rocks have low background levels of chlorine. For example, interferences by nucleogenic ^{36}Cl are minimal, and contamination from atmospheric ^{36}Cl is only a problem in severely weathered material.

The three main sources of *in situ* cosmogenic ^{36}Cl are neutron induced spallation of ^{40}K and ^{40}Ca and neutron activation of ^{35}Cl. A subordinate source is from negative muon capture by ^{40}Ca. The relative importance of these production routes depends on the relative K: Ca: Cl abundances in the target and the degree of shielding by overlying rock. Zreda *et al.* (1991) made an empirical study of spallation production, while Liu *et al.* (1994) studied the depth dependence of the neutron activation reaction. Spallation reactions are caused by fast neutrons, which decrease exponentially with depth. However, activation reactions require (slow) thermal neutrons, which are produced when fast neutrons undergo glancing collisions with substrate atoms. Hence, slow neutrons reach a peak intensity at about 15 cm depth in rocks (Fig. 14.60). The occurrence of multiple routes makes the calculation of total production rates more complex, but it also offers the possibility of greater age control, as will be shown below.

The simplest scenario for cosmogenic dating is the instantaneous transport of rock from below the cosmic ray penetration depth to the surface, followed by exposure without shielding by other material (such as snow cover) and without significant erosion. In this case, all production routes for ^{36}Cl can be summed. However, corrections must be made for the latitude dependence of cosmic ray intensity and the altitude dependence of atmospheric shielding.

A relatively simple scenario is provided by surface-exposure dating of rocks at Meteor Crater, Arizona (Nishiizumi *et al.*, 1991b). Samples were collected from the

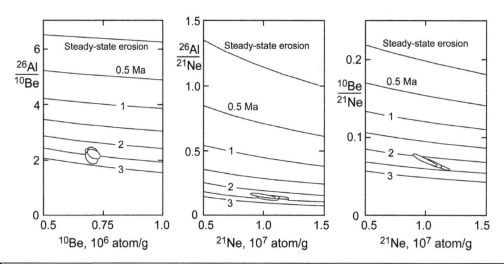

Fig. 14.59 Al–Be–Ne isotope plots comparing burial ages determined by different cosmogenic nuclide pairs. Modified after Balco and Shuster (2009).

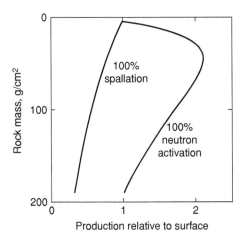

Fig. 14.60 Calculated production profiles for ^{36}Cl by spallation and neutron activation (normalized to equal values at the surface) as a function of rock mass per unit area. After Liu *et al.* (1994).

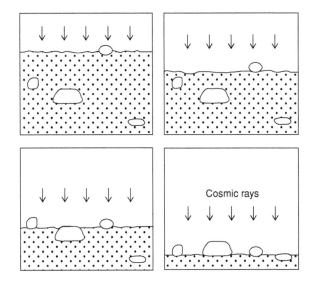

Fig. 14.61 Schematic illustration of the exhumation of clasts from a heterogeneous deposit by erosion of the matrix. After Zreda *et al.* (1994).

upper few cm of large blocks in the ejecta blanket of the impact. The lithology of these blocks shows that they were from strata buried at depths over 10 m before the impact. On the other hand, their large size suggests that they were unroofed of any overlying ash blanket soon after the impact event. The surfaces of the sampled ejecta blocks were found to be coated with 'rock varnish', which takes thousands of years to develop. Hence, erosion was probably negligible in the arid climate of Arizona, so that ^{36}Cl abundances translate directly into an exposure age. Good agreement was found between ^{36}Cl ages and Al–Be ages, with a consensus of ages around 50 ka BP. A few younger ages (e.g. Monument rock) may indicate more recent exposure of these blocks above the ash blanket.

Another simple dating scenario is achieved in relatively young lava flows, which are instantaneously exposed at the surface and have not yet suffered significant weathering. However, in many geological environments, erosion is a significant factor. In this case, a single cosmogenic isotope determination only allows the solution of exposure age at known erosion rate or erosion rate at known exposure age. For example, using a lava flow of known age from Nevada, Shepard *et al.* (1995) were able to estimate the weathering rate based on ^{36}Cl/^{35}Cl ratios.

A more complex scenario arises when both the exposure age and the erosion rate of a deposit are unknown. This applies particularly to rapidly eroding sedimentary deposits. To investigate the constraints which may be applied to such a system, Zreda *et al.* (1994) considered a theoretical model for the erosional exposure of a deposit with buried clasts. These clasts, initially buried within the deposit, are gradually exposed at the surface by erosion of the fine-grained matrix (Fig. 14.61). This model applies to the problem of dating glacial moraines based on the exposure ages of boulders

on the moraine surface, and also to dating meteorite impacts based on blocks exposed on the ejecta blanket. Zreda *et al.* argued that if ^{36}Cl dates are available from several different boulders on a moraine or ejecta blanket, the spread of ages (outside analytical error) could be used to model the exposure history of the deposit.

To demonstrate these principles, Zreda and Phillips (1995) modelled ^{36}Cl/Cl ratios for several boulders buried at depths from zero to 300 g/cm^2 (approximately 1.5 m depth, in soil with a density of 2 g/cm^2). Total ^{36}Cl production was attributed 50% to spallation reactions and 50% to neutron activation. The erosion rate was set so that the deepest boulder just reaches the surface at the present day. The modelling results show a wide range of ^{36}Cl/Cl ratios, both below and slightly above the growth curve for zero depth (Fig 14.62). The growth curves for deeply buried boulders start with slow production, due to the shielding effect of the overlying matrix. As the boulder approaches the surface, the rate of production accelerates, but the total ^{36}Cl inventory remains well below that of a surface sample. On the other hand, boulders which were initially subject to shallow burial can actually show greater total ^{36}Cl production than at the surface, due to the peak of neutron activation production at a depth of 50 g/cm^2.

Zreda and Phillips (1995) modelled the erosion history of ejecta blocks from Meteor Crater, Arizona in a similar way. However, in this case the spread of apparent exposure ages was based on 1000 model points, randomly buried from zero to a chosen maximum depth (Fig. 14.63). The model results were compared with the actual spread of ages in four blocks at Meteor Crater, with a mean age of 49.7 ± 0.9 ka (Monument Rock was excluded). The best fit was obtained

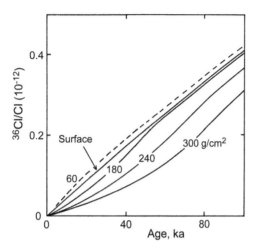

Fig. 14.62 Modelled ^{36}Cl/Cl inventories for rock boulders buried at various depths in an eroding deposit, showing peak cosmogenic production just below the surface. After Zreda and Phillips (1995).

by assuming burial to a maximum depth of 300 g/cm^2 (ca. 1.5 m), removed at 30 g/cm^2/ka (0.15 m/ka), so that all boulders reached the surface within 10 ka. Such modelling cannot yield a unique solution to the erosion process, and the model results should ideally be compared with a larger set of measured ages. Nevertheless, the modelling does suggest that the four blocks yield exposure ages close to the estimated time of impact.

In principle, the use of two spallogenic nuclides (e.g. ^{10}Be and ^{26}Al) allows the deconvolution of exposure ages and erosion rates. However, as noted above, the closely spaced trajectories of the constant erosion and constant exposure saturation lines make the deconvolution weak (e.g. Fig. 14.55). A more powerful application of the ^{36}Cl technique makes use of the neutron activation route to ^{36}Cl, in comparison

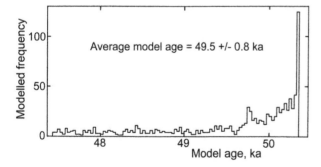

Fig. 14.63 Histogram of model ^{36}Cl exposure ages for 1000 rock clasts exhumed from depths up to ca. 1.5 m in a deposit with a formation age of 50 ka. A few samples over-estimate the real age due to greater cosmogenic production just below the surface. After Zreda and Phillips (1995).

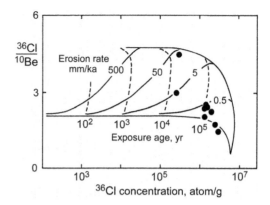

Fig. 14.64 Plot of analysed ^{36}Cl/^{10}Be against ^{36}Cl abundance for boulders from glacial moraines of the Wind River Range, Wyoming. Solid lines indicate erosion rates; dashed lines are isochrons. See text for discussion. After Phillips et al. (1997).

with the purely spallation nuclides (^{10}Be or ^{26}Al). Because spallation and neutron activation yield peak nuclide production at different depths in a geological surface, they should yield a clear resolution of exposure histories involving different rates of erosion (Liu et al., 1994). With this objective in mind, Bierman et al. (1995) described a method for isolating the neutron activation component of ^{36}Cl, by releasing chlorine from fluid inclusions within rock samples. However, the method has not been demonstrated in practice. Alternatively, a crude assessment of the rate of erosion can be obtained by simply ignoring spallogenic ^{36}Cl production.

This approach was taken by Phillips et al. (1997) in a study of glacial deposits from the Wind River Range, Wyoming. Ten boulders from glacial moraines were analysed for both ^{36}Cl and ^{10}Be, and are plotted on a Cl–Be plot (Fig. 14.64). The limits for zero erosion and for infinite age (erosional equilibrium) are given, as in the Al–Be plot. However, the curves are now much further apart, allowing easy resolution between the two models. Most of the data lie below the line corresponding to 0.5 mm of erosion per ka, indicating that the surfaces of these boulders are not undergoing significant loss by erosion. Boulders were sampled from six different moraines, and their ages fall into three distinct groups. The youngest dates from the last glaciation (2 points at 20 ka), the second group dates to the penultimate glaciation (ca. 130–100 ka) and the oldest is >200 ka, assuming rapid exposure of boulders on the moraine surface.

Another application of the ^{36}Cl method was proposed by Stone et al. (1996), using ^{40}Ca spallation as a tool for exposure dating of calcite. The abundance of the target nuclide, along with the relatively low abundance of chlorine in calcite, makes this an analytically favourable method which shows great promise for exposure dating of karst landscapes. For example, Zreda and Noller (1998) and Mitchell et al. (2001) used ^{36}Cl abundances in fault scarps to date the displacement of fault systems in carbonate rocks. In both cases,

^{36}Cl measurements were made at several points down the face of the fault scarp. ^{36}Cl accumulation was then used to determine the date when each sample was first exposed at the surface, allowing a reconstruction of past movement on the fault.

As cosmogenic exposure dating reaches a mature stage of development with wide applicability, it has become increasingly important to develop empirical exposure standards, and to use these standards to test and refine models of cosmogenic nuclide production. This work is being carried out by the CRONUS-Earth collaboration project (Phillips *et al.*, 2016). By establishing standards for each of the three ^{36}Cl production pathways, compositional effects can be disentangled (Marrero *et al.*, 2016). For example, plagioclase from an 18.2 ka basaltic lava at Tabernacle Hill, Utah provides a standard for the Ca pathway. Similarly, the Cl production standard is provided by hornblende separated from boulders on a 13.3 ka glacial moraine at Baboon Lakes, California. Finally, K production standards are provided by K-feldspar from glacial boulders at Huancane, Peru; Skye, Scotland and Baboon Lakes, California. The different latitudes and altitudes of the samples permit cosmogenic production under different conditions to be calibrated, thus promising more reliable exposure age dating into the future.

References

Abreu, J. A., Beer, J., Steinhilber, F., Tobias, S. M. and Weiss, N. O. (2008). For how long will the current grand maximum of solar activity persist? *Geophys. Res. Lett.* **35**, L20109 1–4.

Abreu, J. A., Beer, J., Ferriz-Mas, A., McCracken, K. G. and Steinhilber, F. (2012). Is there a planetary influence on solar activity? *Astron. & Astrophys.* **548**, A88 1–9.

Adkins, J. F. and Boyle, E. A. (1997). Changing atmospheric D^{14}C and the record of deep water paleoventilation ages. *Paleoceanography* **12**, 337–44.

Anderson, E. C. and Libby, W. F. (1951). World-wide distribution of natural radiocarbon. *Phys. Rev.* **81**, 64–9.

Andree, M., Oeschger, H., Broecker, W. *et al.* (1986). Limits on ventilation rates for the deep ocean over the last 12,000 years. *Climate Dynamics* **1**, 53–62.

Andrews, J. N., Davis, S. N., Fabryka-Martin, J. *et al.* (1989). The *in situ* production of radioisotopes in rock matrices with particular reference to the Stripa granite. *Geochim. Cosmochim. Acta* **53**, 1803–15.

Arnold, J. R. (1956). Beryllium-10 produced by cosmic rays. *Science* **124**, 584–5.

Arnold, J. R. (1958). Trace elements and transport rates in the ocean. *2nd UN Conf. on Peaceful Uses of Atomic Energy* **18**, 344–6. IAEA.

Arnold, J. R. and Libby, W. F. (1949). Age determinations by radiocarbon content: Checks with samples of known age. *Science* **110**, 678–80.

Balco, G. and Rovey, C. W. (2008). An isochron method for cosmogenic-nuclide dating of buried soils and sediments. *Amer. J. Sci.* **308**, 1083–114.

Balco, G. and Rovey, C. W. (2010). Absolute chronology for major Pleistocene advances of the Laurentide Ice Sheet. *Geology* **38**, 795–8.

Balco, G. and Shuster, D. L. (2009). ^{26}Al–^{10}Be–^{21}Ne burial dating. *Earth Planet. Sci. Lett.* **286**, 570–5.

Bard, E., Arnold, M., Fairbanks, R. G. and Hamelin, B. (1993). ^{230}Th–^{234}U and ^{14}C ages obtained by mass spectrometry on corals. *Radiocarbon* **35**, 191–9.

Bard, E., Arnold, M., Hamelin, B., Tisnerat-Laborde, N. and Cabioch, G. (1998). Radiocarbon calibration by means of mass spectrometric ^{230}Th/^{234}U and ^{14}C ages of corals: an updated database including samples from Barbados, Mururoa and Tahiti. *Radiocarbon* **40**, 1085–92.

Bard, E., Arnold, M., Mangerud, J. *et al.* (1994). The North Atlantic atmosphere–sea surface ^{14}C gradient during the Younger Dryas climatic event. *Earth Planet. Sci. Lett.* **126**, 275–87.

Bard, E. and Frank, M. (2006). Climate change and solar variability: What's new under the sun?. *Earth Planet. Sci. Lett.* **248**, 1–14.

Bard, E., Hamelin, B., Fairbanks, R. G. and Zindler, A. (1990a). Calibration of the ^{14}C timescale over the past 30,000 years using mass spectrometric U-Th ages from Barbados corals. *Nature* **345**, 405–10.

Bard, E., Hamelin, B., Fairbanks, R. G. *et al.* (1990b). U/Th and ^{14}C ages of corals from Barbados and their use for calibrating the ^{14}C timescale beyond 9000 years B.P. *Nucl. Instr. Meth. in Phys. Res. B* **52**, 461–8.

Bard, E., Raisbeck, G. M., Yiou, F. and Jouzel, J. (1997). Solar modulation of cosmogenic nuclide production over the last millennium: comparison between ^{14}C and ^{10}Be records. *Earth Planet. Sci. Lett.* **150**, 453–62.

Baumgartner, S., Beer, J., Masarik, J. *et al.* (1998). Geomagnetic modulation of the ^{36}Cl flux in the GRIP ice core, Greenland. *Science* **279**, 1330–2.

Baxter, M. S. and Farmer, J. G. (1973). Radiocarbon: short-term variations. *Earth Planet. Sci. Lett.* **20**, 295–9.

Bayes, T. and Price, R. (1763). An essay towards solving a problem in the doctrine of chances. *Phil. Trans.* **53**, 370–418.

Beck, J. W., Richards, D. A., Edwards, R. L. *et al.* (2001). Extremely large variations of atmospheric ^{14}C concentration during the last glacial period. *Science* **292**, 2453–8.

Beer, J., McCracken, K. G., Abreu, J., Heikkilä, U. and Steinhilber, F. (2013). Cosmogenic radionuclides as an extension of the neutron monitor era into the past: potential and limitations. *Space Sci. Rev.* **176**, 89–100.

Beer, J., Andree, M., Oeschger, H. *et al.* (1984). The Camp Century ^{10}Be record: implications for long-term variations of the geomagnetic dipole moment. *Nucl. Instr. Meth. in Phys. Res. B* **5**, 380–4.

Beer, J., Oeschger, H., Finkel, R. C. *et al.* (1985). Accelerator measurements of ^{10}Be: the 11 year solar cycle. *Nucl. Instr. Meth. in Phys. Res. B* **10**, 415–18.

Beer, J., Siegenthaler, U., Bonani, G. *et al.* (1988). Information on past solar activity and geomagnetism from ^{10}Be in the Camp Century ice core. *Nature* **331**, 675–9.

Bentley, H. W., Phillips, F. M., Davis, S. N. *et al.* (1982). Thermonuclear ^{36}Cl pulse in natural water. *Nature* **300**, 737–40.

Bentley, H. W., Phillips, F. M., Davis, S. N. *et al.* (1986). Chlorine 36 dating of very old groundwaters, 1, The Great Artesian Basin, Australia. *Water Resources Res.* **22**, 1991–2002.

Bierman, P., Gillespie, A., Caffee, M. and Elmore, D. (1995). Estimating erosion rates and exposure ages with ^{36}Cl produced by neutron activation. *Geochim. Cosmochim. Acta* **59**, 3779–98.

Broecker, W., Barker, S., Clark, E. *et al.* (2004). Ventilation of the glacial deep Pacific Ocean. *Science* **306**, 1169–72.

Broecker, W., Clark, E. and Barker, S. (2008). Near constancy of the Pacific Ocean surface to mid-depth radiocarbon-age difference over the last 20 kyr. *Earth Planet. Sci. Lett.* **274**, 322–6.

Broecker, W. S. and Denton, G. H. (1989). The role of ocean–atmosphere reorganizations in glacial cycles. *Geochim. Cosmochim. Acta* **53**, 2465–501.

Broecker, W. S., Gerard, R., Ewing, M. and Heezen, B. C. (1960). Natural radiocarbon in the Atlantic Ocean. *J. Geophys. Res.* **65**, 2903–31.

Broecker, W. S. and Peng, T. H. (1982). *Tracers in the Sea*. Lamont-Doherty Geol. Obs. 690 pp.

Brown, E. T., Edmond, J. M., Raisbeck, G. M. *et al*. (1992). Beryllium isotope geochemistry in tropical basins. *Geochim. Cosmochim. Acta* **56**, 1607–24.

Brown, L., Klein, J. and Middleton, R. (1985). Anomalous isotopic concentrations in the sea off Southern California. *Geochim. Cosmochim. Acta* **49**, 153–7.

Brown, L., Klein, J., Middleton, R., Sacks, I. S. and Tera, F. (1982). ^{10}Be in island-arc volcanoes and implications for subduction. *Nature* **299**, 718–20.

Brown, L., Stensland, G. J., Klein, J. and Middleton, R. (1989). Atmospheric deposition of ^{7}Be and ^{10}Be. *Geochim. Cosmochim. Acta* **53**, 135–42.

Bruns, M., Rhein, M., Linick, T. W. and Suess, H. E. (1983). The atmospheric ^{14}C level in the 7th millennium BC. *P.A.C.T. (Physical And Chemical Techniques in Archaeology)* **8**, 511–16.

Bucha, V. and Neustupny, E. (1967). Changes in the Earth's magnetic field and radiocarbon dating. *Nature* **215**, 261–3.

Buck, C. E., Litton, C. D. and Smith, A. F. (1992). Calibration of radiocarbon results pertaining to related archaeological events. *J. Archaeological Sci.* **19**, 497–512.

Burke, A., and Robinson, L. F. (2012). The Southern Ocean's role in carbon exchange during the last deglaciation. *Science* **335**, 557–61.

Campin, J.-M., Fichefet, T. and Duplessy, J.-C. (1999). Problems with using radiocarbon to infer ocean ventilation rates for past and present climates. *Earth Planet. Sci. Lett.* **165**, 17–24.

Chengde, S., Beer, J., Tungsheng, L. *et al*. (1992). ^{10}Be in Chinese loess. *Earth Planet. Sci. Lett.* **109**, 169–77.

Chmeleff, J., von Blanckenburg, F., Kossert, K. and Jakob, D. (2010). Determination of the ^{10}Be half-life by multicollector ICP-MS and liquid scintillation counting. *Nucl. Instr. Meth. in Phys. Res. B* **268**, 192–9.

Cohen, T. J. and Lintz, P. R. (1974). Long term periodicities in the sunspot cycle. *Nature* 250, 398–400.

Craig, H. (1954). Carbon-13 in plants and the relationships between carbon-13 and carbon-14 variations in nature. *J. Geol.* **62**, 115–49.

Craig, H. (1957). Isotopic standards for carbon and oxygen and correction factors for mass-spectrometric analysis of carbon dioxide. *Geochim. Cosmochim. Acta* **12**, 133–49.

Davis, R. and Schaeffer, O. A. (1955). Chlorine-36 in Nature. *Ann. N. Y. Acad. Sci.* **62**, 105–22.

De Jong, A. F. M., Mook, W. G. and Becker, B. (1979). Confirmation of the Suess wiggles: 3200–3700 BC. *Nature* **280**, 48–9.

De Vries, H. (1958). Variation in concentration of radiocarbon with time and location on Earth. *Proc. Konikl. Ned. Akad. Wetenschap B* **61**, 94–102.

De Vries, H. and Barendsen, G. W. (1953). Radiocarbon dating by a proportional counter filled with carbon dioxide. *Physica* **19**, 987–1003.

DeVries, T. and Primeau, F. (2010). An improved method for estimating water-mass ventilation age from radiocarbon data. *Earth and Planetary Science Letters* **295**, 367–78.

Druffel, E. M. (1996). Post-bomb radiocarbon records of surface corals from the tropical Atlantic Ocean. *Radiocarbon* **38**, 563–72.

Eddy, J. A. (1976). The Maunder minimum. *Science* **192**, 1189–202.

Eddy, J. A. (1977). Climate and the changing sun. *Climate Change* **1**, 173–90.

Edwards, R. L., Beck, J. W., Burr, G. S. *et al*. (1993a). A large drop in atmospheric ^{14}C/^{12}C and reduced melting in the Younger Dryas, documented with ^{230}Th ages of corals. *Science* **260**, 962–8.

Edwards, C. M. H., Morris, J. D. and Thirlwall, M. F. (1993b). Separating mantle from slab signatures in arc lavas using B/Be and radiogenic isotope systematics. *Nature* **362**, 530–3.

Elmore, D., Gove, H. E., Ferraro, R. *et al*. (1980). Determination of ^{129}I using tandem accelerator mass spectrometry. *Nature* **286**, 138–40.

Elmore, D., Tubbs, L. E., Newman, D. *et al*. (1982). ^{36}Cl bomb pulse measured in a shallow ice core from Dye 3, Greenland. *Nature* **300**, 735–7.

Elsasser, W., Ney, E. P. and Winckler, J. R. (1956). Cosmic-ray intensity and geomagnetism. *Nature* **178**, 1226–7.

Evans, J. C., Rancitelli, L. A. and Reeves, J. H. (1979). ^{26}Al content of Antarctic meteorites: implications for terrestrial ages and bombardment history. *Proc. 10th Lunar Planet. Sci. Conf.*, 1061–72.

Evans, J. C. and Reeves, J. H. (1987). ^{26}Al survey of Antarctic meteorites. *Earth Planet. Sci. Lett.* **82**, 223–30.

Fabryka-Martin, J., Bentley, H., Elmore, D. and Airey, P. L. (1985). Natural iodine-129 as an environmental tracer. *Geochim. Cosmochim. Acta* **49**, 337–47.

Fabryka-Martin, J., Davis, S. N. and Elmore, D. (1987). Applications of ^{129}I and ^{36}Cl in hydrology. *Nucl. Instr. Meth. in Phys. Res. B* **29**, 361–71.

Fabryka-Martin, J., Davis, S. N., Elmore, D. and Kubik, P. W. (1989). In situ production and migration of ^{129}I in the Stripa granite, Sweden. *Geochim. Cosmochim. Acta* **53**, 1817–23.

Fan, C. Y., Chen, T. M., Yun, S. X. and Dai, K. M. (1986). Radiocarbon activity variation in dated tree rings grown in Mackenzie delta. *Radiocarbon* **28**, 300–5.

Fehn, U. (2012). Tracing crustal fluids: Applications of natural ^{129}I and ^{36}Cl. *Ann. Rev. Earth Planet. Sci.* **40**, 45.

Fehn, U., Holdren, G. R., Elmore, D. *et al*. (1986). Determination of natural and anthropogenic ^{129}I in marine sediments. *Geophys. Res. Lett.* **13**, 137–9.

Fehn, U., Moran, J. E., Snyder, G. T. and Muramatsu, Y. (2007a). The initial ^{129}I/I ratio and the presence of 'old' iodine in continental margins. *Nucl. Instr. Meth. in Phys. Res. B* **259**, 496–502.

Fehn, U., Snyder, G. T. and Muramatsu, Y. (2007b). Iodine as a tracer of organic material: ^{129}I results from gas hydrate systems and fore arc fluids. *J. Geochem. Explor.* **95**, 66–80.

Ferguson, C. W. (1970). Dendrochronology of Bristlecone pine, Pinus aristata. Establishment of a 7484-year chronology in the White Mountains of eastern-central California, USA. In: I. U. Olsson (Ed.) *Radiocarbon Variations and Absolute Chronology, Proc. 12th Nobel Symp.* Wiley, pp. 571–93.

Ferguson, C. W. and Graybill, D. A. (1983). Dendrochronology of Bristlecone pine: a progress report. *Radiocarbon* **25**, 287–8.

Feulner, G. and Rahmstorf, S. (2010). On the effect of a new grand minimum of solar activity on the future climate on Earth. *Geophys. Res. Lett.* **37**, L05707, 1–5.

Fink, D., Middleton, R., Klein, J. and Sharma, P. (1990). ^{41}Ca measurement by accelerator mass spectrometry and applications. *Nucl. Instr. Meth. in Phys. Res. B* **47**, 79–96.

Fleitmann, D., Burns, S. J., Mudelsee, M. *et al*. (2003). Holocene forcing of the Indian monsoon recorded in a stalagmite from southern Oman. *Science* **300**, 1737–9.

Forbush, S. E. (1954). Worldwide cosmic-ray variations, 1937–1952. *J. Geophys. Res.* **59**, 525–42.

Frank, M., Schwarz, B., Baumann, S. *et al*. (1997). A 200 ka record of cosmogenic radionuclide production rate and geomagnetic field intensity from ^{10}Be in globally stacked deep-sea sediments. *Earth Planet. Sci. Lett.* **149**, 121–9.

Galbraith, E. D., Jaccard, S. L., Pedersen, T. F. *et al*. (2007). Carbon dioxide release from the North Pacific abyss during the last deglaciation. *Nature* **449**, 890–3.

Godwin, H. (1962). Half-life of radiocarbon. *Nature* **195**, 984.

Goldberg, E. D. and Arrhenius, G. O. S. (1958). Chemistry of Pacific pelagic sediments. *Geochim. Cosmochim. Acta* **13**, 153–212.

Goldstein, S. J., Lea, D. W., Chakraborty, S., Kashgarian, M. and Murrell, M. T. (2001). Uranium-series and radiocarbon geochronology of deep-sea corals: implications for Southern Ocean ventilation rates and the ocean carbon cycle. *Earth Planet. Sci. Lett.* **193**, 167–82.

Goslar, T., Arnold, M., Bard, E. *et al.* (1995). High concentration of atmospheric ^{14}C during the Younger Dryas cold episode. *Nature* **377**, 414–17.

Gove, H. E. (1987). Tandem-accelerator mass-spectrometry measurements of ^{36}Cl, ^{129}I and osmium isotopes in diverse natural samples. *Phil. Trans. Roy. Soc. Lond. A* **323**, 103–19.

Graf, T., Kohl, C. P., Marti, K. and Nishiizumi, K. (1991). Cosmic-ray produced neon in Antarctic rocks. *Geophys. Res. Lett.* **18**, 203–6.

Granger, D. E., Fabel, D. and Palmer, A. N. (2001). Pliocene–Pleistocene incision of the Green river, Kentucky, determined from radioactive decay of cosmogenic ^{26}Al and ^{10}Be in Mammoth Cave sediments. *GSA Bulletin* **113**, 825–36.

Granger, D. E., Gibbon, R. J., Kuman, K. *et al.* (2015). New cosmogenic burial ages for Sterkfontein Member 2 Australopithecus and Member 5 Oldowan. *Nature* **522**, 85–8.

Granger, D. E. and Muzikar, P. F. (2001). Dating sediment burial with in situ-produced cosmogenic nuclides: theory, techniques, and limitations. *Earth Planet. Sci. Lett.* **188**, 269–81.

Gray, L. J., Beer, J., Geller, M. *et al.* (2010). Solar influences on climate. *Rev. Geophys.* **48**, RG4001, 1–53.

Guyodo, Y. and Valet, J. P. (1996). Relative variations in geomagnetic intensity from sedimentary records: the past 200,000 years. *Earth Planet. Sci. Lett.* **143**, 23–36.

Hammer, C. U., Clausen, H. B. and Tauber, H. (1986). Ice-core dating of the Pleistocene/Holocene boundary applied to a calibration of the ^{14}C time scale. *Radiocarbon* **28**, 284–91.

Heisinger, B., Lal, D., Jull, A. T. *et al.* (2002). Production of selected cosmogenic radionuclides by muons: 1. Fast muons. *Earth Planet. Sci. Lett.* **200**, 345–55.

Henning, W. (1987). Accelerator mass spectrometry of heavy elements: ^{36}Cl to ^{205}Pb. *Phil. Trans. Roy. Soc. Lond. A* **323**, 87–99.

Hillam, J., Groves, C. M., Brown, D. M. *et al.* (1990). Dendrochronology of the English Neolithic. *Antiquity* **64**, 210–20.

Hoffmann, D. L., Beck, J. W., Richards, D. A. *et al.* (2010). Towards radiocarbon calibration beyond 28 ka using speleothems from the Bahamas. *Earth Planet. Sci. Lett.* **289**, 1–10.

Hofmann, H. J., Beer, J., Bonani, G. *et al.* (1987). ^{10}Be half-life and AMS-standards. *Nucl. Instr. Meth. in Phys. Res. B* **29**, 32–6.

Hua, Q., Barbetti, M., Fink, D. *et al.* (2009). Atmospheric ^{14}C variations derived from tree rings during the early Younger Dryas. *Quaternary Sci. Rev.* **28**, 2982–90.

Huber, B. (1970). Dendrochronology of central Europe. In: I. U. Olsson (Ed.) *Radiocarbon Variations and Absolute Chronology, Proc. 12th Nobel Symp.* Wiley, pp. 233–5.

Hughen, K. A., Overpeck, J. T., Lehman, S. J. *et al.* (1998). Deglacial changes in ocean circulation from an extended radiocarbon calibration. *Nature* **391**, 65–8.

Jull, A. T. and Burr, G. S. (2006). Accelerator mass spectrometry: Is the future bigger or smaller? *Earth Planet. Sci. Lett.* **243**, 305–25.

Kamen, M. D. (1963). Early history of carbon-14. *Science* **140**, 584–90.

Keigwin, L. D. and Lehman, S. J. (2015). Radiocarbon evidence for a possible abyssal front near 3.1 km in the glacial equatorial Pacific Ocean. *Earth Planet. Sci. Lett.* **425**, 93–104.

Kelly, P. M. and Wigley, T. M. L. (1992). Solar cycle length, greenhouse forcing and global climate. *Nature* **360**, 328–30.

Key, R., Quay, P. D., Jones, G. A. *et al.* (1996). WOCE AMS radiocarbon I: Pacific Ocean results (P6, P16, P17). *Radiocarbon* **38**, 425–518.

Kieser, W. E., Beukens, R. P., Kilius, L. R., Lee, H. W. and Litherland, A. E. (1986). Isotrace radiocarbon analysis – equipment and procedures. *Nucl. Instr. Meth. in Phys. Res. B* **15**, 718–21.

Kitagawa, H. and van der Plicht, J. (1998). Atmospheric radiocarbon calibration to 45,000 yr B.P.: Late glacial fluctuations and cosmogenic isotope production. *Science* **279**, 1187–90.

Klein, J., Fink, D., Middleton, R., Nishiizumi, K. and Arnold, J. (1991). Determination of the half-life of ^{41}Ca from measurements of Antarctic meteorites. *Earth Planet. Sci. Lett.* **103**, 79–83.

Kober, F., Ivy-Ochs, S., Schlunegger, F. *et al.* (2007). Denudation rates and a topography-driven rainfall threshold in northern Chile: multiple cosmogenic nuclide data and sediment yield budgets. *Geomorphology* **83**, 97–120.

Kok, Y. S. (1999). Climatic influence in NRM and ^{10}Be-derived geomagnetic paleointensity data. *Earth Planet. Sci. Lett.* **166**, 105–19.

Korschinek, G., Bergmaier, A., Faestermann, T. *et al.* (2010). A new value for the half-life of ^{10}Be by heavy-ion elastic recoil detection and liquid scintillation counting. *Nucl. Instr. Meth. in Phys. Res. B* **268**, 187–91.

Krivova, N. A., Vieira, L. E. A. and Solanki, S. K. (2010). Reconstruction of solar spectral irradiance since the Maunder minimum. *J. Geophys. Res.: Space Phys.* **115**, A12112, 1–11.

Kromer, B., Friedrich, M., Hughen, K. A. *et al.* (2004). Late glacial ^{14}C ages from a floating, 1382-ring pine chronology. *Radiocarbon* **46**, 1203–9.

Ku, T. L., Wang, L., Luo, S. and Southon, J. R. (1995). ^{26}Al in seawater and ^{26}Al/^{10}Be as paleo-flux tracer. *Geophys. Res. Lett.* **22**, 2163–6.

Ku, T. L., Kusakabe, M., Measures, C. I. *et al.* (1990). Beryllium isotope distribution in the western North Atlantic: a comparison to the Pacific. *Deep-Sea Res.* **37**, 795–808.

Kusakabe, M., Ku, T. L., Southon, J. R. *et al.* (1987). The distribution of ^{10}Be and ^{9}Be in ocean water. *Nucl. Instr. Meth. in Phys. Res. B* **29**, 306–10.

Labrie, D. and Reid, J. (1981). Radiocarbon dating by infrared laser spectroscopy. *Appl. Phys.* **24**, 381–6.

Laj, C., Kissel, C., Mazaud, A. *et al.*. (2002). Geomagnetic field intensity, North Atlantic Deep Water circulation and atmospheric Δ 14 C during the last 50 kyr. *Earth Planet. Sci. Lett.* **200**, 177–90.

Laken, B. A., Pallé, E., Čalogović, J. and Dunne, E. M. (2012). A cosmic ray-climate link and cloud observations. *J. Space Weather Space Climate* **2**, A18.

Lal, D. (1988). *In situ*-produced cosmogenic isotopes in terrestrial rocks. *Ann. Rev. Earth Planet. Sci.* **16**, 355–88.

Lal, D. (1991). Cosmic ray labeling of erosion surfaces: in situ nuclide production rates and erosion models. *Earth Planet. Sci. Lett.* **104**, 424–39.

Lal, D. (2007). Recycling of cosmogenic nuclides after their removal from the atmosphere; special case of appreciable transport of ^{10}Be to polar regions by aeolian dust. *Earth Planet. Sci. Lett.* **264**, 177–87.

Lal, D. and Peters, B. (1967). Cosmic-ray produced radioactivity on the Earth. In: *Handbook of Physics 46/2.* Springer, pp. 551–612.

Lao, Y., Anderson, R. F., Broecker, W. S. *et al.* (1992). Increased production of cosmogenic ^{10}Be during the Last Glacial Maximum. *Nature* **357**, 576–8.

Lean, J., Skumanich, A. and White, O. (1992). Estimating the Sun's radiative output during the Maunder Minimum. *Geophys. Res. Lett.* **19**, 1591–4.

Lee, H. W., Galindo-Uribarri, A., Chang, K. H., Kilius, L. R. and Litherland, A. E. (1984). The ^{12}CH$_2^{+2}$ molecule and radiocarbon dating by accelerator mass spectrometry. *Nucl. Instrum. Meth. in Phys. Res. B* **5**, 208–10.

Libby, W. F. (1952). *Radiocarbon Dating.* University of Chicago Press, 124 pp.

Libby, W. F. (1970). Ruminations on radiocarbon dating. In: I. U. Olsson (Ed.) *Radiocarbon Variations and Absolute Chronology, Proc. 12th Nobel Symp.* Wiley, pp. 629–40.

Litherland, A. E. (1987). Fundamentals of accelerator mass spectrometry. *Phil. Trans. Roy. Soc. Lond. A* **323**, 5–21.

Litherland, A. E., Zhao, X. L. and Kieser, W. E. (2011). Mass spectrometry with accelerators. *Mass Spec. Rev.* **30**, 1037–72.

Liu, B., Phillips, F. M., Fabryka-Martin, J. T., Fowler, M. M. and Stone, W. D. (1994). Cosmogenic ^{36}Cl accumulation in unstable landforms 1. effects of the thermal neutron distribution. *Water Resour. Res.* **30**, 3115–25.

Lockwood, M. (2010). Solar change and climate: an update in the light of the current exceptional solar minimum. *Proc. Roy. Soc. A* **466**, 303–29.

Mahara, Y., Habermehl, M. A., Hasegawa, T. *et al.* (2009). Groundwater dating by estimation of groundwater flow velocity and dissolved ^{4}He accumulation rate calibrated by ^{36}Cl in the Great Artesian Basin, Australia. *Earth Planet. Sci. Lett.* **287**, 43–56.

Mangini, A., Segl, M., Bonani, G. *et al.* (1984). Mass-spectrometric ^{10}Be dating of deep-sea sediments applying the Zurich tandem accelerator. *Nucl. Instr. Meth. in Phys. Res. B* **5**, 353–8.

Mann, M. E. and Park, J. (1994). Global-scale modes of surface temperature variability on interannual to century timescales. *J. Geophys. Res.: Atm.* **99**, 25819–33.

Marchitto, T. M., Lehman, S. J., Ortiz, J. D., Flückiger, J. and van Geen, A. (2007). Marine radiocarbon evidence for the mechanism of deglacial atmospheric CO_2 rise. *Science*, **316**, 1456–9.

Marrero, S. M., Phillips, F. M., Caffee, M. W. and Gosse, J. C. (2016). CRONUS-Earth cosmogenic ^{36}Cl calibration. *Quaternary Geochron.* **31**, 199–219.

Maunder, E. W. (1922). The sun and sun-spots, 1820–1920. *Monthly Notices Roy. Astron. Soc.* **82**, 534–43.

Maycock, A. C., Ineson, S., Gray, L. J. *et al.* (2015). Possible impacts of a future grand solar minimum on climate: Stratospheric and global circulation changes. *J. Geophys. Res.: Atm.* **120**, 9043–58.

Mazaud, A., Laj, C., Bard, E., Arnold, M. and Tric, E. (1991). Geomagnetic field control of ^{14}C production over the last 80 ka: implications for the radiocarbon time-scale. *Geophys Res. Lett.* **18**, 1885–8.

McCorkell, R., Fireman, E. L. and Langway, C. C. (1967). Aluminium-26 and Beryllium-10 in Greenland Ice. *Science* **158**, 1690–2.

McCracken, K. G., Beer, J. and Steinhilber, F. (2014). Evidence for planetary forcing of the cosmic ray intensity and solar activity throughout the past 9400 years. *Solar Phys.* **289**, 3207–29.

Measures, C. I. and Edmond, J. M. (1982). Beryllium in the water column of the central North Pacific. *Nature* **297**, 51–3.

Merrill, J. R., Lyden, E. F. X., Honda, M. and Arnold, J. R. (1960). The sedimentary geochemistry of the beryllium isotopes. *Geochim. Cosmochim. Acta* **18**, 108–29.

Middleton, R. and Klein, J. (1987). ^{26}Al: measurement and applications. *Phil. Trans. Roy. Soc. Lond. A* **323**, 121–43.

Mitchell, S. G., Matmon, A., Bierman, P. R. *et al.* (2001). Displacement history of a limestone normal fault scarp, northern Israel, from cosmogenic ^{36}Cl. *J. Geophys. Res.* **106**, 4247–64.

Monaghan, M. C., Klein, J. and Measures, C. I. (1988). The origin of ^{10}Be in island-arc volcanic rocks. *Earth Planet. Sci. Lett.* **89**, 288–98.

Monnin, E., Indermühle, A., Dällenbach, A. *et al.* (2001). Atmospheric CO_2 concentrations over the last glacial termination. *Science* **291**, 112–14.

Mook, W. G. and Streurman, H. J. (1983). Physical and chemical aspects of radiocarbon dating. *P.A.C.T. (Physical And Chemical Techniques in Archaeology)* **8**, 31–55.

Moran, J. E., Fehn, U. and Teng, R. T. (1998). Variations in ^{129}I/^{127}I ratios in recent marine sediments: evidence for a fossil organic component. *Chem. Geol.* **152**, 193–203.

Morris, J. D., Leeman, W. P. and Tera, F. (1990). The subducted component in island arc lavas: constraints from Be isotopes and B-Be systematics. *Nature* **344**, 31–6.

Morris, J. D. and Tera, F. (1989). ^{10}Be and ^{9}Be in mineral separates and whole-rocks from island arcs: implications for sediment subduction. *Geochim. Cosmochim. Acta* **53**, 3197–206.

Naylor, J. C. and Smith, A. F. M. (1988). An archaeological inference problem. *J. Amer. Stat. Assoc.* **83**, 588–95.

Neff, U., Burns, S. J., Mangini, A. *et al.* (2001). Strong coherence between solar variability and the monsoon in Oman between 9 and 6 kyr ago. *Nature* **411**, 290–3.

Nishiizumi, K., Arnold, J. R., Elmore, D *et al.* (1979). Measurements of ^{36}Cl in Antarctic meteorites and Antarctic ice using a van de Graaff accelerator. *Earth Planet. Sci. Lett.* **45**, 285–92.

Nishiizumi, K., Elmore, D. and Kubik, P. W. (1989a). Update on terrestrial ages of Antarctic meteorites. *Earth Planet. Sci. Lett.* **93**, 299–313.

Nishiizumi, K., Kohl, C. P., Arnold, J. R. *et al.* (1991a). Cosmic-ray produced ^{10}Be and ^{26}Al in Antarctic rocks: exposure and erosion history. *Earth Planet. Sci. Lett.* **104**, 440–54.

Nishiizumi, K., Kohl, C. P., Shoemaker, J. R. *et al.* (1991b). *In situ* ^{10}Be and ^{26}Al exposure ages at Meteor Crater, Arizona. *Geochim. Cosmochim. Acta* **55**, 2699–703.

Nishiizumi, K., Lal, D., Klein, J., Middleton, R. and Arnold, J. R. (1986). Production of ^{10}Be and ^{26}Al by cosmic rays in terrestrial quartz *in situ* and implications for erosion rates. *Nature* **319**, 134–6.

Nishiizumi, K., Winterer, E. L., Kohl, C. P. *et al.* (1989b). Cosmic ray production rates of ^{10}Be and ^{26}Al in quartz from glacially polished rocks. *J. Geophys. Res.* **94**, 17 907–15.

Nydal, R. (2000). Radiocarbon in the ocean. *Radiocarbon* **42**, 81–98.

Oeschger, H., Houtermans, J., Loosli, H. and Wahlen, M. (1970). The constancy of cosmic radiation from isotope studies in meteorites and on the Earth. In: Olsson, I. U. (Ed.) *Radiocarbon Variations and Absolute Chronology, Proc. 12th Nobel Symp.* Wiley, pp. 471–98.

Oktay, S. D., Santschi, P. H., Moran, J. E. and Sharma, P. (2000). The ^{129}I bomb pulse recorded in Mississippi River Delta sediments: results from isotopes of I, Pu, Cs, Pb, and C. *Geochim. Cosmochim. Acta* **64**, 989–96.

Olsson, I. U., El-Daoushy, M. F. A. F., Abdel-Mageed, A. I. and Klasson, M. (1974). A comparison of different methods for pretreatment of bones. *Geol. Foren. Stockh. Forhandl.* **96**, 171–81.

Olsson, I. U. (1980). ^{14}C in extractives from wood. *Radiocarbon* **22**, 515–24.

Ostlund, H. G. and Rooth, C. G. H. (1990). The North Atlantic tritium and radiocarbon transients 1972–1983. *J. Geophys. Res.* **95**, 20 147–65.

Pavich, M. J., Brown, L., Valette-Silver, J. N., Klein, J. and Middleton, R. (1985). ^{10}Be analysis of a Quaternary weathering profile in the Virginia Piedmont. *Geology* **13**, 39–41.

Pearson, G. W., Pilcher, J. R., Baillie, M. G. L. and Hillam, J. (1977). Absolute radiocarbon dating using a low altitude European tree-ring calibration. *Nature* **270**, 25–8.

Phillips, F. M., Argento, D. C., Balco, G. *et al.* (2016). The CRONUS-Earth project: a synthesis. *Quaternary Geochron.* **31**, 119–54.

Phillips, F. M., Leavy, B. D., Jannik, N. O., Elmore, D. and Kubik, P. W. (1986). The accumulation of cosmogenic chlorine-36 in rocks: a method for surface exposure dating. *Science* **231**, 41–3.

Phillips, F. M., Mattick, J. L. and Duval, T. A. (1988). Chlorine 36 and tritium from nuclear weapons fallout as tracers for long-term liquid and vapour movement in desert soils. *Water Resour. Res.* **24**, 1877–91.

Phillips, F. M., Zreda, M. G., Gosse, J. C. *et al.* (1997). Cosmogenic ^{36}Cl and ^{10}Be ages of Quaternary glacial and fluvial deposits of the Wind River Range, Wyoming. *GSA Bull.* **109**, 1453–63.

Plummer, M. A., Phillips, F. M., Fabryka-Martin, J. *et al.* (1997). Chlorine-36 in fossil rat urine: an archive of cosmogenic nuclide deposition during the past 40,000 years. *Science* **277**, 538–41.

Purser, K. H., Liebert, R. B., Litherland, A. E. *et al.* (1977). An attempt to detect stable N-ions from a sputter ion source and some implications of the results for the design of tandems for ultra-sensitive carbon analysis. *Rev. Phys. Appl.* **12**, 1487–92.

Raisbeck, G. M., Yiou, F., Fruneau, M., Lieuvin, M. and Loiseaux, J. M. (1978). Measurements of ^{10}Be in 1,000- and 5,000-year-old Antarctic ice. *Nature* **275**, 731–3.

Raisbeck, G. M., Yiou, F., Fruneau, M. *et al.* (1979). Deposition rate and seasonal variations in precipitation of cosmogenic ^{10}Be. *Nature* **282**, 279–80.

Raisbeck, G. M., Yiou, F., Fruneau, M. *et al.* (1981a). Cosmogenic ^{10}Be concentrations in Antarctic ice during the past 30,000 years. *Nature* **292**, 825–6.

Raisbeck, G. M., Yiou, F., Fruneau, M. *et al.* (1980). ^{10}Be concentration and residence time in the deep ocean. *Earth Planet. Sci. Lett.* **51**, 275–8.

Raisbeck, G. M., Yiou, F. and Zhou, S. Z. (1994). Palaeointensity puzzle. *Nature* **371**, 207–8.

Ralph, E. K. (1971). Carbon-14 dating. In: Michael, H. N. and Ralph, E. K. (Eds) *Dating Techniques for the Archaeologist.* M.I.T. Press, pp. 1–48.

Ralph, E. K. and Michael, H. N. (1974). Twenty-five years of radiocarbon dating. *Amer. Scient.* **62**, 553–60.

Ramsey, C. B. (1995). Radiocarbon calibration and analysis of stratigraphy; the OxCal program. *Radiocarbon* **37**, 425–30.

Reimer, P. J., Bard, E., Bayliss, A. *et al.* (2013). IntCal13 and Marine13 radiocarbon age calibration curves 0–50,000 years cal BP. *Radiocarbon* **55**, 1869–87.

Reyss, J. L., Yokoyama, Y. and Guichard, F. (1981). Production cross sections of ^{26}Al, ^{22}Na, ^{7}Be from argon and of ^{10}Be, ^{7}Be from nitrogen: implications for the production rates of ^{26}Al and ^{10}Be in the atmosphere. *Earth Planet. Sci. Lett.* **53**, 203–10.

Roberts, M. L., Burton, J. R., Elder, K. L. *et al.* (2010). A high-performance ^{14}C accelerator mass spectrometry system. *Radiocarbon* **52**, 226–35.

Robinson, C., Raisbeck, G. M., Yiou, F., Lehman, B. and Laj, C. (1995). The relationship between ^{10}Be and geomagnetic field strength records in central North Atlantic sediments during the last 80 ka. *Earth Planet. Sci. Lett.* **136**, 551–7.

Scafetta, N. (2014). Discussion on the spectral coherence between planetary, solar and climate oscillations: a reply to some critiques. *Astrophys. Space Sci.* **354**, 275–99.

Scafetta, N. (2016). High resolution coherence analysis between planetary and climate oscillations. *Advances in Space Res.* **57**, 2121–35.

Scafetta, N. and Willson, R. C. (2013). Planetary harmonics in the historical Hungarian aurora record (1523–1960). *Planet. Space Sci.* **78**, 38–44.

Schlesinger, M. E. and Ramankutty, N. (1992). Implications for global warming of intercycle solar irradiance variations. *Nature* **360**, 330–3.

Shackleton, N. J., Duplessy, J.-C., Arnold, M. *et al.* (1988). Radiocarbon age of last glacial Pacific deep water. *Nature* **335**, 708–11.

Sharma, M. (2002). Variations in solar magnetic activity during the last 200 000 years: is there a Sun–climate connection? *Earth Planet. Sci. Lett.* **199**, 459–72.

Shepard, M. K., Arvidson, R. E., Caffee, M., Finkel, R. and Harris, L. (1995). Cosmogenic exposure ages of basalt flows: Lunar Crater volcanic field, Nevada. *Geology* **23**, 21–4.

Siegenthaler, U. and Sarmiento, J. L. (1993). Atmospheric carbon dioxide and the ocean. *Nature* **365**, 119–25.

Sigmarsson, O., Condomines, M., Morris, J. D. and Harmon, R. S. (1990). Uranium and ^{10}Be enrichments by fluids in Andean arc magmas. *Nature* **346**, 163–5.

Sikes, E. L., Samson, C. R., Guilderson, T. P. and Howard, W. R. (2000). Old radiocarbon ages in the southwest Pacific Ocean during the last glacial period and deglaciation. *Nature* **405**, 555–9.

Simpson, J. A. (1951). Neutrons produced in the atmosphere by cosmic radiations. *Phys. Rev.* **83**, 1175–88.

Skinner, L. C., Fallon, S., Waelbroeck, C., Michel, E. and Barker, S. (2010). Ventilation of the deep Southern Ocean and deglacial CO_2 rise. *Science* **328**, 1147–51.

Skinner, L., McCave, I. N., Carter, L. *et al.* (2015). Reduced ventilation and enhanced magnitude of the deep Pacific carbon pool during the last glacial period. *Earth Planet. Sci. Lett.* **411**, 45–52.

Snyder, G., Aldahan, A. and Possnert, G. (2010). Global distribution and long-term fate of anthropogenic ^{129}I in marine and surface water reservoirs. *Geochem. Geophys. Geosys.* **11**, Q04010, 1–19.

Snyder, G. T. and Fehn, U. (2002). Origin of iodine in volcanic fluids: ^{129}I results from the Central American Volcanic Arc. *Geochim. Cosmochim. Acta* **66**, 3827–38.

Snyder, G. T., Fehn, U. and Goff, F. (2002). Iodine isotope ratios and halide concentrations in fluids of the Satsuma–Iwojima volcano, Japan. *Earth,Planets Space* **54**, 265–73.

Steinhilber, F., Abreu, J. A., Beer, J. *et al.* (2012). 9,400 years of cosmic radiation and solar activity from ice cores and tree rings. *Proc. Nat. Acad. Sci.* **109**, 5967–71.

Steinhilber, F., Beer, J. and Fröhlich, C. (2009). Total solar irradiance during the Holocene. *Geophys. Res. Lett.* **36** L19704, 1–5.

Stensland, G. J., Brown, L., Klein, J. and Middleton, R. (1983). Beryllium-10 in rain. *EOS* **64**, 283 (abs.).

Stocker, T. F., Qin, D., Plattner, G. K. *et al.* Eds.(2014). IPCC, Climate Change 2013: The Physical Science Basis. *Contribution of Working Group I to the Fifth Assessment Report of the Intergovernmental Panel on Climate Change.* Cambridge University Press.

Stone, J. O., Allan, G. L., Fifield, L. K. and Cresswell, R. G. (1996). Cosmogenic chlorine-36 from calcium spallation. *Geochim. Cosmochim. Acta* **60**, 679–92.

Stuiver, M. (1961). Variations in radiocarbon concentration and sunspot activity. *J. Geophys. Res.* **66**, 273–6.

Stuiver, M. (1965). Carbon-14 content of 18th- and 19th-century wood: variations correlated with sunspot activity. *Science* **149**, 533–4.

Stuiver, M., Braziunas, T. F., Becker, B. and Kromer, B. (1991). Climatic, solar, oceanic and geomagnetic influences on late-glacial and Holocene atmospheric ^{14}C/^{12}C change. *Quat. Res.* **35**, 1–24.

Stuiver, M. and Pearson, G. W. (1986). High-precision radiocarbon timescale calibration from the present to 500 BC. *Radiocarbon* **28**, 805–38.

Stuiver, M. and Quay, P. D. (1981). Atmospheric ^{14}C changes resulting from fossil fuel CO_2 release and cosmic-ray flux variability. *Earth Planet. Sci. Lett.* **53**, 349–62.

Stuiver, M., Quay, P. D. and Ostlund, H. G. (1983). Abyssal water carbon-14 distribution and the age of the world oceans. *Science* **219**, 849–51.

Suess, H. E. (1955). Radiocarbon concentrations in modern wood. *Science* **122**, 415–17.

Suess, H. E. (1965). Secular variations of the cosmic-ray-produced carbon 14 in the atmosphere and their interpretations. *J. Geophys. Res.* **70**, 5937–52.

Suess, H. E. (1970). Bristlecone-pine calibration of the radiocarbon timescale 5200 B.C. to the present. In: I. U. Olsson (Ed.) *Radiocarbon Variations and Absolute Chronology, Proc. 12th Nobel Symp.* Wiley, pp. 303–11.

Suess, H. E. and Strahm, C. (1970). The Neolithic of Auvernier, Switzerland. *Antiquity* **44**, 91–9.

Sun, B. and Bradley, R. S. (2002). Solar influences on cosmic rays and cloud formation: A reassessment. *J. Geophys. Res.: Atm.* **107** AAC5, 1–12.

Suter, M., Jacob, S. W. A. and Synal, H. A. (2000). Tandem AMS at sub-MeV energies–Status and prospects. *Nucl. Instr. Meth. in Phys. Res. B* **172**, 144–51.

Svensmark, H. and Friis-Christensen, E. (1997). Variation of cosmic ray flux and global cloud coverage – a missing link in solar–climate relationships. *J. Atm. Solar–Terrest. Phys.* **59**, 1225–32.

Tanaka, S. and Inoue, T. (1979). [10]Be dating of North Pacific sediment cores up to 2.5 million years B.P. *Earth Planet. Sci. Lett.* **45**, 181–7.

Tauber, H. (1970). The Scandinavian varve chronology and C-14 dating. In: Olsson, I. U. (Ed.) *Radiocarbon Variations and Absolute Chronology, Proc. 12th Nobel Symp.* Wiley, pp. 173–96.

Tera, F., Brown, L., Morris, J. *et al.* (1986). Sediment incorporation in island-arc magmas: inferences from [10]Be. *Geochim. Cosmochim. Acta* **50**, 535–50.

Thellier, E. O. (1941). Sur la verification d'une methode permettant de determiner l'intensite du champ magnetique terrestre dans le passe. *Compte Rendu Acad. Sci. Paris* **212**, 281.

Tobias, S. M. (1996). Grand minimia in nonlinear dynamos. *Astron. Astrophys* **307**, L21–24.

Torgersen, T., Habermehl, M. A., Phillips, F. M. *et al.* (1991). Chlorine 36 dating of very old groundwater 3. Further studies in the Great Artesian Basin, Australia. *Water. Resources. Res.* **27**, 3201–13.

Turekian, K. K., Cochran, J. K., Krishnaswami, S. *et al.* (1979). The measurement of [10]Be in manganese nodules using a tandem van de Graaff accelerator. *Geophys. Res. Lett.* **6**, 417–20.

Ullman, W. J. and Aller, R. C. (1980). Dissolved iodine flux from estuarine sediments and implications for the enrichment of iodine at the sediment water interface. *Geochim. Cosmochim. Acta* **44**, 1177–84.

Usoskin, I. G., Solanki, S. K. and Kovaltsov, G. A. (2007). Grand minima and maxima of solar activity: new observational constraints. *Astron. Astrophys.* **471**, 301–9.

Vieira, L. E. A., Solanki, S. K., Krivova, N. A. and Usoskin, I. (2011). Evolution of the solar irradiance during the Holocene. *Astron. Astrophys.* **531**, A6 1–20.

von Blankenburg, F., O'Nions, R. K., Belshaw, N. S., Gibb, A. and Hein, J. R. (1996). Global distribution of beryllium isotopes in deep ocean water as derived from Fe-Mn crusts. *Earth Planet. Sci. Lett.* **141**, 213–26.

Wagner, G., Laj, C., Beer, J. *et al.* (2001). Reconstruction of the paleoaccumulation rate of central Greenland during the last 75 ka using the cosmogenic radionuclides [36]Cl and [10]Be and geomagnetic field intensity data. *Earth Planet. Sci. Lett.* **193**, 515–21.

Wang, L., Ku, T. L., Luo, S., Southon, J. R. and Kusakabe, M. (1996). [26]Al-[10]Be systematics in deep-sea sediments. *Geochim. Cosmochim. Acta* **60**, 109–19.

Wolfli, W. (1987). Advances in accelerator mass spectrometry. *Nucl. Instr. Meth. in Phys. Res. B* **29**, 1–13.

Yamazaki, T. and Oda, H. (2002). Orbital influence on Earth's magnetic field: 100,000-year periodicity in inclination. *Science* **295**, 2435–8.

Yiou, F., Raisbeck, G. M., Bourles, D., Lorius, C. and Barkov, N. I. (1985). [10]Be in ice at Vostok Antarctica during the last climatic cycle. *Nature* **316**, 616–17.

Yokoyama, Y., Guichard, F., Reyss, J. L., Van, N. H. (1978). Oceanic residence times of dissolved beryllium and aluminium deduced from cosmogenic tracers [10]Be and [26]Al. *Science* **201**, 1016–17.

You, C. F., Lee, T. and Li, Y. H. (1989). The partition of Be between soil and water. *Chem. Geol.* **77**, 105–18.

Zreda, M. and Noller, J. S. (1998). Ages of prehistoric earthquakes revealed by cosmogenic chlorine-36 in a bedrock fault scarp at Hebgen Lake. *Science* **282**, 1097–9.

Zreda, M. G., Phillips, F. M., Elmore, D. *et al.* (1991). Cosmogenic chlorine-36 production rates in terrestrial rocks. *Earth Planet. Sci. Lett.* **105**, 94–109.

Zreda, M. G. and Phillips, F. M. (1995). Surface exposure dating by cosmogenic chlorine-36 accumulation. In: Beck. C. (Ed.) *Dating in Exposed and Surface Contexts*, University of New Mexico Press, pp. 161–83.

Zreda, M. G., Phillips, F. M. and Elmore, D. (1994). Cosmogenic [36]Cl accumulation in unstable landforms 2. Simulations and measurements on eroding moraines. *Water Resour. Res.* **30**, 3127–36.

Chapter 15

Extinct Radionuclides

15.1 Introduction

The 'extinct' radionuclides may in the past have seemed like an esoteric field with little connection to the world of terrestrial geology. This changed over the past couple of decades, as new analytical methods produced evidence that challenged long-held beliefs about the origins and composition of the Earth. Many 'terrestrial' geochemists, with their empirical models of Earth reservoirs, may not have been directly affected by this cosmochemical revolution. However, a survey of extinct nuclide systems provides an opportunity to better understand why the Earth evolved as it did, and how it is related to other solar system objects.

15.1.1 Nuclide Production and Decay

An extinct radionuclide is understood to be one that was formed by a process of stellar nucleosynthesis prior to the coalescence of the solar system, and which has subsequently decayed away to zero. Most extinct nuclides were very short-lived, but a few have half-lives in the millions of years range. These may have persisted in solar system materials at high enough concentrations to generate observable variations in the isotopic composition of daughter products. These parent–daughter pairs are of interest to cosmochemists because they can provide information about the origins of the solar system and its early history.

The production rate of an arbitrary solar system nuclide as a function of time is shown schematically in Fig. 15.1. After the Big Bang at 13.8 Ga (Bennett *et al.*, 2013), nucleosynthetic production in stars proceeded at a rate p which may have been steady or very variable, depending on the process. However, it is thought that prior to condensation of the solar nebula, much or all solar system matter was out of nucleosynthetic 'circulation' for a period of time in some kind of interstellar cloud. This time between last nucleosynthesis ('star death') and major condensation ('glob formation') is termed delta (Fig. 15.1).

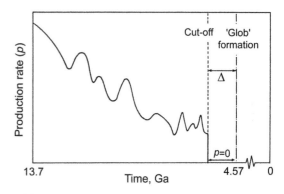

Fig. 15.1 Schematic illustration of the variation in production rate (p) of a given nuclide between the Big Bang and the termination of nucleosynthesis, followed by a period 'Δ' prior to solar system coalescence. After Wasserburg (1985).

Determination of Δ for various extinct nuclides may reveal information about the process which led to solar system coalescence, and this is therefore a major goal of isotope cosmochemistry. A related goal is the dating of early solar system events, and progress in this field necessitates a more precise definition of the beginning of the solar system than the poetic term 'glob formation' (Wasserburg, 1985). A useful way to approach the issue is to proceed backwards in time, noting some of the major events that we seek to date. However, we must first discuss the material that is used for these studies.

15.1.2 Celestial Objects and Ages

Most of the material that survives from the early history of the solar system is preserved in meteorites of various types. Individual meteorites are named as far as possible after their find sites, and some, such as the Allende carbonaceous chondrite and the Canyon Diablo iron meteorite have become very well known. However, numbered suites from desert regions such as the Sahara (NW Africa) and Antarctica (Allan Hills, Asuka, Elephant Moraine, Lewis Cliff, MacAlpine

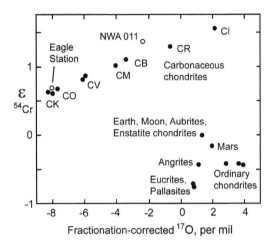

Fig. 15.2 Stable isotope plot showing proposed groupings for different classes of meteorites, along with Earth, Moon and Mars, according to their assembly in different regions of the solar system (see Appendix 2 for meteorite abbreviations). Modified after Trinquier et al. (2007).

Hills, Queen Alexandra Range) have made instant recognition of meteorite names more difficult. In addition, individual meteorites are grouped into 'clans' that are often named after less well-known bodies. Therefore, to assist the reader, a general classification of meteorite types is given in Appendix 2, including most of the named meteorites referred to in this chapter.

Appendix 2 revises the traditional meteorite classification (e.g. Krot et al., 2005) in favour of a new scheme proposed by Warren (2011). This uses stable isotope signatures (Section 15.2) to distinguish volatile-rich meteorites that originated in the outer solar system (mostly carbonaceous chondrites) from other meteorite classes originating from the inner solar system. The latter have stable isotope signatures similar to the terrestrial planets (Fig. 15.2). There are two exceptions (Eagle Station and NWA 011), representing bodies that would previously have been grouped with other achondrites and pallasites, but must now be grouped as unusual bodies that originated in the outer solar system but underwent planetary processing.

Over the past few years, great progress has been made in dating early solar system events. Probably the most recent of these events is the giant impact that is believed to have formed the Earth–Moon system, which may have rendered both of these bodies largely molten a few tens of Ma after the origin of the solar system. Gravitational kinetic energy was clearly important in melting such large planetary bodies, but some of the extinct nuclides (e.g. ^{26}Al) were responsible for a much earlier melting episode in small planetessimals, causing their differentiation into iron cores, ultramafic mantles and mafic crustal rocks. Collisions between some of these bodies broke them into the iron and

achondrite meteorites that we know today, but the differentiation process can now be precisely dated to only a few Ma after the origin of the solar system.

The 'HED' clan represents the most common type of achondrite, and its members are attributed to the basaltic crust of an asteroid. Within the HED clan, the Eucrite group is generally volcanic, the Diogenites are generally plutonic and the Howardites form breccias of the other two types. The HED clan is often identified with the asteroid Vesta, but has also been attributed to a destroyed planetessimal whose core forms the IIIAB group of iron meteorites (Wasson, 2013). The Angrites (named after Angra dos Reis) represent another important group of basaltic achondrites similar to the eucrites. Finally, the primitive achondrites (e.g. Acapulco) and the ultramafic achondrites (ureilites) may represent fragments of variably differentiated planetessimal mantles.

In contrast to achondrites, the chondritic meteorites represent cold agglomerations of material that escaped thermal differentiation. However, according to the 'onion skin' model (Dodd, 1969; Taylor et al., 1987; Elkins-Tanton et al., 2011), many chondrites might actually represent partially buried material from small planetessimals that melted at their centres, but were too small to undergo full differentiation into core, mantle and crust. This context may explain chondrite alteration patterns signified by the numerical classification: 3 = pristine; 1–2 = aqueous alteration; 4–7 = thermal metamorphism (e.g. Weisberg et al., 2006).

Chondrites are largely made of small spherical objects, around 1 mm in diameter, which contain high-temperature minerals. They seem to have condensed directly from an incandescent gas, although most of them were probably remelted shortly afterwards to form molten droplets. Of the two main types, the smaller but more abundant 'chondrules' are dominantly composed of fairly coarse-grained olivine or ortho-pyroxene (enstatite). The rarer but larger 'calcium–aluminium inclusions' (CAIs) are composed of a variety of finer-grained refractory minerals such as spinel, hibonite, melilite, anorthite and perovskite.

The CAIs and chondrules must have formed in a hot region of the solar nebula, but seem to have been suddenly transported outwards by an energetic stellar wind emanating from the developing disc (Salmeron and Ireland, 2012; Brennecka et al., 2013). CAIs are the oldest known relics of this type of heating, and they appear to date the initiation of solar nuclear reactions. However, their extinct nuclide signatures imply that they formed only a short time after the last *pre-solar* nuclear event.

The Pb–Pb age of CAIs (4568 Ma) is generally deemed as the formation age of the solar system. However, this result (and all other meteorite ages) may need to be adjusted downwards by 4–5 Ma due to errors in the ^{235}U half-life (Sections 5.1, 5.3.2). The only materials in the solar system thought to be older than CAIs are nanoscopic grains of even more refractory materials such as Si-C, which are believed to be pre-solar relics.

Table 15.1	Some important extinct radionuclide systems.				
Parent	Daughter	Decay Mode	Mean life Ma	Half-life Ma	Lambda yr^{-1}
^{146}Sm	^{144}Nd	Alpha	149	103	6.7×10^{-9}
^{244}Pu	Various	Fission	119	82	8.4×10^{-9}
^{129}I	^{129}Xe	Beta	23	16	4.3×10^{-8}
^{247}Cm	^{235}U	$3\alpha, 2\beta$	22.5	15.6	4.4×10^{-8}
^{182}Hf	^{182}W	2β	13	8.9	7.7×10^{-8}
^{107}Pd	^{107}Ag	β	9.4	6.5	1.1×10^{-7}
^{53}Mn	^{53}Cr	β	5.3	3.7	1.9×10^{-7}
^{60}Fe	^{60}Ni	2β	3.8	2.6	2.6×10^{-7}
^{26}Al	^{26}Mg	β	1.05	0.7	9.5×10^{-7}
^{36}Cl	^{36}Ar	β	0.43	0.3	2.3×10^{-6}
^{41}Ca	^{41}K	β	0.15	0.1	6.7×10^{-6}

It should also be noted that meteorites and their parent bodies have been subject to cosmic ray bombardment over the life of the solar system. This may have caused nuclear transformations, which should be excluded as a mechanism for generating daughter-product anomalies before these are attributed to extinct radioactivities.

15.1.3 Parent–Daughter Pairs

Most of the scientifically important extinct radionuclides with half-lives over 10^5 years are shown in Table 15.1 (in order of decreasing stability) and will be discussed below. Mean lives $(1/\lambda)$ are quoted in addition to half-lives because they are helpful in understanding the production history of extinct nuclides.

The key to 'detection' of an extinct nuclide is to find material that is strongly depleted in the daughter element. This causes the very small isotopic variations due to decay of the parent to be detectable. These variations must also be visible above the 'background' variation of isotopic abundances in solar system material, caused by heterogeneous mixing of species synthesized over the lifetime of the galaxy.

15.2 Stable Isotopes

It may seem strange to begin a chapter on extinct radionuclides with a discussion of stable isotope abundances. However, most stable nuclides are ultimately nucleogenic, in the sense that they were made by nucleosynthetic processes in stars (Section 1.2). The focus here is specifically on variations in isotopic abundance that do *not* result from mass fractionation processes in nature, the latter being outside the scope of this book. After correcting for such fractionation effects, residual isotopic variations are typically attributed to incomplete mixing of different nucleosynthetic components during the condensation of the solar system. These heterogeneities provide tools for classifying objects and determining the genetic relationships between them.

The first clear signs of stable isotope variations that could not be explained by mass-dependent fractionation were deviations of ^{17}O abundances from the terrestrial ^{18}O/^{16}O mass fractionation line (Clayton *et al.*, 1973). The largest deviations were seen in carbonaceous chondrites, but later work showed that most meteorites, with the exception of enstatite chondrites and enstatite achondrites (Aubrites) lie off the terrestrial mass fractionation line (Clayton, 1993).

The advent of ICP–MS has hugely expanded the number of elements whose isotopes are available for study. Some of the most significant results come from fractionation-normalized chromium and titanium isotope data (Trinquier *et al.*, 2007; Leya *et al.*, 2008). These new tracers (Fig. 15.3) greatly strengthen the oxygen isotope evidence, since they yield a complete separation between outer and inner solar

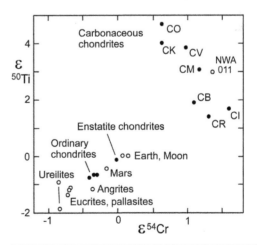

Fig. 15.3 Stable Ti versus Cr isotope plot for different classes of meteorites, along with Earth, Moon and Mars, derived from different regions of the solar system (see Appendix 2 for meteorite names and abbreviations). After Warren (2011).

system signatures, whereas their ^{17}O signatures overlap (Fig. 15.2).

15.2.1 Cosmic Building Blocks of the Earth

The discovery that the Earth and Moon have identical isotope ratios for several stable isotope systems (Figs. 15.2, 15.3) creates problems for the giant impact model of the origin of the Moon (Wiechert *et al.*, 2001). In the 'conventional' version of this model (Fig. 15.4a), the Earth–Moon system was created by the collision between approximately Venus-sized and Mars-sized objects. In this model, most of the impactor's core is added to the Earth's core, but the impactor's mantle forms most of the Moon.

For this model to explain the identical stable isotope signatures of Earth and Moon, the two colliding objects must have had very similar isotope signatures, implying that they grew in similar solar system orbits. However, various lines of evidence suggest that the giant impact occurred more than 50 Ma after the origin of the solar system. This evidence comes from the Pb–Pb age of the Earth (Section 5.3.1), its xenon isotope systematics (Section 11.5) and siderophile element abundances (Jacobson *et al.*, 2014). This implies that the Earth and its impactor (generally named Theia) grew in relatively close proximity to one another for over 50 Ma without colliding.

To avoid this problem, two alternative oblique collision scenarios have been proposed (Fig. 15.4b, c). One involves a 'total annihilation' collision between two equally sized bodies (Canup, 2013), whereby the Earth and Moon are both formed from the well mixed products. The alternative (Ćuk and Stewart, 2012) involves a much smaller impactor hitting a fast counter-spinning earth, whereby a huge lump of the Earth's mantle is excavated to form the Moon. Both of these scenarios involve rather specific conditions of collision which might seem to be unlikely coincidences. They also generate a system with much more angular momentum than the present Earth–Moon system, which must then be lost by resonance between an elliptical lunar orbit and Earth's orbit around the Sun. However, these are the only models that can satisfy tungsten isotope evidence (Section 15.6.6).

Aside from the tungsten isotope evidence, constraints on giant impact models from other isotope systems may now be somewhat relaxed. By showing that enstatite chondrites and enstatite achondrites (Aubrites) have similar fractionation-normalized signatures to the Earth–Moon system (Fig. 15.3),

the new stable isotope data imply that a relatively large volume of the inner solar system experienced similar conditions. This region was termed the 'inner disc uniform reservoir' by Dauphas *et al.* (2014). Its conditions were somewhat more reducing than those experienced by the ordinary (olivine) chondrites, but much less reducing than the region where carbonaceous chondrites grew.

This model represents a modified version of the 'enstatite chondrite' model for the origin of the Earth, developed by Javoy (1995). The new model does not see the Earth as derived directly *from* enstatite chondrites, but it does imply a close genetic relationship between them. This model has been supported by stable molybdenum, nickel and zirconium isotope evidence (Dauphas *et al.*, 2002; Burkhardt *et al.*, 2011; Steele *et al.*, 2012; Akram *et al.*, 2015), and also goes a long way to solving the '^{142}Nd conundrum' (Section 15.7).

15.2.2 Solar System Isotope Heterogeneity

Moving beyond the Earth–Moon system, stable isotope data are also important for studying the nucleosynthetic origins of heterogeneity in the solar nebula. Apart from oxygen, most of the stable isotope systems cited above are part of the 'iron group' in terms of their nucleosynthesis (Section 1.2.2). This implies that these elements and their isotope signatures were produced in the early stages of a supernova explosion. Therefore, fractionation-corrected isotope heterogeneity of elements such as chromium and titanium implies heterogeneous mixing of more than one supernova source in the solar system. A major objective is therefore to identify these sources, and if possible to date them using extinct (or extant) nuclide chronometers.

Typically, the lightest or heaviest isotopes of an element are of greatest interest, since they may have nucleosynthetic sources different from the intermediate isotopes. Titanium, with five isotopes, is of particular interest because its lightest and heaviest isotopes, ^{46}Ti and ^{50}Ti, form two independent tracers, while the three intermediate isotopes can be used to monitor and correct for mass fractionation effects (Trinquier *et al.*, 2009). On a titanium three-isotope plot (normalized against ^{47}Ti), data from various meteorite types and CAIs form a linear array interpreted as a two-component mixing line (Fig. 15.5).

In this diagram, carbonaceous chondrites have the greatest internal heterogeneity, with bulk samples lying on a mixing line between chondrules and CAIs. Because the two axes in Fig. 15.5 represent isotope ratios of the same element, mixing lines are linear, and therefore point towards the mixing end-members. In contrast, bivariate plots involving isotopes of different elements normally form hyperbolic mixing lines that do not point towards the end-members, and may therefore be harder to interpret (Dauphas *et al.*, 2014).

Other stable isotopes of particular interest for monitoring distinct nucleosynthetic sources of solar system materials are the heaviest isotope of calcium, ^{48}Ca (Dauphas *et al.*, 2014) and the lightest isotope of strontium, ^{84}Sr (Paton *et al.*, 2013; Yokoyama *et al.*, 2015). Because these isotopes are both

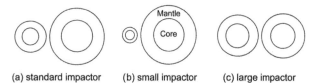

(a) standard impactor   (b) small impactor   (c) large impactor

Fig. 15.4 Alternative giant impact scenarios for the formation of the Earth–Moon system. After Halliday (2012).

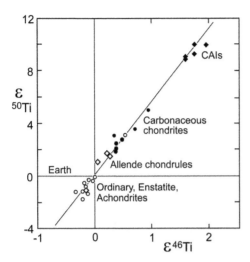

Fig. 15.5 Fractionation-corrected titanium three isotope plot normalized against ^{47}Ti, showing a two-component mixing line. After Trinquier *et al.* (2009).

isolated from the *s*-process pathway, supernova induced *r*- and *p*-process synthesis is required to produce them. As with titanium, the greatest anomalies in these nuclides are shown by CAIs. This has important implications for understanding the signatures of extinct nuclide systems in CAIs, such as Al–Mg (Section 15.5).

15.3 Extant Actinides

While extinct nuclide systems can give age information about the period shortly before or after solar system condensation, the extant actinides (^{235}U, ^{238}U and ^{232}Th) are important sources of information about nucleosynthetic processes over the whole life of the Galaxy (Fig. 15.1).

Based on the now well-established 13.8 Ga age of the Universe, the age of the Galaxy at the time of solar system condensation was ca. 9 Ga, which is of the same order as the lifetime of a typical star like the Sun (Section 1.2). However, large stars have much shorter lifetimes, which may even be less than 1 Ma in duration. Therefore, since the solar system coalesced from the debris of 'dead' stars, it is theoretically possible that any given atom in the solar system could have been processed through only one previous star, or through numerous previous stars. This indeterminacy leads to uncertainty in the production rate of solar system nuclides over the life of the Galaxy, but the long half-lives of the extant actinides allows a rough calculation of their production rates.

The gulf of unstable nuclides between the end of the *s*-process nucleosynthetic ladder and the actinide elements (Section 1.2.2) means that these nuclides can only be generated by the *r*-process. The seed nuclei for this process are clearly the nuclei at the top of the *s*-process ladder, but

these nuclides, especially Pb, have small neutron capture cross-sections, creating a nucleosynthetic barrier. As a result, the production *ratios* of the actinides are constrained to be close to unity. This factor is critical in using them to model *r*-process production rates over the lifetime of the Galaxy.

We begin with the abundances of these nuclides in carbonaceous chondrites. Normalizing to ^{235}U = 1, ^{238}U and ^{232}Th have present day abundances of ca. 138 and 520 respectively. Correcting for decay to initial abundances at 4.55 Ga and re-normalizing to ^{235}U = 1, we obtain lower relative abundances of 3.45 and 8.18 for ^{238}U and ^{232}Th respectively, but these are still higher than estimated production ratios (relative to ^{235}U) of 0.66 and 1.27 respectively (e.g. Broecker, 1986). Therefore, we can use these differences, along with the different half-lives of the three nuclides, to test alternative end-member production models.

If all uranium formation was attributed to a single supernova event, we could calculate the apparent timing of this event by the subsequent decay of short-lived ^{235}U (half-life ca. 700 Ma) relative to longer-lived ^{238}U (half-life ca. 4500 Ma). The calculation based on uranium isotopes alone may be somewhat more reliable than that involving thorium, because this avoids chemical fractionation that could have occurred during solar system coalescence. The model leads to a calculated Δ value for the nucleosynthetic event of 2.1 Ga before solar system coalescence. However, evidence for the presence of the short-lived actinide ^{244}Pu in the early solar system rules out a model with such a large Δ value.

The other extreme model involves 'continuous' supernova production through the life of the galaxy. Taken to its limit this is impossible, since each supernova event terminates the evolution of a star, and the scattered debris must be incorporated in a new star before nucleosynthesis can continue. However, for nuclides with half-lives of hundreds of Ma, a supernova frequency as low as one per 100 Ma in the production history of an element will be a close approximation to continuous production. Under this model, the abundance of an unstable nuclide builds up until it reaches a level where the rate of synthesis is equalled by the rate of decay. This point of saturation is reached sooner in short-lived relative to long-lived nuclides (Fig. 15.6).

The growth curves in Fig. 15.6 can be presented in the form of isotope ratios (Fig. 15.7) of ^{238}U and ^{232}Th against ^{235}U. The time at which the curves intersect the primordial solar system composition (calculated above) yields a crude estimate for the age of the galaxy at solar system coalescence. Adding 4.55 Ga, we obtain estimates for the age of the Galaxy of 16.5 and 13.5 Ga from Figs. 15.7a and 15.7b respectively. Given the many assumptions made, these estimates for the age of the Galaxy agree surprisingly well with the estimated 13.8 Ga age of the Universe, and therefore provide strong support for the 'continuous supernova' model. Under this model, the extant actinides provide a very poor constraint on the value of Δ. However, that is the role of the *extinct* nuclides.

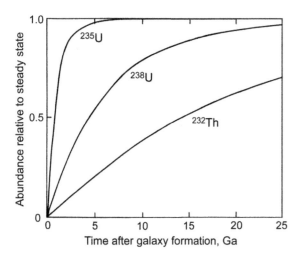

Fig. 15.6 Contrasting rates of approach of U and Th isotopes to a steady state abundance for a model of 'continuous' supernova actinide production. After Broecker (1986).

In between the two extreme models described above, there is an infinite number of intermediate models in which discrete production events of variable size occur at variable intervals. These are termed 'granular' models. Ideally, we would like to use the actinide data to put an upper limit on the 'granularity' of these models. However, the indeterminacy of the system prevents the application of precise limits. Many workers have used rather arbitrary models involving a combination of 'continuous' production and late 'discrete'

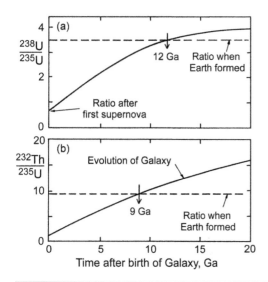

Fig. 15.7 Growth curves for $^{238}U/^{235}U$ and $^{232}Th/^{235}U$ in a 'continuous' supernova production model, relative to the primordial solar system value. After Broecker (1986).

events. However, Trivedi (1977) suggested that it was more reasonable to assume supernova events at regular intervals. He proposed a simple model in which 50 supernova spikes of equal size were equally spaced with an interval of 140 Ma (assuming a galactic age at solar system formation of 7 Ga). Relative to a theoretical 'stable' actinide, this means that the most recent supernova products would undergo 50-fold dilution by isotopically 'cold' material generated in previous events. This model represents a useful yardstick for comparison with the dilution factors for extinct nuclides discussed below.

15.4 Iodine–Xenon

Xenon isotopes are the products of two extinct nuclides, ^{129}I and ^{244}Pu, but the former, with a 16 Ma half-life, has been the most useful in applying constraints to early solar system evolution. The more volatile character of xenon relative to its parent iodine makes radiogenic xenon isotope anomalies theoretically the easiest extinct nuclide signature to detect. The first search for ^{129}Xe anomalies was unsuccessful (Wasserburg and Hayden, 1955), but five years later they were demonstrated by Reynolds (1960), making this the first extinct nuclide to be 'found'.

15.4.1 The Xe–Xe Correlation Diagram

If the excess ^{129}Xe signatures discovered by Reynolds (1960) were due to the decay of once-live ^{129}I in meteorites, it would be expected that stable ^{127}I would still be present. Therefore, to test this model, it was necessary to look for correlations between the abundance of ^{127}I and excess ^{129}Xe in each sample. To do this by chemical analysis would have been laborious and inaccurate. However, Jeffrey and Reynolds (1961) conceived of an elegant means of measuring the ratio $^{129}Xe_{excess}/^{127}I$ in a single mass spectrometric analysis. By irradiating whole-rock samples of a meteorite with slow neutrons in a reactor, it was possible to generate the stable isotope ^{128}Xe from ^{127}I by the following n, γ and β decay reactions:

$$^{127}I + n \rightarrow \,^{128}I + \gamma$$
$$^{128}I \rightarrow \,^{128}Xe + \beta + \bar{\nu}$$

Hence, the I/Xe ratio could be determined by isotopic analysis of xenon alone.

Jeffery and Reynolds made a further technical advance in their method of sample analysis. Rather than one-step outgassing of xenon from each meteorite sample by melting it (to produce a single data point), they outgassed the sample in a series of increasing temperature steps, admitting each successive gas-release separately to the mass spectrometer for analysis. A very similar neutron irradiation and 'step-heating' mass spectrometric method was applied to K–Ar geochronology five years later, revolutionizing it to the ^{40}Ar–^{39}Ar method (Section 10.2).

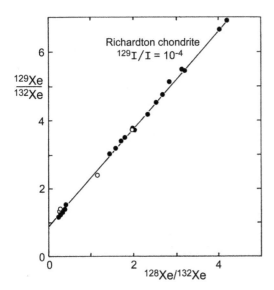

Fig. 15.8 Xe–Xe plot for stepwise degassed samples of the Richardton ordinary chondrite, showing the line of 'iso-concentration' of ^{129}I. Solid and open symbols indicate gas fractions released above and below 1100 °C respectively. After Hohenberg et al. (1967).

It is convenient to display the Xe isotope data on a plot somewhat analogous to an Ar–Ar isochron diagram (Section 10.2.3). Jeffrey and Reynolds demonstrated a correlation between ^{129}Xe and ^{128}Xe abundance in the Richardton chondrite, ratioing both of these against the non-radiogenic isotope ^{132}Xe (e.g. Fig. 15.8). If the efficiency of the activation process is calibrated, the excess ^{128}Xe abundance translates into the abundance of the non-radioactive iodine isotope, ^{127}I. Similarly, because every ^{129}I atom has by now been converted to ^{129}Xe by radioactive decay, the excess ^{129}Xe abundance translates into the ^{129}I abundance at the time when meteorite components were isolated from a common reservoir. Hence, the slope of any array of data points observed in this diagram (^{129}Xe$_{excess}$/^{128}Xe$_{excess}$) has no direct age significance. Instead, it indicates the initial ^{129}I/^{127}I ratio when the meteorite cooled to the point where its minerals became closed to diffusional loss of xenon into space. Meanwhile, the intercept of the correlation line on the y axis represents the initial ^{129}Xe/^{132}Xe ratio of the sample before it became a closed system able to record the decay of live ^{129}I present in the sample.

The slope of the array for Richardton (Fig. 15.8) corresponds to an initial ^{129}I/^{127}I ratio of 1×10^{-4}. Subsequent work on a wide variety of meteorites has confirmed this value with only small variations. This suggests that ^{129}I was widely distributed through the solar nebula (e.g. Podosek, 1970; Wasserburg et al., 1977; Niemeyer, 1979).

A very different explanation was proposed by Clayton (1975), who argued that most isotopic anomalies observed in meteorites, including the Xe–Xe pseudochron, were inherited from pre-solar dust grains. This argument was rejected by most other cosmochemists, and conclusive proof against it was eventually provided by pseudochrons in differentiated meteorites (see below). Excess ^{129}Xe is also found in terrestrial rocks and magmas, but in the Earth, ^{129}Xe abundances are not correlated with ^{127}I abundances. This observation is attributed to the outgassing of noble gases from a deep Earth reservoir which once contained 'live' ^{129}I (Section 11.5.1).

15.4.2 The Determination of 'Delta'

The ^{129}I/^{127}I ratio calculated from meteorite studies can be used as a 'model age' chronometer to measure the time interval (Δ) between last nucleosynthesis and coalescence of the solar nebula. However, as with model ages in general, there are several major assumptions which must be made in any attempt to calculate a Δ value. In particular, estimates must be made of the following quantities:

(1) the ratio of ^{129}I/^{127}I originally produced by nucleosynthesis;
(2) the rate of nucleosynthesis over time, prior to the Δ period; and
(3) in a granular model, any dilution of the last addition of radioactively 'hot' iodine by 'cold' iodine from earlier events.

These questions are best examined by comparing some extreme solutions which were summarized by Wasserburg (1985).

It has traditionally been assumed that iodine is generated by the r-process (Fig. 1.7). Production ratios (p_{127}/p_{129}) can only be determined by theoretical calculation; hence, there are large uncertainties. However, they are generally assumed to be near unity. For example, values which have been used in the literature are unity (Wasserburg et al., 1960; Wasserburg, 1985), 1.3 (Cameron, 1962; quoted by Hohenberg, 1969) and 2.9 $+1/-2$ (Seeger et al., 1965; quoted by Schramm and Wasserburg, 1970).

The model originally conceived by Reynolds (1960) involved synthesis of iodine with a ^{129}I/^{127}I ratio of unity in a single event. This would decay to a ratio of 10^{-4} over ca. 12 half-lives, yielding a maximum Δ value of ca. 200 Ma (Fig. 15.9a). However, if all iodine was generated in a single supernova (i.e. zero dilution of 'hot' supernova iodine by 'cold' ^{127}I), then all other r-process elements would have to have formed at this time, which is incompatible with actinide evidence (as well as some short-lived extinct nuclides).

The other extreme model assumes more or less constant supernova activity throughout the lifetime of the Galaxy. If their products were kept mixed, then r-process production of solar system material might be regarded as relatively constant (Dicke, 1969). Under this model, the total number of atoms of stable ^{127}I after time T (at the termination of nucleosynthesis, Fig. 15.1) is defined as

$$n_T{}^{127} = p^{127} T \qquad [15.1]$$

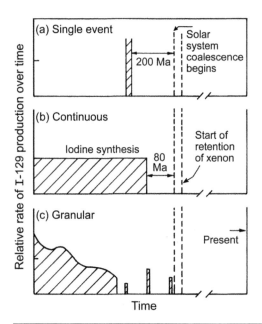

Fig. 15.9 Schematic illustration of iodine production models and consequent calculations. (a) Single supernova event yielding maximum value of Δ (ca. 200 Ma); (b) constant 'continuous' production followed by a period Δ of ca. 80 Ma; (c) complex variation in production rate ('granular model') yielding indeterminate Δ value. After Wasserburg and Papanastassiou (1982).

where p is the average production rate. Similarly, the total number of ^{129}I atoms after time T can be approximated as

$$n_T{}^{129} = p^{129}/\lambda = p^{129} \cdot \text{mean life} \qquad [15.2]$$

Dividing [15.2] by [15.1] we obtain

$$\left(\frac{n^{129}}{n^{127}}\right)_T = \frac{p_{129}}{p_{127}} \cdot \frac{\text{mean life}}{T} \qquad [15.3]$$

Therefore, assuming a production ratio of unity in a model where iodine is formed by frequent and well-mixed supernovae over a period of 10 Ga, the ^{129}I/^{127}I ratio when nucleosynthesis is interrupted is 2.3×10^{-3}. This would take nearly five half-lives to decay to a value of 1×10^{-4}, yielding a Δ value of ca. 80 Ma (Fig. 15.9b). However, this model has a major conceptual problem. It is very sensitive to the contributions of iodine from 'late' supernovae near the end of the nucleosynthetic period. If these form a significant fraction of the total iodine budget, they are apt to de-stabilize the smooth growth model, giving rise to a 'granular' model (Wasserburg and Papanastassiou, 1982), as illustrated in Fig 15.9c.

In reality, consideration of the rate of supernova occurrence in the whole galaxy (about one every 100 years) relative to the size of the galaxy, suggests that in our corner of the galaxy, a granular model is almost inevitable, in relation to the comparatively short half-life of ^{129}I. In that case, the

most critical quantity is the dilution factor for the last addition of hot iodine (^{129}I/^{127}I $= 1$) by cold or nearly cold iodine from earlier events (^{129}I/^{127}I ~ 0). A dilution factor of 100 has been proposed by Cameron and Truran (1977). Coupled with a production ratio of unity, this would yield a Δ value of ca. 110 Ma. However, as the dilution factor approaches 10^4, Δ can approach zero. Not until the review of Wasserburg (1985) was it explicitly pointed out that such 'extreme' solutions are possible. In fact, some evidence from much shorted-lived extinct nuclides suggested that Δ really was approximately zero (see below).

In addition to uncertainties about the dilution factor, there are also questions about the site of nucleosynthetic production of ^{129}I. Because ^{129}I is not greatly separated from the s-process nucleosynthetic pathway, there remains a possibility that ^{129}I might be produced under less extreme conditions than the r-process, such as the n-process (Section 1.2.2). Furthermore, it must not be forgotten that ^{127}I is certainly generated by the s-process, so that uncertainties about the relative s- and r-process contributions to total iodine production are also an issue. As a consequence, ^{129}I cannot place tight constraints on the relative timing of nucleosynthesis and solar system condensation (contrary to early claims). Nevertheless, it may provide useful constraints in combination with other systems.

15.4.3 Pu–Xe

^{244}Pu has a half-life of 82 Ma. The clearest evidence of extinct ^{244}Pu in meteorite materials is provided by fission products, most notably a large excess abundance of 132, 134 and 136 xenon, which has been matched to the signature of laboratory Pu fission products. ^{244}Pu is always compared to the abundance of other actinide elements in drawing conclusions about solar system origins. It is most conveniently ratioed against ^{238}U, but this involves elemental as well as isotopic abundances, and the former are susceptible to chemical fractionation after condensation of the nebula. Whole-rock analysis of different meteorites has yielded a large range of values, but the best consensus is for a chondritic value of around 0.007 (Fig. 15.10).

Plutonium data alone are not able to apply tight constraints to Δ, for the same reasons that were given for iodine. However, if we assume that ^{129}I and ^{244}Pu were added at the same time, we can use the data together to constrain Δ. For example, a continuous model yields an equilibrium ^{244}Pu/^{238}U ratio of 0.018, which can decay to 0.007 after a Δ period of ca. 100 Ma. This is in reasonable agreement with the continuous model for ^{129}I, which yields a Δ value of 80 Ma. Similarly, if we assume a granular model in which both elements undergo 50-fold dilution of the last r-process addition with cold material, both plutonium and iodine yield Δ values of ca. 130 Ma. Both of these results argue against very late addition of very dilute r-process material (which was able to explain the iodine data alone). This conclusion is supported by evidence from the Cm–U system, which yields a Δ value of ca. 160 Ma after the last r-process

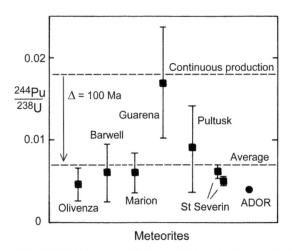

Fig. 15.10 Summary of Pu/U ratios for whole-rock samples of various meteorites. ADOR = Angra dos Reis. After Hagee *et al.* (1990).

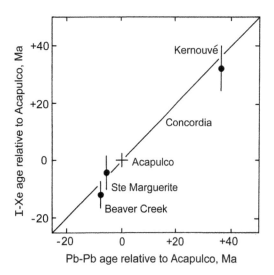

Fig. 15.11 Correlation diagram to test the concordancy of I–Xe and Pb–Pb ages on apatite separates from ordinary chondrites, relative to the achondrite Acapulco. After Brazzle *et al.* (1999).

addition to the solar system (Section 15.8). On the other hand, Hf–W evidence appears to contradict this conclusion, since it suggests late *r*-process addition to the solar nebula. This problem is discussed in Section 15.10.

15.4.4 I–Xe Chronology

In addition to providing evidence about the timing of nucleosynthesis, ^{129}I may also provide information about early solar system evolution. By assuming that 'hot' and 'cold' iodine sources were homogenized in the solar nebula, Podosek (1970) used the slightly variable initial ^{129}I/^{127}I ratios of 0.7 to 1.3×10^{-4} in different meteorites to calculate their relative cooling times. It is not necessary to know the iodine isotope production ratio or the dilution factor of 'hot' by 'cold' iodine in the solar nebula for this calculation. However, to interpret the isotopic variations in terms of cooling times, it is necessary to assume initial ^{129}I/^{127}I homogeneity in the nebula. Crabb *et al.* (1982) argued instead that the variations in initial ratio are due to imperfect mixing of iodine from different sources (variable dilution factors).

To solve this uncertainty it was necessary to test the accuracy of I–Xe ages by comparison with absolute dating methods. Until recently this was not possible, because absolute dating methods (which measure ages back from the present) were not sufficiently precise to measure age differences of ca. 1 Ma in materials 4570 Ma old. However, the availability of new high-precision Pb–Pb ages (Section 5.3.2) has allowed relative I–Xe ages to be anchored against absolute Pb–Pb ages and tested for their reliability.

Nichols *et al.* (1994) began this work by dating phosphate mineral separates (apatite) from the primitive achondrite Acapulco, whose Pb–Pb age is 4557 ± 2 Ma (Gopel *et al.*, 1994). Brazzle *et al.* (1999) continued this work by obtaining step heating I–Xe ages on apatite separates from three ordinary chondrites dated by Pb–Pb. The results (Fig. 15.11) showed excellent concordance between the two methods, demonstrating that the relative I–Xe ages date real events. Since iodine is a volatile element, Brazzle *et al.* inferred that it could not have been incorporated into the refractory minerals that were first crystallized from the nebula. Therefore, it must have been adsorbed onto cooling surfaces and subsequently redistributed into secondary minerals (e.g. apatite) during very early post-formational metamorphism. It is this latter event that I–Xe and Pb–Pb analysis of apatite is thought to be dating.

I–Xe dating played an important role as the first system to show concordance with Pb–Pb dating, and hence to confirm the model of relative homogeneity of late 'hot' addition to the solar nebula. However, because of its relatively long half-life and susceptibility to thermal resetting, it was inevitable that I–Xe dating of early solar system objects would be eclipsed by other methods with more immobile chemistry and shorter half-lives, as discussed below.

15.5 Al–Mg

The nuclide ^{26}Al has a half-life of 0.72 Ma and decays to ^{26}Mg. The discovery of extinct ^{26}Al was a much more difficult task than ^{129}I, and was only made possible by the fall of the Allende CV3 carbonaceous chondrite in February 1969. More than a ton of material was collected, consisting of an agglomerate of abundant chondrules and fine-grained debris, and also containing relatively abundant CAIs.

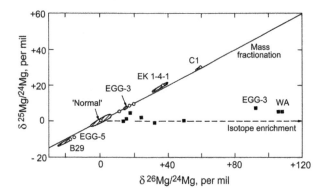

Fig. 15.12 $^{25}Mg/^{24}Mg$ versus $^{26}Mg/^{24}Mg$ isotope diagram showing deviations of Allende inclusions from the normal solar system value, in parts per mil (δ). These may be due to mass fractionation in the solar nebula (open symbols) or decay of extinct ^{26}Al (solid symbols). After Wasserburg and Papanastassiou (1982).

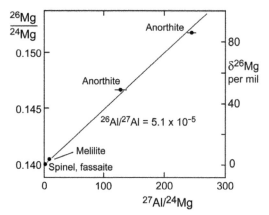

Fig. 15.13 Plot of δMg against Al/Mg for the Allende WA inclusion showing best-fit line of constant initial $^{26}Al/^{27}Al$ ratio with a value of 5.1×10^{-5}. After Lee et al. (1977).

15.5.1 ^{26}Al in the Allende Meteorite

Analysis of Mg isotope ratios in minerals separated from Allende inclusions demonstrates mass-fractionation-dependent variations in $^{25}Mg/^{24}Mg$ ratio. These are normally of the order of a few parts per mil (Fig. 15.12), but some inclusions (e.g. EK1–4-1 and C1) show larger effects. These were termed FUN samples (showing fractionation and unknown nuclear anomalies) by Wasserburg et al. (1977). Because ^{27}Al was always a nuclide of comparatively low abundance, it was at first very difficult to demonstrate radiogenic ^{26}Mg abundances outside error of mass fractionation processes. The first such evidence for radiogenic ^{26}Mg was demonstrated by Lee et al. (1976). Subsequently, larger ^{26}Mg anomalies were found by very careful hand picking of inclusion- and alteration-free plagioclase grains from the Allende WA inclusion. Because these grains have very high Al/Mg ratios, they provide the best chance of finding ^{26}Mg anomalies. This search revealed an excess ^{26}Mg abundance of 97 parts per mil in one anorthite grain (Lee et al., 1977).

It is most convenient to display Al–Mg data on an isotope ratio versus element ratio plot somewhat analogous to an isochron diagram (Fig. 15.13). Hence, $^{26}Mg/^{24}Mg$ is plotted against an Al/Mg ratio. In a conventional isochron diagram, the ratio plotted on the abscissa would be $^{26}Al/^{24}Mg$. However, since every ^{26}Al atom has decayed to ^{26}Mg at the present day, the $^{27}Al/^{24}Mg$ ratio is plotted instead. Therefore, the slope of any array of data points in this diagram has no direct age significance, but indicates the initial $^{26}Al/^{27}Al$ ratio in the sample suite at the time when sub-systems were isolated from a common reservoir. Hence, these arrays are often called pseudochrons.

Lee et al. (1977) found a $^{26}Mg/^{27}Al$ correlation in separated minerals from the Allende WA inclusion, yielding a well-defined initial $^{26}Al/^{27}Al$ ratio of $5.1 \pm 0.6 \times 10^{-5}$. Well-correlated data were also found for the Allende EGG-3 inclusion, yielding a similar $^{26}Al/^{27}Al$ ratio of 4.9×10^{-5} (Armstrong et al., 1984). This consensus of $^{26}Al/^{27}Al$ ratios around 5×10^{-5} has largely been confirmed in subsequent work, and has thus been termed the 'canonical' solar system initial ratio of this tracer. In contrast, minerals from different parts of inclusion USNM 3529–26 (Armstrong et al., 1984) gave lower initial Al/Al ratios, but with a higher value in the core (3.8×10^{-5}) than the rim (2.3×10^{-5}). This spatial variation is best explained by Mg loss, particularly from the margins of the inclusion, during a later metamorphic event. The event may have resulted from the heat output of ^{26}Al decay itself.

15.5.2 Determination of Delta

The great significance of ^{26}Al for cosmochemistry is its short half-life of 0.72 Ma. Because this is only 4% of the ^{129}I half-life, its presence in the early solar system constrains a nucleosynthetic event to have occurred much more imminently before the coalescence of the solar system. Classical nucleosynthetic models (e.g. Arnett, 1969; Truran and Cameron, 1978) attribute ^{26}Al to explosive carbon burning in the envelope of a supernova, and predict a $^{26}Al/^{27}Al$ production ratio of ca. 10^{-3}. It would take only 3 Ma for this ratio to decay to the value of 5×10^{-5} found in several meteorite samples. Therefore, Lee et al. (1976) suggested that the observed anomalies were due to late addition to the solar nebula of freshly synthesized material from a nearby nova or supernova explosion. Cameron and Truran (1977) pointed out that such an explosion in the vicinity of a condensing solar nebula was a very unlikely coincidence unless the supernova itself triggered the collapse of an interstellar cloud to form the solar system. This model is illustrated in Fig. 15.14.

For a time, the 'supernova trigger' model for solar system coalescence was widely accepted. However, spectral data from the High Energy Astronomical Observatory satellite

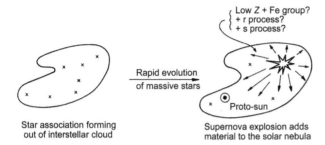

Star association forming
out of interstellar cloud

Supernova explosion adds
material to the solar nebula

Fig. 15.14 Schematic illustration of a model in which solar system collapse is promoted by a supernova which also seeds it with short-lived nuclides. After Wasserburg (1985).

(HEAO 3) revealed a gamma line due to ^{26}Al decay from a diffuse galactic source (Mahoney *et al.*, 1984). An average galactic ^{26}Al/^{27}Al ratio of ca. 10^{-5} was determined, remarkably close to that deduced for the early solar system from Allende inclusions. This high ^{26}Al abundance in the galaxy means that supernovae, which are rare, cannot be the principal source. Indeed, recent experimental studies (Champagne *et al.*, 1983) suggest that asymtotic giant branch (AGB) stars ('red giants') can generate aluminium with a 26/27 production ratio of unity. Nevertheless, the isotopic data on Allende inclusions still imply late injection of ^{26}Al into the pre-solar cloud. For example, Wasserburg (1985) suggested that this injection could be supplied by rapidly evolving stars within the interstellar cloud itself, followed by rapid condensation on a ca. 1 Ma timescale (Wasserburg, 1985).

In contrast, Clayton (1979) repeated his earlier arguments (Section 15.4.1) that extinct nuclide signatures in meteorites were inherited from pre-solar dust grains. If this model were true for ^{26}Al, then both Al isotope and Al/Mg ratios would have to be inherited intact from these pre-solar materials. This would in turn imply that many meteorite mineral phases in CAIs are pre-solar. Wasserburg (1985) contested this argument on mineralogical grounds, believing that most or all of the analysed meteorite phases crystallized within the solar system. Evidence supporting Wasserburg was provided by the discovery of ^{26}Mg excesses in a clast from the Semarkona ordinary chondrite (Hutcheon and Hutchison, 1989). Hutcheon and Hutchison argued that the mineralogy and REE chemistry of the clast were the result of igneous processes, implying a planetary, rather than nebular origin.

All of the samples described above (inclusions and clasts) are now generally thought to have formed during solar system condensation. However, some evidence has been found for the preservation of pre-solar *grains* in the Murchison carbonaceous chondrite. These grains are composed of silicon carbide, and have exotic rare gas signatures which match the abundance patterns expected in carbon-burning red giants (Lewis *et al.*, 1990). This means that they may have escaped significant heating during solar system coalescence. Zinner *et al.* (1991) found large excesses of ^{26}Mg in some of these grains, equivalent to initial ^{26}Al/^{27}Al ratios from 10^{-5} up to

nearly unity. These ratios probably date from the time of expulsion of the grains, in the solar wind of a red giant, into interstellar space. Therefore, one model suggests that the solar wind from such a star (rather than a supernova) triggered the collapse of a giant molecular cloud to form the solar system (Nuth, 1991).

A totally different explanation of ^{26}Al signatures that has been proposed at various times (e.g. Lee, 1978) is their generation by spallation reactions caused by intense solar radiation. For example, such irradiation may have occurred if the sun went through a 'T Tauri' stage early in its evolution. However, Shu *et al.* (1997) argued that the greatest radiation intensity was reached at an earlier stage in the Sun's evolution, as a protostar still 'embedded' in the nebular disc. Lee *et al.* (1998) used this model to explain the presence of several of the very short-lived species whose traces are observed in meteorites. This model is discussed further during an examination of the spallogenic nuclide ^{10}Be (Section 15.9.1). However, the relative consistency of ^{26}Al/^{27}Al ratios that have been found in many different CAIs (the 'canonical' value) argues against such a late source for ^{26}Al.

15.5.3 Al–Mg Early Nebular Chronometry

The use of ^{26}Al as a dating tool for early solar system objects is intimately bound up with the question of the homogeneity of its distribution in different solar system objects. This question has been investigated by further study of CAIs, and also by the analysis of other solar system objects.

Russell *et al.* (1996) found Mg isotope anomalies in CAIs from the unmetamorphosed ordinary chondrites Semarkona and Moorabie. These gave initial ^{26}Al/^{27}Al ratios of 5×10^{-5}, exactly the same as found in inclusions from carbonaceous chondrites. Russell *et al.* also found the first evidence for extinct ^{26}Al in Al-rich *chondrules* (as opposed to inclusions) from the ordinary chondrites Inman and Chainpur. The chondrules gave lower initial ^{26}Al/^{27}Al ratios than the inclusions, around 1×10^{-5}. If the decrease from a typical CAI value of 5×10^{-5} to a chondrule value of 1×10^{-5} is attributed to ^{26}Al decay in a homogeneous reservoir, the data imply a period of ca. 2 Ma between CAI and chondrule formation.

The occurrence of ^{26}Al was finally extended to achondrites by Srinivasan *et al.* (1999). These workers found a very small ^{26}Al signal (^{26}Al/^{27}Al $= 7.5 \times 10^{-7}$) in the eucrite Piplia Kalan, showing that this nuclide survived into the period of differentiation of the eucrite parent body and could have provided a heat source for the melting of such bodies. The ^{26}Al signal was actually larger than expected, and suggested that planetary differentiation occurred as early as 4 Ma after the formation of CAIs, assuming a homogeneous distribution of the nuclide. Subsequent work extended the ^{26}Al signal to objects as young as 5 Ma after CAIs in the D'Orbigny angrite and the Dar al Gani ultramafic achondrite (ureilite). Data are summarized in Nyquist *et al.* (2009).

These objects also yield coherent decay patterns when plotted against other extinct nuclide systems such as Hf–W

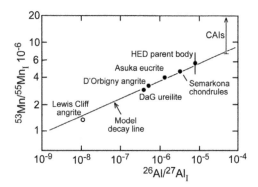

Fig. 15.15 Comparison of Al–Mg and Mn–Cr ages on a variety of meteorites. The $^{26}Al/^{27}Al$ ratio for Lewis Cliff was determined indirectly (see text). After Nyquist et al. (2009).

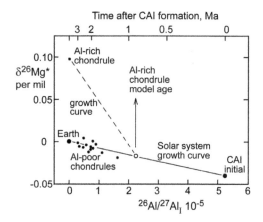

Fig. 15.16 Mg isotope evolution diagram for Allende chondrules relative to the solar system growth curve, showing model age calculation for an Al-rich chondrule. After Villeneuve et al. (2009).

and Mn–Cr (Nyquist et al., 2009). This is demonstrated in Fig. 15.15 by analysed isotope ratios that fall very close to the log–log decay curve calculated from the relative half-lives of the two species. However, the ^{26}Al abundance in Lewis Cliff 86010 is predicted from its Pb–Pb age, since it is too low for accurate measurement. In this figure, the ^{53}Mn data for CAIs yield more complex signatures, which are discussed below. Relationships between Mg–Al and absolute Pb–Pb ages are also less coherent. However, the small range of these ages is at the limit of resolution of the Pb–Pb method, so the errors may be in this method rather than the extinct nuclide system (Kita et al., 2015).

15.5.4 Testing the 'Canonical' Model

There has been a continuing effort over the past few years to test the validity of the 'canonical' CAI model. Claims for supra-canonical $^{26}Al/^{27}Al$ ratios were made by Young et al. (2005) based on in situ laser ablation MC–ICP–MS of CAIs, and by Bizzarro et al. (2005) and Thrane et al. (2006) using solution MC–ICP–MS analysis of bulk CAI fragments. However, subsequent work failed to reproduce these findings. The supra-canonical conventional data were attributed to analytical error (Jacobsen et al., 2008), while the supra-canonical laser ablation data may result from analytical artefacts.

On the other hand, several studies have identified sub-canonical $^{26}Al/^{27}Al$ ratios in CAI, implying the possibility of isotopic heterogeneity in the solar nebula (e.g. Liu et al., 2012; Holst et al., 2013; Makide et al., 2013; Schiller et al., 2015). Some of these cases seem well founded. For example, Holst et al. (2013) determined a precise $^{26}Al/^{27}Al$ ratio of $2.9 \pm 0.2 \times 10^{-6}$ in a variety of separated minerals from an Allende FUN CAI. However, FUN CAIs differ from canonical CAIs in many isotopic tracers, and it has been suggested that they actually predate canonical CAIs (Sahijpal et al., 1998). Hence they may have escaped later ^{26}Al additions from a nearby star that triggered the collapse of the nebula.

In other cases, apparent sub-canonical $^{26}Al/^{27}Al$ ratios may be due to previously unrecognized thermal reprocessing of older material. This would be analogous to the distur-

bance of conventional isochron systems by thermal events (e.g. Podosek et al., 1991). These issues can be examined by placing $^{26}Al/^{27}Al$ pseudochrons in the context of initial Mg isotope abundances using a ^{26}Mg evolution plot (Villeneuve et al., 2009). The y-axis variable, $\delta^{26}Mg^*$, represents fractionation-corrected Mg isotope data, in the form of radiogenic ^{26}Mg abundances relative to terrestrial standards (Fig. 15.16). Standards were processed through the same chemical separation to exclude isotope fractionation during chemical extraction. The x axis is plotted on a log scale, due to the short half-life of ^{26}Al. The closed-system evolution of a reservoir then yields a linear growth curve.

The solar system growth line is drawn from the CAI initial ratio to the present day terrestrial value. The CAI initial ratio of Jacobsen et al. (2008) had a relatively large error of +/−0.03 per mil. However, Schiller et al. (2010) showed that bulk samples from all major chondrite classes (CI, CM, CO, CV, CK, CR, CH, CB, and ordinary and enstatite chondrites) all lay on the CAI pseudochron. Hence, by combining the precise slope of the CAI pseudochron with the low Mg/Al ratios of bulk chondrites, Schiller et al. (2010) determined a precise solar system initial $\delta^{26}Mg$ ratio of -0.034 ± 0.0016 per mil.

Most Allende chondrules yield initial $^{26}Al/^{27}Al$ ratios of between 0.3 and 1.6×10^{-5}, and initial $\delta^{26}Mg$ values within error of a ^{26}Mg growth line (Fig. 15.16). This implies homogenization ages 1.2 to 3 Ma after CAIs. In contrast, one Al-rich chondrule had a lower initial $^{26}Al/^{27}Al$ ratio of less than 10^{-6}, implying a younger homogenization age at least 5 Ma after CAI. However, this sample has a much higher $\delta^{26}Mg$ value. Using the bulk Mg/Al ratio of the chondrule, its Mg isotope growth curve can be projected back to the solar system growth curve, yielding an Al/Mg model age for this chondrule less than 1 Ma after CAI. This approach is analogous to the determination of Nd model ages on metamorphosed rocks (Section 4.3.3).

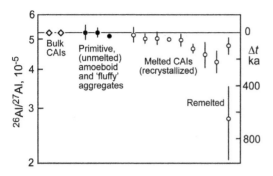

Fig. 15.17 Plot of $^{26}Al/^{27}Al$ ratios for CAIs grouped according to textural type. After MacPherson *et al.* (2012).

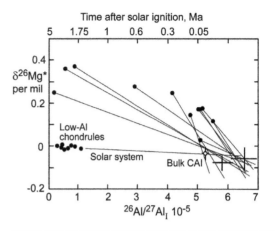

Fig. 15.18 Mg isotope evolution diagram showing growth lines for the precursors of various Efremovka and Vigarano inclusions. Error bars are shown for two inclusions with δ $^{26}Mg*$ below zero, and two points with very large errors are omitted. Modified after Mishra and Chaudisson (2014a).

Another way of studying thermal resetting is the analysis of material whose textural and mineralogical features are indicative of lesser amounts of thermal processing. For example, MacPherson *et al.* (2010, 2012) studied rare inclusions from the CV3 chondrites Vigarano and Leoville that appeared to have escaped melting. These inclusions have irregular outlines and fine-grained inter-grown mineral textures, forming amoeboid and 'fluffy' aggregates. This marks these inclusions as direct condensates from the gas phase, unlike the melt-crystallization textures of most CAIs. Hence, these unusual inclusions are argued to represent the precursors of typical CAIs. Ion microprobe analysis of their minerals yielded Al–Mg pseudochrons with canonical $^{26}Al/^{27}Al$ ratios of 5.2×10^{-5}, consistent with previous bulk CAI pseudochrons (Jacobsen *et al.*, 2008). In contrast, CAIs with melt-crystallization textures yielded mineral pseudochrons with slightly lower $^{26}Al/^{27}Al$ ratios, indicative of open-system behaviour (Fig. 15.17).

Further work on the solar system initial Mg isotope ratio was based on the analysis of a variety of inclusions from Efremovka and Vigarano (Mishra and Chaudisson, 2014a). Figure 15.18 shows calculated initial ratios for each object, along with its Mg isotope growth curve that can be projected back to predict the isotopic composition of the precursor. Most of the data for individual inclusions converge towards a composition slightly richer in ^{26}Al but depleted in ^{26}Mg relative to the canonical bulk CAI value. The new estimate for the precursor composition is $^{26}Al/^{27}Al = 5.6 \pm 0.4 \times 10^{-5}$ and $\delta^{26}Mg* = -0.052 \pm 0.013$ per mil. This suggests that CAI precursors grew from a homogenized gaseous reservoir only a few tens of thousands of years older than CAIs themselves, when the sun first ignited.

15.6 Short-Lived Species in Planetary Differentiation

There are several extinct nuclide systems with relatively short half-lives in which parent–daughter fractionation was caused by processes of planetary accretion or differentia-

tion. As such, these systems are particularly important for studying the early evolution of the solar system, including the development of planetessimals and the Earth–Moon system.

The first of these systems to be studied (Pd–Ag) is characterized by a siderophile parent nuclide, and is therefore particularly applicable to iron meteorites. The second system to be studied (Mn–Cr) is fractionated by the greater volatility of the parent during condensation and accretion. Finally, the last two systems (Hf–W and Fe–Ni) both have more siderophile *daughter* nuclides. Therefore, the build-up of the daughter product in lithophile phases can be used to date core formation and planetary differentiation.

15.6.1 Pd–Ag

^{107}Pd decays by β emission to ^{107}Ag with a half-life of 6.5 Ma. The strong partition of Pd into the metal phases of iron meteorites means that these have high Pd/Ag ratios suitable for detecting the signal of extinct ^{107}Pd. The first successful measurement of this signal was made by Kelly and Wasserburg (1978) on the Santa Clara IVB iron meteorite. Hence, they deduced that only 10 Ma might have elapsed between the last nucleosynthetic event and the coalescence and differentiation of iron-cored small planets. Since ^{107}Pd versus Ag correlations were observed in bodies which had clearly been melted since the accretion of the solar system, they provided some of the earliest proof for the existence of 'live' ^{107}Pd in the early solar system that could not have been inherited from pre-solar grains.

Subsequent work revealed radiogenic ^{107}Ag in several other iron meteorites, of which the best data are from the IVA iron Gibeon (Chen and Wasserburg, 1984, 1990). These results yield a precise initial $^{107}Pd/^{108}Pd$ ratio of $2.31 \pm 0.06 \times 10^{-5}$ (Fig. 15.19). The development of MC-ICP-MS

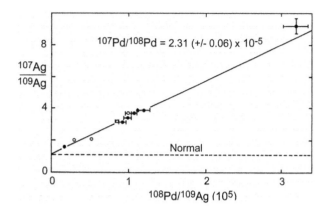

Fig. 15.19 Pd–Ag pseudochron diagram showing evidence of extinct ^{107}Pd in iron meteorites. The best-fit line to seven metal samples from the Gibeon meteorite (•) yields an initial ^{107}Pd/^{108}Pd ratio of 2.3×10^{-5}. Open symbols denote other group IVA iron meteorites. After Wasserburg (1985).

analysis allowed the detection of a radiogenic ^{107}Ag signal in carbonaceous and ordinary chondrites (Schonbachler *et al.*, 2008). An attempt was made to use these data to estimate an improved initial ^{107}Pd/^{108}Pd ratio for the solar system, but scatter of the data prevented an accurate determination.

The Pd–Ag data for Gibeon have now been supported by analysis of Muonionalusta, the oldest well-dated iron meteorite, with a Pb/Pb age of 4565.3 ± 0.1 Ma on the sulphide phase troilite (Blichert-Toft *et al.*, 2010). This age suggests that the core of the IVA iron parent body had not only formed, but cooled below the Pb blocking temperature within 2 to 3 Ma of CAIs. Pd–Ag data for Muonionalusta are somewhat scattered, as seen in other bodies, but give a best fit initial ^{107}Pd/^{108}Pd ratio of $2.15 \pm 0.3 \times 10^{-5}$ (Horan *et al.*, 2012). This value is within error of the more precise value for Gibeon, which is believed to come from the same parent body. Based on the Pb–Pb age of Muonionalusta, this implies a primordial ^{107}Pd/^{108}Pd ratio in the solar nebula of around 2.8×10^{-5}.

In contrast to the simple cooling history of these magmatic iron meteorites, the non-magmatic irons (group IAB, including Canyon Diablo) have a more complex history, probably involving a later collisional disruption/heating event. This was investigated using Pd–Ag systematics by Theis *et al.* (2013). Again, the data were somewhat scattered, with a troilite-rich group including Canyon Diablo giving the lowest initial ^{107}Pd/^{108}Pd ratios at around 0.8×10^{-5}. Using a solar system initial of 2.8×10^{-5} implies a Pd–Ag closure age of ca. 12 Ma after CAIs for the Canyon Diablo group, in general agreement with Hf–W evidence (Schultz *et al.*, 2009).

15.6.2 Mn–Cr

^{53}Mn decays to ^{53}Cr with a half-life of 3.7 Ma. For several reasons this extinct nuclide reinforces the evidence from

^{107}Pd for the early history of planetary differentiation. Birck and Allegre (1985, 1988) found correlated variations in ^{53}Cr, ^{52}Cr and Mn/Cr ratio in several meteorites, including the carbonaceous chondrites Allende and Murchison, the enstatite chondrite Indarch and the pallasite Eagle Station. Since the mixed silicate–iron mineralogy of the latter indicates a high temperature origin, Mn–Cr isotope correlations must result from *in situ* decay of ^{53}Mn in differentiated planetary bodies. Hence ^{53}Mn provides additional evidence for nucleosynthetic processes immediately before coalescence of the solar system.

^{53}Mn is part of the iron group of elements (Section 1.2) which are thought to be synthesized in large stars shortly before a supernova explosion. Hence, it was argued (e.g. Rotaru *et al.*, 1992) that a supernova briefly pre-dated solar system condensation, as initially proposed on the basis of ^{26}Al. From the 3.7 Ma half-life of ^{53}Mn, it would follow that such an event occurred ca. 20 Ma before planetary differentiation. However, there are other possible routes for the synthesis of ^{53}Mn (Birck and Allegre, 1985) and the late-supernova model therefore remains in doubt.

Although its nucleosynthetic significance is unclear, the 3.7 Ma half-life of ^{53}Mn gives it great potential as a chronometer of early solar system evolution. To test this potential, Lugmair and Shukolykov (1998) made a Mn–Cr study of a large group of achondrites, beginning with the angrites Lewis Cliff 86010 (LEW) and Angra dos Reis (ADOR). These bodies show evidence of rapid cooling, despite relatively young ages relative to other angrites. Based on a Pb–Pb age of 4557 Ma, their initial ^{53}Mn/^{55}Mn ratio of 1.25×10^{-6} was used to anchor the Mn–Cr method in the same way that the I–Xe method was anchored.

Lugmair and Shukolykov (1998) also analysed a suite of eight eucrites, attributed to a single parent body (the HED body). Two of these (Chervony Kut and Juvinas) gave good mineral pseudochrons, with Mn–Cr ages of ca. 4563 and 4562 Ma using the angrite Pb–Pb anchor. In addition, the complete suite gave an excellent whole-rock pseudochron with a ^{53}Mn/^{55}Mn ratio of 4.7×10^{-6} (revised to 4.2×10^{-6} by Trinquier *et al.*, 2008). This is equivalent to an age of 4564.5 Ma, interpreted as the time of differentiation of the parent body, ca. 3 Ma after CAIs. This supports Pd–Ag ages in pointing to extremely early formation and differentiation of planetessimals.

Subsequent work by Glavin *et al.* (2004) provided a second anchor for Mn–Cr dating in the form of the angrite D'Orbigny, which is unique in containing quenched glass. The excellent mineral–glass pseudochron yields a very precise ^{53}Mn/^{55}Mn ratio of $3.24 \pm 0.04 \times 10^{-6}$ (Fig. 15.20). This implies a Mn–Cr age 5 Ma older than the LEW–ADOR anchor. Pb–Pb dating corrected for variable uranium compositions (Section 5.1.2) yields ages of 4556.6 (LEW–ADOR) and 4563.4 (D'Orbigny), representing an age difference of 6.8 ± 0.5 Ma. The 2 Ma discrepancy in relative ages implies differences in the blocking temperatures of the Mn–Cr and Pb–Pb systems, but these are not yet fully understood.

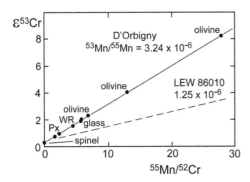

Fig. 15.20 Mn–Cr pseudochron for separated minerals from the quenched angrite D'Orbigny, in comparison with the LEW 86010 anchor. After Glavin et al. (2004).

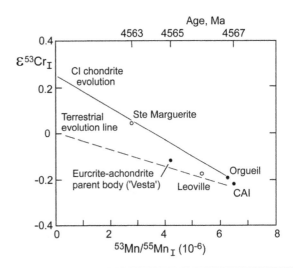

Fig. 15.21 Cr isotope evolution diagram for the solar system based on data for the Earth, the HED parent body, and chondrite leachates and whole-rocks. After Trinquier et al. (2008).

In contrast to the achondrites, establishing the Mn–Cr ages of chondrites and CAIs proved more difficult. Mineral pseudochrons for the ordinary chondrites Chainpur and Bishunpur gave internal pseudochrons with a $^{53}Mn/^{55}Mn$ ratio of 9.4×10^{-6} (Nyquist et al., 2001) implying an age of ca. 4568 Ma, which was surprisingly old. In addition, preliminary data for CAIs implied ratios even higher. However, Yin et al. (2007) obtained a much lower $^{53}Mn/^{55}Mn$ ratio of 5.1 $\times 10^{-6}$ for six chondrules from Chainpur. The lower value was supported by Trinquier et al. (2008), who obtained a $^{53}Mn/^{55}Mn$ ratio of 6.5×10^{-6} for bulk chondrite samples, and Gopel et al. (2015), who determined a $^{53}Mn/^{55}Mn$ ratio of 6.2×10^{-6} for a composite set of analyses of whole-rock chondrites.

Trinquier et al. (2008) also obtained consistent ^{53}Cr initial ratios, which they plotted on a Cr isotope evolution diagram against $^{53}Mn/^{55}Mn$ ages (Fig. 15.21). The different slopes of the Cr isotope evolution lines for carbonaceous chondrites, the HED parent body and the Earth are attributed to volatile-driven Mn/Cr fractionation in the solar nebula. In addition to the eucrite and chondrite whole-rocks, leaching experiments for three individual chondrites (Orgueil, Leoville and Ste Marguerite) gave ages and initial ratios on the chondritic evolution line, attributed to Mn/Cr closure after different cooling histories. However, leaching experiments for other carbonaceous chondrites gave more scattered data. This is not surprising, because variation in stable ^{54}Cr isotope abundances (Fig. 15.2) implies residual nucleosynthetic Cr isotope heterogeneity in solar system objects.

This heterogeneity was explored by Yamakawa and Yin (2014) using leaching experiments on the CM chondrites Murchison and Sutter's Mill (Fig. 15.22). Leachate fractions defined a steep negative correlation which lay on the same trend as insoluble organic residues from Allende and Murchison (Qin et al., 2010). Unlike early results for FUN inclusions (Papanastassiou, 1986), these results imply that nucleosynthetic heterogeneity of stable ^{54}Cr is much greater than for ^{53}Cr.

15.6.3 Fe–Ni

^{60}Fe decays to ^{60}Ni (via ^{60}Co) with a half-life of 2.6 Ma (Rugel et al., 2009; Wallner et al., 2015). Using TIMS analysis, Shukolyukov and Lugmair (1993a, b) found ^{60}Ni excesses as high as 50 ε units in the eucrite Chervony Kut, facilitated by the extremely high Fe/Ni ratios in this meteorite (up to 350 000 in one sample). These ratios are attributed to preferential nickel partition into the cores of differentiated planetesimals. Whole-rock samples of Chervony Kut displayed a correlation between $^{60}Ni/^{58}Ni$ and Fe/Ni ratio, indicating that ^{60}Fe was still alive at the time of differentiation, with an initial $^{60}Fe/^{56}Fe$ ratio of $3.9 \pm 0.6 \times 10^{-9}$. This result was confirmed by MC-ICP-MS analysis of whole-rock samples of

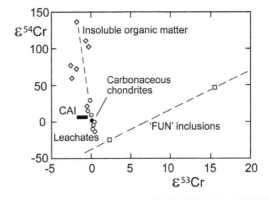

Fig. 15.22 Cr isotope plot for components of carbonaceous chondrites showing possible mixing lines defined by various meteorite components. (●) = bulk chondrites; (○) = leachates. After Yamakawa and Yin (2014).

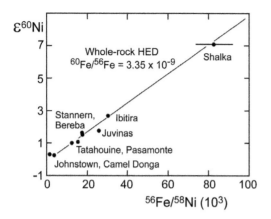

Fig. 15.23 Fe–Ni pseudochron for whole-rock samples of HED meteorites. After Tang and Dauphas (2012).

other HED meteorites (Tang and Dauphas, 2012), yielding a $^{60}Fe/^{56}Fe$ ratio of $3.5 \pm 0.3 \times 10^{-9}$ (Fig. 15.23). On the other hand, whole-rock angrites gave a slightly lower ratio of $3.1 \pm 0.8 \times 10^{-9}$, consistent with their younger Pb–Pb ages (Quitte et al., 2010).

In contrast to these relatively coherent whole-rock data, separated minerals and acid leaches from the same meteorites gave very scattered results, attributed by Shukolyukov and Lugmair to open-system behaviour of nickel during thermal metamorphism after ^{60}Fe became extinct. For example, leachates of a sulphide-rich patch of Chervony Kut plotted far above the whole-rock isochron, with an $\varepsilon^{60}Ni$ value of 50, equivalent to an initial $^{60}Fe/^{56}Fe$ ratio around 10^{-7}. Even higher initial $^{60}Fe/^{56}Fe$ ratios from 2 to 7×10^{-7} were obtained by SIMS analysis of chondrules in carbonaceous chondrites (open circles in Fig. 15.24; Mishra and Goswami, 2014; Mishra and Chaussidon, 2014b). On the other hand,

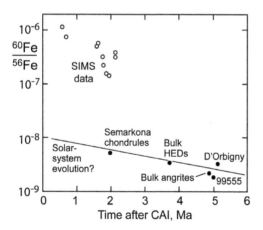

Fig. 15.24 ^{60}Fe decay plot for a variety of chondrites and achondrites relative to absolute ages from Pb–Pb dating. After Tang and Dauphas (2015).

Chen et al. (2009) and Moynier et al. (2011) determined very low initial $^{60}Fe/^{56}Fe$ ratios ($<4 \times 10^{-9}$) for separated metal and sulphide phases from a variety of iron meteorites.

This creates a severe dichotomy in the interpretation of Fe–Ni data between the different data sets. However, it seems more logical to attribute the mineral data rather than the whole-rocks to open-system behaviour of nickel. For example, using the whole-rock data, Tang and Dauphas (2015) plotted a coherent Ni isotope evolution line for several achondrites, along with a less precise $^{60}Fe/^{56}Fe$ ratio of $5.4 \pm 3.3 \times 10^{-9}$ for the Semarkona ordinary chondrite (based on a chondrule pseudochron). Together, these data point to a homogeneous distribution of ^{60}Fe in the solar nebula, with an initial $^{60}Fe/^{56}Fe$ ratio at the time of CAI formation of around 1×10^{-8} (Fig. 15.24).

Unlike several other nuclides, most notably ^{10}Be, but also possibly including ^{26}Al and ^{53}Mn, there is no suitable seed nuclide from which the neutron-rich ^{60}Fe nuclide could have been produced by cosmogenic reactions (Section 15.9). This makes this nuclide important for assessing additions of freshly synthesized material shortly before condensation of the solar nebula. However, the alternative initial $^{60}Fe/^{56}Fe$ ratios proposed above must first be compared with the galactic $^{60}Fe/^{56}Fe$ ratio.

Tang and Dauphas (2012) estimated a $^{60}Fe/^{56}Fe$ ratio in the present day Galaxy of 2.8×10^{-7} from gamma emission lines (as for ^{26}Al). If the galactic value was the same at the time of solar system origins, it implies that the solar nebula was out of nucleosynthetic circulation for over 10 Ma before its coalescence, and hence was *not* seeded with hot ^{60}Fe at the same time as ^{26}Al addition immediately before condensation. This may imply that the late ^{26}Al addition was from the solar wind of a red giant (AGB star) rather than a supernova (Young, 2014). However, until the discrepancies in Fe–Ni measurement are resolved, this conclusion remains tentative.

It is also important to examine the evidence for variations of initial nickel isotope ratio (as opposed to $^{60}Fe/^{56}Fe$) in the nebula. These measurements have also been controversial, due to difficulties in resolving the effects of radiogenic production, nucleosynthetic heterogeneity, and mass fractionation. For example, Bizzarro et al. (2007) claimed to observe variations of $\varepsilon^{60}Ni$ and ^{62}Ni in different meteorite types, but Regelous et al. (2008) and Dauphas et al. (2008) attributed these variations to isobaric interferences on the ^{61}Ni isotope used as a fractionation monitor.

Regelous et al. (2008) observed small −15 ppm ^{60}Ni anomalies in Group IVB iron meteorites relative to terrestrial standards. These anomalies were confirmed in subsequent data (Steele et al., 2011; Tang and Dauphas, 2014) and could theoretically be explained by lower in-growth of ^{60}Ni from live ^{60}Fe, due to Fe/Ni ratios below chondritic. However, a detailed comparison of isotopic variation in all nickel isotopes by Steele et al. (2012) suggested that the negative ^{60}Ni signals were correlated with stable ^{58}Ni when a more reliable fractionation normalization was adopted using the heavy nickel isotopes.

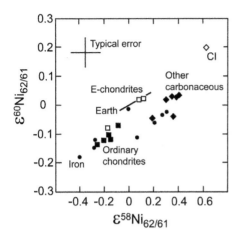

Fig. 15.25 Nickel three-isotope plot showing correlations in isotope abundance for samples with Fe/Ni ratios too low to generate significant changes in the radiogenic ^{60}Ni signal. After Steele et al. (2012).

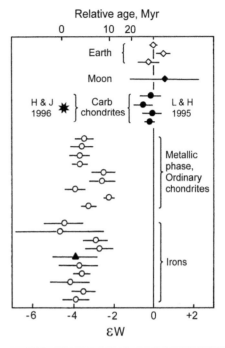

Fig. 15.26 Tungsten isotope values for different solar system objects relative to a terrestrial tungsten standard. Open and closed symbols = measured data of Harper and Jacobsen (1996) and Lee and Halliday (1996) respectively. Star = alternative *modelled* composition of carbonaceous chondrites. Modified after Lee and Halliday (1996).

The resulting correlation on a nickel three-isotope plot (Fig. 15.25) shows that these ^{60}Ni deficits are due to small nucleosynthetic stable isotope variations, as shown previously in Section 15.2. The observed Fe/Ni ratios in these samples would only generate variations of about 1 ppm in radiogenic ^{60}Ni for a ^{60}Fe/^{56}Fe ratio of 3×10^{-9}, which is below analytical resolution. On the other hand, nucleosynthetic effects of around 10 ppm (0.1 ε) are too small to significantly affect the Fe–Ni pseudochron for samples with high Fe/Ni ratios (Fig. 15.25). The consequence is that the initial nickel isotopic heterogeneity is effectively decoupled from the use of Fe–Ni as a chronometer.

15.6.4 Hf–W

^{182}Hf decays to ^{182}W (tungsten) by double β decay with a half-life of 8.9 Ma. Norman and Schramm (1983) proposed the Hf–W system as an *r*-process chronometer, but its application was delayed by the technical difficulties of tungsten isotope analysis (whose high ionization potential is similar to osmium, Section 8.1).

In a sense, this system is a mirror image of Pd–Ag, since for Hf–W the daughter product is siderophile and the parent lithophile. This fact, coupled with its reasonably long half-life, promised a way of measuring the age of core formation in the earth. Thus, if the Earth's core separated while ^{182}Hf was still alive (less than 60 Ma after CAIs), the resulting Hf/W-enriched mantle should have developed a W isotope ratio more radiogenic than chondrites (which did not undergo core formation). This represents a two-stage Hf–W model for the Earth. On the other hand, if the Earth's core separated after Hf was extinct, the mantle would have the same W isotope ratio as chondrites (one-stage model).

Unfortunately, the low W abundance of chondrites made them difficult to analyse, so early work by N-TIMS focussed

on comparing the W isotope composition of the Earth with iron meteorites (Harper et al., 1991; Harper and Jacobsen, 1996). The results showed significant depletions of ^{182}W in iron meteorites relative to terrestrial rocks, which were verified by Lee and Halliday (1995) using MC–ICP–MS (Fig. 15.26). Thus, sawn blocks from iron meteorites (as well as metal phases from ordinary chondrites) had εW values clustering around −4 relative to the Silicate Earth.

In principle, such variations could be explained by cosmogenic production as meteorites travelled through space. However, the negative ^{182}W anomaly in iron meteorites should be more difficult to produce in this way than an enriched signature. Therefore, the logical inference was that the planetessimal cores represented by iron meteorites segregated much earlier than the Earth's core. However, dating the actual time of terrestrial core formation requires comparison with the bulk composition of the solar system.

Harper and Jacobsen (1996) did not have ^{182}W data on chondrites to use as a benchmark for solar system W isotope evolution (their paper was accepted for publication before they knew of the results of Lee and Halliday, 1995). Instead, they estimated the initial solar system ^{182}Hf/^{180}Hf value from a predicted supernova production ratio of 2×10^{-5}. This implied that the present day chondritic tungsten

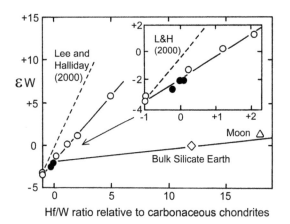

Fig. 15.27 Hf–W pseudochron for metal–silicate separates from two chondrites (open symbols) and two carbonaceous chondrites (closed symbols). Inset shows the lower part of the chondrite pseudochron. After Yin *et al.* (2002).

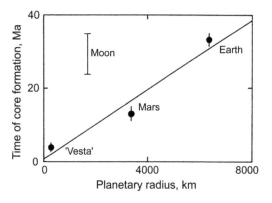

Fig. 15.28 Plot of estimated times of core formation for solar system bodies, compared to their measured or estimated radii. After Kleine *et al.* (2002).

composition, and hence the Bulk Earth, was essentially the same as iron meteorites (star labelled 'H & J' in Fig. 15.26). Based on this assumption, the measured 4 ε unit excess in the Silicate Earth implied that terrestrial Hf/W fractionation occurred while ^{182}Hf was still alive, due to very early core formation. In contrast, Lee and Halliday (1995) made direct measurements of W isotopes in two bulk carbonaceous chondrite samples (Allende and Murchison), obtaining radiogenic ε^{182}W values similar to the Silicate Earth (points labelled 'L & H' in Fig. 15.26). This led to the opposite conclusion from Harper and Jacobsen (1996), namely that the Earth's core segregated relatively *late*, after ^{182}Hf had become extinct.

Lee and Halliday (2000) attempted to validate their relatively radiogenic bulk chondrite data by determining internal Hf–W pseudochrons on three ordinary chondrites. The pseudochrons appeared to support the earlier data, with slopes indicating an initial ^{182}Hf/^{180}Hf ratio of ca. 2×10^{-4}. However, subsequent work by three different research groups showed the chondrite data of Lee and Halliday to be erroneous (Kleine *et al.*, 2002; Schoenberg *et al.*, 2002; Yin *et al.*, 2002). The latter two groups obtained metal-silicate pseudochrons for four different ordinary chondrites (Dhurmsala, Dalgety Downs, Forest Vale and Ste Marguerite), with initial ^{182}Hf/^{180}Hf ratios of ca. 1×10^{-4} and ε^{182}W values of around −2. Both groups also found that whole-rock carbonaceous chondrites plotted on the metal–silicate pseudochrons (Fig. 15.27). Finally, Yin *et al.* determined a CAI composition lying on the same chondrite pseudochron, suggesting that a ^{182}Hf/^{180}Hf ratio of around 1×10^{-4} is also the best estimate for the initial composition of the solar nebula.

Estimating Hf/W = 12 × chondritic in the Bulk Silicate Earth (BSE) at an εW value of zero, a two-point pseudochron can be drawn between the BSE and the average chondritic composition (ε^{182}W = −2). This pseudochron yields an initial ^{182}Hf/^{180}Hf ratio for the BSE at the average time of core formation. The calculated value of 1.1×10^{-5}, when coupled with a starting composition in the solar nebula of 1×10^{-4}, led to an estimated age of terrestrial core formation of ca. 30 Ma after CAI. This assumes that core formation occurred in a single event, corresponding to a simple two-stage model. An estimated lunar composition also plotted near the same pseudochron (Fig. 15.27), suggested that the Moon is approximately the same age as terrestrial core formation.

Similar calculations were made for the time of core formation in the asteroid Vesta and for Mars, based on the inferred tungsten compositions of the silicate fractions of these bodies. These were determined by analysing the compositions of eucrites (usually attributed to Vesta) and SNC meteorites (attributed to fragments ejected from Mars during impact events). Based on these analyses, Kleine *et al.* (2002) argued that the time of core formation in Vesta, Mars and the Earth was consistent with their planetary radii (Fig. 15.28).

When the estimated formation age of the Moon is plotted in Fig. 15.28, it falls far off the correlation line, since it has the same apparent Hf–W age as the Earth's core, despite the Moon's small radius. This is consistent with the giant impact model for the origin of the Moon, involving a collision between the proto-Earth and an impactor (often called Theia) around the size of Mars (e.g. Hartman, 1986). Therefore, when speaking of the age of the Moon, we are estimating the time between CAI formation and this giant impact.

15.6.5 Hf–W Solar System Chronometry

The Hf–W system has developed as one of the most powerful methods for dating early solar system evolution. This is because its half-life and chemistry are applicable to studying the origins of a range of different solar system objects. For example, the refractory nature of both elements ensures good closed-system behaviour except at the highest temperatures (Touboul *et al.*, 2009), whereas the strong partition of the daughter into the metal phase allows it to date very early planetessimal differentiation processes (Kruijer *et al.*, 2014b).

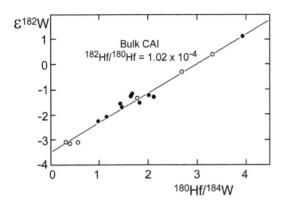

Fig. 15.29 Hf–W pseudochron for CAI from CV3 carbonaceous chondrites Allende, NWA 6717 and NWA 6870. (●) = coarse-grained; (○) = fine-grained. Data were fractionation-normalized using $^{186}W/^{183}W$ ratios corrected for nucleosynthetic effects. After Kruijer et al. (2014a).

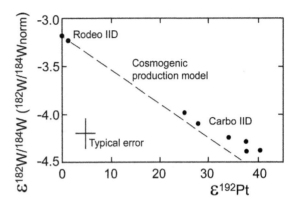

Fig. 15.30 Plot of $\varepsilon^{182}W/^{184}W$ against $\varepsilon^{192}Pt$, demonstrating good agreement of measured data with a cosmogenic neutron irradiation model (dashed line). After Kruijer et al. (2013).

As usual, chronometry begins with the analysis of CAIs, which were the focus of major studies by Kleine et al. (2005a), Burkhardt et al. (2008) and Kruijer et al. (2014a). The slope of the CAI Hf–W pseudochron then forms the basis for relative dating of achondrites, whereas the intercept (εW) anchors the dating of (Hf-free) iron meteorites. Although the results have varied slightly, the most recent CAI data (Fig. 15.29) are close to the mean of the earlier data.

The newest study also took account of varying nucleosynthetic W isotope abundances in CAIs. This is an important issue, because CAIs may be less homogenized than later solar system objects. This was done by comparing coarse-grained and fine-grained CAIs, which have different susceptibility to this problem, allowing quantitative corrections to be developed. The work also compared the effects of nucleosynthetic anomalies on alternative W isotope mass-fractionation correction procedures. The result was a solar system initial $^{182}Hf/^{180}Hf$ ratio of 1.02 ± 0.04 and initial εW = −3.5 ± 0.1. The significance of this result for determination of the Δ period for ^{182}Hf is discussed in Section 15.10.

Core formation in the iron meteorite parent bodies can be dated from their W isotope compositions relative to the CAI initial. However, cosmogenic isotope production turned out to be more significant than expected in iron meteorites, which were originally thought to be immune to these effects. This problem became apparent when Kleine et al. (2005a) determined the first high-precision Hf–W pseudochron for Allende CAIs, yielding an initial εW value of −3.5 ± 0.2. Relative to this value, the measured εW values of several magmatic iron meteorites (with εW values from −3.5 to −4) implied improbable 'negative' ages for iron meteorite differentiation relative to CAIs.

Rejecting this interpretation, Kleine et al. proposed that εW values in these meteorites had been reduced by ^{182}W 'burn-out' due to neutron activation by galactic cosmic ray gen-

erated thermal neutrons. Since many iron meteorites have exposure ages of hundreds of Ma, this could lead to significant ^{182}W burn-out. The fact that these same bodies have younger Pt-Ag and Mn-Cr ages relative to CAI supports this interpretation (Schersten et al., 2006). Previous experience with terrestrial cosmogenic isotopes implies substantial penetration depths for such cosmogenic neutrons. This was confirmed by the modelling of Qin et al. (2008), which suggested a peak burn-out penetration depth of around 70 cm for ^{182}W in iron meteorites. In contrast, non-magmatic IAB iron meteorites apparently do not display this problem (Schulz et al., 2009).

It might be anticipated that ^{182}W burn-out would be accompanied by changes in the abundance of other W isotopes, allowing the effect to be monitored and corrected. However, it seems that normalization for analytical mass fractionation largely cancels out these effects, preventing internal correction of ^{182}W burn-out (Kleine et al., 2005a). One response to this problem was to use rare gas isotopes such as ^3He and ^{21}Ne as exposure indicators, either to correct for cosmogenic effects (Markowski et al., 2006) or to identify favourable meteorites with exposure ages of less than 50 Ma (Kruijer et al., 2012). For example, Muonionalusta (IVA iron group) is estimated to have an exposure age <6 Ma.

A problem with this approach is that deeply buried samples from large iron meteorites can be shielded from rare gas cosmogenic effects, but still be susceptible to ^{182}W burnout. Therefore, Kruijer et al. (2013) used platinum isotopes to monitor and correct for cosmogenic ^{182}W burn-out. They demonstrated a strong correlation between $^{182}W/^{184}W$ and two Pt isotope monitors (Fig. 15.30). The correlation was particularly important for IID iron meteorites, since these meteorites previously had the earliest apparent ages of 0.7 +/− 2.7 Ma before CAIs (Qin et al., 2008). The W–Pt cosmogenic exposure line was anchored by the Rodeo IID iron, and the corrected data for Carbo were in very good agreement

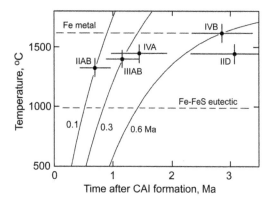

Fig. 15.31 Ages of various iron meteorite groups relative to CAI, based on exposure-corrected Hf–W data, relative to estimated core melting temperatures. Numbers on curves are accretion ages. After Kruijer et al. (2014b).

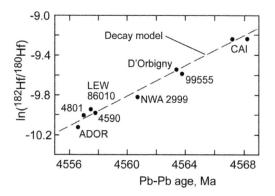

Fig. 15.32 Plot of log initial $^{182}Hf/^{180}Hf$ ratio against Pb–Pb ages (corrected for variable U isotope ratios). Dashed line is the Hf decay curve based on a 8.9 Ma half-life. After Kleine et al. (2012).

with a predicted correlation based on neutron cross-sections (Fig. 15.30).

Application of this method to additional magmatic iron meteorite groups by Kruijer et al. (2014b) gave apparent core segregation ages of 0.7 to 3 Ma after CAI, assuming a single core forming event (Fig. 15.31). Taking into account the chemistry of the iron meteorite groups, their variable sulphur contents can explain some of these variations, due to their different melting temperatures (movement up each curved line in Fig. 15.31). The high apparent age of the sulphur-rich IID group (Fig. 15.31) can be attributed to episodic core formation, during which higher Hf/W ratios in the unsegregated planetessimal caused more rapid W isotope evolution. However, it is necessary to postulate a later accretion age for the sulphur-poor IVB group. The 1.5 Ma core segregation age of the IVA meteorite Muonionalusta compares well with its Pb–Pb age of 4565.3 ± 0.1 (2 Ma after CAI) determined by Blichert-Toft et al. (2010). The difference can be attributed to the lower closure temperature of the U–Pb system relative to Hf–W.

While this work was going on, other studies were aimed at Hf–W dating of the silicate fraction of differentiated planetessimals. Angrite meteorites were of particular interest because they cooled quickly, yielding precise Pb–Pb ages. 'Quenched' angrites (e.g. D'Orbigny and Sahara 99555) yield Pb–Pb ages of around 4563–4 Ma, dating early primary differentiation, whereas metamorphic angrites (e.g. Angra dos Reis, Lewis Cliff 86010) yield Pb–Pb ages of around 4557–8 Ma, probably dating a later collisional heating event (these ages were corrected for variable U isotope ratios, as discussed in Section 5.1.2).

Three detailed studies of this problem (Markowski et al., 2007; Kleine et al., 2009, 2012) gave coherent Hf–W ages, yielding a good correlation with Pb–Pb ages (using D'Orbigny as an anchor). This is demonstrated in Fig. 15.32 by plotting the initial $^{182}Hf/^{180}Hf$ ratio (derived from pseudochron slopes) against Pb–Pb ages. The log scale provides an excellent

fit to the slope of a predicted decay curve based on a half-life of 8.9 Ma (Kleine et al., 2012), therefore demonstrating homogeneous ^{182}Hf distribution in the solar nebula. This correlation appears better than a corresponding one between Al–Mg and Pb–Pb ages (Section 15.5.3). However, the longer half-life of ^{182}Hf makes the correlation less sensitive to possible small discrepancies in the Pb–Pb ages of these meteorites (Section 5.3.2).

15.6.6 Revisiting the Giant Impact Model

Several factors require a re-examination of the age of terrestrial core formation and the giant Moon-forming impact. One of these is that Hf–W determinations of the age of core formation have been shown to be much more model dependent than initially realized (e.g. Allegre et al., 2008). Realistic models of core formation recognize that this process occurred over a significant period of time (perhaps tens of millions of years) in response to progressive addition of impactor material to the terrestrial nucleus. However, there are two end-member models that can describe the accretion process (e.g. Kleine et al., 2009).

In the 'equilibration' model, each impactor is largely vaporized by the collision, so that its internal Hf–W systematics are homogenized to a chondritic composition. In that case, late additions to the Earth would equilibrate with Earth's mantle and lower the evolving (radiogenic) W isotope composition of the mantle back towards the chondritic value. This type of evolution is demonstrated by the zig-zag 'proto-Earth mantle' line in Fig. 15.33. In this model, the 30 Ma estimate for core formation is the average age of these additions.

Subsequent work by Konig et al. (2011) suggested that the Hf/W ratio of the silicate Earth had been underestimated. Using new estimates from terrestrial mantle models, it was inferred that the Earth has a Hf/W ratio 20% higher than the previous estimate. This increases the apparent formation age of the Earth's core to 38 Ma after CAI. This is an

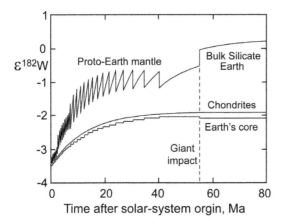

Fig. 15.33 Model of terrestrial εW against time showing a possible evolution path involving equilibration of early accreted material to the silicate mantle, followed by a late giant impact involving minimal equilibration of the impactor core with the silicate Earth. After Halliday (2004).

effective *minimum* age of core formation in models involving late accretion. However, Hf–W dating of the *last* core addition event, corresponding to the giant Moon-forming impact, is highly model dependent.

For example, if a late addition is fully differentiated, and its core is accreted directly to the Earth's core without interacting with the mantle, the impactor's mantle can either raise or lower the W isotope composition of the Earth's mantle, depending on the impactor's own core separation age. In the most likely scenario of early core formation in the impactor, accretion of its mantle would significantly increase the εW composition of Earth's mantle (Fig. 15.33). However, this model can only fit the W isotope composition of the Earth if previous impacts reached much higher equilibration factors, thus holding down the earlier W evolution of the mantle (e.g. Halliday, 2004). A representative example of this type of W isotope evolution model in shown in Fig. 15.33.

Better constraints on the giant impact model come from the W isotope composition of the Moon itself. Early work by Lee *et al.* (1997) claimed W isotope variations amongst different lunar rocks, implying lunar differentiation while ^{182}Hf was still alive. However, it soon became apparent (Lee *et al.*, 2002) that the relatively Ta-rich silicate samples analysed in the early work were affected by cosmogenic W isotope production from thermal neutron activation and β decay of Ta on the Moon's surface. Analysis of Ta-poor metal samples avoided this problem, showing minimal variation in W isotope composition (Kleine *et al.*, 2005b). However, even these samples were susceptible to a smaller cosmogenic W burn-out problem, as seen in iron meteorites.

One of the samples analysed by Lee *et al.* (1997) was a ferroan anorthosite with minimal exposure history, which gave a non-terrestrial signature. However, when this sample was re-analysed by Touboul *et al.* (2009), the new data were indistinguishable from terrestrial. These findings were surprising, because it was not expected that the giant impact would have given rise to similar W isotope signatures in the terrestrial and lunar mantles. Even if the Earth and Moon accreted within the same region of the solar system, thereby inheriting similar stable isotope signatures, it would be anticipated that their accretion and core-separation histories would have generated distinct W isotope compositions (e.g. Fig. 15.33). Therefore, if the terrestrial and lunar mantles inherited the same W isotope composition, the inference is that they were homogenized by the collision itself (Halliday, 2012).

Two alternative models that yield this result were shown in Fig. 15.4. A collision between two equally sized objects leads to homogenization between them. Alternatively, the collision of a small impactor with a fast counter-spinning Earth derives the Moon from Earth's mantle. Presently there is no evidence to distinguish between these models, and they both seem to require rather specific collision conditions. However, this requirement should be balanced against the unlikely coincidence that the Earth and Theia both developed independently with similar W isotope ratios.

More recent work has revealed a small (ca. 20–30 ppm) excess in the ^{182}W isotope abundance of the Moon, relative to Earth (Kruijer *et al.*, 2015; Touboul *et al.*, 2015). However, this discrepancy is now attributed to different accretion rates of late cosmic veneer to each body. It is assumed that the giant impact generated identical W isotope compositions in the mantles of the Earth and Moon, but that the greater gravitational pull of the Earth caused greater subsequent accretion of material with chondritic W isotope signatures, thus lowering the terrestrial W isotope composition by a greater amount. This brings models of mantle W isotopes into line with mantle osmium (Section 8.3.1), and has been supported by recent osmium isotope studies of lunar rocks (Day and Walker, 2015).

15.7 The Sm–Nd System

^{146}Sm is the longest lived of the extinct nuclides, with a half-life of 103 Ma (a proposal by Kinoshita *et al.*, 2012 to revise this value was refuted by Marks *et al.*, 2014). The extant lifetime of a few hundred million years would have been long enough to reach into the period of early differentiation of the silicate Earth. Therefore, if strong Sm/Nd fractionation occurred during early crust formation, variations in the abundance of the daughter product, ^{142}Nd, might be detected in Early Archean rocks. However, the relatively long half-life also means that ^{142}Nd abundance variations are in the low ppm range for most materials, placing extreme demands on analytical precision.

15.7.1 Early Work

Small ^{142}Nd isotopic variations were detected in early studies of the achondrite, Angra dos Reis (Lugmair and Marti, 1977). These were confirmed in more detailed studies of the

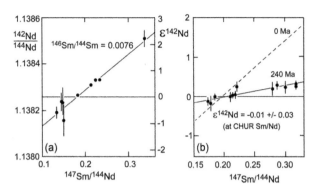

Fig. 15.34 [142]Nd pseudochrons for (a) mineral and whole-rock samples of the Lewis Cliff 86010 angrite; (b) lunar rocks yielding an estimated age of 240 Ma after CAI for crystallization of the lunar magma ocean. Modified after Nyquist et al. (1994, 1995).

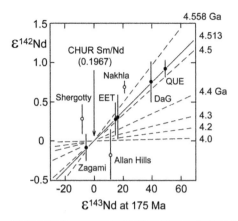

Fig. 15.35 [142]Nd pseudochron against [143]Nd/[144]Nd for Martian meteorites, yielding an estimated age of 4.513 Ga for Martian mantle differentiation. After Borg et al. (2003)

Lewis Cliff 86010 angrite (Lugmair and Galer, 1992; Nyquist et al., 1994) which produced a strong pseudochron array (Fig. 15.34a). Based on a Pb–Pb age of 10 Ma after CAI, the pseudochron defines a solar system initial ^{146}Sm/^{144}Nd ratio of 0.008, which can be used to anchor ages for younger objects.

Large ^{142}Nd anomalies were found in the FUN Allende inclusions EKE141 and C1 (McCulloch and Wasserburg, 1978). These inclusions show a wide variety of isotopic anomalies, so the Nd isotope variations were not particularly surprising. However, these results have become important for understanding much more recent work, to be discussed below.

Much smaller ^{142}Nd variations were observed in lunar rocks (Nyquist et al., 1995), and these samples have also suffered more serious cosmogenic neutron activation effects than most meteorites, due to their long exposure on the surface of the Moon. However, Nyquist et al. showed that these effects could be accurately corrected, since ^{149}Sm has a much larger neutron cross-section than other Sm and Nd isotopes, making it a sensitive radiation monitor. After small cosmogenic corrections, lunar ε^{142}Nd data formed a good pseudochron array against ^{147}Sm/^{144}Nd, giving an age of 240 ± 50 Ma after CAI (Fig. 15.34b). This was interpreted as the crystallization age of the lunar magma ocean. The pseudochron crossed the chondritic Sm/Nd value near ε = 0 (relative to terrestrial standards), as expected from the common origin of the Earth–Moon system.

Similar results were obtained on shergottite meteorites, which (along with Chassigny and the nakhlites) are interpreted as fragments of Mars that were ejected by impact processes and travelled through space to the Earth. Several studies reported ^{142}Nd analyses for individual shergottites, and these were summarized by Borg et al. (2003). Because shergottites represent Martian igneous rocks of relatively young age, their ε^{142}Nd values cannot be plotted directly against

^{147}Sm/^{144}Nd ratios to generate a ^{142}Nd pseudochron. However, because variations of ^{147}Sm/^{144}Nd ratio in the Martian mantle gave rise to time-integrated variations in ^{143}Nd/^{144}Nd, the latter can be used as a proxy for ^{147}Sm/^{144}Nd.

In Fig. 15.35, ε^{142}Nd is plotted against the calculated initial ^{143}Nd/^{144}Nd for each Martian rock. The good correlation observed for several shergottites (excluding the type example) suggests that at least part of the Martian mantle had a simple two-stage history. This could be represented by a brief Martian magma ocean, followed by global differentiation. The slope of the pseudochron was used to determine an age of 54 ± 30 Ma after CAI, which has been supported by more recent studies (32 ± 7 Ma by Debaille et al., 2007; 40 ± 18 Ma by Caro et al., 2008 and 63 ± 6 Ma by Borg et al., 2016).

The first claim for an old terrestrial rock with anomalous ^{142}Nd (relative to terrestrial standards) was made by Harper and Jacobsen (1992), based on analysis of a sample from Isua, western Greenland. The anomaly was at the limits of analytical precision of the instruments available at that time (such as the Finnigan MAT 262), but was verified in another Isua sample by Sharma et al. (1996). However, progress in the field was revolutionized a few years later by the development of the Finnigan Triton TIMS instrument, which improved analytical reproducibility to the 5 ppm level, allowing routine accurate analysis of ^{142}Nd (Section 2.4.3). Nevertheless, there was one unfortunate 'innovation' in the design of the Triton instrument, whereby a small electromagnet was placed in the ion collimation source. This caused small mass-dependent biases in Nd analysis at the 10 ppm level (Caro et al., 2006). The effect was discovered by its relatively larger effect on ^{150}Nd signals, and was initially corrected by double-normalizing ^{142}Nd abundances using the ^{150}Nd signal. After the cause was discovered and the magnet removed, the effect disappeared and analytical accuracy came into line with precision, potentially reaching as low as ± 2 ppm (Caro et al., 2006).

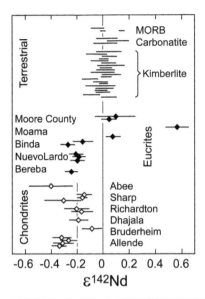

Fig. 15.36 Comparison of ^{142}Nd data in different meteorite classes with terrestrial data. After Boyet and Carlson (2005).

15.7.2 Chondrites and the Bulk Earth

Chondritic meteorites have been used as a reference for the Bulk Earth composition since the earliest Sm–Nd isotope studies (Section 4.1.1), because lithophile elements are not expected to be fractionated by core segregation. Therefore the discovery of negative ^{142}Nd anomalies in chondrites and basaltic eucrites (relative to terrestrial standards) came as a major surprise (Boyet and Carlson, 2005). In fact, Nyquist *et al.* (1995) had previously detected hints of such an anomaly, but the new data of Boyet and Carlson (2005) showed a robust deviation of chondritic ^{142}Nd from the terrestrial value, with a large negative anomaly (−30 ppm) in the Allende carbonaceous chondrite, and smaller negative anomalies (averaging −20 ppm) in several ordinary chondrites (Fig. 15.36).

Because negative ^{142}Nd anomalies were observed for several different types of chondrites, Boyet and Carlson attributed the positive ^{142}Nd signal in terrestrial rocks to a deviation of the Earth's mantle from a chondritic composition, caused by early terrestrial differentiation. This would imply that most of the mantle is Nd depleted relative to Bulk Earth, which should be balanced by the existence of a hidden Nd-enriched reservoir, presumably located at the base of the mantle.

The relatively rapid decay of ^{146}Sm in the early Earth places very strict limits on the time of this hypothetical fractionation event: the later the event is placed, the larger its magnitude required to explain the ^{142}Nd data. Furthermore, a global Sm/Nd fractionation event would also affect the long-lived ^{147}Sm/^{143}Nd system. Thus, Boyet and Carlson calculated that a 20 ppm enrichment of terrestrial ^{142}Nd relative to chondrites would require a fractionation event within 30 Ma of CAI, otherwise it would produce a mantle reser-

voir more depleted than the MORB source. This is problematical, because mantle differentiation must have occurred after the proposed Moon-forming giant impact, which would have remixed any earlier mantle differentiation. This therefore provides a very narrow window for the timing of these events.

15.7.3 The ^{142}Nd Conundrum

Notwithstanding these difficulties, Boyet and Carlson (2006) searched for any signs of an early Nd-enriched reservoir, which might be expected to cause positive ^{142}Nd anomalies in rocks derived from deep mantle plumes. However, no such signal was discovered. Similar investigations of modern terrestrial reservoirs by other workers (Andreasen *et al.*, 2008; Jackson and Carlson, 2012) also failed to reveal any signs of a hidden enriched reservoir. Therefore, in response to these problems, three additional models have been proposed to explain the ^{142}Nd–^{143}Nd isotope systematics of the Earth.

The first model, proposed by Andreasen and Sharma (2006) is that ^{142}Nd abundance variations in the Earth and chondrites do not reflect time-integrated effects of Sm/Nd fractionation, but nucleosynthetic abundance variations inherited by the early solar system. It was shown in Fig. 1.8 (Section 1.2.2) that Sm and Nd isotopes are the products of three different nucleosynthetic processes, *s*, *r* and *p*. Therefore, variable contributions of these processes to different parts of the solar nebula could in principle cause Sm and Nd isotopic heterogeneities.

Most Sm and Nd isotopes are produced by both the *r* and *s* processes, but the heaviest Sm and Nd isotopes (including ^{148}Nd and ^{150}Nd) lie off the *s*-process pathway, and are therefore *r*-only nuclides. On the other hand, ^{142}Nd itself is shielded from *r*-process production, and is therefore almost entirely an *s*-process product. Finally, the two lightest Sm isotopes, ^{144}Sm and ^{146}Sm, lie on the low-mass side of the *s*-process pathway and are also shielded from *r*-process production. Hence, these nuclides can only be made by the *p*-process. Because ^{142}Nd has two components, a non-radiogenic *s*-only component and the radiogenic product of a *p*-process parent, its abundance is particularly sensitive to mixing of different nucleosynthetic sources.

These complexities were borne out when Andreasen and Sharma (2006) found distinct ranges of ε^{142}Nd in carbonaceous and ordinary chondrites, suggesting that these bodies might sample different nucleosynthetic components. This observation was later extended by the analysis of enstatite chondrites (Gannoun *et al.*, 2011). To assess the effect of variable *p*-process synthesis of its ^{146}Sm parent nuclide, ^{142}Nd data are plotted in Fig. 15.37 against another *p*-process nuclide, ^{144}Sm. A weak positive correlation between ε^{142}Nd and ε^{144}Sm is observed in bulk meteorite samples, which can be explained by the effect of variable *p*-process ^{146}Sm abundances on that of its ^{142}Nd daughter (Gannoun *et al.*, 2011). If E-chondrite analyses with large ^{144}Sm errors are excluded (open symbols in Fig. 15.37) then the meteorite array passes close to the Earth–Moon system, suggesting that

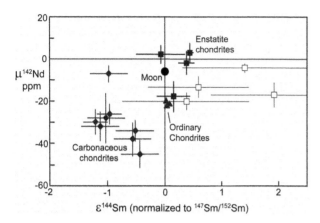

Fig. 15.37 Nd–Sm isotope plot showing the composition of different whole-rock chondrite analyses relative to terrestrial standards. Open symbols denote analyses with large uncertainties in ^{144}Sm. Modified after Gannoun et al. (2011).

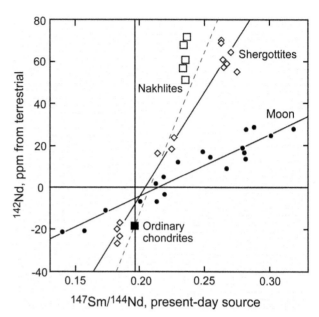

Fig. 15.38 ^{142}Nd pseudochron diagram showing a regression for Shergottite meteorites that intersects with the terrestrial ^{142}Nd composition to the right of the chondritic Sm/Nd value. Lunar data available at that time are also shown. Modified after Caro et al. (2008).

heterogeneities in p-process components might explain the differences between the Earth and chondrites.

This work implies that mixing between distinct nucleosynthetic components could explain the variable signatures of different chondrite groups, but it does not explain why the Earth lies at the extreme upper end of the chondrite array. To address this question, it is important to compare the Earth with other major bodies such as the Moon and Mars. However, the data of Borg et al. (2003) and Debaille et al. (2007) were not of sufficient precision to see small variations in ^{142}Nd. Therefore, Caro et al. (2008) presented new high-precision ^{142}Nd data for 16 Martian meteorites. The ^{147}Sm/^{144}Nd ratios for the Martian mantle were deduced from initial^{143}Nd/^{144}Nd data, yielding the pseudochron plot in Fig. 15.38.

The Martian data in Fig. 15.38 show considerable scatter, indicative of open-system behaviour in the Martian mantle source. However, this would have had most effect on the ^{147}Sm/^{144}Nd system, since ^{142}Nd was extinct by the time the igneous rocks were formed. Therefore, Caro et al. (2008) argued that the most depleted (highest) Sm/Nd ratios were most reliable, leading to the regression line shown. However, unlike the earlier work (Fig. 15.35), the new shergottite pseudochron array did *not* pass through the terrestrial ^{142}Nd composition ($\varepsilon = 0$) at a chondritic Sm/Nd ratio (CHUR), but at a super-chondritic Sm/Nd ratio. In addition, the compilation of lunar data available at that time suggested that the Moon might also intersect the terrestrial ^{142}Nd composition at the same super-chondritic Sm/Nd ratio as Mars. Therefore Caro et al. suggested that all inner solar system bodies shared a supra-chondritic Sm/Nd value that had elevated their bulk ^{142}Nd compositions above the chondritic value, due to some form of elemental Sm/Nd fractionation in the solar nebula. This implied that a super-chondritic Earth model (SCHEM) should also define the Sm/Nd ratio of the Bulk Earth.

15.7.4 SCHEM or Chondritic Moon

The supra-chondritic Earth model has profound implications for understanding the geochemical evolution of the Earth (see Sections 4.2.1 and 6.2.3). However, strong challenges to the model have come from comparative Nd–Hf isotope studies of solar system objects (Bouvier et al., 2008; Sprung et al., 2013). To compare the CHUR and SCHEM models, Bouvier et al. first needed to refine the CHUR composition, based on Sm/Nd and Lu/Hf analysis of less thermally disturbed (petrographic type 1–3) meteorites. A variety of solar system bodies were then plotted relative to the new CHUR composition (Fig. 15.39).

Both the revised CHUR and SCHEM points lie on the Nd–Hf mantle array of modern oceanic volcanics (Fig. 15.39). However, it is difficult to make definitive arguments about the Bulk Earth Sm/Nd ratio from terrestrial rocks because of the complexities of geochemical cycling in the Earth. In fact this was the reason for the importance of the CHUR reference value in the first place (Section 6.2.1). The Martian array also falls close to both the CHUR and SCHEM points. However, lunar rocks plot in a fan-shaped array whose focus is at the CHUR point, along with eucrite meteorites. Hence these data are more consistent with the CHUR model than SCHEM.

A potential problem with the lunar data in Fig. 15.39 is that these rocks had a long residence on the surface of the Moon, where they were subject to cosmogenic isotope production. Sprung et al. (2013) showed that this has a

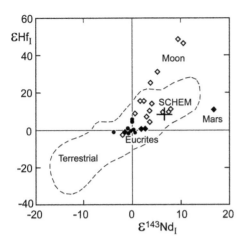

Fig. 15.39 Hf versus Nd isotope plot for a variety of solar system objects: ($\Diamond$) = Moon; ($\blacklozenge$) = Mars; ($\bullet$) = eucrites; (+) = SCHEM. After Bouvier *et al.* (2008).

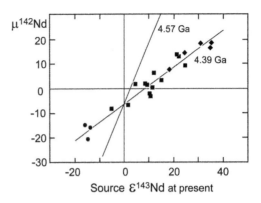

Fig. 15.41 Nd pseudochron plot for lunar samples showing a compilation of high-precision data. ($\blacklozenge$) = High Ti; ($\blacksquare$) = low Ti; ($\bullet$) = KREEP samples. After McLeod *et al.* (2014).

significant effect on Hf isotope analyses of lunar basalts. However, after making corrections for cosmogenic disturbance, the revised data (Fig. 15.40) still defined the same overall pattern shown by Bouvier *et al.* (2008). The data for the low-Ti basalts are a particular problem for the SCHEM model, since they have Nd isotope signatures considerably less radiogenic than SCHEM for a corresponding Hf signature. In other words, the SCHEM source is too incompatible-element-depleted to produce these Nd isotope signatures. Sprung *et al.* (2013) also tested the ability of mixing models to generate the low-Ti lunar basalt source from SCHEM, but this was difficult, even using complex mixing models such as lunar mantle metasomatism.

This evidence re-opened the problem of explaining the ^{142}Nd signatures of the Earth and Moon, and specifically,

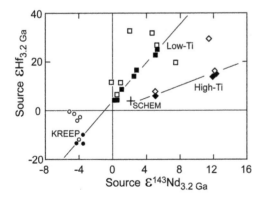

Fig. 15.40 Hf–Nd isotope plot for lunar basalts. Open symbols = raw data; solid = corrected for cosmogenic production on the lunar surface. Modified after Sprung *et al.* (2013).

the evidence cited by Caro *et al.* (2008) that the lunar ^{142}Nd pseudochron intersects the terrestrial ($\varepsilon = 0$) composition at a supra-chondritic Sm/Nd ratio. However, the earlier lunar data were not of high enough precision to clearly test this assertion. Therefore, a new set of high-precision lunar ^{142}Nd data were measured by McLeod *et al.* (2014). This showed that the main lunar basalt types all plot on a single pseudochron (Fig. 15.41) whose slope (assuming a simple two-stage model) yields a revised age for differentiation of the lunar magma ocean of 4.39 Ga (180 Ma after CAI). More importantly, the data yield improved precision on the initial ^{142}Nd ratio of the Moon for a chondritic Sm/Nd ratio. This value falls at around −6 to −7, depending on the proxy that is used for the Sm/Nd ratio of lunar basalt source reservoirs. For example, Dickin (2016) obtained the lowest scatter on the pseudochron by plotting the ^{142}Nd data of McLeod against initial ε^{143}Nd values at 3.3 Ga, leading to a ^{142}Nd value for the Bulk Moon of −5.9 ± 1.5 ppm (2σ) relative to terrestrial standards.

In view of the evidence cited in Section 15.6.6 that the Earth and Moon have almost identical tungsten isotope signatures, despite the unpredictable nature of W isotope growth, it is concluded that the Bulk Earth and the Bulk Moon also share the same Nd composition, with a chondritic Sm/Nd ratio, but negative ^{142}Nd. This signature can be explained by the enstatite chondrite Earth model (Section 15.2), which suggests that the Earth–Moon system was formed in a similar part of the solar nebula to enstatite chondrites.

If the Earth had an initial ^{142}Nd similar to the Bulk Moon, then the +6 ppm ^{142}Nd composition of the accessible mantle must be due to Sm/Nd fractionation in the early Earth, as proposed by Boyet and Carlson (2005). However, since the required magnitude of this fractionation is approximately three times less than the 20 ppm offset relative to carbonaceous and ordinary chondrites originally proposed, this avoids the extreme aspects of the Boyet and Carlson model. The two models are compared in Fig. 15.42. In the

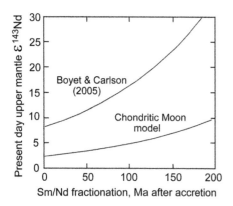

Fig. 15.42 Comparison of modelled Nd evolution paths for the depleted (accessible) mantle of Boyet and Carlson (2005) compared with a 'chondritic Moon' model (Dickin, 2016).

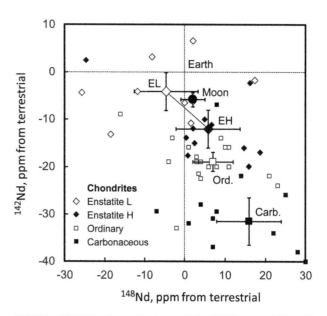

Fig. 15.43 Plot of ppm deviations of $^{142}Nd/^{144}Nd$ versus $^{148}Nd/^{144}Nd$ from terrestrial standards for individual chondrites, means for each chondrite group, and the Moon. Error bars are 2 SEM on means. Data from Gannoun et al. (2011), Boyet and Gannoun (2013), Burkhardt et al. (2016).

model of Boyet and Carlson (2005), the Sm/Nd fractionation event must occur within 30 Ma of CAI in order that the composition of the depleted mantle does not exceed the $\varepsilon^{143}Nd$ value of the MORB source. However, as noted above, this creates extreme difficulties, because it is before the giant Moon-forming impact event.

The new model allows Sm/Nd fractionation to be delayed until 150 Ma after CAI, consistent with the expected crystallization of the global terrestrial magma ocean several tens of Ma after the giant impact. The effect of this differentiation event is still to generate a composition for the accessible mantle that is close to the FOZO component (Section 6.5.1). Hence this model still has dramatic consequences for global geochemical modelling of the Earth (as discussed in several sections of this book). However, the hidden enriched reservoir generated by this model will have a much less extreme ^{142}Nd composition than that predicted by Boyet and Carlson (2005), which explains why this signature has not been detected within the analytical noise of ^{142}Nd analyses of ocean island plume sources.

15.7.5 ^{142}Nd, Core Sulphide and E-Chondrites

Wohlers and Wood (2015) argued that a sulphide phase in the core could also enhance the partition of lanthanide and actinide elements into the core, and hence fractionate Sm/Nd ratios. Such a process during Earth accretion would affect the $^{146}Sm-^{142}Nd$ extinct nuclide system, and might therefore explain the offset of terrestrial $^{142}Nd/^{144}Nd$ ratios from chondrites, the conundrum discussed above. However, the close agreement between the tungsten isotope compositions of the Earth and Moon strongly suggests that the Moon sampled the mantle of the early Earth during the giant Moon-forming impact, placing tight constraints on the timing of core partition processes.

Numerical models of the collision (e.g. Ćuk and Stewart, 2012) typically suggest that the core of the impactor

descended within an hour or so to join the Earth's core. Because of the rapidity of this process, it is not feasible that significant lithophile isotope partition into Earth's core could occur after the impact. Therefore, the distinct ^{142}Nd signatures of the Moon and the accessible Earth must be attributed to Sm/Nd fractionation in the mantle, as proposed by Dickin (2016).

Additional evidence that the Moon sampled the primordial ^{142}Nd of the early Earth comes from recent ^{142}Nd data for Martian meteorites (Borg et al., 2016). These data yield a better Martian pseudochron with a precise ^{142}Nd value of -6 ± 2 ppm (relative to terrestrial standards) at a chondritic Sm/Nd value. This is identical to the ^{142}Nd signature calculated above for the Moon, and provides evidence for a uniform ^{142}Nd composition in the terrestrial planets.

It is possible that limited Sm/Nd fractionation due to core partition could have occurred during the original accretion of both the Earth and Mars. However, the extent of this partitioning may be constrained by comparing lunar and meteorite ^{142}Nd signatures. Chondrites have variable $^{142}Nd/^{144}Nd$ ratios that correlate negatively with $^{148}Nd/^{144}Nd$ (Fig. 15.43). This is best explained by nucleosynthetic heterogeneities inherited from the solar nebula (Gannoun et al., 2011; Boyet and Gannoun, 2013; Burkhardt et al., 2016). A tie-line between the average compositions of EH and EL chondrites passes about 9 ppm below the terrestrial point, but only 3 ppm below the Bulk Moon $^{142}Nd/^{144}Nd$ ratio (assuming chondritic Sm/Nd). On this plot, ε $^{148}Nd/^{144}Nd$ for the

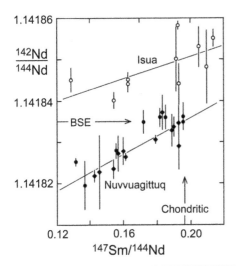

Fig. 15.44 ^{142}Nd isochron diagram for amphibolites from Isua (open symbols) and Nuvvuagittuq (solid symbols). Modified after Roth *et al.* (2013).

Moon is expected to be zero, since the Moon should have the same composition as the Earth for non-radiogenic Nd. Its measured ε ^{148}Nd/^{144}Nd value of +2 ppm is within analytical error of zero.

Because analytical uncertainties for ^{142}Nd/^{144}Nd analysis of chondrites are larger than for lunar rocks, the lunar composition in Fig. 15.43 lies well within two sigma error limits of E-chondrites. Therefore, there is no requirement that Sm/Nd fractionation occurred during accretion to allow the Moon's ^{142}Nd signature to deviate from the E-chondrite tie-line. It is possible that limited Sm/Nd fractionation did occur, but this would be about four times less than that modelled by Wohlers and Wood (2015). This shows that if small ^{142}Nd discrepancies are substantiated between the Moon, Mars and E-chondrites, then these can easily be accommodated by small core-partition effects during the growth of the proto-Earth, Theia and Mars.

15.7.6 ^{142}Nd in the Archean Earth

Eoarchean (>3.5 Ga) terrestrial rocks can potentially also shed light on the Earth's very early differentiation history, but because the Earth is much more geologically complex than the Moon or Mars, the data are difficult to interpret. ^{142}Nd signatures respectively more and less radiogenic than terrestrial standards have been obtained from the Isua and Nuvvuagittuq supracrustal sequences of Greenland and northern Quebec (Fig. 15.44). These deviations from the terrestrial standard value are typically attributed to early terrestrial differentiation, but the age of the fractionation event has been calculated in two different ways.

Harper and Jacobsen (1992) and Caro *et al.* (2003) used combined ^{142}Nd–^{143}Nd data to estimate a very early differentiation age of 4.46 ± 0.1 Ga (ca. 100 Ma after solar sys-

tem initial) for Isua supracrustals. On the other hand, O'Neil *et al.* (2008) used a Sm/Nd pseudochron (Fig. 15.45) to calculate a much younger age of 4.28 Ga (ca. 300 Ma after solar system initial) for the Nuvvuagittuq Supracrustal Belt (using the standard 103 Ma ^{146}Sm half-life).

O'Neil *et al.* (2008) interpreted their pseudochron as an isochron that dated the actual age of the Nuvvuagittuq supracrustals, despite the fact that ^{143}Nd isochron ages and U–Pb ages for these rocks cluster at around 3.75 Ga. A comparison with lunar basalt data confirms that the interpretation of O'Neil *et al.* is problematic. Lunar basalts give a very good correlation between ^{142}Nd and measured ^{147}Sm/^{142}Nd ratios (McLeod *et al.*, 2014). This correlation is not as strong as the lunar ^{142}Nd–^{143}Nd correlation (Fig. 15.41), but it is much stronger than the Nuvvuagittuq correlation line in Fig. 15.44 (Roth *et al.*, 2013). However, there is no suggestion that the lunar ^{142}Nd pseudochron is dating the actual age of lunar basalts. Instead, it is interpreted as a (lunar) mantle isochron, probably dating the crystallization of the lunar magma ocean.

A best-fit regression to the Isua data in Fig. 15.44 does not pass close to the Bulk Earth composition, ruling out its interpretation as a valid ^{146}Sm/^{142}Nd mantle isochron. On the other hand, the regression through the mafic Nuvvuagittuq samples *does* pass close to the estimated Bulk Silicate Earth composition, making it a possible mantle isochron candidate. This allows an estimate to be made of the ^{142}Nd signature of the accessible mantle at the regression age of the isochron (4.28 Ga). In view of the relatively low slope of the regression (compared to Mars and the Moon), it is insensitive to the Sm/Nd ratio of the source, and cannot constrain this source as chondritic or super-chondritic. However, it does constrain the ^{142}Nd signature of the source, suggesting that by 4.28 Ga, this was the same as modern terrestrial standards. As expected, this confirms that the terrestrial mantle Sm/Nd fractionation event must have occurred very early in Earth history.

15.8 The Curium–Uranium–(Nd) System

^{247}Cm (curium) decays to ^{235}U with a half-life of 16 Ma. The significance of this species is that like other actinides it is only formed by the *r*-process, but unlike them it has a comparatively short half-life. Hence, the existence of ^{247}Cm in the early solar system would indicate relatively late additions of *r*-process material to the nebula (Blake and Schramm, 1976).

The clear detection of extinct ^{247}Cm is particularly difficult, because its daughter product ^{235}U is itself a long-lived extinct nuclide. However, whereas the overall uranium isotope ratio in solar system materials speaks of long-term nucleosynthesis during the life of the Galaxy, small *variations* in the relative abundance of the two uranium isotopes in different solar system materials (indicative of the decay of ^{247}Cm) could be used to estimate Δ for r-process nucleosynthesis.

The search for ^{247}Cm therefore consists of looking for very small variations in uranium isotope composition. Unfortunately, ^{235}U is normally used as an enriched isotope in U abundance determinations by isotope dilution, and hence most laboratories are susceptible to artificial perturbations in this ratio. Perhaps for this reason, all early claims for excess ^{235}U signatures were suspect (e.g. Arden, 1977). In an attempt to conclusively resolve this problem, Chen and Wasserburg (1981) made a very careful investigation of U isotope compositions in Allende CAIs and in the phosphate phase whitlockite from the St Severin ordinary chondrite. The data were corrected for mass fractionation during analysis using a double 233–236 uranium spike (Section 2.5.2), but all of the measured ^{238}U/^{235}U data were within $\pm 0.4\%$ (4 per mil) of terrestrial standards.

To convert this result into a maximum ^{247}Cm/^{235}U ratio for the solar nebula, the data should be plotted on a pseudochron diagram. However, there is no stable isotope of curium to allow the translation of ^{238}U/^{235}U variations into a ^{247}Cm/^{235}U ratio. Nevertheless, Chen and Wasserburg argued that rare earth elements (REE) represent a reasonable proxy for curium. Whitlockite is normally enriched in REE, but in the absence of measured REE/U ratios, they estimated a Cm/U enrichment factor of 2 in the phosphate phase, and hence estimated a maximum initial ^{247}Cm/^{235}U ratio of 4×10^{-3} in the early solar system. If the nucleosynthetic model of 'continuous' supernova production is applied to ^{247}Cm, it implies a 'steady state' ^{247}Cm/^{235}U ratio of 8×10^{-2} in the Galaxy. Decay to a ratio of 4×10^{-3} then takes just over four half-lives, leading to a minimum Δ value of ca. 70 Ma. Hence, Chen and Wasserburg argued that this places a lower limit on the date of last r-process addition to the solar nebula.

This issue was revisited by Stirling et al. (2005, 2006) using MC–ICP–MS analysis to achieve a ca. 20-fold improvement in the precision of uranium isotope analysis to ± 2 epsilon units (0.2 per mil). They analysed a variety of meteorite whole-rocks, separated minerals from the Tafassasset carbonaceous chondrite, and leachates from Allende and Murchison. Nd concentration data were also obtained as a proxy for stable curium. They found ^{144}Nd/^{238}U ratios as high as 32 in bulk chondrites and 73 in leachates (Nd/U = 300), relative to an estimated cosmic Nd/U ratio of 56 (Newsome, 1995). However, they did not observe any deviations of uranium isotope ratio outside error of a terrestrial in-house standard (Fig. 15.45).

The error limits of the regression in Fig. 15.45 were used to determine the maximum possible ratio for ^{235}U$_{excess}$/^{144}Nd at the present, equal to 2.6×10^{-8}. However, to determine the initial abundance of the curium parent against Nd, the present day abundance of the daughter nuclide must be age-corrected for radioactive decay during the life of the solar system. Hence, multiplying by $e^{\lambda \cdot 235t}$ ($= 90$), yields a ^{247}Cm/^{144}Nd ratio of 2.4×10^{-6}. Assuming that the analysed samples are representative of bulk solar system, the cosmic Nd/U ratio can be used to determine a maximum initial ^{247}Cm/^{235}U ratio

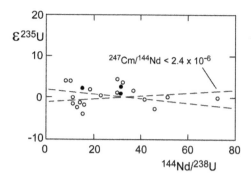

Fig. 15.45 Curium–Uranium–Nd pseudochron plot for bulk samples (●) and leachates (○) of Allende and Murchison. Dashed lines are error limits on the regression. (Low-precision data are omitted.) Modified after Stirling et al. (2006).

of 8×10^{-5}. This implies a Δ value of 170 Ma using the steady state galactic ^{247}Cm/^{235}U ratio quoted above. On the other hand, starting from a fresh supernova production ratio of 0.56 implies a longer value of 230 Ma.

In contrast to Stirling et al., Brennecka et al. (2010) succeeded in finding anomalous uranium isotope ratios in Allende CAIs, and showed that these were correlated with Nd/U (Fig. 15.46). It is generally supposed that CAIs are the most favourable material for this search because they contain anomalies of the short-lived species ^{26}Al and ^{41}Ca. However, it is actually the chemistry of CAIs that makes them a favourable target, since they have Nd/U ratios up to 2000 (^{144}Nd/^{238}U = 500). In these samples, Brennecka observed a positive correlation between ^{235}U and Nd/U ratio, which yields a negative slope when the data are plotted in the form of ^{238}U/^{235}U versus Nd/U ratio (Fig. 15.46). The slope of this correlation implies an initial solar nebula ^{247}Cm/^{235}U ratio of $1.1 \pm 0.2 \times 10^{-4}$.

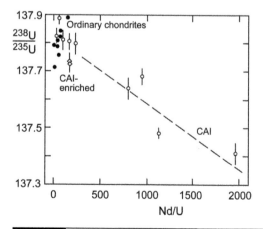

Fig. 15.46 Plot of uranium isotope ratio against Nd/U ratio to test for the presence of ^{247}Cm in the early solar system. Modified after Goldmann et al. (2015).

In contrast, whole-rock analysis of ordinary chondrites (Goldmann *et al.*, 2015) yields a steep correlation line in the opposite sense (Fig. 15.46). This weak reverse correlation could be caused by a variety of isotopic or chemical fractionation processes in the nebula, or even by primary nucleosynthetic uranium production heterogeneity. This noisy background means that the signal of extinct ^{247}Cm is only visible in samples with very high Nd/U ratios (hence explaining the negative result obtained by Stirling *et al.*, 2006). Therefore, the initial ^{247}Cm/^{235}U ratio of 1.1×10^{-4} determined from CAI represents the most reliable value to date, leading to a Δ value of ca. 150 Ma. This will be compared with other Δ values in Section 15.10.

Surprisingly, Brennecka *et al.* (2010) also found that the SRM950a uranium standard previously used to define a terrestrial ^{238}U/^{235}U ratio of 137.88 was anomalous relative to most terrestrial rocks and meteorites, which have ratios near 137.8 (Fig. 15.46). Since the anomalous ratio has been used in most U–Pb and Pb–Pb dating studies, this necessitates a correction of ca. 1 Ma to most meteorite Pb–Pb ages (and lesser corrections to younger terrestrial ages). This affects the absolute ages used to anchor extinct nuclide dating systems (Brennecka and Wadhwa, 2012), and is discussed further in Section 5.1.2.

15.9 Spallogenic Extinct Nuclides

The significance of spallogenic extinct nuclides is that they are best explained by particle interactions in the early solar system rather than inherited nucleosynthetic signatures. If such species are detected, it raises the question of whether other extinct nuclide signatures could be explained by spallation rather nucleosynthetic processes. The most important example is ^{10}Be, whose significance for this argument will be examined.

15.9.1 Be-10

^{10}Be decays to ^{10}B with a half-life of 1.4 Ma (Section 14.4). ^{10}Be is one of the class of nuclides that cannot be produced by stellar nucleosynthesis because it is unstable in stars. Production of ^{10}Be is attributed to spallation reactions involving cosmic rays, termed the 'x process' (Section 1.2.2). Hence, evidence for extinct ^{10}Be in the early solar system would imply that other extinct nuclides might likewise have had spallogenic origins. Therefore, the discovery of 'live' ^{10}Be in the early solar system (McKeegan *et al.*, 2000) raised questions about the canonical model of late incorporation of hot nucleosynthetic material into the solar nebula.

McKeegan *et al.* (2000) observed ^{10}B/^{11}B variations that were positively correlated with Be/B ratios in melilite grains from an Allende CAI. The slope of the array indicated an initial ^{10}Be/^{9}Be ratio of 9.5×10^{-4} at the time of CAI crystallization, which was much too high to be explained by cosmogenic production by exposure to galactic cosmic rays. Therefore, McKeegan suggested that the ^{10}Be signal resulted

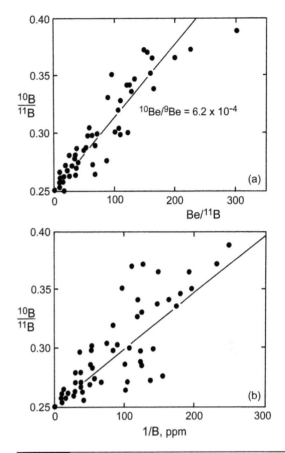

Fig. 15.47 Plots of boron isotope ratio against (a) Be/B ratio and (b) 1/B ratio to evaluate pseudochron versus mixing models for aggregate data from six CAIs. Error bars are omitted for simplicity. Modified after Sugiura *et al.* (2001).

from very intense radiation of the solar nebula early in its history. This raises the possibility that the other very short-lived species such as ^{26}Al, ^{41}K, and ^{53}Mn could also have been produced in this way. Such models have been proposed before (e.g. Lee, 1978), but had been largely discarded due to the success of the 'late addition' model. The discovery of ^{10}Be in the solar nebula requires a re-examination of these models.

Further evidence for the existence of extinct ^{10}Be in the early solar system was provided by Sugiura *et al.* (2001) from the analysis of CAIs from Allende and Efremovka. Each of four inclusions from Efremovka and two from Allende showed positive correlations between ^{10}B/^{11}B and Be/B in analysed melilite grains. The slopes of the arrays varied slightly, but all fell within error of an average ^{10}Be/^{9}Be ratio of 6.2×10^{-4} (Fig. 15.47a). Al–Mg isotope measurements were made in coexisting anorthite from the same inclusions; however, these were scattered, and fell below the 'canonical' ^{26}Al/^{27}Al ratio of 5×10^{-5} previously observed in many CAIs.

This raised the alternative possibility that the Be–B array could be a mixing line.

To test the possibility of a mixing line, Sugiura et al. compared the data on the Be–B pseudochron diagram with a plot of $^{10}B/^{11}B$ against $1/B$ concentration (Fig. 15.47b). A good correlation in the latter diagram would suggest that the data are primarily a younger mixing line between a spallogenic ^{10}B-enriched component and common boron. However, since the pseudochron shows less scatter than the mixing line, the preferred interpretation is that the pseudochron results from a very early nebular homogenization event after spallation reactions were complete, but before ^{10}Be decay (Srinivasan and Chaussidon, 2013).

In order to understand the effect that spallogenic production could have had on other nuclides, it is important to constrain the timing of the irradiation process. However, even ignoring cosmogenic production over the life of the solar system, there are at least three possible production routes: irradiation of a molecular cloud *before* condensation; irradiation of the gas phase of the cloud *during* condensation; irradiation of very early solid phases *after* condensation; or some combination of these processes. Unfortunately, resolving these alternatives has proven difficult.

Chaussidon et al. (2006) claimed to have found correlations between $^7Li/^6Li$ ratios (attributed to ^7Be decay) and Be/Li ratios. Because the half-life of ^7Be is only 53 days, this would imply that ^7Be (and hence ^{10}Be) was produced *in situ*, by irradiation of solids. However, subsequent work suggested that the measured $^7Li/^6Li$ variations resulted from inaccurate correction of cosmogenic production (Leya, 2011). When the corrections were improved, the correlation disappeared.

Another approach to constraining the timing of extinct nuclide production is to examine the isotopic evolution of product nuclides over time. To do this, Srinivasan and Chaussidon (2013) plotted a boron isotope evolution diagram using data on carbonaceous chondrites from several other studies (Fig. 15.48). If the initial $^{10}Be/^9Be$ ratios determined from each pseudochron are attributed age significance, the $^{10}B/^{11}B$ initial ratios appear to show a significant increase with time, which could be produced by closed-system evolution with $^9Be/^{11}B$ ratios between 2 and 10. However, the solar nebula has an inferred $^9Be/^{11}B$ ratio (based on chondrites) of only 0.03, which is far too low to generate this increase.

The more refractory chemistry of Be would have caused early condensed solids to be markedly enriched in $^9Be/^{11}B$ ratio compared with the nebula. Decay of earlier spallogenic ^9Be that was incorporated into these solids could then explain the growth of $^{10}B/^{11}B$ ratios with time, starting from a primordial point represented by hibonite grains from the Isheyevo carbonaceous chondrite (Gounelle et al., 2013). However, this model would require that the Be–B systems in CAI were re-set over a period of up to 3 Ma (Fig. 15.48), in order to generate the Be–B pseudochrons. But if this type of disturbance had happened, CAIs could not have generated the canonical signatures widely seen in the Al–Mg system (which has a similar blocking temperature to the Be–B system).

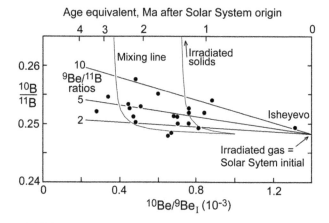

Fig. 15.48 Plot showing alternative growth and mixing models to explain the initial Be isotope ratios of CAIs, based on several Be–B pseudochron diagrams (mostly for Allende and Efremovka). After Srinivasan and Chaussidon (2013).

An alternative model attributes the observed initial ratios to mixing between components derived from irradiated solid and gaseous phases (curved lines in Fig. 15.48). Because the solids have much higher $^9Be/^{11}B$ ratios, they develop much higher $^{10}B/^{11}B$ ratios than the gas phase. In this case the initial $^{10}Be/^9Be$ ratios have no age significance, but reflect different proportions of material from solid and gaseous origins that were mixed very early in the nebula, before the flash-melting events that created most CAI pseudochrons. In this model, the gaseous component could have been irradiated in the molecular cloud shortly before solar system condensation, but this component would be a minority of the total ^{10}B inventory.

15.9.2 Ca–K

^{41}Ca decays to ^{41}K with a half-life of only 0.1 Ma. Hence, if ^{41}K excesses were found in solar system material they would imply a very late addition of nucleosynthetically 'hot' material to the solar nebula. Such anomalies might be expected in CAIs displaying ^{26}Al signatures. Begemann and Stegmann (1976) sought ^{41}Ca signatures in Allende samples, and believed they had found them. However, subsequent work by Hutcheon et al. (1984) attributed this signal to $(^{40}Ca^{42}Ca)^{++}$ dimers, creating a peak which could not be resolved in mass from ^{41}K.

In subsequent work by Srinivasan et al. (1994, 1996), special precautions were taken to resolve the ^{41}K signal from interfering Ca-based molecular ions. This was demonstrated by observing constant $^{41}K/^{39}K$ ratios in terrestrial samples with Ca/K variations spanning nine orders of magnitude. An ion microprobe was used to analyse pyroxene in CAIs from the Efremovka CV3 carbonaceous chondrite. These inclusions were regarded as an ideal place to search for ^{41}K anomalies because of their unusually fresh petrography

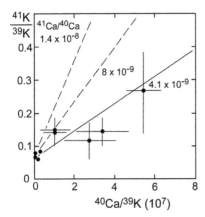

Fig. 15.49 K–Ca pseudochron diagram for pyroxenes in Allende CAIs, indicating a $^{41}Ca/^{40}Ca$ ratio of $4.1 \pm 2 \times 10^{-9}$ in the solar nebula at the time of their crystallization. After Ito et al. (2006).

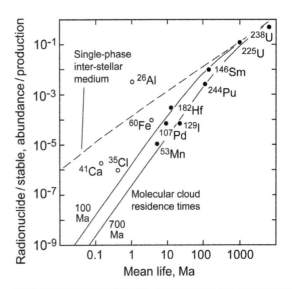

Fig. 15.50 Plots of abundance/production ratios for extinct nuclides as a function of their mean life. Nuclides denoted by filled circles provide an approximate fit to a galactic supernova mixing model (solid lines), whereas open circles, most notably ^{26}Al, show excess abundance. Modified after Young (2014).

and the known presence of ^{26}Al anomalies. The pristine nature of the samples was very important, because ^{41}K anomalies are only visible at extreme Ca/K ratios above 10^5, which would be compromised by re-mobilization of common potassium.

The result of this work was a good Ca–K pseudochron with a slope yielding an initial $^{41}Ca/^{40}Ca$ ratio of 1.4×10^{-8}. However, subsequent studies have indicated that this result was probably biased by inadequate correction of Ca isobaric interferences. Thus, a study of Allende CAIs by Ito et al. (2006) yielded a $^{41}Ca/^{40}Ca$ ratio of $4.1 \pm 2 \times 10^{-9}$ in inclusions with canonical ^{26}Al signatures (Fig. 15.49). Further work on Efremovka CAIs gave slightly lower initial $^{41}Ca/^{40}Ca$ ratios on two inclusions with sub-canonical ^{26}Al, attributed to open-system behaviour during thermal events up to 155 ka after CAI formation (Liu et al., 2012). However, when these data were back-corrected to estimated initial ratios at the time of formation of canonical CAIs, they gave $^{41}Ca/^{40}Ca$ values of ca. 4.1 and 4.6×10^{-9}, close to the result of Ito et al. (2006).

Srinivasan et al. (1996) attributed their $^{41}K/^{39}K$ signal to the presence of live ^{41}Ca in the early solar nebula, rather than recent cosmogenic production in the Efremovka meteorite or ancient bombardment of the nebula by an early active sun. However, the discovery of $^{10}Be/^{9}Be$ signals in the same CAI that previously yielded K isotope anomalies led Srinivasan and Chaussidon (2013) to reconsider this argument. Ironically, the large size of the spallogenic $^{10}Be/^{9}Be$ signals in Efremovka CAIs implies that spallogenic $^{41}K/^{39}K$ anomalies should also be observed, of greater magnitude than the $^{41}K/^{39}K$ ratios actually discovered. This may imply that ^{41}Ca should be moved back into the category of absent species, since it would no longer be seen as the product of pre-solar nucleosynthesis. In contrast, spallogenic production is thought to have had only minor effects on $^{26}Al/^{27}Al$ abundances in CAI.

15.10 Conclusions

To achieve a realistic model for the formation and early evolution of the solar nebula, it is necessary to integrate the results from as many as possible of the extinct nuclide systems discussed above. A useful way of integrating these data is to plot their abundances relative to their half-lives, or preferably, their mean lives. Several authors have presented different variations of this plot, including Wasserburg et al. (1996), Jacobsen (2005) and Huss et al. (2009). The version presented here is modified after Young (2014).

In Fig. 15.50, the y axis plots the abundance of each extinct nuclide relative to a stable nuclide (usually of the same element), divided by the production ratio of the two nuclides. For a simple steady state (single-stage) production model, as described in Section 15.3, the interstellar medium is considered to be continually seeded by supernova ejecta, such that the steady state abundance of each nuclide depends on its mean life. However, the initial solar system abundances of most short-lived extinct nuclides lie well below the single-stage model (dashed line), showing that they were out of nucleosynthetic production for a certain period before solar system coalescence.

As described in various sections above, one approach to this problem has been to propose an ad hoc Δ period for each nuclide between last nucleosynthesis and solar system collapse. However, to avoid these types of granular models, Clayton (1983) proposed a galactic mixing model, in which supernova ejecta were not added directly to the molecular cloud that would form the solar nebula, but *indirectly* via a hot gaseous inter-stellar medium, which mixed gradually with the molecular cloud. The two solid curves in Fig. 15.50 represent the range of mixing times between these reservoirs that can explain extinct nuclide abundances using a steady state supernova model.

This model represents an advance on numerous *ad hoc* models with different Δ periods for different nuclides that have previously been invoked. However, some nuclides lie above the abundances predicted by the mixing model. One of these nuclides (^{36}Cl) that was plotted far above the mixing model by Young (2014) is shown only slightly above it in Fig. 15.50, due to new measurements by Turner *et al.* (2013). Another of the aberrant nuclides (^{41}Ca) was argued above to very likely be the product of spallogenic nuclide production from the early sun. This leaves ^{26}Al as the one nuclide that shows an excess abundance more than three orders of magnitude above its predicted abundance from the supernova mixing model. However, many recent authors, including Young (2014) have attributed this excess to the solar wind of a red giant (AGB star). Such a source could also explain smaller excess abundances of other nuclides such as ^{36}Cl, ^{60}Fe and ^{182}Hf.

The case of ^{182}Hf is particularly interesting because it is separated from the s-processes pathway by the short-lived isotope ^{181}Hf, with a half-life of 42 days (Appendix 1). Such a half-life is long enough to be bridged by the s-process neutron flux. However, previous modelling of the nuclear stability of ^{181}Hf suggested that its half-life was reduced to only 30 hours at the 300 million K core temperature of an AGB star (the s-process stellar site). In contrast, a recent reassessment of pertinent experimental work by Lugaro *et al.* (2014) suggested that the stability of ^{181}Hf is not temperature dependant. Using the 42 day half-life means that the s-process yield of ^{182}Hf could be a factor of five times greater than previously predicted, explaining its excess abundance in Fig. 15.50 relative to the galactic mixing model.

The super-position of an AGB wind onto the supernova mixing model brings extinct nuclide abundances in meteorites into line with normal galactic evolution processes. Molecular clouds are expected to be subjected to the solar winds of AGB stars, and this provides a simple explanation for the compression of a molecular cloud that led to the formation of the solar nebula.

References

Akram W., Schönbächler M., Bisterzo S. and Gallino R. (2015). Zirconium isotope evidence for the heterogeneous distribution of s-process materials in the solar system. *Geochim. Cosmochim. Acta* **165**, 484–500.

Allegre C. J., Manhes G. and Gopel C. (2008). The major differentiation of the Earth at 4.45 Ga. *Earth Planet. Sci. Lett.* **267**, 386–98.

Andreasen R. and Sharma M. (2006). Solar nebula heterogeneity in p-process samarium and neodymium isotopes. *Science* **314**, 806–9.

Andreasen R., Sharma M., Subbarao K. V. and Viladkar S. G. (2008). Where on Earth is the enriched Hadean reservoir? *Earth Planet. Sci. Lett.* **266**, 14–28.

Arden J. W. (1977). Isotopic composition of uranium in chondritic meteorites. *Nature* **269**, 788–9.

Armstrong J. T., Hutcheon I. D., and Wasserburg G. J. (1984). Disturbed Mg isotopic systematics in Allende CAI. In: *Lunar Planet. Sci.* **XV**, Lunar Planet. Inst., 15 (abstract).

Arnett W. D. (1969). Explosive nucleosynthesis in stars. *Astrophys. J.* **157**, 1369–80.

Begemann F., and Stegmann W. (1976). Implications from the absence of ^{41}K anomaly in a Allende inclusion. *Nature* **259**, 549–50.

Bennett C. L., Larson D., Weiland J. L. *et al.* (2013). Nine-year Wilkinson Microwave Anisotropy Probe (WMAP) observations: final maps and results. *Astrophy. J. Supp. Ser.* **208**, 20.

Birck J. L. and Allegre C. J. (1985). Evidence for the presence of ^{53}Mn in the early solar-system. *Geophys. Res. Lett.* **12**, 745–8.

Birck J. L. and Allegre C. J. (1988). Manganese–chromium isotope systematics and the development of the early solar-system. *Nature* **351**, 579–84.

Bizzarro M., Baker J. A., Haack H. and Lundgaard K. L. (2005). Rapid timescales for accretion and melting of differentiated planetesimals inferred from ^{26}Al–^{26}Mg chronometry. *Astrophys. J. Lett.* **632**, L41.

Bizzarro M., Ulfbeck D., Trinquier A. *et al.* (2007). Evidence for a late supernova injection of ^{60}Fe into the protoplanetary disk. *Science* **316**, 1178–81.

Blake J. B. and Schramm, D. N. (1976). A possible alternative to the r-process. *Astrophys. J.* **209**, 846–9.

Blichert-Toft, J., Moynier, F., Lee, C. T. A., Telouk, P. and Albarede, F. (2010). The early formation of the IVA iron meteorite parent body. *Earth Planet. Sci. Lett.* **296**, 469–80.

Borg, L. E., Brennecka, G. A. and Symes, S. J. K. (2016). Accretion timescale and impact history of Mars deduced from the isotopic systematics of martian meteorites. *Geochim. Cosmochim. Acta* **175**, 150–67.

Borg, L. E., Nyquist, L. E., Wiesmann, H., Shih, C. Y. and Reese, Y. (2003). The age of Dar al Gani 476 and the differentiation history of the Martian meteorites inferred from their radiogenic isotopic systematics. *Geochim. Cosmochim. Acta* **67**, 3519–36.

Bouvier, A., Vervoort, J. D. and Patchett, P. J. (2008). The Lu-Hf and Sm-Nd isotopic composition of CHUR: constraints from unequilibrated chondrites and implications for the bulk composition of terrestrial planets. *Earth Planet. Sci. Lett.* **273**, 48–57.

Boyet, M. and Carlson, R.W. (2005). ^{142}Nd Evidence for early (>4.53 Ga) global differentiation of the silicate Earth. *Science* **309**, 576–81.

Boyet, M. and Carlson, R. W. (2006). A new geochemical model for the Earth's mantle inferred from ^{146}Sm-^{142}Nd systematics. *Earth Planet. Sci. Lett.* **250**, 254–68.

Boyet, M. and Gannoun, A. (2013). Nucleosynthetic Nd isotope anomalies in primitive enstatite chondrites. *Geochim. Cosmochim. Acta* **121**, 652–66.

Brazzle, R. H., Pravdivtseva, O. V., Meshik, A. P. and Hohenberg, C. M. (1999). Verification and interpretation of the I Xe chronometer. *Geochim. Cosmochim. Acta* **63**, 739–60.

Brennecka, G. A., Borg, L. E. and Wadhwa, M. (2013). Evidence for supernova injection into the solar nebula and the decoupling of r-process nucleosynthesis. *Proc. Nat. Acad. Sci.* **110**, 17 241–6.

Brennecka, G. A. and Wadhwa, M. (2012). Uranium isotope compositions of the basaltic angrite meteorites and the chronological implications for the early Solar System. *Proc. Nat. Acad. Sci.* **109**, 9299–303.

Brennecka, G. A., Weyer, S., Wadhwa, M. *et al.* (2010). ^{238}U/^{235}U variations in meteorites: Extant ^{247}Cm and implications for Pb-Pb dating. *Science*, **327**, 449–51.

Broecker, W. (1986). *How to Build a Habitable Planet*. Eldigio Press, Columbia University, 291 pp.

Burkhardt, C., Borg, L. E., Brennecka, G. A. *et al.* (2016). A nucleosynthetic origin for the Earth's anomalous ^{142}Nd composition. *Nature* **537**, 394–8.

Burkhardt, C., Kleine, T., Bourdon, B. *et al.* (2008). Hf–W mineral isochron for Ca, Al-rich inclusions: age of the solar system and the timing of core formation in planetesimals. *Geochim. Cosmochim. Acta* **72**, 6177–97.

Burkhardt, C., Kleine, T., Oberli, F. *et al.* (2011). Molybdenum isotope anomalies in meteorites: constraints on solar nebula evolution and origin of the Earth. *Earth Planet. Sci. Lett.* **312**, 390–400.

Cameron, A. G. W. (1962). The formation of the sun and the planets. *Icarus* **1**, 13–69.

Cameron, A. G. W. and Truran, J. W. (1977). The supernova trigger for formation of the solar-system. *Icarus* **30**, 447–61.

Canup, R. (2013). Lunar conspiracies. *Nature* **504**, 27–9.

Caro, G., Bourdon, B., Birck, J. L. and Moorbath, S. (2003). ^{146}Sm–^{142}Nd evidence from Isua metamorphosed sediments for early differentiation of the Earth's mantle. *Nature* **423**, 428–32.

Caro, G., Bourdon, B., Birck, J. and Moorbath, S. (2006). High-precision ^{142}Nd/^{144}Nd measurements in terrestrial rocks: Constraints on the early differentiation of the Earth's mantle. *Geochim. Cosmochim. Acta* **70**, 164–91.

Caro, G., Bourdon, B., Halliday, A. N. and Quitte, G. (2008). Super-chondritic Sm/Nd ratios in Mars, the Earth and the Moon. *Nature* **452**, 336–9.

Champagne, A. E., Howard, A. J., and Parker, P. D. (1983). Nucleosynthesis of ^{26}Al at low stellar temperatures. *Astrophys. J.* **269**, 686–8.

Chaussidon, M., Robert, F. and McKeegan, K.D. (2006). Li and B isotopic variations in an Allende CAI: Evidence for the in situ decay of short-lived ^{10}Be and for the possible presence of the short-lived nuclide ^{7}Be in the early solar system. *Geochim. Cosmochim. Acta* **70**, 224–45.

Chen, J. H., Papanastassiou, D. A. and Wasserburg, G. J. (2009). A search for nickel isotopic anomalies in iron meteorites and chondrites. *Geochim. Cosmochim. Acta* **73**, 1461–71.

Chen, J. H. and Wasserburg, G. J. (1981). The isotopic composition of uranium and lead in Allende inclusions and meteorite phosphates. *Earth Planet. Sci. Lett.* **52**, 1–15.

Chen, J. H. and Wasserburg, G. J. (1984). The origin of excess ^{107}Ag in Gibeon (IVA) and other iron meteorites. In: *Lunar Planet. Sci.* **XV**, Lunar Planet. Inst., 144 (abstract).

Chen, J. H. and Wasserburg, G. J. (1990). The isotopic composition of Ag in meteorites and the presence of ^{107}Pd in proto-planets. *Geochim. Cosmochim. Acta* **54**, 1729–43.

Clayton, D. D. (1975). Extinct radioactivities: trapped residuals of presolar grains. *Astrophys. J.* **199**, 765–9.

Clayton, D. D. (1979). Supernovae and the origin of the solar-system. *Space Sci. Rev.* **24**, 147–226.

Clayton, D. D. (1983). Extinct radioactivities – A three-phase mixing model. *Astrophys. J.* **268**, 381–4.

Clayton, R. N. (1993). Oxygen isotopes in meteorites. *Ann. Rev. Earth Planet. Sci.* **21**, 115–49.

Clayton, R. N., Grossman, L. and Mayeda, T. K. (1973). A component of primitive nuclear composition in carbonaceous meteorites. *Science* **182**, 485–8.

Crabb, J., Lewis, R. S. and Anders, E. (1982). Extinct ^{129}I in C3 chondrites. *Geochim. Cosmochim. Acta* **46**, 2511–26.

Ćuk, M. and Stewart, S. T. (2012). Making the Moon from a fast-spinning Earth: a giant impact followed by resonant despinning. *Science* **338**, 1047–52.

Dauphas, N., Chen, J. H., Zhang, J. *et al.* (2014). Calcium-48 isotopic anomalies in bulk chondrites and achondrites: evidence for a uniform isotopic reservoir in the inner protoplanetary disk. *Earth Planet. Sci. Lett.* **407**, 96–108.

Dauphas, N., Cook, D. L., Sacarabany, A. *et al.* (2008). Iron 60 evidence for early injection and efficient mixing of stellar debris in the protosolar nebula. *Astrophys. J.* **686**, 560.

Dauphas, N., Marty, B. and Reisberg, L. (2002). Inference on terrestrial genesis from molybdenum isotope systematics. *Geophys. Res. Lett.* **29** (6) 1–3.

Day, J. M. and Walker, R. J. (2015). Highly siderophile element depletion in the Moon. *Earth Planet. Sci. Lett.* **423**, 114–24.

Debaille, V., Brandon, A. D., Yin, Q. Z. and Jacobsen, B. (2007). Coupled ^{142}Nd–^{143}Nd evidence for a protracted magma ocean in Mars. *Nature* **450**, 525–8.

Dicke, R. H. (1969). The age of the galaxy from the decay of uranium. *Astrophys. J.* **155**, 123–34.

Dickin, A. P. (2016). The Chondritic Moon: a solution to the ^{142} d conundrum and implications for terrestrial mantle evolution. *Geol. Mag.* **153**, 548–55.

Dodd, R. T. (1969). Metamorphism of the ordinary chondrites: a review. *Geochim. Cosmochim. Acta* **33**, 161–203.

Elkins-Tanton, L. T., Weiss, B. P. and Zuber, M. T. (2011). Chondrites as samples of differentiated planetesimals. *Earth Planet. Sci. Lett.* **305**, 1–10.

Gannoun, A., Boyet, M., Rizo, H. and El Goresy, A. (2011). ^{146}Sm–^{142}Nd systematics measured in enstatite chondrites reveals a heterogeneous distribution of ^{142}Nd in the solar nebula. *Proc. Nat. Acad. Sci.* **108**, 7693–7.

Glavin, D. P., Kubny, A., Jagoutz, E. and Lugmair, G. W. (2004). Mn-Cr isotope systematics of the D'Orbigny angrite. *Meteoritics Planet. Sci.* **39**, 693–700.

Goldmann, A., Brennecka, G., Noordmann, J., Weyer, S. and Wadhwa, M. (2015). The uranium isotopic composition of the Earth and the Solar System. *Geochim. Cosmochim. Acta* **148**, 145–58.

Gopel, C., Birck, J. L., Galy, A., Barrat, J. A. and Zanda, B. (2015). Mn-Cr systematics in primitive meteorites: Insights from mineral separation and partial dissolution. *Geochim. Cosmochim. Acta* **156**, 1–24.

Gopel, C., Manhes, G. and Allegre, C. J. (1994). U-Pb systematics of phosphates from equilibrated ordinary chondrites. *Earth Planet. Sci. Lett.* **121**, 153–71.

Gounelle, M., Chaussidon, M. and Rollion-Bard, C. (2013). Variable and extreme irradiation conditions in the early solar system inferred from the initial abundance of ^{10}Be in Isheyevo CAIs. *Astrophys. J. Lett.* **763**, L33.

Hagee, B., Bernatowicz, T. J., Podosek, F. A. *et al.* (1990). Actinide abundances in ordinary chondrites. *Geochim. Cosmochim. Acta* **54**, 2847–58.

Halliday, A. N. (2004). Mixing, volatile loss and compositional change during impact-driven accretion of the Earth. *Nature* **427**, 505–9.

Halliday, A. N. (2012). The origin of the Moon. *Science* **338**, 4–41.

Harper, C. L. and Jacobsen, S. B. (1992). Evidence from coupled ^{147}Sm–^{143}Nd and ^{146}Sm–^{142}Nd systematics for very early (4.5-Gyr) differentiation of the Earth's mantle. *Nature* **360**, 728–32.

Harper, C. L. and Jacobsen, S. B. (1996). Evidence for ^{182}Hf in the early solar system and constraints on the timescale of terrestrial accretion and core formation. *Geochim. Cosmochim. Acta* **60**, 1131–53.

Harper, C. L., Volkening, J., Heumann, K. G., Shih, C.-Y. and Weismann, H. (1991). ^{182}Hf-^{182}W: new cosmochronometric constraints on terrestrial accretion, core formation, the astrophysical site of the r-process, and the origin of the solar system. *Lunar Planet. Sci.* **XXII**, 515–16.

Hartman, W. K. (1986). Moon origin: the impact-trigger hypothesis. In: Hartman, W. K., Philips, R. J. and Taylor, G. J. (Eds) *Origin of the Moon*, Lunar Planet. Institute, pp. 579–608.

Hohenberg, C. M. (1969). Radioisotopes and the history of nucleosynthesis in the galaxy. *Science* **166**, 212–15.

Hohenberg, C. M., Podosek, F. A. and Reynolds, J. H. (1967). Xenon–iodine dating: sharp isochronism in chondrites. *Science* **156**, 233–6.

Holst, J. C., Olsen, M. B., Paton, C. et al. (2013). ^{182}Hf–^{182}W age dating of a ^{26}Al-poor inclusion and implications for the origin of short-lived radioisotopes in the early Solar System. *Proc. Nat. Acad. Sci.* **110**, 8819–23.

Horan, M. F., Carlson, R. W. and Blichert-Toft, J. (2012). Pd-Ag chronology of volatile depletion, crystallization and shock in the Muonionalusta IVA iron meteorite and implications for its parent body. *Earth Planet. Sci. Lett.* **351**, 215–22.

Huss, G. R., Meyer, B. S., Srinivasan, G., Goswami, J. N. and Sahijpal, S. (2009). Stellar sources of the short-lived radionuclides in the early solar system. *Geochim. Cosmochim. Acta* **73**, 4922–45.

Hutcheon, I. D., Armstrong, J. T., and Wasserburg, G. J. (1984). Excess in ^{41}K in Allende CAZ: confirmation of a hint. In: *Lunar Planet. Sci.* **XV**, Lunar Planet. Inst., 387–8 (abstract).

Hutcheon, I. D. and Hutchison, R. (1989). Evidence from the Semarkona ordinary chondrite for ^{26}Al heating of small planets. *Nature* **337**, 238–41.

Ito, M., Nagasawa, H. and Yurimoto, H. (2006). A study of Mg and K isotopes in Allende CAIs: Implications to the time scale for the multiple heating processes. *Meteoritics Planet. Sci.* **41**, 1871–81.

Jackson, M. G. and Carlson, R. W. (2012). Homogeneous superchondritic ^{142}Nd/^{144}Nd in the mid-ocean ridge basalt and ocean island basalt mantle. *Geochem. Geophys. Geosys.* **13**(6) 1–10.

Jacobsen, B., Yin, Q. Z., Moynier, F. et al. (2008). ^{26}Al–^{26}Mg and ^{207}Pb–^{206}Pb systematics of Allende CAIs: Canonical solar initial ^{26}Al/^{27}Al ratio reinstated. *Earth Planet. Sci. Lett.* **272**, 353–64.

Jacobsen, S. B. (2005). The Hf-W isotopic system and the origin of the Earth and Moon. *Ann. Rev. Earth Planet. Sci.* **33**, 531–70.

Jacobson, S. A., Morbidelli, A., Raymond, et al. (2014). Highly siderophile elements in the Earth's mantle as a clock for the Moon-forming impact. *Nature* **508**, 84–7.

Javoy, M. (1995). The integral enstatite chondrite model of the Earth. *Geophys. Res. Lett.* **22**, 2219–22.

Jeffrey, P. M., and Reynolds, J. H. (1961). Origin of excess ^{129}Xe in stone meteorites. *J. Geophys. Res.* **66**, 3582–3.

Kelly, W. R., and Wasserburg, G. J. (1978). Evidence for the existence of ^{107}Pd in the early solar-system. *Geophys. Res. Lett.* **5**, 1079–82.

Kinoshita, N., Paul, M., Kashiv, Y. et al. (2012). A shorter ^{146}Sm half-life measured and implications for ^{146}Sm–^{142}Nd chronology in the solar system. *Science* **335**, 1614–17.

Kita, N. T., Tenner, T. J., Ushikubo, T. et al. (2015). Why do U–Pb ages of chondrules and CAIs have more spread than their ^{26}Al ages? *78th Ann. Meet. Meteoritical Soc.* **1856**, 5360.

Kleine, T., Hans, U., Irving, A. J. and Bourdon, B. (2012). Chronology of the angrite parent body and implications for core formation in protoplanets. *Geochim. Cosmochim. Acta* **84**, 186–203.

Kleine, T., Mezger, K., Palme, H., Scherer, E. and Munker, C. (2005a). Early core formation in asteroids and late accretion of chondrite parent bodies: Evidence from ^{182}Hf–^{182}W in CAIs, metal-rich chondrites, and iron meteorites. *Geochim. Cosmochim. Acta* **69**, 5805–18.

Kleine, T., Munker, C., Mezger, K. and Palme, H. (2002). Rapid accretion and early core formation on asteroids and the terrestrial planets from Hf-W chronometry. *Nature* **418**, 952–5.

Kleine, T., Palme, H., Mezger, K. and Halliday, A.N. (2005b). Hf-W chronometry of lunar metals and the age and early differentiation of the Moon. *Science* **310**, 1671–4.

Kleine, T., Touboul, M., Bourdon, B. et al. (2009). Hf-W chronology of the accretion and early evolution of asteroids and terrestrial planets. *Geochim. Cosmochim. Acta* **73**, 5150–88.

Konig, S., Munker, C., Hohl, S. et al. (2011). The Earth's tungsten budget during mantle melting and crust formation. *Geochim. Cosmochim. Acta* **75**, 2119–36.

Krot, A. N., Keil, K., Goodrich, C. A., Scott, E. R. D. and Weisberg, M. K. (2005). Classification of meteorites. In: Davis, A. M. (Ed.) *Meteorites, Comets and Planets: Treatise on Geochemistry*, vol. 1. Elsevier.

Kruijer, T. S., Fischer-Godde, M., Kleine, T. et al. (2013). Neutron capture on Pt isotopes in iron meteorites and the Hf-W chronology of core formation in planetessimals. *Earth Planet. Sci. Lett.* **361**, 162–72.

Kruijer, T. S., Kleine, T., Fischer-Gödde, M., Burkhardt, C. and Wieler, R. (2014a). Nucleosynthetic W isotope anomalies and the Hf-W chronometry of Ca-Al-rich inclusions. *Earth Planet. Sci. Lett.* **403**, 317–27.

Kruijer, T. S., Kleine, T., Fischer-Godde, M. and Sprung, P. (2015). Lunar tungsten isotopic evidence for the late veneer. *Nature* **520**, 534–7.

Kruijer, T. S., Sprung, P., Kleine, T. et al. (2012). Hf-W chronometry of core formation in planetesimals inferred from weakly irradiated iron meteorites. *Geochim. Cosmochim. Acta* **99**, 287–304.

Kruijer, T. S., Touboul, M., Fischer-Godde, M. et al. (2014b). Protracted core formation and rapid accretion of protoplanets. *Science* **344**, 1150–4.

Lee, D.-C. and Halliday, A. N. (1995). Hafnium–tungsten chronometry and the timing of terrestrial core formation. *Nature* **378**, 771–4.

Lee, D.-C. and Halliday, A. N. (1996). Hf-W isotopic evidence for rapid accretion and differentiation in the early solar system. *Science* **274**, 1876–9.

Lee, D.-C. and Halliday, A. N. (2000). Hf-W internal isochrons for ordinary chondrites and the initial ^{182}Hf–^{180}Hf of the solar system. *Chem. Geol.* **169**, 35–43.

Lee, D. C., Halliday, A. N., Leya, I., Wieler, R. and Wiechert, U. (2002). Cosmogenic tungsten and the origin and earliest differentiation of the Moon. *Earth Planet. Sci. Lett.* **198**, 267–74.

Lee, D.-C., Halliday, A. N., Snyder, G. A. and Taylor, L. A. (1997). Age and origin of the Moon. *Science* **278**, 1098–103.

Lee, T., Papanastassiou, D. A. and Wasserburg, G. J. (1976). Demonstration of ^{26}Mg excess in Allende and evidence for ^{26}Al. *Geophys. Res. Lett.* **3**, 109–13.

Lee, T., Papanastassiou, D. A. and Wasserburg, G. J. (1977). Aluminum-26 in the early solar-system: Fossil or fuel? *Astrophys. J. Lett.* **211**, L107–10.

Lee, T. (1978). A local proton irradiation model for isotopic anomalies in the solar system. *Astrophys. J.* **224**, 217–26.

Lee, T., Shu, F. H., Shang, H., Glassgold, A. E. and Rehm, K. E. (1998). Protostellar cosmic rays and extinct radioactivities in meteorites. *Astrophys. J.* **506**, 898–912.

Lewis, R. S., Amari, S. and Anders, E. (1990). Meteoritic silicon carbide: pristine material from carbon stars. *Nature* **348**, 293–8.

Leya, I. (2011). Cosmogenic effects on ^{7}Li/^{6}Li, ^{10}B/^{11}B, and ^{182}W/^{184}W in CAIs from carbonaceous chondrites. *Geochim. Cosmochim. Acta* **75**, 1507–18.

Leya, I., Schonbachler, M., Wiechert, U., Krahenbuhl, U. and Halliday, A. N. (2008). Titanium isotopes and the radial heterogeneity of the solar system. *Earth Planet. Sci. Lett.* **266**, 233–44.

Liu, M. C., Chaussidon, M., Gopel, C. and Lee, T. (2012). A heterogeneous solar nebula as sampled by CM hibonite grains. *Earth Planet. Sci. Lett.* **327**, 75–83.

Lugaro, M., Heger, A., Osrin, D. et al. (2014). Stellar origin of the ^{182}Hf cosmochronometer and the presolar history of solar system matter. Science 345, 650–3.

Lugmair, G. W. and Galer, S. J. G. (1992). Age and isotopic relationships among the angrites Lewis Cliff 86010 and Angra dos Reis. Geochim. Cosmochim. Acta 56, 1673–94.

Lugmair, G. W. and Marti, K. (1977). Sm–Nd–Pu timepieces in the Angra dos Reis meteorite. Earth Planet. Sci. Lett. 35, 273–84.

Lugmair, G. W. and Shukolyukov, A. (1998). Early solar system timescales according to ^{53}Mn–^{53}Cr systematics. Geochim. Cosmochim. Acta 62, 2863–86.

MacPherson, G. J., Bullock, E. S., Janney, P. E. et al. (2010). Early solar nebula condensates with canonical, not supracanonical, initial ^{26}Al/^{27}Al ratios. Astrophys. J. Lett. 711, L117–21.

MacPherson, G. J., Kita, N. T., Ushikubo, T., Bullock, E. S. and Davis, A. M. (2012). Well-resolved variations in the formation ages for Ca–Al-rich inclusions in the early Solar System. Earth Planet. Sci. Lett. 331, 43–54.

Mahoney, W. A., Ling, J. C., Wheaton, W. A. and Jacobsen, A. S. (1984). HEAO-3 discovery of ^{26}Al in the interstellar medium. Astrophys. J. 286, 578–85.

Makide, K., Nagashima, K., Krot, A. N. et al. (2013). Heterogeneous distribution of ^{26}Al at the birth of the Solar System: Evidence from corundum-bearing refractory inclusions in carbonaceous chondrites. Geochim. Cosmochim. Acta 110, 190–215.

Markowski, A., Quitte, G., Halliday, A. N. and Kleine, T. (2006). Tungsten isotopic compositions of iron meteorites: chronological constraints vs. cosmogenic effects. Earth Planet. Sci. Lett. 242, 1–15.

Markowski, A., Quitte, G., Kleine, T. et al. (2007). Hafnium–tungsten chronometry of angrites and the earliest evolution of planetary objects. Earth Planet. Sci. Lett. 262, 214–29.

Marks, N. E., Borg, L. E., Hutcheon, I. D., Jacobsen, B. and Clayton, R. N. (2014). Samarium–neodymium chronology and rubidium–strontium systematics of an Allende calcium–aluminum-rich inclusion with implications for ^{146}Sm half-life. Earth Planet. Sci. Lett. 405, 15–24.

McCulloch, M. T. and Wasserburg, G. J. (1978). Barium and neodymium isotopic anomalies in the Allende meteorite. Astrophys. J. 220, L15–19.

McKeegan, K. D., Chaussidon, M. and Robert, F. (2000). Incorporation of short-lived ^{10}Be in a calcium aluminum-rich inclusion from the Allende meteorite. Science 289, 1334–7.

McLeod, C. L., Brandon, A. D. and Armytage, R. M. (2014). Constraints on the formation age and evolution of the Moon from ^{142}Nd–^{143}Nd systematics of Apollo 12 basalts. Earth Planet. Sci. Lett. 396, 179–89.

Mishra, R. K. and Chaussidon, M. (2014a). Timing and extent of Mg and Al isotopic homogenization in the early inner Solar System. Earth Planet. Sci. Lett. 390, 318–26.

Mishra, R. K. and Chaussidon, M. (2014b). Fossil records of high level of ^{60}Fe in chondrules from unequilibrated chondrites. Earth Planet. Sci. Lett. 398, 90–100.

Mishra, R. K. and Goswami, J. N. (2014). Fe–Ni and Al–Mg isotope records in UOC chondrules: Plausible stellar source of ^{60}Fe and other short-lived nuclides in the early Solar System. Geochim. Cosmochim. Acta 132, 440–57.

Moynier, F., Blichert-Toft, J., Wang, K., Herzog, G. F. and Albarede, F. (2011). The elusive ^{60}Fe in the solar nebula. Astrophys. J. 741 (71), 1–6.

Newsome, H. E. (1995). Composition of the solar system, planets, meteorites, and major terrestrial reservoirs. Global Earth Physics. American Geophysical Union, pp. 159–189.

Nichols, R. H., Hohenberg, C. M., Kehm, K., Kim, Y. and Marti, K. (1994). I–Xe studies of the Acapulco meteorite: absolute I–Xe ages of individual phosphate grains and the Bjurbole standard. Geochim. Cosmochim. Acta 58, 2553–61.

Niemeyer, S. (1979). I-Xe dating of silicate and troilite from IAB iron meteorites. Geochim. Cosmochim. Acta 43, 843–60.

Norman, E. B. and Schramm, D. N. (1983). ^{182}Hf chronometer for the early solar system. Nature 304, 515–17.

Nuth, J. (1991). Small grains of truth. Nature 349, 18–19.

Nyquist, L. E., Bansal, B., Wiesmann, H. and Shih, C. Y. (1994). Neodymium, strontium and chromium isotopic studies of the LEW86010 and Angra dos Reis meteorites and the chronology of the angrite parent body. Meteoritics 29, 872–85.

Nyquist, L. E., Kleine, T., Shih, C. Y. and Reese, Y. D. (2009). The distribution of short-lived radioisotopes in the early solar system and the chronology of asteroid accretion, differentiation, and secondary mineralization. Geochim. Cosmochim. Acta 73, 5115–36.

Nyquist, L., Lindstrom, D., Mittlefehldt, D. et al. (2001). Manganese-chromium formation intervals for chondrules from Bishunpur and Chainpur meteorites. Meteoritics Planet. Sci. 36, 911–38.

Nyquist, L. E., Wiesmann, H., Bansal, B. et al. (1995). ^{146}Sm–^{142}Nd formation interval for the lunar mantle. Geochim. Cosmochim. Acta 59, 2817–37.

O'Neil, J., Carlson, R. W., Francis, D. and Stevenson, R. K. (2008). Neodymium-142 evidence for Hadean mafic crust. Science 321, 1828–31.

Papanastassiou, D. A. (1986). Chromium isotopic anomalies in the Allende meteorite. Astrophys. J. 308, L27–30.

Paton, C., Schiller, M. and Bizzarro, M. (2013). Identification of an ^{84}Sr-depleted carrier in primitive meteorites and implications for thermal processing in the solar protoplanetary disk. Astrophys. J. Lett. 763 (L40), 1–6.

Podosek, F. A. (1970). Dating of meteorites by the high-temperature release of iodine-correlated Xe-129. Geochim. Cosmochim. Acta 34, 341–65.

Podosek, F. A., Zinner, E. K., Macpherson, G. J. et al. (1991). Correlated study of initial ^{87}Sr/^{86}Sr and Al–Mg isotopic systematics and petrologic properties in a suite of refractory inclusions from the Allende meteorite. Geochim. Cosmochim. Acta 55, 1083–110.

Qin, L., Alexander, C. M. D., Carlson, R. W., Horan, M. F. and Yokoyama, T. (2010). Contributors to chromium isotope variation of meteorites. Geochim. Cosmochim. Acta 74, 1122–45.

Qin, L., Dauphas, N., Wadhwa, M., Masarik, J. and Janney, P. E. (2008). Rapid accretion and differentiation of iron meteorite parent bodies inferred from ^{182}Hf–^{182}W chronometry and thermal modeling. Earth Planet. Sci. Lett. 273, 94–104.

Quitte, G., Markowski, A., Latkoczy, C., Gabriel, A. and Pack, A. (2010). Iron-60 heterogeneity and incomplete isotope mixing in the early solar system. Astrophys. J. 720, 1215–24.

Regelous, M., Elliott, T. and Coath, C. D. (2008). Nickel isotope heterogeneity in the early Solar System. Earth Planet. Sci. Lett. 272, 330–38.

Reynolds, J. H. (1960). Determination of the age of the elements. Phys. Rev. Lett. 4, 8–9.

Rotaru, M., Birck, J. L. and Allegre, C. J. (1992). Clues to early solar-system history from chromium isotopes in carbonaceous chondrites. Nature 358, 465–70.

Roth, A. S., Bourdon, B., Mojzsis, S. J. et al. (2013). Inherited ^{142}Nd anomalies in Eoarchean protoliths. Earth Planet. Sci. Lett. 361, 50–7.

Rugel, G., Faestermann, T., Knie, K. et al. (2009). New measurement of the ^{60}Fe half-life. Phys. Rev. Lett. 103 (7), 072502.

Russell, S. S., Srinivasan, G., Huss, G. R., Wasserburg, G. J. and MacPherson, G. J. (1996). Evidence for widespread ^{26}Al in the Solar Nebula and constraints for nebular time scales. Science 273, 757–62.

Sahijpal, S., Goswami, J. N., Davis, A. M., Grossman, L. and Lewis, R. S. (1998). A stellar origin for the short-lived nuclides in the early solar system. Nature 391, 559–61.

Salmeron, R. and Ireland, T. R. (2012). Formation of chondrules in magnetic winds blowing through the proto-asteroid belt. *Earth Planet. Sci. Lett.* **327**, 61–7.

Schersten, A., Elliott, T., Hawkesworth, C., Russell, S. and Masarik, J. (2006). Hf–W evidence for rapid differentiation of iron meteorite parent bodies. *Earth Planet. Sci. Lett.* **241**, 530–42.

Schiller, M., Baker, J. A. and Bizzarro, M. (2010). ^{26}Al–^{26}Mg dating of asteroidal magmatism in the young Solar System. *Geochim. Cosmochim. Acta* **74**, 4844–64.

Schiller, M., Connelly, J. N., Glad, A. C., Mikouchi, T. and Bizzarro, M. (2015). Early accretion of protoplanets inferred from a reduced inner solar system ^{26}Al inventory. *Earth Planet. Sci. Lett.* **420**, 45–54.

Schoenberg, R., Kamber, B. S., Collerson, K. D. and Eugster, O. (2002). New W-isotope evidence for rapid terrestrial accretion and very early core formation. *Geochim. Cosmochim. Acta* **66**, 3151–60.

Schonbachler, M., Carlson, R. W., Horan, M. F., Mock, T. D. and Hauri, E. H. (2008). Silver isotope variations in chondrites: volatile depletion and the initial ^{107}Pd abundance of the solar system. *Geochim. Cosmochim. Acta* **72**, 5330–41.

Schramm, D. N. and Wasserburg, G. J. (1970). Nucleochronologies and the mean age of the elements. *Astrophys. J.* **162**, 57–69.

Schulz, T., Munker, C., Palme, H. and Mezger, K. (2009). Hf–W chronometry of the IAB iron meteorite parent body. *Earth Planet. Sci. Lett.* **280**, 185–93.

Seeger, P. A., Fowler, W. A. and Clayton, D. D. (1965). Nucleosynthesis of heavy elements by neutron capture. *Astrophys. J. Supp.* **11**, 121–66

Sharma, M., Papanastassiou, D. A., Wasserburg, G. J. and Dymek, R. F. (1996). The issue of the terrestrial record of Sm-146. *Geochim. Cosmochim. Acta* **60**, 2037–47.

Shu, F. H., Shang, H., Glassgold, A. E. and Lee, T. (1997). X-rays and fluctuating X-winds from protostars. *Science* **277**, 1475–9.

Shukolyukov, A. and Lugmair, G. W. (1993a). Live iron-60 in the early solar system. *Science* **259**, 1138–42.

Shukolyukov, A. and Lugmair, G. W. (1993b). ^{60}Fe in eucrites. *Earth Planet. Sci. Lett.* **119**, 159–66.

Sprung, P., Kleine, T. and Scherer, E. E. (2013). Isotopic evidence for chondritic Lu/Hf and Sm/Nd of the Moon. *Earth Planet. Sci. Lett.* **380**, 77–87.

Srinivasan, G. and Chaussidon, M. (2013). Constraints on ^{10}Be and ^{41}Ca distribution in the early solar system from ^{26}Al and ^{10}Be studies of Efremovka CAIs. *Earth Planet. Sci. Lett.* **374**, 11–23.

Srinivasan, G., Goswami, J. N. and Bhandari, N. (1999). ^{26}Al in eucrite Piplia Kalan: plausible heat source and formation chronology. *Science* **284**, 1348–50.

Srinivasan, G., Sahijpal, S., Ulyanov, A. A. and Goswami, J. N. (1996). Ion microprobe studies of Efremovka CAIs: II. Potassium isotope composition and ^{41}Ca in the early solar system. *Geochim. Cosmochim. Acta* **60**, 1823–35.

Srinivasan, G., Ulyanov, A. A. and Goswami, J. N. (1994). ^{41}Ca in the early solar system. *Astrophys. J. Lett.* **431**, L67–70.

Steele, R. C., Coath, C. D., Regelous, M., Russell, S. and Elliott, T. (2012). Neutron-poor nickel isotope anomalies in meteorites. *Astrophys. J.* **758** (59), 1–21.

Steele, R. C., Elliott, T., Coath, C. D. and Regelous, M. (2011). Confirmation of mass-independent Ni isotopic variability in iron meteorites. *Geochim. Cosmochim. Acta* **75**, 7906–25.

Stirling, C. H., Halliday, A. N. Porcelli, D. (2005). In search of live ^{247}Cm in the early solar system. *Geochim. Cosmochim. Acta* **69**, 1059–71.

Stirling, C. H., Halliday, A. N., Potter, E. K., Andersen, M. B. and Zanda, B. (2006). A low initial abundance of ^{247}Cm in the early solar system and implications for r-process nucleosynthesis. *Earth Planet. Sci. Lett.* **251**, 386–97.

Sugiura, N., Shuzou, Y. and Ulyanov, A. (2001). Beryllium–boron and aluminium–magnesium chronology of calcium–aluminium-rich inclusions in CV chondrites. *Meteoritics Planet. Sci.* **36**, 1397–408.

Tang, H. and Dauphas, N. (2012). Abundance, distribution, and origin of ^{60}Fe in the solar protoplanetary disk. *Earth Planet. Sci. Lett.* **359**, 248–63.

Tang, H. and Dauphas, N. (2014). ^{60}Fe–^{60}Ni chronology of core formation in Mars. *Earth Planet. Sci. Lett.* **390**, 264–74.

Tang, H. and Dauphas, N. (2015). Low ^{60}Fe abundance in Semarkona and Sahara 99555. *Astrophys. J.* **802** (22), 1–9.

Taylor, G. J., Maggiore, P., Scott, E. R., Rubin, A. E. and Keil, K. (1987). Original structures, and fragmentation and reassembly histories of asteroids: Evidence from meteorites. *Icarus* **69**, 1–13.

Theis, K. J., Schonbachler, M., Benedix, G. K. et al. (2013). Palladium–silver chronology of IAB iron meteorites. *Earth Planet. Sci. Lett.* **361**, 402–11.

Thrane, K., Bizzarro, M. and Baker, J. A. (2006). Extremely brief formation interval for refractory inclusions and uniform distribution of ^{26}Al in the early solar system. *Astrophys. J. Lett.* **646** (2), L159.

Touboul, M., Kleine, T., Bourdon, B. et al. (2009). Hf–W thermochronometry: II. Accretion and thermal history of the acapulcoite–lodranite parent body. *Earth Planet. Sci. Lett.* **284**, 168–78.

Touboul, M., Puchtel, I. S. and Walker, R. J. (2015). Tungsten isotopic evidence for disproportional late accretion to the Earth and Moon. *Nature* **520**, 530–3.

Trinquier, A., Birck, J. L. and Allegre, C. J. (2007). Widespread ^{54}Cr heterogeneity in the inner solar system. *Astrophys. J.* **655** (2), 1179.

Trinquier, A., Birck, J. L., Allegre, C. J., Gopel, C. and Ulfbeck, D. (2008). ^{53}Mn–^{53}Cr systematics of the early Solar System revisited. *Geochim. Cosmochim. Acta* **72**, 5146–63.

Trinquier, A., Elliott, T., Ulfbeck, D. et al. (2009). Origin of nucleosynthetic isotope heterogeneity in the solar protoplanetary disk. *Science* **324**, 374–6.

Trivedi, B. M. P. (1977). A new approach to nucleocosmochronology. *Astrophys. J.* **215**, 877–84.

Truran, J. W., and Cameron, A. G. W. (1978). ^{26}Al production in explosive carbon burning. *Astrophys. J.* **219**, 226–9.

Turner, G., Crowther, S. A., Burgess, R. et al. (2013). Short lived ^{36}Cl and its decay products ^{36}Ar and ^{36}S in the early solar system. *Geochim. Cosmochim. Acta* **123**, 358–67.

Villeneuve, J., Chaussidon, M. and Libourel, G. (2009). Homogeneous distribution of ^{26}Al in the solar system from the Mg isotopic composition of chondrules. *Science* **325**, 985–8.

Wallner, A., Bichler, M., Buczak, K. et al. (2015). Settling the half-life of ^{60}Fe: fundamental for a versatile astrophysical chronometer. *Phys. Rev. Lett.* **114** (4), 041101.

Warren, P. H. (2011). Stable-isotopic anomalies and the accretionary assemblage of the Earth and Mars: a subordinate role for carbonaceous chondrites. *Earth Planet. Sci. Lett.* **311**, 93–100.

Wasserburg, G. J. (1985). Short-lived nuclei in the early solar-system. In: Black, D. C. and Matthews, M. S. (Eds) *Protostars and Planets.* University of Arizona Press, pp. 703–37.

Wasserburg, G. J., Busso, M. and Gallino, R. (1996). Abundances of actinides and short-lived non-actinides in the interstellar medium: diverse supernova sources for the r-process. *Astrophys. J. Lett.* **431**, L109–13.

Wasserburg, G. J., Fowler, W. A. and Hoyle, F. (1960). Duration of nucleosynthesis. *Phys. Rev. Lett.* **4**, 112–14.

Wasserburg, G. J. and Hayden, R. J. (1955). Time interval between nucleogenesis and the formation of meteorites. *Nature* **176**, 130–1.

Wasserburg, G. J., Lee, T. and Papanastassiou, D. A. (1977). Correlated O and Mg isotopic anomalies in Allende inclusions: II magnesium. *Geophys. Res. Lett.* **4**, 299–302.

Wasserburg, G. J. and Papanastassiou, D. A. (1982). Some short-lived nuclides in the early solar-system – a connection with the placental ISM. In: Barnes, C. A., Clayton, D. D. and Schramm, D. N. (Eds), *Essays in Nuclear Astrophysics.* Cambridge University Press, pp. 77–140.

Wasson, J. T. (2013). Vesta and extensively melted asteroids: Why HED meteorites are probably not from Vesta. *Earth Planet. Sci. Lett.* **381**, 138–46.

Weisberg, M. K., McCoy, T. J. and Krot, A. N. (2006). Systematics and evaluation of meteorite classification. In: Lauretta, D. S. and McSween, H. Y. (Eds.), *Meteorites and The Early Solar System II.* University of Arizona Press, pp. 19–52.

Wiechert, U., Halliday, A. N., Lee, D. C. *et al.* (2001). Oxygen isotopes and the Moon-forming giant impact. *Science* **294**, 345–8.

Wohlers, A. and Wood, B. J. (2015). A Mercury-like component of early Earth yields uranium in the core and high mantle ^{142}Nd. *Nature* **520**, 337–40.

Yamakawa, A. and Yin, Q. Z. (2014). Chromium isotopic systematics of the Sutter's Mill carbonaceous chondrite: Implications for isotopic heterogeneities of the early solar system. *Meteoritics Planet. Sci.* **49**, 2118–27.

Yin, Q. Z., Jacobsen, B., Moynier, F. and Hutcheon, I. D. (2007). Toward consistent chronology in the early solar system: high-resolution ^{53}Mn–^{53}Cr chronometry for chondrules. *Astrophys. J. Lett.* **662** (1), L43.

Yin, Q., Jacobsen, S. B., Yamashita, K. *et al.* (2002). A short timescale for terrestrial planet formation from Hf–W chronometry of meteorites. *Nature* **418**, 949–52.

Yokoyama, T., Fukami, Y., Okui, W., Ito, N. and Yamazaki, H. (2015). Nucleosynthetic strontium isotope anomalies in carbonaceous chondrites. *Earth Planet. Sci. Lett.* **416**, 46–55.

Young, E. D. (2014). Inheritance of solar short-and long-lived radionuclides from molecular clouds and the unexceptional nature of the solar system. *Earth Planet. Sci. Lett.* **392**, 16–27.

Young, E.E., Simon, J.I., Galy, A. *et al.* (2005). Supra-canonical ^{26}Al/^{27}Al and the residence time of the CAIs in the solar protoplanetary disk. *Science* **308**, 223–7.

Zinner, E., Amari, S., Anders, E. and Lewis, R. (1991). Large amounts of extinct ^{26}Al in interstellar grains from the Murchison meteorite. *Nature* **349**, 51–4.

Chapter 16

Fission-Track Dating

'Fission tracks' are not strictly speaking radiogenic nuclides, but they are the damage tracks left by fission products, which represent a special kind of radiogenic nuclide. As such, the abundance of fission tracks in geological materials increases over time in the same way as radiogenic isotopes. Fission-track dating was first developed as a dating tool for general application. However, the susceptibility of fission tracks to thermal resetting, originally a disadvantage, has been put to very good use as a measure of cooling, uplift or burial processes.

16.1 Track Formation

The spontaneous fission of ^{238}U releases about 200 MeV of energy, much of which is transferred to the two product nuclides as kinetic energy. They travel about 7 μm in opposite directions, leaving a single trail of damage through the medium which is about 15 μm long. Fission fragment tracks were originally observed in cloud chambers and photographic emulsions (e.g. Corson and Thornton, 1939). Subsequently, Silk and Barnes (1959) produced artificial tracks in muscovite by irradiating uranium-coated flakes in a reactor to cause induced fission of ^{235}U. The resulting fragment tracks were observed at high magnification under the electron microscope.

'Fission tracks' (Fleischer et al., 1964) are only found in insulating materials. Fleischer et al. (1965a) proposed that the passage of the charged fission fragment causes ionization of atoms along its path by stripping away electrons (Fig. 16.1a). The positively charged ions then repel each other, creating a cylindrical zone of disordered structure (Fig. 16.1b). This, in turn, causes relaxation stress in the surrounding matrix. It is the resulting 100 Å (10 nm)-wide zone of strain (Fig. 16.1c) which is actually seen under the electron microscope. Conductors do not display fission tracks because the free movement of electrons in their lattice structure neutralizes the charged damage zone.

The ability to generate tracks depends on the mass of the ionizing particle and the density of the medium. In muscovite, the lowest mass particle which can generate tracks by irradiation is about 30 atomic mass units (a.m.u.). Fission fragments, with masses of ca. 90 and 135 a.m.u. respectively, are well above this threshold, so that they always generate tracks. On the other hand, α particles, the major product of uranium decay, are so far below the critical mass that they cannot create tracks. Neither can they cause track erasure (Fleischer et al., 1965b).

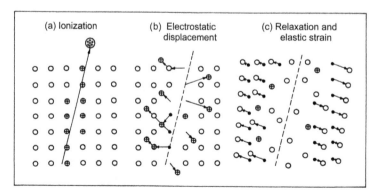

Fig. 16.1 Schematic illustration of the process of formation of a fission track in a crystalline insulating solid. After Fleischer et al. (1975).

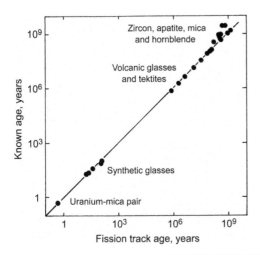

Fig. 16.2 A comparison of specimen ages determined by fission-track analysis with those from historical or other radiometric sources. After Fleischer *et al.* (1965a).

Price and Walker (1962a) showed that when irradiated material was abraded to expose fission tracks at the surface, the damage zone could be preferentially dissolved by mineral acids, leading initially to a very fine channel only 2.5 nm wide. However, this could be enlarged by further chemical etching to yield a wide pit which was observable under the optical microscope. Price and Walker (1962b) first discovered 'fossil' fission tracks in minerals, created by the spontaneous fission of dispersed uranium atoms. They went on to suggest (Price and Walker, 1963) that their density could be used as a dating tool for geological materials up to a billion years old. This was verified by Fleischer *et al.* (1965a) who obtained dates on artificial and natural glasses and minerals which were in agreement with other methods (Fig. 16.2).

Price and Walker (1963) demonstrated that spontaneous fission of ^{238}U was the only significant source of tracks in most natural materials. Induced fission of ^{235}U by natural thermal neutrons can be ignored, as can cosmic ray induced fission of uranium. Spallation recoils induced by cosmic rays could, in principle, generate tracks in geological material exposed at the surface for very long time periods. This is the principal source of tracks in meteorites (e.g. Lal *et al.*, 1969), but atmospheric shielding reduces their abundance to negligible levels in terrestrial rocks (Fleischer *et al.*, 1975). Therefore the total production of spontaneous fission tracks (F_s) per unit volume of rock can be derived from the general decay equation [1.9]:

$$F_s = \frac{\lambda_{\text{fission}}}{\lambda_\alpha} \, ^{238}\text{U} \left(e^{\lambda_\alpha t} - 1\right) \qquad [16.1]$$

The value of the ^{238}U fission decay constant recommended by the International Union of Pure and Applied Chemistry (Holden and Hoffman, 2000) is $8.5 \pm 0.1 \times 10^{-17}$ yr^{-1} ($t_{1/2} = 8.4 \times 10^{15}$ yr). This value ignores erroneous data from before

1984 and has more recently been verified by Yoshioka *et al.* (2005). Although it has a larger uncertainty than many other decay constants, it will be seen below that this uncertainty need not enter into geological age determinations. The fissiogenic decay constant is over a million times lower than the α decay constant of ^{238}U, so it can be ignored in determining the isotopic abundance of uranium through time.

After polishing and etching a surface of the material to be dated, a fraction q of the total tracks will be visible at the surface. Therefore the measured spontaneous fission-track density, ρ_s, will be $q \times F_s$:

$$\rho_s = q . \frac{\lambda_{\text{fission}}}{\lambda_\alpha} \, ^{238}\text{U} \left(e^{\lambda_\alpha t} - 1\right) \qquad [16.2]$$

Price and Walker recognized that the most effective way of measuring the uranium concentration was to irradiate the sample with neutrons in a reactor, and thereby produce artificial tracks by the induced fission of ^{235}U. Based on equation [16.2], the induced track density will be

$$\rho_i = q \, ^{235}\text{U} \, \phi \sigma \qquad [16.3]$$

where ϕ is the thermal neutron flux per unit volume and σ is the cross-section of ^{235}U for induced fission by thermal neutrons. If the sample material, including uranium concentration and etching procedure, is identical for these two experiments, then the ratio of track densities can be used to solve for t, so that q goes out of the equation and the uranium concentrations are replaced by the ^{238}U/^{235}U isotope ratio only:

$$\frac{\rho_s}{\rho_i} = \frac{\lambda_{\text{fission}}}{\lambda_\alpha} . \frac{137.8}{\phi \sigma} \left(e^{\lambda_\alpha t} - 1\right) \qquad [16.4]$$

This can be rearranged to yield an equation in terms of t:

$$t = \frac{1}{\lambda_\alpha} \ln \left[1 + \frac{\rho_s}{\rho_i} . \frac{\lambda_\alpha}{\lambda_{\text{fission}}} . \frac{\phi \sigma}{137.8}\right] \qquad [16.5]$$

It is possible to determine ϕ and σ directly by using flux monitors such as iron wire or copper foil. However, these types of flux monitor may not respond to reactor conditions in exactly the same way as geological material. Therefore, an alternative procedure is to do a fission-track analysis of a standard material with known uranium concentration. Fleischer *et al.* (1965a) used fragments of glass microscope slides to calibrate the Brookhaven graphite reactor in this way. However, this does not avoid the uncertainty of the ^{238}U fission decay constant.

To eliminate both the flux term and the decay constant term, many workers started to use minerals dated by K–Ar as internal standards for the irradiation. Fleischer and Hart (1972) formalized this system into the 'zeta calibration'. A sample of known age is used to calculate ζ by rearranging equation [16.4] and dividing both sides by the track density ρ_d in a given glass dosimeter:

$$\zeta = \frac{\phi \sigma}{137.8 \, \lambda_{\text{fission}}} = \frac{e^{\lambda_\alpha t} - 1}{\rho_d \lambda_\alpha (\rho_s / \rho_i) \rho_d} \qquad [16.6]$$

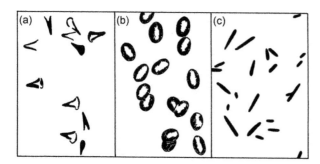

Fig. 16.3 Drawings of etched fission tracks induced by the same source (^{252}Cf) in different materials: (a) K feldspar; (b) soda-lime glass; (c) Lexan polycarbonate. Width of each field is 40 μm. From photographs by Fleischer et al. (1968).

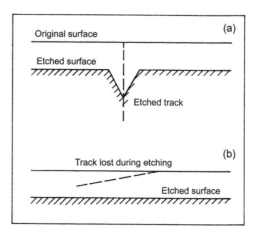

Fig. 16.4 Schematic illustration of the progress of track etching: (a) perpendicular to surface; (b) tangential. After Fleischer and Price (1964b).

To date an unknown sample, the age equation [16.5] is now modified by substitution of ζ:

$$t = \frac{1}{\lambda_\alpha} \ln\left[1 + \frac{\zeta \, \lambda_\alpha \, \rho_s \, \rho_d}{\rho_i}\right] \qquad [16.7]$$

This method transfers uncertainties about the fission decay constant and the physics of the irradiation process into an age determination of the geological reference material. Use of such material was recommended for all fission-track dating studies by a working group of the IUGS Subcommission on Geochronology (Hurford, 1990). One of the most well known of these standards is the 28 Ma-old Fish Canyon Tuff, Colorado (Naeser et al., 1981; Hurford and Green, 1983).

With the determination of a more reliable value for the decay constant, there have been suggestions from some workers that 'standardless' fission-track dating is now feasible (e.g. Jonckheere and Ratschbacher, 2015). This approach attempts to replace the zeta calibration with absolute measurements of the radiation flux used to create induced tracks, or direct measurement of the U content of the sample (Section 16.3.4). However, these alternatives are unlikely to improve the accuracy of fission-track dating, because errors in fission-track analysis are invariably larger than those of other dating methods, such as U–Pb or Ar–Ar, used to determine the absolute ages of fission-track standards.

16.2 Track Etching

Several different types of geological material are suitable for the determination of fission-track ages. Fleischer and Price (1964a) tested them with different acid or alkali leaching solutions to determine the most effective for track observation. The precise progress of the etching process depends on the composition of the matrix and the nature, concentration and temperature of the acid. This can give rise to a surprising variation in the appearance of etched tracks in different materials (e.g. Fig. 16.3), and may affect the accuracy of track counting. These problems were discussed by Fleischer and Price (1964b) in an assessment of the fission-track dating of glass.

The geometry of an etched track depends on the rate of etching down the axis of the track (from its intersection with the surface), relative to the general rate of attack on the polished surface (Fig. 16.4a). One problem in accurate track counting is to distinguish etched tracks from other features. For example, track pits in glass are at first pointed, but with increased etching time they round out. The optimal etching time is then a compromise between the need to make large enough pits to count quickly, and the tendency for large round-bottomed pits to be confused with etched porosity. However, this is not such a problem in mineral phases.

Another source of uncertainty for both glass and mineral dating is caused by tracks which barely register in the etched surface. For example, tracks which are almost tangential to the surface may be completely erased by etching (Fig. 16.4b). Other tracks may not have intersected the original polished surface, but are exposed by the general attack of the surface during etching. These discrepancies will average out statistically if large numbers of tracks are counted with identical spatial geometry (see below), but may cause large errors when spatial geometry varies. A more detailed discussion of track formation and track etching is given by Fleischer et al. (1975).

Fleischer and Price (1964a) estimated the dating range of fission-track analysis with different types of material. Using the criterion that dates of reasonable precision can only be determined when the track density is at least 100 per cm^2, the lower end of the dating range can be estimated for different types of material according to uranium content (Fig. 16.5).

16.3 Counting Techniques

Close examination must now be given to the assumptions involved in fission-track dating. The first of these, noted

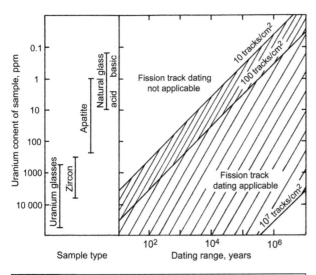

Fig. 16.5 Diagram to show the dating range for fission-track analysis of different kinds of geological material according to uranium content. After Wagner (1978).

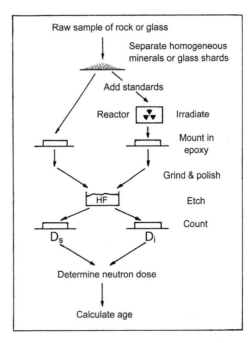

Fig. 16.6 Schematic illustration of the population method of fission-track analysis. After Naeser and Naeser (1984).

above, is that the induced track count is performed on identical material to the spontaneous track count. Several different experimental methods are available which attempt to reach this ideal. Different approaches may be best for different types of sample material.

16.3.1 Population Method

This expression was coined by Naeser (1979a), but was effectively the method adopted by the earliest workers (e.g. Price and Walker, 1963). The term refers to the fact that spontaneous and induced tracks are counted in different splits or sub-populations of material, which are nevertheless assumed to sample the same population. This depends on the material having a homogeneous distribution of uranium between the two splits. The method has proved particularly successful for dating glass and apatite, but unsuccessful for sphene and zircon, where uranium distribution is very variable both within and between grains.

To apply the population method, the sample is separated into two splits (Fig. 16.6). One is irradiated with thermal neutrons along with the standard (flux monitor). Both spontaneous and induced tracks are to be registered under spherical (4π) spatial geometry. Therefore, after irradiation of the induced-track split, both splits are mounted in epoxy, ground, polished and etched under identical conditions. This reveals an internal surface of the material and also removes any extraneous superficial tracks generated by uranium-bearing dust particles. Track densities are counted in both splits. The induced-track density is calculated by subtracting the spontaneous-track density (un-irradiated sample) from the total track density (irradiated sample).

The population method should be statistically tested by counting track densities in numerous grains or glass shards in each split. Alternatively, if a large piece of glass or mineral is available it can be cut or cleaved so that the two faces to be counted are nearly identical sections through the sample. The latter method was adopted by Price and Walker (1963) in their analyses of muscovite. Price and Walker took the extra precaution of irradiating the split for spontaneous-track counting in a cadmium box (which screens out thermal neutrons) so that both splits should be treated as nearly identically as possible prior to etching. However, this precaution has now been dispensed with.

In the analysis of apatite, pre-irradiation heating of the induced-track split has been found advantageous to erase all spontaneous tracks by thermal annealing (see below). This allows the induced-track density to be determined directly in the irradiated split. However, this procedure may be problematical in dating glass because it may affect the etching properties of the irradiated split, leading to systematic track counting errors.

16.3.2 External Detector Method

In this technique (Fleischer et al., 1965a), the uranium content of the material to be dated is determined by inducing counts in an external detector rather than in the sample material itself. The sample is ground, polished, etched and counted, after which a sheet of detector material is placed in intimate contact with the etched surface. This must be done with absolute cleanliness to exclude uranium-bearing dust grains (see above). The external detector is commonly a low-uranium mica or a plastic such as lexan. After irradiation, the external detector is removed from the sample, etched and counted (Fig. 16.7).

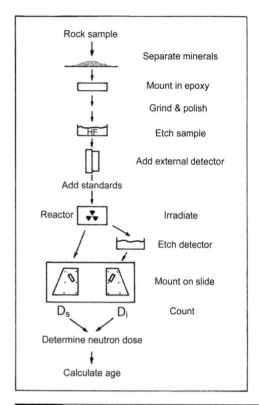

Fig. 16.7 Schematic illustration of the external detector method of fission-track analysis, as described by Naeser (1979a). In this version, the counting of spontaneous tracks is performed after irradiation, unlike the sequence described in the text. After Naeser and Naeser (1984).

etched plane (spherical or 4π geometry). In contrast, tracks induced in the external detector come out from the surface of the analysed material and are therefore generated with approximately one-half the frequency (hemi-spherical or 2π geometry). Reimer *et al.* (1970) questioned whether the efficiency of induced-track formation is exactly 50%, or whether small biases are introduced. However, subsequent experiments (discussed by Hurford and Green, 1982) showed that in most cases the ideal efficiency of 50% is achieved.

16.3.3 Re-Etching and Re-Polishing

The re-etching technique, described by Price and Walker (1963), is similar to the external detector method in that a sample is irradiated *after* polishing, etching and counting of spontaneous tracks. However, the sample itself is now re-etched and re-counted to determine the induced-track density by subtraction. As for the external method, induced tracks are formed with only 50% efficiency (2π geometry). The disadvantage of this method is that spontaneous-track pits will be unduly enlarged after the second etch, and may obscure some induced tracks. It is consequently less popular than the external method.

The re-polishing technique (Naeser *et al.*, 1989) is an improvement on the re-etching method, and yields results similar to the 'mirror image' population method (Price and Walker, 1963). The sample is polished, etched and counted for spontaneous-track density. After irradiation it is re-polished to a depth of at least 20 μm to reveal a new internal face with 4π track geometry. This is then etched and counted to determine the induced-track density by subtraction. The method has the advantage that both spontaneous and induced tracks are recorded under identical geometry, and spontaneous tracks are not over-enlarged by double etching. Also, surface contamination during irradiation is not a problem. The spontaneous and induced tracks are not generated by exactly the same sample material, but the two etched surfaces are so close together that uranium inhomogeneity in the grain as a whole is unlikely to significantly bias the data. A disadvantage compared with the normal population method is that the two etching steps are performed separately, and may therefore vary slightly in efficiency.

The advantage of the external detector method is that both spontaneous and induced tracks are generated by the same sample material. Hence, it is suited to the analysis of material with a very heterogeneous distribution of uranium. The main disadvantage of the method is that the spontaneous and induced tracks are recorded under different spatial geometry conditions (Fig. 16.8). Spontaneous tracks are generated in the interior of the rock, and can therefore be formed by uranium atoms both above and below the

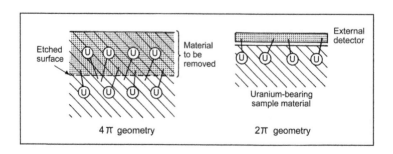

Fig. 16.8 Schematic illustration of the difference between 4π and 2π geometry in track formation.

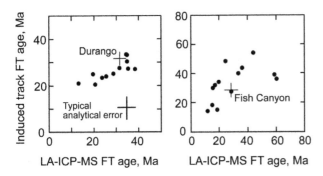

Fig. 16.9 Comparison of fission-track ages using LA–ICP–MS and induced-track counting, on individual apatite grains from the Durango and Fish Canyon age standards. Error estimates omitted for clarity. Modified after Hasebe *et al.* (2004).

16.3.4 LA–ICP–MS

With the widespread availability of *in situ* trace element analysis by laser ablation ICP–MS, this has become a viable alternative to induced-track analysis as a means of determining the U content of fission-track dating materials. U determination in zircon by LA–ICP–MS has now become a routine technique for U–Pb dating (Section 2.7.1), but application to U measurements in apatite has only more recently been explored (Kimura *et al.*, 2000). Nevertheless, these results were promising, showing good agreement in U determinations by *in situ* LA–ICP–MS, solution ICP–MS, and other methods.

A direct comparison of LA–ICP–MS and induced-track counting for fission-track dating was made by Hasebe *et al.* (2004). Results for the Durango and Fish Canyon dating standards are shown in Fig. 16.9. They suggest that ICP–MS data are comparable in quality to induced-track counting, although these data also illustrate the heterogeneity of individual grain fission-track ages, possibly due to the heterogeneous distribution of uranium. The large errors emphasize the fact that the modern application of fission-track ages is in support of thermal history analysis using track lengths, rather than as a dating method in its own right.

16.3.5 Automated Track Counting

Classically, all of the track counting methods described above have been based on human observation and counting of etched tracks by optical microscopy. With the development of artificial intelligence methods of pattern recognition, attempts have been made to automate the laborious process of track counting (e.g. Petford *et al.*, 1993). However, it has been shown in practice that manual correction of automated track counting is necessary in order to achieve accurate age results (e.g. Enkelmann *et al.*, 2012).

A major development of this approach was achieved by Gleadow *et al.* (2009), based on a combination of transmitted and reflected light microscopy. The reflected light image creates a relatively standard response to each etched track,

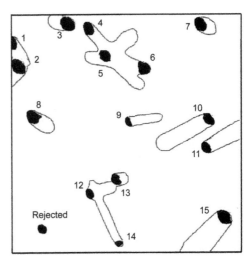

Fig. 16.10 Automated counting of tracks (#1–15) in a ca. 50 μm square of muscovite, based on coincidence of reflected light (black) and transmitted light (white) image outlines. After Gleadow *et al.* (2009).

based on its creation of a low-reflectance pit on the surface. This is then verified as a genuine fission track by comparison with the transmitted light image. This generates a more variable response, depending on the dip of each track relative to the etched surface, but it allows spurious reflected light responses not caused by tracks to be rejected (Fig. 16.10). Gleadow *et al.* tested the method on apatite grains from the 280 Ma Harcourt granodiorite standard, along with a muscovite external detector. The muscovite gave an excellent correlation between automated and manual track counting, whereas the apatite showed a very good (but slightly inferior) correlation.

The approach was tested on Fish Canyon apatite grains by Enkelmann *et al.* (2012). Unfortunately, the agreement between automated track-density measurements was much lower than for manual counting. This suggests that the automated method still has some way to go to reach the reliability of manual counting for spontaneous tracks (as opposed to more easily counted induced tracks in an external detector).

16.4 Detrital Populations

An advantage of the external detector method of fission-track counting is the ability to determine a separate age from each grain of the population (this also applies to the less widely used re-polishing method). This capability is useful if a heterogeneous age population is suspected in the sample, as in the case of sedimentary rocks with mixed provenance (e.g. Hurford and Carter, 1991). However, the scatter of analysis points generated by these kinds of samples can be a challenge when it comes to data presentation.

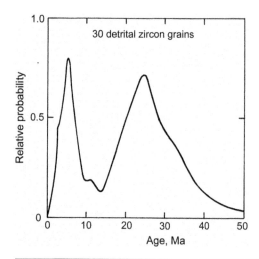

Fig. 16.11 Plot of probability density as a function of age for fission-track data on detrital zircons from the re-worked El Ocote tephra from Mexico. After Kowallis et al. (1986).

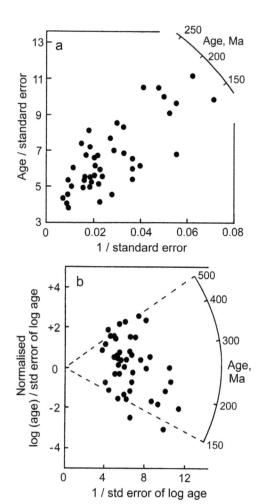

Fig. 16.12 Variations on the radial plot for presentation of single grain fission-track data on heterogeneous samples: (a) raw data; (b) normalized. Modified after Galbraith (1988).

Fission-track results for individual detrital grains may be presented in histogram form. However, a more quantitative age estimate is possible if errors are assigned to each individual grain determination, so that the data can be presented as a probability density function (Hurford et al., 1984). This function is simply the summation of the Poisson age distributions for each of the individual grain determinations. Figure 16.11 shows such a plot for zircons from the re-worked El Ocote tephra deposit in Mexico, which displays a bimodal age distribution (Kowallis et al., 1986). The younger peak places a maximum age on the time of sedimentary re-working, and is in agreement with the estimated biostratigraphic age of associated fossil material. These results show that application of the population method to fission-track dating of this tephra would yield a meaningless average of the two age populations.

A problem with the probability density plot is that individual data points cannot be distinguished, so that some important but small components in the data distribution can be buried under the other data. To avoid this problem, Galbraith (1988) introduced a kind of isochron diagram for the presentation of fission-track data measured on individual grains of a heterogeneous sample. This 'radial' plot is designed for fission-track data sets with a high degree of scatter, either due to mixed detrital ages or variable cooling ages.

This diagram differs from other isochron plots used in geology because the two variables plotted are the apparent fission-track age of each grain, and the standard error of each grain age (σ). These quantities are plotted in the form of age/σ against $1/\sigma$ (Fig. 16.12a). In this plot, the slope of an array indicates the average age of the suite of grains analysed, and this age can be indicated on a calibrated arc. In practice, Galbraith argued, it is more convenient to normalize the average slope to a horizontal, and plot the y axis on a log scale from $+2$ to -2 (Fig. 16.12b). The age of any individual point

is then determined by projecting from the zero point on the y axis, through the data point, to the calibrated arc of ages (on a log scale).

An alternative data presentation of this type was proposed by Walter (1989). He suggested that additional assessments could be made of the quality of detrital fission-track ages if the raw data (spontaneous- versus induced-track densities) were plotted for each grain. This also yields an isochron diagram (Fig. 16.13) where the slope of each correlation line is proportional to age. These lines should pass through the origin, corresponding to a grain with zero uranium content. The linearity of each correlation line can be used to assess the influence of systematic analytical errors or geological disturbance on the reliability of the best-fit ages. However, this presentation has not been as popular as the isochron diagram of Galbraith, which has found wide application to complex fission-track data sets, including partially annealed systems as well as detrital systems.

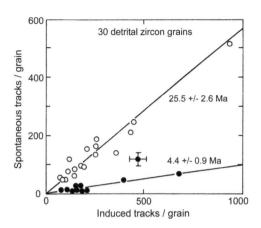

Fig. 16.13 Spontaneous- versus induced-track isochron diagram, showing data for individual zircon grains from the El Ocote tephra. After Walter (1989).

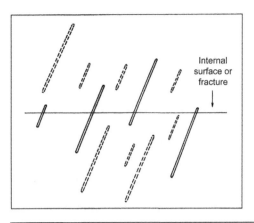

Fig. 16.14 Schematic illustration of the effect of track shortening on the observed density of etched tracks. Short and long tracks are of equal abundance, but the latter have a higher probability of becoming etched. After Laslett et al. (1982).

With the development of single-grain U–Pb dating of detrital zircons by *in situ* analysis (Sections 2.7.1, 5.2.6), it became possible to combine fission-track cooling ages of zircons with U–Pb crystallization ages (Carter and Moss, 1999). Alternatively, U–Pb dating can be combined with U–He dating (Rahl *et al.*, 2003). The combination of these dating methods for 'double dating' of detrital zircons allows 'first-cycle' erosion of source rocks to be distinguished from zircons recycled within the sedimentary system (Reiners *et al.*, 2005). Since U–Pb and U–He dating are both mass spectrometric methods, it is perhaps understandable that these methods form an attractive combination, although the function of U–He thermochronometry (Section 10.5) is very similar to fission-track dating.

16.5 Track Annealing

From the very beginning of fission-track studies (Silk and Barnes, 1959) it has been known that fission tracks can fade under certain conditions. This was first seen as a result of electron bombardment during microscopy. However, elevated temperatures are the most important cause of track fading or 'annealing'. During this process the displaced ions within the damaged track lose their charge and return to their normal lattice positions, after which the track is no longer susceptible to preferential acid attack.

Following experiments on track annealing in mica, Fleischer *et al.* (1964) claimed that track annealing progressed by the accumulated 'healing up' of short segments at random points along the length of tracks. However, subsequent work on other materials (e.g. on glass by Storzer and Wagner, 1969) has shown that the healing process occurs principally at the ends of each track, causing a regular and progressive shortening. As the length of tracks is diminished by healing, they have a smaller probability of intersecting the free surface during the etching treatment. Hence, fewer tracks become

etched and the apparent track density decreases (Fig. 16.14). This correlation between track length and track density is termed the 'random line segment model' (Fleischer *et al.*, 1975).

Early studies showed that different materials have different degrees of resistance to fission-track annealing (Fleischer and Price, 1964a). In addition, however, a temperature–time relationship is found for the annealing process. The higher the temperature, the shorter the time required for complete annealing of tracks in any given material. To examine this behaviour, Fleischer and Price (1964b) performed laboratory annealing experiments on the mineral indochinite and found that annealing obeyed an Arrhenius/Boltzmann's law relation:

$$t = A \, e^{E/kT} \qquad [16.8]$$

where t is the time for track fading, A is a constant, E is the activation energy, k is Boltzmann's constant and T is absolute temperature. Much of the work since this time has been devoted to determining accurate Boltzmann relation annealing curves for different materials, both by laboratory and well-constrained geological studies.

Detailed laboratory experiments were performed on apatite and sphene by Naeser and Faul (1969) and on tektite glass by Storzer and Wagner (1969). These studies showed that annealing is a progressive process. Different degrees of track annealing in different materials each define their own Boltzmann's relation lines when shown on Arrhenius plots of time against reciprocal temperature (Fig. 16.15). The fan of annealing lines in Fig. 16.15 is evidence for the existence of a range of activation energies for track annealing within a single type of material. This implies that as annealing progresses (as measured by the fraction of tracks lost) it also becomes progressively more difficult (Storzer and Wagner, 1969). Hence, when comparing the annealing properties of different minerals it is necessary to compare equal fractions of track loss, such as 50% (Fig. 16.15).

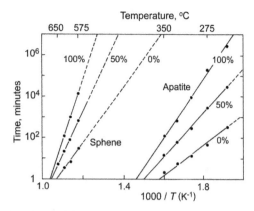

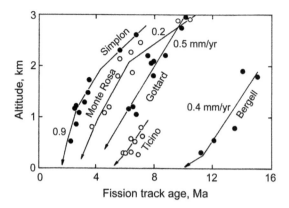

Following this line of investigation, Storzer and Poupeau (1973) compared laboratory annealing rates (in the same material) for freshly induced tracks and spontaneous tracks which had been partially annealed in nature. They found that as the temperature was raised, the fresh tracks were initially lost at a much higher rate, but that at a certain 'plateau' temperature the rates of annealing became equal.

Storzer and Poupeau argued that if both spontaneous and induced tracks were subjected to a heat treatment before counting then fission-track ages could be corrected for partial annealing in the environment. Track counting must be by the population method; therefore, the sample must have a uniform distribution of uranium. After irradiation of the induced-track sample, track-counting analysis is performed by stepwise annealing of both spontaneous- and induced-track samples in the laboratory. After each heating step a new surface of both samples is polished, etched and counted.

Results from this procedure are shown in Fig. 16.16 for a North American tektite. Above a certain threshold temperature (ca. 100 °C), induced tracks start to fade, but spontaneous tracks are resistant. Therefore the apparent fission-

track age increases rapidly with temperature. However, as laboratory heating approaches the temperature at which annealing occurred in the environment, spontaneous tracks also start to fade, and the apparent age therefore reaches a plateau (Fig. 16.16). Storzer and Wagner (1982) argued that this 'plateau-annealing' technique can yield corrected fission-track ages in glasses with a precision of $\pm$ 10% (2σ).

16.6 Uplift and Subsidence Rates

Wagner and Reimer (1972) demonstrated the usefulness of apatite fission-track ages for tectonic studies by applying them to Alpine uplift rates. Subsequently, Wagner et al. (1977) developed this technique by measuring apatite fission-track ages over a 3000 m range of vertical relief in the Central Alps. Fission-track ages on Alpine apatites do not conform to metamorphic isograds or terrane boundaries, but display a strong correlation with topographic relief (Fig. 16.17). They clearly represent cooling ages from Alpine metamorphism due to tectonic uplift, and can be used directly to calculate apparent uplift rates over the last few million years.

If a 'freezing-in' or 'blocking' temperature could be calculated for the Alpine apatites then the uplift rates in Fig. 16.17 could be converted into cooling rates. Two problems are faced in this task. The first is that the laboratory experiments show that blocking occurs over a range of temperatures. The second is that this range is itself dependent on the cooling rate. Hence, the argument is to some extent circular. Wagner et al. (1977) estimated a temperature for 50% track retention of 100–120 °C, half-way between 0% and 100% annealing temperatures of 60 and 180 °C, at a cooling rate of ca. 20 °C/Ma estimated from Rb–Sr biotite ages.

By combining apatite fission-track data with biotite K–Ar and Rb–Sr, muscovite K–Ar, muscovite Rb–Sr and monazite U–Pb data, Wagner et al. were able to calculate cooling rates for different regions of the Alps over the last 35 Ma

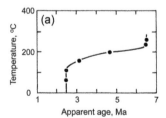

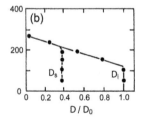

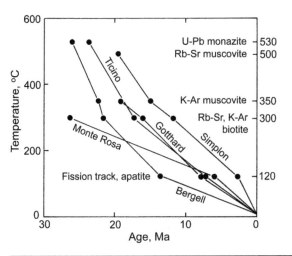

Fig. 16.18 Proposed cooling history for different regions of the Alps, based on time since closure of different radiogenic mineral systems. After Wagner et al. (1977).

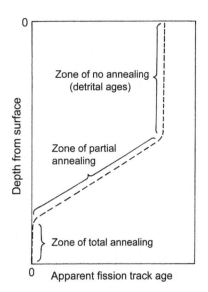

Fig. 16.19 Schematic illustration of the variation of apparent fission-track age with depth in bore-hole samples from a sedimentary basin. After Naeser (1979b); Naeser et al. (1989).

(Fig. 16.18). These results suggest that cooling in the Central Alps (Ticino and Gotthard areas) has been relatively uniform, while that in the East (Bergell) has slowed and the West (Simplon and Monte Rosa) has speeded up in the last few Ma. These conclusions are consistent with Fig. 16.17, and suggest that the Alps have undergone differential geographic uplift through time.

The idea of using a vertical traverse of apatite fission-track ages to deduce tectonic histories was applied by Naeser (1979b) to bore-hole studies of sedimentary basins. Naeser proposed that in sedimentary sequences which are at their maximum burial temperature, apparent fission-track ages would show a relationship with burial depth similar to Fig. 16.19. At shallow depths, burial heating is insignificant and fission-track ages reflect the sediment source (detrital ages). As depth of burial increases, apatites undergo increased thermal annealing, and display decreasing apparent fission-track ages, until they finally reach a total annealing zone with zero apparent age. The interval between zero and total annealing is called the partial annealing zone (PAZ).

The upper and lower temperature bounds of the PAZ will depend on the age of the sedimentary basin. Naeser (1981) collected fission-track age data from sedimentary basins with different burial rates. By making geological estimates of the effective burial (annealing) time in each basin, Naeser was able to make geological determinations of the Boltzmann lines for thermal annealing in apatite. These were confirmed by Gleadow and Duddy (1981) in a study of bore-hole data from the Otway Basin of Victoria, SE Australia. This basin is particularly suitable for fission-track studies of burial rates because the basinal sediments were derived from an active volcanic province. Hence, the sediments entering the basin were essentially of zero age, with very little older provenance.

The effective annealing time at present day down-hole temperatures was estimated from the burial history of the basin, suggesting that peak temperatures have been maintained for ca. 30 Ma. Using these estimates, annealing properties were determined for the Otway Basin apatites (Fig. 16.20), which were consistent with other bore-hole data (Naeser, 1981). In addition, the Boltzmann line for 50% annealing was consistent with laboratory annealing experiments. However, the temperature *interval* between 0 and 100% annealing was narrower than that predicted by the divergence of Arrhenius relation annealing lines from the laboratory data of Naeser and Faul (1969).

A complicating factor in the analysis of track fading in apatite is the discovery that annealing temperature is compositionally dependent (Green et al., 1985). Fission-track analyses were performed on individual apatite grains from a single horizon in an Otway drill hole with a present day temperature of 92 °C. These conditions result from progressive burial over the last 120 Ma. Chlor-apatite grains were found to give results near the depositional age, whereas fluor-apatites gave ages as low as zero (Fig. 16.21). Hence, when laboratory and geological annealing processes are compared, it is important that the material in the two types of experiment is as near compositionally identical as possible.

In the above discussion, geologically well-known thermal histories were used to calibrate the annealing behaviour of apatite tracks. Given this background, fission-track data can then be used to study geologically unknown basins. This evidence is especially pertinent to oil fields, because fission-track annealing occurs over the same temperature range as hydrocarbon maturation (Gleadow et al., 1983). For example, Briggs et al. (1981) used this approach to compare the

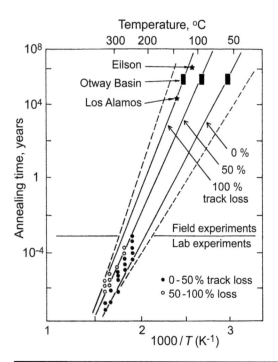

Fig. 16.20 Arrhenius plot for fission-track annealing in apatite from Otway Basin bore-holes (blocks), other bore-holes (stars) and laboratory experiments (spots and dashed lines). After Gleadow and Duddy (1981).

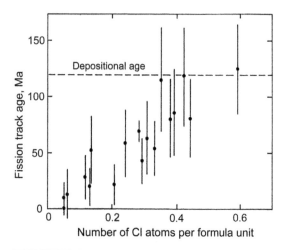

Fig. 16.21 Plot of measured fission-track ages in individual apatites from Otway Group sandstones, Australia, to show compositional dependence of track annealing. After Green et al. (1985).

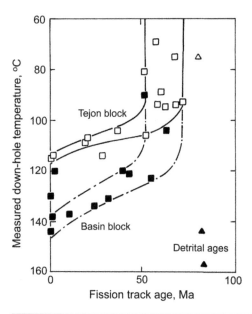

Fig. 16.22 A comparison between fission-track ages in bore-holes from the Basin and Tejon blocks of the San Joaquin Valley, California. Apatite data (■, □) give thermal history information while zircon data (▲, △) yield provenance ages. After Naeser et al. (1989).

thermal histories of two sedimentary basins of the Tejon oil field, in the San Joaquin Valley of California. This oil field is divided into two parts by the seismically active White Wolf fault. One part, the Basin Block, was a Late Tertiary depocen-tre which underwent strong subsidence. The other, Tejon Block, was less depressed. Fission-track analysis of apatite from bore-holes reveals the different geological history of the two blocks (Fig. 16.22).

Naeser et al. (1989) used these data, along with Boltzmann annealing lines from other geological locations, to calculate the thermal histories of the two blocks. Given geological evidence that the present down-hole temperatures represent peak values, the temperatures necessary for total annealing can be used to calculate effective heating times of ca. 1 Ma and 10 Ma for the Basin and Tejon blocks respectively (Fig. 16.23). These results are consistent with geological evidence for much more rapid burial of the Basin block, and do not require a perturbation in geothermal gradient.

16.7 Track Length Measurements

Because annealing initially causes shortening of tracks rather than their complete erasure, the use of etched track densities to chart the progress of annealing is an indirect approach. Under the 'random line segment model' of Fleischer et al. (1975) there should be a linear relationship between track density and average track length. However, when we only count track densities, we ignore information about the variation in track length on either side of the mean value. This *variation* in track length can yield additional information about the cooling history of a sample,

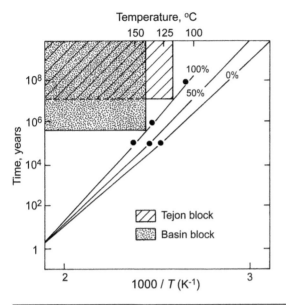

Fig. 16.23 Use of the Arrhenius plot to calculate effective heating times for the total annealing horizon in bore-holes from the Tejon and Basin blocks, San Joaquin Valley, California. After Naeser et al. (1989).

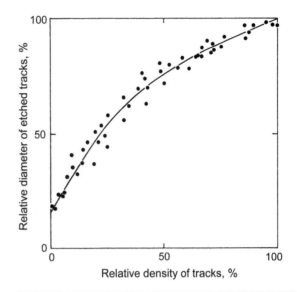

Fig. 16.24 Plot of mean pit diameter against pit density in variably annealed tektite glasses, relative to the original pit size and density before annealing. After Storzer and Wagner (1969).

because the longer the residence time of a sample in the partial annealing zone, the larger will be the variation in track length around the mean value. Therefore, fission-track data can be used more effectively to study the thermal history of a sample if the apparent age determined from track density is augmented by study of the length of etched tracks.

Track length studies were first made in dating micas (Maurette et al., 1964; Bigazzi, 1967), and were applied in detail to tektite glasses by Storzer and Wagner (1969). However, the etching of fission tracks in glasses tends to yield circular pits because of the smaller difference in structure, and hence etching rate, between tracks and the free surface. Consequently, the progress of annealing in glasses is accompanied by a decrease in the diameter rather than length of etched tracks. Nevertheless, Storzer and Wagner showed that pit diameter was correlated with pit density in variably annealed tektite glasses (Fig. 16.24), and were able to use the measurements of pit diameter to correct fission-track ages for the effects of annealing.

In mineral samples, track length data can be collected in two different ways. One way is to measure the apparent length of tracks which intersect the etched surface. Because these tracks intersect the surface, they are typically cut in half, hence the term 'semi-tracks' (e.g. Galbraith et al., 1990). Measurements of these tracks are often called 'projected track lengths' (Dakowski, 1978), a term that must not be confused with 'c-axis projection' (Section 16.7.4). The alternative approach is to measure the length of 'confined tracks' that do not reach the surface, but become etched due to indirect contact with the surface through other etched tracks

or etched cleavage traces (Bhandari et al., 1971; Green, 1981; Laslett et al. 1982, 1984). Apatite is by far the most important mineral for track length studies due to its widespread occurrence and relative chemical homogeneity. Hence, most of the following examples will be for apatite.

All track length measurements are subject to bias. These biases can be corrected to some extent by comparison between the lengths of spontaneous and induced tracks. However, it is critical to understand these biases in order to optimize the quality of the data. Longer projected tracks are more likely to be etched because they intersect the surface, but longer confined tracks are also more likely to be etched by intersection with secondary etching channels.

16.7.1 Projected Tracks (Semi-Tracks)

The apparent length of projected tracks is biased from the true length distribution by two additional factors. Firstly, truncation by the surface reduces apparent length. Secondly, tracks undergo visual foreshortening to an extent which depends on their angle to the surface. Together, these biases cause projected track length distributions to be much more complex than confined tracks, as shown in comparison studies (Fig. 16.25) by Laslett et al. (1982) and Gleadow et al. (1986). These authors argued that while projected track lengths yield only subtle indications of different thermal histories, confined tracks gave much clearer diagnostic indicators of thermal history (Section 16.7.5).

Wagner (1988) argued that useful thermal history information can be extracted from projected length distributions of (variably annealed) spontaneous tracks if these are ratioed against the length of projected induced tracks (in a population experiment). In theory this cancels out the bias involved

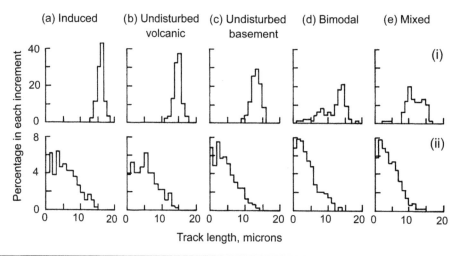

Fig. 16.25 Histograms of track length (as a percentage of total sample) for apatites with different types of thermal history (see text). Top row: horizontal confined tracks; bottom row: projected tracks. After Gleadow et al. (1986).

in using projected track measurements, but in practice this method has the effect of degrading the statistical quality of the data. This was demonstrated by Laslett et al. (1994) in a comparison of the usefulness of confined and projected track lengths for thermal history analysis (Fig. 16.26).

Confined track lengths bear a simpler relationship to the true track length distribution, but projected tracks (semi-tracks) are more numerous. Therefore, Laslett et al. performed simulations to test the ability of 2000 projected tracks and 100 confined tracks to recover the true track length variation in a mixed population with lengths of 14.5 and 10 μm. The proportion of shorter tracks (p) was varied in different simulations from 20% to 80%. Results showed that when the shorter tracks made up 60 to 80% of the population, both methods could recover the length of these tracks and the correct value of p with similar error bars (Fig. 16.26). However, when the proportion of short tracks fell below 50%, the projected track data were seriously compromised.

Laslett et al. concluded from this experiment that a relatively small number of confined track measurements can more reliably recover the true track length distribution than a large number of projected track lengths. Hence, it is now generally agreed that confined tracks are much superior to projected tracks for thermal history analysis, and the measurement of projected track lengths has largely been abandoned.

16.7.2 Confined Tracks

Confined tracks do not reach the general etched surface, but become etched by the penetration of acid down a channel inside the mineral which intersects the track (Fig. 16.27). The two most common types are termed 'Track-IN-Track' (TINT) and 'Track-IN-CLEavage' (TINCLE) respectively (Lal et al., 1969; Fleischer et al., 1975). However, some workers (e.g. Carlson et al., 1999; Donelick et al., 2005) have argued that the only

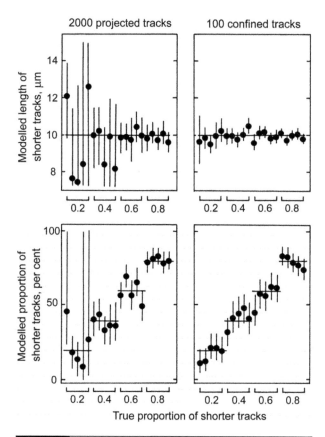

Fig. 16.26 Comparison of the success of projected and confined tracks to recover the length and the proportion of a population of short tracks amongst a population of longer tracks. Error bars indicate 95% confidence limits, except where they are truncated by the limits of the graphs. After Laslett et al. (1994).

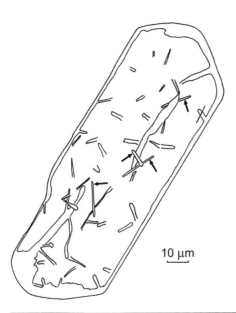

Fig. 16.27 Line drawing of high-contrast features in an etched apatite grain, viewed under dry (non-immersion) conditions. Four confined tracks are visible (arrowed). From a photograph by Gleadow et al. (1986).

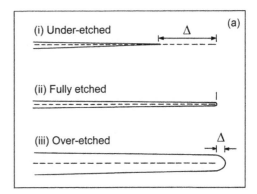

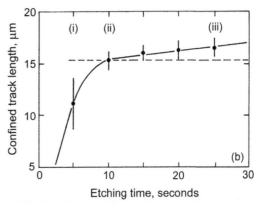

Fig. 16.28 Illustration of the progress of etching: (a) for a single track; (b) average track length in Durango apatite (Mexico) as a function of etching time in 5M HNO_3 at 21°C. After Laslett et al. (1984).

reliable type of confined track for thermal history analysis is the TINT type. The track-in-cleavage type is less reliable, because the track was open to percolation of fluids during the geological lifetime of the sample. In contrast, TINT tracks form isolated pairs or groups that are only open to chemical attack (etching) once the sample is cut open.

Restricting confined track analysis to TINT examples clearly limits the number of tracks that can be counted. However, Donelick and Miller (1991) proposed a way for overcoming this limitation (for minerals with low U content) by specifically inducing the creation of additional tracks perpendicular to the etched surface. This was achieved by placing the sample against a planar [252]Cf source before etching. The vertical induced tracks did not interfere with the process of horizontal track counting, but, by providing many more access points for the etching solution, they were able to increase the number of etched horizontal confined tracks up to 20-fold. Further testing of this method by Jonckheere et al. (2007) has shown that it does not introduce any extra biases to the measurement of confined track lengths.

Measurement of confined tracks which are horizontal (parallel to the etched surface) leads to the minimum bias from true track length, and this is therefore the only type of confined length measurement now used. These tracks must be counted using fixed selection criteria to exclude the possibility of subjective bias. Laslett et al. (1994) recommended the 'bright reflection' criterion, which exploits the property of etched tracks at a low angle to the horizontal (less than about 15%) to show a bright image in *reflected* light. In contrast, a criterion which requires tracks to be in focus along

their entire length is unsuitable because it causes a higher rejection rate for long tracks than for short tracks.

Another analytical variable that must be carefully controlled is the etching time. It is necessary that some tracks become over-etched to ensure that others are not under-etched. Since tracks have effectively zero width before etching, over-etched tracks can be recognized by their non-zero width, which is almost the same as the excess length (2 × Δ, Fig. 16.28a). The optimum etching time is determined by experiments on incremental etching (Fig. 16.28b). The standard etching time of 20 seconds used by Laslett et al. (1984) leads on average to 1 μm of over-etching.

16.7.3 Track Widths

Confined tracks are subject to much smaller bias than projected tracks, but are still subject to crystallographic bias due to the composition of the mineral and the orientation of tracks in the mineral. Such bias can affect the progress of track annealing and track etching differently, and was first investigated by Green and Durrani (1977).

Green and Durrani observed that un-annealed tracks gave rise to circular pits for apatite sections cut parallel to the

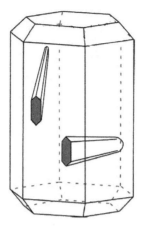

Fig. 16.29 Diagrammatic illustration of etched tracks in apatite showing the apparent width of etched tracks and their cross-sectional dimensions ('negative crystal' shape) on a face parallel to the long axis of the crystal (c). Modified after Gleadow et al. (2002).

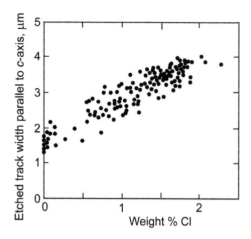

Fig. 16.30 Plot of D_{PAR} (the long axis of the track cross-section) as a function of the chlorine content of a fluorapatite. Modified after Burtner et al. (1994).

(0001) plane, commonly known as 'basal' sections. In contrast, sections cut parallel to the long axis of the crystal revealed elongated pits after etching.

Typical etching conditions for apatite are 5% nitric acid (0.8 M) for 50 secs (Green and Durrani, 1977) or 30% nitric (5.5 M) for 20 secs at 21 °C (Donelick et al., 2005). Detailed examination of etched pits (track cross-sections) shows that they have the form of a 'negative crystal' (Gleadow and Seiler, 2015), forming a cavity which is normally elongated in the c axis direction, and whose shape can be described by its 'aspect ratio'. For example, an unusual hydroxy-apatite from Renfrew, Canada has only slightly elongated cross-sections with an aspect ratio below 2 (Ketcham, 2003). In contrast, Durango apatite (Mexico), has the more typical elongated track cross-section, with an aspect ratio of around 5 (Durango is a fluorapatite with a small Cl component).

The effect of this phenomenon on the overall appearance of etched confined tracks is shown in Fig. 16.29. Etched tracks perpendicular to the c axis are easier to see because they are up to five times wider than tracks aligned along the c axis. This will typically lead to counting bias, because observers tend to overlook the much narrower tracks parallel to the c axis.

The width of a track cross-section along the c axis (i.e. the long axis of the cross-section) was called D_{PAR} (diameter parallel to c) by Burtner et al. (1994). It was also referred to as (the long axis of) the etch-pit diameter by Carlson et al. (1999). Burtner et al. showed that this measurement (which is related to the aspect ratio) was strongly dependent on the composition of the apatite (Fig. 16.30). Hence, they argued that measurement of this parameter should be useful in characterizing the chemical affinity of unknown apatite grains. This is critical in interpreting track-length

data because the annealing behaviour of apatites is also dependent on their chemistry. However, Ketcham et al. (1999) showed that etch-pit diameter (D_{PAR}) is not actually a very good predictor of the chemical dependence of annealing (referred to by them as the 'kinetic factor'). Instead, the best predictor (Carlson et al., 1999) was a complex combination of chemical factors, including the chlorine, hydroxide, Mn and Fe content of apatites (relative to typical fluorapatites).

16.7.4 c Axis Projection

Green and Durrani (1977) showed that crystallographic orientation effects in apatite also lead to biased track lengths due to annealing. This was explored in more detail by Laslett et al. (1984), who showed that for a given intensity of annealing, tracks are progressively shortened as the angle to the c axis increases (Fig. 16.31). In the early thermal-history modelling based on track lengths (Section 16.7.5), variable track lengths at different angles to the c axis were simply averaged (Green et al., 1989). These authors recognized that this was a simplification of the real distribution, but argued that this did not significantly degrade the resulting thermal history models. On the other hand, Galbraith and Laslett (1988) argued that it was important to routinely measure confined track orientations as well as their lengths in order to verify relationships such as those presented in Fig. 16.34.

Donelick (1991) argued that the length of confined tracks parallel (and perpendicular) to c was a more fundamental property of annealed tracks, and therefore preferable to the use of average track lengths. This presents a problem, since the number of tracks counted at these two extreme orientations is generally less than for intermediate directions. However, Donelick showed that an elliptical function could be fitted to the distribution of etched confined track lengths plotted in polar coordinates (radially) as a function of their angle to the c axis (Fig. 16.32). This allows the length on the

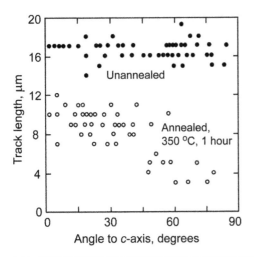

Fig. 16.31 Effect of crystallographic orientation in causing increased annealing of tracks with increasing angle to the c axis of apatite. After Laslett et al. (1984).

c axis to be recovered, leading to the concept of 'c axis projection' (Donelick et al., 1999). This term should not be confused with the analysis of projected track lengths ('semi-tracks'), which is now largely discontinued (Section 16.7.1).

The effectiveness of c axis projection was demonstrated by Donelick et al. (1999) using a large data set of induced tracks in apatite that were subjected to annealing experiments in the laboratory (Carlson et al., 1999). With zero annealing (Fig. 16.32, curve 1), the tracks define a circular distribution (track lengths the same in every direction). However, even in this case, it should be noted that fewer tracks with alignments

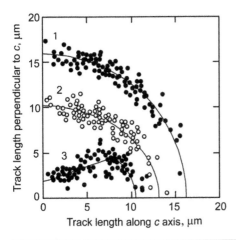

Fig. 16.32 Polar coordinate (radial) track-length plot, showing elliptical distributions of annealed track lengths as a function of angle to the c axis of apatite crystals. Symbol types distinguish three annealing experiments. Modified after Donelick et al. (1999).

close to the c axis were counted. This is partly explained by the greater difficulty in observing these narrow tracks in reflected light, but may also be caused by incomplete etching of the ends of these tracks, causing the observer to reject them as abnormal (Ketcham, 2003).

As annealing progresses (curve 2), the distribution becomes elliptical as track lengths undergo more shortening at high angles to the c axis than along the axis. Finally, at a certain degree of annealing, the curve 'collapses' as tracks at high angles to the c axis start to heal in the middle of the track, causing non-linear shortening (curve 3). Donelick et al. called this behaviour 'rapid track shortening' and showed that their elliptical model could identify this change in behaviour, and to some extent recover the projected c axis lengths, even of such strongly shortened track sets.

One consequence of this approach is that it focuses on the lengths of the tracks in the direction most resistant to annealing. This may seem to run counter to the objective of using confined track lengths to study thermal histories by means of track annealing. However, Donelick et al. (2005) argued that consistency of analysis is more important than maximum sensitivity. The latter does not confer advantage if it is subject to greater observer bias.

Ketcham et al. (1999) argued that observational bias in the determination of mean track lengths is likely because the mean is strongly influenced by the relative numbers of tracks documented at each end of the length spectrum. These extreme tracks are also the hardest to identify and measure reliably, due to their large variation in width. In highly annealed apatites, the mean is also strongly biased by tracks showing rapid shortening due to segmentation. Finally, Ketcham et al. argued that mean track lengths are more susceptible to bias when used with the ^{252}Cf irradiation technique to enhance TINT frequencies. This technique is necessary for generating good track length data sets for many apatite samples because their U contents are too low to generate adequate numbers of TINTs.

The benefits of c axis projection in compensating for variations in observation style were tested by Ketcham et al. (2007) using large data sets collected by Barbarand et al. (2003a, b). This comparison showed that c axis projection greatly improved the consistency of observations by different analysts at high degrees of annealing, which are the most critical cases. The quality of data for less annealed tracks was also improved, but to a lesser extent (Fig. 16.33).

16.7.5 Forward and Inverse Modelling

The objective of track length determinations (in combination with fission track ages derived from track densities) is to recover thermal histories, especially those involving extensive periods in the partial-annealing zone. Forward modelling is the most straightforward method, in which a known or estimated thermal history is used to predict the distribution of track lengths. This prediction is then tested against measured track-length data. Reverse modelling is more

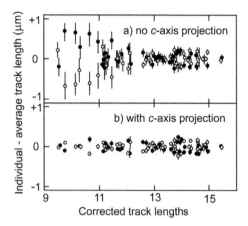

Fig. 16.33 Comparison of the scatter of confined track length measurements as a function of track length for two different analysts: (a) using average track lengths; (b) using c axis projection. After Ketcham et al. (2007).

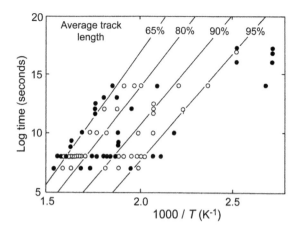

Fig. 16.34 Interpretation of laboratory annealing data showing gently fanning Boltzmann lines for different degrees of track shortening that are consistent with field data from the Otway basin. After Laslett et al. (1987).

difficult because it attempts to back-predict thermal history data directly from a measured track-length data set. This can only work when the parameters have previously been subject to detailed testing based on forward modelling. Examples of these approaches for fission-track densities were shown in Section 16.6. They will now be discussed specifically using track-length data.

Examples of *qualitative* forward models were shown in Fig. 16.25, in a comparison of semi-track and confined-track data (Gleadow et al., 1986). In this plot, confined track distributions are divided into five main types which yield overall qualitative information about their thermal history (Fig. 16.25a–e). Induced tracks (Fig. 16.25a) are the longest and most uniform type (16 ± 1 μm, based on several different sample types). Tracks in undisturbed volcanics and rapidly cooled shallow intrusions are also uniform within a single sample (Fig. 16.25b), but there is some variation between sample means (ca. 14–15.5 μm). This can be attributed to limited annealing at near-ambient temperatures over periods of tens of Ma. Tracks in undisturbed basement apatite (Fig. 16.25c) are somewhat shorter (means of 12–14 μm), with a skewed distribution attributed to slow cooling from regional metamorphism. Finally, bimodal and mixed distributions (Fig. 16.25 d and e) are attributed to various types of two-stage thermal history, in which pre-existing tracks were partially erased by a thermal event between initial cooling and the present.

In order to use confined track length analysis to make *quantitative* interpretations of thermal histories, it is first necessary to define Arrhenius (temperature–time) relationships for given fractions of track shortening. These are analogous to the Arrhenius relationships for track density (e.g. Fig. 16.20). A set of experimental data on the annealing of induced tracks in the Durango apatite standard was presented by Green et al. (1986). These data describe the progressive shortening of tracks for increased temperatures and durations of annealing under constant temperature (isothermal conditions). However, the laboratory experiments are conducted over a restricted range of annealing periods that are obviously much shorter than geological time periods. Therefore, the laboratory data must be greatly extrapolated to use them to model geological thermal histories. This led to divergence of opinion on whether Arrhenius equations yielding parallel or fanning Boltzmann lines should be used.

Based on the laboratory data, Green et al. (1985) argued that Boltzmann lines for different percentages of track annealing did not have a fan-shaped distribution, but were parallel. However, re-examination of the data set of Green et al. (1985) by Laslett et al. (1987) suggested that the data were more consistent with fanning Boltzmann lines (Fig. 16.34). These lines are not as divergent as those suggested by early experiments on track densities (Fig. 16.20), but the mildly fanning lines are in better agreement with drill-hole data from the Otway basin, Australia.

These isothermal (constant temperature) models were extended to more complex thermal histories involving variable temperature conditions by Duddy et al. (1988), and applied to geological timescales by Green et al. (1989). In their approach, a predicted temperature–time curve is divided into intervals (e.g. 1 Ma each), and after each interval the degree of shortening of existing tracks is calculated. At a constant elevated temperature, track shortening occurs rapidly at first, because the track-ends are least energetically stable, but the rate subsequently slows considerably. In addition to the annealing of old tracks, new track formation is simulated at 10 Ma intervals. The example shown in Fig. 16.35 simulates very steady slow cooling from 100 °C to 20 °C over 200 Ma. Tracks formed in each time interval define

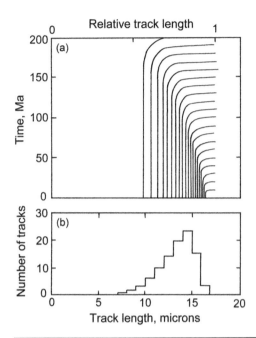

Fig. 16.35 Model for fission-track formation and annealing in a thermal history defined by very slow cooling. (a) Progress of track formation and shortening; (b) resulting distribution of track length data. Modified after Green *et al.* (1989).

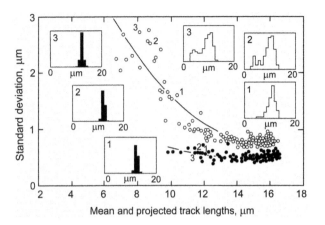

Fig. 16.36 Comparison of track lengths, standard deviations and within-grain distributions for partially annealed tracks. (○) = mean lengths; (●) = (c axis) projected lengths. Within-grain length distributions are for numbered points in each data set. Modified after Ketcham (2005).

an evolution path of reduced track length against time (Fig. 16.35a). The sum of these evolution paths at the present day forms a skewed histogram of track length distribution (Fig. 16.35b) indicative of such a slowly cooled thermal history.

Unfortunately, subsequent work on the reproducibility of track length distributions has shown that distributions such as that modelled in Fig. 16.35 are susceptible to bias during the counting of spontaneous tracks in geological samples. For example, Ketcham (2005) showed that 'raw' track length distributions within individual grains could be highly skewed, and yet nevertheless yield symmetrical distributions of c axis projected lengths. This is demonstrated in Fig. 16.36 using the experimental data of Carlson *et al.* (1999). As a result of these biases, the overall distributions of mean and (c axis) projected track lengths will be different (Fig. 16.37). Projected track-length distributions show more subtle skew (tailing effects) for slowly cooled samples (Fig. 16.37b). However, they show sharper resolution of double heating episodes (Fig. 16.37c).

In addition to analytical uncertainties, track development and shortening are processes that are subject to random noise. Therefore, inverse modelling of thermal histories can only generate possible models with probability estimates. In practice, thermal histories chosen at random (the Monte Carlo method) are used to calculate track length distributions by forward modelling (e.g. Corrigan, 1991). The results are then tested against the observed (or simulated)

track length data. After a few hundred iterations it is possible to map out a range of possible thermal histories which are consistent with the data set. Within this range of possibilities, the highest probability density defines an optimum, but not unique, thermal history (e.g. Fig. 16.38). Relatively well-constrained thermal histories can be projected back to the last temperature maximum, but beyond this time the thermal history is very poorly constrained. This is indicated by the fanning of the probability contours at ages over 60 Ma in the example shown. Several software packages are now available for the inverse modelling of thermal histories from fission-track data, as summarized by Ketcham (2005).

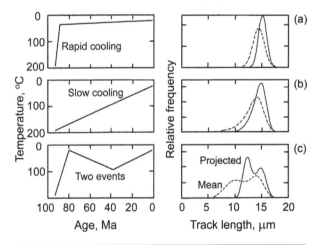

Fig. 16.37 Three different thermal histories (left) and forward-modelled track-length distributions (right). Dashed = mean lengths; solid = (c axis) projected lengths. After Ketcham (2005).

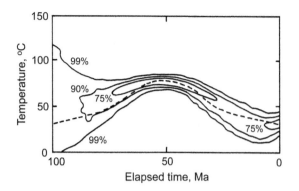

Fig. 16.38 Plot of probability density for modelled thermal histories of a sedimentary basin, based on fission-track data. In this test case the actual thermal history is known, and is shown for reference (dashed line). After Corrigan (1991).

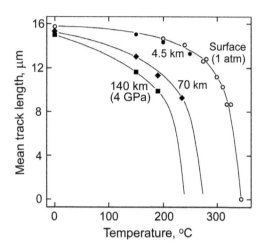

Fig. 16.39 Plot of mean track length against temperature for an experimental run time of 10 hours at different pressures. After Schmidt et al. (2014).

16.8 Pressure Effects

Experiments to test for the pressure dependence of fission-track annealing were performed during the early work of Fleischer et al. (1965b), but there was little evidence for any pressure dependence. This view was supported by fission-track evidence from deep boreholes, which did not show any deviation from the annealing rate expected from the measured down-hole temperatures. Strong control on this behaviour is provided by boreholes in shield regions with low geothermal gradients, where any pressure dependence would be more visible relative to purely thermal annealing (Wagner et al., 1997).

In contrast to the conventional view, surprising new data were obtained by Wendt et al. (2002) which suggested a strong pressure dependence of apatite fission-track ages. Even more surprising was the evidence that pressure appeared to *reduce* the extent of annealing, affecting both average track length and track density. The effects were observed at both 100 MPa and 300 MPa (1 and 3 kbar, equivalent to ca. 3.5 and 10.5 km depth), and implied a dramatic pressure effect on thermal annealing in natural systems.

To test these findings, Donelick et al. (2003) attempted to replicate the experimental work, but were not able to reproduce the results of Wendt et al. (2002). However, Donelick et al., pointed out that Wendt et al. had used different heating and temperature sensing equipment on their experiments at ambient and 100/300 MPa experiments. This is a critical point, because laboratory annealing experiments must be conducted at high temperatures (in this case 250 °C) to complete them in a reasonable period of time. This makes the experiments very sensitive to the temperature conditions of the runs, so that small errors in the measured temperature could easily explain the aberrant results. In addition, Kohn et al. (2003) argued that the results of Wendt et al. were also biased by the use of partially annealed spontaneous tracks, which have a complex response to renewed annealing procedures compared with the response of pristine induced tracks.

Subsequent experiments at much higher pressure (2–4 GPa, equal to 20–40 kb) showed some pressure dependence of annealing (Schmidt et al., 2014). However, the effect was in the opposite sense to that proposed by Wendt et al. (2002), such that increasing pressure caused enhanced annealing (Fig. 16.39). Schmidt et al. also showed that when this pressure dependence is extrapolated to 150 MPa (5 km burial depth), it would have a negligible effect relative to temperature in controlling annealing rates.

References

Barbarand, J., Carter, A., Wood, I. and Hurford, T. (2003a). Compositional and structural control of fission-track annealing in apatite. *Chem. Geol.* **198**, 107–37.

Barbarand, J., Hurford, T. and Carter, A. (2003b). Variation in apatite fission-track length measurement: implications for thermal history modelling. *Chem. Geol.* **198**, 77–106.

Bhandari, N. Bhat, S. G., Rajogopalan, G., Tamhane, A. S. and Venkatavaradan, V. S. (1971). Fission fragment lengths in apatite: recordable track lengths. *Earth Planet. Sci. Lett.* **13**, 191–9.

Bigazzi, G. (1967). Length of fission tracks and age of muscovite samples. *Earth Planet. Sci. Lett.* **3**, 434–8.

Briggs, N. D., Naeser, C. W. and McCulloh, T. H. (1981). Thermal history of sedimentary basins by fission-track dating. *Nucl. Tracks* **5**, 235–7 (abstract).

Burtner, R. L., Nigrini, A. and Donelick, R. A. (1994). Thermochronology of Lower Cretaceous source rocks in the Idaho–Wyoming thrust belt. *AAPG Bull.* **78**, 1613–36.

Carlson, W. D., Donelick, R. A. and Ketcham, R. A. (1999). Variability of apatite fission-track annealing kinetics: I. Experimental results. *Amer. Mineral.* **84**, 1213–23.

Carter, A. and Moss, S. J. (1999). Combined detrital-zircon fission-track and U–Pb dating: A new approach to understanding hinterland evolution. *Geology* **27**, 235–8.

Corrigan, J. (1991). Inversion of apatite fission track data for thermal history information. *J. Geophys. Res.* **96**, 10 347–60.

Corson, D. R. and Thornton, R. L. (1939). Disintegration of uranium. *Phys. Rev.* **55**, 509.

Dakowski, M. (1978). Length distributions of fission tracks in thick crystals. *Nucl. Track Det.* **2**, 181–9.

Donelick, R. A. (1991). Crystallographic orientation dependence of mean etchable fission track length in apatite: An empirical model and experimental observations. *Amer. Mineral.* **76**, 83–91.

Donelick, R., Farley, K., O'Sullivan, P. and Asimow, P. (2003). Experimental evidence concerning the pressure dependence of He diffusion and fission-track annealing kinetics in apatite. *On Track* **26**, 19–21.

Donelick, R. A., Ketcham, R. A. and Carlson, W. D. (1999). Variability of apatite fission-track annealing kinetics: II. Crystallographic orientation effects. *Amer. Mineral.* **84**, 1224–34.

Donelick, R. A. and Miller, D. S. (1991). Enhanced TINT fission track densities in low spontaneous track density apatites using ^{252}Cf-derived fission fragment tracks: A model and experimental observations. *Int. J. Rad. App. Instrum.* **18**, 301–7.

Donelick, R. A., O'Sullivan, P. B. and Ketcham, R. A. (2005). Apatite fission-track analysis. *Rev. Mineral. Geochem.* **58**, 49–94.

Duddy, I. R., Green, P. F. and Laslett, G. M. (1988). Thermal annealing of fission tracks in apatite 3. Variable temperature behaviour. *Chem. Geol. (Isot. Geosci. Sect.)* **73**, 25–38.

Enkelmann, E., Ehlers, T. A., Buck, G. and Schatz, A. K. (2012). Advantages and challenges of automated apatite fission track counting. *Chem. Geol.* **322**, 278–89.

Fleischer, R. L. and Hart, H. R. (1972). Fission track dating: techniques and problems. In: Bishop, W., Miller, J. and Cole, S. (Eds) *Calibration of Hominoid Evolution*. Scottish Academic Press, pp. 135–70.

Fleischer, R. L. and Price, P. B. (1964a). Techniques for geological dating of minerals by chemical etching of fission fragment tracks. *Geochim. Cosmochim. Acta* **28**, 1705–14.

Fleischer, R. L. and Price, P. B. (1964b). Glass dating by fission fragment tracks. *J. Geophys. Res.* **69**, 331–9.

Fleischer, R. L., Price, P. B., Symes, E. M. and Miller, D. S. (1964). Fission track ages and track-annealing behaviour of some micas. *Science* **143**, 349–51.

Fleischer, R. L., Price, P. B. and Walker, R. M. (1965a). Tracks of charged particles in solids. *Science* **149**, 383–93.

Fleischer, R. L., Price, P. B. and Walker, R. M. (1965b). Effects of temperature, pressure, and ionization on the formation and stability of fission tracks in minerals and glasses. *J. Geophys. Res.* **70**, 1497–502.

Fleischer, R. L., Price, P. B. and Walker, R. M. (1968). Charged particle tracks: tools for geochronology and meteor studies. In: Hamilton, E. and Farquhar, R. M. (Eds) *Radiometric Dating for Geologists*. Wiley Interscience, pp. 417–35.

Fleischer, R. L., Price, P. B. and Walker, R. M. (1975). *Nuclear Tracks in Solids*. University of California Press, 605 pp.

Galbraith, R. F. (1988). Graphical display of estimates having differing standard errors. *Tectonometrics* **30**, 271–81.

Galbraith, R. F. and Laslett, G. M. (1988). Some calculations relevant to thermal annealing of fission tracks in apatite. *Proc. Roy. Soc. Lond. A.***419**, 305–21.

Galbraith, R. F., Laslett, G. M., Green, P. F. and Duddy, I. R. (1990). Apatite fission track analysis: geological thermal history analysis based on a three-dimensional random process of linear radiation damage. *Phil. Trans. Roy. Soc. Lond. A* **332**, 419–38.

Gleadow, A. J., Belton, D. X., Kohn, B. P. and Brown, R. W. (2002). Fission track dating of phosphate minerals and the thermochronology of apatite. *Rev. Mineral. Geochem.* **48**, 579–630.

Gleadow, A. J. W. and Duddy, I. R. (1981). A natural long-term track annealing experiment for apatite. *Nucl. Tracks* **5**, 169–74.

Gleadow, A. J. W., Duddy, I. R., Green, P. F. and Lovering, J. F. (1986). Confined fission track lengths in apatite: a diagnostic tool for thermal history analysis. *Contrib. Mineral. Petrol.* **94**, 405–15.

Gleadow, A. J. W., Duddy, I. R. and Lovering, J. F. (1983). Apatite fission-track analysis as a paleotemperature indicator for hydrocarbon exploration. *Aust. Petrol. Explor. Soc. J.* **23**, 93–102.

Gleadow, A. J., Gleadow, S. J., Belton, D. X. *et al.* (2009). Coincidence mapping – a key strategy for the automatic counting of fission tracks in natural minerals. *Geol. Soc. Lond. Spec. Pub.* **324**, 25–36.

Gleadow, A. J. and Seiler, C. (2015). Fission track dating and thermochronology. In: Rink, W. J. and Thompson, J. W. (Eds) *Encyclopedia of Scientific Dating Methods. Springer*, pp. 285–96.

Green, P. F. (1981). 'Track-in track' length measurements in annealed apatites. *Nucl. Tracks* **5**, 121–8.

Green, P. F., Duddy, I. R., Gleadow, A. J. W. and Tingate, P. R. (1985). Fission-track annealing in apatite: track length measurements and the form of the Arrhenius plot. *Nucl. Tracks* **10**, 323–8.

Green, P. F., Duddy, I. R., Gleadow, A. J. W., Tingate, P. R. and Laslett, G. M. (1986). Thermal annealing of fission tracks in apatite. 1. *A qualitative description. Chem. Geol. (Isot. Geosci. Sect.)* **59**, 237–53.

Green, P. F., Duddy, I. R., Laslett, G. M. *et al.* (1989). Thermal annealing of fission tracks in apatite 4. Quantitative modelling techniques and extension to geological timescales. *Chem. Geol. (Isot. Geosci. Sect.)* **79**, 155–82.

Green, P. F. and Durrani, S. A. (1977). Annealing studies of tracks in crystals. *Nucl. Track Det.* **1**, 33–9.

Hasebe, N., Barbarand, J., Jarvis, K., Carter, A. and Hurford, A. J. (2004). Apatite fission-track chronometry using laser ablation ICP–MS. *Chem. Geol.* **207**, 135–45.

Holden, N. E. and Hoffman, D. C. (2000). Spontaneous fission half-lives for ground-state nuclide (Technical report). *Pure Appl. Chem.* **72**, 1525–62.

Hurford, A. J. (1990). Standardization of fission track calibration: recommendation by the Fission Track Working Group of the I.U.G.S. Subcommission on Geochronology. *Chem. Geol. (Isot. Geosci. Sect.)* **80**, 171–8.

Hurford, A. J. and Carter, A. (1991). The role of fission track dating in discrimination of provenance. In: Morton, A. C., Todd, S. P. and Haughton, P. D. W. (Eds) *Developments in Sedimentary Provenance Studies. Geol. Soc. Spec. Pub.* **57**, pp. 67–78.

Hurford, A. J., Fitch, F. J. and Clarke, A. (1984). Resolution of the age structure of the detrital zircon populations of two Lower Cretaceous sandstones from the Weald of England by fission track dating. *Geol. Mag.* **121**, 269–77.

Hurford, A. J. and Green, P. F. (1982). A users' guide to fission track dating calibration. *Earth Planet. Sci. Lett.* **59**, 343–54.

Hurford, A. J. and Green, P. F. (1983). The ζ age calibration of fission-track dating. *Isot. Geosci.* **1**, 285–317.

Jonckheere, R., Enkelmann, E., Min, M., Trautmann, C. and Ratschbacher, L. (2007). Confined fission tracks in ion-irradiated and step-etched prismatic sections of Durango apatite. *Chem. Geol.* **242**, 202–17.

Jonckheere, R. and Ratschbacher, L. (2015). Standardless fission-track dating of the Durango apatite age standard. *Chem. Geol.* **417**, 44–57.

Ketcham, R. A. (2003). Observations on the relationship between crystallographic orientation and biasing in apatite fission-track measurements. *Amer. Mineral.* **88**, 817–29.

Ketcham, R. A. (2005). Forward and inverse modeling of low-temperature thermochronometry data. *Rev. Mineral. Geochem.* **58**, 275–314.

Ketcham, R. A., Carter, A., Donelick, R. A., Barbarand, J. and Hurford, A. J. (2007). Improved modeling of fission-track annealing in apatite. *Amer. Mineral.* **92**, 799–810.

Ketcham, R. A., Donelick, R. A. and Carlson, W. D. (1999). Variability of apatite fission-track annealing kinetics: III. Extrapolation to geological time scales. *Amer. Mineral.* **84**, 1235–55.

Kimura, J. I., Danhara, T. and Iwano, H. (2000). A preliminary report on trace element determinations in zircon and apatite crystals using excimer laser ablation-inductively coupled plasma mass spectrometry (ExLA-ICPMS). *Fission Track News Lett.* **13**, 11–20.

Kohn, B. P., Belton, D. X., Brown, R. W. *et al.* (2003). Comment on: "Experimental evidence for the pressure dependence of fission track annealing in apatite" by A. S. Wendt *et al.* [Earth Planet. Sci. Lett. 201 (2002) 593–607]. *Earth Planet. Sci. Lett.* **215**, 299–306.

Kowallis, B. J., Heaton, J. S. and Bringhurst, K. (1986). Fission-track dating of volcanically derived sedimentary rocks. *Geology* **14**, 19–22.

Lal, D., Rajan, R. S. and Tamhane, A. S. (1969). Chemical composition of nuclei of $Z > 22$ in cosmic rays using meteoritic minerals as detectors. *Nature* **221**, 33–7.

Laslett, G. M., Galbraith, R. F. and Green, P. F. (1994). The analysis of projected fission track lengths. *Rad. Meas.* **23**, 103–23.

Laslett, G. M., Gleadow, A. J. W. and Duddy, I. R. (1984). The relationship between fission track length and track density in apatite. *Nucl. Tracks* **9**, 29–37.

Laslett, G. M., Green, P. F., Duddy, I. R. and Gleadow, A. J. W. (1987). Thermal annealing of fission tracks in apatite, 2. A quantitative analysis. *Chem. Geol. (Isot. Geosci. Sect.)* **65**, 1–13.

Laslett, G. M., Kendall, W. S., Gleadow, A. J. W. and Duddy, I. R. (1982). Bias in measurement of fission-track length distributions. *Nucl. Tracks* **6**, 79–85.

Maurette, M., Pellas, P. and Walker, R. M. (1964). Etude des traces fission fossiles dans le mica. *Bull. Soc. Franc. Miner. Cryst.* **87**, 6–17.

Naeser, C. W. (1979a). Fission-track dating and geological annealing of fission tracks. In: Jager, E. and Hunziker, J. C. (Eds) *Lectures in Isotope Geology*. Springer-Verlag, pp. 154–69.

Naeser, C. W. (1979b). Thermal history of sedimentary basins: Fission-track dating of subsurface rocks. In: Scholle, P. A., and Schluger, P. R. (Eds) *Aspects of Diagenesis. Soc. Econ. Paleontol. Mineral. Spec. Pub.* **26**, pp. 109–12.

Naeser, C. W. (1981). The fading of fission tracks in the geologic environment – data from deep drill holes. *Nucl. Tracks.* **5**, 248–50 (abs).

Naeser, C. W. and Faul, H. (1969). Fission track annealing in apatite and sphene. *J. Geophys. Res.* **74**, 705–10.

Naeser, C. W., Zimmermann, R. A. and Cebula, G. T. (1981). Fission-track dating of apatite and zircon: an inter-laboratory comparison. *Nucl. Tracks* **5**, 65–72.

Naeser, N. D. and Naeser, C. W. (1984). Fission-track dating. In: Mahaney, W. C. (Ed.), *Quaternary Dating Methods. Developments in Paleontology and Stratigraphy* **7**. Elsevier, pp. 87–100.

Naeser, N. D., Naeser, C. W. and McCulloh, T. H. (1989). The application of fission-track dating to the depositional and thermal history of rocks in sedimentary basins. In: Naeser, N. D. and McCulloh, T. H. (Eds) *Thermal History of Sedimentary Basins*. Springer-Verlag, pp. 157–80.

Petford, N., Miller, J. A. and Briggs, J. (1993). The automated counting of fission tracks in an external detector by image analysis. *Comput. Geosci.* **19**, 585–91.

Price, P. B. and Walker, R. M. (1962a). Chemical etching of charged particle tracks in solids. *J. Appl. Phys.* **33**, 3407–12.

Price, P. B. and Walker, R. M. (1962b). Observation of fossil particle tracks in natural micas. *Nature* **196**, 732–4.

Price, P. B. and Walker, R. M. (1963). Fossil tracks of charged particles in mica and the age of minerals. *J. Geophys. Res.* **68**, 4847–62.

Rahl, J. M., Reiners, P. W., Campbell, I. H., Nicolescu, S. and Allen, C. M. (2003). Combined single-grain (U-Th)/He and U/Pb dating of detrital zircons from the Navajo Sandstone, Utah. *Geology* **31**, 761–4.

Reimer, G. M., Storzer, D. and Wagner, G. A. (1970). Geometry factor in fission track counting. *Earth Planet. Sci. Lett.* **9**, 401–4.

Reiners, P. W., Campbell, I. H., Nicolescu, S. *et al.* (2005). (U-Th)/(He-Pb) double dating of detrital zircons. *Amer. J. Sci.* **305**, 259–311.

Schmidt, J. S., Lelarge, M. L. M. V., Conceicao, R. V. and Balzaretti, N. M. (2014). Experimental evidence regarding the pressure dependence of fission track annealing in apatite. *Earth Planet. Sci. Lett.* **390**, 1–7.

Silk, E. C. H. and Barnes, R. S. (1959). Examination of fission fragment tracks with an electron microscope. *Phil. Mag.* **4**, 970–2.

Storzer, D. and Poupeau, G. (1973). Ages plateaux de mineraux et verres par la methode des traces de fission. *C. R. Acad. Sci. Paris* **276**, 137–9.

Storzer, D. and Wagner, G. A. (1969). Correction of thermally lowered fission track ages of tektites. *Earth Planet. Sci. Lett.* **5**, 463–8.

Storzer, D. and Wagner, G. A. (1982). The application of fission track dating in stratigraphy: a critical review. In: Odin, G. S. (Ed.) *Numerical Dating in Stratigraphy*. Wiley, pp. 199–221.

Wagner, G. A. (1978). Archaeological applications of fission-track dating. *Nucl. Track Det.* **2**, 51–63.

Wagner, G. A. (1988). Apatite fission-track geochrono-thermometer to 60°C: projected length studies. *Chem. Geol. (Isot. Geosci. Sect.)* **72**, 145–53.

Wagner, G. A., Coyle, D. A., Duyster, J. *et al.* (1997). Post-Variscan thermal and tectonic evolution of the KTB site and its surroundings. *J. Geophys. Res.* **102** (B8), 18 221–32.

Wagner, G. A. and Reimer, G. M. (1972). Fission-track tectonics: the tectonic interpretation of fission track apatite ages. *Earth Planet. Sci. Lett.* **14**, 263–8.

Wagner, G. A., Reimer, G. M. and Jager, E. (1977). Cooling ages derived by apatite fission-track, mica Rb-Sr and K-Ar dating: the uplift and cooling history of the Central Alps. *Mem. Inst. Geol. Min. Univ. Padova* **30**, 1–27.

Walter, R. C. (1989). Application and limitation of fission-track geochronology to Quaternary tephras. *Quat. Int.* **1**, 35–46.

Wendt, A. S., Vidal, O. and Chadderton, L. T. (2002). Experimental evidence for the pressure dependence of fission track annealing in apatite. *Earth Planet. Sci. Lett.* **201**, 593–607.

Yoshioka, T., Tsuruta, T., Iwano, H. and Danhara, T. (2005). Spontaneous fission decay constant of ^{238}U determined by SSNTD method using CR-39 and DAP plates. *Nucl. Instrum. Meth. Phys. Res. A* **555**, 386–95.

Chart of the Nuclides

The chart is broken into six sections. The atomic number, Z, is plotted on the ordinate against the neutron number, N, on the abscissa. Mass numbers, A, are ringed. Stable and naturally occurring unstable nuclides are shown, along with a few extinct nuclides of cosmochemical interest. The abundance of stable nuclides is quoted in per cent. (Unstable nuclides with half-lives over 10^{12} yr are included in this category.) Half-lives of naturally occurring unstable nuclides are quoted in seconds, minutes, hours, days and years (s, m, h, d, yr). Data mainly from Lide, D. R. (Ed.) *CRC Handbook of Chemistry and Physics* 1994–5 (75th Edn), CRC Press.

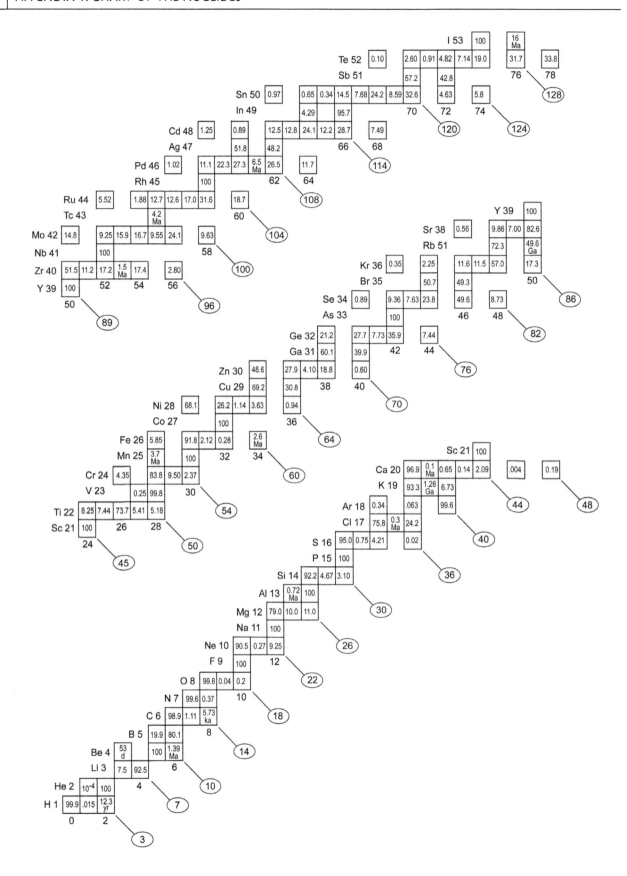

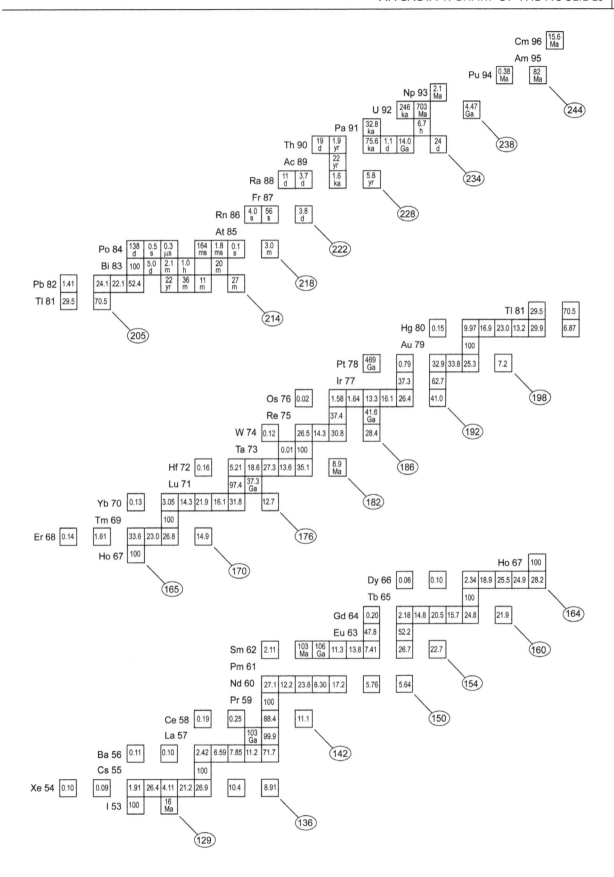

Meteorite Types

1 Outer Solar System Objects

Carbonaceous Chondrite Class
'Chondrites' lacking chondrules (formerly called C1)
    CI (Ivuna) group
        Orgueil
Mini-chondrule clan
    CM (Mighei) group
        Murchison
    CO (Ornans) group
CV-CK chondrite clan
    CV (Vigarano) group
        Allende, Efremovka, Mokoia, NWAfrica 6717, 6670
    CK (Karoonda) group
CR chondrite clan
    CR (Renazzo) group
        Tafassasset
    CB/CH (Bencubbin / Allan Hills 85085) group
        Isheyevo

Anomalous Achondrites
NW Africa 011

Anomalous Pllasites (Stoney-iron meteorites)
Eagle Station

2 Inner Solar System Objects

Ordinary Chondrite Class
H (high iron) group
    Ste Marguerite, Richardton, Kernouve, Beaver Creek, Forest Vale
L (low iron) group
    Bruderheim, Inman, Que (Queen Alex. Range) 97008, Dalgetty Downs
LL (low iron, low metal) group
    Bjurbole, Bishunpur, Chainpur, Semarkona, St Severin, Dhurmsala

Enstatite Chondrite Class
EH (high iron) chondrite group
    Indarch, Kota Kota, Qingzhen, ALHA 77295, Sahara 97158
EL (low iron) chondrite group

Miscellaneous Chondrites
R (Rumuruti) group
K (Kakangari) grouplet

Primitive Achondrites
Acapulco group
Lodranite group
Winonaite group

Asteroidal Achondrites
Basaltic achondrites (including the HED clan, possibly from asteroid 4 Vesta)
    Howardite group
    Eucrite group
        Juvinas, Piplia Kalan, Chervony Kut, Asuka 881394
    Diogenite group
    Angrite group (Angra dos Reis)
        D'Orbigny, Sahara 99555, Lewis Cliff 86010
Enstatite achondrites
    Aubrite group
        Shallowater
Ultramafic achondrites
    Ureilite group
        Dar al Gani 165
    Brachinite group

Lunar Meteorites
Martian meteorites (sometimes called 'SNC meteorites')
Shergottite group
    Shergotty, Zagami, Dar al Gani 476, Que 94201, EET 79001
Nakhlite group

Chassignite group
Other Martian meteorites, e.g., ALH 84001

Pallasites (stoney-iron meteorites)
Main group pallasites
Pyroxene pallasite grouplet
Mesosiderite group

Magmatic Iron Class
Iron I clan
  IC group
Iron II clan
  IIAB, IIC, II D, II F, II G groups
    Rodeo, Carbo

Iron III clan
  IIIAB, IIIE, III F groups
Iron IV clan
  IVA, IVB groups
    Gibeon, Muonionalusta, Santa Clara

Non-magmatic or Primitive Iron Class
I AB clan (formerly groups IAB and IIICD)
  IAB main group
    Canyon Diablo, Nantan
  Udei Station grouplet
  Pitts grouplet
  IIE iron meteorite group

Index

AAA treatment, 365
Abitibi, 80, 86
Abouchami, 128, 129
abrasion, 100, 101, 104, 105, 106, 107, 108,
    197, 207, 257, 258, 312
Abreu, 377, 378
abundance sensitivity, 24, 333
Acapulco, 220, 415
Acasta, 86, 87
accelerator mass spectrometry, 374, 378
achondrite, 45, 67, 68, 72, 113, 185, 220, 408,
    415, 417, 427
Ackerman, 168
Adkins, 374
advection, 308, 382, 383, 384, 385
AFC, 171, 172, 175
Africa, 9, 50, 84, 89, 90, 91, 93, 144, 159, 167,
    168, 169, 179, 181, 197, 198, 262, 316, 383,
    397, 407
Aftalion, 108
AGB star. See red giant
age spectrum plot, 246, 247, 249, 265, 266
Ahrens, 102, 103
Akram, 410
Alard, 203
Albarede, 29, 57, 88, 89, 114, 115, 218, 219,
    228, 230, 231
Alberta, 285
Aldrich, 14, 275, 290
Aleinikoff, 46, 105
Aleutians, 162, 389
Alexander, 91
Allan Hills, 218, 219, 395, 396, 407
Allegre, 58, 75, 83, 84, 114, 117, 138, 139,
    140, 145, 146, 147, 153, 155, 194, 196, 197,
    199, 200, 201, 278, 281, 286, 291, 296, 327,
    334, 335, 338, 339, 343, 344, 426
Allen, 31
Allende, 27, 45, 117, 118, 218, 407, 415, 416,
    417, 418, 420, 421, 424, 425, 428, 429, 434,
    435, 436, 437
alpha decay, 8
Alpher, 119
Alps, 47, 48, 71, 220, 452, 453
Alvarez, 274, 275
Amakawa, 234, 235
Amazon, 57, 60, 79, 88
Amelin, 117, 118, 219, 220, 227, 228
Amitsoq, 86, 87, 121, 122, 123, 124, 185, 186,
    226, 227, 229, 232
AMOC, 317, 318, 319, 320
AMS, 363, 378, 379, 380, 390, 393, 394, 395,
    396, 397, 398
analytical blank, 100, 126, 292
Anders, 247
Andersen, 148
Anderson, 136, 277, 281, 364

Andes, 60, 355, 356, 358, 389, 397, 398
Ando, 56, 61
Andreasen, 429
Andree, 374
Andrews, 392
Angra dos Reis, 45, 408, 415, 420, 426, 427
angrite, 67, 68, 219, 417, 420, 421, 428
annealing, 31, 105, 106, 111, 112, 269, 447,
    451, 452, 453, 454, 455, 457, 458, 459, 460,
    461, 462
anorthite, 408, 416, 435
Antarctic, 88, 90, 93, 128, 129, 218, 269, 270,
    317, 319, 372, 373, 375, 376, 383, 385, 394,
    395
Antarctica, 55, 110, 129, 250, 256, 336, 348,
    375, 377, 384, 385, 391, 396, 407
anthropogenic, 125, 126, 127, 276, 320, 371,
    372, 373, 376, 378, 391, 392, 393, 394
anthropogenic Pb, 126, 127, 320
anti-coincidence, 364
apatite, 45, 220, 267, 268, 269, 415, 447, 449,
    451, 452, 453, 454, 455, 457, 458, 459, 460,
    462
Apollo, 43, 275
Appalachians, 49, 125
Appel, 114
aragonite, 62
Arden, 434
argon loss, 244, 246, 247, 248, 249, 251, 252,
    255, 256, 259, 261, 262, 263, 264, 265
Arizona, 81, 113, 262, 398, 399
Armstrong, 56, 57, 61, 82, 83, 84, 85, 121,
    281, 416
Armytage, 299
Arnaud, 266
Arndt, 76, 77, 78
Arnett, 416
Arnold, 364, 380, 383, 387
Arrhenius, 47, 198, 250, 259, 260, 261, 263,
    265, 266, 269, 451, 452, 453, 454, 455, 460
Artesian Basin, 392, 393
Ascension, 134, 135, 151
Asmerom, 53, 354
assimilation, 44, 171, 172, 174, 175, 176, 181,
    190, 202, 283, 293, 389
Aston, 13, 22
astrochronology, 101, 102
Atkinson, 3
Atlantic, 88, 89, 91, 92, 93, 94, 127, 128, 129,
    136, 137, 138, 139, 145, 147, 149, 150, 151,
    154, 157, 160, 177, 181, 182, 202, 203, 212,
    222, 230, 231, 234, 235, 277, 283, 284, 293,
    310, 311, 315, 317, 318, 319, 343, 344, 347,
    348, 371, 373, 374, 375, 383
Atlantic Ocean, 88, 89, 92, 93, 94, 127, 128,
    212, 230, 231, 234, 317, 318, 319, 371, 373,
    383

atmospheric contamination, 241, 242, 243,
    244, 246, 274, 285, 286, 287, 288, 289, 290,
    291, 292, 293, 296
atmospheric fractionation, 286
Augland, 107
Australia, 49, 50, 69, 78, 109, 111, 179, 180,
    189, 202, 227, 233, 269, 300, 322, 392, 453,
    454, 460
Australopithecus, 397
Awwiller, 79, 80
Azores, 135, 136, 140, 141, 148, 151, 154, 202,
    278, 347, 348, 353

BABI, 44, 45, 185
Bachmann, 257
Bacon, 89
baddeleyite, 105, 111, 112
Baffin Island, 278, 279, 281, 283
Bahamas, 323, 371, 372
Baksi, 256, 257
Balco, 396, 397, 398
Ballentine, 288
Barbados, 311, 370
Barbarand, 459
Barberton, 80, 252
Bard, 311, 370, 374, 377, 384, 385
Barling, 154
Bar-Matthews, 324
Barovich, 15, 81
Basin and Range, 354
Basu, 58
Bateman, 10
Bateman equation, 321
Batiza, 138
Bau, 231
Baumgartner, 391
Baxter, 368
Bayes, 369
Bayesian statistics, 369
Bayon, 231
Beattie, 347
Beck, 58, 371, 372
Becker, 60
Beckinsale, 240, 257
Beer, 384, 385, 386, 387, 391
Beerling, 212
Begemann, 11, 41, 436
Bell, 158, 180, 228, 230
Bellot, 233
Belshaw, 31
Ben Othman, 161
Ben Vuirich, 108
Bender, 311
Bennett, 75, 86, 87, 199, 200, 202, 226, 407
Benoit, 320
Bentley, 391, 392
Berger, 248, 249, 259, 260, 261

Bermuda, 94, 318, 319
Bernatowicz, 300
Bertram, 89
beta decay, 7
Bethe, 3
Bezard, 160
Bhandari, 455
Bierman, 400
Bigazzi, 455
Bijwaard, 137
biotite, 42, 46, 47, 48, 173, 198, 207, 235, 244, 248, 249, 253, 255, 257, 259, 260, 261, 262, 264, 265, 452
bioturbation, 323, 369, 374, 375, 376
Birck, 44, 420
Bischoff, 325, 326
Bishop Tuff, 341
Bizimis, 222, 223
Bizzarro, 219, 220, 227, 228, 418, 422
Bjurbole, 246, 247
Black, 36, 287, 433
Blake, 6, 433
blank levels, 16
Blichert-Toft, 29, 114, 118, 148, 218, 219, 222, 225, 420, 426
blocking temperature, 47, 48, 111, 112, 198, 220, 221, 251, 259, 260, 262, 265, 269, 420, 436
Blum, 60, 61
Bock, 80
Boehnke, 67
Bohm, 319, 320
Boltzmann, 451, 453, 454, 460
bomb signature, 372
Borg, 428, 430, 432
Bouhifd, 280
Bourdon, 326, 346, 347, 348, 353, 354, 358, 359
Bouvier, 68, 218, 220, 227, 228, 229, 430, 431
Bowring, 83, 86
box models, 90, 141, 142, 143, 144, 280, 281
Boyet, 29, 143, 144, 155, 429, 431, 432
Brahmaputra, 58, 60
Branca, 337
Brand, 53
Brandon, 11, 184, 185, 198, 199, 205, 206
Brannon, 49
Brass, 57
Brazzle, 415
Brenan, 198
Brennecka, 101, 117, 118, 408, 434, 435
Brereton, 245
Brett, 181
Briggs, 453
bright reflection, 457
Brinkman, 40
Briquet, 171, 172
bristlecone pine, 367, 368
Broecker, 373, 374, 375, 393, 411, 412
Brooks, 32, 33, 43, 44
Bros, 80

Brown, 47, 381, 383, 387
Bruderheim, 248
Bruland, 320
Brunhes, 255, 256, 259
Bruns, 368
Bucha, 366, 367
Buchan, 259
Buck, 369
Bulk Earth, 72, 73, 113, 114, 119, 125, 140, 141, 143, 145, 147, 148, 149, 151, 155, 168, 179, 199, 200, 233, 234, 424, 429, 430, 431, 433
Bulk Silicate Earth, 74, 114, 144, 149, 150, 152, 198, 199, 290, 424, 433
Bunge, 139
Burbidge, 3, 4
Burgoyne, 18
burial ages, 396
Burke, 53, 54, 144, 159, 375
Burkhardt, 410, 425, 432
Burnham, 181
burn-out, 425, 427
Burrows, 4
Burtner, 458
Burton, 71, 91, 92, 93, 129, 149, 150, 202, 214, 312, 313
Bushveld, 43, 207, 208, 209, 233
Bushveld complex, 43
Butler, 295
Butterfield, 58

Cabral, 157
Cadogan, 250
Caffee, 297, 298, 300
CAI, 408, 417, 418, 419, 422, 424, 425, 426, 428, 429, 431, 432, 435, 436, 437
CAIs, 116, 117, 118, 408, 410, 411, 415, 417, 418, 419, 420, 421, 423, 425, 434, 435, 436, 437
California, 187, 189, 214, 254, 335, 340, 367, 374, 375, 401, 454, 455
Cameron, 17, 413, 414
Cameroon, 181, 182
Campbell, 31, 143
Campin, 373
Canada, 49, 77, 83, 86, 100, 187, 198, 209, 260, 285, 297, 458
Canadian Shield, 73, 82
Canaries, 135, 148, 151, 204, 234
Canup, 410
Canyon Diablo, 113, 114, 117, 118, 119, 120, 121, 407, 420
Capaldi, 335, 341
Cape Hatteras, 316, 393
Cape Verde, 148, 234
Carbo, 425
carbonaceous chondrites, 2, 68, 116, 218, 219, 275, 288, 294, 408, 409, 410, 411, 417, 420, 421, 422, 423, 424, 425, 436
carbonatite, 158, 180, 181, 234, 287, 341, 342
Cariaco Basin, 214, 370, 371

Caribbean, 92, 160, 214, 233, 314, 370, 371
Carl, 115
Carlson, 29, 116, 144, 183, 184, 185, 429, 431, 456, 458, 459, 461
Caro, 26, 29, 143, 155, 236, 428, 430, 431, 433
Carter, 451
Cartigny, 181
Casse, 5
Cassidy, 15
Castillo, 158, 159
Catanzaro, 15, 40
Catchen, 10
Catlos, 112
Cattel, 86
cerium anomalies, 234
Cerling, 277
Cerrai, 334, 335
Chabaux, 312, 353, 357
Chainpur, 417, 421
chalcophile, 194, 197
Chamberlain, 185
Champagne, 417
Channell, 256
Chapman, 70, 105
Chappell, 189
charge stripping, 378
Charlier, 335, 336, 340
Chase, 85, 146, 158, 317
Chassigny, 299, 428
Chaudhuri, 58
Chaudisson, 419
Chaussidon, 181, 422, 436, 437
Chauvel, 69, 158
Chauvenet, 25
Chen, 15, 25, 117, 135, 308, 310, 419, 422, 434
Cheng, 11, 29, 308, 309, 327, 328
Chengde, 387
Cherdyntsev, 309
Chervony Kut, 420, 421, 422
Chesley, 185
Chester, 89
Chiu, 327
Chmeleff, 380, 395
chondrite, 45, 68, 117, 162, 196, 197, 198, 199, 218, 220, 226, 233, 289, 293, 294, 295, 300, 407, 408, 410, 413, 415, 417, 418, 420, 421, 422, 424, 429, 430, 431, 432, 433, 434, 436
chondrules, 116, 117, 118, 247, 408, 410, 415, 417, 418, 421
Chow, 126, 127, 128
Christensen, 29, 49, 141
Christiansen, 185
chromitite, 207, 208, 209
CHUR, 31, 68, 69, 72, 73, 74, 81, 82, 140, 141, 143, 198, 218, 227, 228, 229, 230, 430
Circumpolar, 319, 375
Claoue-Long, 69
Clark, 47

Clarke, 275, 277
Class, 135, 204, 279, 281
Clauer, 51, 58
Clayton, 27, 409, 413, 417, 437
Clemens, 55, 56
Cliff, 47
climate change, 94, 372, 376, 378
closure temperature. *See* blocking
    temperature
cobalt dating, 92
Cochran, 316
Coe, 256
Coggon, 59, 208
Cohen, 134, 138, 168, 198, 209, 210, 211, 212,
    333, 344, 377
Colin, 287, 288
Collerson, 148, 158
Collins, 190
Colorado, 74, 226, 235, 236, 244, 354, 446
Columbia, 72, 183, 184, 185, 336, 388
Columbia River, 72, 183, 184, 185, 388
Comore, 135, 353
Compston, 27, 28, 42, 45, 50, 69, 109, 110
concordia, 35, 41, 69, 83, 99, 102, 103, 104,
    107, 108, 110, 111, 115, 116, 257, 258, 326,
    327, 328, 329
Condomines, 335, 337, 338, 339, 340, 341,
    342, 343, 348, 349
confined tracks, 456, 457
conformable lead, 120
Connelly, 118
continental run-off, 56, 57, 90
conventional ages, 365
Coogan, 58, 59, 60
cooling rate, 47, 259, 260, 262, 269, 452
Copeland, 111
core–mantle boundary, 137, 144, 154, 159,
    206, 280
Corfu, 104
correlated errors, 32
Corrigan, 461, 462
Corson, 444
cosmic abundance, 2
cosmic rays, 2, 5, 219, 276, 363, 364, 366, 376,
    380, 395, 435, 445
Cotte, 23
Cougar Point, 31
Cowan, 9, 101
Cowie, 51
Cox, 242, 254
cpx, 70, 136, 149, 156, 168, 169, 170, 222,
    223, 347, 353
Crabb, 294, 415
Craig, 127, 276, 285, 292, 366
Creaser, 194, 195, 198
Crock, 14, 15
CRONUS, 401
Croudace, 14
Crowley, 31
Crozaz, 320
Crumpler, 25

crustal contamination, 44, 69, 119, 135, 160,
    167, 171, 172, 173, 174, 175, 176, 178, 179,
    180, 183, 188, 190, 205, 208, 209, 233, 341
crustal formation, 73, 74, 75, 76, 77, 78, 80,
    82, 83, 122, 123, 124, 125, 187, 228
crustal residence age, 79, 83
Cuk, 410, 432
Cullers, 80
Culshaw, 78
Cumming, 120

D, 139, 281
D'Orbigny, 420, 421, 426
Dahl, 111, 112
Dakowski, 455
Dalrymple, 241, 242, 243, 245, 246, 247, 256
Daly, 24
Daly detector, 24
Damon, 242, 290
Danebury, 369
Das, 106
Dasch, 52, 128, 134, 135
dating gap, 306
daughter-deficiency, 306, 309, 321
daughter-excess, 306
Dauphas, 410, 422
David, 15, 107
Davidson, 160, 178
Davies, 139
Davis, 32, 34, 41, 57, 58, 59, 104, 106, 187,
    398
Dawson, 17, 168, 169, 179, 180
Day, 427
de Bievre, 309
De Jong, 368
de Vries, 364, 366, 368, 384
Debaille, 428, 430
DeBievre, 27
decay chains, 10, 99, 161, 285, 306, 307, 341
decay constant, 10, 40
decay equation, 41, 99, 118, 119, 198, 218,
    232, 235, 240, 245, 310, 313, 321, 341,
    445
deep sea corals, 375
defects, 104, 262, 264
deglaciation, 62, 93, 94, 318, 319, 323, 324,
    373, 374, 375, 376
Deino, 256
Del Moro, 47
DeLaeter, 101
Delavault, 157
Dempster, 13
dendrochronology, 11, 367, 369, 370, 371,
    374, 376
Deng, 318
DePaolo, 17, 58, 68, 69, 72, 73, 74, 75, 77, 82,
    85, 140, 143, 171, 172, 179, 183, 184, 185,
    188, 226, 228, 235, 236
depleted mantle, 72, 73, 74, 75, 77, 80, 82,
    84, 85, 86, 88, 142, 143, 144, 145, 150, 152,
    154, 155, 158, 161, 183, 184, 185, 186, 189,

200, 222, 226, 227, 228, 229, 230, 278, 356,
    432
Derry, 53
detrital zircons, 35, 78, 79, 227, 228, 450, 451
Devils Hole, 323, 324
DeVries, 374
Dewey, 159
DeWolf, 77, 112, 125
Dhuime, 75, 187, 228
Dia, 55, 84
diamond, 11, 167, 168, 181, 276, 296, 300
Dicke, 413
Dickin, 15, 73, 74, 75, 77, 124, 143, 155, 176,
    177, 197, 209, 210, 233, 234, 431, 432
Dietz, 209
diffusivity, 136, 259, 261, 263, 264, 268, 269,
    274, 277, 278, 281
discordia, 34, 35, 102, 103, 104, 106, 107,
    108, 109, 111, 112, 115, 116, 327, 328
disequilibrium melting, 136, 174, 182, 222,
    351
Dixon, 218
DMM, 144, 152, 153, 154, 168, 204, 228, 283
Dodd, 242, 408
Dodson, 25, 27, 35, 47, 48, 185, 259, 269
Doe, 114, 120, 121, 125
Doell, 255
Dome C, 256, 384, 385
Donelick, 456, 457, 458, 459, 462
Dosso, 15, 140, 141
double focussing, 18, 24, 109
double spiking, 27
double-spike, 101
Doucelance, 234
Douglass, 157
Drake Passage, 129, 375
Dreyfus, 256
Druffel, 372
DSDP, 54, 55, 255
Duce, 128
Duchene, 220
Duddy, 460
Dunoyer, 50
Dupal, 159
Dupre, 134, 145, 150, 159, 160
Durango, 268, 269, 449, 457, 458, 460
Durrani, 457
dynamic melting, 344, 345, 346, 351, 352,
    353, 359, 360
Dziewonski, 144, 159

Eagle Station, 408, 420
East African Rift, 158, 180, 341
East Pacific Rise, 57, 128, 138, 213, 214, 289,
    336, 337, 344, 346, 349, 350, 351
Eberhardt, 19, 20
Eddy, 376, 384
Edmond, 57
Edwards, 29, 308, 309, 322, 323, 354, 370, 389
Efremovka, 117, 419, 435, 436, 437
Eiler, 156, 157

Eisele, 225
Elderfield, 58, 89, 234
Eldridge, 181
electron capture, 8, 11, 218, 232, 240, 257, 258, 300
Elkins, 348, 352, 408
Ellam, 160, 161
Elliott, 135, 141, 147, 148, 281, 345, 356
Elmore, 391, 393
Elsasser, 366
EMI, 152, 153, 154, 156, 157, 158, 159, 168, 180, 181, 204, 223, 225, 289
EMII, 152, 154, 155, 156, 159, 168, 204, 223
Endt, 257
English, 61
Enkelmann, 449
enstatite, 143, 155, 199, 287, 288, 289, 293, 294, 408, 409, 410, 418, 420, 429, 431
enstatite chondrites, 143, 155, 199, 287, 288, 289, 293, 294, 409, 410, 418, 429, 431
equiline, 334, 337, 338, 342, 343, 344, 345, 349, 350, 354, 355, 356
Erlank, 169, 170
erosion island, 396, 397, 398
errorchron, 33, 34, 36, 48, 70, 86, 87, 123
Esat, 311
ESR dating, 328
Esser, 212, 250
etching, 445, 446, 447, 448, 451, 455, 457, 458, 459
Etna, 335, 337, 339, 341
eucrite, 67, 218, 219, 408, 417, 421, 430
Eugster, 15, 294
Evans, 394, 395
Excel, 35
excess argon, 242, 243, 244, 249, 250, 255, 290
exposure ages, 269, 270, 277, 285, 394, 395, 399, 400, 425
extended geometry, 23
external detector method, 448, 449

Fabryka-Martin, 393, 394
Faeroes, 137
Faggart, 209
Fairbairn, 45
FAMOUS, 343, 344
Fanale, 290
Faraday, 18, 22, 23, 24, 25, 26, 29, 30, 194, 275, 308, 309, 379
Faraday bucket, 24
Farley, 152, 268, 269, 283, 284, 285, 287, 292, 293
Farquhar, 157
Farrell, 56
Faul, 13
Faure, 56, 134
Fehn, 393, 394
Fe–Mn crusts, 91, 92, 128, 129, 211, 230, 231, 312, 313
Fe–Mn nodules, 88, 230, 235

Feng, 30
Ferguson, 367
Fermi, 7
Feulner, 378
Fick's first law, 261
Field, 45
Fietzke, 62
Fink, 395
Finland, 226, 232
Fish Canyon, 101, 256, 257, 258, 446, 449
Fish Canyon tuff, 101, 256, 258
Fisher, 29, 31, 32, 125, 279, 290, 292
fissiogenic, 274, 295, 296, 297, 298, 301
fission product, 9, 274, 393, 394
Fitton, 182
FitzGerald, 266
Fleischer, 444, 445, 446, 447, 451, 454, 456, 462
Fleitmann, 376
flicker, 18
Fleitmann, 376
flicker, 18
Florida, 55
Flower, 135
fluid inclusions, 48, 49, 250, 283, 292, 300, 400
fluorapatite, 458
flux-monitor, 447
Flynn, 40
Foland, 251, 265
foram, 54, 55, 59, 61, 92, 94, 284, 374, 376
Forbush, 366
Fornari, 350
Forsyth, 346
forward modelling, 459
Foster, 111
FOZO, 152, 153, 154, 155, 159, 204, 278, 279, 432
fractional correction, 55
fractionation processes, 19
Francalanci, 178
Frank, 92, 129, 386
Fraser, 180
Frei, 91
French, 112
Froude, 109, 110
Fujimaki, 222
FUN inclusions, 421

Gaber, 261
Galapagos, 62, 151, 279, 288, 289, 293, 349
Galbraith, 375, 450, 455, 458
Gale, 27
galena, 99, 114, 118, 119, 120, 121, 124, 145, 158
Galer, 28, 67, 85, 114, 147, 148, 150, 345, 346, 428
Gallup, 311
Ganges, 58, 60, 212
Gannoun, 202, 429, 430, 432
Garapic, 279

Garner, 240
Garnero, 144
garnet, 67, 70, 71, 111, 112, 161, 167, 168, 169, 170, 201, 220, 221, 222, 347, 348, 349, 352, 353, 354
Garnet, 70, 112
Gast, 52, 134, 135, 145, 342
Gatti, 195
Geiss, 275
Geissel, 2
General Circulation Model, 373, 374
Genesis Mission, 275
Gentry, 105
geochron, 113, 114, 121, 145, 151, 281
GEOSECS, 372, 373, 374, 375
Gerling, 276
Gherardi, 318, 319
giant impact, 115, 144, 155, 299, 408, 410, 424, 427, 429, 432
giant impact model, 424, 427
Gibeon, 419, 420
Gibson, 178, 179
Giletti, 261
Gill, 345, 348, 355, 357
Gillson, 21
glacial cycles, 11, 55, 56, 61, 92, 94, 213, 214, 255, 284, 285, 322, 323, 386
glacial maximum, 93, 94, 317, 318, 372, 375, 376, 385
glauconite, 50, 51, 52, 251
Glavin, 420, 421
Gleadow, 449, 453, 454, 455, 456, 457, 458, 460
global warming, 371, 376, 378
Godfrey, 230
Godwin, 365
gold standard, 35, 107, 257
Goldberg, 88, 234, 309, 315, 320, 382
Goldmann, 434, 435
Goldrich, 103
Goldstein, 57, 74, 75, 76, 79, 84, 85, 88, 129, 279, 333, 375
Gopalan, 236
Gopel, 415, 421
Goslar, 370
Goswami, 418
Gounelle, 436
Graf, 397
Graham, 185, 222, 223
Granger, 396, 397
Grant, 52, 324
Gray, 44, 45, 189, 376
Green, 73, 396, 446, 453, 454, 455, 457, 458, 460, 461
Greenland, 84, 86, 87, 91, 114, 121, 122, 123, 124, 129, 185, 186, 187, 202, 227, 229, 230, 232, 371, 373, 377, 378, 380, 384, 385, 391, 428, 433
Gregoire, 170
Grenville, 73, 74, 77, 78, 100, 124, 125, 209, 219, 260, 261

Grenville Province, 73, 74, 77, 100, 124, 125, 260
Griffin, 70, 124
Grimberg, 275
Grove, 112
Grove and Harrison, 112
Grun, 328, 329
Guitreau, 76, 77
Gurenko, 181
Gussone, 62
Guyodo, 371, 372

Habfast, 21
Hagee, 415
Haines, 121
Hale, 252
Haleakala, 256, 276
half-life, 9, 10, 40, 41, 67, 68, 72, 99, 118, 122, 143, 194, 195, 196, 197, 198, 207, 218, 219, 220, 232, 240, 244, 256, 257, 268, 299, 300, 306, 307, 308, 309, 310, 314, 319, 320, 321, 326, 328, 333, 336, 338, 340, 342, 343, 345, 349, 350, 353, 363, 364, 365, 367, 380, 381, 384, 387, 389, 390, 393, 394, 395, 408, 411, 412, 414, 415, 416, 418, 419, 420, 421, 423, 424, 426, 427, 433, 435, 436, 438
Halliday, 18, 19, 30, 45, 108, 115, 182, 294, 301, 410, 427
Hamelin, 28, 127, 222
Hamilton, 68, 70, 84, 86
Hammer, 370
Hammouda, 173
Hanan, 152, 222, 278
Handley, 225
Hanes, 261
Hanna, 7
Hansen, 2, 43
Hanson, 81
Hanyu, 278, 289
Harding County, 286, 288, 294, 295, 297, 300
Harmon, 156
harmonic oscillator, 19
Harper, 260, 287, 423, 428, 433
Harquahala, 81
Harris, 51, 135, 136
Harrison, 126, 227, 228, 259, 262, 265
Hart, 136, 149, 151, 152, 153, 154, 155, 156, 159, 161, 168, 183, 184, 199, 207, 208, 242, 244, 278, 279, 290, 291, 296
Harte, 167
Hartman, 424
Harvey, 58, 149, 203
harzburgite, 138, 205, 208, 296
Hasebe, 449
Haskin, 73
Hastie, 85
Hattori, 199
Hauri, 149, 153, 154, 155, 156, 204
Hawaii, 135, 151, 155, 157, 204, 223, 243, 277, 279, 281, 310, 342, 346, 353, 381

Hawkesworth, 140, 141, 160, 161, 168, 169, 170, 171, 179, 355
HDEHP, 15
Heaman, 112
heat flow, 280, 282
Hebrides, 124
HED parent body, 421
Heinrich, 318, 319, 324, 374
Heirtzler, 11
Heisinger, 397
Heizler, 265
Hekla, 350
helium burning, 4
helium paradox, 279
Hellstrom, 326
Hemming, 181
Hemond, 335, 350, 355
Henderson, 55, 56, 61, 62, 127, 310, 312, 313, 323
Hennecke, 295
Henning, 393
Hensley, 11
Herr, 218
Hertzsprung–Russell, 3
Hess, 54, 58
HIBA, 15
Hiess, 101, 229
Hillam, 369
Hilton, 277, 279, 282, 283
Himalayas, 60, 111, 265
HIMU, 152, 153, 154, 155, 156, 157, 158, 159, 168, 180, 181, 204, 223, 278, 283, 348
Hinton, 109, 110
Hirata, 30, 195
Hirt, 194, 196, 198
Hiyagon, 278, 286, 287
Hodell, 52, 55, 61
Hodges, 262, 263
Hoefs, 188
Hoffmann, 230, 371
Hofmann, 47, 136, 137, 141, 149, 150, 158, 210, 380, 396
Hohenberg, 413
Holden, 227, 445
Holland, 294
Holmes, 103, 118, 119, 235
Holst, 418
Honda, 286
Hooker, 17
Hooper, 185
Hopp, 289
Horan, 209, 420
Horie, 88
Horn, 30
hornblende, 220, 242, 244, 248, 249, 252, 253, 254, 257, 260, 261, 264, 335, 340, 401
Horwitz, 14
hot atom effect, 309, 311, 323, 335
Houk, 17, 21
House, 269
Houtermans, 113, 118, 119

Hua, 371
Huang, 282, 359
Huber, 368
Hughen, 371
Hulett, 181
Hungary, 282, 325
Hunziker, 47
Huon Pine, 371
Huppert, 175
Hurford, 446, 448, 449, 450
Hurley, 51, 82, 83, 188, 268
Huss, 288, 437
Hutcheon, 417, 436
Hutton, 10
hydrofluoric acid, 13, 104
hydrogen-burning, 4
hydrothermal exchange, 56

Iben, 3
ice core, 256, 384, 385, 391
Iceland, 136, 137, 145, 148, 151, 155, 156, 204, 222, 277, 278, 279, 287, 293, 296, 297, 298, 346, 348, 350, 353
Ickert, 181, 341
ICP-MS, 15, 17, 18, 19, 21, 22, 25, 28, 29, 30, 31, 35, 111, 178, 194, 195, 218, 220, 222, 224, 225, 227, 230, 309, 389, 409, 418, 419, 423, 434, 449
IDP, 283, 284, 285
Illinois, 381
illite, 50, 51
Imbrie, 322
impact event, 55, 212, 399, 432
Indian Ocean, 92, 93, 128, 129, 154, 159, 212, 214, 225, 342, 348, 376, 389
Ingram, 16, 91
inherited argon, 243, 244, 246, 252, 267, 268, 290
insolation, 101, 255, 322, 323, 324
Intcal, 372, 377
inter-planetary dust, 283
ion exchange, 14
ion gun, 109, 379
Ion optics, 22
Ireland, 206, 319
iron meteorites, 44, 113, 118, 196, 197, 206, 395, 408, 419, 420, 422, 423, 424, 425, 427
irradiation, 219, 220, 244, 245, 246, 250, 251, 252, 264, 266, 287, 290, 412, 417, 425, 436, 444, 445, 446, 447, 448, 452, 459
island-arc, 159, 160, 161, 188, 224, 236
isobaric interference, 14, 28, 31, 109, 233, 380, 390, 393
isochron diagram, 42, 43, 44, 46, 48, 51, 67, 69, 87, 100, 113, 116, 117, 119, 120, 121, 122, 123, 124, 169, 170, 176, 182, 189, 196, 198, 199, 200, 201, 203, 222, 223, 232, 243, 244, 246, 247, 249, 252, 253, 321, 324, 325, 326, 334, 335, 336, 339, 340, 341, 343, 350, 355, 356, 357, 397, 413, 416, 433, 450, 451
isochron regression, 32

isomer, 8, 219
isoplot, 35
isotope dilution, 14, 16, 19, 26, 27, 32, 43, 125, 194, 195, 196, 240, 241, 242, 257, 268, 308, 434
isotope mapping, 75
Israel, 324, 329
Isua, 84, 86, 87, 91, 110, 114, 124, 230, 428, 433
Italy, 47, 161, 335
Ito, 437
IUGS Subcommission on Geochronology, 11, 446
Ivanovich, 308, 328

Jack Hills, 111, 227, 228, 229
Jackson, 155, 205, 279, 281, 283, 429
Jacobsen, 68, 83, 90, 91, 124, 141, 142, 410, 418, 419, 424, 437
Jacobson, 60
Jaffey, 11, 100, 107, 120
Jager, 47, 100
Jakes, 355
Jan Mayen, 315
Japan, 149, 286, 370
Japanese, 205, 206, 388, 389
Java, 225
Javoy, 293, 410
Jeandel, 90
Jeffrey, 276, 412, 413
Jego, 162
Jenner, 149, 150
Johannesson, 90
Johnson, 18, 223, 255
Joly, 333
Jonas, 328
Jonckheere, 446, 457
Jones, 58, 128, 169, 175
Jonkers, 319
Joplin, 120
Jordan, 167, 200
Juan de Fuca, 58, 59, 138
Juan Fernandez, 292
Jull, 379
Juvinas, 44, 67, 68, 72, 148, 149, 420

Kaapvaal, 167, 168, 170
Kaczor, 173
Kalsbeek, 36
Kambalda, 69, 85
Kamber, 87, 91, 123, 148, 234
Kamen, 363
Kamenetsky, 180, 181
Kaneoka, 250, 279, 291
kappa conundrum, 122, 147, 148
Kapusta, 251
Kauffman, 336
Kaufman, 53, 310, 321, 322, 326
Kaufmann, 327
Keay, 190
Keigwin, 318, 375, 376
Kelemen, 158, 352

Kelley, 162, 250, 254, 264, 377
Kellogg, 139, 140, 143, 281, 296
Kelly, 419
Kemp, 78, 190, 228, 229, 230
Kennedy, 285
Kentucky, 396, 397
Kenyon, 138, 139
Kerguelen, 140, 141, 151, 154
Kerr, 176, 177
Ketcham, 458, 459, 461
Keto, 80, 90, 91
Key, 373
K-feldspar, 42, 45, 46, 50, 235, 244, 249, 250, 251, 254, 257, 260, 261, 263, 265, 266, 267, 268, 269, 401
Khlapin, 321
Kick'em Jenny, 359
Kieser, 379
Kigoshi, 334, 335
Kilauea, 206, 242, 285, 286, 287, 354
Kilburn Hole, 201, 202
Kille, 176
Kim, 90
kimberlite, 168, 169, 170, 179, 180, 181, 200, 222, 249, 262, 276
Kimberly, 169, 170
Kimura, 449
kinetic factor, 458
Kinoshita, 427
Kirk, 197
Kirkley, 181
Kita, 118, 418
Kitagawa, 370
Klein, 395
Kleine, 424, 425, 426, 427
Klemm, 129, 211
Klotzli, 30
Knesel, 173, 174
Kober, 103, 104, 105, 106, 397, 398
Kohn, 462
Koide, 320
Kok, 386
Kokfelt, 346
Kola Peninsula, 180, 287
komatiite, 69, 80, 167, 168, 252, 300
Konig, 426
Koolau, 156, 157, 204, 205, 206, 225
Korenaga, 144
Korschinek, 380, 395
Kosler, 30
Kossert, 41
Kowallis, 450
Krabbenhoft, 62
Kramers, 122, 123, 147, 148, 169
KREEP, 431
Krishnaswamy, 320
Krivova, 377
Krogh, 13, 34, 104, 105, 112, 209
Kromer, 371
Krot, 408
Kruijer, 424, 425, 426, 427
K-T boundary, 55, 211, 259, 284

Ku, 309, 310, 314, 315, 324, 327, 383, 384
Kubler, 50
Kuiper, 256, 259
Kulp, 290
Kumari, 122
Kunz, 276, 293, 297
Kuritani, 15
Kuroda, 295
Kurz, 24, 275, 276, 277, 278, 279, 289
Kusakabe, 383
Kyser, 286

Labanieh, 160
Labidi, 157
Labrie, 379
Lacan, 90
Lachlan fold belt, 189, 190
Lagos, 115
Lahaye, 80, 81
Laj, 372
Laken, 376
Lal, 276, 277, 378, 381, 385, 396, 445, 456
Lambert, 69, 208, 209
Landwehr, 347
Langmuir, 344
Lanphere, 46, 241, 244, 249
Lao, 385
Larderello, 47
laser, 13, 29, 30, 31, 79, 111, 178, 195, 208, 227, 228, 229, 252, 253, 254, 255, 257, 261, 262, 263, 264, 267, 287, 309, 379, 418, 449
laser ablation, 13, 29, 30, 31, 79, 111, 178, 195, 208, 227, 228, 229, 252, 253, 254, 264, 267, 418, 449
Laslett, 451, 455, 456, 457, 458, 459, 460
Lassiter, 201, 203, 204, 205
LaTourrette, 347
Lau Basin, 283
Laurentia, 125
Lay, 144, 159
Layer, 253, 261
Lean, 377
Lear, 59
Lederer, 7, 8
Lee, 29, 230, 247, 253, 254, 263, 264, 267, 379, 416, 417, 423, 424, 427, 435
Lesser Antilles, 160, 161, 357, 358, 359
Leucite Hills, 223, 224, 226
Levasseur, 212, 213, 214
Lewis, 67, 68, 407, 417, 418, 420, 426, 428
Lewis Cliff, 67, 68, 407, 418, 420, 426, 428
Lewisian, 70, 81, 176, 177, 233
Leya, 409, 436
lherzolite, 138, 139, 149, 167, 169, 201, 202, 205
Li, 2, 4, 5, 17, 276, 308, 323, 380, 436
Libby, 363, 364, 365
LILE, 161
Lindner, 195, 196
linear regression, 32
Ling, 91, 92, 129

Lippold, 318
Litherland, 378, 379, 380
lithophile, 48, 74, 135, 145, 146, 152, 155, 156, 161, 185, 194, 197, 200, 204, 209, 218, 244, 278, 279, 281, 282, 283, 289, 299, 419, 423, 429, 432
Liu, 157, 204, 398, 399, 400, 418, 437
LLSVPs, 144, 159
Lo, 264
Lockwood, 378
loess, 128, 129, 231, 387
Loihi, 152, 156, 157, 206, 277, 278, 286, 287, 289, 291, 292, 296, 297
Long Valley, 340, 341
Lopez Martinez, 252
Lovera, 263, 264, 265, 266, 267, 268, 269
Luais, 29
Luck, 194, 196, 197, 198, 199, 200
Ludwig, 26, 32, 33, 34, 35, 100, 107, 111, 115, 309, 310
Lugaro, 438
Lugmair, 11, 17, 67, 68, 73, 257, 420, 421, 422, 427
Lundstrom, 349, 351, 352
Luo, 29, 317, 326
Lupton, 275, 277, 278

Maaloe, 167
Maas, 180, 181
Macfarlane, 198
MacLeod, 55
MacPherson, 419
magma chamber evolution, 336, 337, 338, 340
magnetic remanence, 251
magnetic reversal, 254, 255
magnetic sector, 22
magnetite, 278, 340
Mahaffy, 275
Mahara, 392
Mahoney, 417
Maier, 200
Maine, 80, 265, 266
Makide, 418
Makishima, 233, 234
Malaviarach, 149, 150
Mammoth Cave, 396, 397
Mamyrin, 276, 277
Mangaia, 154, 156, 204, 348
Mangini, 316, 383
Manhes, 114
Manitoba, 112
Mankinen, 255, 256
Mann, 376
mantle array, 140, 141, 145, 169, 188, 221, 222, 224, 225, 227, 231, 234, 348, 355, 356, 430
mantle heterogeneity, 43, 134, 135, 137, 138, 140, 169
mantle upwelling, 345, 346, 352, 354
mantle viscosity, 139, 280
Manton, 16

marble cake mantle, 138, 139, 201, 202, 349, 350
Marcantonio, 204, 208, 284, 285
Marchitto, 374, 375
Marechal, 21
Marianas, 162, 355, 356
Marillo-Sialer, 31
Markowski, 425, 426
Marks, 427
Marrero, 401
Mars, 155, 299, 408, 409, 410, 424, 428, 430, 431, 432, 433
Marshall, 235, 236
Martin, 54, 204, 208, 213, 391
Martinique, 160, 161, 233
Marty, 285, 286, 288, 289
mass fractionation, 20, 21, 22, 26, 27, 28, 29, 30, 31, 32, 33, 62, 68, 101, 117, 128, 156, 181, 235, 236, 274, 281, 286, 288, 289, 294, 299, 300, 301, 409, 410, 416, 422, 425, 434
mass spectrometer, 13
Mathews, 135
Matsuda, 278, 280
Mattauch, 8
Mattinson, 11, 31, 100, 101, 105, 106, 107, 108, 111
Mauna Kea, 157, 204, 205, 342, 354
Maunder, 376, 377, 378, 384, 385
Maurette, 455
Mauritania, 50, 51
Maury, 173
Maycock, 378
Mazaud, 370
McArthur, 55
McCandless, 197, 207
McCorkell, 380
McCracken, 378
McCulloch, 69, 73, 74, 78, 161, 162, 180, 189, 428
McDermott, 161, 328, 329, 333, 355, 356
McDougall, 243, 250, 254, 261, 263, 264
McDowell, 257, 268
MC–ICP–MS, 28
McIntyre, 32, 33, 34, 35
McKeegan, 435
McKenzie, 138, 156, 178, 344, 345
McLeod, 431, 433
McManus, 318, 319
McMullen, 41
MDD model, 267, 268
Meadows, 309
mean life, 409
Measures, 383
Medina-Elizalde, 311
Megrue, 252
Meijer, 149
Meisel, 201, 202
memory effects, 25
Menzies, 168, 169
Merensky Reef, 207
Merrihue, 245, 246, 247, 277
Merrill, 382

metamict, 104
metamorphism, 45, 46, 47, 50, 51, 70, 71, 72, 73, 78, 79, 81, 99, 112, 121, 122, 124, 176, 177, 186, 187, 220, 244, 248, 264, 408, 415, 422, 452, 460
metasomatism, 141, 157, 158, 161, 167, 168, 169, 170, 200, 201, 207, 355, 356, 357, 358, 359, 431
meteor crater, 113, 398, 399
Mexico, 178, 201, 202, 256, 268, 269, 295, 354, 391, 450, 457, 458
Meyer, 6
Mezger, 112
Michard, 88, 158
micromass, 19
Mid Atlantic Ridge, 136, 137, 145, 149, 177, 202, 203, 222, 277, 284, 315, 343
Middleton, 395, 396
Milankovich, 101, 255, 378, 386
Milankovitch, 322, 323
Mildowski, 80
Miller, 90
Min, 257, 258, 259
Minster, 41
Mishra, 419, 422
missing xenon, 299, 301
Mississippi, 48, 79, 88, 212, 393
Mitchell, 245, 400
mixing hyperbola, 189, 293
model age, 42, 44, 45, 51, 52, 72, 73, 75, 76, 77, 78, 79, 80, 81, 82, 83, 99, 118, 123, 124, 200, 228, 335, 337, 339, 340, 413, 418
Mojave Desert, 364, 365
Mokadem, 56, 61, 62
Molasse Basin, 282
molybdenite, 196, 197, 209
Monaghan, 388
monazite, 103, 111, 112, 219, 220, 452
Monnin, 375
Monte Carlo method, 461
Montel, 112
Mook, 364, 365, 367
Moon, 74, 85, 115, 143, 155, 229, 288, 289, 294, 299, 408, 409, 410, 419, 424, 426, 427, 428, 429, 430, 431, 432, 433
Moorbath, 70, 83, 86, 87, 122, 123, 124, 174, 175, 176, 185, 186
Moore, 78
moraine, 61, 399, 400, 401
Moran, 393
MORB, 73, 74, 75, 85, 122, 128, 134, 135, 136, 138, 140, 141, 142, 143, 144, 145, 146, 147, 148, 149, 150, 151, 152, 153, 155, 156, 157, 158, 159, 160, 162, 182, 183, 184, 185, 202, 203, 221, 222, 223, 224, 225, 226, 228, 230, 234, 235, 236, 276, 277, 278, 279, 281, 282, 285, 286, 288, 289, 291, 292, 293, 295, 296, 297, 298, 300, 336, 337, 342, 343, 344, 345, 347, 348, 349, 350, 351, 352, 353, 354, 355, 356, 429, 432
Moreira, 275, 276, 286, 287, 288, 289, 296, 297

Morgan, 136, 137, 178, 197, 199, 209, 210
Mork, 70, 71
Morocco, 139, 234, 256
Morris, 388, 389, 390
Morrison, 177, 178, 276
Morton, 51, 52, 57
Mount Narryer, 106, 109, 110
Mousterian, 325
Moynier, 422
MSWD, 33, 34, 35, 36, 48, 70, 86, 87, 117, 123, 197, 198, 207, 208, 219
Mukhopadhyay, 284, 287, 296, 297, 298, 300
Muller, 285
multi-dynamic analysis, 25
multi-path diffusion, 263
multiple collector, 18, 23, 25, 195
Murakami, 144
Murchison, 218, 417, 420, 421, 424, 434
Murphy, 285
Murthy, 169
muscovite, 42, 47, 173, 197, 261, 262, 263, 267, 444, 447, 449, 452
Mussett, 242

NADW, 88, 93, 94, 129, 317, 318, 319, 373, 383
Naeser, 269, 446, 447, 448, 451, 452, 453, 454, 455
Nagao, 286
Nakai, 48, 49, 232
Naldrett, 207, 209, 210
Nantan, 114, 118
Naudet, 9
Naumenko, 257
Naylor, 369
NBS, 32, 33, 62, 100
Nd paradox, 89
Neal, 140
Nebel, 41
Neff, 312, 376
Negre, 319
Nelson, 74, 75, 78, 79, 81, 82, 180, 236
neon-A, 288
neon-B, 287, 288
Nepal, 61
Neumann, 40, 41
Neustupny, 366
neutron activation, 245, 295, 326, 372, 391, 398, 399, 400, 425, 427, 428
neutron capture, 5, 9, 10, 244, 411
Nevada, 49, 188, 267, 323, 391, 392, 399
New Mexico, 288, 391
New York State, 80
New Zealand, 243, 335, 336, 340, 371
Newman, 342, 344
Newsome, 146, 434
Nichols, 415
Nicolaysen, 32, 42
Niedermann, 289, 293
Niemeyer, 413

Nier, 13, 22, 113, 277
Nier-Johnson geometry, 18
Nir-El, 11, 219
Nishiizumi, 395, 396, 398
NIST, 62, 100
Niu, 18
Nixon, 179
Noble, 20
Noe-Nygaard, 137
Noril'sk, 198, 199, 206
Normade, 207
Norman, 219, 232, 423
North Atlantic, 56, 88, 89, 92, 93, 127, 129, 148, 150, 159, 202, 222, 231, 256, 277, 278, 312, 316, 317, 318, 319, 371, 373, 386
Norway, 47, 70, 71, 112, 124
Nu Instruments, 19, 31
nuclear tests, 364, 366, 372, 373, 391
nucleosynthesis, 2, 4
nuclide chart, 1
Nuk, 122, 123, 124, 185, 186
Nunes, 68
Nuth, 417
Nutman, 85, 87, 187
Nuuvuagittuq, 106, 107
Nuvvuagittuq, 433
Nydal, 372
Nyquist, 417, 418, 421, 428, 429

O'Hara, 135, 158, 343
O'Neil, 433
O'Neill, 150
O'Nions, 6, 20, 68, 84, 88, 90, 91, 128, 129, 135, 140, 141, 142, 147, 280, 345
Obert, 311, 312
ocean conveyer belt, 231, 373, 374, 383
oceanographic tracer, 52, 316
Odin, 51
ODP, 59
Oeschger, 366
OIB, 134, 135, 136, 140, 141, 143, 144, 145, 146, 148, 149, 150, 151, 152, 153, 154, 155, 156, 157, 158, 159, 168, 180, 202, 203, 204, 205, 206, 221, 223, 224, 225, 234, 278, 279, 281, 282, 286, 287, 289, 291, 293, 296, 297, 298, 300, 342, 343, 346, 348, 349, 351, 353, 354, 355, 356
Oklo, 8, 9, 10
Oktay, 393
Oldoinyo Lengai, 180, 341, 342
olivine, 44, 136, 156, 157, 158, 173, 181, 183, 202, 204, 277, 292, 335, 408, 410
Olsen, 159
Olsson, 365, 366
onion skin model, 408
Onstott, 250, 251, 262, 264
Ontario, 77, 78, 86, 125, 197, 209, 253, 260, 261, 262, 391
open-system behaviour, 50, 52, 68, 91, 100, 197, 209, 308, 310, 311, 324, 327, 328, 419, 422, 430

ordinary chondrites, 199, 201, 219, 415, 417, 420, 421, 423, 424, 429, 431, 435
ore deposits, 48
Orgueil, 421
Orinoco, 160, 387
Osmond, 309
Ostlund, 372
Ott, 288, 293, 299, 300
Otway, 269, 453, 454, 460
Oversby, 120, 333, 342
Owen, 294, 295, 299
Oxburgh, 213, 214, 280, 282
OxCal, 369
Ozima, 276, 283, 284, 286, 295, 296, 297, 299

Pacific, 57, 58, 88, 89, 90, 91, 92, 93, 127, 128, 129, 138, 144, 147, 148, 150, 151, 153, 154, 155, 157, 159, 161, 182, 210, 211, 213, 214, 230, 231, 234, 235, 255, 277, 278, 283, 284, 285, 289, 310, 312, 336, 337, 344, 346, 349, 350, 351, 352, 371, 373, 374, 375, 376, 382, 383, 384, 386
Pacific Ocean, 58, 90, 92, 127, 128, 231, 235, 277, 371, 375, 376, 382, 383, 386
Pakistan, 60, 61
pallasites, 408
Palmer, 57, 60, 90
Panama Gateway, 92
Pankhurst, 108, 186
Panov, 6
Panthalassan Ocean, 90
Papanastassiou, 43, 44, 45, 421
Papua New Guinea, 90, 213, 370
Paquay, 212, 214
Parana Basin, 179
Parrish, 13, 28, 111
Parsons, 267
partial annealing zone, 453
Patchett, 14, 15, 17, 27, 76, 83, 218, 219, 220, 221, 223, 224, 226, 227
path of stability, 2, 7
Paton, 410
Patterson, 113, 114, 125, 127, 128, 292, 296
Pavich, 387
Pavlov, 180
Pb loss, 102, 103, 106, 110, 111, 112, 114, 115, 116, 327
Pb paradox, 114, 115, 120, 121, 146, 147, 149, 150
PDB, 93, 366, 367
Pearce, 62, 89, 160
Pearson, 195, 368
Pegram, 210, 211, 212, 213
pelagic sediment, 114, 204, 224, 225, 315, 357
Penokean orogeny, 77
Pepin, 275, 288, 293, 299
peridotite, 138, 139, 149, 150, 155, 156, 157, 158, 159, 167, 168, 169, 170, 200, 201, 203, 204, 206, 223, 347, 349, 352, 354, 359
Peron, 288

perovskite, 144, 408
Peru, 401
Peterman, 52, 56
Petford, 449
Peto, 296, 297, 298
Pettke, 49, 50
Peucker-Ehrenbrink, 211
PFE, 13
PGE, 194, 200, 206, 207, 208, 210
phenocryst, 156, 178, 335, 336
Philippines, 356
Phillips, 262, 391, 398, 400, 401
Phinney, 286, 295, 297
phlogopite, 135, 136, 168, 169, 170, 173, 262
phosphate, 15, 44, 89, 90, 91, 111, 116, 220, 415, 434
phosphoric acid, 16
Picciotto, 314
Pickett, 326, 353, 358
Pickles, 264
Pidgeon, 50, 70, 108
Piepgras, 88, 89
Pierce, 25
Pierson-Wickmann, 61
Pikes Peak, 74, 226, 235, 236
Pin, 15
Piotrowski, 93, 94, 230
Pitcairn, 154, 156, 204, 225, 279, 289
plagioclase, 45, 68, 71, 81, 149, 156, 157, 171, 172, 173, 176, 178, 189, 250, 251, 260, 261, 340, 401, 416
plasma source mass spectrometry, 17
plateau age, 247, 248, 250, 252, 253, 259, 262, 265
plum pudding mantle, 138
Plumbotectonics, 121, 122, 148
Plummer, 391, 392
Podosek, 413, 415, 418
Pollington, 71
Polve, 138
Popp, 53
Porcelli, 280, 281, 289, 296
Poreda, 275, 289, 296
porosity, 344, 345, 347, 351, 352, 353, 446
positron, 8, 240, 257
Potter, 311
Potts, 17, 109, 306
Powell, 28, 36
p-process, 6
PREMA, 152, 154, 155
Premo, 68
Price, 369, 445, 447, 448
primordial helium, 155, 275, 277, 278, 279, 280, 281, 282, 283, 286
Prinzhofer, 138
probability density, 35, 450, 461, 462
probability of fit, 34, 35
projected track lengths, 455, 456, 458, 461
Provost, 46
Przybylowicz, 324, 325

pseudochron, 358, 359, 413, 418, 420, 421, 422, 423, 424, 425, 426, 428, 430, 431, 432, 433, 434, 435, 436, 437
pseudo-isochron, 44, 176, 189
pseudo-isochron', 44
Puchtel, 199
Pujol, 300, 301
PUM, 201, 202, 203
Purdy, 47
Purser, 379
Pyle, 341, 342
Pyrenees, 201
pyroxenite, 138, 139, 159, 207, 348, 349
pyrrhotite, 197, 198

Qin, 351, 421, 425
Quebec, 106, 433
Quitte, 422

R/R_A, 157, 274, 276, 277, 278, 279, 282, 283
radioactive decay, 6
Raffenach, 9
Rahaman, 58
Rahl, 451
Raisbeck, 380, 382, 384, 385, 386
Ralph, 363, 364
Ramsey, 369
Raquin, 287, 293
Rausch, 59
Ravizza, 198, 210, 211
Rayleigh fractionation, 19, 338, 343
Raymo, 60
Read, 167
Reagan, 336
recoil, 8, 103, 250, 251, 267, 309, 311, 323
red giant, 3, 4, 5, 417, 422, 438
redwood, 368
REE, 14, 15, 16, 29, 72, 73, 79, 80, 88, 91, 141, 149, 160, 161, 218, 222, 232, 233, 234, 235, 417, 434
Reeves, 5
Regelous, 422
regression, 32, 315, 325
Rehkamper, 15
Reid, 335, 340
Reimer, 372, 448
Reiners, 269, 451
Reisberg, 201, 204
Renne, 11, 256, 257, 258, 259
residence time, 52, 56, 88, 89, 127, 129, 143, 147, 148, 150, 176, 212, 213, 214, 230, 231, 258, 281, 283, 296, 309, 310, 313, 316, 317, 319, 320, 338, 341, 343, 344, 372, 380, 382, 383, 384, 387, 455
Reunion, 151, 289, 343
Rex, 248
Reykjanes Ridge, 136, 145, 277, 296
Reymer, 76, 77
Reynolds, 91, 92, 128, 295, 299, 412, 413
Reyss, 383
rhenium-depletion age, 203

Rhine Graben, 282
Richard, 15, 140
Richardson, 148, 167
Richardton, 220, 413
Richter, 101, 138, 149, 167, 264, 265, 266
Rickli, 231
Ringwood, 2, 137
Rink, 328, 329
Rio Grande Rift, 354
Rison, 286
Rivera, 257, 259
Robert, 11
Roberts, 94, 253, 314, 379
Robinson, 386
Roddick, 16, 243, 244
Rogers, 108
Rohling, 323
Ronda, 201
Rose, 8
Rosholt, 100, 113, 309, 317
Ross, 2
Rotaru, 420
Rotenberg, 41
Roth, 433
Roy-Barman, 202, 326
r-process, 5
Rubin, 336, 337, 351
Rugel, 421
Ruiz, 185, 336
Rundberg, 54
Russ, 194
Russell, 20, 21, 103, 119, 120, 121, 235, 417
Russia, 82, 287
Russo, 348
Rutberg, 93, 94
Rutherford, 10, 274

Sackett, 317
Sahara, 89, 407, 426
Sahijpal, 418
Salmeron, 408
Salt Lake Crater, 222, 223
Salters, 221, 222, 223, 224, 225, 226
Samoa, 140, 141, 151, 154, 156, 278, 283, 348
Samson, 84
Sanborn, 220
Sanchez-Cabeza, 321
sanidine, 255, 256, 257, 258
Santschi, 320
Sao Miguel, 135, 136, 140, 141, 151
Sarda, 277, 286
Sasada, 297
Sato, 232, 233
Savage, 281
Scafetta, 378
Schaefer, 336
Scharer, 111
SCHEM, 143, 430, 431
Scherer, 219, 220, 221
Schersten, 206, 425
Schiano, 202

Schiller, 418
Schilling, 136, 137
Schimizu, 91
Schmidt, 462
Schmitz, 257, 284
Schoenberg, 207, 424
Schoene, 30, 100, 101, 107
Scholten, 315, 316
Scholz, 311
Schonbachler, 420
Schramm, 413, 433
Schreiner, 42, 43
Schubert, 77
Schwarcz, 324, 325, 328
Schwartzman, 290
Science Citation Index, 75
Scoates, 207
Scotia arc, 160
Scotland, 44, 70, 81, 108, 124, 174, 175, 177, 233, 401
Scott, 317
seawater, 53, 55, 57, 59, 61
seawater array, 230, 231
secular equilibrium, 161, 306, 307, 308, 309, 310, 312, 313, 315, 321, 322, 324, 327, 328, 333, 334, 337, 338, 340, 341, 349, 350, 358, 392
sedimentation rate, 283, 284, 285, 313, 314, 315, 316, 320, 374, 382, 383, 388
Seeger, 6, 413
Selby, 197
Semarkona, 417, 422
Seta, 281
Shackleton, 255, 374
shales, 50, 51, 80, 198, 210, 224
Sharma, 212, 213, 214, 386, 428
Sharpe, 207
Shaw, 90
Shearer, 137
Sheldon, 277
Shen, 25
Shepard, 399
Sheppard, 48
shergottite, 428, 430
Shirahata, 320
Shirey, 195
Shlyakhter, 10
SHRIMP, 30, 31, 35, 46, 86, 87, 105, 109, 111, 256
Shu, 417
Shukolyukov, 420, 421, 422
Siddall, 90
siderophile, 114, 146, 194, 199, 206, 289, 299, 410, 419, 423
Siegenthaler, 372
Sierra Nevada, 173, 187, 188
Sigmarsson, 348, 349, 350, 355, 356, 357, 358, 389
Sikes, 374
silica gel, 19
Silk, 444, 451

Silver, 74, 103, 104, 226
Simon, 170, 257, 341
Simpson, 364
Sims, 346, 348, 349, 350, 351, 352, 353
SIMS, 109, 194, 335, 336, 340, 422
Siqueiros Fracture Zone, 349, 350, 352
Skellefte, 76, 77
Skinner, 374, 375
Skiold, 76
Skovgaard, 204
Skye, 174, 175, 176, 177, 233, 401
slab-derived fluids, 160, 161, 162, 168, 205, 206, 341, 355, 358, 359, 390
Slave Province, 86
Smart, 187
Smith, 85, 111, 169, 179, 251
Smoliar, 11, 196, 197
Smyth, 13
SNC meteorites, 424
Snyder, 393
Soderlund, 219
soil profile, 387, 391
solar cycles, 376
solar irradiance, 377, 378
solar nebula, 29, 40, 44, 45, 62, 68, 118, 119, 197, 219, 294, 407, 408, 410, 413, 415, 416, 418, 420, 421, 422, 424, 426, 429, 430, 431, 432, 434, 435, 436, 437, 438
solar wind, 275, 286, 287, 366, 378, 384, 391, 417, 422, 438
Solari, 31, 35
SOPITA, 153, 159
South America, 129, 179, 375
South Pole, 320, 384, 385
spallation, 5, 269, 274, 285, 295, 363, 380, 383, 390, 393, 394, 396, 398, 399, 400, 417, 435, 436
SPECMAP, 322, 323, 386
speleothem, 323, 324, 328, 371, 372, 376
Spell, 256
sphalerite, 48, 49, 50
sphene, 101, 111, 112, 232, 447, 451, 452
Spiegelman, 351
spike, 26
spinel, 167, 168, 201, 202, 206, 221, 223, 347, 352, 353, 354, 359, 408
spontaneous fission, 4, 7, 8, 297, 444, 445
Spooner, 56, 57
s-process, 5
Sprung, 430, 431
Srinivasan, 301, 417, 436, 437
St Helena, 140, 146, 151, 182, 204
Stacey, 103, 120, 121, 123
Stanton, 119, 120
Starkey, 279
static analysis, 25
Staudacher, 278, 287, 291, 292, 293, 295, 296
Staudigel, 90, 153, 159
Ste Marguerite, 258, 421, 424
Steele, 410, 422, 423
Steiger, 11, 41, 100, 240, 255, 256, 257, 258

Steinhilber, 377
stellar evolution, 3
Stensland, 381
step heating, 246, 247, 249, 251, 253, 257, 261, 262, 263, 265, 267, 276, 291, 415
Steuber, 59
Stewart, 410, 432
Stichel, 90
Stille, 80
Stillwater, 68, 69, 208, 209, 290
stilpnomelane, 47, 252
Stirling, 309, 434
Stocker, 376
Stoll, 61
Stone, 269, 400
Stordal, 88
Storzer, 451, 452, 455
Stracke, 154, 155, 158, 223, 352, 353
Stromboli, 178, 335, 341
Stroncik, 293
Strutt, 268
Stuiver, 366, 368, 370, 373
Sturges, 127
subcontinental lithosphere, 135, 156, 158, 201, 223, 224, 354
subduction barrier, 278, 287, 295
subduction zone, 159, 160, 359, 389, 393
subduction-related magmas, 162, 205, 341, 352, 355, 356, 358, 359
Sudbury, 106, 197, 198, 209, 210
Suess, 366, 367, 368
Suganuma, 256
Sugiura, 435, 436
Suigetsu, 370, 371
Sun, 43, 136, 145, 151, 158, 376
sunspot, 366, 368, 376, 377, 384
Superior Province, 83, 125
supernova, 4, 5, 6, 9, 410, 411, 412, 413, 414, 416, 417, 420, 422, 423, 434, 437, 438
Suter, 379
Suzuki, 197, 205, 206
Svensmark, 376
Sweden, 76, 392, 394

Tachikawa, 89
Taddeucci, 335
Tafassasset, 434
Tahiti, 324
Takayanagi, 283, 284
Tanaka, 15, 232, 233, 234, 382
tandem accelerator, 378, 379
Tang, 288, 422
Tanguy, 337
Tasmania, 371
Tatsumoto, 43, 120, 125, 126, 134, 145, 146, 218, 219, 220, 221
Tauber, 370
Taupo Volcanic Zone, 335, 340
Tauxe, 255, 256
Taylor, 84, 122, 124, 183, 186, 188, 256, 408
Tazoe, 235

T_{DM}, 74, 75, 76, 77, 78, 81, 82, 228
Tejon Block, 454
tektite, 247, 451, 452, 455
Temora, 31
Tennessee, 48, 49
Tepley, 178
Tera, 111, 115, 116, 388, 389
Theia, 410, 424, 427, 433
Theis, 420
Thellier, 366
thermal ionization, 16
thermochronometry, 251, 260, 261, 263, 265, 266, 267, 277, 451
thermohaline circulation, 373
Thermo-Scientific, 19
Thirlwall, 17, 20, 21, 26, 28, 31, 170, 171, 175, 176
Thoennessen, 2
Thomas, 324, 359
Thompkins, 14
Thompson, 13, 173, 174, 177, 311, 344
Thrane, 219, 418
Thurber, 310
Thurston, 83
Tibet, 60
tidal forces, 378
Tilton, 45, 103, 112, 180
time-interpolation, 25
TIMS, 13, 14, 15, 17, 18, 19, 20, 21, 22, 24, 25, 28, 29, 30, 56, 101, 106, 107, 194, 218, 230, 233, 241, 308, 309, 335, 336, 340, 421, 423, 428
TINCLE, 456
TINT, 456, 457, 459
titanite. *See* sphene
Titterington, 32
Tobias, 378
Todt, 28
Tolstikhin, 85, 139, 140, 143, 144, 297, 298, 299
Tommasini, 173
Tonga, 355, 356, 357
Torgersen, 392, 393
Touboul, 424, 427
transposed paleo-isochron, 124
tree rings, 11, 365, 368, 370, 384, 385
Tremblay, 269, 270
Tretbar, 49
Trieloff, 258, 287, 293, 296, 297
Trinquier, 25, 408, 409, 410, 411, 420, 421
Tristan, 140, 151, 152, 278
Trivedi, 412
Troodos, 58, 59
Truran, 416
Tubuai, 155, 156, 348
Tucker, 112, 297, 298
Turcotte, 134, 138
Turekian, 290, 382
Turner, 85, 245, 246, 247, 248, 250, 252, 253, 259, 265, 298, 348, 356, 357, 358, 438
two-reservoir model, 279, 280, 281

Ullman, 393
United States, 74, 75, 82, 320
Usoskin, 378
U-uptake, 328

Valbracht, 293
Valley, 111
van Breemen, 104
van de Flierdt, 230
van der Hilst, 137, 139
van Keken, 278, 280
Vance, 21, 22, 29, 57, 61, 71, 92
Vanuatu, 322
varved sediment, 370
Veeh, 322
Veizer, 52, 53, 54
Venezuela, 259
ventilation ages, 318, 374, 375
Vermeesch, 35, 36
Verschure, 47
Vervoort, 224, 225, 226, 227, 229, 230
Vesta, 408, 424
Vesuvius, 335, 341
VG Elemental, 19
Vidal, 146
Vieira, 377
Vigarano, 419
Vikan gneiss complex, 124
Villa, 11, 41, 250
Villemant, 311
Villeneuve, 418
Vincent, 240
Virginia, 387
Vlastelic, 129
Volkening, 194
Volpe, 337, 351
volume diffusion, 47, 247, 248, 250, 261, 262, 263, 264, 266, 267, 351
von Blankenburg, 128, 129, 383
von Quadt, 31
VonderHaar, 91

Wadhwa, 435
Wagner, 385, 392, 447, 452, 453, 455, 462
Walczyk, 194
Walder, 18, 19, 28, 29, 30
Wales, 80
Walker, 196, 198, 199, 200, 201, 206, 209, 445
Wallner, 421
Walter, 317, 450, 451
Walvis Ridge, 151, 226
Wanajo, 6
Wang, 384
Warren, 408, 409
Wartho, 267
Wasserburg, 20, 27, 44, 68, 69, 72, 73, 74, 88, 111, 115, 116, 117, 140, 142, 184, 185, 296, 407, 412, 413, 414, 416, 417, 420, 434, 437
Wasson, 408
Watson, 108

weathering, 60, 61, 62, 73, 91, 99, 113, 134, 160, 212, 214, 227, 230, 231, 247, 309, 386, 387, 388, 396, 397, 399
Weaver, 154, 156, 158
Weinberg, 2
Weis, 140
Weisberg, 408
Weiss, 158
well-gases, 288, 300
Wen, 144
Wendlandt, 179
Wendt, 34, 115, 462
West, 85
Wetherill, 46, 102, 103, 285, 295
White, 3, 21, 28, 134, 136, 137, 140, 141, 150, 151, 158, 159, 160, 161, 189, 223, 224, 230, 454
Whitehouse, 70, 81, 87, 88, 124
Wickman, 52
Widom, 135, 206
Wiechert, 410
wiggle matching, 368, 371
Wilde, 111
Willbold, 233
Williams, 109, 111, 336, 341, 342, 345
Wilson, 76, 340, 378
Wimpenny, 220
Winckler, 285
Wind River, 61, 400
Winograd, 323
Wisconsin, 49, 50, 78
Witwatersrand, 197, 199
Wohlers, 432, 433
Wolf, 269
Wolff, 185
Wolfli, 390
Wood, 115, 347
Woodhead, 28, 31, 156, 228
Woodhouse, 213
Workman, 143, 144, 156
Wotzlaw, 102
Wright, 261, 262
Wu, 126
Wust, 88
Wyllie, 159
Wyoming, 61, 69, 99, 113, 400

XRD, 50, 51

Yamakawa, 421
Yamazaki, 386
Yang, 317
Yangtze, 60
Yellowstone, 185
Yin, 421, 424
Yiou, 385
Yokochi, 287
Yokoyama, 182, 307, 382, 410
Yokoyama, 287
York, 32, 33, 34, 35, 219, 247, 249, 251, 253, 259, 260, 261
Yoshioka, 445

You, 387
Young, 422, 437
Younger Dryas, 93, 94, 370, 371, 373, 375
Yu, 316, 317

Zadnik, 276
Zahnle, 301
Zartman, 121, 122, 148
Zeh, 228
Zeitler, 268, 274
Zhang, 280

Zhao, 231
Zimbabwe, 102
Zimmermann, 231
Zindler, 138, 146, 151, 152, 153, 155, 168, 223
Zinner, 417
zircon, 13, 16, 28, 30, 31, 32, 35, 36, 67, 69, 70, 76, 77, 78, 86, 87, 88, 100, 101, 102, 103, 104, 105, 107, 108, 109, 110, 111, 112, 116, 148, 190, 197, 207, 220, 221, 224, 226, 227, 228, 229, 230, 231, 233, 254, 257, 258,

262, 327, 335, 336, 340, 341, 447, 449, 451, 454
Zolnai, 78
Zreda, 398, 399, 400
α counting, 308, 309, 322, 323, 326, 342, 351
α decay, 8, 67, 198, 274, 445
β decay, 8, 11, 40, 195, 232, 233, 240, 257, 258, 259, 285, 300, 412, 423, 427
ε notation, 43, 71, 72
μ value, 114, 115, 119, 120, 121, 122, 145, 146, 150, 162, 281